Epigenetic Cancer Therapy

Epigenetic Cancer Therapy

Edited by

Steven G. Gray

ELSEVIER

AMSTERDAM • BOSTON • HEIDELBERG • LONDON
NEW YORK • OXFORD • PARIS • SAN DIEGO
SAN FRANCISCO • SINGAPORE • SYDNEY • TOKYO
Academic Press is an imprint of Elsevier

Academic Press is an imprint of Elsevier
125, London Wall, EC2Y 5AS
525 B Street, Suite 1800, San Diego, CA 92101-4495, USA
225 Wyman Street, Waltham, MA 02451, USA
The Boulevard, Langford Lane, Kidlington, Oxford OX5 1GB, UK

Notices
Knowledge and best practice in this field are constantly changing. As new research and experience
broaden our understanding, changes in research methods, professional practices, or medical treatment
may become necessary.

Practitioners and researchers must always rely on their own experience and knowledge in evaluating and
using any information, methods, compounds, or experiments described herein. In using such information or
methods they should be mindful of their own safety and the safety of others, including parties for whom they
have a professional responsibility.

To the fullest extent of the law, neither the Publisher nor the authors, contributors, or editors, assume any liability
for any injury and/or damage to persons or property as a matter of products liability, negligence or otherwise,
or from any use or operation of any methods, products, instructions, or ideas contained in the material herein.

ISBN: 978-0-12-800206-3

Library of Congress Cataloging-in-Publication Data
A catalog record for this book is available from the Library of Congress

British Library Cataloguing-in-Publication Data
A catalogue record for this book is available from the British Library

For information on all Academic Press publications
visit our website at http://store.elsevier.com/

Publisher: Mica Haley
Acquisition Editor: Catherine Van Der Laan
Editorial Project Manager: Lisa Eppich
Production Project Manager: Melissa Read
Designer: Mark Rogers

Working together
to grow libraries in
developing countries

www.elsevier.com • www.bookaid.org

Contents

CHAPTER 7   Mining the Epigenetic Landscape: Surface Mining or Deep Underground 141

Viren Amin, Vitor Onuchic, and Aleksandar Milosavljevic

PART 2   EPIGENETICS AND CANCER

CHAPTER 8   Development of Epigenetic Targeted Therapies in Hematological Malignancies: From Serendipity to Synthetic Lethality 169

Thomas Prebet and Steven D. Gore

PART 3   TARGETING ABERRANT EPIGENETICS

List of Contributors

Bryce K. Allen
Center for Therapeutic Innovation, The Miami Project to Cure Paralysis, Department of Psychiatry and Behavioral Sciences, University of Miami Miller School of Medicine, Miami, FL, USA

Donat Alpar
Centre for Evolution and Cancer, The Institute of Cancer Research, London, UK

Viren Amin
Epigenome Center, Baylor College of Medicine, Houston, Texas, USA

Fazila Asmar
Department of Hematology, Rigshospitalet, Copenhagen, Denmark

Nagi G. Ayad
Center for Therapeutic Innovation, The Miami Project to Cure Paralysis, Department of Psychiatry and Behavioral Sciences, University of Miami Miller School of Medicine, Miami, FL, USA

Anne-Marie Baird
Genome Stability Laboratory, Cancer and Ageing Research Program, Institute of Health and Biomedical Innovation, Queensland University of Technology, Woolloongabba, Queensland, Australia; Thoracic Oncology Research Group, Institute of Molecular Medicine, St. James's Hospital, Dublin, Ireland

Louise J. Barber
Centre for Evolution and Cancer, The Institute of Cancer Research, London, UK

Becky A.S. Bibby
Cancer Biology and Therapeutics Lab, School of Biological, Biomedical and Environmental Sciences, University of Hull, Yorkshire, UK

Emma Bolderson
Genome Stability Laboratory, Cancer and Ageing Research Program, Institute of Health and Biomedical Innovation, Queensland University of Technology, Woolloongabba, Queensland, Australia

Philippe Bouvet
Laboratoire Joliot-Curie, Ecole Normale Supérieure de Lyon, Université de Lyon, Lyon, France

Frédéric Catez
Centre de Recherche en Cancérologie de Lyon, Centre Léon Bérard, Lyon, France; Université de Lyon, Lyon, France

Leandro Cerchietti
Hematology and Oncology Division, Medicine Department, Weill Cornell Medical College of Cornell University, New York, NY, USA

Snehajyoti Chatterjee
Transcription and Disease Laboratory, Molecular Biology and Genetics Unit, Jawaharlal Nehru Centre for Advanced Scientific Research, Bangalore, Karnataka, India; Laboratoire de Neurosciences Cognitives et Adaptatives, Université de Strasbourg-CNRS, GDR CNRS, Strasbourg, France

Taiping Chen
Department of Molecular Carcinogenesis and the Center for Cancer Epigenetics, The University of Texas MD Anderson Cancer Center, Smithville, Texas, USA; The University of Texas Graduate School of Biomedical Sciences at Houston, Houston, Texas, USA

Andreas I. Constantinou
Laboratory of Cancer Biology and Chemoprevention, Department of Biological Sciences, School of Pure and Applied Sciences, University of Cyprus, Nicosia, Cyprus

Stuart J. Conway
Department of Chemistry, Chemistry Research Laboratory, University of Oxford, Oxford, UK

Dashyant Dhanak
Janssen Research & Development, Spring House, PA, USA

Jean-Jacques Diaz
Centre de Recherche en Cancérologie de Lyon, Centre Léon Bérard, Lyon, France; Université de Lyon, Lyon, France

Marc Diederich
College of Pharmacy, Seoul National University, Gwanak-gu, Seoul, Korea

Xinmin Fan
Department of Pathology, The Shenzhen University School of Medicine, Shenzhen, Guangdong, People's Republic of China

Brendan Ffrench
Department of Pathology, Coombe Women's and Infant's University Hospital, Dublin, Ireland

Michael F. Gallagher
Department of Histopathology, University of Dublin, Trinity College, Trinity Centre, St James Hospital, Dublin, Ireland

Marco Gerlinger
Centre for Evolution and Cancer, The Institute of Cancer Research, London, UK

Steven D. Gore
Department of Internal Medicine (Hematology), Yale Cancer Center, New Haven, CT, USA

Steven G. Gray
HOPE Directorate, St. James's Hospital, Dublin, Ireland; Thoracic Oncology Research Group, Institute of Molecular Medicine, St. James's Hospital, Dublin, Ireland

Kirsten Grønbæk
Department of Hematology, Rigshospitalet, Copenhagen, Denmark

David S. Hewings
Department of Chemistry, Chemistry Research Laboratory, University of Oxford, Oxford, UK

Holger Heyn
Cancer Epigenetics and Biology Program, Bellvitge Biomedical Research Institute, Barcelona, Catalonia, Spain

Zhe Jin
Department of Pathology, The Shenzhen University School of Medicine, Shenzhen, Guangdong, People's Republic of China; Shenzhen Key Laboratory of Micromolecule Innovatal Drugs, The Shenzhen University School of Medicine, Shenzhen, Guangdong, People's Republic of China; Shenzhen Key Laboratory of Translational Medicine of Tumor, The Shenzhen University School of Medicine, Shenzhen, Guangdong, People's Republic of China

Stephanie Kaypee
Transcription and Disease Laboratory, Molecular Biology and Genetics Unit, Jawaharlal Nehru Centre for Advanced Scientific Research, Bangalore, Karnataka, India

Yutaka Kondo
Department of Epigenomics, Nagoya City University Graduate School of Medical Sciences, Mizuho-cho, Mizuho-ku, Nagoya, Japan

Tapas K. Kundu
Transcription and Disease Laboratory, Molecular Biology and Genetics Unit, Jawaharlal Nehru Centre for Advanced Scientific Research, Bangalore, Karnataka, India

Florian Laforêts
Centre de Recherche en Cancérologie de Lyon, Centre Léon Bérard, Lyon, France; Université de Lyon, Lyon, France

Matthew W. Lawless
Experimental Medicine, UCD School of Medicine and Medical Science, Mater Misericordiae University Hospital, Catherine McAuley Centre, Dublin, Ireland

Stephen G. Maher
Cancer Biology and Therapeutics Lab, School of Biological, Biomedical and Environmental Sciences, University of Hull, Yorkshire, UK

Somnath Mandal
Transcription and Disease Laboratory, Molecular Biology and Genetics Unit, Jawaharlal Nehru Centre for Advanced Scientific Research, Bangalore, Karnataka, India; Department of Biochemistry, Faculty of Agriculture, Uttar Banga Krishi Viswavidyalaya, Pundibari, Cooch-Behar, West Bengal, India

Virginie Marcel
Centre de Recherche en Cancérologie de Lyon, Centre Léon Bérard, Lyon, France; Université de Lyon, Lyon, France

Aleksandar Milosavljevic
Epigenome Center, Baylor College of Medicine, Houston, Texas, USA

Hannah L. Moody
Cancer Biology and Therapeutics Lab, School of Biological, Biomedical and Environmental Sciences, University of Hull, Yorkshire, UK; Hull York Medical School, Yorkshire, UK

Atsushi Natsume
Department of Neurosurgery, Nagoya University Graduate School of Medicine, Nagoya, Japan

Thomas B. Nicholson
Developmental and Molecular Pathways, Novartis Institutes for BioMedical Research, Cambridge, MA, USA

Kenneth J. O'Byrne
Genome Stability Laboratory, Cancer and Ageing Research Program, Institute of Health and Biomedical Innovation, Queensland University of Technology, Woolloongabba, Queensland, Australia; Thoracic Oncology Research Group, Institute of Molecular Medicine, St. James's Hospital, Dublin, Ireland

Shane O'Grady
Experimental Medicine, UCD School of Medicine and Medical Science, Mater Misericordiae University Hospital, Catherine McAuley Centre, Dublin, Ireland

John J. O'Leary
Department of Pathology, Coombe Women's and Infant's University Hospital, Dublin, Ireland

Fumiharu Ohka
Department of Epigenomics, Nagoya City University Graduate School of Medical Sciences, Mizuho-cho, Mizuho-ku, Nagoya, Japan; Department of Neurosurgery, Nagoya University Graduate School of Medicine, Nagoya, Japan

Vitor Onuchic
Epigenome Center, Baylor College of Medicine, Houston, Texas, USA

Vineet Pande
Janssen Research & Development, Beerse, Belgium

Clara Penas
Center for Therapeutic Innovation, The Miami Project to Cure Paralysis, Department of Psychiatry and Behavioral Sciences, University of Miami Miller School of Medicine, Miami, FL, USA

Li-Xia Peng
State Key Laboratory of Oncology in South China, Sun Yat-sen University Cancer Center, Guangzhou, Guangdong, People's Republic of China

Benet Pera
Hematology and Oncology Division, Medicine Department, Weill Cornell Medical College of Cornell University, New York, NY, USA

Antoinette Sabrina Perry
Prostate Molecular Oncology, Institute of Molecular Medicine, Trinity College Dublin, Dublin, Ireland

Olga Piskareva
Cancer Genetics, Molecular and Cellular Therapeutics, Royal College of Surgeons in Ireland, Dublin, Ireland and Children's Research Centre Our Lady's Children's Hospital, Crumlin, Dublin, Ireland

Chara A. Pitta
Laboratory of Cancer Biology and Chemoprevention, Department of Biological Sciences, School of Pure and Applied Sciences, University of Cyprus, Nicosia, Cyprus

David J. Pocalyko
Janssen Research & Development, Spring House, PA, USA

Thomas Prebet
Department of Internal Medicine (Hematology), Yale Cancer Center, New Haven, CT, USA

Chao-Nan Qian
State Key Laboratory of Oncology in South China, Sun Yat-sen University Cancer Center, Guangzhou, Guangdong, People's Republic of China

Glen Reid
Asbestos Diseases Research Institute, University of Sydney, New South Wales, Australia

Derek J. Richard
Genome Stability Laboratory, Cancer and Ageing Research Program, Institute of Health and Biomedical Innovation, Queensland University of Technology, Woolloongabba, Queensland, Australia

Timothy P.C. Rooney
Department of Chemistry, Chemistry Research Laboratory, University of Oxford, Oxford, UK

Michael Schnekenburger
Laboratoire de Biologie Moléculaire et Cellulaire du Cancer, Hôpital Kirchberg, Luxembourg, Luxembourg

Alexandra Søgaard
Department of Hematology, Rigshospitalet, Copenhagen, Denmark

Peter Staller
EpiTherapeutics ApS, Copenhagen, Denmark

Raymond L. Stallings
Cancer Genetics, Molecular and Cellular Therapeutics, Royal College of Surgeons in Ireland, Dublin, Ireland and Children's Research Centre Our Lady's Children's Hospital, Crumlin, Dublin, Ireland

Vasileios Stathias
Center for Therapeutic Innovation, The Miami Project to Cure Paralysis, Department of Psychiatry and Behavioral Sciences, University of Miami Miller School of Medicine, Miami, FL, USA

Gabriel Therizols
Centre de Recherche en Cancérologie de Lyon, Centre Léon Bérard, Lyon, France; Université de Lyon, Lyon, France

Nicolas Veland
Department of Molecular Carcinogenesis and the Center for Cancer Epigenetics, The University of Texas MD Anderson Cancer Center, Smithville, Texas, USA; The University of Texas Graduate School of Biomedical Sciences at Houston, Houston, Texas, USA

Casey M. Wright
Asbestos Diseases Research Institute, The University of Sydney, Concord, NSW, Australia

Xiaojing Zhang
Department of Pathology, The Shenzhen University School of Medicine, Shenzhen, Guangdong, People's Republic of China; Shenzhen Key Laboratory of Translational Medicine of Tumor, The Shenzhen University School of Medicine, Shenzhen, Guangdong, People's Republic of China

Wei Zhu
Department of Oncology, First Affiliated Hospital of Nanjing Medical University, Nanjing, People's Republic of China

Jiaqi Qian
Department of Oncology, First Affiliated Hospital of Nanjing Medical University, Nanjing, People's Republic of China

INTRODUCTION

1

Steven G. Gray[1,2]

*[1]Thoracic Oncology Research Group, Institute of Molecular Medicine,
St. James's Hospital, Dublin, Ireland, [2]HOPE Directorate, St. James's Hospital, Dublin, Ireland*

CHAPTER OUTLINE

Epigenetics has come a long way from its earliest incarnations. But how do we now define this phenomenon? Classically, Waddington in 1942 used the phrase "epigenetic landscape" to describe conceptually how genes might interact with their surroundings to produce a phenotype [1]. A more recent commonly consensus definition describes epigenetics as a "stably heritable phenotype resulting from changes in a chromosome without alterations in the DNA sequence" [2]. Another common definition describes epigenetics as "chromatin-based events that regulate DNA-templated processes" [3].

Critically, it is now well established that aberrant epigenetics are a common feature in cancer [4]. There are currently four known mechanisms that underpin epigenetic regulation. They are: DNA CpG methylation, histone posttranslational modifications (PTMs), histone variants, and noncoding RNA (ncRNA). All four work individually or together to elicit what is often described as the "Epigenetic code." The epigenetic code is encompassed within the cell as the "epigenome," and has led to several large consortium-based mapping programs (e.g., the NIH Roadmap Epigenomics Program or the International Human Epigenetics Consortium) working to create epigenetic "roadmaps" or "blueprints" of human health and disease including cancer [5].

Given the scale of epigenetic cancer therapy today, this volume has been designed to have subsections on key areas as follows.

S.G. Gray (Ed): Epigenetic Cancer Therapy. DOI: http://dx.doi.org/10.1016/B978-0-12-800206-3.00001-X

1 INTRODUCTION TO THE AREA (KEY CONCEPTS)

In beginning the formulation for this volume it was felt that as this book is aimed not only at students but also at scientists, medics, and researchers, it would be critical to have a significant introductory section, allowing the reader to become acquainted with the known mechanisms by which epigenetic regulation of gene expression is achieved and how epigenetic dysregulation is a common feature in cancer [3,4].

As a conceptual concept the notion of an "epigenetic code" has simplified the way we can describe the mechanistic elements that the cell uses to recognize and elicit epigenetic or aberrant epigenetic regulation in a cancer cell. Generally speaking there are three "types" of regulatory protein known as "readers," "writers," or "erasers" of the epigenetic code. These are most generally associated with proteins that interact with histones through histone PTMs, although some have suggested a fourth category "remodeler" [6,7], whilst proteins such as DNA methyltransferases which modify DNA have also been classified as "writers" [8].

DNA methylation was one of the first described epigenetic events, and the suggestion that methylases could act as oncogenic agents was first postulated in 1964 by Srinivasan and Borek [9]. Drugs that target DNA methylation were rapidly developed but it was not until 2004/2006 that US Food and Drug Administration (FDA) approval was obtained for the use of 5-azacytidine (Vidaza) and 5-aza-2-deoxy-cytidine (Dacogen) for the treatment of myelodysplastic syndrome [10–12]. Over the next 50 years our understanding of DNA methylation has dramatically changed with the recent identification of additional methylation states such as hydroxymethylation (5-hmC) [13]. In this book, the roles of DNA methylation and hydroxymethylation in cancer are discussed in detail, how mutations in epigenetic regulators are involved with the fidelity of DNA methylation maintenance, how DNA methylation may be a driving process during tumorigenesis, and discuss the current available methodologies by which researchers can investigate the landscape of DNA methylation and hydroxymethylation in the laboratory setting.

We have known about histone posttranslational mechanisms such as histone acetylation since the 1960s [14], but the notion of a "Histone Code" was not formulated until the turn of the century by Jenuwein and Allis [15]. The histone code basically involves mechanisms by which cells mark histones using PTMs as "marks" in order to regulate gene expression. Various proteins which recognize and act on these "marks" can collectively be described as "readers," "writers," or "erasers" of this code. Readers are introduced to the key concepts surrounding these three classes of proteins, with discussions on the key elements in the complex interplay of epigenetic proteins in the regulation of gene transcription, and how defects in these systems can contribute to cancer initiation and progression. ncRNAs have also emerged as elements with important roles in epigenetic regulation. One class of ncRNAs are called microRNAs, or miRNAs, whilst another are long noncoding RNAs (lncRNAs) [16,17]. The roles of both in cancer are discussed in detail, their biogenesis, mechanism of action and their role in cancer initiation, promotion and progression, and their potential as epigenetic anticancer therapeutics.

Ribosomal RNA (rRNA) may seem a strange bedfellow with respect to epigenetics but unsurprisingly, a significant body of work has now shown that not only are the genes for rRNAs epigenetically regulated [18] but the RNA itself can be methylated. This is a new concept within the framework of epigenetics but additional data is now linking methylation of tRNAs [19] and epigenetic heritability [20]. The role of epigenetic modification of RNA, whilst in its infancy, demonstrates the importance of elucidating the functional role of RNA methylation with regard to cancer.

The emergence of next-generation sequencing and the landmark epigenome mapping projects have lead to an explosion of data [5,21–23]. A significant challenge now facing scientists is how to interrogate this wealth of data effectively. To this end a discussion on some of the most widely used epigenomic profiling methods, comparing their similarities and differences, as well as their strengths and weaknesses is welcome. From epigenome-wide association studies (EWAS), the scope of epigenomic profiling has extended significantly, through annotation of the epigenome itself. Analysis of such large datasets typically involve (i) enrichment analysis, allowing for interpretation of genomic and epigenomic variability, and (ii) sample clustering and classification. Case studies are provided to illustrate the potential types of analysis possible with large-scale epigenomic datasets.

2 EPIGENETICS AND CANCER

Having discussed the basics, the following section deals with a series of actual cancer settings. No one cancer is the same and so individual chapters on important cancers discuss the roles of aberrant epigenetics within particular tumor types, and describe the recent advances in our knowledge regarding the potential role of epigenetic targeting agents in these cancers. Finally the key currently identified potential epigenetic targets/biomarkers for therapy have been discussed in detail for each cancer type.

3 TARGETING ABERRANT EPIGENETICS

Since the development and approval of demethylating agents such as Dacogen and Vidaza, and the emergence of histone deacetylase inhibitors such as Vorinostat and Romidepsin for the treatment of myelodysplastic syndrome and cutaneous T-cell lymphoma, new regulatory approvals for agents targeting these proteins have not emerged. An earlier estimate indicated that prior to 2013, over 490 clinical trials had been conducted with HDACi with very limited success [24]. However, greater improvements are beginning to emerge for HDACi therapy. For instance, Entinostat has recently received "breakthrough therapy FDA designation" for breast cancer when added to exemestane in postmenopausal women with ER+ metastatic breast cancer whose cancer had progressed after treatment with a nonsteroidal aromatase inhibitor [25]. Likewise Panobinostat has recently achieved priority review at the FDA for the treatment of multiple myeloma when combined with bortezomib based on results from the PANORAMA1 trial (NCT01023308) [26] which showed a significant interim PFS (11.99 vs. 8.08 months). It must be noted that a similar trial utilizing Vorinostat and bortezomib did not see as significant a PFS (7.63 vs. 6.83 months) [27]. Within lung cancer a phase II NSCLC study of carboplatin and paclitaxel with randomization to the HDAC inhibitor Vorinostat, a superior response rate was demonstrated in the Vorinostat arm albeit with only a trend toward an improvement in survival [28]. This section includes a section on the current clinical trials being conducted using epigenetic therapies (and including the design, and implementation of said trials). In addition, the issue of dosing regimens for currently licensed epigenetic targeting agents has become very important and has been discussed in some detail.

The rest of this section is aimed at introducing the reader to various aspects with respect to the development of new and emerging epigenetic therapeutics themselves. The potential importance of nutritional intervention strategies and the identification and role of natural compounds that inhibit the epigenetic machinery are discussed. Within the last few years, focus has shifted to other members of the epigenetic regulatory machinery and discussions on the new developments in targeting lysine methyltransferases, lysine demethylases, lysine acetyltransferases, and bromodomain inhibitors round off this section.

4 ISSUES TO OVERCOME/AREAS OF CONCERN

One of the major drawbacks encountered with epigenetic therapies has been the issue of ineffectively low concentrations within the context of solid tumors [24]. But these are not the only issues of concern. Section 4 of *Epigenetic Cancer Therapy* highlights to the reader some of the other current emerging issues for epigenetic targeting of cancer.

One such critical issue with respect to epigenetic targeting concerns the actual compositional makeup of the tumor itself. It is now becoming apparent that intratumoral heterogeneity (ITH) is a major issue affecting epigenetic profiles within different regions of tumors themselves [29–31]. Obviously, this has major implications for both standard therapies as well as epigenetic therapies, and ITH is discussed in detail, providing insights into the mechanisms underpinning how ITH is generated via new mutations, the implications such heterogeneity has on clinical outcome, and places an emphasis on the potential role of epigenetic therapeutic interventions to improve treatment efficacy.

Heterogeneity is one issue, another emerging concern relates to the ability of cancer cells to repair DNA damage. It will come as no surprise that as DNA is compacted into chromatin, DNA damage and hence the repair of DNA damage by default involves epigenetics. As many standard chemotherapy regimens involve the use of DNA damaging agents (e.g., cisplatin), resistance to chemotherapy may therefore involve epigenetic regulation of DNA repair pathways. How a cell responds to, and deals with, genomic instability involves chromatin remodeling and highlights the critical importance of epigenetics to correct DNA repair and cell survival following DNA damage. The current clinical evidence and importance of additional epigenetic events in the development of resistance to cisplatin are discussed with descriptions on how epigenetic targeting therapies may prove to play an important role in the way cisplatin-based chemotherapy regimens may be managed in the future.

It is now clear that epigenetics plays central roles in the differentiation of embryonic stem cells [32], and a similar role for epigenetics has emerged for the differentiation of cancer stem cells (CSCs) during carcinogenesis [33]. CSCs are described as "a small subset of cancerous population responsible for tumor initiation and growth, which also possess the characteristic properties of quiescence, indefinite self-renewal, intrinsic resistance to chemo- and radiotherapy and capability to give rise to differentiated progeny" [34]. Data supporting this comes from seminal studies such as the development of resistance to gefitinib or erlotinib [35]. The role of "stemness" within the context of both epigenetics and cancer is therefore critical to future epigenetic cancer therapy.

5 FUTURE DIRECTIONS: TRANSLATION TO THE CLINIC

In the final section, the reader is introduced to how the field of epigenetic therapy may evolve in the near future, particularly how we may achieve personalized epigenetic therapy. Two emerging areas have been defined. One is the issue of chemosensitivity testing. Evidence for the power of such strategies came from predictive modeling of anticancer drug sensitivity within the Cancer Cell Line Encyclopedia, a "compilation of gene expression, chromosomal copy number, and massively parallel sequencing data from 947 human cancer cell lines" [36], and recently recapitulated using a patient *ex vivo* platform [37]. The issue of chemosensitivity testing will become critically important in the future for selecting the right drug or epigenetic therapy tailored to suit the patient. The other future direction in personalized therapy is the potential utility of epigenetic profiling of the patient for predictive or prognostic value. This may invoke the notion of epigenetic biomarkers to complement current strategies for diagnosis, prognosis, and prediction of drug response and assisting with therapeutic decision making.

The wealth of data emerging regarding both the aberrant epigenetics underpinning cancer combined with the exciting new developments with respect to therapeutically targeting cancers through inhibition of the epigenetic regulatory machinery has thrust epigenetics to the forefront of cancer research. By providing this comprehensive review of how epigenetic cancer therapy is evolving, I hope that readers of this work will identify or gain benefit in their studies for the treatment of cancer. I would also like to thank all of the contributors for the effort they have put in and for their time and patience.

REFERENCES

[1] Waddington CH. The epigenotype. Int J Epidemiol 2012;41(1):10–13.
[2] Berger SL, Kouzarides T, Shiekhattar R, Shilatifard A. An operational definition of epigenetics. Genes Dev 2009;23(7):781–3.
[3] Dawson MA, Kouzarides T. Cancer epigenetics: from mechanism to therapy. Cell 2012;150(1):12–27.
[4] Baylin SB, Jones PA. A decade of exploring the cancer epigenome—biological and translational implications. Nat Rev Cancer 2011;11(10):726–34.
[5] Kundaje A, Meuleman W, Ernst J, Bilenky M, Yen A, Heravi-Moussavi A, et al. Integrative analysis of 111 reference human epigenomes. Nature 2015;518(7539):317–30.
[6] Simo-Riudalbas L, Esteller M. Targeting the histone orthography of cancer: drugs for writers, erasers and readers. Br J Pharmacol 2015;172(11):2716–32.
[7] Wang Q, Huang J, Sun H, Liu J, Wang J, Qin Q, et al. CR Cistrome: a ChIP-Seq database for chromatin regulators and histone modification linkages in human and mouse. Nucleic Acids Res 2014;42(Database issue): D450–8.
[8] Moore LD, Le T, Fan G. DNA methylation and its basic function. Neuropsychopharmacology 2013;38(1):23–38.
[9] Srinivasan PR, Borek E. Enzymatic alteration of nucleic acid structure. Science 1964;145(3632):548–53.
[10] Kaminskas E, Farrell A, Abraham S, Baird A, Hsieh LS, Lee SL, et al. Approval summary: azacitidine for treatment of myelodysplastic syndrome subtypes. Clin Cancer Res 2005;11(10):3604–8.
[11] Issa JP, Kantarjian HM, Kirkpatrick P. Azacitidine. Nat Rev Drug Discov 2005;4(4):275–6.
[12] Gore SD, Jones C, Kirkpatrick P. Decitabine. Nat Rev Drug Discov 2006;5(11):891–2.
[13] Kroeze LI, van der Reijden BA, Jansen JH. 5-Hydroxymethylcytosine: an epigenetic mark frequently deregulated in cancer. Biochim Biophys Acta 2015;1855(2):144–54.
[14] Verdin E, Ott M. 50 years of protein acetylation: from gene regulation to epigenetics, metabolism and beyond. Nat Rev Mol Cell Biol 2015;16(4):258–64.
[15] Jenuwein T, Allis CD. Translating the histone code. Science 2001;293(5532):1074–80.
[16] Gupta RA, Shah N, Wang KC, Kim J, Horlings HM, Wong DJ, et al. Long non-coding RNA HOTAIR reprograms chromatin state to promote cancer metastasis. Nature 2010;464(7291):1071–6.
[17] Bonasio R, Shiekhattar R. Regulation of transcription by long noncoding RNAs. Annu Rev Genet 2014;48:433–55.
[18] Bierhoff H, Postepska-Igielska A, Grummt I. Noisy silence: non-coding RNA and heterochromatin formation at repetitive elements. Epigenetics 2014;9(1):53–61.

[19] Goll MG, Kirpekar F, Maggert KA, Yoder JA, Hsieh CL, Zhang X, et al. Methylation of tRNAAsp by the DNA methyltransferase homolog Dnmt2. Science 2006;311(5759):395–8.

[20] Kiani J, Grandjean V, Liebers R, Tuorto F, Ghanbarian H, Lyko F, et al. RNA-mediated epigenetic heredity requires the cytosine methyltransferase Dnmt2. PLoS Genet 2013;9(5):e1003498.

[21] Leung D, Jung I, Rajagopal N, Schmitt A, Selvaraj S, Lee AY, et al. Integrative analysis of haplotype-resolved epigenomes across human tissues. Nature 2015;518(7539):350–4.

[22] Amin V, Harris RA, Onuchic V, Jackson AR, Charnecki T, Paithankar S, et al. Epigenomic footprints across 111 reference epigenomes reveal tissue-specific epigenetic regulation of lincRNAs. Nat Commun 2015;6:6370.

[23] Elliott G, Hong C, Xing X, Zhou X, Li D, Coarfa C, et al. Intermediate DNA methylation is a conserved signature of genome regulation. Nat Commun 2015;6:6363.

[24] Gryder BE, Sodji QH, Oyelere AK. Targeted cancer therapy: giving histone deacetylase inhibitors all they need to succeed. Future Med Chem 2012;4(4):505–24.

[25] Yardley DA, Ismail-Khan RR, Melichar B, Lichinitser M, Munster PN, Klein PM, et al. Randomized phase II, double-blind, placebo-controlled study of exemestane with or without entinostat in postmenopausal women with locally recurrent or metastatic estrogen receptor-positive breast cancer progressing on treatment with a nonsteroidal aromatase inhibitor. J Clin Oncol 2013;31(17):2128–35.

[26] San-Miguel JF, Hungria VT, Yoon SS, Beksac M, Dimopoulos MA, Elghandour A, et al. Panobinostat plus bortezomib and dexamethasone versus placebo plus bortezomib and dexamethasone in patients with relapsed or relapsed and refractory multiple myeloma: a multicentre, randomised, double-blind phase 3 trial. Lancet Oncol 2014;15(11):1195–206.

[27] Dimopoulos M, Siegel DS, Lonial S, Qi J, Hajek R, Facon T, et al. Vorinostat or placebo in combination with bortezomib in patients with multiple myeloma (VANTAGE 088): a multicentre, randomised, double-blind study. Lancet Oncol 2013;14(11):1129–40.

[28] Ramalingam SS, Maitland ML, Frankel P, Argiris AE, Koczywas M, Gitlitz B, et al. Carboplatin and Paclitaxel in combination with either vorinostat or placebo for first-line therapy of advanced non-small-cell lung cancer. J Clin Oncol 2010;28(1):56–62.

[29] Easwaran H, Tsai HC, Baylin SB. Cancer epigenetics: tumor heterogeneity, plasticity of stem-like states, and drug resistance. Mol Cell 2014;54(5):716–27.

[30] Swanton C, Beck S. Epigenetic noise fuels cancer evolution. Cancer Cell 2014;26(6):775–6.

[31] Landau DA, Clement K, Ziller MJ, Boyle P, Fan J, Gu H, et al. Locally disordered methylation forms the basis of intratumor methylome variation in chronic lymphocytic leukemia. Cancer Cell 2014;26(6):813–25.

[32] Dixon JR, Jung I, Selvaraj S, Shen Y, Antosiewicz-Bourget JE, Lee AY, et al. Chromatin architecture reorganization during stem cell differentiation. Nature 2015;518(7539):331–6.

[33] Shukla S, Meeran SM. Epigenetics of cancer stem cells: pathways and therapeutics. Biochim Biophys Acta 2014;1840(12):3494–502.

[34] O'Flaherty JD, Barr M, Fennell D, Richard D, Reynolds J, O'Leary J, et al. The cancer stem-cell hypothesis: its emerging role in lung cancer biology and its relevance for future therapy. J Thorac Oncol 2012;7(12):1880–90.

[35] Sharma SV, Lee DY, Li B, Quinlan MP, Takahashi F, Maheswaran S, et al. A chromatin-mediated reversible drug-tolerant state in cancer cell subpopulations. Cell 2010;141(1):69–80.

[36] Barretina J, Caponigro G, Stransky N, Venkatesan K, Margolin AA, Kim S, et al. The Cancer Cell Line Encyclopedia enables predictive modelling of anticancer drug sensitivity. Nature 2012;483(7391):603–7.

[37] Majumder B, Baraneedharan U, Thiyagarajan S, Radhakrishnan P, Narasimhan H, Dhandapani M, et al. Predicting clinical response to anticancer drugs using an *ex vivo* platform that captures tumour heterogeneity. Nat Commun 2015;6:6169.

INTRODUCTION AND KEY CONCEPTS

DNA METHYLATION AND HYDROXYMETHYLATION IN CANCER

2

Fazila Asmar, Alexandra Søgaard, and Kirsten Grønbæk

Department of Hematology, Rigshospitalet, Copenhagen, Denmark

CHAPTER OUTLINE

S.G. Gray (Ed): Epigenetic Cancer Therapy. DOI: http://dx.doi.org/10.1016/B978-0-12-800206-3.00002-1

1 INTRODUCTION

Cancer is characterized by abnormal proliferation of cells without the presence of growth signals due to accumulation of mitotically heritable genetic and epigenetic aberrations causing deregulation of, e.g., cell cycling, DNA damage response, differentiation, and apoptosis. In the recent years, genome-wide analyses have been used to characterize the genetic landscapes of many cancers. One of the most frequent classes of genes found to be mutated is the epigenetic regulators, providing a link between genetic alterations and epigenetic changes in cancer. Mutations have been detected in the DNA methyltransferases, enzymes involved in DNA demethylation, histone modifiers such as the histone methyl- and acetyl transferases, and in the core histones. In addition, multiple studies have reported aberrant DNA and histone methylation patterns in cancer.

The recognition of the mutuality between genetic and epigenetic aberrations in cancer and the reversible features of epigenetic changes provide a rationale for combining epigenetic therapy with conventional chemotherapy. Detection and functional characterization of epigenetic marks and regulators, and correlation of these findings to the pathogenesis and clinical outcome of specific diseases, form basis for the development of novel and targeted treatment modalities. The focus of this chapter is to review aberrant DNA methylation and hydroxymethylation patterns in cancer, mutations in epigenetic regulators involved in DNA methylation maintenance fidelity, and the driving events in the process of DNA methylation in tumorigenesis. In order to understand the aberrant DNA modifications in cancer, an overview of DNA modifications in normal cells is provided as well.

2 EPIGENETICS

Traditionally, epigenetic marks have been broadly classified into three groups: direct modification of the DNA (primarily cytosine methylation and hydroxymethylation), posttranslational modifications of histone proteins, and positioning of nucleosomes along the DNA. These together make up what is referred to as the epigenome. With the advent of genome-wide studies, our understanding of the epigenome is rapidly growing. The epigenome is involved in most cellular functions, including transcription, replication, and DNA repair [1,2].

The epigenetic modifications gracefully combine forces to direct, regulate, and orchestrate cellular fate. Failure of the epigenome to function properly can result in inappropriate activation/inhibition of genes and have been shown to be initiators and drivers in cancer, right alongside and in combination with genetic aberrations [1,2].

2.1 CHROMATIN STRUCTURE

Interpretation of the concept of epigenetics requires an understanding of the chromatin structure. Chromatin is organized in repeating units of nucleosomes, each of which is a complex of 146 bp of double-stranded DNA wrapped around an octamer protein structure consisting of two subunits of each of the histone proteins H2A, H2B, H3, and H4 [3]. Histone H1 localizes to internucleosomal DNA and is also named the linker histone. Additional histone variants that can be incorporated into nucleosomes are also reported in eukaryotes, such as the H2A.Z found in nucleosomes bordering the nucleosome-depleted regions (NDRs) at the transcription start sites (TSS) of active genes [4].

Numerous nonhistone proteins such as transcription factors, polymerases, and other enzymes bind to internucleosomal DNA and NDRs. DNA in mammalian cells consists of approximately 3 billion base pairs. Due to spatial organization and gene regulatory function, nucleosomes are folded in a complex manner to eventually form a chromosome. The structure of chromatin in mammalian cells changes dynamically, enforced by transcriptional needs. It may exist in a condensed, transcriptionally silent form, called heterochromatin, or in less condensed chromatin, named euchromatin, with a "beads-on-a-string" conformation that is accessible for the transcriptional machinery. DNA modifications, posttranslational histone modifications, and nucleosome remodeling operate in a dynamic fashion to change the chromatin structure [5].

2.2 DNA METHYLATION IN CELLULAR HOMEOSTASIS

DNA methylation provides a mechanism of stable, yet reversible, gene silencing and plays an important role in regulating not only gene expression but also chromatin architecture and chromosome stability [5]. In mammalian DNA, methylation takes place at the carbon-5 position of cytosines (5-methylcytosines, 5mC) preceding guanines, so-called CpG (cytosine–phosphate–guanine) dinucleotides, where "p" refers to the phosphodiester bond in the DNA backbone. 5mC is also referred to the fifth base.

2.2.1 Genomic distribution of DNA methylation

In order to understand the role of DNA methylation in cellular function, one must take into consideration the CpG distribution within the genome. CpG sites are rare (1% of all dinucleotides), which is believed to be caused by the spontaneous deamination of 5mC into thymine [6]. Interestingly, the CpG distribution genome wide is nonrandom. Large areas of the genome are only sparsely punctuated by CpG sites, and these are in turn heavily methylated. These CpG poor oceans are interrupted by short, CpG-rich regions termed "CpG islands" [7]. These islands are defined as >0.5 kb stretches of DNA with a G + C content ≥55%, and an observed:expected frequency of at least 0.6. Based on the deamination rates of 5mC, it can be speculated how CpG islands exist. While they are most likely maintained through evolution, one explanation may be that CpG islands are rarely methylated, or only transiently methylated in the germline, hence avoiding conversion into thymine [8].

CpG islands preferentially locate to the promoter/5' region of genes and 60% of human promoters have associated CpG islands. Unmethylated islands at promoters correspond to either active transcription or a poised state, where genes can be expressed if the appropriate cellular cues are present. It should be noted that CpG methylation within the gene body is correlated to active transcription. It is believed that the body methylation inhibits spurious initiation of transcription within the gene body [9]. In contrast, methylated CpG islands in gene promoters correspond to a silenced state (mechanism described later in detail). Promoter CpG islands tend to remain unmethylated during development and in normal somatic tissues, except for a few (~6%) that become methylated in a tissue-specific manner during early development [10]. For example, developmentally important genes may be tissue-specifically methylated in the somatic, differentiated tissue. Furthermore, X-chromosome inactivation and genomic imprinting are normal methylation events at specific CpG islands coordinated during development [5]. Lastly, CpG islands covering repetitive elements assumed to have evolved from parasitic elements, are also highly methylated and help maintain chromosomal stability by inhibiting the transposition of these elements [11].

Recently, two new classes of CpG distributions have been defined. Regions with lower CpG density that lie within close proximity (~2 kb) of CpG islands, demarking areas between oceans and islands, are

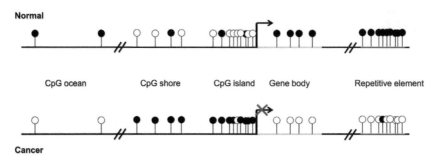

FIGURE 2.1

Methylation of CpG islands in gene promoters relates to transcriptional activity. Methylation of CpGs in the promoters of genes relates to transcriptional activity, demonstrated by the observation that several proteins bind to methylated CpGs, but not to unmethylated CpGs. One such family of proteins is the highly conserved methyl-CpG-binding domain (MBD). When the MBD proteins (MBD1, MBD2, MBD4, and MeCP2) bind to methylated CpGs, chromatin remodelers, histone deacetylases, and methylases are recruited to the methylated DNA leading to a closed chromatin structure and transcriptional repression.

termed CpG island shores (Figure 2.1) [12]. Together, these publications showed that (i) shore methylation also correlated with gene silencing, (ii) most tissue-specific methylation does not occur at islands, but at CpG island shores, and (iii) differentially methylated shore regions could sufficiently distinguish different somatic tissues. Sequences (~2 kb) that border the shores are referred to as shelves.

The final class has been termed "canyons," and are regions (≥3.5 kb) distinct from islands and shores, but contain them with their boundaries of low methylation that span conserved domains frequently containing transcription factors [13]. Their relevance will be discussed in reference to DNA methylation and cancer.

Lastly, it should be mentioned that promoter and enhancer regions with CpG-rich regions that do not meet CpG island criteria, or categorized into any of the above, also show an inverse correlation between DNA methylation status and gene expression [14] (Box 2.1).

2.2.2 Functional role of DNA methylation

The functional consequence of DNA methylation at CpG islands/shores in gene promoters/enhancers is the inhibition of gene expression, while unmethylated promoter regions are permissive for transcription. Transcriptional inhibition occurs as a consequence of numerous factors. First, the methyl group itself can sterically block the binding of transcription factors to the promoter [15]. Second, methyl-CpG-binding proteins are recruited to methylated DNA, and these in turn bind chromatin-remodeling complexes that further compact the area, making it inaccessible (Figure 2.2). In a similar fashion, unmethylated CpG sites bind different proteins, which recruit histone methyltransferases (HMTs) that mark the chromatin with active marks [15]. Recently, it has been shown that DNA methylation also affects and directs RNA splicing [16].

2.2.3 DNA methyltransferases

DNA methylation is catalyzed by a group of enzymes collectively named DNA methyltransferases (DNMTs). DNMTs covalently modify the carbon-5 position of cytosine residues, using *S*-adenosyl

---

BOX 2.1 THE DNA METHYLATION LANDSCAPE AND RELATED TERMS

CpG islands	• Short CpG-rich areas defined as >0.5 kb stretches of DNA with a G + C content ≥55%, and an observed:expected frequency of at least 0.6
	• CpG islands are typically located in the promoter region, 5' to the TSS
CpG shores	• Regions with lower CpG density that lie within the 2 kb up- and downstream of a CpG island
CpG shelves	• Shelves are defined as the 2 kb outside of a shore
CpG canyon	• Regions of low methylation more than 3.5 kb distant from islands and shores, frequently containing transcription factors
CpG ocean	• Regions with low methylation and not characterized in any of the above
Gene body	• Gene body is defined as the entire gene from transcription start site to the end of transcript
Gene desert	• Regions with very few, if any genes in a 500 kb region

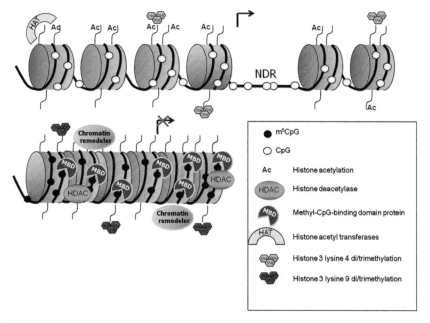

FIGURE 2.2

Distribution of CpG sites within the genome. The top panel illustrates an overview of the CpG distribution and transcriptional effects in normal cells, while the bottom panel illustrates alterations in cancer. CpG islands are typically located in the promoter region, 5' to the TSS, and are regions of high CpG density. These remain unmethylated in normal cells, but become aberrantly methylated in cancer. Bordering CpG islands are CpG shores. These have a lower CpG density and can cover promoter and/or enhancer elements. Regions of lower CpG density are named CpG oceans and comprise most of the genome. Repetitive elements and gene bodies have higher densities of CpGs and are methylated in normal cells. Unmethylated CpG sites are indicated as white lollipops, while methylated sites are denoted as black lollipops.

methionine as a methyl donor. The mechanisms are reviewed elsewhere [17], but in a simplified model, DNMTs are divided into maintenance (DNMT1) and *de novo* methyltransferases (DNMT3A and DNMT3B). During development, the *de novo* DNMTs are highly expressed and establish the methylation patterns independently of replication. Through differentiation, they are downregulated, and DNMT1 takes over and ensures that DNA methylation patterns are copied and stably inherited to daughter cells. Hence, DNMT1 is highly expressed during S phase and has a strong affinity (30–40×) toward hemi-methylated DNA. However, the distinction is not as clear cut as stated above; studies have shown that the *de novo* DNMTs are also required for maintenance methylation in human embryonic stem cells (ESCs). DNMT1 itself is not sufficient for maintaining DNA methylation, since a gradual loss of methylation occurs in subsequent cell divisions in DNMT3A and DNMT3B knockout ESCs [18]. In a revised model, it is suggested that the three enzymes cooperate to maintain DNA methylation at densely methylated regions, repetitive elements, and imprinted genes, and that the cooperativity of these three enzymes may be a mechanism to ensure that the fidelity of methylation patterns is maintained [18].

2.2.4 Recruitment of DNMTs

An intriguing question is how DNA methylation is targeted to specific sites in the genome. In plants, RNA interference is a dominant mechanism, but only few examples have been observed in humans. There are multiple theories, and these are reviewed elsewhere, but the most convincing studies show that other epigenetic factors (e.g., histone modifications) recruit the *de novo* DNMTs to specific genes, and that the underlying DNA sequence also guides DNMTs [19,20]. Conversely, CpG islands may be protected from methylation through R-loop formations coupled with GC strand asymmetry and through active histone marks in the vicinity directly blocking DNMT access to the DNA [21].

2.3 DNA DEMETHYLATION

Until recently, methylation of cytosines was thought to be an irreversible modification. Since 2009, several studies have revealed that the Ten–eleven translocation (TET) family of α-ketoglutarate (α-KG) and Fe^{2+}-dependent hydroxylases oxidize 5mC to 5-hydroxymethylcytosine (5hmC) (Figure 2.3), and that 5hmC is a normal constituent of mammalian DNA particularly abundant in ESCs and cerebellar Purkinje neurons. 5hmC is, however, less abundant than 5mC (0.1–0.3% of all bases) [22–24].

Several studies support that 5hmC has a direct role in DNA demethylation, since it is converted to 5-formylcytosine (5fC) and 5-carboxylcytosine (5caC), that is excised by thymine-DNA glycosylase (TDG) and base excision repair (BER) (Figure 2.4) [24]. The cofactor, α-KG, is converted from isocitrate by isocitrate dehydrogenases (cytosolic IDH1 and mitochondrial IDH2), both important components of the citric acid cycle.

5hmC is also passively involved in demethylation, since DNMT1 has a much lower affinity for 5hmC, and the content of methylated cytosines can thereby be diluted during DNA replication [25]. Recently, an *in vitro* study suggests that the *de novo* DNMTs DNMT3A and DNMT3B also are involved in conversion of 5hmC to unmodified cytosine [26]. However, the significance of this finding needs to be demonstrated *in vivo*.

The regulatory functions of TET1 and 5hmC in ESCs have been confirmed by several studies using affinity-based approaches. Unique genomic distribution patterns of TET1 and 5hmC were mapped to TSSs and promoters, as well as gene bodies of actively transcribed genes [27,28]. 5hmC was enriched

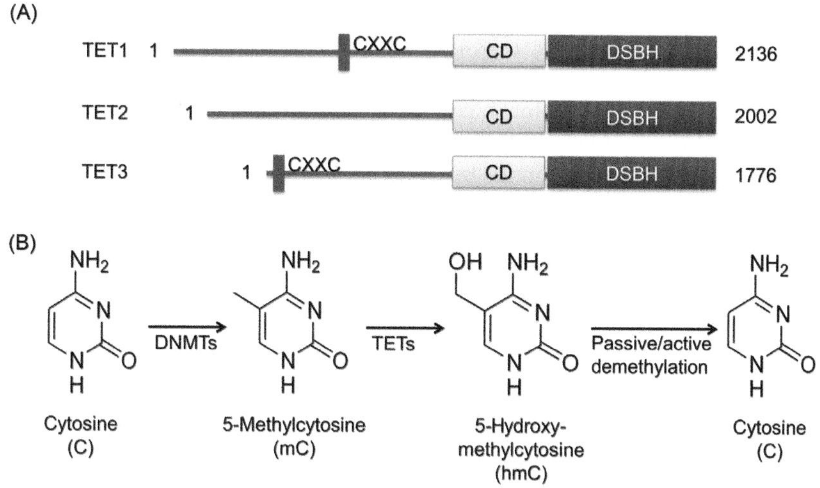

(A)

TET1   1 ———————CXXC——— CD | DSBH   2136

TET2   1 ——————————— CD | DSBH   2002

TET3   1 CXXC——————— CD | DSBH   1776

(B)

Cytosine
(C) →(DNMTs)→ 5-Methylcytosine
(mC) →(TETs)→ 5-Hydroxy-methylcytosine
(hmC) →(Passive/active demethylation)→ Cytosine
(C)

FIGURE 2.3

Schematic diagrams of the TET proteins. (A) All three TET enzymes have the highly conserved cysteine rich region (Cys) and the double-stranded β-helix (DSBH) fold of the α-KG and Fe(II)-dependent dioxygenase domain. TET1 and TET3 also contain the DNA binding CXXC zing finger. (B) All three enzymes oxidize the methyl group at the fifth carbon position of cytosines that is incorporated by the DNMTs.

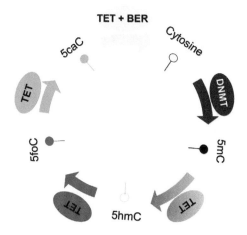

FIGURE 2.4

5mC hydroxylation promotes active DNA demethylation. TET proteins can further oxidize 5hmC to 5fC and 5caC, which can be removed by TDG and BER.

Figure modified from Shen et al., 2013 [2].

at CpG islands with low to medium GC content, at promoters with intermediate CpG density, at promoters with the bivalent histone marks, H3K4me3 and H3K27me3, and at intergenic *cis*-regulatory elements such as active enhancers and transcription factor binding sites [27,28]. Tet-assisted bisulfite and oxidative bisulfite sequencing enabling quantitative sequencing at base resolution level (described in Section 5.1) revealed the distribution of 5hmC around and not within transcription factor binding sites in ESCs [29] and at 5′ splice sites in the brain suggesting a role in the regulation of splicing [30]. Furthermore, in human brain, 5hmC was enriched at poised enhancers and was negatively correlated with the H3K27me3 and H3K9me3 enriched genomic regions. Interestingly, increased levels of hmC were observed in adult brain relative to fetal brain [30].

3 DNA METHYLATION PATTERNS IN CANCER

Cancer cells have been characterized by global hypomethylation together with focal, *de novo* promoter hypermethylation. These observations were made more than three decades ago, where initial studies showed hypomethylation of repetitive elements in both cell lines and primary tumors, which was contrasted by studies showing hypermethylation of promoter CpG islands, including those of classic tumor suppressor genes [2]. While these observations still hold true, recent epigenome-wide studies have shown that alternations in the DNA methylome of cancer cells are far more intricate. The underlying initiating mechanisms of cancer-specific methylation changes are still largely unclear, but it is apparent that they occur early in tumorigenesis and contribute to both cancer initiation and progression [31].

3.1 HYPERMETHYLATION IN CANCER

As mentioned above, cancer methylomes exhibit focal regions of hypermethylation. Early studies were focused on cancer-specific methylation of CpG islands in gene promoters causing silencing of the associated gene. The list of aberrantly methylated genes in cancer is steadily growing, including hundreds of genes affecting major cellular pathways, including cell cycle control (*p15*^{INK4B}, cyclin-dependent kinase 4 inhibitor B; *p16*^{INK4A}, cyclin-dependent kinase 4 inhibitor A; *RB*, retinoblastoma tumor suppressor 2), apoptosis (*TMS1*, target of methylation-induced silencing; *DAPK1*, death-associated protein kinase 1; *SFRP1*, secreted frizzled-related protein 1), and DNA repair (*MGMT*, O-6-methylguanine-DNA methyltransferase; *BRCA1*, breast cancer 1, early onset; *hMLH1*, human mutL homolog 1) [1]. Silencing of genes in DNA repair pathways will further propagate the carcinogenic state by allowing cells to accumulate additional genetic lesions. Also, silencing of transcription factors will indirectly silence or downregulate a large number of other genes.

Worth mentioning is also the mutagenic properties of 5mC, since the spontaneous deamination of methylated cytosine to thymine described previously, may result in point mutations if remained impaired [32]. More than 30% of all germline and almost half of all somatic *TP53* (tumor protein p53) mutations occur at methylated CpGs. Recent sequencing studies showed that 70% of tier 1 mutations in acute myeloid leukemia (AML) comprised 5mC-T transitions [33]. Furthermore, 5mC may also be involved in the mutagenic effect of exogenous carcinogens from cigarette smoke by promoting formation of adducts on the subsequent guanine and thus the conversion of guanine to thymine [32].

Genome-wide studies of methylation and corresponding gene expression patterns in tumors have brought up certain discrepancies between methylation status and gene expression. Interestingly, recent studies show that hypermethylation events at CpG island shores and CpG-rich distal regions (e.g., enhancers) may also be cancer-specific alterations, which in some studies correlate more closely with gene expression [12,34,35]. Moreover, a comparison between gene expression and DNA methylation levels in chronic lymphocytic leukemia (CLL) cases showed a significant positive correlation between differentially methylated CpGs only within gene bodies and gene expression [36]. Aberrant DNA methylation within gene bodies may have an indirect effect on gene expression or isoform expression by altering RNA splicing [16]. Although most hypermethylated loci in cancer are affected on an individual basis, entire methylated subchromosomal domains have also recently been reported [37]. It is apparent that methylation patterns are tumor specific and can be used as biomarkers to stratify tumors into subtypes according to their distinct methylation profiles.

Initially, the causal relevance for epigenetic alterations in cancer was questioned, and it was suggested that these were merely passengers and not drivers of carcinogenesis. This notion has however been disproven based on a number of observations. First, hypermethylation of tumor suppressors serves as an alternative mechanism to mutation in Knudson's two hit hypothesis [32]. As seen for both *BRCA1* in breast cancer and *CDKN2A* (cyclin-dependent kinase inhibitor 2A) in lung cancer, methylation of the promoter is mutually exclusive to any mutational or structural inactivation events [38,39]. Second, Carvalho et al. showed that DNA methylation is a driver of tumorigenesis [40]. These data suggest that cancer cells become addicted to epigenetic alterations and these are essential for cancer cell survival. Moreover, tumors show hypermethylation as an early event in carcinogenesis, which is supported by the finding of the so-called "field effect" where adjacent normal tissues also harbor altered DNA methylation patterns [41].

Issa et al. demonstrated that there was a distinct subset of colorectal cancers (CRCs) with extensive hypermethylation of a subset of CpG islands that remained unmethylated in other colorectal tumors, a phenomenon termed "CpG island methylator phenotype" (CIMP) [42]. The efforts of the cancer genome atlas (TCGA) network have identified CIMPs in breast, endometrial, glioblastomas, and AMLs, but not in serous ovarian, squamous lung or renal kidney tumors [43]. Finally, regions subjected to cancer-associated DNA methylation changes comprise short interspersed or clustered regions as well as long blocks in so-called long-range epigenetic silencing [37]. Such phenotypes highly indicate that methylation events in cancer are not random and occur through coordinated mechanisms.

The question remains, why some regions become methylated and others do not? It is commonly accepted that genetic changes in cancer occur randomly and are maintained through selection. A similar model has been proposed for epigenetic alterations; hypermethylation events are random, stochastic events that then are selected for because they are advantageous for cell survival. This would explain how cancers have been stratified in subtypes according to their distinct methylation profiles. While this is still believed to hold true, a number of recent studies have added to the complexity of cancer methylome establishment.

An initial study indicated that *de novo* methylation in cancer, as during development, may partially be determined by an instructive mechanism that recognizes specifically marked regions in the genome [44]. This group utilized DNA methyl-specific antibodies coupled with microarray analysis to investigate the genome-wide *de novo* methylation found in colon and prostate cancer cells. The authors found that only ~15% of the genes methylated in cancer samples were actively transcribed in normal tissue and that these were already inactivated by methylation in precancerous tissue. Collectively, these observations led the authors to suggest that much of the *de novo* methylation observed in cancer is not necessarily the result of growth selection, but may instead occur in an instructive manner. The following year, three groups reported that genes that become aberrantly methylated in cancer are Polycomb group targets in the ESC [45–47]. Their data suggested that cancer cells target *de novo* methylation by taking advantage of a pre-existing epigenetic repression program, namely the PRC2 (polycomb repressive complex 2)-mediated H3K27me3 mark, i.e., genes that are already silenced are consistently targeted for cancer-specific methylation. This would explain why genes that do not necessarily confer a growth advantage become methylated. Notably, many CIMP loci are known polycomb targets [47]. Thus, the cancer cell epigenome is in part determined by cell of origin, as well as passenger events at genes that are not required for that particular cancer [48]. Finally, in some instances, fusion proteins can misdirect DNMTs to genes, thereby causing their silencing [49].

3.2 HYPOMETHYLATION IN CANCER

In addition to regions of hypermethylation, cancer cells display marked losses of DNA methylation genome wide (20–60% less 5mC). This hypomethylation occurs at multiple genomic sites including CpG islands at repetitive regions and transposable elements, CpG poor promoters, CpG island shores, introns and in gene deserts (typically same area as CpG oceans) [12]. The consequence of DNA hypomethylation at repetitive regions is genomic instability that in turn promotes chromosomal rearrangements and copy number changes. Demethylation of transposable elements also increases genomic instability, and their transposition can in turn inactivate other genes [11]. Although rare, hypomethylation occurring at promoters of known oncogenes results in their expression and further exacerbation of the carcinogenic state [5]. Hypomethylation events can also cause loss of imprinting (LOI). The most common example is *IGF2* (insulin-like growth factor 2), where LOI at the paternal allele has been reported in a large number of cancers including breast, liver, lung, and colon cancer [50]. Finally, demethylation events at enhancers may affect transcriptional rate, while demethylation of gene bodies may affect RNA splicing [51].

3.3 METHODS FOR 5MC DETECTION

A plethora of methods for DNA methylation analysis have evolved over the past decade ranging from analyses restricted to specific loci to genome-wide approaches and at single base pair resolution. Since there is such a variety of advanced DNA methylation profiling techniques that are reviewed in detail elsewhere [52], it is within the scope of this chapter to provide a general overview of the different approaches.

DNA methylation information is not preserved during PCR (polymerase chain reaction) amplification, cloning in bacteria, or revealed by hybridization. Thus, most DNA methylation analysis techniques necessitate methylation-dependent treatment of DNA prior to amplification or hybridization. Three major DNA modifying methods are in use: (i) bisulfite conversion, (ii) restriction enzyme digestion, and (iii) affinity enrichment. These DNA treatment methods are followed by various read-out analytical techniques, such as PCR amplification with methylation independent primer (MIP) or methylation-specific primer (MSP), probe hybridization, and sequencing. Analysis of DNA methylation is further complicated by the uneven distribution of methylation across the genome, and importantly, none of the methods apart from the affinity-based methods described in the current section distinguish mC from hmCs [52].

- (Ad 1) Bisulfite treatment changes cytosine (C) to uracil (U), but the conversion of 5mC/hmC to U is much slower than the conversion of nonmethylated C to U. This is exploited in various time restricted assays, in which 5mC/hmC remain unconverted and are read as C, while nonmethylated Cs are converted to Us. Bisulfite sequencing has been viewed as the golden standard for DNA methylation analysis, this is however being challenged by the discovery of hmC, since mC/hmC are indistinguishable. In addition to the reduced complexity of the bisulfite-converted DNA, the development of primers and technology has been somewhat challenging.

For locus-specific DNA methylation detection, most bisulfite-based methods use MIP or MSP PCR primers. Genome-wide DNA methylation detection of bisulfite-converted DNA is done by array hybridization or sequencing. The latest version of Illumina bead arrays allows interrogation of > 485,000 CpG sites at single base resolution. By next-generation sequencing (NGS), a higher CpG coverage and information of allele-specific methylation are obtained, however, the bioinformatic analyses are challenging. By reduced representation bisulfite sequencing (RRBS), the genomic DNA is digested by methylation insensitive restriction enzyme, *Msp*I (an isoschizomer of *Hpa*II which cleaves both meth/unmeth *Hpa*II sites), and will therefore mainly cover methylation in CpG-rich areas such as CpG islands.

- (Ad 2) Methylation of DNA protects against endonuclease digestion, so the patterns of restriction by such enzymes can provide a read-out of DNA methylation. Commonly used restriction enzymes are the *Hpa*II (a restriction endonuclease that cleaves at the sequence C^CG_G) and *Msp*I. CpG methylation prevents *Hpa*II, but not *Msp*I, from DNA cleavage. Methylation-sensitive endonuclease restriction followed by PCR across the restriction site is a very sensitive technique, however, false-positive results caused by incomplete digestion may be observed. The endonuclease restriction method has been coupled to array-based analysis, such as methylated CpG island amplification with microarray hybridization (MCAM), differential methylation hybridization (DMH), or *Hpa*II tiny fragment enrichment by ligation-mediated PCR (HELP, *Hpa*II tiny fragment enrichment by ligation-mediated PCR). Recently, the endonuclease restriction method has been coupled to NGS techniques, which enable allele-specific DNA methylation analysis, cover more of the genome, and avoid hybridization biases. These methods include Methyl-, HELP-, and MCA-seq, amongst others.

- (Ad 3) Genome-wide assessment of DNA methylation by affinity enrichment of methylated regions uses antibodies with specificity for 5 mC or methyl-CpG-binding proteins coupled with array hybridization (MeDIP-ChIP) or NGS (MeDIP-seq). These assays have been widely used, however, the resolution is confined to the size of the sonicated immunoprecipitated DNA fragments, and the methods are laborious and need bioinformatic adjustment for the uneven distribution of CpG sites in the genome.

4 ABERRATIONS OF ENZYMES INVOLVED IN DNA METHYLATION HOMEOSTASIS IN CANCER

The underlying mechanisms that direct DNA methylation to specific gene promoters in cancer are largely unknown. They may involve failed fidelity of DNA maintenance methylation caused by aberrant expression or mutations of the enzymes involved in the homeostasis of CpG methylation, which will be elaborated in the following sections.

4.1 DNA METHYLTRANSFERASE

Overexpression of the DNMTs has been correlated with unfavorable prognostic outcome in several cancers. For example, in diffuse large B-cell lymphoma (DLBCL) overexpression of DNMT3B evaluated by immunohistochemistry (IHC) was significantly correlated to advanced clinical stage, overall and progression-free survival, and promoter hypermethylation of specific genes was observed [53]. Whether the overexpression of DNMT3B is specific to DLBCL or a consequence of proliferation is not clarified. Another study showed that EZH2 (enhancer of Zeste homolog 2) may directly recruit DNMTs to promoters [54]. However, a direct interaction between EZH2 and DNMT has not been consistently substantiated.

Based on tumors showing overexpression of DNMTs, especially *de novo* DNMTs, it was suggested that DNMTs function as oncogenes by causing aberrant hypermethylation of tumor suppressor genes. However, inactivating mutations in DNMT3A are found to correlate with poor prognosis in myeloid malignancies, and deletion of DNMT3A promotes tumor progression in a lung cancer mouse model, indicating its tumor suppressor function [55]. This correlates well with the recent finding that DNMT3A is essential for hematopoietic stem cell differentiation, where deletion of the gene caused both hyper- and hypomethylation events at promoters [56]. Interestingly, methylation differences between DNMT3A wild-type and mutant AML patient samples are limited [57,58]. Of the 182 genomic regions that were affected, no correlation with local changes in gene expression was found. Another group reported that in mice, DNMT3A-deficient tumors had altered, mainly loss of DNA methylation within gene bodies [59]. Jeong et al. showed that methylation at the edges of CpG canyons diminished in the absence of DNMT3A-null mice hematopoietic stem cells, and genes that are typically dysregulated in human leukemias are enriched for canyon-associated genes [13]. These findings suggest that DNMT3A may maintain methylation at the boundaries of CpG canyons.

Finally, microRNAs of the miR-29 family have been shown to be involved in the regulation of DNA methylation by targeting the DNMTs, DNMT3A/3B and DNMT1 [60].

It is thus evident that DNMT3A deregulation is important in hematopoietic malignancies, and future studies are likely to uncover the mechanisms involved in DNMT-mediated tumor progression.

Drugs interfering with DNMT activity are already approved for clinical use, and several new drugs are currently in preclinical and clinical trials. The only class currently used routinely in the clinics are the nucleoside analogs, 5-aza-2'-deoxycytidine (5-Aza-CdR, decitabine) and 5-azacytidine (5-Aza-CR, azacytidine). Both drugs are initially phosphorylated by intracellular kinases [61]. 5-Aza-CR is incorporated preferentially into RNA, however, approximately 20% is converted by ribonucleotide reductase, and the phosphorylated forms are incorporated into DNA during replication. When incorporated into DNA, the drugs form a covalent bond with the DNMT, thereby trapping the enzyme and preventing it from further methyltransferase activity. This results in a passive demethylation of DNA in the subsequent cell cycles [61].

Azacytidine and decitabine are approved by the FDA (US Food and Drug Administration) for treatment of myelodysplastic syndrome (MDS), chronic myelomonocytic leukemia (CMML) with 10–29% blasts in bone marrow, and AML with 20–30% blasts. Whereas, EMA (European Medicines Agency) has approved the use of 5-Aza-CR for treatment of higher risk MDS, CMML with 10–29% blasts without myeloproliferative disorder, and AML with 20–30% blasts and multilineage dysplasia. Furthermore, Aza-CdR is approved by EMA for treatment of AML patients above 65 years who are not candidates for standard induction therapy.

4.2 TET PROTEINS

The TET1 and TET2 proteins have both been implicated in cancer. *TET1* was first identified as an MLL translocation partner in rare cases of AML and acute lymphoid leukemia (ALL) carrying *t*(10:11) (q22;q23). *TET2* is located on chromosome 4q24, a region that is commonly deleted or involved in chromosomal rearrangements in myeloid malignancies. *TET2* point mutations and deletions are frequent in a variety of myeloid cancers (10–25%), such as myeloproliferative neoplasms, systemic mastocytosis, CMML, MDS, and AML [62]. The highest frequency is reported in CMML (35–50%). Early studies have shown that tumors in patients with concomitant myeloid and lymphoid malignancies have deletion of chromosome 4q24 [63], and recent studies show that *TET2* is also involved in lymphomagenesis [64–66]. *TET2* mutations are frequently detected in T-cell lymphoma, with the highest frequency observed in peripheral T-cell lymphoma and angioimmunoblastic T-cell lymphoma (AITL) (38% and 47%, respectively) [65]. The detection rate of *TET2* mutations in DLBCL varies in the published studies from 2% to 12%. Recently, targeted sequencing of Mantle cell lymphoma revealed *TET2* mutations with a frequency of 4%. The observations that lymphoma-associated *TET2* mutations are found in common hematopoietic progenitors of the same patients [63], and that *Tet2*-deficient mice in addition to myeloid expansion develop increased proliferation of lymphoid cells [64], suggest that *TET2* mutations may be early events in lymphomagenesis.

The effect of *TET2* mutations on DNA methylation patterns has been investigated in CMML, AML, and DLBCL with somewhat divergent findings. In CMML, *TET2* mutations were associated with global hypermethylation as measured by the LINE-1 (long interspersed element 1) assay; however, DNA methylation levels of 10 promoter CpG islands that are frequently hypermethylated in myeloid leukemia did not differ in *TET2* mutated versus wild-type samples [67]. In AML, Figueroa et al. found correlation of *TET2* mutations with a hypermethylation signature that overlapped considerably with the methylation signature observed in *IDH1/IDH2* mutant AML [68]. Ko et al. showed a strong correlation between *TET2* mutations and low 5hmC levels (details given in Section 5.2) and reported 2510 differentially hypomethylated regions in "5hmC low AML samples" (22 *TET2* mutated, 7 *TET2* wild-type) in comparison with "5hmC high AML samples" (22 *TET2* wild-type, 2 *TET2* mutated) [69].

In DLBCL, distinct *TET2* mutations were associated with a hypermethylation signature including genes involved in hematopoietic development and cancer [66]. Notably, 53.4% of the "*TET2* methylation signature genes" carried the bivalent H3K4me3/H3K27me3 silencing mark in human ESCs. The significant overlap between polycomb target genes and DNA promoter hypermethylated genes questions whether TET2-directed DNA methylation occurs at genes that are already silenced by polycomb. *TET2* mutations have mainly been reported in hematological malignancies. In solid tumors, somatic alterations of *TET2* have so far only been reported in metastatic tissue of castration-resistant prostate cancer [70]. Furthermore, an increasing number of studies find downregulation of the TET proteins in a variety of solid tumors (specified in Section 5.2).

4.3 ISOCITRATE DEHYDROGENASES

IDH1 missense mutation at R132 and *IDH2* missense mutation at R172 and R140 are reported in diffuse and anaplastic gliomas, secondary glioblastomas (60–90%), AML (15–30%), AITLs (20–45%), chondrosarcoma (50%), cholangiocarcinoma (15%), and melanoma (10%) [71]. *IDH1* mutations at R132 have been detected in metastatic lesions of melanoma, whereas the *IDH2* mutations (P158T and G171D) have been detected in primary melanomas [72]. Additional somatic *IDH1* variants (G70D, G123R, I130M, H133Q, A134D, V71I, and V178I) variants have been detected in thyroid carcinomas (18%) [73].

The *IDH1* missense mutations at R132 and *IDH2* missense mutations at R172 or R140 reside at the enzymes' catalytic domains and alter the activity of the enzymes so that α-KG is reduced to D-2-hydroxygluterate (2-HG) that is a competitive inhibitor of α-KG (Figure 2.5) [71]. Enzymes that

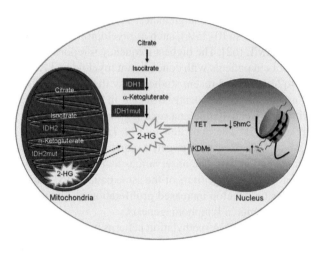

FIGURE 2.5

IDH1/2 mutations result in synthesis of 2-HG, a competitive inhibitor of α-ketoglutarate. α-KG is converted from isocitrate by IDH1 and IDH2, both important components of the citric acid cycle. *IDH1/2* missense mutations alter the activity of the enzymes so that α-KG is reduced to 2-HG resulting in impairment of enzymes that require α-KG as a cofactor, such as the TET proteins and the JMJC family of histone demethylases (KDMs).

require α-KG as a cofactor, such as the TET proteins and the Jumonji family (JMJC family) of histone demethylases, may thus have impairment of their enzymatic activity. Consistently, a significant increase in promoter hypermethylation has been observed in AMLs [68] and gliomas [74] carrying *IDH1/2* mutations. Interestingly, *IDH1/2* mutations were present in 98% of glioma-CIMP positive tumors and associated with a distinct transcriptome, while none of the glioma-CIMP negative tumors had *IDH1/2* mutations. Furthermore, immortalized primary human astrocytes expressing mutant IDH1 (R132H) resulted in production of 2-HG and hypermethylation of a substantial number of genes [74].

Since α-KG is also a cofactor for the JMJC family, *IDH* mutations are also associated with impairment of histone demethylation, resulting in differentiation blockage due to repression of lineage-specific differentiation genes [68,75]. Conditional knock-in mice engineered to express the IDH1 (R132H) mutation in the myeloid compartment showed increased numbers of hematopoietic progenitors, developed splenomegaly, anemia, and extramedullary hematopoiesis. The cells of myeloid lineage had changes in DNA methylation comparable to those observed in human *IDH1/2* mutant AML as well as hypermethylated histones [76].

Treatment of cancers carrying *IDH* mutations could thus potentially benefit from differentiation cancer therapy. Interestingly, treatment with a small molecule, AGI-6780, that selectively inhibits the tumor-associated mutant IDH2/R140Q led to differentiation of an erythroleukemia cell line, TF-1, and primary human AML cells *in vitro* [77]. Other small molecule selective inhibitors of the mutant IDH1/R132H, such as AGI-5198 and ML309, lower the 2-HG levels, delay growth, and promote differentiation of glioma cells in an *in vivo* mouse tumor graft model [78,79]. In addition, it was recently shown that inhibition of the bromodomain-containing protein Brd4 led to differentiation and apoptosis of *IDH2* mutant AML [80].

Finally, the observations that *TET2* and *IDH* mutations are present in both lymphoid and myeloid malignancies suggest that these epigenetic defects may at least in some instances arise in hematopoietic progenitor cells. It can be speculated that these mutations may prime the malignant clone by accumulation of methylation errors that affect the programming of genes involved in hematopoietic differentiation. Most likely, secondary genetic or epigenetic events direct the clone toward the lymphoid or the myeloid compartment, respectively.

This is not only interesting from a biological viewpoint, but may revise the existing theory of how lymphoma develop, and could potentially have significance for future clinical and therapeutic decision making.

4.4 SUCCINATE DEHYDROGENASES

Gastrointestinal stromal tumors (GIST) without *KIT* (v-kit Hardy-Zuckerman 4 feline sarcoma viral oncogene homolog) mutations frequently harbor mutations of succinate dehydrogenase (SDH) subunit genes *SDHA*, *SDHB*, *SDHC*, or *SDHD* [81]. SDH catalyzes the conversion of succinate to fumarate in the citric acid cycle downstream of the IDH2. The subsequent accumulation of succinate in SDH-deficient cells also leads to inhibition of α-KG-dependent dioxygenases, such as the TET proteins and the JMJDs [81]. Interestingly, two independent studies have shown a correlation between SDH mutations in GIST, a hypermethylation signature, and decreased levels of hydroxymethylation [82,83]. *SDHB*-deficient mice had increased 5mC, decreased 5hmC levels, and increased histone methylation in neuroendocrine chromaffin cells compared to *SDHB* wild-type mice. Interestingly, an enhanced migratory capacity of chromaffin cells with *SDHB* loss was observed *in vitro*, which was reversible by treatment with DNMT inhibitor. Germline *SDH* mutations have also been reported in pheochromocytoma–paraganglioma syndrome as well as in sporadic cases [84].

The discoveries of the mutations in the metabolic enzymes, IDH and SDH, and the effect of these mutations on the epigenome, strongly support a connection between metabolic disruption and epigenetic aberrations in cancer. This opens new opportunities for treatment of cancers carrying mutations in the enzymes of the citric acid cycle by targeting DNA and histone methylation.

5 DNA HYDROXYMETHYLATION IN CANCER

The study of DNA hydroxymethylation in cancer and normal cells has been rather challenging, since DNA modifications are not preserved during PCR amplification and the standard methods used in the analysis of DNA methylation such as bisulfite- or restriction enzyme-based techniques are incapable of discriminating 5hmC and 5mC [85,86]. Given the very low 5hmC content in mammalian DNA (as low as 0.009% of cytosine in 293T cells) [24], highly sensitive detection methods are required for quantitative evaluation of 5hmC content. There are a steadily increasing number of methods available for specific detection of 5hmC either at single base resolution or genome wide. Since these methods have been developed recently and taken in use for evaluation of 5hmC in cancer, a brief review of the methods will be given followed by an overview of the methods for 5mC detection as well.

5.1 METHODS FOR 5HMC DETECTION

Affinity-based enrichment techniques such as 5hmC DNA immunoprecipitation (hMeDIP), single molecule real-time sequencing (SMRT), and chemical methods such as mass spectrometry can specifically detect 5hmC without preceding manipulations of the DNA, while other methods rely on manipulation of DNA by glucosylation or oxidation in combination with bisulfite conversion [87]. Intriguingly, a recent study describes a new method using an innovative optical detection approach for investigation of 5hmC levels in DNA [88]. Some of these methods are reviewed in the separate sections below.

5.1.1 Affinity-based enrichment approach

The anti-5hmC has been commercially available in the recent years. An obstacle with regard to hMeDIP is that the anti-5hmC antibody reacts only with DNA containing high densities of 5hmC. This has been overcome by an initial bisulfite treatment of the DNA resulting in conversion of 5hmC to cytosine 5-methylenesulfonate (CMS) and use of an anti-CMS antibody instead [69]. The activity of the anti-CMS antibody does not rely on the 5hmC density. Additionally, affinity-based enrichment can be combined with selective chemical labeling of 5hmC such as glucosylation, periodate oxidation, or biotinylation followed by deep sequencing.

5.1.2 Chemical methods

Chemical methods using liquid chromatography–mass spectrometry (LC–MS) can specifically and quantitatively detect 5hmC simultaneously with 5mC. For instance, reversed phase liquid chromatography (RPLC) with MS has been used for quantification of 5hmC in lung squamous cell cancers, adenocarcinomas, and astrocytomas [89]. However, there are technical weaknesses related to RPLC which diminish the detection capability, and another technique corresponding to RPLC, the hydrophilic interaction liquid chromatography (HILIC), with better resolution of polar compounds, has been developed [90].

5.1.3 Quantitative methods at single base resolution

Methods that require initial experimental modifications of DNA, such as glucosylation, oxidative or Tet-assisted bisulfite sequencing, followed by PCR, are described below.

The first method takes advantage of restriction enzymes' incapability of cleaving glucosylated 5hmC. Glucosylation of 5hmC is mediated by treatment of DNA with glucosyl transferase. The DNA is then subjected to restriction enzymes and PCR. Glucosylated 5hmC protects from cleavage by restriction enzymes such as *MspI*, which otherwise cleaves the DNA sequence, CCGG, when modified either by methylation or hydroxymethylation [91].

By oxidative bisulfite sequencing, 5hmC is converted to 5fC upon treatment with potassium perruthenate ($KRuO_4$). Given that 5fC undergoes deamination to uracil by bisulfite treatment, quantitative measurement of 5hmC is enabled by comparing bisulfite-treated DNA with bisulfite-treated DNA that is pretreated by $KRuO_4$ [87].

Lastly, the Tet-assisted bisulfite sequencing method is based on a chemical modification of the DNA that allows specific recognition of 5hmC, combined with conversion of 5mC to uracil during bisulfite treatment. This is performed by glucosylation of 5hmC followed by treatment of this DNA with recombinant Tet dioxygenase leading to conversion of 5mC to 5caC. As a result, only 5hmC will remain unconverted by the subsequent bisulfite treatment [29].

These methods are relatively feasible, although limitations may be incomplete enzymatic reaction, incomplete bisulfite conversion, and that 5mC cannot be measured simultaneously.

5.1.4 Third-generation sequencing

The measurement of 5hmC and 5mC by SMRT (Pacific Blues) is based on the specific polymerase kinetics signature of each base modification in a strand-specific manner and at base resolution. The technology is now accessible; however, it still needs improvement [92]. Other upcoming third-generation sequencing approaches are reviewed elsewhere [52].

5.2 DNA HYDROXYMETHYLATION PATTERNS IN CANCER

The first study to investigate 5hmC levels in cancer was performed in AML by Ko et al. in order to investigate the functional consequence of *TET2* mutations. The authors found relatively lower levels of 5hmC in bone marrow samples from patients with *TET2* mutations comparing to *TET2* wild type. This was assessed by IHC and dot blot assays as well as quantitative affinity enrichment of the 5hmC bisulfite-converted product, CMS [69].

Recent studies have shown global loss of 5hmC in a variety of human solid tumors (breast, colon, gastric, liver, lung, melanoma, and prostate cancer) compared with the normal surrounding matched tissue demonstrated by IHC and dot blot assays [93,94]. The decrease of 5hmC is associated with downregulation of the TETs, e.g., a significant reduction of 5hmC was found in 72.7% of CRCs and 75% of gastric cancers compared to the normal counterpart tissues, and the reduction of 5hmC was correlated to downregulation of TET1. In melanoma, reduction of 5hmC was correlated to downregulation of TET2 and IDH2. In breast and liver cancers, 5hmC and the expression levels of all three TETs were significantly reduced compared to matched normal tissues [94]. In most immortalized tumor cell cultures, 5hmC levels are reduced, and in *in vitro* experiments oncogene-induced cellular transformation was linked to downregulation of TET1 [93–95]. The underlying mechanisms for low expression levels of the TETs are not further characterized; however, one mechanism may be translational inhibition by miR-22 as reported in MDS [95].

Depletion of 5hmC in a variety of cancers has also been confirmed by the more specific and quantitative mass spectrometry methods. In hepatocellular carcinoma (HCC) quantification of both 5mC and 5hmC in DNA was performed by capillary hydrophilic interaction liquid chromatography and mass spectrometry. A four- to fivefold lower 5hmC content was found in HCC compared to normal tumor-adjacent tissues and a significant correlation between 5hmC levels and tumor stage was seen [90]. A sixfold reduction was observed in CRC using an isotype-based LC–MS method [96]. In lung squamous cell cancers, two- to fivefold lower 5hmC compared to normal-matched tissue was detected by RPLC–MS. In astrocytomas, a strong depletion of 5hmC was observed; in some tumors a reduction of more than 30 fold was detected compared to normal brain tissue. There was no correlation between the levels of 5mC and 5hmC or with tumor stage or patient survival. No difference in 5hmC levels was detected in *IDH1* mutated versus *IDH1* wild-type gliomas [89]. The latter finding is in conflict with the observation from another study, which reported significant downregulation of 5hmC levels in *IDH1*-mutant glioblastomas as assessed by IHC [97]. This discrepancy is most likely attributed to the different methods used. IHC is semiquantitative and caution must therefore be taken when assessing 5mC or 5hmC by IHC. The levels of TET proteins have not been investigated in the aforementioned studies.

One study that used selective glucosylation of 5hmC followed by enzymatic digestion showed that decreased 5hmC levels of the *LZTS1* (leucine zipper, putative tumor suppressor 1) gene promoter correlated to low mRNA expression of LZTS1 in breast cancer [98].

Interestingly, a recently published report suggests that the loss of 5hmC is replication dependent in mouse preimplantation embryos [99]. Since low 5hmC levels are also reported in liver adenomas as compared with normal liver tissue [94], one may speculate whether the loss of 5hmC recently documented in several cancers results from a replication-dependent passive process.

Finally, a recent study has reported that all cytosine derivatives are able to induce CT transition mutation in *Escherichia coli* cells, and thus may have potential mutagenic properties if not repaired. The cytosine derivatives did not influence DNA replication [100]. These findings, however, need to be validated in mammalian cells.

6 CONCLUSION

While DNA methylation has been implicated in numerous biological processes and aberrant DNA methylation patterns are considered as a hallmark of cancer, the biological significance of 5hmC in cancer remains elusive. DNA hydroxymethylation is an intermediate in the demethylation of cytosine, but has a proposed function in transcriptional regulation as well. Global as well as locus-specific loss of DNA hydroxymethylation has been reported in several cancers in comparison with normal counterpart tissue. The depletion of DNA hydroxymethylation in cancer could be a potential biomarker for early detection and prognosis of distinct cancers.

Furthermore, emerging evidence based on high throughput and genome-wide analysis of DNA methylation points toward that DNA hyper- and hypomethylation is not that tightly limited to specific genomic boundaries as previously recognized, but found throughout the genome. Hypermethylation events at CpG island shores and CpG-rich distal regions (e.g., enhancers) may also be cancer-specific alterations, which in some studies correlate more closely with gene expression. In addition, the underlying mechanisms that direct DNA methylation to specific gene promoters in cancer are largely unknown. They may involve failed fidelity of DNA maintenance methylation caused by aberrant expression or mutations of the enzymes involved in the homeostasis of CpG methylation.

An obstacle in studies of DNA modifications is the requirement of experimental treatment of the DNA in order to preserve the information on covalent DNA modifications during enrichment. Future studies using third-generation sequencing techniques will allow detection of DNA modifications without manipulation and may thus contribute further to the genomic mapping of DNA modifications in cancer.

REFERENCES

[1] You JS, Jones PA. Cancer genetics and epigenetics: two sides of the same coin? Cancer Cell 2012;22(1):9–20.
[2] Shen H, Laird PW. Interplay between the cancer genome and epigenome. Cell 2013;153(1):38–55.
[3] Luger K, Mäder AW, Richmond RK, Sargent DF, Richmond TJ. Crystal structure of the nucleosome core particle at 2.8A resolution. Nature 1997;389:251–60.
[4] Kelly TK, Miranda TB, Liang G, Berman BP, Lin JC, Tanay A, et al. H2A.Z maintenance during mitosis reveals nucleosome shifting on mitotically silenced genes. Mol Cell 2010;39(6):901–11.
[5] Sharma S, Kelly TK, Jones PA. Epigenetics in cancer. Carcinogenesis 2010;31(1):27–36.
[6] Jones PA. Functions of DNA methylation: islands, start sites, gene bodies and beyond. Nat Rev Genet 2012;13(7):484–92.
[7] Bird A, Taggart M, Frommer M, Miller OJ, Macleod D. A fraction of the mouse genome that is derived from islands of nonmethylated, CpG-rich DNA. Cell 1985;40(1):91–9.
[8] Smallwood SA, Tomizawa S-I, Krueger F, Ruf N, Carli N, Segonds-Pichon A, et al. Dynamic CpG island methylation landscape in oocytes and preimplantation embryos. Nat Genet 2011;43(8):811–14.

[9] Zilberman D, Gehring M, Tran RK, Ballinger T, Henikoff S. Genome-wide analysis of *Arabidopsis thaliana* DNA methylation uncovers an interdependence between methylation and transcription. Nat Genet 2007;39(1):61–9.

[10] Straussman R, Nejman D, Roberts D, Steinfeld I, Blum B, Benvenisty N, et al. Developmental programming of CpG island methylation profiles in the human genome. Nat Struct Mol Biol 2009;16(5):564–71.

[11] Suzuki MM, Bird A. DNA methylation landscapes: provocative insights from epigenomics. Nat Rev Genet 2008;9(6):465–76.

[12] Doi A, Park I-H, Wen B, Murakami P, Aryee MJ, Irizarry R, et al. Differential methylation of tissue- and cancer-specific CpG island shores distinguishes human induced pluripotent stem cells, embryonic stem cells and fibroblasts. Nat Genet 2009;41(12):1350–3.

[13] Jeong M, Sun D, Luo M, Huang Y, Challen GA, Rodriguez B, et al. Large conserved domains of low DNA methylation maintained by Dnmt3a. Nat Genet 2014;46(1):17–23.

[14] Han H, Cortez CC, Yang X, Nichols PW, Jones PA, Liang G. DNA methylation directly silences genes with non-CpG island promoters and establishes a nucleosome occupied promoter. Hum Mol Genet 2011:1–12.

[15] Deaton AM, Bird A. CpG islands and the regulation of transcription. Genes Dev 2011;25:1010–22.

[16] Gelfman S, Cohen N, Yearim A, Ast G. DNA-methylation effect on cotranscriptional splicing is dependent on GC architecture of the exon-intron structure. Genome Res 2013;23(5):789–99.

[17] Jones PA, Liang G. Rethinking how DNA methylation patterns are maintained. Nat Rev Genet 2009;10(11):805–11.

[18] Liang G, Chan MF, Tomigahara Y, Tsai YC, Gonzales FA, Li E, et al. Cooperativity between DNA methyltransferases in the maintenance methylation of repetitive elements. Mol Cell Biol 2002;22(2):480–91.

[19] Cedar H, Bergman Y. Linking DNA methylation and histone modification: patterns and paradigms. Nat Rev Genet 2009;10(5):295–304.

[20] Lienert F, Wirbelauer C, Som I, Dean A, Mohn F, Schübeler D. Identification of genetic elements that autonomously determine DNA methylation states. Nat Genet 2011;43(11):1091–7.

[21] Ginno PA, Lott PL, Christensen HC, Korf I, Chédin F. R-loop formation is a distinctive characteristic of unmethylated human CpG island promoters. Mol Cell 2012;45(6):814–25.

[22] Kriaucionis S, Heintz N. The nuclear DNA base 5-hydroxymethylcytosine is present in Purkinje neurons and the brain. Science 2009;324(5929):929–30.

[23] Tahiliani M, Koh KP, Shen Y, Pastor WA, Bandukwala H, Brudno Y, et al. Conversion of 5-methylcytosine to 5-hydroxymethylcytosine in mammalian DNA by MLL partner TET1. Science 2009;324(5929):930–5.

[24] Ito S, Shen L, Dai Q, Wu SC, Collins LB, Swenberg JA, et al. Tet proteins can convert 5-methylcytosine to 5-formylcytosine and 5-carboxylcytosine. Science 2011;333(6047):1300–3.

[25] Inoue A, Zhang Y. Replication-dependent loss of 5-hydroxymethylcytosine in mouse preimplantation embryos. Science 2011;334(6053):194.

[26] Chen C-C, Wang K-Y, Shen C-KJ. The mammalian de novo DNA methyltransferases DNMT3A and DNMT3B are also DNA 5-hydroxymethylcytosine dehydroxymethylases. J Biol Chem 2012;287(40):33116–21.

[27] Williams K, Christensen J, Pedersen MT, Johansen JV, Cloos PAC, Rappsilber J, et al. TET1 and hydroxymethylcytosine in transcription and DNA methylation fidelity. Nature 2011;473(7347):343–8.

[28] Pastor WA, Pape UJ, Huang Y, Henderson HR, Lister R, Ko M, et al. Genome-wide mapping of 5-hydroxymethylcytosine in embryonic stem cells. Nature 2011;473(7347):394–7.

[29] Yu M, Hon GC, Szulwach KE, Song CX, Zhang L, Kim A, et al. Base-resolution analysis of 5-hydroxymethylcytosine in the mammalian genome. Cell 2012;149:1368–80.

[30] Wen L, Li X, Yan L, Tan Y, Li R, Zhao Y, et al. Whole-genome analysis of 5-hydroxymethylcytosine and 5-methylcytosine at base resolution in the human brain. Genome Biol 2014;15:R49.

[31] Feinberg AP, Ohlsson R, Henikoff S. The epigenetic progenitor origin of human cancer. Nat Rev Genet 2006;7(1):21–33.

[32] Grønbæk K, Hother C, Jones PA. Epigenetic changes in cancer. APMIS 2007;115:1039–59.

[33] Walter MJ, Shen D, Ding L, Shao J, Koboldt DC, Chen K, et al. Clonal architecture of secondary acute myeloid leukemia. N Engl J Med 2012;366:1090–8.

[34] Aran D, Hellman A. DNA methylation of transcriptional enhancers and cancer predisposition. Cell 2013;154(1):11–13.

[35] Aran D, Sabato S, Hellman A. DNA methylation of distal regulatory sites characterizes dysregulation of cancer genes. Genome Biol 2013;14(3):R21.

[36] Kulis M, Heath S, Bibikova M, Queirós AC, Navarro A, Clot G, et al. Epigenomic analysis detects widespread gene-body DNA hypomethylation in chronic lymphocytic leukemia. Nat Genet 2012;44(11):1236–42.

[37] Coolen MW, Stirzaker C, Song JZ, Statham AL, Kassir Z, Moreno CS, et al. Consolidation of the cancer genome into domains of repressive chromatin by long-range epigenetic silencing (LRES) reduces transcriptional plasticity. Nat Cell Biol 2010;12:235–46.

[38] Network C Genome Atlas Research Integrated genomic analyses of ovarian carcinoma. Nature 2011;474(7353): 609–15.

[39] Network C Genome Atlas Research Comprehensive genomic characterization of squamous cell lung cancers. Nature 2012;489(7417):519–25.

[40] De Carvalho DD, Sharma S, You JS, Su S-F, Taberlay PC, Kelly TK, et al. DNA methylation screening identifies driver epigenetic events of cancer cell survival. Cancer Cell 2012;21(5):655–67.

[41] Taby R, Issa JJ. Cancer epigenetics. CA Cancer J Clin 2010;60(6):376–92.

[42] Toyota M, Ahuja N, Ohe-Toyota M, Herman JG, Baylin SB, Issa JP. CpG island methylator phenotype in colorectal cancer. Proc Natl Acad Sci USA 1999;96(15):8681–6.

[43] Weisenberger DJ. Characterizing DNA methylation alterations from the cancer genome atlas. J Clin Invest 2014;124:17–23.

[44] Keshet I, Schlesinger Y, Farkash S, Rand E, Hecht M, Segal E, et al. Evidence for an instructive mechanism of de novo methylation in cancer cells. Nat Genet 2006;38(2):149–53.

[45] Ohm JE, McGarvey KM, Yu X, Cheng L, Schuebel KE, Cope L, et al. A stem cell-like chromatin pattern may predispose tumor suppressor genes to DNA hypermethylation and heritable silencing. Nat Genet 2007;39(2):237–42.

[46] Schlesinger Y, Straussman R, Keshet I, Farkash S, Hecht M, Zimmerman J, et al. Polycomb-mediated methylation on Lys27 of histone H3 pre-marks genes for *de novo* methylation in cancer. Nat Genet 2007;39(2):232–6.

[47] Widschwendter M, Fiegl H, Egle D, Mueller-Holzner E, Spizzo G, Marth C, et al. Epigenetic stem cell signature in cancer. Nat Genet 2007;39(2):157–8.

[48] Sproul D, Kitchen RR, Nestor CE, Dixon JM, Sims AH, Harrison DJ, et al. Tissue of origin determines cancer-associated CpG island promoter hypermethylation patterns. Genome Biol 2012;13(10):R84.

[49] Di Croce L, Raker VA, Corsaro M, Fazi F, Fanelli M, Faretta M, et al. Methyltransferase recruitment and DNA hypermethylation of target promoters by an oncogenic transcription factor. Science 2002;295(5557):1079–82.

[50] Ito Y, Koessler T, Ibrahim AEK, Rai S, Vowler SL, Abu-Amero S, et al. Somatically acquired hypomethylation of IGF2 in breast and colorectal cancer. Hum Mol Genet 2008;17(17):2633–43.

[51] Ehrlich M, Lacey M. DNA hypomethylation and hemimethylation in cancer. Adv Exp Med Biol 2013:31–56.

[52] Kristensen LS, Treppendahl MB, Grønbæk K. Analysis of epigenetic modifications of DNA in human cells. Curr Protoc Hum Genet 2013 Unit20.2.

[53] Amara K, Ziadi S, Hachana M, Soltani N, Korbi S, Trimeche M. DNA methyltransferase DNMT3b protein overexpression as a prognostic factor in patients with diffuse large B-cell lymphomas. Cancer Sci 2010;101(7):1722–30.

[54] Viré E, Brenner C, Deplus R, Blanchon L, Fraga M, Didelot C, et al. The Polycomb group protein EZH2 directly controls DNA methylation. Nature 2006;439(7078):871–4.

[55] Chen B-F, Chan W-Y. The *de novo* DNA methyltransferase DNMT3A in development and cancer. Epigenetics 2014;9(5):1–9.

[56] Challen GA, Sun D, Jeong M, Luo M, Jelinek J, Berg JS, et al. Dnmt3a is essential for hematopoietic stem cell differentiation. Nat Genet 2012;44(1):23–31.

[57] Ley TJ, Ding L, Walter MJ, McLellan MD, Lamprecht T, Larson DE, et al. DNMT3A mutations in acute myeloid leukemia. N Engl J Med 2010;363(25):2424–33.

[58] Yan X-J, Xu J, Gu Z-H, Pan C-M, Lu G, Shen Y, et al. Exome sequencing identifies somatic mutations of DNA methyltransferase gene DNMT3A in acute monocytic leukemia. Nat Genet 2011;43(4):309–15.

[59] Raddatz G, Gao Q, Bender S, Jaenisch R, Lyko F. Dnmt3a protects active chromosome domains against cancer-associated hypomethylation. PLoS Genet 2012;8(12):e1003146.

[60] Garzon R, Liu S, Fabbri M, Liu Z, Heaphy CEA, Callegari E, et al. MicroRNA-29b induces global DNA hypomethylation and tumor suppressor gene reexpression in acute myeloid leukemia by targeting directly DNMT3A and 3B and indirectly DNMT1. Blood 2009;113(25):6411–18.

[61] Cortez CC, Jones PA. Chromatin, cancer and drug therapies. Mutat Res 2008;647(1–2):44–51.

[62] Solary E, Bernard OA, Tefferi A, Fuks F, Vainchenker W. The Ten–Eleven Translocation-2 (TET2) gene in hematopoiesis and hematopoietic diseases. Leukemia 2014;28(3):485–96.

[63] Viguié F, Aboura A, Bouscary D, Ramond S, Delmer A, Tachdjian G, et al. Common 4q24 deletion in four cases of hematopoietic malignancy: early stem cell involvement? Leukemia 2005;19(8):1411–15.

[64] Quivoron C, Couronné L, Della Valle V, Lopez CK, Plo I, Wagner-Ballon O, et al. TET2 inactivation results in pleiotropic hematopoietic abnormalities in mouse and is a recurrent event during human lymphomagenesis. Cancer Cell 2011;20(1):25–38.

[65] Lemonnier F, Couronné L, Parrens M, Jaïs J-P, Travert M, Lamant L, et al. Recurrent TET2 mutations in peripheral T-cell lymphomas correlate with TFH-like features and adverse clinical parameters. Blood 2012;120(7):1466–9.

[66] Asmar F, Punj V, Christensen J, Pedersen MT, Pedersen A, Nielsen AB, et al. Genome-wide profiling identifies a DNA methylation signature that associates with TET2 mutations in diffuse large B-cell lymphoma. Haematologica 2013;98:1912–20.

[67] Yamazaki J, Taby R, Vasanthakumar A, Macrae T, Ostler KR, Shen L, et al. Effects of TET2 mutations on DNA methylation in chronic myelomonocytic leukemia. Epigenetics 2012;7(2):201–7.

[68] Figueroa ME, Abdel-Wahab O, Lu C, Ward PS, Patel J, Shih A, et al. Leukemic IDH1 and IDH2 mutations result in a hypermethylation phenotype, disrupt TET2 function, and impair hematopoietic differentiation. Cancer Cell 2010;18(6):553–67.

[69] Ko M, Huang Y, Jankowska AM, Pape UJ, Tahiliani M, Bandukwala HS, et al. Impaired hydroxylation of 5-methylcytosine in myeloid cancers with mutant TET2. Nature 2010;468(7325):839–43.

[70] Nickerson ML, Im KM, Misner KJ, Tan W, Lou H, Gold B, et al. Somatic alterations contributing to metastasis of a castration-resistant prostate cancer. Hum Mutat 2013;34:1231–41.

[71] Cairns RA, Mak TW. Oncogenic isocitrate dehydrogenase mutations: mechanisms, models, and clinical opportunities. Cancer Discov 2013;3:730–41.

[72] Shibata T, Kokubu A, Miyamoto M, Sasajima Y, Yamazaki N. Mutant IDH1 confers an *in vivo* growth in a melanoma cell line with BRAF mutation. Am J Pathol 2011;178(3):1395–402.

[73] Hemerly JP, Bastos AU, Cerutti JM. Identification of several novel non-p.R132 IDH1 variants in thyroid carcinomas. Eur J Endocrinol 2010;163:747–55.

[74] Turcan S, Rohle D, Goenka A, Walsh LA, Fang F, Yilmaz E, et al. IDH1 mutation is sufficient to establish the glioma hypermethylator phenotype. Nature 2012;483:479–83.

[75] Lu C, Ward PS, Kapoor GS, Rohle D, Turcan S, Abdel-Wahab O, et al. IDH mutation impairs histone demethylation and results in a block to cell differentiation. Nature 2012;483:474–8.

[76] Sasaki M, Knobbe CB, Munger JC, Lind EF, Brenner D, Brüstle A, et al. IDH1(R132H) mutation increases murine haematopoietic progenitors and alters epigenetics. Nature 2012;488:656–9.

[77] Wang F, Travins J, DeLaBarre B, Penard-Lacronique V, Schalm S, Hansen E, et al. Targeted inhibition of mutant IDH2 in leukemia cells induces cellular differentiation. Science 2013;340:622–6.

[78] Popovici-Muller J, Saunders JO, Salituro FG, Travins JM, Yan S, Zhao F, et al. Discovery of the first potent inhibitors of mutant IDH1 that lower tumor 2-HG *in vivo*. ACS Med Chem Lett 2012;3:850–5.

[79] Rohle D, Popovici-Muller J, Palaskas N, Turcan S, Grommes C, Campos C, et al. An inhibitor of mutant IDH1 delays growth and promotes differentiation of glioma cells. Science 2013;340:626–30.

[80] Chen C, Liu Y, Lu C, Cross JR, Morris JP, Shroff AS, et al. Cancer-associated IDH2 mutants drive an acute myeloid leukemia that is susceptible to Brd4 inhibition. Genes Dev 2013;27:1974–85.

[81] Janeway KA, Kim SY, Lodish M, Nosé V, Rustin P, Gaal J, et al. Defects in succinate dehydrogenase in gastrointestinal stromal tumors lacking KIT and PDGFRA mutations. Proc Natl Acad Sci USA 2011;108:314–18.

[82] Killian JK, Kim SY, Miettinen M, Smith C, Merino M, Tsokos M, et al. Succinate dehydrogenase mutation underlies global epigenomic divergence in gastrointestinal stromal tumor. Cancer Discov 2013;3:648–57.

[83] Mason EF, Hornick JL. Succinate dehydrogenase deficiency is associated with decreased 5-hydroxymethylcytosine production in gastrointestinal stromal tumors: implications for mechanisms of tumorigenesis. Mod Pathol 2013;26:1492–7.

[84] Van Nederveen FH, Gaal J, Favier J, Korpershoek E, Oldenburg RA, de Bruyn EM, et al. An immunohistochemical procedure to detect patients with paraganglioma and phaeochromocytoma with germline SDHB, SDHC, or SDHD gene mutations: a retrospective and prospective analysis. Lancet Oncol 2009;10:764–71.

[85] Nestor C, Ruzov A, Meehan R, Dunican D. Enzymatic approaches and bisulfite sequencing cannot distinguish between 5-methylcytosine and 5-hydroxymethylcytosine in DNA. Biotechniques 2010;48:317–19.

[86] Huang Y, Pastor WA, Shen Y, Tahiliani M, Liu DR, Rao A. The behaviour of 5-hydroxymethylcytosine in bisulfite sequencing. PLoS One 2010;5(1):e8888.

[87] Booth MJ, Branco MR, Ficz G, Oxley D, Krueger F, Reik W, et al. Quantitative sequencing of 5-methylcytosine and 5-hydroxymethylcytosine at single-base resolution. Science 2012;336:934–7.

[88] Michaeli Y, Shahal T, Torchinsky D, Grunwald A, Hoch R, Ebenstein Y. Optical detection of epigenetic marks: sensitive quantification and direct imaging of individual hydroxymethylcytosine bases. Chem Commun (Camb) 2013;49:8599–601.

[89] Jin S-G, Jiang Y, Qiu R, Rauch TA, Wang Y, Schackert G, et al. 5-hydroxymethylcytosine is strongly depleted in human cancers but its levels do not correlate with IDH1 mutations. Cancer Res 2011;71(24):7360–5.

[90] Chen M-L, Shen F, Huang W, Qi J-H, Wang Y, Feng Y-Q, et al. Quantification of 5-methylcytosine and 5-hydroxymethylcytosine in genomic DNA from hepatocellular carcinoma tissues by capillary hydrophilic-interaction liquid chromatography/quadrupole time-of-flight mass spectrometry. Clin Chem 2013;59(5):824–32.

[91] Davis T, Vaisvila R. High sensitivity 5-hydroxymethylcytosine detection in Balb/C brain tissue. J Vis Exp 2011;48:2661.

[92] Song C-X, Clark TA, Lu X-Y, Kislyuk A, Dai Q, Turner SW, et al. Sensitive and specific single-molecule sequencing of 5-hydroxymethylcytosine. Nat. Methods. 2011;9(1):75–7.

[93] Haffner MC, Chaux A, Meeker AK, Esopi DM, Gerber J, Pellakuru LG, et al. Global 5-hydroxymethylcytosine content is significantly reduced in tissue stem/progenitor cell compartments and in human cancers. Oncotarget 2011;2(8):627–37.

[94] Yang H, Liu Y, Bai F, Zhang J-Y, Ma S-H, Liu J, et al. Tumor development is associated with decrease of TET gene expression and 5-methylcytosine hydroxylation. Oncogene 2013;32(5):663–9.

[95] Song SJ, Ito K, Ala U, Kats L, Webster K, Sun SM, et al. The oncogenic microRNA miR-22 targets the TET2 tumor suppressor to promote hematopoietic stem cell self-renewal and transformation. Cell Stem Cell 2013;13:87–101.

[96] Zhang L-T, Zhang L-J, Zhang J-J, Ye X-X, Xie A-M, Chen L-Y, et al. Quantification of the sixth DNA base 5-hydroxymethylcytosine in colorectal cancer tissue and C-26 cell line. Bioanalysis 2013;5:839–45.

[97] Xu W, Yang H, Liu Y, Yang Y, Wang P, Kim S-H, et al. Oncometabolite 2-hydroxyglutarate is a competitive inhibitor of α-ketoglutarate-dependent dioxygenases. Cancer Cell 2011;19(1):17–30.

[98] Wielscher M, Liou W, Pulverer W, Singer CF, Rappaport-fuerhauser C, Kandioler D, et al. Cytosine 5-hydroxymethylation of the LZTS1 gene is reduced. Transl Onol 2013;6:715–21.

[99] Inoue A, Zhang Y. Replication-dependent loss of 5-hydroxymethylcytosine in mouse preimplantation embryos. Science 2011;334:194.

[100] Xing XW, Liu YL, Vargas M, Wang Y, Feng YQ, Zhou X, et al. Mutagenic and cytotoxic properties of oxidation products of 5-methylcytosine revealed by next-generation sequencing. PLoS One 2013;8:e72993.

WRITERS, READERS, AND ERASERS OF EPIGENETIC MARKS

3

Thomas B. Nicholson[1], Nicolas Veland[2], and Taiping Chen[2]

[1]*Developmental and Molecular Pathways, Novartis Institutes for BioMedical Research, Cambridge, MA, USA,*
[2]*Department of Molecular Carcinogenesis and the Center for Cancer Epigenetics, The University of Texas MD Anderson Cancer Center, Smithville, Texas, USA; The University of Texas Graduate School of Biomedical Sciences at Houston, Houston, Texas, USA*

CHAPTER OUTLINE

S.G. Gray (Ed): Epigenetic Cancer Therapy. DOI: http://dx.doi.org/10.1016/B978-0-12-800206-3.00003-3

1 INTRODUCTION

The cellular nucleus contains DNA that is complexed with histones to form chromatin, which serves a critical role in the modulation of gene expression. Chromatin consists of DNA wrapped around an octamer of two proteins each of histones H2A, H2B, H3, and H4, creating a compact central structure that has N-terminal tails that project out from this core and are the targets of many posttranslational modifications (PTMs). These modifications, in combination with direct modification of DNA, combine to form the epigenetic code. One requisite for this code is that it has the flexibility and adaptability to be modified as an organism develops from a fertilized egg to an adult organism, as well as to respond to its environment as appropriate. At least four different types of DNA cytosine modifications—methylation, hydroxymethylation, formylation, and carboxylation—and dozens of different varieties of histone modifications, including methylation, acetylation, phosphorylation, sumoylation, ubiquitylation, and ADP-ribosylation, have been identified to date [1]. A wide variety of enzymes have evolved that modify specific amino acids and nucleotide bases (so-called writers), along with enzymes to remove these marks ("erasers"). Cells also express a diverse complement of proteins that bind chromatin modifications ("readers"), providing the ability to sense the chromatin state of a given locus. Because these proteins are intimately involved in ensuring the desired gene expression, it is not surprising that recent research has identified a widespread contribution of these factors to cancer development, and it is hoped that these enzymes may serve as druggable targets. This chapter, which will focus on the epigenetic processes of methylation and acetylation, will introduce the key concepts around the classes of proteins involved in these modifications, and offer insights into why they are thought to represent attractive drug targets in oncology.

2 WRITERS

Epigenetic modifications arise in a variety of forms; this review will focus on the methylation of DNA and histones, and the acetylation of histones. Methylation and acetylation of chromatin components can serve a variety of functions, with both positive and negative effects on transcription, combining to ensure the desired gene expression pattern for any particular cell. Multiple proteins mediate the deposition and removal of the methyl and acetyl marks, providing a mechanism to ensure specific gene expression modulation. The end result is that the epigenetic status of chromatin is complex, as in humans it has been estimated that over 50 different chromatin states occur [2].

2.1 DNA METHYLTRANSFERASES

DNA is methylated on cytosine residues, predominantly in the context of cytosine–guanine (CpG) dinucleotides, forming 5-methylcytosine (5mC). Due to deamination reactions that convert 5mCs to thymines, CpG dinucleotides are underrepresented in the mammalian genome on average, although they are enriched in certain regions, known as CpG islands, that are components of the promoters of many genes [3]. Promoter methylation generally correlates with gene silencing, through the binding of proteins such as MeCP2 (see later) and the recruitment of various factors that block RNA Pol II transcription. Conversely, methylation of DNA in gene bodies has been correlated with gene expression, although the exact role of this modification remains under debate [4]. DNA methylation is mediated by the DNA methyltransferase (DNMT) family, which consists of DNMT1, DNMT3A, and DNMT3B, as well as the nonenzymatically active DNMT3L. DNMT1 functions as the maintenance methyltransferase, ensuring faithful propagation of DNA methylation patterns during DNA replication by copying the methylation pattern from the parental strand onto the daughter strand [5,6]. UHRF1 (ubiquitin-like with PHD and ring finger domains 1, also known as Np95 and ICBP90) plays an essential role in recruiting DNMT1 to hemi-methylated CpG sites [7–9]. DNMT3A and DNMT3B, conversely, act primarily as *de novo* DNA methyltransferases that establish methylation patterns during gametogenesis and early embryogenesis [10,11]. DNMT3L serves as an accessory for *de novo* methylation by DNMT3A (and perhaps DNMT3B as well) and is essential for the establishment of genomic imprinting during gametogenesis [12,13]. There is evidence that DNMT3L recruits DNMT3A to chromatin regions where lysine 4 of histone H3 (H3K4) is unmethylated [14].

2.2 HISTONE LYSINE METHYLTRANSFERASES

The methylation of histone lysines occurs on multiple residues, and results in a diverse set of transcriptional effects, leading to both up- and downregulation of transcription depending on the modification. Lysines can be modified with one (monomethyl, me1), two (dimethyl, me2), or three (trimethyl, me3) methyl groups; the number of methyl groups, in addition to the location of modification, determines the resulting effect. Due to the extensive complexity inherent in this system, this section will outline a few well-characterized modifications and their downstream effects. A detailed understanding of these marks has come from genome-wide analyses, employing techniques such as ChIP-Seq [15]. Methylation of residues K4, K36, and K79 of histone H3 are considered marks of active chromatin, while histone H3K9, H3K27, and H4K20 methylation is linked to repressed chromatin. Methylation of H3K4 generally occurs at active genes, with the trimethyl mark being found at the promoter of transcriptionally active genes, while H3K4me1 identifies active and poised enhancers [16]. The gene bodies of active genes contain multiple marks, including H2BK5me1, H3K36me3, and H4K20me1 [17]. H3K79 methylation, particularly di- and tri-methyl marks at the 5′ end of genes, correlates with the level of transcription, while the monomethyl mark is present at transcriptional activator–binding sites [18]. Conversely, modification of promoters with H3K9me3 or H3K27me3 occurs at inactive genes [19]. H4K20me2 has been linked to DNA repair, while trimethylation of this residue is associated with constitutive heterochromatin [20]. A unique situation occurs in stem cells, where a subset of promoters contains both H3K4me3 and H3K27me3. It is generally considered that these bivalent promoters are poised for differentiation signals, to be rapidly activated or repressed during stem cell differentiation [21].

The enzymatic addition of methyl marks to histone lysine residues is mediated by multiple redundant proteins, which are generically referred to as lysine methyltransferases (KMTs) (Table 3.1) [22]. There are two KMT subgroups, which are divided based on the presence or absence of the Su(var)3-9, Enhancer-of-zeste and Trithorax (SET) domain: DOT1L/KMT4 lacks this domain, while all other characterized KMTs contain this domain. KMT proteins in general exhibit a high degree of enzymatic specificity, modifying specific lysine residues in the histone tail with defined number(s) of methyl residues (from 1 to 3). For example, the enzymes SUV39H1/KMT1A and SUV39H2/KMT1B exclusively convert H3K9me1 to H3K9me3, while the related enzyme G9a/KMT1C produces H3K9me2 [23]. DOT1L is the only enzyme known to modify H3K79 in humans, adding the mono-, di-, and tri-methyl marks. Methylation by DOT1L plays a critical role in a variety of biological processes, such as

Table 3.1 The Human Lysine Methyltransferase Family

Enzyme	Other names	Substrate	Cancer implication	References
KMT1A	SUV39H1	H3K9	7.4% overexpress (uterine, ovarian, prostate, sarcoma, melanoma); 4% mutated (colon); 3.6% deleted (head and neck, stomach, lung)	[244, 245, 246]
KMT1B	SUV39H2	H3K9	5.5% overexpress (bladder, ovarian, uterine, breast, sarcoma, stomach, liver, glioma, melanoma); 4.2% mutated (colon, lung, uterine)	[244, 245, 246]
KMT1C	G9a/EHMT2	H3K9	8.3% mutated (cervical, pancreatic, melanoma, kidney, stomach, bladder, colon, head and neck); 5.5% overexpress (ovarian, melanoma, lung, liver); 3.6% deleted (prostate)	[244, 245, 246]
KMT1D	EuHMTase/ GLP/EHMT1	H3K9	6.6% mutated (lung, pancreatic, uterine, bladder, melanoma, colon, esophageal, head and neck); 4% overexpress (ovarian, sarcoma, glioma); 3% deleted (ACyC)	[244, 245]
KMT1E	ESET/ SETDB1	H3K9	18% overexpress (lung, bladder, uterine, breast, liver, ovarian, sarcoma, ACyC, melanoma); 5.6% mutated (colon, uterine, stomach, lung)	[244, 245, 246]
KMT1F	CLL8/ SETDB2	H3K9	21.3% deleted (prostate, bladder, sarcoma); 4.6% mutated (uterine, melanoma, lung)	[244, 245]
KMT2A	MLL1	H3K4	23.1% mutated (bladder, melanoma, stomach, lung, colon, pancreatic, liver); 6.4% overexpress (AML, uterine, prostate, ovarian, ALL)	[244, 245, 246]
KMT2B	MLL2	H3K4	12% mutated (uterine, stomach, bladder, melanoma, colon, lung, glioblastoma, medulloblastoma, leukemias); 8% overexpress (ovarian, sarcoma)	[244, 245, 246]
KMT2C	MLL3	H3K4	25% mutated (melanoma, cervical, bladder, lung, stomach, kidney, uterine, breast, pancreatic, head and neck, prostate); 8.7% overexpress (ovarian); 3% deleted (AML, sarcoma)	[244, 245, 246]
KMT2D	MLL4	H3K4	26% mutated (bladder, melanoma, lung, stomach, head and neck, cervical, colon, uterine, pancreatic, prostate, medulloblastoma); 7% overexpress (B-cell lymphoma)	[244, 245]
KMT2E	MLL5	H3K4	7% mutated (uterine, lung, stomach, colon, melanoma, bladder, breast); 4.3% overexpress (head and neck, lung, ovarian, uterine, glioblastoma, sarcoma)	[244, 245]
KMT2F	SET1A	H3K4	7.6% mutated (melanoma, lung, colon, head and neck, pancreatic, uterine, bladder); 5% overexpress (breast, uterine)	[244, 245]
KMT2G	SET1B	H3K4	11% mutated (colon, uterine, prostate); 3.3% overexpress (prostate, ovarian, uterine, glioma)	[244, 245]
KMT2H	ASH1L	H3K4	16.4% overexpress (liver, lung, breast, ovarian, uterine, bladder, ACC); 10% mutated (melanoma, stomach, uterine, bladder, lung, colon, pancreatic)	[244, 245]

(Continued)

Table 3.1 (Continued)

Enzyme	Other names	Substrate	Cancer implication	References
KMT3A	SETD2	H3K36	19.2% mutated (bladder, kidney, lung, uterine, colon, pancreatic, melanoma, cervical, stomach); 7% deleted (B-cell lymphoma, kidney)	[244, 245]
KMT3B	NSD1	H3K36	11% mutated (head and neck, uterine, bladder, lung, melanoma, colon, stomach, ACC); 6.5% overexpress (kidney, uterine, pancreatic, ovarian, sarcoma, liver, ACC, AML)	[244, 245, 246]
KMT3C	SMYD2	H3K36	13.3% overexpress (liver, breast, lung, B-cell lymphoma, uterine, ovarian, melanoma, esophageal); 2% mutated (uterine, lung)	[244, 245, 246]
KMT3D	SMYD1	H3K4	6.6.% mutated (melanoma, lung, uterine, colon, head and neck); 4.7% overexpress (sarcoma)	[244, 245]
KMT3E	SMYD3	H3K4	14.7% overexpress (breast, liver, ovarian, lung, melanoma, uterine, colon); 7% deleted (B-cell lymphoma); 2.8% mutated (colon, uterine)	[244, 245, 246]
KMT4	DOT1L	H3K79	11% mutated (colon, bladder, melanoma, lung, stomach); 5.5% deleted (uterine, ovarian, ACyC); 3.8% overexpress (glioma, sarcoma, ACC, glioblastoma)	[244, 245]
KMT5A	SET7/8 (SETD8)	H4K20	3.3% deleted (prostate, lung); 3% mutated (colon, bladder); 2.4% overexpress (sarcoma, ovarian, uterine)	[244, 245]
KMT5B	SUV4-20H1	H4K20	8.2% overexpress (head and neck, prostate, bladder, stomach, breast, melanoma, ovarian, liver); 7.7% mutated (bladder, uterine, stomach, colon, lung)	[244, 245]
KMT5C	SUV4-20H2	H4K20	6.3% overexpress (bladder, uterine, ACyC, ovarian, breast, ACC); 4.2% mutated (colon, stomach); 3.8% deleted (glioma)	[244, 245]
KMT6A	EZH2	H3K27	9.6% overexpress (ovarian, melanoma, prostate, glioblastoma, uterine, breast, bladder, lung, pancreatic, colon); 5% mutated (B-cell lymphoma, uterine, melanoma, head and neck, lung); 2.7% deleted (AML, sarcoma, lung)	[244, 245, 246]
KMT6B	EZH1	H3K27	3.3% mutated (uterine, stomach, melanoma, lung, colon); 2% deleted (prostate, ACC); 1.4% overexpress (breast, stomach)	[244, 245]
KMT7	SET7/9 (SETD7)	H3K4	3.3% overexpress (prostate, ovarian, sarcoma, stomach); 2.7% mutated (uterine, colon, lung, melanoma, esophageal); 2% deleted (glioma, glioblastoma)	[244, 245]
KMT8	RIZ1/PRDM2	H3K9	11% mutated (colon, stomach, pancreatic, uterine, melanoma, lung, prostate, bladder); 4% overexpress (sarcoma, bladder, pancreatic, glioblastoma, ovarian); 2% deleted (liver)	[244, 245]

(Continued)

Table 3.1 (Continued)

Enzyme	Other names	Substrate	Cancer implication	References
	WHSC1/ NSD2	H3K27	18% overexpress (uterine, bladder, ovarian, sarcoma, multiple myeloma, neuroblastoma); 9.7% mutated (colon, melanoma, stomach, bladder, uterine, lung)	[244, 245, 246]
	WHSC1L1/ NSD3	H3K4/ H3K27	13.6% overexpress (lung, breast, uterine, bladder, head and neck, ovarian, sarcoma, AML); 6.6% deleted (prostate, liver); 5.6% mutated (colon, uterine, melanoma)	[244, 245, 246]
	PRDM9	H3K4	31% mutated (lung, melanoma, head and neck, colon, esophageal); 10% overexpress (ovarian, sarcoma, prostate, lung, cervical, bladder)	[244, 245]
	SETD3	H3K36	8.3% mutated (colon, lung); 3.4% overexpress (sarcoma, head and neck, ovarian); 3% deleted (bladder)	[244, 245]

maintenance of telomere lengths, regulation of imprinted gene expression, and cell cycle progression [24–26]. Furthermore, the specificity of enzymes can be modified by the complex within which they are located. For instance, EZH2/KMT6 is a component of a least two different complexes, Polycomb Repressive Complex 2 (PRC2) and PRC3, where the complex components differ and the enzymatic activity toward nucleosome substrates is affected [27].

2.3 PROTEIN ARGININE METHYLTRANSFERASES

Arginine methylation is a widespread PTM that occurs on a variety of substrates [28], including multiple histone arginine residues such as H2AR11, H2AR29, H3R2, H3R8, H3R17, H3R26, and H4R3. This methylation is mediated by a family of nine proteins, collectively referred to as the protein arginine methyltransferases (PRMTs) (Table 3.2). These proteins are subdivided into different classes (Type I, II, or III), based on the enzyme's end product. All three classes initially generate ω-N^G-monomethylarginine (MMA), via the addition of one methyl group to one of the two terminal nitrogen atoms of the arginine amino acid. The Type III class, of which PRMT7 is the only example, does not modify the arginine further. Type I enzymes, including PRMT1, PRMT2, PRMT3, PRMT4/ CARM1, PRMT6, and PRMT8, then methylate the same arginine with the further addition of a methyl group to the same nitrogen atom previously methylated, generating ω-N^G, N^G-asymmetric dimethylarginine (ADMA). PRMT5 is the primary Type II enzyme, which methylates the other, unmodified terminal nitrogen atom, generating ω-N^G, N'^G-symmetric dimethylarginine (SDMA). PRMT9/FBXO11 has yet to be fully characterized.

The methylation of H4R3, one of the best characterized histone arginine modifications, was the first to be demonstrated *in vivo*, with PRMT1 responsible for the majority of this methylation [34,35]. In the absence of this ADMA mark, decreased histone acetylation and increased levels of H3K9 and H3K27 methylation result, which combine to decrease chromatin transcription [36]. PRMT5, which methylates H4R3 with SDMA, causes transcriptional repression, in part through the recruitment of

Table 3.2 The Human Protein Arginine Methyltransferase Family

Enzyme	Other common names	Modification	Cancer implication	References
PRMT1		ADMA	up to 5.5% overexpression (uterine, bladder, pancreas, breast, lung); up to 2.8% mutated (cervical, uterine, melanoma); other alterations (MLL)	[244, 245, 29, 247, 30, 31]
PRMT2		ADMA	up to 3.5% overexpression (ovarian, bladder, AML, sarcoma, breast); up to 2.8% mutated (colon, uterine, lung); up to 1.4% deletion (stomach, colon)	[244, 245, 29]
PRMT3		ADMA	up to 3.8% mutated (uterine, lung, stomach, colon, melanoma); up to 2% overexpression (bladder, stomach, sarcoma); regulated by DAL-1/4 tumor suppressor gene	[244, 245, 29, 248]
PRMT4	CARM1	ADMA	up to 11.3% overexpression (uterine, ovarian, ACyC, ACC, sarcoma, glioma, colon, breast, prostate, lung, liver); up to 2% mutated (uterine, bladder, lung, head and neck, melanoma)	[244, 245, 29, 249, 250, 251, 32]
PRMT5		SDMA	up to 4% mutated (uterine, head and neck, bladder, colon, melanoma, leukemia, lymphoma); up to 3% overexpression (ovarian, lung, sarcoma, glioma, breast, liver, glioblastoma)	[244, 245, 246, 33, 252, 253, 254]
PRMT6		ADMA	up to 3.6% overexpression (prostate, melanoma, sarcoma, bladder, lung); up to 1.6% mutated (lung, uterine, liver)	[244, 245, 246, 30, 255, 256]
PRMT7		MMA/ SDMA[a]	up to 6.6% deletion (prostate, ovarian, breast, AML); up to 4% mutated (bladder, lung, uterine, stomach, melanoma, colon)	[244, 245, 246, 257]
PRMT8		ADMA	up to 11% overexpression (ovarian, uterine, bladder, glioma, sarcoma, head and neck, breast, glioblastoma); up to 7% mutated (lung, melanoma, colon)	[244, 245, 258]
PRMT9	FBXO11	SDMA	up to 3.8% mutated (uterine, stomach, colon, bladder, lung, esophageal, liver, glioblastoma); up to 3.6% overexpression (uterine, prostate, ovarian, sarcoma, breast, lung)	[244, 245]

ACyC, adenoid cystic carcinoma. ACC, adrenocortical carcinoma.
[a]Literature reports have suggested either MMA and SDMA activity for PRMT7, further clarification will be required.

DNMT3A and DNA methylation [37]. PRMT5 can also methylate H2AR3 and H3R8; the latter modification along with H4R3 symmetrical methylation serves to repress, for example, RB family genes [33]. Reducing PRMT5 levels in lymphoma cells results in reexpression of RBL2, activation of RB1, and induced expression of several pro-apoptotic genes via PRC2 inhibition, which ultimately leads to decreased cell proliferation [38]. Similar to H4R3 methylation, modification of H3R2 has different biological effects depending on whether the modification is SDMA or ADMA. At this amino acid, ADMA addition by PRMT4 and PRMT6 results in repression of transcription, while the presence of SDMA or MMA serves as an activation signal [39]. PRMT4/CARM1 not only methylates multiple histone arginine residues, including H3R2, H3R17, H3R26, H3R128, H3R129, H3R131, and H3R134, but it also cooperates with acetyltransferases to initiate chromatin modification [40]. PRMT7, which has two histone substrates, H2AR3 and H4R3, has an apparent role in regulating the expression of DNA repair enzymes [41].

2.4 HISTONE ACETYLTRANSFERASES

Acetylation of lysine ε-amino groups results in the neutralization of its positive charge, and leads to a decrease in association with the negatively charged DNA phosphate backbone. This reduces chromatin compaction, rendering the modified locus more accessible to transcription factors and thereby increasing gene expression [42]. As such, it is perhaps not surprising that genome-wide analyses have demonstrated that the lysine acetylation mark [43,44], as well as the histone acetyltransferases (HATs) that generate this mark [45], is commonly localized at promoters and enhancers of expressed genes. Numerous lysine residues in the tails of histone H2A, H2B, H3, and H4 are acetylated. These modifications are involved in more than modulating gene transcription; for example, H3K56ac plays a role in DNA damage response [43,44,46] and DNA replication [47]. Due to the relatively limited number of these enzymes, a major mechanism that is employed to control their activity and substrate specificity is through the formation of multi-protein complexes [48]. The human GNAT family (see following paragraph) has been found to be a part of at least three different complexes, namely, the PCAF, STAGA, and TFTC complexes [49]. These complexes play critical roles in determining protein activity, as the associated factors help to determine the location in the chromatin that will be modified [50].

HATs, which are also known as lysine acetyltransferases (KATs) (Table 3.3), are grouped into families based on sequence conservation, with the best characterized being the GNAT (GCN5-related N-acetyl transferase), the MYST (MOZ, YBF2/SAS3, SAS2, TIP60), and the p300/CBP families. The GNAT HATs (GCN5/KAT2A; PCAF/KAT2B) belong to a larger superfamily, which consists of proteins with a variety of N-acetyltransferase enzymatic activities, including toward aminoglycoside, serotonin, and glucosamine-6-phosphate substrates [51]. While GCN5 and PCAF are related proteins that serve as coactivators for many transcription factors [52], they have distinct physiological functions as demonstrated by the different phenotypes of the respective knockout mice [53,54]. The MYST family of enzymes, which consists of five proteins in mammals (MOF/MYST1/KAT8, HBO1/MYST2/KAT7, MOZ/MYST3/KAT6A, MORF/MYST4/KAT6B, and TIP60/KAT5), all contain a MYST domain that consists of acetyl-CoA-binding and zinc finger (ZnF) motifs. These proteins have diverse properties, being involved in proliferation, differentiation, and apoptosis [55]. The members of the third family, CBP/KAT3A and p300/KAT3B, have several regions of homology, including the catalytic domain and bromodomain [56]. CBP and p300 are acetyltransferases and coactivators of transcription [57,58], with many shared histone substrates but also divergent preferences for different lysine residues [59].

Table 3.3 The Human Histone Acetyltransferase Family

Enzyme	Other names	Histone substrate	Cancer implication	References
KAT1	HAT1	H2AK5, H4K5, H4K12	6% overexpress (ovarian, uterine, head and neck, lung, sarcoma, kidney); 2% mutated (uterine, lung, stomach); 3.7% deleted (prostate)	[244, 245]
KAT2A	GCN5	H3K9, H3K14, H3K56, H4K5, H4K8, H4K12, H4K16, H4K91	7.7.% mutated (bladder, uterine, colon); 2% overexpress (stomach); 2.3% deleted (ACC, prostate)	[244, 245]
KAT2B	PCAF	H3K9, H3K14	6.3% overexpress (bladder, sarcoma, ovarian); 4.8% mutated (lung, melanoma, uterine, liver); 3.6% deleted (kidney)	[244, 245]
KAT3A	CBP	H3K9, H3K14, H3K56, H4K5, H4K8, H4K12, H4K16, H4K91	15% mutated (bladder, cervical, lung, colon, stomach, uterine, melanoma, head and neck, ACyC); 5% overexpress (breast)	[244, 245]
KAT3B	P300	H3K14, H3K18, H3K27, H3K56, H4K5, H4K8, H4K12, H4K16	15.7% mutated (bladder, colon, melanoma, uterine, lung, stomach, head and neck, pancreatic, kidney, glioblastoma)	[244, 245]
KAT4	TAF1	H3K14	14.6% mutated (uterine, lung, cervical, colon, melanoma, bladder, pancreatic, stomach, breast); 8% overexpress (prostate)	[244, 245]
KAT5	TIP60	H2AK5, H4K5, H4K8, H4K12, H4K16	4.8% mutated (pancreatic, uterine, colon, head and neck); 4.3% overexpress (head and neck, bladder, ovarian)	[244, 245]
KAT6A	MYST3	H3K9, H3K14	16.4% overexpress (uterine, bladder, lung, breast, prostate); 9% mutated (melanoma, lung, stomach, esophageal)	[244, 245]
KAT6B	MYST4	H4K5, H4K8, H4K12, H4K16	10% mutated (melanoma, colon, bladder, lung, uterine, prostate); 7% overexpress (uterine, ovarian, prostate)	[244, 245]
KAT7	MYST2	H3K14, H4K5, H4K8, H4K12	7% overexpress (breast, kidney, lung, melanoma, prostate, sarcoma); 4% mutated (colon, uterine, lung, head and neck)	[244, 245]
KAT8	MYST1	H4K16	5.5% overexpress (uterine, breast, bladder); 3.4% mutated (lung, stomach, colon, esophageal)	[244, 245]
KAT9	ELP3	H3K9, H3K18	11.5% deleted (bladder, prostate, uterine, ovarian, lung, colon, liver, breast); 2% overexpress (sarcoma)	[244, 245]
KAT12	GTF3C4	H3K14	5.6% mutated (colon, stomach, uterine, lung, melanoma); 3.3% deleted (ACyC); 2% overexpress (sarcoma, head and neck)	[244, 245]
KAT13A	NCOA1	H3K14	6.3% mutated (bladder, uterine, stomach, lung, colon, breast, melanoma); 7.7% overexpress (bladder, uterine, liver)	[244, 245]

(Continued)

Table 3.3 (Continued)

Enzyme	Other names	Histone substrate	Cancer implication	References
KAT13B	NCOA3	H3K14	16.4% overexpress (uterine, colon, ovarian, stomach, lung, breast, sarcoma); 5.8% mutated (melanoma)	[244, 245]
KAT13D	CLOCK	H3K14	7% mutated (pancreatic, uterine, colon, melanoma, head and neck); 5.6% overexpress (lung, glioblastoma, ACyC, sarcoma)	[244, 245]
	CDY1		-	[244, 245]
	CDY2 (CDY2A)		-	[244, 245]
	CDYL		10% overexpress (ovarian, melanoma, uterine, liver, bladder, sarcoma); 5.6% mutated (colon, melanoma, lung, stomach)	[244, 245]
	MGEA5	H4K8, H3K14	6.6% deleted (prostate); 3.3% mutated (uterine, bladder, stomach, lung, melanoma, colon, pancreatic)	[244, 245]
	NAT10		6.7% mutated (uterine, colon, melanoma, stomach); 4% overexpress (ovarian, breast, head and neck, bladder, sarcoma)	[244, 245]

Furthermore, point mutations in these proteins have differing effects on enzymatic activity, and CBP but not p300 requires its PHD finger for activity [60]; knockout mouse models also exhibit different phenotypes [61–63]. CBP and p300 acetylate H3K56, which leads to incorporation of this histone into chromatin following DNA damage [43]. Interestingly, recent studies have demonstrated cooperativity between different KAT family members. Thus, activating PPARγ initially causes acetylation of H3K18 and H3K27 by the KAT3 family members, which is subsequently followed by H3K9 acetylation by KAT2 proteins in conjunction with transcriptional elongation [64].

3 READERS

In order to mediate their effects, epigenetic modifications need to be recognized by other cellular factors. A wide variety of protein domains have thus evolved to bind these modifications; these proteins are known as "readers." Due to the important role these readers play, a variety of techniques have been developed to understand protein/methyl-mark interaction, such as peptide-based enrichment [65]. This method, which exploits the fact that most histone-binding proteins recognize short, linear amino acid stretches, provides an unbiased method for studying binding activity. This assay is particularly powerful as it allows for peptides with defined, combinatorial modifications to dissect the interaction of a reader with a variety of histone modifications. Further enhancements have allowed the binding domains to be spotted in an array, followed by incubation with a fluorescently labeled peptide mimicking a histone fragment with different modification(s) and the high-throughput analysis of protein binding to particular methyl marks. While most readers positively interact with a particular mark, a subset of proteins, including those with PHD fingers, ADD (ATRX-DNMT3-DNMT3L) domains, and WD40 domains, binds to histone tails that are unmodified at specific locations, with this binding decreased upon histone modification. The role of unmodified histone binding is still being explored, and as such the following sections will focus on the specific recognition of epigenetic posttranslational marks.

3.1 METHYL-CPG-BINDING PROTEINS

Methylated DNA is recognized by methyl-CpG-binding proteins (MBPs), which bind to the methylated DNA and initiate the silencing of chromatin through the recruitment of other factors. There are three distinct families that bind to methylated DNA, including the methyl-CpG-binding domain (MBD) family, ZnF proteins, and SET and RING finger-associated (SRA) domain proteins [66]. The ability of flanking DNA sequences to modulate binding to specific CpG dinucleotides suggests a mechanism for specific genome recognition by these proteins [67]. The MBD proteins include MeCP2 and MBD1-4; MeCP2, MBD1, and MBD2 bind to methylated DNA, and, through their transcriptional repression domains (TRD), interact with various repressor complexes to silence the bound locus. MBD3 contains mutations in its MBD domain and as such is unable to bind DNA on its own; rather, it localizes to genomic regions as part of the NuRD complex [68]. MBD4 binds to regions of base pair mismatch, and through its glycosylase activity enhances DNA mismatch repair [69]. The ZnF domain–containing proteins that are known to interact with methylated DNA are ZBTB33/Kaiso, ZBTB4, and ZBTB38. Despite an apparent lack of specificity for methylated DNA *in vitro*, these proteins repress transcription in a DNA methylation–dependent manner [66]. Recent ChIP-seq analysis, however, suggests that ZBTB33/Kaiso preferentially binds unmethylated regulatory regions in the genome in cells [70]. Thus, the significance of ZBTB proteins as DNA methylation "readers" remains to be determined. Another ZnF protein, ZFP57, recognizes a methylated hexanucleotide (TGCCGC) present in imprinting control regions (ICRs) and is essential for protecting DNA methylation imprints against the wave of demethylation that erases most DNA methylation marks in the mammalian genome during early embryogenesis [71]. The SRA protein UHRF1 binds to both hemimethylated DNA and DNMT1 and is critical for recruiting DNMT1 to hemimethylated sites during DNA replication [72–74]. In general, methyl-DNA-binding proteins serve to reinforce the transcriptional-repressive activity of DNA methylation. Recently, proteins that recognize the other cytosine modifications, notably 5-hydroxymethylcytosine (5hmC) and 5-formylcytosine (5fC), have been identified. It has been reported that MBD3, MeCP2, and UHRF1 interact with 5hmC in addition to 5mC [75–77]. Subsequent to these observations, screens have been performed which identified several proteins, including transcription factors, DNA damage repair proteins, and epigenetic proteins, that exhibit enhanced binding to 5hmC or to 5fC [78,79]. These proteins appear to show specificity for the type of modification, and the specific binding partners can change with, for example, differentiation [79]; there will likely be many novel functions for these DNA modifications discovered in the future.

3.2 HISTONE METHYLATION–BINDING DOMAINS

Methylated histone lysines are bound by several domains, including ankyrin, ADD, bromo-adjacent homology (BAH), chromo-barrel, chromodomain, double chromodomain (DCD), malignant brain tumor (MBT), plant homeodomain (PHD), PWWP, Tudor, tandem Tudor domain (TTD), and WD40 domains [80]. The substantial number and variety of domains that recognize methylated lysines demonstrate the importance this modification has in the regulation of gene expression. Each domain has specific binding characteristics, specifically recognizing certain lysines with a particular methylation status. For example, PHD fingers interact with H3K4me3, a mark of active transcription, while WD40 domains bind to multiple trimethylated lysines associated with repressive marks. In contrast, the presence of a particular methyl mark can interfere with protein binding in some cases. For example, the ADD domains of DNMT3A and DNMT3L interact with histone H3, but H3K4 methylation can disrupt this interaction [14,81]. Similarly, the ADD domain of ATRX binds to histone tails that contain the H3K9me3 mark, but only when the H3K4me3/me2 mark is not present [82]. The ligand specificity of the various domains is determined by two mechanisms. On the one hand, the conformation of the binding pocket provides for either steric hindrance, inhibiting interaction with certain ligands, or specific domain-ligand binding to enhance interaction. Secondly, the amino acids surrounding the modified lysine can interact with the reader protein, which can positively or negatively affect binding.

Methylation of arginine similarly has dual effects, causing either enhanced binding by proteins that recognize the methylarginine, or reducing the interaction of proteins that specifically bind to unmodified arginines. Tudor domains are one of the best-characterized domains that recognize methylated arginines. For example, the protein TDRD3 serves as a transcriptional coactivator, and is recruited to regions of the genome based on histone arginine methylation [83]. PHD finger domains, conversely, interact with unmodified arginine; the binding and suppression, by UHRF1, of target genes requires an interaction with unmethylated H3R2 [84]. WD40 repeats and chromodomains have also been implicated in recognition of unmodified arginine residues [85].

3.3 HISTONE ACETYLATION–BINDING DOMAINS

While acetylation of histones generally loosens chromatin structure, proteins also specifically bind to this modification through bromodomains [86] and tandem PHD domains [87]. In contrast to methylation readers, these binding domains are rather non-specific, with individual domains often able to bind to multiple modified residues. Bromodomains adopt a similar secondary structure, despite differences in primary amino acid sequence, which creates a deep, hydrophobic pocket that can bind acetylated lysines but will not bind charged, unmodified lysine residues [80]. Humans express 61 proteins that contain bromodomains, which are involved in a variety of pathways and include epigenetic modifiers, chromatin remodelers, and transcriptional activators [88]. Structural studies indicated that binding by these domains is not only dependent on the acetyl mark but may involve other histone modifications in the same region, in particular acetylation and phosphorylation [89]. TRIM family bromodomain-containing proteins recognize not only acetylated lysines, but the specific histone context of these modifications [85]. Multiple binding domains can be present in a single protein, such as TAF1, which contains a double bromodomain where the two domains are positioned in a manner to optimize binding to H4 tails containing multiple acetylated lysine residues [90]. Alternatively, the single bromodomain in BET-family proteins, as represented by the mouse Brdt protein, binds histones with at least two acetylation marks [91]. Thus, while bromodomains are known to serve as binding partners for acetylated histone marks, a wide variety of mechanisms have evolved to allow for specific interaction between particular proteins and histone marks.

While bromodomains are the best characterized acetyl-lysine-binding domains, other domains can bind this epigenetic mark. The double PHD finger protein DPF3b has a unique binding property, whereby both domains function together to bind to H3K14ac mark [87]. This binding, interestingly, can be inhibited by methylation of H3K4, demonstrating an interplay between these two marks. Similarly, the tandem PHD finger present in the acetyltransferase MOZ binds to histone tails that contain acetylated H3K14, but only in the situation where H3R2 is unmodified [92]. Additionally, recent data has demonstrated that a double pleckstrin homology (PH) domain located in the yeast protein Rtt106 exhibits increased binding to the H3K56ac mark [93], an interaction which promotes new nucleosome assembly [94].

4 ERASERS

In order to modify the epigenome, a class of proteins known as "erasers," which oppose the activity of the writers, is required. These enzymes catalyze the removal of epigenetic marks from specific residues. The removal of an epigenetic mark relieves its effect on transcription, and the result can modulate gene expression. These enzymes ensure that the cell has the capability to rapidly modify its chromatin in order to respond to changes in differentiation status or external stimuli.

4.1 PROTEINS INVOLVED IN DNA DEMETHYLATION

DNA methylation, which for many years was thought to be a semi-permanent modification of the genome that was only passively removed by dilution upon DNA replication, has recently been demonstrated to be actively reversed by the ten-eleven translocation (TET) family of proteins [95]. The TET proteins (TET1, TET2, and TET3) oxidize 5mC to 5hmC, with further oxidation possible to produce 5fC and 5-carboxylcytosine (5caC) [96,97]. TET1 is involved in the erasure of imprinting marks in primordial germ cells [98]. TET2 has pleiotropic roles during hematopoiesis through, at least in part, modulation of DNA methylation, and TET2 mutations are frequently found in hematopoietic malignancies [99]. TET3, meanwhile, is essential for the active erasure of paternal DNA methylation marks in the zygote, an important reprogramming event during preimplantation development [100]. There exist multiple mechanisms for the further removal of 5hmC following TET-mediated conversion of 5mC [101]. These include dilution as a result of DNA replication, as seen in preimplantation embryos [102]; the removal of 5fC and 5caC by thymine-DNA glycosylase followed by base excision repair [103,104]; and, there are recent suggestions that DNMT3A and DNMT3B are capable of converting 5hmC to cytosine, at least *in vitro* [105].

In addition to being intermediates of DNA demethylation, 5hmC, 5fC, and 5caC may also function as epigenetic marks. It is worth noting that various studies have established the 5hmC mark as being actively regulated in the genome. Notably, in mouse ES cells 5hmC is preferentially located in euchromatic regions, including promoter, exons, and transcription start sites [101]. Furthermore, as described above, several 5hmC-binding proteins (readers) have been identified. Data in ES cells has indicated that 5fC is enriched at poised enhancer elements, with greater binding of the acetyltransferase p300 at these regions [106]. However, given the low levels of these modifications (especially 5fC and 5caC) in most cell types, their significance as meaningful epigenetic marks remains to be established.

4.2 HISTONE DEMETHYLASES

Prior to the characterization of the first histone lysine demethylase (KDM), lysine-specific demethylase 1 (LSD1, also known as KDM1A) [107], it was believed that histone lysine methylation was not actively removed. Subsequently, numerous enzymes have been identified that demethylate lysine residues, encompassing two families with distinct enzymatic mechanisms (Table 3.4). The flavin adenine dinucleotide (FAD)-dependent amine oxidase family comprises two members, LSD1/KDM1A and LSD2/

Table 3.4 The Human Lysine Demethylase Family

Enzyme	Other names	Histone substrate	Cancer implication	References
KDM1A	LSD1	H3K4me1/2, H3K9me1/2	7.7% overexpress (bladder, colon, breast, prostate, neuroblastoma, sarcoma, lung, ovarian, liver); 4.8% mutated (pancreatic, bladder, lung, melanoma, uterine)	[244, 245, 259, 246]
KDM1B	AOF1	H3K4me1/2	14.4% overexpress (bladder, ovarian, melanoma, uterine, lung, liver, breast); 5.4% mutated (uterine, lung, stomach, melanoma, colon)	[244, 245, 259]
KDM2A	FBXL11/ JHDM1A	H3K36me1/2	9.4% mutated (pancreatic, bladder, stomach, uterine, melanoma, lung, ovarian, prostate, colon); 9% overexpress (head and neck, stomach, breast, bladder, melanoma, ovarian, prostate, sarcoma, liver)	[244, 245]
KDM2B	FBXL10/ JHDM1B	H3K36me1/2	6.9% mutated (colon, melanoma, lung, stomach, uterine, kidney, esophageal, head and neck); 2.6% overexpress (AML, bladder, pancreatic, uterine, sarcoma)	[244, 245, 259]
KDM3A	JHDM2a/ JMJD1A	H3K9me1/2	3% mutated (prostate, bladder, stomach, uterine, colon, head and neck, lung, breast, melanoma); 5.5% overexpress (colon, prostate, kidney, liver, sarcoma, lung, ovarian)	[244, 245, 259]
KDM3B	JHDM2b/ JMJD1B	H3K9me1/2	11.5% mutated (bladder, melanoma, uterine, colon, stomach, AML, ovarian, lung, liver, prostate, head and neck); 4.3% overexpress (kidney, sarcoma); 2.7% deleted (AML)	[244, 245]
KDM3C	JMJD1C	H3K9me1/2	8.3% mutated (colon, stomach, uterine, prostate, bladder, melanoma, lung, head and neck, breast, liver, kidney, esophageal); 6.6% overexpress (prostate, ovarian, breast)	[244, 245]
KDM4A	JMJD2A/ JHDM3A	H3K9, H3K36me2/3	7% overexpress (ovarian, bladder, uterine, sarcoma, breast, melanoma, head and neck, liver); 3% mutated (bladder, uterine, stomach, lung, melanoma, colon, head and neck)	[244, 245, 259]
KDM4B	JMJD2B	H3K9, H3K36me2/3	10% mutated (colon, uterine, stomach, lung, melanoma, breast, head and neck, liver); 5% overexpress (sarcoma, ACyC, glioma, breast, bladder, ovarian, colon, glioblastoma); 5% deleted (ACyC, uterine, stomach, ovarian, bladder)	[244, 245, 259, 246]
KDM4C	JMJD2C/ GASC1	H3K9, H3K36me2/3	11.5% deleted (bladder, lung, stomach, sarcoma, ovarian, uterine, melanoma, prostate, glioma, glioblastoma); 5% overexpress (breast, medulloblastoma, ACyC, head and neck, bladder, lung, stomach, sarcoma, ovarian, uterine, colon); 3.8% mutated (uterine, stomach, lung, melanoma, colon); translocated in lymphoma	[244, 245, 259, 246]

(Continued)

Table 3.4 (Continued)

Enzyme	Other names	Histone substrate	Cancer implication	References
KDM4D	JMJD2D	H3K9me2/3	6.4% overexpress (ovarian, lung, prostate, head and neck, breast, glioblastoma); 2.5% mutated (uterine, lung, melanoma, colon, esophageal, bladder, stomach); 1.8% deleted (head and neck, uterine, melanoma)	[244, 245]
KDM5A	JARID1A/ RBP2	H3K4me2/3	12% overexpress (ovarian, bladder, lung, uterine, head and neck, breast, glioma, sarcoma, glioblastoma); 7% mutated (bladder, uterine, lung, colon, stomach, melanoma, cervical, prostate, breast, head and neck, liver); deleted (melanoma); translocated (acute leukemia)	[244, 245, 259, 246]
KDM5B	JARID1B/ PLU-1	H3K4me2/3	14% overexpress (liver, breast, melanoma, lung, ovarian, uterine, B-cell lymphoma, pancreatic, prostate, bladder, stomach, sarcoma); 11.5% mutated (bladder, melanoma, lung, uterine, colon, pancreatic, stomach, head and neck)	[244, 245, 259, 246]
KDM5C	JARID1C/ SMCX	H3K4me2/3	9.7% mutated (colon, kidney, uterine, stomach, lung, breast, melanoma, pancreatic); 9.8% overexpressed (prostate, sarcoma, B-cell lymphoma, ovarian, uterine); 4% deleted (head and neck)	[244, 245, 259, 246]
KDM5D	JARID1D/ SMCY	H3K4me2/3	21.3% deleted (prostate); 4% mutated (colon, lung, melanoma, bladder)	[244, 245, 259]
KDM6A	UTX	H3K27me2/3	32% mutated (bladder, medulloblastoma, multiple myeloma, head and neck, ACyC, stomach, uterine, lung, kidney, breast, colon, AML, CML, prostate, pancreatic); 5.5% overexpress (uterine, prostate, sarcoma, ovarian); 5% deleted (head and neck, stomach, prostate)	[244, 245, 259, 246]
KDM6B	JMJD3	H3K27me2/3	7.7% deleted (bladder, prostate, lung, liver); 5.6% mutated (colon, melanoma, stomach, lung, uterine, breast, kidney, ACC, ACyC); overexpressed in lung, liver, and Hodgkin's lymphoma	[244, 245, 259, 246]
KDM6C	UTY	None identified	16.4% deleted (prostate); 7% mutated (lung, bladder, colon)	[244, 245]
KDM7A	JMJD1D	H3K9me2, H3K27me2, H4K20me1	11% overexpress (uterine, melanoma, ovarian, glioblastoma, prostate, lung, stomach, glioma, sarcoma); 5% mutated (melanoma, ACC, stomach, head and neck, lung, uterine); 2.7% deleted (AML)	[244, 245]

(Continued)

Table 3.4 (Continued)

Enzyme	Other names	Histone substrate	Cancer implication	References
KDM7B	PHF8	H3K9me1/2, H3K27me2, H4K20me1	5% mutated (uterine, stomach, colon, lung, melanoma); 5% overexpress (uterine, prostate, sarcoma, ovarian, glioma); 3.6% deleted (head and neck)	[244, 245]
KDM7C	PHF2	H3K9me2	11% mutated (breast, colon, melanoma, uterine, stomach, lung); 3% overexpress (sarcoma, head and neck)	[244, 245, 259]
KDM8	JMJD5	h3k36me2	4.3% overexpress (breast, bladder, B-cell lymphoma); 3% mutated (uterine, lung, melanoma, stomach)	[244, 245]

KDM1B, which are capable of demethylating H3K4me1/2, but not H3K4me3 [107,108]. The jumonji C (JmjC)–domain-containing, iron- and α–ketoglutarate-dependent dioxygenase family has over 30 members in humans. Unlike the KDM1 family, these enzymes are capable of removing tri-methyl marks (in addition to mono- and di-methyl marks), as they do not require a free electron pair due to their mechanism of action. The JmjC-containing family members have been subdivided into five separate groups (JHDM1/KDM2, JMJD1/JHDM2/KDM3, JMJD2/JHDM3/KDM4, JARID1/KDM5, and UTX/JMJD3/KDM6), with each subfamily exhibiting activity for specific lysine residue(s). There remain some histone methyl marks (e.g., H3K79) for which no demethylase has been identified to date [109].

For arginine methylation, there have been no conclusive reports identifying a demethylase so far. The enzyme peptidylarginine deiminase 4 (PAD4) converts methylarginine residues to citrulline [110]. However, PAD4 also acts on unmodified arginines, and cannot demethylate dimethylated arginines, so this activity is not considered that of a *bona fide* demethylase. It has been reported that JMJD6 serves as an arginine demethylase [111], although more recent data suggested that JMJD6 serves instead as a lysyl-5 hydroxylase [112]. Thus, to date no equivalent to the KDM family has been identified for methylated arginines.

4.3 HISTONE DEACETYLASES (HDACS)

The removal of the histone acetylation mark has a negative effect on transcription, as it leads to increased chromatin compaction, while also removing binding sites for various proteins. Histone deacetylases (HDACs), one of the families that remove this mark, comprise a group of proteins that is divided into four distinct classes (I, IIa, IIb, and IV) based on homology with yeast proteins and structure and function (Table 3.5) [113]. The mammalian HDAC family consists of 11 proteins, each containing a conserved deacetylase domain that requires a zinc atom for activity. HDACs lack DNA-binding motifs and as such are recruited to DNA through their presence in large multi-protein complexes, which contain subunits that bind to specific DNA regions to target HDAC activity and thus gene repression [114]. For example, the methylated DNA–binding protein MeCP2 binds to methylated DNA, and recruits an HDAC complex to deacetylate histones and silence the locus [115,116]. HDAC1, 2, 3, and 8 are the members of the class I family, and in general show widespread tissue expression and nuclear localization [117]. Class

I HDACs are present in several complexes, including the Sin3, NuRD, and CoREST complexes [118], which have a variety of cellular effects. Interestingly, these complexes also contain lysine demethylase enzymes, allowing for the coordination of the removal of multiple histone marks. The class I HDACs also all contain motifs in a C-terminal tail that allow for phosphorylation, providing another mechanism to modulate their function [119]. Class IIa enzymes, HDAC4, 5, 7, and 9, contain an N-terminal extension that results in them also being responsive to signaling pathways, through phosphorylation by numerous kinases including CaMK I, CaMK IV, PKD, and MARK1 [120]. Class IIa enzymes, which contain both nuclear export and nuclear localization signals, shuttle between the cytoplasm and nucleus [121]. For example, phosphorylation of HDAC4 and HDAC5 results in their export from the nucleus, and relieves their inhibitory effects on the MEF2 transcription factor [122,123]. The recruitment of these enzymes to chromatin is mediated by a variety of transcription factors, including GATA, Forkhead, and hypoxia-inducible factors (HIFs) [120]. HDAC6 and 10 are the class IIb enzymes, which are cytoplasmic and nuclear proteins with a variety of substrates [124]. HDAC11, meanwhile, is the only class IV enzyme, and shows homology to class I and class II enzymes [125]. HDAC11 regulates the expression of IL-10 in antigen-presenting cells, suggesting it has a role in immunity [126].

Sirtuins, which are sometimes referred to as the class III HDAC family, consist of seven family members, six of which actively remove lysine acetylation (Table 3.5). In contrast to the other classes of deacetylases, sirtuins function through a NAD^+ pathway to mediate their enzymatic reaction [127]. Sirtuin activity is regulated by several mechanisms, including protein expression level, complex formation, and the levels of cofactor NAD^+ [128]. The requirement for NAD^+ links the activity of these enzymes to cellular metabolism, and they have been suggested to play key roles in organism longevity [128]. The sirtuin family has a variety of activities; with current data indicating that SIRT1-3 and SIRT5-7 function as deacetylases, SIRT4 and SIRT6 have ADP-ribosyltransferase function, and SIRT5 also acts as a desuccinylase and demalonylase [129]. These enzymes have varied tissue expression and cellular localization, further pointing to specialization of function. SIRT1, SIRT6, and SIRT7 are localized to different regions of the nucleus, while SIRT3, SIRT4, and SIRT5 are present in the mitochondria, and SIRT2 is found in the cytoplasm [130]. The three nuclear sirtuins have different localizations, with SIRT1 excluded from nucleoli, SIRT6 present throughout the nucleus, and SIRT7 predominantly in the nucleolus [130]. SIRT1 has strong activity against multiple histone targets, including residues in histones H1, H3, and H4, as well as a long list of other proteins [131,132]. SIRT1 knockout mice exhibit a wide variety of background strain-dependent negative phenotypes, with the most severe being death in the first month of life [131]. SIRT6 was first characterized as a H3K9 deacetylase, affecting telomere structure [133], while it also appears to negatively affect the transcriptional activation of several transcription factors, including NF-kappaB [134] and HIF1α [135]. SIRT7, meanwhile, plays a role in RNA polymerase I transcription, as expected from its nucleolar localization [136]. SIRT2 is predominantly cytoplasmic, where it deacetylates multiple proteins including tubulin [137]; during the G^2 to M transition it apparently is involved in chromatin condensation [138].

5 INTERACTIONS BETWEEN THE VARIOUS COMPONENTS

Early studies of histone modifications dealt with their effect in isolation, however, chromatin can have multiple modifications at a given locus, which combine to determine the activity of that region. Thus, recent experiments have demonstrated a coordination of various epigenetic marks that combine to affect

Table 3.5 The Human Histone Deacetylase Family

Enzyme	Class	Cancer implication	References
HDAC1	I	4% overexpress (ovarian, sarcoma, lung); 1.7% mutated (uterine, liver)	[244, 245]
HDAC2	I	10% deleted (prostate, B-cell lymphoma, ACyC, melanoma); 4.2% mutated (colon, lung, melanoma, uterine); 4.7% overexpress (sarcoma)	[244, 245]
HDAC3	I	4.8% overexpress (kidney, pancreatic, sarcoma); 3.8% mutated (uterine, lung, melanoma)	[244, 245]
HDAC4	II A	8.3% mutated (colon, bladder, lung, melanoma, uterine, stomach); 5.6% deleted (cervical, glioma, prostate, bladder, sarcoma); 2.3% overexpress (ovarian)	[244, 245]
HDAC5	II A	6.6% deleted (prostate, ACC); 4% mutated (colon, prostate, uterine); 3.6% overexpress (uterine)	[244, 245]
HDAC6	II B	7.3% overexpress (uterine, ovarian, cervical, prostate, sarcoma); 7% mutated (lung, uterine, stomach); 3.6% deleted (head and neck)	[244, 245]
HDAC7	II A	16.7% deletion (ACyC); 7% overexpress (B-cell lymphoma, sarcoma, uterine, ACC); 4% mutated (colon, uterine, stomach)	[244, 245]
HDAC8	I	8% overexpress (prostate, ovarian); 2.4% deletion (sarcoma)	[244, 245]
HDAC9	II A	15.7% mutated (melanoma, lung, colon, uterine, head and neck); 8.2% overexpress (prostate, bladder, pancreatic, sarcoma); 7% deleted (B-cell lymphoma)	[244, 245]
HDAC10	II B	10.3% deleted (ovarian, bladder, cervical, pancreatic, stomach); 3.3% mutated (melanoma); 2.4% overexpress (liver, sarcoma)	[244, 245]
HDAC11	IV	10.2% overexpress (bladder, uterine, sarcoma, ovarian); 3.6% deleted (kidney); 3.3% mutated (prostate, melanoma, stomach, uterine)	[244, 245]
SIRT1	III	6.6% overexpress (prostate, ovarian, uterine); 3% mutated (kidney, colon, pancreatic, stomach, lung, uterine)	[244, 245]
SIRT2	III	22% overexpress (bladder, uterine, ovarian, ACyC, cervical, sarcoma, lung, breast, stomach, colon); 2.5% mutated (melanoma, uterine)	[244, 245]
SIRT3	III	3.3% overexpress (ACyC); 3% deleted (glioma, sarcoma, ovarian, breast, glioblastoma, bladder, uterine); 2.8% mutated (colon, melanoma, uterine, head and neck)	[244, 245]
SIRT4	III	6.6% overexpress (prostate, ACC, ovarian, sarcoma, stomach, glioblastoma); 2.8% mutated (colon); 1.6% deleted (lung)	[244, 245]
SIRT5	III	11% overexpress (ovarian, melanoma, bladder, uterine, liver, breast, lung); 2% mutated (uterine, stomach)	[244, 245]
SIRT6	III	3.6% deleted (uterine, ovarian, stomach); 3% overexpress (sarcoma, glioma); 1.7% mutated (lung)	[244, 245]
SIRT7	III	8.5% overexpress (liver, ovarian, bladder, uterine, breast, melanoma, prostate, sarcoma, kidney, ACC, ACyC); 2% mutated (uterine)	[244, 245]

gene expression. These effects are not limited to the same type of modification, as growing evidence indicates crosstalk occurs between histone modifications and DNA methylation. This includes the demonstration that defective histone demethylation, due to knockout of Kdm1b/Lsd2, results in abnormal DNA methylation and germline imprinting in mice [108]. Furthermore, Kdm1a/Lsd1 knockout mice exhibit severe developmental defects, in large part due to a loss of demethylation of Dnmt1, which causes its degradation and results in a loss of DNA methylation [139]. Reports have indicated that methylation of DNA-methylation-sensing proteins, arginines alters their activity, suggesting a crosstalk between arginine methylation and DNA methylation [140]. The deposition of a specific histone modification can also affect the subsequent addition of other epigenetic marks, including interaction between marks on different tails. This has in particular been demonstrated in the case of H3K4 and H3K79 methylation, which requires monoubiquitylation of H2B for their addition [141]. Asymmetric methylation of H3R2 blocks the methylation of the H3K4 by Set1 in yeast [142], reinforcing the transcriptionally repressive effect of this mark. Conversely, the addition of SDMA to H3R2 leads to binding of WDR5, which recruits various coactivator complexes and ensures the associated chromatin is not silenced [143]. Coordinated histone modification can also be achieved via the interaction of multiple proteins, which guides different enzymatic activities to the same locus. Recent results have demonstrated that the PRC2 complex is recruited, in stem cells, to specific DNA regions based on DNA sequence and transcription factor occupancy [144]. Two components of this complex are the protein EZH2, which mediates repressive chromatin formation through the deposition of H3K27 methylation, and DNMT enzymes, which methylate DNA to inhibit gene transcription, allowing for the coordination of repressive histone and DNA marks [145]. Signaling pathways can also lead to the coordinated activation of multiple histone modifications. Stimulation of cells with estrogen leads to multiple histone modifications, beginning with acetylation of H3K18, followed by acetylation of H3K23, and then recruitment of PRMT4/CARM1 to methylate H3R17 [40]. Thus, coordination of multiple components enables a specific gene expression pattern in a given cell, so it is perhaps not surprising that multiple enzymes and marks work together.

6 EPIGENETICS AND CANCER

A recent analysis identified approximately 40 epigenetic regulators that exhibit some form of alteration in cancer, demonstrating the important emerging role these proteins play in oncology [1]. These findings open up new avenues of pursuit for the understanding of cancer initiation and progression as well as for treatment. With the increased use of whole genome sequencing and genome-wide association studies, more associations between epigenetics and cancer are certain to be identified. Not only are epigenetic proteins directly altered in various cancers, but other mutations may impinge on the proper function of these enzymes. For example, recurrent mutations in the enzymes IDH1 and IDH2 alter their enzymatic activity such that these enzymes now produce a molecule (2-hydroxyglutarate) that may inhibit several dioxygenases, including TET2 [146] and KDM4C [147]. This indicates that not only are epigenetic factors directly altered in cancer, but mutations in other pathways can impinge on the regulation of gene expression.

6.1 DNA METHYLATION AND CANCER

The role that epigenetics plays in cancer was initially demonstrated by observations of altered DNA methylation [148]. While global hypomethylation is commonly observed in tumors, potential tumor suppressors, including genes encoding proteins involved in the cell cycle, signal transduction, and transcription, are often hypermethylated, resulting in their suppression [149]. The global hypomethylation is thought to result in increased expression of pro-tumorigenic proteins along with impaired genome stability and imprinting [150]. Somatic heterozygous mutations in DNMT3A are found in ~20% of patients with acute myeloid leukemia (AML) and occur, at lower frequencies, in other hematological malignancies [151,152]. While the mechanism by which heterozygous mutations of DNMT3A contributes to leukemogenesis is not fully understood, recent evidence suggests that at least some of the mutations exhibit dominant-negative effects by inhibiting the function of the wild-type DNMT3A allele [153–155]. Genetic alterations in TET1 and TET2 have also been identified in cancer. In leukemia, translocations that lead to gene fusions commonly occur with the mixed lineage leukemia (MLL) gene [156,157]. One such MLL fusion partner in AML is TET1, resulting in increased 5hmC levels and enhanced expression of MLL target genes [158,159]. TET2 is frequently mutated in AML and other hematological malignancies [99]. Many tumor tissues exhibit decreased 5hmC levels compared to healthy controls; however, in low-grade gliomas there is an increase in 5hmC, suggesting that there may be a cell of origin effect on the role of TET proteins in cancer [159]. Conversely, recent data has demonstrated that 5hmC levels are decreased in melanoma, and can be used for both prognosis and diagnosis [160].

6.2 HISTONE METHYLATION AND CANCER

A wide range of mutations, affecting numerous histone modifying pathways and occurring in a variety of cancers, have been identified for proteins that affect histone methylation levels [161]. Histone methylation pathways are commonly altered in cancer, including methylation of histone H3K4 (via the MLL methyltransferase family, as well as KDM5A and KDM5C), H3K27 (EZH2 methyltransferase and UTX demethylase), H3K36 (NSD methyltransferase family), and H3K79 (DOT1L) [162]. In the majority of situations, cancer cells exhibit haploinsufficiency of the associated genes, although EZH2-activating mutations have been described [163,164]. Wild-type EZH2 methylates (mono, di, and tri) H3K27, which serves as a transcriptional repressive mark; the enzyme is most efficient at catalyzing the monomethyl step of the reaction. However, certain lymphomas carry heterozygous mutations (of residues Tyr641 or Ala677) that render the enzyme more efficient at catalyzing the addition of the trimethyl mark [165]. Thus, heterozygous mutant cells exhibit enhanced efficiency at depositing all stages of this repressive mark, likely resulting in the silencing of genes that negatively affect tumor growth. In an alternative mechanism to enhance EZH2 activity, approximately half of all prostate cancers contain a translocation resulting in the generation of a TMPRSS2-ERG fusion protein. ERG is a transcription factor that can increase the transcription of EZH2, leading to increased silencing of EZH2 target genes, which thereby contributes to cancer progression [166]. DOT1L is a key component in the increased transcription mediated by MLL fusion proteins in leukemia, via the abnormal deposition of H3K79 methyl marks [167]. It is required for immortalization in cells containing MLL-fusion proteins, and for leukemia formation [168–170].

Lysine demethylases also have diverse roles in cancer development. Some examples include the demonstration that LSD1/KDM1A overexpression has been correlated with poor prognosis in breast and prostate cancer [171,172]. In pancreatic cancer, it has been shown that cooperation of KDM1A and HIF1α regulates glycolysis, contributing to cancer progression [173]. KDM2B has been suggested, based on expression levels and insertional mutagenesis, to function as a tumor suppressor [161], while the closely related KDM2A is overexpressed in lung cancer and promotes its development, through effects on the ERK1/2 pathway and/or HDAC3 [174,175]. A novel role for the enzyme KDM4A was recently demonstrated, as increased protein expression led to copy number increase of particular regions of the genome [176]. Results, meanwhile, have indicated that KDM4B acts as a cofactor for the estrogen receptor (ER), and decreasing its expression in breast cancer cells affects their tumorigenic properties [177]. KDM5B is commonly overexpressed in certain cancers, and depletion of this protein in cancer cell lines can increase tumor suppressor expression and decrease proliferation [161]. While most demethylases to date have been found to have altered expression in cancer, UTX/KDM6A was one of the first demethylases to demonstrate point mutations in cancer, with loss-of-function mutations observed in multiple types of cancer [178].

Overexpression, but not mutation, of PRMT genes is a common feature in cancer [29]. In fact, the only family member that to date has demonstrated appreciable mutation rates in cancer is PRMT8, where mutations have been identified in ovarian, skin, and colon cancer [29]. All other PRMT family members are overexpressed in tumors, and emerging mechanistic investigations are exploring their contribution to cancer. PRMT1 is a component of an MLL complex that also exhibits acetyltransferase activity and is a requisite component for cellular transformation driven by this fusion protein [31]. PRMT2 is upregulated in breast cancer, with isoforms identified in these tumors that increase ERα activity [179]. PRMT4/CARM1 has widespread roles in cancer, including an ability to serve as a coactivator in ERα-dependent breast cancer [180,181], which may explain its observed overexpression in a subset of breast cancers [32]. Furthermore, at the molecular level, PRMT4 regulates transcription of the Cyclin E1 gene, through histone methylation [182], and directly methylates the SWI/SNF subunit BAF155, enhancing tumorigenesis [183]. Altered miRNA expression patterns lead to PRMT5 overexpression in hematological malignancies, and this causes downregulation of RB family proteins [33]. PRMT5 also directly methylates the tumor suppressors p53 [184] and PDCD4 [185], affecting the activity of both proteins. PRMT6 expression is increased in a variety of cancers [30], perhaps in part due to its negative effects on the expression of tumor suppressors [186].

6.3 HISTONE ACETYLATION AND CANCER

Because histone acetylation plays a role in chromatin compaction, it is perhaps not surprising that proteins that control the balance of this mark have been implicated in human cancer. Decreased acetylation has been associated with tumorigenesis, invasion, and metastasis [187,188]. Global loss of monoacetylated H4K16, along with trimethylated H4K20, is a common feature of cancer cells [187]; the H4K16ac decrease occurs in multiple cancer types, and correlates with sensitivity to chemotherapy [189]. The deposition of this mark is predominantly mediated by MOF/MYST1/KAT8, for which decreased expression has been observed in multiple cancers [190,191]. Acetylation mediated by CBP and p300 has also been implicated in cancer, [162] with p300 proposed to have both tumor suppressive and promoting activities, depending on the cellular context [192]. Both somatic mutations and deletions in acetyltransferases, predominantly involving CBP, have been identified in approximately 40% of certain lymphomas [193] and 18% of relapsed leukemia [194]. Germline mutations in CBP result in Rubinstein–Taybi syndrome, which includes many developmental abnormalities and an increased risk of cancer [195]. Mutations in the p300 gene were also noted in a variety of solid tumors, with the alterations typically resulting in truncated proteins or occurring in key residues [196]. Some common translocations involving KATs have been characterized, including *MLL-CBP* [197] and *MOZ-TIF2*

[198,199]. The MOZ-TIF2 fusion protein is sufficient for cellular transformation [200,201], through interactions with the transcription factor PU.1 that enhance macrophage colony-stimulating factor receptor (CSF1R) protein levels [202]. Similarly, GCN5 and PCAF are members of a complex that act as coactivators of various oncogenes, including the E2F transcription factors, p53, and BRCA1/2 [203]. Histone acetylation readers can also be altered in cancer. NUT midline cancer occurs upon fusions of the *NUT* gene with other activators; the predominant fusion partner is bromodomain-containing protein 4 (*BRD4*) [204]. This fusion protein retains the bromodomains of BRD4, targeting the protein to acetylated histones and leading to MYC expression [205]. The protein Twist, when acetylated, interacts with BRD4 and results in increased expression of WNT5A, which promotes tumorigenicity in breast cancer [206]. There have also been suggestions that BRD4 enhances NF-κB activity, thereby promoting proliferation of cancer cells [207].

HDAC proteins also have a role in cancer development, but to date few somatic mutations in this family have been identified [1]. The most common alteration in HDACs in cancer is overexpression, which has been observed in a wide variety of cancers [208]. However, 43% of colon cancers with microsatellite instability contain a frameshift mutation in HDAC2, with a corresponding loss of protein expression [209,210]. Another mechanism that has often been observed in hematological malignancies is the mistargeting of HDAC-containing complexes to specific genomic regions by fusion proteins, such as AML1-ETO [211]. In acute promyelocytic leukemia, which is driven by the fusion proteins RAR-PML and RAR-PLZF, the fusion proteins recruit an HDAC-containing complex to specific genomic loci [212]. The majority of sirtuin proteins have also been associated with cancer [131]. SIRT1 is overexpressed in a variety of tumors [129], and is often correlated with poorer prognosis [213–215]. Conversely, mouse model studies have suggested that Sirt1 may serve as a tumor suppressor, although to date no obvious deletions or point mutations have been noted in human cancer samples [216]. SIRT2, similar to SIRT1, exhibits functions that may consist of both tumor suppressor and oncogenic role(s) [131]. Thus, SIRT2 shows decreased expression levels in some cancers, while increased levels in others. SIRT2-knockout mice develop cancer, with males showing hepatocellular carcinoma and females breast cancer, predominantly, suggesting that SIRT2 functions as a tumor suppressor [217]. SIRT7 deacetylation of H3K18ac has been implicated in cellular transformation [218]. SIRT6, meanwhile, has been demonstrated to function as a tumor suppressor by controlling glycolysis [219]; its expression is commonly downregulated in human tumors [131].

7 EPIGENETIC PROTEINS AS THERAPEUTIC TARGETS

Due to the increasing evidence for the role of epigenetic factors in human cancer, there is a growing interest in exploiting these factors therapeutically. Despite the promise of epigenetic therapy, however, to date limited therapeutic modalities have been approved for treatment. HDAC inhibitors, such as SAHA (vorinostat), are currently approved for treatment of cutaneous T-cell lymphoma [220], while therapeutic activity in other tumor types is being assessed. Another HDAC inhibitor, romidepsin, has shown promising activity against peripheral T-cell lymphoma, resulting in FDA approval [221]. While HDAC inhibitors have to date been among the more successful epigenetic-based therapies, substantial work has been initiated to develop newer molecules that will expand the utility of these treatments [222]. The nucleoside analogues azacytidine and decitabine, which serve as DNMT inhibitors, are also being studied in the clinic and have demonstrated promising results in hematological malignancies [223].

Newer generation inhibitors, such as zebularine, which has been found to affect cancer cell proliferation more than normal cells [224], are being developed in order to expand the indication for this category of molecules.

For histone methylation, early drug discovery has revolved around a few well-characterized targets, such as EZH2 and DOT1L, where recently specific inhibitors of both have been identified [225]. Small molecule inhibitors of EZH2 decrease cellular proliferation, and cause cell cycle arrest and apoptosis, of lymphoma cells carrying an EZH2-activating mutation [226,227]. Similarly, inhibitors of DOT1L decrease H3K79 methylation and selectively killing leukemia cells containing MLL translocations, but not cells with wild-type MLL [228]. The inhibitors of DOT1L and EZH2 have recently entered Phase I trials, examining tolerability and safety in patients with leukemia and lymphoma [225]. The ability to achieve relatively specific inhibition of EZH2, including a 35-fold specificity over the closely related EZH1 [229], offers hope that other KMTs can also be specifically targeted in the future. Other targeted molecules, including one that specifically blocks the methyltransferase activity of MLL1 and affects MLL-driven cell survival and proliferation, have also been described [230]. Similarly, the recent identification of some small molecules that demonstrate selectivity between PRTM1 and PRMT4 [231] suggests that specific inhibitors of arginine methyltransferases may be achievable.

Recently, a group of pharmaceutical and academics entities have partnered, as part of the Structural Genomics Consortium, with the goal of determining crystal structures for medically relevant protein targets, with an emphasis on epigenetic modifiers [232]. The resulting data should enhance the ability to rationally design small molecule inhibitors of these proteins. The crystal structure of the histone demethylase KDM1A has allowed for a greater understanding of the functional site of the protein, including the fact that multiple amino acids from the histone tail appear to enter the active site, which may serve to improve its specificity, and the lysine methylation status likely does not affect binding [233]. Using structure-based analysis, the first specific inhibitors for the H3K27 demethylases UTX/KDM6A and JMJD3/KDM6B were identified, and subsequently showed biological activity in primary human macrophages [234]. The similarity in active sites amongst many of these family members will pose challenges for the identification of specific inhibitors, but these efforts will be greatly aided by improved structural understanding of the proteins.

The use of epigenetic inhibitors may also provide a mechanism to target other protein families, especially transcription factors, that have so far proved challenging to target therapeutically. One example that has been widely explored recently is the MYC family of transcription factors, which are frequently mutated in cancer but have proven refractory to drug targeting to date. Inhibitors of the bromodomain of the bromodomain and extraterminal (BET) domain-containing family reduce transcription from a variety of natural and modified (translocated, amplified) MYC loci, resulting in cell cycle arrest and apoptosis of leukemia and lymphoma cell lines [235]. An unbiased RNAi screen of chromatin modifying enzymes in an AML mouse model demonstrated that the protein BRD4 was critical for cancer progression [236]. Mechanistic studies, involving RNAi reagents and small molecule inhibitors, showed that this effect was mediated through alteration of MYC activity. Notably, treatment with the small molecule JQ1, a potent and selective inhibitor of the bromodomain of BRD4 that blocks its ability to bind acetylated lysines, results in differentiation and decreased proliferation of BRD4-dependent cells [237]. MLL fusion proteins are present in complexes with BET family members [238], and inhibition of the BET-containing proteins, using JQ1, affected complex interaction with DNA, resulting in decreased transcription of several known oncogenes [239]. Further studies involving various cancer cell models have demonstrated that BET inhibitors affect cancer cell growth, and can lead to silencing of MYC [240].

Studies of combined epigenetic therapies are being initiated, based on the belief that inhibiting multiple related pathways may have a stronger therapeutic benefit than modifying any single pathway. In one example, it was hypothesized that inhibiting the repressive effects of both DNA methylation and histone deacetylation could cooperate for greater benefit. In a study of non-small cell lung cancer patients with recurrent, previously treated disease, 4 out of 19 patients showed major responses when treated with a combination of DNMT and HDAC inhibitors prior to other treatment [241]. In cell culture, combined treatment of an EZH2 inhibitor and an HDAC inhibitor exhibited better activity than either inhibitor alone in a human AML cell line [242]. Considering the relatively recent appreciation for the role the majority of these enzymes play in oncogenesis, substantial progress in many different avenues has been made to identify small molecule therapeutics, which will hopefully lead to improved patient outcomes in the near future.

8 CONCLUSION AND FUTURE OPPORTUNITIES

A complex network, involving a variety of proteins, is required in order to maintain the desired gene expression programs; defects in these pathways can result in cancer. While a great deal has already been learned about epigenetic factors, the relatively recent identification of many of the proteins, the expansion of novel techniques for studying chromatin dynamics, and the continued exploration of their various biological properties suggest that new and exciting discoveries remain to be made in this field. Emerging evidence is demonstrating that a wide variety of human malignancies coopt these pathways in order to provide a selective growth advantage. These observations, in combination with an ever expanding dataset derived from whole genome sequencing of patient tumors to identify key driver mutations, should allow for personalized treatment of patients as the field matures. It is tempting to equate the writer/reader/eraser epigenetic field to that of the kinase/phosphatase field of several decades ago, where the discovery of Imatinib to target BCR-ABL driven CML began the field of personalized medicine in oncology [243]. While there have been complications due to resistance developing to these targeted therapies, they offer an exciting path forward for treating patients. Epigenetic modifications, which also involve the posttranslational addition and removal of a small molecule (methylation/acetylation versus phosphorylation), could be the next field that is exploited to benefit cancer patients.

REFERENCES

[1] Dawson Mark A, Kouzarides T. Cancer epigenetics: from mechanism to therapy. Cell 2012;150(1):12–27.

[2] Ernst J, Kellis M. Discovery and characterization of chromatin states for systematic annotation of the human genome. Nat Biotechnol 2010;28(8):817–25.

[3] Ehrlich M, Gama-Sosa MA, Huang L-H, Midgett RM, Kuo KC, McCune RA, et al. Amount and distribution of 5-methylcytosine in human DNA from different types of tissues or cells. Nucleic Acids Res 1982;10(8):2709–21.

[4] Shenker N, Flanagan JM. Intragenic DNA methylation: implications of this epigenetic mechanism for cancer research. Br J Cancer 2012;106(2):248–53.

[5] Lei H, Oh SP, Okano M, Juttermann R, Goss KA, Jaenisch R, et al. De novo DNA cytosine methyltransferase activities in mouse embryonic stem cells. Development 1996;122(10):3195–205.

[6] Li E, Bestor TH, Jaenisch R. Targeted mutation of the DNA methyltransferase gene results in embryonic lethality. Cell 1992;69(6):915–26.

[7] Bostick M, Kim JK, Esteve PO, Clark A, Pradhan S, Jacobsen SE. UHRF1 plays a role in maintaining DNA methylation in mammalian cells. Science 2007;317(5845):1760–4.

[8] Sharif J, Muto M, Takebayashi S, Suetake I, Iwamatsu A, Endo TA, et al. The SRA protein Np95 mediates epigenetic inheritance by recruiting Dnmt1 to methylated DNA. Nature 2007;450(7171):908–12.

[9] Nishiyama A, Yamaguchi L, Sharif J, Johmura Y, Kawamura T, Nakanishi K, et al. Uhrf1-dependent H3K23 ubiquitylation couples maintenance DNA methylation and replication. Nature 2013;502(7470):249–53.

[10] Kaneda M, Okano M, Hata K, Sado T, Tsujimoto N, Li E, et al. Essential role for de novo DNA methyltransferase Dnmt3a in paternal and maternal imprinting. Nature 2004;429(6994):900–3.

[11] Okano M, Bell DW, Haber DA, Li E. DNA methyltransferases Dnmt3a and Dnmt3b are essential for de novo methylation and mammalian development. Cell 1999;99(3):247–57.

[12] Hata K, Okano M, Lei H, Li E. Dnmt3L cooperates with the Dnmt3 family of de novo DNA methyltransferases to establish maternal imprints in mice. Development 2002;129(8):1983–93.

[13] Bourc'his D, Xu GL, Lin CS, Bollman B, Bestor TH. Dnmt3L and the establishment of maternal genomic imprints. Science 2001;294(5551):2536–9.

[14] Ooi SK, Qiu C, Bernstein E, Li K, Jia D, Yang Z, et al. DNMT3L connects unmethylated lysine 4 of histone H3 to de novo methylation of DNA. Nature 2007;448(7154):714–17.

[15] Mikkelsen TS, Ku M, Jaffe DB, Issac B, Lieberman E, Giannoukos G, et al. Genome-wide maps of chromatin state in pluripotent and lineage-committed cells. Nature 2007;448(7153):553–60.

[16] Hon GC, Hawkins RD, Ren B. Predictive chromatin signatures in the mammalian genome. Hum Mol Genet 2009;18(R2):R195–201.

[17] Hon G, Wang W, Ren B. Discovery and annotation of functional chromatin signatures in the human genome. PLoS Comput Biol 2009;5(11):e1000566.

[18] Steger DJ, Lefterova MI, Ying L, Stonestrom AJ, Schupp M, Zhuo D, et al. DOT1L/KMT4 recruitment and H3K79 methylation are ubiquitously coupled with gene transcription in mammalian cells. Mol Cell Biol 2008;28(8):2825–39.

[19] Kimura H. Histone modifications for human epigenome analysis. J Hum Genet 2013;58(7):439–45.

[20] Balakrishnan L, Milavetz B. Decoding the histone H4 lysine 20 methylation mark. Crit Rev Biochem Mol Biol 2010;45(5):440–52.

[21] Bernstein BE, Mikkelsen TS, Xie X, Kamal M, Huebert DJ, Cuff J, et al. A bivalent chromatin structure marks key developmental genes in embryonic stem cells. Cell 2006;125(2):315–26.

[22] Allis CD, Berger SL, Cote J, Dent S, Jenuwien T, Kouzarides T, et al. New Nomenclature for Chromatin-Modifying Enzymes. Cell 2007;131(4):633–6.

[23] Shinkai Y, Tachibana M. H3K9 methyltransferase G9a and the related molecule GLP. Genes Dev 2011;25(8):781–8.

[24] Kim W, Choi M, Kim JE. The histone methyltransferase Dot1/DOT1L as a critical regulator of the cell cycle. Cell Cycle 2014;13:5.

[25] Singh P, Han L, Rivas GE, Lee DH, Nicholson TB, Larson GP, et al. Allele-specific H3K79 Di- versus trimethylation distinguishes opposite parental alleles at imprinted regions. Mol Cell Biol 2010;30(11):2693–707.

[26] Jones B, Su H, Bhat A, Lei H, Bajko J, Hevi S, et al. The histone H3K79 methyltransferase Dot1L is essential for mammalian development and heterochromatin structure. PLoS Genet 2008;4(9):e1000190.

[27] Kuzmichev A, Jenuwein T, Tempst P, Reinberg D. Different EZH2-containing complexes target methylation of histone h1 or nucleosomal histone H3. Mol Cell 2004;14(2):183–93.

[28] Boisvert FM, Cote J, Boulanger MC, Richard S. A proteomic analysis of arginine-methylated protein complexes. Mol Cell Proteomics 2003;2(12):1319–30.

[29] Yang Y. Bedford MT. Protein arginine methyltransferases and cancer. Nat Rev Cancer 2013;13(1):37–50.

[30] Yoshimatsu M, Toyokawa G, Hayami S, Unoki M, Tsunoda T, Field HI, et al. Dysregulation of PRMT1 and PRMT6, Type I arginine methyltransferases, is involved in various types of human cancers. Int J Cancer 2011;128(3):562–73.

[31] Cheung N, Chan LC, Thompson A, Cleary ML, So CW. Protein arginine-methyltransferase-dependent oncogenesis. Nat Cell Biol 2007;9(10):1208–15.

[32] Cheng H, Qin Y, Fan H, Su P, Zhang X, Zhang H, et al. Overexpression of CARM1 in breast cancer is correlated with poorly characterized clinicopathologic parameters and molecular subtypes. Diagn Pathol 2013;8:129.

[33] Wang L, Pal S, Sif S. Protein arginine methyltransferase 5 suppresses the transcription of the RB family of tumor suppressors in leukemia and lymphoma cells. Mol Cell Biol 2008;28(20):6262–77.

[34] Wang H, Huang ZQ, Xia L, Feng Q, Erdjument-Bromage H, Strahl BD, et al. Methylation of histone H4 at arginine 3 facilitating transcriptional activation by nuclear hormone receptor. Science 2001;293(5531):853–7.

[35] Strahl BD, Briggs SD, Brame CJ, Caldwell JA, Koh SS, Ma H, et al. Methylation of histone H4 at arginine 3 occurs in vivo and is mediated by the nuclear receptor coactivator PRMT1. Curr Biol 2001;11(12):996–1000.

[36] Huang S, Litt M, Felsenfeld G. Methylation of histone H4 by arginine methyltransferase PRMT1 is essential in vivo for many subsequent histone modifications. Genes Dev 2005;19(16):1885–93.

[37] Zhao Q, Rank G, Tan YT, Li H, Moritz RL, Simpson RJ, et al. PRMT5-mediated methylation of histone H4R3 recruits DNMT3A, coupling histone and DNA methylation in gene silencing. Nat Struct Mol Biol 2009;16(3):304–11.

[38] Chung J, Karkhanis V, Tae S, Yan F, Smith P, Ayers LW, et al. Protein arginine methyltransferase 5 (PRMT5) inhibition induces lymphoma cell death through reactivation of the retinoblastoma tumor suppressor pathway and polycomb repressor complex 2 (PRC2) silencing. J Biol Chem 2013;288(49):35534–47.

[39] Molina-Serrano D, Schiza V, Kirmizis A. Cross-talk among epigenetic modifications: lessons from histone arginine methylation. Biochem Soc Transact 2013;41(3):751–9.

[40] Daujat S, Bauer UM, Shah V, Turner B, Berger S, Kouzarides T. Crosstalk between CARM1 methylation and CBP acetylation on histone H3. Curr Biol 2002;12(24):2090–7.

[41] Karkhanis V, Wang L, Tae S, Hu YJ, Imbalzano AN, Sif S. Protein arginine methyltransferase 7 regulates cellular response to DNA damage by methylating promoter histones H2A and H4 of the polymerase delta catalytic subunit gene, POLD1. J Biol Chem 2012;287(35):29801–14.

[42] Shahbazian MD, Grunstein M. Functions of site-specific histone acetylation and deacetylation. Annu Rev Biochem 2007;76:75–100.

[43] Das C, Lucia MS, Hansen KC, Tyler JK. CBP/p300-mediated acetylation of histone H3 on lysine 56. Nature 2009;459(7243):113–17.

[44] Yuan J, Pu M, Zhang Z, Lou Z. Histone H3-K56 acetylation is important for genomic stability in mammals. Cell Cycle 2009;8(11):1747–53.

[45] Wang Z, Zang C, Cui K, Schones DE, Barski A, Peng W, et al. Genome-wide mapping of HATs and HDACs reveals distinct functions in active and inactive genes. Cell 2009;138(5):1019–31.

[46] Vempati RK, Jayani RS, Notani D, Sengupta A, Galande S, Haldar D. p300-mediated acetylation of histone H3 lysine 56 functions in DNA damage response in mammals. J Biol Chem 2010;285(37):28553–64.

[47] Clemente-Ruiz M, Gonzalez-Prieto R, Prado F. Histone H3K56 acetylation, CAF1, and Rtt106 coordinate nucleosome assembly and stability of advancing replication forks. PLoS Genet 2011;7(11):e1002376.

[48] Berndsen CE, Denu JM. Catalysis and substrate selection by histone/protein lysine acetyltransferases. Curr Opin Struct Biol 2008;18(6):682–9.

[49] Lee KK, Workman JL. Histone acetyltransferase complexes: one size doesn't fit all. Nat Rev Mol Cell Biol 2007;8(4):284–95.

[50] Lalonde ME, Cheng X, Cote J. Histone target selection within chromatin: an exemplary case of teamwork. Genes Dev 2014;28(10):1029–41.

[51] Vetting MW, SdC LP, Yu M, Hegde SS, Magnet S, Roderick SL, et al. Structure and functions of the GNAT superfamily of acetyltransferases. Arch Biochem Biophys 2005;433(1):212–26.

[52] Nagy Z, Tora L. Distinct GCN5/PCAF-containing complexes function as co-activators and are involved in transcription factor and global histone acetylation. Oncogene 2007;26(37):5341–57.

[53] Yamauchi T, Yamauchi J, Kuwata T, Tamura T, Yamashita T, Bae N, et al. Distinct but overlapping roles of histone acetylase PCAF and of the closely related PCAF-B/GCN5 in mouse embryogenesis. Proc Natl Acad Sci USA 2000;97(21):11303–6.

[54] Xu W, Edmondson DG, Evrard YA, Wakamiya M, Behringer RR, Roth SY. Loss of Gcn5l2 leads to increased apoptosis and mesodermal defects during mouse development. Nat Genet 2000;26(2):229–32.

[55] Thomas T, Voss AK. The diverse biological roles of MYST histone acetyltransferase family proteins. Cell Cycle 2007;6(6):696–704.

[56] Kalkhoven E. CBP and p300: HATs for different occasions. Biochem Pharmacol 2004;68(6):1145–55.

[57] Ogryzko VV, Schiltz RL, Russanova V, Howard BH, Nakatani Y. The transcriptional coactivators p300 and CBP are histone acetyltransferases. Cell 1996;87(5):953–9.

[58] Chakravarti D, LaMorte VJ, Nelson MC, Nakajima T, Schulman IG, Juguilon H, et al. Role of CBP/P300 in nuclear receptor signalling. Nature 1996;383(6595):99–103.

[59] Henry RA, Kuo YM, Andrews AJ. Differences in specificity and selectivity between CBP and p300 acetylation of histone H3 and H3/H4. Biochemistry 2013;52(34):5746–59.

[60] Bordoli L, Husser S, Luthi U, Netsch M, Osmani H, Eckner R. Functional analysis of the p300 acetyltransferase domain: the PHD finger of p300 but not of CBP is dispensable for enzymatic activity. Nucleic Acids Res 2001;29(21):4462–71.

[61] Tanaka Y, Naruse I, Maekawa T, Masuya H, Shiroishi T, Ishii S. Abnormal skeletal patterning in embryos lacking a single Cbp allele: a partial similarity with Rubinstein–Taybi syndrome. Proc Natl Acad Sci 1997;94(19):10215–20.

[62] Yao TP, Oh SP, Fuchs M, Zhou ND, Ch'ng LE, Newsome D, et al. Gene dosage-dependent embryonic development and proliferation defects in mice lacking the transcriptional integrator p300. Cell 1998;93(3):361–72.

[63] Kasper LH, Fukuyama T, Biesen MA, Boussouar F, Tong C, de Pauw A, et al. Conditional knockout mice reveal distinct functions for the global transcriptional coactivators CBP and p300 in T-cell development. Mol Cell Biol 2006;26(3):789–809.

[64] Jin Q, Yu LR, Wang L, Zhang Z, Kasper LH, Lee JE, et al. Distinct roles of GCN5/PCAF-mediated H3K9ac and CBP/p300-mediated H3K18/27ac in nuclear receptor transactivation. EMBO J 2011;30(2):249–62.

[65] Nikolov M, Fischle W. Systematic analysis of histone modification readout. Mol Biosyst 2013;9(2):182–94.

[66] Moore LD, Le T, Fan G. DNA methylation and its basic function. Neuropsychopharmacology 2013;38(1):23–38.

[67] Sasai N, Nakao M, Defossez PA. Sequence-specific recognition of methylated DNA by human zinc-finger proteins. Nucleic Acids Res 2010;38(15):5015–22.

[68] Zhang Y, Ng HH, Erdjument-Bromage H, Tempst P, Bird A, Reinberg D. Analysis of the NuRD subunits reveals a histone deacetylase core complex and a connection with DNA methylation. Genes Dev 1999;13(15):1924–35.

[69] Hendrich B, Hardeland U, Ng HH, Jiricny J, Bird A. The thymine glycosylase MBD4 can bind to the product of deamination at methylated CpG sites. Nature 1999;401(6750):301–4.

[70] Blattler A, Yao L, Wang Y, Ye Z, Jin VX, Farnham PJ. ZBTB33 binds unmethylated regions of the genome associated with actively expressed genes. Epigenetics Chromatin 2013;6(1):13.

[71] Quenneville S, Verde G, Corsinotti A, Kapopoulou A, Jakobsson J, Offner S, et al. In embryonic stem cells, ZFP57/KAP1 recognize a methylated hexanucleotide to affect chromatin and DNA methylation of imprinting control regions. Mol Cell 2011;44(3):361–72.

[72] Hashimoto H, Horton JR, Zhang X, Bostick M, Jacobsen SE, Cheng X. The SRA domain of UHRF1 flips 5-methylcytosine out of the DNA helix. Nature 2008;455(7214):826–9.

[73] Avvakumov GV, Walker JR, Xue S, Li Y, Duan S, Bronner C, et al. Structural basis for recognition of hemi-methylated DNA by the SRA domain of human UHRF1. Nature 2008;455(7214):822–5.

[74] Arita K, Ariyoshi M, Tochio H, Nakamura Y, Shirakawa M. Recognition of hemi-methylated DNA by the SRA protein UHRF1 by a base-flipping mechanism. Nature 2008;455(7214):818–21.

[75] Frauer C, Hoffmann T, Bultmann S, Casa V, Cardoso MC, Antes I, et al. Recognition of 5-hydroxymethylcytosine by the Uhrf1 SRA domain. PLoS One 2011;6(6):e21306.

[76] Mellen M, Ayata P, Dewell S, Kriaucionis S, Heintz N. MeCP2 binds to 5hmC enriched within active genes and accessible chromatin in the nervous system. Cell 2012;151(7):1417–30.

[77] Yildirim O, Li R, Hung JH, Chen PB, Dong X, Ee LS, et al. Mbd3/NURD complex regulates expression of 5-hydroxymethylcytosine marked genes in embryonic stem cells. Cell 2011;147(7):1498–510.

[78] Iurlaro M, Ficz G, Oxley D, Raiber EA, Bachman M, Booth MJ, et al. A screen for hydroxymethylcytosine and formylcytosine binding proteins suggests functions in transcription and chromatin regulation. Genome Biol 2013;14(10):R119.

[79] Spruijt CG, Gnerlich F, Smits AH, Pfaffeneder T, Jansen PW, Bauer C, et al. Dynamic readers for 5-(hydroxy) methylcytosine and its oxidized derivatives. Cell 2013;152(5):1146–59.

[80] Musselman CA, Lalonde M-E, Cote J, Kutateladze TG. Perceiving the epigenetic landscape through histone readers. Nat Struct Mol Biol 2012;19(12):1218–27.

[81] Otani J, Nankumo T, Arita K, Inamoto S, Ariyoshi M, Shirakawa M. Structural basis for recognition of H3K4 methylation status by the DNA methyltransferase 3A ATRX-DNMT3-DNMT3L domain. EMBO Rep 2009;10(11):1235–41.

[82] Dhayalan A, Tamas R, Bock I, Tattermusch A, Dimitrova E, Kudithipudi S, et al. The ATRX-ADD domain binds to H3 tail peptides and reads the combined methylation state of K4 and K9. Hum Mol Genet 2011;20(11):2195–203.

[83] Yang Y, Lu Y, Espejo A, Wu J, Xu W, Liang S, et al. TDRD3 is an effector molecule for arginine-methylated histone marks. Mol Cell 2010;40(6):1016–23.

[84] Rajakumara E, Wang Z, Ma H, Hu L, Chen H, Lin Y, et al. PHD finger recognition of unmodified histone H3R2 links UHRF1 to regulation of euchromatic gene expression. Mol Cell 2011;43(2):275–84.

[85] Patel DJ, Wang Z. Readout of epigenetic modifications. Annu Rev Biochem 2013;82:81–118.

[86] Dhalluin C, Carlson JE, Zeng L, He C, Aggarwal AK, Zhou MM. Structure and ligand of a histone acetyl-transferase bromodomain. Nature 1999;399(6735):491–6.

[87] Zeng L, Zhang Q, Li S, Plotnikov AN, Walsh MJ, Zhou MM. Mechanism and regulation of acetylated histone binding by the tandem PHD finger of DPF3b. Nature 2010;466(7303):258–62.

[88] Filippakopoulos P, Knapp S. Targeting bromodomains: epigenetic readers of lysine acetylation. Nat Rev Drug Discov 2014;13(5):337–56.

[89] Filippakopoulos P, Picaud S, Mangos M, Keates T, Lambert JP, Barsyte-Lovejoy D, et al. Histone recognition and large-scale structural analysis of the human bromodomain family. Cell 2012;149(1):214–31.

[90] Jacobson RH, Ladurner AG, King DS, Tjian R. Structure and function of a human TAFII250 double bromo-domain module. Science 2000;288(5470):1422–5.

[91] Moriniere J, Rousseaux S, Steuerwald U, Soler-Lopez M, Curtet S, Vitte AL, et al. Cooperative binding of two acetylation marks on a histone tail by a single bromodomain. Nature 2009;461(7264):664–8.

[92] Qiu Y, Liu L, Zhao C, Han C, Li F, Zhang J, et al. Combinatorial readout of unmodified H3R2 and acetylated H3K14 by the tandem PHD finger of MOZ reveals a regulatory mechanism for HOXA9 transcription. Genes Dev 2012;26(12):1376–91.

[93] Su D, Hu Q, Li Q, Thompson JR, Cui G, Fazly A, et al. Structural basis for recognition of H3K56-acetylated histone H3-H4 by the chaperone Rtt106. Nature 2012;483(7387):104–7.

[94] Fazly A, Li Q, Hu Q, Mer G, Horazdovsky B, Zhang Z. Histone chaperone Rtt106 promotes nucleosome formation using (H3-H4)2 tetramers. J Biol Chem 2012;287(14):10753–60.

[95] Tahiliani M, Koh KP, Shen Y, Pastor WA, Bandukwala H, Brudno Y, et al. Conversion of 5-methylcytosine to 5-hydroxymethylcytosine in mammalian DNA by MLL partner TET1. Science 2009;324(5929):930–5.

[96] He YF, Li BZ, Li Z, Liu P, Wang Y, Tang Q, et al. Tet-mediated formation of 5-carboxylcytosine and its excision by TDG in mammalian DNA. Science 2011;333(6047):1303–7.

[97] Ito S, Shen L, Dai Q, Wu SC, Collins LB, Swenberg JA, et al. Tet proteins can convert 5-methylcytosine to 5-formylcytosine and 5-carboxylcytosine. Science 2011;333(6047):1300–3.

[98] Yamaguchi S, Shen L, Liu Y, Sendler D, Zhang Y. Role of Tet1 in erasure of genomic imprinting. Nature 2013;504(7480):460–4.

[99] Solary E, Bernard OA, Tefferi A, Fuks F, Vainchenker W. The Ten-Eleven Translocation-2 (TET2) gene in hematopoiesis and hematopoietic diseases. Leukemia 2014;28(3):485–96.

[100] Gu TP, Guo F, Yang H, Wu HP, Xu GF, Liu W, et al. The role of Tet3 DNA dioxygenase in epigenetic reprogramming by oocytes. Nature 2011;477(7366):606–10.

[101] Shen L, Zhang Y. 5-Hydroxymethylcytosine: generation, fate, and genomic distribution. Curr Opin Cell Biol 2013;25(3):289–96.

[102] Reik W, Dean W, Walter J. Epigenetic reprogramming in mammalian development. Science 2001;293(5532):1089–93.

[103] He Y-F, Li B-Z, Li Z, Liu P, Wang Y, Tang Q, et al. Tet-mediated formation of 5-carboxylcytosine and its excision by TDG in mammalian DNA. Science 2011;333(6047):1303–7.

[104] Maiti A, Drohat AC, Thymine DNA. Glycosylase can rapidly excise 5-formylcytosine and 5-carboxylcytosine: potential implications for active demethylation of CpG sites. J Biol Chem 2011;286(41):35334–8.

[105] Chen CC, Wang KY, Shen CK. The mammalian de novo DNA methyltransferases DNMT3A and DNMT3B are also DNA 5-hydroxymethylcytosine dehydroxymethylases. J Biol Chem 2012;287(40):33116–21.

[106] Song CX, Szulwach KE, Dai Q, Fu Y, Mao SQ, Lin L, et al. Genome-wide profiling of 5-formylcytosine reveals its roles in epigenetic priming. Cell 2013;153(3):678–91.

[107] Shi Y, Lan F, Matson C, Mulligan P, Whetstine JR, Cole PA, et al. Histone demethylation mediated by the nuclear amine oxidase homolog LSD1. Cell 2004;119(7):941–53.

[108] Ciccone DN, Su H, Hevi S, Gay F, Lei H, Bajko J, et al. KDM1B is a histone H3K4 demethylase required to establish maternal genomic imprints. Nature 2009;461(7262):415–18.

[109] Black JC, Van Rechem C, Whetstine JR. Histone lysine methylation dynamics: establishment, regulation, and biological impact. Mol Cell 2012;48(4):491–507.

[110] Wang Y, Wysocka J, Sayegh J, Lee Y-H, Perlin JR, Leonelli L, et al. Human PAD4 regulates histone arginine methylation levels via demethylimination. Science 2004;306(5694):279–83.

[111] Chang B, Chen Y, Zhao Y, Bruick RK. JMJD6 is a histone arginine demethylase. Science 2007;318(5849):444–7.

[112] Webby CJ, Wolf A, Gromak N, Dreger M, Kramer H, Kessler B, et al. Jmjd6 catalyses lysyl-hydroxylation of U2AF65, a protein associated with RNA splicing. Science 2009;325(5936):90–3.

[113] Gregoretti IV, Lee YM, Goodson HV. Molecular evolution of the histone deacetylase family: functional implications of phylogenetic analysis. J Mol Biol 2004;338(1):17–31.

[114] Glass CK, Rosenfeld MG. The coregulator exchange in transcriptional functions of nuclear receptors. Genes Dev 2000;14(2):121–41.

[115] Jones PL, Veenstra GJ, Wade PA, Vermaak D, Kass SU, Landsberger N, et al. Methylated DNA and MeCP2 recruit histone deacetylase to repress transcription. Nat Genet 1998;19(2):187–91.

[116] Nan X, Ng HH, Johnson CA, Laherty CD, Turner BM, Eisenman RN, et al. Transcriptional repression by the methyl-CpG-binding protein MeCP2 involves a histone deacetylase complex. Nature 1998;393(6683):386–9.

[117] Reichert N, Choukrallah MA, Matthias P. Multiple roles of class I HDACs in proliferation, differentiation, and development. Cell Mol Life Sci 2012;69(13):2173–87.

[118] Hayakawa T, Nakayama J. Physiological roles of class I HDAC complex and histone demethylase. J Biomed Biotechnol 2011;2011:129383.

[119] Yang XJ, Seto E. The Rpd3/Hda1 family of lysine deacetylases: from bacteria and yeast to mice and men. Nat Rev Mol Cell Biol 2008;9(3):206–18.

[120] Clocchiatti A, Florean C, Brancolini C. Class IIa HDACs: from important roles in differentiation to possible implications in tumourigenesis. J Cell Mol Med 2011;15(9):1833–46.

[121] Wang AH, Yang XJ. Histone deacetylase 4 possesses intrinsic nuclear import and export signals. Mol Cell Biol 2001;21(17):5992–6005.

[122] McKinsey TA, Zhang CL, Olson EN. Identification of a signal-responsive nuclear export sequence in class II histone deacetylases. Mol Cell Biol 2001;21(18):6312–21.

[123] McKinsey TA, Zhang CL, Lu J, Olson EN. Signal-dependent nuclear export of a histone deacetylase regulates muscle differentiation. Nature 2000;408(6808):106–11.

[124] Yang XJ, Gregoire S. Class II histone deacetylases: from sequence to function, regulation, and clinical implication. Mol Cell Biol 2005;25(8):2873–84.

[125] Gao L, Cueto MA, Asselbergs F, Atadja P. Cloning and functional characterization of HDAC11, a novel member of the human histone deacetylase family. J Biol Chem 2002;277(28):25748–55.

[126] Villagra A, Cheng F, Wang HW, Suarez I, Glozak M, Maurin M, et al. The histone deacetylase HDAC11 regulates the expression of interleukin 10 and immune tolerance. Nat Immunol 2009;10(1):92–100.

[127] Sauve AA, Youn DY. Sirtuins: NAD(+)-dependent deacetylase mechanism and regulation. Curr Opin Chem Biol 2012;16(5-6):535–43.

[128] Houtkooper RH, Pirinen E, Auwerx J. Sirtuins as regulators of metabolism and healthspan. Nat Rev Mol Cell Biol 2012;13(4):225–38.

[129] Roth M, Chen WY. Sorting out functions of sirtuins in cancer. Oncogene 2014;33(13):1609–20.

[130] Michishita E, Park JY, Burneskis JM, Barrett JC, Horikawa I. Evolutionarily conserved and nonconserved cellular localizations and functions of human SIRT proteins. Mol Biol Cell 2005;16(10):4623–35.

[131] Yuan H, Su L, Chen WY. The emerging and diverse roles of sirtuins in cancer: a clinical perspective. OncoTargets Ther 2013;6:1399–416.

[132] Vaquero A, Scher M, Lee D, Erdjument-Bromage H, Tempst P, Reinberg D. Human SirT1 interacts with histone H1 and promotes formation of facultative heterochromatin. Mol Cell 2004;16(1):93–105.

[133] Michishita E, McCord RA, Berber E, Kioi M, Padilla-Nash H, Damian M, et al. SIRT6 is a histone H3 lysine 9 deacetylase that modulates telomeric chromatin. Nature 2008;452(7186):492–6.

[134] Kawahara TL, Michishita E, Adler AS, Damian M, Berber E, Lin M, et al. SIRT6 links histone H3 lysine 9 deacetylation to NF-kappaB-dependent gene expression and organismal life span. Cell 2009;136(1):62–74.

[135] Zhong L, D'Urso A, Toiber D, Sebastian C, Henry RE, Vadysirisack DD, et al. The histone deacetylase Sirt6 regulates glucose homeostasis via Hif1alpha. Cell 2010;140(2):280–93.

[136] Ford E, Voit R, Liszt G, Magin C, Grummt I, Guarente L. Mammalian Sir2 homolog SIRT7 is an activator of RNA polymerase I transcription. Genes Dev 2006;20(9):1075–80.

[137] North BJ, Marshall BL, Borra MT, Denu JM, Verdin E. The human Sir2 ortholog, SIRT2, is an NAD + -dependent tubulin deacetylase. Mol Cell 2003;11(2):437–44.

[138] Vaquero A, Scher MB, Lee DH, Sutton A, Cheng HL, Alt FW, et al. SirT2 is a histone deacetylase with preference for histone H4 Lys 16 during mitosis. Genes Dev 2006;20(10):1256–61.

[139] Wang J, Hevi S, Kurash JK, Lei H, Gay F, Bajko J, et al. The lysine demethylase LSD1 (KDM1) is required for maintenance of global DNA methylation. Nat Genet 2009;41(1):125–9.

[140] Tan CP, Nakielny S. Control of the DNA methylation system component MBD2 by protein arginine methylation. Mol Cell Biol 2006;26(19):7224–35.

[141] Nakanishi S, Lee JS, Gardner KE, Gardner JM, Takahashi YH, Chandrasekharan MB, et al. Histone H2BK123 monoubiquitination is the critical determinant for H3K4 and H3K79 trimethylation by COMPASS and Dot1. J Cell Biol 2009;186(3):371–7.

[142] Kirmizis A, Santos-Rosa H, Penkett CJ, Singer MA, Vermeulen M, Mann M, et al. Arginine methylation at histone H3R2 controls deposition of H3K4 trimethylation. Nature 2007;449(7164):928–32.

[143] Migliori V, Muller J, Phalke S, Low D, Bezzi M, Mok WC, et al. Symmetric dimethylation of H3R2 is a newly identified histone mark that supports euchromatin maintenance. Nat Struct Mol Biol 2012;19(2): 136–44.

[144] Mendenhall EM, Koche RP, Truong T, Zhou VW, Issac B, Chi AS, et al. GC-rich sequence elements recruit PRC2 in mammalian ES cells. PLoS Genet 2010;6(12):e1001244.

[145] Vire E, Brenner C, Deplus R, Blanchon L, Fraga M, Didelot C, et al. The Polycomb group protein EZH2 directly controls DNA methylation. Nature 2006;439(7078):871–4.

[146] Losman JA, Looper RE, Koivunen P, Lee S, Schneider RK, McMahon C, et al. (R)-2-hydroxyglutarate is sufficient to promote leukemogenesis and its effects are reversible. Science 2013;339(6127):1621–5.

[147] Lu C, Ward PS, Kapoor GS, Rohle D, Turcan S, Abdel-Wahab O, et al. IDH mutation impairs histone demethylation and results in a block to cell differentiation. Nature 2012;483(7390):474–8.

[148] Feinberg AP, Tycko B. The history of cancer epigenetics. Nat Rev Cancer 2004;4(2):143–53.

[149] Esteller M. Epigenetic gene silencing in cancer: the DNA hypermethylome. Hum Mol Genet 2007;16 Spec No 1:R50–9.

[150] Esteller M. Epigenetics in cancer. N Engl J Med 2008;358(11):1148–59.

[151] Ley TJ, Ding L, Walter MJ, McLellan MD, Lamprecht T, Larson DE, et al. DNMT3A mutations in acute myeloid leukemia. N Engl J Med 2010;363(25):2424–33.

[152] Yan XJ, Xu J, Gu ZH, Pan CM, Lu G, Shen Y, et al. Exome sequencing identifies somatic mutations of DNA methyltransferase gene DNMT3A in acute monocytic leukemia. Nat Genet 2011;43(4):309–15.

[153] Kim SJ, Zhao H, Hardikar S, Singh AK, Goodell MA, Chen T. A DNMT3A mutation common in AML exhibits dominant-negative effects in murine ES cells. Blood 2013;122(25):4086–9.

[154] Xu J, Wang YY, Dai YJ, Zhang W, Zhang WN, Xiong SM, et al. DNMT3A Arg882 mutation drives chronic myelomonocytic leukemia through disturbing gene expression/DNA methylation in hematopoietic cells. Proc Natl Acad Sci USA 2014;111(7):2620–5.

[155] Russler-Germain DA, Spencer DH, Young MA, Lamprecht TL, Miller CA, Fulton R, et al. The R882H DNMT3A mutation associated with aml dominantly inhibits wild-type DNMT3A by blocking its ability to form active tetramers. Cancer Cell 2014;25(4):442–54.

[156] Cimino G, Moir DT, Canaani O, Williams K, Crist WM, Katzav S, et al. Cloning of ALL-1, the locus involved in leukemias with the t(4;11)(q21;q23), t(9;11)(p22;q23), and t(11;19)(q23;p13) chromosome translocations. Cancer Res 1991;51(24):6712–14.

[157] Ziemin-van der Poel S, McCabe NR, Gill HJ, Espinosa R, Patel Y, Harden A, et al. Identification of a gene, MLL, that spans the breakpoint in 11q23 translocations associated with human leukemias. Proc Natl Acad Sci 1991;88(23):10735–9.

[158] Huang H, Jiang X, Li Z, Li Y, Song CX, He C, et al. TET1 plays an essential oncogenic role in MLL-rearranged leukemia. Proc Natl Acad Sci USA 2013;110(29):11994–9.

[159] Ye C, Li L. 5-Hydroxymethylcytosine: a new insight into epigenetics in cancer. Cancer Biol Ther 2014;15(1):10–15.

[160] Lian CG, Xu Y, Ceol C, Wu F, Larson A, Dresser K, et al. Loss of 5-hydroxymethylcytosine is an epigenetic hallmark of melanoma. Cell 2012;150(6):1135–46.

[161] Varier RA, Timmers HT. Histone lysine methylation and demethylation pathways in cancer. Biochim Biophys Acta 2011;1815(1):75–89.

[162] Garraway LA, Lander ES. Lessons from the cancer genome. Cell 2013;153(1):17–37.

[163] Yap DB, Chu J, Berg T, Schapira M, Cheng SW, Moradian A, et al. Somatic mutations at EZH2 Y641 act dominantly through a mechanism of selectively altered PRC2 catalytic activity, to increase H3K27 trimethylation. Blood 2011;117(8):2451–9.

[164] McCabe MT, Graves AP, Ganji G, Diaz E, Halsey WS, Jiang Y, et al. Mutation of A677 in histone methyl-transferase EZH2 in human B-cell lymphoma promotes hypertrimethylation of histone H3 on lysine 27 (H3K27). Proc Natl Acad Sci USA 2012;109(8):2989–94.

[165] Sneeringer CJ, Scott MP, Kuntz KW, Knutson SK, Pollock RM, Richon VM, et al. Coordinated activities of wild-type plus mutant EZH2 drive tumor-associated hypertrimethylation of lysine 27 on histone H3 (H3K27) in human B-cell lymphomas. Proc Natl Acad Sci USA 2010;107(49):20980–5.

[166] Yu J, Mani RS, Cao Q, Brenner CJ, Cao X, Wang X, et al. An integrated network of androgen receptor, poly-comb, and TMPRSS2-ERG gene fusions in prostate cancer progression. Cancer Cell 2010;17(5):443–54.

[167] Bernt KM, Zhu N, Sinha AU, Vempati S, Faber J, Krivtsov AV, et al. MLL-rearranged leukemia is dependent on aberrant H3K79 methylation by DOT1L. Cancer Cell 2011;20(1):66–78.

[168] Chang MJ, Wu H, Achille NJ, Reisenauer MR, Chou CW, Zeleznik-Le NJ, et al. Histone H3 lysine 79 meth-yltransferase Dot1 is required for immortalization by MLL oncogenes. Cancer Res 2010;70(24):10234–42.

[169] Nguyen AT, Taranova O, He J, Zhang Y. DOT1L, the H3K79 methyltransferase, is required for MLL-AF9-mediated leukemogenesis. Blood 2011;117(25):6912–22.

[170] Deshpande AJ, Chen L, Fazio M, Sinha AU, Bernt KM, Banka D, et al. Leukemic transformation by the MLL-AF6 fusion oncogene requires the H3K79 methyltransferase Dot1l. Blood 2013;121(13):2533–41.

[171] Kahl P, Gullotti L, Heukamp LC, Wolf S, Friedrichs N, Vorreuther R, et al. Androgen receptor coactivators lysine-specific histone demethylase 1 and four and a half LIM domain protein 2 predict risk of prostate cancer recurrence. Cancer Res 2006;66(23):11341–7.

[172] Lim S, Janzer A, Becker A, Zimmer A, Schule R, Buettner R, et al. Lysine-specific demethylase 1 (LSD1) is highly expressed in ER-negative breast cancers and a biomarker predicting aggressive biology. Carcinogenesis 2010;31(3):512–20.

[173] Qin Y, Zhu W, Xu W, Zhang B, Shi S, Ji S, et al. LSD1 sustains pancreatic cancer growth via maintaining HIF1alpha-dependent glycolytic process. Cancer Lett 2014;347(2):225–32.

[174] Dhar SS, Alam H, Li N, Wagner KW, Chung J, Ahn YW, et al. Transcriptional repression of histone deacety-lase 3 by the histone demethylase KDM2A is coupled to tumorigenicity of lung cancer cells. J Biol Chem 2014;289(11):7483–96.

[175] Wagner KW, Alam H, Dhar SS, Giri U, Li N, Wei Y, et al. KDM2A promotes lung tumorigenesis by epige-netically enhancing ERK1/2 signaling. J Clin Invest 2013;123(12):5231–46.

[176] Black JC, Manning AL, Van Rechem C, Kim J, Ladd B, Cho J, et al. KDM4A lysine demethylase induces site-specific copy gain and rereplication of regions amplified in tumors. Cell 2013;154(3):541–55.

[177] Shi L, Sun L, Li Q, Liang J, Yu W, Yi X, et al. Histone demethylase JMJD2B coordinates H3K4/H3K9 methylation and promotes hormonally responsive breast carcinogenesis. Proc Natl Acad Sci USA 2011;108(18):7541–6.

[178] van Haaften G, Dalgliesh GL, Davies H, Chen L, Bignell G, Greenman C, et al. Somatic mutations of the histone H3K27 demethylase gene UTX in human cancer. Nat Genet 2009;41(5):521–3.

[179] Zhong J, Cao RX, Zu XY, Hong T, Yang J, Liu L, et al. Identification and characterization of novel spliced variants of PRMT2 in breast carcinoma. FEBS J 2012;279(2):316–35.

[180] Al-Dhaheri M, Wu J, Skliris GP, Li J, Higashimato K, Wang Y, et al. CARM1 is an important determinant of ERalpha-dependent breast cancer cell differentiation and proliferation in breast cancer cells. Cancer Res 2011;71(6):2118–28.

[181] Frietze S, Lupien M, Silver PA, Brown M. CARM1 regulates estrogen-stimulated breast cancer growth through up-regulation of E2F1. Cancer Res 2008;68(1):301–6.

[182] El Messaoudi S, Fabbrizio E, Rodriguez C, Chuchana P, Fauquier L, Cheng D, et al. Coactivator-associated arginine methyltransferase 1 (CARM1) is a positive regulator of the Cyclin E1 gene. Proc Natl Acad Sci USA 2006;103(36):13351–6.

[183] Wang L, Zhao Z, Meyer MB, Saha S, Yu M, Guo A, et al. CARM1 methylates chromatin remodeling factor BAF155 to enhance tumor progression and metastasis. Cancer Cell 2014;25(1):21–36.

[184] Jansson M, Durant ST, Cho EC, Sheahan S, Edelmann M, Kessler B, et al. Arginine methylation regulates the p53 response. Nat Cell Biol. 2008;10(12):1431–9.

[185] Powers MA, Fay MM, Factor RE, Welm AL, Ullman KS. Protein arginine methyltransferase 5 accelerates tumor growth by arginine methylation of the tumor suppressor programmed cell death 4. Cancer Res 2011;71(16):5579–87.

[186] Stein C, Riedl S, Ruthnick D, Notzold RR, Bauer UM. The arginine methyltransferase PRMT6 regulates cell proliferation and senescence through transcriptional repression of tumor suppressor genes. Nucleic Acids Res 2012;40(19):9522–33.

[187] Fraga MF, Ballestar E, Villar-Garea A, Boix-Chornet M, Espada J, Schotta G, et al. Loss of acetylation at Lys16 and trimethylation at Lys20 of histone H4 is a common hallmark of human cancer. Nat Genet 2005;37(4):391–400.

[188] Yasui W, Oue N, Ono S, Mitani Y, Ito R, Nakayama H. Histone acetylation and gastrointestinal carcinogenesis. Ann N Y Acad Sci 2003;983:220–31.

[189] Fullgrabe J, Kavanagh E, Joseph B. Histone onco-modifications. Oncogene 2011;30(31):3391–403.

[190] Cao L, Zhu L, Yang J, Su J, Ni J, Du Y, et al. Correlation of low expression of hMOF with clinicopathological features of colorectal carcinoma, gastric cancer and renal cell carcinoma. Int J Oncol 2014;44(4):1207–14.

[191] Pfister S, Rea S, Taipale M, Mendrzyk F, Straub B, Ittrich C, et al. The histone acetyltransferase hMOF is frequently downregulated in primary breast carcinoma and medulloblastoma and constitutes a biomarker for clinical outcome in medulloblastoma. Int J Cancer 2008;122(6):1207–13.

[192] Dekker FJ, Haisma HJ. Histone acetyl transferases as emerging drug targets. Drug Discov Today 2009;14(19-20):942–8.

[193] Pasqualucci L, Dominguez-Sola D, Chiarenza A, Fabbri G, Grunn A, Trifonov V, et al. Inactivating mutations of acetyltransferase genes in B-cell lymphoma. Nature 2011;471(7337):189–95.

[194] Mullighan CG, Zhang J, Kasper LH, Lerach S, Payne-Turner D, Phillips LA, et al. CREBBP mutations in relapsed acute lymphoblastic leukaemia. Nature 2011;471(7337):235–9.

[195] Petrij F, Giles RH, Dauwerse HG, Saris JJ, Hennekam RC, Masuno M, et al. Rubinstein-Taybi syndrome caused by mutations in the transcriptional co-activator CBP. Nature 1995;376(6538):348–51.

[196] Iyer NG, Ozdag H, Caldas C. p300/CBP and cancer. Oncogene 2004;23(24):4225–31.

[197] Taki T, Sako M, Tsuchida M, Hayashi Y. The t(11;16)(q23;p13) translocation in myelodysplastic syndrome fuses the MLL gene to the CBP gene. Blood 1997;89(11):3945–50.

[198] Carapeti M, Aguiar RC, Goldman JM, Cross NC. A novel fusion between MOZ and the nuclear receptor coactivator TIF2 in acute myeloid leukemia. Blood 1998;91(9):3127–33.

[199] Liang J, Prouty L, Williams BJ, Dayton MA, Blanchard KL. Acute mixed lineage leukemia with an inv(8)(p11q13) resulting in fusion of the genes for MOZ and TIF2. Blood 1998;92(6):2118–22.

[200] Deguchi K, Ayton PM, Carapeti M, Kutok JL, Snyder CS, Williams IR, et al. MOZ-TIF2-induced acute myeloid leukemia requires the MOZ nucleosome binding motif and TIF2-mediated recruitment of CBP. Cancer Cell 2003;3(3):259–71.

[201] Huntly BJ, Shigematsu H, Deguchi K, Lee BH, Mizuno S, Duclos N, et al. MOZ-TIF2, but not BCR-ABL, confers properties of leukemic stem cells to committed murine hematopoietic progenitors. Cancer Cell 2004;6(6):587–96.

[202] Aikawa Y, Katsumoto T, Zhang P, Shima H, Shino M, Terui K, et al. PU.1-mediated upregulation of CSF1R is crucial for leukemia stem cell potential induced by MOZ-TIF2. Nat Med 2010;16(5):580–5. 1p following 5.

[203] Nagy Z, Tora L. Distinct GCN5//PCAF-containing complexes function as co-activators and are involved in transcription factor and global histone acetylation. Oncogene 2007;26(37):5341–57.

[204] French CA, Miyoshi I, Kubonishi I, Grier HE, Perez-Atayde AR, Fletcher JA. BRD4-NUT fusion oncogene: a novel mechanism in aggressive carcinoma. Cancer Res 2003;63(2):304–7.

[205] Grayson AR, Walsh EM, Cameron MJ, Godec J, Ashworth T, Ambrose JM, et al. MYC, a downstream target of BRD-NUT, is necessary and sufficient for the blockade of differentiation in NUT midline carcinoma. Oncogene 2014;33(13):1736–42.

[206] Shi J, Wang Y, Zeng L, Wu Y, Deng J, Zhang Q, et al. Disrupting the interaction of BRD4 with diacetylated twist suppresses tumorigenesis in basal-like breast cancer. Cancer Cell 2014;25(2):210–25.

[207] Zou Z, Huang B, Wu X, Zhang H, Qi J, Bradner J, et al. Brd4 maintains constitutively active NF-kappaB in cancer cells by binding to acetylated RelA. Oncogene 2014;33(18):2395–404.

[208] Hagelkruys A, Sawicka A, Rennmayr M, Seiser C. The biology of HDAC in cancer: the nuclear and epigenetic components. Yao T-P, Seto E, editors. Histone Deacetylases: The Biology and Clinical Implication. Berlin Heidelberg: Springer; 2011. p. 13–37.

[209] Ropero S, Fraga MF, Ballestar E, Hamelin R, Yamamoto H, Boix-Chornet M, et al. A truncating mutation of HDAC2 in human cancers confers resistance to histone deacetylase inhibition. Nat Genet 2006;38(5):566–9.

[210] Hanigan CL, Van Engeland M, De Bruine AP, Wouters KA, Weijenberg MP, Eshleman JR, et al. An inactivating mutation in HDAC2 leads to dysregulation of apoptosis mediated by APAF1. Gastroenterology 2008;135(5): 1654–64 e2.

[211] Gelmetti V, Zhang J, Fanelli M, Minucci S, Pelicci PG, Lazar MA. Aberrant recruitment of the nuclear receptor corepressor-histone deacetylase complex by the acute myeloid leukemia fusion partner ETO. Mol Cell Biol 1998;18(12):7185–91.

[212] Lin RJ, Sternsdorf T, Tini M, Evans RM. Transcriptional regulation in acute promyelocytic leukemia. Oncogene 2001;20(49):7204–15.

[213] Jang KY, Hwang SH, Kwon KS, Kim KR, Choi HN, Lee NR, et al. SIRT1 expression is associated with poor prognosis of diffuse large B-cell lymphoma. Am J Surg Pathol 2008;32(10):1523–31.

[214] Cha EJ, Noh SJ, Kwon KS, Kim CY, Park BH, Park HS, et al. Expression of DBC1 and SIRT1 is associated with poor prognosis of gastric carcinoma. Clin Cancer Res 2009;15(13):4453–9.

[215] Chen HC, Jeng YM, Yuan RH, Hsu HC, Chen YL. SIRT1 promotes tumorigenesis and resistance to chemotherapy in hepatocellular carcinoma and its expression predicts poor prognosis. Ann Surg Oncol 2012;19(6):2011–19.

[216] Roth M, Chen WY. Sorting out functions of sirtuins in cancer. Oncogene 2013.

[217] Kim HS, Vassilopoulos A, Wang RH, Lahusen T, Xiao Z, Xu X, et al. SIRT2 maintains genome integrity and suppresses tumorigenesis through regulating APC/C activity. Cancer Cell 2011;20(4):487–99.

[218] Barber MF, Michishita-Kioi E, Xi Y, Tasselli L, Kioi M, Moqtaderi Z, et al. SIRT7 links H3K18 deacetylation to maintenance of oncogenic transformation. Nature 2012;487(7405):114–18.

[219] Sebastian C, Zwaans BM, Silberman DM, Gymrek M, Goren A, Zhong L, et al. The histone deacetylase SIRT6 is a tumor suppressor that controls cancer metabolism. Cell 2012;151(6):1185–99.

[220] Mann BS, Johnson JR, He K, Sridhara R, Abraham S, Booth BP, et al. Vorinostat for treatment of cutaneous manifestations of advanced primary cutaneous T-cell lymphoma. Clin Cancer Res 2007;13(8):2318–22.

[221] Coiffier B, Pro B, Prince HM, Foss F, Sokol L, Greenwood M, et al. Results from a pivotal, open-label, phase II study of romidepsin in relapsed or refractory peripheral T-cell lymphoma after prior systemic therapy. J Clin Oncol 2012;30(6):631–6.

[222] Thaler F, Minucci S. Next generation histone deacetylase inhibitors: the answer to the search for optimized epigenetic therapies? Exp Opin Drug Discov 2011;6(4):393–404.

[223] Gnyszka A, Jastrzebski Z, Flis S. DNA methyltransferase inhibitors and their emerging role in epigenetic therapy of cancer. Anticancer Res 2013;33(8):2989–96.

[224] Cheng JC, Yoo CB, Weisenberger DJ, Chuang J, Wozniak C, Liang G, et al. Preferential response of cancer cells to zebularine. Cancer Cell 2004;6(2):151–8.

[225] Copeland RA. Molecular pathways: protein methyltransferases in cancer. Clin Cancer Res 2013;19(23):6344–50.

[226] Qi W, Chan H, Teng L, Li L, Chuai S, Zhang R, et al. Selective inhibition of Ezh2 by a small molecule inhibitor blocks tumor cells proliferation. Proc Natl Acad Sci 2012;109(52):21360–5.

[227] McCabe MT, Ott HM, Ganji G, Korenchuk S, Thompson C, Van Aller GS, et al. EZH2 inhibition as a therapeutic strategy for lymphoma with EZH2-activating mutations. Nature 2012;492(7427):108–12.

[228] Daigle SR, Olhava EJ, Therkelsen CA, Majer CR, Sneeringer CJ, Song J, et al. Selective killing of mixed lineage leukemia cells by a potent small-molecule DOT1L inhibitor. Cancer Cell 2011;20(1):53–65.

[229] Knutson SK, Warholic NM, Wigle TJ, Klaus CR, Allain CJ, Raimondi A, et al. Durable tumor regression in genetically altered malignant rhabdoid tumors by inhibition of methyltransferase EZH2. Proc Natl Acad Sci 2013;110(19):7922–7.

[230] Cao F, Townsend EC, Karatas H, Xu J, Li L, Lee S, et al. Targeting MLL1 H3K4 methyltransferase activity in mixed-lineage leukemia. Mol Cell 2014;53(2):247–61.

[231] Dowden J, Pike RA, Parry RV, Hong W, Muhsen UA, Ward SG. Small molecule inhibitors that discriminate between protein arginine N-methyltransferases PRMT1 and CARM1. Org Biomol Chem 2011;9(22):7814–21.

[232] Williamson AR. Creating a structural genomics consortium. Nat Struct Mol Biol 2000;7:953.

[233] Stavropoulos P, Blobel G, Hoelz A. Crystal structure and mechanism of human lysine-specific demethylase-1. Nat Struct Mol Biol 2006;13(7):626–32.

[234] Kruidenier L, Chung C-W, Cheng Z, Liddle J, Che K, Joberty G, et al. A selective jumonji H3K27 demethylase inhibitor modulates the proinflammatory macrophage response. Nature 2012;488(7411):404–8.

[235] Mertz JA, Conery AR, Bryant BM, Sandy P, Balasubramanian S, Mele DA, et al. Targeting MYC dependence in cancer by inhibiting BET bromodomains. Proc Natl Acad Sci 2011;108(40):16669–74.

[236] Zuber J, Shi J, Wang E, Rappaport AR, Herrmann H, Sison EA, et al. RNAi screen identifies Brd4 as a therapeutic target in acute myeloid leukaemia. Nature 2011;478(7370):524–8.

[237] Filippakopoulos P, Qi J, Picaud S, Shen Y, Smith WB, Fedorov O, et al. Selective inhibition of BET bromodomains. Nature 2010;468(7327):1067–73.

[238] Dawson MA, Prinjha RK, Dittmann A, Giotopoulos G, Bantscheff M, Chan W-I, et al. Inhibition of BET recruitment to chromatin as an effective treatment for MLL-fusion leukaemia. Nature 2011;478(7370):529–33.

[239] Delmore Jake E, Issa Ghayas C, Lemieux Madeleine E, Rahl Peter B, Shi J, Jacobs Hannah M, et al. BET bromodomain inhibition as a therapeutic strategy to target c-Myc. Cell 2011;146(6):904–17.

[240] Helin K, Dhanak D. Chromatin proteins and modifications as drug targets. Nature 2013;502(7472):480–8.

[241] Juergens RA, Wrangle J, Vendetti FP, Murphy SC, Zhao M, Coleman B, et al. Combination epigenetic therapy has efficacy in patients with refractory advanced non–small cell lung cancer. Cancer Discov 2011.

[242] Fiskus W, Wang Y, Sreekumar A, Buckley KM, Shi H, Jillella A, et al. Combined epigenetic therapy with the histone methyltransferase EZH2 inhibitor 3-deazaneplanocin A and the histone deacetylase inhibitor panobinostat against human AML cells. Blood 2009;114(13):2733–43.

[243] Garber K. STI571 Revolution: can the newer targeted drugs measure up? J Natl Cancer Inst 2001;93(13):970–3.

[244] Gao J, Aksoy BA, Dogrusoz U, Dresdner G, Gross B, Sumer SO, et al. Integrative analysis of complex cancer genomics and clinical profiles using the cBioPortal. Sci Signal 2013;6(269):pl1.

[245] Cerami E, Gao J, Dogrusoz U, Gross BE, Sumer SO, Aksoy BA, et al. The cBio cancer genomics portal: an open platform for exploring multidimensional cancer genomics data. Cancer Discov 2012;2(5):401–4. Erratum in: Cancer Discov. 2012;2(10):960.

[246] Tian X, Zhang S, Liu HM, Zhang YB, Blair CA, Mercola D, et al. Histone lysine-specific methyltransferases and demethylases in carcinogenesis: new targets for cancer therapy and prevention. Curr Cancer Drug Targets 2013;13(5):558–79.

[247] Elakoum R, Gauchotte G, Oussalah A, Wissler MP, Clément-Duchêne C, Vignaud JM, et al. CARM1 and PRMT1 are dysregulated in lung cancer without hierarchical features. Biochimie 2014;97:210–8.

[248] Singh V, Miranda TB, Jiang W, Frankel A, Roemer ME, Robb VA, et al. DAL-1/4.1B tumor suppressor interacts with protein arginine N-methyltransferase 3 (PRMT3) and inhibits its ability to methylate substrates in vitro and in vivo. Oncogene 2004;23(47):7761–71.

[249] Kim YR, Lee BK, Park RY, Nguyen NT, Bae JA, Kwon DD, et al. Differential CARM1 expression in prostate and colorectal cancers. BMC Cancer 2010;10:197.

[250] Habashy HO, Rakha EA, Ellis IO, Powe DG. The oestrogen receptor coactivator CARM1 has an oncogenic effect and is associated with poor prognosis in breast cancer. Breast Cancer Res Treat 2013;140(2):307–16.

[251] Osada S, Suzuki S, Yoshimi C, Matsumoto M, Shirai T, Takahashi S, et al. Elevated expression of coactivator-associated arginine methyltransferase 1 is associated with early hepatocarcinogenesis. Oncol Rep 2013;30(4):1669–74.

[252] Bao X, Zhao S, Liu T, Liu Y, Liu Y, Yang X. Overexpression of PRMT5 promotes tumor cell growth and is associated with poor disease prognosis in epithelial ovarian cancer. J Histochem Cytochem 2013;61(3):206–17.

[253] Shilo K, Wu X, Sharma S, Welliver M, Duan W, Villalona-Calero M, et al. Cellular localization of protein arginine methyltransferase-5 correlates with grade of lung tumors. Diagn Pathol 2013;8:201.

[254] Yan F, Alinari L, Lustberg ME, Martin LK, Cordero-Nieves HM, Banasavadi-Siddegowda Y, et al. Genetic validation of the protein arginine methyltransferase PRMT5 as a candidate therapeutic target in glioblastoma. Cancer Res 2014;74(6):1752–65.

[255] Vieira FQ, Costa-Pinheiro P, Ramalho-Carvalho J, Pereira A, Menezes FD, Antunes L, et al. Deregulated expression of selected histone methylases and demethylases in prostate carcinoma. Endocr Relat Cancer 2013;21(1):51–61.

[256] Meerzaman DM, Yan C, Chen QR, Edmonson MN, Schaefer CF, Clifford RJ, et al. Genome-wide transcriptional sequencing identifies novel mutations in metabolic genes in human hepatocellular carcinoma. Cancer Genomics Proteomics 2014;11(1):1–12.

[257] Thomassen M, Jochumsen KM, Mogensen O, Tan Q, Kruse TA. Gene expression meta-analysis identifies chromosomal regions involved in ovarian cancer survival. Genes Chromosomes Cancer 2009;48(8):711–24.

[258] Liu Y, Melin BS, Rajaraman P, Wang Z, Linet M, Shete S, et al. Insight in glioma susceptibility through an analysis of 6p22.3, 12p13.33-12.1, 17q22-23.2 and 18q23 SNP genotypes in familial and non-familial glioma. Hum Genet 2012;131(9):1507–17.

[259] Højfeldt JW, Agger K, Helin K. Histone lysine demethylases as targets for anticancer therapy. Nat Rev Drug Discov 2013;12(12):917–30.

MicroRNAs AND CANCER

4

Stephen G. Maher[1], Becky A.S. Bibby[1], Hannah L. Moody[1,2], and Glen Reid[3]

[1]Cancer Biology and Therapeutics Lab, School of Biological, Biomedical and Environmental Sciences, University of Hull, Yorkshire, UK, [2]Hull York Medical School, Yorkshire, UK, [3]Asbestos Diseases Research Institute, University of Sydney, New South Wales, Australia

CHAPTER OUTLINE

1 miRNA BIOGENESIS AND FUNCTIONALITY

microRNAs (miRNA) are a species of the non-protein-coding RNA family, represented by short, single-stranded RNA approximately 18–22 nucleotides in length [1]. miRNAs are regulators of gene expression at a posttranscriptional level [1]. In combination with an RNA-induced silencing complex (RISC), miRNAs specifically target complementary messenger RNA (mRNA) transcripts, typically via imperfect complementary base pairing, and repress their translation [1,2]. Although miRNAs are generally regarded as regulators of gene expression at the posttranscriptional level, newer miRNA functions are being identified. The role of miRNA as bystanders and effectors within the epigenetic landscape of the cell is becoming apparent [3]. In diseases, such as cancer, the complex relationship between miRNA and epigenetics is uncovering a new understanding of cancer cell biology.

S.G. Gray (Ed): Epigenetic Cancer Therapy. DOI: http://dx.doi.org/10.1016/B978-0-12-800206-3.00004-5

In 1993, Lee and colleagues identified the first miRNA in *Caenorhabditis elegans* [4]. The *lin-4* gene encodes two small non-protein-coding RNA transcripts approximately 22 and 61 nucleotides in length, and is essential for post-embryonic development in *C. elegans* [4]. The sequences of the RNA transcripts have complementarity to the 3'-UTR (3' untranslated region) of the *lin-14* mRNA, and bind to the mRNA via antisense RNA:RNA interactions, repressing mRNA translation, which results in downregulation of LIN-14 protein levels [4–6]. Several years later a second miRNA was identified in *C. elegans*. The *let-7* gene encodes a 21-nucleotide RNA transcript with complementarity to the 3' UTR of five heterochronic genes involved in normal development: *lin-14*, *lin-28*, *lin-41*, *lin-42*, and *daf-12* [7,8]. Expression of *lin-4* and *let-7* are essential to post-embryonic development and developmental timing in *C. elegans*. In 2001, numerous small non-protein-coding RNAs were identified and were collectively termed "miRNA" [9–11], and included 21 novel human miRNAs, miR-1 to miR-33 [9]. Intensified research efforts identified additional miRNAs in mammals, fish, worms, flies, and plants [1]. To date approximately 1600 human miRNA precursor sequences have been identified, but the functional roles of most miRNA are still unknown.

The genes encoding miRNA are located throughout the genome as individual genes, polycistrons, or within introns of pre-mRNA [1]. miRNA genes located within the introns of pre-mRNA are ideally located for translational repression of their mRNA host [1]. Polycistronic miRNA genes encode a cluster of miRNA precursors, which are transcribed as a single transcript then processed into individual mature miRNAs that may have related or non-related functions [1]. miRNA genes are transcribed in the nucleus by RNA polymerase II or III, most often RNA polymerase II, producing a single-stranded RNA transcript 1–7 kb in length [12,13] (Figure 4.1).

The primary miRNA (pri-miRNA) transcript folds into an imperfect hairpin structure as a result of Watson–Crick base pairing and is processed in the nucleus by Drosha and DGCR [14,15]. The 5' and 3' ends of the hairpin structure are asymmetrically cleaved producing a precursor-miRNA (pre-miRNA), which is exported to the cytoplasm via exportin-5 in the nuclear membrane [16]. In the cytoplasm, Dicer and TRBP cleave the loop structure off the hairpin, forming the miRNA duplex, termed "miR-5p/-3p" (formerly known as miRNA-miRNA*) [17,18]. The relative thermodynamic stability of the strands determines which arm of the duplex will be incorporated into the RISC as the mature miRNA, while the passenger strand (miRNA*) is subsequently degraded [17,19]. Argonaute (Ago) proteins constitute the major functional element of the RISC. In mammals, there are four Ago proteins, all of which are capable of repressing mRNA translation as part of the RISC, but only Ago2 is able to directly cleave the mRNA target [20]. The piwi-argonaute-zwille (PAZ) domain of the Ago protein binds the 2-nucleotide overhang at the 3' end of the mature miRNA strand [19]. The duplex then unwinds and the 5' end of the mature miRNA strand is bound by the Ago MID (middle) domain [1]. Exposed bases of the mature miRNA within the RISC bind to target mRNA sequences via complementary base pairing [17]. The mRNA may be regulated by an RNAi mechanism involving direct cleavage and degradation by the RISC, or the translation of the mRNA can be repressed, as is most frequently the case in mammals [1].

Within the target mRNA, the "seed site" is a sequence of approximately seven nucleotides, and is essential for miRNA binding [21–23]. These sequences are often highly conserved between species [21–23]. The mRNA seed site is frequently, but not exclusively, located in the 3'-UTR. The "seed region" at the 5' end of the miRNA binds the seed site in the mRNA [21,24]. The 5'-UTR and coding sequence of the mRNA can also contain "seed sites"; the miRNA RISC complex can potentially bind to any region of the mRNA [24,25]. The general consensus is that the 3'-UTR is the most accessible region of

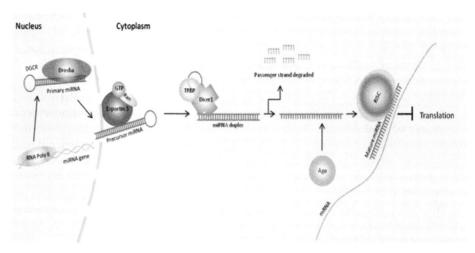

FIGURE 4.1

miRNA biogenesis. miRNA genes are transcribed in the nucleus by RNA polymerase II to produce a long, single-stranded RNA transcript. The primary-miRNA (pri-miRNA) folds into a hairpin structure and is processed by Drosha and DGCR, whereby the 5′ and 3′ ends of the hairpin structure are asymmetrically cleaved to produce the precursor-miRNA (pre-miRNA). The pre-miRNA is exported from the nucleus to the cytoplasm via exportin-5 and Ran-GTP on the nuclear membrane and is further processed in the cytoplasm by Dicer and TRBP. The loop of the hairpin structure is cleaved to produce the miRNA duplex. Within the miRNA duplex one arm is the passenger strand and the other is the mature (guide) miRNA. The passenger strand is degraded while the mature strand complexes with Argonaute (Ago) proteins. The Ago proteins constitute the major functional element of the RISC. The exposed bases of the mature miRNA bind to complementary target mRNA sequences, typically via imperfect complementary base pairing. Subsequently, translation of the mRNA is repressed, thereby downregulating expression of the target at protein level.

the mRNA, as the RISC has less competition binding to the mRNA furthest away from the ribosome and translational machinery [1,24,26]. Furthermore, regulated target mRNAs generally have longer 3′-UTR compared to ubiquitously expressed genes which tend to have shorter 3′-UTRs that are depleted of miRNA-binding sites [27]. A mature miRNA within the RISC guides the complex toward complementary mRNA targets, stringent seed sites have perfect Watson–Crick base pairing between the mRNA and miRNA [24]. However, the RISC can also tolerate G:U wobble and mismatch binding between the miRNA seed region and mRNA seed site [24]. In mammals, miRNA:mRNA binding is generally the result of imperfect complementary base pairing, while in contrast, near-perfect complementary base pairing is most common in plants [1]. The imperfect nature of miRNA target binding enables a single miRNA to target multiple mRNA targets, hence there is a degree of redundancy between miRNAs, conversely a single mRNA can be targeted by multiple miRNAs [28]. This variation and flexibility in miRNA:mRNA binding can make it difficult to predict mRNA targets using bioinformatic tools [24].

Once incorporated into the RISC, the mature miRNA downregulates target gene expression at the posttranscriptional level. There are a range of mechanisms by which this may occur and these are broadly divided into two categories: translational repression and mRNA degradation [29]. In metazoans, mRNA targets are typically translationally repressed by their regulating miRNA; the imperfect base pairing between the miRNA and mRNA is generally associated with translational repression, as opposed to mRNA cleavage [1]. Experimental evidence in mammals has demonstrated that the levels of target mRNA remain unchanged upon miRNA targeting, but a decrease in protein expression is observed [1,4]. However, studies also demonstrate that miRNA binding to targets can frequently result in degradation, with the miRNA promoting translational quiescence, followed by degradation of the target as a secondary consequence. The exact mechanism of translational repression is unclear; the RISC complex may repress translation at the initiation or post-initiation stage, or both [29]. Alternatively, mRNA targets can be guided by the RISC into processing (P)-bodies, sequestering them from ribosomes and the translational machinery [19,29]. Endonucleases can subsequently enter P-bodies and degrade the sequestered mRNAs, or these mRNAs can later be released back into the cytoplasm for translation if protein levels decrease below the requirements of the cell, thus demonstrating miRNA-mediated repression is reversible [30,31]. Near-perfect base pairing between the miRNA:mRNA is associated with direct cleavage of the mRNA by the RISC [1]. miRNAs are also destabilized by the gradual shortening of the poly A tail, resulting in mRNA degradation by progressive decay catalyzed by the exosome or degradation by endonucleases [32].

Generally, miRNAs are recognized as functioning to downregulate gene expression via translational repression. However, there is evidence of further functional roles of miRNAs, including links between miRNAs and epigenetics [3]. The expression of miRNA genes is regulated by their epigenetic status and miRNAs are known to have specific epigenetic functions [3]. Although miRNAs are processed and function in the cytoplasm, there is evidence that mature miRNAs are associated with Ago proteins found in the nucleus [33,34]. The miRNAs in the nucleus are reported to have epigenetic functions, such as modulating mRNA splicing and targeting gene promoters to activate or repress transcription [33,34].

miRNAs account for approximately 1% of the genome and are estimated to regulate approximately 30% of genes [1]. The imperfect nature of miRNA target binding enables a single miRNA to target multiple mRNA targets, therefore miRNAs essentially regulate all cellular pathways. As different cell types have specialized functions and express a specific set of genes related to the function of the cell, this is reflected in tissue-specific miRNA expression profiles [35]. Disrupting the highly complex miRNA regulatory network within the cell can induce abnormal cell behavior and disease initiation or progression [36], and as such, dysregulated miRNA expression is a common feature in human diseases, especially cancer.

2 miRNAs IN CANCER BIOLOGY

miRNAs are involved in all pathways and cellular processes within the cell, hence it is not surprising that miRNA dysregulation is viewed as a fundamental feature of cancer and is considered instrumental in the acquisition of the hallmarks of cancer, such as invasion, angiogenesis, and evasion of apoptosis. Tumors most often have a reduced level of mature miRNA, due to the loss of genetic material, alterations to the machinery associated with biogenesis, and epigenetic silencing [37]. It is proposed that cancer-associated miRNAs either have an oncogenic or tumor-suppressive activity [38]. Again, this relates to tissue type and location of the cancer. The link between miRNA and cancer was first established over a decade ago, when Calin et al. [39] reported the deletion or downregulation of miR-15a and miR-16-1 encoded at the 13q14 loci in a majority of B-cell chronic lymphocytic leukemia cases. The same group later reported that the alterations in expression of miRNA in various cancers could be critical to the understanding of cancer pathophysiology [40].

Interestingly, Calin et al. [40] revealed that many miRNAs are encoded at fragile sites and within common breakpoint regions in the genome, thus increasing their susceptibility to mutation and deletion.

As miRNAs play a significant role in the regulation of many aspects of cellular machinery, it has been suggested that the deregulation of these small noncoding molecules could profoundly and substantially affect the cell and its progression through the cell cycle. It is considered that during carcinogenesis, miRNAs become either upregulated or downregulated; however, this can vary depending on the tissue of origin. The frequently altered miRNA, miR-31, is disrupted in opposing directions in a wide range of tumor types [41], for example, in mesothelioma it has been established that miR-31 is significantly downregulated, which effects the expression of PPP6C, a pro-survival phosphatase [42]. Conversely, the same miRNA within the colorectal cancer microenvironment is highly upregulated across all stages of the disease [43].

A cluster of miRNA associated with tumor formation was discovered by He et al. [44]. The multiple component miRNA polycistron miR-17~92 was found to be amplified in B-cell lymphomas in both cell line studies and samples of tumor tissue [45]. Further investigation by Suarez et al. [45] concluded that the cluster carries out pleiotropic functions and modifies postnatal angiogenesis in response to vascular factors, such as vascular endothelial growth factor (VEGF). The cluster has also been shown to promote carcinogenesis by altering cell cycle phase distribution, as in Sylvestre et al. [46]. MiR-17~92 is activated by members of the E2F family, which stimulate a number of S phase genes, including thymidine kinase. E2F1, E2F2, and E2F3 are all modulated by the miR-17~92 cluster, via their 3'-UTR binding sites. In addition, overexpression of a member of the cluster, miR-20a, decreased apoptosis in a prostate cancer cell line whilst the inhibition of miR-20a produced an increase in cell death. The miR-20a anti-apoptotic properties may elucidate some of the oncogenic capabilities of the miR-17-92 cluster. The study suggested that the autoregulation between E2F1-3 and miR-20a may contribute to the regulation of apoptotic events and proliferation [46].

Recently, Gao et al. [47] investigated miR-184 for regulatory functions within hepatocellular carcinoma (HCC). It had previously been established that miR-122 had a significant role within HCC [48]; however, miR-184 was novel in the research into both the development and onset of HCC. Using the inositol polyphosphate phosphatase-like 1 (INPPL1) insulin regulator as a recognized target, miR-184 was found to be central in HCC cell proliferative activity, and that silencing of miR-184 leads to the overexpression of INPPL1. The miRNA was also allied to the inhibition of caspase 3 and 7, suggesting a role in the evasion of apoptosis [47].

In pancreatic cancer, miR-106a is highly expressed in tumor tissue and in four cell lines, one of which, SW-1990, is a highly invasive line [49]. Interestingly, results have indicated that the highest expression of miR-106a was present in the most invasive line. In cells transfected with a miR-106a mimic, tumor cell growth was stimulated, whereas an miR-106a inhibitor decreased cell viability [49]. Mace et al. [50] investigated miRNA in relation to pancreatic tumor cells under hypoxia; findings suggested that miR-21 was induced via hypoxia inducible factor (HIF)-1α upregulation, and that miR-21 overexpression promoted the evasion of apoptosis in a hypoxic environment.

Let-7a, a tumor suppressor miRNA, is lost in malignant melanoma, where it has been demonstrated to regulate integrin β_3 expression [51]. The integrin β_3 subunit $\alpha v \beta_3$ family of adhesion receptors is involved in the transition from dysplastic nevi to tumorigenic melanomas, and overexpression has been linked to increased cellular motility [52]. MiR-143 and miR-145, located at the 5q33 fragile site, represent tumor-suppressive miRNA [53]. The miR-143~145 cluster is downregulated in many cancers, suggesting a "protective" role for the miRNA. miR-143 and miR-145 are recurrently coordinately downregulated in endometrial cancer, with a connection made between downregulation of miR-143/~145 and overexpression of DNA methyltransferase (DNMT) 3B [54]. The DNMT group contributes to the coordination of mRNA expression in normal tissues and overexpression in many tumors [55]. In addition, miR-29 has been identified to target DNMT2A and DNMT3B, which have been found to be upregulated in lung carcinoma [56]. The miR-29 family has been identified as being upregulated in induced and replicative senescence, and functions to inhibit DNA synthesis and repress the *B-Myb* oncogene in combination with Rb; this may be beneficial as the tumorigenic cell may become senescent as a result [57,58].

The well-established hallmarks of cancer are integrally linked with miRNA expression, and indeed hundreds of miRNAs have been found to be novel regulators of these distinctive carcinogenic hallmarks. miR-519 has demonstrated the ability to inhibit proliferation in cervical, colon, and ovarian cell lines, through one of its targets, HuR, an RNA-binding protein [59]. Furthermore, the RhoA pathway can be modulated through miR-146a in prostate cancer [60]. This results in the downregulation of the serine/threonine protein kinase ROCK1, leading to dysregulation of the actin cytoskeleton [60], and alterations in cellular motility. The expression of miR-34a correlates with p53 expression, and so can be termed a regulator of apoptosis [61]. The cluster of miR-290 can directly regulate the DNMT expression in Dicer1-null cells, which indirectly affects telomere integrity and length, thus implying significance in the regulation of replicative potential [62]. An example relating to angiogenesis regulation is that of miR-378, which binds to 3'-UTR region of VEGF, which promotes neovascularization [63]. Other miRNAs, such as miR-10b and miR-23a, are implicated in the regulation of invasion and genomic stability, respectively [64].

miRNAs are novel therapeutic targets and promising cancer biomarkers with potential applications in diagnostics, prognostic, tumor staging, patient response to treatment, and determination of developmental lineages and clinical subtypes [65]. Interestingly, miRNA treatment in the form of a nucleic acid–modified DNA phosphorothioate antisense oligonucleotide has already entered human clinical trials in treatment of disease, specifically viral infection. miR-122, an abundant liver-expressed miRNA, essential for the replication of the hepatitis C virus (HCV), is sequestered by the oligonucleotide (as will be discussed later) and is bound in a duplex that inhibits its function [66]. Results thus far have shown promise, with long-standing dose-dependent decreases in viral titer levels without evidence of acquired resistance [66].

In summary thus far, miRNAs function as regulators of gene expression in all cellular pathways, and aberrant miRNA expression is associated with cancer initiation, promotion, and progression. miRNAs globally downregulated in cancer and tissue-specific expression profiles have been determined for several cancer types. However, the functional roles of many miRNAs are yet to be determined. miRNAs are involved in all cancer-associated pathways and the acquisition of the hallmarks of cancer, and thus represent promising cancer biomarkers, and in translational therapeutics, which will be discussed later. More specifically, miRNAs may play a fundamental role in the epigenetic changes observed within carcinogenesis, and understanding the dysregulation of these small, noncoding molecules could be imperative in our understanding of cancer biology [67].

3 miRNA: AN EPIGENETIC PERSPECTIVE

Generally, miRNAs are regarded as regulators of gene expression at the posttranscriptional level. However, miRNAs are directly and indirectly linked with epigenetics [3]. The indirect link involves the regulation of miRNA expression via epigenetic modifications, namely DNA methylation and chromatin remodeling, which ultimately determine the accessibility of miRNA genes for transcription [68]. Conversely, miRNAs are directly linked to epigenetics in one of two ways, firstly epi-miRNAs (epigenetic-miRNA) control the expression of the epigenetic machinery via translational repression [3]. Secondly, miRNAs have emerging roles as functional epigenetic components that modulate gene expression at the transcriptional level by interacting with gene promoter regions and the transcriptional machinery, to positively or negatively regulate gene expression [69]. Epigenetics and miRNAs are entwined in a complex circuit and modulate the expression of vast numbers of genes in the genome (Figure 4.2).

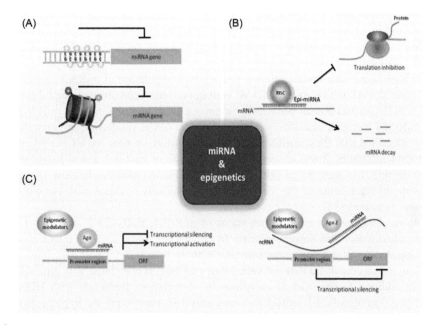

FIGURE 4.2

Epigenetics and miRNA. (A) **Epigenetic alteration of miRNA expression.** Both DNA methylation (*top*) and histone modification (*bottom*) are adopted in the regulation of miRNA gene expression within many different cancers. Susceptibility to methylation has been allied with miRNA gene proximity to CpG islands. (B) **Epigenetic-miRNA.** Epi-miRNAs are miRNAs that specifically regulate the expression of genes encoding epigenetic machinery. The RISC in complex with the epi-miRNA is guided to complementary mRNA targets, which are subsequently degraded or translationally repressed. (C) **miRNA with epigenetic function.** miRNAs are functional components of the epigenetic machinery. In collaboration with Ago proteins and components of the epigenetic machinery, miRNAs are capable of inducing transcriptional activation and silencing. The miRNA interacts with the promoter region of genes and controls gene expression much like a transcription factor. Alternatively, the miRNA may recruit additional epigenetic machinery, which alter the epigenetic status of the gene (*left*). An alternative mechanism involves noncoding RNA (ncRNA) produced from the promoter region of a gene which serves as the miRNA target. The ncRNA intermediate recruits the complementary miRNA to the promoter region of the gene. Additional epigenetic machinery and Ago2 are subsequently recruited to silence transcription (*right*).

3.1 EPIGENETIC ALTERATION OF miRNA EXPRESSION

Analysis of miRNA genes suggests that almost half of the sequences are associated with CpG islands, and thus are liable to epigenetic regulation [70,71]. Histone modification controls DNA accessibility by manipulating chromatin states in order for gene expression to be controlled throughout various developmental stages of life [72]. The loci of a number of miRNA genes being so closely located to CpG islands reveals how they would be subject to DNA methylation; this is a major route by which miRNA expression of some primary miRNA transcripts is regulated [73], but it has limited impact on the control of mature miRNA levels. DNA methylation decreases in many tissue types during aging, and it is suggested that DNMT3b activity is increased with age, which feeds back into examples such as the miR-143~145 cluster, as discussed previously [54,74]. As cancer risk and incidence increases with age, it is interesting that miRNAs could have a role not only within carcinogenesis, but also in earlier stages where cells age and become more prone to mutation.

There have been significant studies relating to epigenetic regulation of miRNA expression in various cancers. Using DNMT inhibitors and HDAC inhibitors, it has been observed that there are a range of effects on miRNA expression in renal cell carcinoma [75]. Upregulation of miR-9-1, miR-642, miR-95, and miR-184 has been identified via microarray and qPCR validation, and upon further investigation it was identified that these miRNAs are located within, or near to, a CpG island [75]. Post addition of DNMT inhibitors, it was observed that expression of miR-9-1 and miR-642 was restored, indicating that the miRNAs were downregulated by a DNA methylation mechanism. Overall, the inhibition of DNA methylation initiated the reexpression of previously silenced miRNAs with recognized tumor suppressor functions [75].

Within colorectal cancer, the methylation status of miR-9-1, miR-129-2, and miR-137 has been analyzed in matched tumor and normal (disease-free) adjacent tissue samples. DNA methylation of miR-137 was apparent in all primary tumor samples, with the matched normal tissue having 23% miR-137 methylation; it was concluded that colorectal mucosal defects correlated with anomalous miRNA methylation. In addition to the evident variations in methylation, treatment with HDAC inhibitor restored the expression of miR-9-1, miR-129-2, and miR-137, supporting the hypothesis that miRNA silencing is induced by histone modifications. Both DNA hypermethylation and histone modification have been illustrated as being contributors to the transcriptional downregulation of specific miRNAs in colorectal cancer [76].

Epithelial ovarian cancer is also subject to epigenetic control of miRNA expression [77]. Three genomic loci, on chromosome 14, 19, and X, harbor 25 miRNAs significantly downregulated in ovarian cancer. Using ovarian cancer cell lines, it has been established that treatment with DNMT inhibitors and HDAC inhibitors restored expression of miR-34b, a tumor suppressor miRNA regulated by p53 [77]. Additionally, epigenetic mechanisms may also alter expression of miRNA genes in prostate cancer. For example, miR-205, miR-21, and miR-196b have been established to be downregulated through promoter methylation [78]. Conversely, miR-615 has been identified to be activated through epigenetic means in prostate cancer cell lines [79].

The prevalence of miRNAs that illustrate varying expression due to epigenetic regulation suggests that there is a complex relationship underpinning the regulation of miRNAs that mainly represses the promoter region of tumor suppressor miRNA genes; however, there is evidence of the recruitment of miRNA "activators", as in prostate carcinogenesis. It has been proposed, as in Hulf et al., that changes at the pri-miRNA transcript level compared with alterations in mature miRNA expression can allow for the detection of functionally relevant miRNAs that are epigenetically regulated [79]. The study of epigenetic mechanisms governing miRNA expression may provide further answers as to how the cancer cell has a selective advantage over normal tissue, and aid in the development of both novel diagnostics and therapeutics. In another dimension to this concept, certain miRNAs, termed epi-miRNAs, can themselves modulate the epigenetic machinery, which alludes to a circular feedback loop, demonstrating an even greater depth of complexity in miRNA biology [70].

3.2 EPI-miRNA

Epi-miRNAs modulate the expression of genes encoding the epigenetic machinery, such as DNMTs, HDACs, and polycomb group (PcG) proteins [3]. The genes encoding epi-miRNA are also subject to epigenetic regulation via DNA methylation and chromatin remodeling. Thus, complex feedback loops exist between the epigenetic machinery and miRNA. Dysregulation of these intricate networks within normal cells can contribute toward the initiation and promotion of cancer and may provide novel therapeutic targets.

The miR-29 family members were the first epi-miRNAs to be identified [56]. The *de novo* DNMTs, DNMT3A and DNMT3B, are translationally repressed by the miR-29 family (miR-29a, miR-29b, and miR-29c). Binding sites within the 3′-UTRs of DNMT3A and DNMT3B were identified and validated in lung cancer cell lines, and miR-29 expression was inversely correlated with DNMT3A and DNMT3B in lung cancer tissues. In addition to the *de novo* DNMTs, miR-29 was shown to indirectly repress DNMT1 in acute myeloid leukemia [80]. Ectopic expression of miR-29 in cancer cell lines resulted in global DNA hypomethylation, reexpression of tumor suppressor genes, and inhibition of tumorigenesis is both *in vitro* and *in vivo* [56,80]. However, an additional layer of complexity has recently developed. As well as regulating DNMTs, the miR-29 family also regulates the DNA demethylation machinery [81]. The tet methylcytosine dioxygenase (TET) protein family, in conjunction with thymine DNA glycosylase (TDG), modulates DNA demethylation [81]. The TET1 and TDG genes are both targets of miR-29 translational repression in lung cancer cell lines [81]. These recent developments suggest miR-29 is responsible for maintaining the existing DNA methylation status in normal cells by modulating the activity of both DNMTs and demethylases. Restoring endogenous miR-29 expression has the potential to repress *de novo* methylation, maintain cellular methylation status, and release hypermethylated tumor suppressor genes.

Other epi-miRNAs known to regulate the DNA methylation machinery include miR-152, miR-301, and miR-148 [82]. Interestingly, miR-148 expression is regulated by methylation and is itself a regulator of the *DNMT3B* gene [3,82]. The miR-148 binding site is located within the protein-coding region of the DNMT3B splice variants DNMT3B1, DNMT3B2, and DNMT3B4. The binding site is conserved between these splice variants; however, the binding site is absent in the *DNMT3B3* gene which is therefore resistant to miR-148 repression [82]. The abundance of DNMT3B splice variants is modulated by miR-148 and expression of miR-148 is regulated by DNA methylation, again suggesting a complex regulatory feedback mechanisms exist between miRNA and the epigenetic machinery.

PcG proteins mediate gene silencing through histone modifications and chromatin remodeling. The PRC1 and PRC2 complexes are multidomain protein complexes consisting of PcG proteins. Expression of the PcG proteins is regulated by miRNA and the PRC complexes in turn modulate miRNA expression [83]. Enhancer of Zeste Homolog 2 (EZH2) is a histone methyltransferase, and the catalytic subunit of the PRC2 complex. EZH2 contributes to epigenetic silencing via tri-methylation of the core histone H3 lysine 27 (H3K27me3). Overexpression of EZH2 in solid tumors is associated with enhanced tumorigenesis and metastasis [83]. In a prostate cancer study, miR-101 was inversely correlated with EZH2 protein levels, and miR-101 expression decreased during cancer progression [84]. Hence, miR-101 was identified as a tumor suppressor in prostate cancer and was also downregulated in bladder, lung, breast, and colorectal cancer [83,85–87]. Furthermore, DNMT3B has been identified as a target of miR-101 repression [88]. Restoring miR-101 may attenuate the oncogenic effects of EZH2 overexpression and return the epigenetic status of tumor cells to a more normal state.

These studies suggest complex feedback loops exist between epi-miRNAs and the proteins of the epigenetic machinery. The expression levels of epi-miRNAs are regulated by epigenetic modifications, such as DNA methylation status and histone modifications. In addition, the proteins of the epigenetic machinery are themselves regulated by the epi-miRNAs [3]. In cancer these networks are dysregulated and as a result multiple cellular pathways are in turn dysregulated.

3.3 miRNA WITH EPIGENETIC FUNCTIONS

New evidence is emerging to suggest miRNAs also function to regulate gene expression at the transcriptional level within the nucleus [34]. High levels of miRNA and Ago proteins are reported to exist within the nucleus [34,89]. Emerging evidence suggests miRNAs in collaboration with Ago proteins and other noncoding RNA species regulate gene transcription [34]. miRNAs in the nucleus may function like transcription factors to initiate gene expression, while, alternatively, miRNA binding at the promoter region may induce changes in chromatin structure or recruit other RNA species or proteins to induce transcription activation [69,90].

The first direct epigenetic modification induced by a miRNA was initially identified via a bioinformatics approach [90]. Gene promoter regions were scanned for complementarity to miR-373, thereby identifying 80% sequence complementarity between miR-373 and the promoter regions of the genes encoding E-Cadherin and CSDC2 (cold shock domain-containing protein C2) [90]. Ectopic expression of pre-miR-373 and mature miR-373 both induced target expression *in vitro*. The mechanism by which miR-373 is able to regulate transcription initiation remains unclear. Other miRNAs demonstrated to positively regulate gene expression include miR-744, miR-1186, and miR-466d-3p [91]. These miRNAs have high promoter complementarity with the *CCNB1* gene. Ectopic expression of miR-744, miR-1186, and miR-466d-3p induces cyclinB1 expression from the *CCNB1* gene *in vitro*, and silencing endogenous miR-744 decreases cyclin B1 levels [91]. Furthermore, miR-744 increases RNA polymerase II and histone H3 lysine 4 tri-methylation (H3K4me3) at the CCNB1 transcriptional start site [91]. In addition, Ago1 is localized to the promoter region and is proposed to recruit chromatin-modifying proteins to activate transcription [91]. *In vivo*, miR-747 and miR-1186 overexpression initially promotes tumor growth through *CCNB1* gene expression, however, long-term exposure results in chromosomal instability and inhibits tumor growth [91].

Alternatively, miRNAs can epigenetically silence transcriptional activation of genes [69]. MiR-320 is encoded within the promoter region of the cell cycle gene *POLR3D* in the antisense orientation, and mature miR-320 has inherent complementarity to its target [92]. Acting in a *cis* orientation, miR-320 is transcribed from the promoter region and processed to the mature form, which silences the *POLR3D* gene by recruiting Ago1 and EZH2, thereby inducing epigenetic transcriptional gene silencing [92]. In another study, miRNA mimics have been employed to determine the mechanism by which miR-423-5p silences transcription of the human progesterone receptor (PR) [93]. Despite low complementarity between miR-423-5p and the promoter region of the PR gene, the miR-423-5p mimic silences transcription, acting in a *trans* orientation [93]. Noncoding RNA (ncRNA) transcribed from the PR promoter served as the miR-423-5p target, effectively recruiting miR-423-5p to the promoter region of the PR gene. In contrast to other studies, Ago2, as opposed to Ago1, is localized to the promoter suggesting an alternative mechanism for transcriptional gene silencing depending on the sequence complementarities between the miRNA and promoter region [93]. In addition, decreased localization of RNA polymerase II at the promoter region and increased histone H3 lysine 9 dimethylation (H3K9me2) induced epigenetic transcriptional silencing [93].

Most often, miRNAs are described as regulators of gene expression via translational repression of mRNA targets. However, research is uncovering new and diverse miRNA functions. In the context of epigenetics, not only is the expression of certain miRNAs modulated by the epigenetic status of their encoding genes, but miRNAs themselves also have specific epigenetic functions and are able to modulate the epigenetic landscape within the cell [3]. Undoubtedly, miRNA and epigenetics are intertwined and dysregulation of the miRNA epigenetic circuit is both cause and consequence in cancer initiation and progression [3]. The extensive feedback loops and collaborations that exist between miRNAs and epigenetics suggest that a fragile balance must be maintained for the normal function of a cell.

4 miRNA EPIGENETIC THERAPY

The goal in developing miRNA–modulating therapeutic drugs is to return miRNA expression to levels found in non-diseased cells. The attraction of this strategy is that it has the potential to restore control over many coregulated pathways with a single agent. miRNAs are dysregulated in most cancers and because expression can be up- or downregulated, miRNA-based therapy thus depends on inhibition or replacement of the target miRNA and can be achieved in a variety of ways (Table 4.1). In essence, miRNA inhibition borrows heavily from antisense technology, whereas miRNA replacement mirrors the gene therapy techniques previously developed for RNAi-based therapeutic approaches. In addition, there are several miRNA-specific approaches, such as sponges and small-molecule modulators of miRNA biogenesis, which have been developed to alter miRNA levels in cancer cells. Nucleic acid–based methods offer the most specific methods, and effective control of miRNA levels can be achieved in each case but, as for antisense and gene therapy, the translation of *in vitro* results into *in vivo* efficacy is heavily dependent on delivery.

4.1 miRNA INHIBITION IN CANCER

The overexpression of miRNAs targeting tumor suppressor genes in cancer led to the coining of the term oncomiRs. In order to reduce overexpressed oncogenic miRNAs to the appropriate levels, several overlapping strategies can be employed, all of which are based on sequence-specific inhibitors that lead to sequestration or induced cleavage of the target. In a broad sense, miRNA inhibitors can be grouped into synthetic antisense oligonucleotides (so-called antagomiRs or antimiRs) and expressed decoys (known as sponges). Direct miRNA inhibitors are essentially antisense oligonucleotides (AS-ODN) complimentary to the mature miRNA target. These AS-ODNs are generally chemically modified to improve activity, stability (resistance to nucleases), and delivery [94]. Common modifications are those

Table 4.1 Methods for Modulating miRNA Expression Levels *In Vivo*

Inhibition		
Method	**Description**	**Mechanism**
AntimiRs	Single-stranded AS-ODN complimentary to an miR target, modified by LNA substitutions; usually 13–22 nucleotides in length	High-affinity duplex formation, leading to RNaseH-mediated degradation of target miRNA
Tiny LNAs	Similar to antimiRs in composition, but much shorter (8 nucleotides); target seed sequence of individual miRNAs or families	As above, but may target multiple miRNA
AntagomiRs	Single-stranded AS-ODN complimentary to the miR target, modified by 2′ O-Me groups; usually 22 or 23 nucleotides in length	Bind single miRNA target, but mechanism not fully understood
Sponges	Expressed transcripts with multiple tandem miRNA binding sites; requires viral/nonviral delivery vehicle	Act as decoys for target miRNA(s), reducing regulation (and increasing expression) of true mRNA targets
Replacement		
Method	**Description**	**Mechanism**
miRNA mimic	Double-stranded synthetic RNAs, with or without modification, of sequence related to mature miRNAs; requires delivery vehicle	Mimic post-dicer miRNA activity via incorporation into RISC and regulation of target mRNAs
Expression constructs	Plasmid or virus-encoded expression construct; former requires nonviral delivery vehicle	Expressed miRNAs enter miRNA processing pathway in the nucleus; exported into the cytoplasm and incorporated into RISC
Small molecules	Transcriptional activators (decitabine) or processing modulators (enoxacin)	Drugs that non specifically alter expression or RNA processing, leading to upregulation of miRNA levels

AS-ODN, antisense oligonucleotides.

altering the sugar and/or backbone, and in addition, the placement of these modifications can further alter the properties of the AS-ODN (Figure 4.3). Typically, these AS-ODN inhibitors are administered without delivery vehicle when used *in vivo*, although conjugation to cholesterol and other functional groups has been used to improve uptake [95].

Typical sugar modifications include 2′ O-Me and locked nucleic acid (LNA) substitutions, both of which increase binding affinity and nuclease resistance of the AS-ODN. Inhibitors with 2′ O-Me modification (consisting of a substitution of the 2′ hydroxyl with a methoxy moiety) were one of the first to be shown to inhibit miRNA activity [95]. These inhibitors were originally termed "antagomiRs." With only moderate affinity for the cognate sequence, antagomiR needs to be used at relatively high concentrations. In the case of LNA substitutions, the 2′ O,4′ C-methylene bridge causes the nucleotide to be locked in an RNA-mimicking conformation. The resultant inhibitors, called antimiRs, have greatly increased affinity for target miRNAs compared with an unmodified AS-ODN, equivalent to an increase in melting temperature of 2°C to 8°C per incorporated LNA. In addition, this increased affinity enables LNA-based

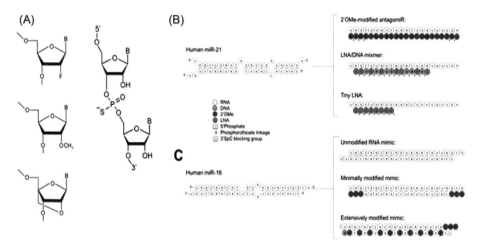

FIGURE 4.3

Common chemical modification strategies for synthetic miRNA inhibitors and mimics. (A) Modified sugars commonly used in synthetic oligonucleotides include 2′ F (*top left*), 2′ O-Me (*middle left*), and LNA ribose modifications, all of which can impart resistance to nucleases and/or prevent off-target effects and immune stimulation. Introducing phosphorothioate (PS) linkages (*right*) between nucleotides also provides nuclease resistance. (B) Common types of miRNA inhibitors are shown, using the human miR-21 sequence as target. AntagomiRs (*top*) consist of a fully 2′ O-Me-modified oligo, with terminal PS; LNA/DNA mixmers (*middle*) are, as the name suggests, a mix of LNA-modified ribose and deoxyribose nucleotides, and are usually around 15–22 nucleotides; tiny LNAs (*bottom*) are fully LNA-modified oligos specific for the seed sequence, and have a minimum length of eight nucleotides. Both mixmers and tiny LNAs have fully PS-modified backbones. (C) Examples of mimics of the human miR-16 sequence are shown. An unmodified double-stranded RNA (*top*), with fully complementary sense strand is active, but can give rise to off-target effects and stimulation of the innate immune system. More common are minimally (*middle*) or extensively (*bottom*) modified mimics. (NB: All examples are illustrative only and have not necessarily been tested for activity.)

antimiRs of much shorter length to maintain activity. This has been exploited in the case of so-called Tiny LNAs, 8-nucleotide long AS-ODN, that inhibit either single miRNAs or miRNA families via interaction with the seed sequence in the miRNA target [96]. Finally, changes in the phosphodiester backbone also impart desirable properties to AS-ODN inhibitors. Most commonly used is the phosphorothioate modification, consisting of replacement of a non-bridging oxygen atom with sulfur [94]. See Figure 4.3.

Now that they have become a very common reagent in cancer research, the use of miRNA inhibitors *in vivo* with the aim of developing anticancer therapies is also increasing. Examples of oncogenic miRNAs that have been targeted by inhibitors include the well-characterized oncomiRs miR-21, miR-155, the miR-17~92 cluster, and others (Table 4.2). One of the first oncogenic miRNAs identified, miR-21, is commonly upregulated in a range of cancer types and it has been linked to inhibition of a range of tumor suppressor genes [125] and drug resistance [126]. While experimental inhibition of miR-21 via genetic knockout or transfection of inhibitors prior to implantation leads to tumor growth inhibition and has revealed details of the function of miR-21 (and other miRNAs), these methods are of limited clinical relevance. Several groups, however, have employed miR-21 inhibitors of various types using

Table 4.2 Studies Demonstrating miRNA Modulation in Preclinical Cancer Models *In Vivo*

Inhibitor			
miRNA	**Model**	**Administration**	**Reference**
miR-21	Tongue SCC (scxg)	AS-ODN, i.t. (LF2000)	[97]
	Glioblastoma (scxg)	LNA/DNA AS-ODN, i.t. (LF2000)	[98]
	Breast, colon cancer (scxg)	AS-ODN, i.p.	[99]
	Breast cancer (otxg)	8-mer LNA, i.v.	[96]
	Glioblastoma (scxg)	AS-ODN, i.t. (LF2000)	[100]
	Breast cancer (scxg)	AS-ODN, i.t.	[101]
	Multiple myeloma (scxg)	AS-ODN (RNA Lancer II)	[102]
miR-17-5p	Neuroblastoma (scxg)	AS-ODN, i.t.	[103]
miR-17~92	Medulloblastoma (otag)	AS-ODN, i.p.	[104]
miR-19a	Breast cancer (scxg)	LNA/DNA AS-ODN, i.v.	[105]
miR-155	Breast cancer (scxg)	2′O-Me, i.t. (LF2000)	[106]
	Lymphoma (otxg)	8-mer LNA, i.v.	[107]
miR-10b	Breast cancer (otxg)		[108]
miR-380-5p	Neuroblastoma (otxg)	AS-ODN, i.p.	[109]
miR-181b	Breast, colon cancer (scxg)	AS-ODN, i.p.	[99]
	Breast cancer (otxg)	AS-ODN, i.t.	[110]
miR-222	Breast cancer (otxg)	AS-ODN, i.t.	[110]
miR-9	Hodgkin lymphoma (otxg)	8-mer LNA, i.v.	[111]
miR-135b	Metastatic lung cancer (otxg)	AS-ODN, i.v.	[112]
miR-335	Astrocytoma (scxg)	AS-ODN, i.t.	[113]

Mimic			
miRNA	**Model**	**Administration**	**References**
Let-7	Lung cancer (scxg)	i.t. (siPORT)	[114]
	Lung cancer (scxg)	i.v. (NLE)	[115]
miR-34a	Lung cancer (scxg)	i.t.; i.v. (RNA Lancer II)	[116]
	Lung cancer (scxg)	i.v. (NLE)	[115]
	Prostate cancer (sc/otxg)	i.v. (NLE)	[117]
	Neuroblastoma (otxg)	i.v. (NP)	[118]
miR-15a	Multiple myeloma	i.v. (RNA Lancer II)	[119]
miR-16	Metastatic prostate cancer	i.v. (atelocollagen)	[120]
	Multiple myeloma	i.v. (RNA Lancer II)	[119]
	Mesothelioma (scxg)	i.v. (minicells)	[121]
miR-143	Colon cancer (scxg)	i.v. (Lipo Trust)	[122]
miR-145	Colon cancer (scxg)	i.t.; i.v. (PEI)	[123]
	Colon cancer (scxg)	i.v. (Lipo Trust)	[122]
miR-33a	Colon cancer (scxg)	i.t.; i.v. (PEI)	[123]
miR-122	HCC (scxg)	i.t. (LNP-DP1)	[124]

AS-ODN, antisense oligonucleotides; i.p., intraperitoneal; i.t, intratumoral; i.v., intravenous; LNA, locked nucleic acid; ot, orthotopic; sc, subcutaneous; SCC, squamous cell carcinoma; xg, xenograft.

various routes of administration in preclinical xenograft models (Table 4.2). For example, direct intratumoral injection of antimiR was able to inhibit growth of glioblastoma [98,100], multiple myeloma [102], and tongue squamous cell carcinoma tumors [97], and to affect expression of miR-21 and its targets. In addition, miR-21 antimiR has been administered systemically via intraperitoneal injection in mice with subcutaneous breast or colon cancer xenografts, leading to inhibition of miR-21 levels and inhibition of tumor growth [99]. This effect was enhanced by coadministration of an miR-181b-1 antimiR. Tiny LNAs specific for miR-21 were effectively delivered to breast cancer xenografts after systemic injection, leading to inhibition of miR-21 activity, as demonstrated by derepression of a luciferase reporter, although no effect on tumor growth was observed in this aggressive model [96]. There is also evidence that the inhibition of miR-21 following intratumoral injection has anti-angiogenic effects, further hampering tumor growth [101].

Further studies have reported silencing of additional cancer-associated miRNAs with systemic antisense treatment in tumor models in mice (Table 4.2). Intratumoral injections of a cholesterol-conjugated antagomiR of miR-17-5p resulted in striking inhibition of xenografts derived from *MYCN*-amplified neuroblastoma cells [103]. This correlated with an increase in the miR-17-5p targets p21 and BIM, and an induction of apoptosis in the tumors. In a breast cancer model, miR-19a was able to reduce tumor growth, and antimiR treatment was synergistic in effect when combined with taxol [105]. Employing a tiny LNA approach, allograft tumors in a model of medulloblastoma were treated by intravenous injection of 8-mer LNA AS-ODN targeting either the miR-17 or miR-19 seed sequences [104]. This led to a specific reduction in levels of the miRNAs sharing the miR-17 or miR-19 seed sequence, and a decrease in tumor growth. In line with its pro-metastatic role, systemic intravenous administration of a miR-10b antagomiR led to a reduction in breast cancer metastases with minimal effects on the primary tumor [108]. This treatment reduced miR-10b levels (with a resultant increase in Hoxd10 expression) in the primary tumor and significantly reduced numbers of lung metastases. In a neuroblastoma model, inhibition of miR-380-5p was shown to arrest cell growth in a p53-dependent manner, and intraperitoneal injection with anti-miR-380 led to a greatly reduced tumor mass [109].

The oncomiR miR-155 has also been silenced *in vivo*. Intratumoral injection of anti-miR-155 was used to inhibit miR-155 in subcutaneous breast cancer xenografts, leading to increased caspase-3 expression and reduced tumor growth [106]. Using the tiny LNA antimiR strategy, miR-155 was inhibited in a multiple myeloma model, with the antimiR found in the bone marrow and spleen [107]. Interestingly, and although not a cancer-related study, it was recently shown that tiny LNAs targeting the seed sequence could effectively inhibit miR-33a/b in nonhuman primates [127], paving the way for similar approaches targeting cancer-associated miRNA families in patients.

In addition to miRNA-specific-antisense-based approaches, miRNA decoys—known as sponges— have also been used to inhibit oncomiR function. These sponges consist of expression constructs in which the RNA has multiple miRNA target sites, which "mop up" the miRNA thereby limiting regulation of *bona fide* targets [128]. Such a strategy has been used to demonstrate that the metastasis-reducing effects of miR-10b inhibition are tumor cell specific, and not due to effects on the host microenvironment [108]. Similarly, a miR-9 sponge reduced the levels of miR-9, and subsequent development of metastases, by half in the same model [108]. Despite their utility in experimental systems *in vitro* and *in vivo*, the application of sponges in a clinical setting will face greater hurdles, as they require access to the nucleus for expression. Nevertheless, long-term depletion of miR-122 was achieved in the mouse liver using an AAV-driven sponge construct [129], suggesting that such a strategy may be useful in HCC.

4.2 miRNA REPLACEMENT IN CANCER

A number of miRNA families with tumor suppressor activity are downregulated in cancer. These tend to be those most conserved throughout evolution (e.g., let-7, miR-15/16, miR-34, and miR-200) and the pathways these miRNAs control are essential pathways in development and cell proliferation. Restoring the miRNA levels thus enables control over these growth-promoting pathways to be re-exerted. Experimentally, this can be achieved using miRNA expression vectors (either plasmid or virus encoded) but therapeutic applications have generally employed synthetic miRNA mimics complexed with a variety of liposomal and/or nanoparticle-based delivery vehicles (see Table 4.2).

In early experiments, mimics were designed with structures (and sequences) equivalent to the endogenous form produced by Dicer processing. As the mature miRNA guide strand and its complementary passenger strand are not fully homologous, the duplex formed includes the bulges and mismatches commonly found in miRNA secondary structure. In contrast, more recent designs have used artificial passenger strands with complete homology [130]. Although this introduces a nonnatural sequence into the cells, off-target effects can be avoided through chemical modification (see below). Furthermore, in common with small interfering RNAs (siRNAs), mimic duplexes are effective with overhangs, or as blunt-ended double-stranded RNAs.

In addition to duplex structure, chemical modification of miRNA mimics has also borrowed heavily from strategies used for siRNA. As is the case for AS-ODN, mimics are also susceptible to degradation, and similar modifications affording protection from nucleases. A detailed discussion of these approaches is beyond the scope of the current review, but in brief, 2′ O-Me and LNA substitutions can be included without loss of miRNA function (Figure 4.3, and reviewed for siRNA [131] and miRNA [130]). It should also be kept in mind that many studies using mimics *in vivo* use one of the various proprietary modification strategies found in off-the-shelf reagents. An important point to note is that modifications of the passenger strands of miRNA mimics are tolerated more readily than those of the mature miRNA (guide) strand. Typically, modifications in and around the seed sequence of the guide strand are avoided. Also, as the goal with miRNA mimics is generally to prevent misleading of the passenger strand, this can be terminally modified with 2′ O-Me to prevent loading, while at the same time avoiding phosphorothioate and LNA modifications will permit nuclease degradation of the passenger strand following duplex unwinding and the loading of the mature miRNA into the RISC [130].

As in the case of inhibitors, synthetic (or expressed) mimics of tumor suppressor miRNAs can be introduced into tumor cells before implantation to investigate effects on tumor growth but again, this is of limited clinical relevance. Here, we will consider only mimics that have been used as a therapeutic approach in preclinical tumor models. The earliest reports aimed to restore levels of the let-7 and miR-34 family. Intratumoral injection of let-7b with siPORT transfection agent was able to significantly reduce tumor burden in a subcutaneous lung cancer xenograft model [114]; inhibition correlated with reduced Ki-67 and increased caspase-3 staining. Systemic administration in an autochthonous lung cancer model reduced the number of lesions and the tumor burden [115]. The first study of miR-34a mimics as a therapeutic approach focused on a lung cancer model, and demonstrated that both intratumoral and intravenous administrations of lipid-formulated mimic could inhibit xenograft growth, with inhibition of Bcl-2 and c-Met involved [116]. Of relevance to therapeutic applications, the mimics also had effects in cancer cells with normal miR-34a expression [116]. The same miR-34a formulation was later shown to inhibit autochthonous lung tumors [115], and to reduce the growth of orthotopic prostate xenografts and the incidence of lung metastases derived from these tumors [117]. As the miR-34a mimic was able to inhibit the function of the cancer stem cell population [117], this underlines the importance of this miRNA and its relevance as a therapeutic target.

Like miR-34, the miR-15/16 family is also associated with a tumor suppressor phenotype, through regulation of cell cycle and apoptosis genes. The first study to target miR-16 as a therapeutic approach used atelocollagen to systemically deliver a miR-16 mimic in a metastatic prostate cancer model [120]. This reduced overall tumor burden by decreasing the number of metastases. Similarly, in a model of multiple myeloma, treatment with either miR-15a or miR-16 formulated in neutral lipid emulsion was able to greatly inhibit tumor formation following intravenous injection of tumor cells [119]. This significantly prolonged survival of mice in both treatment groups when compared with controls. Finally, restoring miR-16 levels in a malignant pleural mesothelioma model using a miR-16 mimic encapsulated in minicells was able to strongly inhibit tumor growth [121]. This effect was dose-dependent and involved induction of apoptosis. As predicted targets of the miR-15/16 family include VEGF, inhibition of angiogenesis may play a role in tumor growth inhibition, as suggested by the reduced number of vessels in miR-15a-treated tumors [119].

4.3 SMALL-MOLECULE-BASED miRNA MODULATION

A number of drugs have been shown to have effects on miRNA expression, although in most cases these are not miRNA-specific effects. Like protein-coding genes, many miRNAs are epigenetically silenced in cancer, and this can be reversed by demethylating agents [68]. For example, decitabine was shown to produce upregulation of the expression of epigenetically silenced miRNAs, and was used to identify three miRNAs commonly silenced in metastatic cancer [132]. Small molecules that modulate RNA processing can also affect levels of mature miRNAs. The antibacterial fluoroquinolone enoxacin binds the protein TARRNA-binding protein 2, a partner of Dicer, and increases the levels of a number of mature miRNAs, including miR-16, and leads to growth inhibition in a range of cancer cells [133]. Likewise, curcumin derivatives have been shown to alter the expression of miR-21 [134,135] and other selected miRNAs [135] both *in vitro* and *in vivo*, possibly due to inhibition of NF-kB signaling [135]. Similarly, other agents, such as kinase inhibitors, which alter transcription factor expression, also contribute to miRNA regulation but are beyond the scope of this review.

5 FUTURE PERSPECTIVES

The rapid progress that has been made in understanding the functions of miRNAs in the biology of cancer has led a number of companies to develop inhibitors or mimics toward clinical use. At the moment, however, an inhibitor of miR-122, *Miravirsen*, is the only one currently in clinical trial. Developed by Santaris Pharma as a treatment for HCV infection, this is an LNA-based antimiR that sequesters miR-122, which is, as previously mentioned, an essential factor for HCV replication in the liver. In completed clinical trials, it has proved to be well tolerated and effective in reducing HCV titers [66]. Other companies, such as Regulus Therapeutics, have inhibitors of miRNAs in preclinical development. In contrast, miRNA mimics as a treatment for cancer are in (or near) clinical trial. The first miRNA replacement therapy to reach the clinic—MRX34—is based on the miR-34a mimic studies described above, and is being developed by Mirna Therapeutics [136]. The lipid-formulated mimic is currently undergoing an open-label multicenter Phase I clinical trial, and was reported to have a manageable toxicity profile in interim safety data [137]. The current dose-escalation trial aims to identify maximum tolerated dose for planned Phase 2 trial, in which efficacy will be assessed.

In summary, targeting aberrant miRNA expression has great promise for cancer therapy. Optimism in the field is tempered somewhat by the continuing difficulties in achieving controlled delivery, and the various off-target effects related to stimulation of the innate immune system and cell-type specific (opposing) roles of miRNAs in different cells. Furthermore, it should be kept in mind that miRNAs with oncogenic functions in one cancer type can be tumor suppressive in another [138], and also that dysregulated miRNAs expression in cancer can be changed in the opposite direction in other diseases (e.g., miR-15 [139] and miR-34 [140]). Nevertheless, results from current clinical trials are eagerly awaited, and will provide the stimulus for further studies.

ACKNOWLEDGMENTS

SGM is supported by grants from the Health Research Board, Ireland (HRA/POR/2012/82), and the Cancer and Polio Research Fund, UK. BASB and HLM are funded by scholarships from the University of Hull and the Hull York Medical School, respectively.

REFERENCES

[1] Bartel DP. MicroRNAs: genomics, biogenesis, mechanism, and function. Cell 2004;116(2):281–97.
[2] He L, Hannon GJ. MicroRNAs: small RNAs with a big role in gene regulation. Nat Rev Genet 2004;5(7):522–31.
[3] Malumbres M. miRNAs and cancer: an epigenetics view. Mol Aspects Med. 2013;34(4):863–74.
[4] Lee RC, Feinbaum RL, Ambros V. The *C. elegans* heterochronic gene lin-4 encodes small RNAs with antisense complementarity to lin-14. Cell 1993;75(5):843–54.
[5] Wightman B, Burglin TR, Gatto J, Arasu P, Ruvkun G. Negative regulatory sequences in the lin-14 3′-untranslated region are necessary to generate a temporal switch during *Caenorhabditis elegans* development. Genes Dev 1991;5(10):1813–24.
[6] Wightman B, Ha I, Ruvkun G. Posttranscriptional regulation of the heterochronic gene lin-14 by lin-4 mediates temporal pattern formation in *C. elegans*. Cell 1993;75(5):855–62.
[7] Reinhart BJ, Slack FJ, Basson M, et al. The 21-nucleotide let-7 RNA regulates developmental timing in *Caenorhabditis elegans*. Nature 2000;403(6772):901–6.
[8] Slack FJ, Basson M, Liu Z, Ambros V, Horvitz HR, Ruvkun G. The lin-41 RBCC gene acts in the *C. elegans* heterochronic pathway between the let-7 regulatory RNA and the LIN-29 transcription factor. Mol cell 2000;5(4):659–69.
[9] Lagos-Quintana M, Rauhut R, Lendeckel W, Tuschl T. Identification of novel genes coding for small expressed RNAs. Science 2001;294(5543):853–8.
[10] Lau NC, Lim LP, Weinstein EG, Bartel DP. An abundant class of tiny RNAs with probable regulatory roles in *Caenorhabditis elegans*. Science 2001;294(5543):858–62.
[11] Lee RC, Ambros V. An extensive class of small RNAs in *Caenorhabditis elegans*. Science 2001; 294(5543):862–4.
[12] Borchert GM, Lanier W, Davidson BL. RNA polymerase III transcribes human microRNAs. Nat Struct Mol Biol 2006;13(12):1097–101.

[13] Lee Y, Kim M, Han J, et al. MicroRNA genes are transcribed by RNA polymerase II. EMBO J 2004;23(20):4051–60.

[14] Han J, Lee Y, Yeom KH, Kim YK, Jin H, Kim VN. The Drosha-DGCR8 complex in primary microRNA processing. Genes Dev 2004;18(24):3016–27.

[15] Lee Y, Jeon K, Lee JT, Kim S, Kim VN. MicroRNA maturation: stepwise processing and subcellular localization. EMBO J 2002;21(17):4663–70.

[16] Bohnsack MT, Czaplinski K, Gorlich D. Exportin 5 is a RanGTP-dependent dsRNA-binding protein that mediates nuclear export of pre-miRNAs. RNA 2004;10(2):185–91.

[17] Schwarz DS, Hutvagner G, Du T, Xu Z, Aronin N, Zamore PD. Asymmetry in the assembly of the RNAi enzyme complex. Cell 2003;115(2):199–208.

[18] Ketting RF, Fischer SE, Bernstein E, Sijen T, Hannon GJ, Plasterk RH. Dicer functions in RNA interference and in synthesis of small RNA involved in developmental timing in *C. elegans*. Genes Dev 2001;15(20):2654–9.

[19] Peters L, Meister G. Argonaute proteins: mediators of RNA silencing. Mol Cell 2007;26(5):611–23.

[20] Meister G, Landthaler M, Patkaniowska A, Dorsett Y, Teng G, Tuschl T. Human Argonaute2 mediates RNA cleavage targeted by miRNAs and siRNAs. Mol Cell 2004;15(2):185–97.

[21] Lewis BP, Shih IH, Jones-Rhoades MW, Bartel DP, Burge CB. Prediction of mammalian microRNA targets. Cell 2003;115(7):787–98.

[22] Pasquinelli AE, Reinhart BJ, Slack F, et al. Conservation of the sequence and temporal expression of let-7 heterochronic regulatory RNA. Nature 2000;408(6808):86–9.

[23] Friedman RC, Farh KK, Burge CB, Bartel DP. Most mammalian mRNAs are conserved targets of microRNAs. Genome Res 2009;19(1):92–105.

[24] Saito T, Saetrom P. MicroRNAs--targeting and target prediction. N Biotechnol 2010;27(3):243–9.

[25] Lytle JR, Yario TA, Steitz JA. Target mRNAs are repressed as efficiently by microRNA-binding sites in the 5′ UTR as in the 3′ UTR. Proc Natl Acad Sci USA 2007;104(23):9667–72.

[26] Grimson A, Farh KK, Johnston WK, Garrett-Engele P, Lim LP, Bartel DP. MicroRNA targeting specificity in mammals: determinants beyond seed pairing. Mol cell 2007;27(1):91–105.

[27] Stark A, Brennecke J, Bushati N, Russell RB, Cohen SM. Animal MicroRNAs confer robustness to gene expression and have a significant impact on 3′UTR evolution. Cell 2005;123(6):1133–46.

[28] Lim LP, Lau NC, Garrett-Engele P, et al. Microarray analysis shows that some microRNAs downregulate large numbers of target mRNAs. Nature 2005;433(7027):769–73.

[29] Filipowicz W, Bhattacharyya SN, Sonenberg N. Mechanisms of post-transcriptional regulation by microRNAs: are the answers in sight? Nat Rev Genet 2008;9(2):102–14.

[30] Bhattacharyya SN, Habermacher R, Martine U, Closs EI, Filipowicz W. Relief of microRNA-mediated translational repression in human cells subjected to stress. Cell 2006;125(6):1111–24.

[31] Brengues M, Teixeira D, Parker R. Movement of eukaryotic mRNAs between polysomes and cytoplasmic processing bodies. Science 2005;310(5747):486–9.

[32] Parker R, Song H. The enzymes and control of eukaryotic mRNA turnover. Nat Struct Mol Biol 2004;11(2):121–7.

[33] Younger ST, Pertsemlidis A, Corey DR. Predicting potential miRNA target sites within gene promoters. Bioorg Med Chem Lett 2009;19(14):3791–4.

[34] Huang V, Li LC. miRNA goes nuclear. RNA Biol 2012;9(3):269–73.

[35] Lu J, Getz G, Miska EA, et al. MicroRNA expression profiles classify human cancers. Nature 2005;435(7043):834–8.

[36] Li M, Li J, Ding X, He M, Cheng SY. microRNA and cancer. AAPS J 2010;12(3):309–17.

[37] Jansson MD, Lund AH. MicroRNA and cancer. Mol Oncol 2012;6(6):590–610.

[38] Vandenboom Ii TG, Li Y, Philip PA, Sarkar FH. MicroRNA and cancer: tiny molecules with major implications. Curr Genomics 2008;9(2):97–109.

[39] Calin GA, Dumitru CD, Shimizu M, et al. Frequent deletions and down-regulation of micro-RNA genes miR15 and miR16 at 13q14 in chronic lymphocytic leukemia. Proc Natl Acad Sci USA 2002;99(24):15524–9.

[40] Calin GA, Croce CM. MicroRNA-cancer connection: the beginning of a new tale. Cancer Res 2006;66(15):7390–4.

[41] Laurila EM, Kallioniemi A. The diverse role of miR-31 in regulating cancer associated phenotypes. Genes Chromosomes Cancer 2013;52(12):1103–13.

[42] Ivanov SV, Goparaju CM, Lopez P, et al. Pro-tumorigenic effects of miR-31 loss in mesothelioma. J Biol Chem 2010;285(30):22809–17.

[43] Cekaite L, Rantala JK, Bruun J, et al. MiR-9, -31, and -182 Deregulation promote proliferation and tumor cell survival in colon cancer. Neoplasia 2012;14(9) 868-IN21.

[44] He L, Thomson JM, Hemann MT, et al. A microRNA polycistron as a potential human oncogene. Nature 2005;435(7043):828–33.

[45] Suarez Y, Fernandez-Hernando C, Yu J, et al. Dicer-dependent endothelial microRNAs are necessary for postnatal angiogenesis. Proc Natl Acad Sci USA 2008;105(37):14082–7.

[46] Sylvestre Y, De Guire V, Querido E, et al. An E2F/miR-20a autoregulatory feedback loop. J Biol Chem 2007;282(4):2135–43.

[47] Gao B, Gao K, Li L, Huang Z, Lin L. miR-184 functions as an oncogenic regulator in hepatocellular carcinoma (HCC). Biomed Pharmacother 2014;68(2):143–8.

[48] Tsai WC, Hsu PW, Lai TC, et al. MicroRNA-122, a tumor suppressor microRNA that regulates intrahepatic metastasis of hepatocellular carcinoma. Hepatology 2009;49(5):1571–82.

[49] Li P, Xu Q, Zhang D, et al. Upregulated miR-106a plays an oncogenic role in pancreatic cancer. FEBS Lett 2014;588(5):705–12.

[50] Mace TA, Collins AL, Wojcik SE, Croce CM, Lesinski GB, Bloomston M. Hypoxia induces the overexpression of microRNA-21 in pancreatic cancer cells. J Surg Res 2013;184(2):855–60.

[51] Muller DW, Bosserhoff AK. Integrin [beta]3 expression is regulated by let-7a miRNA in malignant melanoma. Oncogene 2008;27(52):6698–706.

[52] Dang D, Bamburg JR, Ramos DM. αvβ3 integrin and cofilin modulate K1735 melanoma cell invasion. Exp Cell Res 2006;312(4):468–77.

[53] Koturbash I, Zemp FJ, Pogribny I, Kovalchuk O. Small molecules with big effects: the role of the microRNAome in cancer and carcinogenesis. Mut Res/Genet Toxicol Environ Mutagen 2011;722(2):94–105.

[54] Zhang X, Dong Y, Ti H, et al. Down-regulation of miR-145 and miR-143 might be associated with DNA methyltransferase 3B overexpression and worse prognosis in endometrioid carcinomas. Hum Pathol 2013;44(11):2571–80.

[55] Robertson KD, Uzvolgyi E, Liang G, et al. The human DNA methyltransferases (DNMTs) 1, 3a and 3b: coordinate mRNA expression in normal tissues and overexpression in tumors. Nucleic Acids Res 1999;27(11):2291–8.

[56] Fabbri M, Garzon R, Cimmino A, et al. MicroRNA-29 family reverts aberrant methylation in lung cancer by targeting DNA methyltransferases 3A and 3B. Proc Natl Acad Sci USA 2007;104(40):15805–10.

[57] Martinez I, Cazalla D, Almstead LL, Steitz JA, DiMaio D. miR-29 and miR-30 regulate B-Myb expression during cellular senescence. Proc Natl Acad Sci USA 2011;108(2):522–7.

[58] Wang Y, Zhang X, Li H, Yu J, Ren X. The role of miRNA-29 family in cancer. Eur J Cell Biol 2013;92(3):123–8.

[59] Abdelmohsen K, Srikantan S, Kuwano Y, Gorospe M. miR-519 reduces cell proliferation by lowering RNA-binding protein HuR levels. Proc Natl Acad Sci USA 2008;105(51):20297–302.

[60] Zhou X, Wei M, Wang W. MicroRNA-340 suppresses osteosarcoma tumor growth and metastasis by directly targeting ROCK1. Biochem Biophys Res Commun 2013;437(4):653–8.

[61] Chen X, Zhang Y, Yan J, Sadiq R, Chen T. miR-34a suppresses mutagenesis by inducing apoptosis in human lymphoblastoid TK6 cells. Mut Res/Genet Toxicol Environ Mutagen 2013;758(1–2):35–40.

[62] Benetti R, Gonzalo S, Jaco I, et al. A mammalian microRNA cluster controls DNA methylation and telomere recombination via Rbl2-dependent regulation of DNA methyltransferases. Nat Struct Mol Biol 2008;15(3):268–79.

[63] Hua Z, Lv Q, Ye W, et al. miRNA-directed regulation of VEGF and other angiogenic factors under hypoxia. PLoS One 2006;1(1):e116.

[64] Ruan K, Fang X, Ouyang G. MicroRNAs: novel regulators in the hallmarks of human cancer. Cancer Lett 2009;285(2):116–26.

[65] Schwarzenbach H, Nishida N, Calin GA, Pantel K. Clinical relevance of circulating cell-free microRNAs in cancer. Nat Rev Clin Oncol 2014;11(3):145–56.

[66] Janssen HL, Reesink HW, Lawitz EJ, et al. Treatment of HCV infection by targeting microRNA. N Engl J Med 2013;368(18):1685–94.

[67] Croce CM. Causes and consequences of microRNA dysregulation in cancer. Nat Rev Genet 2009;10(10):704–14.

[68] Baer C, Claus R, Plass C. Genome-wide epigenetic regulation of miRNAs in cancer. Cancer Res 2013;73(2):473–7.

[69] Li LC. Chromatin remodeling by the small RNA machinery in mammalian cells. Epigenetics 2013;9(1).

[70] Iorio MV, Croce CM. MicroRNAs in cancer: small molecules with a huge impact. J Clin Oncol 2009;27(34):5848–56.

[71] Weber B, Stresemann C, Brueckner B, Lyko F. Methylation of human microRNA genes in normal and neoplastic cells. Cell Cycle 2007;6(9):1001–5.

[72] Wang Z, Yao H, Lin S, et al. Transcriptional and epigenetic regulation of human microRNAs. Cancer Lett 2013;331(1):1–10.

[73] Han L, Witmer PD, Casey E, Valle D, Sukumar S. DNA methylation regulates MicroRNA expression. Cancer Biol Ther 2007;6(8):1284–8.

[74] Huidobro C, Fernandez AF, Fraga MF. Aging epigenetics: causes and consequences. Mol Aspects Med 2013;34(4):765–81.

[75] Schiffgen M, Schmidt DH, von Rücker A, Müller SC, Ellinger J. Epigenetic regulation of microRNA expression in renal cell carcinoma. Biochem Biophys Res Commun 2013;436(1):79–84.

[76] Bandres E, Agirre X, Bitarte N, et al. Epigenetic regulation of microRNA expression in colorectal cancer. Int J Cancer 2009;125(11):2737–43.

[77] Zhang L, Volinia S, Bonome T, et al. Genomic and epigenetic alterations deregulate microRNA expression in human epithelial ovarian cancer. Proc Natl Acad Sci USA 2008;105(19):7004–9.

[78] Jerónimo C, Bastian PJ, Bjartell A, et al. Epigenetics in prostate cancer: biologic and clinical relevance. Eur Urol 2011;60(4):753–66.

[79] Hulf T, Sibbritt T, Wiklund E, et al. Discovery pipeline for epigenetically deregulated miRNAs in cancer: integration of primary miRNA transcription. BMC Genomics 2011;12(1):54.

[80] Garzon R, Liu S, Fabbri M, et al. MicroRNA-29b induces global DNA hypomethylation and tumor suppressor gene reexpression in acute myeloid leukemia by targeting directly DNMT3A and 3B and indirectly DNMT1. Blood 2009;113(25):6411–18.

[81] Morita S, Horii T, Kimura M, Ochiya T, Tajima S, Hatada I. miR-29 represses the activities of DNA methyltransferases and DNA demethylases. Int J Mol Sci 2013;14(7):14647–58.

[82] Duursma AM, Kedde M, Schrier M, le Sage C, Agami R. miR-148 targets human DNMT3b protein coding region. RNA 2008;14(5):872–7.

[83] Cao Q, Mani RS, Ateeq B, et al. Coordinated regulation of polycomb group complexes through microRNAs in cancer. Cancer Cell 2011;20(2):187–99.

[84] Varambally S, Cao Q, Mani RS, et al. Genomic loss of microRNA-101 leads to overexpression of histone methyltransferase EZH2 in cancer. Science 2008;322(5908):1695–9.

[85] Friedman JM, Liang G, Liu CC, et al. The putative tumor suppressor microRNA-101 modulates the cancer epigenome by repressing the polycomb group protein EZH2. Cancer Res 2009;69(6):2623–9.

[86] Kleer CG, Cao Q, Varambally S, et al. EZH2 is a marker of aggressive breast cancer and promotes neoplastic transformation of breast epithelial cells. Proc Natl Acad Sci USA 2003;100(20):11606–11.

[87] Varambally S, Dhanasekaran SM, Zhou M, et al. The polycomb group protein EZH2 is involved in progression of prostate cancer. Nature 2002;419(6907):624–9.

[88] Wei X, Xiang T, Ren G, et al. miR-101 is down-regulated by the hepatitis B virus x protein and induces aberrant DNA methylation by targeting DNA methyltransferase 3A. Cell Signal 2013;25(2):439–46.

[89] Hwang HW, Wentzel EA, Mendell JT. A hexanucleotide element directs microRNA nuclear import. Science 2007;315(5808):97–100.

[90] Place RF, Li LC, Pookot D, Noonan EJ, Dahiya R. MicroRNA-373 induces expression of genes with complementary promoter sequences. Proc Natl Acad Sci USA 2008;105(5):1608–13.

[91] Huang V, Place RF, Portnoy V, et al. Upregulation of cyclin B1 by miRNA and its implications in cancer. Nucleic Acids Res 2012;40(4):1695–707.

[92] Kim DH, Saetrom P, Snove Jr. O, Rossi JJ. MicroRNA-directed transcriptional gene silencing in mammalian cells. Proc Natl Acad Sci USA 2008;105(42):16230–5.

[93] Younger ST, Corey DR. Transcriptional gene silencing in mammalian cells by miRNA mimics that target gene promoters. Nucleic Acids Res 2011;39(13):5682–91.

[94] Hydbring P, Badalian-Very G. Clinical applications of microRNAs. F1000Research 2013;2:136.

[95] Stenvang J, Petri A, Lindow M, Obad S, Kauppinen S. Inhibition of microRNA function by antimiR oligonucleotides. Silence 2012;3(1):1.

[96] Obad S, dos Santos CO, Petri A, et al. Silencing of microRNA families by seed-targeting tiny LNAs. Nat Genet 2011;43(4):371–8.

[97] Li J, Huang H, Sun L, et al. MiR-21 indicates poor prognosis in tongue squamous cell carcinomas as an apoptosis inhibitor. Clin Cancer Res 2009;15(12):3998–4008.

[98] Zhou X, Ren Y, Moore L, et al. Downregulation of miR-21 inhibits EGFR pathway and suppresses the growth of human glioblastoma cells independent of PTEN status. Lab Invest 2010;90(2):144–55.

[99] Iliopoulos D, Jaeger SA, Hirsch HA, Bulyk ML, Struhl K. STAT3 activation of miR-21 and miR-181b-1 via PTEN and CYLD are part of the epigenetic switch linking inflammation to cancer. Mol Cell 2010;39(4):493–506.

[100] Wang YY, Sun G, Luo H, et al. MiR-21 modulates hTERT through a STAT3-dependent manner on glioblastoma cell growth. CNS Neurosci Ther 2012;18(9):722–8.

[101] Zhao D, Tu Y, Wan L, et al. In vivo monitoring of angiogenesis inhibition via down-regulation of mir-21 in a VEGFR2-luc murine breast cancer model using bioluminescent imaging. PLoS One 2013;8(8):e71472.

[102] Leone E, Morelli E, Di Martino MT, et al. Targeting miR-21 inhibits in vitro and in vivo multiple myeloma cell growth. Clin Cancer Res 2013;19(8):2096–106.

[103] Fontana L, Fiori ME, Albini S, et al. Antagomir-17-5p abolishes the growth of therapy-resistant neuroblastoma through p21 and BIM. PLoS One 2008;3(5):e2236.

[104] Murphy BL, Obad S, Bihannic L, et al. Silencing of the miR-17~92 cluster family inhibits medulloblastoma progression. Cancer Res 2013;73(23):7068–78.

[105] Liang Z, Li Y, Huang K, Wagar N, Shim H. Regulation of miR-19 to breast cancer chemoresistance through targeting PTEN. Pharm Res 2011;28(12):3091–100.

[106] Zheng SR, Guo GL, Zhai Q, Zou ZY, Zhang W. Effects of miR-155 antisense oligonucleotide on breast carcinoma cell line MDA-MB-157 and implanted tumors. Asian Pac J Cancer Prev 2013;14(4):2361–6.

[107] Zhang Y, Roccaro AM, Rombaoa C, et al. LNA-mediated anti-miR-155 silencing in low-grade B-cell lymphomas. Blood 2012;120(8):1678–86.

[108] Ma L, Reinhardt F, Pan E, et al. Therapeutic silencing of miR-10b inhibits metastasis in a mouse mammary tumor model. Nat Biotechnol 2010;28(4):341–7.

[109] Swarbrick A, Woods SL, Shaw A, et al. miR-380-5p represses p53 to control cellular survival and is associated with poor outcome in MYCN-amplified neuroblastoma. Nat Med 2010;16(10):1134–40.

[110] Lu Y, Roy S, Nuovo G, et al. Anti-microRNA-222 (anti-miR-222) and -181B suppress growth of tamoxifen-resistant xenografts in mouse by targeting TIMP3 protein and modulating mitogenic signal. J Biol Chem 2011;286(49):42292–302.

[111] Leucci E, Zriwil A, Gregersen LH, et al. Inhibition of miR-9 de-represses HuR and DICER1 and impairs Hodgkin lymphoma tumour outgrowth in vivo. Oncogene 2012;31(49):5081–9.

[112] Lin CW, Chang YL, Chang YC, et al. MicroRNA-135b promotes lung cancer metastasis by regulating multiple targets in the Hippo pathway and LZTS1. Nat Commun 2013;4:1877.

[113] Shu M, Zheng X, Wu S, et al. Targeting oncogenic miR-335 inhibits growth and invasion of malignant astrocytoma cells. Mol Cancer 2011;10:59.

[114] Trang P, Medina PP, Wiggins JF, et al. Regression of murine lung tumors by the let-7 microRNA. Oncogene 2010;29(11):1580–7.

[115] Trang P, Wiggins JF, Daige CL, et al. Systemic delivery of tumor suppressor microRNA mimics using a neutral lipid emulsion inhibits lung tumors in mice. Mol Ther 2011;19(6):1116–22.

[116] Wiggins JF, Ruffino L, Kelnar K, et al. Development of a lung cancer therapeutic based on the tumor suppressor microRNA-34. Cancer Res 2010;70(14):5923–30.

[117] Liu C, Kelnar K, Liu B, et al. The microRNA miR-34a inhibits prostate cancer stem cells and metastasis by directly repressing CD44. Nat Med 2011;17(2):211–15.

[118] Tivnan A, Orr WS, Gubala V, et al. Inhibition of neuroblastoma tumor growth by targeted delivery of microRNA-34a using anti-disialoganglioside GD2 coated nanoparticles. PLoS One 2012;7(5):e38129.

[119] Sun CY, She XM, Qin Y, et al. miR-15a and miR-16 affect the angiogenesis of multiple myeloma by targeting VEGF. Carcinogenesis 2013;34(2):426–35.

[120] Takeshita F, Patrawala L, Osaki M, et al. Systemic delivery of synthetic microRNA-16 inhibits the growth of metastatic prostate tumors via downregulation of multiple cell-cycle genes. Mol Ther 2010;18(1):181–7.

[121] Reid G, Pel ME, Kirschner MB, et al. Restoring expression of miR-16: a novel approach to therapy for malignant pleural mesothelioma. Ann Oncol 2013;24(12):3128–35.

[122] Akao Y, Nakagawa Y, Hirata I, et al. Role of anti-oncomirs miR-143 and -145 in human colorectal tumors. Cancer Gene Ther 2010;17(6):398–408.

[123] Ibrahim AF, Weirauch U, Thomas M, Grunweller A, Hartmann RK, Aigner A. MicroRNA replacement therapy for miR-145 and miR-33a is efficacious in a model of colon carcinoma. Cancer Res 2011;71(15):5214–24.

[124] Hsu SH, Yu B, Wang X, et al. Cationic lipid nanoparticles for therapeutic delivery of siRNA and miRNA to murine liver tumor. Nanomedicine 2013;9(8):1169–80.

[125] Pan X, Wang ZX, Wang R. MicroRNA-21: a novel therapeutic target in human cancer. Cancer Biol Ther 2010;10(12):1224–32.

[126] Hong L, Han Y, Zhang Y, et al. MicroRNA-21: a therapeutic target for reversing drug resistance in cancer. Expert Opin Ther Targets 2013;17(9):1073–80.

[127] Rottiers V, Obad S, Petri A, et al. Pharmacological inhibition of a microRNA family in nonhuman primates by a seed-targeting 8-mer antimiR. Sci Transl Med 2013;5(212) 212ra162.

[128] Ebert MS, Neilson JR, Sharp PA. MicroRNA sponges: competitive inhibitors of small RNAs in mammalian cells. Nat Methods 2007;4(9):721–6.

[129] Xie J, Ameres SL, Friedline R, et al. Long-term, efficient inhibition of microRNA function in mice using rAAV vectors. Nat Methods 2012;9(4):403–9.

[130] Henry JC, Azevedo-Pouly AC, Schmittgen TD. MicroRNA replacement therapy for cancer. Pharm Res 2011;28(12):3030–42.

[131] Rettig GR, Behlke MA. Progress toward in vivo use of siRNAs-II. Mol Ther 2012;20(3):483–512.

[132] Lujambio A, Calin GA, Villanueva A, et al. A microRNA DNA methylation signature for human cancer metastasis. Proc Natl Acad Sci USA 2008;105(36):13556–61.

[133] Melo S, Villanueva A, Moutinho C, et al. Small molecule enoxacin is a cancer-specific growth inhibitor that acts by enhancing TAR RNA-binding protein 2-mediated microRNA processing. Proc Natl Acad Sci USA 2011;108(11):4394–9.

[134] Roy S, Yu Y, Padhye SB, Sarkar FH, Majumdar AP. Difluorinated-curcumin (CDF) restores PTEN expression in colon cancer cells by down-regulating miR-21. PLoS One 2013;8(7):e68543.

[135] Yang CH, Yue J, Sims M, Pfeffer LM. The curcumin analog EF24 targets NF-kappaB and miRNA-21, and has potent anticancer activity in vitro and in vivo. PLoS One 2013;8(8):e71130.

[136] Bader AG. miR-34—a microRNA replacement therapy is headed to the clinic. Front Genet 2012;3:120.

[137] Beg MS, Borad M, Sachdev J, Hong DS, Smith S, Bader A, Stoudemire J, Kim S, Brenner A. Multicenter phase I study of MRX34, a first-in-class microRNA miR-34 mimic liposomal injection. AACR Annual Meeting. San Diego, April 5–9; 2014.

[138] Ling H, Fabbri M, Calin GA. MicroRNAs and other non-coding RNAs as targets for anticancer drug development. Nat Rev Drug Discov 2013;12(11):847–65.

[139] Hullinger TG, Montgomery RL, Seto AG, et al. Inhibition of miR-15 protects against cardiac ischemic injury. Circ Res 2012;110(1):71–81.

[140] Bernardo BC, Gao XM, Winbanks CE, et al. Therapeutic inhibition of the miR-34 family attenuates pathological cardiac remodeling and improves heart function. Proc Natl Acad Sci USA 2012;109(43):17615–20.

LONG NONCODING RNAs AND CANCER

5

Casey M. Wright

Asbestos Diseases Research Institute, The University of Sydney, Concord, NSW, Australia

CHAPTER OUTLINE

S.G. Gray (Ed): Epigenetic Cancer Therapy. DOI: http://dx.doi.org/10.1016/B978-0-12-800206-3.00005-7

1 INTRODUCTION

Since completion of the Human Genome Project, several genome-wide analyses investigating the eurokaryotic transcriptome have been performed. These studies have revealed that eukaryotic genomes are extensively transcribed, with only a small portion coding for proteins. The remaining RNAs, lacking protein coding potential, are together referred to as noncoding RNAs and incorporate several subgroups including longer noncoding RNAs (lncRNAs) and smaller noncoding RNAs such as small nuclear RNAs (snRNAs), small nucleolar RNAs (snoRNA), RNase P and mitochondrial RNA processing (MRP) RNAs, signal recognition particles (SRP) RNAs, telomerase RNAs, piwi RNAs (piRNAs), and microRNAs (miRNAs) [1]. For the purposes of this chapter, we will focus on the role and function of lncRNAs.

lncRNAs are RNA sequences >200 nucleotides in length that lack open reading frames. These lncRNAs accumulate differently in the nucleus and cytoplasm, and can be either polyadenylated or nonpolyadenylated [2,3]. Unlike smaller noncoding RNAs such as miRNAs, which are highly conserved, lncRNAs are poorly conserved and regulate gene expression in several different ways [4–7]. It is estimated that the human genome encodes ~15,000–17,000 lncRNAs, with only 2% of these having a proven function [8–10]. Despite this, there is compelling evidence that the large number of noncoding RNAs located throughout the genome have functionality, as demonstrated by their conservation of promoters [11], splice junctions [9], sequence [9,10,12], genomic position [13,14], secondary structure [15], expression [16], tissue- and cell-specific expression patterns, specific subcellular localization [17–19], altered expression and splicing, and association with particular chromatin signatures [20]. Interestingly, many of these transcripts do not share sequence homology with each other and are often spliced, capped, and polyadenylated [21]. In addition, evidence is rapidly emerging to suggest that lncRNAs have critical roles in epigenetic regulation of gene expression.

Although only a small fraction of lncRNAs have been functionally characterized, there is much debate as to whether the majority of identified lncRNAs are transcriptional noise or in fact have meaningful biological functions [22,23]. Despite this controversy, it is clear that those lncRNAs to which functions have been ascribed play important roles in the regulation of multiple biological processes including transcription, chromatin remodeling and histone modification, modulation of alternative splicing, generation of endo-siRNAs (small interfering RNAs), modulation of protein activity, structural/organization roles, and protein localization [24] (Figure 5.1). As we discover more about their functions, lncRNAs will impact greatly on our knowledge and understanding of gene regulation in human disease. In the following sections, we will explore the classification and nomenclature of lncRNAs, their functional activities, their role in epigenetic regulation and transcriptional control of gene expression, their role in human disease, potential as diagnostic, prognostic and predictive markers, and potential strategies for therapeutic targeting of lncRNAs.

Mechanisms of lncRNA function.

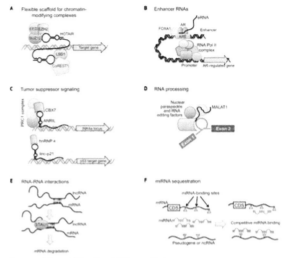

Prensner J R , and Chinnaiyan A M Cancer Discovery
2011;1:391-407

FIGURE 5.1

Mechanisms of lncRNA function. (A) lncRNAs may serve as a scaffolding base for the coordination of epigenetic or histone-modifying complexes. (B) Enhancer RNAs (eRNAs) transcribed from gene enhancers may facilitate hormone signaling. (C) lncRNAs may directly impact tumor suppressor signaling either by transcriptional regulation of tumor suppressor genes through epigenetic silencing (e.g., *ANRIL*, **upper**) or by mediating activation of tumor suppressor target genes (e.g., *linc-p21*, **lower**). (D) *MALAT1* and *NEAT2* lncRNAs may be integral components of the nuclear paraspeckle and contribute to posttranscriptional processing of mRNAs. (E) Gene expression regulation may occur through direct lncRNA–mRNA interactions, which arise from hybridization of homologous sequences and can serve as a signaling for STAU1-mediated degradation of the mRNA. (F) RNA molecules, including mRNAs, pseudogenes, and ncRNAs, can serve as molecular sponges for miRNAs. This generates an environment of competitive binding of miRNAs to achieve gene expression control based upon the degree of miRNA binding to each transcript. The colored triangles represent different miRNA binding sites in a transcript. CDS, coding sequence. *Reproduced from Ref. [25].*

2 CLASSIFICATION AND NOMENCLATURE OF lncRNAs

2.1 CLASSIFICATION

In recent times, there has been a concerted effort to characterize and compile lncRNAs in public databases such as lncRNAdb [8], NRED [26], NONCODE [27–30], LNCipedia [31], ncFANS [32], and Starbase [33,34] (Table 5.1). These databases are assisting the field of lncRNA research, by systematically cataloguing all reported lncRNAs and providing annotation based on experimental data, structural

Table 5.1 Public lncRNA Databases

Database	Website (reference)	Species	Comments	References
lncRNAdb	www.lncrnadb.org	Multiple	Annotation, expression, function, conservation, sequence, literature	[8]
NONCODE	www.noncode.org	Multiple	Annotation, literature	[27–30]
LNCipedia	http://www.lncipedia.org/	Human	Annotation, sequence, secondary structure, locus conservation, protein coding potential	[31]
NRED	http://nred.matticklab.com/cgi-bin/ncrnadb.pl	Human, mouse	Annotated expression data from various sources	[26]
STARBASE	http://starbase.sysu.edu.cn/mirLncRNA.php	Human, mouse	Predicted miRNA–lncRNA interactions, miRNA–mRNA interactions	[33,34]
lncRNome	http://genome.igib.res.in/lncRNome/	Human	Provides information on disease association, chromosomal location, biological functions, protein–lncRNA interactions, genomic variations	[35]
Chip Base	http://deepbase.sysu.edu.cn/chipbase/	–	Transcription factor binding maps, expression profiles, transcriptional regulation from ChIP-Seq data	[36]
LncRNA disease	http://202.38.126.151/hmdd/html/tools/lncrnadisease.html	Human	Experimentally supported lncRNA disease associations	[37]

information, expression, conservation, function, subcellular localization, and interactions with other noncoding RNAs such as miRNAs. Generating a comprehensive classification system for lncRNAs that is universally accepted among the research community is difficult. A survey of the literature has found that lncRNAs can be broadly grouped into five main categories based on their proximity to known protein coding genes: (i) sense lncRNAs—often considered transcript variants of protein coding genes, (ii) antisense lncRNAs—RNA molecules transcribed from the antisense strand of protein coding genes, (iii) bidirectional—transcripts orientated head to head with a protein coding gene, (iv) intronic lncRNAs—RNA overlapping with introns from annotated coding genes, and (v) intergenic—lncRNAs located in between annotated protein coding genes. Each of these categories can also be further subclassified on the basis of their individual features including (i) genomic location and context, (ii) effects exerted on DNA sequences, (iii) functional mechanism, and (iv) targeting mechanisms (as reviewed in [38]). In contrast miRNAs, a family of smaller and highly conserved noncoding RNAs, lncRNAs have yet to be thoroughly categorized, with many functions yet to be determined. The aforementioned features will be explored further in subsequent sections (Figure 5.2).

2.2 NOMENCLATURE

The HUGO Nomenclature Committee (HGNC), responsible for cataloguing and assigning standardized nomenclature to human genes, has undertaken a project to correctly annotate and characterize lncRNAs in a systematic manner [21]. In cases where lncRNAs are located antisense to a protein

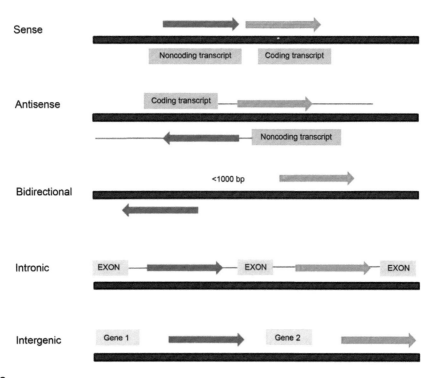

FIGURE 5.2

Categories of lncRNAs. lncRNAs can be broadly grouped into five main categories based on their proximity to known protein coding genes: (i) sense lncRNAs—often considered transcript variants of protein coding genes, (ii) antisense lncRNAs—RNA molecules transcribed from the antisense strand of protein coding genes, (iii) bidirectional—transcripts orientated head to head with a protein coding gene, (iv) intronic lncRNAs—RNA overlapping with introns from annotated coding genes, and (v) intergenic—lncRNAs located in between annotated protein coding genes.

coding gene, they are generally labeled using the HGNC approved gene symbol, with a suffix –AS for "antisense," if they are intronic, they are symbolized with the suffix –IT for "intronic" and for host genes that contain smaller lncRNAs, they contain the suffix –HG for "host gene" [21]. In cases where the function is unknown or they are not located close to protein coding genes, the lncRNAs are prefixed with the symbol "LINC" for long intergenic nonprotein coding RNA [21] (Table 5.2). This work has been essential for ongoing development of a standardized nomenclature for lncRNAs, which is now being integrated with the public lncRNA databases.

3 MECHANISMS OF lncRNA FUNCTION

lncRNAs have been found to control every level of gene expression including posttranscriptional gene regulation, transcription factor activity, and chromatin structure [39]. Wang et al. suggested the presence

Table 5.2 Nomenclature for Naming of lncRNAs

Type of lncRNA	Suffix/prefix	Examples
Antisense	–AS	*CDKN2B-AS1*
Intronic	–IT	*SPRY4-IT1*
Host gene	–HG	*SNHG1*
Long intergenic nonprotein coding RNA	LINC–	*LINC-P21, LINC-UBC1*

of four molecular functions for lncRNAs—as signals, decoys, guides, and scaffolds [39]. Gutschner et al. built on this structure to suggest additional molecular functions of lncRNAs as (i) regulators of gene expression, (ii) miRNA sponges [40], (iii) modulators of protein activity and localization, (iv) endo-siRNAs that target RNAs for degradation, (v) regulators of alternative splicing, (vi) scaffolds, and finally (vii) important controllers of chromatin remodeling and histone modifications [41] (Figure 5.3).

3.1 lncRNAs AS DECOYS AND GUIDES

In addition to the aforementioned functions, lncRNAs can act as decoys or guides that bind proteins and direction localization of protein targets to highly specific targets [39]. In these circumstances, lncRNAs can act in *trans* (distant genes) or *cis* (neighboring genes) to direct changes in gene expression by binding to RNA:DNA heteroduplexes, RNA:DNA:DNA triplexes, recognizing distinctive chromatin features (reviewed in [42]) and guiding chromatin changes co-transcriptionally or as targets for smaller regulatory RNAs. Trithorax and polycomb repressor complex (PcGs) proteins are involved in this process, with lncRNA acting as tethers to bind these effector molecules [42]. One example of this is the binding of lincRNAs to chromatin-modifying complexes. This binding helps to facilitate the guiding of chromatin-modifying complexes to specific locations in the genome, in turn altering the epigenetic state of certain cell types [43].

3.2 lncRNAs AS SCAFFOLDS

By acting as molecular scaffolds, lncRNAs can help to bring together several different factors including chromatin and histone-modifying complexes such as the polycomb repressor complex. In this role, lncRNAs are the organizers and associate with proteins to form ribonucleic particles (RNPs), thereby acting as versatile protein assembly platforms. These scaffolds have unique properties including tight RNA–protein bridges (if scaffolds are tight) and flexible rotation (if RNA contains regions that have hinges) [44]. The scaffold function can also be exploited under stressful conditions to regulate protein activity, e.g., the ncRNA 7SK can negatively regulate the transcription factor P-TEFb by sequestering it in an RNP complex [44]. In addition to this, some lncRNAs like *NEAT*1 (nuclear enriched abundant transcript 1) have been implicated in nuclear organization of small bodies known as paraspeckles [17]. Paraspeckles are small nuclear structures that localize to the edge of SC-35 domains and are observed through interphase. Their precise function is not entirely understood; however, it has been shown that the lncRNA *NEAT1* is necessary for their formation and binds two critical proteins PSP1 and p54 [17]. Both of these proteins are involved in RNA and DNA binding, and are implicated in transcriptional regulation, pre-mRNA (messenger RNA) splicing, and nuclear retention of RNA [17]. In RNA inhibition

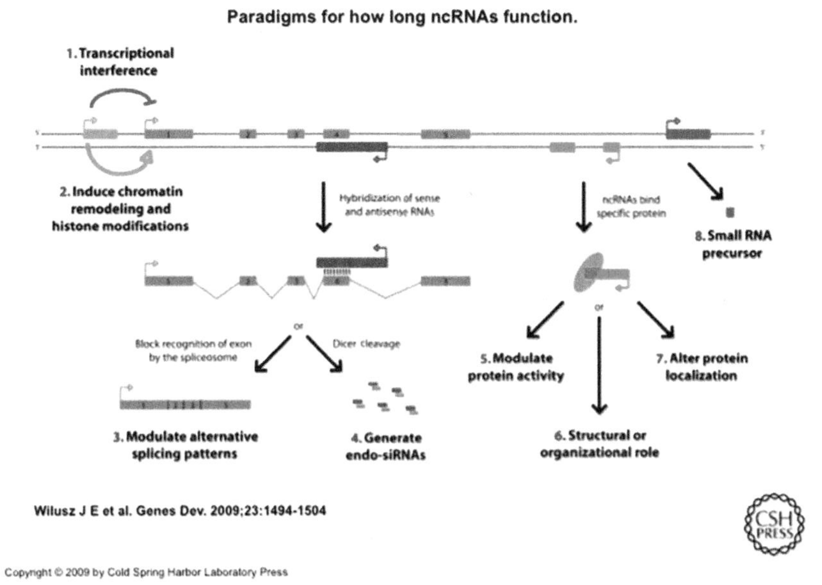

Paradigms for how long ncRNAs function.

Wilusz J E et al. Genes Dev. 2009;23:1494-1504

FIGURE 5.3

Paradigms for how long ncRNAs function. Recent studies have identified a variety of regulatory paradigms for how long ncRNAs function, many of which are highlighted here. Transcription from an upstream noncoding promoter (orange) can negatively (1) or positively (2) affect expression of the downstream gene (blue) by inhibiting RNA polymerase II recruitment or inducing chromatin remodeling, respectively. (3) An antisense transcript (purple) is able to hybridize to the overlapping sense transcript (blue) and block recognition of the splice sites by the spliceosome, thus resulting in an alternatively spliced transcript. (4) Alternatively, hybridization of the sense and antisense transcripts can allow Dicer to generate endogenous siRNAs. By binding to specific protein partners, a noncoding transcript (green) can modulate the activity of the protein (5), serve as a structural component that allows a larger RNA–protein complex to form (6), or alter where the protein localizes in the cell (7). (8) Long ncRNAs (pink) can be processed to yield small RNAs, such as miRNAs, piRNAs, and other less well-characterized classes of small transcripts. (For interpretation of the references to color in this figure legend, the reader is referred to the web version of this book.)

Reproduced from Ref. [24].

experiments, knock down of *NEAT1* has been shown to result in significantly less paraspeckles being formed, providing further evidence of their critical architectural role [17]. It is interesting to speculate on how other lncRNAs may serve as molecular scaffolds, to tether other molecular complexes such as transcription factors and RNA polymerases.

3.3 lncRNAs AS SIGNALING MOLECULES

lncRNAs exhibit cell-type specific lncRNA expression patterns and respond to different stimuli in a multitude of ways. Many lncRNAs possess some sort of regulatory function and can function at the initiation, elongation, or termination stage of transcription. Examples of this include *KCNQ1OT1* (KCNQ1 opposite strand/antisense transcript 1), which is important for epigenetic regulation [45], and *AIR*, which is important in repressing several imprinted genes on the paternal allele. lncRNAs can also respond to environmental cues including induction of DNA damage through *PANDA* (promoter of CDKN1A antisense DNA damage activated RNA) and *lincRNA-p21*. *PANDA* can interact with the transcription factor NF-YA (nuclear transcription factor Y-alpha) to limit the expression of proapoptopic genes, thereby enabling cell cycle arrest [46]. It also participates in the p53 signaling pathway and is situated close to a p53 binding site within the *CDKN1A* locus. Hung et al. showed that siRNA-mediated knockdown of p53 resulted in inhibition of *PANDA* expression by 70% following DNA damage [42]. In addition, a gain-of-function p53 mutant was found to preserve the ability to induce *PANDA*, while complementation experiments in wild-type H1299 lung carcinoma cells restored *PANDA* expression following DNA damage [42]. This work demonstrates that lncRNAs can interact with important cancer signaling molecules and can affect DNA damage responses.

lncRNAs have also been implicated in pluripotency and cell reprogramming [10], recruitment of RNA polymerase to promoter regions (enhancer RNAs) [39], and mediating RNA decay. Using loss of function-based studies, Guttman et al. showed that knockdown of long intergenic noncoding RNAs (lincRNAs) has major effects on gene expression [47]. In mouse embryonic stem (ES) cells, knockdown of 26 lincRNAs was found to result in loss of the pluripotent state, in the process repressing lineage specific ES expression [47] (see Ref. [47] for further details). These lincRNAs were also shown to physically interact with chromatin, suggesting that loss of these lincRNAs results in loss of signaling pathways that affect RNA transcription. The idea of signaling lncRNAs acting in mediation of RNA decay was first described by Gong and Maquat [48]. They showed that staufen 1 (*STAU1*), an important protein binding double-stranded RNA, can bind alu elements in a cytoplasmic, polyadenylated lncRNA termed half-STAU1-binding site RNA (½-sbsRNA). *STAU1* binding to mRNAs is therefore facilitated through transactivation by the lncRNA (½-sbsRNA) [48]. This provides evidence that lncRNAs are important regulators of RNA-mediated decay and critical for binding proteins involved in double-stranded RNA. In addition to acting as signaling molecules, lncRNAs can also act as miRNA sponges, which act to sequester miRNAs involved in regulation of oncogenes and tumor suppressor genes.

3.4 miRNA SEQUESTRATION

There is evidence to suggest that lncRNAs can act as molecular sponges for miRNAs. lncRNAs can interact with miRNAs to form extensive regulatory networks that protect target RNAs from repression [40]. It has been hypothesized that lncRNAs contain miRNA response elements (MRE) and act to titrate miRNAs, making them less available for binding to their target mRNAs, resulting in decreased miRNA expression and activity [40]. This complex relationship is dependent on the concentration of lncRNAs and miRNAs available (i.e., expression), the number of miRNAs the lncRNA is capable of binding (i.e., how many it can sequester), and the specific nucleotide composition they recognize (as not all MRE bind the same miRNA) [40]. One example of this repression was demonstrated by Braconi et al., who showed that the tissue-specific increase in expression of the maternally expressed gene 3 (*MEG3*) was accompanied by overexpression of miR-29a in hepatocellular carcinoma (HCC) [49]. Similarly, depletion of the lincRNA linc-MD1 in muscle differentiation was found to be accompanied by reduced levels of two predicted targets of miR-133 and miR-135, *MAML1* and *MEF2C* [50], suggesting linc-MD1 may act as a decoy for miR-133 and miR-135. Finally, mutations in the *PTENP1/ PTEN* region have been shown to disrupt miRNA binding, negating the protective effect of the lncRNA transcript, *PTENP1* [51]. These studies highlight the interdependence observed between lncRNAs and miRNAs, and suggest that lncRNAs and miRNAs work together in complex regulatory networks to activate or suppress gene expression. In summary, lncRNAs have a variety of functions and act in a myriad of ways to control gene transcription.

3.5 lncRNAs AND EPIGENETIC REGULATION

Chromatin remodeling and histone modifications including methylation, acetylation, and phosphory-lation are critical for the regulation of gene expression and chromatin conformation. Disruption of the fragile histone acetylation and deacetylation balance can contribute to tumorigenesis. Disruption of histones and chromatins can occur in a variety of ways, mostly via the polycomb (PcG) and trithorax (TrxG) groups of proteins [52,53]. There is increasing evidence to suggest that lncRNAs can interact with these proteins and are critical in the control of chromatin modulation and epigenetic regulation. lincRNAs, antisense transcripts, and other lncRNAs have all been found to regulate chromatin states. Examples of lncRNAs involved in epigenetic regulation include *H19, AIR, KCNQOT1*, and lincRNAs, which are discussed below.

3.5.1 H19

The first studies investigating the role of lncRNAs in epigenetic regulation originated from the studies of genomic imprinting and X chromosome inactivation [54,55]. One of the first lncRNA genes identi-fied was *H19*, an imprinted maternally expressed lncRNA co-expressed with *IGF2* (insulin growth fac-tor 2, an imprinted gene expressed on the paternal allele), that is spliced, polyadenylated, and exported to the nucleus [56]. The *H19/IGF2* locus is regulated by downstream enhancers and a differentially methylated region (DMR) [57]. The methylation-sensitive CTCF (CCCTC-binding factor) transcrip-tion factor, which is important for maintaining chromatin loop structure, binds to the DMR region near *H19* [58]. Recently Ito et al. showed that DNA demethylation of the *H19/IGF2* locus with 5-aza-2'deoxycytidine, reduced *IGF2* expression, increased *H19* expression, and led to a cascade of epige-netic changes in the choriocarcinoma cell line JEG3, including altered chromatin looping structure and gene expression [58]. *H19* is also thought to act in *trans* to regulate the expression of several genes in the imprinted gene network (IGN) [59]. Loss of the *H19* noncoding RNA via deletion also results in reactivation of *IGF2*, indicating an important regulatory role [59]. This lncRNA has been implicated in a myriad of cancer subtypes and has been associated with early tumor recurrence in bladder cancer [60] and increased cell proliferation in gastric cancer [61]. Taken together, this work suggests that *H19* is critical for control of genomic imprinting and when disrupted leads to a series of epigenetic events that lead to altered chromatin structure.

3.5.2 AIR and KCNQ1OT1

Other notable examples of lncRNAs that target chromatin regulators include the paternally imprinted lncRNAs *AIR* and *KCNQ1OT1*. Both of these lncRNAs specifically bind to the histone H3 lysine 9 methylase G9a to mediate transcriptional silencing of *IGFR2* and *KCNQ1* and recruitment of H3K9me3 histones [62]. *KCNQ1OT1* is expressed antisense to a highly conserved and differentially methylated region, and it thought to play a bidirectional role in gene silencing [45,63]. Truncation of this transcript has been shown to result in upregulation of several imprinted genes. *KCNQ1OT1* has also been shown to influence chromatin modifications at the *KCNQ1* locus. Using a chromatin immunoprecipitation (ChIP) and RNA immunoprecipitation (RIP)-based assay, Pandey et al. have demonstrated that the

KCNQ1OT1 ncRNA interacts with chromatin and is thought to be lineage specific [63]. Furthermore, profiling of histone modifications using Nimblegen high-resolution oligonucleotide profiling arrays showed enrichment of the H2K27me3 histone mark [63]. When mice with a 244 bp deletion of the *KCNQ1OT1* promoter were profiled, loss of the H2K27me3 mark was observed across several imprinted regions, suggesting that *KCNQ1OT1* is critical for regulation of these genes [64] via a chromatin-dependent manner.

3.5.3 lincRNAs

In addition to these lncRNAs, it is thought that approximately 20% of lincRNAs can bind to the polycomb repressor complex 2 (PRC2), an important regulator of transcription. It has been proposed that lincRNAs can direct the polycomb repressor complex to specific DNA segments contained within the genome, however, it is not currently understood how this mechanism may work. One example of a lincRNA binding to PCR2 is the linc—upregulated in Bladder Cancer 1 (*linc-UBC1*) ncRNA. He et al. showed that *linc-UBC1* physically associates with PRC2, and when knocked down decreases H3K27 trimethylation [65]. Khalil et al. suggest that there are greater than 3300 lincRNAs present throughout the genome, and that up to 20% of these lincRNAs are bound to PCR2 [43]. They propose that lincRNAs bind to chromatin-modifying complexes and guide them to specific genomic locations. To test the hypothesis that lincRNAs act through the PRC2 pathway, Khalil et al. performed siRNA knockdown experiments targeting five candidate lincRNAs and compared their expression to that in cells transfected with control siRNAs. They then used gene expression microarrays to identify genes that were different between treated and untreated cells [43]. Differential expression analysis identified genes significantly different between treated and control cells, however, this work did not reveal any lincRNAs commonly altered. Furthermore, they did not observe any difference in expression for genes located close to the lincRNA suggesting it does not act by a *cis* mechanism.

These examples highlight the potential functional role of lncRNAs in transcriptional regulation and demonstrate the highly complex and layered network required to control expression of genes via modulation of chromatin structures and histone marks. Future studies are required to elucidate their roles further and to better understand the exact mechanisms underpinning lncRNA expression and chromatin modulation.

4 lncRNAs AND HUMAN DISEASE

The field of lncRNA research is expanding rapidly with many studies now exploring the potential of lncRNAs as diagnostic, prognostic, and predictive markers. Cancer is a multistep process involving acquisition of several genetic and epigenetic changes, of which dysregulation of lncRNAs appears to be an important one. For most types of malignancies, an increase in the number of prognostic and diagnostic tools is desirable to improve management and treatment of patients. In addition to their functional roles in cancer described above, lncRNAs have the potential to serve as molecular markers. So far, very few discovery-based studies of large well-annotated cohorts have been undertaken, and even fewer studies have tested the ability to assess lncRNA expression levels in blood or plasma. Future research efforts should focus on these aspects to determine whether lncRNAs have the potential to be useful biomarkers of disease.

4.1 PROGNOSTIC MARKERS

There are an increasing number of studies relating lncRNA expression to prognostic outcomes including overall survival, disease-free survival, metastasis, tumor size and recurrence (Table 5.3). The imprinted gene *H19* was one of the first lncRNAs found to be associated with cancer. Chen et al. demonstrated a loss of imprinting of H19 in 30.2% of epithelial ovarian cancers and loss of heterozygosity in advanced stage ovarian cancer [87]. In contrast, upregulation of *H19* was found to be associated with invasive potential in bladder cancer [66] and hepatic metastasis [88]. Similarly, *HOTAIR* overexpression is an independent prognostic factor of overall survival and disease-free survival in epithelial ovarian cancer, endometrial carcinoma, glioma, esophageal squamous cell carcinoma (SCC), nonsmall cell lung cancer (NSCLC), and nasopharyngeal carcinoma [69–74]. Several other lncRNAs including *XLOC_010588*, *BANCR, NCRAN, GAS6-AS1, MEG3, LINC-UBC1, LOC285194, PCAT-1, MALAT1,* and *HOTTIP* have

Table 5.3 lncRNAs Associated with Prognosis

Gene	Analysis and major findings	References
H19	High expression of H19 associated with poor overall survival in gastric cancer	[66–68]
	Higher expression associated with disease progression and invasive potential in bladder cancer	
HOTAIR	High expression is an independent prognostic factor of overall survival and disease-free survival in epithelial ovarian cancer	[69–75]
	Correlates with poor prognosis in endometrial carcinoma	
	Strong predictor of survival in glioma	
	Associated with poor prognosis in esophageal SCC	
	Associated with shorter disease-free survival in NSCLC	
	Independent prognostic factor for progression and survival in nasopharyngeal carcinoma	
	Associated with metastasis in breast cancer	
XLOC_010588	Low expression in cervical cancer associated with poorer prognosis and tumor size	[76]
BANCR	Low expression associated with tumor size, advanced pathological stage, metastasis, and shorter overall survival of NSCLC patients	[77]
ncRAN	Low expression associated with worse overall survival in colorectal cancer	[78,79]
	High expression in bladder cancer associated with invasive tumors	
GAS6-AS1	Low expression predicts poor prognosis in NSCLC	[80]
MEG3	Low expression associated with TNM stage, tumor size, and poor prognosis in gastric cancer	[81]
Linc-UBC1	Overexpression associated with poor survival and lymph node metastasis in bladder cancer	[65]
LOC285194	Low expression associated with poor prognosis in colorectal cancer	[82]
PCAT-1	Higher expression associated with poorer overall survival	[83]
MALAT1	Overexpression predicts tumor recurrence in HCC	[84,85]
	Overexpression associated with poor prognosis in NSCLC	
HOTTIP	Associated with overall survival and metastasis formation in HCC	[86]

all been associated with tumor recurrence, overall survival, tumor metastasis, tumor size, and progno-sis in colorectal cancer, NSCLC, gastric cancer, and bladder cancer [65,76–86]. Therefore, lncRNAs clearly have the potential to be useful as prognostic markers. While some of these markers have been validated in several studies, further confirmation is required, before these molecules can be used in the management of patients in a clinical setting.

4.1.1 lncRNAs in the circulation

The release of nucleic acids into the circulation is thought to be a result from a combination of active release in the secreted microvesicles and of dying tumor cells releasing their contents during apoptosis and necrosis. The ability to detect these circulating nucleic acids has significant clinical utility. For a marker to have diagnostic capacity, it should be able to demonstrate high specificity and high sensitivity to avoid false negatives, and should be minimally invasive for the patient. Small noncoding RNAs have been found to be relatively stable in the circulation and are often detected in the serum and plasma of cancer patients. Currently very few lncRNAs have been characterized as potential markers in human body fluids (Table 5.4). lncRNAs are much larger than the smaller noncoding RNAs and it has been suggested that they may be protected by exosome encapsulation, or may form complexes with proteins, much like miRNAs [93,96]. Using deep sequencing, Huang et al. characterized the exosomal RNA profile of three human plasma samples using RNA library protocols from three different manufacturers [97]. While their main aim was to identify exosomal miRNAs, they also detected low levels of lncR-NAs (~3%) [97]. Since their protocols were designed to detect smaller lncRNAs, they suggested that the lncRNA fragments detected were in fact fragments and that the presence of lncRNAs in exosomes may in fact be lncRNAs derived from the cytosol [97]. Despite using suboptimal protocols, Huang et al. demonstrated that multiple lncRNAs can be detected in exosomes, suggesting that this may be a mechanism used by cells to remove degraded lncRNAs [97]. For these reasons, it is possible that lncR-NAs may be detectable in plasma and serum, and as such could serve as good molecular biomarkers.

4.1.2 Prostate cancer antigen 3 (nonprotein coding)

Perhaps one of the best characterized circulating lncRNAs is Prostate cancer antigen 3(*PCA3*). In a cohort of 201 prostate cancer patients, Tinzl et al. showed that *PCA3* was more sensitive and specific for diagnosing prostate cancer, compared to the more commonly used *PSA* (prostate-specific anti-gen) [89]. Its value as a diagnostic marker in prostate cancer has been extensively documented with

Table 5.4 lncRNAs Detected in the Circulation

Gene	Major analysis and findings	References
PCA3	Detectable in urine from prostate cancer patients. Found to have higher sensitivity and specificity compared to the commonly used prostate-specific antigen (PSA)	[89,90]
MALAT1	Detectable in the cellular fraction of blood. Found to exhibit higher expression in NSCLC cases compared to healthy controls	[91,92]
	Detectable in plasma from prostate cancer patients	
H19	Detectable in plasma from gastric cancer patients	[93]
LIPCAR	Detectable in plasma from patients with cardiac remodeling	[94]
HULC	Detectable in blood from HCC patients	[94,95]

many highlighting the advantages of using *PCA3* over *PSA*. In a study of 3073 men undergoing *PCA3* analysis of urine samples, prostate cancer was identified in 1341 (43.6%) of men [98]. Multivariate analysis showed that *PCA3* was significantly associated with prostate cancer generally and high-grade prostate cancer after adjusting for commonly used markers including *PSA*, free *PSA*, family history, abnormal digital rectal examination, body mass index, and prostate volume [98]. Receiver operating characteristic (ROC) analysis also showed that *PCA3* consistently outperformed *PSA* for prediction of prostate cancer (AUC 0.697 vs. 0.599, $P < 0.01$) [98], suggesting that this particular lncRNA is a good biomarker for prostate cancer. This is supported by the Food and Drug Administration's recent approval of PROGENSA's *PCA3* assay to help determine the need for repeat prostate biopsies in men who have previously returned a negative biopsy for prostate cancer [99]. This development of an lncRNA-based molecular test indicates the suitability of lncRNAs as molecular biomarkers in body fluids.

4.1.3 MALAT1

The lncRNA *MALAT1* is associated with poor prognosis in NSCLC. In a recent publication, Weber et al. evaluated the potential for using *MALAT1* as a blood-based biomarker in NSCLC [91]. They found that although *MALAT1* satisfied many of the criteria desirable in a diagnostic marker, it was not sensitive enough to be routinely used, despite showing expression differences between NSCLC and cancer-free controls [91]. In contrast, Ren et al. used plasma from prostate cancer patients to measure expression of *MALAT1*. They were able to show that *MALAT1* lncRNA fragments were both stable and detectable in human plasma and serum, are secreted from tumor cells into the circulation and are poten-tially useful as a diagnostic marker [92]. These studies provide evidence that *MALAT1* is detectable in the circulation and has potential clinical utility.

4.1.4 Highly upregulated in liver cancer

Finally, the highly upregulated in liver cancer (*HULC*) lncRNA has been shown to be upregulated in HCC [95]. In a pilot study of 9 healthy volunteers, 10 liver cirrhosis patients, and 4 HCC patients, Panzitt et al. showed that the levels of HULC were higher in the peripheral blood of patients with HCC as compared to healthy controls or liver cirrhosis patients using conventional PCR [95]. Despite having a small sample size, this pilot study showed the possibility for detecting lncRNA expression in blood.

While there are several publications showing the ability to detect smaller noncoding RNAs such as miRNAs in plasma and serum, the study of lncRNAs in body fluids is still in its infancy. Despite this, the literature supports the hypothesis that lncRNAs are detectable in body fluids and have the potential to serve as diagnostic biomarkers. The detection and identification of lncRNAs can be achieved via RT-qPCR (real-time quantitative polymerase chain reaction) and microarray-based methods. However, there needs to be a consensus regarding standardization of results, similar to that proposed for miRNA detection in the circulation [100]. This is critical to ensure that results are reproducible among study groups that appropriate sensitivities and specificities are achieved, and that appropriate tests can be translated to the clinic.

5 METHODS FOR STUDYING LNCRNA EXPRESSION

5.1 HIGH-THROUGHPUT ANALYSIS FOR DISCOVERY

Several high-throughput methods have been used for the discovery, establishment of function, and prediction of function of lncRNAs including microarrays, transcriptome analysis, immunoprecipitation and computational analyses, and their combinations.

5.1.1 Microarrays

Microarrays have provided a unique opportunity to study lncRNA expression in various diseases. They allow multiple genes to be studied in parallel for a given sample, making it an ideal method of quickly assessing expression of multiple lncRNAs in a single experiment. Two novel microarrays were released in 2008 by Invitrogen (now Life Technologies): the NCode™ Human and Mouse noncoding RNA microarrays. These arrays were created by John Mattick's group at the Institute of Molecular Biosciences, University of Queensland, Australia, based on sequences identified using a specific algorithm. This algorithm scores sequences based on characteristics of protein coding genes including open reading frame length, synonymous/nonsynonymous base substitution rates, and similarity to known proteins. In addition to this, DNA Arraystar has also released a commercial platform, constructed from several public databases including content identified by RefSeq, Gencode, RNAdb 2.0, NRED, and lncRNAdb [43,101]. These platforms have been used in several studies for discovery-based projects.

5.1.2 Transcriptome analysis

Transcriptome analysis using next-generation sequencing technology (RNA-seq) allows unbiased, high-throughput analysis of all transcripts. This technology is not based on *a priori* knowledge of targets and is advantageous for the discovery of new transcripts, as it does not rely on known genomic sequences. Unlike microarrays which often have a limited dynamic range, and rely on hybridization, RNA-seq analysis allows detection of low abundance transcripts and has very low background signals. It therefore has a much larger dynamic range and allows the transcriptome to be sequenced at higher coverage in a high throughput and quantitative manner.

Pauli et al. were one of the first to utilize this technology to identify lncRNAs involved in zebrafish embryogenesis [102]. They identified a set of 1133 embryonic lncRNAs including 184 intronic overlapping lncRNAs, 566 antisense-exonic overlapping lncRNAs, and 397 lincRNAs with no overlap to other genes involved in zebrafish development [102]. Esteve-Codina et al. also used RNA-seq to identify a set of 2047 putative lncRNAs from two phenotypically different pigs [103]. More recently, RNA-seq analysis has been used to identify cancer-associated lncRNAs in prostate cancer [104], long intergenic transcripts involved in adipogenesis [105], and differentially expressed lncRNAs in pancreatic islet cells [106]. In a study of 10 cancer cell lines, Chen et al. used RNA-seq to identify differentially expressed lncRNAs that were correlated with certain cancer types [107]. These studies demonstrate the utility of transcriptome profiling in the identification of lncRNAs involved in human disease.

5.1.3 Computational prediction

Several computational approaches have been used to predict noncoding transcripts based on secondary structure. In general, approaches can be categorized into two methods: (i) methods based on homology and (ii) methods based on common features [108]. Homology-based methods rely on the assumption that sequences are conserved between family members, even though it is known that a high degree of sequence similarity is not maintained for ncRNAs [108]. Software tools that can be used for detecting sequence homology include BLAST [109], FASTA [110], and S SEARCH and BLAT [111]. In addition to sequence homology, assessment of structural features is also used. Algorithms such as RNAfold [112], Mfold [113], and Afold [114] use the minimum folding energy (MFE) to predict secondary structure of RNAs. Recently, machine learning techniques have been used to identify polycomb-associated lncRNAs [115]. Computational methods are often less expensive than profiling studies which require large sample numbers to make meaningful conclusions. Using computational methods to analyze publicly available data will be useful in identifying novel lncRNAs with key roles in cancer based on structural and sequence homology.

5.2 VERIFICATION OF SINGLE CANDIDATES IDENTIFIED FROM HIGH-THROUGHPUT DATA

5.2.1 Northern blots and RT-qPCR

RT-qPCR and northern blot experiments can be used to detect different RNA transcripts and expression patterns between different tissues and cell types. RT-qPCR when combined with reverse transcription can quantify both mRNA and lncRNAs. Northern blotting is limited by its sensitivity when compared with RT-qPCR, but gives an indication of RNA size, and can show the presence of alternative splice products that can be missed using RT-qPCR. Northern blots also require significantly more input than RT-qPCR-based reactions, which can be a limitation for studies where sample is limited.

5.2.2 Fluorescence in situ hybridization

Fluorescence in situ hybridization (FISH) is a cytogenic technique used for the detection and localization of RNA sequences within tissues or cells [116]. It is particularly important for defining the spatial–temporal patterns of gene expression [116]. FISH relies on fluorescent probes that bind to complementary sequences of the lncRNA, mRNA, or miRNA of interest. A series of hybridization steps are performed to achieve signal amplification of the target of interest. This amplification is then viewed using a fluorescent microscope. This technique can be used on formalin-fixed paraffin embedded (FFPE) tissue, frozen tissues, fresh tissues, cells and circulating tumor cells. These characteristics make FISH particularly good for studying lncRNAs, especially given their function is largely unknown.

5.3 STUDYING THE INTERACTIONS BETWEEN lncRNAs AND PROTEIN

5.3.1 RNA immunoprecipitation

RIP is a technique used to study the interactions between RNA and protein, and allows interrogation of RNA binding sites across the genome for a particular protein of interest. In this technique, RNA is immunoprecipitated using an antibody-based approach. Once precipitated, RNAs associated with the protein of interest can be detected using a variety of methods including real-time PCR, microarrays, or sequencing. This technique can easily be applied to lncRNA research especially given many lncRNAs can regulate gene expression via chromatin-modifying complexes and modulation of transcription factor activity. For example, Zhao et al. used RIP to investigate whether the lncRNA *XIST* existed in specific protein complexes [117]. They identified a 1.6 kB ncRNA termed RepA that was contained within the XIST transcript and targeted the polycomb complex, PRC2 [117]. This complex is critical for targeting specific genomic regions for epigenetic silencing of chromatin and is important in stem cell development. RIP has also used in combination with microarrays to analyze the expression of stroke-responsive lncRNAs in relation to the transcription factors Sin3A and coREST [118]. Dharap et al. showed that there was an association between ischemia-induced lncRNAs and the chromatin-modifying complexes Sin3A and coREST [118]. Taken together, these studies show that RIP is an effective methodology for studying the interactions between lncRNAs and protein.

6 STRATEGIES FOR MANIPULATING lncRNA EXPRESSION

Several strategies have been developed to target RNA and protein markers for drug-based discovery, however, it is important to consider the options for analyzing the mechanisms and functions of lncRNAs. Considering our knowledge of the noncoding RNA genome and function is increasing, it may be possible to modulate disease pathways that are controlled by RNAs rather than protein. This could be done using oligonucleotide-based methods.

6.1 OLIGONUCLEOTIDE-BASED METHODS

Oligonucleotide methods such as antisense oligonucleotides (ASOs) and siRNAs are increasingly being used in *in vitro* functional studies of lncRNAs. The simplicity, low cost (compared with traditional drug discovery programs), and high specificity of ASOs make them an attractive method for targeting RNA. Several ASOs are already in clinical trials for the treatment of human disease [119]. Oligonucleotide-based methods involve the use of complementary sequences (oligonucleotides) targeted to a specific RNA sequence resulting in the inhibition of gene expression. Two classes of oligonucleotides exist: (i) oligonucleotides which rely on an RNase-H dependent mechanism to degrade mRNA and (ii) steric blocker oligonucleotides [120]. Often these two mechanisms are used to target mRNA and ultimately protein, however for noncoding RNAs, this is not the case, and hence the RNA transcript itself must be targeted. Future drug discovery efforts will need to focus on this aspect if lncRNAs are to be successfully targeted therapeutically (Table 5.5).

6.2 RNA INTERFERENCE

RNA interference (RNAi) is a naturally occurring process used to regulate gene expression. Gene expression can be altered by either knocking down or reintroducing the target of interest. RNAi often requires a lipid-based system to deliver the oligonucleotides targeting the gene of interest, into the cell. In contrast to mRNAs, which are often exported to the cytoplasm, lncRNAs are often retained in the nucleus of cells and are degraded by exosomes [124]. Studies investigating budding in yeast have demonstrated that transcription is dependent on mechanisms involving the 3' ends of genes [124]. For lncRNAs, these ends are highly heterogeneous, making it difficult for researchers to effectively target lncRNAs. Despite this, RNAi has been used in lncRNA research to demonstrate the functional implications of altered lncRNA expression, often in combination with cell phenotype-based assays. This work has provided several key insights into the functional roles of lncRNAs in human disease. Future studies will need to concentrate on designing methods of effectively targeting nuclear lncRNAs, which are often more difficult to target using traditional RNAi methods.

6.3 TARGETING lncRNAs WITH NATURAL ANTISENSE TRANSCRIPTS

As protein coding genes require a lot of energy to control gene expression at the posttranscriptional level, it may be possible to target these transcripts at the mRNA level by using antisense transcripts. Controlling mRNA expression could therefore save energy that would otherwise be required for degradation of the target mRNA transcript. Targeting natural antisense transcripts (NATs) could block the action of the sense transcript for targeted degradation. Affecting gene expression in this manner would

Table 5.5 Examples of Studies Using Oligonucleotides for Targeting of lncRNAs _In Vivo_

Gene	Cancer type	Major analysis and findings	References
HEIH	Hepatocellular carcinoma	Nude mice were injected with tumor cells in which HEIH was upregulated with an expression construct or downregulated with shRNAs. Growth of tumors with downregulated HEIH significantly less than controls, and growth of tumors with upregulated HEIH significantly increased	[121]
PRNCR1/ PGEM1	Prostate cancer	shRNA-mediated knockdown inhibited tumor growth in a CWR22Rv1 prostate cancer xenograft mouse model	[122]
HOTAIR	Epithelial ovarian cancer	Stable HOTAIR knockdown HEY-A8 cells expressing GFP created and injected into mice. Mice with HOTAIR knockdown had fewer tumors developing in the mesentery compared to negative controls	[73]
PCNA-AS	Hepatocellular carcinoma	Lentivirus expressed PCNA-AS injected into mice with overexpression associated with larger tumor size	[123]

Please note that this is not an extensive list and merely serves to highlight examples of how oligonucleotides can be used in in vivo _models for knockdown of lncRNA expression._

be highly desirable given that _cis_-acting NATs are highly specific and unlike miRNAs do not affect several mRNA transcripts at once. Inhibition of NATs can occur via the use of antagoNATs, small single-stranded oligonucleotides designed to block the interactions between the sense and antisense transcripts, targeting it for degradation and causing transcriptional derepression [125]. The first _in vivo_ demonstration of the efficacy of these molecules was performed by Modarresi et al. [125]. In this study, Modarresi et al. used siRNA to transiently reverse the expression of the targets _BDNF_, _GDNF_, and _EPHB2_ by targeting their antisense transcripts. They suggest that these antisense transcripts may require the _PRC2_ complex, a key player in epigenetic silencing of chromatin, suggesting that they function to regulate expression of their sense transcripts, through chromatin modification. Thus, antago-NATs represent a novel method of targeting lncRNAs therapeutically.

6.4 TARGETING lncRNAs WITH SMALL MOLECULE INHIBITORS

Small molecule inhibitors are another potential avenue for targeting lncRNAs therapeutically, particularly given they have already been used to target proteins, and have attractive pharmacodynamics and pharmacokinetic properties [126]. When trying to find ligands specific for the lncRNA of interest, it is important to consider the RNA secondary structure. There is increasing evidence to suggest that small molecules can trigger RNA conformational changes [127]. Some suggest that regions around the A-form helix are optimal for RNA targeting [128] and can result in changes to secondary structure. RNA contains both major and minor grooves, however unlike DNA, the A-form helix makes binding of small molecules difficult [128]. Despite this, mismatches or unpaired bases can help to widen the major grove, making it more accessible for small molecule binding [128]. It is this aspect of the RNA structure that could be potentially exploited for small molecule targeting of lncRNAs.

One example of how lncRNAs can impact sensitivity to small molecular inhibitors can be seen in prostate cancer. _PCAT-1_ is an lncRNA shown to be upregulated in prostate cancer. Recent studies have shown that high levels of *PCAT-1* are associated with low levels of *BRCA2*, a well-known tumor suppressor gene [129]. Dysregulation of *PCAT-1* expression can result in downstream impairment of homologous recombination, resulting in increased double-stranded DNA breaks. Prensner et al. demonstrated that *PCAT-1* expressing cells increased cell death in response to PARP small molecule inhibitors, changing cell sensitivity by up to fivefold [129]. They concluded that this particular lncRNA has a critical role in posttranscriptionally regulating *BRCA2*. This study is interesting in that it provides evidence that small molecule inhibitors can be useful for the targeting of specific lncRNAs. If these changes can be targeted in an effective manner, this could have great utility in the clinic.

7 CONCLUSIONS

lncRNAs are an emerging area of research and are provoking immense interest in research circles. While very few lncRNAs have been well characterized, there is increasing evidence to suggest that they have important roles in cancer biology, particularly in regulating chromatin structure and epigenetic changes. Their discovery indicates that RNA transcription is far more complex than first thought, involving an intricate network of chromatin, regulatory elements, and other noncoding RNAs. Already, several groups have demonstrated that dysregulation of lncRNA expression can contribute to the development and progression of cancer, and that changes in lncRNA expression can be exploited to find biomarkers. However, determining the best way of targeting these changes for therapeutic intervention is still in its infancy. Future work will need to focus on further characterization of these lncRNAs, particularly with respect to function and mechanisms of action, before their implementation as biomarkers and drug targets in the clinic. To conclude, lncRNAs represent an additional mechanism of epigenetic control that must be considered in the context of epigenetic dysregulation in cancer. Determining their function will be critical in assessing how to best to target epigenetic changes therapeutically.

REFERENCES

[1] Brown JW, Marshall DF, Echeverria M. Intronic noncoding RNAs and splicing. Trends Plant Sci 2008;13(7):335–42.

[2] Kapranov P, Cheng J, Dike S, Nix DA, Duttagupta R, Willingham AT, et al. RNA maps reveal new RNA classes and a possible function for pervasive transcription. Science 2007;316(5830):1484–8.

[3] Kapranov P, St Laurent G, Raz T, Ozsolak F, Reynolds CP, Sorensen PH, et al. The majority of total nuclear-encoded non-ribosomal RNA in a human cell is "dark matter" un-annotated RNA. BMC Biol 2010;8:149.

[4] Mercer TR, Dinger ME, Mattick JS. Long non-coding RNAs: insights into functions. Nat Rev Genet 2009;10(3):155–9.

[5] Bernstein E, Allis CD. RNA meets chromatin. Genes Dev 2005;19(14):1635–55.

[6] Faghihi MA, Wahlestedt C. Regulatory roles of natural antisense transcripts. Nat Rev Mol Cell Biol 2009;10(9):637–43.

[7] Whitehead J, Pandey GK, Kanduri C. Regulation of the mammalian epigenome by long noncoding RNAs. Biochim Biophys Acta 2009;1790(9):936–47.

[8] Amaral PP, Clark MB, Gascoigne DK, Dinger ME, Mattick JS. lncRN0Adb: a reference database for long noncoding RNAs. Nucleic Acids Res 2011;39(Database issue):D146–51.

[9] Ponjavic J, Ponting CP, Lunter G. Functionality or transcriptional noise? Evidence for selection within long noncoding RNAs. Genome Res 2007;17(5):556–65.

[10] Guttman M, Amit I, Garber M, French C, Lin MF, Feldser D, et al. Chromatin signature reveals over a thousand highly conserved large non-coding RNAs in mammals. Nature 2009;458(7235):223–7.

[11] Carninci P, Kasukawa T, Katayama S, Gough J, Frith MC, Maeda N, et al. The transcriptional landscape of the mammalian genome. Science 2005;309(5740):1559–63.

[12] Pang KC, Frith MC, Mattick JS. Rapid evolution of noncoding RNAs: lack of conservation does not mean lack of function. Trends Genet 2006;22(1):1–5.

[13] Dinger ME, Amaral PP, Mercer TR, Pang KC, Bruce SJ, Gardiner BB, et al. Long noncoding RNAs in mouse embryonic stem cell pluripotency and differentiation. Genome Res 2008;18(9):1433–45.

[14] Trinklein ND, Aldred SF, Hartman SJ, Schroeder DI, Otillar RP, Myers RM. An abundance of bidirectional promoters in the human genome. Genome Res 2004;14(1):62–6.

[15] Washietl S, Hofacker IL, Lukasser M, Huttenhofer A, Stadler PF. Mapping of conserved RNA secondary structures predicts thousands of functional noncoding RNAs in the human genome. Nat Biotechnol 2005;23(11):1383–90.

[16] Louro R, El-Jundi T, Nakaya HI, Reis EM, Verjovski-Almeida S. Conserved tissue expression signatures of intronic noncoding RNAs transcribed from human and mouse loci. Genomics 2008;92(1):18–25.

[17] Clemson CM, Hutchinson JN, Sara SA, Ensminger AW, Fox AH, Chess A, et al. An architectural role for a nuclear noncoding RNA: NEAT1 RNA is essential for the structure of paraspeckles. Mol Cell 2009;33(6):717–26.

[18] Souquere S, Beauclair G, Harper F, Fox A, Pierron G. Highly ordered spatial organization of the structural long noncoding NEAT1 RNAs within paraspeckle nuclear bodies. Mol Biol Cell 2010;21(22):4020–7.

[19] Ip JY, Nakagawa S. Long non-coding RNAs in nuclear bodies. Dev Growth Differ 2012;54(1):44–54.

[20] Mattick JS. The genetic signatures of noncoding RNAs. PLoS Genet 2009;5(4):e1000459.

[21] Wright MW, Bruford EA. Naming "junk": human non-protein coding RNA (ncRNA) gene nomenclature. Hum Genomics 2011;5(2):90–8.

[22] Brosius J. Waste not, want not—transcript excess in multicellular eukaryotes. Trends Genet 2005;21(5):287–8.

[23] Chakalova L, Debrand E, Mitchell JA, Osborne CS, Fraser P. Replication and transcription: shaping the landscape of the genome. Nat Rev Genet 2005;6(9):669–77.

[24] Wilusz JE, Sunwoo H, Spector DL. Long noncoding RNAs: functional surprises from the RNA world. Genes Dev 2009;23(13):1494–504.

[25] Prensner JR, Chinnaiyan AM. Cancer Discov 2011;1:391–407.

[26] Dinger ME, Pang KC, Mercer TR, Crowe ML, Grimmond SM, Mattick JS. NRED: a database of long noncoding RNA expression. Nucleic Acids Res 2009;37(Database issue):D122–6.

[27] Xie C, Yuan J, Li H, Li M, Zhao G, Bu D, et al. NONCODEv4: exploring the world of long non-coding RNA genes. Nucleic Acids Res 2014;42(Database issue):D98–D103.

[28] Bu D, Yu K, Sun S, Xie C, Skogerbo G, Miao R, et al. NONCODE v3.0: integrative annotation of long noncoding RNAs. Nucleic Acids Res 2012;40(Database issue):D210–5.

[29] He S, Liu C, Skogerbo G, Zhao H, Wang J, Liu T, et al. NONCODE v2.0: decoding the non-coding. Nucleic Acids Res 2008;36(Database issue):D170–2.

[30] Liu C, Bai B, Skogerbo G, Cai L, Deng W, Zhang Y, et al. NONCODE: an integrated knowledge database of non-coding RNAs. Nucleic Acids Res 2005;33(Database issue):D112–15.

[31] Volders PJ, Helsens K, Wang X, Menten B, Martens L, Gevaert K, et al. LNCipedia: a database for annotated human lncRNA transcript sequences and structures. Nucleic Acids Res 2013;41(Database issue):D246–51.

[32] Liao Q, Xiao H, Bu D, Xie C, Miao R, Luo H, et al. ncFANs: a web server for functional annotation of long non-coding RNAs. Nucleic Acids Res 2011;39(Web Server issue):W118–24.

[33] Li JH, Liu S, Zhou H, Qu LH, Yang JH. starBase v2.0: decoding miRNA-ceRNA, miRNA–ncRNA and protein–RNA interaction networks from large-scale CLIP-Seq data. Nucleic Acids Res 2014;42(Database issue):D92–7.

[34] Yang JH, Li JH, Shao P, Zhou H, Chen YQ, Qu LH. starBase: a database for exploring microRNA–mRNA interaction maps from Argonaute CLIP-Seq and Degradome-Seq data. Nucleic Acids Res 2011;39(Database issue):D202–9.

[35] Bhartiya D, Pal K, Ghosh S, Kapoor S, Jalali S, Panwar B, et al. lncRNome: a comprehensive knowledgebase of human long noncoding RNAs. Database 2013;2013:bat034.

[36] Yang JH, Li JH, Jiang S, Zhou H, Qu LH. ChIPBase: a database for decoding the transcriptional regulation of long non-coding RNA and microRNA genes from ChIP-Seq data. Nucleic Acids Res 2013;41(Database issue):D177–87.

[37] Chen G, Wang Z, Wang D, Qiu C, Liu M, Chen X, et al. LncRNADisease: a database for long-non-coding RNA-associated diseases. Nucleic Acids Res 2013;41(Database issue):D983–6.

[38] Ma L, Bajic VB, Zhang Z. On the classification of long non-coding RNAs. RNA Biol 2013;10(6):925–33.

[39] Wang KC, Chang HY. Molecular mechanisms of long noncoding RNAs. Mol Cell 2011;43(6):904–14.

[40] Salmena L, Poliseno L, Tay Y, Kats L, Pandolfi PP. A ceRNA hypothesis: the Rosetta Stone of a hidden RNA language? Cell 2011;146(3):353–8.

[41] Gutschner T, Diederichs S. The hallmarks of cancer: a long non-coding RNA point of view. RNA Biol 2012;9(6):703–19.

[42] Hung T, Chang HY. Long noncoding RNA in genome regulation: prospects and mechanisms. RNA Biol 2010;7(5):582–5.

[43] Khalil AM, Guttman M, Huarte M, Garber M, Raj A, Rivea Morales D, et al. Many human large intergenic noncoding RNAs associate with chromatin-modifying complexes and affect gene expression. Proc Natl Acad Sci USA 2009;106(28):11667–72.

[44] Hogg JR, Collins K. Structured non-coding RNAs and the RNP Renaissance. Curr Opin Chem Biol 2008;12(6):684–9.

[45] Thakur N, Tiwari VK, Thomassin H, Pandey RR, Kanduri M, Gondor A, et al. An antisense RNA regulates the bidirectional silencing property of the Kcnq1 imprinting control region. Mol Cell Biol 2004;24(18):7855–62.

[46] Hung T, Wang Y, Lin MF, Koegel AK, Kotake Y, Grant GD, et al. Extensive and coordinated transcription of noncoding RNAs within cell-cycle promoters. Nat Genet 2011;43(7):621–9.

[47] Guttman M, Donaghey J, Carey BW, Garber M, Grenier JK, Munson G, et al. lincRNAs act in the circuitry controlling pluripotency and differentiation. Nature 2011;477(7364):295–300.

[48] Gong C, Maquat LE. lncRNAs transactivate STAU1-mediated mRNA decay by duplexing with 3' UTRs via Alu elements. Nature 2011;470(7333):284–8.

[49] Braconi C, Kogure T, Valeri N, Huang N, Nuovo G, Costinean S, et al. microRNA-29 can regulate expression of the long non-coding RNA gene MEG3 in hepatocellular cancer. Oncogene 2011;30(47):4750–6.

[50] Sumazin P, Yang X, Chiu HS, Chung WJ, Iyer A, Llobet-Navas D, et al. An extensive microRNA-mediated network of RNA–RNA interactions regulates established oncogenic pathways in glioblastoma. Cell 2011;147(2):370–81.

[51] Poliseno L, Salmena L, Zhang J, Carver B, Haveman WJ, Pandolfi PP. A coding-independent function of gene and pseudogene mRNAs regulates tumour biology. Nature 2010;465(7301):1033–8.

[52] Simon JA, Kingston RE. Occupying chromatin: polycomb mechanisms for getting to genomic targets, stopping transcriptional traffic, and staying put. Mol Cell 2013;49(5):808–24.

[53] Schuettengruber B, Martinez AM, Iovino N, Cavalli G. Trithorax group proteins: switching genes on and keeping them active. Nat Rev Mol Cell Biol 2011;12(12):799–814.

[54] Yang PK, Kuroda MI. Noncoding RNAs and intranuclear positioning in monoallelic gene expression. Cell 2007;128(4):777–86.

[55] Chaumeil J, Le Baccon P, Wutz A, Heard E. A novel role for Xist RNA in the formation of a repressive nuclear compartment into which genes are recruited when silenced. Genes Dev 2006;20(16):2223–37.

[56] Bartolomei MS, Zemel S, Tilghman SM. Parental imprinting of the mouse H19 gene. Nature 1991;351(6322):153–5.

[57] Monnier P, Martinet C, Pontis J, Stancheva I, Ait-Si-Ali S, Dandolo L. H19 lncRNA controls gene expression of the imprinted gene network by recruiting MBD1. Proc Natl Acad Sci USA 2013;110(51):20693–8.

[58] Ito Y, Nativio R, Murrell A. Induced DNA demethylation can reshape chromatin topology at the IGF2-H19 locus. Nucleic Acids Res 2013;41(10):5290–302.

[59] Gabory A, Jammes H, Dandolo L. The H19 locus: role of an imprinted non-coding RNA in growth and development. Bioessays 2010;32(6):473–80.

[60] Ariel I, Sughayer M, Fellig Y, Pizov G, Ayesh S, Podeh D, et al. The imprinted H19 gene is a marker of early recurrence in human bladder carcinoma. Mol Pathol 2000;53(6):320–3.

[61] Yang F, Bi J, Xue X, Zheng L, Zhi K, Hua J, et al. Up-regulated long non-coding RNA H19 contributes to proliferation of gastric cancer cells. FEBS J 2012;279(17):3159–65.

[62] Rinn JL, Chang HY. Genome regulation by long noncoding RNAs. Annu Rev Biochem 2012;81:145–66.

[63] Pandey RR, Mondal T, Mohammad F, Enroth S, Redrup L, Komorowski J, et al. Kcnq1ot1 antisense noncoding RNA mediates lineage-specific transcriptional silencing through chromatin-level regulation. Mol Cell 2008;32(2):232–46.

[64] Mancini-Dinardo D, Steele SJ, Levorse JM, Ingram RS, Tilghman SM. Elongation of the Kcnq1ot1 transcript is required for genomic imprinting of neighboring genes. Genes Dev 2006;20(10):1268–82.

[65] He W, Cai Q, Sun F, Zhong G, Wang P, Liu H, et al. linc-UBC1 physically associates with polycomb repressive complex 2 (PRC2) and acts as a negative prognostic factor for lymph node metastasis and survival in bladder cancer. Biochim Biophys Acta 2013;1832(10):1528–37.

[66] Ariel I, Lustig O, Schneider T, Pizov G, Sappir M, De-Groot N, et al. The imprinted H19 gene as a tumor marker in bladder carcinoma. Urology 1995;45(2):335–8.

[67] Zhang EB, Han L, Yin DD, Kong R, De W, Chen J. c-Myc-induced, long, noncoding H19 affects cell proliferation and predicts a poor prognosis in patients with gastric cancer. Med Oncol 2014;31(5):914.

[68] Iizuka N, Oka M, Tamesa T, Hamamoto Y, Yamada-Okabe H. Imbalance in expression levels of insulin-like growth factor 2 and H19 transcripts linked to progression of hepatocellular carcinoma. Anticancer Res 2004;24(6):4085–9.

[69] Chen FJ, Sun M, Li SQ, Wu QQ, Ji L, Liu ZL, et al. Upregulation of the long non-coding RNA HOTAIR promotes esophageal squamous cell carcinoma metastasis and poor prognosis. Mol Carcinog 2013;52(11):908–15.

[70] He X, Bao W, Li X, Chen Z, Che Q, Wang H, et al. The long non-coding RNA HOTAIR is upregulated in endometrial carcinoma and correlates with poor prognosis. Int J Mol Med 2014;33(2):325–32.

[71] Nakagawa T, Endo H, Yokoyama M, Abe J, Tamai K, Tanaka N, et al. Large noncoding RNA HOTAIR enhances aggressive biological behavior and is associated with short disease-free survival in human non-small cell lung cancer. Biochem Biophys Res Commun 2013;436(2):319–24.

[72] Nie Y, Liu X, Qu S, Song E, Zou H, Gong C. Long non-coding RNA HOTAIR is an independent prognostic marker for nasopharyngeal carcinoma progression and survival. Cancer Sci 2013;104(4):458–64.

[73] Qiu JJ, Lin YY, Ye LC, Ding JX, Feng WW, Jin HY, et al. Overexpression of long non-coding RNA HOTAIR predicts poor patient prognosis and promotes tumor metastasis in epithelial ovarian cancer. Gynecol Oncol 2014;134(1):121–8.

[74] Zhang JX, Han L, Bao ZS, Wang YY, Chen LY, Yan W, et al. HOTAIR, a cell cycle-associated long noncoding RNA and a strong predictor of survival, is preferentially expressed in classical and mesenchymal glioma. Neuro Oncol 2013;15(12):1595–603.

[75] Gupta RA, Shah N, Wang KC, Kim J, Horlings HM, Wong DJ, et al. Long non-coding RNA HOTAIR reprograms chromatin state to promote cancer metastasis. Nature 2010;464(7291):1071–6.

[76] Liao LM, Sun XY, Liu AW, Wu JB, Cheng XL, Lin JX, et al. Low expression of long noncoding XLOC_010588 indicates a poor prognosis and promotes proliferation through upregulation of c-Myc in cervical cancer. Gynecol Oncol 2014;133(3):616–23.

[77] Sun M, Liu XH, Wang KM, Nie FQ, Kong R, Yang JS, et al. Downregulation of BRAF activated non-coding RNA is associated with poor prognosis for non-small cell lung cancer and promotes metastasis by affecting epithelial-mesenchymal transition. Mol Cancer 2014;13:68.

[78] Qi P, Xu MD, Ni SJ, Shen XH, Wei P, Huang D, et al. Down-regulation of ncRAN, a long non-coding RNA, contributes to colorectal cancer cell migration and invasion and predicts poor overall survival for colorectal cancer patients. Mol Carcinog 2014. http://dx.doi.org/10.1002/mc.22137.

[79] Zhu Y, Yu M, Li Z, Kong C, Bi J, Li J, et al. ncRAN, a newly identified long noncoding RNA, enhances human bladder tumor growth, invasion, and survival. Urology 2011;77(2) 510.e1–5.

[80] Han L, Kong R, Yin DD, Zhang EB, Xu TP, De W, et al. Low expression of long noncoding RNA GAS6-AS1 predicts a poor prognosis in patients with NSCLC. Med Oncol 2013;30(4):694.

[81] Sun M, Xia R, Jin F, Xu T, Liu Z, De W, et al. Downregulated long noncoding RNA MEG3 is associated with poor prognosis and promotes cell proliferation in gastric cancer. Tumour Biol 2014;35(2):1065–73.

[82] Qi P, Xu MD, Ni SJ, Huang D, Wei P, Tan C, et al. Low expression of LOC285194 is associated with poor prognosis in colorectal cancer. J Transl Med 2013;11(1):122.

[83] Ge X, Chen Y, Liao X, Liu D, Li F, Ruan H, et al. Overexpression of long noncoding RNA PCAT-1 is a novel biomarker of poor prognosis in patients with colorectal cancer. Med Oncol 2013;30(2):588.

[84] Lai MC, Yang Z, Zhou L, Zhu QQ, Xie HY, Zhang F, et al. Long non-coding RNA MALAT-1 overexpression predicts tumor recurrence of hepatocellular carcinoma after liver transplantation. Med Oncol 2012;29(3):1810–16.

[85] Schmidt LH, Spieker T, Koschmieder S, Schaffers S, Humberg J, Jungen D, et al. The long noncoding MALAT-1 RNA indicates a poor prognosis in non-small cell lung cancer and induces migration and tumor growth. J Thorac Oncol 2011;6(12):1984–92.

[86] Quagliata L, Matter MS, Piscuoglio S, Arabi L, Ruiz C, Procino A, et al. Long noncoding RNA HOTTIP/HOXA13 expression is associated with disease progression and predicts outcome in hepatocellular carcinoma patients. Hepatology 2014;59(3):911–23.

[87] Chen CL, Ip SM, Cheng D, Wong LC, Ngan HY. Loss of imprinting of the IGF-II and H19 genes in epithelial ovarian cancer. Clin Cancer Res 2000;6(2):474–9.

[88] Fellig Y, Ariel I, Ohana P, Schachter P, Sinelnikov I, Birman T, et al. H19 expression in hepatic metastases from a range of human carcinomas. J Clin Pathol 2005;58(10):1064–8.

[89] Tinzl M, Marberger M, Horvath S, Chypre C. DD3PCA3 RNA analysis in urine—a new perspective for detecting prostate cancer. Eur Urol 2004;46(2):182–6. discussion 7.

[90] de Kok JB, Verhaegh GW, Roelofs RW, Hessels D, Kiemeney LA, Aalders TW, et al. DD3(PCA3), a very sensitive and specific marker to detect prostate tumors. Cancer Res 2002;62(9):2695–8.

[91] Weber DG, Johnen G, Casjens S, Bryk O, Pesch B, Jockel KH, et al. Evaluation of long noncoding RNA MALAT1 as a candidate blood-based biomarker for the diagnosis of non-small cell lung cancer. BMC Res Notes 2013;6(6):518.

[92] Ren S, Wang F, Shen J, Sun Y, Xu W, Lu J, et al. Long non-coding RNA metastasis associated in lung adenocarcinoma transcript 1 derived miniRNA as a novel plasma-based biomarker for diagnosing prostate cancer. Eur J Cancer 2013;49(13):2949–59.

[93] Arita T, Ichikawa D, Konishi H, Komatsu S, Shiozaki A, Shoda K, et al. Circulating long non-coding RNAs in plasma of patients with gastric cancer. Anticancer Res 2013;33(8):3185–93.

[94] Kumarswamy R, Bauters C, Volkmann I, Maury F, Fetisch J, Holzmann A, et al. The circulating long non-coding RNA LIPCAR predicts survival in heart failure patients. Circ Res 2014.

[95] Panzitt K, Tschernatsch MM, Guelly C, Moustafa T, Stradner M, Strohmaier HM, et al. Characterization of HULC, a novel gene with striking up-regulation in hepatocellular carcinoma, as noncoding RNA. Gastroenterology 2007;132(1):330–42.

[96] Iguchi H, Kosaka N, Ochiya T. Secretory microRNAs as a versatile communication tool. Commun Integr Biol 2010;3(5):478–81.

[97] Huang X, Yuan T, Tschannen M, Sun Z, Jacob H, Du M, et al. Characterization of human plasma-derived exosomal RNAs by deep sequencing. BMC Genomics 2013;14:319.

[98] Chevli KK, Duff M, Walter P, Yu C, Capuder B, Elshafei A, et al. Urinary PCA3 as a predictor of prostate cancer in a cohort of 3,073 men undergoing initial prostate biopsy. J Urol 2013;191(6):1743–8.

[99] lncRNA PCA3 has been approved by FDA to help determine need for repeat prostate biopsies Long noncoding RNA Blog 2012 [cited 2014 17/4/2014]. Available from: http://deepbase.sysu.edu.cn/lncrnablog/?p=424.

[100] Kirschner MB, van Zandwijk N, Reid G. Cell-free microRNAs: potential biomarkers in need of standardized reporting. Front Genet 2013;4:56.

[101] Cabili MN, Trapnell C, Goff L, Koziol M, Tazon-Vega B, Regev A, et al. Integrative annotation of human large intergenic noncoding RNAs reveals global properties and specific subclasses. Genes Dev 2011;25(18):1915–27.

[102] Pauli A, Valen E, Lin MF, Garber M, Vastenhouw NL, Levin JZ, et al. Systematic identification of long noncoding RNAs expressed during zebrafish embryogenesis. Genome Res 2012;22(3):577–91.

[103] Esteve-Codina A, Kofler R, Palmieri N, Bussotti G, Notredame C, Perez-Enciso M. Exploring the gonad transcriptome of two extreme male pigs with RNA-seq. BMC Genomics 2011;12:552.

[104] Ren S, Peng Z, Mao JH, Yu Y, Yin C, Gao X, et al. RNA-seq analysis of prostate cancer in the Chinese population identifies recurrent gene fusions, cancer-associated long noncoding RNAs and aberrant alternative splicings. Cell Res 2012;22(5):806–21.

[105] Yi F, Yang F, Liu X, Chen H, Ji T, Jiang L, et al. RNA-seq identified a super-long intergenic transcript functioning in adipogenesis. RNA Biol 2013;10(6):991–1001.

[106] Li B, Bi CL, Lang N, Li YZ, Xu C, Zhang YQ, et al. RNA-seq methods for identifying differentially expressed gene in human pancreatic islet cells treated with pro-inflammatory cytokines. Mol Biol Rep 2014;41(4):1917–25.

[107] Chen G, Yin K, Shi L, Fang Y, Qi Y, Li P, et al. Comparative analysis of human protein-coding and noncoding RNAs between brain and 10 mixed cell lines by RNA-Seq. PLoS One 2011;6(11):e28318.

[108] Wang C, Wei L, Guo M, Zou Q. Computational approaches in detecting non-coding RNA. Curr Genomics 2013;14(6):371–7.

[109] Altschul SF, Gish W, Miller W, Myers EW, Lipman DJ. Basic local alignment search tool. J Mol Biol 1990;215(3):403–10.

[110] Pearson WR. Flexible sequence similarity searching with the FASTA3 program package. Methods Mol Biol 2000;132:185–219.

[111] Kent WJ. BLAT—the BLAST-like alignment tool. Genome Res 2002;12(4):656–64.

[112] Mathews DH, Turner DH. Prediction of RNA secondary structure by free energy minimization. Curr Opin Struct Biol 2006;16(3):270–8.

[113] Hofacker IL. Vienna RNA secondary structure server. Nucleic Acids Res 2003;31(13):3429–31.

[114] Zuker M, Stiegler P. Optimal computer folding of large RNA sequences using thermodynamics and auxiliary information. Nucleic Acids Res 1981;9(1):133–48.

[115] Derrien T, Johnson R, Bussotti G, Tanzer A, Djebali S, Tilgner H, et al. The GENCODE v7 catalog of human long noncoding RNAs: analysis of their gene structure, evolution, and expression. Genome Res 2012;22(9):1775–89.

[116] Yan B, Wang ZH, Guo JT. The research strategies for probing the function of long noncoding RNAs. Genomics 2012;99(2):76–80.

[117] Zhao J, Sun BK, Erwin JA, Song JJ, Lee JT. Polycomb proteins targeted by a short repeat RNA to the mouse X chromosome. Science 2008;322(5902):750–6.

[118] Dharap A, Pokrzywa C, Vemuganti R. Increased binding of stroke-induced long non-coding RNAs to the transcriptional corepressors Sin3A and coREST. ASN Neuro 2013;5(4):283–9.

[119] Agrawal S, Kandimalla ER. Antisense therapeutics: is it as simple as complementary base recognition? Mol Med Today 2000;6(2):72–81.

[120] Dias N, Stein CA. Antisense oligonucleotides: basic concepts and mechanisms. Mol Cancer Ther 2002;1(5):347–55.

[121] Yang F, Zhang L, Huo XS, Yuan JH, Xu D, Yuan SX, et al. Long noncoding RNA high expression in hepatocellular carcinoma facilitates tumor growth through enhancer of zeste homolog 2 in humans. Hepatology 2011;54(5):1679–89.

[122] Yang L, Lin C, Jin C, Yang JC, Tanasa B, Li W, et al. lncRNA-dependent mechanisms of androgen-receptor-regulated gene activation programs. Nature 2013;500(7464):598–602.

[123] Yuan SX, Tao QF, Wang J, Yang F, Liu L, Wang LL, et al. Antisense long non-coding RNA PCNA-AS1 promotes tumor growth by regulating proliferating cell nuclear antigen in hepatocellular carcinoma. Cancer Lett 2014;349(1):87–94.

[124] Shah S, Wittmann S, Kilchert C, Vasiljeva L. lncRNA recruits RNAi and the exosome to dynamically regulate pho1 expression in response to phosphate levels in fission yeast. Genes Dev 2014;28(3):231–44.

[125] Modarresi F, Faghihi MA, Lopez-Toledano MA, Fatemi RP, Magistri M, Brothers SP, et al. Inhibition of natural antisense transcripts *in vivo* results in gene-specific transcriptional upregulation. Nat Biotechnol 2012;30(5):453–9.

[126] Bhartiya D, Kapoor S, Jalali S, Sati S, Kaushik K, Sachidanandan C, et al. Conceptual approaches for lncRNA drug discovery and future strategies. Expert Opin Drug Discov 2012;7(6):503–13.

[127] Stelzer AC, Frank AT, Kratz JD, Swanson MD, Gonzalez-Hernandez MJ, Lee J, et al. Discovery of selective bioactive small molecules by targeting an RNA dynamic ensemble. Nat Chem Biol 2011;7(8):553–9.

[128] Thomas JR, Hergenrother PJ. Targeting RNA with small molecules. Chem Rev 2008;108(4):1171–224.

[129] Prensner JR, Chen W, Iyer MK, Cao Q, Ma T, Han S, et al. PCAT-1, a long noncoding RNA, regulates BRCA2 and controls homologous recombination in cancer. Cancer Res 2014;74(6):1651–60.

RIBOSOMAL RNA METHYLATION AND CANCER

Gabriel Therizols[1,2]**, Florian Laforêts**[1,2]**, Virginie Marcel**[1,2]**, Frédéric Catez**[1,2]**, Philippe Bouvet**[3]**, and Jean-Jacques Diaz**[1,2]

[1]*Centre de Recherche en Cancérologie de Lyon, Centre Léon Bérard, Lyon, France,* [2]*Université de Lyon, Lyon, France,* [3]*Laboratoire Joliot-Curie, Ecole Normale Supérieure de Lyon, Université de Lyon, Lyon, France*

CHAPTER OUTLINE

1 INTRODUCTION

Nucleotides are the building blocks that constitute the RNA biopolymers found within living cells, messenger RNA (mRNA), transfer RNA (tRNA), ribosomal RNA (rRNA), and long and small noncoding RNAs. Nucleotides are made up of a sugar molecule, a ribose, a phosphate group, and different nitrogen-containing bases. The most abundant bases found in RNA are the purines, adenine and guanine, and the pyrimidines, uracil and cytosine.

RNAs are synthesized by RNA polymerases. Identity of nucleotides and their order within the RNA molecules arise from the sequences coded by the genomes. However, many processes occurring co-transcriptionally or posttranscriptionally lead to important modifications of the RNA molecule (Table 6.1), increasing, therefore, the diversity of gene expression. Thanks to the huge development of large scale technologies allowing to have a global view of the differences between a given genome and the corresponding transcriptome it appears that these RNA modifications are regulated and that this regulation represents an important level of gene expression regulation. In addition to the extensively studied processing and modifications of RNA such as splicing, capping, polyadenylation those affecting the chemical nature of the internal sugar and bases of RNA appear more and more crucial for controlling structure–function relationships of all the types of RNAs [3,4]. Since these modifications change the expression message contained within the initial genomic DNA and because they are in some circumstances transmitted to daughter cells during cell division they could be regarded as a part of an epigenetic regulation of gene expression. Interestingly, these epigenetic RNA modifications appear to modify the function of the RNA, providing a key level of gene expression regulation.

The finding that more than the four frequent nitrogen-containing bases can be incorporated within RNA allowed to view these nucleotide polymers as biomolecules exhibiting a more important than expected diversity in their physicochemical and structural features. It appears now very clearly that the variations of these physicochemical and structural features are intensely regulated to provide specific RNA species dedicated to specific cellular functions and that deregulation of these processes is linked to pathological development, notably of cancer.

Among the many modifications of RNA, RNA editing is a controlled enzymatic reaction leading to the transformation of adenine to inosine. Deamination of adenine occurs for all types of RNA. RNA editing is performed by enzymes of the adesosine deaminases acting on RNA (ADARs) family. This A-to-I editing has been conserved during evolution [5] and is widespread in eukaryotes since it affects many thousands of transcripts [6]. Most of these edited transcripts do not code for proteins but rather are involved in regulatory functions of gene expression [7].

Besides this change of the nature of the base, the ribose and all the bases of RNA can be modified by a very large number, probably more than 100, of chemical residues [1,8]. A tentative exhaustive description of base and ribose modifications is given in the RNA modification database [1]. Almost any position of the RNA bases has been found modified associated or not with a modification of the ribose at position 2′ [8]. They are found for every type of RNA nucleotides, adenosine, inosine, cytidine, guanosine, 7-deazaguanosine, and uridine for the three major RNA species tRNA, rRNA, mRNA and also for small nuclear RNA (snRNA) and some small nucleolar RNA (snoRNA). They are found in all primary phylogenetic domains (Archaea, Bacteria, and Eukarya).

Among all these chemical modifications, pseudouridylation and methylation are the most frequent in RNA. These particular modifications represent those that have been clearly shown to play major roles in cancer. Pseudouridylation is a posttranscriptional chemical modification of an RNA base. Pseudouridine (Ψ) is found in various stable RNAs (tRNA, rRNA, snRNA including spliceosomal snRNA and snoRNA) of all organisms [9]. Ψ is 5-ribosyluracil formed by isomerization of uridine. It is catalyzed by an enzymatic complex made of four proteins and the H/ACA class of snoRNA. These enzymatic complexes will be described below. Conversion of uridine into ψ induces a modification of the physiochemical and physical properties of the RNA and, as a consequence, of its functional behavior. Deregulations of RNA Ψ has been associated with several types of diseases such as Alzheimer's and Parkinson's diseases and the X-linked form of dyskeratosis congenital [9].

Table 6.1 Posttranscriptional Modifications of Human rRNA Nucleotides

Residue	Modified residue
Ribose	Am; Gm; Um; Cm:
Uridine	m^3U: ψ: $m^1acp^3\psi$:
Adenosine	m^1A: m^6A: m^6_2A:
Guanosine	m^7G:
Cytidine	m^5C:

The modified and/or added chemical groups are shown in red. Am, 2'-O-methyladenosine; Gm, 2'-O-methylguanosine; Um, 2'-O-methyluridine; Cm, 2'-O-methylcytosine; m^3U, 3-methyluridine; ψ, pseudouridine; $m^1acp^3\Psi$, 1-methyl-3-(3-amino-3-carboxypropyl) pseudouridine; m^1A, N^1-methyladenosine; m^6A, N^6-methyladenosine; m^6_2A, N^6, N^6-dimethyladenosine; m^7G, N^7-methylguanosine; m^5C, 5-methylcytidine. These data are extracted from the RNA Modification Database [1] and 3D Ribosomal Modification Maps [2].

Methylation occurs for the sugar and for the bases of the RNA. This modification could be considered as quite unique in the sense that it affects both DNA and RNA. Particularly, methylation of RNA bases methylation (mN) displays some striking similarities with that of DNA suggesting that this particular modification plays a key role in the epigenetic regulation of RNA activity. Interestingly, ribosomal DNA (rDNA) expression is, like many other genes, regulated epigenetically by methylation of rDNA promoter and rDNA chromatin. Very recently it has been shown that in humans, a part of this epigenetic regulation could be mediated by fibrillarin (FBL), the enzyme responsible for 2'-O ribose methylation (2'-O-m) of rRNA, that represents the most abundant modification of eukaryotic rRNA. Therefore, a prototypic example to illustrate the central role that methylation of nucleic acids could play in gene expression regulation and of its deregulation occurring during the cancer process is provided by the functional consequences resulting from methylation of rDNA genes and chromatin and of their products, the rRNA. In this chapter we will describe, in the context of the cancer process, the function of rDNA and rRNA methylation. More particularly we will describe how rRNAs are methylated, whether rRNA 2'-O-m is regulated and how rRNA 2'-O-m in concert with rRNA Ψ could affect rRNA functions.

2 OVERVIEW OF HUMAN RIBOSOME BIOGENESIS: AN INTENSE ENERGY CONSUMING PROCESS

A human ribosome contains 80 ribosomal proteins (RP) and 4 rRNAs (called 28S, 18S, 5.8S, and 5S) organized in two asymmetric subunits: the large 60S subunit and the small 40S subunit. More than 400 factors (proteins and snoRNA) and the three RNA polymerases are involved in the multiple steps allowing the synthesis, maturation, assembly, and export of the ribosomal components [10]. Most of these steps occur within a specialized nuclear domain, the nucleolus (Figure 6.1). The rate of ribosomal subunits synthesis has been evaluated at 7500 subunits per minutes in a human cancer cell line (HeLa) [11]. This rate of ribosome biogenesis requires approximately 300,000 RP. Indeed, ribosome biogenesis is one of the most energy consuming processes in eukaryotic cells. The limiting step of ribosome biogenesis is the transcription of rDNA genes by RNA polymerase I (RNAPI), which generates a polycistronic RNA precursor of three out of the four rRNA (Figure 6.1). This precursor, called the 47S pre-rRNA, contains the sequences of the 18S, 5.8S, and 28S rRNA, separated by internal transcribed spacer sequences (ITS) and surrounded by external transcribed spacer sequences (ETS). These spacer sequences are removed by endo- and exonucleases at 12 defined sites. This generates many rRNA processing intermediates [12]. In addition, rRNAs are extensively modified. Human rRNA contains 105 2'-O-m, 95 Ψ, and 10 mN (m^7G, m^1A, m^6A, m^6_2A, m^5C, m^3U, m^1acp^3 Ψ). All rRNA posttranscriptional modifications (PTMs) are listed in Refs. [2,13]. Most of rRNA methylations and Ψ are catalyzed co-transcriptionally or during the first processing steps.

3 TRANSCRIPTIONAL AND EPIGENETIC DEREGULATION OF rDNA GENE EXPRESSION IN CANCER

3.1 ORGANIZATION OF rDNA GENES AND THEIR TRANSCRIPTION

Since several decades, the increase in size and number of nucleoli is used by pathologists as a marker of tumor aggressiveness. Because nucleoli structural characteristics are a direct consequence of their

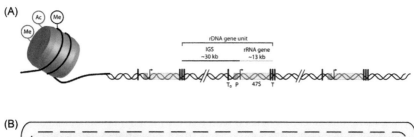

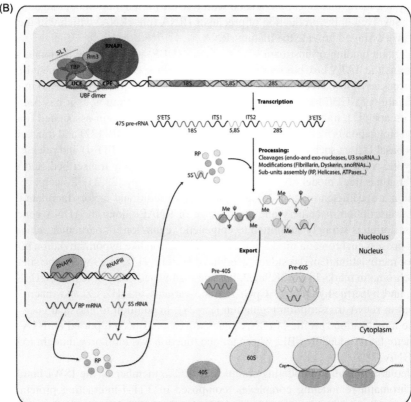

FIGURE 6.1

Overview of ribosome biogenesis and rDNA gene organization. (A) rDNA genes are organized in cluster of tandemly repeated units. One rDNA gene unit of ~43 kb contains an intergenic spacer region (IGS) of ~30 kb and a rRNA coding region of ~13 kb. The promoter region (P) is located upstream the rDNA gene. Several transcription–termination elements are located at the 3′ end of rDNA genes (T) and upstream the promoter (T_0). (B) The rDNA gene promoter consists of two elements (UCE, upstream control element; CPE, core promoter element) that interact with UBF and SL1 to recruit the RNAPI-Rrn3 complex. Inside the nucleolus, RNAPI synthesizes the 47S pre-rRNA, which is processed by a series of cleavages and chemical modifications implicating many non-ribosomal factors, producing mature 28S, 18S, and 5.8S rRNAs. The 5S rRNA is transcribed by RNAPIII, and the RP genes are transcribed by RNAPII. RP and the 5S are imported together in the nucleolus, where they are assembled to rRNA to form the pre-ribosomal subunits pre-40S and pre-60S. The final maturation steps occur in the nucleoplasm and in cytoplasm, to produce functional ribosomes.

activity (i.e., transcription of rDNA genes and assembly of pre-ribosomes), it is crucial to understand how rDNA genes are regulated in normal cells and how they can be deregulated in cancer cells to facilitate ribosome biogenesis and to support high cell growth and proliferative capacity.

In the human genome, about four hundreds of rDNA genes are organized in clusters located on the short arms of the five acrocentric chromosomes (13, 14, 15, 21, and 22) in a telomere to centromere orientation (Figure 6.1). During interphase, nucleoli are formed around these clusters that are then named nucleolar organizer regions (NOR). Assembly of several specific nucleolar complexes containing the basal transcription factor machinery associated with RNAPI for the initiation, elongation, and termination steps are involved in the production of precursor rRNA (pre-rRNA) which will then be subjected to extensive processing to provide the mature rRNA.

The synergistic binding of upstream binding factor (UBF) and SL1, the promoter selectivity factor that contains the TATA box binding protein and RNAPI-specific TBP-associated factors (TAF$_{IS}$) [14–16], is necessary for the recruitment of RNAPI to the transcription start site to initiate pre-rRNA synthesis (Figure 6.1). UBF has additional roles in RNAPI transcription as it has been also involved in promoter escape [17], transcription elongation [18], and in chromatin-associated function [19–21]. Transcription elongation is facilitated by a number of factors like TFIIS [22] and transcription termination requires basal factors such as termination transcription factor I (TTF-I) and release factor (RF).

In eukaryotic cells, only about half of the rDNA genes are highly transcribed, and give rise to the classical "Christmas tree" model observed by electron microscopy [23,24]. To achieve this high level of transcription, a specific chromatin structure and several additional factors facilitate the recruitment of the basal transcription machinery and the passage of RNAPI along the rDNA genes. In addition, RNAPI transcription is strongly submitted to epigenetic regulation. In particular, actively transcribed rDNA genes possess euchromatin marks in the promoter region like hypomethylation of CG DNA base pair targets of methylation (CpG) residues, acetylated H4K12, H3K4me3 while the silent rDNA genes carry heterochromatin marks like H3K9me2, histone methylated at a lysine group (H4K) H4K20me3, H3K27me3, and hypermethylation of CpG. Histone variants like H2A.Z and macroH2A have also been involved in rDNA transcription regulation [25–27]. In addition to classical euchromatin marks, novel histone modification specific of rDNA genes in yeast has been reported on H2A that presents a glutamine methylation added by FBL, whose unique function was until now thought to be restricted to 2′-O-m of rRNA [28].

The nucleolar chromatin remodeling complex NoRC, a member of the ISWI family of the ATP-dependent chromatin remodeling complexes, composed of TTF-I-interacting protein 5 (TIP5) and Snf2H (ATPase) plays a key role in establishing the silencing of rDNA genes by recruitment of histone deacetylase (HDAC) and methyltransferases (HMT), DNA methyltransferase and by shifting the promoter-bound nucleosome into a silent position [29,30]. In contrast, a major nucleolar protein, nucleolin (NCL) which is often overexpressed in cancer [31–37], is associated with the activation of transcription of rDNA genes. NCL interacts with rDNA chromatin and overexpression of NCL in human cells is associated with a decrease of heterochromatin marks on rDNA while euchromatin marks and the rate of RNAPI transcription elongation increase [38]. NCL possesses a histone chaperone activity and facilitates the transcription through nucleosomal structures (FACT activity) [39,40] that probably also contribute to this high transcription rate, without any change in rDNA methylation or in the number of active rDNA genes.

Although the increase of rDNA transcription in liver cancer seems to be related to a lower methylation of the promoter region [41], this does not seem to be a general feature of rDNA activation in cancer as in prostate cancer cells no such correlation could be made [42]. Since active rDNA copy genes are rather unmethylated, we could expect that demethylation of rDNA genes could be associated with an increase of transcription. Indeed, in cells lacking the DNA methyltransferase (DNMT1 or DMT3b) which decreases the level of methylated CpG along the rDNA, a significant increase in the number of actively transcribed rDNA genes is observed, but still about half of the rDNA genes remains silent. Despite this higher number of actively transcribed genes, the global level of rRNA is lower and the cells grow more slowly. This could be explained by a strong reduction of the RNAPI transcription elongation rate and by the accumulation of pre-rRNAs that have not been correctly processed [43]. These findings may provide a new mechanism of action for drugs that inhibit DNA methylation (5-azacytidine and (5-aza-2′-deoxycytidine) that were thought to act through the reactivation of tumor suppressor genes as it is shown now that they act also on ribosome biogenesis through RNAPI transcription and pre-rRNA processing which may contribute to the therapeutic effects of these drugs.

If an alteration of DNA methylation cannot explain the higher rate of RNAPI transcription observed in many cancer cells, multiple factors like products of proto-oncogenes and numerous kinase pathways (phosphoinositide 3-kinase, c-Myc, RAS/ERK, ErbB2, AKT, etc.) have been shown to contribute to the deregulation of RNAPI transcription in cancer cells. Since c-Myc is very often deregulated in cancer cells, its role in rDNA gene transcription has been the focus of many studies. Several nucleolar proteins involved in ribosome biogenesis like NCL and FBL are upregulated when c-Myc is overexpressed and RNAPI transcription can be directly activated by c-Myc. c-Myc binds directly to specific sequences located in rDNA and interacts with SL1. The presence of c-Myc on rDNA coincides with the recruitment of SL1 to the rDNA promoter, an increase of histone acetylation [44], and an increase of the number of active genes through the regulation of UBF [45].

3.2 TARGETING rDNA TRANSCRIPTION FOR ANTICANCER DRUG DEVELOPMENT

As ribosome biogenesis is crucial to support cell growth and proliferation, different drugs that target the production of rRNA have been developed [46]. The compounds CX-3543 and CX-5461 are small molecules identified in a screen for selective inhibitors of RNAPI activity. CX-5461 inhibited RNAPI activity as a consequence of the release of the transcription factor SL1 from rDNA genes promoter. CX-5461 was found to cause a disorganization of the nucleolar structure, leading to the activation of p53-dependent apoptotic signaling and to an inhibition of rDNA transcription. Interestingly, CX-5461 selectively kills B-lymphoma cells while wild type B-cell remains unaffected [47]. In addition, CX-5461 induced autophagy and senescence in various tumor cells, indicating that RNAPI inhibition may lead to different cytotoxic responses. CX-5461 is being evaluated in Phase I clinical trial in hematological malignancies by Senhwa Biosciences. Another drug, CX-3543 selectively disrupts the interaction of NCL with rDNA G-quadruplex in nucleoli resulting in inhibition of RNAPI transcription. It was shown that this compound induces apoptosis in cancer cells [48]. Several clinical trials, Phase I and Phase II were launched to evaluate CX-3543 as anticancer drug in hematological, neuroendocrine, and carcinoid tumors. Additional compounds, AS1411 (G-quartet aptamer) and HB19 (pseudopeptide) that target NCL also show some activity in reducing tumor growth, but it is not clear if this antitumoral activity is through the regulation of NCL nucleolar role in rDNA transcription [49]. Most recently, BMH-21, a compound identified as an inducer of p53 was subsequently found to repress RNAPI transcription [50]. BMH-21 binds GC-rich sequences, in particular within regions of rDNA genes, which result in an accelerated proteasome-dependent degradation of RPA194, one of the major RNAPI subunits. Finally, topoisomerase II promotes topological changes in the rDNA promoter and facilitates the efficiency of binding of the pre-initiation complex. Drugs that inhibit topoisomerase function, therefore, have a repressive effect on RNAPI transcription [51]. Altogether, these observations showed that inhibition of RNAPI activity by different means results in a clear anticancer activity, opening a new field for anticancer drug development. Yet, a highly significant contribution of the studies on CX-5461 and BMH-21 is the demonstration that inhibiting rRNA synthesis provides a way to efficiently and selectively kill cancer cells. This should change our view on the potential of such therapeutic strategy.

4 CHEMICAL MODIFICATIONS OF rRNA

In eukaryotes and archaea, rRNA PTMs occur during rRNA processing and are mediated by snoRNP. snoRNPs are composed of a core set of proteins and a small nucleolar RNA (snoRNA) that guide the enzymatic complex on the nucleotide that will be modified. The target nucleotide is localized through base paring of the snoRNA with the substrate rRNA [52–54]. The snoRNAs are evolutionarily conserved, and share homologues in eukaryotes and archaea [55]. Two types of snoRNPs have been described; the H/ACA box and C/D box snoRNPs, each recruiting a different set of core proteins, and respectively, guiding Ψ and 2′-O-m. Expression of these snoRNP components is frequently altered in cancer.

4.1 THE rRNA PSEUDOURIDYLATION MACHINERY

The RNA component of the pseudouridylation machinery is a H/ACA box snoRNA. The H/ACA box snoRNA possesses two hairpins linked by a hinge region consisting of the box H sequence ANANNA where N is any nucleotide. The box ACA consists of three nucleotides and is localized near the 3′ end [56–58]. The hairpins (one or both of them) contain internal loop sequences complementary to rRNAs that are responsible for guiding the H/ACA snoRNP on the target nucleotide in a site specific manner [53]. The H/ACA snoRNA assembles with four core proteins known in human as Gar1, Nhp2, Nop10, and Dyskerin (DKC) [55–57]. DKC is the enzyme responsible for catalysis of Ψ, although all the proteins of the complex are required for the reaction [56,57,59,60].

4.2 THE rRNA 2′-O-METHYLATION MACHINERY

The RNA component of the 2′-O-methylation machinery is a C/D box snoRNA, which is evolutionarily conserved. It is synthesized through different expression strategies, mostly depending on the species involved. In most vertebrates, and rarely in yeast or plants, snoRNAs are encoded in introns of genes transcribed by RNA polymerase II, and are excised from pre-mRNA [55,61,62]. Interestingly, a majority of these mRNAs encode genes which play essential roles in ribosome biogenesis and function, suggesting coregulation of snoRNA amount and ribosome production [55,61,62]. A few snoRNAs, however, have been reported to be transcribed and excised from introns of genes whose exon sequences do not encode proteins [55,61,62].

A few snoRNA in vertebrates, and most snoRNAs in yeast are expressed independently from their own promoter by RNA polymerase II and less frequently by RNA polymerase III [55,61,62].

Very frequently in plants and also in yeast and trypanosomes, snoRNAs arise from processing of polycistronic transcripts, containing up to nine different snoRNAs. In this case, snoRNAs are excised through a mechanism that does not involve splicing, but are processed by RNA III-like endonucleic activity and subsequently trimmed by exonucleases [55,61,62].

In the case of intronic genes, pre-snoRNAs generally arise from excised and debranched introns, which are then trimmed by exonucleases digesting away non-snoRNA sequences [63]. A second, much less frequent pathway involves the direct excision of pre-snoRNA intronic sequences by endonuclease and generation of mature ends by exonucleases, independently of splicing [62,63]. In the case of mono- and polycistronic genes, snoRNA sequences are also cleaved from the transcript by endonucleases and trimmed by exonucleases [63]. The processing also takes place at least partly in the Cajal Bodies as it has been shown in plants [62]. A large proportion of mammal C/D snoRNA has been reported to be tri-methylated on the 5′-end guanosine when transiting through Cajal Bodies [64–67]. C/D snoRNA maturation, stability, and localization require at least initiation of snoRNP assembly, which occurs co-transcriptionally [56]. snoRNPs are thereafter targeted to the dense fibrillar component of the nucleolus where they catalyze the rRNA 2′-O-m reaction [68].

All C/D box snoRNAs contain the consensus sequences C and D, respectively localized near their 5′ and 3′ ends, which are usually brought together by two short base paring sequences [56,60,61] (Figure 6.2). The internal part of C/D box snoRNA often contains imperfect copies of C and D boxes, respectively called C′ and D′, usually well conserved in archaea species, but less in eukaryotes [56,60,61]. C/D and C′/D′ boxes fold into a "stem-bulge-stem" motif called "kink-turn" (k-turn), but C′/D′ boxes often lack this motif in eukaryotes [56,60]. In archaea, C′/D′ boxes are also believed to fold into a motif related to the k-turn called k-loop [71] (Figure 6.2).

Most C/D snoRNAs possess long sequences (10–20 nucleotides) with complementarity to rRNA localized upstream D and/or D′ boxes, which allow snoRNA to target the snoRNP to the target nucleotide in a site-specific manner [56,60,61]. This methylation site is invariably the complementary to the fifth nucleotide upstream D and/or D′ box [52,54,72,73] (Figure 6.2). Several identified C/D box snoRNAs are known to possess two sequences that target two distinct rRNA methylation sites [13]. However, most known C/D snoRNAs have been mapped to only one rRNA 2′-O-m [13]. It is possible that snoRNA could target other types of RNA, whose sites have not yet been matched to their complementary sequences, or that mutations acquired during evolution either in these sequences or in rRNA have modified the complementarity.

The protein components of the 2′-O-methylation complex are FBL, Nop56, Nop58, and 15.5kD in eukaryotes [56,74,75], and have homologues in archaea species (Figure 6.2).

FBL is the methyltransferase of the complex and possesses conserved methyltransferase structural features [56]. It accumulates mainly in the nucleolus and in Cajal bodies [76]. FBL is encoded by the *Fbl* gene located on the chromosome 19 at position 19q13 in human, and was first described in physarum [77]. FBL is a 321-amino acid–long protein, and both its primary and tertiary structures are highly conserved, even between human and archaea [56]. In addition to being very well conserved, it is also an essential protein. Knock out of FBL is lethal in yeast and tetrahymena [78–80], and its expression is required for early embryonic development in mouse [81].

Nop56 and Nop58 are paralogues, and are both nucleolar proteins. They are respectively encoded by the *Nop56* and *Nop58* genes, located on locus 20p13 of chromosome 20 and locus 2q33.1 on chromosome 2 in human. *Nop56* gene gives rise to several transcripts encoding different protein isoforms, most of which have not been studied. The most studied isoform encompasses a 594-amino acid–long

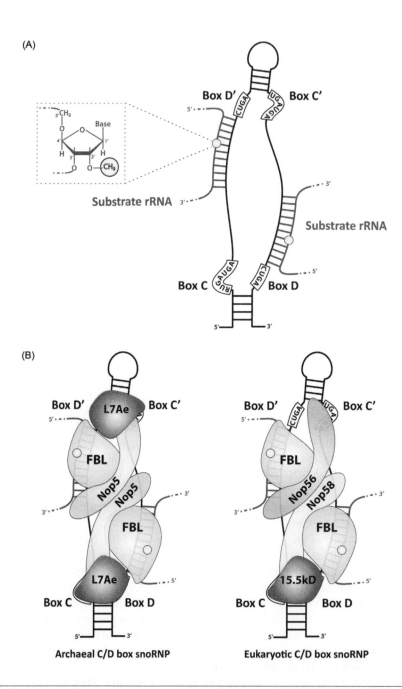

FIGURE 6.2

C/D box snoRNA and snoRNPs organization. (A) Schematic representation of a C/D box snoRNA. Box C and D consensus sequences (shown in white boxes) are, respectively, close to 5′ and 3′ ends, which are

polypeptide. *Nop58* gene also gives rise to several isoforms and its most studied protein isoform contains 529 residues.

15.5kD protein (official name in human: NHP2L1 for "non-histone binding protein 2-like 1") is a 128-amino acid–long protein that belongs to the L7Ae RP family. It is encoded by the *NHP2L1* gene located on the chromosome 22 at position 22q13, from which arises several protein coding isoforms, most of whose functions remain unstudied. In eukaryotes, 15.5kD initiates snoRNP assembly by binding k-turn motif formed by C/D boxes [56,74]. This binding is required to recruit the other core members of the snoRNP [82], all required for rRNA 2'-O-m [56].

No crystal structure data is available concerning the snoRNP complex in eukaryotes, but several archaean structures have been solved. Briefly, it seems that Nop5 (archaean homologue replacing both Nop56/Nop58) and L7Ae (archaean homologue of 15.5kD) hold FBL between them and position it with its catalytic domain facing the substrate ribose [69,70] (Figure 6.2). FBL itself has no site specificity, but the very precise position of the Nop5 and L7Ae on the snoRNA makes sure that FBL does not slide along the snoRNA–rRNA duplex helix and remains facing the fifth nucleotide upstream D or D' box [70] (Figure 6.2). The methyl transfer reaction from the universal methyl donor *S*-adenosyl-methionine (SAM) to the target ribose is an S_N2-type substitution although the precise chemical mechanism underlying this transfer remains unknown. Moreover, it should be noted that C/D snoRNP complexes contain two copies of FBL, one Nop56 and one Nop58 and at least one 15.5kD (Figure 6.2). It is likely that Nop58 and Nop56 form a dimer, as do the two Nop5 copies in archaean sRNP (Figure 6.2). Finally, it is still discussed whether s(no)RNP are formed by a monomer or dimer of snoRNA each containing two copies of the core proteins [69,70].

◄ **FIGURE 6.2** (Continued) brought together by a short base-paring region. The internal part of the snoRNA may contain boxes C' and D', imperfect copies of C and D boxes. C/D box and C'/D' box form a characteristic stem-bulge-stem structure called the kink-turn, allowing the snoRNA to recruit core snoRNP proteins. In archaea, the motif formed by C'/D' box is a variant of the k-turn called the k-loop. C' and D' boxes can be linked by a short stem structure that is not present in all snoRNAs. Upstream D and D' boxes are guide sequences complementary to rRNA able to form base pair with the later, thus allowing the snoRNA to bind the rRNA substrate in a site-specific manner. The 2'-O-ribose-methylation occurs invariably on the fifth nucleotide upstream D or D' box. In eukaryotes, snoRNA may possess only one guide sequence. When two such sequences are present they often guide modifications on two sites located on the same rRNA, but not always. Substrate rRNAs are indicated in red, s(no)RNA in black, methylated residue in yellow. (B) Schematic representation of archaeal and eukaryotic C/D box s(no)RNPs. Right panel, archaeal C/D box sRNP organization, represented using known archaeal C/D sRNP crystal structures [69,70]. Two copies of L7ae (pink) bind motifs formed by box C/D and box C'/D'. This allows recruitment of other core proteins: 2 copies of Nop5 (light blue), homodimerized, each binding one copy of fibrillarin (FBL, green). FBL can interact with rRNA substrate but the site specificity is conferred by Nop5 and L7ae that have a fixed and precise position on the sRNA through interaction with C/D and C'/D' boxes, and which hold FBL between them with its catalytic domain facing the fifth nucleotide upstream D or D' box. Left panel, eukaryotic C/D snoRNP represented using known protein–protein and RNA–protein interaction data. Only one copy of 15.5kD (red) recruited on C/D box (15.5 kD cannot bind C'/D' box in eukaryotes) is present. Two copies of FBL, each interacting with either one copy of Nop56 (orange) or one copy of Nop58 (yellow). Nop56 and Nop58 are likely to form a dimer as do the two copies of Nop5 in archaeal sRNP.

5 FUNCTION OF rRNA METHYLATION

5.1 THE COMPLEX FUNCTIONAL ARCHITECTURE OF THE RIBOSOME

Since the elucidation of prokaryotic ribosome structures early in 2000 [83–85], it appears that the ribosome displays ribozyme activities. Indeed, the rRNA carries out the peptidyl transferase activity and is involved in mRNA decoding and proofreading [86]. The complex network of rRNA and RP interactions highlights the requirement of a highly organized architecture that gives rise to several functional domains connected to each other (Figure 6.3). The large subunit contains the peptidyl transferase center, the peptide exit tunnel, and most of the A, P, and E tRNA binding sites. The small subunit contains the mRNA channel and the mRNA decoding center. Several bridges composed of both rRNA and RP connect the two subunits to each other [89]. During the last 3 years, the resolution of the structure of eukaryotic ribosomes from *Saccharomyces cerevisiae*, *Tetrahymena thermophila*, drosophila, and human have provided important breakthrough in our knowledge regarding the evolution of ribosome structure and function [87,90–92]. The main functional domains are highly conserved in all the life kingdoms. The conserved core ribosome is composed of 34 conserved RP and ~4400 ribonucleotides [93,94]. Eukaryotic ribosomes display additional layers of RP and rRNA externally to the core ribosomes. These differences are thought to reflect the more complex and more regulated mode of translation of eukaryotic cells. This is also true for differences observed between lower and higher eukaryotes [87]. The yeast ribosome contains 79 PR, ~5500 ribonucleotides and has a molecular weight of 3.3 MDa, and the human ribosome is composed of 80 PR, ~7200 rRNA base and has a molecular

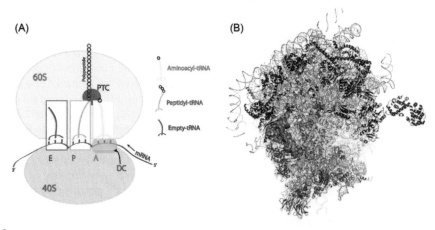

FIGURE 6.3

Functional organization of ribosome. (A) Schematic representation of the different functional domains of the ribosome. The mRNA channel is located on the interface of the 40S subunit. The tRNAs that carry the new amino acid, the nascent polypeptide, or the empty tRNA are, respectively, located in the A, P, and E tRNA binding sites. The mRNA decoding center (DC) is located on the 40S subunit, at the bottom of the A site. The peptidyl transferase center (PTC), where the peptide bond formation occurs, is located in the 60S subunits, between the A and P sites. (B) Structure of the human ribosome fitted with the A, P, and E tRNA from *Thermus thermophilus*. This picture was generated using PyMOL software and the PDB files 3J3A, 3J3B, 3J3D, and 3J3F (for the human ribosome, from [87]) (and 2WDK (for *T. thermophilus* tRNA, from [88]).

weight of 4.3 MDa [94]. The function of species–specifics components is still poorly understood, although some of them are thought to integrate translation regulation signals, such as ribosomal protein of the small subunit (RPS) 6, which is implicated in protein synthesis regulation when phosphorylated in response to various pathways [95]. Furthermore, a growing set of data suggests that in addition to the variations between species, the composition of ribosome could change in different cell types of the same species or under specific conditions. This heterogeneity has been shown, in few examples, to be responsible for selectivity of ribosome recruitment on specific mRNA [96–98].

5.2 REGULATORY ROLE OF rRNA METHYLATION: AN EMERGING CONCEPT

In addition to the composition evolution of the external layers of ribosomes, many rRNA PTMs have been added to nucleotides of the core ribosome. Among the 105 2′-O-m sites, 10 mN sites, and 95 Ψ sites, almost 40% of human rRNA PTM sites are conserved in yeast and nearly 5% is universally conserved (Figure 6.4). The majority of modified nucleotides are part of functionally important rRNA

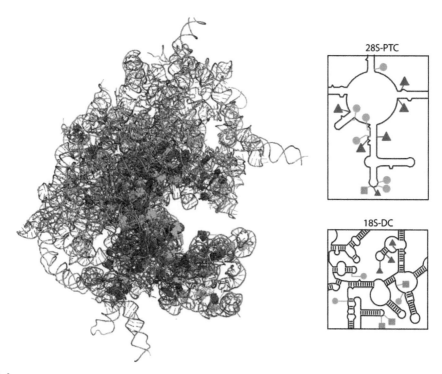

FIGURE 6.4

Location of rRNA posttranscriptional modifications on ribosome. Location of rRNA 2′-O-methylation (green) and pseudouridylations (red) on the structure of human rRNA. rRNA PTMs are particularly enriched in ribosome functional domains. 2D structures of parts of the PTC and DC rRNA regions are shown in the boxes on the right. Red dots: 2′-O-methylations; green triangles: pseudouridylations; orange squares: base methylations. 2D structures were taken from [99]. 3D structure representation was generated using PyMOL software and the PDB files 3J3D and 3J3F from [87]. 2′-O-methylation and pseudouridylation positions were adapted from [2].

domains such as the peptidyl transferase center, the decoding center, the tRNA binding sites, and the subunit interfaces [99].

Although rRNA PTM has been discovered more than 50 years ago [100–103], efficient methods to modulate and quantify the level of modification at a single site only emerged in the last few years [104,105], giving the possibility to study modifications occurring at a single site. Indeed, the level of 2′-O-m rRNA is subject to some heterogeneity. Two studies of our laboratory conducted on breast cancer cell lines have revealed that the levels of multiple 2′-O-m sites significantly varied depending on tumor aggressiveness [106] or FBL expression [107]. Only some sites are affected by an increase in the level of FBL, mainly the sites localized within the 18S rRNA. In the analyzed cell lines, the expression of the C/D box snoRNAs is also changed, but at this stage no correlation has been observed between the level of methylation at a particular site and the level of the corresponding snoRNA. The fact that some sites may be more or less modified under various conditions suggests that different subtypes of ribosome may exist within a cell, each of them displaying a specific 2′-O-m pattern.

Thus, it appears that the rRNA PTM pattern depends on many factors such as the expression level of enzymes that catalyze them or even the snoRNAs that specify the site of modification. A recent study conducted in yeast shows that a site of 18S rRNA (A100) is methylated on 68% of ribosomes in wild type condition [108]. The methylated and unmethylated 18S rRNAs are found in monoribosomes and polyribosomes, demonstrating that 2′-O-m at this site does not preclude dramatically translational activity of the ribosome. This study is the first to precisely quantify the level of 2′-O-m on a given site. In addition, inhibition and overexpression of snoRNA targeting the A100 site induce changes in the level of 2′-O-m to this site.

From several studies performed to elucidate the role of PTM in ribosome function, it appears that PTM rRNA plays a role in optimizing the structure and function of ribosomes. With few exceptions, blocking a single PTM in yeast has no or very small impact on ribosome function, but the inhibition of several PTM modifies the structure of rRNA, fidelity, and efficiency of translation and sensitivity to antibiotics. However, these studies on the role of PTM were mainly focused on the peptidyl transferase center [109,110], the mRNA decoding center [111,112], and the interface between the two subunits, including helix 69 of the 25S rRNA (28S in mammals) [112,113]. Furthermore, addition of 2'-O-m to sites naturally unmethylated of the peptidyl transferase center strongly affects the stability and function of ribosomes [114].

Interestingly, the notion emerges from different studies performed in mammals that the PTMs could be the mediators of regulatory process enabling the ribosome to modulate directly the translational process according to intrinsic as well as external stimuli. Some studies suggest a role of 2'-O-m rRNA in the internal ribosome entry site (IRES)–dependent translation, an alternative translation initiation way particularly activated when the classical cap-dependent initiation is impaired [115–117]. Depletion of L13a RP causes a decrease in the rate of 2'-O-m by an unknown mechanism. Under these conditions, the IRES-dependent translation of p53, p27, and SNAT2 mRNAs is decreased while the overall level of translation is not affected [118]. Inhibition of L13a does not affect the translation of the viral IRES of Hepatitis C virus (HCV) and Cricket paralysis virus (CrPV). Similar results are observed when the methylation is inhibited by treatment with cycloleucine, a competitive inhibitor of the methyl donor SAM. Recruitment of translation pre-initiation complex is not affected by inhibition of 2'-O-m, but the recruitment of 60S subunit is ineffective on SNAT2 IRES [119]. A correlation between the profiles of 2'-O-m rRNA and IRES-dependent translation of insulin growth factor-1 receptor (IGF-1R), c-Myc, fibroblast growth factor (FGF)1, FGF2, vascular endothelial growth factor (VEGF)A, and EMCV mRNAs have also been observed in various breast cancer cell lines [106,107]. In addition to 2′-O-m, rRNA Ψ also influences several aspects of ribosome function. In human cells and in yeast, a reduced level of pseudouridylation mediated by the inactivation of DKC (homologous of the yeast Cbf5p) affects translation fidelity, tRNA binding, and IRES-dependent translation initiation [120–123].

The heterogeneity of rRNA PTM could also play a role in other ribosomal functions. In zebrafish, inhibition of U26, U44, and U78 snoARN alters the 2′-O-m level of their target sites, which are respectively 28S-389, 28S-4593, and 18S-166 (human rRNA numbering). Inhibition of each of these 2′-O-m creates defects in the development of fish embryos [124]. The involved mechanisms are not known; however, it appears that these three PTMs are close to three RP binding sites. The 18S-166 nucleotide contacts RPS6, the 28S-389 nucleotide contacts ribosomal protein of the large subunit (RPL)17, and the nucleotide 28S-4593 contacts RPL3. Interestingly, these three RPs exhibit extraribosomal functions. Notably, RPL3 is involved in cell cycle and apoptosis [125], and RPS6 has a role in growth and cell proliferation [126]. Therefore, it is possible that rRNA PTMs also control the sequestration and/or the release of specific RP with extraribosomal functions.

6 MODIFICATION OF rRNA METHYLATION IN CANCER

The role of rRNA Ψ in pathological process and in cancer has been extensively studied and described in several reviews by the groups of Davide Ruggero and Lorenzo Montanaro [120,121,127–131]. On the contrary, few studies from our group have yet focused their attention on the role of rRNA methylation in pathological processes and particularly in cancer [106,107]. Therefore, at this stage, it is difficult to have a global view of the variations of rRNA 2′-O-m in cancer and more importantly to claim firmly that rRNA 2′-O-m, probably in concert with Ψ, participates to the pathological process. Nevertheless, several studies have described the variations of the expression of components of the methylation complex (proteins and snoRNA) in cancer. Therefore, from these studies, we can anticipate that the rRNA 2′-O-m itself is also very probably modified in cancer although this will have to be demonstrated in the upcoming years.

6.1 ALTERATIONS OF snoRNA EXPRESSION IN CANCER

Changes in rRNA 2′-O-m pattern could rely on modulation of the expression of the snoRNA part of the methylation machinery. snoRNA host genes are known to often encode either RP or proteins involved in ribosome biogenesis within their exons. It is well known that in cancer, several pathways are altered and lead to increase in ribosome biogenesis, as previously described [127,132]. As a result, it is very likely that these same cancer-altered pathways have an impact on snoRNA production and therefore on rRNA 2′-O-m. There are already many studies linking C/D snoRNA with tumor biology, and they have been reported to be involved in tumor development and progression, and also to be potential prognosis biomarkers. For example, low levels of C/D snoRNA SNORD44 encoded within *Gas5* introns, along with RNU43 and RNU48 were associated with poor prognosis in breast cancer and head and neck squamous cell carcinoma [133].

C/D snoRNA U50 copy loss and transcriptional down regulation have been reported in breast primary tumors [134], along with 2 bp deletion which leads to increased homozygosis [135]. Moreover, reexpressing U50 in breast cancer cell lines reduced colony formation [134], suggesting its involvement in tumorigenesis. As in breast cancers, U50 has also been reported to be transcriptionally downregulated and to be mutated in prostate cancers. This mutation has been reported to reduce colony formation, suggesting U50 as a tumor suppressor [135,136]. U50 is also known to be located at breakpoint of chromosomal translocation t(3;6)(q27;q15) in human B-cell lymphoma [135], and it has been shown that U50 levels in proliferation-stimulated lymphocytes isolated from healthy donors are decreased compared to non-stimulated ones. In addition, the site-specific 2′-O-m of C2848 in 28S rRNA guided by U50 was also decreased in stimulated lymphocytes [137]. U50 was also downregulated in colorectal cancer tissues compared with normal tissues, and its level was inversely linked to tumor grade (high tumor grade correlated with lower level of U50). Knockdown and overexpression of U50 expression in one colorectal cancer cell line, respectively, decrease and increase 2′-O-m of C2848 site of the 28S rRNA which modulated ribosome activity in terms of IRES-mediated translation [137], thus reinforcing the link between snoRNA expression, rRNA methylation, ribosome activity, and cancer.

In nonsmall cells lung cancer (NSCLC), six C/D snoRNAs were found to be overexpressed in surgical tissues. These snoRNAs are located at frequently amplified genomic regions in human cancers [138]. SNORD66 and SNORD76 in particular can be found in the two most amplified genomic regions in human solid tumors [135]. Three of these snoRNAs were also overexpressed in patients' plasma, SNORD33, SNORD66, and SNORD76 [138], and could be potential markers for early NSCLC detection, one of the major issues in cancer prognosis, NSCLC being one of the leading cause of death by cancer.

Finally, C/D snoRNA U32a, U33, and U35a have been reported to be involved in oxidative stress response, and suppression of their expression seemed to confer palmitate-induced oxidative stress resistance, although this was independent of rRNA 2′-O-m [135,136], linking further C/D snoRNA to cancer biology. Lastly, the imprinted gene *Rian* encodes at least nine C/D snoRNAs [139], which were found overexpressed 9- to 539-fold in various cancers [135].

6.2 DEREGULATION OF GENE EXPRESSION CODING FOR rRNA METHYLATION COMPLEX PROTEINS

Changes in rRNA 2′-O-m pattern could rely on individual or concomitant modulation of expression of the four proteins assembling the rRNA methylation complex (FBL, Nop56, Nop58, and NHP2L1). While their encoding genes are currently considered as housekeeping genes, several pieces of data revealed they are finely regulated at transcriptional levels by master genes. In particular, a huge literature demonstrates that expression of rRNA methylation complex genes is coordinated by c-Myc. Transcriptomic analyses and chromatin immuno-precipitation (ChIP)–based assays performed in various models revealed a positive correlation between expression of c-Myc and the four rRNA methylation complex genes, among them, *FBL* showing the greatest range of variation[140–147]. Further studies confirmed the positive correlation between FBL and c-Myc in primary human fibroblasts, breast, lymphoma cancer cell lines, and tumor samples [140,142,148] and clearly demonstrated that *FBL* is a classical c-Myc target gene [146]. Another example of positive correlation can be given, Nop58 mRNA levels being increased in response to PDGF stimulation, while the transcription factor and the molecular mechanism involved has not been yet identified [149]. Finally, we recently reported that *FBL* transcription is repressed by the tumor suppressor p53 through its direct DNA-binding to *FBL* intron 1 [107]. The *FBL*-induced expression in response to p53 inactivation in immortalized human mammary epithelial cells was associated with variation in rRNA 2′-O-m pattern [107], supporting the notion that changes in expression of rRNA methylation complex proteins is sufficient to modify rRNA 2′-O-m pattern. Owing to the regulation of genes encoding rRNA methylation complex proteins by key master genes involved in tumorigenesis, gene expression of rRNA methylation complex proteins can be deregulated in numerous cancers. In melanoma, higher Nop58 mRNA levels were observed in meta-static lesions than in primary lesions [150]. Overexpression of *FBL* has been reported in different cancers, including squamous cell cervical carcinoma, prostatic, and breast cancers [141,146,148]. In breast tumors, increased *FBL* expression was associated with expression of mutant p53 [107]. Moreover, high expression levels of *FBL* were independently associated with poor breast cancer patient's outcome

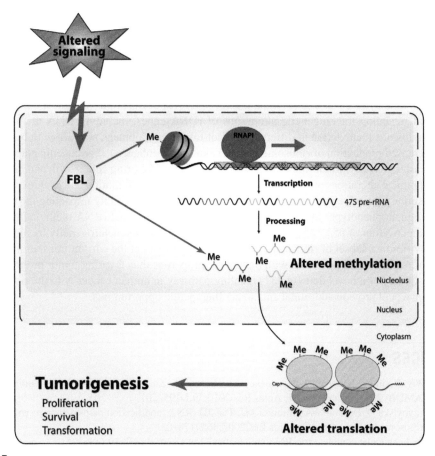

FIGURE 6.5

Model of the impact of rRNA methylation on the cancerous phenotype. In cancer cells, FBL is overexpressed as a consequence of altered signaling, leading to increased methylation of the rDNA chromatin on the one hand, and a highly methylated rRNA on the other hand. The resulting consequence of the altered rRNA methylation pattern is an alteration of the translational control that promotes expression of pro-oncogenic proteins and tumorigenesis. Me represents a CH_3 group.

[107]. Finally, *Nop56* expression has been shown to be associated with Burkitt's lymphoma and, with other c-Myc target genes, allows the molecular diagnosis of this specific subset of lymphoma [144]. Not only the *Nop56* increased expression is associated with c-Myc translocation in Burkitt's lymphoma, but also with the expression of specific c-Myc mutants, which promote *Nop56* transcription by increasing H4 acetylation of *Nop56* promoter [143]. In addition to be deregulated in human cancers, expression of the four genes encoding rRNA methylation complex proteins are altered in other human pathologies, including hypomethylation-associated disorders that involve an allele-specific reduction of *NHP2L1* due to inheritance of promoter hypomethylation [151].

Altogether, these data draw a large picture showing many links between snoRNP biology and cancer, many of which might involve their rRNA methylation function, and modulation of ribosomal activity (Figure 6.5).

7 CONCLUDING REMARKS

Deregulation of ribosomal DNA and RNA methylation during tumorigenesis remains an intriguing field still at its infancy that will highlight the unexpected roles of ribosomes in cancers in the next few decades. Understanding the epigenetic alterations of rDNA genes, including DNA methylation and FBL-induced histone methylation at different steps of tumor development, will allow to better characterize the increased production of ribosomes—that is known to enhance the oncogenic potential of different key master genes by promoting protein synthesis and thus favoring cell growth and proliferation. Moreover, chemical alterations of ribosomes occurring during tumor initiation and progression will add a novel and subtle layer of regulation in the gene expression process directly impacting the translatome and the subsequent phenotype. In particular, FBL-induced alterations of rRNA methylation pattern in response to inactivation of p53, for example, might promote ribosome heterogeneity by modifying the intrinsic activities of a subset of ribosomes that would, in turn, affect the current translational program of a subset of mRNA-coding proteins with oncogenic properties thus favoring tumor progression. The emerging discovery of a novel deregulated signaling pathway in human cancer with the ribosome as a central player would provide additional anticancer therapeutic opportunities.

REFERENCES

[1] Cantara WA, Crain PF, Rozenski J, McCloskey JA, Harris KA, Zhang X, et al. The RNA modification database, RNAMDB: 2011 update. Nucleic Acids Res 2010;39:D195–201.
[2] Piekna-Przybylska D, Decatur WA, Fournier MJ. The 3D rRNA modification maps database: with interactive tools for ribosome analysis. Nucleic Acids Res 2007;36:D178–83.
[3] He C. Grand challenge commentary: RNA epigenetics? Nat Chem Biol 2010;6:863–5.
[4] Liu N, Pan T. RNA epigenetics. Transl Res, 2015;165:28-35.
[5] Jin Y, Zhang W, Li Q. Origins and evolution of ADAR-mediated RNA editing. IUBMB Life 2009;61:572–8.
[6] Avesson L, Barry G. The emerging role of RNA and DNA editing in cancer. Biochim Biophys Acta 2014;1845:308–16.
[7] Wulff B-E, Sakurai M, Nishikura K. Elucidating the inosinome: global approaches to adenosine-to-inosine RNA editing. Nat Rev Genet 2011;12:81–5.

[8]   Behm-Ansmant I, Helm M, Motorin Y. Use of specific chemical reagents for detection of modified nucleo-tides in RNA. J Nucleic Acids 2011;2011:1–17.

[9]   Ge J, Yu Y-T. RNA pseudouridylation: new insights into an old modification. Trends Biochem Sci 2013;38:210–18.

[10] Henras AK, Soudet J, Gérus M, Lebaron S, Caizergues-Ferrer M, Mougin A, et al. The post-transcriptional steps of eukaryotic ribosome biogenesis. Cell Mol Life Sci 2008;65:2334–59.

[11] Lewis JD. Like attracts like: getting RNA processing together in the nucleus. Science 2000;288:1385–9.

[12] Mullineux S-T, Lafontaine DLJ. Mapping the cleavage sites on mammalian pre-rRNAs: where do we stand? Biochimie 2012;94:1521–32.

[13] Lestrade L. snoRNA-LBME-db, a comprehensive database of human H/ACA and C/D box snoRNAs. Nucleic Acids Res 2006;34:D158–62.

[14] Zomerdijk JC, Beckmann H, Comai L, Tjian R. Assembly of transcriptionally active RNA polymerase I initiation factor SL1 from recombinant subunits. Science 1994;266:2015–18.

[15] Comai L, Zomerdijk JC, Beckmann H, Zhou S, Admon A, Tjian R. Reconstitution of transcription factor SL1: exclusive binding of TBP by SL1 or TFIID subunits. Science 1994;266:1966–72.

[16] Heix J, Zomerdijk JC, Ravanpay A, Tjian R, Grummt I. Cloning of murine RNA polymerase I-specific TAF factors: conserved interactions between the subunits of the species-specific transcription initiation factor TIF-IB/SL1. Proc Natl Acad Sci USA 1997;94:1733–8.

[17] Panov KI, Friedrich JK, Russell J, Zomerdijk JC. UBF activates RNA polymerase I transcription by stimulat-ing promoter escape. The EMBO J 2006;25:3310–22.

[18] Stefanovsky V, Langlois F, Gagnon-Kugler T, Rothblum LI, Moss T. Growth factor signaling regulates elon-gation of RNA polymerase I transcription in mammals via UBF phosphorylation and r-chromatin remodel-ing. Mol Cell 2006;21:629–39.

[19] Wright JE, Mais C, Prieto JL, McStay B. A role for upstream binding factor in organizing ribosomal gene chromatin. Biochem Soc Symp 2006;73:77–84.

[20] Chen D, Dundr M, Wang C, Leung A, Lamond A, Misteli T, et al. Condensed mitotic chromatin is accessible to transcription factors and chromatin structural proteins. J Cell Biol 2004;168:41–54.

[21] Mais C, Wright JE, Prieto JL, Raggett SL, McStay B. UBF-binding site arrays form pseudo-NORs and sequester the RNA polymerase I transcription machinery. Genes Dev 2005;19:50–64.

[22] Schnapp G, Graveley BR, Grummt I. TFIIS binds to mouse RNA polymerase I and stimulates transcript elongation and hydrolytic cleavage of nascent rRNA. Mol Gen Genet 1996;252:412–19.

[23] Scheer U, Weisenberger D. The nucleolus. Curr Opin Cell Biol 1994;6:354–9.

[24] Miller OLJ, Beatty BR. Extrachromosomal nucleolar genes in amphibian oocytes. Genetics 1969;61 Suppl:133–Suppl:143.

[25] Németh A, Guibert S, Tiwari VK, Ohlsson R, Längst G. Epigenetic regulation of TTF-I-mediated promoter–terminator interactions of rRNA genes. Embo J 2008;27:1255–65.

[26] Cong R, Das S, Douet J, Wong J, Buschbeck M, Mongelard F, et al. macroH2A1 histone variant represses rDNA transcription. Nucleic Acids Res 2013;42:181–92.

[27] Araya I, Nardocci G, Morales J, Vera M, Molina A, Alvarez M. MacroH2A subtypes contribute antagonisti-cally to the transcriptional regulation of the ribosomal cistron during seasonal acclimatization of the carp fish. Epigenetics Chromatin 2010;3:14.

[28] Tessarz P, Santos-Rosa H, Robson SC, Sylvestersen KB, Nelson CJ, Nielsen ML, et al. Glutamine methyla-tion in histone H2A is an RNA-polymerase-I-dedicated modification. Nature 2014;505:564–8.

[29] Santoro R, Li J, Grummt I. The nucleolar remodeling complex NoRC mediates heterochromatin formation and silencing of ribosomal gene transcription. Nat Genet 2002;32:393–6.

[30] Nemeth A, Strohner R, Grummt I, Langst G. The chromatin remodeling complex NoRC and TTF-I cooperate in the regulation of the mammalian rRNA genes in vivo. Nucleic Acids Res 2004;32:4091–9.

[31] Abdelmohsen K, Gorospe M. RNA-binding protein nucleolin in disease. RNA Biol 2012;9:799–808.

[32] Roussel P, Hernandez-Verdun D. Identification of Ag-NOR proteins, markers of proliferation related to ribosomal gene activity. Exp Cell Res 1994;214:465–72.

[33] Sirri V, Roussel P, Gendron MC, Hernandez-Verdun D. Amount of the two major Ag-NOR proteins, nucleolin, and protein B23 is cell-cycle dependent. Cytometry 1997;28:147–56.

[34] Sirri V, Roussel P, Trere D, Derenzini M, Hernandez-Verdun D. Amount variability of total and individual Ag-NOR proteins in cells stimulated to proliferate. J Histochem Cytochem 1995;43:887–93.

[35] Storck S, Shukla M, Dimitrov S, Bouvet P. Functions of the histone chaperone nucleolin in diseases. Subcell Biochem 2007;41:125–44.

[36] Derenzini M, Sirri V, Pession A, Trere D, Roussel P, Ochs RL, et al. Quantitative changes of the two major AgNOR proteins, nucleolin and protein B23, related to stimulation of rDNA transcription. Exp Cell Res 1995;219:276–82.

[37] Mehes G, Pajor L. Nucleolin and fibrillarin expression in stimulated lymphocytes and differentiating HL-60 cells. A flow cytometric assay. Cell Prolif 1995;28:329–36.

[38] Cong R, Das S, Ugrinova I, Kumar S, Mongelard F, Wong J, et al. Interaction of nucleolin with ribosomal RNA genes and its role in RNA polymerase I transcription. Nucleic Acids Res 2012;40:9441–54.

[39] Angelov D, Bondarenko VA, Almagro S, Menoni H, Mongelard F, Hans F, et al. Nucleolin is a histone chaperone with FACT-like activity and assists remodeling of nucleosomes. The EMBO J 2006;25:1669–79.

[40] Mongelard F, Bouvet P. Nucleolin: a multiFACeTed protein. Trends Cell Biol 2007;17:80–6.

[41] Ghoshal K, Majumder S, Datta J, Motiwala T, Bai S, Sharma SM, et al. Role of Human Ribosomal RNA (rRNA) promoter methylation and of Methyl-CpG-binding protein MBD2 in the suppression of rRNA Gene expression. J Biol Chem 2004;279:6783–93.

[42] Uemura M, Zheng Q, Koh CM, Nelson WG, Yegnasubramanian S, De Marzo AM. Overexpression of ribosomal RNA in prostate cancer is common but not linked to rDNA promoter hypomethylation. Oncogene 2011;31:1254–63.

[43] Gagnon-Kugler T, Langlois F, Stefanovsky V, Lessard F, Moss T. Loss of human ribosomal Gene CpG methylation enhances cryptic RNA polymerase II transcription and disrupts ribosomal RNA processing. Mol Cell 2009;35:414–25.

[44] Grandori C, Gomez-Roman N, Felton-Edkins ZA, Ngouenet C, Galloway DA, Eisenman RN, et al. c-Myc binds to human ribosomal DNA and stimulates transcription of rRNA genes by RNA polymerase I. Nat Cell Biol 2005;7:311–18.

[45] Poortinga G, Wall M, Sanij E, Siwicki K, Ellul J, Brown D, et al. c-MYC coordinately regulates ribosomal gene chromatin remodeling and Pol I availability during granulocyte differentiation. Nucleic Acids Res 2011;39:3267–81.

[46] Pickard AJ, Bierbach U. The cell's nucleolus: an emerging target for chemotherapeutic intervention. Chem Med Chem 2013;8:1441–9.

[47] Bywater MJ, Poortinga G, Sanij E, Hein N, Peck A, Cullinane C, et al. Inhibition of RNA polymerase I as a therapeutic strategy to promote cancer-specific activation of p53. Cancer Cell 2012;22:51–65.

[48] Drygin D, Siddiqui-Jain A, O'Brien S, Schwaebe M, Lin A, Bliesath J, et al. Anticancer activity of CX-3543: a direct inhibitor of rRNA biogenesis. Cancer Res 2009;69:7653–61.

[49] Mongelard F, Bouvet P. AS-1411, a guanosine-rich oligonucleotide aptamer targeting nucleolin for the potential treatment of cancer, including acute myeloid leukemia. Curr Opin Mol Ther 2010;12:107–14.

[50] Peltonen K, Colis L, Liu H, Trivedi R, Moubarek MS, Moore HM, et al. A targeting modality for destruction of RNA polymerase I that possesses anticancer activity. Cancer Cell 2014;25:77–90.

[51] Ray S, Panova T, Miller G, Volkov A, Porter AC, Russell J, et al. Topoisomerase IIalpha promotes activation of RNA polymerase I transcription by facilitating pre-initiation complex formation. Nat Commn 2013;4:1598.

[52] Cavaillé J, Nicoloso M, Bachellerie J-P. Targeted ribose methylation of RNA in vivo directed by tailored antisense RNA guides. Nature 1996;383:732–5.

[53] Ganot P, Bortolin M-L, Kiss T. Site-specific pseudouridine formation in preribosomal RNA is guided by small nucleolar RNAs. Cell 1997;89:799–809.

[54] Kiss-László Z, Henry Y, Bachellerie JP, Caizergues-Ferrer M, Kiss T. Site-specific ribose methylation of preribosomal RNA: a novel function for small nucleolar RNAs. Cell 1996;85:1077–88.

[55] Bratkovič T, Rogelj B. Biology and applications of small nucleolar RNAs. Cell Mol Life Sci 2011;68:3843–51.

[56] Reichow SL, Hamma T, Ferré-D'Amaré AR, Varani G. The structure and function of small nucleolar ribonucleoproteins. Nucleic Acids Res 2007;35:1452–64.

[57] Watkins NJ, Bohnsack MT. The box C/D and H/ACA snoRNPs: key players in the modification, processing and the dynamic folding of ribosomal RNA. WIREs RNA 2012;3:397–414.

[58] Bortolin ML, Ganot P, Kiss T. Elements essential for accumulation and function of small nucleolar RNAs directing site-specific pseudouridylation of ribosomal RNAs. The EMBO J 1999;18:457–69.

[59] Henras A, Henry Y, Bousquet-Antonelli C, Noaillac-Depeyre J, Gélugne JP, Caizergues-Ferrer M. Nhp2p and Nop10p are essential for the function of H/ACA snoRNPs. The EMBO J 1998;17:7078–90.

[60] Henras AK, Dez C, Henry Y. RNA structure and function in C/D and H/ACA s(no)RNPs. Curr Opin Struct Biol 2004;14:335–43.

[61] Tollervey D, Kiss T. Function and synthesis of small nucleolar RNAs. Curr Opin Cell Biol 1997;9: 337–42.

[62] Filipowicz W, Pogacić V. Biogenesis of small nucleolar ribonucleoproteins. Curr Opin Cell Biol 2002;14:319–27.

[63] Tycowski KT, Steitz JA. Non-coding snoRNA host genes in drosophila: expression strategies for modification guide snoRNAs. Eur J Cell Biol 2001;80:119–25.

[64] Speckmann WA, Terns RM, Terns MP. The box C/D motif directs snoRNA 5'-cap hypermethylation. Nucleic Acids Res 2000;28:4467–73.

[65] Verheggen C. Mammalian and yeast U3 snoRNPs are matured in specific and related nuclear compartments. The EMBO J 2002;21:2736–45.

[66] Maxwell ES, Fournier MJ. The small nucleolar RNAs. Annu Rev Biochem 1995;64:897–934.

[67] Mouaikel J, Verheggen C, Bertrand E, Tazi J, Bordonné R. Hypermethylation of the cap structure of both yeast snRNAs and snoRNAs requires a conserved methyltransferase that is localized to the nucleolus. Mol Cell 2002;9:891–901.

[68] Lechertier T, Grob A, Hernandez-Verdun D, Roussel P. Fibrillarin and Nop56 interact before being co-assembled in box C/D snoRNPs. Exp Cell Res 2009;315:928–42.

[69] Lapinaite A, Simon B, Skjaerven L, Rakwalska-Bange M, Gabel F, Carlomagno T. The structure of the box C/D enzyme reveals regulation of RNA methylation. Nature 2013;502:519–23.

[70] Lin J, Lai S, Jia R, Xu A, Zhang L, Lu J, et al. Structural basis for site-specific ribose methylation by box C/D RNA protein complexes. Nature 2011;469:559–63.

[71] Nolivos S, Carpousis AJ, Clouet-d'Orval B. The K-loop, a general feature of the Pyrococcus C/D guide RNAs, is an RNA structural motif related to the K-turn. Nucleic Acids Res 2005;33:6507–14.

[72] Chen C-L, Perasso R, Qu L-H, Amar L. Exploration of pairing constraints identifies a 9 base-pair core within Box C/D snoRNA–rRNA duplexes. J Mol Biol 2007;369:771–83.

[73] Kiss-László Z, Henry Y, Kiss T. Sequence and structural elements of methylation guide snoRNAs essential for site-specific ribose methylation of pre-rRNA. The EMBO J 1998;17:797–807.

[74] Biswas S, Buhrman G, Gagnon K, Mattos C, Brown II BA, Maxwell ES. Comparative analysis of the 15.5kD box C/D snoRNP core protein in the primitive eukaryote *Giardia lamblia* reveals unique structural and functional features. Biochemistry 2011;50:2907–18.

[75] Bower-Phipps KR, Taylor DW, Wang HW, Baserga SJ. The box C/D sRNP dimeric architecture is conserved across domain Archaea. Rna 2012;18:1527–40.

[76] Snaar S, Wiesmeijer K, Jochemsen AG, Tanke HJ, Dirks RW. Mutational analysis of fibrillarin and its mobility in living human cells. J Cell Biol 2000;151:653–62.

[77] Christensen ME, Beyer AL, Walker B, Lestourgeon WM. Identification of NG, NG-dimethylarginine in a nuclear protein from the lower eukaryote *Physarum polycephalum* homologous to the major proteins of mammalian 40S ribonucleoprotein particles. Biochem Biophys Res Commun 1977;74:621–9.

[78] Tollervey D, Lehtonen H, Carmo-Fonseca M, Hurt EC. The small nucleolar RNP protein NOP1 (fibrillarin) is required for pre-rRNA processing in yeast. The EMBO J 1991;10:573–83.

[79] David E, McNeil JB, Basile V, Pearlman RE. An unusual fibrillarin gene and protein: structure and functional implications. Mol Biol Cell 1997;8:1051–61.

[80] Jansen RP, Hurt EC, Kern H, Lehtonen H, Carmo-Fonseca M, Lapeyre B, et al. Evolutionary conservation of the human nucleolar protein fibrillarin and its functional expression in yeast. J Cell Biol 1991;113:715–29.

[81] Newton K, Petfalski E, Tollervey D, Cáceres JF. Fibrillarin is essential for early development and required for accumulation of an intron-encoded small nucleolar RNA in the mouse. Mol Cell Biol 2003;23:8519–27.

[82] Watkins NJ, Dickmanns A, Lührmann R. Conserved stem II of the box C/D motif is essential for nucleolar localization and is required, along with the 15.5K protein, for the hierarchical assembly of the box C/D snoRNP. Mol Cell Biol 2002;22:8342–52.

[83] Wimberly BT, Brodersen DE, Clemons WM, Morgan-Warren RJ, Carter AP, Vonrhein C, et al. Structure of the 30S ribosomal subunit. Nature 2000;407:327–39.

[84] Yusupov MM. Crystal structure of the ribosome at 5.5 A resolution. Science 2001;292:883–96.

[85] Ban N. The complete atomic structure of the large ribosomal subunit at 2.4 A Resolution. Science 2000;289:905–20.

[86] Ramakrishnan V. Ribosome structure and the mechanism of translation. Cell 2002;108:557–72.

[87] Anger AM, Armache J-P, Berninghausen O, Habeck M, Subklewe M, Wilson DN, et al. Structures of the human and Drosophila 80S ribosome. Nature 2013;497:80–5.

[88] Voorhees RM, Weixlbaumer A, Loakes D, Kelley AC, Ramakrishnan V. Insights into substrate stabilization from snapshots of the peptidyl transferase center of the intact 70S ribosome. Nat Struct Mol Biol 2009;16:528–33.

[89] Korostelev A, Trakhanov S, Laurberg M, Noller HF. Crystal structure of a 70S Ribosome-tRNA complex reveals functional interactions and rearrangements. Cell 2006;126:1065–77.

[90] Rabl J, Leibundgut M, Ataide SF, Haag A, Ban N. Crystal structure of the eukaryotic 40S ribosomal subunit in complex with initiation factor 1. Science 2011;331:730–6.

[91] Ben-Shem A, Garreau de Loubresse N, Melnikov S, Jenner L, Yusupova G, Yusupov M. The structure of the eukaryotic ribosome at 3.0 A resolution. Science 2011;334:1524–9.

[92] Klinge S, Voigts-Hoffmann F, Leibundgut M, Arpagaus S, Ban N. Crystal structure of the eukaryotic 60S ribosomal subunit in complex with initiation factor 6. Science 2011;334:941–8.

[93] Wilson DN, Doudna Cate JH. The structure and function of the eukaryotic ribosome. Cold Spring Harb Perspect Biol 2012;4:a011536.

[94] Melnikov S, Ben-Shem A, Garreau de Loubresse N, Jenner L, Yusupova G, Yusupov M. One core, two shells: bacterial and eukaryotic ribosomes. Nat Struct Mol Biol 2012;19:560–7.

[95] Ruvinsky I, Meyuhas O. Ribosomal protein S6 phosphorylation: from protein synthesis to cell size. Trends Biochem Sci 2006;31:342–8.

[96] Xue S, Barna M. Specialized ribosomes: a new frontier in gene regulation and organismal biology. Nat Rev Mol Cell Biol 2012;13:355–69.

[97] Kondrashov N, Pusic A, Stumpf CR, Shimizu K, Hsieh AC, Xue S, et al. Ribosome-mediated specificity in hox mRNA translation and vertebrate tissue patterning. Cell 2011;145:383–97.

[98] Lee AS-Y, Burdeinick-Kerr R, Whelan SPJ. A ribosome-specialized translation initiation pathway is required for cap-dependent translation of vesicular stomatitis virus mRNAs. Proc Natl Acad Sci 2013;110:324–9.

[99] Decatur WA, Fournier MJ. rRNA modifications and ribosome function. Trends Biochem Sci 2002;27:344–51.

[100] Lane BG, Tamaoki T. Methylated bases and sugars in 16-S and 28-S RNA from L cells. Biochim Biophys Acta 1969;179:332–40.

[101] Lane BG, Ofengand J, Gray MW. Pseudouridine and O2'-methylated nucleosides. Significance of their selective occurrence in rRNA domains that function in ribosome-catalyzed synthesis of the peptide bonds in proteins. Biochimie 1995;77:7–15.

[102] Littlefield JW, Dunn DB. The occurrence and distribution of thymine and three methylated-adenine bases in ribonucleic acids from several sources. Biochem J 1958;70:642–51.

[103] Nazar RN, Lo AC, Wildeman AG, Sitz TO. Effect of 2'-O-methylation on the structure of mammalian 5.8S rRNAs and the 5.8S-28S rRNA junction. Nucleic Acids Res 1983;11:5989–6001.

[104] Liang XH, Vickers TA, Guo S, Crooke ST. Efficient and specific knockdown of small non-coding RNAs in mammalian cells and in mice. Nucleic Acids Res 2011;39:e13.

[105] Motorin Y, Muller S, Behm-Ansmant I, Branlant C. Identification of modified residues in RNAs by reverse transcription-based methods. Methods Enzymol 2007;425:21–53.

[106] Belin S, Beghin A, Solano-Gonzàlez E, Bezin L, Brunet-Manquat S, Textoris J, et al. Dysregulation of ribosome biogenesis and translational capacity is associated with tumor progression of human breast cancer cells. PLoS One 2009;4:e7147.

[107] Marcel V, Ghayad SE, Belin S, Therizols G, Morel A-P, Solano-Gonzàlez E, et al. p53 Acts as a safeguard of translational control by regulating fibrillarin and rRNA methylation in cancer. Cancer Cell 2013;24:318–30.

[108] Buchhaupt M, Sharma S, Kellner S, Oswald S, Paetzold M, Peifer C, et al. Partial methylation at Am100 in 18S rRNA of Baker's yeast reveals ribosome heterogeneity on the level of eukaryotic rRNA modification. PLoS One 2014;9:e89640.

[109] Baxter-Roshek JL, Petrov AN, Dinman JD. Optimization of ribosome structure and function by rRNA base modification. PLoS One 2007;2:e174.

[110] King TH, Liu B, McCully RR, Fournier MJ. Ribosome structure and activity are altered in cells lacking snoRNPs that form pseudouridines in the peptidyl transferase center. Mol Cell 2003;11:425–35.

[111] Liang XH, Liu Q, Fournier MJ. Loss of rRNA modifications in the decoding center of the ribosome impairs translation and strongly delays pre-rRNA processing. Rna 2009;15:1716–28.

[112] Baudin-Baillieu A, Fabret C, Liang XH, Piekna-Przybylska D, Fournier MJ, Rousset JP. Nucleotide modifications in three functionally important regions of the *Saccharomyces cerevisiae* ribosome affect translation accuracy. Nucleic Acids Res 2009;37:7665–77.

[113] Liang X-H, Liu Q, Fournier MJ. rRNA Modifications in an intersubunit bridge of the ribosome strongly affect both ribosome biogenesis and activity. Mol Cell 2007;28:965–77.

[114] Liu B, Liang X-H, Piekna-Przybylska D, Liu Q, Fournier MJ. Mis-targeted methylation in rRNA can severely impair ribosome synthesis and activity. RNA Biol 2008;5:249–54.

[115] Sonenberg N, Hinnebusch AG. Regulation of translation initiation in eukaryotes: mechanisms and biological targets. Cell 2009;136:731–45.

[116] Thakor N, Holcik M. IRES-mediated translation of cellular messenger RNA operates in eIF2 - independent manner during stress. Nucleic Acids Res 2012;40:541–52.

[117] Mokrejs M, Masek T, Vopalensky V, Hlubucek P, Delbos P, Pospisek M. IRESite--a tool for the examination of viral and cellular internal ribosome entry sites. Nucleic Acids Res 2009;38:D131–6.

[118] Chaudhuri S, Vyas K, Kapasi P, Komar AA, Dinman JD, Barik S, et al. Human ribosomal protein L13a is dispensable for canonical ribosome function but indispensable for efficient rRNA methylation. Rna 2007;13:2224–37.

[119] Basu A, Das P, Chaudhuri S, Bevilacqua E, Andrews J, Barik S, et al. Requirement of rRNA methylation for 80S ribosome assembly on a cohort of cellular internal ribosome entry sites. Mol Cell Biol 2011;31:4482–99.

[120] Rocchi L, Pacilli A, Sethi R, Penzo M, Schneider RJ, Trere D, et al. Dyskerin depletion increases VEGF mRNA internal ribosome entry site-mediated translation. Nucleic Acids Res 2013;41:8308–18.

[121] Bellodi C, Krasnykh O, Haynes N, Theodoropoulou M, Peng G, Montanaro L, et al. Loss of function of the tumor suppressor DKC1 perturbs p27 translation control and contributes to pituitary tumorigenesis. Cancer Res 2010;70:6026–35.

[122] Yoon A, Peng G, Brandenburger Y, Brandenburg Y, Zollo O, Xu W, et al. Impaired control of IRES-mediated translation in X-linked dyskeratosis congenita. Science 2006;312:902–6.

[123] Jack K, Bellodi C, Landry DM, Niederer RO, Meskauskas A, Musalgaonkar S, et al. rRNA pseudouridylation defects affect ribosomal ligand binding and translational fidelity from yeast to human cells. Mol Cell 2011;44:660–6.

[124] Higa-Nakamine S, Suzuki T, Uechi T, Chakraborty A, Nakajima Y, Nakamura M, et al. Loss of ribosomal RNA modification causes developmental defects in zebrafish. Nucleic Acids Res 2011;40:391–8.

[125] Russo A, Esposito D, Catillo M, Pietropaolo C, Crescenzi E, Russo G. Human rpL3 induces G_1/S arrest or apoptosis by modulating p21waf1/cip1 levels in a p53-independent manner. Cell Cycle 2013;12:76–87.

[126] Warner JR, McIntosh KB. How common are extraribosomal functions of ribosomal proteins? Mol Cell 2009;34:3–11.

[127] Ruggero D, Pandolfi PP. Does the ribosome translate cancer? Nat Rev Cancer 2003;3:179–92.

[128] Stumpf CR, Ruggero D. The cancerous translation apparatus. Curr Opin Genet Dev 2011;21:474–83.

[129] Bellodi C, Kopmar N, Ruggero D. Deregulation of oncogene-induced senescence and p53 translational control in X-linked dyskeratosis congenita. The EMBO J 2010:1–12.

[130] Bellodi C, McMahon M, Contreras A, Juliano D, Kopmar N, Nakamura T, et al. H/ACA small RNA dysfunctions in disease reveal key roles for noncoding RNA modifications in hematopoietic stem cell differentiation. Cell Rep 2013;3:1493–502.

[131] Montanaro L, Calienni M, Bertoni S, Rocchi L, Sansone P, Storci G, et al. Novel dyskerin-mediated mechanism of p53 inactivation through defective mRNA translation. Cancer Res 2010;70:4767–77.

[132] Whittaker S, Martin M, Marais R. All roads lead to the ribosome. Cancer Cell 2010;18:5–6.

[133] Gee HE, Buffa FM, Camps C, Ramachandran A, Leek R, Taylor M, et al. The small-nucleolar RNAs commonly used for microRNA normalisation correlate with tumour pathology and prognosis. Br J Cancer 2011;104:1168–77.

[134] Dong X-Y, Guo P, Boyd J, Sun X, Li Q, Zhou W, et al. Implication of snoRNA U50 in human breast cancer. J Genet Genomics 2009;36:447–54.

[135] Mannoor K, Liao J, Jiang F. Small nucleolar RNAs in cancer. Biochim Biophys Acta 2012;1826:121–8.

[136] Williams GT, Farzaneh F. Are snoRNAs and snoRNA host genes new players in cancer? Nat Rev Cancer 2012;12:84–8.

[137] Pacilli A, Ceccarelli C, Treré D, Montanaro L. SnoRNA U50 levels are regulated by cell proliferation and rRNA transcription. Ijms 2013;14:14923–35.

[138] Liao J, Yu L, Mei Y, Guarnera M, Shen J, Li R, et al. Small nucleolar RNA signatures as biomarkers for non-small-cell lung cancer. Mol Cancer 2010;9:198.

[139] Cavaillé J, Seitz H, Paulsen M, Ferguson-Smith AC, Bachellerie J-P. Identification of tandemly-repeated C/D snoRNA genes at the imprinted human 14q32 domain reminiscent of those at the Prader-Willi/Angelman syndrome region. Hum Mol Genet 2002;11:1527–38.

[140] Schlosser I, Holzel M, Murnseer M, Burtscher H, Weidle UH, Eick D. A role for c-Myc in the regulation of ribosomal RNA processing. Nucleic Acids Res 2003;31:6148–56.

[141] Choi YW, Kim YW, Bae SM, Kwak SY, Chun HJ, Tong SY, et al. Identification of differentially expressed genes using annealing control primer-based GeneFishing in human squamous cell cervical carcinoma. Clin Oncol (R Coll Radiol) 2007;19:308–18.

[142] Coller HA, Grandori C, Tamayo P, Colbert T, Lander ES, Eisenman RN, et al. Expression analysis with oligonucleotide microarrays reveals that MYC regulates genes involved in growth, cell cycle, signaling, and adhesion. Proc Natl Acad Sci USA 2000;97:3260–5.

[143] Cowling VH, Turner SA, Cole MD. Burkitt's lymphoma-associated c-Myc mutations converge on a dramatically altered target gene response and implicate Nol5a/Nop56 in oncogenesis. Oncogene 2014;33(27):3519–27.

[144] Dave SS, Fu K, Wright GW, Lam LT, Kluin P, Boerma EJ, et al. Molecular diagnosis of burkitt's lymphoma. N Engl J Med 2006;354:2431–42.

[145] Fan J, Zeller K, Chen YC, Watkins T, Barnes KC, Becker KG, et al. Time-dependent c-Myc transactomes mapped by array-based nuclear run-on reveal transcriptional modules in human B cells. PLoS One 2010;5:e9691.

[146] Koh CM, Gurel B, Sutcliffe S, Aryee MJ, Schultz D, Iwata T, et al. Alterations in nucleolar structure and gene expression programs in prostatic neoplasia are driven by the MYC oncogene. Am J Pathol 2011;178:1824–34.

[147] Menssen A, Hermeking H. Characterization of the c-MYC-regulated transcriptome by SAGE: identification and analysis of c-MYC target genes. Proc Natl Acad Sci USA 2002;99:6274–9.

[148] Su H, Xu T, Ganapathy S, Shadfan M, Long M, Huang TH-M, et al. Elevated snoRNA biogenesis is essential in breast cancer. Oncogene 2014;33:1348–58.

[149] Nelson SA, Santora KE, LaRochelle WJ. Isolation and characterization of a novel PDGF-induced human gene. Gene 2000;253:87–93.

[150] Nakamoto K, Ito A, Watabe K, Koma Y, Asada H, Yoshikawa K, et al. Increased expression of a nucleolar Nop5/Sik family member in metastatic melanoma cells: evidence for its role in nucleolar sizing and function. Am J Pathol 2001;159:1363–74.

[151] Docherty LE, Rezwan FI, Poole RL, Jagoe H, Lake H, Lockett GA, et al. Genome-wide DNA methylation analysis of patients with imprinting disorders identifies differentially methylated regions associated with novel candidate imprinted genes. J Med Genet 2014;51:229–38.

MINING THE EPIGENETIC LANDSCAPE: SURFACE MINING OR DEEP UNDERGROUND

7

Viren Amin, Vitor Onuchic, and Aleksandar Milosavljevic

Epigenome Center, Baylor College of Medicine, Houston, Texas, USA

CHAPTER OUTLINE

S.G. Gray (Ed): Epigenetic Cancer Therapy. DOI: http://dx.doi.org/10.1016/B978-0-12-800206-3.00007-0

1 INTRODUCTION

Over the past decade, the decreasing cost of epigenomic profiling techniques, as well as the large amounts of resources allocated to epigenomic research, have led to a leap in our understanding of the epigenome. In particular, there has been tremendous progress in our understanding of the role of epigenetics in cancer. Advances over the past decade have indicated the central role of epigenetics in the silencing of tumor suppressors and activation of oncogenes [1]. Furthermore, many mutations that are likely drivers of cancer progression affect epigenetic regulator genes resulting in global changes in the epigenome [2,3]. Based on these discoveries, anti-cancer drugs targeting epigenomic regulators have already been developed and are currently on the market [4,5]. Epigenomic profiles of different cancer samples have also been used to determine cancer subtypes with distinct outcomes and with potentially distinct responses to different treatments. Further, epigenomic alterations have been shown to be important steps in the progression of not only cancer, but other human pathologies. Changes in epigenomic profiles can also give important insights into the mechanism of action of particular drugs.

The potential for new discoveries arising from epigenomic studies is not limited to epigenomic changes associated with diseases. The epigenome can be highly informative in the annotation of the non-coding portion of the genome as different epigenomic marks have been shown to associate with distinct genomic elements and with their level of activity. A better understanding of the non-coding portion of the genome will be invaluable for the interpretation of functional effects of genetic variants that occur in those areas of the genome. Further, the patterns of activation across cell types of different genomic elements, inferred through the epigenome, can indicate the cell type that is most relevant to a particular disease.

The epigenome is composed of many layers, and its function is dependent on the crosstalk between them. Several different assays have been proposed to probe the distinct constituents of the epigenome. In the beginning of this chapter, we will briefly review some of the most commonly used assays that allow us to study the epigenome. Following the description of experimental assays used to probe the epigenome, we will present common study designs and analysis techniques in the field of epigenomics. We also provide use cases that illustrate the types of analyses mentioned in this chapter, with the hope that the reader will be able to apply similar epigenomic analysis in their research area.

2 SUMMARY OF EPIGENOMIC PROFILING METHODS AND THE DATA GENERATED BY LARGE-SCALE EPIGENOMIC PROJECTS

The epigenome of a cell consists of many interdependent layers, including DNA methylation, histone tail modifications, alternative types of histones, chromatin accessibility, gene expression, and three-dimensional chromosomal conformation. Many experimental procedures have been designed to probe each of the elements of the epigenome. Such methods have varying degrees of resolution, throughput and coverage, as well as unique biases that must be accounted for. In this section, we will discuss some of the most widely used epigenomic profiling methods, describe their similarities and differences, as well as their strengths and weaknesses. Our purpose here is not to provide a step-by-step description of the experimental protocols, as there are several reviews that describe them in great detail [6–8]. Instead, the goal is to make the reader aware of the different methods and the types of data generated by each of them (see Table 7.1 for summary). We will also describe in this section publicly available datasets generated using the methods described herein.

Table 7.1 Summary of Epigenomic Methods, Assay Description, Their Advantages and Disadvantages

Methylation	Description	Pros	Cons	References
450K array	Method uses bisulfite conversion followed by detection of methylation status based on differentially labeled Cy3 or Cy5 using bead array containing probe for ~480K individual CpGs	Detection of methylation status at single nucleotide resolution at ~480K individual CpGs	Does not assess closely located CpGs and biological insight can be limited by the selection of CpGs. Cannot distinguish DNA methylation from hydroxymethylation	[9]
MeDIP-seq	Method uses an antibody to immunoprecipitate methylated DNA or hydroxymethylated-specific regions. Sequencing of immunoprecipitated DNA determines methylation or hydroxymethylation status	Comprehensive methylome coverage at fraction of the cost and can specifically detect 5mC and 5hmC based on antibody used	Resolution is 100–300 bp	[10]
MRE-seq	Method uses methyl-sensitive restriction enzyme (MRE) to cut unmethylated cytosine within their recognition sequence. Sequencing of size selected fragments detects unmethylated CpGs that can be used to determine relative methylation status of other CpGs	Method is reliable, inexpensive, and can detect methylation status of >1 million CpGs or 65% of CpG islands at single nucleotide resolution	Only assays CpG islands that contain the restriction enzyme recognition sequence. This only represents 1–2% of the genome and 7% of total CpGs	[11]
RRBS	Method uses methylation-insensitive restriction enzyme to digest DNA and select fragments containing CpG dinucleotides. This is followed by bisulfite treatment and sequencing	Retains the advantage of WGBS while reducing the cost. Provides single base pair resolution, captures 90% of CpG islands, determines methylation status at non-CpG sites, and provides information regarding strand and allele-specific methylation	Fragment selected are biased based on restriction enzyme used and cannot distinguish DNA methylation from hydroxymethylation	[12,13]
WGBS	DNA sonication is followed by bisulfite treatment. Sequencing of bisulfite-treated DNA fragments detects methylation status at single base pair level	Provides single CpG resolution of DNA methylation and provides information regarding strand and allele-specific methylation	Costly since many reads do not contain CpG dinucleotides. Cannot distinguish DNA methylation from hydroxymethylation	[14,15]

(Continued)

Table 7.1 (Continued)

Methylation	Description	Pros	Cons	References
Histone modifications				
ChIP-seq	Method uses a specific antibody to detect histone modification and immunoprecipitate the DNA fragments. Histone-modified chromatin sites are determined by sequencing the immunoprecipitated DNA	Inexpensive, provides higher resolution maps than ChIP-chip and better signal-to-noise ratios	Chromatin maps rely upon availability of specific antibodies	[16]
Chromatin accessibility				
DNase-seq	Method uses DNase I enzyme to digest DNase I hypersensitive sites and detect open chromatin regions by sequencing DNase I digested DNA fragments	Provides high-resolution mapping of open chromatin regions across the genome	DNase I optimal concentration for digestion varies from sample to sample, so each sample needs to be individually titrated	[17]
Gene expression				
RNA-seq	Method uses polyT fractionated or RNA fragments to convert to a library of cDNAs, which are PCR amplified and sequenced. Different libraries need to be constructed for the type of RNA quantified (small RNA, <200 bp, or large RNA, >200 bp) based on size selection	Genome-wide snapshot of transcript level in the sample of coding and noncoding regions	Transcriptomic profile with polyT-based fractionation does not include several non-polyT-based noncoding transcripts. Requires significant depth to detect expression of noncoding RNAs	[18]

2.1 EPIGENOMIC PROFILING METHODS

2.1.1 DNA methylation

Methylation of cytosine bases in the DNA (DNA methylation) is the most widely studied portion of the epigenetic landscape of human cells. In mammalian cells DNA methylation is found almost exclusively at CpG loci (a cytosine followed by a guanine). It has been shown that methylation profiles are deeply involved in the regulation of gene expression, cellular differentiation, and embryonic development. The disruption of such patterns has been associated with several human pathologies, such as cancer, neurodevelopmental disorders, and neurodegenerative disorders. Due to the relevance of methylation profiling for understanding human biology, several techniques have been developed to assess the methylation state of cytosines in a genome-wide fashion.

The most comprehensive method for detecting DNA methylation is whole genome bisulfite sequencing (WGBS). This method, as well as many others, involves treating purified and fragmented DNA with the chemical bisulfite, which converts unmethylated cytosines to uracils while methylated cytosines are protected from bisulfite treatment and are not converted. DNA fragments are then PCR amplified, sequenced and reads are mapped to the reference genome. It is then possible to measure the level of methylation in specific cytosine positions by counting the number of reads with a "C" base call in that position and the number with a "T" (uracils are converted to thymines after PCR amplification) base call for the same position. This technique has genome-wide coverage and single base resolution. However, due to the high depth of sequencing necessary to achieve high accuracy of methylation calls and its genome-wide coverage, this technique can be very costly both in terms of sequencing and in terms of computational time and resources needed to process the data.

Other techniques sacrifice the coverage of all cytosines in the genome for a decreased cost and increased throughput. Examples of such techniques are reduced representation bisulfite sequencing (RRBS) and the currently very popular Infinium Human Methylation 450 Bead Chip (450K array). Both of these techniques also use bisulfite treatment to distinguish between methylated and unmethylated cytosines. However, in RRBS instead of sequencing the whole genome, a combination of restriction enzyme treatment and size selection is applied to DNA in order to enrich for genomic regions with high CpG density before sequencing. In the case of 450K arrays, instead of sequencing a bead chip containing probes designed to assess the methylation status of about 480,000 CpGs distributed throughout the genome is used. Note that these methods also have a single base resolution but do not investigate every cytosine position in the genome.

Also worth mentioning are methods that achieve a decreased cost and do not involve bisulfite treatment. Examples of this type of method are methylated DNA immunoprecipitation sequencing (MeDIP-seq) and methylation-sensitive restriction enzyme sequencing (MRE-seq). In MeDIP-seq, antibodies that recognize methylated CpGs are used to immunoprecipitate and enrich for DNA fragments containing methylated CpGs. After sequencing and mapping of reads, genomic regions with high levels of methylated CpGs will have many reads mapped to them, and regions with low levels of CpG methylation will have few or no reads mapped to them. On the contrary, MRE-seq uses methylation-sensitive restriction enzymes and size selection of DNA fragments in order to enrich for regions with unmethylated CpGs. This makes MeDIP-seq and MRE-seq highly complementary assays and the output of these two methods is often combined to estimate a single base resolution of methylation profile [19,20].

Recent studies have recognized hydroxymethylation of cytosines as an important component of the epigenome. Similar methods as the ones mentioned above have been developed to study that mark [21], but such methods will not be discussed in detail here.

2.1.2 Alternative histones and histone modifications

Histones are found in complexes called nucleosomes. Each nucleosome is comprised of eight histones (usually two copies of H2A, H2B, H3, and H4) bound by 147 bp of DNA. Many chemical modifications can be found in the tails of the histones. These include, but are not limited to acetylation, mono-, di-, or tri-methylation and ubiquitination, and can occur in different amino acids in the tails of the different histones. Further, alternative versions of specific histones are sometimes observed in the nucleosomes. It is clear that the combination of all these epigenetic modifications can lead to very complex patterns that have been shown to associate with cellular differentiation, embryonic development, and regulation of gene expression.

The identification of genomic regions that are associated with modified histones or alternative histones is usually achieved through chromatin immunoprecipitation (ChIP) studies with antibodies for specific modified histones. Early studies focused on chromatin marks in one particular genomic locus, and later looking at a variety of them using microarrays after the immunoprecipitation (ChIP-Chip). However, with the decreasing cost of DNA sequencing, ChIP-seq is currently the most widely-used assay for histone modification profiling. It uses antibodies that specifically bind to histones with a particular modification and co-immunoprecipitates the DNA fragments bound to the modified histone (by cross-linking). The immunoprecipitated fragments are then sequenced and mapped to a reference genome. Regions with large number of mapped reads will then correspond to those where DNA was wrapped around the histones carrying specific marks.

2.1.3 Chromatin accessibility

Another layer of epigenomic information relates to the degree of accessibility of different regions of the genome. The genomic DNA in the nucleus generally highly compacted. However, different regions of the genome can be in more or less compact states. It has been shown that many active genomic regions tend to be in a less compact state, since it must be possible for the transcription machinery to recognize DNA motifs and bind to those regions. As one would expect, different cell types will have distinct profiles of chromatin accessibility, making this epigenomic layer another important determinant of cell type identity.

The most commonly used assay to profile chromatin accessibility genome-wide is called DNase I hypersensitive site sequencing (DNase-seq). It begins by digesting genomic DNA with low levels of DNase I enzyme. Genomic regions that are in a less condensed chromatin state will have a higher rate of digestion by DNAse I. Biotinylated linkers are then bound to the digested ends of DNA and using them such DNA fragments are purified and sequenced. Reads are then mapped to a reference genome. Regions with a high number of mapped reads correspond to those with higher chromatin accessibility.

2.1.4 Gene expression

Transcription of both coding and non-coding genes is closely intertwined with epigenomic states. Many assays have been developed to profile the transcriptome. Not long ago microarrays were the driving force in this field. However, in recent years, array methods have been largely supplanted by sequencing based methods, referred to as RNA-seq. This type of method begins with the extraction of RNA from a population of cells (or possibly from a single cell). Many methods exist for RNA extraction, and the chosen method will depend on the type of RNA species that one is interested in profiling. In case of polyadenylated transcripts, the extraction is followed by reverse transcription of RNA into cDNA, which is then followed by sequencing. Due to the many variations of this method, we chose not go into details about this technique in this chapter, but to point the reader to more in-depth reviews on the topic [22,23].

2.1.5 Chromatin conformation

The assays and data types mentioned so far only provide a one-dimensional view of the epigenomic landscape of a cell. However, it has been shown that the three-dimensional conformation of the genomic DNA in the nucleus is also highly significant for the behavior of a cell [24]. Three-dimensional chromosomal conformation is also highly cell type specific and can be disrupted in disease states.

The three-dimensional architecture of the genome can be profiled through assays categorized as chromosomal conformation capture assays. The Hi-C method leads to genome-wide profiles of long range DNA interactions. Other techniques follow many of the same concepts. The Hi-C method begins with cross-linking of genomic DNA and associated protein to capture native long range interactions. After that, a restriction enzyme digestion step is performed followed by end filling with the addition of a biotinylated base. DNA fragments from interacting genomic regions are then ligated together and purified through their biotin tags. Sequencing using the paired-end approach, the two ends of the same DNA fragment in a Hi-C library will come from different genomic regions that were in nearby at the time of cross-linking. Each end of the fragment is then mapped to the reference genome, providing evidence for the interaction between the specific genomic regions. The number of paired-end reads that map to each pair of genomic regions is used as a measure of the level of interaction between those regions.

2.1.6 Epigenomic projects and publicly available epigenomic data

Several large scale national and international epigenome projects have generated and analyzed a large number of complete epigenomes of normal and cancer cell types. Table 7.2 summarizes these projects and provides a description of their goals. The timelines of these projects are also displayed in Figure 7.1B. Such projects have made a significant contribution to the standardization of protocols, data processing and analytical tools used to mine epigenomic data. Further, they have significantly improved the metadata standards used for epigenomic data, which will be highly significant for the integrative epigenomic analyses soon to come. Finally, large amounts of data are now publicly available. Links to these data resources are provided in Table 7.2. Summary of the types of epigenomic maps generated by these projects are shown in Figure 7.1C. A total of 3436 sequencing-based epigenomic profiles

Table 7.2 Summary of Large-Scale Epigenome Projects

Projects	Description	Data portal	References
Roadmap Epigenomics Project	Generated reference human epigenome profiles across >100 different cell types and tissues	http://www.roadmapepigenomics.org/	[25,26]
International Human Epigenome Consortium (IHEC)	Extends the efforts of the Roadmap Epigenomics Project to generate 1000 reference maps of human epigenome	http://ihec-epigenomes.org	[27,28]
The Cancer Genome Atlas (TCGA)	Perform comprehensive genomic and epigenomic characterization in more than 20 types of human cancers.	http://cancergenome.nih.gov/	[2]
International Cancer Genome Consortium (ICGC)	Extends the TCGA project to generate comprehensive catalogues of genomic and epigenomic characterization of more than 50 different cancer types.	http://dcc.icgc.org	[27]
Encyclopedia of DNA Elements (ENCODE)	Determines functional elements in the human genome by generating transcription factor and maps of chromatin modifications	http://encodeproject.org/ENCODE/	[29]

Adapted from Rivera and Ren [6].

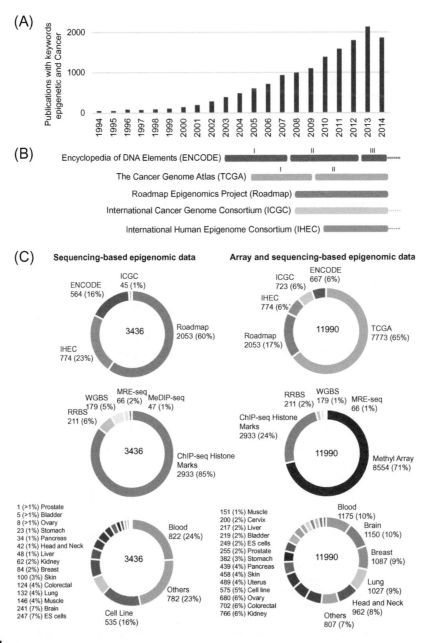

FIGURE 7.1

Summary of publicly available epigenomic datasets generated by large-scale projects. (A) The number of
PubMed entries by year in past two decades on cancer epigenetics. (B) Timeline of the large-scale epigenome
projects. (C) Distribution of the current epigenomic data sets generated by the consortium, by the epigenomic
method and by the cell types.

were generated by these projects. This number significantly increases to 11990 if we include array based epigenomic data mainly contributed by The Cancer Genome Atlas (TCGA) project. Regarding histone modification data, the vast majority of samples contains profiling of H3K27me3, H3K36me3, H3K4me1, H3K4me3, H3K9me3, and H3K27ac, which were primarily generated through the efforts of the Roadmap Epigenomics Project (25). The figure also provides distribution of various cell types for which epigenomic data is available. Also, NCBI Epigenomics portal and Ensembl portal have additional epigenomic dataset generated from other individual projects. With the availability of such data and with the use of proper tools, it possible to mine such data.

3 CURRENT TOOLS AND ANALYSES THAT CAN EXTEND OUR INSIGHTS INTO EPIGENETIC ABERRATIONS IN CANCER

Epigenomic alterations can be the cause or important steps in the progression of cancer. Epigenome-wide association studies (EWAS) attempt to identify epigenomic alterations associated with particular phenotypes. Although the concept may seem simple, many challenges are involved in the appropriate design and analysis of this type of study.

Despite the wide applicability of EWAS and the proven informativeness of the different epigenomic marks, the majority of the studies of this type performed in recent years has considered only DNA methylation. This situation arises from the fact that many layers of the epigenome cannot currently be probed at the scale and with the accuracy necessary for this type of study. However, alteration in such epigenomic marks can still be investigated in studies performed under more controlled conditions.

3.1 EPIGENOME-WIDE ASSOCIATION STUDIES

In the last decade, many studies have focused on finding associations between genomic mutations and particular phenotypes. Many such studies referred to as genome-wide association studies (GWAS), were performed by probing the whole genome of a large number of individuals and looking for loci that associate with the phenotype of interest.

EWAS have recently emerged as a method for understanding the epigenomic variability associated with health, disease, and with the response to environmental factors. Epigenomic changes promise to explain much of the phenotypic differences between individuals that could not be explained by genetic variation alone. When compared to the genome, the epigenome is in principle easier to modify by drugs or even by changes in lifestyle. Therefore, associations identified between diseases and the epigenome have a greater potential of giving rise to new treatments or preventive strategies than those associations found in GWAS. However, despite the great potential of EWAS for human health and disease, there are many challenges that need to be overcome when performing this type of study.

As we mentioned in the previous section, the epigenome is composed of many layers of biological information. However, due to the complexity involved with gathering and properly analyzing other epigenomic data types, DNA methylation is the focus of the vast majority of EWAS. DNA methylation is a more stable epigenomic mark and the assays used to probe it are more quantitative and reproducible than those probing histone marks.

The first tip for designing an EWAS is to have a clear hypothesis in mind. This will help determine what tissue or cell type to profile, possible confounding factors that must be accounted for when selecting individuals to participate in the study, as well as greatly assist in the interpretation of the results of the study.

One must note that when analyzing epigenomic data the appropriate choice of cell type or tissue type is very important. Different cell types have very distinct epigenomic profiles, and the effects of a disease state or an environmental factor over their epigenomes are likely very different. Furthermore, when dealing with complex tissue samples (consisting of a combination of different cell types), one must remember that methylation profiling techniques generally measure the average methylation per site across all the cells in the input sample. Therefore, changes in cell type proportions between different samples will lead to differences in the detected methylation profiles. A clear solution for dealing with this issue is to purify the cell type of interest prior to the analysis. However, such approaches are costly and time consuming, making them unfeasible for most EWAS. Further, in many situations, determining particular surface markers that can be used to purify specific cell types is not a trivial task. In light of the difficulties in solving this problem through purification, a few computational approaches have been proposed to address this issue, but even though such techniques work well for specific situations, correcting the bias that comes from shifting cell type proportions in the general case is still an open problem.

Other sources of bias can also have an effect on the results of EWAS. The epigenomic profile of an individual can be affected by age, gender, exposure to environmental factors, and by the genetic background of that individual. Therefore, the population structure of the subjects involved in an EWAS should be very carefully considered to control for such confounding factors. When performing an EWAS, one should either select a population that is homogeneous in terms of these characteristics, or profile a population large enough to allow stratification or statistical correction for such factors. Further, technical biases can also lead to spurious associations in EWAS. Thus, balanced study designs should be implemented, and samples should be processed under strictly standardized conditions.

After the data has been gathered in an appropriate manner, statistical analyses must be applied to the dataset in order to identify associations between methylation and the phenotype of interest. The most basic way of performing such analysis is by using univariate tests of association, such as t-tests or linear regressions in a site-by-site manner. However, the measurements of DNA methylation at the single site level can be noisy. Therefore, since methylation statuses tend to be similar among nearby CpGs, many researchers tend to perform their association tests focusing on changes in the level of methylation across longer regions of the genome. This approach generally leads to greatly improved specificity. Nonetheless, if a true association does involve only an individual CpG site, such methods will likely miss it. Note that either in a site-by-site approach or approaches focusing on longer regions, multiple hypothesis testing correction is essential. In general false discovery rate based correction is recommended for this type of study.

Despite our best efforts to avoid them, such studies will always find at least a few spurious associations. Therefore, once one has identified regions with statistically significant associations with the phenotype it is important to interpret the possible functional effect of such associations, as well as to validate the identified associations in an independent dataset using a different type of methylation profiling assay. This will guarantee that the results are indeed reproducible and that they are not due to technical biases specific to the assay originally used. For additional discussion about the analysis and interpretation of EWAS, the reader should consult recently published guidelines [30].

3.2 DIFFERENTIAL EPIGENOMIC SIGNALS BEYOND EWAS

The term association studies refers to situations where one is trying to discover the molecular basis of traits commonly found in the population without using model organisms under a carefully controlled environment. Many variables change between the different subjects involved in this type of study, not only the variable of interest. Hence large population sizes and very careful planning are involved in such studies. The use of inherently noisy assays, such as ChIP-seq and DNAse-seq, would make such studies that much more complicated and is thus not commonly done. Nonetheless, in more carefully controlled environments, enrichment-based assays can be a powerful tool, and can lead to important biological discoveries.

As we mentioned above, methylation assays with single nucleotide resolution give us a measure of the percentage of DNA fragments that are methylated at each cytosine position that was covered in the experiment. Conversely, in the case of enrichment-based assays, such as ChIP-seq and DNAse-seq, the signals coming from the assay do not always provide a highly accurate measurement of a real biological variable. Such measurements can sometimes be interpreted only relative to each other or relative to a background distribution, while carefully controlling for many possible sources of bias. For a detailed description of the possible biases affecting enrichment based epigenomic methods and how to adequately handle them see Ref. [31].

Furthermore, the genomic regions where these epigenomic signals occur are not clearly defined *a priori*. Thus, an essential step in any comparative study using this type of data is to define the regions over which such comparisons will take place. The simplest, though commonly applied, approach for determining the genomic regions used in a comparative study is to perform the comparisons over a previously defined set of genomic regions. Gene promoters, enhancers, and gene bodies are common examples of such regions. Many such annotations were gathered and made available by the GENCODE project. A drawback of this type of approach, however, is that changes in the epigenomic profile that occur in regions that were not included in the analysis cannot be identified. Two main techniques have been used to address this problem. The first is to use sliding windows of fixed size that cover the whole genome, and perform the comparisons over these windows. The second and widely adopted approach is to use peak calling algorithms to determine the regions of relatively high signals in such experiments.

Peak calling methods parse epigenomic tracks by identifying regions with significant signal over the background. The background can be defined based on the assumption that read mapping follows a particular probability distribution, but peak calling procedures may become much more accurate when a negative control (usually sequencing of the input library) can be used to estimate the background distribution. An issue that must be considered when applying peak calling algorithms to epigenomic datasets is that different epigenomic marks have distinct patterns of distribution across the genome. Applying the same statistical framework of peak detection to all different classes of epigenomic marks can result in biological misinterpretation [32,33]. Epigenomic marks can possibly be categorized into three classes: Point-source, Broad-source, and Mixed-source [33]. Point-source factors bind to specific positions in the genome. They yield punctate signals and are typically profiles of H3K4me1, H3K4me3, which are active chromatin marks localized at enhancers and transcription start sites. Broad-source factors are associated with large genomic domains. Therefore, they result in broad peak signals and are typically observed in H3K36me3 and H3K9me3 chromatin mark profiles which are involved in chromatin elongation and repression. Finally, mixed-source factors bind at small genomic domains in some parts of the genome and broader domains in others. These are observed for chromatin accessibility and H3K27me3 chromatin mark profiles. Different algorithms are optimized for detecting different types of peak sources. However, detecting broad peaks is still a challenge. Several well-known peak calling tools exists that perform such analysis - examples of such tools are MACS, HOTSPOT, CisGenome, and F-Seq. For more details about peak calling tools and algorithms see elsewhere [34,35].

Once the regions where the comparisons are going to be made are defined, one can analyze the differences in read count over those regions between different conditions. Note that although large differences in the binding signal can be detected with a small number of biological replicates, detecting less extreme changes in the signal may require more replicates. Also, as we mentioned above other sources of bias may confound such comparisons. A number of tools, such as edgeR and DEseq, have been developed to perform such comparisons in a manner that accounts for certain biases.

3.3 ANNOTATING THE GENOME THROUGH EPIGENOMIC PROFILING

Protein coding regions represent only 1% of the genome. Many GWAS performed in the last decade have shown that a large portion of the variants associated with a diverse range of human pathologies actually affect the noncoding portion of the genome, making its understanding essential not only for basic biology but also for medical research. Therefore, a significant effort is now being applied toward the understanding of the noncoding portion of the genome. The study of the epigenome has been invaluable in this pursuit.

Through years of research in epigenetics, studies were able to conclude that different chromatin marks are associated with specific genomic elements and their level of activity. Some examples of such associations are monomethylation of lysine 4 on histone H3 (H3K4me1) marking active enhancer regions and trimethylation of lysine 4 on histone H3 marking active promoter regions. By profiling such marks in a genome-wide fashion, researchers are now able to classify genomic regions into different categories of genomic elements based on their epigenomic profile. Further, by generating epigenomic profiles of different cell types, we can identify genomic elements that display different levels of activity between different cell types. Note, however, that such associations between marks and genomic elements are not absolute, and our knowledge about them is still not complete.

The type of study mentioned above can be performed based on a single epigenomic mark or by analyzing multiple marks simultaneously. Peak calling algorithms, such as the ones described in the previous section, can be used to identify regions with high signal over background for individual chromatin marks. Such peaks can then be used to assign a particular annotation to that genomic region based the mark being profiled. However, it is now clear that the co-occurrence of multiple marks is highly relevant for the activity of multiple genomic elements, making annotations generated by analyzing multiple epigenomic marks simultaneously more advantageous for this purpose.

Most approaches to annotating the genome based on multiple epigenomic marks begin by identifying recurrent patterns of epigenomic signals that consistently appear throughout the genome. To accomplish this, one usually divides the genome into non-overlapping windows of a fixed size and groups them based on their multi-mark epigenomic profile. Spark is one such tool that uses k-means clustering to identify groups of genomic regions with similar patterns of epigenomic signals (Figure 7.2C) [36]. The ChromHMM tool, on the other hand, trains a Hidden Markov Model with a given number of states in a way that each state in the model will be associated with a particular combination of epigenomic signals [37]. The chromatin states can be associated with particular genomic annotations based on the overlap of genomic regions that belong to that state and previously known genomic elements (Figure 7.2D). Previous knowledge of chromatin biology can also inform us about what each of these states represents. Once the possible chromatin states are defined, one can then classify new genomic regions into these different states based on their epigenomic profile.

Projects such as the NIH Epigenomics Roadmap consortium and ENCODE have recently generated and analyzed epigenomic profiles of multiple chromatin marks across a wide range of cell types (Figure 7.2) using the methods described above. In fact, IHEC is currently in the process of generating

(A) Data Slice

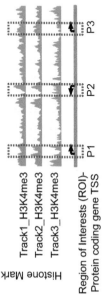

(C) Spark

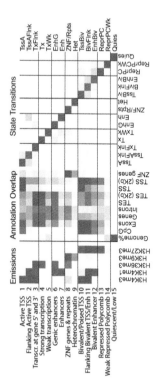

(B) Heatmap/Clustering

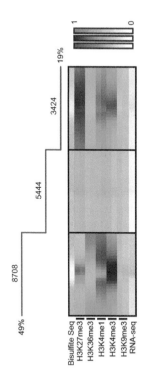

(D) ChromHMM

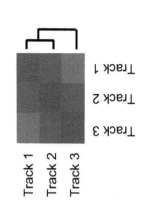

FIGURE 7.2

Common methods used for comparative epigenomic analysis. (A) A "Data Slice" consists of average signals over genomic regions. For illustration, data slice is shown for three tracks over three genomic regions (P1, P2, and P3). Data matrix displays the signal values for each track and region. (B) Heatmap/Clustering tool takes in the data slice and performs unsupervised hierarchical clustering to group samples based on similarity. (C) Spark tool also performs clustering but instead of samples it clusters regions based on different epigenomic patterns. It performs supervised k-means clustering to groups regions. In the illustration, protein coding promoters are grouped into three clusters based on different epigenomic patterns. (D) ChromHMM tool classifies the binned genomic regions into distinct chromatin states based on combinations of histone modification patterns. ChromHMM panel adapted from (40).

1000 complete epigenomes of human cell and tissue types. These projects have led to a leap in our understanding of the noncoding portion of the genome. Determining the number of existing chromatin states is still a challenge, since that number often needs to be adjusted based on the number and identity of histone marks that are profiled. Furthermore, our knowledge of cell-type specific functional regulatory regions is limited by the number of epigenomic profiles available for these different cell types. Nonetheless, as more epigenomic data becomes available across a larger diversity of cell types, our knowledge about the noncoding portion of the genome will rapidly increase.

3.4 INTERPRETATION OF GENOMIC AND EPIGENOMIC VARIABILITY THROUGH ENRICHMENT ANALYSES

Enrichment analyses can be generally referred to as methods that investigate whether a particular set of genes or genomic regions (obtained from epigenomic comparison studies for example) have a statistical significant overlap (is enriched) with sets of previously known biological features.

An example of this type of analysis is the gene set enrichment analysis. Examples of reference gene sets are those made up of genes involved in particular biological processes or activated in response to a particular genetic or experimental perturbation. The Gene Ontology is a great resource, containing a wide diversity of such gene sets. As an example of this type of analysis, imagine one performs a EWAS that identifies a set of genes that, through a change in the methylation of their promoter region, get repressed under certain conditions. Through gene set enrichment analysis, one could then interpret the result of the study by comparing the identified group of genes with sets of genes involved in different biochemical pathways. Through such analysis, one could determine whether any known pathways seem to be specifically involved in the disease being studied. Several tools exist for performing this type of analysis. For instance, DAVID takes a gene list as input and searches through many previously defined gene sets for those that are enriched in the gene list provided as input.

A problem that frequently comes up when trying to apply gene set enrichment analysis to regions defined through epigenomic comparisons is that it is not always straightforward to determine what are the genes affected by particular changes in the epigenome, since most epigenomic variability involves non-coding regions. When epigenomic alterations happen in the promoter regions or regions overlapping a gene, it is reasonably safe to assume that such change somehow affects that gene. However, when the alterations happen in genomic regions that are far from any gene, the interpretation of the effect of such variation is not as straightforward [38]. The GREAT tool tries to address this issue by assigning a gene to each of the input regions based on the linear genomic distance between the gene and the given genomic region [39]. It then performs enrichment analysis using gene sets from multiple biological databases. This approach seems to give good results in most cases, but determining the genes affected by epigenomic alterations in non-coding regions is still a highly relevant and open problem.

The interpretation of genomic and epigenomic variability in the non-coding portion of the genome is, in fact, one of the major objectives of projects such as ENCODE and Roadmap Epigenomics. As we mentioned in the previous section, through epigenomic data analysis, such projects have been able to annotate a large part of the non-coding portion of the genome. Binding sites of many transcription factors in a reasonably large number of human cell types are now known. Further, by profiling different chromatin marks in a wide variety of cell types, it has been possible to identify enhancer regions, as well as regions that are transcribed into non-coding RNAs, and to determine in which cell types such regions are active or inactive based on chromatin state information [40]. Sets of such annotations can also be used in enrichment analyses in a similar fashion as gene sets. A few tools are available to perform this type of analysis, but many of the publications in which they are performed use in-house scripts for them [41].

A common type of enrichment analysis, that uses the aforementioned annotations, is to determine whether a given set of genomic regions is particularly enriched for binding sites of specific transcription factors/epigenetic modifying enzymes or for sequence motifs to which particular transcription factors are known to bind. This can be used to determine whether regions affected by genomic or epigenomic alterations are specifically regulated by that transcription factor. As an example, imagine one wants to determine the functional effect of a particular drug. When comparing the epigenome of cells treated with the drug and of those not treated with the drug, a particular set of genomic regions is determined to display epigenomic changes. By performing enrichment analysis for transcription factors binding sites, one can determine that many of the regions affected by the drug treatment are binding sites for a particular transcription factor. From that result, one could infer that the drug likely acts by affecting that transcription factor, which leads to the changes in the epigenome specifically where that factor binds [42,43].

Similar types of analyses can be performed using other types of annotations. One could, for example, verify enrichment of a set of genomic regions for enhancer regions specifically active in different cell types. Suppose one wants to determine what cell type is affected by a particular disease, and that one has a list of genomic variants associated with that disease that fall in the noncoding portion of the genome. One could verify the enrichment of these genomic variants over enhancer regions that are specifically active in each of the many cell types profiled in the Roadmap Epigenomics Project. If such enrichment were found for enhancers active in a particular cell type, one could then infer that the disease affects that same cell type [44].

Another very similar type of analyses that can be performed using cell type specific regulatory annotation is to prioritize genomic variants identified in cancer. This type of technique is different from enrichment analysis, but can also be used in the interpretation of genomic variants in cancer. Thousands of variants are normally identified by sequencing the cancer genome, and it is difficult to determine which of those are truly involved the progression of the disease. It is reasonable to assume that variants overlapping active regulatory regions are more likely to have functional effects than variants that affect inactive regions of the genome. Such variants can, therefore, be assigned a higher priority for further investigation of their involvement in the cancer progression. Regulatory regions that are specifically active or inactive in different cell types can be inferred from the profiles produced by the Roadmap Epigenomics Project [38].

Note that the types of analyses mentioned here are by no means a comprehensive description of all the possible ways to interpret and make use of epigenomic data. The increasing availability of such datasets, as well as the creativity of individual researchers, will undoubtedly lead to many new applications of this type of analysis.

3.5 SAMPLE CLUSTERING AND CLASSIFICATION

We mentioned above that the Spark tool that clusters genomic regions based on their multimark epigenomic profiles. In this section, we would like to point out that not only genomic regions can be clustered, but samples can also be grouped together based on their epigenomic profiles across a set of genomic regions. Many algorithms are available to cluster objects based on the similarity measures between sets of features associated with them. This very general technique is applied in many areas of science. Typical examples of this type of algorithm are hierarchical clustering and k-means clustering, but many other possibilities exist. The choice of clustering algorithms, the similarity metric used to compare the features associated with each object, and other input parameters may affect the final result of this type of analysis. The choice of such parameters needs to be made on a case by case basis, as no single parameter set can be thought of as the best in general.

This type of analysis has been successfully applied to many different situations, such as for defining molecular subtypes of cancer based on their gene expression, methylation, or other epigenomic profiles [5,45,46]. Such subtypes have been shown to be associated with survival and are even used to determine best treatment options for individual cases. Once groups of samples have been defined, it is possible to determine the group to which a new sample belongs. The problem of determining the cancer subtype of a new cancer sample based on predefined subtypes is generally referred to as classification. As with clustering, this is a very general problem, and many algorithms are available to perform this type of analysis. The main difference between classification and clustering is that in classification the possible classes (groups) that can be assigned to a particular sample are defined *a priori*.

3.6 USE CASES ILLUSTRATING TYPES OF EPIGENOMIC ANALYSES

We will use the epigenome toolset within the Genboree Workbench to illustrate several types of epigenomic analyses. The goal of the following use cases is to give examples of real applications of the techniques described above using publicly available tools hosted in the Genboree Workbench. Step by step instructions for these use cases can be found at www.genboree.org/site/epigenomics_toolset.

3.6.1 Use Case—Inferring the Most Likely Cell Type of Origin for Breast Cancer Cell Lines

Background: Breast epithelial cells undergo dramatic epigenomic alterations during cancer progression. Breast cancer cell lines originate from a specific cancerous cell that is extracted from a breast tumor. That cell is then propagated in culture for a very long time. During its extraction and especially during its growth in culture, the epigenome of such cells can become highly distinct from the original cell that gave rise to that cell line. This happens because culture conditions are highly distinct from the environmental conditions this cell would have found in vivo. The objective of this use case is to determine, despite the large amount of epigenomic differences that arise both during cancer progression and during growth in culture of breast cancer cell lines, whether we can still determine the type of breast epithelial cell from which the cell line originated.

Dataset: In this use we will use the methylation profiles of 7 breast cancer cell lines (4 ER+ and 3 ER-), as well as that of a normal breast cell line, generated using the RRBS technique [47]. We will also use the methylation profiles of three different types of purified breast epithelial cells: luminal epithelial cells, myoepithelial cells, and epithelial stem cells. These purified breast epithelial cells were profiled using the MeDIP-seq technique as part of the Roadmap Epigenomics Project.

Approach: This use case begins by identifying a set of genomic regions that have highly distinct methylation signal across the three types of normal breast epithelial cells. For this, the LIMMA tool on the Genboree Workbench can be used. The idea is that by comparing the methylation signal of breast cancer cell lines against the normal breast cell types over a set of regions for which the three normal cell types are highly different from each other, we will get a more clear picture of what is the normal cell type that is most similar to each of the breast cancer cell lines.

Once the set of regions of interest (ROI) is defined, the next step is to compute the average methylation signal for each sample included in our analysis over these ROIs, generating a data slice (a data matrix with columns being samples and rows being average methylation signals for each ROI). Quantile normalization is then applied to this data slice matrix in order to make the methylation measurements over different samples more comparable. Once that is done we will compute the correlation between the methylation profiles over the chosen ROIs of each of the breast cancer cell lines against each of the normal breast epithelial cells. A hierarchical clustering with Euclidian distance metric will then be performed in order to group breast cancer cell lines according to their similarity to the normal breast epithelial cells, as well as to group the normal breast epithelial cells according to their similarity to breast cancer cell lines. A heat map in which the color corresponds to the level of correlation between a pair of samples will be plotted. The rows and columns of this heatmap will be sorted according to the result of the hierarchical clustering procedure, and dendrograms will be added to the sides of the heat map to indicate the clustering of the samples. Compute Similarity Matrix (heatmap) tool in the Genboree Workbench performs this analysis based on user supplied input of regions of interest and genome-wide methylation profiles.

Result: As can be observed in Figure 7.3A, all breast cancer cell lines as well as the normal breast cell line had their best match being the luminal epithelial cells from breast. This result indicated that despite all the epigenomic changes that occur in breast cancer cell lines, one can still identify its cell type of origin as being a luminal epithelial cell. Further, one can observe that ER+ and ER- breast cancer cell lines, as well as the normal breast cell type all seem to be of luminal epithelial origin.

3.6.2 Use Case—Comparing Methylation Profile of Normal Breast and Breast Cancer Samples Against Immune Cell Reference Epigenomes to Examine Infiltration by Immune Cells

Background: Complex tissue samples are typically constituted by several different cell types, and different samples from the same tissue may have different proportions of constituent cell types. In breast cancer, for example, immune cell infiltration is often observed. True methylation changes between cases and controls in a particular cell type may thus be masked by differences that result from changes in proportions of constituent cell types found in breast cancer. Comparison of breast cancer samples with reference epigenomes from the Roadmap Epigenomics Project provides an approach to help identify shifts in cell type composition in complex tissues.

Dataset: 57 WGBS profiles across 45 different cell types were obtained from the Roadmap Epigenomics Project. We also obtained methylation profiles generated using the Illumina 450K array for 16 breast tissue samples (8 normal breast samples and 8 primary breast tumor samples) [9].

Analysis approach: Instead of comparing the methylation profiles in this analysis over all CpG positions probed in Illumina 450K array, we chose to perform the comparisons over a subset of those CpGs that were previously found to display highly cell type specific patterns of methylation (defined by comparing methylation profiles of all cell types analyzed in the Roadmap Epigenomics Project) [48]. This sub-setting reduces the noise introduced by uninformative CpGs in the comparison. We will begin this use case by slicing both the 450K array breast samples and the reference epigenomes over this same set of CpG positions (calculating the average levels of methylation for each sample over each

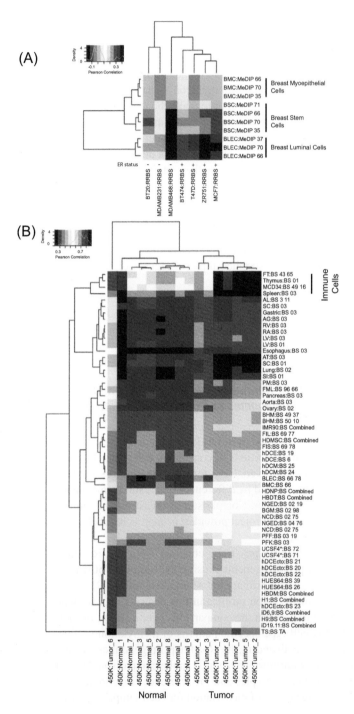

FIGURE 7.3

Use cases involving comparison of methylation profiles. (a) Comparison of 4 ER (+) and 3 ER (-) breast cancer cell lines (RRBS, rows) against the reference epigenomes of normal breast cell types (MeDIP, columns) from the Roadmap Epigenomics Project. Methylation comparisons are made over set of CpG islands to assess the cell type of origin of breast cancer cell lines. (b) Detection of shifts in the cell type composition of complex tissues. Illumina 450K array methylomes of breast tumors and normal mammary tissues (columns) are compared against MeDIP profiles from the Roadmap Epigenomics Project (rows) using the set of lineage-specific marker CpGs from the 450K array as ROIs. Breast tumor samples have much higher similarity to immune cell methylomes than normal breast samples, indicating infiltration of breast tumors by immune cells.

of the chosen CpGs). We will then compute the correlation between the methylation profiles of each of the breast samples against the methylation profiles of the cell/tissue types analyzed in the Roadmap Epigenomics Project. Hierarchical clustering (Euclidian distance) based on this correlation matrix will then be performed in order to group the breast samples based on their similarity to the different cell/tissue types in the Epigenome Atlas. A heat map will be plotted in which the colors indicate the level of correlation between each pair of samples. The rows and columns of the heatmap will be sorted based on the hierarchical clustering, and dendrograms indicating the result of the clustering procedure will be displayed on the sides of the heat map. The whole process described here can be performed using the Compute Similarity Matrix (heatmap) tool in the Genboree Workbench.

Results: We find that 450k array profiles of tumor samples are more epigenetically similar to immune cell samples from the Epigenome Atlas than the normal breast profiles (Figure 7.3B). This indicates, in agreement with previous knowledge of breast cancer biology, an increase in the proportion of immune cells in the tumor samples.

3.6.3 Use Case—Studying Epigenomic Aberrations in Promyelocytic Leukemia to Determine Biological Pathways That Are Affected

Background: Studying regions of epigenomic changes in normal versus disease state can help determine pathways that are involved in disease progression. In particular, changes in the epigenomic landscape identified in the comparison of the epigenomic profile of a tumor with an appropriate normal reference epigenome can be used to indicate biochemical pathways that may be involved in the progression of the disease. To illustrate this we will examine epigenomic differences between HL60, a human myeloid leukemia cell line, its closest matching normal reference epigenome (as a surrogate for a control sample), and a hematopoietic stem cell (CD34+) from the Epigenome Atlas. The idea is to determine whether we can indeed observe the previously reported pattern of dedifferentiation that occurs in leukemia cells. If such pattern is present, we expect to observe that for particular genomic regions the HL60 epigenome should be more similar to the hematopoietic stem cells then to the more differentiated myeloid cell type.

Dataset: In this use case we used epigenomic profiles produced by the Roadmap Epigenomics Project and IHEC. ChIP-seq profiles of multiple histone marks (H3K4me1, H3K4me3, H3K27me3, H3K36me3, and H3K9me3) for three cell types (HL-60, CD14+ cells and CD34+ cells) are used in this analysis. H3K4me1 ChIP-seq profiles of other immune cell types from the Roadmap Epigenomics Project are also included in this use case.

Analysis approach: We begin the use case by comparing the H3K4me1 profile of the HL60 cell line against the H3K4me1 profiles of many immune cell types available in the Epigenome Atlas. This is done in order to identify the cell type that has the most similar epigenome to HL60 in order to use it as a surrogate for a control sample. As regions of interest for this comparison, we used the set of 2.5 million enhancer regions previously annotated by the Roadmap Epigenomics Project. By looking at the heat map in Figure 7.4, we can observe that CD14+ cells had the highest similarity with the HL60 cell line among the immune cells included in the comparison. As in the two previous use cases, the Compute Similarity Matrix (heatmap) tool in the Genboree Workbench was used to perform this comparison.

The next step in this use case is to analyze the pattern of changes in the signals of histone marks across three cell types: HL60, CD14+ cells (its closest matching reference epigenome), and CD34+ cells (hematopoietic stem cells). The idea is to determine whether there is a large number of genomic regions in which the epigenomic profile of the HL60 cell line becomes more similar to the stem cell profile then to the more differentiated CD14+ cell. The Spark tool from the Genboree Workbench was

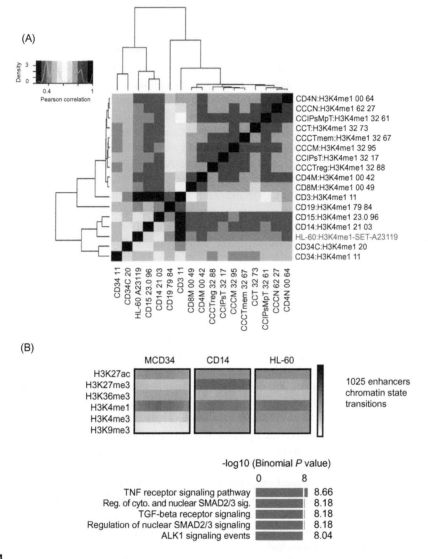

FIGURE 7.4

Use case about the interpretation of epigenomic aberration. (A) Comparing H3K4me1 ChIP-seq profiled in HL-60 (Promyelocytic leukemia) with H3K4me1 ChIP-seq profiles from other samples profiled in the Roadmap Epigenomics Project. Cluster analysis identifies the cell type of origin and the closest reference epigenomes for the promyelocytic leukemia cell line HL-60. HL-60 clustered with the myeloid reference epigenomes (CD14 and CD15) based on H3K4me1 signals over enhancers. (B) Spark tool is used to study epigenomic perturbations at myeloid specific enhancer regions that occurred in HL-60 using reference epigenomes of myeloid cells (CD14 and CD15) and hematopoietic stem cells (CD34). Spark analysis shows changes in 1025 of the ~37,000 myeloid specific enhancer regions. The most prominent change is marked by the reversal of H3K27me3 marks, consistent with known de-differentiation mechanisms in promyelocytic leukemia. GREAT tool was used to analyze pathway enrichment for enhancer regions that underwent epigenomic perturbations in HL-60. Significant pathways (FDR corrected binomial P-value <0.05) are biologically relevant for promyelocytic leukemia.

used to cluster enhancer regions based on their epigenomic signal across 5 different histone marks and the three different cell types. We observe that a particular cluster of enhancer regions (containing 1025 enhancers), displays a pattern of epigenomic signals in which both HL60 and the hematopoietic stem cells show a low level of H3K27me3, while the CD14+ cells show a high level of that same mark.

In order to understand what pathways are likely affected by such changes in the epigenome, the 1025 enhancer regions that displayed the pattern of change mentioned above were included in an enrichment analysis using the GREAT tool. As a result we can observe a high enrichment for particular pathways that are likely affected by the process of dedifferentiation that occurs in leukemia cells.

Results: Comparison of HL60 with the reference epigenomes from the Roadmap Epigenomics Project correctly identifies HL60 being of myeloid origin, since its closest matching reference profile was an immune cell type of the myeloid lineage (Figure 7.4A). Furthermore, Spark analysis shows that for 1025 enhancer regions the signal of H3K27me3 histone mark in HL60 became more similar to the hematopoietic stem cells then to the more differentiated CD14+ cells, highlighting the pattern of dedifferentiation in the leukemia cells. This pattern of reversal of H3K27me3 is consistent with known dedifferentiation mechanisms in promyelocytic leukemia. Finally, GREAT tool shows the associated pathways that are perturbed in HL-60, and are likely involved or affected by the dedifferentiation process in these cancerous cells (Figure 7.4B).

3.6.4 Use Case—Interpreting Genomic Variation in the Context of Global Chromatin State Patterns

Background: Catalogs of genomic variants have been compiled for various cancer types. However, the effect of these variants on the epigenome remains poorly understood. Understanding of the variants in the context of the epigenome may provide possible mechanistic insights that might not be otherwise uncovered. As an example, one may interpret genomic variants through the maps of global chromatin state patterns. Chromatin maps have been generated for several different cell types. Such maps can be used to generate tentative hypothesis to interpret the effect of genomic variation on the epigenome.

Dataset: We used the Cosmic database to obtain somatic variants cataloged for hepatocellular carcinoma. ChIP-seq profiles in Adult liver for multiple histone marks - H3K4me1, H3K4me3, H3K27me3, H3K36me3, and H3K9me3 were obtained from the Roadmap Epigenomics Project.

Analysis approach: We first determine the chromatin state map of the adult liver by running chromHMM tool in the Genboree Workbench. The tool segments the genome in 200 bp bins and based on the combination of chromatin marks assigns each bin to a particular chromatin state. Here we used 15 state chromatin model to run chromHMM. The results from the analysis output a file containing chromatin state information in a 4 column bed-format providing chromosome, start, stop, and chromatin state information for each 200bp bin in the genome. Next, genomic variants cataloged for hepatocellular carcinoma were obtained from the Cosmic database. The hypergeometric test was used to calculate the p-values for the enrichment of genomic variants for a particular chromatin state.

Results: Chromatin state map of Adult liver profiled in Roadmap Epigenomics Project and their enrichment to annotated genomic regions are shown in Figure 7.5A. Genomic variants are associated with weak transcription (Figure 7.5B). This pattern is expected because most variants affect transcribed protein-coding genes that are marked by H3K36me3.

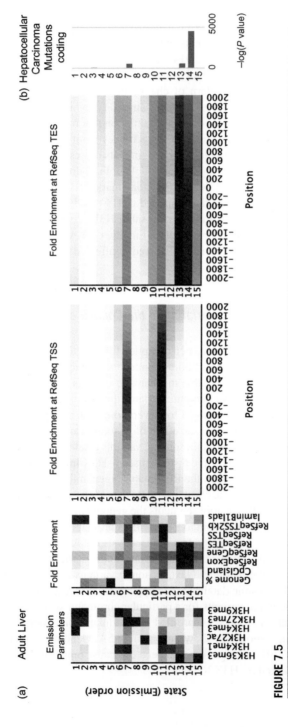

FIGURE 7.5

Interpretation of genomic variants in epigenomic context. (A) Using the Genboree Workbench, chromatin state information for adult liver (profiled by the Roadmap Epigenomics Project) was obtained by running ChromHMM tool. (B) Somatic variants for hepatocellular carcinoma were obtained from the Cosmic database. Using R statistical program, Hypergeometric test was used to calculate the p-value for the enrichment of genomic variants for each chromatin state.

4 CONCLUSION

The past decade has brought significant improvements to high-throughput epigenomic profiling methods. This has led to a leap in our understanding of the cancer epigenome. As the cost to generate genome-wide datasets decreases and the amount of epigenomic data increases, opportunity arises to mine epigenome-wide datasets and detect epigenetic alterations involved in cancer progression. In this chapter, we have introduced various types of analyses that are commonly performed using epigenomic datasets. However, our survey is by no means comprehensive.

So far, DNA methylation has been widely used for the interpretation of epigenetic perturbations in cancer. It is preferred over other layers of the epigenome because of stability and absolute quantitation. However, with proper interpretation, other layers of the epigenome can provide independent information that complements DNA methylation. For instance, histone modifications and DNase hypersensitivity profiles can provide chromatin state information that can be very informative for understanding the dynamics of cancer disease progression.

Emerging methods that map the three-dimensional structure of the genome may also help better understand how chromosomal conformation is maintained and how it gets disrupted in cancer. Also single cell epigenomic profiling methods promise to dissect the epigenome of individual cells, which will likely lead to understanding of the cellular heterogeneity in cancer and its changes during cancer progression. With better understanding of the topographical structure of the mammalian epigenome, allelic epigenomes, and chromatin state maps, we will be able to better understand the ramifications of the genetic mutations and epigenetic abnormalities that work together to cause cancer.

As exciting as this prospect may be, several challenges remain to be addressed in the coming years. As studies increase in scale and epigenomic dimensions, development of new methods and tools for analysis and data visualization will be required. Also, with increasing amount of data from different large-scale projects, efficient data warehousing and linking will be crucial. For instance, integration of EWAS with GWAS can help determine genetic variants that can influence both epigenetics and disease phenotypes. Currently, there are public databases for epigenomic data, structural variation data (dbVar), genotypes and phenotypes data (dbGAP) which also maintains GWAS data, but no such database currently exists for EWAS. Second, in order to facilitate data linking between databases, metadata standards need to be improved with the use of standard ontologies. The integrative analysis that will be made possible in this way will open new perspectives on the epigenetic landscape of cancer.

REFERENCES

[1] Jones PA, Baylin SB. The epigenomics of cancer. Cell 2007;128(4):683–92.
[2] Garraway LA, Lander ES. Lessons from the cancer genome. Cell 2013;153(1):17–37.
[3] Dawson MA, Kouzarides T. Cancer epigenetics: from mechanism to therapy. Cell 2012;150(1):12–27.
[4] Ntziachristos P, Tsirigos A, Welstead G, Trimarchi T, Bakogianni S, Xu L, et al. Contrasting roles of histone 3 lysine 27 demethylases in acute lymphoblastic leukaemia. Nature 2014;514(7523):513–7.
[5] Mack SC, Witt H, Piro RM, Gu L, Zuyderduyn S, Stütz AM. Epigenomic alterations define lethal CIMP-positive ependymomas of infancy.
[6] Rivera CM, Ren B. Mapping human epigenomes. Cell 2013;155(1):39–55.

[7] Nagarajan RP, Fouse SD, Bell RJA, Costello JF. Methods for cancer epigenome analysis. Adv Exp Med Biol 2013;754:313–38.

[8] Bock C, Tomazou EM, Brinkman AB, Müller F, Simmer F, Gu H, et al. Quantitative comparison of genome-wide DNA methylation mapping technologies. Nat Biotechnol 2010;28(10):1106–14.

[9] Dedeurwaerder S, Defrance M, Calonne E, Denis H, Sotiriou C, Fuks F. Evaluation of the infinium methylation 450K technology. Epigenomics 2011;3(6):771–84.

[10] Pomraning KR, Smith KM, Freitag M. Genome-wide high throughput analysis of DNA methylation in eukaryotes. Methods 2009;47(3):142–50.

[11] Ball MP, Li JB, Gao Y, Lee J-H, LeProust EM, Park I-H, et al. Targeted and genome-scale strategies reveal gene-body methylation signatures in human cells. Nat Biotechnol 2009;27(4):361–8.

[12] Meissner A, Mikkelsen TS, Gu H, Wernig M, Hanna J, Sivachenko A, et al. Genome-scale DNA methylation maps of pluripotent and differentiated cells. Nature 2008;454(7205):766–70.

[13] Meissner A, Gnirke A, Bell GW, Ramsahoye B, Lander ES, Jaenisch R. Reduced representation bisulfite sequencing for comparative high-resolution DNA methylation analysis. Nucleic Acids Res 2005;33(18):5868–77.

[14] Laurent L, Wong E, Li G, Huynh T, Tsirigos A, Ong CT, et al. Dynamic changes in the human methylome during differentiation. Genome Res 2010;20(3):320–31.

[15] Lister R, Pelizzola M, Kida YS, Hawkins R, Nery JR, Hon G, et al. Hotspots of aberrant epigenomic reprogramming in human induced pluripotent stem cells. Nature 2011;471(7336):68–73.

[16] Park PJ. ChIP-seq: advantages and challenges of a maturing technology. Nat Rev Genet 2009;10(10):669–80.

[17] Boyle AP, Davis S, Shulha HP, Meltzer P, Margulies EH, Weng Z, et al. High-resolution mapping and characterization of open chromatin across the genome. Cell 2008;132(2):311–22.

[18] Wang Z, Gerstein M, Snyder M. RNA-Seq: a revolutionary tool for transcriptomics. Nat Rev Genet 2009;10(1):57–63.

[19] Harris R, Wang T, Coarfa C, Nagarajan RP, Hong C, Downey SL, et al. Comparison of sequencing-based methods to profile DNA methylation and identification of monoallelic epigenetic modifications. Nat Biotechnol 2010;28(10):1097–105.

[20] Stevens M, Cheng JB, Li D, Xie M, Hong C, Maire CL, et al. Estimating absolute methylation levels at single-CpG resolution from methylation enrichment and restriction enzyme sequencing methods. Genome Res 2013;23(9):1541–53.

[21] Yu M, Hon GC, Szulwach KE, Song C-X, Jin P, Ren B, et al. Tet-assisted bisulfite sequencing of 5-hydroxymethylcytosine. Nat Protoc 2012;7(12):2159–70.

[22] Rapaport F, Khanin R, Liang Y, Pirun M, Krek A, Zumbo P, et al. Comprehensive evaluation of differential gene expression analysis methods for RNA-seq data. Genome Biol 2013;14(9):R95.

[23] Trapnell C, Roberts A, Goff L, Pertea G, Kim D, Kelley DR, et al. Differential gene and transcript expression analysis of RNA-seq experiments with TopHat and Cufflinks. Nat Protoc 2012;7(3):562–78.

[24] Jost D, Carrivain P, Cavalli G, Vaillant C. Modeling epigenome folding: formation and dynamics of topologically associated chromatin domains. Nucleic Acids Res 2014:gku698.

[25] Chadwick LH. The NIH roadmap epigenomics program data resource. Epigenomics 2012;4(3):317–24.

[26] Bernstein BE, Stamatoyannopoulos JA, Costello JF, Ren B, Milosavljevic A, Meissner A, et al. The NIH roadmap epigenomics mapping consortium. Nat Biotechnol 2010;28(10):1045–8.

[27] (Chairperson) TJH, Anderson W, Aretz A, Barker AD, Bell C, Bernabé RR, et al. International network of cancer genome projects. Nature 2010;464(7291):993–8.

[28] Bae J-B. Perspectives of International Human Epigenome Consortium. Genomics Inform 2013;11(1):7–14.

[29] Consortium TEP. An integrated encyclopedia of DNA elements in the human genome. Nature 2012;489(7414):57–74.

[30] Michels KB, Binder AM, Dedeurwaerder S, Epstein CB, Greally JM, Gut I, et al. Recommendations for the design and analysis of epigenome-wide association studies. Nat Methods 2013;10(10):949–55.

[31] Meyer CA, Liu XS. Identifying and mitigating bias in next-generation sequencing methods for chromatin biology. Nat Rev Genet 2014;15(11):709–21.

[32] Pepke S, Wold B, Mortazavi A. Computation for ChIP-seq and RNA-seq studies. Nat Methods 2009;6(11 Suppl.):S22–32.

[33] Landt SG, Marinov GK, Kundaje A, Kheradpour P, Pauli F, Batzoglou S, et al. ChIP-seq guidelines and practices of the ENCODE and modENCODE consortia. Genome Res 2012;22(9):1813–31.

[34] Wilbanks EG, Facciotti MT. Evaluation of algorithm performance in ChIP-seq peak detection. PLoS One 2009;5(7).

[35] Koohy H, Down TA, Spivakov M, Hubbard T. A comparison of peak callers used for DNase-Seq data. PLoS One 2013;9(5).

[36] Nielsen C, Younesy H, O'Geen H, Xu X, Jackson A, Milosavljevic A, et al. Spark: a navigational paradigm for genomic data exploration. Genome Res 2012;22(11):2262–9.

[37] Ernst J, Kellis M. ChromHMM: automating chromatin-state discovery and characterization. Nat Methods 2012;9(3):215–6.

[38] Greenman C, Stephens P, Smith R, Dalgliesh GL, Hunter C, Bignell G, et al. Patterns of somatic mutation in human cancer genomes. Nature 2007;446(7132):153–8.

[39] McLean C, Bristor D, Hiller M, Clarke S, Schaar B, Lowe C, et al. GREAT improves functional interpretation of cis-regulatory regions. Nat Biotechnol 2010;28(5):495–501.

[40] Roadmap Epigenomics Consortium Kundaje A, Meuleman W, Ernst J, Bilenky M, Yen A, et al. Integrative analysis of 111 reference human epigenomes. Nature 2015;518(7539):317–30.

[41] Favorov A, Mularoni L, Cope LM, Medvedeva Y, Mironov AA, Makeev VJ, et al. Exploring massive, genome scale datasets with the GenometriCorr package. PLoS Comput Biol 2012;8(5):e1002529.

[42] Arrowsmith CH, Bountra C, Fish PV, Lee K, Schapira M. Epigenetic protein families: a new frontier for drug discovery. Nat Rev Drug Discov 2012;11(5):384–400.

[43] Conway O'Brien E, Prideaux S, Chevassut T. The epigenetic landscape of acute myeloid leukemia. Adv Hematol 2014;2014:e103175.

[44] Polak P, Karlić R, Koren A, Thurman R, Sandstrom R, Lawrence MS, et al. Cell-of-origin chromatin organization shapes the mutational landscape of cancer. Nature 2015;518(7539):360–4.

[45] Kulis M, Heath S, Bibikova M, Queirós AC, Navarro A, Clot G, et al. Epigenomic analysis detects widespread gene-body DNA hypomethylation in chronic lymphocytic leukemia. Nat Genet 2012;44(11):1236–42.

[46] Salhia B, Kiefer J, Ross JTD, Metapally R, Martinez RA, Johnson KN, et al. Integrated genomic and epigenomic analysis of breast cancer brain metastasis. PLoS One 2014;9(1):e85448.

[47] Weber M, Hellmann I, Stadler MB, Ramos L, Pääbo S, Rebhan M, et al. Distribution, silencing potential and evolutionary impact of promoter DNA methylation in the human genome. Nat Genet 2007;39(4):457–66.

[48] Amin V, Harris RA, Onuchic V, Jackson AR, Charnecki T, Paithankar S, et al. Epigenomic footprints across 111 reference epigenomes reveal tissue-specific epigenetic regulation of lincRNAs. Nat Commun 2015;6:6370.

EPIGENETICS AND CANCER

DEVELOPMENT OF EPIGENETIC TARGETED THERAPIES IN HEMATOLOGICAL MALIGNANCIES: FROM SERENDIPITY TO SYNTHETIC LETHALITY

8

Thomas Prebet and Steven D. Gore

Department of Internal Medicine (Hematology), Yale Cancer Center, New Haven, CT, USA

CHAPTER OUTLINE

S.G. Gray (Ed): Epigenetic Cancer Therapy. DOI: http://dx.doi.org/10.1016/B978-0-12-800206-3.00008-2

1 INTRODUCTION

The revisionist historical retrospectoscope often assigns scientific wisdom to serendipitous findings. For example, many believe that the transformative success of the introduction of retinoic acid into the treatment of acute promyelocytic leukemia originated from the cloning of the breakpoint at *t*(15;17) revealing a fusion gene involving retinoic receptor alpha [1]. In fact, the cloning of the breakpoint postdated the empiric use of tretinoin in China based on successful *in vitro* differentiation of HL60 cells (which are neither derived from an APL patient nor have aberrant retinoic acid receptors) by retinoids. Similarly, given the approval of 5-azacytidine to treat high-risk myelodysplastic syndrome (MDS), and demonstration of improvement of survival of such patients when treated with azacitidine (5AC), one might assume that this therapy was developed following understanding of the dysregulation of the MDS epigenome. As with APL and retinoic acid, the development of azanucleosides was somewhat fortuitous, and in retrospect based on somewhat fallacious scientific concepts. Nonetheless, the successful development of azacitidine and its cousin drug 2′-deoxy-5-azacitidine (decitabine, DAC) established targeting of aberrant epigenomes as a valid concept in cancer therapeutics and provides a variety of lessons, in some way serving as a cautionary tale [2]. This chapter will review the historic development of putative epigenetically targeted therapies in hematologic malignancies, current understanding of epigenomic dysregulation in hematologic malignancies, clinical utilization of azanucleosides and histone deacetylase (HDAC) inhibitors, and current efforts in targeting the epigenome in bone marrow cancers.

2 THE METHYLOME IN HEMATOLOGIC MALIGNANCY
2.1 MYELODYSPLASTIC SYNDROME

Early interest in the epigenome in hematologic malignancy focused on MDS due to early reports of clinical responses in azacitidine trials. In the 1990s, technology permitted examination of gene-by-gene investigation of promoter methylation using techniques such as methylation-specific PCR, COBRA, and others [3]. The promoter regions of a variety of tumor suppressor genes were found to have highly methylated CpG islands, associated with transcriptional silencing. These included CDKN2B, CDH2, SOCS1, and DAPK [4–6]. Methylation of CDKN2B in MDS suggested inferior outcome. CDKN2B methylation also characterizes therapy-related myeloid neoplasm (tMN), associated with abnormalities of chromosome 7 and poor prognosis [7]. Examining methylation of a variety of tumor suppressor genes, Hokland et al. [8] found that hypermethylation of at least one gene decreased survival in lower risk MDS patients. Similar impact on prognosis was found using a score derived from methylation of 24 genes [9]. Methylation also predicted inferior outcome of cytarabine-based chemotherapy [10].

Using an array-based methylation assay enriched for promoter regions (so-called HpaII tiny fragment Enrichment by Ligation-mediated PCR (HELP) assay), MDS samples showed more extensive hypermethylation than *de novo* AML (acute myeloid leukemia) samples or normal CD34$^+$ progenitor cells [11]. Early deep sequencing studies [12] of the MDS genome identified frequent mutations in genes which directly impact DNA methylation: DNMT3A, TET2, and IDH1 and 2. TET2 was the second most commonly mutated gene in a large series of MDS genomes (22%) [13]. Other mutations which may modify the epigenome frequently mutated in MDS include BCOR, EZH2, and ASXL1 [14]. Overall, almost 50% of MDS cases have mutations which impact DNA methylation, with 30%

of cases bearing mutations involving chromatin modification (ASXL1, EZH2, KDM6A, and ATRX). These two classes of mutations are not mutually exclusive [15]. How mutations which impact the epigenome contribute to the pathogenesis of MDS and how the specific changes in the epigenome dictate the phenotype of MDS remain unanswered questions. The frequent changes in the epigenome must be integrated with abnormalities in RNA splicing, since mutations in components of the spliceosome occur in over 60% of MDS cases [15].

2.2 ACUTE MYELOID LEUKEMIA

Perhaps not surprisingly, evolution of our knowledge of epigenetic abnormalities in AML parallel that of MDS; in many studies, AML and MDS cases were investigated concomitantly. While not impacting the methylome directly, some of the earliest and best examples of cancer genomics dictating epigenomic changes came from studies of the so-called core-binding factor leukemias (*t*(8;21), RUNX1-ETO, Inv(16) or *t*(16;16), CBF-MYH11) and APL (*t*(15;17), PML-RARA). In each of these three leukemias, a fusion gene is formed which includes the DNA binding domain of a transcriptional regulator. The fusion gene recruits transcriptional corepressors rather than coactivators, including HDACs rather than histone acetyltransferases, leading to chromatin compaction and transcriptional silencing [16].

As with MDS, mutations in DNMT3, TET2, IDH1, and IDH2 occur in AML [17]. In fact, methylomic patterns subsequently attributed to the latter three mutations (at that time unknown) were discovered as "unknown" genetic subtypes [18]. EZH2, RUNX1, and BCOR are also mutated in AML. Certain mutations in AML lead to aberrancies in histone methylation. MLL (mixed lineage leukemia)-mutated AMLs figure prominently in this category. Other epigenome impacting mutations include ASXL2 and LSD1. An excellent recent review of the mutational spectrum and epigenetic landscape in AML has recently been published [19].

3 HYPOMETHYLATING AGENTS: THE STRANGE BUT TRUE HISTORY OF THE EARLY DEVELOPMENT OF AZANUCLEOSIDES IN MDS

5-Azacitidine and DAC were synthesized in the 1960s in then Czechoslovakia and were conceived of as classic antimetabolites [20]. Based on Phase II data in refractory and relapsed AML as a high dose S phase specific cytotoxic agent, a new drug application for azacitidine was unsuccessfully brought to the FDA in 1982 by its then sponsor, Upjohn (subsequently merged with Pharmacia and eventually acquired by Pfizer).

Manufacture of azacitidine was subsequently contracted by the National Cancer Institute. The 1979 Annual Report of the National Cancer Institute notes "the pharmaceutical production activity has successfully maintained clinical inventory of high quality investigational agents… [including] 33,674 vials of 5-azacytidine." The ongoing supply of azacitidine through the NCI mechanism through two decades in which this nucleoside had no active pharmaceutical sponsor represents an unsung success story of NCI in cancer therapeutics.

The ability of azacitidine to induce terminal differentiation in cell lines was published by Taylor and Jones in 1980 [21]. The same authors demonstrated that azacitidine administration led to the reduction in methylated DNA; the biological significance of cytosine methylation of DNA remained controversial for a subsequent decade. Data supported the association of DNA methylation with transcriptional repression, particularly in the literature examining the mechanism of hemoglobin switching [22].

Dover et al. [23] thus successfully demonstrated the induction of hemoglobin F-containing reticulo-cytes and erythrocytes in patients with sickle cell anemia following administration of azacitidine. In an interesting historical footnote, the so-called Phase I study published by these authors (which involved four patients!) demonstrated the ability to successfully administer this compound orally; a "Holy Grail" which was not again successfully addressed until the beginning of the development of CC486 [24].

A variety of clinical trials in the 1980s focused on the concept of inducing terminal differentiation to malignant cells. Prominent among these studies were trials of low dose cytarabine, retinoids as well as vitamin D [25]. Many of these studies focused on MDS, a myeloid malignancy in which morpho-logically aberrant differentiation of hematopoietic cells in the bone marrow accompanies peripheral blood cytopenias. The concept, which in retrospect seems naïve, was that drugs which could cause blastic cell lines to mature into neutrophils or monocytes could somehow correct the *aberrant* differen-tiation in MDS marrows (which we now know leads to intramedullary apoptosis) and induce effective hematopoiesis [26,27]. It was in this context that Silverman and Holland commenced investigation of azacitidine as "differentiation" therapy for MDS [28]. The ultimate success of these investigations led to the approval of azacitidine for the treatment of MDS and the recognition that azacitidine improves survival for patients with higher risk MDS. While these authors are unsung heroes as the true parents of epigenetically targeted therapies in hematopoietic neoplasms, as with many important discoveries, the result was in part serendipitous.

4 AZACITIDINE: THE BIRTH OF "EPIGENETIC" THERAPIES

As discussed above, the initial experience with azacitidine in myeloid malignancies followed the obser-vation that azacitidine administration led to the expression of hemoglobin F in patients with sickle cell anemia. The idea of reprogramming gene expression, leading to effective cellular terminal dif-ferentiation led to initial trials of azacitidine with the Cancer and Leukemia Group B (CALGB) led by Silverman and Holland. In a series of Phase II trials, hematopoietic responses to azacitidine in patients with MDS were documented. In CALGB 9221, the efficacy of azacitidine was studied in a Phase 3 trial randomized against best supportive care [29,30] (Table 1). In this trial, patients in the supportive care arm could cross over to receive azacitidine if no response had been seen after 4 months. Thus, the majority of patients eventually received azacitidine. This trial demonstrated the efficacy of this drug in higher risk MDS, with approximately 50% of patients developing hematopoietic responses, although only 10% developed complete responses. Because of the crossover design of the study, the impact of azacitidine on survival could not be directly studied. Nonetheless, a landmark analysis suggested that earlier use of azacitidine (i.e., patients on the azacitidine arm) had potentially improved survival com-pared to no azacitidine or later azacitidine (the crossover arm) [29].

Following reworking of the CALGB data for FDA submission, the Pharmion corporation designed a subsequent study to examine the impact of azacitidine on survival of higher risk patients (AZA001) [31]. In this trial, conducted primarily in Europe, patients were randomly assigned to the subsequently FDA-approved dose schedule of azacitidine or to conventional care regimens (CCRs). The latter included best supportive care, low dose cytarabine, and intensive cytarabine plus anthracycline acute leukemia induction therapy. Investigators preselected the CCR to which a patient would be assigned prior to randomization. Thus, an individual was randomized between preselected cytarabine plus anthracycline versus azacitidine, low dose cytarabine versus azacitidine, or best supportive care versus azacitidine.

Table 8.1 Studies of Azacitidine or Decitabine Monotherapy

Study	Refs.	Patients (*n*)	Patient Population	ORR (%)	CR + PR (%)	Median OS
CALGB 9221	[30]	191	MDS 5-Aza 7 days	44	14	NE
AZA-001	[31]	179	Higher risk 5-Aza 7 days	49	29	24 months
E1905 (5-AZA arm)	[6]	75	98 MDS/52 AML 5-Aza 10 days	46	20	18 months
French registry	[101]	282	High risk MDS AZA 7 days	43	17	14 months
Spanish registry	[102]	144	Intermediate-2 + high risk: 59%; 5-Aza 7 days	74	33	NA
Korean registry	[103]	97	MDS 5-Aza 7 days	44	26	26 months
ADOPT	[32]	99	MDS DAC 5 days	51	17	NA
EORTC	[104]	116	MDS high-risk DAC 3 days	34	19	10 months
Japan	[105]	37	MDS DAC 5 days	47	27	>24 months
DEC arm	[33]	74	MDS/AML DAC 5 days	55	34	12 months

ORR, overall response rate; CR, complete remission; PR, partial remission.

The most commonly selected comparator was best supportive care, followed by low dose cytarabine, with a relatively small number randomized between azacitidine and intensive chemotherapy.

Not only did the trial confirm the response rate of the CALGB 9221, AZA001 demonstrated improvement in overall survival (OS) in azacitidine-treated patients, with an improvement in median survival from 15 to 24 months, and a doubling of 2-year survival (26% vs. 52%). While the trial was not powered to investigate survival differences within each CCR assignment stratum, survival benefit was seen across the regimens (even if not statistically significant in the comparison with intensive chemotherapy) and across cytogenetic risk groups.

The kinetics of response to azacitidine (and its congener decitabine) are unique and poorly understood compared to common cytotoxic chemotherapy regimens. The median time to first hematologic response is 2 months, with the median time to best hematologic response 6 months [34]. In fact, 12 cycles of administration are required to develop 95% of hematologic responses. Another unique feature of azacitidine treatment is that complete response does not seem to be required in order for the treated patient to experience a survival benefit. In a multivariate analysis of data from the AZA001 study, achievement of any hematologic improvement (even in one lineage) or better response was associated with improved survival compared to achievement of an equivalent response in the CCR arms [35]. Administration of azacitidine in responding patients through treatment failure is recommended; anecdotal data suggests that responses are rapidly lost upon discontinuation of azacitidine and usually cannot be regained [36].

The prolonged kinetics required for response to azanucleosides have led some to question whether these drugs exert their activity on malignant stem cells rather than more mature progeny. Such a mechanism might require repeated administrations and prolonged time to lead to clonal replacement. *Ex vivo* treatment of human $CD34^+$ progenitor cells with decitabine in combination with an HDAC inhibitor led to expansion of phenotypic stem cells and led to increased numbers of marrow repopulating cells [37,38]. In contrast, such treatment led to the disappearance of stem cells from JAK-2 mutated chronic idiopathic myelofibrosis [39]. Azacitidine administered with valproic acid to patients with MDS and AML reduced but did not eradicate the leukemic stem cell population (measured immunophenotypically) in clinical responders but not nonresponders [40].

In a third somewhat unique feature of azacitidine-based therapy, survival following failure of azacitidine appears to be markedly limited (median 6 months) [41]. Such patients tend not to respond to other forms of treatment and to date there is no standard of care in this situation. The molecular changes underpinning the resistance which manifests following azanucleoside failure has not been rigorously studied. It seems logical that additional mutations may be induced through azanucleoside treatment; alternatively, if azanucleosides suppress the initial clone well, highly resistant subclones may emerge through selection. This phenomenon raises potential concerns regarding the widespread use of azacitidine to treat lower risk MDS patients for the palliation of cytopenias. Long-term follow-up data on such patients are scarce, as are the outcomes of subsequent therapies.

5 IS THERE A CORRELATION BETWEEN METHYLATION AND RESPONSE TO HYPOMETHYLATING AGENTS?

The relationship between the impact of azanucleosides on the methylome and the clinical response of the patient remains murky. Using a variety of assays including methylation-specific PCR, bisulfite pyrophosphate sequencing, methylation microarrays, and mass spectrometry, many investigators have repeatedly demonstrated that administration of azacitidine or decitabine leads to extensive demethylation of the genome [42]. With decitabine, methylation reversal peaks at day 8; with azacitidine at day 15. Prior to the next cycle of treatment, some degree of remethylation has been observed in some [24,43] but not all studies.

Administration of azacitidine or decitabine leads to methylation changes in hundreds of genes [42]. The relationship between changes in methylation and clinical response remains unclear. Using methylation arrays enriched for promoter regions, no relationship between baseline methylation status and clinical response has been found. Similarly, changes in the methylome upon treatment have not predicted clinical response [6]. Examination of methylation reversal of targeted genes, often but not restricted to tumor suppressors, has provided conflicting data [5,6]. Among the challenges faced in correlating methylation dynamics with clinical response is the time lag between proximal molecular events (hours to days) and clinical response (weeks to months). Examination of cells following one or more cycles of treatment suffers from the problems of clonal replacement: as normal cells repopulate the bone marrow, the cellular population surveyed is different from the starting population.

Recent applications of next-generation sequencing (NGS) have enabled more detailed examination of the anatomy of the methylome. Previous studies examining the clinical consequences of methylation reversal focused on promoter regions, due in part to the almost complete lack of knowledge about the significance of cytosine methylation elsewhere in the methylome. Studies have used NGS to examine the impact of decitabine treatment on hematopoietic cells [44]. DNMT inhibitor treatment leads to changes in methylation not only in promoters but also in enhancers and many other intergenic regions. Complete sequencing of the methylome before, and potentially following treatment, will allow definitive analysis of the relationship between methylome anatomy, changes in methylome with DNMTi treatment, and clinical response. Such a study performed in bone marrow samples from patients with chronic myelomonocytic leukemia (CMMoL) has allowed development of a methylome-based predictor of decitabine response; these results require validation and extension to other myeloid neoplasms [106].

Two studies support the importance of azanucleoside uptake and phosphorylation for clinical activity. The ratio of cytidine deaminase to deoxycytidine kinase appeared higher in clinical responders to decitabine than in nonresponders [45]. Similarly, the expression of uridine cytidine kinase was lower in nonresponders to azacitidine than in patients who achieved clinical response [46]. These suggest that the phosphorylation of these nucleosides, and probable incorporation into DNA, is indeed responsible for their clinical activity.

However, the need for uptake into DNA, which is associated with DNMT adduct formation and depletion of the enzyme [42], does not lend certainty to the need for methylation reversal for clinical activity. Exposure to both azanucleosides leads to evidence of double-stranded breaks in DNA [42]. Decitabine has been shown to lead to development of reactive oxygen species and ATM activation [47]. In contrast, the direct DNMT enzyme inhibitor RG108 induces methylation reversal and gene re-expression without DNA damage in cell lines [47,48] and without impact on cell cycle, in contrast to the azanucleosides. This suggests a potentially important disconnect between the methyltransferase inhibition and pharmacodynamics of the azanucleosides.

6 WHAT IS THE OPTIMAL DOSING REGIMEN?

The current schedules of administration of the azanucleosides were empirically derived. The 7-day schedule of azacitidine approved by regulatory agencies was arbitrarily picked for the early CALGB trials, borrowing from experience with sickle cell anemia [23]. Early studies of decitabine used a fairly toxic dosing schedule requiring inpatient hospitalization ($15\,mg/m^2$ Q8h * 9 doses intravenously). Subsequent pharmacodynamically driven studies of decitabine dosing suggested optimal schedules of $15\,mg/m^2$/day * 10 days [43] or $20\,mg/m^2$/day * 10 days [49]. A Bayesian-designed dose finding trial which included neither of those schedules selected $20\,mg/m^2$/day * 5 days for further study (the closest comparator to the Phase I schedules was $10\,mg/m^2$/day * 10 days); this schedule underwent further Phase II dosing in MDS [32] and received regulatory approval.

The question of prolonged administration of lower daily doses of azanucleosides remains intriguing based on the presumed mechanism of action of these drugs. If methylation reversal is required for clinical activity, then treated cells must continue to undergo cell cycle once the nucleoside is incorporated into DNA in order to generate strands of DNA which are demethylated [50]. Because both nucleosides have cell cycle inhibitory activity, lower dosing to minimize cell cycle arrest, and prolonged administration, to allow as many cells as possible to traverse S phase would seem to offer therapeutic advantage.

In a community-practiced base scheduling study in which only peripheral blood improvement was used as an endpoint for MDS patients, a 7-day schedule (incorporating a 2-day break for weekends) at the approved dosage was compared to a 10-day schedule of a lower dosage, also interrupting administration for the weekend. While no overall advantage was seen for the longer dosing, in patients with thrombocytopenia (thus presumably more severe disease), only patients receiving the prolonged dosing developed responses in the erythroid lineage [51]. A cooperative group Phase 2 study in high-risk MDS examined azacitidine $50\,mg/m^2$/day for 10 days (no weekend breaks), aiming to increase the CR plus PR plus trilineage improvement rates from 15% (Aza 001) to 30%. This endpoint was met; however, because the control was historic, confirmation would be required in a randomized trial [6]. Among 25 patients with MDS treated with intermittent low dose decitabine ($0.2\,mg/kg$/dose twice to three times weekly), 11 patients responded with hematologic improvement or better, with complete responses in 4 [52].

7 HDAC INHIBITORS

7.1 HDAC INHIBITORS FOR AML AND MDS: A DEAD END?

The effects of HDACi monotherapy for AML and MDS patients are deceiving [53–59]. Overall response ranges between 0% and 15% and are mostly limited to hematological improvement or transient blast clearance (Table 2). Most of those studies were Phase I dose-escalation studies conducted in heavily pretreated patients. Studies of belinostat in this population have not been published. *In vitro* studies have highlighted that resistance to HDACi could be associated with overexpression of HSP90 [60] and expression of genes implicated in reactive oxygen species pathway [55]. Data from sodium valproate studies suggested that low risk patients were more likely to respond than higher risk patients [53,61]. This population has been poorly represented in other studies of HDACi monotherapy and may represent reasonable subjects for such trials. A new (third) generation of HDACi is currently being tested. These new drugs have been chemically optimized to increase their affinity to one or several HDAC. Results are still preliminary but some compounds seems to show a better efficacy than the prior generation.

The more common limiting toxicities [62,63] were fatigue and gastrointestinal toxicities (vomiting, diarrhea). Neutropenia and thrombocytopenia are frequent in advanced phase AML/MDS patients. Early concern about problematic QT prolongation with romidepsin has been allayed. Neurologic toxicity including dizziness and confusion has been seen with sodium valproate, sodium phenylbutyrate, and the benzamide derivatives (Entinostat, MGCD0103).

Table 8.2 Clinical Studies Using HDAC Inhibitors as Single Agents

Study	Schedule	N	ORR	HI + MLFS	CR + CRi
[107]	Phenylbutyrate Ph1	27	15%	15%	0%
[57]	Vorinostat Ph 1 monotherapy	41	17%	7%	10%
[55]	Vorinostat Ph2 AML monotherapy	37	0%/9%*	0%/5%*	0%/5%*
[63]	MS 275 (entinostat) Ph 1 monotherapy	38	0%	NE	0%
[58]	Depsipeptide Ph1 monotherapy	10	0%	0%	0%
[59]	Depsipeptide Ph1 monotherapy	12	8%	0%	8%
[57]	MGCD0103 (mocetinostat) Ph 1 monotherapy	23	13%	13%	0%
[61]	VPA ± ATRA	55	16%	16%	0%

*2 arm study: arm A with continuous 400 mg regimen and arm B with 200 mg t.i.d. regimen for 14 days over a 21 days period. Results are reported armA/ arm B.
HI: hematological improvement; MLFS: marrow leukemia free state.

7.2 COMBINATIONS APPROACHES BASED ON HDACi EXCLUDING COMBINATION WITH HYPOMETHYLATING AGENTS

In vitro studies have shown that HDACi could have synergistic effects with a large number of treatments used against cancer including, but not limited to, radiation therapy, anthracyclines [64], fludarabine, proteasome inhibitors [65], AKT inhibitors [66], and antiangiogenic agents [67]. These synergistic effects rely on both epigenetic and nonhistone-mediated effects of HDACi according to the tested combination (i.e., inhibition of HDAC5 for angiogenesis vs. inhibition on HSP90 and tubulin for proteasome inhibitors). The schedule of administration appears important in such combinations as suggested by *in vitro* studies with conventional chemotherapy drugs [64,68]. In clinical practice, ongoing clinical trials with such combinations are not restricted to AML and MDS. The most mature data come from the combination of chemotherapy and HDACi in AML: Garcia-Manero et al. [69] reported a high frequency of response (80% ORR) of combination of vorinostat and an intensive induction regimen (idarubicin + high dose cytarabine) in high-risk patients including adverse cytogenetics, secondary AML, and FLT3-mutated patients. Based on *in vitro* data, vorinostat was given at high doses and prior to chemotherapy for putative "priming" before the addition of chemotherapy. The combination of Belinostat and idarubicin has been studied [70,71] in AML and MDS patients and the response rate was modest (18%).

7.3 HDAC INHIBITORS FOR LYMPHOID MALIGNANCIES

There is little to no evidence of an activity of hypomethylating agents in either lymphoma or myeloma. In contrast, HDAC inhibitors show significant activity in lymphomas. Two HDAC inhibitors are currently registered in cutaneous T-cell lymphomas. Vorinostat was the first in class HDAC registered based on the results of a Phase II study in advanced CTCL in 2006. 74 patients were treated. Tolerance profile was similar to what is observed in myeloid malignancies and an ORR of 30% was observed. Time to progression remains limited with a median of 202 days [72]. This study was supported by an additional Phase II study of 33 pts [73]. More recently, Romidepsin was also registered in this indication supported by two phase II trials [74,75]. Response rate was 34% in both studies and time to progression ranged between 8 and 9 months. Panobinostat has also been tested with very similar results but is not yet registered [76]. Interestingly, the underlying physiopathological rational remains a matter of debate. In myeloma, there is a strong rational for using HDAC inhibitors mostly for their nonepigenetic properties of cytoplasmic protein acetylation and proteasome targeting. There is also strong evidence of synergy with proteasome inhibitors. Several second-generation HDAC have been or are being evaluated in myeloma. Vorinostat was recently evaluated in combination with bortezomib (a proteasome inhibitor) in a large Phase III randomized trial [77] versus placebo for patients with relapsed/refractory myeloma (380 patients per arm of treatment). Progression-free survival (PFS) was the primary endpoint and was significantly prolonged of 1 month as compared to placebo (7.6 vs. 6.8 months, $P = 0.01$). However, the clinical significance of the benefit remains limited, taking in consideration the related side effects of vorinostat. Further development of vorinostat in this setting is on hold. Panobinostat was also tested in the same settings [78], median PFS for the experimental arm was prolonged as compared to placebo (12 m vs. 8.1 m, $P < 0.0001$). Several other agents are currently developed.

8 COMBINATION THERAPIES

8.1 AZA/HDAC COMBINATIONS

Combinations of azanucleosides with HDAC inhibitors have been proposed to increase the re-expression of methylated genes putatively important to the biological and clinical activities of the azanucleosides [79]. This concept was derived from *in vitro* studies linking silencing of gene expression silencing due to methylation of CpG islands in gene promoter regions to heterochromatin formation due in part to HDAC recruitment. Cameron et al. published the classic demonstration of such synergy [80]. This synergistic gene re-expression was strictly dependent on sequence specificity, requiring exposure to a DNMT inhibitor prior to the HDAC inhibitor. The synergy was also demonstrated using pharmacodynamically submaximal concentrations of DNMT inhibitors.

Combinations which have been tested clinically utilizing this concept are listed in Table 3. The combinations were feasible from a tolerability point of view, and several (azacitidine/entinostat; azacitidine/Vorinostat) appeared quite promising clinically. Three combinations have now been tested in randomized Phase II trials in patients with MDS and/or AML: azacitidine with entinostat overlapping beginning on day 3 of a 10-day regimen [6], azacitidine with overlapping Vorinostat [81], and decitabine plus valproic acid [33] (overlapping). In none of the three studies was the combination more effective in inducting remissions or in remission duration. In the study with entinostat, methylation reversal was inhibited in the combination compared to azacitidine alone, suggesting that the cell cycle inhibitory activity of the HDAC inhibitor was pharmacodynamically antagonistic to the azanucleoside. A comparative study of a truly sequential DNMTi/HDACi combination has not been performed to date. While pharmacodynamic antagonism of the overlapping schedules may explain the failure of these clinical trials, consideration must be given to the possibility that the current clinical dosage of azanucleosides provide maximal gene re-expression and that synergy (discovered *in vitro* based on submaximal concentrations of DNMT inhibitor) is not possible to achieve *in vivo*. The recently published very low dose decitabine schedule may represent a better platform in which to explore these combinations [52].

Table 8.3 Combined Therapies Based on Hypomethylating Agents

	Combination agent	Demethylating agent	Phase	N	ORR	CR + CRi	OS
[30]	Vorinostat	5-aza 55 or 75 mg/m^2 d1–7	1–2	28	72%	42%	21 months
[6]	Entinostat	5-aza (50 d1–10)	2 RAND	68	44%*	14%*	14 months
[57]	MGCD0103	5-aza (75 d1–7)	1	12	30%	25%	unk
[108]	Lenalidomide	5 aza	1–2	36	72%	44%	14 months
[109]	Vorinostat	5 aza	2	30	30%	27%	7 months
[110]	Etanercept	5 aza	1	32	72%	30%	>12 months
[33]	Valproic acid	Decitabine	2 RAND	149	55%	34%	12

8.2 OTHER AZANUCLEOSIDE COMBINATIONS

Azacitidine and decitabine have been combined with a variety of other agents. Some of these combinations have been based on preclinical modeling; some have been empiric. The empiric combination of the thalidomide analogue lenalidomide with azacitidine appeared to have remarkable activity in the Phase I/II setting [82] in patients with MDS. Tested in a randomized Phase II setting against azacitidine alone, the combination failed to show enhanced activity. While the final report of this study has not been published, preliminary data suggest that patients receiving the combination received less total therapy than patients receiving azacitidine monotherapy (this was true with azacitidine-vorinostat arm of the same study). This suggests that while it is possible that the combination possessed increased activity, any positive effects were outweighed by increased toxicity of the regimen [83].

Azacitidine was combined with the anti-CD33 immunotoxin gemtuzumab ozogamycin in elderly patients with AML [84]. This was based on the premise that azacitidine would upregulate CD33 expression on blasts and inhibit mdr1. Response rate was modest and response duration was limited. Azacitidine has been combined with low dose cytarabine [85] in AML and high-risk MDS with minimal effect.

Tyrosine kinase inhibitors have been combined with azanucleosides for the treatment of chronic myeloid leukemia [86] and myeloproliferative neoplasms in blast phase in a case series [87]. Azacitidine plus sorafenib successfully induced remissions in approximately half of relapsed patients with AML characterized by an flt-3 mutation [88].

9 AZANUCLEOSIDES AS IMMUNOMODULATORS

The ability of azanucleosides to induce regulatory T cells may point to a potential role for these compounds in adoptive immunotherapy and allogeneic stem cell transplantation. Both azacitidine and decitabine can induce FOXP3 expression, associated with a Treg phenotype, in $CD4^+$ $CD25^-$ T cells [89,90]. Cells so generated could act as suppressor cells functionally. Azacitidine, but not decitabine, protected mice receiving mismatched T cells from graft versus host disease.

The administration of azanucleosides may lead to expression of antigens which may serve as therapeutic targets. For example, MDS patients treated with azacitidine or decitabine demonstrated modest upregulation of PD-1 and PD-L1, PD-L2, and CTLA-4 on peripheral blood mononuclear cells, suggesting that this treatment may sensitize such cells to immune checkpoint inhibitors [91].

Another neoantigen whose expression is induced by DNMT inhibitors is the cancer testis antigen NY-ESO-1. Decitabine induces expression of NY-ESO-1 in leukemia cell lines [92]. Similar results were obtained with the decitabine prodrug SGI-110 in both cell lines and in AML-bearing xenograft models [93]. The MAGE A antigen was also induced in the latter study. These data suggest the possibility of utilizing azanucleosides prior to the administration of vaccines under development for cancer testis antigens.

10 TARGETED THERAPEUTICS WITH MAJOR EPIGENETIC IMPACT: THE NEXT GENERATION

While early generations of epigenetically targeted agents demonstrate that such approaches have significant anticancer potential, the optimistic future includes targeted agents with far greater specificity.

This new generation of epigenetic therapies are more likely to be called real "targeted" therapies as they have been developed to interfere with mutations and aberrations disrupting epigenetic regulation (i.e., epigenetic writers, readers, and erasers) and not less specifically on epigenetic marks. Even if mutations in epigenetic related genes are frequent in hematological diseases, the more frequent (TET2 and ASXL1) are not the one for which drugs are currently available. For example, IDH1 and 2 mutations are present in a total of 5–15% of the mutations, EZH2 mutations are usually described in less than 5% of the cases [12].

An increasing number of compounds have been successfully tested *in vitro* with but only a limited number have entered clinical development program. Phase I studies of IDH1, IDH2, DOT1L, EZH2, Brd2/4 inhibitors are currently accruing or have just been completed. In AMLs with IDH2 mutations, IDH2 inhibition with AG-221 [94] is tolerable with mainly gastrointestinal side effects. AG-221 leads to significant clinical responses (20 objective responses on 32 evaluated patients) and *in vivo* evidence of cell differentiation, a feature usually limited to acute promyelocytic leukemia treated with ATRA. Encouraging preliminary results are also observed with AG-120 for AML patients with IDH1 mutation [95] with 4 CRs among 17 treated patients and a pattern of toxicities similar to AG-221. OTX-015, a BRD2/4 inhibitor, is currently undergoing evaluation for both AML and lymphoid diseases. The most frequent side effects are gastrointestinal toxicity, hyperglycemia, and fatigue. Inclusion into the OTX-015 trial was not limited to the presence of any specific mutation or alteration. In the AML subgroup (32 patients) [96], 4 patients experienced some degree of clinical response (including 1 CR) and most of the patients had an acceptable tolerance of the drug. In the lymphoid disease subgroup [97], some activity was documented in diffuse large B-cell lymphomas (4 responses on 18 patients) whereas no sign of activity was observed in myeloma. EZ7438 is an EZH2 inhibitor used in advanced phase lymphoma [98] (follicular and DLBCL); Phase I data showed acceptable toxicity and significant response in 4 of the 10 evaluable patients. There are no available data on modulation of EZH2 for AML in clinical settings, the type of mutation (and its consequences) differing from what is observed in lymphomas. DOT1L inhibitors have shown compelling evidence of activity in MLL driven leukemia, EP5676 is a potent DOT1L inhibitor currently tested in adults and children leukemia with MLL leukemias. Safety profile is acceptable and clinical responses were observed in 8 of 34 MLL-driven leukemias including CR [99].

11 CONCLUSION

In the present chapter, we have seen evidence that our scientific knowledge of epigenetics could translate into clinical benefit for our patients. However, even if benefit could be striking for some patients there is still a long road to go and a significant number of patients have little to no option of response and cure. We know from a large and increasing number of studies that cancer is a heterogeneous set of diseases with different subtypes in each histology and different clones for each individual disease [100]. This landscape is able to change with time and with the exposure to the selective pressure of our current therapies. The ultimate goal will be to personalize each patient's treatment to the extent possible. Conversely, it means that we should be able to have ways to select the best treatment or the best combination. With the exception of mutation-driven drugs like IDH1, IDH2, EZH2 inhibitors, we currently do not have any strong biological or clinical correlate able to predict response to these drugs; potentially further methylome profiling will assist in the case of DNMT inhibitors. Despite much translational effort to date, none of the potential biomarker candidates has been validated. The significant "grey area" linking the epigenetic effect of drugs and outcome of treatment requires resolution if further progress is to be made. Even if the clinical efficacy of demethylating agents in AML/MDS or HDACi in CTCL is demonstrated, there is no strong evidence linking demethylation or deacetylation with response to therapy. The example of HDACi is the most striking as we know that the ability of drugs to deacetylase cytosolic proteins through HDAC6 inhibition (FLT3 is a good example) may be the mainstay of the biological effect with little to no impact of nuclear histone modifications. This last issue could be perceived as negative, which may be true with regard to potential association with side effects occurrence for example. However, this pleiotropy or "lack of targeting" may also be seen as useful considering the high risk of development of resistance associated with truly targeted agents. The field of clinical development of epidrugs is quickly evolving and we can expect more answers and potential breakthroughs will be available in the coming years.

REFERENCES

[1] Ablain J, de The H. Retinoic acid signaling in cancer: the parable of acute promyelocytic leukemia. Int J Cancer 2014;135(10):2262–72.

[2] Esteller M. Epigenetics in cancer. N Engl J Med 2008;358(11):1148–59.

[3] Plongthongkum N, Diep DH, Zhang K. Advances in the profiling of DNA modifications: cytosine methylation and beyond. Nat Rev Genet 2014;15(10):647–61.

[4] Deneberg S, Grovdal M, Karimi M, et al. Gene-specific and global methylation patterns predict outcome in patients with acute myeloid leukemia. Leukemia 2010;24(5):932–41.

[5] Raj K, John A, Ho A, et al. CDKN2B methylation status and isolated chromosome 7 abnormalities predict responses to treatment with 5-azacytidine. Leukemia 2007;21(9):1937–44.

[6] Prebet T, Sun Z, Figueroa ME, et al. Prolonged administration of azacitidine with or without entinostat increases rate of hematologic normalization for myelodysplastic syndrome and acute myeloid leukemia with myelodysplasia-related changes: results of the US Leukemia Intergroup Trial E1905. J Clin Oncol 2014.

[7] Christiansen DH, Andersen MK, Pedersen-Bjergaard J. Methylation of p15INK4B is common, is associated with deletion of genes on chromosome arm 7q and predicts a poor prognosis in therapy-related myelodysplasia and acute myeloid leukemia. Leukemia 2003;17(9):1813–19.

[8] Aggerholm A, Holm MS, Guldberg P, Olesen LH, Hokland P. Promoter hypermethylation of p15INK4B, HIC1, CDH1, and ER is frequent in myelodysplastic syndrome and predicts poor prognosis in early-stage patients. Eur J Haematol 2006;76(1):23–32.

[9] Shen L, Kantarjian H, Guo Y, et al. DNA methylation predicts survival and response to therapy in patients with myelodysplastic syndromes. J Clin Oncol 2010;28(4):605–13.

[10] Grovdal M, Khan R, Aggerholm A, et al. Negative effect of DNA hypermethylation on the outcome of intensive chemotherapy in older patients with high-risk myelodysplastic syndromes and acute myeloid leukemia following myelodysplastic syndrome. Clin Cancer Res 2007;13(23):7107–12.

[11] Figueroa ME, Skrabanek L, Li Y, et al. MDS and secondary AML display unique patterns and abundance of aberrant DNA methylation. Blood 2009;114(16):3448–58.

[12] Bejar R, Stevenson K, Abdel-Wahab O, et al. Clinical effect of point mutations in myelodysplastic syndromes. N Engl J Med 2011;364(26):2496–506.

[13] Papaemmanuil E, Gerstung M, Malcovati L, et al. Clinical and biological implications of driver mutations in myelodysplastic syndromes. Blood 2013;122(22):3616–27.

[14] Walter MJ, Shen D, Shao J, et al. Clonal diversity of recurrently mutated genes in myelodysplastic syndromes. Leukemia 2013;27(6):1275–82.

[15] Haferlach T, Nagata Y, Grossmann V, et al. Landscape of genetic lesions in 944 patients with myelodysplastic syndromes. Leukemia 2014;28(2):241–7.

[16] Mrozek K, Marcucci G, Paschka P, Bloomfield CD. Advances in molecular genetics and treatment of core-binding factor acute myeloid leukemia. Curr Opin Oncol 2008;20(6):711–18.

[17] Patel JP, Gonen M, Figueroa ME, et al. Prognostic relevance of integrated genetic profiling in acute myeloid leukemia. N Engl J Med 2012;366(12):1079–89.

[18] Figueroa ME, Lugthart S, Li Y, et al. DNA methylation signatures identify biologically distinct subtypes in acute myeloid leukemia. Cancer Cell 2010;17(1):13–27.

[19] Mazzarella L, Riva L, Luzi L, Ronchini C, Pelicci PG. The genomic and epigenomic landscapes of AML. Semin Hematol 2014;51(4):259–72.

[20] Troetel WM, Weiss AJ, Stambaugh JE, Laucius JF, Manthei RW. Absorption, distribution, and excretion of 5-azacytidine (NSC-102816) in man. Cancer Chemother Rep 1972;56(3):405–11.

[21] Jones PA, Taylor SM. Cellular differentiation, cytidine analogs and DNA methylation. Cell 1980;20(1):85–93.

[22] Sankaran VG, Xu J, Orkin SH. Advances in the understanding of haemoglobin switching. Br J Haematol 2010;149(2):181–94.

[23] Dover GJ, Charache SH, Boyer SH, Talbot Jr. CC, Smith KD. 5-Azacytidine increases fetal hemoglobin production in a patient with sickle cell disease. Prog Clin Biol Res 1983;134:475–88.

[24] Garcia-Manero G, Gore SD, Cogle C, et al. Phase I study of oral azacitidine in myelodysplastic syndromes, chronic myelomonocytic leukemia, and acute myeloid leukemia. J Clin Oncol 2011;29(18):2521–7.

[25] Chomienne C, Chedeville A, Balitrand N, De Cremoux P, Abita JP, Degos L. Discrepancy between *in vitro* and *in vivo* passaged U-937 human leukemic cells: tumorigenicity and sensitivity to differentiating drugs. In vivo 1988;2(3–4):281–8.

[26] Tilly H, Castaigne S, Bordessoule D, et al. Low-dose cytosine arabinoside treatment for acute nonlymphocytic leukemia in elderly patients. Cancer 1985;55(8):1633–6.

[27] Chomienne C, Najean Y, Degos L, et al. Present results of the treatment of myelodysplastic syndromes with low-dose cytosine arabinoside. Preliminary results of a cooperative protocol (38 patients) and review of the literature. Acta Haematol 1987;78(Suppl. 1):109–15.

[28] Silverman LR, Holland JF, Weinberg RS, et al. Effects of treatment with 5-azacytidine on the *in vivo* and *in vitro* hematopoiesis in patients with myelodysplastic syndromes. Leukemia 1993;7(Suppl. 1):21–9.

[29] Silverman LR, Demakos EP, Peterson BL, et al. Randomized controlled trial of azacitidine in patients with the myelodysplastic syndrome: a study of the cancer and leukemia group B. J Clin Oncol 2002;20(10):2429–40.

[30] Silverman LR, McKenzie DR, Peterson BL, et al. Further analysis of trials with azacitidine in patients with myelodysplastic syndrome: studies 8421, 8921, and 9221 by the Cancer and Leukemia Group B. J Clin Oncol 2006;24(24):3895–903.

[31] Fenaux P, Mufti GJ, Hellstrom-Lindberg E, et al. Efficacy of azacitidine compared with that of conventional care regimens in the treatment of higher-risk myelodysplastic syndromes: a randomised, open-label, phase III study. Lancet Oncol 2009;10(3):223–32.

[32] Steensma DP, Baer MR, Slack JL, et al. Multicenter study of decitabine administered daily for 5 days every 4 weeks to adults with myelodysplastic syndromes: the alternative dosing for outpatient treatment (ADOPT) trial. J Clin Oncol 2009;27(23):3842–8.

[33] Issa JP, Garcia-Manero G, Huang X, et al. Results of phase 2 randomized study of low-dose decitabine with or without valproic acid in patients with myelodysplastic syndrome and acute myelogenous leukemia. Cancer 2015;121(4):556–61.

[34] Silverman LR, Fenaux P, Mufti GJ, et al. Continued azacitidine therapy beyond time of first response improves quality of response in patients with higher-risk myelodysplastic syndromes. Cancer 2011;117(12):2697–702.

[35] Gore SD, Fenaux P, Santini V, et al. A multivariate analysis of the relationship between response and survival among patients with higher-risk myelodysplastic syndromes treated within azacitidine or conventional care regimens in the randomized AZA-001 trial. Haematologica 2013;98(7):1067–72.

[36] Voso MT, Breccia M, Lunghi M, et al. Rapid loss of response after withdrawal of treatment with azacitidine: a case series in patients with higher-risk myelodysplastic syndromes or chronic myelomonocytic leukemia. Eur J Haematol 2013;90(4):345–8.

[37] Milhem M, Mahmud N, Lavelle D, et al. Modification of hematopoietic stem cell fate by 5-aza 2′deoxycytidine and trichostatin A. Blood 2004;103(11):4102–10.

[38] Araki H, Yoshinaga K, Boccuni P, Zhao Y, Hoffman R, Mahmud N. Chromatin-modifying agents permit human hematopoietic stem cells to undergo multiple cell divisions while retaining their repopulating potential. Blood 2007;109(8):3570–8.

[39] Hemavathy K, Wang JC. Epigenetic modifications: new therapeutic targets in primary myelofibrosis. Curr Stem Cell Res Ther 2009;4(4):281–6.

[40] Craddock C, Quek L, Goardon N, et al. Azacitidine fails to eradicate leukemic stem/progenitor cell populations in patients with acute myeloid leukemia and myelodysplasia. Leukemia 2012.

[41] Prebet T, Gore SD, Esterni B, et al. Outcome of high-risk myelodysplastic syndrome after azacitidine treatment failure. J Clin Oncol 2011;29(24):3322–7.

[42] Hollenbach PW, Nguyen AN, Brady H, et al. A comparison of azacitidine and decitabine activities in acute myeloid leukemia cell lines. PLoS One 2010;5(2):e9001.

[43] Issa JP, Garcia-Manero G, Giles FJ, et al. Phase 1 study of low-dose prolonged exposure schedules of the hypomethylating agent 5-aza-2′-deoxycytidine (decitabine) in hematopoietic malignancies. Blood 2004;103(5):1635–40.

[44] Klco JM, Spencer DH, Lamprecht TL, et al. Genomic impact of transient low-dose decitabine treatment on primary AML cells. Blood 2013;121(9):1633–43.

[45] Qin T, Castoro R, El Ahdab S, et al. Mechanisms of resistance to decitabine in the myelodysplastic syndrome. PLoS One 2011;6(8):e23372.

[46] Valencia A, Masala E, Rossi A, et al. Expression of nucleoside-metabolizing enzymes in myelodysplastic syndromes and modulation of response to azacitidine. Leukemia 2014;28(3):621–8.

[47] Fandy TE, Herman JG, Kerns P, et al. Early epigenetic changes and DNA damage do not predict clinical response in an overlapping schedule of 5-azacytidine and entinostat in patients with myeloid malignancies. Blood 2009;114(13):2764–73.

[48] Wong YF, Micklem CN, Taguchi M, et al. Longitudinal analysis of DNA methylation in CD34+ hematopoietic progenitors in myelodysplastic syndrome. Stem Cells Transl Med 2014;3(10):1188–98.

[49] Blum W, Garzon R, Klisovic RB, et al. Clinical response and miR-29b predictive significance in older AML patients treated with a 10-day schedule of decitabine. Proc Natl Acad Sci USA 2010;107(16):7473–8.

[50] Jasielec J, Saloura V, Godley LA. The mechanistic role of DNA methylation in myeloid leukemogenesis. Leukemia 2014;28(9):1765–73.

[51] Lyons RM, Cosgriff TM, Modi SS, et al. Hematologic response to three alternative dosing schedules of azacitidine in patients with myelodysplastic syndromes. J Clin Oncol 2009;27(11):1850–6.

[52] Saunthararajah Y, Sekeres M, Advani A, et al. Evaluation of noncytotoxic DNMT1-depleting therapy in patients with myelodysplastic syndromes. J Clin Invest 2015;125(3):1043–55.

[53] Kuendgen A, Gattermann N. Valproic acid for the treatment of myeloid malignancies. Cancer 2007;110(5):943–54.

[54] Gore SD, Weng LJ, Figg WD, et al. Impact of prolonged infusions of the putative differentiating agent sodium phenylbutyrate on myelodysplastic syndromes and acute myeloid leukemia. Clin Cancer Res 2002;8(4):963–70.

[55] Schaefer EW, Loaiza-Bonilla A, Juckett M, et al. A phase 2 study of vorinostat in acute myeloid leukemia. Haematologica 2009;94(10):1375–82.

[56] Gimsing P, Hansen M, Knudsen LM, et al. A phase I clinical trial of the histone deacetylase inhibitor belinostat in patients with advanced hematological neoplasia. Eur J Haematol 2008;81(3):170–6.

[57] Garcia-Manero G, Assouline S, Cortes J, et al. Phase 1 study of the oral isotype specific histone deacetylase inhibitor MGCD0103 in leukemia. Blood 2008;112(4):981–9.

[58] Byrd JC, Marcucci G, Parthun MR, et al. A phase 1 and pharmacodynamic study of depsipeptide (FK228) in chronic lymphocytic leukemia and acute myeloid leukemia. Blood 2005;105(3):959–67.

[59] Klimek VM, Fircanis S, Maslak P, et al. Tolerability, pharmacodynamics, and pharmacokinetics studies of depsipeptide (romidepsin) in patients with acute myelogenous leukemia or advanced myelodysplastic syndromes. Clin Cancer Res 2008;14(3):826–32.

[60] Fiskus W, Rao R, Fernandez P, et al. Molecular and biologic characterization and drug-sensitivity of pan histone deacetylase inhibitor resistant acute myeloid leukemia cells. Blood 2008;112(7):2896–905.

[61] Kuendgen A, Schmid M, Schlenk R, et al. The histone deacetylase (HDAC) inhibitor valproic acid as monotherapy or in combination with all-trans retinoic acid in patients with acute myeloid leukemia. Cancer 2006;106(1):112–19.

[62] Kelly WK, O'Connor OA, Krug LM, et al. Phase I study of an oral histone deacetylase inhibitor, suberoylanilide hydroxamic acid, in patients with advanced cancer. J Clin Oncol 2005;23(17):3923–31.

[63] Gojo I, Jiemjit A, Trepel JB, et al. Phase 1 and pharmacologic study of MS-275, a histone deacetylase inhibitor, in adults with refractory and relapsed acute leukemias. Blood 2007;109(7):2781–90.

[64] Sanchez-Gonzalez B, Yang H, Bueso-Ramos C, et al. Antileukemia activity of the combination of an anthracycline with a histone deacetylase inhibitor. Blood 2006;108(4):1174–82.

[65] Miller CP, Ban K, Dujka ME, et al. NPI-0052, a novel proteasome inhibitor, induces caspase-8 and ROS-dependent apoptosis alone and in combination with HDAC inhibitors in leukemia cells. Blood 2007;110(1):267–77.

[66] Mahalingam D, Medina EC, Esquivel 2nd JA, et al. Vorinostat enhances the activity of temsirolimus in renal cell carcinoma through suppression of survivin levels. Clin Cancer Res 2010;16(1):141–53.

[67] Urbich C, Rossig L, Kaluza D, et al. HDAC5 is a repressor of angiogenesis and determines the angiogenic gene expression pattern of endothelial cells. Blood 2009;113(22):5669–79.

[68] Qin T, Youssef EM, Jelinek J, et al. Effect of cytarabine and decitabine in combination in human leukemic cell lines. Clin Cancer Res 2007;13(14):4225–32.

[69] Garcia-Manero G, Tambaro F, Nebiyou Bekele B, et al. Phase II study of vorinostat in combination with idarubicin (Ida) and cytarabine (ara-C) as front line therapy in acute myelogenous leukemia (AML) or higher risk myelodysplastic syndrome (MDS). *ASH abstract book* 2009.

[70] Gimsing P. Belinostat: a new broad acting antineoplastic histone deacetylase inhibitor. Expert Opin Investig Drugs 2009;18(4):501–8.

[71] Schlenk R, Sohlbach K, Hutter M, et al. Interim results of a phase I/II clinical trial of belinostat in combination with idarubicin in patients with AML not suitable for standard intensive therapy. *ASH abstract book* 2008.

[72] Olsen EA, Kim YH, Kuzel TM, et al. Phase IIb multicenter trial of vorinostat in patients with persistent, progressive, or treatment refractory cutaneous T-cell lymphoma. J Clin Oncol 2007;25(21):3109–15.

[73] Duvic M, Talpur R, Ni X, et al. Phase 2 trial of oral vorinostat (suberoylanilide hydroxamic acid, SAHA) for refractory cutaneous T-cell lymphoma (CTCL). Blood 2007;109(1):31–9.

[74] Piekarz RL, Frye R, Turner M, et al. Phase II multi-institutional trial of the histone deacetylase inhibitor romidepsin as monotherapy for patients with cutaneous T-cell lymphoma. J Clin Oncol 2009;27(32):5410–17.

[75] Whittaker SJ, Demierre MF, Kim EJ, et al. Final results from a multicenter, international, pivotal study of romidepsin in refractory cutaneous T-cell lymphoma. J Clin Oncol 2010;28(29):4485–91.

[76] Duvic M, Dummer R, Becker JC, et al. Panobinostat activity in both bexarotene-exposed and -naive patients with refractory cutaneous T-cell lymphoma: results of a phase II trial. Eur J Cancer 2013;49(2):386–94.

[77] Dimopoulos M, Siegel DS, Lonial S, et al. Vorinostat or placebo in combination with bortezomib in patients with multiple myeloma (VANTAGE 088): a multicentre, randomised, double-blind study. Lancet Oncol 2013;14(11):1129–40.

[78] San-Miguel JF, Hungria VT, Yoon SS, et al. Panobinostat plus bortezomib and dexamethasone versus placebo plus bortezomib and dexamethasone in patients with relapsed or relapsed and refractory multiple myeloma: a multicentre, randomised, double-blind phase 3 trial. Lancet Oncol 2014;15(11):1195–206.

[79] Gore SD, Baylin S, Sugar E, et al. Combined DNA methyltransferase and histone deacetylase inhibition in the treatment of myeloid neoplasms. Cancer Res 2006;66(12):6361–9.

[80] Cameron EE, Bachman KE, Myohanen S, Herman JG, Baylin SB. Synergy of demethylation and histone deacetylase inhibition in the re-expression of genes silenced in cancer. Nat Genet 1999;21(1):103–7.

[81] Silverman L, Verma A, Odchimar-Reissig R, Cozza A, Najfeld V, Licht JD, et al. A phase I/II study of vorinostat, an oral histone deacetylase inhibitor, in combination with azacitidine in patients with the myelodysplastic syndrome (MDS) and acute myeloid leukemia (AML). Initial results of the phase I trial: A New York Cancer Consortium. J Clin Oncol 2008;26(suppl.) Abstract 7000.

[82] Sekeres MA, O'Keefe C, List AF, et al. Demonstration of additional benefit in adding lenalidomide to azacitidine in patients with higher-risk myelodysplastic syndromes. Am J Hematol 2011;86(1):102–3.

[83] Sekeres M, Othus M, List A, et al. A randomized phase II study of azacitidine combined with lenalidomide or with vorinostat vs. azacitidine monotherapy in higher-risk myelodysplastic syndromes (MDS) and chronic myelomonocytic leukemia (CMML): North American Intergroup Study SWOG S1117. *asH Annual Meeting Abstracts 2014*; 2014.

[84] Nand S, Godwin J, Smith S, et al. Hydroxyurea, azacitidine and gemtuzumab ozogamicin therapy in patients with previously untreated non-M3 acute myeloid leukemia and high-risk myelodysplastic syndromes in the elderly: results from a pilot trial. Leuk Lymphoma 2008;49(11):2141–7.

[85] Borthakur G, Huang X, Kantarjian H, et al. Report of a phase 1/2 study of a combination of azacitidine and cytarabine in acute myelogenous leukemia and high-risk myelodysplastic syndromes. Leuk Lymphoma 2010;51(1):73–8.

[86] Oki Y, Kantarjian HM, Gharibyan V, et al. Phase II study of low-dose decitabine in combination with imatinib mesylate in patients with accelerated or myeloid blastic phase of chronic myelogenous leukemia. Cancer 2007;109(5):899–906.

[87] Mwirigi A, Galli S, Keohane C, et al. Combination therapy with ruxolitinib plus 5-azacytidine or continuous infusion of low dose cytarabine is feasible in patients with blast-phase myeloproliferative neoplasms. Br J Haematol 2014;167(5):714–16.

[88] Ravandi F, Alattar ML, Grunwald MR, et al. Phase 2 study of azacytidine plus sorafenib in patients with acute myeloid leukemia and FLT-3 internal tandem duplication mutation. Blood 2013;121(23):4655–62.

[89] Sanchez-Abarca LI, Gutierrez-Cosio S, Santamaria C, et al. Immunomodulatory effect of 5-azacytidine (5-azaC): potential role in the transplantation setting. Blood 2010;115(1):107–21.

[90] Choi J, Ritchey J, Prior JL, et al. *In vivo* administration of hypomethylating agents mitigate graft-versus-host disease without sacrificing graft-versus-leukemia. Blood 2010;116(1):129–39.

[91] Yang H, Bueso-Ramos C, DiNardo C, et al. Expression of PD-L1, PD-L2, PD-1 and CTLA4 in myelodysplastic syndromes is enhanced by treatment with hypomethylating agents. Leukemia 2014;28(6):1280–8.

[92] Almstedt M, Blagitko-Dorfs N, Duque-Afonso J, et al. The DNA demethylating agent 5-aza-2′-deoxycytidine induces expression of NY-ESO-1 and other cancer/testis antigens in myeloid leukemia cells. Leuk Res 2010;34(7):899–905.

[93] Srivastava P, Paluch BE, Matsuzaki J, et al. Immunomodulatory action of SGI-110, a hypomethylating agent, in acute myeloid leukemia cells and xenografts. Leuk Res 2014;38(11):1332–41.

[94] Stein E, Altman JK, collins R, et al. AG-221, an oral, selective, first-in-class, potent inhibitor of the IDH2 mutant metabolic enzyme, induces durable remissions in a phase I study in patients with IDH2 mutation positive advanced hematologic malignancies. *ASH Annual Meeting Abstracts 2014*; 2014.

[95] Pollyea DA. Clinical safety and activity in a phase I trial of AG-120, a first in class, selective, potent inhibitor of the IDH1-mutant protein, in patients with IDH1 mutant positive advanced haematologic malignancies. In: 26th EORTC-NCI-AACR Symposium on Molecular Targets and Cancer Therapeutics, Barcelona, Spain; 2014.

[96] Dombret H, Preudhomme C, Berthon C, et al. A phase 1 study of the BET-bromodomain inhibitor OTX015 in patients with advanced acute leukemia. *asH Annual Meeting Abstracts 2014*; 2014.

[97] thieblemont C, Stathis A, Inghirami G, et al. A phase 1 study of the BET-bromodomain inhibitor OTX015 in patients with non-leukemic hematologic malignancies. *asH Annual Meeting Abstracts 2014*; 2014.

[98] Study of E7438 (EZH2 Histone Methyl Transferase [HMT] Inhibitor) as a single agent in subjects with advanced solid tumors or with B cell lymphomas. 2014. clinicaltrials.gov.

[99] Stein E, Garcia-Manero G, Rizzieri DA, et al. The DOT1L inhibitor EPZ-5676: Safety and activity in relapsed/refractory patients with MLL-rearranged leukemia. *ash Annual Meeting Abstracts 2014*; 2014.

[100] Welch JS, Ley TJ, Link DC, et al. The origin and evolution of mutations in acute myeloid leukemia. Cell 2012;150(2):264–78.

[101] Itzykson R, Thépot S, Quesnel B, Dreyfus F, Beyne-Rauzy O, Turlure P, et al. Prognostic factors for response and overall survival in 282 patients with higher-risk myelodysplastic syndromes treated with azacitidine. Blood 2011;117(2):403–11.

[102] Garcia R., de Miguel D., Bailen A., González J., Sanz G., Falantes J., et al. (2009). Different clinical results with the use of different dosing schedules of azacitidine in patients with myelodysplastic syndrome managed in community-based practice: effectiveness and safety data from the Spanish Azacitidine Compassionate Use Registry. Blood (ASH Annual Meeting Abstracts) 114: abstract 2773.

[103] Lee JH, Choi Y, Kim SD, Kim DY, Lee JH, Lee KH, et al. Comparison of 7-day azacitidine and 5-day decitabine for treating myelodysplastic syndrome. Ann Hematol 2013;92(7):889–97.

[104] Lübbert M, Suciu S, Baila L, Rüter BH, Platzbecker U, Giagounidis A, et al. Low-dose decitabine versus best supportive care in elderly patients with intermediate- or high-risk myelodysplastic syndrome (MDS) ineligible for intensive chemotherapy: final results of the randomized phase III study of the European Organisation for Research and Treatment of Cancer Leukemia Group and the German MDS Study Group. J Clin Oncol 2011;29(15):1987–96.

[105] Oki Y, Kondo Y, Yamamoto K, Ogura M, Kasai M, Kobayashi Y, et al. Phase I/II study of decitabine in patients with myelodysplastic syndrome: a multi-center study in Japan. Cancer Sci 2012;103(10):1839–47.

[106] Figueroa, et al. Specific molecular signatures predict decitabine response in chronic myelomonocytic leukemia. J Clin Invest 2015;125(5):1857–72.

[107] Gore SD, Weng LJ, Zhai S, Figg WD, Donehower RC, Dover GJ, et al. Impact of the putative differentiating agent sodium phenylbutyrate on myelodysplastic syndromes and acute myeloid leukemia. Clin Cancer Res 2001;7(8):2330–9.

[108] Sekeres MA, List AF, Cuthbertson D, Paquette R, Ganetzky R, Latham D, et al. Phase I combination trial of lenalidomide and azacitidine in patients with higher-risk myelodysplastic syndromes. J Clin Oncol 2010;28(13):2253–8.

[109] Garcia-Manero G, Estey EH, Jabbour E, Borthakur G, Kadia T, Naqvi K, et al. Final Report of a Phase II Study of 5-Azacitidine and Vorinostat in Patients (pts) with Newly Diagnosed Myelodysplastic Syndrome (MDS) or Acute Myelogenous Leukemia (AML) Not Eligible for Clinical Trials Because Poor Performance and Presence of Other Comorbidities *Blood* (ASH Annual Meeting Abstracts) 2011 118: Abstract 608.

[110] Scott BL, Ramakrishnan A, Storer B, Becker PS, Petersdorf S, Estey EH, et al. Prolonged responses in patients with MDS and CMML treated with azacitidine and etanercept. Br J Haematol 2010;148(6):944–7.

EPIGENETIC THERAPY IN LUNG CANCER AND MESOTHELIOMA

Anne-Marie Baird[1,2], Derek Richard[2], Kenneth J. O'Byrne[1,2], and Steven G. Gray[1,3]

[1]*Thoracic Oncology Research Group, Institute of Molecular Medicine, St. James's Hospital, Dublin, Ireland,*
[2]*Genome Stability Laboratory, Cancer Research and Ageing Program, Institute of Health and Biomedical Innovation,*
Queensland University of Technology, Woolloongabba, Queensland, Australia, [3]*HOPE Directorate,*
St. James's Hospital, Dublin, Ireland

CHAPTER OUTLINE

S.G. Gray (Ed): Epigenetic Cancer Therapy. DOI: http://dx.doi.org/10.1016/B978-0-12-800206-3.00009-4

1 INTRODUCTION

Thoracic malignancies present a considerable global health burden with the incidence and mortality of both lung cancer and malignant pleural mesothelioma (MPM) increasing year on year. Survival rates are poor and treatment options are limited in these cancers. Several epigenetic modifications have been associated with the development of both of these diseases with alterations discriminating between MPM and adenocarcinoma (AC) of the lung. In addition, studies have suggested that epigenetic agents are effective in altering the cellular characteristics of lung and MPM cells in terms of proliferation and migration. Furthermore, it has been demonstrated that epigenetic therapy can alter a pathologically relevant gene expression profile, with one that is more associated with comparative normal tissue. Therefore agents, which target the epi-genomes of lung cancer and MPM, may provide a substantial therapeutic improvement when used in combination with current therapy or indeed benefit when used as a single treatment modality.

2 OVERVIEW OF LUNG CANCER

Lung cancer is responsible for more cancer-related deaths worldwide than any other cancer type, accounting for approximately 1.59 million deaths annually [1]. This translates as one in five of all cancer deaths are attributable to lung cancer [1]. In the majority of countries, lung cancer kills more females than breast cancer and more males than prostate cancer. Stages of disease, gender, and histology are significant prognostic factors for survival [2]. For patients suitable for surgery, the completeness of resection is also an important factor in survival rates [2]. While, treatment options have improved somewhat, 5-year survival rates are dismal at approximately between 10% and 15% [3].

Lung cancer is subdivided into two main types with approximately 85% of patients presenting with nonsmall cell lung cancer (NSCLC) and the remainder with small cell lung cancer (SCLC). NSCLC is further divided into a number of subcategories with AC and squamous cell carcinoma (SCC) accounting for the most frequently diagnosed histological types. Chemotherapy is a common treatment option in lung cancer, as the vast majority of patients present with advanced disease, where surgery for curative intent is not possible. SCLC tends to be more aggressive than NSCLC and is classed as limited or extensive disease. Treatment is usually chemotherapy combined with radiotherapy.

In the absence of specific treatable mutations [4], most advanced NSCLC patients are treated with platinum-based doublet chemotherapy, the specifics of which vary depending on subtype [5]. Current treatable mutations consist of epidermal growth factor receptor (EGFR) and the echinoderm microtubule-associated protein-like 4: anaplastic lymphoma kinase fusion protein (EML4-ALK), which account for 14% and 7% of all NSCLC patients, respectively (Source: Lung cancer Mutation Consortium). EGFR-mutated patients receive specific EGFR tyrosine kinase inhibitors (TKIs) such as Erlotinib or Gefitinib, while EML4-ALK positive patients receive Crizotinib [6]. Usually, these mutations are mutually exclusive and tend to be found in light or never smokers with AC histology.

Despite these recent advances, 5-year survival rates in lung cancer patients remain poor. This is due in part to intrinsic and acquired resistance to both targeted and non-targeted therapies, the heterogeneous nature of this disease and the lack of effective screening protocols. Consequently, there is an urgent need to identify new avenues for effective therapeutic intervention in lung cancer. Epigenetic therapy may be the way forward, given the dysfunctional epigenetic background in this disease.

3 LUNG CANCER AS AN EPIGENETIC DISEASE

Lung cancer can arise in a flawed epigenetic environment, leading to the activation of proto-oncogenes and the silencing of tumor suppressor genes (TSGs), through changes in the epigenetic machinery such as histone modifications, DNA methylation, and miRNA regulation among others [7]. The relationship between the alteration in epigenetic marks and lung cancer is outlined in Figure 9.1.

A number of studies have generated promoter methylation signatures in cohorts of lung cancer patients. More than 80 genes have been reported as hypermethylated in this cancer, including *RARβ* (retinoic acid receptor), *p16*, *RASSF1A* (Ras association domain family 1 isoform A), *MGMT* (O^6-alkylguanine DNA alkyltransferase), *FHIT* (fragile histidine triad), and *DAPK* (death-associated protein kinase) to name but a few [8–10]. In a multivariate model, the promoter methylation of *p16*, *CDH13* (T-cadherin), *RASSF1A*, and *APC* (adenomatous polyposis coli) in stage I NSCLC patients was associated with tumor recurrence, independently of other characteristics such as stage, age, and smoking history [11].

Changes in gene expression may also be linked to histone modifications such as acetylation and methylation. In AC, the main epigenetic network alterations appear to be linked to cell cycle regulation, cellular development, and death compared with DNA replication, recombination, and repair pathways in SCC [12].

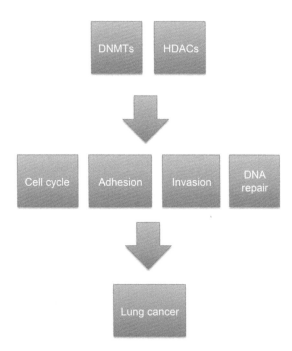

FIGURE 9.1

The link between epigenetic alterations and lung cancer. As discussed throughout the text, alterations in levels of HDACs and DNMTs can potentiate differential gene expression in a number of pathways. These pathways lead to dysfunctional cell cycle regulation, adhesion, invasion, and DNA repair processes which contribute to the lung cancer phenotype.

Although the majority of epigenetic changes are associated with TSG and transcription factors, recent studies have determined that chemokine mediators are also affected. Silencing of *CXCL5, 12,* and *14* through promoter methylation is common in a large number of NSCLC samples [13]. CXCL14 can promote events affecting over 1000 genes such as those involved in inhibition of cell cycle and pro-apoptotic gene regulation. In murine models, reexpression of CXCL14 reduced tumor growth [13].

To further underline the importance of epigenetic issues in NSCLC, studies have shown that small nucleotide polymorphisms (SNPs) in epigenetic modifiers can significantly predict outcomes in early stage resected NSCLC patients [14]. Genes examined included methyl-CpG-binding domain proteins, histone lysine methyltransferases, and DNA methyltransferases [14]. Summaries of epigenetic associated changes are outlined in Table 9.1.

Table 9.1 Epigenetic Modifiers in Lung Cancer

Modifier	Level in lung	Significance	Reference
DNA Methyltransferases			
DNMT1	Elevated	Poor prognosis	[15–17,19]
DNMT3A	Elevated	Cell cycle[a]	[15,17,18]
DNMT3B	Elevated	Poor prognosis	[15,17,19]
Lysine Methyltransferases			
KMT6/EZH2	Elevated	Poor prognosis	[12,20,21]
MMSET	Elevated	Proliferation[a]	[26,27]
SMYD3		Growth[a]	[28]
WHSC1L1/NSD3	Elevated	Proliferation[a]	[30–32]
KMT3C/SYMD2		Bioenergetics[a]	[33]
KMT1C/G9a	Elevated	Poor prognosis	[34]
KMT4/DOT1L	Elevated	Proliferation[a]	[35]
KMT1B/SUV39H2		Increased risk[b]	[36]
KMT8/RIZ1		Decreased risk[b]	[37]
Lysine Acetyltransferases			
KAT3A	Elevated	Reduced overall survival	[39,40]
KAT3B	Elevated	Reduced overall survival	[39]
KAT8/hMOF	Elevated	Migration[a]	[42]
KAT2A/hGCN5	Elevated	Growth[a]	[43]
KAT5/TIP60		DNA repair[a]	[44]
SRC-3	Elevated	Poor overall survival	[46]
Lysine demethylases			
KDM2A/JHDM1a	Elevated	Poor prognosis	[47]
KDM5A/RBP2	Elevated	EMT[a]	[50,51]
KDM1/LSD1	Elevated	Shorter survival	[52]
KDM4A/JMJD22A	Elevated	OIS[a]	[55]

(Continued)

Table 9.1 (Continued)

Modifier	Level in lung	Significance	Reference
KDM5B/JAIRD1B	Elevated	Cell cycle[a]	[56]
KDM3A	Elevated	Proliferation[a]	[57]
Arginine Methyltransferases			
PRMT1	Elevated	Proliferation[a]	[58]
PRMT5	Elevated	Growth[a]	[59,60]
PRMT6	Elevated	Proliferation[a]	[58]
Histone Deacetylases			
HDAC1	Elevated	Stage of disease	[62]
HDAC2	Elevated	Growth[a]	[64]
HDAC3	Elevated	Poor 5-year survival	[65]
HDAC4	Reduced	Poor prognosis	[70]
HDAC5	Reduced	Poor prognosis	[70]
HDAC6	Reduced	Poor prognosis	[70]
HDAC7	Reduced	Poor prognosis	[70]
HDAC8	Reduced	Growth[a]	[69,70]
HDAC9	Reduced	Poor prognosis	[70]
HDAC10	Reduced	Poor prognosis	[70]

[a]*In cell line studies.*
[b]*Gene polymorphism.*

3.1 DNA METHYLTRANSFERASES

All three DNA methyltransferases, DNMT1, DNMT3A, and DNMT3B, are increased in lung cancer tumors [15–17]. High levels of DNMT1 had a trend for a poorer prognosis particularly with SCC histology. As with DNMT3A, high levels of DNMT1 and DNMT3B are significantly associated with methylation status promoter changes in common lung cancer associated genes such as *FHIT*, *p16*, and *RARβ* [15]. Mouse double minute 2 (MDM2) homolog promotes transcriptional suppression of the Rb/E2F pathway, resulting in increased levels of DNMT3A. This overexpression of DNMT3A resulted in promoter hypermethylation of *RARβ*, *FHIT*, and *RASSF1A* [18]. Treatment with an MDM2 antagonist (Nutlin-3) resulted in reduced DNMT3A levels with a parallel decrease in the levels of hypermethylation genes in xenograft models [18]. In other cohorts of NSCLC patients, high levels of DNMT1/3B were associated with hypermethylation of *H-cadherin*, *GSTP1* (glutathione S-transferase pi), and *RIZ* (retinoblastoma protein interacting zinc finger) in addition to those listed above [16]. Elevated levels of DNMT1 were evident in patients with all stages of NSCLC disease and correlated with significantly reduced survival [16]. Abnormal levels of DNMT1 have also been significantly associated with an increased risk of death in NSCLC [19]. High levels of DNMT3B were linked to poor prognosis when patients were stratified by age (>65 years old) [19]. Conversely, raised levels of methyl-CpG-binding domain protein 2 (MBD2) were associated in a reduced risk of death in male patients and the SCLC cohort [19].

3.2 LYSINE METHYLTRANSFERASES

Histone lysine methyltransferases are also modified in lung cancer with significantly higher levels of KMT6/EZH2 (enhancer of zeste homolog 2) in SCC compared with AC patient tumors [12]. In an IHC study of NSCLC, positive KMT6/EZH2 expression correlated with larger tumor size, significantly shorter overall survival [20], and poor prognosis [21]. However, in other cohorts of patients an increase in H3K27me3 via KMT6/EZH2 correlated with better outcomes in NSCLC [22] and certain polymorphisms in this gene are associated with reduced lung cancer risk [23]. The effect of KMT6/EZH2 may vary depending on ethnicity. NSCLC cell line studies have indicated that KMT6/EZH2 plays a role in NSCLC invasion [20] and growth [24]. The compound 3-Deazaneplanocin A (DZNep) can inhibit KMT6/EZH2 resulting in apoptosis of NSCLC cells [24]. Another methyltransferase that is coupled with KMT6/EZH2 function is multiple myeloma SET domain (MMSET) also known as NSD2 or WHSC1 [25]. MMSET expression is increased in SCLC and NSCLC patient samples compared with normal lung tissue [26,27]. Knockdown of MMSET reduced the proliferative capacity of NSCLC cells and was also determined to regulate a number of proteins in the WNT pathway [27].

Ras is a common mutation in NSCLC tumors and recent data has shown that SMYD3 methylates MAP3K2, which in turn promotes MAPK signaling [28]. In murine lung AC models, abolishing SMYD3 activity resulted in tumor growth inhibition in response to oncogenic Ras [28]. Furthermore, methylation restricts the negative regulation of MAPK signaling by blocking PP2Aphosphatase complex binding to MAP3K2 [28]. However, a polymorphism in *SYMD3* is not linked to lung cancer risk [29].

WHSC1L1/NSD3 is overexpressed in tumor tissues [30], lung cell lines [31] and can promote the amplification of the 8p11-12 amplicon (region which contains a number of ongogenic drivers) in SCLC [31]. The Twist pathway is regulated by WHSC1L1 in lung cancer and the reduction of WHSC1L1 levels reduced tumor metastasis *in vivo* [32]. A reduction in proliferation was observed in lung cancer cells where WHSC1L1 was knocked down [30].

Other methyltransferases with an involvement in lung cancer include: KMT3C/SYMD2, which may be important in lung cancer tumor bioenergetics [33]; increased KMT1C (G9a) levels are linked to poor prognosis and is involved in invasion and metastases through the repression of epithelial cell adhesion molecule (Ep-CAM) [34]; KMT4 (DOT1L) is upregulated in patient tumor tissues and lung cancer cell lines whereby it promotes increased cellular proliferative capacity [35]; KMT1B/SUV39H2 polymorphisms are linked to an increased risk of lung cancer [36]; KMT8/RIZ1 polymorphisms are associated with a decreased risk of NSCLC [37].

3.3 LYSINE ACETYLTRANSFERASES

A number of lysine acetyltransferases are deregulated lung cancer. It has been demonstrated that a decrease in KAT3B levels through a CDK1 and ERK1/2 mechanism leads to increases in c-myc, MMP9, and catenin among others, therefore augmenting lung tumorigenesis [38]. In SCLC, high levels of KAT3A (CBP) and KAT3B (P300) protein were associated with significantly shorter overall survival and disease-free survival [39]. These proteins were independent prognosticators of poor overall survival in operable SCLC patients [39]. KAT3A levels are also higher in NSCLC patient tumors compared with normal [40] and it is found mutated in a small subset of tumors [41]. KAT8 (hMOF) expression is significantly elevated in NSCLC tissue compared with normal and in cell lines it promotes migration and proliferation [42]. KAT2A (hGCN5) is also highly expressed in NSCLC where it promotes tumor growth through an E2F1-cyclin E1/D1 mechanism [43]. E2F1 proteins are also regulated through KAT5/TIP60 and play a role in the repair of cisplatin induced DNA breaks [44]. Indeed KAT5 and a number of other epigenetic modifiers may play an important role in chemoresistance in NSCLC [45]. Steroid receptor coactivator-3 (SRC-3) is also upregulated in NSCLC tumors, where it significantly correlates with poor disease-free and overall survival [46]. Knockdown studies indicated an involvement in tumor growth and proliferation [46].

3.4 LYSINE DEMETHYLASES

It has been demonstrated that histone demethylases such as KDM2A/JHDM1a are upregulated in NSCLC cell lines, with increased levels associated with poor prognosis in patients [47]. Targets of KDM2A include dual-specificity phosphoprotein phosphatase (*DUSP3*), which exerts an effect on extracellular signal-regulated kinase (ERK1/2) signaling [47]. This signaling pathway is involved in epithelial-mesenchymal transition (EMT) in lung [48]. KDM2A can also promote tumorigenesis through the elevation of *CDK6*, *NEK7* (NIMA-related kinase 7), *NANOS1* (Nanos Homolog 1 (Drosophila)), and *RAPH1* (Ras association and pleckstrin homology domains 1) genes [49].

KDM5A (RBP2) is also overexpressed in lung tissues, where it has been shown to repress the expression of p27 [50]. KDM5A is implicated in carcinogenesis and metastasis of lung cancer cells in an *in vivo* murine model possibly through p27, cyclin D1, and ITGB1 (integrin beta-1) mechanisms as it can bind to these promoters [50]. In addition, KDM5A can promote EMT in NSCLC through the repression of E-cadherin, the promotion of N-cadherin and SNAIL through AKT (protein kinase B) signaling [51].

KDM1 (LSD1) is upregulated in NSCLC compared with normal tissue and is associated with shorter survival [52]. Its overexpression is linked to increases in migration, invasion, proliferation, and EMT [52]. Knockdown of KDM1 inhibited the growth of lung cancer cell lines [53] and the inhibition of KDM1 increases methylation at H3K4 [54].

Other demethylases with an involvement in lung cancer include KDM4A (JMJD22A), which is overexpressed in lung cancer, and can reduce p53 oncogene-induced senescence through the downregulation of CHD5 [55]; KDM5B (JARID1B) is elevated in NSCLC and target genes include E2F1/E2F2 [56]; KDM3A expression is increased in cancer tissue compared with normal, particularly in AC samples. Reduced levels of KDM3A were associated with a parallel decrease in proliferation capacity [57].

4 ADDITIONAL EPIGENETIC REGULATORY PROTEINS

A number of other histone modification enzymes are dysregulated in lung cancer such as histone chaperones (ASF1B), arginine methyltransferases (PRMT), sumoylation (SAE1), and chromatin modifiers (CHAF1A) [12]. These may contribute to the overexpression of EZH2 [12]. Arginine methyltransferases such as PRMT1 and PRMT 6 are upregulated in NSCLC and suppression of these reduced the proliferative capacity of lung cancer cells [58]. PRMT5 is a plausible oncoprotein as it can upregulate the PI3K/AKT pathway [59]. It is overexpressed in lung cancer tissues [59,60] and knockdown in NSCLC cell lines stunted cell growth [59,60]. Moreover, silencing PRMT5 in the A549 cell line inhibited tumor growth in mouse studies [60]. In other studies, cells with reduced levels of this PRMT displayed increased sensitivity to TRAIL-induced apoptosis [61].

4.1 HISTONE DEACETYLASES

HDAC levels are also important in lung cancer pathogenesis. HDAC1 levels increase with advanced stage of lung cancer and are associated with metastasis-associated protein 1 (MTA1) levels and they can interact together [62]. MTA1 can promote the invasion and migration of lung cancer cells [63]. Depletion of HDAC2 resulted in reduced *in vivo* lung tumor growth through increased apoptosis [64]. Five-year disease free survival was poor in AC patients with high levels of HDAC3 compared with individuals with low levels of HDAC3 [65]. Reduction of HDAC6 can also reduce xenograft tumor growth and enhances sensitivity to cisplatin in lung cancer cell lines [66]. In addition, it has been demonstrated than knockdown of HDAC8 can inhibit the growth of lung cancer cells [67]. An SNP in HDAC9 may increase susceptibility to lung cancer [68] and HDAC9 levels are reduced in cancer tissue compared with normal [69]. Reactivation of HDAC9 led to a decrease in lung cancer cellular growth [69]. Levels of Class II HDACs 4, 5, 6, 7, 9, and 10 are reduced in NSCLC patient samples and are significantly correlated with poor prognosis [70]. Of all HDACs examined HDAC10 was the strongest predictor of prognosis [70].

4.2 OTHER HISTONE MARKS

A number of histone epigenetic marks such as H2AK5Ac, H2BK12Ac, H3K9Ac, H4K8Ac, and H3K4diMe were examined in a large cohort of NSCLC patients [71]. In this group of patients, the expression of H3K4diMe was significantly associated with stage of disease. In addition, levels of H3K4diMe, H2AK5Ac, and H3K9Ac influenced overall and disease-free survival in resected stage I NSCLC patients [71]. Other histone H4 marks are also associated with survival in NSCLC. Carcinoma samples had hypoacetylation of K12 and K16 with concomitant hyperacetylation on K5 and K8 when compared with normal samples. A loss of the H4K20Me3 marker in cancer tissues was correlated with significantly reduced survival [72].

4.3 miRNA AND lncRNA IN LUNG CANCER

MicroRNAs (miRNAs) and long non-coding RNAs (lncRNAs) are short non-coding RNA of about 20–22 and >200 nucleotides in length, respectively. They serve as important regulatory mediators in gene expression. miR-29b can target and regulate DNMTs with levels of DNMTs increasing as mirR-29b decreases in a urethane of lung cancer [73]. The family of miR29 (a/b/c) can target DNMT1, 3A, and 3B and are downregulated in NSCLC [74]. miRNA-124a is also heavily methylated in lung cancer cell lines and tumor tissue. Hypermethylaton of miR124a correlated with changes in cell cycle mediators CDK6 and Rb [75]. Let-7a-3 is involved in the regulation of DNMTs and is found hypomethylated in a number of ACs of the lung. DNMT1 and DNMT3B can affect the methylation state of this gene, with reexpression promoting anchorage independent growth in lung cell lines [76]. miR-212 is also reduced in lung cancer through histone modifications and the more reduced levels are associated with a worse TNM stage [77].

Recent studies suggest that lncRNAs are important in the regulation of critical cellular processes. EZH2 expression can result in the silencing of SPRY4 intronic transcript 1 (SPRY4-IT1) in NSCLC [78]. A low level of SPRY4-IT1 was a predictor of poor prognosis in NSCLC. Reexpression of this lncRNA resulted in the inhibition of tumor growth *in vivo* [78]. The KDM, JMJD1A, can induce the expression of another lncRNA, MALAT1 [79]. This lncRNA promotes migration [79]. MALAT1 is a critical mediator of metastasis in lung cancer [79,80]. An overview of lncRNA in NSCLC is given in Yang et al. [81].

4.4 SMOKING CAN ALSO AFFECT THE LUNG EPI-GENOME

Lung tumors isolated from individuals with a smoking history have increased levels of DNMTs [15]. Occurrence for methylation of TNFRSF10C (TNF receptor superfamily member), BHLHB5, and BOLL were significantly higher in ACs from never smokers than smokers [82]. Other genes with significantly dissimilar methylation status between smokers and non-smokers were *CPEB1* (cytoplasmic polyadenylation element-binding protein 1), *CST6* (cystatin-M), *EMILIN2* (elastin microfibril interface-located protein 2), *LAYN* (layilin), and *MARVELD3* (tight junction associated protein) in lung ACs. Knockdown of *MARVELD3* significantly reduced anchorage-independent growth of lung cancer cells [83]. *In vitro* and *in vivo* studies using nicotine-derived nitrosamine ketone (NNK), a chemical found in many tobacco products, causes methylation status gene changes. NNK can lengthen DNMT1 protein stability through AKT signaling [84] as well as promoting the hypermethylation of lung cancer–associated genes such as *RARβ*, *DAPK*, and *p16* [84,85].

5 EPIGENETIC TARGETING OF LUNG CANCER

Given the number of epigenetic changes in lung cancer, it is reasonable to explore the benefit of epigenetic-targeted drugs as a therapeutic option in this disease. There are a number of pan HDAC inhibitors (HDi), in addition to more specific second- and third-generation HDi that have been developed or are in the developmental stages. The FDA has previously approved HDi, vorinostat, and romidepsin, in the treatment of cutaneous T-cell lymphoma and more recently Belinostat for relapsed or refractory peripheral T-cell lymphoma (PTCL). DNA methyltransferase inhibitors (DNMTi) agents have also been established, two of which have FDA approval in myelodysplastic syndrome: 5-azacytidine (AZA) and 5-aza-2′-deoxycytidine (DAC). A number of epigenetic agents currently in trials for lung cancer are outlined in Table 9.2.

5.1 HISTONE DEACETYLASE INHIBITORS

A large amount of data has been generated from *in vitro* and *in vivo* studies and clinical trials in combination with chemotherapy agents, usually in a refractory population of heavily pretreated patients.

Mocetinostat (MGCD0103) is an orally available benzamide, which inhibits a specific profile of HDAC isoforms that includes Class I (HDAC1, 2, 3, and 8) and Class IV (HDAC 11) (Source: Methylgene information leaflet). A Phase I trial was completed in patients with advanced cancer, five of which had lung cancer. No objective tumor responses were observed but one lung cancer patient had disease stabilization after seven cycles of drug. However, WBCs (white blood cells) isolated from patients showed inhibition of HDAC activity with a concomitant induction of H3 histone acetylation [86].

A number of Phase I studies have been conducted with HDi in conjunction with topoisomerase inhibitors. Topoisomerases are involved in regulating DNA supercoiling and inhibitors can be divided into two classes (Topo I and Topo II) depending on their exact mode of action. A number of these inhibitors are approved for cancer treatment such as topotecan (Topo I inhibitor) in lung cancer. Combination

Table 9.2 Current Epigenetic Targeted Drugs in Clinical Trials for Lung Cancer

Drug	Additional Drugs	NCT
Vorinostat	Cisplatin, pemetrexed, radiation	NCT01059552
	Iressa	NCT02151721
	Paclitaxel, carboplatin	NCT01249443
Panobinostat	Pemetrexed	NCT00907179
	Sorafenib	NCT01005797
	Cisplatin, pemetrexed	NCT01336842
Decitabine	Genistein	NCT01628471
Romidepsin	5-Azacitidine	NCT01537744
Azacitidine and Entinostat		NCT01935947
		NCT01886573
Azacitidine	CC-223, erlotinib	NCT01545947
Vidaza		NCT02009436
Azacitidine and Entinostat	Nivolumab	NCT01928576
Azacitidine	Carboplatin or ABI-007	NCT01478685
Valproic Acid	Chemoradiotherapy	NCT01203735
Belinostat	Cisplatin, etoposide	NCT00926640

of valproic acid (VPA) and epirubicin (Topo II inhibitor) was completed in advanced solid tumors. Patients were pretreated with VPA 48 hours prior to epirubicin. This cohort included four patients with SCLC, three of which showed responses. One patient had a partial response and two had stable disease. There was a significant correlation between histone H4 acetylation and VPA dose [87].

In SCLC cell lines, the combination of mocetinostat and vorinostat with Topo inhibitors (amribicin, epirubicin, etoposide, and topotecan) resulted in increased cytotoxic effects [88]. Laboratory studies have also combined camptothecin (Topo I inhibitor) with (a) sodium butyrate (NaB), (b) suberoylani-lide hydroxamic acid (SAHA), or (c) trichostatin A (TSA). It was found that the treatment schedule was critical for the optimal effects as initial dosing with camptothecin followed by HDi was the most effective in lung cell lines [89]. There was also synergy observed with vorinostat and Topo I inhibitor (Topotecan) in SCLC. The mechanism of which is thought to be through ROS generation and apoptosis. In terms of dosing, simultaneous or sequential treatment with these agents produced synergistic effects in both Topo I sensitive and resistant SCLC cell lines. Mechanistically this was through increased cas-pase-dependant apoptosis and oxidative injury [90]. *SULF2* (heparan sulfate 6-O-endosulfatase) meth-ylation is associated with better overall survival in NSCLC. The methylation of this gene can sensitize lung cells to camptotecin and topotecan treatment via an ISG15-related mechanism (interferon-induced 17 kDa) [91]. The success of HDi with Topo inhibitors is possibly due to the interaction of HDAC1/2 with Topo II complexes. It has been shown that Topo II complexes have HDAC activity and equally complexes with HDAC1 or HDAC2 have Topo II capabilities [92].

A Phase II trial was undertaken in treatment naïve advanced NSCLC patients with vorinostat or pla-cebo in combination with carboplatin and paclitaxel [93]. The response rate was 34% in the vorinostat and chemotherapy cohort, compared with 12.5% among patients who received placebo and chemotherapy. One patient experienced a complete response in the vorinostat treatment arm. Further analysis based on histologies indicated that 33% of squamous patients experienced a partial response with vorinostat compared with no patients in the placebo cohort. In the non-squamous group, 34% of patients achieved objective responses compared with 16% in the placebo group. Furthermore the 1-year overall survival rate was better with vorinostat at 51% compared with 33% in the placebo [93]. However, in a pretreated relapsed NSCLC cohort, vorinostat resulted in no objective antitumor responses, although 57% of patients achieved stable disease (range 1.4–19.4 months) [94]. A Phase I study of vorinostat in combination with bortezomib in patients with advanced malignancies, two of which had NSCLC, resulted in one SCC patient experiencing a partial response lasting 8 months [95]. Using this combination in a cohort of advanced NSCLC patients as a third-line therapy resulted in stable disease for 27.8% of participants [96].

Phase I studies of entinostat (Ms-275) in advanced or refractory solid tumors showed variations in response. In one study with four NSCLC patients, one patient had stable disease for 9 months [97], while in another study (two NSCLC cases) no patients had objective responses [98]. However, histone acetylation in peripheral blood mononuclear cells (PBMCs) was observed in both studies [97,98]. There has been evidence of increased disease stabilization in additional cohorts of patients with entinostat, ranging from 45 days to 10 months, again with changes to acetylated histone levels in PBMCs [99]. A Phase II trial in advanced NSCLC patients with erlotinib in combination with entinostat did not show any overall benefit. However, in a cohort of patients with high E-cadherin levels, the combination treatment significantly improved overall survival compared with erlotinib alone (9.4 vs. 5.4 months, respectively) with a trend for increased progression free survival [100]. Entinostat is primarily an HDAC1 selective inhibitor [101].

In a Phase I trial with refractory-solid tumors (three NSCLC patients), romidepsin (depsipeptide FK228; DP) treatment resulted in one partial response and eight with stable disease [102]. However, an additional study with romidepsin resulted in no objective responses (lung cancer), although posttreatment biopsies showed increased acetylation of histone H4 and improved p21 expression [103]. Romidepsin treatment can cause alterations to global histone acetylation levels and decrease in global levels of repressive H3K9Me3 in cell line models [104]. In SCLC cell line models, romidepsin produced synergistic effects with both cisplatin and eptoposide when added simultaneously [105].

A Phase II trial of pivaloyloxymethyl butyrate (pivanex, AN-9) resulted in three partial responses and 14 patients with stable disease (>12 weeks) in a cohort of refractory NSCLC patients. Median survival in the cohort was 6.2 months and 26% of patients were alive at 1 year [106].

Other compounds have been used in conjunction with HDi such as silibinin. Silibinin is a flavonolignan with anticancer efficacy in NSCLC [107]. TSA and SAHA were used in combination with silibinin. Silibinin reduced levels of HDAC 1–3 and in combination with either HDi displayed enhanced cytotoxic affects [108]. Treatment also correlated with upregulated p21 expression in NSCLC cells. In a xenograft model, the combination of silibinin and SAHA produced an enhanced significant effect on tumor growth inhibition than compared with silibinin and TSA or either drug alone [109].

5.2 DNA METHYLTRANSFERASE INHIBITORS

DNMTi has also been used in the clinical and laboratory setting, with the mode of delivery affecting the overall efficacy of the drugs. Aerosolized AZA suppressed growth and also altered the epigenome in an orthotopic model of lung cancer. Due to the aerosolized nature, the half-life of the drug was improved and effects were evident at lower doses than what was required via systemic therapy [110]. There was an elevation in the expression of Bik and demethylation of important lung cancer–associated genes such as *CXCL5* and *FOX* (forkhead box), HOX (homeobox), and GATA transcription factor families [110]. In other studies, intratracheal dosing of AZA was three times more effective in prolonging mice survival in orthotopic NSCLC xenografts compared with IV administration by up to 40 days [111]. Intratracheal administration increased the therapeutic index by 75-fold and produced a better toxicity profile [111]. Combination of AZA with silibinin restored E-cadherin expression in NSCLC cells with a concomitant decrease in cellular migration and invasive potential possibly through a zinc finger E-box-binding homeobox 1 (Zeb1) mechanism [108].

5.3 COMBINATION OF HDAC AND DNMT INHIBITORS

Combination of HDi and DNMTi can have a more profound anticancer effect than when used as single agents. Entinostat and vidaza treatment resulted in the reduction of KRAS/P53 mutant lung ACs in rats [112]. An increase in pro-apoptotic genes and *p21* was observed, in addition to other key genes involved in cell cycle and DNA damage and *EZH2* [112]. A clinical trial of AZA with NaB did not result in any clinical benefit; however, the acetylation status of histone H3 and H4 was altered [113]. When combined with VPA in a Phase I study, stable disease was observed in some patients with changes in DNA methylation and histone acetylation [114].

The other FDA-approved DNMTi, DAC, can interact synergistically with depsipeptide to reduce NSCLC cellular proliferation by abolishing the removal of incorporated abases [115]. In addition, in a murine model of tobacco induced lung cancer, a combination of NaB and DAC prevented lung cancer growth. On its own DAC decreased tumors by 30%, however, this was increased > 50% when combined with NaB [116]. In Phase I trials, DAC and VPA treatment in advanced NSCLC, resulted in the reactivation of epigenetically silenced genes, however, clinical benefit was limited by toxicities [117]. An additional trial with single agent DAC resulted in stable disease in three NSCLC and one SCLC patients with reexpression of melanoma-associated antigen 3 (MAGE3), p16, and NY-ESO-1 (cancer-testis antigen) [118]. It is interesting to note that knockdown of KMT6, KDM1, or KDM5B augmented DAC-mediated activation of cancer-testis genes in lung cancer cells (*NY-ESO-1*, *MAGE-A1*, and *MAGE-A3*) [119]. The most promising study so far utilized DAC with entinostat in a Phase I/II trial in recurrent metastatic NSCLC cohort. A complete and partial response was observed in one patient. The methylation pattern was also altered in APC, RASSF1A, CDH13, and p16 [120]. The demethylation of these genes was associated improved progression free and overall survival. Of interest was that four patients when on to achieve major objective responses with other anticancer therapies after the end of this study [120]. The results obtained from this trial lead to two further trials being initiated based on the idea of "priming" to subsequent therapy. The first Phase II involves low dose azacitidine and enionstat prior to chemotherapy in advanced NSCLC patients (NCT01935947). The second Phase II study also involves both of these drugs given prior to nivolumab (NCT01928576). Nivolumab is an immune-modulatory agent, which blocks PD-L1 (programmed death-ligand 1) binding to PD-1 (programmed cell death 1). Both studies are currently recruiting patients.

SGI-110, a second-generation DNMTi, which is resistant to deamination by cytidine deaminase was used in combination with entinostat. It reduced lung cancer growth in a rat model by 36% (as a single agent) versus 56% (combined with entinostat) [121]. In microarray studies SGI-110 resulted in methylation changes to over 300 gene promoters including 18 cancer-testis antigen genes. Tumors isolated from treated rats had changes in *p21*, *Bik*, and *EZH2* target genes. *EZH2* genes were significantly increased in treated tumors compared with vehicle [121].

6 OVERVIEW OF MALIGNANT PLEURAL MESOTHELIOMA

MPM arises in the pleural cavity in the lungs, from the mesothelial cells. It is an aggressive inflammatory cancer, which has been associated with asbestos exposure since the early 1960s [122]. The lag period between exposure and the development of MPM is significant, anywhere between 20 and 40 years [123]. Some success has come from using mesothelin-related protein as a screening marker [124], however, survival is dismal with most patients dying within 1 year of diagnosis [123,125]. Conservative estimates have determined that 43,000 people die from this disease each year [126]. In the United Kingdom, it is projected that between 1968 and 2050 approximately 90,000 deaths will be attributed to mesothelioma [127].

MPM is the most common form of mesothelioma accounting for over 80% of all cases. Much less frequently diagnosed cases include mesothelioma occurring in the pericardium and peritoneum. MPM consists of three main histological subtypes; epithelioid (60%) is the most common followed by biphasic (30%) and sarcomatoid (10%).

Symptoms tend to be nondescript and most patients present with late disease where chemotherapy is the only treatment option [128]. While no chemotherapy regimen has proven highly successful, the current gold standard of care is cisplatin combined with pemetrexed. This regimen is non-curative and results in a response rate of approximately 40% [129]. Despite this, almost all patients are dead within 2 years. There is currently no standard second line therapy for MPM.

There is some controversy regarding surgery in MPM. Two large trials were undertaken termed MARS (mesothelioma and radical surgery). The first study determined that it was feasible to conduct a randomized trial in patients to receive extrapleural pneumonectomy (EPP) or no surgery [130]. In the follow-on trial from this, the results suggested that "radical surgery in the form of EPP within trimodal (neoadjuvant chemotherapy, EPP, and postoperative radiotherapy) therapy offered no patient benefit." [131]. There were a large number of mortalities in the EPP group [131]. However, other studies do not agree with this interpretation of the MARS study [132]. One study using EPP in trimodality demonstrated a 23-month median survival rate [133].

7 MALIGNANT PLEURAL MESOTHELIOMA AS AN EPIGENETIC DISEASE

A number of epigenetic events are observed in MPM. DNMT1/DNMT3A/DNMT3B are all overexpressed in MPM and may contribute to carcinogenesis, as targeting these with antisense oligos result in MPM growth inhibition [134]. Some studies have shown specific methylation profiles in MPM compared with normal pleura samples [135] and lung AC [136]. Comparing MPM samples with normal pleura demonstrated a unique methylation profile. DNMT3B was significantly reduced in tumors with a concomitant increase in methylation of RASSF1 [135]. A number of genes, which are differentially expressed, are involved in chronic inflammation, which is a hallmark of MPM. Of particular interest was HOXA9, which is involved in regulating some nuclear factor kappa B (NF-κB) responses [135]. Furthermore silencing of SOC was observed, which may have resulted in chronic activation of signal transducer and activator of transcription (STAT) further perpetuating the inflammatory phenotype. The difference in methylation status between MPM and normal samples was an independent predictor of patient survival [135]. Other genes that are significantly changed through methylation are *ESR1* (estrogen receptor 1), *SLC6A20* (solute carrier family 6), and *SYK* (spleen tyrosine kinase) [137]. In addition, WNT inhibitory factor 1 (WIF-1) and serum response factor (SFR (1, 2, 4)) promoter methylation was observed in a high number of mesothelioma tissue samples [138]. MPM patients with methylation of *TMS1* (target of methylation-induced silencing 1) or *HIC-1* (hypermethylated in cancer 1) had significantly reduced overall survival [139].

Furthermore, in a large cohort of MPM patient samples there were differences observed in the methylation frequencies in certain genes, some of which are also common in lung cancer. High incidences of hypermethylation at the promoter region of E-cadherin and FHIT were observed at 71.4% and 78%, respectively [140]. Values of between 14% and 20% were determined for *APC1A*, *RASSF1A*, and *DAPK*; and values of 28–55% for *p16*, *APC1B*, *p14*, and *RARβ* (55.8%) [140]. The number of genes methylated significantly affected the overall survival; for example, patients with methylation of *RARβ* with either *DAPK* or *RASSF1A* had a shorter overall survival compared with patients who had only one or no epigenetic alteration. This was also the case for patients with double or triple methylations in these genes compared to those with one or no methylation [140].

As is the case with lung, lysine methyltranferases are also important in MPM. Epigenetic facilitators of stem cell pluripotency are the polycomb group (PcG) of proteins with EZH2 and EED being part of the polycomb repressor complex (PRC-2). EZH2 was overexpressed in MPM patient samples compared with normal pleura [141]. This increase was associated with decreased survival. Knockdown of these genes or treatment with DZnep resulted in decreased H3K27Me3 levels [91]. Functionally, there was a significant inhibition in cellular proliferation and migration of MPM cells. DZNep treatment reduced tumor size by 50% and decreased H3K27Me3 levels within the RASSF1A, HIC-1, and p21 promoters. H3K27Me3 levels within the promoters of RASSFIA and HIC-1 were markedly decreased in cells exhibiting knockdown of EZH2 or EED [141]. In an additional study, patients with low methylation profiles lived significantly longer compared with those with higher levels of methylated genes [136]. Three genes were specifically methylated in MPM compared with AC of the lung: transmembrane protein 30B (TMEM30B), Kazal-type serine peptidase inhibitor domain 1 (KAZALD1), and mitogen-activated protein kinase 13 (MAPK13) [136].

Asbestos can also change the epi-epigenome with methylation of *p16*, *CDKN2B*, and *RASSF1* significantly associated with asbestos exposure [135]. The methylation of MT1A was also correlated with asbestos burden [135,137]. Increased asbestos burden was associated with increased hypermethylation of cell cycle genes such as *APC*, *CCND2* in addition to those mentioned above [142]. Figure 9.2 outlines the association between epigenetic alterations and MPM.

7.1 miRNAS AND lncRNAs IN MESOTHELIOMA

Zinc finger of the cerebellum (ZIC1), a possible TSG, is downregulated in MPM cell lines through promoter methylation [143]. The low level of this gene was associated with high levels of miR23a and miR27a. miR23a is correlated to shorter patient survival [143]. miR-34 is also dysregulated in MPM through promoter methylation silencing [144]. Reexpression of miR-34a/b through DAC treatment resulted in cellular arrest, and the reduction of migration and invasion in an MPM cell line [144]. hsa-miR-29c* analysis in MPM tumor tissues showed that its expression was an independent predictive factor for survival [145]. Induction of this miR in MPM cell lines resulted in decreased migration and invasion. The expression of this miR is regulated through DNMTs [145].

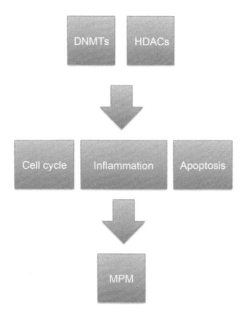

FIGURE 9.2

The link between epigenetic alterations and MPM. As detailed throughout the text, alterations in epigenetic modifiers such as HDACs and DNMT result in aberrant expression of many genes. In MPM, a number of these gene changes promote a chronic inflammatory state, a decrease in cell cycle inhibitors in addition to the promotion of anti-apoptotic mediators.

lncRNA are dysregulated in MPM and expression patterns can distinguish from normal (MET-5A) mesothelial cells and MPM tissue and are associated with nodal stage [146]. GAS5 is reduced in MPM cell lines compared with normal mesothelial cells and is associated with cell cycle length [147].

8 EPIGENETIC TARGETING IN MALIGNANT PLEURAL MESOTHELIOMA

The current standard of care for MPM patients is pemetrexed and cisplatin. One study has shown that VPA in combination with these drugs provided improved apoptosis in MPM cell lines and in cells from patient biopsies possibly through Bid and cytochrome c release and the production of reactive oxygen species. VPA on its own or in combination with theses chemotherapy drugs increased the acetylation of histone H3. In a mouse model of MPM, a combination of all three drugs resulted in tumor growth inhibition [148]. In a Phase II trial of VPA with doxorubicin resulted in seven partial responses in MPM patients [149]. In MPM cell lines, VPA alone or in combination with lovastatin induced autophagy and reduced cellular invasion [150]. In a Phase I study of SAHA in advanced cancer (13 MPM cases), resulted in two partial responses with increased acetylation in PBMCs [151]. In a Phase II study, PDX101 (bellinostat) was not effective as a monotherapy with no objective responses [152]. In MPM animal models, panobinostat treatment significantly reduced tumor growth compared with untreated

Table 9.3 Epigenetic Targeted Drugs in Clinical Trials for Mesothelioma

Drug	Additional drugs	NCT
DAC and Depsipeptide	Celecoxib	NCT00037817
Belinostat		NCT00365053
Belinostat	5-Fluorouracil	NCT00413322
LBH589	Dextromethorphan	NCT00535951
DAC		NCT00019825
SAHA		NCT00128102
Valproate	Doxorubicin	NCT00634205
Azacitidine	Temozolomide	NCT00629343
Entinostat		NCT00020579

mice, through the upregulation of p21, caspase 3/7 and the downregulation of BCL-2 and BCL-XL [153]. One of the biggest mesothelioma trials to date was VANTAGE 14. This was a Phase III study using vorinostat or placebo in a cohort of 661 patients who had previously been treated with chemotherapy. Vorinostat did not improve survival compared with placebo, with median survivals of 31 weeks and 27 weeks, respectively. However, progression free survival was improved but not in a clinically relevant way (Source: European Multidisciplinary Cancer Congress). A summary of other clinical trials undertaken in mesothelioma is outlined in Table 9.3.

Natural compounds can possess inherent HDi activity such as curcumin, which can induce autophagy in MPM cell lines [154]. In addition, curcumin can inhibit MPM cellular growth in both *in vitro* and *in vivo* models, through the activation of p38 kinase, caspases 3/9 with a concomitant increase of Bax and PARP cleavage. In addition, pretreatment with curcumin improved cisplatin efficacy [155].

In MPM cell lines, DAC induced senescence possibly through an increase in β galactosidase and also increased the phosphorylation γH2AX [156]. In addition, DAC treatment reduced MPM cell survival through an upregulation of p21 levels [157]. DAC in combination with VPA demonstrated synergistic effects in reducing MPM cellular survival and induced tumor antigen expression, thus increasing cell killing through CD8[+] cytotoxic T cells. This combination *in vivo* inhibited tumor growth and potentiated the immune response [158]. DAC can stimulate the expression of a range of antigens in MPM cell lines such as MAGE-1, 2, 3, and 4, NY-ESO-1, and synovial sarcoma X (SSX-2) [159]. In a Phase I trial (six MPM cases), DAC resulted in no responses in MPM patients, however, there was a reexpression of NY-ESO-1, MAGE3, and p16 [118]. DAC can enhance the therapeutic benefit of anti-IL-13Rα2 in MPM through increased growth inhibition. In *in vivo* models combination of both drugs significantly prolonged the survival of mice with MPM xenografts compares with either drug alone [160].

Cells grown in 3D culture are thought to more truly represent the physiological nature of tumors. MPM cells are usually resistant to reagents that promote apoptosis and this is evident in 3D culture. In 3D culture models, it was established that the resistance of MPM cells to apoptotic agents such as bortezomib is through BCL-2 dependant mechanisms [161]. Treatment of this 3D model with vorinostat overcame the resistance to bortezomib through the upregulation of NOXA. In addition, vorinostat enhanced the response of 3D-cultured cells to cisplatin and pemetrexed [162].

TNF levels are higher in MPM patients compared with controls, which would indicate the possibility of TRAIL (tumor necrosis factor–related apoptosis-inducing ligand) and SMAC compounds as putative treatment options. However, most MPM cells demonstrate resistance to these agents. SAHA treatment can sensitize MPM cells to the affect of TRAIL, with a downregulation of BCL-XL [163]. NaB can also reduce levels of the BCL–XL expression in MPM, leading to increased cell death [164]. MPM cells are also resistant to SMAC mimetic compounds. SAHA can sensitize cells to the effects of these compounds by decreasing the levels of the caspase 8 inhibitor, FLIP [165]. In mesothelioma cell lines, LBH589 increases the sensitivity to TRAIL, possibly through the degradation of XIAP [166].

9 CONCLUSION

Although HDi and DNMTi have shown promising effects in cell line and animal models, this has not translated through to the clinical setting. There are numerous reasons and issues that must be addressed. Intra-tumoral heterogeneity is a factor which can affect the efficacy of the majority of treatments [167] including epigenetic modifiers. Identification of robust biomarkers and adequate profiling of the tumor may help in overcoming this factor. The mode of delivery of the drug needs to be improved and in doing so will hopefully improve the drug therapeutic index, while decreasing the toxicity profiles. Studies using aerosolized forms of drug have proven beneficial. The next issue is the scheduling of treatment, as it may be better to pretreat with epigenetic therapies prior to the beginning of chemotherapeutic agents compared with using drugs simultaneously. Most studies have shown improved efficacy with current chemotherapy protocols, so the way forward may be combinations of drugs and indeed combinations of HDi and DNMTi with chemotherapy or other targeted agents. In addition, most clinical trials have been performed in heavily pretreated patients and this may be skewing trial results. If possible, future trials should be undertaken in treatment naïve patients. Furthermore, patients should undergo promoter methylation profiling and/or DNMT and/or HDAC expression level screens as a means to stratify for epigenetic therapies. It will be interesting to determine the HDAC expression levels in patients enrolled in the VANTAGE 14 trial and analyze specific responses to therapy within high and low level expression groups.

In the interim, epigenetic profiling of patients may prove beneficial as a screening tool for high-risk MPM and lung cancer patients, given the increased promoter methylation alterations as disease progresses as well as being a marker of asbestos burden. A number of studies have been carried out in recent times, particularly in sputum samples of high-risk lung cancer cohorts. The aberrant methylation of the promoter regions of *p16* and/or *MGMT* in sputum was detected in all patients up to 3 years before a clinical diagnosis of lung cancer [168]. Belinksy et al. demonstrated that the methylation of six of these genes in sputum (*p16, MGMT, DAPK, RASSF1A, PAX5β,* and *GATA5*) was associated with > 50% increased lung cancer risk [169]. An encouraging recent study has refined the number of methylated genes further. Wrangle et al. [170] have shown that the methylation of one or any of these three genes: *CDO1* (cysteine dioxygenase type 1), *HOXA9*, and *TAC1* (protachykinin-1) in lung tissue had 100% specificity and 83–99% sensitivity for NSCLC. Indeed, as these screening markers are validated in the clinical setting, the true power of epigenetic therapy may be at the precancer stage, and prevent high-risk patients from developing these diseases.

REFERENCES

[1] Ferlay J.S.I., Ervik M., Dikshit R., Eser S., Mathers C., Rebelo M., Parkin D.M., Forman D, Bray F. GLOBOCAN 2012 v1.0, Cancer incidence and mortality worldwide: IARC CancerBase No. 11. 2013 [accessed 10.05.14].

[2] Pfannschmidt J, Muley T, Bulzebruck H, Hoffmann H, Dienemann H. Prognostic assessment after surgical resection for non-small cell lung cancer: experiences in 2083 patients. Lung Cancer 2007;55(3):371–7.

[3] Molina JR, Yang P, Cassivi SD, Schild SE, Adjei AA. Non-small cell lung cancer: epidemiology, risk factors, treatment, and survivorship. Mayo Clin Proc 2008;83(5):584–94.

[4] Lindeman NI, Cagle PT, Beasley MB, et al. Molecular testing guideline for selection of lung cancer patients for EGFR and ALK tyrosine kinase inhibitors: guideline from the college of American Pathologists, International Association for the Study of Lung Cancer, and Association for Molecular Pathology. J Thorac Oncol 2013;8(7):823–59.

[5] Scagliotti GV, Parikh P, von Pawel J, et al. Phase III study comparing cisplatin plus gemcitabine with cisplatin plus pemetrexed in chemotherapy-naive patients with advanced-stage non-small-cell lung cancer. J Clin Oncol 2008;26(21):3543–51.

[6] Gridelli C, de Marinis F, Cappuzzo F, et al. Treatment of advanced non-small-cell lung cancer with epidermal growth factor receptor (EGFR) mutation or ALK gene rearrangement: results of an international expert panel meeting of the italian association of thoracic oncology. Clin Lung Cancer 2014;15(3):173–81.

[7] Baylin SB, Jones PA. A decade of exploring the cancer epigenome-biological and translational implications. Nat Rev Cancer 2011;11(10):726–34.

[8] Sato M, Shames DS, Gazdar AF, Minna JD. A translational view of the molecular pathogenesis of lung cancer. J Thorac Oncol 2007;2(4):327–43.

[9] Esteller M, Sanchez-Cespedes M, Rosell R, Sidransky D, Baylin SB, Herman JG. Detection of aberrant promoter hypermethylation of tumor suppressor genes in serum DNA from non-small cell lung cancer patients. Cancer Res 1999;59(1):67–70.

[10] Li W, Deng J, Jiang P, Zeng X, Hu S, Tang J. Methylation of the RASSF1A and RARbeta genes as a candidate biomarker for lung cancer. Exp Ther Med 2012;3(6):1067–71.

[11] Brock MV, Hooker CM, Ota-Machida E, et al. DNA methylation markers and early recurrence in stage I lung cancer. N Engl J Med 2008;358(11):1118–28.

[12] Lockwood WW, Wilson IM, Coe BP, et al. Divergent genomic and epigenomic landscapes of lung cancer subtypes underscore the selection of different oncogenic pathways during tumor development. PLoS One 2012;7(5):e37775.

[13] Tessema M, Klinge DM, Yingling CM, Do K, Van Neste L, Belinsky SA. Re-expression of CXCL14, a common target for epigenetic silencing in lung cancer, induces tumor necrosis. Oncogene 2010;29(37):5159–70.

[14] Wagner KW, Ye Y, Lin J, Vaporciyan AA, Roth JA, Wu X. Genetic variations in epigenetic genes are predictors of recurrence in stage I or II non-small cell lung cancer patients. Clin Cancer Res 2012;18(2):585–92.

[15] Lin RK, Hsu HS, Chang JW, Chen CY, Chen JT, Wang YC. Alteration of DNA methyltransferases contributes to 5'CpG methylation and poor prognosis in lung cancer. Lung Cancer 2007;55(2):205–13.

[16] Kim H, Kwon YM, Kim JS, et al. Elevated mRNA levels of DNA methyltransferase-1 as an independent prognostic factor in primary nonsmall cell lung cancer. Cancer 2006;107(5):1042–9.

[17] Vallbohmer D, Brabender J, Yang D, et al. DNA methyltransferases messenger RNA expression and aberrant methylation of CpG islands in non-small-cell lung cancer: association and prognostic value. Clin Lung Cancer 2006;8(1):39–44.

[18] Tang YA, Lin RK, Tsai YT, et al. MDM2 overexpression deregulates the transcriptional control of RB/E2F leading to DNA methyltransferase 3A overexpression in lung cancer. Clin Cancer Res 2012;18(16):4325–33.

[19] Xing J, Stewart DJ, Gu J, Lu C, Spitz MR, Wu X. Expression of methylation-related genes is associated with overall survival in patients with non-small cell lung cancer. Br J Cancer 2008;98(10):1716–22.

[20] Huqun Ishikawa R, Zhang J, et al. Enhancer of zeste homolog 2 is a novel prognostic biomarker in nonsmall cell lung cancer. Cancer 2012;118(6):1599–606.

[21] Kikuchi J, Kinoshita I, Shimizu Y, et al. Distinctive expression of the polycomb group proteins Bmi1 polycomb ring finger oncogene and enhancer of zeste homolog 2 in nonsmall cell lung cancers and their clinical and clinicopathologic significance. Cancer 2010;116(12):3015–24.

[22] Chen X, Song N, Matsumoto K, et al. High expression of trimethylated histone H3 at lysine 27 predicts better prognosis in non-small cell lung cancer. Int J Oncol 2013;43(5):1467–80.

[23] Yoon KA, Gil HJ, Han J, Park J, Lee JS. Genetic polymorphisms in the polycomb group gene EZH2 and the risk of lung cancer. J Thorac Oncol 2010;5(1):10–16.

[24] Kikuchi J, Takashina T, Kinoshita I, et al. Epigenetic therapy with 3-deazaneplanocin A, an inhibitor of the histone methyltransferase EZH2, inhibits growth of non-small cell lung cancer cells. Lung Cancer 2012;78(2):138–43.

[25] Asangani IA, Ateeq B, Cao Q, et al. Characterization of the EZH2-MMSET histone methyltransferase regulatory axis in cancer. Mol Cell 2013;49(1):80–93.

[26] Hudlebusch HR, Santoni-Rugiu E, Simon R, et al. The histone methyltransferase and putative oncoprotein MMSET is overexpressed in a large variety of human tumors. Clin Cancer Res 2011;17(9):2919–33.

[27] Toyokawa G, Cho HS, Masuda K, et al. Histone lysine methyltransferase wolf-hirschhorn syndrome candidate 1 is involved in human carcinogenesis through regulation of the Wnt pathway. Neoplasia (New York, NY) 2011;13(10):887–98.

[28] Mazur PK, Reynoird N, Khatri P, et al. SMYD3 links lysine methylation of MAP3K2 to Ras-driven cancer. Nature 2014;510(7504):283–7.

[29] Barlesi F, Giaccone G, Gallegos-Ruiz MI, et al. Genotype analysis of the VNTR polymorphism in the SMYD3 histone methyltransferase gene: lack of correlation with the level of histone H3 methylation in NSCLC tissues or with the risk of NSCLC. Int J Cancer 2008;122(6):1441–2.

[30] Kang D, Cho HS, Toyokawa G, et al. The histone methyltransferase wolf-hirschhorn syndrome candidate 1-like 1 (WHSC1L1) is involved in human carcinogenesis. Genes Chromosomes Cancer 2013;52(2):126–39.

[31] Mahmood SF, Gruel N, Nicolle R, et al. PPAPDC1B and WHSC1L1 are common drivers of the 8p11-12 amplicon, not only in breast tumors but also in pancreatic adenocarcinomas and lung tumors. Am J Pathol 2013;183(5):1634–44.

[32] Kuo CH, Chen KF, Chou SH, et al. Lung tumor-associated dendritic cell-derived resistin promoted cancer progression by increasing Wolf-Hirschhorn syndrome candidate 1/Twist pathway. Carcinogenesis 2013;34(11):2600–9.

[33] O'Byrne KJ, Baird AM, Kilmartin L, Leonard J, Sacevich C, Gray SG. Epigenetic regulation of glucose transporters in non-small cell lung cancer. Cancers 2011;3(2):1550–65.

[34] Chen MW, Hua KT, Kao HJ, et al. H3K9 histone methyltransferase G9a promotes lung cancer invasion and metastasis by silencing the cell adhesion molecule Ep-CAM. Cancer Res 2010;70(20):7830–40.

[35] Kim W, Kim R, Park G, Park JW, Kim JE. Deficiency of H3K79 histone methyltransferase Dot1-like protein (DOT1L) inhibits cell proliferation. J Biol Chem 2012;287(8):5588–99.

[36] Yoon KA, Hwangbo B, Kim IJ, et al. Novel polymorphisms in the SUV39H2 histone methyltransferase and the risk of lung cancer. Carcinogenesis 2006;27(11):2217–22.

[37] Yoon KA, Park S, Hwangbo B, et al. Genetic polymorphisms in the Rb-binding zinc finger gene RIZ and the risk of lung cancer. Carcinogenesis 2007;28(9):1971–7.

[38] Wang SA, Hung CY, Chuang JY, Chang WC, Hsu TI, Hung JJ. Phosphorylation of p300 increases its protein degradation to enhance the lung cancer progression. Biochim Biophys Acta 2014;1843(6):1135–49.

[39] Gao Y, Geng J, Hong X, et al. Expression of p300 and CBP is associated with poor prognosis in small cell lung cancer. Int J Clin Exp Pathol 2014;7(2):760–7.

[40] Gorgoulis VG, Zacharatos P, Mariatos G, et al. Transcription factor E2F-1 acts as a growth-promoting factor and is associated with adverse prognosis in non-small cell lung carcinomas. J Pathol 2002;198(2):142–56.

[41] Kishimoto M, Kohno T, Okudela K, et al. Mutations and deletions of the CBP gene in human lung cancer. Clin Cancer Res 2005;11(2 Pt 1):512–19.

[42] Zhao L, Wang DL, Liu Y, Chen S, Sun FL. Histone acetyltransferase hMOF promotes S phase entry and tumorigenesis in lung cancer. Cell Signal 2013;25(8):1689–98.

[43] Chen L, Wei T, Si X, et al. Lysine acetyltransferase GCN5 potentiates the growth of non-small cell lung cancer via promotion of E2F1, cyclin D1, and cyclin E1 expression. J Biol Chem 2013;288(20):14510–21.

[44] Van Den Broeck A, Nissou D, Brambilla E, Eymin B, Gazzeri S. Activation of a Tip60/E2F1/ERCC1 network in human lung adenocarcinoma cells exposed to cisplatin. Carcinogenesis 2012;33(2):320–5.

[45] O'Byrne KJ, Barr MP, Gray SG. The role of epigenetics in resistance to Cisplatin chemotherapy in lung cancer. Cancers 2011;3(1):1426–53.

[46] Cai D, Shames DS, Raso MG, et al. Steroid receptor coactivator-3 expression in lung cancer and its role in the regulation of cancer cell survival and proliferation. Cancer Res 2010;70(16):6477–85.

[47] Wagner KW, Alam H, Dhar SS, et al. KDM2A promotes lung tumorigenesis by epigenetically enhancing ERK1/2 signaling. J Clin Invest 2013;123(12):5231–46.

[48] Buonato JM, Lazzara MJ. ERK1/2 blockade prevents epithelial-mesenchymal transition in lung cancer cells and promotes their sensitivity to EGFR inhibition. Cancer Res 2014;74(1):309–19.

[49] Dhar SS, Alam H, Li N, et al. Transcriptional repression of histone deacetylase 3 by the histone demethylase KDM2A is coupled to tumorigenicity of lung cancer cells. J Biol Chem 2014;289(11):7483–96.

[50] Teng YC, Lee CF, Li YS, et al. Histone demethylase RBP2 promotes lung tumorigenesis and cancer metastasis. Cancer Res 2013;73(15):4711–21.

[51] Wang S, Wang Y, Wu H, Hu L. RBP2 induces epithelial-mesenchymal transition in non-small cell lung cancer. PLoS One 2013;8(12):e84735.

[52] Lv T, Yuan D, Miao X, et al. Over-expression of LSD1 promotes proliferation, migration and invasion in non-small cell lung cancer. PLoS One 2012;7(4):e35065.

[53] Hayami S, Kelly JD, Cho HS, et al. Overexpression of LSD1 contributes to human carcinogenesis through chromatin regulation in various cancers. Int J Cancer 2011;128(3):574–86.

[54] Sharma SK, Wu Y, Steinbergs N, et al. (Bis)urea and (bis)thiourea inhibitors of lysine-specific demethylase 1 as epigenetic modulators. J Med Chem 2010;53(14):5197–212.

[55] Mallette FA, Richard S. JMJD2A promotes cellular transformation by blocking cellular senescence through transcriptional repression of the tumor suppressor CHD5. Cell Rep 2012;2(5):1233–43.

[56] Hayami S, Yoshimatsu M, Veerakumarasivam A, et al. Overexpression of the JmjC histone demethylase KDM5B in human carcinogenesis: involvement in the proliferation of cancer cells through the E2F/RB pathway. Mol Cancer 2010;9:59.

[57] Cho HS, Toyokawa G, Daigo Y, et al. The JmjC domain-containing histone demethylase KDM3A is a positive regulator of the G1/S transition in cancer cells via transcriptional regulation of the HOXA1 gene. Int J Cancer 2012;131(3):E179–89.

[58] Yoshimatsu M, Toyokawa G, Hayami S, et al. Dysregulation of PRMT1 and PRMT6, Type I arginine methyltransferases, is involved in various types of human cancers. Int J Cancer 2011;128(3):562–73.

[59] Wei TY, Juan CC, Hisa JY, et al. Protein arginine methyltransferase 5 is a potential oncoprotein that upregulates G1 cyclins/cyclin-dependent kinases and the phosphoinositide 3-kinase/AKT signaling cascade. Cancer Sci 2012;103(9):1640–50.

[60] Gu Z, Gao S, Zhang F, et al. Protein arginine methyltransferase 5 is essential for growth of lung cancer cells. Biochem J 2012;446(2):235–41.

[61] Tanaka H, Hoshikawa Y, Oh-hara T, et al. PRMT5, a novel TRAIL receptor-binding protein, inhibits TRAIL-induced apoptosis via nuclear factor-kappaB activation. Mol Cancer Res 2009;7(4):557–69.

[62] Sasaki H, Moriyama S, Nakashima Y, et al. Histone deacetylase 1 mRNA expression in lung cancer. Lung Cancer 2004;46(2):171–8.

[63] Li Y, Chao Y, Fang Y, et al. MTA1 promotes the invasion and migration of non-small cell lung cancer cells by downregulating miR-125b. J Exp Clin Cancer Res 2013;32:33.

[64] Jung KH, Noh JH, Kim JK, et al. HDAC2 overexpression confers oncogenic potential to human lung cancer cells by deregulating expression of apoptosis and cell cycle proteins. J Cell Biochem 2012;113(6): 2167–77.

[65] Minamiya Y, Ono T, Saito H, et al. Strong expression of HDAC3 correlates with a poor prognosis in patients with adenocarcinoma of the lung. Tumour Biol 2010;31(5):533–9.

[66] Wang L, Xiang S, Williams KA, et al. Depletion of HDAC6 enhances cisplatin-induced DNA damage and apoptosis in non-small cell lung cancer cells. PLoS One 2012;7(9):e44265.

[67] Vannini A, Volpari C, Filocamo G, et al. Crystal structure of a eukaryotic zinc-dependent histone deacetylase, human HDAC8, complexed with a hydroxamic acid inhibitor. Proc Natl Acad Sci USA 2004;101(42):15064–9.

[68] Lai LC, Tsai MH, Chen PC, et al. SNP rs10248565 in HDAC9 as a novel genomic aberration biomarker of lung adenocarcinoma in non-smoking women. J Biomed Sci 2014;21:24.

[69] Okudela K, Mitsui H, Suzuki T, et al. Expression of HDAC9 in lung cancer--potential role in lung carcinogenesis. Int J Clin Exp Pathol 2014;7(1):213–20.

[70] Osada H, Tatematsu Y, Saito H, Yatabe Y, Mitsudomi T, Takahashi T. Reduced expression of class II histone deacetylase genes is associated with poor prognosis in lung cancer patients. Int J Cancer 2004;112(1):26–32.

[71] Barlesi F, Giaccone G, Gallegos-Ruiz MI, et al. Global histone modifications predict prognosis of resected non small-cell lung cancer. J Clin Oncol 2007;25(28):4358–64.

[72] Van Den Broeck A, Brambilla E, Moro-Sibilot D, et al. Loss of histone H4K20 trimethylation occurs in preneoplasia and influences prognosis of non-small cell lung cancer. Clin Cancer Res 2008;14(22):7237–45.

[73] Pandey M, Sultana S, Gupta KP. Involvement of epigenetics and microRNA-29b in the urethane induced inception and establishment of mouse lung tumors. Exp Mol Pathol 2014;96(1):61–70.

[74] Fabbri M, Garzon R, Cimmino A, et al. MicroRNA-29 family reverts aberrant methylation in lung cancer by targeting DNA methyltransferases 3A and 3B. Proc Natl Acad Sci USA 2007;104(40):15805–10.

[75] Lujambio A, Ropero S, Ballestar E, et al. Genetic unmasking of an epigenetically silenced microRNA in human cancer cells. Cancer Res 2007;67(4):1424–9.

[76] Brueckner B, Stresemann C, Kuner R, et al. The human let-7a-3 locus contains an epigenetically regulated microRNA gene with oncogenic function. Cancer Res 2007;67(4):1419–23.

[77] Incoronato M, Urso L, Portela A, et al. Epigenetic regulation of miR-212 expression in lung cancer. PLoS One 2011;6(11):e27722.

[78] Sun M, Liu XH, Lu KH, et al. EZH2-mediated epigenetic suppression of long noncoding RNA SPRY4-IT1 promotes NSCLC cell proliferation and metastasis by affecting the epithelial-mesenchymal transition. Cell Death Dis 2014;5:e1298.

[79] Tee AE, Ling D, Nelson C, et al. The histone demethylase JMJD1A induces cell migration and invasion by up-regulating the expression of the long noncoding RNA MALAT1. Oncotarget 2014;5(7):1793–804.

[80] Gutschner T, Hammerle M, Eissmann M, et al. The noncoding RNA MALAT1 is a critical regulator of the metastasis phenotype of lung cancer cells. Cancer Res 2013;73(3):1180–9.

[81] Yang J, Lin J, Liu T, et al. Analysis of lncRNA expression profiles in non-small cell lung cancers (NSCLC) and their clinical subtypes. Lung Cancer 2014.

[82] Tessema M, Yu YY, Stidley CA, et al. Concomitant promoter methylation of multiple genes in lung adenocarcinomas from current, former and never smokers. Carcinogenesis 2009;30(7):1132–8.

[83] Tessema M, Yingling CM, Liu Y, et al. Genome-wide unmasking of epigenetically silenced genes in lung adenocarcinoma from smokers and never smokers. Carcinogenesis 2014.

[84] Lin RK, Hsieh YS, Lin P, et al. The tobacco-specific carcinogen NNK induces DNA methyltransferase 1 accumulation and tumor suppressor gene hypermethylation in mice and lung cancer patients. J Clin Invest 2010;120(2):521–32.

[85] Belinsky SA. Silencing of genes by promoter hypermethylation: key event in rodent and human lung cancer. Carcinogenesis 2005;26(9):1481–7.

[86] Siu LL, Pili R, Duran I, et al. Phase I study of MGCD0103 given as a three-times-per-week oral dose in patients with advanced solid tumors. J Clin Oncol 2008;26(12):1940–7.

[87] Munster P, Marchion D, Bicaku E, et al. Phase I trial of histone deacetylase inhibition by valproic acid followed by the topoisomerase II inhibitor epirubicin in advanced solid tumors: a clinical and translational study. J Clin Oncol 2007;25(15):1979–85.

[88] Gray J, Cubitt CL, Zhang S, Chiappori A. Combination of HDAC and topoisomerase inhibitors in small cell lung cancer. Cancer Biol Ther 2012;13(8):614–22.

[89] Bevins RL, Zimmer SG. It's about time: scheduling alters effect of histone deacetylase inhibitors on camptothecin-treated cells. Cancer Res 2005;65(15):6957–66.

[90] Bruzzese F, Rocco M, Castelli S, Di Gennaro E, Desideri A, Budillon A. Synergistic antitumor effect between vorinostat and topotecan in small cell lung cancer cells is mediated by generation of reactive oxygen species and DNA damage-induced apoptosis. Mol Cancer Ther 2009;8(11):3075–87.

[91] Tessema M, Yingling CM, Thomas CL, et al. SULF2 methylation is prognostic for lung cancer survival and increases sensitivity to topoisomerase-I inhibitors via induction of ISG15. Oncogene 2012;31(37):4107–16.

[92] Tsai SC, Valkov N, Yang WM, Gump J, Sullivan D, Seto E. Histone deacetylase interacts directly with DNA topoisomerase II. Nat Genet 2000;26(3):349–53.

[93] Ramalingam SS, Maitland ML, Frankel P, et al. Carboplatin and paclitaxel in combination with either vorinostat or placebo for first-line therapy of advanced non-small-cell lung cancer. J Clin Oncol 2010;28(1):56–62.

[94] Traynor AM, Dubey S, Eickhoff JC, et al. Vorinostat (NSC# 701852) in patients with relapsed non-small cell lung cancer: a wisconsin oncology network phase II study. J Thorac Oncol 2009;4(4):522–6.

[95] Schelman WR, Traynor AM, Holen KD, et al. A phase I study of vorinostat in combination with bortezomib in patients with advanced malignancies. Invest New Drugs 2013;31(6):1539–46.

[96] Hoang T, Campbell TC, Zhang C, et al. Vorinostat and bortezomib as third-line therapy in patients with advanced non-small cell lung cancer: a wisconsin oncology network phase II study. Invest New Drugs 2014;32(1):195–9.

[97] Ryan QC, Headlee D, Acharya M, et al. Phase I and pharmacokinetic study of Ms-275, a histone deacetylase inhibitor, in patients with advanced and refractory solid tumors or lymphoma. J Clin Oncol 2005;23(17):3912–22.

[98] Kummar S, Gutierrez M, Gardner ER, et al. Phase I trial of Ms-275, a histone deacetylase inhibitor, administered weekly in refractory solid tumors and lymphoid malignancies. Clin Cancer Res 2007;13(18 Pt 1):5411–17.

[99] Gore L, Rothenberg ML, O'Bryant CL, et al. A phase I and pharmacokinetic study of the oral histone deacetylase inhibitor, Ms-275, in patients with refractory solid tumors and lymphomas. Clin Cancer Res 2008;14(14):4517–25.

[100] Witta SE, Jotte RM, Konduri K, et al. Randomized phase II trial of erlotinib with and without entinostat in patients with advanced non-small-cell lung cancer who progressed on prior chemotherapy. J Clin Oncol 2012;30(18):2248–55.

[101] Hu E, Dul E, Sung CM, et al. Identification of novel isoform-selective inhibitors within class I histone deacetylases. J Pharmacol Exp Ther 2003;307(2):720–8.

[102] Sandor V, Bakke S, Robey RW, et al. Phase I trial of the histone deacetylase inhibitor, depsipeptide (FR901228, NSC 630176), in patients with refractory neoplasms. Clin Cancer Res 2002;8(3):718–28.

[103] Schrump DS, Fischette MR, Nguyen DM, et al. Clinical and molecular responses in lung cancer patients receiving romidepsin. Clin Cancer Res 2008;14(1):188–98.

[104] Chang J, Varghese DS, Gillam MC, et al. Differential response of cancer cells to HDAC inhibitors trichostatin A and depsipeptide. Br J Cancer 2012;106(1):116–25.

[105] Luchenko VL, Salcido CD, Zhang Y, et al. Schedule-dependent synergy of histone deacetylase inhibitors with DNA damaging agents in small cell lung cancer. Cell Cycle 2011;10(18):3119–28.

[106] Reid T, Valone F, Lipera W, et al. Phase II trial of the histone deacetylase inhibitor pivaloyloxymethyl butyrate (Pivanex, AN-9) in advanced non-small cell lung cancer. Lung Cancer 2004;45(3):381–6.

[107] Mateen S, Raina K, Agarwal R. Chemopreventive and anti-cancer efficacy of silibinin against growth and progression of lung cancer. Nutr Cancer 2013;65(Suppl. 1):3–11.

[108] Mateen S, Raina K, Agarwal C, Chan D, Agarwal R. Silibinin synergizes with histone deacetylase and DNA methyltransferase inhibitors in upregulating E-cadherin expression together with inhibition of migration and invasion of human non-small cell lung cancer cells. J Pharmacol Exp Ther 2013;345(2):206–14.

[109] Mateen S, Raina K, Jain AK, Agarwal C, Chan D, Agarwal R. Epigenetic modifications and p21-cyclin B1 nexus in anticancer effect of histone deacetylase inhibitors in combination with silibinin on non-small cell lung cancer cells. Epigenetics 2012;7(10):1161–72.

[110] Reed MD, Tellez CS, Grimes MJ, et al. Aerosolised 5-azacytidine suppresses tumour growth and reprogrammes the epigenome in an orthotopic lung cancer model. Br J Cancer 2013;109(7):1775–81.

[111] Mahesh S, Saxena A, Qiu X, Perez-Soler R, Zou Y. Intratracheally administered 5-azacytidine is effective against orthotopic human lung cancer xenograft models and devoid of important systemic toxicity. Clin Lung Cancer 2010;11(6):405–11.

[112] Belinsky SA, Grimes MJ, Picchi MA, et al. Combination therapy with vidaza and entinostat suppresses tumor growth and reprograms the epigenome in an orthotopic lung cancer model. Cancer Res 2011;71(2):454–62.

[113] Lin J, Gilbert J, Rudek MA, et al. A phase I dose-finding study of 5-azacytidine in combination with sodium phenylbutyrate in patients with refractory solid tumors. Clin Cancer Res 2009;15(19):6241–9.

[114] Braiteh F, Soriano AO, Garcia-Manero G, et al. Phase I study of epigenetic modulation with 5-azacytidine and valproic acid in patients with advanced cancers. Clin Cancer Res 2008;14(19):6296–301.

[115] Chai G, Li L, Zhou W, et al. HDAC inhibitors act with 5-aza-2'-deoxycytidine to inhibit cell proliferation by suppressing removal of incorporated abases in lung cancer cells. PLoS One 2008;3(6):e2445.

[116] Belinsky SA, Klinge DM, Stidley CA, et al. Inhibition of DNA methylation and histone deacetylation prevents murine lung cancer. Cancer Res 2003;63(21):7089–93.

[117] Chu BF, Karpenko MJ, Liu Z, et al. Phase I study of 5-aza-2'-deoxycytidine in combination with valproic acid in non-small-cell lung cancer. Cancer Chemother Pharmacol 2013;71(1):115–21.

[118] Schrump DS, Fischette MR, Nguyen DM, et al. Phase I study of decitabine-mediated gene expression in patients with cancers involving the lungs, esophagus, or pleura. Clin Cancer Res 2006;12(19):5777–85.

[119] Rao M, Chinnasamy N, Hong JA, et al. Inhibition of histone lysine methylation enhances cancer-testis antigen expression in lung cancer cells: implications for adoptive immunotherapy of cancer. Cancer Res 2011;71(12):4192–204.

[120] Juergens RA, Wrangle J, Vendetti FP, et al. Combination epigenetic therapy has efficacy in patients with refractory advanced non-small cell lung cancer. Cancer Discov 2011;1(7):598–607.

[121] Tellez CS, Grimes MJ, Picchi MA, et al. SGI-110 and entinostat therapy reduces lung tumor burden and reprograms the epigenome. Int J Cancer 2014.

[122] Wagner JC, Sleggs CA, Marchand P. Diffuse pleural mesothelioma and asbestos exposure in the North Western Cape Province. Br J Ind Med 1960;17:260–71.

[123] Robinson BW, Musk AW, Lake RA. Malignant mesothelioma. Lancet 2005;366(9483):397–408.

[124] Robinson BW, Creaney J, Lake R, et al. Mesothelin-family proteins and diagnosis of mesothelioma. Lancet 2003;362(9396):1612–16.

[125] Peto J, Hodgson JT, Matthews FE, Jones JR. Continuing increase in mesothelioma mortality in Britain. Lancet 1995;345(8949):535–9.

[126] Driscoll T, Nelson DI, Steenland K, et al. The global burden of disease due to occupational carcinogens. Am J Ind Med 2005;48(6):419–31.

[127] Hodgson JT, McElvenny DM, Darnton AJ, Price MJ, Peto J. The expected burden of mesothelioma mortality in Great Britain from 2002 to 2050. Br J Cancer 2005;92(3):587–93.

[128] Fennell DA, Gaudino G, O'Byrne KJ, Mutti L, van Meerbeeck J. Advances in the systemic therapy of malignant pleural mesothelioma. Nat Clin Pract Oncol 2008;5(3):136–47.

[129] Vogelzang NJ, Rusthoven JJ, Symanowski J, et al. Phase III study of pemetrexed in combination with cisplatin versus cisplatin alone in patients with malignant pleural mesothelioma. J Clin Oncol 2003;21(14):2636–44.

[130] Treasure T, Waller D, Tan C, et al. The mesothelioma and radical surgery randomized controlled trial: the mars feasibility study. J Thorac Oncol 2009;4(10):1254–8.

[131] Treasure T, Lang-Lazdunski L, Waller D, et al. Extra-pleural pneumonectomy versus no extra-pleural pneumonectomy for patients with malignant pleural mesothelioma: clinical outcomes of the Mesothelioma and Radical Surgery (MARS) randomised feasibility study. Lancet Oncol 2011;12(8):763–72.

[132] Weder W, Stahel RA, Baas P, et al. The MARS feasibility trial: conclusions not supported by data. Lancet Oncol 2011;12(12):1093–4. author reply 4–5.

[133] Weder W, Stahel RA, Bernhard J, et al. Multicenter trial of neo-adjuvant chemotherapy followed by extra-pleural pneumonectomy in malignant pleural mesothelioma. Ann Oncol 2007;18(7):1196–202.

[134] Kassis ES, Zhao M, Hong JA, Chen GA, Nguyen DM, Schrump DS. Depletion of DNA methyltransferase 1 and/or DNA methyltransferase 3b mediates growth arrest and apoptosis in lung and esophageal cancer and malignant pleural mesothelioma cells. J Thorac Cardiovasc Surg 2006;131(2):298–306.

[135] Christensen BC, Houseman EA, Godleski JJ, et al. Epigenetic profiles distinguish pleural mesothelioma from normal pleura and predict lung asbestos burden and clinical outcome. Cancer Res 2009;69(1):227–34.

[136] Goto Y, Shinjo K, Kondo Y, et al. Epigenetic profiles distinguish malignant pleural mesothelioma from lung adenocarcinoma. Cancer Res 2009;69(23):9073–82.

[137] Tsou JA, Galler JS, Wali A, et al. DNA methylation profile of 28 potential marker loci in malignant mesothelioma. Lung Cancer 2007;58(2):220–30.

[138] Kohno H, Amatya VJ, Takeshima Y, et al. Aberrant promoter methylation of WIF-1 and SFRP1, 2, 4 genes in mesothelioma. Oncol Rep 2010;24(2):423–31.

[139] Suzuki M, Toyooka S, Shivapurkar N, et al. Aberrant methylation profile of human malignant mesotheliomas and its relationship to SV40 infection. Oncogene 2005;24(7):1302–8.

[140] Fischer JR, Ohnmacht U, Rieger N, et al. Promoter methylation of RASSF1A, RARbeta and DAPK predict poor prognosis of patients with malignant mesothelioma. Lung Cancer 2006;54(1):109–16.

[141] Kemp CD, Rao M, Xi S, et al. Polycomb repressor complex-2 is a novel target for mesothelioma therapy. Clin Cancer Res 2012;18(1):77–90.

[142] Christensen BC, Godleski JJ, Marsit CJ, et al. Asbestos exposure predicts cell cycle control gene promoter methylation in pleural mesothelioma. Carcinogenesis 2008;29(8):1555–9.

[143] Cheng YY, Kirschner MB, Cheng NC, et al. ZIC1 is silenced and has tumor suppressor function in malignant pleural mesothelioma. J Thorac Oncol 2013;8(10):1317–28.

[144] Kubo T, Toyooka S, Tsukuda K, et al. Epigenetic silencing of microRNA-34b/c plays an important role in the pathogenesis of malignant pleural mesothelioma. Clin Cancer Res 2011;17(15):4965–74.

[145] Pass HI, Goparaju C, Ivanov S, et al. hsa-miR-29c* is linked to the prognosis of malignant pleural mesothelioma. Cancer Res 2010;70(5):1916–24.

[146] Wright CM, Kirschner MB, Cheng YY, et al. Long non coding RNAs (lncRNAs) are dysregulated in Malignant Pleural Mesothelioma (MPM). PLoS One 2013;8(8):e70940.

[147] Renganathan A, Kresoja-Rakic J, Echeverry N, et al. GAS5 long non-coding RNA in malignant pleural mesothelioma. Mol Cancer 2014;13(1):119.

[148] Vandermeers F, Hubert P, Delvenne P, et al. Valproate, in combination with pemetrexed and cisplatin, provides additional efficacy to the treatment of malignant mesothelioma. Clin Cancer Res 2009;15(8):2818–28.

[149] Scherpereel A, Berghmans T, Lafitte JJ, et al. Valproate-doxorubicin: promising therapy for progressing mesothelioma. A phase II study. Eur Respir J 2011;37(1):129–35.

[150] Yamauchi Y, Izumi Y, Asakura K, et al. Lovastatin and valproic acid additively attenuate cell invasion in ACC-MESO-1 cells. Biochem Biophys Res Commun 2011;410(2):328–32.

[151] Krug LM, Curley T, Schwartz L, et al. Potential role of histone deacetylase inhibitors in mesothelioma: clinical experience with suberoylanilide hydroxamic acid. Clin Lung Cancer 2006;7(4):257–61.

[152] Ramalingam SS, Belani CP, Ruel C, et al. Phase II study of belinostat (PXD101), a histone deacetylase inhibitor, for second line therapy of advanced malignant pleural mesothelioma. J Thorac Oncol 2009;4(1):97–101.

[153] Crisanti MC, Wallace AF, Kapoor V, et al. The HDAC inhibitor panobinostat (LBH589) inhibits mesothelioma and lung cancer cells in vitro and in vivo with particular efficacy for small cell lung cancer. Mol Cancer Ther 2009;8(8):2221–31.

[154] Yamauchi Y, Izumi Y, Asakura K, Hayashi Y, Nomori H. Curcumin induces autophagy in ACC-MESO-1 cells. Phytother Res 2012;26(12):1779–83.

[155] Wang Y, Rishi AK, Wu W, et al. Curcumin suppresses growth of mesothelioma cells in vitro and in vivo, in part, by stimulating apoptosis. Mol Cell Biochem 2011;357(1–2):83–94.

[156] Amatori S, Bagaloni I, Viti D, Fanelli M. Premature senescence induced by DNA demethylating agent (Decitabine) as therapeutic option for malignant pleural mesothelioma. Lung Cancer 2011;71(1):113–15.

[157] Amatori S, Papalini F, Lazzarini R, et al. Decitabine, differently from DNMT1 silencing, exerts its antiproliferative activity through p21 upregulation in malignant pleural mesothelioma (MPM) cells. Lung Cancer 2009;66(2):184–90.

[158] Leclercq S, Gueugnon F, Boutin B, et al. A 5-aza-2′-deoxycytidine/valproate combination induces cytotoxic T-cell response against mesothelioma. Eur Respir J 2011;38(5):1105–16.

[159] Sigalotti L, Coral S, Altomonte M, et al. Cancer testis antigens expression in mesothelioma: role of DNA methylation and bioimmunotherapeutic implications. Br J Cancer 2002;86(6):979–82.

[160] Takenouchi M, Hirai S, Sakuragi N, Yagita H, Hamada H, Kato K. Epigenetic modulation enhances the therapeutic effect of anti-IL-13R(alpha)2 antibody in human mesothelioma xenografts. Clin Cancer Res 2011;17(9):2819–29.

[161] Barbone D, Ryan JA, Kolhatkar N, et al. The Bcl-2 repertoire of mesothelioma spheroids underlies acquired apoptotic multicellular resistance. Cell Death Dis 2011;2:e174.

[162] Barbone D, Cheung P, Battula S, et al. Vorinostat eliminates multicellular resistance of mesothelioma 3D spheroids via restoration of Noxa expression. PLoS One 2012;7(12):e52753.

[163] Neuzil J, Swettenham E, Gellert N. Sensitization of mesothelioma to TRAIL apoptosis by inhibition of histone deacetylase: role of Bcl-xL down-regulation. Biochem Biophys Res Commun 2004;314(1):186–91.

[164] Cao XX, Mohuiddin I, Ece F, McConkey DJ, Smythe WR. Histone deacetylase inhibitor downregulation of bcl-xl gene expression leads to apoptotic cell death in mesothelioma. Am J Respir Cell Mol Biol 2001;25(5):562–8.

[165] Crawford N, Stasik I, Holohan C, et al. SAHA overcomes FLIP-mediated inhibition of SMAC mimetic-induced apoptosis in mesothelioma. Cell Death Dis 2013;4:e733.

[166] Symanowski J, Vogelzang N, Zawel L, Atadja P, Pass H, Sharma S. A histone deacetylase inhibitor LBH589 downregulates XIAP in mesothelioma cell lines which is likely responsible for increased apoptosis with TRAIL. J Thorac Oncol 2009;4(2):149–60.

[167] Lavi O, Greene JM, Levy D, Gottesman MM. The role of cell density and intratumoral heterogeneity in multidrug resistance. Cancer Res 2013;73(24):7168–75.

[168] Palmisano WA, Divine KK, Saccomanno G, et al. Predicting lung cancer by detecting aberrant promoter methylation in sputum. Cancer Res 2000;60(21):5954–8.

[169] Belinsky SA, Liechty KC, Gentry FD, et al. Promoter hypermethylation of multiple genes in sputum precedes lung cancer incidence in a high-risk cohort. Cancer Res 2006;66(6):3338–44.

[170] Wrangle J, Machida EO, Danilova L, et al. Functional identification of cancer-specific methylation of CDO1, HOXA9, and TAC1 for the diagnosis of lung cancer. Clin Cancer Res 2014;20(7):1856–64.

BREAST CANCER EPIGENETICS 10

Chara A. Pitta and Andreas I. Constantinou

*Laboratory of Cancer Biology and Chemoprevention, Department of Biological Sciences,
School of Pure and Applied Sciences, University of Cyprus, Nicosia, Cyprus*

CHAPTER OUTLINE

1 INTRODUCTION

Breast cancer is the most common cancer among women in the United States. Approximately 200,000 women and 2000 men are diagnosed with breast cancer each year (http://www.cancer.org). There are well-understood genetic alterations related to breast carcinogenesis, including specific gene amplifications, deletions, point mutations, chromosome rearrangements, and aneuploidy. In addition to these vastly described mutations, epigenetic alterations resulting in aberrant gene expression are key contributors to breast tumorigenesis. These modifications are quite attractive as targets for preventative care and therapeutics because of their potential for reversal [1–3].

Epigenetic modifications can be defined as constant molecular alterations of a cellular phenotype, such as the gene expression profile of a cell, that are heritable during somatic cell divisions (and sometimes germ line transmissions) but do not engage alterations of the DNA sequence. Epigenetic mechanisms transmit genomic adaption to the environment, ultimately contributing toward a given phenotype. Epigenetic events are mediated by a number of molecular mechanisms including DNA methylation, histone modifications, polycomb/trithorax protein complexes, and small noncoding or

S.G. Gray (Ed): Epigenetic Cancer Therapy. DOI: http://dx.doi.org/10.1016/B978-0-12-800206-3.00010-0

antisense RNAs. These different alterations are strongly interconnected and play a significant role in normal growth and development [4]. Potential medical care for breast cancer patients will likely depend upon a better understanding of the significance of epigenetic modifications in tumorigenesis.

2 DNA METHYLATION AND BREAST CANCER

DNA methylation is a covalent chemical addition of a methyl (CH_3) group at the carbon 5 position of the cytosine ring. Most often cytosine methylation occurs in the sequence context 5'-CG-3', also called the CpG dinucleotide. However in some instances it involves the CpA and CpT dinucleotides [5]. The human genome is not consistently methylated as it includes regions of unmethylated segments that are flanked by methylated regions [6]. CpG islands ranging from 0.5 to 5 kb are distributed on the average every 100 kb. Approximately, half of the humanness contains CpG islands [7], and these are present on both housekeeping genes and genes with tissue-specific patterns of expression [8].

Methylation of DNA is mediated by enzymes collectively known as DNA methyltransferases (DNMTs). The terminal catalytic domain of DNMTs transfers methyl groups onto cytosine residues within the DNA [9]. In mammals, five members of the DNMT protein family have been well characterized (Dnmt1, 2, 3a, 3b, 3L). Only Dnmt1, Dnmt3a, and Dnmt3b are shown to possess catalytic methyltransferase activity. Dnmt1 strongly prefers to hemimethylate over unmethylated DNA and is targeting the replication foci, as shown by colocalization with the proliferating cell nuclear antigen (PCNA). Thus, DNMT1 is engaged into copying the parental DNA methylation pattern onto the newly synthesized DNA daughter strand and is regarded as a maintenance methyltransferase. Dnmt3a and Dnmt3b, the two other catalytic active members, are *de novo* methyltransferases as they exhibit increased methyltransferase activity toward unmethylated over hemimethylated DNA [9].

DNMTs are ubiquitously expressed at discrete levels in normal human tissues [9]. However, they are overexpressed in various tumor types, such as leukemia, colorectal cancer, prostate cancer, ovarian cancer, endometrial cancer, and breast cancer [10–16]. In a series of 130 primary unilateral nonmetastatic breast carcinomas, *DNMT1*, DNMT*3a*, and DNMT*3b* mRNA levels were correlated positively with each other, suggesting a common regulation pathway. *DNMT3b* overexpression status was being observed in 30% of the 130 patients; in contrast, only 5.4% and 3.1% of the patients' tumors overexpressed *DNMT1* and *DNMT3a*, respectively [10]. High expression of DNMT3b was determined to be correlated with higher histological grade, lack of estrogen receptor-α (ER-α) and presence of the proliferation marker Ki67, pointing to a prospective involvement of DNMT3b in breast tumor progression and aggressiveness [10]. An association of high DNMT3b expression and reduced disease-free survival time was also detected in a group of patients receiving adjuvant hormone therapy.

One can visualize two possible scenarios with respect of gene promoter CpG island methylation that may affect the transcriptional expression of genes and leading to carcinogenesis: (i) normally unmethylated CpG islands may become abnormally methylated resulting in gene silencing of tumor suppressor genes and (ii) CpG islands that are normally methylated may become unmethylated, resulting in unsilencing of genes such as oncogenes or retrotransposons [17].

Besides tumor suppressor genes, other classes of genes that are silenced by DNA hypermethylation are those that suppress tumor invasion and metastasis, DNA repair genes, hormone receptor, cellular homeostasis, cell adhesion genes, as well as genes that inhibit cell cycle progression, angiogenesis, and cell survival. As of the present, more than 100 genes have been described to be hypermethylated in breast cancer cell lines and breast tumors [18,19]. Specific examples include, cyclin D2 (CCND2) which is an imperative regulator of the cell cycle and when it is overexpressed it inhibits the transition between G1 and S phases. CCND2 is frequently methylated in breast cancer and is also methylated in ductal carcinomas *in situ* (DCIS), suggesting that this is an event during tumorigenesis [20]. p16ink4A/CDKN2A is another cell cycle regulator attracting a lot of interest and is frequently methylated in many human cancers including breast cancer [21]. Methylation of CDKN2A in human mammary epithelial cells (HMECs) leads to escape from senescence, telomere crisis, and chromosomal abnormalities similar to those taking place during neoplasia [22,23]. CDH3 is a cell adhesion molecule regularly silenced in breast carcinomas by DNA methylation and this silencing might be significant for tumor cell invasion and metastasis [22]. HIN1 is an inhibitor of cell growth, migration, and invasion that is commonly silenced by DNA methylation in breast cancer [24,25].

Breast cancer genes critical in the development of the disease are frequently not expressed due to promoter hypermethylation; among them are the ER-α and ER-β genes and BRCA1 [26]. The steroid receptor, *ER* and *PR*, genes have long been associated with breast cancer. Decreased ER-α and ER-β expression has been extensively reported in various cancer types including breast, ovary, colon, and prostate [27–33]. *ER-α* and *ER-β* promoter regions contain CpG islands, making them susceptible to epigenetic modifications. Importantly, loss of ER-β expression was found to be associated with the transition from *in situ* to invasive ductal carcinomas, and this silencing was mediated by reversible promoter DNA hypermethylation in ER-β negative breast cancer cell lines [34]. It has also been suggested that expression of different ER-β isoforms in breast cancer may also be regulated by DNA methylation [35].

The BRCA1 gene located on chromosome 17q21 is one of the most commonly altered genes in breast cancer, and often its protein product is decreased or absent. DNA methylation has been suggested as one of the causes of its inactivation [26]. BRCA1 methylation has been demonstrated in sporadic breast cancers but it is not a repeated event [36,37], and it is possible that BRCA1 methylation is most common in rare subtypes of basal-like origin [38]. Taking into account its important role in familial breast cancer and the fact that no BRCA1 mutations are evident in sporadic breast cancers, DNA methylation became attractive mechanism for BRCA1 silencing in breast cancer tumors. Sporadic and inherited breast tumors have overall similar methylation profiles but BRCA1 tumors have reduced methylation of certain nonfamilial genes and have a phenotype correlated with basal-like carcinomas [37].

The above reports clearly demonstrate that gene silencing by hypermethylation of CpG island gene promoters is a common mechanism of breast carcinogenesis and consequently it has a great potential for cancer prevention and therapy. Furthermore, the methylation status of these genes could serve as markers for breast cancer prognosis and response to epigenetic therapy.

High-resolution analysis of DNA hypomethylation in breast cancer demonstrated a large number of hypomethylated sites with around 1500 regions hypomethylated in a cancer-specific manner [39,40]. It is possible that many of these regions include genes or regulatory sequences that play crucial roles in tumorigenesis. Among the genes that are hypomethylated in primary breast tumors are the genes for endonucleases FEN1 [41], cadherin CDH3 [42], *N*-acetyltransferase NAT1 [43], breast cancer-specific gene 1 BCSG1 [44], and urokinase-type plasminogen activator PLAU [45]. The only imprinted genes that have been described to be hypomethylated in breast cancer so far is the insulin-like growth factor II IGF2 gene and ARH1 [46].

Although breast cancer can be segregated into distinct histological subclasses, there is no evidence for large differences in DNA methylation patterns between ductal and lobular breast cancers [47,48]. Regardless, distinct epigenetic profiles were evident in breast tumors classified based on their hormone receptor status [49]. The candidate genes examined were typically less methylated in ER-negative tumors compared to the ER-positive tumors. Mutations in the p53 gene have also been associated to differential methylation patterns in breast cancer where tumors with p53 mutations are typically hypomethylated compared with tumors with wild p53 [49]. Since mutations in p53 and loss of ER expression are directly linked to the basal-like expression subtype, it is likely that tumors with a basal-like phenotype are hypomethylated in cancer in a specific manner.

It is becoming clear from the previous studies that during breast tumor carcinogenesis and tumor development, deregulation of normal cell signaling, which is maintained by the epigenetic machinery, may result in abnormal gene silencing/unsilencing.

3 THE ROLE OF HISTONE DEACETYLASES IN BREAST CANCER

The connection between acetylation and transcription had long been assumed. Acetylation neutralizes the positive charge of the histone lysine residues, unfolding the chromatin and enabling greater accessibility of the transcription machinery [50]. Consequently, histone acetylation is associated with gene activation, while the exclusion of acetyl groups from histones stimulates chromatin condensation and transcriptional repression [50].

Histone acetylation is catalyzed by histone acetyltransferases (HATs). These enzymes transfer acetyl groups from acetyl CoA to the ε-amino group of lysine residue. The human coactivator p300/CBP has HAT activity which is closely related with its effect on transcription. p300/CBP interacts with numerous DNA-binding regulatory proteins integrating and transducing signals controlling the cell cycle, differentiation, DNA repair, and apoptosis. Histone deacetylases (HDACs) stimulate the removal of acetyl group from the acetylated residues, releasing an acetate molecule [51,52]. Eighteen human HDACs have been recognized, identified, and grouped into four classes on the basis of their homology with yeast proteins. These are: class I (HDAC1, 2, 3 and 8), class II (HDAC4, 5, 6, 7, 9, 10), class III that are also called sirtuins and are homologous with Sir2 (SIRT1, 2, 3, 4, 5, 6, and 7), and class IV (HDAC11) which is homologous to class I and class II enzymes. The class III group is considered an atypical category of its own, as it is NAD^+ dependent, whereas the other groups require Zn^{2+} as a cofactor. Because of their gene regulatory significance and their differential expression at both the mRNA and protein levels, HDACs attracted strong interest among cancer researchers including breast cancer [50].

Normal mammary glands exhibit nuclear positivity for HDAC1 in the luminal epithelium but not in the basal epithelium. HDAC1 protein expression was observed in 40% and HDAC3 in 44% of breast cancer cases. HDAC1 mRNA protein levels and HDAC3 protein levels were found to be elevated in ER and progesterone receptor (PR) positive tumors [53,54]. HDAC1 and HDAC3 expression was correlated to steroid hormone receptor, Her2/neu and proliferation status of breast cancer tumors [53,54]. This was also partially confirmed in patients with invasive breast cancers since HDAC1 protein expression levels were found to predict disease free survival (DFS) but not overall survival (OS); survival differences were especially striking in the group of patients with small tumors of all differentiation types. mRNA expression of HDAC6 has been reported to be mainly prominent in small, low-grade, ER and PR positive tumors in comparison to larger high-grade hormone receptor negative cancers [55]. These studies point out that the targeting of HDACs is an active area of investigation with promise to find applications in breast cancer prognosis and targeted therapeutics.

4 HISTONE MODIFICATIONS

Histone modifications operate in a synchronized and orderly fashion and directly affect chromatin function and structure. Histone modifications and histone-modifying complexes play a crucial role in chromatin-based processes including DNA repair, DNA replication, and gene transcription [56]. In a histone octamer, positively charged histone tails extrude from the central domain of the nucleosome and bind the negatively charged DNA, through charge interactions, or arbitrate interactions between nucleosomes contributing to chromatin compaction [57]. Over 60 different residues in histones have been found to be modified, using either specific antibodies or mass spectrometry for their detection. Methylation at lysines or arginines may take one of three different forms, mono-, di- or trimethyl for the lysines and mono- or di- for the arginine residues, and this makes the interpretation of their regulatory consequences even more complicated.

Considering the essential role of histone modifications, it is not surprising that aberrant histone modifications are often associated with the cancer tissues. Fundamentally, all these abnormalities have been demonstrated to occur at individual gene promoters due to inappropriate targeting of one or more histone-modifying enzymes. Promoter-specific epigenetic patterns result in improper suppression or activation of genes that could eventually lead to cellular transformation, carcinogenesis, or cancer progression. The use of the chromatin immunoprecipitation (ChIP) technique [58,59] that allows the determination of the sites of histones that interact with specific DNA sequences provided valuable information regarding histone modifications and transcriptional regulation. For instance, it was determined that silencing of the CDKN2A locus (which encodes the p16INK4A and p14ARF tumor suppressors) is correlated with hypermethylation of mK9H3 and hypomethylation of meK4H3 [60]. Lower levels of histone acK16H4 and meK20H3 were found in hematological malignancies and colorectal adenocarcinomas compared to normal tissues [61].

Breast cancer is a heterogeneous disease, ranging from premalignant hyperplasia to invasive and metastatic carcinomas. Assessment of a well-characterized series of human breast carcinomas demonstrated that there was a strong correlation between global histone modifications status, tumor biomarker phenotype, and clinical outcome [62]. High levels of global histone acetylation and methylation were detected almost exclusively in luminal-like breast tumors and this was correlated with a favorable prognosis [62]. To the contrary, lower levels of histone acetylation and methylation (such as H3K18ac, H4K12ac, H3K4me2, H4K20me3, and H4R3me2) were detected in basal carcinomas and this was associated with unfavorable response [62]. Moreover, this analysis also showed that the majority of breast cancer cases have low or absent H4K16ac, suggesting that this modification may represent an early sign of breast cancer [63]. In addition to variations in the global levels of histone modifications, breast cancer also appears to exhibit gene-specific histone alterations. ChIP analysis of breast cancer tissues from cancer patients revealed that centromeric satellite (SAT2) levels on H3K9me3 and H4K20me3 are upregulated in breast cancer and downregulated in colorectal cancer compared with healthy patients [64]. We have shown previously that in the breast cancer tamoxifen-resistant cell line MCF-7/TAMR, ER-β is silenced due to the recruitment of enzymes producing histone modifications that create an environment of heterochromatin around the promoter that makes the gene inaccessible to the transcriptional machinery [3]. The 200 region of the ER-β promoter is where most of the changes occurred. Repressive chromatin marks like 2meK9H3, 3meK9H3, 2meK27H3, and 3meK27H3 were observed in this region [3]. This is consistent with previous findings with other genes associated with growth control showing that 3meK27H3 is associated with gene silencing and frequently represses genes [65–67]. Uncharacteristic expression of a number of histone-modifying enzymes has also been associated with breast cancer prognosis. Overexpression of Enhancer of Zeste (EZH2), which is part of polycomb repressive complex 2 (PRC2) and methylates histone 3 lysine 27 (H3K27), was shown to correlate with breast cancer aggressiveness and poor patient prognosis [68,69].

There is increasing evidence that histone modifications control dynamic processes that affect nucleosomes. Modifications associated with heterochromatin or euchromatin play an important role in chromatin remodeling and as a result in the regulation of gene expression. It is clear that histone modifications are perhaps the most important in the above regulatory schemes and understanding the mechanisms governing their modifications and controlling chromatin remodeling will lead to the development of new efficacious and specific drugs against breast cancer.

5 THE ROLE OF MicroRNAs IN BREAST CANCER

MicroRNAs (miRNAs) have been demonstrated to play a crucial role in the regulation of a wide range of biological and pathological processes. Until recently, there was little knowledge regarding their roles in normal breast physiology. Their role in the development of the normal mammary gland was realized in a comparative study that examined miRNA expression in juveniles and adults, throughout pregnancy, lactation, and involution [70]. Seven groups of miRNAs were identified on the basis of their temporal patterns of expression, connecting several families of miRNAs to different phases of the breast cycle [70]. Fascinatingly, several miRNAs (including miR-25 and the miR-17e92 cluster) associated with aggressive basal breast cancer are greatly expressed during puberty and gestation, the periods of peak physiological ductal and alveolar proliferation, signifying that they may form part of a physiological programme of proliferation and growth subverted by tumors [70]. Furthermore, miRNAs (such as let-7) that are known to be associated with the terminally differentiated phenotype increase throughout puberty and gestation, correlating with an increase in the proportion of mature epithelium. Increased levels of miR-29 were observed during postlactational involution, with a corresponding reduction in the expression of its predicted mRNA targets, which are enriched for genes connected to focal adhesion, suggesting a role for miR-29 in breast remodeling.

Current large-scale profiling approaches have shown that miRNAs are globally downregulated in several cancer types, including breast cancer. The first study describing genome-wide profiling of miRNAs in breast cancer recognized 29 differentially expressed candidate genes, of which 15 predictive miRNAs were able to differentiate between breast cancer and normal breast tissue [71]. Additionally, different associations between downregulation of certain miRNAs and clinicopathological features, such as ER/PgR positivity, tumor size, lymph node status, and the expression of p53, were demonstrated. At an early stage in breast cancer progression, Let-7 expression is lost while its continued expression is associated with low-grade, ER-positive, luminal A tumors [72,73]. Let-7 expression is downregulated in subpopulations of murine breast cells with stem-like properties [74] and during epithelial-mesenchyme transition (EMT) [75]. Furthermore, loss of let-7 family members expression was correlated with clinical features, such as PgR status (let-7c), a positive lymph node status (let-7f-1, let-7a-3, and let-7a-2), or an elevated proliferation index (let-7c and let-7d) [71]. We will briefly describe below key miRNAs that seem to be important in breast cancer.

miR-34 (a family of three miRNAs) is transcriptionally upregulated by p53 and influences a range of genes involved in cell proliferation and apoptosis, including Bcl-2 [76,77]. In a recent study, 25%

of breast cancer cell lines were found to have methylated miR-34a promoters [78]. In cell lines derived from ER-/PR-/HER2- tumors, miR-34 levels are especially low, which may reflect the higher incidence of p53 mutations in this subtype [79].

Early in the development of the field, miR-125a and miR-125b were identified as being downregulated in breast tumors [71]. miR-125 inhibits cell growth and favors apoptosis in breast cancer cell lines via targeting of the mRNA encoding the RNA-stabilizing protein HuR [80]. It also inhibits the growth of HER2-dependent SKBR3 cells when artificially overexpressed, via the targeting of HER2 and HER3 [81].

miR-205 normally is expressed in basal epithelium [72] and is considered a ubiquitous marker of basal cells in stratified epithelia. Its expression is generally lost early in breast cancer progression, but persistent expression is linked to the presence of basal immunohistochemical markers, within ER/PR/HER2-negative tumors [72]. miR-205 is correlated with stem cell behavior in a murine mammary gland cell line [74], although in the mature breast it is mostly expressed in differentiated basal myoepithelial cells.

The miR-10 family consists of miR-10a and miR-10b. miR-10a is overexpressed in breast cancer and their expression is correlated with prognosis in ER-positive breast tumors [82, 83]. It has also been demonstrated that miR-10b overexpression initiates invasion and metastasis in a murine xenograft model of breast cancer via its targeting of HOXD10 [84].

One of the first characterized oncogenic miRNAs was the miR-21 which is upregulated in numerous tumors, a finding initially made in glioblastoma [87] and in breast cancer [88]. The overexpression of miR-21 in breast tumors was confirmed by *in situ* hybridization [72]. Upregulation of miR-21 is also observed in intratumoral fibroblasts although the implication of this is not known. The expression of miR-21 in breast tumors is correlated with advanced stage metastasis and is characterized by poor survival independently of grade and stage [89,90]. As expected, knockdown of miR-21 reduces MCF-7-derived tumor growth in xenograft models via decreased proliferation and increased apoptosis possibly due to Bcl-2 targeting [91]. Other targets of miR-21 include the tumor suppressor PTEN in hepatocellular carcinoma and breast cancer [89].

The miR-17e92 cluster is one of three clusters, each transcribed as a single unit. They include 15 mature miRNAs, containing the miR-17, miR-18, miR-19, miR-20, miR-25miR-92, miR-106, and miR-363 families, representing four distinct seed sequences between them. The miR-17e92 cluster was found to be increased in lymphomas [92], although it seems to be deleted in breast tumors [77]. The miR-17e92 cluster inhibits breast cancer cell proliferation in cell culture, by targeting directly cyclin D1 [83]. Other members of the cluster were initially reported as being overexpressed in solid tumors including lung and breast [71,85]. Additionally, in breast cells miR-17 has been demonstrated to limit anchorage-independent growth by targeting the oncogene AIB1, which amplifies the trans activating effects of the estrogen receptor [86].

There is now rising evidence that signatures of miRNA expression may be utilized in the future as tumor biomarkers for diagnosis and patient risk stratification. Since the realization that hypermethylation is a crucial mechanism of miRNA silencing, it became recognized that deregulated miRNAs may also serve as novel targets for anticancer therapy. However, a lot needs to be done regarding miRNA-based therapies. It is as yet unclear what is the best protocol to manipulate miRNA-based mechanisms. Definitely, it is possible to deliver miRNAs or "anti-miRNAs" into experimental animals and humans and some of these therapies are now being tested in human trials [93]. The innovation of miRNA-specific maturation pathways holds open the possibility of small molecule-targeted therapies. At this stage, it may even be probable to target-specific miRNAs or miRNA:mRNA hybrids directly. However, in a diverse disease as breast cancer, any such therapy must be very carefully tailored to a specific tumor to avoid unwanted complications.

6 BREAST CANCER EPIGENETIC TREATMENT

Current treatment protocols of breast cancer require multidisciplinary therapies. The state-of-the-art treatment and depending on the type of breast cancer usually consists of a combination of surgery, radiation, cytotoxic chemotherapy, and molecularly targeted endocrine therapy. The last several years, considerable effort has been put into developing targeted therapeutic approaches. Presently, new treatment strategies focusing on epigenetic alterations are more promising than existing treatments because of the reversibility of epigenetic modifications. The establishment and maintenance of epigenetic alterations are based on the operations of DNMTs and HDACs, which have become the primary targets for epigenetic therapy [94]. Epigenetic therapies utilizing the inhibitors of these enzymes have shown antitumorigenic effects on several malignant conditions [95]. Epigenetic changes leading to breast cancer can find applications in breast cancer prevention, treatment, and diagnostics. At present, breast cancer treatments are focused on reversing aberrant DNA methylation and histone acetylation of tumor suppressor genes. Combinations of epigenetic-targeted therapies with conventional chemotherapeutic drugs seem promising in resensitizing drug-resistant tumors. Also, it seems that combinations of epigenetic drug treatments with conventional chemotherapeutics could potentially work synergistically, increasing in this manner the therapeutic index.

6.1 DNMTS AND HDACS INHIBITORS

The commonly utilized drugs targeting methylation are azacytidine (5-azacytidine), decitabine (5-aza-2'-deoxycytidine, 5-aza-dc), fazarabine (1-β-D-arabinofurasonyl-5-azacytosine), and dihydro-5-azacytidine [96]. These are derivatives of deoxycytidine with some modifications at the fifth position in the pyrimidine ring. Zebularine and antisense oligodeoxynucleotides also target DNA methylation [97].

5-Azacytidine was originally developed and tested as a nucleoside antimetabolite with clinical specificity for acute myelogenous leukemia [98,99]. Because it could be activated to the nucleoside triphosphate and incorporated into both DNA and RNA, 5-azacytidine treatment of cells led to inhibition of DNA, RNA, and protein synthesis [100]. 5-aza-dc is only incorporated to DNA [101] and is at least 10-fold more cytotoxic than 5-azacytidine for cultured cells and animals [102]. At dose levels low enough to avoid triggering cell death, incorporation of 5-aza or 5-aza-2′-dc into DNA of cultured cells leads to rapid loss of DNMT enzyme because it becomes irreversibly bound to DNA [103].

Preclinical studies with 5-aza-2'-dc have demonstrated that it reverses methylation in a number of cell lines and in cells from human leukemia patients [104,105]. Clinical trials in solid tumors illustrated response rates of less than 10% [96]. Nevertheless, trials on hematologic malignancies have been more successful. A randomized study evaluated the combination of amsacrine and etoposide with the same two agents plus 5-aza-dc for the therapy of induction-resistant childhood acute myeloid leukemia. The inclusive response rate was higher with the three drug regimen [106]. In patients with myelodysplastic syndrome, demethylation of p15 cyclin-dependent kinase inhibitor gene and restoration of normal p15 protein expression levels after treatment with 5-aza-dc supports pharmacologic demethylation as a potential mechanism that results in hematologic response [107].

HDAC inhibitors played important roles in the functional research of HDACs. These can be grouped into several structural classes including hydroxamates, cyclic peptides, aliphatic acids, and benzamides [108]. Interestingly, none of these compounds were initially selected for their ability to inhibit HDACs activity; they were rather discovered on the basis of their phenotypic effects on transformed cells. For instance, Trichostatin A (TSA) [7-[4-(dimethylamino)phenyl]-N-hydroxy-4,6-dimethyl-7-oxo-(2E,4E,6R)-2,4 heptadienamide)], a well-known HDAC inhibitor, was initially isolated as an antifungal agent. However, it has been demonstrated to be potent anticancer agent that stimulates differentiation of murine erythroleukemia (MEL) cells and identified as an HDAC molecular target [109]. It has also been shown that TSA has the ability to arrest cells in the G_1 and G_2 phases of the cell cycle, provoke differentiation, and revert the transformed morphology of cells in culture [109]. Suberoylanilide hydroxamic acid (SAHA) [N-hydroxy-N-phenyloctanediamide, vorinostat, Zolinza®] is a synthetic hydroxamic acid, which is structurally correlated to the natural product TSA. It is produced by selected strains of *Streptomyces platensis*, *Streptomyces hygroscopicus* Y-50, or *Streptomyces sioyaensis*. Hydroxamic acids have a high affinity to biometals, including Fe(III), Ni(II), and Zn(II) [110]. The synthesis of SAHA was first reported in 1996 when it was determined to induce MEL cell differentiation [111]. SAHA illustrates strong inhibition, at nanomolar concentrations, for both class I and class II HDACs via coordination to the catalytic Zn(II) cofactor [112].

HDAC inhibitors stimulate the increase in the status of acetylation of histones and transcription factors that results to both increased and decreased expression of a limited number of genes (2–5% of the expressed genes) [113,114]. It has been demonstrated that HDAC inhibitors stimulate the expression of p21*WAF1* which is associated with an increase in the acetylation of histones within the p21*WAF1* promoter region [115,116]. These data indicate that p21*WAF1* is a direct target gene of HDAC inhibitors. It has also been shown that SAHA caused specific modifications in the acetylation and methylation of lysines in histones H3 and H4 within the p21*WAF1* promoter [115]. Among the genes being repressed by HDAC inhibitors are *cyclin D1*, *ErbB2*, and *thymidylate synthase*.

Tamoxifen is an ER-α antagonist classified as a nonsteroidal SERM, widely used in breast cancer chemoprevention and chemotherapy [117,118]. Tamoxifen and its bioactive metabolite 4-hydroxytamoxifen (4-OHT) control tumor growth by inhibiting proliferation and inducing apoptosis in several types of ER-positive breast cancers [117,118]. However, it is well established that although antiestrogen therapy targeting ERs is one of the most successful first-line adjuvant therapies in women with breast cancer, more than 30% of treated patients will develop resistance associated with relapse within 5 years of tamoxifen treatment [119]. Various studies attempted to decipher the causes of tamoxifen resistance [120,121]. Chang et al. [122] investigated the proposed mechanism of the role of methylation of ER-α and ER-β effects on tamoxifen resistance in breast cancer. They found that the CpG methylation rate of ER-β gene is lower in tamoxifen-resistant tumors than controls. Among the methylated tumors, the tamoxifen-resistant tumors showed denser methylation of the ER-α and ER-β genes than controls. This result suggests that hypermethylation of the ER-β gene is involved in the development of tamoxifen resistance [122]. For more effective endocrine therapy, it is important to identify patients most likely to respond to tamoxifen and those that are likely to acquire resistance. We report below approaches taken to overcome tamoxifen resistance by combining HDAC and DNMT inhibitors with tamoxifen.

DNA hypermethylation of promoter regions and aberrant histone modification patterns are involved in the regulation of ER expression [2,123,124]. In the ER-negative and PR-negative MDA-MB-435 cell line, ER-α was re-expressed by treatment with 5-aza-dc or TSA. In fact their combination provided a synergistic effect [125]. Moreover, based on the induction of PR and pS2 gene expression, it has been suggested that the re-expression of ER-α in these cells leads to a functional protein production [126]. In a recent study, we demonstrated that pretreatment of MCF-7/TAM-R that do not normally express ER-β with 5-aza-dc/TSA results in the re-expression of ER-β. This finding compliments and extends a previous report where the use of DNMT inhibitors reactivated the expression of ER-β in MDA-MB-435 and SK-BR-3 breast cancer cells [3]. These results showed that hypermethylation of the ER-β promoter region is involved in ER-β silencing and suggest that the loss of ER-α expression is not the only mechanism driving acquired tamoxifen resistance [127–131].

6.2 NUTRITION AND BREAST CANCER EPIGENETICS

Natural products and dietary constituents with chemopreventive potential have recently been determined to have an impact on DNA methylation, histone modifications, and miRNA expression [132]. These natural chemopreventive agents counteract cancer-related epigenetic alterations by influencing the activity or expression of DNMTs and histone-modifying enzymes. Natural chemopreventive agents employing epigenetic modifications have recently been reviewed by Gerhausen [132]. The list includes selenium, folate, retinoic acid, Epigallocatechin-3-gallate (EGCG) butyrate, curcumin, genistein, soy isoflavones, resveratrol, nordihydroguaiareticacid (NDGA), dihydrocoumarin, lycopene, anacardic acid, garcinol, indol-3-carbinol (I3C), diindolylmethane (DIM), phenylethylisothiocyanate (PEITC), sulforaphane, phenylhexylisothiocyanate(PHI), diallyldisulfide (DADS) and its metabolite allylmercaptan (AM) and cambinol [132]. Their epigenetic effects were mainly based on *in vitro* assays. Results from animal models or human intervention studies are not available. Furthermore, most studies have focused on single candidate genes or mechanisms. Using, novel technologies such as next-generation sequencing, future research has the potential to explore nutriepigenomics at a genome-wide level to comprehend better the significance of epigenetic mechanisms for gene regulation in cancer chemoprevention [132]

We will limit our discussion below to a small number of natural agents with epigenetic impact. Folate and B vitamins have a potential impact on DNA hypomethylation. They affect the so-called one-carbon metabolism which supplies methyl groups for methylation reactions. Folate is an essential factor for the maintenance of DNA biosynthesis and DNA repair, and folate depletion results to global DNA hypomethylation, genomic instability, and chromosomal damage [132]. As a vital micronutrient, folate needs to be taken up from dietary sources, such as citrus fruits, dark green vegetables, whole grains, and dried beans. Epidemiological studies have found an association between low folate levels and cancer risk including colorectal, breast, ovarian, pancreatic, brain, lung, and cervical cancers [133–135]. Subsequently, the relationship between folate status, DNA methylation, and cancer risk has been studied in numerous rodent carcinogenesis models and in human intervention studies.

The effects of genistein and the tomato-derived carotenoid lycopene on DNA methylation in breast cancer cells have also been investigated [136]. Reactivation of glutathione S-transferase P (GSTP1) mRNA expression has been achieved by a single application of lycopene and it was associated with reduced promoter methylation in MDA-MB-468 cells [137]. However, genistein was weakly effective and only after repetitive treatments. Retinoic acid receptor beta (RARb) and HIN1 promoter methylation was unaffected by these treatments. Treatment of MCF7 cells with a series of dietary polyphenols, including ellagic acid, protocatechuic acid, sinapic acid, syringic acid, rosmarinic acid, betanin, and phloretin did not result to demethylation and re-expression of GSTP1, RASSF1A, and HIN1, although all of these compounds at the same concentrations inhibited DNMT activity in vitro by 20–88% [137].

Nordihydroguaiaretic acid (NDGA) was examined in RKO and T47D breast cancer cell lines. p16 promoter demethylation and reactivation was associated with reduced cyclin D1 expression and RB phosphorylation, G1 cell cycle arrest, increased senescence [138].

PTEN is hypermethylated in MCF-7 and MDA-MB-231 breast cancer cell lines. It has been examined whether PTEN silencing could be reversed in these cell lines after incubation with the chemopreventive agents vitamin D3, and resveratrol alone and in combination with nucleoside analogues such as 2-chloro-2' deoxyadenosine (2CdA), 9-b-D-arabinosyl-2-fluoroadenine (F-ara-A), and 5-aza-dc [139]. In MCF-7 cells with a methylation level of about 30% at the PTEN promoter, demethylation and re-expression of PTEN were observed after incubation with the two natural products. This was correlated with downregulation of DNMT1 and upregulation of p21 after incubation with vitamin D3 and resveratrol. The effects were robustly enhanced upon coincubation with 2CdA and F-ara-A. In the highly invasive MDA-MB-231 cells, PTEN promoter was >90% methylated. Only vitamin D3 treatment was able to inhibit methylation and to enhance concomitant expression of PTEN, whereas the combined treatment with nucleoside analogues did not enhance efficacy [139].

From these studies, it is clear that natural products have the ability to introduce epigenetic changes. Their effects on global DNA methylation, histone modifications, tumor suppressor genes silenced by promoter methylation, and miRNAs deregulated during carcinogenesis have a possible impact on several mechanisms related for cancer chemoprevention. Furthermore, cell cycle progression, signal transduction mediated by nuclear receptors and transcription factors, cellular differentiation, apoptosis, and senescence are known to be modulated by natural products. *In vivo* studies that reveal the functional relevance of epigenetic mechanisms for chemopreventive efficacy by natural products are still limited. Future studies should focus on identifying the best strategies for chemopreventive intervention, taking into consideration the knowledge that has been gathered so far from the *in vitro* systems regarding the epigenetic mode of action of natural products.

7 CONCLUSION

Our knowledge on hypermethylated DNA sequences encoding either proteins or miRNAs has significantly increased over the last decade. Yet, the precise mechanisms initiating hypermethylation during tumor development and progression are still not clear. Apparently, cancer cells acquire a complex pattern of genetic as well as epigenetic lesions, which most intriguingly may even become interconnected. Conversely, structural aberrations in the DNMT3b gene may be partly responsible for increased Dnmt3b expression and consequently hypermethylation of critical genes in human tumors. Adding complexity to this, such specific genetic–epigenetic relations may not necessarily be found in all tumor types; BRCA1 methylation, for instance, is observed in breast and ovarian cancers only, but it is almost absent in other cancer types. Further investigations are required in order to unravel the question of how the hypermethylome is established and maintained in a cancer cell.

Epigenetic alterations are clearly implicated in breast cancer initiation and progression. Early studies focused on single genes important in prognosis and prediction, but newer genome-wide methods are identifying many genes whose regulation is epigenetically altered during breast cancer progression. Use of epigenetic alterations as a means of screening tumors or adjacent tissues may help clinicians to improve prognosis and provide more effective treatment options for breast cancer patients. Furthermore, analysis of histologically normal tumor margins for epigenetic modifications and field cancerization will increase the capability to remove all "precancerous" tissues. Anticancer drugs targeting DNA methylation and histone modifications in breast cancer are already available. We expect that newer versions of these drugs are likely to become an important component of future clinical treatment protocols. It is therefore becoming apparent that the recognition and identification of specific epigenetic modifications will lead to significant advances for the prognosis and treatment of the disease.

ACKNOWLEDGMENT

This work was supported by funding from the Cyprus Research Promotion Foundation (ΥΓΕΙΑ/ΒΙΟΣ/0311(ΒΕ)/14).

REFERENCES

[1] Campan M, Weisenberger DJ, Laird PW. DNA methylation profiles of female steroid hormone-driven human malignancies. Curr Top Microbiol Immunol 2006;310:141–78.

[2] Sharma D, Blum J, Yang X, Beaulieu N, Macleod AR, Davidson NE. Release of methyl CpG binding proteins and histone deacetylase 1 from the estrogen receptor alpha (ER) promoter upon reactivation in ER-negative human breast cancer cells. Mol Endocrinol 2005;19:1740–51.

[3] Pitta CA, Papageorgis P, Charalambous C, Constantinou AI. Reversal of ER-beta silencing by chromatin modifying agents overrides acquired tamoxifen resistance. Cancer Lett 2013;337:167–76.

[4] Bannister AJ, Kouzarides T. Regulation of chromatin by histone modifications. Cell Res 2011;21:381–95.

[5] Ramsahoye BH, Biniszkiewicz D, Lyko F, Clark V, Bird AP, Jaenisch R. Non-CpG methylation is prevalent in embryonic stem cells and may be mediated by DNA methyltransferase 3a. Proc Natl Acad Sci USA 2000;97:5237–42.

[6] Bird AP. CpG-rich islands and the function of DNA methylation. Nature 1986;321:209–13.

[7] Antequera F, Bird A. CpG islands. EXS 1993;64:169–85.

[8] Singal R, Ginder GD. DNA methylation. Blood 1999;93:4059–70.

[9] Robertson KD, Uzvolgyi E, Liang G, Talmadge C, Sumegi J, Gonzales FA, et al. The human DNA methyltransferases (DNMTs) 1, 3a and 3b: coordinate mRNA expression in normal tissues and overexpression in tumors. Nucleic Acids Res 1999;27:2291–8.

[10] Girault I, Tozlu S, Lidereau R, Bieche I. Expression analysis of DNA methyltransferases 1, 3A, and 3B in sporadic breast carcinomas. Clin Cancer Res 2003;9:4415–22.

[11] Issa JP, Vertino PM, Wu J, Sazawal S, Celano P, Nelkin BD, et al. Increased cytosine DNA-methyltransferase activity during colon cancer progression. J Natl Cancer Inst 1993;85:1235–40.

[12] Mizuno S, Chijiwa T, Okamura T, Akashi K, Fukumaki Y, Niho Y, et al. Expression of DNA methyltransferases DNMT1, 3A, and 3B in normal hematopoiesis and in acute and chronic myelogenous leukemia. Blood 2001;97:1172–9.

[13] Eads CA, Danenberg KD, Kawakami K, Saltz LB, Danenberg PV, Laird PW. CpG island hypermethylation in human colorectal tumors is not associated with DNA methyltransferase overexpression. Cancer Res 1999;59:2302–6.

[14] Patra SK, Patra A, Zhao H, Dahiya R. DNA methyltransferase and demethylase in human prostate cancer. Mol Carcinog 2002;33:163–71.

[15] Ahluwalia A, Hurteau JA, Bigsby RM, Nephew KP. DNA methylation in ovarian cancer. II. Expression of DNA methyltransferases in ovarian cancer cell lines and normal ovarian epithelial cells. Gynecol Oncol 2001;82:299–304.

[16] Jin F, Dowdy SC, Xiong Y, Eberhardt NL, Podratz KC, Jiang SW. Up-regulation of DNA methyltransferase 3B expression in endometrial cancers. Gynecol Oncol 2005;96:531–8.

[17] Jovanovic J, Ronneberg JA, Tost J, Kristensen V. The epigenetics of breast cancer. Mol Oncol 2010;4:242–54.

[18] Hinshelwood RA, Clark SJ. Breast cancer epigenetics: normal human mammary epithelial cells as a model system. J Mol Med (Berl) 2008;86:1315–28.

[19] Widschwendter M, Jones PA. DNA methylation and breast carcinogenesis. Oncogene 2002;21:5462–82.

[20] Evron E, Dooley WC, Umbricht CB, Rosenthal D, Sacchi N, Gabrielson E, et al. Detection of breast cancer cells in ductal lavage fluid by methylation-specific PCR. Lancet 2001;357:1335–6.

[21] Herman JG, Merlo A, Mao L, Lapidus RG, Issa JP, Davidson NE, et al. Inactivation of the CDKN2/p16/MTS1 gene is frequently associated with aberrant DNA methylation in all common human cancers. Cancer Res 1995;55:4525–30.

[22] Graff JR, Herman JG, Lapidus RG, Chopra H, Xu R, Jarrard DF, et al. E-cadherin expression is silenced by DNA hypermethylation in human breast and prostate carcinomas. Cancer Res 1995;55:5195–9.

[23] Romanov SR, Kozakiewicz BK, Holst CR, Stampfer MR, Haupt LM, Tlsty TD. Normal human mammary epithelial cells spontaneously escape senescence and acquire genomic changes. Nature 2001;409:633–7.

[24] Krop I, Parker MT, Bloushtain-Qimron N, Porter D, Gelman R, Sasaki H, et al. HIN-1, an inhibitor of cell growth, invasion, and AKT activation. Cancer Res 2005;65:9659–69.

[25] Krop IE, Sgroi D, Porter DA, Lunetta KL, LeVangie R, Seth P, et al. HIN-1, a putative cytokine highly expressed in normal but not cancerous mammary epithelial cells. Proc Natl Acad Sci USA 2001;98:9796–801.

[26] Yang X, Yan L, Davidson NE. DNA methylation in breast cancer. Endocr Relat Cancer 2001;8:115–27.

[27] Fixemer T, Remberger K, Bonkhoff H. Differential expression of the estrogen receptor beta (ERbeta) in human prostate tissue, premalignant changes, and in primary, metastatic, and recurrent prostatic adenocarcinoma. Prostate 2003;54:79–87.

[28] Foley EF, Jazaeri AA, Shupnik MA, Jazaeri O, Rice LW. Selective loss of estrogen receptor beta in malignant human colon. Cancer Res 2000;60:245–8.

[29] Roger P, Sahla ME, Makela S, Gustafsson JA, Baldet P, Rochefort H. Decreased expression of estrogen receptor beta protein in proliferative preinvasive mammary tumors. Cancer Res 2001;61:2537–41.

[30] Iwao K, Miyoshi Y, Egawa C, Ikeda N, Noguchi S. Quantitative analysis of estrogen receptor-beta mRNA and its variants in human breast cancers. Int J Cancer 2000;88:733–6.

[31] Lau KM, Mok SC, Ho SM. Expression of human estrogen receptor-alpha and -beta, progesterone receptor, and androgen receptor mRNA in normal and malignant ovarian epithelial cells. Proc Natl Acad Sci USA 1999;96:5722–7.

[32] Linja MJ, Savinainen KJ, Tammela TL, Isola JJ, Visakorpi T. Expression of ERalpha and ERbeta in prostate cancer. Prostate 2003;55:180–6.

[33] Sharma D, Saxena NK, Davidson NE, Vertino PM. Restoration of tamoxifen sensitivity in estrogen receptor-negative breast cancer cells: tamoxifen-bound reactivated ER recruits distinctive corepressor complexes. Cancer Res 2006;66:6370–8.

[34] Skliris GP, Munot K, Bell SM, Carder PJ, Lane S, Horgan K, et al. Reduced expression of oestrogen receptor beta in invasive breast cancer and its re-expression using DNA methyl transferase inhibitors in a cell line model. J Pathol 2003;201:213–20.

[35] Zhao C, Lam EW, Sunters A, Enmark E, De Bella MT, Coombes RC, et al. Expression of estrogen receptor beta isoforms in normal breast epithelial cells and breast cancer: regulation by methylation. Oncogene 2003;22:7600–6.

[36] Dobrovic A, Simpfendorfer D. Methylation of the BRCA1 gene in sporadic breast cancer. Cancer Res 1997;57:3347–50.

[37] Esteller M, Silva JM, Dominguez G, Bonilla F, Matias-Guiu X, Lerma E, et al. Promoter hypermethylation and BRCA1 inactivation in sporadic breast and ovarian tumors. J Natl Cancer Inst 2000;92:564–9.

[38] Turner NC, Reis-Filho JS, Russell AM, Springall RJ, Ryder K, Steele D, et al. BRCA1 dysfunction in sporadic basal-like breast cancer. Oncogene 2007;26:2126–32.

[39] Novak P, Jensen T, Oshiro MM, Watts GS, Kim CJ, Futscher BW. Agglomerative epigenetic aberrations are a common event in human breast cancer. Cancer Res 2008;68:8616–25.

[40] Shann YJ, Cheng C, Chiao CH, Chen DT, Li PH, Hsu MT. Genome-wide mapping and characterization of hypomethylated sites in human tissues and breast cancer cell lines. Genome Res 2008;18:791–801.

[41] Singh P, Yang M, Dai H, Yu D, Huang Q, Tan W, et al. Overexpression and hypomethylation of flap endonuclease 1 gene in breast and other cancers. Mol Cancer Res 2008;6:1710–17.

[42] Paredes J, Albergaria A, Oliveira JT, Jeronimo C, Milanezi F, Schmitt FC. P-cadherin overexpression is an indicator of clinical outcome in invasive breast carcinomas and is associated with CDH3 promoter hypomethylation. Clin Cancer Res 2005;11:5869–77.

[43] Kim SJ, Kang HS, Chang HL, Jung YC, Sim HB, Lee KS, et al. Promoter hypomethylation of the N-acetyltransferase 1 gene in breast cancer. Oncol Rep 2008;19:663–8.

[44] Gupta A, Godwin AK, Vanderveer L, Lu A, Liu J. Hypomethylation of the synuclein gamma gene CpG island promotes its aberrant expression in breast carcinoma and ovarian carcinoma. Cancer Res 2003;63:664–73.

[45] Pakneshan P, Szyf M, Farias-Eisner R, Rabbani SA. Reversal of the hypomethylation status of urokinase (uPA) promoter blocks breast cancer growth and metastasis. J Biol Chem 2004;279:31735–44.

[46] Yuan J, Luo RZ, Fujii S, Wang L, Hu W, Andreeff M, et al. Aberrant methylation and silencing of ARHI, an imprinted tumor suppressor gene in which the function is lost in breast cancers. Cancer Res 2003;63:4174–80.

[47] Bae YK, Brown A, Garrett E, Bornman D, Fackler MJ, Sukumar S, et al. Hypermethylation in histologically distinct classes of breast cancer. Clin Cancer Res 2004;10:5998–6005.

[48] Fackler MJ, McVeigh M, Evron E, Garrett E, Mehrotra J, Polyak K, et al. DNA methylation of RASSF1A, HIN-1, RAR-beta, Cyclin D2 and Twist in in situ and invasive lobular breast carcinoma. Int J Cancer 2003;107:970–5.

[49] Feng W, Shen L, Wen S, Rosen DG, Jelinek J, Hu X, et al. Correlation between CpG methylation profiles and hormone receptor status in breast cancers. Breast Cancer Res 2007;9:R57.

[50] Haberland M, Montgomery RL, Olson EN. The many roles of histone deacetylases in development and physiology: implications for disease and therapy. Nat Rev Genet 2009;10:32–42.

[51] Bosch-Presegue L, Vaquero A. The dual role of sirtuins in cancer. Genes Cancer 2011;2:648–62.

[52] Peng L, Seto E. Deacetylation of nonhistone proteins by HDACs and the implications in cancer. Handb Exp Pharmacol 2011;206:39–56.

[53] Zhang Z, Yamashita H, Toyama T, Sugiura H, Ando Y, Mita K, et al. Quantitation of HDAC1 mRNA expression in invasive carcinoma of the breast. Breast Cancer Res Treat 2005;94:11–16.

[54] Krusche CA, Wulfing P, Kersting C, Vloet A, Bocker W, Kiesel L, et al. Histone deacetylase-1 and -3 protein expression in human breast cancer: a tissue microarray analysis. Breast Cancer Res Treat 2005;90:15–23.

[55] Saji S, Kawakami M, Hayashi S, Yoshida N, Hirose M, Horiguchi S, et al. Significance of HDAC6 regulation via estrogen signaling for cell motility and prognosis in estrogen receptor-positive breast cancer. Oncogene 2005;24:4531–9.

[56] Kouzarides T. Chromatin modifications and their function. Cell 2007;128:693–705.

[57] Luger K, Rechsteiner TJ, Flaus AJ, Waye MM, Richmond TJ. Characterization of nucleosome core particles containing histone proteins made in bacteria. J Mol Biol 1997;272:301–11.

[58] Dedon PC, Soults JA, Allis CD, Gorovsky MA. A simplified formaldehyde fixation and immunoprecipitation technique for studying protein–DNA interactions. Anal Biochem 1991;197:83–90.

[59] Dedon PC, Soults JA, Allis CD, Gorovsky MA. Formaldehyde cross-linking and immunoprecipitation demonstrate developmental changes in H1 association with transcriptionally active genes. Mol Cell Biol 1991;11:1729–33.

[60] Nguyen CT, Weisenberger DJ, Velicescu M, Gonzales FA, Lin JC, Liang G, et al. Histone H3-lysine 9 methylation is associated with aberrant gene silencing in cancer cells and is rapidly reversed by 5-aza-2'-deoxycytidine. Cancer Res 2002;62:6456–61.

[61] Fraga MF, Ballestar E, Villar-Garea A, Boix-Chornet M, Espada J, Schotta G, et al. Loss of acetylation at Lys16 and trimethylation at Lys20 of histone H4 is a common hallmark of human cancer. Nat Genet 2005;37:391–400.

[62] Chervona Y, Costa M. Histone modifications and cancer: biomarkers of prognosis? Am J Cancer Res 2012;2:589–97.

[63] Elsheikh SE, Green AR, Rakha EA, Powe DG, Ahmed RA, Collins HM, et al. Global histone modifications in breast cancer correlate with tumor phenotypes, prognostic factors, and patient outcome. Cancer Res 2009;69:3802–9.

[64] Leszinski G, Gezer U, Siegele B, Stoetzer O, Holdenrieder S. Relevance of histone marks H3K9me3 and H4K20me3 in cancer. Anticancer Res 2012;32:2199–205.

[65] Cao R, Wang L, Wang H, Xia L, Erdjument-Bromage H, Tempst P, et al. Role of histone H3 lysine 27 methylation in Polycomb-group silencing. Science 2002;298:1039–43.

[66] Kuzmichev A, Nishioka K, Erdjument-Bromage H, Tempst P, Reinberg D. Histone methyltransferase activity associated with a human multiprotein complex containing the enhancer of Zeste protein. Genes Dev 2002;16:2893–905.

[67] Kirmizis A, Bartley SM, Kuzmichev A, Margueron R, Reinberg D, Green R, et al. Silencing of human polycomb target genes is associated with methylation of histone H3 Lys 27. Genes Dev 2004;18:1592–605.

[68] Kleer CG, Cao Q, Varambally S, Shen R, Ota I, Tomlins SA, et al. EZH2 is a marker of aggressive breast cancer and promotes neoplastic transformation of breast epithelial cells. Proc Natl Acad Sci USA 2003;100:11606–11.

[69] Pietersen AM, Horlings HM, Hauptmann M, Langerod A, Ajouaou A, Cornelissen-Steijger P, et al. EZH2 and BMI1 inversely correlate with prognosis and TP53 mutation in breast cancer. Breast Cancer Res 2008;10:R109.

[70] Avril-Sassen S, Goldstein LD, Stingl J, Blenkiron C, Le Quesne J, Spiteri I, et al. Characterisation of microRNA expression in post-natal mouse mammary gland development. BMC Genomics 2009;10:548.

[71] Lu J, Getz G, Miska EA, Alvarez-Saavedra E, Lamb J, Peck D, et al. MicroRNA expression profiles classify human cancers. Nature 2005;435:834–8.

[72] Sempere LF, Christensen M, Silahtaroglu A, Bak M, Heath CV, Schwartz G, et al. Altered microRNA expression confined to specific epithelial cell subpopulations in breast cancer. Cancer Res 2007;67:11612–20.

[73] Blenkiron C, Goldstein LD, Thorne NP, Spiteri I, Chin SF, Dunning MJ, et al. MicroRNA expression profiling of human breast cancer identifies new markers of tumor subtype. Genome Biol 2007;8:R214.

[74] Ibarra I, Erlich Y, Muthuswamy SK, Sachidanandam R, Hannon GJ. A role for microRNAs in maintenance of mouse mammary epithelial progenitor cells. Genes Dev 2007;21:3238–43.

[75] Dangi-Garimella S, Yun J, Eves EM, Newman M, Erkeland SJ, Hammond SM, et al. Raf kinase inhibitory protein suppresses a metastasis signalling cascade involving LIN28 and let-7. EMBO J 2009;28:347–58.

[76] Bommer GT, Gerin I, Feng Y, Kaczorowski AJ, Kuick R, Love RE, et al. p53-mediated activation of miRNA34 candidate tumor-suppressor genes. Curr Biol 2007;17:1298–307.

[77] Zhang L, Huang J, Yang N, Greshock J, Megraw MS, Giannakakis A, et al. microRNAs exhibit high frequency genomic alterations in human cancer. Proc Natl Acad Sci USA 2006;103:9136–41.

[78] Lodygin D, Tarasov V, Epanchintsev A, Berking C, Knyazeva T, Korner H, et al. Inactivation of miR-34a by aberrant CpG methylation in multiple types of cancer. Cell Cycle 2008;7:2591–600.

[79] Kato M, Paranjape T, Muller RU, Nallur S, Gillespie E, Keane K, et al. The mir-34 microRNA is required for the DNA damage response *in vivo* in *C. elegans* and *in vitro* in human breast cancer cells. Oncogene 2009;28:2419–24.

[80] Guo X, Wu Y, Hartley RS. MicroRNA-125a represses cell growth by targeting HuR in breast cancer. RNA Biol 2009;6:575–83.

[81] Scott GK, Goga A, Bhaumik D, Berger CE, Sullivan CS, Benz CC. Coordinate suppression of ERBB2 and ERBB3 by enforced expression of micro-RNA miR-125a or miR-125b. J Biol Chem 2007;282:1479–86.

[82] Hoppe R, Achinger-Kawecka J, Winter S, Fritz P, Lo WY, Schroth W, et al. Increased expression of miR-126 and miR-10a predict prolonged relapse-free time of primary oestrogen receptor-positive breast cancer following tamoxifen treatment. Eur J Cancer 2013;49:3598–608.

[83] Parrella P, Barbano R, Pasculli B, Fontana A, Copetti M, Valori VM, et al. Evaluation of microRNA-10b prognostic significance in a prospective cohort of breast cancer patients. Mol Cancer 2014;13:142.

[84] Ma L, Teruya-Feldstein J, Weinberg RA. Tumour invasion and metastasis initiated by microRNA-10b in breast cancer. Nature 2007;449:682–8.

[85] Volinia S, Calin GA, Liu CG, Ambs S, Cimmino A, Petrocca F, et al. A microRNA expression signature of human solid tumors defines cancer gene targets. Proc Natl Acad Sci USA 2006;103:2257–61.

[86] Hossain A, Kuo MT, Saunders GF. Mir-17-5p regulates breast cancer cell proliferation by inhibiting translation of AIB1 mRNA. Mol Cell Biol 2006;26:8191–201.

[87] Chan JA, Krichevsky AM, Kosik KS. MicroRNA-21 is an antiapoptotic factor in human glioblastoma cells. Cancer Res 2005;65:6029–33.

[88] Iorio MV, Ferracin M, Liu CG, Veronese A, Spizzo R, Sabbioni S, et al. MicroRNA gene expression deregulation in human breast cancer. Cancer Res 2005;65:7065–70.

[89] Huang GL, Zhang XH, Guo GL, Huang KT, Yang KY, Shen X, et al. Clinical significance of miR-21 expression in breast cancer: SYBR-Green I-based real-time RT-PCR study of invasive ductal carcinoma. Oncol Rep 2009;21:673–9.

[90] Yan LX, Huang XF, Shao Q, Huang MY, Deng L, Wu QL, et al. MicroRNA miR-21 overexpression in human breast cancer is associated with advanced clinical stage, lymph node metastasis and patient poor prognosis. RNA 2008;14:2348–60.

[91] Si ML, Zhu S, Wu H, Lu Z, Wu F, Mo YY. miR-21-mediated tumor growth. Oncogene 2007;26:2799–803.

[92] Ota A, Tagawa H, Karnan S, Tsuzuki S, Karpas A, Kira S, et al. Identification and characterization of a novel gene, C13orf25, as a target for 13q31-q32 amplification in malignant lymphoma. Cancer Res 2004;64:3087–95.

[93] Corella D, Sorli JV, Estruch R, Coltell O, Ortega-Azorin C, Portoles O, et al. MicroRNA-410 regulated lipoprotein lipase variant rs13702 is associated with stroke incidence and modulated by diet in the randomized controlled PREDIMED trial. Am J Clin Nutr 2014;100:719–31.

[94] Esteller M. Cancer epigenomics: DNA methylomes and histone-modification maps. Nat Rev Genet 2007;8:286–98.

[95] Egger G, Liang G, Aparicio A, Jones PA. Epigenetics in human disease and prospects for epigenetic therapy. Nature 2004;429:457–63.

[96] Goffin J, Eisenhauer E. DNA methyltransferase inhibitors-state of the art. Ann Oncol 2002;13:1699–716.

[97] Cheng JC, Matsen CB, Gonzales FA, Ye W, Greer S, Marquez VE, et al. Inhibition of DNA methylation and reactivation of silenced genes by zebularine. J Natl Cancer Inst 2003;95:399–409.

[98] Cihak A. Biological effects of 5-azacytidine in eukaryotes. Oncology 1974;30:405–22.

[99] Sorm F, Piskala A, Cihak A, Vesely J. 5-Azacytidine, a new, highly effective cancerostatic. Experientia 1964;20:202–3.

[100] Cihak A, Vesely J. Effects of 5-aza-2'-deoxycytidine on DNA synthesis in mouse lymphatic tissues. Neoplasma 1978;25:385–93.

[101] Li LH, Olin EJ, Fraser TJ, Bhuyan BK. Phase specificity of 5-azacytidine against mammalian cells in tissue culture. Cancer Res 1970;30:2770–5.

[102] Flatau E, Gonzales FA, Michalowsky LA, Jones PA. DNA methylation in 5-aza-2'-deoxycytidine-resistant variants of C3H 10T1/2 C18 cells. Mol Cell Biol 1984;4:2098–102.

[103] Christman JK, Schneiderman N, Acs G. Interaction of DNA methyltransferase and other non-histone proteins isolated from friend erythroleukemia cell nuclei with 5-azacytosine residues in DNA. Prog Clin Biol Res 1985;198:105–18.

[104] Momparler RL, Bouchard J, Onetto N, Rivard GE. 5-aza-2'-deoxycytidine therapy in patients with acute leukemia inhibits DNA methylation. Leuk Res 1984;8:181–5.

[105] Momparler RL, Momparler LF, Samson J. Comparison of the antileukemic activity of 5-AZA-2'-deoxycytidine, 1-beta-D-arabinofuranosylcytosine and 5-azacytidine against L1210 leukemia. Leuk Res 1984;8:1043–9.

[106] Steuber CP, Krischer J, Holbrook T, Camitta B, Land V, Sexauer C, et al. Therapy of refractory or recurrent childhood acute myeloid leukemia using amsacrine and etoposide with or without azacitidine: a Pediatric Oncology Group randomized phase II study. J Clin Oncol 1996;14:1521–5.

[107] Daskalakis M, Nguyen TT, Nguyen C, Guldberg P, Kohler G, Wijermans P, et al. Demethylation of a hypermethylated P15/INK4B gene in patients with myelodysplastic syndrome by 5-aza-2'-deoxycytidine (decitabine) treatment. Blood 2002;100:2957–64.

[108] Miller TA, Witter DJ, Belvedere S. Histone deacetylase inhibitors. J Med Chem 2003;46:5097–116.

[109] Yoshida M, Horinouchi S, Beppu T. Trichostatin A and trapoxin: novel chemical probes for the role of histone acetylation in chromatin structure and function. Bioessays 1995;17:423–30.

[110] Codd R, Braich N, Liu J, Soe CZ, Pakchung AA. Zn(II)-dependent histone deacetylase inhibitors: suberoylanilide hydroxamic acid and trichostatin A. Int J Biochem Cell Biol 2009;41:736–9.

[111] Thomas M, Rivault F, Tranoy-Opalinski I, Roche J, Gesson JP, Papot S. Synthesis and biological evaluation of the suberoylanilide hydroxamic acid (SAHA) beta-glucuronide and beta-galactoside for application in selective prodrug chemotherapy. Bioorg Med Chem Lett 2007;17:983–6.

[112] Richon VM, Emiliani S, Verdin E, Webb Y, Breslow R, Rifkind RA, et al. A class of hybrid polar inducers of transformed cell differentiation inhibits histone deacetylases. Proc Natl Acad Sci USA 1998;95:3003–7.

[113] Butler LM, Zhou X, Xu WS, Scher HI, Rifkind RA, Marks PA, et al. The histone deacetylase inhibitor SAHA arrests cancer cell growth, up-regulates thioredoxin-binding protein-2, and down-regulates thioredoxin. Proc Natl Acad Sci USA 2002;99:11700–5.

[114] Glaser KB, Li J, Staver MJ, Wei RQ, Albert DH, Davidsen SK. Role of class I and class II histone deacetylases in carcinoma cells using siRNA. Biochem Biophys Res Commun 2003;310:529–36.

[115] Gui CY, Ngo L, Xu WS, Richon VM, Marks PA. Histone deacetylase (HDAC) inhibitor activation of p21WAF1 involves changes in promoter-associated proteins, including HDAC1. Proc Natl Acad Sci USA 2004;101:1241–6.

[116] Richon VM, Zhou X, Secrist JP, Cordon-Cardo C, Kelly WK, Drobnjak M, et al. Histone deacetylase inhibitors: assays to assess effectiveness *in vitro* and *in vivo*. Methods Enzymol 2004;376:199–205.

[117] Lykkesfeldt AE, Larsen JK, Christensen IJ, Briand P. Effects of the antioestrogen tamoxifen on the cell cycle kinetics of the human breast cancer cell line, MCF-7. Br J Cancer 1984;49:717–22.

[118] Kumar R, Mandal M, Lipton A, Harvey H, Thompson CB. Overexpression of HER2 modulates bcl-2, bcl-XL, and tamoxifen-induced apoptosis in human MCF-7 breast cancer cells. Clin Cancer Res 1996;2:1215–19.

[119] Esslimani-Sahla M, Simony-Lafontaine J, Kramar A, Lavaill R, Mollevi C, Warner M, et al. Estrogen receptor beta (ER beta) level but not its ER beta cx variant helps to predict tamoxifen resistance in breast cancer. Clin Cancer Res 2004;10:5769–76.

[120] Sweeney EE, McDaniel RE, Maximov PY, Fan P, Jordan VC. Models and mechanisms of acquired antihormone resistance in breast cancer: significant clinical progress despite limitations. Horm Mol Biol Clin Investig 2012;9:143–63.

[121] Clarke R, Liu MC, Bouker KB, Gu Z, Lee RY, Zhu Y, et al. Antiestrogen resistance in breast cancer and the role of estrogen receptor signaling. Oncogene 2003;22:7316–39.

[122] Chang HG, Kim SJ, Chung KW, Noh DY, Kwon Y, Lee ES, et al. Tamoxifen-resistant breast cancers show less frequent methylation of the estrogen receptor beta but not the estrogen receptor alpha gene. J Mol Med (Berl) 2005;83:132–9.

[123] Duong V, Licznar A, Margueron R, Boulle N, Busson M, Lacroix M, et al. ERalpha and ERbeta expression and transcriptional activity are differentially regulated by HDAC inhibitors. Oncogene 2006;25:1799–806.

[124] Furst RW, Kliem H, Meyer HH, Ulbrich SE. A differentially methylated single CpG-site is correlated with estrogen receptor alpha transcription. J Steroid Biochem Mol Biol 2012;130:96–104.

[125] Yang X, Ferguson AT, Nass SJ, Phillips DL, Butash KA, Wang SM, et al. Transcriptional activation of estrogen receptor alpha in human breast cancer cells by histone deacetylase inhibition. Cancer Res 2000;60:6890–4.

[126] Fan J, Yin WJ, Lu JS, Wang L, Wu J, Wu FY, et al. ER alpha negative breast cancer cells restore response to endocrine therapy by combination treatment with both HDAC inhibitor and DNMT inhibitor. J Cancer Res Clin Oncol 2008;134:883–90.

[127] Riggins RB, Schrecengost RS, Guerrero MS, Bouton AH. Pathways to tamoxifen resistance. Cancer Lett 2007;256:1–24.

[128] Arpino G, Green SJ, Allred DC, Lew D, Martino S, Osborne CK, et al. HER-2 amplification, HER-1 expression, and tamoxifen response in estrogen receptor-positive metastatic breast cancer: a southwest oncology group study. Clin Cancer Res 2004;10:5670–6.

[129] Gutierrez MC, Detre S, Johnston S, Mohsin SK, Shou J, Allred DC, et al. Molecular changes in tamoxifen-resistant breast cancer: relationship between estrogen receptor, HER-2, and p38 mitogen-activated protein kinase. J Clin Oncol 2005;23:2469–76.

[130] Cannings E, Kirkegaard T, Tovey SM, Dunne B, Cooke TG, Bartlett JM. Bad expression predicts outcome in patients treated with tamoxifen. Breast Cancer Res Treat 2007;102:173–9.

[131] Planas-Silva MD, Bruggeman RD, Grenko RT, Smith JS. Overexpression of c-Myc and Bcl-2 during progression and distant metastasis of hormone-treated breast cancer. Exp Mol Pathol 2007;82:85–90.

[132] Gerhauser C. Cancer chemoprevention and nutriepigenetics: state of the art and future challenges. Top Curr Chem 2013;329:73–132.

[133] Kim YI. Nutritional epigenetics: impact of folate deficiency on DNA methylation and colon cancer susceptibility. J Nutr 2005;135:2703–9.

[134] Lamprecht SA, Lipkin M. Chemoprevention of colon cancer by calcium, vitamin D and folate: molecular mechanisms. Nat Rev Cancer 2003;3:601–14.

[135] Huang S. Histone methyltransferases, diet nutrients and tumour suppressors. Nat Rev Cancer 2002;2:469–76.

[136] King-Batoon A, Leszczynska JM, Klein CB. Modulation of gene methylation by genistein or lycopene in breast cancer cells. Environ Mol Mutagen 2008;49:36–45.

[137] Paluszczak J, Krajka-Kuzniak V, Baer-Dubowska W. The effect of dietary polyphenols on the epigenetic regulation of gene expression in MCF7 breast cancer cells. Toxicol Lett 2010;192:119–25.

[138] Cui Y, Lu C, Liu L, Sun D, Yao N, Tan S, et al. Reactivation of methylation-silenced tumor suppressor gene p16INK4a by nordihydroguaiaretic acid and its implication in G1 cell cycle arrest. Life Sci 2008;82:247–55.

[139] Stefanska B, Salame P, Bednarek A, Fabianowska-Majewska K. Comparative effects of retinoic acid, vitamin D and resveratrol alone and in combination with adenosine analogues on methylation and expression of phosphatase and tensin homologue tumour suppressor gene in breast cancer cells. Br J Nutr 2012;107:781–90.

THERAPEUTIC APPLICATIONS OF THE PROSTATE CANCER EPIGENOME

11

Antoinette Sabrina Perry

Prostate Molecular Oncology, Institute of Molecular Medicine, Trinity College Dublin, Dublin, Ireland

CHAPTER OUTLINE

1 INTRODUCTION TO PROSTATE CANCER

Prostate cancer (PCa) is the second most common cancer in men. An estimated 1.1 million men worldwide were diagnosed in 2012, accounting for 15% of male cancers, with almost 70% of cases (759,000) occurring in more developed regions [1]. The incidence of the disease varies more than

S.G. Gray (Ed): Epigenetic Cancer Therapy. DOI: http://dx.doi.org/10.1016/B978-0-12-800206-3.00011-2

25-fold worldwide; the rates are highest in Northern America, Australia/New Zealand, and in Northern and Western Europe [2]. Incidence rates remain low in Asian populations pointing toward dietary and genetic risk factors. With an estimated 307,000 deaths in 2012, PCa is the fifth leading cause of male cancer-related deaths, accounting for 6.6% of fatalities, with rates highest in developing regions [1,3].

The marked variation in incidence can largely be attributed to the practice of prostate-specific antigen (PSA) testing and subsequent biopsy in asymptomatic men, which has become commonplace in certain parts of the world. PCa is a remarkably heterogeneous disease whose etiology is not well understood; most tumors have a very slow natural trajectory and pose little likelihood of clinical manifestation, deemed indolent in nature [4]. Three large clinical trials have reported on the significant impact of PSA screening on the over-treatment of such indolent tumors [5–7]. Nevertheless, a proportion of prostate tumors are highly aggressive and are associated with the lethal form of the disease. Current prognostic tools at time of diagnosis include a combination of tumor grade (measured using the Gleason scale), PSA level, and clinical stage. Identifying molecular correlates to discern between indolent and aggressive prostate tumors at an early stage while potentially curable is one of the largest unmet needs in this field and is the subject of active research.

1.1 CLINICAL MANAGEMENT AND TREATMENT OF PROSTATE CANCER

Given the spectrum of PCa behavior, treatment options range accordingly. Localized disease has a number of successful therapeutic options, including radical prostatectomy, external beam radiotherapy, and active surveillance for low-risk disease; all of which offer close to 100% 10-year disease-specific survival rates [8,9].

Approximately 20–25% of patients develop a recurrence following radical local therapy. The primary intervention for disease recurrence and metastasis is androgen deprivation therapies. Reducing circulating androgens (by luteinizing hormone-releasing hormone (LHRH) analogues) and/or antiandrogens (which block androgen signaling by targeting the ligand binding domain of the androgen receptor (AR)) instigate a favorable response in almost all men, measured by a decline in serum PSA, improvement in disease-related symptoms and radiographic imaging. However, the positive response is short lived and the disease invariably progresses to the lethal castration-resistant phase (CRPC) [10]. Despite administration of androgen-depleting therapies, continued AR signaling is a common feature of CRPC, attributed to *AR* gene amplification, mutation, overexpression, or increased androgen biosynthesis in prostate tumors [11]. Progression to castration resistance is invariably accompanied by an increase in the level of PSA, the archetypical AR transactivated gene, indicating that the AR is aberrantly activated under castrate conditions [12]. This apparent "addiction" of CRPC to AR signaling does however present opportunities to develop novel antihormonal therapeutic agents.

New drug discovery and development has yielded two newly approved agents. The androgen biosynthesis inhibitor Abiraterone targets the CYP17A1 enzyme, inhibiting residual androgen synthesis [13]. Abiraterone in combination with prednisone improved survival of CRPC patients by 3.9 months in a Phase III study [14,15]. The novel AR antagonist enzalutamide achieved significant serum PSA responses in more than 50% of chemotherapy-naive CRPC patients in Phase I/II studies [16,17]. Resistance to these "second-generation" AR antagonists is emerging, however, through missense *AR* mutations [18]. Yet, the continued reliance on AR signaling warrants the development of further next-generation AR antagonists. The emergence, however, of AR-independent clones (e.g., neuroendocrine, small cell morphology) at any stage during disease progression poses a major clinical challenge, as these cells by definition will not respond to AR inhibition. This chapter will provide a synopsis of research activity into epigenetic therapies as viable alternatives or adjuncts to conventional therapeutics for CRPC.

2 A SNAPSHOT OF THE PROSTATE CANCER EPIGENOME

2.1 THE PROSTATE CANCER METHYLOME

PCa has an unusually low mutation frequency [19]. By contrast, promoter hypermethylation of tumor suppressor genes and genes with important regulatory functions is widespread affecting virtually all cellular pathways [20–24]. Table 11.1 provides a comprehensive summary of genes methylated in PCa.

There is overwhelming evidence to support a role for promoter hypermethylation of individual genes, in particular the intracellular detoxification enzyme *GSTP1*, at the earliest stages of PCa initiation [25–27]. Promoter hypermethylation and silencing of *GSTP1* is a molecular hallmark of PCa, observed in more than 90% of tumors and in ~75% of preinvasive high grade prostatic intraepithelial neoplastic (HGPIN) lesions [25,26,28]. Since these early studies, quantitative assessments of gene panels have demonstrated an "epigenetic catastrophe" during tumor initiation, with rampant hypermethylation at specific gene loci. *De novo* hypermethylation events in PCa precursor cells presumably generate polyclonal populations of cells that are prone to acquiring further (epi)genetic hits. There is also strong evidence to support clonal inheritance of methylation during metastatic progression [20]. However, recent advances in whole methylome analyses have identified distinct subsets of genes that become hypermethylated in late stage PCa, during which epigenetic changes become markedly more heterogeneous [21,29].

De novo losses in DNA methylation (or hypomethylation) are approximately four times more common than hypermethylated alterations in PCa metastases and notably show no enrichment at promoters or with any particular gene ontology or cancer-related gene sets [30]. These observations support previous work that showed extensive losses in 5mC, particularly at regions of low CpG density, accompanied by focal gains in DNA methylation at promoter CpG islands (CGIs) in PCa metastases [31]. There remains very limited evidence of *de novo* promoter hypomethylation and subsequent proto-oncogene activation in PCa [32–34]. Global analysis of DNA methylation in parallel with gene expression in PCa metastases showed in fact that promoter hypomethylation principally has a negligible effect on gene expression [30]. A review of the literature suggests that there are substantial differences in the timing and extent of DNA hypomethylation between different cancer types, which could relate to the proliferative index of the tissue and the metabolism of one carbon. This would explain why tumors with a protracted natural history (such as prostate) do not exhibit global losses in 5mC until more advanced stages of disease.

Global DNA methylation analysis of multiple [3–6] metastatic deposits and [1–2] matched normal tissues from 13 individuals revealed that PCa metastases have monoclonal origins and display subsequent clonal expansion within individuals [30]. Relatively little tumor heterogeneity is observed between metastases within subjects, compared to the significant tumor heterogeneity between subjects, when examining genetic (copy number) and epigenetic (DNA methylation) events. The authors hypothesize that genetic and epigenetic changes may therefore develop through parallel clonal evolutionary pathways. Importantly, DNA methylation alterations that are associated with phenotypic changes in

Table 11.1 Genes Hypermethylated in Prostate Cancer

Gene Symbol	Total PCa Samples (n)[a]	References (n)	Methylation Frequency (%)
Bioenergetics			
ABCB1	477	4	63
FHIT	101	1	15
Cell cycle regulation			
CDKN2A	609	10	32
CDKN1A	48	3	8
CDKN1B	48	2	10
CCND2	219	2	69
SFN	103	2	92
CDC6	78	1	67
CDKN1C	41	1	56
Cell signaling			
DKK3	41	1	68
GREM1	50	1	20
EDNRB	387	6	48
CAV1	22	1	91
DLC1	27	22	48
IGFBP7	0	1	Cell lines
RARB	933	11	50
TGFBR2	67	1	100
SFRP1	41	1	83
CTNNB1	78	1	9
Cell–cell/cell–matrix interactions			
CDH1	663	9	32
CD44	603	7	33
TIMP3	447	5	33
CDH13	280	2	46
LAMA3	101	1	44

Gene Symbol	Total PCa Samples (n)[a]	References (n)	Methylation Frequency (%)
Tumor suppressor			
APC	941	12	55
RB1	48	2	8
PRDM2	47	1	43
Gene expression			
AR	209	3	21
ESR1	214	5	55
ESR2	61	3	87
PGR	0	1	Cell lines
NEUROG1	179	1	34
NKX3-1	40	1	83
PCQAP	179	1	42
SRY	0	1	Cell lines
HLTF	179	1	16
RBMY1A1	0	1	Cell lines
NAB2	78	1	92
HIC1	73	2	100
RUNX 3	273	3	16
DAZ1	0	1	Cell lines
Nervous system			
PRIMA1	62	1	45
TMEFF2	50	1	38
Programmed cell death			
RASSF1	1050	14	62
DAPK1	277	3	16
PYCARD	303	3	52
TNFRSF10C	179	2	65
BCL2	179	1	52

Gene			
PDLIM4	62	1	98
CVY	0	1	Cell lines
SVIL	62	1	83
TIMP2	0	1	Cell lines
THBS1	179	1	25
LAMC2	101	1	41
LAMB3	101	1	18
DNA recombination and repair			
WRN	20	1	20
DDB2	41	1	83
Hormone activity			
INHA	0	1	Cell lines
Immunity			
SCGB3A1	21	2	95
Metabolism			
CYP1A1	30	1	37
FOLH1	0	1	Cell lines
MME	21	1	14
PTGS2	481	6	61
HPGD	41	1	73

Gene			
TNFRSF10D	179	1	25
FAS	32	1	13
Intracellular transport/detoxification			
GSTP1	2001	31	72
MGMT	443	6	17
S100A2	152	2	98
RBP1	215	2	36
MT1G	121	1	24
TUSC3	0	1	Cell lines
GPX3	41	1	93
NULL (RHAG)	118	1	81
Unknown function			
RARRES1	50	1	52
S100A6	27	1	52
HPP1	179	1	36
LOC728412	0	1	Cell lines
RBMY 1H	0	1	Cell lines
PRY	0	1	Cell lines

[a]Combined number of total tumor cases examined across studies.
Data extracted from PubMeth (www.pubmeth.org) as of April 2014.

gene expression showed the greatest tendency to be maintained during metastatic dissemination, which may be due to selection for these phenotypes. In addition, promoter-associated CGIs were more frequently hypermethylated than nonpromoter CGIs.

2.2 HISTONE MODIFICATIONS, VARIANTS, AND EPIGENETIC ENZYMES IN PROSTATE CANCER

Global levels of histone modifications are significantly altered in PCa and may be useful in predicting disease outcome. For example, global decreases in histone acetylation correspond with increases in tumor grade and disease recurrence [35]. Increases in H3K4me2 and H3K18ac are correlated with tumor recurrence [36]; H3K4me1 and H3K4me2/3 are significantly increased in CRPC [37], and H3K4me1, H3K9me2/3, HSK9me3 and acetylation at H3 and H4 are all significantly reduced in PCa compared with nonmalignant tissue [37].

In turn, the overexpression of histone-modifying enzymes has also been linked with PCa progression, with many shown to be independent prognostic markers (Figure 11.1). For example, several classes of histone deacetylases (HDACs) are overexpressed in aggressive and CRPC [38–40]. So too are the lysine demethylases (KDMs), which also offer prognostic value [41,42].

Of course in addition to posttranslational modifications, replacement histone variants can also modify the genomic landscape and have been shown to play a role in cancer. The most noteworthy variant with respect to PCa is the core histone variant H2A.Z [43,44]. It operates as both a transcriptional repressor and activator, influenced by its posttranslational modifications (particularly acetylation), existence of H2A.Z-H2B heterotypic dimers and presence of multiple variants, namely H2A.Z1 and H2A.Z2. A global reorganization of H2A.Z is observed in PCa LNCaP cells: enrichment of acetylated H2A.Z at transcriptional start sites in conjunction with oncogene activation (such as *KLK2* and *ERBB2*), and an overall reduction in H2A.Z levels. Conversely, H2A.Z becomes deacetylated at the transcriptional start sites of tumor suppressor genes (*CAV1*, *CAV2*, and *RND3*) [44]. The authors also demonstrated an anti-correlation between acetylated H2A.Z and promoter H3K27me3 and DNA methylation, suggesting that acetylation of H2A.Z is an important epigenetic mark involved in gene dysregulation in PCa. Indeed increased expression of the H2A.Z gene (*H2AFZ*) is reported in a CRPC xenograft model [45] as well as HGPIN and primary tumors [46]. Sirtium 1, a class III HDAC, reduces histone variant H2A.Z levels via proteasomal degradation; the two show reciprocal levels of gene expression in PCa [46].

Enhancer of zeste homolog 2 (EZH2), a subunit of Polycomb repressive complex 2 (PRC2), silences gene expression through its lysine methyltransferase (KMT) activity on H3K27. EZH2 also acts as a molecular scaffold, attracting DNMTs (DNA methyltransferases) to gene promoters [47], thus bridging two distinct modes of epigenetic gene regulation: nucleosome remodeling and DNA methylation. EZH2 is frequently overexpressed in many cancers including prostate; its overexpression represses tumor and metastases suppressor genes, is associated with elevated proliferation, and increases tumor aggressiveness [48]. Elevated EZH2 is also linked with poor survival, making it a relevant therapeutic target. It was recently shown that the expression of the *EZH2* gene is stimulated by the ERG transcription factor, which of course is elevated in approximately half of all prostate tumors, due to fusion with the androgen-regulated *TMPRSS2* [49]. In a remarkable twist, the oncogenic properties of EZH2 have in fact been attributed to a PRC2-independent transcriptional activating function of the enzyme, discussed further in the following section [50].

A genome-wide systematic analysis of the functional significance of 615 epigenetic proteins in PCa cells revealed that subsets of enzymes influence different cancer cell phenotypes [51]. Several KDMs were highly expressed and found to mainly impact cell proliferation, whereas HDACs were primarily involved in regulating AR expression.

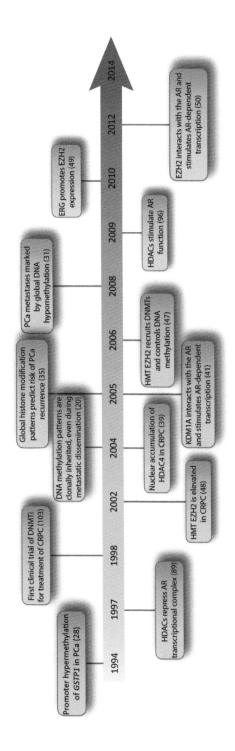

FIGURE 11.1

A snapshot of prostate cancer epigenomics research over the past 20 years.

2.3 NONCODING RNAs IN PROSTATE CANCER

microRNAs remain the best characterized class of small noncoding RNAs (ncRNAs) and tend to be preferentially downregulated during PCa progression and metastatic dissemination [52–57]. One of the most noteworthy tumor suppressor microRNA in PCa is miR-205 [58]. DNA hypermethylation at the *MIR-205* locus is associated with reduced miR-205 expression in PCa cell lines and localized PCa [59]. Functionally miR-205 counteracts epithelial-to-mesenchymal transition (EMT) and the mitogen-activated protein kinase (MAPK) and AR signaling pathways, by targeting a number of mRNAs, including the *AR* itself [60–63].

Several studies have focused on miRNAs involved in the transformation of hormone-sensitive PCa to the lethal castration-resistant phenotype. A gain-of-function analysis of 1,129 microRNAs combined with AR protein quantification in a panel of PCa cell lines identified 13 microRNAs that target *AR* mRNA (miR-135b, miR-185, miR-297, miR-299-3p, miR-34a, miR-34c, miR-371-3p, miR-421, miR-449a, miR-449b, miR-634, miR-654-5p, and miR-9); several of these also inhibited androgen-induced proliferation. Analysis in clinical specimens confirmed a negative correlation with miR-34a and miR-34c expression and AR levels [64]. In addition to the *AR*, *MYC* and cell adhesion and stem cell marker *CD44* have been identified and validated as direct and functional targets of miR-34a [65,66].

Deep sequencing of the PCa transcriptome reveals that almost a quarter of abundant transcripts are either long ncRNAs (lncRNAs) or transcripts that remain as yet unannotated [67], reviewed in full [68]. Recently, two lncRNAs, *PCa Noncoding RNA1 (PRNCR1)* and *PCa gene expression marker 1 (PCGEM1)*, have been identified, which are both involved in controlling AR-mediated gene transcription [69]. These lncRNAs localize to distal androgen response elements (AREs) and facilitate enhancer-promoter looping and three-dimensional gene activation. *PRNCR1* binds to the acetylated carboxy-terminal domain of the AR and recruits DOT1L, which methylates the AR at K349 of its amino terminal domain [19]. This activation by DOT1L is required for the AR to associate with *PCGEM1*. *PCGEM1* mediates the recruitment of PYGO2, which can bind to H3K4me3 at gene promoters, enabling the AR to engage with its target genes. The importance of *PCGEM1* and *PRNCR1* in CRPC is evident from the observation that constitutively active truncated AR isoforms (even in the absence of ligand binding) require both lncRNAs to activate transcription [69].

Other lncRNAs with a particularly well-characterized role in PCa are those involved in alternative splicing, such as *MALAT1* [70,71], those involved in antisense gene silencing, such as *CDKN2B-AS1* (which plays a role in silencing the *CDKN2A/CDKN2B* tumor suppressor locus through heterochromatin formation, by increasing dimethylation of H3K9 and decreasing dimethylation of H3K4, at gene promoter) [72], those involved in antagonizing transcriptional regulators, such as *SCHLAP1* [67,73] and finally, those involved in repressing DNA repair, e.g., *PCAT1*, which represses expression of BRCA2, leading to downstream impairment of homologous recombination repair of double-stranded DNA breaks [74]. High *PCAT1* expression is also inversely correlated with expression of *EZH2* [48].

3 EPIGENETIC INFLUENCES ON THE ANDROGEN SIGNALING AXIS

The development, differentiation, and growth of the prostate rely on androgens and their stimulation of the AR. The reliance on AR signaling is retained in PCa. One cannot consider epigenetic therapies in PCa in isolation without also considering their influence on this pivotal signaling axis in the prostate cell (Figure 11.2).

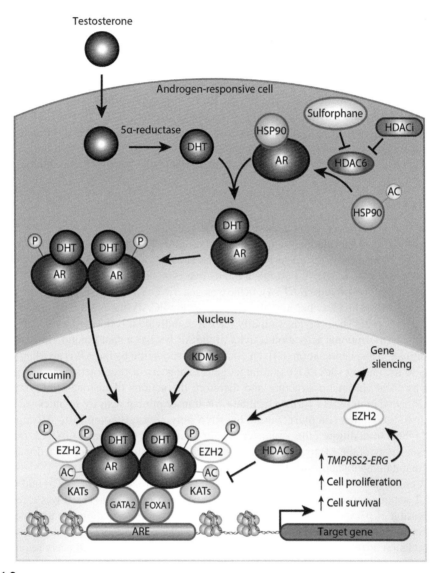

FIGURE 11.2

The role for epigenetic therapies in modifying AR signaling. Free testosterone enters prostate cells and is converted to the more potent DHT by the enzyme 5α reductase. Binding of DHT to the AR induces dissociation from heat shock proteins (HSPs) and receptor phosphorylation. The AR dimerizes and translocates to the nucleus. The AR engages with a host of coactivators including lysine acetyltransferases (KAT3A (CBP) and KAT3B (p300)) and pioneer factors FOXA1 and GATA2 to bind to the ARE and drive expression of its target genes. Activation of target genes leads to biological responses including proliferation and cell survival and the production of PSA (*KLK3*) and *TMPRSS2*.

There are three sources of androgenic steroids: the testes, which synthesize testosterone from adrenal precursors; the adrenal glands, which secrete the weakly androgenic precursors androstenedione and dehydroepiandrosterone (DHEA), and diet [75]. Within prostate cells, the enzyme 5α reductase converts testosterone into the more potent dihydrotestosterone (DHT), which has a high affinity for binding to the AR. Ligand-activated AR functions as a ligand-dependent transcription factor, by homodimerizing and translocating to the nucleus, where it promotes transcription of a multitude of target genes (including *KLK3* (PSA) and *TMRPSS2*), leading to androgen-dependent cell growth [76]. AR-mediated transcriptional regulation is dependent on AR activity, which can be dysregulated through a variety of mechanisms including mutations in the *AR* gene, increased expression of the AR protein, promiscuous activation of the AR (e.g., through growth factors), or aberrant behavior of AR coregulators [12,77].

3.1 ACETYLATION

AR coregulators (-activators or -repressors) modify AR gene transactivation and prostate cell survival through acetylation, phosphorylation, ubiquitination, and sumoylation [78–80]. Acetylation in particular appears to be intimately involved in regulating the activity of the AR. Lysine acetyltransferases (KATs), such as p300 (KAT3B), TIP60 (TAT interactive protein 60, KAT5) [81], and CBP (cAMP response element-binding protein, KAT3A), behave as generic coactivators of many transcription factors including the AR and p53. Other notable coactivators are PCAF (the p300-CBP-associated factor, KAT2B) and ARD1 [82]. The acetylases modify the AR at individual lysine residues. This acetylation is critical for the transcriptional activation activity of the AR but has a dual function in also precluding its polyubiquitination and degradation [83]. Of course, transactivation by the AR is also largely dependent on access to its binding sites (AREs) at target gene enhancers or promoters, which is governed to some extent by local chromatin structure and therefore the actions of histone-modifying enzymes. Following AR acetylation, KATs further facilitate AR transcriptional activity by nucleosome remodeling via histone acetylation and by recruiting the RNA polymerase II complex to AR-regulated promoters [79,84–86]. In addition, other "pioneer" factors, namely, FOXA1 and GATA2, modulate the transcriptional activity of the AR by promoting an open chromatin structure, which facilitates AR binding [87] and maintaining nucleosome depletion at AREs, in a locus-specific manner [88].

Many AR coactivators, such as the acetylases p300 (KAT3B) and TIP60 (KAT5) and the pioneer factor GATA2, act in a concerted effort to promote PCa progression [83,89–91]. It was recently shown that the expression of p300 (KAT3B) is correlated with AKT phosphorylation (a surrogate for PTEN inactivation and a prevalent feature of metastatic PCa). Mechanistically, PTEN inactivation increases AR phosphorylation at serine 81, which promotes p300 (KAT3B) binding and acetylation of the AR [83]. Thus, p300 (KAT3B) behaves as a bona fide oncogenic factor in PCa. Similarly, by promoting transcriptional activation activity of the AR, these coactivators promote CRPC [92].

Further evidence in support of the positive effects of acetylation on AR activity comes from multiple studies, which demonstrate that HDACs (1–3, 6, and Sirt1) suppress AR activity by catalyzing AR deacetylation [81,85,93,94] or indeed sumoylation (by HDAC4) [95]. The situation is complicated, however, by evidence of HDACs exerting pro-AR signaling effects. HDAC-1 and -3 are required for the activation of approximately 50% of AR target genes, including *TMPRSS2*, the transcriptional driver of the ETS fusion gene, implicated in approximately 50–70% of prostate tumors [96,97]. A further example is HDAC6, a cytoplasmic nonhistone protein deacetylase whose substrates include the AR chaperone protein, HSP90. HDAC6-mediated deacetylation of HSP90 enhances its binding to the AR, thus lessening degradation of client proteins including the AR [98,99].

3.2 METHYLATION

Lysine methylation and demethylation are involved in regulating AR transcriptional activation, at the level of histones, the AR protein itself the interplay between the two. EZH2 (KMT6) has been shown to methylate the AR, at lysine 630 and 632, which potentiates its transactivation function. Again, this has been mechanistically linked with AKT hyperactivity/PTEN loss, which mediates EZH2 (KMT6) phosphorylation at serine 21, stimulating AR methylation, through the SET domain of EZH2 (KMT6) [50].

KDM1A interacts with the AR and promotes androgen-dependent transcription of target genes by demethylating repressive histone marks (mono- and dimethylation at H3K9) [41]. Several members of the jumonji domain containing family of KDMs (KDM4A, KDM4B, KDM4C and KDM4D) also function as AR coactivators [100]. KDM1A and KDM4C colocalize and cooperatively stimulate AR-dependent gene transcription in the prostate microenvironment by removing methyl groups (me1, me2, and me3) from H3K9 [101]. KDM4B was identified as the first androgen-regulated KDM [100]. It influences AR transcriptional activation via its histone demethylation activity and also by directly modulating ubiquitination of the AR. Knockdown of KDM4B almost completely depletes AR protein in the LNCaP cell line. Furthermore, KDM4B expression in clinical specimens positively correlates with clinical grade, implying that KDM4B may be a viable therapeutic target for PCa.

4 DRUGGING THE METHYLOME FOR THE TREATMENT OF CRPC

The fact that DNA methylation patterns are clonally maintained and could therefore serve as driver events makes them attractive therapeutic targets. In addition unlike genetic alterations, DNA methylation can be dynamic and reversible. Genes silenced by promoter hypermethylation are typically wild type in their genetic code, thus normal genetic function might be restored by removing the DNA methyl marks.

Many preclinical studies of nucleoside analogue DNMTi have validated their potential use in restoring gene expression of numerous hypermethylated loci in PCa cell lines. For example, in 2000, the expression of the *AR* gene was restored in DU145 PCa cell lines upon treatment with 5,6-dihydro-5-azacytidine [102].

Over the last decade, there have been a handful of clinical trials assessing the efficacy of DNMTi for treatment of CRPC (Table 11.2). However, as has been the case for other solid tumors, these therapies have been marred by immediate cytotoxic and off-target effects. As early as 1998, a Phase II clinical trial investigated intravenous administration of decitabine for treatment of CRPC, in men following complete androgen blockade and flutamide withdrawal. Only 12 men were evaluable, of which 2 showed a time of progression of >10 weeks [103]. More recently, a Phase II clinical trial of vidaza in a larger number of CRPC patients increased PSA doubling time $\geq$3 months in 55.8% of patients [104].

There is some evidence to suggest that DNMTi may be useful in amplifying the effectiveness of conventional therapies (chemotherapeutic agents or androgen withdrawal/blockade) in men with advanced PCa. Treatment of DU145 PCa cell lines with docetaxel and either 5-Aza-CdR or 5-Aza-CR enhanced docetaxel sensitivity through hypomethylation and reactivation of DNA damage response gene GADD45A [105]. Treatment with decitabine and either paclitaxel or cisplatinum caused synergistic growth suppression in prostate cell lines through enhanced induction of apoptosis and G_2/M cell cycle arrest [106,107]. In the TRAMP (transgenic adenocarcinoma of mouse prostate) mouse model of PCa, combined castration and decitabine significantly improved survival over either single treatment ($P < 0.05$) and reduced the presence of malignant disease ($P < 0.0001$) [108]. In preclinical cell

Table 11.2 Epigenetic Drug Trials for Prostate Cancer

Drug	Clinical Trial Identifier	Study Design	Phase	Status	Study Objectives	Description	Outcome Measures	Outcome	Reference
Vidaza	NCT00384839	Open-label	Phase II	Completed	To find out the effects of vidaza on patients with PCa	Injectable suspension of vidaza (75 mg/m²) days 1–5 every 28-day cycle. Maximum 12 cycles	**Primary:** 1) To determine whether vidaza can restore hormone responsiveness in CRPC	34/36 men enrolled were evaluable. 80.6% had metastatic disease. PSA-DT ≥3 months was attained in 19 patients (55.8%). Median progression-free survival was 12.4 weeks	[104]
Decitabine			Phase II	Completed	To find out the effects of decitabine on patients with CRPC	I.v. decitabine (75 mg/m²) every 8 h for 3 doses, repeated every 5–8 weeks	**Primary:** 1) To determine whether vidaza can restore hormone responsiveness in CRPC	12/14 men enrolled were evaluable. 2/12 men had stable disease and delayed time to progression ≥10 weeks	[103]
Vidaza, docetaxel, and prednisone	NCT00503984	Nonrandomized, open-label	Phase I/II	Ongoing, not recruiting	To study the side effects and best dose of azacitidine and docetaxel when given together with prednisone and to see how well they work in treating patients with CRPC who did not respond to hormone therapy	Phase I: i.v. vidaza over 30 min, days 1–5 of each 3 weekly cycle, docetaxel: i.v. over 1 h on day 6 of each 3 weekly cycle, prednisone (5 mg) twice a day from days 1–21 of each cycle	**Primary:** Phase I: MTD of vidaza and docetaxel combination Phase II: Response, defined as PSA response or complete or partial response, by RECIST criteria		

Agent	Trial	Design	Phase	Status	Objective	Dosing schedule	Outcomes	Results	References
						Phase II: As above without prednisone	**Secondary:** Phase I: Toxicity Phase II: Duration of response, progression-free survival, overall survival, *GADD45A* methylation and expression		
SB939	NCT01075308	Nonrandomized, open-label	Phase II	Ongoing, not recruiting	To determine the efficacy and progression-free survival of patients with recurrent or metastatic CRPC in response to HDAC inhibitor SB939	Oral SB939 once daily on days 1, 3, 5, 8, 10, 12, 15, 17, and 19 Treatment repeats every 4 weeks for up to 12 courses in the absence of disease progression or unacceptable toxicity	**Primary:** 1) PSA response 2) PFS **Secondary:** 1) Objective response rate 2) Duration of response 3) Safety 4) Enumeration of CTCs 5) Analysis of *TMPRSS2-ERG* fusion and *PTEN* deletion in CTCs		
Panobinostat	NCT00667862	Open-label	Phase II	Completed	To characterize the safety, tolerability, and efficacy of i.v. panobinostat as a single agent treatment in patients with CRPC	IV panobinostat ($20\,mg/m^2$) on days 1 and 8 of a 21-day cycle	**Primary:** 1) PFS at 24 weeks **Secondary:** 1) Safety 2) Tolerability 3) Proportion of patients with a PSA decline	Of the 35 patients enrolled, 4 (11.4%) were progression-free at 24 weeks; 0 exhibited a PSA decline $\geq 50\%$. I.v. panobinostat did not show sufficient level of clinical activity to pursue further activity as a single agent in CRPC	[129]

Table 11.2 (Continued)

Drug	Clinical Trial Identifier	Study Design	Phase	Status	Study Objectives	Description	Outcome Measures	Outcome	Reference
Panobinostat	NCT00493766	Nonrandomized, open-label	Phase I	Terminated	To define the MTD, toxicity, activity, and pharmacokinetics of oral panobinostat, alone in combination with docetaxel in patients with CRPC	Arm A: Oral panobinostat (20 mg) on days 1, 3, and 5 for 2 consecutive weeks, followed by 1-week break Arm B: Oral panobinostat (15 mg) administered on the same schedule in combination with docetaxel (75 mg/m²) every 21 days	**Primary:** 1) MTD and DLT of escalating doses of panobinostat 2) MTD and DLT of escalating doses of panobinostat in combination with standard dose of docetaxel and daily prednisone **Secondary:** 1) Safety and tolerability of combination (acute and chronic toxicities) 2) Single- and multidose pharmacokinetics of the combination	16 patients were enrolled, 8 in each arm. Grade 3 toxicities were observed. Arm A: All patients developed progressive disease. Arm B: 5 patients had a PSA decline ≥50%. Docetaxel had no apparent effect on the pharmacokinetics of panobinostat	[131]
Panobinostat combined with Bicalutamide (Casodex)	NCT00878436	Randomized, open-label	Phase 1 & P	Recruiting	To test whether panobinostat enhances efficacy of second line hormonal therapy in CRPC	Arm A: 21-day treatment cycle of Bicalutamide 50 mg P.O. daily, continuously, with the addition of 40 mg	**Primary:** 1) The proportion of patients free of progression and without symptomatic deterioration by 9 months of therapy		

Vorinostat combined with Temsirolimus	NCT01174199	Open-label	Phase I	Recruiting	To determine the efficacy and survival of patients with chemoresistant mCRPC in response to HDAC inhibitor Vorinostat in combination with mTOR inhibitor Temsirolimus	Panobinostat 3 times/week (120 mg/ week) for 2 consecutive weeks with 1-week rest Arm B: 21-day treatment cycle of Bicalutamide 50 mg P.O. daily, continuously, with the addition of 20 mg Panobinostat 3 times/ week (60 mg/ week) for 2 consecutive weeks with 1-week rest Oral vorinostat once daily on days 1–14 and temsirolimus IV on days 1, 8, and 15. Courses repeat every 21 days in the absence of disease progression or unacceptable toxicity	**Secondary:** 1) Time to PSA progression (up to 2 years) 2) Proportion of patients that achieve a PSA decline of ≥50% by 9 months of therapy **Primary:** 1) Frequencies of DLT and toxicity (within 4 years) 2) Adverse events (within 4 years) **Secondary:** 1) Median survival, median PFS, and frequency of deaths (within 4 years) 2) PSA response 3) Changes in expression of bone remodeling markers and angiogenesis-related gene and protein expression 4) Changes in tumor metabolism as assessed by PET/CT scan

Table 11.2 (Continued)

Drug	Clinical Trial Identifier	Study Design	Phase	Status	Study Objectives	Description	Outcome Measures	Outcome	Reference
Romidepsin	NCT00106418	Open-label, nonrandomized	Phase II	Completed	To evaluate the activity and tolerability of romidepsin in CRPC with rising PSA	Intravenous romidepsin infusion ($13\,mg/m^2$) on days 1, 8, and 15 every 28-day cycle	**Primary:** 1) No evidence of radiological progression at 6 months	Of the 35 patients enrolled, 2 achieved a confirmed radiological partial response lasting ≥ 6 months and a PSA decline $\geq 50\%$; 11 were discontinued due to toxicities. At the selected dose and schedule, romidepsin demonstrated minimal antitumor activity in chemonaive patients with CRPC	[128]
Vorinostat	NCT00330161	Open-label	Phase II	Completed	To evaluate vorinostat in patients with advanced PCa who have progressed on 1 prior chemotherapy	Oral vorinostat once daily on days 1–21. Treatment repeats every 21 days for at least 4 courses in the absence of disease progression or unacceptable toxicity.	**Primary:** 1) Proportion of patients who do not demonstrate disease progression (at 6 months) **Secondary:** 1) Incidence of toxicity 2) Rate of PSA decline 3) Progression-free survival 4) Median survival 5) Objective response rate	Of the 29 patients enrolled, 27 were evaluable. All 27 evaluable patients were off therapy before the 6 month time point	

Vorinostat combined with androgen deprivation therapy	NCT00589472	Open-label	Phase II	Ongoing, not recruiting	To determine the efficacy of neoadjuvant androgen deprivation therapy and vorinostat followed by radical prostatectomy for the treatment of localized PCa	Bicalutamide PO QD for 1 month and leuprolide acetate IM or goserelin acetate SC once a month until surgery, vorinostat PO QD beginning on the first day of androgen depletion therapy and continuing for up to 8 weeks or until the day of surgery (open or laparoscopic radical prostatectomy). Patients with positive surgical margins undergo immediate adjuvant external beam radiotherapy	**Primary:** 1) Pathologic complete response at the time of surgery **Secondary:** 1) Postradical prostatectomy peripheral blood PSA, testosterone, DHT, DHEA, and DHEA-S levels (up to 1 year) 2) Levels of testosterone, androstenedion, androstenediole, DHT, DHEA, and DHEA-S in prostate tissue (up to 1 year) 3) Gene and protein expression analysis, including *AR* target genes, *PSA* and *TMPRSS2*
Sipuleucel-T with/without Tasquinimod	NCT02159950	Randomized, open-label	Phase II	Not yet recruiting	To determine how well sipuleucel-T works with/ without tasquinimod in treating patients with mCRPC	Arm I: sipuleucel-T i.v. over 60 min (day 4). Arm II: sipuleucel-T i.v. over 60 min (day 4) with daily, oral tasquinimod (day 14+).	**Primary:** 1) Immune response 2) PFS 3) OS 4) PSA response

Table 11.2 (Continued)

Drug	Clinical Trial Identifier	Study Design	Phase	Status	Study Objectives	Description	Outcome Measures	Outcome	Reference
						Patients continue on tasquinimod treatment after day 42 until disease progression Both arms: Treatment repeats every 2 weeks for 3 courses in the absence of disease progression or unacceptable toxicity.	**Secondary:** 1) Assessment of multiple immune responses		
Tasquinimod	NCT02057666	Randomized, double-bind, placebo-controlled	Phase III	Recruiting	To confirm the effect of tasquinimod in delaying disease progression or death as compared with placebo in chemo-naïve Asian men with mCRPC	Drug: 0.25 mg/day, titrated through 0.5 mg/day (from day 15) to a maximum of 1 mg/day (from day 29)	**Primary:** 1) Radiological PFS (up to 3 years) **Secondary:** 1) OS (up to 3 years) 2) Symptomatic PFS 3) QoL		

| Tasquinimod | NCT01234311 | Randomized, double-bind, placebo-controlled | Phase III | Ongoing, not recruiting | To confirm the effect of tasquinimod in delaying disease progression or death as compared with placebo in asymptomatic/mildly symptomatic men with mCRPC | Group A: Tasquinimod 0.25, 0.5, or 1 mg/day; n=800 Group B: Placebo; n=400 | **Primary:** 1) Radiological PFS (up to 3 years) |
| Tasquinimod | NCT01732549 | Randomized, double-bind, placebo-controlled | Phase II | Recruiting | Proof-of-concept study of maintenance therapy with tasquinimod in patients with mCRPC who are not progressing after a first-line Docetaxel-Based chemotherapy | Drug: 0.25 mg/day, escalated to 0.5 or 1 mg/day until disease progression | **Primary:** 1) Radiological PFS (up to 3.5 years) **Secondary:** 1) OS 2) PFS on next-line therapy 3) symptomatic PFS 4) Adverse events |

All data obtained from clinicaltrials.gov and associated publications, where available. Abbreviations: NCT, National Clinical Trials identifier; CRPC, castration-resistant prostate cancer; mCRPC, metastatic CRPC; CTC, circulating tumor cell; MTD, maximum tolerated dose; DLT, dose.

line and xenograft models, vidaza has been shown to restore AR expression and enhance the apoptotic effects of antiandrogen bicalutamide [109].

4.1 NEW CLASSES OF DNMT INHIBITORS

The inherent toxicity of DNMTi has steered research toward compounds that target DNMTs more directly, rather than intercalating into the DNA double helix. For example, procainamide (derived from local anesthetic procaine) inhibits DNMT1 by binding to CpG-dense DNA, thus perturbing interactions between the enzyme and its target CpG sites [110]. The drug reverses *GSTP1* hypermethylation and restores its expression in LNCaP PCa cells propagated *in vitro* and *in vivo* as xenograft tumors [111]. Disulfiram, a thiol-reactive compound used to treat alcoholism, possesses intrinsic DNMTi properties. In PCa xenograft and cell line models, the drug globally reduced 5mC levels, reactivated *RARβ2* and *APC* tumor suppressor genes and functionally reduced proliferation and induced apoptosis [112].

5 HDAC INHIBITORS FOR THE TREATMENT OF CRPC

HDAC inhibitors (HDACi) are the other major class of epigenome modulating agents under investigation for the treatment of CRPC. HDAC inhibition causes histones to become hyperacetylated and chromatin to adopt an open conformation conducive to interaction with the transcriptional apparatus. Interestingly, it has been reported that less than 10% of genes are influenced directly by HDACi [113]. Of course, HDACs interact with numerous nonhistone proteins, e.g., the AR, thus implying more pleiotropic mechanisms of HDACi activity (see Section 5.1). The strong attraction for developing HDACi is their selective action on tumor cells, inducing apoptosis, growth arrest, and autophagy, among other phenotypic effects (reviewed elsewhere).

Multiple dose-dependent mechanisms of action have been observed for HDACi, both epigenetic and cytotoxic. There is a wealth of preclinical information showing that multiple HDACi exert effective antiproliferative and proapoptotic effects in PCa cell lines and xenograft models. Valproic acid (VPA, a short chain fatty acid inhibitor) inhibits growth of PCa cells *in vitro* and reduces tumor xenograft growth in athymic nude mice by modulating multiple pathways including cell cycle arrest, apoptosis, angiogenesis, and senescence through its effects on HDACI1 acetylation [114,115]. The more potent hydroxamate HDACi inhibits HDAC activity at low micro/nanomolar ranges through binding to the catalytic pocket of the enzyme. Both TSA (trichostatin A) and SAHA (suberoylanilide hydroxamic acid, vorinostat) induce cell death and inhibit growth in PCa cell lines through inhibiting *AR* gene expression [116,117]. SAHA also suppresses tumor growth and volume with little toxicity in mice transplanted with CWR22 human prostate tumors [118]. PXD101 (belinostat) is a potent pan HDACi and retards tumor growth in xenograft models [119]. LBH589 causes growth arrest in PCa cell lines via inactivation of HDAC6 [120]. Micromolar concentrations of the benzamide HDACi MS-275 (entinostat) causes growth arrest of PC3 and LNCaP cell lines and induces cell death in DU145 cells. *In vivo*, MS-275 inhibits the growth of DU145, LNCaP, and PC3 in subcutaneous xenografts. Molecular analysis shows that treatment increases histone H3 acetylation and CDKN1A (p21) expression in tumors. In the TRAMP mouse model, long-term treatment with MS-275 slowed tumor progression with significant reduction in cell proliferation [121].

5.1 SYNERGISTIC ACTIVITY

HDACi have been used with demethylating agents to synergistically activate expression of methylated genes. For example, cotreatment of AR-negative cell line DU145 with 5-Aza-CR and TSA is more effective in restoring functional expression of the *AR* gene and its downstream targets, compared with either agent alone [102]. HDACi have also been shown to synergize with gamma radiation to kill tumor cells *in vitro*. In mouse models, MS-275 and VPA enhances the radiosensitivity of PCa DU145 and PC3 xenografts [122,123]. Other studies support a combination of HDACi with conventional chemotherapeutic agents or antiandrogens. Combined treatment of an HDACi (depsipeptide, SAHA, sodium butyrate, or TSA) with TNFα-related apoptosis-inducing ligand (TRAIL) leads to enhanced cytotoxicity and synergistic apoptosis in PCa cell lines LNCaP and DU145 [124–126]. Combined treatment of PC3 cells with panobinostat and the dual PI3K-mTOR inhibitor, BEZ235 significantly attenuates DNA damage repair protein ATM and increases antitumor activity over either agent alone [127].

A number of clinical trials with HDACi have been instigated, however, the response rate is generally poor (Table 11.2) [128–130]. Phase I studies assessing HDACi (panobinostat) with chemotherapy have reported feasibility of the combination [131]. A phase I/II study is currently ongoing assessing the efficacy of HDACi panobinostat in combination with antiandrogen bicalutamide (Table 11.2). The pan activity of most HDACi results in the activation of unwanted prosurvival genes. This could be counteracted by developing more selective inhibitors. A notable example is the oral antiangiogenic drug tasquinimod, which has been shown to double progression-free survival time in men with metastatic CRPC [132]. Tasquinimod mediates its anticancer effects, in part by specifically inactivating HDAC4. Tasquinimod binds allosterically to HDAC4 and locks it into a conformation that prevents formation of the HDAC4/NCoR/HDAC3 complex, thus inhibiting deacetylation of histones and other "client" proteins (such as transcription factors), that are required for cancer cell survival and angiogenic response [133]. Currently, there are a number of phase II/III clinical trials assessing the efficacy of tasquinimod for the treatment of metastatic CRPC (Table 11.2).

6 TARGETING AR SIGNALING BY EPIGENETIC DRUGS

Traditionally, HDACi have been demonstrated to potentiate AR transcription by enabling the activity of KATs on the AR signaling axis. Antiandrogens such as bicalutamide recruit corepressors (i.e., NCoR or SMRT) that complex with HDACs and inhibit AR-transactivation. However, several lines of evidence actually support a negative effect of HDACi on AR signaling [96]. Certain HDACi result in hyperacetylation of HSP90, dissociation from the AR and subsequent AR degradation [98,99] (Figure 11.2). In addition, HDACi (TSA, SAHA, and LBH589) block transcriptional activation of HDAC-dependent AR target genes, predominantly by interfering with the assembly of the RNA polymerase II complex, and also by inhibiting transcription of the *AR* gene [96,116]. This apparent paradoxical situation must surely be driven by the severe molecular (and thus clinical) heterogeneity of the disease. To illustrate this point, CBP (KAT3A), which behaves as a *bona fide* AR coactivator is lost in a proportion of PCa and correlates with PTEN loss [134]. Concomitant loss of PTEN and CBP (KAT3A) induces preinvasive HGPIN and low grade tumor formation; panobinostat treatment in these $CPB^{-/-}$; $PTEN^{+/-}$ mouse models diminish PIN lesions [134].

An exciting thought is of course the possibility of combinatorially suppressing AR and AR coactivator activity. This could be achieved by developing inhibitors of AR coactivators, in addition to conventional agents that deplete androgens or compete for binding with the AR. A small number of KAT inhibitors (KATi) have been reported. In LNCaP and PC3 PCa cell line models, a synthetic inhibitor of TIP60, namely NU9056, inhibited proliferation and induced apoptosis in a concentration- and time-dependent manner (IC50: 2 μM). Reduced levels of the AR, PSA, and p53 were observed in response to treatment [135]. Similarly, knockdown of p300 (KAT3B) by siRNA or a synthetic small molecule (C646) induces apoptosis in androgen-dependent and castration-resistant cell lines [136].

The bromodomain and extraterminal (BET) family of chromatin readers modulate gene expression by binding to acetylated histone tails and subsequently activating RNA polymerase II-driven transcriptional elongation. Small molecule BET inhibitors block binding of BET proteins to acetylated chromatin marks, thus preventing transcriptional activation of BET target genes [137].

BRD4 plays an important role in transcription by RNA polymerase II. BRD4 is known to interact with a number of transcription factors, including the AR. It was recently shown that this interaction can be hindered in AR signaling competent CRPC cell lines, by JQ1, a small molecule BET inhibitor [138]. Phenotypically, JQ1 causes cell cycle arrest and apoptosis. Similar to AR antagonist (MDV3100), JQ1 disrupts AR recruitment to target gene loci. In contrast to MDV3100, however, JQ1 appears to function downstream of the AR, thus offering promise in the context of acquired castration resistance, mediated by AR amplification or mutation. In addition to modulating AR signaling, another promising line of research surrounds the potent effect of BRD4 inhibition on downregulation of MYC [137,138].

7 CHEMOPREVENTION AND NEUTRACEUTICAL THERAPIES

The chronic nature of latent PCa makes it appealing to consider therapeutic and lifestyle interventions that could reduce the risk of disease progression into clinically apparent disease and/or that could prevent the formation of lethal metastases. Autopsy studies reveal the presence of precursor lesions and microscopic tumor foci in up to 29% of men aged in their 30s and 40s [139]. There is substantial evidence in support of an "epigenetic catastrophe" at the earliest stages of prostate carcinogenesis, positioning adverse epigenetic changes as seminal events in PCa initiation [140,141]. Does the almost universal epigenetic inactivation of *GSTP1* in PCa suggest that preventing this singular event could represent an effective means for PCa prevention [142]. 5-Aza-CdR reduces the appearance of preinvasive intestinal polyps in APC⁻ mice [143]. Similarly, 5-Aza-CR has chemopreventative properties in TRAMP mice, where treatment prevents hypermethylation and silencing of tumor suppressor genes, reduces the development of tumors, and improves survival time [144].

There is considerable epidemiological evidence to support a role for dietary factors as chemopreventative agents against PCa. Elucidating the molecular mechanisms by which the "active" compounds contained in these dietary constituents mediate their anticancer activity has lagged behind. This section attempts to summarize our current knowledge of how dietary agents may serve as therapeutic modalities by restoring epigenomics normality.

7.1 ISOTHIOCYANATES

A high intake of cruciferous vegetables is associated with a reduced risk of PCa in epidemiological studies [145]. There is preliminary preclinical evidence that sulforaphane, derived from glucoraphanin found in a number of cruciferous vegetables, may prevent and induce regression of PCa and other malignancies, although through mechanisms that are poorly understood. Sulforaphane treatment of PCa cells *in vitro* inhibits HDAC activity and induces caspase-dependent apoptosis [146]. Similarly, sulforaphane administration inhibits PCa progression and pulmonary metastasis in TRAMP mice, by reducing cell proliferation and augmenting natural killer cell lytic activity [147]. More specifically, sulforaphane inhibits HDAC6 enzymatic activity, enhances HSP90 acetylation and AR degradation by the proteasome, ultimately reducing AR occupancy at target gene enhancers and reducing expression of target genes (*KLK3* and *TMPRSS2*) [99]. Sulforaphane treatment of benign and malignant prostate cells in culture also inhibits DNMT expression and exerts broad changes on DNA methylation patterns [148], although the functional consequences of these observations remain to be elucidated. Other studies suggest that the chemopreventative efficacy of sulforaphane against late-stage poorly differentiated PCa can be augmented by pharmacologic inhibition of autophagy [149]. There are several ongoing clinical trials assessing the preventative properties of sulforaphane (Table 11.3).

7.2 CURCUMIN

The isoflavone curcumin, a component of the spice turmeric, long used in Asian cooking, has numerous medicinal and anticancer properties (reviewed elsewhere). Curcumin has radiosensitizing properties in PCa cells *in vitro* [150], and in combination with phenethyl isothiocyanate (PEITC) significantly reduces the growth of PC3 xenografts in mice [151]. Mechanistically, curcumin represses histone H4 acetylation at AREs and subsequent AR recruitment by impinging on KAT3A and KAT3B (p300 and CPB) and pioneer factor (GATA2 and FOXA1) occupancy at these enhancer elements. These effects were observed in xenograft models mimicking both androgen sensitive and CRPC [92]. HDACi (TSA and SAHA) reversed the inhibitory effects of curcumin on cell survival, thereby implying that the ability of curcumin to attenuate AR activity is dependent on alteration of histone acetylation. The findings of this preclinical study are of strong clinical relevance because HAT coactivators and pioneer factors support activity of the AR in a castrate-resistant environment. Consonantly, curcumin could be used to alter the chromatin landscape and act in concert with hormone deprivation. There are a number of ongoing clinical trials assessing the efficacy of curcumin in synergistic effect with radiotherapy or taxanes to improve survival in men with CRPC. However, epigenetic alterations are not listed as outcome measures for any of these trials.

7.3 PHYTOESTROGENS

Plasma and serum phytoestrogen levels of Japanese men are at least 10-fold higher than Caucasian men in the United Kingdom. Soy isoflavones, e.g., genistein, classified as phytoestrogens, act as both estrogen agonists and antagonists by differentially binding to the estrogen receptor alpha or beta and/or altering enzymes involved in hormone metabolism [152]. Epidemiological studies link an increased intake of dietary genistein with reduced rates of PCa metastasis and mortality [153].

Genistein appears to have many disparate biologic effects. Using both *in vitro* and *in vivo* models, genistein impedes cell detachment [154] and inhibits cell invasion [155]. Studies in PCa cell lines show that genistein treatment blocks activation of the p38 MAPK MMP2 proinvasion pathway [155]. Subsequently, genistein was found to exert its inhibitory effects on cell invasion by directly inhibiting the activity of p38 MAPK activator MEK4 at nanomolar concentrations in PCa cell lines [156].

Table 11.3 Neutraceutical Epigenetic Drug Trials for Prostate Cancer

Drug	Clinical Trial Identifier	Study Design	Phase	Status	Study Objective	Description	Outcome measures
Sulforaphane (SFN) glucosinolate	NCT01265953	Randomized, double-blind	Chemoprevention	Recruiting	To identify mechanisms by which dietary compounds, such as those found in cruciferous vegetables, decrease PCa risk	Four weeks SFN glucosinolate capsules: 250 mg of broccoli seed extract (30 mg SFN glucosinolate), 8 capsules (4 capsules B.I.D.) daily in subjects at risk of PCa undergoing prostate biopsy	**Primary:** 1) Presence of SFN and its metabolites (SFN-Cys, SFN-NAC) 2) Expression of acetylated H3 and H4, and absolute histone levels **Secondary:** 1) Methylation of *GSTP1, AR, sigma14-3-3, P21* and global 5mC levels. 2) Cell proliferation (Ki-67 expression) and apoptosis (TUNEL assay) Analyses will be carried out on peripheral blood plasma, urine, and prostate biopsy cores following supplementation with SFN or placebo
Sulforaphane (SFN) glucosinolate	NCT01950143	Randomized, double-blind	Not provided	Recruiting (invite-only)	To determine whether a 12-month diet rich in SFN will prevent PCa progression in men diagnosed with low- and	Group A: 2 portions of standard broccoli soup/week for 12 months Group B: 2 portions of glucoraphanin-enriched broccoli soup/week for 12 months	**Primary:** 1) Global transcriptome expression analysis (baseline and 12 months)

Agent	NCT identifier	Study design	Phase	Status	Aim	Intervention	Outcomes
Sulforaphane (SFN) glucosinolate	NCT00946309	Randomized, double-blind	Phase II	Recruiting (invite-only)	intermediate-risk prostate cancer on active surveillance. To study the biological *in vivo* effects of SFN supplementation on normal prostate tissue	Group C: 2 portions of glucoraphanin-extra enriched broccoli soup/week for 12 months. Experimental: 100µmol sulforaphane, every other day for 5 weeks. Placebo: 250mg microcrystalline cellulose NF, every other day for 6 weeks	**Secondary:** 1) Metabolite concentration (baseline and 12 months) **Primary:** 1) Gene expression of Phase II enzymes 2) DNA and lipid oxidation 3) DHT levels. All assessed at baseline and 5 weeks
Sulforaphane (SFN) glucosinolate	NCT01228084	Open-label, single group assignment	Phase II	Completed	To study the effects of sulforaphane in patients with biochemical recurrence of PCa	Experimental: Four 50µmol capsules Sulforaphane taken once daily from Week 1 Day 1 to Week 20 Day 7	**Primary:** 1) Proportion of patients who achieve a 50% decline in PSA **Secondary:** 1) % change in PSA from baseline to end of study (20 weeks) 2) Toxicities 3) Half-life of SFN in blood in relation to patient's GSTM1 genotype

All data obtained from clinicaltrials.gov and associated publications, where available. Abbreviations: NCT, National Clinical Trials identifer; CRPC, castration-resistant prostate cancer; mCRPC, metastatic CRPC; CTC, circulating tumor cell; DLT, dose limiting toxicity; I.V., intravenous; MTD, maximum tolerated dose; PFS, progression-free survival; OS, overall survival; QoL, quality of life.

Evidence on the effects of genistein on the epigenome is conflicting. In androgen-dependent and -independent cell lines, genistein dose dependently inhibits DNMT activity and reactivates methylation-associated silenced genes such as *GSTP1* and *RASSF1A* [157,158]. However, feeding studies with genistein increased DNA methylation *in vivo* [159]. Genistein-rich diets fed to mice have also been shown to suppress the trimethylation of H3K9 and the phosphorylation of H3S10 and increase methylation of Wnt signaling-related genes *SFRP2*, *SFRP5*, and *WNT5A* genes, suppressing their expression and maintain normal levels of Wnt signaling [160].

The significance of these studies is that they demonstrate prevention of epigenetic aberrations and restoration to normal physiological states by commonly consumed dietary constituents in parts of the world that experience some of the lowest incidence of PCa. Thus, one may speculate that inhibiting hypermethylation-induced inactivation of key tumor suppressor genes or avoiding hyperstimulation of AR signaling by dietary molecules could afford a chemopreventative effect against PCa. It is possible that long-term consumption of polyphenols, isoflavones, and isothiocyanates with dietary DNMT/HDACi may have a cumulative effect over a man's lifetime, providing protection against PCa development [151,161].

8 FUTURE DIRECTIONS

In the past, epigenetic research has concentrated around promoter CGIs because of their innate association with gene expression. It follows naturally that epigenetic cancer therapies have focused largely around reactivating tumor suppressor genes and restoring their functional products. However, the ever-growing surge in epigenome-wide capabilities coupled with Next-Gen sequencing offers the delectable opportunity to explore epigenomic therapeutics beyond the promoter CGI. Elucidating the full range of molecular mechanisms involved in the therapeutic responsiveness to epigenetic drugs is needed before their successful clinical application for the treatment of CRPC.

Clinical trial results have for the most part not supported the use of conventional DNMTi and HDACi as effective single agent CRPC therapies. However, in recent years, evidence from breast and lung tumors supports the use of "low-dose" epigenetic therapies to "reprogram" tumor cells and sensitize them to additional therapeutic regimes, while ameliorating the intolerable side effects observed at higher doses. Indeed, pretreatment of DU145 tumors with low-dose FK228 enhances their sensitivity to gemcitabine and docetaxel *in vivo*, while LBH589 restores androgen sensitivity and in combination with bicalutamide synergistically inhibits cell growth and induces caspase 3/7 activation [162,163]. Consonantly, an ongoing Phase II clinical trial is assessing a low dose of panobinostat to enhance second line hormonal therapy with bicalutamide, with a view to preventing metastatic progression (Table 11.2).

Other potentially exciting avenues are novel mechanisms of perturbing AR signaling, which is the cornerstone of CRPC. Worthy of mention are possible EZH2 (KMT6) inhibitors [164,165]. The identification of KMT6 phosphorylation as an inducer of EZH2-mediated gene activation via interaction with the AR opens new therapeutic avenues targeting the activation function of KMT6, without affecting H3K27me3. The possibility of pharmacologically inhibiting KDMs has also been demonstrated [166]. In PCa, KDM1A remains the most promising target for inhibition, with several substrate analogues and inhibitors in development [167]. However, a degree of caution must be applied when interpreting findings from *in vitro* cell culture models as clinical realities. This is particularly pertinent, in view of recent findings that showed how cellular context is crucially important in the functional genomic positioning of the AR [168,169]. The study by Sharma et al., which demonstrated a tissue-specific transcriptional network (not observed in cultured cells), is a stark reminder of the need to utilize clinical material in order to shape our understanding of CRPC.

As we enter an age of personalized medicine and cancer care, what does the future hold for epigenetic therapies for CRPC? The answer must lie in advancing our understanding of the molecular pathobiology of the disease to enable patient stratification and individualized therapies.

ACKNOWLEDGMENTS

Sincere thanks to Colm O'Rourke for help with table formatting and Alexandra Tuzova for illustrations. Dr Perry's current research program is funded by the Irish Cancer Society, Movember, and the Prostate Cancer Foundation.

REFERENCES

[1] Bray F, Ren JS, Masuyer E, Ferlay J. Global estimates of cancer prevalence for 27 sites in the adult population in 2008. Int J Cancer 2013;132(5):1133–45.
[2] Siegel R, Ma J, Zou Z, Jemal A. Cancer statistics, 2014. CA Cancer J Clin 2014;64(1):9–29.
[3] Ferlay J, Steliarova-Foucher E, Lortet-Tieulent J, et al. Cancer incidence and mortality patterns in Europe: estimates for 40 countries in 2012. Eur J Cancer 2013;49(6):1374–403.
[4] Carter HB, Partin AW, Walsh PC, et al. Gleason score 6 adenocarcinoma: should it be labeled as cancer? J Clin Oncol 2012;30(35):4294–6.
[5] Hugosson J, Carlsson S, Aus G, et al. Mortality results from the Goteborg randomised population-based prostate-cancer screening trial. Lancet Oncol 2010;11(8):725–32.
[6] Schroder FH, Hugosson J, Roobol MJ, et al. Screening and prostate-cancer mortality in a randomized European study. N Engl J Med 2009;360(13):1320–8.
[7] Andriole GL, Crawford ED, Grubb 3rd RL, et al. Mortality results from a randomized prostate-cancer screening trial. N Engl J Med 2009;360(13):1310–19.
[8] Dall'Era MA, Albertsen PC, Bangma C, et al. Active surveillance for prostate cancer: a systematic review of the literature. Eur Urol 2012;62(6):976–83.
[9] Bill-Axelson A, Holmberg L, Ruutu M, et al. Radical prostatectomy versus watchful waiting in early prostate cancer. N Engl J Med 2011;364(18):1708–17.
[10] Scher HI, Sawyers CL. Biology of progressive, castration-resistant prostate cancer: directed therapies targeting the androgen-receptor signaling axis. J Clin Oncol 2005;23(32):8253–61.
[11] Chen CD, Welsbie DS, Tran C, et al. Molecular determinants of resistance to antiandrogen therapy. Nat Med 2004;10(1):33–9.
[12] Feldman BJ, Feldman D. The development of androgen-independent prostate cancer. Nat Rev Cancer 2001;1(1):34–45.
[13] Reid AH, Attard G, Danila DC, et al. Significant and sustained antitumor activity in post-docetaxel, castration-resistant prostate cancer with the CYP17 inhibitor abiraterone acetate. J Clin Oncol 2010;28(9):1489–95.
[14] de Bono JS, Logothetis CJ, Molina A, et al. Abiraterone and increased survival in metastatic prostate cancer. N Engl J Med 2011;364(21):1995–2005.

[15] Danila DC, Morris MJ, de Bono JS, et al. Phase II multicenter study of abiraterone acetate plus prednisone therapy in patients with docetaxel-treated castration-resistant prostate cancer. J Clin Oncol 2010; 28(9):1496–501.

[16] Tran C, Ouk S, Clegg NJ, et al. Development of a second-generation antiandrogen for treatment of advanced prostate cancer. Science 2009;324(5928):787–90.

[17] Scher HI, Beer TM, Higano CS, et al. Antitumour activity of MDV3100 in castration-resistant prostate cancer: a phase 1–2 study. Lancet 2010;375(9724):1437–46.

[18] Nelson WG, Yegnasubramanian S. Resistance emerges to second-generation antiandrogens in prostate cancer. Cancer Discov 2013;3(9):971–4.

[19] Kan Z, Jaiswal BS, Stinson J, et al. Diverse somatic mutation patterns and pathway alterations in human cancers. Nature 2010;466(7308):869–73.

[20] Yegnasubramanian SKJ, Gonzalgo ML, Zahurak M, Piantadosi S, Walsh PC, Bova GS, et al. Hypermethylation of CpG islands in primary and metastatic human prostate cancer. Cancer Res 2004;64(6):1975–86.

[21] Mahapatra S, Klee EW, Young CY, et al. Global methylation profiling for risk prediction of prostate cancer. Clin Cancer Res 2012;18(10):2882–95.

[22] Kobayashi Y, Absher DM, Gulzar ZG, et al. DNA methylation profiling reveals novel biomarkers and important roles for DNA methyltransferases in prostate cancer. Genome Res 2011;21(7):1017–27.

[23] Perry AS, O'Hurley G, Raheem OA, et al. Gene expression and epigenetic discovery screen reveal methylation of SFRP2 in prostate cancer. Int J Cancer 2013;132(8):1771–80.

[24] Sullivan L, Murphy TM, Barrett C, et al. IGFBP7 promoter methylation and gene expression analysis in prostate cancer. J Urol 2012;188(4):1354–60.

[25] Brooks JD, Weinstein M, Lin X, et al. CG island methylation changes near the GSTP1 gene in prostatic intraepithelial neoplasia. Cancer Epidemiol Biomarkers Prev 1998;7(6):531–6.

[26] Kang GH, Lee S, Lee HJ, Hwang KS. Aberrant CpG island hypermethylation of multiple genes in prostate cancer and prostatic intraepithelial neoplasia. J Pathol 2004;202(2):233–40.

[27] Perry AS, Loftus B, Moroose R, et al. *In silico* mining identifies IGFBP3 as a novel target of methylation in prostate cancer. Br J Cancer 2007;96(10):1587–94.

[28] Lee WH, Morton RA, Epstein JI, et al. Cytidine methylation of regulatory sequences near the pi-class glutathione S-transferase gene accompanies human prostatic carcinogenesis. Proc Natl Acad Sci USA 1994;91(24):11733–7.

[29] Kim JH, Dhanasekaran SM, Prensner JR, et al. Deep sequencing reveals distinct patterns of DNA methylation in prostate cancer. Genome Res 2011;21(7):1028–41.

[30] Aryee MJ, Liu W, Engelmann JC, et al. DNA methylation alterations exhibit intraindividual stability and interindividual heterogeneity in prostate cancer metastases. Sci Transl Med 2013;5(169) 169ra10.

[31] Yegnasubramanian S, Haffner MC, Zhang Y, et al. DNA hypomethylation arises later in prostate cancer progression than CpG island hypermethylation and contributes to metastatic tumor heterogeneity. Cancer Res 2008;68(21):8954–67.

[32] Ogishima T, Shiina H, Breault JE, et al. Increased heparanase expression is caused by promoter hypomethylation and up-regulation of transcriptional factor early growth response-1 in human prostate cancer. Clin Cancer Res 2005;11(3):1028–36.

[33] Wang Q, Williamson M, Bott S, et al. Hypomethylation of WNT5A, CRIP1 and S100P in prostate cancer. Oncogene 2007;26(45):6560–5.

[34] Shukeir N, Pakneshan P, Chen G, Szyf M, Rabbani SA. Alteration of the methylation status of tumor-promoting genes decreases prostate cancer cell invasiveness and tumorigenesis *in vitro* and *in vivo*. Cancer Res 2006;66(18):9202–10.

[35] Seligson DB, Horvath S, Shi T, et al. Global histone modification patterns predict risk of prostate cancer recurrence. Nature 2005;435(7046):1262–6.

[36] Bianco-Miotto T, Chiam K, Buchanan G, et al. Global levels of specific histone modifications and an epigenetic gene signature predict prostate cancer progression and development. Cancer Epidemiol Biomarkers Prev 2010;19(10):2611–22.

[37] Ellinger J, Kahl P, von der Gathen J, et al. Global levels of histone modifications predict prostate cancer recurrence. Prostate 2010;70(1):61–9.

[38] Weichert W, Roske A, Gekeler V, et al. Histone deacetylases 1, 2 and 3 are highly expressed in prostate cancer and HDAC2 expression is associated with shorter PSA relapse time after radical prostatectomy. Br J Cancer 2008;98(3):604–10.

[39] Halkidou K, Cook S, Leung HY, Neal DE, Robson CN. Nuclear accumulation of histone deacetylase 4 (HDAC4) coincides with the loss of androgen sensitivity in hormone refractory cancer of the prostate. Eur Urol 2004;45(3):382–9. author reply 9.

[40] Huffman DM, Grizzle WE, Bamman MM, et al. SIRT1 is significantly elevated in mouse and human prostate cancer. Cancer Res 2007;67(14):6612–8.

[41] Metzger E, Wissmann M, Yin N, et al. LSD1 demethylates repressive histone marks to promote androgen-receptor-dependent transcription. Nature 2005;437(7057):436–9.

[42] Kahl P, Gullotti L, Heukamp LC, et al. Androgen receptor coactivators lysine-specific histone demethylase 1 and 4 and a half LIM domain protein 2 predict risk of prostate cancer recurrence. Cancer Res 2006;66(23):11341–7.

[43] Dryhurst D, Ausio J. Histone H2A.Z deregulation in prostate cancer. Cause or effect? Cancer Metastasis Rev 2014;33(2–3):429–39.

[44] Valdes-Mora F, Song JZ, Statham AL, et al. Acetylation of H2A.Z is a key epigenetic modification associated with gene deregulation and epigenetic remodeling in cancer. Genome Res 2012;22(2):307–21.

[45] Dryhurst D, McMullen B, Fazli L, Rennie PS, Ausio J. Histone H2A.Z prepares the prostate specific antigen (PSA) gene for androgen receptor-mediated transcription and is upregulated in a model of prostate cancer progression. Cancer Lett 2012;315(1):38–47.

[46] Baptista T, Graca I, Sousa EJ, et al. Regulation of histone H2A.Z expression is mediated by sirtuin 1 in prostate cancer. Oncotarget 2013;4(10):1673–85.

[47] Vire E, Brenner C, Deplus R, et al. The Polycomb group protein EZH2 directly controls DNA methylation. Nature 2006;439(7078):871–4.

[48] Varambally S, Dhanasekaran SM, Zhou M, et al. The polycomb group protein EZH2 is involved in progression of prostate cancer. Nature 2002;419(6907):624–9.

[49] Yu J, Yu J, Mani RS, et al. An integrated network of androgen receptor, polycomb, and TMPRSS2-ERG gene fusions in prostate cancer progression. Cancer Cell 2010;17(5):443–54.

[50] Xu K, Wu ZJ, Groner AC, et al. EZH2 oncogenic activity in castration-resistant prostate cancer cells is Polycomb-independent. Science 2012;338(6113):1465–9.

[51] Bjorkman M, Ostling P, Harma V, et al. Systematic knockdown of epigenetic enzymes identifies a novel histone demethylase PHF8 overexpressed in prostate cancer with an impact on cell proliferation, migration and invasion. Oncogene 2012;31(29):3444–56.

[52] Porkka KP, Pfeiffer MJ, Waltering KK, Vessella RL, Tammela TL, Visakorpi T. MicroRNA expression profiling in prostate cancer. Cancer Res 2007;67(13):6130–5.

[53] Ozen M, Creighton CJ, Ozdemir M, Ittmann M. Widespread deregulation of microRNA expression in human prostate cancer. Oncogene 2008;27(12):1788–93.

[54] Ambs S, Prueitt RL, Yi M, et al. Genomic profiling of microRNA and messenger RNA reveals deregulated microRNA expression in prostate cancer. Cancer Res 2008;68(15):6162–70.

[55] Tong AW, Fulgham P, Jay C, et al. MicroRNA profile analysis of human prostate cancers. Cancer Gene Ther 2009;16(3):206–16.

[56] Szczyrba J, Loprich E, Wach S, et al. The microRNA profile of prostate carcinoma obtained by deep sequencing. Mol Cancer Res 2010;8(4):529–38.

[57] Martens-Uzunova ES, Jalava SE, Dits NF, et al. Diagnostic and prognostic signatures from the small non-coding RNA transcriptome in prostate cancer. Oncogene 2012;31(8):978–91.

[58] Bolton EM, Tuzova AV, Walsh AL, Lynch T, Perry AS. Noncoding RNAs in prostate cancer: the long and the short of it. Clin Cancer Res 2014;20(1):35–43.

[59] Hulf T, Sibbritt T, Wiklund ED, et al. Epigenetic-induced repression of microRNA-205 is associated with MED1 activation and a poorer prognosis in localized prostate cancer. Oncogene 2013;32(23):2891–9.

[60] Boll K, Reiche K, Kasack K, et al. MiR-130a, miR-203 and miR-205 jointly repress key oncogenic pathways and are downregulated in prostate carcinoma. Oncogene 2013;32(3):277–85.

[61] Hagman Z, Haflidadottir BS, Ceder JA, et al. miR-205 negatively regulates the androgen receptor and is associated with adverse outcome of prostate cancer patients. Br J Cancer 2013;108(8):1668–76.

[62] Gregory PA, Bert AG, Paterson EL, et al. The miR-200 family and miR-205 regulate epithelial to mesenchymal transition by targeting ZEB1 and SIP1. Nat Cell Biol 2008;10(5):593–601.

[63] Gandellini P, Folini M, Longoni N, et al. miR-205 Exerts tumor-suppressive functions in human prostate through down-regulation of protein kinase Cepsilon. Cancer Res 2009;69(6):2287–95.

[64] Ostling P, Leivonen SK, Aakula A, et al. Systematic analysis of microRNAs targeting the androgen receptor in prostate cancer cells. Cancer Res 2011;71(5):1956–67.

[65] Yamamura S, Saini S, Majid S, et al. MicroRNA-34a modulates c-Myc transcriptional complexes to suppress malignancy in human prostate cancer cells. PLoS One 2012;7(1):e29722.

[66] Liu C, Kelnar K, Liu B, et al. The microRNA miR-34a inhibits prostate cancer stem cells and metastasis by directly repressing CD44. Nat Med 2011;17(2):211–15.

[67] Prensner JR, Iyer MK, Balbin OA, et al. Transcriptome sequencing across a prostate cancer cohort identifies PCAT-1, an unannotated lincRNA implicated in disease progression. Nat Biotechnol 2011;29(8):742–9.

[68] Walsh AL, Tuzova AV, Bolton EM, Lynch TH, Perry AS. Long noncoding RNAs and prostate carcinogenesis: the missing "linc"? Trends Mol Med 2014;20(8):428–36.

[69] Yang L, Lin C, Jin C, et al. lncRNA-dependent mechanisms of androgen-receptor-regulated gene activation programs. Nature 2013;500(7464):598–602.

[70] Bernard D, Prasanth KV, Tripathi V, et al. A long nuclear-retained non-coding RNA regulates synaptogenesis by modulating gene expression. EMBO J 2010;29(18):3082–93.

[71] Ren S, Liu Y, Xu W, et al. Long noncoding RNA MALAT-1 is a new potential therapeutic target for castration resistant prostate cancer. J Urol 2013;190(6):2278–87.

[72] Yu W, Gius D, Onyango P, et al. Epigenetic silencing of tumour suppressor gene p15 by its antisense RNA. Nature 2008;451(7175):202–6.

[73] Prensner JR, Iyer MK, Sahu A, et al. The long noncoding RNA SChLAP1 promotes aggressive prostate cancer and antagonizes the SWI/SNF complex. Nat Genet 2013;45(11):1392–8.

[74] Prensner JR, Chen W, Iyer MK, et al. PCAT-1, a long noncoding RNA, regulates BRCA2 and controls homologous recombination in cancer. Cancer Res 2014.

[75] Dart DA, Brooke GN, Sita-Lumsden A, Waxman J, Bevan CL. Reducing prohibitin increases histone acetylation, and promotes androgen independence in prostate tumours by increasing androgen receptor activation by adrenal androgens. Oncogene 2012;31(43):4588–98.

[76] Takayama K, Tsutsumi S, Katayama S, et al. Integration of cap analysis of gene expression and chromatin immunoprecipitation analysis on array reveals genome-wide androgen receptor signaling in prostate cancer cells. Oncogene 2011;30(5):619–30.

[77] Isaacs JT, Isaacs WB. Androgen receptor outwits prostate cancer drugs. Nat Med 2004;10(1):26–7.

[78] Heemers HV, Tindall DJ. Androgen receptor (AR) coregulators: a diversity of functions converging on and regulating the AR transcriptional complex. Endocr Rev 2007;28(7):778–808.

[79] Fu M, Rao M, Wang C, et al. Acetylation of androgen receptor enhances coactivator binding and promotes prostate cancer cell growth. Mol Cell Biol 2003;23(23):8563–75.

[80] Coffey K, Robson CN. Regulation of the androgen receptor by post-translational modifications. J Endocrinol 2012;215(2):221–37.

[81] Gaughan L, Logan IR, Cook S, Neal DE, Robson CN. Tip60 and histone deacetylase 1 regulate androgen receptor activity through changes to the acetylation status of the receptor. J Biol Chem 2002;277(29):25904–13.

[82] Wang Z, Wang Z, Guo J, et al. Inactivation of androgen-induced regulator ARD1 inhibits androgen receptor acetylation and prostate tumorigenesis. Proc Natl Acad Sci USA 2012;109(8):3053–8.

[83] Zhong J, Ding L, Bohrer LR, et al. p300 acetyltransferase regulates androgen receptor degradation and PTEN-deficient prostate tumorigenesis. Cancer Res 2014;74(6):1870–80.

[84] Fu M, Wang C, Reutens AT, et al. p300 and p300/cAMP-response element-binding protein-associated factor acetylate the androgen receptor at sites governing hormone-dependent transactivation. J Biol Chem 2000;275(27):20853–60.

[85] Shang Y, Myers M, Brown M. Formation of the androgen receptor transcription complex. Mol Cell 2002;9(3):601–10.

[86] Ianculescu I, Wu DY, Siegmund KD, Stallcup MR. Selective roles for cAMP response element-binding protein binding protein and p300 protein as coregulators for androgen-regulated gene expression in advanced prostate cancer cells. J Biol Chem 2012;287(6):4000–13.

[87] Lupien M, Eeckhoute J, Meyer CA, et al. FoxA1 translates epigenetic signatures into enhancer-driven lineage-specific transcription. Cell 2008;132(6):958–70.

[88] Andreu-Vieyra C, Lai J, Berman BP, et al. Dynamic nucleosome-depleted regions at androgen receptor enhancers in the absence of ligand in prostate cancer cells. Mol Cell Biol 2011;31(23):4648–62.

[89] Halkidou K, Gnanapragasam VJ, Mehta PB, et al. Expression of Tip60, an androgen receptor coactivator, and its role in prostate cancer development. Oncogene 2003;22(16):2466–77.

[90] Debes JD, Sebo TJ, Lohse CM, Murphy LM, Haugen DA, Tindall DJ. p300 in prostate cancer proliferation and progression. Cancer Res 2003;63(22):7638–40.

[91] Bohm M, Locke WJ, Sutherland RL, Kench JG, Henshall SM. A role for GATA-2 in transition to an aggressive phenotype in prostate cancer through modulation of key androgen-regulated genes. Oncogene 2009;28(43):3847–56.

[92] Shah S, Prasad S, Knudsen KE. Targeting pioneering factor and hormone receptor cooperative pathways to suppress tumor progression. Cancer Res 2012;72(5):1248–59.

[93] Nagy L, Kao HY, Chakravarti D, et al. Nuclear receptor repression mediated by a complex containing SMRT, mSin3A, and histone deacetylase. Cell 1997;89(3):373–80.

[94] Dai Y, Ngo D, Forman LW, Qin DC, Jacob J, Faller DV. Sirt1 is required for antagonist-induced transcriptional repression of androgen-responsive genes by the androgen receptor. Mol Endocrinol 2007.

[95] Yang Y, Tse AK, Li P, et al. Inhibition of androgen receptor activity by histone deacetylase 4 through receptor SUMOylation. Oncogene 2011;30(19):2207–18.

[96] Welsbie DS, Xu J, Chen Y, et al. Histone deacetylases are required for androgen receptor function in hormone-sensitive and castrate-resistant prostate cancer. Cancer Res 2009;69(3):958–66.

[97] Kumar-Sinha C, Tomlins SA, Chinnaiyan AM. Recurrent gene fusions in prostate cancer. Nat Rev Cancer 2008;8(7):497–511.

[98] Kovacs JJ, Murphy PJ, Gaillard S, et al. HDAC6 regulates Hsp90 acetylation and chaperone-dependent activation of glucocorticoid receptor. Mol Cell 2005;18(5):601–7.

[99] Gibbs A, Schwartzman J, Deng V, Alumkal J. Sulforaphane destabilizes the androgen receptor in prostate cancer cells by inactivating histone deacetylase 6. Proc Natl Acad Sci USA 2009;106(39):16663–8.

[100] Coffey K, Rogerson L, Ryan-Munden C, et al. The lysine demethylase, KDM4B, is a key molecule in androgen receptor signalling and turnover. Nucleic Acids Res 2013;41(8):4433–46.

[101] Wissmann M, Yin N, Muller JM, et al. Cooperative demethylation by JMJD2C and LSD1 promotes androgen receptor-dependent gene expression. Nat Cell Biol 2007;9(3):347–53.

[102] Nakayama TWM, Suzuki H, Toyota M, Sekita N, Hirokawa Y, Mizokami A, et al. Epigenetic regulation of androgen receptor gene expression in human prostate cancers. Lab Invest 2000;80:1789–96.

[103] Thibault A, Figg WD, Bergan RC, et al. A phase II study of 5-aza-2′deoxycytidine (decitabine) in hormone independent metastatic (D2) prostate cancer. Tumori 1998;84(1):87–9.

[104] Sonpavde G, Aparicio AM, Zhan F, et al. Azacitidine favorably modulates PSA kinetics correlating with plasma DNA LINE-1 hypomethylation in men with chemonaive castration-resistant prostate cancer. Urol Oncol 2011;29(6):682–9.

[105] Ramachandran K, Gopisetty G, Gordian E, et al. Methylation-mediated repression of GADD45alpha in prostate cancer and its role as a potential therapeutic target. Cancer Res 2009;69(4):1527–35.

[106] Shang D, Liu Y, Liu Q, et al. Synergy of 5-aza-2′-deoxycytidine (DAC) and paclitaxel in both androgen-dependent and -independent prostate cancer cell lines. Cancer Lett 2009;278(1):82–7.

[107] Fang X, Zheng C, Liu Z, Ekman P, Xu D. Enhanced sensitivity of prostate cancer DU145 cells to cisplatinum by 5-aza-2′-deoxycytidine. Oncol Rep 2004;12(3):523–6.

[108] Zorn CS, Wojno KJ, McCabe MT, Kuefer R, Gschwend JE, Day ML. 5-Aza-2′-deoxycytidine delays androgen-independent disease and improves survival in the transgenic adenocarcinoma of the mouse prostate mouse model of prostate cancer. Clin Cancer Res 2007;13(7):2136–43.

[109] Gravina GL, Marampon F, Di Staso M, et al. 5-Azacitidine restores and amplifies the bicalutamide response on preclinical models of androgen receptor expressing or deficient prostate tumors. Prostate 2010;70(11):1166–78.

[110] Villar-Garea A, Fraga MF, Espada J, Esteller M. Procaine is a DNA-demethylating agent with growth-inhibitory effects in human cancer cells. Cancer Res 2003;63(16):4984–9.

[111] Lin X, Asgari K, Putzi M, et al. Reversal of GSTP1 CpG island hypermethylation and reactivation of pi-class glutathione S-transferase (GSTP1) expression in human prostate cancer cells by treatment with procainamide. Cancer Res 2001;61(24):8611–16.

[112] Lin J, Haffner MC, Zhang Y, et al. Disulfiram is a DNA demethylating agent and inhibits prostate cancer cell growth. Prostate 2011;71(4):333–43.

[113] Peart MJ, Smyth GK, van Laar RK, et al. Identification and functional significance of genes regulated by structurally different histone deacetylase inhibitors. Proc Natl Acad Sci USA 2005;102(10):3697–702.

[114] Shabbeer S, Kortenhorst MS, Kachhap S, Galloway N, Rodriguez R, Carducci MA. Multiple molecular pathways explain the anti-proliferative effect of valproic acid on prostate cancer cells *in vitro* and *in vivo*. Prostate 2007;67(10):1099–110.

[115] Xia Q, Sung J, Chowdhury W, et al. Chronic administration of valproic acid inhibits prostate cancer cell growth *in vitro* and *in vivo*. Cancer Res 2006;66(14):7237–44.

[116] Rokhlin OW, Glover RB, Guseva NV, Taghiyev AF, Kohlgraf KG, Cohen MB. Mechanisms of cell death induced by histone deacetylase inhibitors in androgen receptor-positive prostate cancer cells. Mol Cancer Res 2006;4(2):113–23.

[117] Marrocco DL, Tilley WD, Bianco-Miotto T, et al. Suberoylanilide hydroxamic acid (vorinostat) represses androgen receptor expression and acts synergistically with an androgen receptor antagonist to inhibit prostate cancer cell proliferation. Mol Cancer Ther 2007;6(1):51–60.

[118] Butler LM, Agus DB, Scher HI, et al. Suberoylanilide hydroxamic acid, an inhibitor of histone deacetylase, suppresses the growth of prostate cancer cells *in vitro* and *in vivo*. Cancer Res 2000;60(18):5165–70.

[119] Gravina GL, Marampon F, Muzi P, et al. PXD101 potentiates hormonal therapy and prevents the onset of castration-resistant phenotype modulating androgen receptor, HSP90, and CRM1 in preclinical models of prostate cancer. Endocr Relat Cancer 2013;20(3):321–37.

[120] Chuang MJ, Wu ST, Tang SH, et al. The HDAC inhibitor LBH589 induces ERK-dependent prometaphase arrest in prostate cancer via HDAC6 inactivation and down-regulation. PLoS One 2013;8(9):e73401.

[121] Qian DZ, Wei YF, Wang X, Kato Y, Cheng L, Pili R. Antitumor activity of the histone deacetylase inhibitor MS-275 in prostate cancer models. Prostate 2007;67(11):1182–93.

[122] Camphausen K, Scott T, Sproull M, Tofilon PJ. Enhancement of xenograft tumor radiosensitivity by the histone deacetylase inhibitor MS-275 and correlation with histone hyperacetylation. Clin Cancer Res 2004;10(18 Pt 1):6066–71.

[123] Annicotte JS, Iankova I, Miard S, et al. Peroxisome proliferator-activated receptor gamma regulates E-cadherin expression and inhibits growth and invasion of prostate cancer. Mol Cell Biol 2006;26(20):7561–74.

[124] Lakshmikanthan V, Kaddour-Djebbar I, Lewis RW, Kumar MV. SAHA-sensitized prostate cancer cells to TNFalpha-related apoptosis-inducing ligand (TRAIL): mechanisms leading to synergistic apoptosis. Int J Cancer 2006;119(1):221–8.

[125] VanOosten RL, Earel Jr. JK, Griffith TS. Histone deacetylase inhibitors enhance Ad5-TRAIL killing of TRAIL-resistant prostate tumor cells through increased caspase-2 activity. Apoptosis 2007;12(3):561–71.

[126] Vanoosten RL, Moore JM, Ludwig AT, Griffith TS. Depsipeptide (FR901228) enhances the cytotoxic activity of TRAIL by redistributing TRAIL receptor to membrane lipid rafts. Mol Ther 2005;11(4):542–52.

[127] Ellis L, Ku SY, Ramakrishnan S, et al. Combinatorial antitumor effect of HDAC and the PI3K-Akt-mTOR pathway inhibition in a Pten deficient model of prostate cancer. Oncotarget 2013;4(12):2225–36.

[128] Molife LR, Attard G, Fong PC, et al. Phase II, two-stage, single-arm trial of the histone deacetylase inhibitor (HDACi) romidepsin in metastatic castration-resistant prostate cancer (CRPC). Ann Oncol 2010;21(1):109–13.

[129] Rathkopf DE, Picus J, Hussain A, et al. A phase 2 study of intravenous panobinostat in patients with castration-resistant prostate cancer. Cancer Chemother Pharmacol 2013;72(3):537–44.

[130] Bradley D, Rathkopf D, Dunn R, et al. Vorinostat in advanced prostate cancer patients progressing on prior chemotherapy (National Cancer Institute Trial 6862): trial results and interleukin-6 analysis: a study by the Department of Defense Prostate Cancer Clinical Trial Consortium and University of Chicago Phase 2 Consortium. Cancer 2009;115(23):5541–9.

[131] Rathkopf D, Wong BY, Ross RW, et al. A phase I study of oral panobinostat alone and in combination with docetaxel in patients with castration-resistant prostate cancer. Cancer Chemother Pharmacol 2010;66(1):181–9.

[132] Pili R, Haggman M, Stadler WM, et al. Phase II randomized, double-blind, placebo-controlled study of tasquinimod in men with minimally symptomatic metastatic castrate-resistant prostate cancer. J Clin Oncol 2011;29(30):4022–8.

[133] Isaacs JT, Antony L, Dalrymple SL, et al. Tasquinimod is an allosteric modulator of HDAC4 survival signaling within the compromised cancer microenvironment. Cancer Res 2013;73(4):1386–99.

[134] Ding L, Chen S, Liu P, et al. CBP loss cooperates with PTEN haploinsufficiency to drive prostate cancer: implications for epigenetic therapy. Cancer Res 2014;74(7):2050–61.

[135] Coffey K, Blackburn TJ, Cook S, et al. Characterisation of a Tip60 specific inhibitor, NU9056, in prostate cancer. PLoS One 2012;7(10):e45539.

[136] Santer FR, Hoschele PP, Oh SJ, et al. Inhibition of the acetyltransferases p300 and CBP reveals a targetable function for p300 in the survival and invasion pathways of prostate cancer cell lines. Mol Cancer Ther 2011;10(9):1644–55.

[137] Wyce A, Degenhardt Y, Bai Y, et al. Inhibition of BET bromodomain proteins as a therapeutic approach in prostate cancer. Oncotarget 2013;4(12):2419–29.

[138] Asangani IA, Dommeti VL, Wang X, et al. Therapeutic targeting of BET bromodomain proteins in castration-resistant prostate cancer. Nature 2014;510(7504):278–82.

[139] Sakr WA, Haas GP, Cassin BF, Pontes JE, Crissman JD. The frequency of carcinoma and intraepithelial neoplasia of the prostate in young male patients. J Urol 1993;150(2 Pt 1):379–85.

[140] Baylin SB, Ohm JE. Epigenetic gene silencing in cancer—a mechanism for early oncogenic pathway addiction? Nat Rev Cancer 2006;6(2):107–16.

[141] Feinberg AP, Ohlsson R, Henikoff S. The epigenetic progenitor origin of human cancer. Nat Rev Genet 2006;7(1):21–33.

[142] Nelson WG, De Marzo AM, Deweese TL, et al. Preneoplastic prostate lesions: an opportunity for prostate cancer prevention. Ann N Y Acad Sci 2001;952:135–44.

[143] Laird PW, Jackson-Grusby L, Fazeli A, et al. Suppression of intestinal neoplasia by DNA hypomethylation. Cell 1995;81(2):197–205.

[144] McCabe MT, Low JA, Daignault S, Imperiale MJ, Wojno KJ, Day ML. Inhibition of DNA methyltransferase activity prevents tumorigenesis in a mouse model of prostate cancer. Cancer Res 2006;66(1):385–92.

[145] Giovannucci E, Rimm EB, Liu Y, Stampfer MJ, Willett WC. A prospective study of cruciferous vegetables and prostate cancer. Cancer Epidemiol Biomarkers Prev 2003;12(12):1403–9.

[146] Myzak MC, Hardin K, Wang R, Dashwood RH, Ho E. Sulforaphane inhibits histone deacetylase activity in BPH-1, LnCaP and PC-3 prostate epithelial cells. Carcinogenesis 2006;27(4):811–19.

[147] Singh SV, Warin R, Xiao D, et al. Sulforaphane inhibits prostate carcinogenesis and pulmonary metastasis in TRAMP mice in association with increased cytotoxicity of natural killer cells. Cancer Res 2009;69(5):2117–25.

[148] Wong CP, Hsu A, Buchanan A, et al. Effects of sulforaphane and 3,3′-diindolylmethane on genome-wide promoter methylation in normal prostate epithelial cells and prostate cancer cells. PLoS One 2014;9(1):e86787.

[149] Vyas AR, Hahm ER, Arlotti JA, et al. Chemoprevention of prostate cancer by d,l-sulforaphane is augmented by pharmacological inhibition of autophagy. Cancer Res 2013;73(19):5985–95.

[150] Chendil D, Ranga RS, Meigooni D, Sathishkumar S, Ahmed MM. Curcumin confers radiosensitizing effect in prostate cancer cell line PC-3. Oncogene 2004;23(8):1599–607.

[151] Khor TO, Keum YS, Lin W, et al. Combined inhibitory effects of curcumin and phenethyl isothiocyanate on the growth of human PC-3 prostate xenografts in immunodeficient mice. Cancer Res 2006;66(2):613–21.

[152] Ho E, Beaver LM, Williams DE, Dashwood RH. Dietary factors and epigenetic regulation for prostate cancer prevention. Adv Nutr 2011;2(6):497–510.

[153] Severson RK, Nomura AM, Grove JS, Stemmermann GN. A prospective study of demographics, diet, and prostate cancer among men of Japanese ancestry in Hawaii. Cancer Res 1989;49(7):1857–60.

[154] Lakshman M, Xu L, Ananthanarayanan V, et al. Dietary genistein inhibits metastasis of human prostate cancer in mice. Cancer Res 2008;68(6):2024–32.

[155] Huang X, Chen S, Xu L, et al. Genistein inhibits p38 map kinase activation, matrix metalloproteinase type 2, and cell invasion in human prostate epithelial cells. Cancer Res 2005;65(8):3470–8.

[156] Xu L, Ding Y, Catalona WJ, et al. MEK4 function, genistein treatment, and invasion of human prostate cancer cells. J Natl Cancer Inst 2009;101(16):1141–55.

[157] Fang MZ, Chen D, Sun Y, Jin Z, Christman JK, Yang CS. Reversal of hypermethylation and reactivation of p16INK4a, RARbeta, and MGMT genes by genistein and other isoflavones from soy. Clin Cancer Res 2005;11(19 Pt 1):7033–41.

[158] Majid S, Dar AA, Shahryari V, et al. Genistein reverses hypermethylation and induces active histone modifications in tumor suppressor gene B-cell translocation gene 3 in prostate cancer. Cancer 2010;116(1):66–76.

[159] Day JK, Bauer AM, DesBordes C, et al. Genistein alters methylation patterns in mice. J Nutr 2002;132 (8 Suppl.):2419S–2423SS.

[160] Zhang Y, Li Q, Chen H. DNA methylation and histone modifications of Wnt genes by genistein during colon cancer development. Carcinogenesis 2013;34(8):1756–63.

[161] Fang M, Chen D, Yang CS. Dietary polyphenols may affect DNA methylation. J Nutr 2007;137 (1 Suppl.):223S–228SS.

[162] Kanzaki M, Kakinuma H, Kumazawa T, et al. Low concentrations of the histone deacetylase inhibitor, depsipeptide, enhance the effects of gemcitabine and docetaxel in hormone refractory prostate cancer cells. Oncol Rep 2007;17(4):761–7.

[163] Liu X, Gomez-Pinillos A, Johnson EM, Ferrari AC. Induction of bicalutamide sensitivity in prostate cancer cells by an epigenetic Puralpha-mediated decrease in androgen receptor levels. Prostate 2010;70(2):179–89.

[164] Knutson SK, Wigle TJ, Warholic NM, et al. A selective inhibitor of EZH2 blocks H3K27 methylation and kills mutant lymphoma cells. Nat Chem Biol 2012;8(11):890–6.

[165] McCabe MT, Ott HM, Ganji G, et al. EZH2 inhibition as a therapeutic strategy for lymphoma with EZH2-activating mutations. Nature 2012;492(7427):108–12.

[166] Kruidenier L, Chung CW, Cheng Z, et al. A selective jumonji H3K27 demethylase inhibitor modulates the proinflammatory macrophage response. Nature 2012;488(7411):404–8.

[167] O'Rourke CJ, Knabben V, Bolton E, et al. Manipulating the epigenome for the treatment of urological malignancies. Pharmacol Ther 2013;138(2):185–96.

[168] Sharma NL, Massie CE, Ramos-Montoya A, et al. The androgen receptor induces a distinct transcriptional program in castration-resistant prostate cancer in man. Cancer cell 2013;23(1):35–47.

[169] Roth JE, Peer CJ, Price DK, Figg WD. The androgen receptor transcriptional program in castration-resistant prostate cancer: cell lines vs. tissue samples. Cancer Biol Ther 2014;15(1):16–18.

LIVER CANCER (HEPATOCELLULAR CARCINOMA)

12

Shane O'Grady and Matthew W. Lawless

*Experimental Medicine, UCD School of Medicine and Medical Science, Mater Misericordiae University Hospital,
Catherine McAuley Centre, Dublin, Ireland*

CHAPTER OUTLINE

1 LIVER CANCER: EPIDEMIOLOGY AND RISK FACTORS

Liver cancer is relatively rare in developed countries, accounting for 1.7% of new cancer cases in the United States and 1% of new cancer cases within the United Kingdom, making it the 15th most common cancer in males and the 19th most common in females [1]. Outside of the developed world, the

S.G. Gray (Ed): Epigenetic Cancer Therapy. DOI: http://dx.doi.org/10.1016/B978-0-12-800206-3.00012-4

statistics paints a very different picture: worldwide, liver cancer is the third greatest cause of cancer deaths, with an estimated 696,000 deaths in 2008 [2]. The vast majority (almost 85%) of new liver cases occur in less developed countries [3].

Liver cancer is rare before the age of 40 (except in high risk areas where it can strike as early as the 20s) and most commonly afflicts people in their 60s or 70s [4,5]. Hepatocellular carcinoma (HCC) is the most common form of liver cancer, accounting for 85% of liver cancers [6]. It is also the most fatal: the median survival rate is 11 months from time of diagnosis and less than 20% is expected to survive 3 or more years [7].

The risk factors of HCC are relatively well defined, unlike many other cancers. By far the greatest cause is chronic infection with the hepatitis B or C virus (HBC and HCV). Together, HBV and HCV account for just under 29.5% of infection-related cancers, almost 5% of total cancer cases [8]. Development of virally induced liver cancer is a multistep process. Approximately 80% of those infected with HCV will develop a chronic infection. Of those chronically infected, 10% to 20% will progress to liver cirrhosis and 1–2.5% will develop HCC [9]. In the United States, HCV now causes more deaths per year than HIV [10]. Approximately 160 million people (over 2% of the world population) are infected with HCV and chronic infection accounts for almost 20% of new HCC cases each year [9]. HBV is even more widespread, with over 2 billion cases worldwide [11].

While chronic hepatitis underlies the majority of HCC cases, recognition of other risk factors is still important. In particular, alcohol consumption, genetics, and consumption of aflatoxin-contaminated food are also key promoters of liver cancer.

In recent years, obesity and its associated state of chronic low-grade inflammation have emerged as highly important factors in the pathology of many cancers. Liver cancer is no exception, with multiple studies demonstrating a clear association between weight gain (especially increases in the metabolically active visceral fat) and increased prevalence of liver cancer. Park et al. demonstrated increased levels of interleukin-6 (IL-6), tumor necrosis factor (TNF) and activation of signal transducer and activator of transcription 3 (STAT3) as key mediators of obesity induced carcinogenesis in HCC patients (Figure 12.1) [12].

2 CURRENT TREATMENT OF HCC

Surgery offers the best hope of cure in most cases of HCC. Multiple surgical options are available, including resection of the tumor, transplant, and ethanol injections. Surgical resection is highly effective, with a 5-year survival rate of between 26% and 57%. Unfortunately, it is only suitable in 5–15% [13] and up to 70% of patients will suffer a recurrence within 5 years [14]. Liver transplants offer the possibility to cure both liver cancer and the underlying cirrhosis, greatly reducing the likelihood of recurrence. Careful selection of the most appropriate patients has allowed transplantation to achieve a 5-year survival rate of 70%, with less than 15% recurrence [6]. The disadvantages of transplantation, however, are obvious. In addition to the relatively low number of patients eligible for the procedure, long waiting lists and issues with donor–recipient tissue compatibility greatly reduce the number of patients who can be treated each year.

Apart from surgery, therapeutic options for HCC patients are extremely limited, leading to a very poor prognosis for patients with late stage HCC. Chemotherapy is generally of limited value in HCC, due to a high level of chemoresistance in liver cancers. Considering that detoxification of endogenous

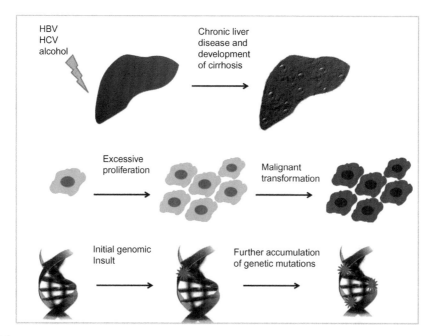

FIGURE 12.1

Histopathological progression of HCC. Onset of HCC can be studied on a multitiered basis. Liver damage can result from a range of hepatotoxic agents including chronic hepatitis infection or long-term alcohol abuse. Sustained liver damage can eventually lead to development of cirrhosis. At a cellular level, liver cirrhosis promotes increased rates of hepatocyte proliferation in response to liver damage possibly leading to a subsequent transformation to malignant cells. At a genetic level, an initial genetic heterozygosity (such as loss of p53) may follow from the increased proliferation and highly stressed environment within a cirrhotic liver and predispose hepatocytes to further genetic damage, possibly culminating in oncogenesis.

molecules is one of the livers primary physiological functions, it is of little surprise that the liver is so resistant to traditional chemotherapy agents. Hypoxia, genetic mutations accumulated during oncogenesis and activation of survival signaling pathways also contribute to HCC chemoresistance [15].

The kinase inhibitor sorafenib is one of the few exceptions and has now become the standard of care for late stage HCC. Sorafenib blocks a range of kinases but it is believed that kinases involved in regulation of growth and proliferation (e.g., members of the rat sarcoma-mitogen-activated protein kinases (Ras-MAPK pathway) and those involved in angiongenesis (e.g., vascular endothelial growth factor receptors (VEGF-R)) are the most clinically relevant. A Phase III trial found a 3-month increase in median survival (10.7 vs. 7.9 months) in sorafenib treated versus non-treated HCC patients [16] and multiple trials are now ongoing, aiming to combine sorafenib with other chemotherapeutic agents such as erlotininb and doxorubicin, or as an adjuvant therapy in combination with surgical resection [17].

3 EPIGENETICS

Clearly, there is a great need for development of effective therapies, especially agents that will be active in late-stage patients. One promising possibility is targeting the epigenetic background of liver tumors. Epigenetics refers to regulation of gene expression levels without alteration of DNA sequence. Eukaryotic cells utilize multiple epigenetic methods, including methylation of CpG DNA motifs, posttranslational modification (PTM) of histone proteins and mi-RNA-mediated suppression of mRNA translation. Aberrations in epigenetic control mechanisms have been implicated in many human diseases including Alzheimer's, diabetes, multiple sclerosis, rheumatoid arthritis, Huntington's, Parkinson's, Fragile X syndrome, and many forms of cancer [18]. Epigenetic alteration can promote tumor development in two key ways, increased suppression of anticancer genes or loss of suppression of pro-cancer genes. The evidence implicating epigenetic silencing as a key driver of oncogenesis is extensive. The most commonly silenced genes in cancers are those with roles in limiting cell growth and proliferation including genes involved in maintenance of genomic stability. For example, methylation of the DNA repair gene methylguanine DNA methyltransferase (*MGMT*) promoter is a common alteration in gliomas, colorectal carcinomas, nonsmall cell lung carcinomas, lymphomas, and head and neck carcinomas [19].

Epigenetic modifications in HCC were analyzed by Calvisi et al., who found that HCC development was linked to a global increase in DNA methylation levels, with much of this increase clustering in the promoter regions of important cell growth regulators and angiogenic inhibitors [20]. Perhaps most importantly, they found epigenetic silencing of multiple inhibitors of the Ras signaling pathway, leading to significantly increased Ras activity, a common driver of tumor development.

4 HISTONE MODIFICATION

With the knowledge that epigenetics plays an important role in both initiating and promoting tumor development, significant interest has arisen recently in the possibility of targeting components of the epigenetic pathway as a potential therapy, specifically histones. Histones consist of an octameric complex of eight proteins, two each of H2A, H2B, H3, and H4, with approximately 146 bp of associated DNA [21]. Primarily functioning as packaging proteins to condense the 3.6 billion base pairs comprising the human genome down to a size small enough to fit inside the nucleus, histones also serve as mediators of gene regulation. In general, the more tightly wound around histone cores DNA is, the less access will be available to the transcription factors, polymerases, and other components necessary for gene expression. Central to histones role in gene regulation is their C-terminal tail domains, which contain a large number of sites for PTMs. Histone tails are acted upon by several families of enzymes to yield a variety of PTMs, including acetylation, methylation, sumoylation, ubiquitination, and phosphorylation, each with its own regulatory effect on chromatin–DNA interactions. Histones can be simultaneously modified on several different sites, leading to a huge number of possible permutations of PTMS which may be read by the cells' transcriptional machinery as a "histone code" [22]. The idea of a regulatory code within patterns of histone modifications has been an exciting concept for oncologists and has led to some important developments, including the creation of novel research tools, such as the online databases human histone modification database (HHMD)[23] as well as generating a great deal of interest in the feasibility of histones as a therapeutic target.

The best characterized histone PTM is acetylation. Acetyl groups are added to lysine residues by histone acetyltransferase (HAT) enzymes. Acetylation of lysine residues neutralizes the positively charged amino acid and reduces attractive force between neighboring histones [21] and between the histone and its associated DNA [24]. In this way, acetylation of histones is usually considered to have an enhancing effect on gene expression.

The activity of HATs is counteracted by a second group of enzymes, histone deacetylases (HDACs). The first HDAC was discovered in 1996 by Taunton et al., who used the microbial deacetylase inhibitor trapoxin to purify and identify HDAC1, a mammalian homologue of the yeast repressor protein [25]. The total number of HDACs identified in humans now stands at 18 proteins, grouped into two broad families : Zn^{2+}-dependent "classic HDACs" (subdivided into class I, II, and IV proteins) and the Zn^{2+} independent, nicotinamide adenine dinucleotide (NAD)–dependent class III HDACs (also known as sirtuins, based on homology to the *Saccharomyces cerevisiae* Sir2p protein) (Figure 12.2) [26].

Gene knockout studies underlie the indispensable role of HDACs in regulating essential cell processes. Knockout of *HDAC1* in a mouse model causes profound disruption of embryonic stem cell proliferation and retardation of growth, resulting in embryonic lethality by E9.5 [27]. Mice deficient in HDAC2 fare slightly better and are able to survive throughout gestation, although they succumb to a myriad of cardiac defects shortly after birth [28]. Embryonic or perinatal lethality has also been demonstrated in mice deficient in HDAC3, HDAC4, and HDAC7 [29].

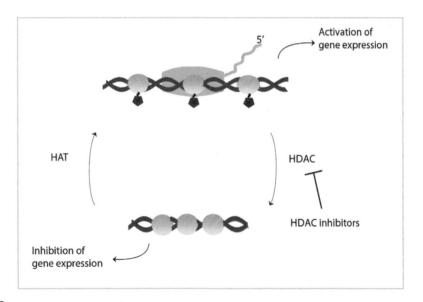

FIGURE 12.2

Histone acetyltransferase and deacetylase interactions dynamically regulate gene expression. Addition of acetyl groups to histone tail groups loosens DNA–histone interaction grants to transcription factors and polymerases. Loss of acetyl groups through the activity of HDACs leads to a greater association between DNA and its accompanying histones and tends to decrease levels of gene expression. In this, gene expression can be altered in a dynamic fashion through alteration of acetylation levels.

5 CHARACTERIZATION OF CLASSIC HDACs

Class I HDACs (HDAC1, HDAC2, HDAC3, and HDAC8) are ubiquitously expressed nuclear proteins with close homology to the yeast RPD3. They typically associate with other regulatory factors to form large complexes such as Sin3 and NuRD, often containing members of the histone demethylase family that remove methyl groups from histone targets in preparation for acetylation [30].

Class II HDACs differ from those of class I in two key ways. Unlike class I HDACs, which are for the most part ubiquitously expressed, class II HDACs possess tissue-specific patterns of expression. They also can be found in both the nuclear and cytoplasmic compartments of a cell, continuously moving between the two, in contrast to the predominantly nuclear class I HDACs. Class II HDACs are split into two subgroups: class IIa (HDAC4, HDAC5, HDAC7, and HDAC9) and class IIb (HDAC6 and HDAC10). Class IIa HDACs are unusual in that, as previously mentioned, they possess very little native deacetylase activity. They are usually associated with class I HDACs (often HDAC3) within repressor complexes [31].

So far, only one class IV HDAC has been characterized. HDAC11 is a predominantly nuclear protein, usually found in a complex with HDAC6. Its expression is restricted to kidney, heart, brain, skeletal muscle, and testis [32].

6 HDACS AND CANCER

Given what we now know about the important role HDACs play in control of gene expression and the extreme deregulation of transcription exhibited by tumors, it is of no surprise that HDACs have been shown to be important players in many aspects of cancer biology, including initial oncogenic transformation, growth, angiogenesis, metastasis, and response to therapeutic intervention. Alteration in acetylation status appears to be an extremely common event in cancer: Fraga et al. analyzed histone modifications in a panel of cancer cell lines and found an almost universal loss of acetylation of Lys16 and (trimethylation of Lys20) at the H4 histone [33]. Loss of H4 acetylation has also been implicated in tumor invasion and metastasis in gastrointestinal cancers [34].

Structural mutations in *HDAC* genes are relatively rare [35], and thus far only identified in colorectal cancer (HDAC2 truncation mutation, leading to loss of sensitivity to HDAC inhibitors [36]) and breast cancer (HDAC4 mutation, relevance unknown [37]). Therefore, aberrant deacetylase activity in human cancers seems to be related more to altered expression levels rather than mutations within *HDAC* gene sequences.

The role of HDACs in HCC has been subject to a great deal of analysis, with multiple links now apparent. A C > T single-nucleotide polymorphism (SNP) within the HDAC10 promoter was found to be associated with greater occurrence and earlier age of onset in patients with chronic HBV infections [38]. SNPs within the HDAC1 and HDAC3 genes have also been linked with greater risk of recurrence after transplant [39]. HDAC1 is essential for hepatocyte epithelial–mesenchymal transition (EMT) and migration in response to transforming growth factor beta (TGF-β) and plays an important role in fibrosis [40].

Rikimaru et al. demonstrated that high HDAC1 expression correlates with poor prognosis, greater incidence of hepatic portal vein invasion, and significantly decreased survival rates after surgical resection (95.5% for low HDAC1 patients vs. 60% for high HDAC1 patients at 3 years) [41]. In contrast,

Quint et al. found a significant association between poor prognosis and increased expression of HDAC2 but not HDAC1 [42] and Wu et al. found that high expression of HDAC2 and HDAC3 but not HDAC1 is a marker of tumor recurrence after treatment with a liver transplant [43]. The disparity between these studies may reflect a difference in the etiology of their case patients: the Asian cohorts of Rikimarus and Wus studies had a higher incidence of virally induced HCC than the European patients of Quints study. Evidently, further work is required to pin down the exact contribution to HDAC expression levels to the survival of HCC patients.

7 HDAC INHIBITORS

The important contribution of HDACs to the development and survival of many forms of cancer has lead oncologists to pursue HDAC inhibitors as a potential source of therapy. HDAC inhibitors are a diverse group of molecules, consisting of short-chain fatty acids, hydroxamates, carboxylates, benzamides, electrophilic ketones, and cyclic peptides [44]. Despite this varied chemical background, most HDAC inhibitors follow a generalized structural motif of a zinc chelating group connected to a cap via a short-chain linking section. The inhibitory activity of HDAC inhibitors resides within the zinc binding region which chelates HDACAs the catalytically active zinc ion of class I, II, and IV HDACs and attenuates their activity. Class III HDACs do not possess a catalytic zinc ion and are completely unaffected by most HDAC inhibitors [30].

8 ANTICANCER EFFECTS OF HDAC INHIBITORS
8.1 APOPTOSIS

Dysregulation of apoptosis is considered one of the hallmarks of cancer [45] and induction of apoptosis is an extremely common mechanism of action in anticancer agents. HDAC inhibitors are potent apoptotic agents, capable of initiating both the intrinsic mitochondrial pathway and the extrinsic death receptor pathway. In one of the first studies of HDAC inhibitors in HCC, Herold et al. demonstrated that treatment of HepG2 hepatocytes with trichostatin A (TSA) leads to increased levels of the pro-apoptotic factor bax and a decrease in anti-apoptotic B-cell lymphoma 2 (Bcl-2) [46]. TSA treatment was also associated with increases in both the expression and activation of caspases 3. Yamamoto also found increased levels of apoptosis in a panel of six HCC cell lines following treatment with sodium butyrate [47]. Yamashita showed an increased level of apoptosis in Huh-7 but could not detect this effect in HepG2 cells [48]. All of this points toward apoptosis as an important effector of HDAC activity within HCC.

Interestingly, HDAC inhibition seems to have differing effects on levels of apoptosis in hepatocytes depending on whether they are of a normal or malignant phenotype. Papeleu et al. found a lack of apoptosis inducing activity and even demonstrated an improvement in hepatocyte function following TSA treatment [49]. Lack of pro-apoptotic effect in nonmalignant hepatocytes is highly promising from a drug development point of view as the liver is primary organ responsible for drug detoxification and is often adversely affected by anticancer drugs: 85% of cancer patients receiving chemotherapy will develop liver steatosis [50].

8.2 CELL CYCLE ARREST

Cell cycle arrest was the first identified effect of HDAC inhibitors on cancer cells. HDAC inhibitors are capable of causing a cell cycle arrest in a broad range of cells, including numerous forms of cancer and both cancerous and noncancerous cells [51]. The type of arrest seems to be dependent on the concentration: blocking deacetylase activity in tumors generally produces a G1 arrest when used in low concentrations but higher concentrations tend to be cytotoxic with surviving cells displaying both G1/S and G2/M arrests [52,53]. The cyclin-dependent kinase inhibitor p21 is commonly implicated in HDAC-mediated cell cycle arrest [53,54]. Treatment of tumor cells with HDAC inhibitors results in increased acetylation of histones associated with both promoter and coding regions of the p21 gene and a resulting increase in CDK-interacting protein 1 (p21) expression levels [55]. Interestingly, HDAC-mediated induction of p21 has been reported to act independently of tumor suppressor protein 53 (p53) [56]: a useful feature for an anticancer agent considering how often p53 is mutated in transformed cancerous cells.

HDAC inhibition has been shown as a potent cell cycle arrestor in liver cancer cell lines. TSA induces cell cycle arrest in both HepG2 and Huh-7 cells [48]. Cell cycle arrest is often associated with increased levels of cells with a differentiated hepatocyte phenotype [47,48,57].

8.3 ANGIOGENESIS

Activation of angiogenic pathways is a critical step in tumor development. Hypoxia-inducible factor 1-alpha (HIF-1α) acts as a master regulator of oxygen homeostasis and is a primary target for therapies aiming to decrease angiogenesis. A potential role for HDACs in regulation of angiogenesis was first discovered by Kim et al., who observed increased expression of HDAC1, HDAC2, and HDAC3 in both normal and transformed cells under hypoxic conditions [58].

This increase in HDAC levels was associated with decrease of the tumor suppressors p53 and Von Hippel–Lindau (VHL) tumor as well as the HIF-1α inhibitor factor inhibiting HIF (FIH) [59] and promotion of pro-angiogenic transcription. Treatment with TSA normalizes expression of p53, VHL, HIF-1α and reduces formation of new blood vessels in a chorioallantoic membrane (CAM) assay [58]. Several HDACs have also been shown to interact directly with HIF-1α: HDAC1 and HDAC3 directly bind to HIF-1α and increase its stability [60].

9 ENDOPLASMIC RETICULUM STRESS

An emerging effect of several clinically relevant HDAC inhibitors is their interaction with the endoplasmic reticulum (ER) and in particular and modulation of its levels of "stress." Most commonly, ER stress is induced when the ERs folding capacity is overwhelmed, leading to an accumulation of unfolded proteins. The ER will usually attempt to clear any backlog of unfolded proteins by increasing its processing capacity via upregulation of chaperone proteins and reducing its workload by decreasing transcriptional and translational activity [61].

This response has been dubbed the unfolded protein response (UPR) and is coordinated by the chaperone protein 78 kDa glucose-regulated protein (Grp78/BiP) and a small set of transmembrane stress sensors within the ER, inositol-requiring 1 alpha (IRE1α), double-strand RNA-activated protein kinase-like ER kinase (PERK), and activating transcription factor 6 (ATF6). These ER stress sensors are powerful signaling agents capable of causing significant changes in cellular activity. Their activity is regulated primarily by Grp78, a highly conserved member of the heat shock protein (Hsp70) family which primarily functions as a chaperone by binding unfolded client proteins via hydrophobic patches to prevent aggregation [62]. Under normal physiological conditions, ER stress sensors are kept in an inactive state through Grp78 binding. Increased levels of unfolded proteins sequesters Grp78 remove this inhibitory effect. Grp78 has a higher affinity for unfolded proteins than for IRE1α, PERK, or ATAF6, allowing even small fluctuations in the level of unfolded proteins to activate ER stress pathways [61].

Together IRE1α, PERK, and ATF6 coordinate a cytoprotective transcriptional response characterized by increased protein folding capacity within the ER and a global decrease in protein production. If these measures fail to resolve ER stress, UPR signaling can switch from a protective to an apoptosis promoting effect. Expression of the transcription factor CCAAT-enhancer-binding protein homologous protein (CHOP) is strongly upregulated during periods of ER stress [63] and seems to be a strong promoter of cell death. Overexpression of CHOP causes apoptosis in transformed cell lines [64] while $CHOP^{-/-}$ cells exhibit significantly less apoptosis when exposed to ER stress–inducing agents [65]. CHOP alters the expression of several Bcl-2 proteins, upregulating B-cell lymphoma 2 interacting mediator of cell death (Bim) [66] and p53 upregulated modulator of apoptosis (PUMA) [67] and down-regulating Bcl-2 [68], shifting the balance toward a pro-apoptotic state. CHOP can also increase the level of oxidative stress within a cell via activation of the ER oxidase enzyme-1 alpha (ERO1α) [69]. Activation of c-Jun N-terminal kinase (JNK) signaling pathways by IRE1α may also contribute to induction of apoptosis [70,71].

10 ER STRESS AND CANCER

ER stress and cancer are closely linked: the high rate of proliferation seen in transformed cells requires synthesis of large quantities of proteins, placing an increased workload on the ER. The often hypoxic, nutrient-starved and acidic extracellular conditions within a tumor are also strong inducers of ER stress [72]. The constant ER stress generated by tumor microenvironment places a strong selective pressure to adapt to stressful conditions, often by altering expression levels of UPR-associated genes.

The UPR pathway seems to play an important role in pathogenesis of HCC. Shuda et al. demonstrated upregulation of Grp78, XBP-1, and ATF6 in surgically resected tissue from HCC patients, correlating with histological grade [73]. Localization of Grp78 to the cytoplasm and ATF6 to the nucleus (indicating activation of UPR) was found to be associated with poor differentiation status. Expression of Grp78 also positively correlates with tumor invasiveness [74,75] and is a significant factor in the development of sorafenib resistance [76]. Grp78 SNPs both within the promoter region [77] and the *Grp78* gene itself [78] have been linked with an increased risk of developing HCC.

11 ER STRESS AND HDACs

HDACs are important regulators of UPR pathways and many studies have shown a synergistic effect on tumor cytotoxicity between inhibition of HDAC activity and inducement of a state of ER stress. HDAC inhibitors can induce ER stress via multiple paths, especially through alteration of components of the UPR signaling pathway. Inhibition of HDAC6 is an especially promising target due to its important role in regulating cellular response to ER stress through altering acetylation levels of heat shock protein 90 (Hsp90) [79]. HDAC6 senses aggregations of misfolded proteins and deacetylates Hsp90 in response to coordinate proteasomal degradation. Aggregation of misfolded proteins is a highly cytotoxic event [80] and any inhibition of a cell's capacity for their removal is likely to adversely affect cell survival. Acetylation of Hsp90 decreases its binding to client proteins and co-chaperones [81] and promotes aggregation of misfolded proteins. Inhibition of Hsp90 also reduces production of cancer promoting oncoproteins, many of which require Hsp90-mediated folding including RAF proto-oncogene serine/ threonine-protein kinase (Raf-1) [82].

Liu et al. recently showed that in transformed hepatocytes, induction of autophagy in response to ER stress is an important driver of cell death. Following vorinostat treatment, Liu et al. found increased levels of PERK activation and eukaryotic translation initiation factor 2A (eIF2α) phosphorylation in a panel of HCC cell lines, as well as downregulation of protein kinase B/mammalian target of rapamycin (Akt/mTOR) leading to parallel induction of both apoptosis and autophagic cell death [83]. Autophagy is a lysosomal protein-degradation pathway which is increasingly being seen as an important factor in determining tumor response to therapy. Autophagy can either confer a survival advantage to tumor cells via increasing stress tolerance or induce autophagic cell death (Figure 12.3) [84].

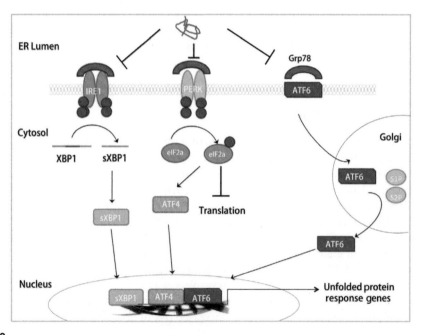

FIGURE 12.3

ER stress and activation of UPR signaling pathways. Accumulation of unfolded proteins sequesters Grp78 and removes its inhibitory effect on UPR signaling components. IRE1 and PERK are activated through dimerization and auto-phosphorylation while ATF6 translocates to the Golgi apparatus and undergoes proteolysis by S1P and S2P. Activated IRE1 promotes splicing of the XBP1mRNA allowing for translation of the XBP1 protein. PERK's primary function is to phosphorylate the vital translation factor eIF2a, promoting a general repression of translation, except for a small group of proteins including ATF4, whose expression is upregulated. The combined effect of the three signaling pathways increases production of chaperone proteins and other factors involved in clearance of misfolded proteins.

12 HDAC INHIBITORS IN TREATMENT OF HCC

Due to the paucity of effective therapies for HCC, HDAC inhibitors have been the subject of a flurry of studies investigating their potential as effective treatments in liver cancer. Multiple important targets and mechanisms of action have now been identified for HDAC inhibitors within malignant hepatocytes. So far, only two HDAC inhibitors (vorinostat and romidepsin) have received FDA approval, both for the treatment of cutaneous T-cell lymphoma.

Initial preclinical studies delivered promising results. Belinostat was found to inhibit growth of PLC/PRF/5, Hep3B, and HepG2 cells in a dose-dependent manner [85]. This was followed by a multicentre Phase I/II study where belinostat was administered to patients with unresectable HCC, the first study to test epigenetic therapy in patients with advanced HCC [86]. Belinostat stabilized tumor in 45.2% of patients with another 2.4% achieving partial response. Immunohistochemical staining also revealed that patients with high levels of the DNA repair factor HR23B had a disease stabilization rate of 58% versus 14% for those with low HR23B histoscores ($P = 0.036$). Subsequent studies have suggested that the role of HR23B in determining response to HDAC inhibitors may stem from interactions between HDAC6 with HR23B and a subsequent regulatory effect on both apoptosis and autophagy [87].

Although successful in certain cases, trials of HDAC in solid cancers have overall been disappointing [88] with most agents failing to achieve significant results when administered in a single drug therapy. Because of this, there has been a shift toward testing HDAC inhibitors as part of a combination therapy. The most commonly used combination in HCC is an HDAC inhibitor and the kinase inhibitor sorafenib.

Lachenmayer et al. reported aberrant HDAC gene expression profiles in 230 samples taken from HCC patients with significant upregulation of HDAC1, 2, 4, 5, and 11 and downregulation of HDAC6 and HDAC7 compared to normal liver [89]. Administration of panobinostat achieved a number of antitumor effects including increased apoptosis and hyperacetylation of histone H3 and Hsp90 in Huh7 cells, both as a single agent and when coadministered with sorafenib, as well as a significant decrease in tumor volume and improved survival in an *in-vivo* murine xenograft model. Activation of the UPR pathway is essential to panobinostat-mediated cytotoxicity, with increased levels of CHOP and JNK activation in response to panobinostat treatment leading to activation of caspases and apoptosis [90].

Chen et al. recently showed that coadministration of the broad spectrum HDAC inhibitor MPT0E028 acted synergistically with sorafenib both *in vitro* and *in vivo* to reduce cell viability and activate apoptotic pathways [91]. Further observations of increased extracellular-signal-regulated kinase (Erk) signaling in response to MPT0E028, an effect that was abrogated in a concentration-dependent manner by sorafenib led to the hypothesis that low concentrations of MPT0E028 may increase the sensitivity of liver cancer cells to sorafenib through increased dependence on Erk signaling.

Interestingly, panobinostat seems to act on more than one epigenetic pathway. 0.1 µM of panobinostat significantly inhibited DNA methyltransferases (DNMTs) in HepG2 and Hep3B cells with a corresponding decrease in methylation status of the tumor suppressor genes RASSF1A (Ras association domain family 1 isoform A) and adenomatous polyposis coli (APC) and a rise in their expression at both mRNA and protein levels [92].

The ongoing stage 2 SHELTER trial aims to evaluate the combination of sorafenib with resminostat in sorafenib-resistant HCC, with preliminary data demonstrating good pharmacokinetics [93] and a re-sensitization to sorafenib [94]. A currently recruiting Phase I trial also aims to evaluate vorinostat in combination with sorafenib in patients with advanced HCC [95]. Recent results have also raised the possibility of giving HDAC inhibitors to high-risk patients as a preventative measure. HDAC inhibitors exert a number of useful protective effects on hepatocytes. In a "resistant hepatocyte" model of HCC in rats, pretreatment with a butyric acid prodrug reduced development of preneoplastic lesions in response to the chemical tumor promotersdiethylnitrosamine and 2-acetylaminofluorene [96]. Treatment with vorinostat greatly inhibits the carcinogenic action of the HCV oncoprotein pre-S2 mutant large HBV surface antigen through increasing expression of thioredoxin-binding protein 2 and preventing degradation of the cyclin kinase inhibitor p27Kip1 [97]. TSA reduces expression of important fibrosis inducing genes in stellate cells including the extracellular matrix (ECM) factors collagen 1 and 3 [98], actin-related protein 2 (Arp2) and Arp3 and RhoA [99] leading to decreased levels of cell migration and proliferation, reduction in fibrosis and an overall protective effect. Fibrosis of the liver is a common step in the development of HCC, making the anti-fibrotic effect of HDAC inhibitors a potentially useful preventative treatment (Table 12.1).

Table 12.1 Overview of Selected Demonstrated Anticancer Effects of HDAC Inhibitors

Drug	Effect	Mechanism	Ref
Vorinostat	Increased efficacy of sorafenib	FLIP suppression and CD95 Activation	[106]
	Increased efficacy of 5-fluorouracil and irinotecan	Bcl-2 downregulation caspase 3 activation	[107]
	Growth inhibition	Decreased expression of RCN1, ANXA3, and HSP27	[108]
	Suppression of Hep-B-induced oncogenesis	Recovery of cell cycle checkpoint	[97]
Valproic acid	Reduced fibrosis	Prevention of hepatic stellate cell activation	[109]
	Increased sensitivity to retinoids	Increased expression of expression of RAR-beta and p21	[110]
Sodium butyrate	Increased efficacy of p53 gene therapy	Increased necrosis and reduced angiogenesis	[111]
	Reduced cell migration	Inhibition of epithelial to mesenchymal transition	[112]
TSA	Reduced inflammation during sepsis	Reduced MAPK activation	[113]
	Inhibition of cancer stem cell self-renewal	Suppression of HDAC3 activity	[114]
	Growth inhibition	Upregulation of miR-449	[115]
Panobinostat	Increased efficacy of sorafenib	Altered expression levels of CDH1 and BIRC5	[105]
	Induction of apoptosis	ER stress signaling	[116]
MS-275	Growth inhibition	Increased expression of IGFBP-3	[117]
Tributyrin	Chemoprevention	Normalized p53 signaling	[118]
AR-42	Sensitization to radiation therapy	Inhibition of Ku70 activity	[119]

13 DNA METHYLATION IN HCC

DNA methylation is a central epigenetic mechanism, whereby such abnormal expression patterns are present in cancers that are associated with aberrant gene regulation. These alterations have been suggested as a cause of the formation of cancerous features with genetic abnormalities. It has been reported that a bidirectional alteration of methylation in cancer occurs noting suppression in the overall level of DNA methylation associated with enhanced proto-oncogenes. In addition, hypermethylated status associated with a negative control of cell growth and genomic stability, leading to transcription silencing. Therefore the establishment of a DNA methylation profile of changes in HCC is paramount for the understanding of the mechanisms and discovery of better diagnostic approaches in cancer clinics. Genome-wide methylation arrays have been investigated to identify genes that are methylated in HCC. Data obtained from such studies have suggested that particular genes may indeed be appropriate as biomarkers of early HCC diagnosis. Further studies expanding such concepts with larger patient numbers are underway. The use of DNA methylation profiles and data as a surrogate clinical biomarker is an intensive area of clinical HCC research. Indeed, reports have identified etiological factors such as HCV infection that can lead to abnormal DNA methylation expression patterns to be found in cancerous tissue. Furthermore, specific DNA methytransferases such as DNMT1, DNMT3A, and DMT3B were also revealed to be upregulated in liver cancer. While the exact mechanisms and greater research into these findings are required, they do represent a strong indication into this important area for HCC [100].

14 OTHER EPIGENETIC REGULATORY PROTEINS

Interestingly, more recently for a more complete picture of various epigenetic interactions at a cellular level consideration must also be given to other epigenetic regulatory proteins that modify the histone code other that histone deacetylases including lysine demethylases, HATs, and lysine methytransferases. These epigenetic regulatory proteins have been studied, in particular in association with clinicopathological features and prognostic value in HCC. Hence the reported findings indicate that the expression level of such epigenetic regulatory proteins such as lysine specific demethylase 1 can clearly indicate tumor progression and predict poor prognosis [101].

15 ncRNAs

Due to the problems with current detection and effective options for liver cancer patients, the identification of new approaches in the context of biochemical tools for HCC is a priority. It is now understood that most genes in the human genome are lacking protein coding ability. Therefore genes frequently transcribed without protein coding ability are defined as noncoding RNAs. Today, the role of micro-RNAs (miRNAs) in the context of the onset and progression of HCC including invasion, metastasis, and apoptosis was reviewed in great detail by two expert groups in this area [102,103]. The impact miRNAs have made to the understanding of HCC has been dramatic, the knowledge of long noncoding RNAs (LncRNAs) is still at an early stage nonetheless is providing a deep understanding and avenues of research for HCC. LncRNAs (>200 base pairs) are expressed in several species, initially considered as a transcriptional background noise in the genomic regulation. The LncRNA highly upregulated in liver cancer (HULC) and its aberrant expression were revealed to be the first liver-specific LncRNA identified to be associated with HCC [104,105]. Noncoding RNAs (ncRNAs) are known to alter cancer through the targeting of oncogenes via mechanisms such as DNA methylation, histone modification, and chromatin modeling. These cellular mechanisms are showing great promise for the first time in the potential identification of epigenetic agents in previously incurable cancers. ncRNAs identification in association with new epigenetic treatments of HCC may act as ideal biomarkers guiding predication of response and categorization of HCC patients to these or more refined generations of future epigenetic agents.

16 CONCLUSION

HDAC inhibitors are an exciting development in the treatment of HCC. Furthermore, it is reported that lysine-specific demethylase inhibitors are proving efficient in providing a novel approach in cancer onset via regulation of epigenetic modification. Indeed, such findings highlight that genes controlling epigenetic programs which are requirements for maintaining chromatin structure and cell identity also include genes that can drive cancer development. While there is still some way to go before they will begin benefitting patients much progress has been made and most importantly clear signposts now exist on a previously unexplored road. Nonetheless, the advent of powerful tools in the form of epigenetic agents that have the potential to alter cancer has been clearly indicated. Like all early stage treatments and discoveries this rough diamond needs further refinement and more clarity in order to reach its full potential. This is against the backdrop of an ever increasing rise in HCC cases that is occurring hand in hand with limited treatment options and low survival rates.

REFERENCES

[1] UK, C.R. *Liver Cancer Incidence Statistics*. 2012 06/09/12 22/04/2013]; Available from: http://www.cancer-researchuk.org/cancer-info/cancerstats/types/liver/incidence/.

[2] Ferlay J., S.H., Bray F., Forman D., Mathers C., Parkin D.M. *Cancer incidence and mortality worldwide: IARC CancerBase No. 10*. 2008; Available from: http://globocan.iarc.fr.

[3] Jemal A, Center MM, DeSantis C, Ward EM, et al. Global patterns of cancer incidence and mortality rates and trends. Cancer Epidemiol Biomarkers Prev 2010;19(8):1893–907.

[4] Bosch FX, Ribes J, Díaz M, Cléries R, et al. Primary liver cancer: worldwide incidence and trends. Gastroenterology 2004;127(5):S5–S16.

[5] Ananthakrishnan A, Gogineni V, Saeian K. Epidemiology of primary and secondary liver cancers. Semin Intervent Radiol 2006;23(01):47–63.

[6] Llovet JM, Burroughs A, Bruix J. Hepatocellular carcinoma. Lancet 2003;362(9399):1907–17.

[7] Greten TF, Papendorf F, Bleck JS, Kirchhoff T, Wohlberedt T, Kubicka S, et al. Survival rate in patients with hepatocellular carcinoma: a retrospective analysis of 389 patients. Br J Cancer 2005;92(10):1862–8.

[8] de Martel C, Ferlay J, Franceschi S, Vignat J, Bray F, Forman D, et al. Global burden of cancers attributable to infections in 2008: a review and synthetic analysis. Lancet Oncol 2012;13(6):607–15.

[9] Lavanchy D. Evolving epidemiology of hepatitis C virus. Clin Microbiol Infect 2011;17(2):107–15.

[10] Ly KN, Xing J, Klevens RM, Jiles RB, Ward JW, Holmberg SD, et al. The increasing burden of mortality from viral hepatitis in the United States between 1999 and 2007. Ann Intern Med 2012;156(4):271–8.

[11] EASL jury. *EASL INTERNATIONAL CONSENSUS CONFERENCE ON HEPATITIS B13–14 September, 2002 Geneva, Switzerland: Consensus statement (Short version)*, 2003. J Hepatol 38(4), 533–540.

[12] Park EJ, Lee JH, Yu GY, He G, Ali SR, Holzer RG, et al. Dietary and genetic obesity promote liver inflammation and tumorigenesis by enhancing IL-6 and TNF expression. Cell 2010;140(2):197–208.

[13] El-Serag HB, Marrero JA, Rudolph L, Reddy KR, et al. Diagnosis and treatment of hepatocellular carcinoma. Gastroenterology 2008;134(6):1752–63.

[14] Villanueva A, Llovet JM. Targeted therapies for hepatocellular carcinoma. Gastroenterology 2011;140(5):1410–26.

[15] Asghar U, Meyer T. Are there opportunities for chemotherapy in the treatment of hepatocellular cancer? J Hepatol 2012;56(3):686–95.

[16] Llovet JM, Ricci S, Mazzaferro V, Hilgard P, Gane E, Blanc JF, et al. Sorafenib in advanced hepatocellular carcinoma. N Engl J Med 2008;359(4):378–90.

[17] FornerA, Llovet JM,Bruix J. Hepatocellular carcinoma. Lancet 379(9822), 1245–1255.

[18] Portela A, Esteller M. Epigenetic modifications and human disease. Nat Biotechnol 2010;28(10):1057–68.

[19] Esteller M, Hamilton SR, Burger PC, Baylin SB, Herman JG, et al. Inactivation of the DNA repair gene O6-methylguanine-DNA methyltransferase by promoter hypermethylation is a common event in primary human neoplasia. Cancer Res 1999;59(4):793–7.

[20] Calvisi DF, Ladu S, Gorden A, Farina M, Lee JS, Conner EA, et al. Mechanistic and prognostic significance of aberrant methylation in the molecular pathogenesis of human hepatocellular carcinoma. J Clin Invest 2007;117(9):2713–22.

[21] Luger K, Mäder AW, Richmond RK, Sargent DF, Richmond TJ, et al. Crystal structure of the nucleosome core particle at 2.8A resolution. Nature 1997;389(6648):251–60.

[22] Strahl BD, Allis CD. The language of covalent histone modifications. Nature 2000;403(6765):41–5.

[23] Zhang Y, Lv J, Liu H, Zhu J, Su J, Wu Q, et al. HHMD: the human histone modification database. Nucleic Acids Res 2010;38(suppl 1):D149–54.

[24] Hong L, Schroth GP, Matthews HR, Yau P, Bradbury EM. Studies of the DNA binding properties of histone H4 amino terminus. Thermal denaturation studies reveal that acetylation markedly reduces the binding constant of the H4 "tail" to DNA. J Biol Chem 1993;268(1):305–14.

[25] Taunton J, Hassig CA, Schreiber SL. A mammalian histone deacetylase related to the yeast transcriptional regulator Rpd3p. Science 1996;272(5260):408–11.

[26] Khan O, La Thangue NB. HDAC inhibitors in cancer biology: emerging mechanisms and clinical applications. Immunol Cell Biol 2012;90(1):85–94.

[27] Lagger G, O'Carroll D, Rembold M, Khier H, Tischler J, Weitzer G, et al. Essential function of histone deacetylase 1 in proliferation control and CDK inhibitor repression. EMBO J 2002;21(11):2672–81.

[28] Montgomery RL, Davis CA, Potthoff MJ, Haberland M, Fielitz J, Qi X, et al. Histone deacetylases 1 and 2 redundantly regulate cardiac morphogenesis, growth, and contractility. Genes Dev 2007;21(14):1790–802.

[29] Wagner JM, Hackanson B, Lübbert M, Jung M. Histone deacetylase (HDAC) inhibitors in recent clinical trials for cancer therapy. Clin Epigenet 2010;1(3–4):117–36.

[30] Imai S, Armstrong CM, Kaeberlein M, Guarente L. Transcriptional silencing and longevity protein Sir2 is an NAD-dependent histone deacetylase. Nature 2000;403(6771):795–800.

[31] Fischle W, Dequiedt F, Hendzel MJ, Guenther MG, Lazar MA, Voelter W, et al. Enzymatic activity associated with Class II HDACs is dependent on a multiprotein complex containing HDAC3 and SMRT/N-CoR. Mol Cell 2002;9(1):45–57.

[32] Gao L, Cueto MA, Asselbergs F, Atadja P, et al. Cloning and functional characterization of HDAC11, a novel member of the human histone deacetylase family. J Biol Chem 2002;277(28):25748–55.

[33] Fraga MF, Ballestar E, Villar-Garea A, Boix-Chornet M, Espada J, Schotta G, et al. Loss of acetylation at Lys16 and trimethylation at Lys20 of histone H4 is a common hallmark of human cancer. Nat Genet 2005;37(4):391–400.

[34] Yasui W, Oue N, Ono S, Mitani Y, Ito R, Nakayama H. Histone acetylation and gastrointestinal carcinogenesis. Ann N Y Acad Sci 2003;983(1):220–31.

[35] Özdağ H, Teschendorff AE, Ahmed AA, Hyland SJ, Blenkiron C, Bobrow L, et al. Differential expression of selected histone modifier genes in human solid cancers. BMC Genomics 2006;7(1):90.

[36] Ropero S, Fraga MF, Ballestar E, Hamelin R, Yamamoto H, Boix-Chornet M, et al. A truncating mutation of HDAC2 in human cancers confers resistance to histone deacetylase inhibition. Nat Genet 2006;38(5): 566–9.

[37] Sjöblom T, Jones S, Wood LD, Parsons DW, Lin J, Barber TD, et al. The consensus coding sequences of human breast and colorectal cancers. Science 2006;314(5797):268–74.

[38] Park BL, Kim YJ, Cheong HS, Lee SO, Han CS, Yoon JH, et al. HDAC10 promoter polymorphism associated with development of HCC among chronic HBV patients. Biochem Biophys Res Commun 2007;363(3):776–81.

[39] Yang Z, Zhou L, Wu LM, Xie HY, Zhang F, Zheng SS. Combination of polymorphisms within the HDAC1 and HDAC3 gene predict tumor recurrence in hepatocellular carcinoma patients that have undergone transplant therapy. Clin Chem Lab Med 2010;48(12):1785–91.

[40] Lei W, Zhang K, Pan X, Hu Y, Wang D, Yuan X, et al. Histone deacetylase 1 is required for transforming growth factor-β1-induced epithelial–mesenchymal transition. Int J Biochem Cell Biol 2010;42(9):1489–97.

[41] Rikimaru T, Taketomi A, Yamashita Y, Shirabe K, Hamatsu T, Shimada M, et al. Clinical significance of histone deacetylase 1 expression in patients with hepatocellular carcinoma. Oncology 2007;72(1–2): 69–74.

[42] Quint K, Agaimy A, Di Fazio P, Montalbano R, Steindorf C, Jung R, et al. Clinical significance of histone deacetylases 1, 2, 3, and 7: HDAC2 is an independent predictor of survival in HCC. Virchows Archiv 2011;459(2):129–39.

[43] Wu LM, Yang Z, Zhou L, Zhang F, Xie HY, Feng XW, et al. Identification of histone deacetylase 3 as a biomarker for tumor recurrence following liver transplantation in HBV-associated hepatocellular carcinoma. PLoS One 2010;5(12):e14460.

[44] Miller TA, Witter DJ, Belvedere S. Histone deacetylase inhibitors. J Med Chem 2003;46(24):5097–116.

[45] Hanahan D, Weinberg RA. The hallmarks of cancer. Cell 2000;100(1):57–70.

[46] Herold C, Ganslmayer M, Ocker M, Hermann M, Geerts A, Hahn EG, et al. The histone-deacetylase inhibitor Trichostatin A blocks proliferation and triggers apoptotic programs in hepatoma cells. J Hepatol 2002;36(2):233–40.

[47] Yamamoto H, Fujimoto J, Okamoto E, Furuyama J, Tamaoki T, Hashimoto-Tamaoki T. Suppression of growth of hepatocellular carcinoma by sodium butyrate in vitro and in vivo. Int J Cancer 1998;76(6):897–902.

[48] Yamashita Y, Shimada M, Harimoto N, Rikimaru T, Shirabe K, Tanaka S, et al. Histone deacetylase inhibitor trichostatin a induces cell-cycle arrest/apoptosis and hepatocyte differentiation in human hepatoma cells. Int J Cancer 2003;103(5):572–6.

[49] Papeleu P, Loyer P, Vanhaecke T, Elaut G, Geerts A, Guguen-Guillouzo C, et al. Trichostatin A induces differential cell cycle arrests but does not induce apoptosis in primary cultures of mitogen-stimulated rat hepatocytes. J Hepatol 2003;39(3):374–82.

[50] Ramadori G, Cameron S. Effects of systemic chemotherapy on the liver. Ann Hepatol 2010;9(2):133–43.

[51] Qiu L, Burgess A, Fairlie DP, Leonard H, Parsons PG, Gabrielli BG. Histone deacetylase inhibitors trigger a G2 checkpoint in normal cells that is defective in tumor cells. Mol Biol Cell 2000;11(6):2069–83.

[52] Marks PA, Dokmanovic M. Histone deacetylase inhibitors: discovery and development as anticancer agents. Expert Opin Investig Drugs 2005;14(12):1497–511.

[53] Richon VM, Sandhoff TW, Rifkind RA, Marks PA, Histone deacetylase inhibitor selectively induces p21WAF1 expression and gene-associated histone acetylation. Proc Natl Acad Sci USA 2000;97(18): 10014–19.

[54] Sowa Y, Orita T, Minamikawa S, Nakano K, Mizuno T, Nomura H, et al. Histone deacetylase inhibitor activates the WAF1/Cip1 gene promoter through the Sp1 sites. Biochem Biophys Res Commun 1997;241(1):142–50.

[55] Gui CY, Ngo L, Xu WS, Richon VM, Marks PA, Histone deacetylase (HDAC) inhibitor activation of p21WAF1 involves changes in promoter-associated proteins, including HDAC1. Proc Natl Acad Sci USA 2004;101(5):1241–6.

[56] Sambucetti LC, Fischer DD, Zabludoff S, Kwon PO, Chamberlin H, Trogani N, et al. Histone deacetylase inhibition selectively alters the activity and expression of cell cycle proteins leading to specific chromatin acetylation and antiproliferative effects. J Biol Chem 1999;274(49):34940–7.

[57] Jiang W, Guo Q, Wu J, Guo B, Wang Y, Zhao S, et al. Dual effects of sodium butyrate on hepatocellular carcinoma cells. Mol Biol Rep 2012;39(5):6235–42.

[58] Kim MS, Kwon HJ, Lee YM, Baek JH, Jang JE, Lee SW, et al. Histone deacetylases induce angiogenesis by negative regulation of tumor suppressor genes. Nat Med 2001;7(4):437–43.

[59] Mahon PC, Hirota K, Semenza GL. FIH-1: a novel protein that interacts with HIF-1α and VHL to mediate repression of HIF-1 transcriptional activity. Genes Dev 2001;15(20):2675–86.

[60] Kim SH, Jeong JW, Park JA, Lee JW, Seo JH, Jung BK, et al. Regulation of the HIF-1alpha stability by histone deacetylases. Oncol Rep 2007;17(3):647–51.

[61] Schröder M, Kaufman RJ. ER stress and the unfolded protein response. Mutat Res/Fundam Mol Mech Mutagen 2005;569(1–2):29–63.

[62] Quinones QJ, de Ridder GG, Pizzo SV. GRP78: a chaperone with diverse roles beyond the endoplasmic reticulum. Histol Histopathol 2008;23(11):1409–16.

[63] Okada T, Yoshida H, Akazawa R, Negishi M, Mori K. Distinct roles of activating transcription factor 6 (ATF6) and double-strandedRNA-activated protein kinase-like endoplasmic reticulum kinase (PERK) in transcription during the mammalian unfolded protein response. Biochem J 2002;366(2):585–94.

[64] Maytin EV, Ubeda M, Lin JC, Habener JF. Stress-inducible transcription factor CHOP/gadd153 induces apoptosis in mammalian cells via p38 kinase-dependent and -independent mechanisms. Exp Cell Res 2001;267(2):193–204.

[65] Zinszner H, Kuroda M, Wang X, Batchvarova N, Lightfoot RT, Remotti H, et al. CHOP is implicated in programmed cell death in response to impaired function of the endoplasmic reticulum. Genes Dev 1998;12(7):982–95.

[66] Puthalakath H, O'Reilly LA, Gunn P, Lee L, Kelly PN, Huntington ND, et al. ER stress triggers apoptosis by activating BH3-only protein Bim. Cell 2007;129(7):1337–49.

[67] Galehdar Z, Swan P, Fuerth B, Callaghan SM, Park DS, Cregan SP. Neuronal apoptosis induced by endoplasmic reticulum stress is regulated by ATF4–CHOP-mediated induction of the Bcl-2 homology 3-only member PUMA. J Neurosci 2010;30(50):16938–48.

[68] McCullough KD, Martindale JL, Klotz LO, Aw TY, Holbrook NJ. Gadd153 sensitizes cells to endoplasmic reticulum stress by down-regulating Bcl2 and perturbing the cellular redox state. Mol Cell Biol 2001;21(4):1249–59.

[69] Marciniak SJ, Yun CY, Oyadomari S, Novoa I, Zhang Y, Jungreis R, et al. CHOP induces death by promoting protein synthesis and oxidation in the stressed endoplasmic reticulum. Genes Dev 2004;18(24):3066–77.

[70] Srinivasan S, Ohsugi M, Liu Z, Fatrai S, Bernal-Mizrachi E, Permutt MA, et al. Endoplasmic reticulum stress–induced apoptosis is partly mediated by reduced insulin signaling through phosphatidylinositol 3-kinase/Akt and increased glycogen synthase kinase-3β in mouse insulinoma cells. Diabetes 2005;54(4):968–75.

[71] Lee H, Park MT, Choi BH, Oh ET, Song MJ, Lee J, et al. Endoplasmic reticulum stress-induced JNK activation is a critical event leading to mitochondria-mediated cell death caused by β-lapachone treatment. PLoS One 2011;6(6):e21533.

[72] Lee AS. GRP78 induction in cancer: therapeutic and prognostic implications. Cancer Res 2007;67(8):3496–9.

[73] Shuda M, Kondoh N, Imazeki N, Tanaka K, Okada T, Mori K, et al. Activation of the ATF6, XBP1 and grp78 genes in human hepatocellular carcinoma: a possible involvement of the ER stress pathway in hepatocarcinogenesis. J Hepatol 2003;38(5):605–14.

[74] Su R, Li Z, Li H, Song H, Bao C, Wei J, et al. Grp78 promotes the invasion of hepatocellular carcinoma. BMC Cancer 2010;10(1):20.

[75] Li H, Song H, Luo J, Liang J, Zhao S, Su R, et al. Knockdown of glucose-regulated protein 78 decreases the invasion, metalloproteinase expression and ECM degradation in hepatocellular carcinoma cells. J Exp Clin Cancer Res 2012;31:39.

[76] Chiou JF, Tai CJ, Huang MT, Wei PL, Wang YH, An J, et al. Glucose-regulated protein 78 is a novel contributor to acquisition of resistance to sorafenib in hepatocellular carcinoma. Ann Surg Oncol 2010;17(2): 603–12.

[77] Zhu X, Zhang J, Fan W, Wang F, Yao H, Wang Z, et al. The rs391957 variant cis-regulating oncogene GRP78 expression contributes to the risk of hepatocellular carcinoma. Carcinogenesis 2013.

[78] Zhu X, Chen MS, Tian LW, Li DP, Xu PL, Lin MC, et al. Single nucleotide polymorphism of rs430397 in the fifth intron of GRP78 gene and clinical relevance of primary hepatocellular carcinoma in Han Chinese: Risk and prognosis. Int J Cancer 2009;125(6):1352–7.

[79] Kovacs JJ, Murphy PJ, Gaillard S, Zhao X, Wu JT, Nicchitta CV, et al. HDAC6 regulates Hsp90 acetylation and chaperone-dependent activation of glucocorticoid receptor. Mol Cell 2005;18(5):601–7.

[80] Bucciantini M, Giannoni E, Chiti F, Baroni F, Formigli L, Zurdo J, et al. Inherent toxicity of aggregates implies a common mechanism for protein misfolding diseases. Nature 2002;416(6880):507–11.

[81] Scroggins BT, Robzyk K, Wang D, Marcu MG, Tsutsumi S, Beebe K, et al. An acetylation site in the middle domain of Hsp90 regulates chaperone function. Mol Cell 2007;25(1):151–9.

[82] Blagosklonny MV. Hsp-90-associated oncoproteins: multiple targets of geldanamycin and its analogs. Leukemia 2002;16(4):455–62.

[83] Liu YL, Yang PM, Shun CT, Wu MS, Weng JR, Chen CC, et al. Autophagy potentiates the anti-cancer effects of the histone deacetylase inhibitors in hepatocellular carcinoma. Autophagy 2010;6(8):1057–65.

[84] Yang ZJ, Chee CE, Huang S, Sinicrope FA. The role of autophagy in cancer: therapeutic implications. Mol Cancer Ther 2011;10(9):1533–1541.

[85] Ma BB, Sung F, Tao Q, Poon FF, Lui VW, Yeo W, et al. The preclinical activity of the histone deacetylase inhibitor PXD101 (belinostat) in hepatocellular carcinoma cell lines. Invest New Drugs 2010;28(2):107–14.

[86] Yeo W, Chung HC, Chan SL, Wang LZ, Lim R, Picus J, et al. Epigenetic therapy using belinostat for patients with unresectable hepatocellular carcinoma: a multicenter phase I/II study with biomarker and pharmaco-kinetic analysis of tumors from patients in the Mayo Phase II Consortium and the Cancer Therapeutics Research Group. J Clin Oncol 2012;30(27):3361–7.

[87] New M, Olzscha H, Liu G, Khan O, Stimson L, McGouran J, et al. A regulatory circuit that involves HR23B and HDAC6 governs the biological response to HDAC inhibitors. Cell Death Differ 2013;20(10):1306–16.

[88] Kim HJ, Bae SC. Histone deacetylase inhibitors: molecular mechanisms of action and clinical trials as anti-cancer drugs. Am J Transl Res 2011;3(2):166–79.

[89] Lachenmayer A, Toffanin S, Cabellos L, Alsinet C, Hoshida Y, Villanueva A, et al. Combination therapy for hepatocellular carcinoma: additive preclinical efficacy of the HDAC inhibitor panobinostat with sorafenib. J Hepatol 2012;56(6):1343–50.

[90] Montalbano R, Waldegger P, Quint K, Jabari S, Neureiter D, Illig R, et al. Endoplasmic reticulum stress plays a pivotal role in cell death mediated by the pan-deacetylase inhibitor panobinostat in human hepato-cellular cancer cells. Transl Oncol 2013;6(2):143–57.

[91] Chen CH, Chen MC, Wang JC, Tsai AC, Chen CS, Liou JP, et al. Synergistic interaction between the HDAC inhibitor, MPT0E028, and sorafenib in liver cancer cells in vitro and in vivo. Clin Cancer Res 2014;20(5):1274–87.

[92] Zopf S, Ocker M, Neureiter D, Alinger B, Gahr S, Neurath MF, et al. Inhibition of DNA methyltransferase activity and expression by treatment with the pan-deacetylase inhibitor panobinostat in hepatocellular car-cinoma cell lines. BMC Cancer 2012;12(1):386.

[93] Bitzer M, Horger M, Ganten T, Ebert MP, Siveke J, Woerns MA, et al. Clinical update on the SHELTER study: a phase I/II trial of the HDAC inhibitor resminostat in patients with sorafenib-resistant hepatocellular carcinoma (HCC). J Clin Oncol 2011;29(Suppl 4) Abstr 275.

[94] Bitzer M, Horger M, Ganten T, Siveke J, Woerns MA, Dollinger MM, et al. Investigation of the HDAC inhibitor resminostat in patients with sorafenib-resistant hepatocellular carcinoma (HCC): clinical data from the phase I/II SHELTER study. J Clin Oncol 2012;30(Suppl 4) Abstr 262.

[95] Clinicaltrials.gov. *Sorafenib and Vorinostat in Treating Patients With Advanced Liver Cancer (NCT01075113)*. Available from: http://clinicaltrials.gov/ct2/show/NCT01075113.

[96] Kuroiwa-Trzmielina J, de Conti A, Scolastici C, Pereira D, Horst MA, Purgatto E, et al. Chemoprevention of rat hepatocarcinogenesis with histone deacetylase inhibitors: efficacy of tributyrin, a butyric acid prodrug. Int J Cancer 2009;124(11):2520–7.

[97] Hsieh YH, Su JJ, Yen CJ, Tsai TF, Tsai HW, Tsai HN, et al. Histone deacetylase inhibitor suberoylanilide hydroxamic acid suppresses the pro-oncogenic effects induced by hepatitis B virus pre-S2 mutant oncoprotein and represents a potential chemopreventive agent in high-risk chronic HBV patients. Carcinogenesis 2013;34(2):475–85.

[98] Niki T, Rombouts K, De Bleser P, De Smet K, Rogiers V, Schuppan D, et al. A histone deacetylase inhibitor, trichostatin A, suppresses myofibroblastic differentiation of rat hepatic stellate cells in primary culture. Hepatology 1999;29(3):858–67.

[99] Rombouts K, Knittel T, Machesky L, Braet F, Wielant A, Hellemans K, et al. Actin filament formation, reorganization and migration are impaired in hepatic stellate cells under influence of trichostatin A, a histone deacetylase inhibitor. J Hepatol 2002;37(6):788–96.

[100] Mah WC, Lee CGL. DNA methylation: potential biomarker in hepatocellular carcinoma. Biomarker Res 2014;2(5) 2050-7771-2-5.

[101] Zhao ZK, Yu HF, Whang DR, Dong P, Chen L, Wi WG, et al. Overexpression of lysine specific demethylase 1 predicts worse prognosis in primary hepatocellular carcinoma patients. World J Gastroenterol 2012;18(45):6651–6.

[102] Greene CM, Varley RB, Lawless MW. MicroRNAs and liver cancer assocaited with iron overload: therapeutic targets unreavelled. World J Gastroenterol 2013;19(32):5212–26.

[103] Giordano S, Columbano A. MicroRNAs: new tools for diagnosis, prognosis and therapy in hepatocellular carcinoma? Hepatology 2013;57(2):840–7.

[104] Zhao J, Lawless MW. Long noncoding RNAs and their role in the liver cancer axis. Nat Rev Gastroenterol Hepatol 2013:87-c1.

[105] Panzitt K, Tschernatsch MM, Guelly C, Moustafa T, Stradner M, Strohnmaier HMet al. Characterization of HULC; a novel gene with striking up-regulation in hepatocellular carcinomea, as noncoding RNA. Gastroenterology 132:330–342.

[106] Zhang G, Park MA, Mitchell C, Hamed H, Rahmani M, Martin AP, et al. Vorinostat and Sorafenib Synergistically Kill Tumor Cells via FLIP Suppression and CD95 Activation. Clin Cancer Res 2008;14(17): 5385–99.

[107] Ocker M, Alajati A, Ganslmayer M, Zopf S, Lüders M, Neureiter D, et al. The histone-deacetylase inhibitor SAHA potentiates proapoptotic effects of 5-fluorouracil and irinotecan in hepatoma cells. J Cancer Res Clin Oncol 2005;131(6):385–94.

[108] Tong A, Zhang H, Li Z, Gou L, Wang Z, Wei H, et al. Proteomic analysis of liver cancer cells treated with suberonylanilide hydroxamic acid. Cancer Chemother Pharmacol 2008;61(5):791–802.

[109] Mannaerts I, Nuytten NR, Rogiers V, Vanderkerken K, van Grunsven LA, Geerts A, et al. Chronic administration of valproic acid inhibits activation of mouse hepatic stellate cells in vitro and in vivo. Hepatology 2010;51(2):603–14.

[110] Tatebe H, Shimizu M, Shirakami Y, Sakai H, Yasuda Y, Tsurumi H, et al. Acyclic retinoid synergises with valproic acid to inhibit growth in human hepatocellular carcinoma cells. Cancer Lett 2009;285(2): 210–17.

[111] Takimoto R, Kato J, Terui T, Takada K, Kuroiwa G, Wu J, et al. Augmentation of antitumor effects of p53 Gene therapy by combination with HDAC inhibitor. Cancer Biol Ther 2005;4(4):427–34.

[112] Wang HG, Huang XD, Shen P, Li LR, Xue HT, Ji GZ, et al. Anticancer effects of sodium butyrate on hepatocellular carcinoma cells in vitro. Int J Mol Med 2013;31(4):967–74.

[113] Finkelstein RA, Li Y, Liu B, Shuja F, Fukudome E, Velmahos GC, et al. Treatment with histone deacetylase inhibitor attenuates MAP kinase mediated liver injury in a lethal model of septic shock. J Surg Res 2010;163(1):146–54.

[114] Liu C, Liu L, Shan J, Shen J, Xu Y, Zhang Q, et al. Histone deacetylase 3 participates in self-renewal of liver cancer stem cells through histone modification. Cancer Lett 2013;339(1):60–9.

[115] Buurman R, Gürlevik E, Schäffer V, Eilers M, Sandbothe M, Kreipe H, et al. Histone deacetylases activate hepatocyte growth factor signaling by repressing microRNA-449 in hepatocellular carcinoma cells. Gastroenterology 2012;143(3):811–20. e15.

[116] Yoshida M. Chemical genetics enables an approach to life phenomena and developments in drug discovery. Riken Res 2008;3:10.

[117] Lin WH, Martin JL, Marsh DJ, Jack MM, Baxter RC, Lin WH. Involvement of insulin-like growth factor-binding protein-3 in the effects of histone deacetylase inhibitor Ms-275 in hepatoma cells. J Biol Chem 2011;286(34):29540–7.

[118] de Conti A, Tryndyak V, Koturbash I, Heidor R, Kuroiwa-Trzmielina J, Ong TP, et al. The chemopreventive activity of the butyric acid prodrug tributyrin in experimental rat hepatocarcinogenesis is associated with p53 acetylation and activation of the p53 apoptotic signaling pathway. Carcinogenesis 2013;34(8):1900–6.

[119] Lu YS, Chou CH, Tzen KY, Gao M, Cheng AL, Kulp SK, et al. Radiosensitizing effect of a phenylbutyrate-derived histone deacetylase inhibitor in hepatocellular carcinoma. Int J Radiat Oncol Biol Phys 2012;83(2):e181–9.

NEUROBLASTOMA

13

Olga Piskareva and Raymond L. Stallings

Cancer Genetics, Molecular and Cellular Therapeutics, Royal College of Surgeons in Ireland, Dublin, Ireland and Children's Research Centre Our Lady's Children's Hospital, Crumlin, Dublin, Ireland

CHAPTER OUTLINE

S.G. Gray (Ed): Epigenetic Cancer Therapy. DOI: http://dx.doi.org/10.1016/B978-0-12-800206-3.00013-6
© 2015 Elsevier Inc. All rights reserved.

1 NEUROBLASTOMA

Neuroblastoma is a highly malignant pediatric cancer, originating from precursor or immature cells of the sympathetic nervous system during embryonic development or early postnatal life. It is the most common solid extracranial malignancy of childhood and the most common malignant tumor in infants. The overall incidence of neuroblastoma is 6–7 cases per million and approximately 15% of all childhood cancer deaths can be attributed to the disease [1–4].

Neuroblastoma displays significant heterogeneity between patients both in disease progression and in clinical outcome. Some tumors will spontaneously regress or mature even without therapy, developing into a benign ganglioneuroma. Others display a very aggressive, malignant behavior that is poorly responsive to current intensive, multimodal therapy. At the time of neuroblastoma diagnosis, more than half of the patients will already have metastatic disease. Metastatic sites most commonly include bone and bone marrow, less frequently lymph nodes and liver, and rarely lung, brain or skin.

Neuroblastoma tumors can be stratified into groups according to risk factors which include age at diagnosis, tumor localization and size, histopathologic classification, DNA content, chromosomal abnormalities, and *MYCN* status (extensively reviewed by [2]). Of note, *MYCN* oncogene amplification occurred in approximately 20% of neuroblastomas and is the most reliable genetic marker of a poor prognosis. This marker is currently used in combination with clinical characteristics to stratify newly diagnosed patients into "risk groups" based on the likelihood of their developing an aggressive or less-severe form of neuroblastoma.

Among the molecular features contributing to neuroblastoma, heterogeneity and stratification are whole chromosome gains and a large number of recurrent large-scale chromosome imbalances, such as loss of heterozygosity at chromosome arms 1p, 3p, 14q, and 11q, unbalanced gain of 1q, 11p, and 17q and numerous mutations in key genes such as ALK, PHOX2B, and PTPRD (reviewed in [2,3]).

Growing evidence indicates that features such as epigenetic changes and miRNA expression are highly correlated with clinical behavior, providing insight into the molecular basis of clinical heterogeneity and offering better defined prognostic signatures [3,5]. These advances may eventually be included to risk assessment and stratification in order to improve treatment regimens and prediction of patient outcome.

These risk factors categorize neuroblastoma into low-, intermediate-, and high-risk groups, which predict the overall survival rate. Half of all neuroblastoma diagnoses are classified as high risk, requiring immediate treatment. The 5-year survival rate for children with low-risk neuroblastoma is higher than 95%, intermediate-risk neuroblastoma is 80% to 90%. Long-term survival for children with high-risk neuroblastoma is very poor, about 30% to 50%.

Standard of care therapy may include chemotherapy, surgery, myeloablative radiation, and a restorative autologous stem cell transplant (comprising the patient's bone marrow cells harvested and stored prior to treatment). Unfortunately, at least 20% of high-risk neuroblastoma cases will be refractory (resistant) to one or the other type of therapy. After remission more than one-third of children will subsequently relapse. These are the highest risk cases. One of the greatest concerns in high-risk neuroblastoma is the persistence of minimal residual disease (MRD) after chemotherapy or radiation which often results in recurrence.

In this chapter, we will discuss the key epigenetic factors contributing to neuroblastoma, the significance of such in relation to improved understanding of neuroblastoma predisposition and development of new therapies.

2 EPIGENETIC CHANGES

The hallmark of cancer is dysregulated gene expression. To date, it has been shown that both genetic and epigenetic factors impact gene expression and contribute to cancer development. Epigenetic alterations are defined as those heritable changes in gene expression that do not result from direct changes in DNA sequence. Mechanisms of epigenetic regulation most commonly include DNA methylation, modification of histones, and changes in microRNA (miRNA) expression.

2.1 DNA METHYLATION

Altered DNA methylation patterns are widely observed in development and progression of various types of cancer including neuroblastoma. Methylation pattern detection has become an enormously important area of study (see Chapter 2 for more details). DNA methylation triggers the binding of methylated DNA-specific binding proteins to CpG sites, attracting histone-modifying enzymes that, in turn, focally establish a silenced chromatin state. Aberrant hypermethylation of CpGs within gene promoters can lead to gene inactivation, while genome wide or global hypomethylation affects mostly repetitive DNA sequences. These alterations are considered to be the most frequent cancer-related epigenetic changes, while the epigenetic silencing is a common mechanism for loss of tumor suppressor gene function rather than mutation as will be discussed in the following sections.

2.1.1 Aberrant hypermethylation of CpGs within gene promoters

The first DNA methylation studies in neuroblastoma discovered that epigenetic inactivation of caspase 8 (*CASP8*) and RAS-association domain family 1 isoform A (*RASSF1A*) are important in the development and progression of disease [6–9]. The promoter of tumor suppressor *RASSF1A*, mapped to 3p21.3 was *de novo* hypermethylated in 55–100% neuroblastoma tumors as reviewed by Decock [5]. The DNA methylation status was correlated with poor prognosis, suggesting that the epigenetic silencing of this tumor suppressor gene could contribute to aberrations of RAS signal pathways observed in neuroblastomas and progression of this disease.

Another commonly methylated and inactivated gene in neuroblastoma is *CASP8*, encoding a cysteine protease involved in the tumor necrosis factor–related apoptosis pathway [6]. Inactivation of the *CASP8* gene was caused by deletion or hypermethylation in neuroblastoma cell lines. However, methylation of its promoter region was the primary mechanism for its silencing in neuroblastoma tumor samples. Interestingly, complete inactivation of *CASP8* was observed predominantly in neuroblastomas with amplified *MYCN*. A consistent CpG island promoter methylation of *CASP8* has been detected in 14–91% neuroblastoma tumors [reviewed in [5]]. For example, in a cohort of 70 neuroblastoma tumor samples *CASP8* displayed 56% hypermethylation which was correlated to poor outcome [10]. Applying clustering of a limited number of hypermethylated genes, *CASP8* was found to be methylated in 77% of the neuroblastoma cell lines investigated, further supporting the importance of the methylation status of this gene *in vitro* [11].

Advances in genome-wide methylation discrimination technologies, such as DNA methylation promoter assay after affinity-based capture, reexpression analysis after treatment with 5-aza-2'-deoxycytidine (DAC), methylation microarray after bisulfate treatment and next-generation sequencing, made it possible to identify nearly 80 different DNA methylation candidate genes associated with patient survival in neuroblastoma and risk factors, such as *MYCN* amplification, patient age, and tumor stage (reviewed in [5]). All these genes are involved in fundamental biological processes, namely, tumor suppression, cell cycle control, DNA damage response, cell migration, apoptosis, etc. Noticeably, a significant correlation between *RASSF1A* and *CASP8* methylation in neuroblastoma was demonstrated [7,9], suggesting that a subset of neuroblastoma may have a CpG island methylator phenotype (CIMP) as described in other cancers [12]. Some of these genes could be included into development of CIMP for neuroblastoma: however, further extensive studies are required.

2.1.2 Genome-wide aberrant hypermethylation of CpGs islands assessment

The first genome-wide assessment of DNA methylation patterns in neuroblastoma, ganglioneuroma, and ganglioneuroblastoma tumors was carried out by Buckley et al. [13]. This study identified recurrent large-scale blocks of contiguously hypermethylated promoter/CpG island sites ($n = 70$) in the tumors. The highest number of large-scale DNA methylation blocks (eight blocks in total) was demonstrated for chromosome 19, which may be explained by the greater gene density of this chromosome. The study recognized a total of 63 methylation blocks that are potentially disease related. In contrast, DNA methylation analysis of three human chromosomes in normal tissue identified a significant correlation for methylation of regions only over distances 1000 bp, suggesting that larger methylated regions may be disease specific. Of note, a significant overrepresentation of methylated blocks toward telomeric ends (31% of the blocks occurring <2 Mb from telomere) was identified (Figure 13.1).

2.2 HISTONE MODIFICATIONS

Lysine-rich tails of core histones (H2A, H2B, H3, and H4) overhang from the nucleosome providing sites for reversible modifications that modify chromatin structure and modulate gene expression (see Chapter 3 for more details). The most characterized histone modifications in cancer are acetylation/deacetylation and methylation/demethylation.

Relatively limited information is available concerning histone modification events in neuroblastoma. Two separate DNA methylation studies focused on DNA methylation status of *PTGER2* [14] and *NSD1* [15] using chromatin immunoprecipitation (ChIP) assays have detected and discussed histone modifications in neuroblastoma.

The *PTGER2* gene, mapping to 14q22, encodes a receptor for prostaglandin E2, which has different biologic activities in a wide range of tissues. Dissection of *PTGER2* promoter region revealed segments of different susceptibility to methylation in neuroblastoma cells [14]. Hypermethylation patterns were concordant with loss of PTGER2 protein expression. Subsequent experimental validation of histone modification status of promoter region of *PTGER2* by ChIP suggested that around its promoter region, histone H3 and H4 are deacetylated and histone H9 is di- and tri-methylated in neuroblastoma cells lacking expression of this gene. The methylation status of *PTGER2* correlated with the aggressiveness of neuroblastoma tumors.

NSD1 gene mapped to 5q35 encodes a histone methyl transferase. CpG island hypermethylation of *NSD1* leads to reduced levels of NSD1 protein in neuroblastoma and glioma [15]. This epigenetic silencing was associated with global reduced levels of trimethylated histone H4 and H3 and poor survival.

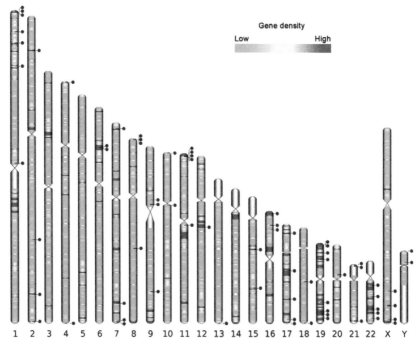

FIGURE 13.1

Identification of genome-wide methylation blocks in neuroblastoma. Regions of consecutive hypermethylation are highlighted in blue dots across chromosomes. Gene density across each chromosome is also depicted.

This figure is the original figure published by Buckley et al. [13] and is reproduced here with permission of the authors and their agreement with the publisher.

2.3 HISTONE DEACETYLASE EXPRESSION

Histone deacetylases (HDACs) are a family of protein deacetylating enzymes that remove acetyl groups from lysine residues of histone and non-histone proteins (see Chapter 3 for more details as well as [16]). Their enzyme activity can be inhibited by small molecule compounds termed HDAC inhibitors. HDACIs are now recognized as a promising mode of anticancer treatment with clinical trials under way (Table 13.1).

The first study investigating expression of all HDAC family members at mRNA level in neuroblastoma was conducted on a large cohort of 251 tumors [17]. Increasing expression of HDAC8 mRNA showed significant correlation with disease progression, known clinical and molecular risk factors, and poor clinical outcome. Individual HDAC family members control different cellular pathways in cancer; for example, suppression of differentiation (HDAC8 [17], HDAC5 [18], HDAC1/HDAC2 [19]) and apoptosis (HDAC2 [17]), mediating *MYCN* function (HDAC2 [20], HDAC5 [18]) and multidrug resistance (HDAC1 [21]). This data may explain the different biological effects observed with nonselective HDAC inhibitors on cancer cells.

Table 13.1 Epigenetic Targeting Therapeutics in Clinical Development

NCT number	Title	Recruitment	Conditions	Age groups	Phases	Enrollment
NCT01048086	90Y DOTA/Retinoic Acid for Neuroblastoma and Neuroendocrine Tumor (NET)	Active, not recruiting	Neuroblastoma/neuroendocrine tumor	Child/adult	Phase 2	70
NCT01183416	High-Dose 3F8/GM-CSF Immunotherapy Plus 13-Cis-Retinoic Acid for Consolidation of First Remission after Myeloablative Therapy and Autologous Stem Cell Transplantation	Active, not recruiting	Neuroblastoma	Child/adult/senior	Phase 2	43
NCT01592045	Ch14.18 Pharmacokinetic Study in High-Risk Neuroblastoma	Active, not recruiting	Neuroblastoma	Child	Phase 1/ Phase 2	28
NCT00499616	Combination Chemotherapy and Surgery with or without Isotretinoin in Treating Young Patients with Neuroblastoma	Active, not recruiting	Neuroblastoma	Child	Phase 3	464
NCT01183429	3F8/GM-CSF Immunotherapy Plus 13-Cis-Retinoic Acid for Consolidation of First Remission after Non-Myeloablative Therapy in Patients with High-Risk Neuroblastoma	Recruiting	Neuroblastoma	Child/adult/senior	Phase 2	58
NCT01183884	3F8/GM-CSF Immunotherapy Plus 13-Cis-Retinoic Acid for Consolidation of Second or Greater Remission of High-Risk Neuroblastoma	Recruiting	Neuroblastoma	Child/adult/senior	Phase 2	63

NCT Number	Title	Status	Condition	Age	Phase	N
NCT01183897	3F8/GM-CSF Immunotherapy Plus 13-Cis-Retinoic Acid for Primary Refractory Neuroblastoma in Bone Marrow	Recruiting	Neuroblastoma	Child/adult/senior	Phase 2	53
NCT01526603	High-Dose Chemotherapy and Autologous Transplant for Neuroblastoma	Recruiting	Neuroblastoma	Child/adult		20
NCT01728155	European Low- and Intermediate-Risk Neuroblastoma Protocol	Recruiting	Low and intermediate pediatric neuroblastoma and neonatal suprarenal masses	Child/adult	Phase 3	685
NCT01857934	Therapy for Children with Advanced Stage Neuroblastoma	Recruiting	Neuroblastoma	Child/adult	Phase 2	42
NCT01701479	Long-Term Continuous Infusion Ch14.18/CHO Plus s.c. Aldesleukin (IL-2)	Recruiting	Neuroblastoma	Child/adult	Phase 1/ Phase 2	60
NCT01208454	Vorinostat and Isotretinoin in Treating Patients with High-Risk Refractory or Recurrent Neuroblastoma	Recruiting	Disseminated neuroblastoma/localized unresectable neuroblastoma/ recurrent neuroblastoma/ regional neuroblastoma/stage 4s neuroblastoma	Child/adult	Phase 1	54
NCT00939965	Isotretinoin in Treating Young Patients with High-Risk Neuroblastoma	Recruiting	Neuroblastoma	Child/adult		75

(Continued)

Table 13.1 (Continued)

NCT number	Title	Recruitment	Conditions	Age groups	Phases	Enrollment
NCT00026312	Isotretinoin with or without Monoclonal Antibody Ch14.18, Aldesleukin, and Sargramostim Following Stem Cell Transplant in Treating Patients with Neuroblastoma	Recruiting	Disseminated neuroblastoma/localized resectable neuroblastoma/ localized unresectable neuroblastoma/regional neuroblastoma/stage 4s neuroblastoma	Child/adult	Phase 3	1660
NCT01404702	Pilot Study of Zoledronic Acid and Interleukin-2 for Refractory Pediatric Neuroblastoma	Recruiting	Neuroblastoma	Child/adult	Phase 1	6
NCT01295762	Immunomonitoring of Children with Neuroblastoma	Recruiting	Neuroblastoma	Child/adult		30
NCT00793845	Tandem High-Dose Chemotherapy and Autologous Stem Cell Rescue in Patients with High-Risk Neuroblastoma	Recruiting	Neuroblastoma	Child/adult/senior	Phase 2	40
NCT01208454	Vorinostat and Isotretinoin in Treating Patients with High-Risk Refractory or Recurrent Neuroblastoma	Recruiting	Disseminated neuroblastoma/ localized unresectable neuroblastoma/recurrent neuroblastoma/regional neuroblastoma/stage 4s neuroblastoma	Child/adult	Phase 1	54

NCT Number	Title	Status	Condition	Age	Phase	Enrollment
NCT01019850	N2007-03: Vorinostat and 131-I MIBG in Treating Patients with Resistant or Relapsed Neuroblastoma	Active, not recruiting	Neuroblastoma	Child/adult	Phase 1	42
NCT02035137	131I-MIBG Alone vs. 131I-MIBG with Vincristine and Irinotecan VS131I-MIBG with Vorinistat	Not yet recruiting	Neuroblastoma	Child/adult	Phase 2	105
NCT01294670	Clinical Study of Vorinostat in Combination with Etoposide in Pediatric Patients <21 Years at Diagnosis with Refractory Solid Tumors	Recruiting	Solid tumors/relapsed/ refractory sarcomas	Child/adult	Phase 1/ Phase 2	50
NCT01422499	Vorinostat in Children	Recruiting	Relapsed solid tumors, lymphoma or leukemia	Child/adult	Phase 1/ Phase 2	50
NCT01829971	A Multicenter Phase I Study of MRX34, MicroRNA miR-RX34 Liposome Injectable Suspension	Recruiting	Primary liver cancer/solid tumors/lymphoma	Adult/senior	Phase 1	48
NCT01241162	Decitabine Followed by a Cancer Antigen Vaccine for Patients with Neuroblastoma and Sarcoma	Recruiting	Neuroblastoma/Ewings sarcoma/osteogenic sarcoma/ rhabdomyosarcoma/synovial sarcoma	Child	Phase 1	15

DNA methylation in neuroblastoma displays a complex pattern, yet to be fully elucidated. Integrated epigenetic events, such as DNA methylation and histone modification are involved in epigenetic gene silencing. Almost 80 differentially DNA methylated genes were validated using various detection techniques. Some of them have been shown to have the prognostic potential as epigenetic biomarkers. Importantly, epigenetic changes are reversible; hence DNA methylating agents may be able to restore functionality of epigenetically inactivated genes with potential tumor growth inhibitory effects.

2.4 miRNA

Epigenetic inactivation of miRNAs with tumor-suppressor activities is now recognized as a major hallmark of neuroblastoma tumors. miRNAs are a class of small, noncoding RNAs that appear to function in gene regulation. MiRNAs are known to regulate oncogenes, tumor suppressor genes, genes involved in cell cycle control, cell migration, differentiation, development, apoptosis, and angiogenesis (reviewed in Chapter 4). MiRNA expression in tumors, and particularly in neuroblastoma has been observed to be up- or downregulated in high versus low risk tumors, supporting their complex dual role as either "oncomirs" or tumor suppressors, respectively [22]. Remarkably, select miRNA signatures can classify multiple cancers more accurately than data from ~16,000 mRNAs [23]. Similarly to protein encoding gene expression, the activity of miRNAs is also under epigenetic regulation. Therapeutic targeting of miRNA in neuroblastoma is currently being explored [24–26].

2.4.1 miRNA expression patterns

The first miRNA expression profiling study of primary neuroblastoma tumors was published by Chen and Stallings in 2007 [27], demonstrating that many miRNAs are differentially expressed in different genomic subtypes of neuroblastoma. Importantly, those miRNA profiles were correlated with the clinical outcome. Distinguished miRNA expression patterns were found between *MYCN* amplified and other tumor subtypes, leading to the hypothesis that overexpression of MYCN results in the down regulation of a large set of miRNAs that have anti-proliferative or pro-apoptotic effects and the upregulation of a smaller set of miRNAs that promote tumor growth. This observation has been validated further in the two independent miRNA expression profiling studies involving larger neuroblastoma tumor cohort and miRNA loci [28]. The hypothesis was confirmed by ChIP studies demonstrated that MYCN binds in close proximity to certain miRNA loci [29]. Further support for the concept that MYCN contributes to aggressive disease pathogenesis through the regulation of miRNAs came from the study by Mestdagh et al. [30]. These authors demonstrated that miRNAs that are upregulated by MYCN/MYC signaling directly downregulate a large set of protein-coding genes that are significantly associated with patient survival. Thus, MYCN directly regulates expression of some miRNAs [28,30,31].

Other factors that significantly alter miRNA expression in neuroblastoma are the recurrent large-scale chromosomal imbalances, including loss of 1p, 3p, 11q, and 14q, along with gain of 1q and 17q [28]. Bray et al. used a machine learning algorithm to explore combinations of miRNAs that are either overexpressed or underexpressed in unfavorable tumors. This study identified a 15-miRNA signature predictive of OS and event-free survival EFS, as illustrated in Figure 13.2, as reproduced from Bray et al. [28]. Notably, this signature is independent of MYCN copy number and the status of other prognostic markers, thus can be used for advanced stratification of neuroblastomas.

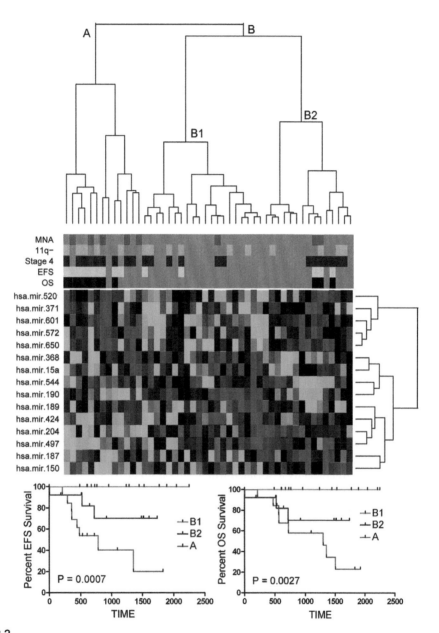

FIGURE 13.2

Hierarchical cluster analysis using a 15-miRNA expression signature divides neuroblastoma patients into two major groups, A and B, which significantly differ in survival. Patients with tumors in group A have significantly worse event-free and overall survival than patients from group B. The analysis represents an independent validation of the signature. Red indicates high miRNA expression and green low expression on the heat map.

This figure is the original figure published by Bray et al. [28] and is reproduced here with permission of the authors and their agreement with the publisher.

Buckley et al. [32] split tumors with 11q deletions into two distinct subtypes using the 15-miRNA signature originally developed by Bray [28]. These subtypes differed significantly in clinical outcome and the overall frequency of large-scale genomic imbalances, with the poor survival group having more imbalances. This is consistent with a study by Fisher et al. where 11q-tumors were subdivided on the basis of mRNA expression profiling into two clinically distinct groups [33]. However, Buckley's study also discovered cases where miRNA expression was inversely related to the genomic imbalance. Some miRNAs were underexpressed in spite of mapping to a region of DNA copy number gain. This strongly suggests that alternative mechanisms can in some instances neutralize the effects of DNA dosage. Importantly, an miRNA expression signature predictive of clinical outcome for 11q-tumors was identified for neuroblastoma, emphasizing the potential for miRNA mediated diagnostics and therapeutics (Figure 13.3).

2.4.2 Individual miRNAs

miRNAs regulate important genes involved in neuroblastoma pathogenesis, playing a complex role as either "oncomirs" or tumor suppressors [22]. One of the tumor suppressor miRNAs in neuroblastoma is miR-34a [27,34–36]. Welch et al. first demonstrated that miR-34a, which maps to a region on distal chromosome 1p, is expressed at lower levels in tumors with 1p deletion, and that ectopic over expression of this miRNA in neuroblastoma cell lines leads to the arrest of cell proliferation and the induction of a caspase-mediated apoptotic pathway [34]. miR-34a directly targets a number of genes in neuroblastoma, including the *E2F3* and *MYCN* transcription factors [34–36] as well as other genes involved in cell proliferation or apoptosis such as *BCL2* [37], *CCND1* and *CDK6* [38], Notch1 [39], survivin [40] and CD44 [41], and many others.

Other miRNAs can also act as tumor suppressors in neuroblastoma, such as let-7 and miR-101 which directly regulate MYCN expression [42], or pro-apoptotic miR-184 [43], the anti-invasive miR-335 [44], miR-542-5p [45], and several differentiation-related miRNA [46,47].

Oncogenic potential of the miR-17-5p-92 polycistronic cluster (miR-17-5p, -18a, -19a, -20a, and -92) was provided by Fontana et al. [25]. This functional study in neuroblastoma reported that the cluster is directly upregulated by MYCN showing that some members of this cluster act in an oncogenic, growth promoting manner *in vitro* and *in vivo*. The upregulation of this cluster is correlated with its host gene, *MIRHG1*, which is significantly associated with neuroblastoma patient survival [48]. The experimentally validated targets of miR-17-5p are the p21 tumor suppressor gene (*CDKN1A*) responsible for enhanced cell proliferation and tumorigenesis; and pro-apoptotic gene, *BIM* [25]. The effects of the miR-17-5p-92 cluster appear to be quite extensive, as Chayka et al. [26] demonstrated direct targeting of the clusterin (CLU) gene mRNA by members of this polycistron.

Other miRNAs can contribute to cisplatin sensitivity in neuroblastoma, reexpression of miR-204 [49] increase, whereas miR-21 [50] reduces sensitivity to the drug. The list of specific miRNAs validated as contributors to neuroblastoma pathogenesis is continually expanding, with some examples summarized/illustrated in Figure 13.4.

In reviewing even the most comprehensive miRNA expression profiling studies of neuroblastoma, it is important to remember that some miRNAs contributing to disease pathogenesis might not be annotated yet in the Sanger miRNA Registry (http://microrna.sanger.ac.uk/sequences/) and therefore might not be included in polymerase chain reaction or microarray-based assays. Afanasyeva et al. [51] identified several novel miRNAs expressed in neuroblastoma following the cloning and sequencing of small RNAs from neuroblastoma tumors and cell lines, while Schulte et al. [52] more recently used an unbiased next-generation sequencing approach to make statistically significant comparisons between

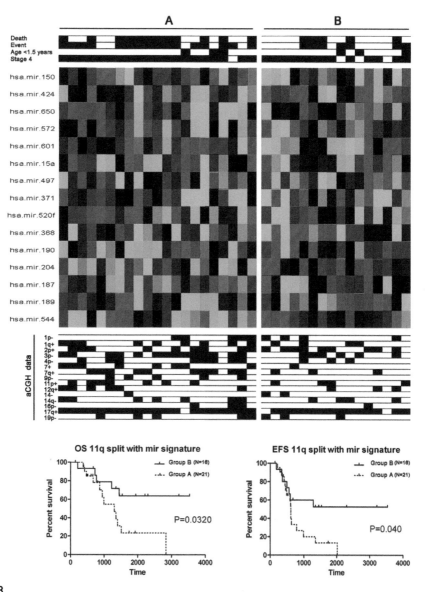

FIGURE 13.3

Analysis of neuroblastoma tumors possessing deletion of chromosome 11q using the same miRNA expression signature described in Figure 13.2. The 11q-tumors could be split by the unsupervised *k*-means partitional algorithm into two groups, A and B, on the basis of miRNA expression. Groups A and B differed significantly in event-free and overall survival, as well as in the frequency of segmental chromosome imbalances. The authors concluded that two distinct biological subtypes that differ in miRNA expression, clinical outcome, and frequency of segmental chromosomal imbalances. *This figure is the original figure published by Buckley et al. [32] and is reproduced here with permission of the authors and their agreement with the publisher.*

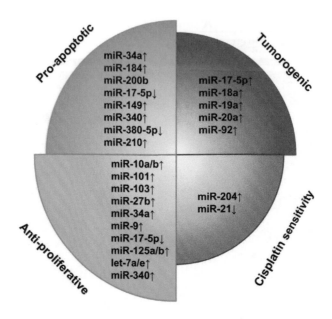

FIGURE 13.4

Summary of miRNA functions in neuroblastoma.

small RNA sequences in unfavorable and favorable tumor subtypes. In addition to identifying novel miRNAs, Schulte et al. [52] also provided information on the clinical relevance of these miRNAs and on the absolute quantification of minor versus major forms of miRNAs. The inclusion of these novel miRNAs in large-scale expression profiling studies will be necessary for a comprehensive understanding of the contribution of miRNAs in neuroblastoma tumorigenesis.

2.4.3 Epigenetic control of miRNA expression

miRNA expression in neuroblastoma is under control of multiple mechanisms, including aberrations in DNA copy number, altered transcriptional activators/repressors, aberrant DNA methylation, or defects of the proteins involved in the miRNA biogenesis machinery and in the posttranscriptional regulation of miRNA expression. The deep understanding of the mechanisms involved in miRNA regulation is required, not only to better understand the role miRNAs play in the development of the disease, but also may help us to identify new therapeutic targets.

Hypermethylation of CpG islands is associated with specific miRNAs. A recent study, reviewing the methylation data available from several different neoplasms, discovered that miRNAs displayed a higher magnitude of methylation in comparison to protein coding genes, with about 11.6% of all known miRNAs being methylated [53]. Therefore, this epigenetic mechanism has been proposed as one of the mechanisms by which the miRNA is selectively downregulated in tumors. In cases where the miRNA is positioned in the coding region of a gene, methylation may simultaneously suppress expression of both the protein-coding gene and also its embedded miRNA. However, very little is known in this area in relation to neuroblastoma disease. In a recent study, Das et al. hypothesized that DNA methylation could be responsible for the dysregulation of miRNA expression in neuroblastoma [54]. In-depth analysis of DNA methylation patterns together with miRNA and mRNA expression profiles in

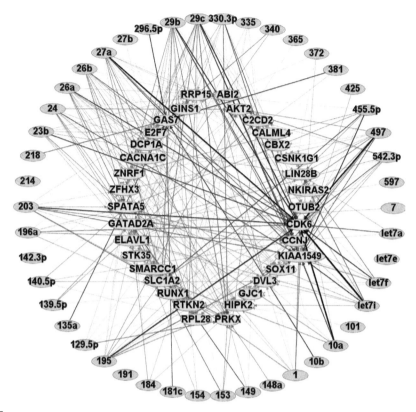

FIGURE 13.5

A radial bipartite graph showing mRNAs (inner circle) that are most redundantly targeted by the epigenetically silenced miRNAs (outer circle) and which show significant correlations with poor patient survival when upregulated in tumors as identified in the recent study [54]. Evidence for cooperativity and redundant targeting of a large set of mRNAs by the miRNA panel was detected. The links between the miRNAs and the mRNA targets are displayed, with the thickness of each line being proportional to both inverse expression correlation between the miRNA and mRNA and the over-representation of predicted miRNA binding sites. *This figure is adapted from the original figure published by Das et al. [54] and is reproduced here with permission of the authors and their agreement with the publisher.*

neuroblastoma samples resulted in the identification of a large set of epigenetically regulated miRNAs with significantly enriched target sites in the 3′-UTRs of genes overexpressed in unfavorable tumor subtypes. Remarkably, a high proportion of both the methylated miRNAs (42%) and their associated mRNA targets (56% of the highly redundantly targeted mRNAs) were highly associated with poor clinical outcome when under- and overexpressed in tumors, respectively. The list of potential epigenetically regulated miRNAs consisted of well-characterized tumor suppressor miRNAs in neuroblastoma such as some of the let-7 miRNAs, miR-29c, miR-101, miR-335, and miR-184. Importantly, many of the genes targeted by this miRNA panel are known to play oncogenic roles in neuroblastoma, such as AKT2, LIN28B, and CDK6, suggesting that epigenetic silencing of miRNAs could contribute to the overexpression of oncogenes in neuroblastoma (Figure 13.5).

The epigenetically modulated miRNAs target genes involved in cell differentiation. Differentiation is important in the treatment of some cancers. 13-*cis*-retinoic acid is currently used as part of the treatment regimen for high-risk neuroblastoma patients [55], and induces some neuroblastoma cell lines to differentiate, leading to profound changes in mRNA and miRNA expression [46]. Thus, the discovery and validation of specific genetic targets causing neuroblastoma cells to differentiate could be of potential therapeutic benefit. In a study by Das et al., DNA methylation changes were compared in SKNBE ATRA-treated versus untreated cells using methylated DNA immunoprecipitation applied to microarrays [56]. The authors identified a total of 402 gene promoters demethylated following all-*trans* retinoic acid (ATRA) treatment, while only 88 genes became hypermethylated. The demethylation events were explained in part by the downregulation of the methyltransferases DNMT1 and DNMT3 along with the upregulation of endogenous miRNAs targeting them, such as miR-152 and miR-26a/b.

A number of reports have validated individual miRNAs that play a major role in neuroblastoma cell differentiation, a highly complex and poorly understood process. Among them are miR-9 [57–59], miR-10ab [46,60], miR-103 [57,61], miR-107 [61], miR-124a [47], miR-125a [58], miR-125b [47,58], miR-128 [62], miR-18a/19a [63], and miR -184 [43]. Le et al. reported that ectopic overexpression of both miR-124a and miR-125b induces neurite outgrowth in SH-SY5Y cells [47]. Importantly, miR-125b directly bound and repressed the expression of 10 genes with seed matches in their 3′ UTRs. These genes are key contributors of neuronal differentiation. The targeting of these key genes by miR-125b launched a complex cascade of transcriptional alterations of downstream players for neuronal differentiation. Discovery of specific miRNA involved in differentiation is being recognized as a potential therapeutic effector of epigenetic therapeutics Figure 13.6).

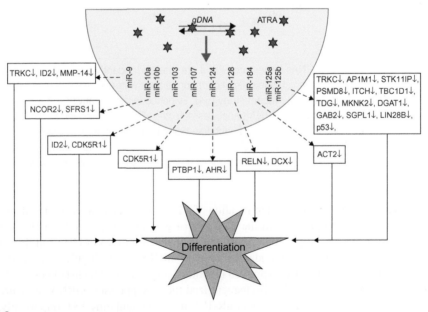

FIGURE 13.6

ATRA regulated miRNA in neuroblastoma.

2.5 LONG NONCODING RNAS

2.5.1 Long noncoding RNAs

Long noncoding RNAs (lncRNAs) comprise another class of regulatory ncRNAs. LncRNAs are commonly defined as 200–100,000 nt long mRNA-like transcripts lacking protein-coding features such as open-reading frames and exert their functional role as RNA transcripts (extensively reviewed in [64]). LncRNAs have altered expression in human cancers, including neuroblastoma. The first lncRNA expression profiling study of neuroblastoma tumors was published by Scaruffi et al. [65], demonstrating that the expression signature of 28 lncRNAs representing a subclass of transcribed ultraconserved regions (T-UCR) lncRNA is associated with survival in high risk, stage 4 neuroblastomas. Later, altered expression of seven T-UCRs was found in MYCN-amplified tumors in comparison with non-MYCN amplified [66]. Of these seven, three (uc.350, uc.379, and uc.460) were proven to be regulated by transcription factor MYCN. Linc00467 is another lncRNAs regulated by this transcription factor [67]. These data indicate that differential lncRNA expression correlates with clinic pathological parameters such as MYCN amplification and can be used for patient stratification.

Differentially expressed in neuroblastoma (DEIN) noncoding transcript (mapping to 4q33–34) is highly expressed in stage 4S tumors compared with localized stage 1, 3, and 4 tumors [68]. Interestingly, DEIN expression displayed no significant association with the common neuroblastoma aberrations of MYCN amplification, 1p/3p/11q deletion or 17q gain, but was found to be significantly associated with event-free survival, although not an independent prognostic marker.

Genetic aberrations such as chromosomal gains and losses can contribute to the over- or underexpression, respectively, of transcripts encoded in these regions. The gain of chromosome arm 17q is a predominant unfavorable prognostic factor in neuroblastoma [2]. The overexpression of 2.3 kb RNA ncRAN (noncoding RNA expressed in aggressive neuroblastoma) mapped to 17q is present in neuroblastomas with partial gain of 17q, but interestingly not present in those with whole chromosome 17 gain [69]. The phenomenon could be partially explained by genomic imprinting when the chromosome gain/amplification happens in a parent-of-origin-dependent manner. The oncogenic potential of this transcript was demonstrated by siRNA knockdown of ncRAN in SH-SY5Y neuroblastoma cells, and overexpression in NIH3T3 mouse fibroblast cells, resulting in significantly inhibited cell growth and an increase in anchorage-dependent cell growth, respectively [69].

2.5.2 Epigenetic control of lncRNA expression

So far little is known about epigenetic modulation of lncRNA expression. Comparing mechanisms of the regulation of miRNA expression, one would expect to see similar regulatory mechanisms involved in lncRNA expression as well. Recent studies demonstrated that ATRA induces profound changes in mRNA and lncRNA expression in some neuroblastoma cell lines [70,71]. In a study by Watters et al. [70], a transcribed ultra-conserved lncRNA, T-UC300A, was determined to play a role in proliferation, invasion, and inhibition of differentiation in neuroblastoma cell lines prior to ATRA treatment. Remarkably, another study discriminated 19 lncRNA classes and evaluated expression dynamics for each individual class during 120 hours after a single stimulation by retinoic acid. The study proposed potential architecture-dependent regulatory mechanisms of co-expression of lncRNAs and neighboring proteins [71]. The growing evidence suggests that these lncRNAs can play important roles in both normal and pathological physiology and represent a challenging new area of research in both normal development and disease.

Together, these data highlight the fact that miRNAs and lncRNA are an important part of cellular processes. Importantly, miRNA and lncRNA expression patterns correlate with patient outcome, hence making them attractive candidates for prognostic biomarkers for neuroblastoma. There is significant experimental evidence that dysregulation of tumor suppressor miRNA expression in neuroblastoma cells leads to loss of control of normal gene function. Investigations of individual miRNA contributions to a differentiated phenotype may have a great impact for novel therapies and revision of existed in light of clinical development of epigenetic therapeutic agents. The deep understanding of the molecular mechanisms regulating lncRNA expression is required, not only to better understand the role lncRNA's play in the development of the disease, but also may help us to identify new therapeutic targets.

3 EPIGENETIC TARGETING AGENTS

Mechanisms of epigenetic regulation most commonly include DNA methylation, modification of histones, and changes in miRNA expression. Epigenetic changes are reversible; hence the epigenetic status of the cancer genome can be restored to a nonmalignant state through epigenetic reprogramming through pharmacological manipulations resulting in tumor growth termination and elimination.

3.1 DNA METHYLATION INHIBITOR

Decitabine (5-aza-2-deoxycytidine) is a cytosine analogue that inhibits DNA methyltransferases, reverses methylation, and can reactivate silenced genes [72,73]. It has shown therapeutic activity in patients with different cancers (reviewed by [72]).

A Phase I clinical trial was conducted through The Children's Oncology Group (COG) in which decitabine (5-aza-2 = -deoxycytidine), was given, together with doxorubicin and cyclophosphamide, to children with relapsed/refractory solid tumors [74]. Low-dose decitabine in combination with doxorubicin/cyclophosphamide displayed tolerable toxicity in children. Unfortunately, doses of decitabine capable of producing therapeutical effects were not well tolerated with this combination. Alternative combinatory strategies of demethylating agents with non-cytotoxic, biologically targeted agents such as histone deacetylase inhibitors should attract more attention. At present, decitabine is being tested in clinical trials for neuroblastoma followed by cancer-antigen vaccination.

3.2 INHIBITORS OF HISTONE MODIFICATION ENZYMES

Histones provide a compact structure of DNA. These proteins together with the DNA form the major components of chromatin. Changes in histones can affect chromatin structure. They can be modified by methylation and acetylation, mediated by histone acetyl transferases and deacetylases, and histone methyltransferases. Each of these processes alters histone function, which, in turn alters the structure of chromatin and, therefore, the accessibility of DNA to transcription factors.

Histone deacetylase inhibitors (HDACIs) inhibit the growth of neuroblastoma *in vitro* [75] and *in vivo* [76,77].

Vorinostat (suberoylanilide hydroxamic acid or SAHA) is a HDAIC (reviewed in Richon, 2006). Vorinostat inhibits deacetylation of HDACs through binding to its active site and leads to an accumulation of hyperacetylated histones and transcription factors which, in turn results in activation of expression of cyclin-dependent kinase p21, followed by G1 cell–cycle arrest (thereby effecting cell-cycle arrest) and/or apoptosis. Hyperacetylation of tumor suppressor p53, alpha tubulin, and heat shock protein 90 produces additional anti-proliferative effects. There are several clinical trials on vorinostat in neuroblastoma summarized in Table 13.1.

HDACIs have a great potential to work synergistically with ATRA [76] or radiation [77] treatment. The combination administration of low dose ATRA with HDACI (100 mg/kg) resulted in growth inhibition comparable with that achieved with 200 mg/kg CBHA on human neuroblastoma xenografts of the SMS-KCN-69n cell line in mice. Synergy between ATRA and HDACIs was confirmed by combining non-inhibitory doses of CBHA (50 mg/kg) and ATRA (2.5 mg/kg) and achieving a statistically significant 52% reduction in final tumor size compared with control [76].

A recent preclinical study suggests that vorinistat can also function as a radiosensitizer in neuroblastoma [77]. The study demonstrated that combination of radiation and vorinostat significantly increased γ-H2AX expression, resulting in decreased tumor size compared to single modality alone in a murine metastatic neuroblastoma model. Thus, vorinostat potentiates anti-neoplastic effects of radiation in neuroblastoma possibly due to inhibitory effects on DNA repair enzymes. The COG is currently testing the safety and efficacy of vorinostat with isotretinoin for the treatment of high-risk refractory or recurrent neuroblastoma. Vorinostat is being assessed in combination with other agents including MBIG therapy, etoposide, and 13-*cis*-retinoic acid (isotretinoin). The favorable toxicity profiles reported in these ongoing trials make HDAC inhibitors promising agents for cancer therapy in children.

3.3 DIFFERENTIATION THERAPEUTICS

Epigenetic changes trigger the differentiation process during embryo development. Some cancer cells undergo differentiation under such stimuli like ATRA. Differentiation agents reprogram cancer cells toward a nonmalignant phenotype, and consequently may suppress tumor growth and development. The process of differentiation is particularly important in neuroblastoma tumors, as patients with more differentiated tumors have a better clinical outcome [78].

ATRA is an active metabolite of vitamin A under the family retinoid [79]. Emerging functional studies identified that 13-*cis*-retinoic acid inhibits the proliferation and induces markers of apoptosis and differentiation in neuroblastoma cell lines [47,57–59,61,62,80]. ATRA has been shown to induce differentiation and has anticancer activities in a number of other types of cancer cells *in vitro* and in *ex vivo* [81].

13-*cis*-retinoic acid is a recognized component of the treatment of high-risk neuroblastoma, regardless of discouraging results of early Phase II trials conducted with low-dose 13-*cis*-retinoic acid. This trial showed limited clinical benefit in patients with recurrent disease [82,83]. However, more recent randomized trials demonstrated improved survival in patients treated with high dose, pulsed 13-*cis*-retinoic acid after myeloablative chemotherapy [84,85]. Other strategies are currently under development to achieve pharmacological efficacy of retinoid therapy. Hence, retinoid therapy may be more effective in combination with other pharmacologic options, such as HDAC inhibitors [76,86].

Synthetic atypical retinoids are an attractive alternative, as their mechanism of action is different from the classic retinoids, and might be effective in treating retinoic acid–resistant neuroblastomas [87].

Neuroblastoma patients, however, are not universally responsive to retinoic acid treatment due to various reasons. To achieve pharmacologically efficacious 13-*cis*-retinoic acid levels, it is important to administrate adequate dose and optimal schedules due to different individual metabolisms or/and acquired resistance mechanisms [88].

Tumor genotype can critically contribute to sensitivity to ATRA. Mutations in the neurofibromatosis type 1 (NF1) tumor suppressor gene were determined in neuroblastoma cell lines [89] and in primary tumors [90,91] and affect response on ATRA treatment in neuroblastoma patients [92]. Loss of NF1 activates signaling in the RAS–MEK pathway, resulting in downregulation of ZNF423, which is a critical transcriptional co-activator of the retinoic acid receptors [93,92]. Inhibition of MEK signaling downstream of NF1 restores sensitivity to retinoic acid. This may represent an alternative approach for the treatment protocol to overcome RA resistance in NF1-deficient neuroblastomas.

There is a remarkable potential in a combinational therapy, as it can reduce toxicity of single agent therapy. Activation of epigenetically silenced miRNA through demethylation or differentiation agents may provide increased therapeutic efficiency in the management of neuroblastoma. Importantly, these agents efficiently modulate miRNA reexpression at low doses [54], and as such make them less toxic and easily tolerated while achieving therapeutic efficacy. This advantage makes this area extremely important in neuroblastoma treatment development, and management.

4 miRNA-BASED THERAPEUTICS

The widespread involvement of miRNAs in various human pathogenic diseases opens a new avenue for the study and development of new therapeutic strategies. The majority of small molecule inhibitor therapeutics target a single oncogene resulting in a modest induction of therapeutic response. In contrast, a single miRNA is able to repress many oncogenes simultaneously across various cell pathways, hence providing a strong rationale for developing miRNA-based cancer therapeutics. The principles for developing miRNA-based therapies are exactly the same as for other targeted therapies that take the path from drug target to drug. Target identification and validation are key steps in the selection of potential miRNAs. Importantly, despite the recent discovery of miRNAs, several candidates have already progressed into product and clinical development. Currently, there are several companies developing miRNA therapeutics for different human pathological conditions, including cancer (extensively reviewed in [94,95]).

There are two tactics to developing miRNA-based therapeutics: miRNA antagonists and miRNA mimics. miRNA antagonists aim to inhibit endogenous miRNAs showing a gain-of-function in diseased tissues; while miRNA mimics aim to restore a loss of function observed in healthy cells. The first approach is also known as "miRNA-knockdown therapy," the second—"miRNA replacement therapy." miRNAs under investigation are those that not only yield satisfactory efficacy in various disease models, but also sufficient experimental data that allow a precise placement of the miRNA into disease-related pathways. Because the miRNA mimic therapeutics is in the pipeline for anticancer therapy rather than anti-miRNA based, we will mainly focus on miRNA mimics and briefly outline prospective anti-miRNA candidates in neuroblastoma.

4.1 miRNA REPLACEMENT THERAPY

The reintroduction of miRNA mimics is expected to restore pathways required for normal cellular function and blocks those that drive the disease. miRNA mimics are ideally to be well tolerated in normal tissues. miRNA mimics are designed and synthesized to have the same sequence as the naturally occurring equivalent, so are expected to target the same set of genes. Since most normal cells already express the miRNA with tumor suppressor function, administration of miRNA mimics to normal tissue is unlikely to induce adverse events as the cellular pathways affected by the mimic are already activated or inactivated by the endogenous miRNA. A subset of the miRNAs which have shown therapeutic promise in neuroblastoma is outlined below and schematized on Figure 13.4.

Proof of concept for miRNA replacement therapy has been demonstrated by mimics of tumor suppressor miRNAs that stimulate anti-oncogenic pathways, apoptosis and eventually lead to an abolition of tumor cells [24,96–99]. To date most targeted cancer therapeutics tackle a gain-of-function, so miRNA replacement therapy provides a new exciting opportunity to exploit tumor suppressors. At present, the most advanced candidates for cancer are mimics of the let-7 and miR-34 miRNAs that target a broad spectrum of solid tumors.

A strategic target downregulated by let-7 is KRAS [100], an oncogene that is frequently mutated in different cancer types, and has escaped many previous attempts for therapeutic intervention [101]. In neuroblastoma, let-7 family directly targets *MYCN*, *DKK3*, and its re-expression is anti-proliferative [42,102]. Therapeutic delivery of let-7 causes an inhibition of tumor growth in human nonsmall cell lung cancer xenografts and the KRAS-G12D transgenic mouse model [98].

The tumor suppressor function of miR-34a, which is lost or expressed at reduced levels in most solid and hematologic malignancies, has been validated by numerous *in vitro* and *in vivo* studies. It functions within the p53 pathway and inhibits cancer cell growth by repressing *MYC/MYCN*, *MET*, *BCL2*, and other oncogenes. Therapeutic delivery of an miR-34 mimic blocked tumor growth in murine models of lung [99], prostate cancer [41], and neuroblastoma [24]. miR-34a inhibits prostate cancer stem cells and metastasis by directly repressing *CD44* [41]. The therapeutic efficacy is anti-proliferative and pro-apoptotic in lung tumor cells with a specific repression of *CDK4*, *MET*, and *BCL2* [99]. In addition, therapeutic delivery of miR-34 did not induce an elevation of cytokines or liver and kidney enzymes in serum, suggesting that treatment is well tolerated and the anti-oncogenic effects are mediated by a specific mechanism of miR-34.

Recently, tumor suppressor potential of miR-34a was successfully explored in a mouse xenograft model of neuroblastoma [24]. The miR-34a delivery system consisted of silica nanoparticles conjugated to a disialoganglioside GD2 (GD2) antibody. Neuroblastoma tumors express high levels of the cell surface antigen GD2, providing a target for tumor-specific delivery. Thus, targeted delivery is to provide a higher concentration of miR-34a to the tumor site, potentially increasing the efficacy of this mode of treatment. Encapsulated miRNA was administered to a well characterized murine orthotopic xenograft disease model. The targeted delivery of tumor suppressor miR-34a significantly inhibited tumor growth (Figure 13.7). Immunohistochemical staining of tumors treated with miR-34a revealed that multiple mechanisms, including increased apoptosis, and decreased angiogenesis, were responsible for the anti-tumorigenic effects of this miRNA. This study demonstrates the viability of a targeted delivery miRNA-based therapy for neuroblastoma.

Commercial MRX34 is one of the first clinical miRNA replacement therapeutics [103]. The therapeutic activity of MRX34 has been tested in a survival study using an orthotopic mouse model of hepatocellular carcinoma. Liposomes bearing mimics of miR-34 were administered into mice with pre-developed human tumors by intravenous tail vein injections. This study has demonstrated that all animals treated with MRX34 stayed alive in comparison with negative controls. Importantly, the histopathologic and molecular analyses of these animals did not find any evidence of remaining tumor cells. In autumn 2013, MRX34 therapeutic was transferred to the Phase I of clinical trials for patients with primary liver cancer or those with liver metastasis from other cancers (clinical trial identifier, NCT01829971).

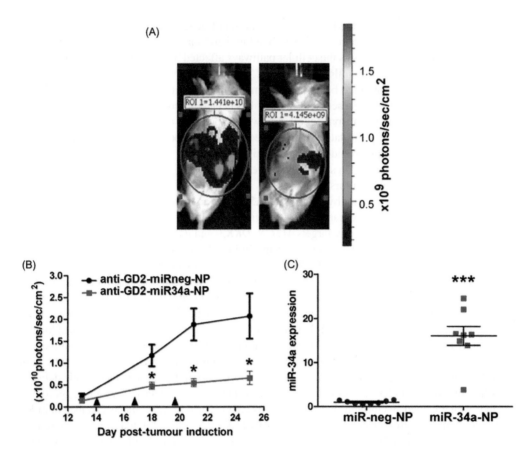

FIGURE 13.7

Anti-neuroblastoma effect of anti-GD2 conjugated nanoparticles bearing miR-34a *in vivo*. (A) Bioluminescent images representative of mice bearing NB1691luc tumors treated with anti-GD2-miRneg-NP (left) or anti-GD2-miR34a-NP (right). (B) Tumor growth curves from mice bearing NB1691luc tumors treated with anti-GD2-miRneg-NP (black line) or anti-GD2-miR34a-NP (red line). Time points for systemic administration of nanoparticles are indicated by the symbol ▲. Differences in tumor growth between mice injected with anti-GD2-miR34a-NP versus anti-GD2-miRneg-NP were statistically significant for both models. (C) Mature miR-34a transcript levels were significantly higher in anti-GD2-miR34a-NP-treated tumors relative to anti-GD2-miRneg-NP-treated control tumors in both. *This figure is modified from the original figure published by Tivnan et al. [24] and is reproduced here with permission of the authors and their agreement with the publisher.*

4.2 miRNA KNOCKDOWN THERAPY

MiRNA antagonists are synthetic molecules that inhibit endogenous miRNAs with a gain-of-function in diseased tissues. This concept is similar to other inhibitory therapeutics that target a single gene product such as small molecule inhibitors and short interfering RNAs (siRNAs). The inhibitory synthetic molecule is a chemically modified miRNA passenger strand (anti-miR or antagomiR) that binds with high affinity to the active miRNA strand. This binding is irreversible, so the new miRNA duplex is unable to be processed by RISC and/or degraded. As a main concern, the antagonist can potentially non-specifically bind to other RNAs, which could result in unwanted side effects. To date, there is no miRNA antagonist-based therapeutic in clinical trials for cancer conditions, while reviewed for other human pathological conditions elsewhere (reviewed in [95]). Some miRNAs of the miR-17-5p-92 polycistronic cluster are the perspective candidates for miRNA knockdown/antagonist in various types of cancer [104], including neuroblastoma [25]. Most impressively, antagonization of miR-17-5p significantly inhibits tumor growth in a mouse xenograft model of the disease, illustrating the potential for miRNA antagonist mediated therapy of neuroblastoma [25,26].

miRNA functional studies suggest the enormous potential of miRNA based therapy. Through in-depth validation of a single miRNA targets, it is feasible to select candidates that simultaneously restore non-malignant cell function through direct targeting of oncogenes. Advancing delivery of these miRNA-based therapeutics in animal models will facilitate clinical trials for miRNA therapeutics for treatment of neuroblastoma in the near future.

REFERENCES

[1] Maris JM, Hogarty MD, Bagatell R, Cohn SL. Neuroblastoma. Lancet 2007:2106–20.

[2] Davidoff A.M. Neuroblastoma. Semin Pediatr Surg [Internet]. 2012 Feb [cited 2014 Jan 21];21(1):2–14.

[3] Domingo-Fernandez R, Watters K, Piskareva O, Stallings RL, Bray I. The role of genetic and epigenetic alterations in neuroblastoma disease pathogenesis. Pediatr Surg Int [Internet] 2013;29(2):101–19.

[4] The William Guy Forbeck Research Foundation: The Neuroblastoma Research Landscape. 2012. Available from: <http://wgfrf.org/the-neuroblastoma-research-landscape-2012>.

[5] Decock A, Ongenaert M, Vandesompele J, Speleman F. Neuroblastoma epigenetics: from candidate gene approaches to genome-wide screenings. Epigenetics [Internet]. 2011 Aug [cited 2014 Feb 12];6(8):962–70.

[6] Teitz T, Wei T, Valentine MB, Vanin EF, Grenet J, Valentine VA, et al. Caspase 8 is deleted or silenced preferentially in childhood neuroblastomas with amplification of MYCN. Nat Med [Internet]. 2000 May;6(5):529–35.

[7] Astuti D, Agathanggelou A, Honorio S, Dallol A, Martinsson T, Kogner P, et al. RASSF1A promoter region CpG island hypermethylation in phaeochromocytomas and neuroblastoma tumours. Oncogene [Internet]. 2001 Nov 8;20(51):7573–7.

[8] Yang Q, Zage P, Kagan D, Tian Y, Seshadri R, Salwen HR, et al. Association of epigenetic inactivation of RASSF1A with poor outcome in human neuroblastoma association of epigenetic inactivation of RASSF1A with poor outcome in human neuroblastoma. Clin Cancer Res 2004;8493–500.

[9] Lázcoz P, Muñoz J, Nistal M, Pestaña A, Encío I, Castresana J.S. Frequent promoter hypermethylation of RASSF1A and CASP8 in neuroblastoma. BMC Cancer [Internet]. 2006 Jan [cited 2014 Feb 18];6:254.

[10] Yang Q, Kiernan CM, Tian Y, Salwen HR, Chlenski A, Brumback BA, et al. Methylation of CASP8, DCR2, and HIN-1 in neuroblastoma is associated with poor outcome. Clin Cancer Res 2007;13:3191–7.

[11] Van Noesel M.M, van Bezouw S, Voûte P.A, Herman J.G, Pieters R, Versteeg R. Clustering of hyper-methylated genes in neuroblastoma. Genes Chromosomes Cancer [Internet]. 2003 Nov [cited 2014 Feb 13];38(3):226–33.

[12] Hughes L.A, Melotte V, de Schrijver J, de Maat M, Smit V.T, Bovée J.V, et al. The CpG island methylator phenotype: what's in a name? Cancer Res [Internet]. 2013 Oct 1 [cited 2014 Jan 29];73(19):5858–68.

[13] Buckley P.G, Das S, Bryan K, Watters K.M, Alcock L, Koster J, et al. Genome-wide DNA methylation analysis of neuroblastic tumors reveals clinically relevant epigenetic events and large-scale epigenomic alterations localized to telomeric regions. Int J Cancer [Internet]. 2011 May 15 [cited 2014 Feb 13];128(10):2296–305.

[14] Sugino Y, Misawa A, Inoue J, Kitagawa M, Hosoi H, Sugimoto T, et al. Epigenetic silencing of prostaglandin E receptor 2 (PTGER2) is associated with progression of neuroblastomas. Oncogene [Internet]. 2007 Nov 22 [cited 2014 Feb 18];26(53):7401–13.

[15] Berdasco M, Ropero S, Setien F, Fraga M.F, Lapunzina P, Losson R, et al. Epigenetic inactivation of the Sotos overgrowth syndrome gene histone methyltransferase NSD1 in human neuroblastoma and glioma. Proc Natl Acad Sci USA [Internet]. 2009 Dec 22;106(51):21830–5.

[16] Witt O, Deubzer HE, Lodrini M, Milde T, Oehme I. Targeting histone deacetylases in neuroblastoma. Curr Pharm Des 2009;15:436–47.

[17] Oehme I, Deubzer HE, Wegener D, Pickert D, Linke J-P, Hero B, et al. Histone deacetylase 8 in neuroblastoma tumorigenesis. Clin Cancer Res 2009;15:91–9.

[18] Sun Y, Liu P.Y, Scarlett CJ, Malyukova A, Liu B, Marshall G.M, et al. Histone deacetylase 5 blocks neuroblastoma cell differentiation by interacting with N-Myc. Oncogene [Internet]. 2013;1–8.

[19] Frumm SM, Fan ZP, Ross KN, Duvall JR, Gupta S, Verplank L, et al. Selective HDAC1/HDAC2 inhibitors induce neuroblastoma differentiation. Chem Biol 2013;20:713–25.

[20] Lodrini M, Oehme I, Schroeder C, Milde T, Schier MC, Kopp-Schneider A, et al. MYCN and HDAC2 cooperate to repress miR-183 signaling in neuroblastoma. Nucleic Acids Res [Internet]. 2013;41:6018–33.

[21] Keshelava N, Davicioni E, Wan Z, Ji L, Sposto R, Triche TJ, et al. Histone deacetylase 1 gene expression and sensitization of multidrug-resistant neuroblastoma cell lines to cytotoxic agents by depsipeptide. J Natl Cancer Inst 2007;99:1107–19.

[22] Croce CM. Causes and consequences of microRNA dysregulation in cancer. Nat Rev Genet [Internet]. Nature Publishing Group; 2009 Oct [cited 2014 Jan 22];10(10):704–14.

[23] Lu J, Getz G, Miska EA, Alvarez-Saavedra E, Lamb J, Peck D, et al. MicroRNA expression profiles classify human cancers. Nature [Internet]. 2005 Jun 9 [cited 2014 Jan 21];435(7043):834–8.

[24] Tivnan A, Orr WS, Gubala V, Nooney R, Williams DE, McDonagh C, et al. Inhibition of neuroblastoma tumor growth by targeted delivery of microRNA-34a using anti-disialoganglioside GD2 coated nanoparticles. PLoS One 2012:e38129.

[25] Fontana L, Fiori M.E, Albini S, Cifaldi L, Giovinazzi S, Forloni M, et al. Antagomir-17-5p abolishes the growth of therapy-resistant neuroblastoma through p21 and BIM. PLoS One [Internet]. 2008 Jan [cited 2014 Jan 22];3(5):e2236.

[26] Chayka O, Corvetta D, Dews M, Caccamo A.E, Piotrowska I, Santilli G, et al. Clusterin, a haploinsufficient tumor suppressor gene in neuroblastomas. J Natl Cancer Inst [Internet]. 2009 May 6 [cited 2014 Feb 18];101(9):663–77.

[27] Chen Y, Stallings R.L. Differential patterns of microRNA expression in neuroblastoma are correlated with prognosis, differentiation, and apoptosis. Cancer Res [Internet]. 2007 Feb 1 [cited 2014 Feb 13];67(3):976–83.

[28] Bray I, Bryan K, Prenter S, Buckley P.G, Foley N.H, Murphy D.M, et al. Widespread dysregulation of MiRNAs by MYCN amplification and chromosomal imbalances in neuroblastoma: association of miRNA expression with survival. PLoS One [Internet]. 2009 Jan [cited 2014 Feb 13];4(11):e7850.

[29] Murphy D.M, Buckley P.G, Bryan K, Das S, Alcock L, Foley N.H, et al. Global MYCN transcription factor binding analysis in neuroblastoma reveals association with distinct E-box motifs and regions of DNA hypermethylation. PLoS One [Internet]. 2009 Jan [cited 2014 Feb 11];4(12):e8154.

[30] Mestdagh P, Fredlund E, Pattyn F, Schulte J.H, Muth D, Vermeulen J, et al. MYCN/c-MYC-induced microRNAs repress coding gene networks associated with poor outcome in MYCN/c-MYC-activated tumors. Oncogene [Internet]. Nature Publishing Group; 2010 Mar 4 [cited 2014 Feb 18];29(9):1394–404.

[31] Schulte JH, Horn S, Otto T, Samans B, Heukamp L.C, Eilers U-C, et al. MYCN regulates oncogenic MicroRNAs in neuroblastoma. Int J Cancer [Internet]. 2008 Feb 1 [cited 2014 Feb 18];122(3):699–704.

[32] Buckley P.G, Alcock L, Bryan K, Bray I, Schulte J.H, Schramm A, et al. Chromosomal and microRNA expression patterns reveal biologically distinct subgroups of 11q- neuroblastoma. Clin Cancer Res [Internet]. 2010 Jun 1 [cited 2014 Feb 13];16(11):2971–8.

[33] Fischer M, Oberthuer A, Brors B, Kahlert Y, Skowron M, Voth H, et al. Differential expression of neuronal genes defines subtypes of disseminated neuroblastoma with favorable and unfavorable outcome. Clin Cancer Res [Internet]. 2006 Sep 1 [cited 2014 Feb 18];12(17):5118–28.

[34] Welch C, Chen Y, Stallings R.L. MicroRNA-34a functions as a potential tumor suppressor by inducing apoptosis in neuroblastoma cells. Oncogene [Internet]. 2007 Jul 26 [cited 2014 Jan 30];26(34):5017–22.

[35] Cole K.A, Attiyeh E.F, Mosse Y.P, Laquaglia M.J, Diskin S.J, Brodeur G.M, et al. A functional screen identifies miR-34a as a candidate neuroblastoma tumor suppressor gene. Mol Cancer Res [Internet]. 2008 May [cited 2014 Jan 27];6(5):735–42.

[36] Wei J.S, Song Y.K, Durinck S, Chen Q-R, Cheuk A.T, Tsang P, et al. The MYCN oncogene is a direct target of miR-34a. Oncogene [Internet]. 2008 Sep 4 [cited 2014 Feb 5];27(39):5204–13.

[37] Bommer G.T, Gerin I, Feng Y, Kaczorowski A.J, Kuick R, Love R.E, et al. p53-mediated activation of miRNA34 candidate tumor-suppressor genes. Curr Biol [Internet]. 2007 Aug 7 [cited 2014 Jan 27];17(15): 1298–307.

[38] Sun F, Fu H, Liu Q, Tie Y, Zhu J, Xing R, et al. Downregulation of CCND1 and CDK6 by miR-34a induces cell cycle arrest. FEBS Lett [Internet]. 2008 Apr 30 [cited 2014 Feb 18];582(10):1564–8.

[39] Pang R.T.K., Leung C.O.N, Ye T-M, Liu W, Chiu P.C.N, Lam K.K.W, et al. MicroRNA-34a suppresses invasion through downregulation of Notch1 and Jagged1 in cervical carcinoma and choriocarcinoma cells. Carcinogenesis [Internet]. 2010 Jun [cited 2014 Feb 18];31(6):1037–44.

[40] Cao W, Fan R, Wang L, Cheng S, Li H, Jiang J, et al. Expression and regulatory function of miRNA-34a in targeting survivin in gastric cancer cells. Tumour Biol [Internet]. 2013 Apr [cited 2014 Feb 5];34(2):963–71.

[41] Liu C, Kelnar K, Liu B, Chen X, Calhoun-Davis T, Li H, et al. The microRNA miR-34a inhibits prostate cancer stem cells and metastasis by directly repressing CD44. Nat Med [Internet]. Nature Publishing Group; 2011 Feb [cited 2014 Jan 24];17(2):211–15.

[42] Buechner J, Tømte E, Haug BH, Henriksen JR, Løkke C, Flægstad T, et al. Tumour-suppressor microRNAs let-7 and mir-101 target the proto-oncogene MYCN and inhibit cell proliferation in MYCN-amplified neuroblastoma. Br J Cancer [Internet]. 2011 Jul 12 [cited 2014 Feb 18];105(2):296–303.

[43] Foley N.H, Bray I.M, Tivnan A, Bryan K, Murphy D.M, Buckley P.G, et al. MicroRNA-184 inhibits neuroblastoma cell survival through targeting the serine/threonine kinase AKT2. Mol Cancer [Internet]. 2010 Jan;9:83.

[44] Lynch J, Fay J, Meehan M, Bryan K, Watters K.M, Murphy D.M, et al. MiRNA-335 suppresses neuroblastoma cell invasiveness by direct targeting of multiple genes from the non-canonical TGF-β signalling pathway. Carcinogenesis [Internet]. 2012 May [cited 2014 Feb 18];33(5):976–85.

[45] Bray I, Tivnan A, Bryan K, Foley N.H, Watters K.M, Tracey L, et al. MicroRNA-542-5p as a novel tumor suppressor in neuroblastoma. Cancer Lett [Internet]. 2011 Apr 1 [cited 2014 Feb 13];303(1):56–64.

[46] Foley N.H, Bray I, Watters K.M, Das S, Bryan K, Bernas T, et al. MicroRNAs 10a and 10b are potent inducers of neuroblastoma cell differentiation through targeting of nuclear receptor corepressor 2. Cell Death Differ [Internet]. Nature Publishing Group; 2011 Jul [cited 2014 Feb 13];18(7):1089–98.

[47] Le M.T.N, Xie H, Zhou B, Chia PH, Rizk P, Um M, et al. MicroRNA-125b promotes neuronal differentiation in human cells by repressing multiple targets. Mol Cell Biol [Internet]. 2009 Oct [cited 2014 Jan 21];29(19):5290–305.

[48] Wei JS, Johansson P, Chen Q-R, Song YK, Durinck S, Wen X, et al. microRNA profiling identifies cancer-specific and prognostic signatures in pediatric malignancies. Clin Cancer Res [Internet]. 2009 Sep 1 [cited 2014 Feb 18];15(17):5560–8.

[49] Ryan J., Tivnan A., Fay J., Bryan K., Meehan M., Creevey L., et al. MicroRNA-204 increases sensitivity of neuroblastoma cells to cisplatin and is associated with a favourable clinical outcome. Br J Cancer [Internet]. 2012 Sep 4 [cited 2014 Feb 18];107(6):967–76.

[50] Chen Y, Tsai YH, Fang Y, Tseng SH. Micro-RNA-21 regulates the sensitivity to cisplatin in human neuroblastoma cells. J Pediatr Surg 2012;47:1797–805.

[51] Afanasyeva E.A, Mestdagh P, Kumps C, Vandesompele J, Ehemann V, Theissen J, et al. MicroRNA miR-885-5p targets CDK2 and MCM5, activates p53 and inhibits proliferation and survival. Cell Death Differ [Internet]. Nature Publishing Group; 2011 Jun [cited 2014 Feb 18];18(6):974–84.

[52] Schulte JH, Marschall T, Martin M, Rosenstiel P, Mestdagh P, Schlierf S, et al. Deep sequencing reveals differential expression of microRNAs in favorable versus unfavorable neuroblastoma. Nucleic Acids Res [Internet]. 2010 Sep [cited 2014 Jan 21];38(17):5919–28.

[53] Kunej T, Godnic I, Ferdin J, Horvat S, Dovc P, Calin G.A. Epigenetic regulation of microRNAs in cancer: an integrated review of literature. Mutat Res [Internet]. Elsevier B.V.; 2011 Dec 1 [cited 2014 Feb 19];717 (1–2):77–84.

[54] Das S, Bryan K, Buckley PG, Piskareva O, Bray IM, Foley N, et al. Modulation of neuroblastoma disease pathogenesis by an extensive network of epigenetically regulated microRNAs. Oncogene 2012.

[55] Reynolds C.P, Matthay K.K, Villablanca J.G, Maurer B.J. Retinoid therapy of high-risk neuroblastoma. Cancer Lett [Internet]. 2003 Jul [cited 2014 Feb 19];197(1–2):185–92.

[56] Das S, Foley N, Bryan K, Watters K.M, Bray I, Murphy D.M, et al. MicroRNA mediates DNA demethylation events triggered by retinoic acid during neuroblastoma cell differentiation. Cancer Res [Internet]. 2010 Oct 15 [cited 2014 Feb 8];70(20):7874–81.

[57] Annibali D, Gioia U, Savino M, Laneve P, Caffarelli E, Nasi S. A new module in neural differentiation control: two microRNAs upregulated by retinoic acid, miR-9 and -103, target the differentiation inhibitor ID2. PLoS One [Internet]. 2012 Jan [cited 2014 Feb 19];7(7):e40269.

[58] Laneve P, Di Marcotullio L, Gioia U, Fiori M.E, Ferretti E, Gulino A, et al. The interplay between microR-NAs and the neurotrophin receptor tropomyosin-related kinase C controls proliferation of human neuroblastoma cells. Proc Natl Acad Sci USA [Internet]. 2007 May 8;104(19):7957–62.

[59] Zhang H, Qi M, Li S, Qi T, Mei H, Huang K, et al. microRNA-9 targets matrix metalloproteinase 14 to inhibit invasion, metastasis, and angiogenesis of neuroblastoma cells. Mol Cancer Ther [Internet]. 2012 Jul [cited 2014 Feb 18];11(7):1454–66.

[60] Meseguer S, Mudduluru G, Escamilla JM, Allgayer H, Barettino D. MicroRNAs-10a and -10b contribute to retinoic acid-induced differentiation of neuroblastoma cells and target the alternative splicing regulatory factor SFRS1 (SF2/ASF). J Biol Chem 2011;286:4150–64.

[61] Moncini S, Salvi A, Zuccotti P, Viero G, Quattrone A, Barlati S, et al. The role of miR-103 and miR-107 in regulation of CDK5R1 expression and in cellular migration. PLoS One [Internet]. 2011 Jan [cited 2014 Feb 19];6(5):e20038.

[62] Evangelisti C, Florian M.C, Massimi I, Dominici C, Giannini G, Galardi S, et al. MiR-128 up-regulation inhibits Reelin and DCX expression and reduces neuroblastoma cell motility and invasiveness. FASEB J [Internet]. 2009 Dec [cited 2014 Feb 19];23(12):4276–87.

[63] Lovén J, Zinin N, Wahlström T, Müller I, Brodin P, Fredlund E, et al. MYCN-regulated microRNAs repress estrogen receptor-alpha (ESR1) expression and neuronal differentiation in human neuroblastoma. Proc Natl Acad Sci USA [Internet]. 2010 Jan 26 [cited 2014 Feb 18];107(4):1553–8.

[64] Rinn JL, Chang HY. Genome regulation by long noncoding RNAs. Annu Rev Biochem 2012:145–66.

[65] Scaruffi P, Stigliani S, Moretti S, Coco S, De Vecchi C, Valdora F, et al. Transcribed-Ultra Conserved Region expression is associated with outcome in high-risk neuroblastoma. BMC Cancer 2009;9:441.

[66] Mestdagh P, Fredlund E, Pattyn F, Rihani A, Van Maerken T, Vermeulen J, et al. An integrative genomics screen uncovers ncRNA T-UCR functions in neuroblastoma tumours. Oncogene [Internet]. 2010 Jun 17 [cited 2014 Jan 27];29(24):3583–3592.

[67] Atmadibrata B, Liu P.Y, Sokolowski N, Zhang L, Wong M, Tee A.E, et al. The novel long noncoding RNA linc00467 promotes cell survival but is down-regulated by N-Myc. PLoS One [Internet]. 2014 Jan [cited 2014 Mar 25];9(2):e88112.

[68] Voth H, Oberthuer A, Simon T, Kahlert Y, Berthold F, Fischer M. Identification of DEIN, a novel gene with high expression levels in stage IVS neuroblastoma. Mol Cancer Res 2007;5:1276–84.

[69] Yu M, Ohira M, Li Y, Niizuma H, Oo ML, Zhu Y, et al. High expression of ncRAN, a novel non-coding RNA mapped to chromosome 17q25.1, is associated with poor prognosis in neuroblastoma. Int J Oncol 2009;34:931–8.

[70] Watters K.M, Bryan K, Foley N.H, Meehan M, Stallings R.L. Expressional alterations in functional ultra-conserved non-coding RNAs in response to all-trans retinoic acid--induced differentiation in neuroblastoma cells. BMC Cancer [Internet]. 2013 Jan;13:184.

[71] Batagov A.O, Yarmishyn A.A, Jenjaroenpun P, Tan J.Z, Nishida Y, Kurochkin I.V. Role of genomic architecture in the expression dynamics of long noncoding RNAs during differentiation of human neuroblastoma cells. BMC Syst Biol [Internet]. BioMed Central Ltd; 2013 Jan [cited 2014 Mar 25];7(Suppl. 3):S11.

[72] Oki Y, Aoki E, Issa J-P.J. Decitabine--bedside to bench. Crit Rev Oncol Hematol [Internet]. 2007 Feb [cited 2014 Feb 6];61(2):140–52.

[73] Stresemann C, Lyko F. Modes of action of the DNA methyltransferase inhibitors azacytidine and decitabine. Int J Cancer [Internet]. 2008 Jul 1 [cited 2014 Jan 26];123(1):8–13.

[74] George R.E, Lahti J.M, Adamson P.C, Zhu K, Finkelstein D, Ingle A.M, et al. Phase I study of decitabine with doxorubicin and cyclophosphamide in children with neuroblastoma and other solid tumors: a Children's Oncology Group Study. 2010;(March):629–38.

[75] Glick RD, Swendeman SL, Coffey DC, Rifkind RA, Marks PA, Richon VM, et al. Hybrid polar histone deacetylase inhibitor induces apoptosis and CD95/CD95 ligand expression in human neuroblastoma. Cancer Res 1999;59:4392–9.

[76] Coffey DC, Kutko MC, Glick RD, Butler LM, Heller G, Rifkind RA, et al. The histone deacetylase inhibitor, CBHA, inhibits growth of human neuroblastoma xenografts in vivo, alone and synergistically with all-trans retinoic acid. Cancer Res 2001;61:3591–4.

[77] Mueller S, Yang X, Sottero T.L, Gragg A, Prasad G, Polley M-Y, et al. Cooperation of the HDAC inhibitor vorinostat and radiation in metastatic neuroblastoma: efficacy and underlying mechanisms. Cancer Lett [Internet]. Elsevier Ireland Ltd; 2011 Jul 28 [cited 2014 Feb 19];306(2):223–9.

[78] Goto S, Umehara S, Gerbing R.B, Stram D.O,., Brodeur G.M, et al. Histopathology (International Neuroblastoma Pathology Classification) and MYCN status in patients with peripheral neuroblastic tumors a report from the Children's Cancer Group. Cancer [Internet]. 2001;92(10):2699–708.

[79] Lovat PE, Irving H, Annicchiarico-Petruzzelli M, Bernassola F, Malcolm AJ, Pearson ADJ, et al. Retinoids in neuroblastoma therapy: distinct biological properties of 9-cis-and all-trans-retinoic acid. Eur J Cancer 1997:2075–80.

[80] Foley N.H, Bray I, Watters K.M, Das S, Bryan K, Bernas T, et al. MicroRNAs 10a and 10b are potent inducers of neuroblastoma cell differentiation through targeting of nuclear receptor corepressor 2. Cell Death Differ [Internet]. Nature Publishing Group; 2011 Jul [cited 2014 Feb 13];18(7):1089–98.

[81] Reynolds CP, Lemons RS. Retinoid therapy of childhood cancer. Hematol Oncol Clin North Am 2001;15:867–910.

[82] Adamson P.C, Matthay K.K, Brien M.O, Reaman G.H, Sato J.K, Balis F.M. A phase 2 trial of all-trans-retinoic acid in combination with interferon-alpha2a in children with recurrent neuroblastoma or Wilms tumor: A Pediatric Oncology Branch, NCI and Children's Oncology Group Study. 2007;(May 2006):661–5.

[83] Finklestein JZ, Krailo MD, Lenarsky C, Ladisch S, Blair GK, Reynolds CP, et al. 13-cis-retinoic acid (NSC 122758) in the treatment of children with metastatic neuroblastoma unresponsive to conventional chemotherapy: report from the Childrens Cancer Study Group. Med Pediatr Oncol 1992;20:307–11.

[84] Matthay K.K, Reynolds C.P, Seeger R.C, Shimada H, Adkins E.S, Haas-Kogan D, et al. Long-term results for children with high-risk neuroblastoma treated on a randomized trial of myeloablative therapy followed by 13-cis-retinoic acid: a children's oncology group study. J Clin Oncol [Internet]. 2009 Mar 1 [cited 2014 Feb 19];27(7):1007–13.

[85] Park J.R, Villablanca J.G, London W.B, Gerbing R.B, Haas-kogan D, Adkins E.S, et al. Outcome of high-risk stage 3 neuroblastoma with myeloablative therapy and 13-cis-retinoic acid: a report from the Children's Oncology Group. 2009; (August 2008):44–50.

[86] Hahn CK, Ross KN, Warrington IM, Mazitschek R, Kanegai CM, Wright RD, et al. Expression-based screening identifies the combination of histone deacetylase inhibitors and retinoids for neuroblastoma differentiation. Proc Natl Acad Sci USA 2008;105:9751–6.

[87] Villablanca J.G, Krailo M.D, Ames M.M, Reid J.M, Reaman G.H, Reynolds C.P, et al. Phase I trial of oral fenretinide in children with high-risk solid tumors: a report from the Children's Oncology Group (CCG 09709). J Clin Oncol [Internet]. 2006 Jul 20 [cited 2014 Feb 19];24(21):3423–30.

[88] Veal G.J, Cole M, Errington J, Pearson A.D, Foot A.B, Whyman G, et al. Pharmacokinetics and metabolism of 13-cis-retinoic acid (isotretinoin) in children with high-risk neuroblastoma-a study of the United Kingdom Children's Cancer Study Group. Br J Cancer [Internet]. 2007 Feb 12 [cited 2014 Feb 19];96(3):424–31.

[89] The I, Murthy A.E, Hannigan G.E, Jacoby L.B, Menon A.G, Gusella J.F, et al. Neurofibromatosis type 1 gene mutations in neuroblastoma. Nat Genet [Internet]. 1993;3:62–6.

[90] Legius E, Marchuk D, Collins F, Glover T. Somatic deletion of the neurofibromatosis type 1 gene in a neurofibrosarcoma supports a tumour suppressor gene hypothesis. Nat Genet [Internet]. 1993 [cited 2014 Feb 20];3:122–6.

[91] Origone P, Defferrari R, Mazzocco K, Lo Cunsolo C, De Bernardi B, Tonini GP. Homozygous inactivation of NF1 gene in a patient with familial NF1 and disseminated neuroblastoma. Am J Med Genet A [Internet]. 2003 May 1 [cited 2014 Feb 20];118A(4):309–13.

[92] Hölzel M, Huang S, Koster J, Øra I, Lakeman A, Caron H, et al. NF1 is a tumor suppressor in neuroblastoma that determines retinoic acid response and disease outcome. Cell 2010;142:218–29.

[93] Huang S, Laoukili J, Epping MT, Koster J, Hölzel M, Westerman BA, et al. ZNF423 is critically required for retinoic acid-induced differentiation and is a marker of neuroblastoma outcome. Cancer Cell 2009;15:328–40.

[94] Broderick J.A, Zamore P.D. MicroRNA therapeutics. Gene Ther [Internet]. Nature Publishing Group; 2011 Dec [cited 2014 Jan 28];18(12):1104–10.

[95] Van Rooij E, Purcell A.L, Levin A.A. Developing microRNA therapeutics. Circ Res [Internet]. 2012 Feb 3 [cited 2014 Jan 21];110(3):496–507.

[96] Kota J, Chivukula R.R, O'Donnell K.A, Wentzel E.A, Montgomery C.L, Hwang H-W, et al. Therapeutic microRNA delivery suppresses tumorigenesis in a murine liver cancer model. Cell [Internet]. Elsevier Ltd.; 2009 Jun 12 [cited 2014 Jan 25];137(6):1005–17.

[97] Takeshita F, Patrawala L, Osaki M, Takahashi R, Yamamoto Y, Kosaka N, et al. Systemic delivery of synthetic microRNA-16 inhibits the growth of metastatic prostate tumors via downregulation of multiple cell-cycle genes. Mol Ther 2010;18:181–7.

[98] Trang P, Medina PP, Wiggins JF, Ruffino L, Kelnar K, Omotola M, et al. Regression of murine lung tumors by the let-7 microRNA. Oncogene 2010;29:1580–7.

[99] Wiggins J.F, Ruffino L, Kelnar K, Omotola M, Patrawala L, Brown D, et al. Development of a lung cancer therapeutic based on the tumor suppressor microRNA-34. Cancer Res [Internet]. 2010 Jul 15 [cited 2014 Feb 3];70(14):5923–30.

[100] Johnson SM, Grosshans H, Shingara J, Byrom M, Jarvis R, Cheng A, et al. RAS is regulated by the let-7 microRNA family. Cell 2005;120:635–47.

[101] Chetty R, Govender D. Gene of the month: KRAS. J Clin Pathol [Internet]. 2013 Jul [cited 2014 Mar 3];66(7):548–50.

[102] Haug BH, Henriksen JR, Buechner J, Geerts D, Tømte E, Kogner P, et al. MYCN-regulated miRNA-92 inhibits secretion of the tumor suppressor DICKKOPF-3 (DKK3) in neuroblastoma. Carcinogenesis 2011;32:1005–12.

[103] Bader A.G. miR-34-a microRNA replacement therapy is headed to the clinic. Front Genet [Internet]. 2012 Jan [cited 2014 Feb 19];3(July):120.

[104] He L, Thomson JM, Hemann MT, Hernando-Monge E, Mu D, Goodson S, et al. A microRNA polycistron as a potential human oncogene. Nature 2005;435:828–33.

THE EPIGENETICS OF MEDULLOBLASTOMA

14

Clara Penas, Vasileios Stathias, Bryce K. Allen, and Nagi G. Ayad

Center for Therapeutic Innovation, The Miami Project to Cure Paralysis, Department of Psychiatry and Behavioral Sciences, University of Miami Miller School of Medicine, Miami, FL, USA

CHAPTER OUTLINE

1 INTRODUCTION

Medulloblastoma is the most common malignant pediatric brain tumor. Although surgical resection and radiotherapy are effective at eliminating some forms of medulloblastoma, more aggressive tumors do not respond to any treatment. In addition, patients who respond to classical treatment often suffer cognitive problems. Thus, new drugs need to be developed to effectively treat medulloblastoma. Many current drug discovery efforts are directed at identifying epigenetic enzymes dysregulated in various cancers including medulloblastoma. Since recent sequencing and expression studies have shown that various epigenetic enzymes are either mutated or differentially expressed in medulloblastoma, validating these targets in preclinical models has become especially important. After target validation, small molecules that bind to each epigenetic enzyme are tested/screened in the hopes of reducing tumor growth. Further refinement of lead compounds then ensures safety as well as neurodevelopmental tolerance.

S.G. Gray (Ed): Epigenetic Cancer Therapy. DOI: http://dx.doi.org/10.1016/B978-0-12-800206-3.00014-8

In addition to targeting epigenetic enzymes, many groups attempt to modulate the expression of oncogenes and tumor suppressor genes through microRNAs. microRNAs are small RNAs, which can regulate the expression of multiple genes at once, thus providing an opportunity to eliminate or at least reduce expression of several medulloblastoma oncogenes or drivers simultaneously. This is especially important since many studies have demonstrated resistance of medulloblastoma and other cancers to single target modulation. What this means is that inhibition of one target in medulloblastoma with a small molecule may select for tumor subpopulations that can evade growth control by bypassing the requirement for the small molecule target. These select tumor cells then proliferate and are resistant to the initial therapy. However, if multiple targets are modulated by the therapy, it becomes more difficult for the tumor cells to evade growth control. Thus, strategies to increase or decrease microRNAs that control multiple oncogenes or tumor suppressor proteins in medulloblastoma may be particularly attractive. Similar strategies that achieve inhibition of several pathways via a combination of small molecule epigenetic enzyme inhibitors and microRNA modulation may be equally beneficial in medulloblastoma treatment. We will highlight that understanding epigenetic enzymes–microRNA pathway modulation can yield effective combination therapies for medulloblastoma. We will also discuss how kinases and ubiquitin ligases interact with epigenetic pathways, thus providing possible combination therapies with kinases, ubiquitin ligases, and epigenetic enzymes. These studies are providing a systems level understanding required to design effective clinical trials for medulloblastoma treatment.

2 MEDULLOBLASTOMA SUBTYPES

Medulloblastoma includes four groups (Sonic Hedgehog or SHH, Wingless or WNT, Group 3 and Group 4), which show differences in karyotype, histology, and prognosis [1,2]. WNT tumors show activated WNT pathway signaling and respond to treatment [3]. SHH tumors show hedgehog pathway activation (Figure 14.1) and are associated with an intermediate prognosis [4]. Group 3 tumors have an amplification of the C-MYC oncogene and are associated with poor prognosis. Similarly, Group 4 tumors are molecularly less well characterized and also present great clinical challenges [4–6].

Although there are no conclusive human studies, data from mouse studies suggest that human medulloblastoma subgroups have different cells of origin. SHH-dependent medulloblastoma may have multiple cells of origin including cerebellar granule precursor cells, neural stem cells [7,8], and cochlear nuclei of the lower rhombic lip [9]. Group 3 medulloblastoma arise from CD133+ NSCs [10], WNT-driven tumors originate from progenitor cells of the dorsal brainstem [11], while the cell of origin for Group 4 tumors has yet to be determined.

Although human medulloblastoma has been grouped into four or more groups, large heterogeneity exists at the DNA level. A recent study highlighted that 77% of the mutated genes in medulloblastoma are unique to a single case, underscoring the genetic heterogeneity of this cancer [12]. Within the cohort of 125 patients studied, only 8 genes were somatically altered in more than 3% of all patients. These include CTNNB1 (catenin (cadherin-associated protein) beta 1, 12%), DDX3X (DEAD (Asp–Glu–Ala–Asp) box helicase 3 X-linked, 8%), PTCH1 (patched 1, 6%), SMARCA4 (SWI/SNF related, matrix-associated actin-dependent regulator of chromatin, subfamily A, 5%), MLL2 (5%), TP53 (tumor protein P53, 4%), KDM6A (lysine (K)-specific demethylase 6A, 4%), and CTDNEP1 (CTD nuclear envelope phosphatase 1, 3%). However, a commonly recurring theme across all medulloblastoma is alterations in genes involved in chromatin modification. For example, the same study described a significant number of alterations in several chromatin modifiers such as MLL2, SMARCA4, and

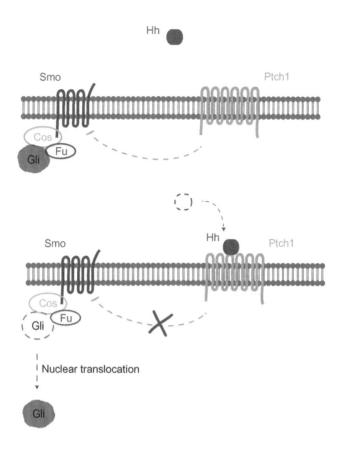

FIGURE 14.1

Components of the SHH pathway. In the absence of Hh ligand, Patched (Ptch) catalytically inhibits the seven-transmembrane receptor-like protein Smoothened (Smo), thus inhibiting its activation. However, Hh ligand binding to Ptch causes Ptch inactivation, disinhibiting Smo leading to the transduction of the Hh signal to the cytoplasm. Hh signaling then promotes the disassociation of Gli1 from scaffolding proteins Fused (Fu) and Costal (Cos) [3a,3b,3c]. Smo activation keeps Gli1 from being cleaved and translocating to the nucleus where it normally acts as a transcription factor stimulating expression of target genes involved in cell proliferation.

KDM6A. Therefore, this and other studies described herein underscore the importance of epigenetics research in medulloblastoma.

3 EPIGENETIC ENZYMES HDACs AND HATs AS TARGETS IN MEDULLOBLASTOMA

Before sequencing studies suggested that epigenetic enzymes and regulators might be attractive therapeutic targets in medulloblastoma, many studies focused on the role of histone deacetylases (HDACs) [13–21]. Histone acetyltransferases (HATs) attach acetyl groups to lysine residues on histone proteins, while HDACs remove those modifications. In mammals, HDACs include the conventional HDACs

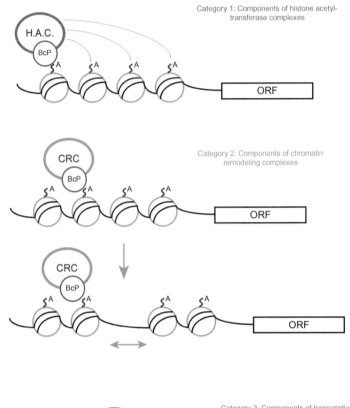

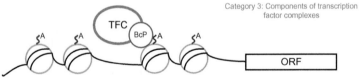

FIGURE 14.2

Complexes. **Category 1: Components of HAT complexes.** The BRD-containing protein (BcP) binds to acetylated H3 and anchors the histone acetyltransferase complex (HAC) to nucleosomes, allowing it to acetylate adjacent nucleosomes (3). **Category 2: Components of chromatin remodeling complexes.** The chromatin remodeling complex (CRC) binds to acetylated nucleosomes through BcP and remodels the promoter region to allow formation of the preinitiation complex and ultimately transcription (3). **Category 3: Components of transcription factor complexes.** BcP binds to an acetylated histone and recruits a transcription factor complex (TFC) (3).

(HDAC1–11) and sirturins (sirturin 1–7), which have other enzymatic functions. Histones normally bind DNA molecules via their positively charged lysine and arginine tails. Histone-DNA binding initiates DNA compaction and transcriptional silencing. Histone tail acetylation reduces this positive charge, attenuates DNA binding and compaction, and allows transcription (Figure 14.2). Histone deacetylation has the opposite effect, thus allowing histone-DNA binding and reducing transcription.

HDAC inhibitors counteract the activity of HDACs and allow transcription to occur. All classic HDACs [22] and some sirtuins are widely expressed in the vertebrate brain where they serve important functions during development and throughout adulthood. In addition, increased expression or activity of these enzymes in medulloblastoma potentiates tumor growth.

HDAC5 and HDAC9 are found significantly upregulated in high-risk medulloblastoma relative to low-risk medulloblastoma, and their expression is associated with poor survival [18]. HDAC5 is localized predominantly in the nucleus, whereas HDAC9 is mostly cytoplasmic in these patient samples. siRNA-mediated knockdown of HDAC5 or HDAC9 results in decreased medulloblastoma cell growth and viability. However, little is known about the physiologic function of HDAC5 and HDAC9 in normal cells and in development, although several studies suggest a role in differentiation [23–25] and cell cycle regulation. In fact, HDAC5 represses the transcription of cyclin D3, a known cell cycle potentiator [26].

HDAC1 and HDAC3 have been correlated with poor clinical outcome in gastrointestinal system cancers [27]. These HDACs have been found in medulloblastoma cells to be part of a repressive complex containing INSM1 (insulinoma-associated 1) along with cyclin D1 both *in vivo* and *in vitro* [28]. These complexes greatly enhance the repressive activity of INSM1 on the neuroD/b2 (neuronal differentiation factor/beta 2) promoter, a gene involved in differentiation [28].

Sirtuin 1 (SIRT1), a class III HDAC, regulates differentiation of neuronal stem cells. It is overexpressed in medulloblastoma samples, and its expression is correlated with human medulloblastoma formation [29]. Indeed, siRNA-mediated inhibition of SIRT1 expression induced G1 arrest and apoptosis in a dose-related fashion in a medulloblastoma cell line. Thus, SIRT1 may be a potential therapeutic target in medulloblastoma.

In mammals, 19 HATs have been described and are separated into five families (p300/CBP, p300/cAMP response element-binding binding protein; MYST, K(lysine) acetyltransferase 8; GNAT, glycine-*N*-Acyltransferase-Like 1; NCOA, nuclear receptor coactivator; and transcription-related HATs), which play a crucial role in brain development and disease [30]. By contrast to HDACs, the role of HATs in medulloblastoma is less studied, although there have been some notable advances. For instance, the HAT human MOF (hMOF, human K(Lysine) Acetyltransferase 8) is required for H4K16 acetylation and is downregulated in medulloblastoma compared to nontransformed control tissues. Downregulation of hMOF protein expression along with H4K16 acetylation is associated with lower survival rates identifying hMOF as an independent prognostic marker for clinical outcome [31].

HATs also regulate the SHH pathway, a main driver of medulloblastoma. The p300/CBP-associated factor, or PCAF (lysine (K)-Acetyltransferase 2B) acetyltransferase, is downregulated in medulloblastoma and glioblastoma (GBM) cells, leading to decreased proliferation and increased apoptosis. Indeed, PCAF interacts with Gli1, the downstream effector in the SHH pathway, and its loss reduces the levels of H3K9 acetylation on Hh target gene promoters [32]. Moreover, PCAF silencing reduces the tumor-forming potential of neural stem cells *in vivo* [32]. By contrast, the HAT coactivator p300 is a unique HAT that acetylates Gli2 at a conserved lysine K757, thereby inhibiting Hh target gene expression. Mechanistically, acetylation at K757 prevents Gli2–chromatin association [33]. Further, Gli2 acetylation is attenuated by Hh agonists, thus providing a therapeutic avenue for drug discovery.

4 HDAC INHIBITORS AS MEDULLOBLASTOMA THERAPEUTICS

The major classes of HDAC inhibitors include hydroxamic acid-based benzamide derivatives, cyclic peptides, and aliphatic acids. Hydroxamic acid-based pan-HDAC inhibitors include suberoylanilide

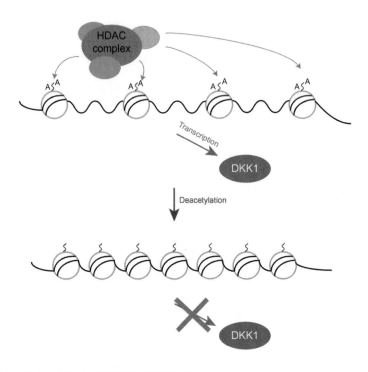

FIGURE 14.3

Histone acetylation regulates DKK1 expression in medulloblastoma. The HDAC deacetylates the histones producing the repression of DKK1 gene expression. In medulloblastoma cells, HDAC inhibition induces acetylation of histones and upregulation of DKK1 expression.

hydroxamic acid (SAHA), trichrostatin A (TSA), and PXD-101. The inhibitors MS-275 (HDAC 1/2/3) and MGCD0103 (class I) are synthetic benzamide derivatives. The synthetic peptide depsipeptide is a cyclic peptide product of prodrug type that inhibits HDAC1/2 selectively. Aliphatic acids include valproic acid and sodium phenylbutyrate, however, this class has less inhibitory potency on the class I/IIa HDACs [34].

Several HDAC inhibitors have been found to decrease medulloblastoma tumor growth. For instance, the HDAC inhibitor valproic acid induces expression of the cyclin-dependent kinase inhibitor $p21^{Cip1}$, which may be one of the mechanisms through which it induces senescence in medulloblastoma cell lines [35]. Similarly, the HDAC inhibitor sodium butyrate was shown to reduce growth of medulloblastoma cell lines and stem cells [36]. Other studies suggest that HDAC inhibitors reduce proliferation via inhibiting expression of repressor element 1-silencing transcription factor or REST [21]. Yet other studies suggest that the treatment of medulloblastoma cells with the HDAC inhibitor trichostatin A (TSA) increased the expression of the negative regulators of the WNT pathway, Dickkopf 1 (DKK1) (Figure 14.3) and DKK3 [37,38]. However, the likelihood that any one transcriptional target is responsible for the efficacy of HDAC inhibitors is low. It is more likely that there are multiple targets whose modulation after HDAC inhibition contributes to medulloblastoma cell death.

In addition to modifying histones, HDACs can also regulate other proteins such as tubulin. For instance, Lee et al. demonstrated that curcumin reduces medulloblastoma cell growth *in vitro* and *in vivo* via HDAC regulation and tubulin modification [17]. Lower *in vivo* growth was associated with a G_2/M arrest and reduced expression of HDAC4. Further, tubulin acetylation was increased after curcumin treatment. Thus, HDACs have multiple nonhistone targets such as tubulin, whose inhibition can be additive or synergistic with histone modification after HDAC inhibitor administration. Future studies may be directed at inhibiting nonhistone HDAC targets via HDAC inhibition and other strategies such as combining microtubule depolymerization inhibitors and HDAC inhibitors as they would both target tubulin dynamics and induce mitotic catastrophe.

Important studies revealed that in addition to single agent therapies, combinations with HDAC inhibitors and other drugs show additive beneficial effects in medulloblastoma. For instance, treating mice bearing medulloblastoma in the flank or intracranially with the HDAC inhibitor SAHA along with retinoic acid reduced tumor growth more profoundly than with either compound alone [39]. Similarly, combining the HDAC inhibitor valproic acid with the kinase inhibitor sorafenib, which targets Raf, VEGF (vascular endothelial growth factor), and PDGF (platelet-endothelial growth factor) receptors [40], induced radiosensitization of medulloblastoma cells [41]. Although discussing all reported combinations of HDAC inhibitors and other agents is not feasible, it is likely that future therapies will include HDAC drug combinations for the treatment of medulloblastoma.

5 METHYLTRANSFERASES IN MEDULLOBLASTOMA

Methylation directly impacts chromatin structure and modulates gene transcription. Although the various methylation mechanisms are complex, hypo- and hypermethylation of DNA are implicated in many diseases. The most common methyltransferases involved in epigenetic regulation are histone methyltransferases (e.g., MLL1–5 (lysine (K)-Specific Methyltransferase 1-5), EZH1–2 (enhancer of zeste homolog 1-2 (Drosophila)), EHMT1–2 (euchromatic histone–lysine *N*-methyltransferase 1), SETD1–7 (SET domain containing 1–7), among others), and DNA methyltransferases (DNMT1, DNMT2, and DNMT3).

Histone methyltransferases can either suppress or activate transcription via histone modification. Recent studies have shown that the expression of several histone methyltransferases is altered in medulloblastoma. Among these, the H3K27 and H3K4 modifiers KDM6a and ZMYM3 (zinc finger MYM-type 3) are deleted or inactivated in medulloblastoma. Interestingly, KDM6a inactivation presents an interesting therapeutic possibility as KDM6a activity is counteracted by EZH2 (KDM6a is a methylase while EZH2 is a methyltransferase). H3K27me3 modification is written by the polycomb complex containing EZH2, which is amplified in medulloblastoma [42–47]. Highly specific small molecule inhibitors of EZH2 have been identified that potently and selectively kill cancer cells [48–50]. Importantly, EZH2 inactivation may provide yet another means of inhibiting multiple pathways as EZH2 has been shown to function as a modifier of the ubiquitin proteasome pathway by creating methyldegrons, which are degradation motifs important for recognition by ubiquitin ligases [51].

In addition to EZH2, frequent mutations in the histone lysine methyltransferase gene *MLL2* and its homolog *MLL3* may also play important roles in various types of human medulloblastomas [51a,51b,12,44]. MLL2 and MLL3 are methyltransferases that add a monomethyl moiety to H3K4. Intriguingly, while the loss of MLL2 in normal cells most likely predisposes cells to transformation, its activity may be beneficial to the proliferation of already transformed cells [52]. Thus, its potential as a therapeutic target remains to be determined.

Apart from histone methylation, DNA methylation is another major epigenetic means of modulating gene expression. Cancer cell DNA is generally hypomethylated relative to nontransformed cell DNA although it displays pervasive hypermethylation in the promoter regions of a subset of genes [53]. This DNA hypermethylation is correlated with transcriptional repression, indicating that epigenetic silencing of tumor suppressor genes may be an early step in the process of carcinogenesis [54]. In fact, there are common regions of cancer-specific methylation changes in primary medulloblastomas in critical developmental regulatory pathways, including SHH, WNT, RAR, and bone morphogenetic protein (BMP). For example, one of the commonly methylated loci is the *PTCH1-1C* promoter, a negative regulator of the SHH pathway that is methylated in both primary patient samples and human medulloblastoma cell lines. This suggests that expression of *PTCH1-1C* inhibits Shh signaling and cellular growth in normal cerebellar development, whereas in a subset of medulloblastomas *PTCH1* is inactivated by transcriptional silencing through DNA methylation of the *PTCH1-1C* promoter [55].

Similarly, methylation of promoters driving expression of tumor suppressor proteins has been shown to be altered in medulloblastoma. One of the main examples of this is the RASSF1A (Ras association domain family 1) promoter. Harada et al. [56] and Lusher et al. [57] examined the methylation status of multiple genes in primary pediatric tumors including medulloblastoma and found that the most frequently methylated gene was RASSF1A (Ras association domain family 1, 40% of all tumors). RASSF1A is a tumor suppressor, which negatively regulates the Ras pathway implicated in proliferation of multiple cancer types. RASSF1A is not an enzyme but is thought to be a scaffolding protein for multiple tumor suppressor complexes. Some of the known downstream effectors of RASSF1A regulate microtubule dynamics. These include Cdc20 (cell division cycle 20), MAP18 (microtubule-associated protein), and C19ORF5. Other effectors such as p120^{E4F} and JNK regulate cell cycle progression [58]. Thus, reduced RASSF1A expression in medulloblastoma is likely to lead to aberrant cell proliferation and tumor growth. Importantly, epigenetic enzymes or factors that may induce re-expression of RASSF1A in medulloblastoma may be attractive drug targets.

6 EPIGENETIC READERS IN MEDULLOBLASTOMA

Recent studies suggest that epigenetic readers may be particularly attractive drug targets. Bromodomain (BRD) and extra terminal (BET) protein inhibitors have been shown to reduce medulloblastoma *in vitro* and *in vivo*. BET proteins are histone readers that bind acetylated histones and recruit transcriptional and translational complexes. One BET protein, BRD4 has emerged as a target in multiple cancers including medulloblastoma [59–61]. Other BET BRD proteins include BRD2, BRD3, and BRDT. BRDT expression is restricted to testis, and thus small molecule inhibitors targeting various cancers that are not testis specific are directed against BRD2, BRD3, and BRD4. Our recent studies and those from other laboratories suggest that small molecule BRD2/BRD4 inhibitors reduce GBM and medulloblastoma cell growth [59,60,62,63]. These small molecules are acetylated histone mimics and thus can compete with the BRD2– or BRD4–histone interaction. As these proteins recruit transcriptional and translational complexes, disabling them to detect acetylated histones can sharply reduce the expression of oncogenic drivers. As stated earlier, one particularly aggressive form of medulloblastoma is driven by overexpression of the C-MYC oncogene. As BET BRD proteins have been shown to control *C-MYC* levels, it is possible that BET BRD protein inhibitors will be effective against this form of medulloblastoma in humans.

However, given the likelihood that cancer cells will become resistant to the inhibition of any one target, it is beneficial to simultaneously modulate two targets (Figure 14.4). Three recent reports have

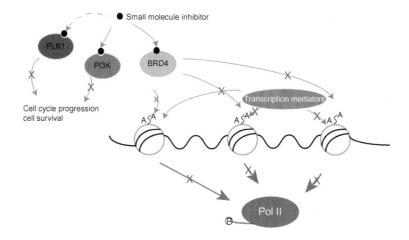

FIGURE 14.4

Polypharmacological small molecule inhibitors can target multiple pathways such as epigenetic readers and kinases. Inhibition of BRD-containing protein 4 reduces the protein's ability to read acetylated (A) lysine residues on histones, interfering with its recruitment of transcription mediators stimulating DNA polymerase II (Pol II) phosphorylation and activation. This same inhibitor can also be designed to inhibit kinases (e.g., polo-like kinase 1 (PLK1) and phosphoinositide 3-kinase (PI3K)) involved in cell cycle progression and survival and antiapoptotic mechanisms. This approach is relevant for cancers, which exhibit drug resistance.

shown that this is feasible for BRD proteins and kinases [64–66]. One group demonstrated that the PI3-kinase inhibitor LY294002 is a BET BRD protein inhibitor while two groups identified the Polo-kinase inhibitor BI-2536 as a potent BET BRD protein inhibitor (Figure 14.4). These results are especially important in medulloblastoma as prior studies have shown that PI3K or Polo-kinase inhibition is effective against reducing medulloblastoma cell growth [67–69].

7 ESTABLISHED KINASES MAY ALSO BE EPIGENETIC REGULATORS IN MEDULLOBLASTOMA

Importantly, well-established kinases may also function as epigenetic modifiers in medulloblastoma and thus provide an opportunity to target both kinase and epigenetic pathways pharmacologically. An example of this is the tyrosine kinase Wee1, which has recently been implicated in epigenetics and medulloblastoma [70–72]. Wee1 is a kinase known to limit mitotic entry via phosphorylation of the mitotic promoting factor CDK1 (cyclin-dependent kinase 1)-Cyclin B1 [73,74]. Wee1 activity ensures that DNA replication is complete before initiating mitosis. Indeed, Wee1 is considered a checkpoint kinase, which couples the DNA damage pathway and mitotic entry [75] (Figure 14.5). However, recent studies suggest that Wee1 plays another role as an epigenetic modifier since it controls histone transcription by phosphorylating Histone H2b at tyrosine 37 (Figure 14.5). This modification is important since cells have to maintain the correct ratio of DNA to histones. Thus, by initiating Histone H2b phosphorylation at the end of S phases, Wee1 is helping to decrease histone transcription [71,72]. Reducing histone levels is then permissive for mitotic entry as the correct DNA/histone ratio is maintained.

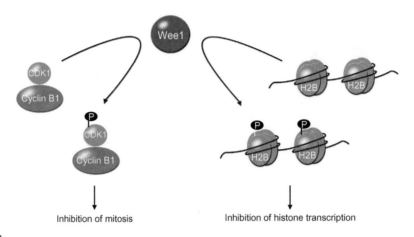

Inhibition of mitosis   Inhibition of histone transcription

FIGURE 14.5

Wee1 has a dual role during cell proliferation. First, Wee1 phosphorylates CDK1 preventing entry into mitosis to allow repair of damaged DNA. Second, it reduces histone gene transcription by phosphorylating H2B in nucleosomes.

8 THE INTERSECTION OF THE UBIQUITIN PROTEASOME PATHWAY AND THE EPIGENETIC PATHWAY IN MEDULLOBLASTOMA

Our studies and those from other laboratories demonstrated that Wee1 is targeted for degradation via the ubiquitin proteasome pathway, thus suggesting that the kinase, ubiquitin, and epigenetic pathways may intersect at Wee1 [76–81]. However, other examples of the intersection of these pathways have been described and may be important in medulloblastoma. We have already mentioned that Ezh2 may be an interesting target in medulloblastoma and functions in both the epigenetic pathway by modulating methylation of histones and the ubiquitin pathway by creating methyldegrons. However, Ezh2 activity is also modulated by the cell cycle regulator CDK1, which stimulates Ezh2 binding to the polycomb recognition complex as well as long noncoding RNAs required for its localization to genomic loci [82–87]. Future studies will delineate whether Ezh2 modulation of the ubiquitin pathway is also controlled by CDK1 activity.

A third example of ubiquitin ligase and epigenetic pathway intersection involves acetylation of the transcriptional regulator Gli1. Canettieri et al. [88] demonstrated that Gli1 is normally acetylated, which represses its activity as a transcription factor. By contrast, deacetylated Gli1 is able to induce transcription. This deacetylation is mediated by HDAC1 and HDAC2. However, SHH responsive cells have a mechanism to degrade these HDACs and turn off this system. A ubiquitin ligase complex containing Cul3 (cullin 3) and a SHH antagonist named REN (renin) is able to induce polyubiquitination and proteasomal degradation of the HDACs. Further, REN is frequently deleted in medulloblastoma given that its gene is located on Chromosome 17p13.1, a region frequently deleted in medulloblastoma. This suggests that REN deletion is one of the mechanisms through which the SHH pathway is constitutively activated in medulloblastoma.

KCASH2 and KCASH3 are two proteins, which share a number of features with REN. These include a BTB domain required for the formation of a Cul3 ubiquitin ligase complex and HDAC1 ubiquitination and degradation capability, which suppresses acetylation-dependent Hh/Gli signaling.

KCASH2 and 3 are expressed in the cerebellum, whereas epigenetic silencing and allelic deletion are observed in human medulloblastoma. Rescuing KCASHs expression reduces Hedgehog-dependent medulloblastoma growth suggesting that loss of members of this novel family of native HDAC inhibitors is crucial in sustaining Hh pathway-mediated tumorigenesis [14].

9 MicroRNAs AS MEDULLOBLASTOMA TARGETS

In addition to well-established kinases, ubiquitin ligases, and epigenetic enzyme targets in medulloblastoma, many groups have identified microRNAs as important medulloblastoma regulators. One of the first reports identified miR-124 as a CDK6 regulator, which is a cell cycle protein that is overexpressed in medulloblastoma [89]. CDK6 expression is associated with poor prognostic outcome in medulloblastoma, thus mechanisms controlling its levels are important for fully understanding its role in tumor progression [90]. Early studies suggested that CDK6 overexpression could not merely be explained by amplification of the CDK6 genomic region, which motivated the search for other factors, such as microRNAs, that could regulate CDK6 expression in medulloblastoma. microRNAs are small RNAs (18–22 nucleotides) with well-established roles in posttranscriptional regulation of gene expression. MicroRNAs can bind to complementary regions of mRNAs and repress translation. In the case of a microRNA controlling CDK6 levels, the prediction would be that it normally acts as a tumor suppressor, reducing CDK6 protein expression in untransformed cells. By contrast, during the process of transformation in medulloblastoma, the levels of that particular microRNA would be predicted to be lower, thereby increasing CDK6 levels in medulloblastoma patients. This is what was observed for miR-124, namely that it was lower in medulloblastoma relative to control tissue. Further, overexpression of miR-124 in human medulloblastoma cells reduced CDK6 expression and decreased proliferation suggesting that upregulating miR-124 levels may be a viable therapeutic strategy [91].

In addition to miR-124, three other microRNAs have been shown to affect CDK6 levels in medulloblastoma [92–94]. The first, microRNA-34a, is a transcriptional target of p53, which is downregulated in multiple cancer cell lines, including medulloblastoma. miR-34a levels are lower in human medulloblastoma and its overexpression reduces proliferation and survival of medulloblastoma and GBM cells but has no effect on proliferation of normal cells such as astrocytes [93]. In addition, miR-34a overexpression reduces levels of Notch-1, Notch-2, and c-MET (MNNG HOS transforming gene), putative miR-34a target genes and drivers of medulloblastoma cell proliferation. Collectively, these studies suggest that miR-34a upregulation might be beneficial in human medulloblastoma since it would potentially downregulate multiple drivers of medulloblastoma growth including CDK6. Similarly, overexpression of another microRNA, miR-129, reduced CDK6 levels as suggested by Wu et al. [92]. Although the authors performed their studies in human lung adenocarcinoma lines, the prediction is that overexpression of miR-129 in human medulloblastoma cells would induce a similar G1 arrest and apoptosis. In addition to miR-129, overexpression of miR-218 has been shown to reduce medulloblastoma cell proliferation [94]. Venkataraman et al. [94] overexpressed miR-218 in medulloblastoma cells and subsequently performed cross-linking and RNA-sequencing to identify 618 miR-218 putative targets. Among those that were later confirmed by overexpression studies were CDK6, RICTOR (RPTOR Independent Companion Of MTOR, Complex 2), and Cathepsin B. Future studies will likely reveal other downregulated microRNAs in medulloblastoma that normally repress CDK6 expression and could potentially be attractive targets for upregulation in medulloblastoma.

MicroRNAs regulating medulloblastoma drivers aside from CDK6 include those potentially controlling the Group 3 medulloblastoma oncogene C-MYC. Lv et al. [95] examined the 3' untranslated region of C-MYC mRNA and screened 48 medulloblastoma samples for 9 microRNAs that could potentially target C-MYC mRNA. They observed correlation between C-MYC overexpression and miR-512-2 gene deletion. Further, miR-512-2 knockdown increased C-MYC levels while its overexpression decreased C-MYC levels. In addition to miR-512-2, miR-33b, miR135a-1, miR-135a-2, miR-135b, miR-186, miR-200b, miR-512-2, miR-548d1, miR-548d02, miR-33b, deletion, amplification, and a point mutation were observed in human medulloblastoma. These studies suggest that networks of microRNAs control oncogenic drivers in medulloblastoma.

MicroRNA networks are also likely to regulate components of the SHH pathway in SHH-dependent medulloblastoma. Ferretti et al. [96] performed high-throughput microRNA analysis for different types of medulloblastoma and identified signatures associated with the SHH pathway. Among these, the microRNAs miR-125b and miR-326 were identified as suppressors of Smoothened expression while miR-324-5p targeted Gli1, one of the main transcription factors in the SHH pathway [96]. Interestingly, reduced miR-324-5p levels in medulloblastoma were associated with chromosome 17p deletion in humans, thus providing a genetic basis for miR-324-5p downregulation in medulloblastoma. Further, greater cell growth was observed after miR-324-5p depletion since this microRNA normally antagonizes SHH-induced proliferation of cerebellar granule cell progenitors, which can give rise to medulloblastoma [96].

Multiple components of the SHH pathway are regulated by microRNAs. Friggi-Grelin et al. [97] showed that multiple microRNAs regulate the SHH pathway via control of Smoothened levels, the kinesin-like protein Costal, and the kinase Sufu [97]. Similarly, Luo et al. tested expression of 90 different miRNAs in wild-type and $Ptch^{-/-}$ mouse embryonic fibroblasts and found that the miR-154 cluster 6 regulating SHH components is downregulated in medulloblastoma [98].

In addition to microRNAs downregulated in SHH medulloblastoma, a group of microRNAs are increased in the SHH subgroup [99]. Uziel et al. [99] compared microRNA levels in cerebellar granule cell progenitors relative to medulloblastoma samples isolated from $Ink4c^{-/-}$; $Ptch1^{+/-}$ and $Ink4c^{-/-}$; $p53^{-/-}$ mice and found that 26 microRNAs were overexpressed in the tumor samples. Among these, nine were from the miR-17–92 cluster, which had been previously implicated as oncogenes and tumor suppressor genes in various cancers. Importantly, three microRNAs from this cluster, miR-92, miR-19a, and miR-20, were overexpressed in SHH medulloblastoma subtypes but not other subtypes. Further proof of a direct link with the SHH pathway came from overexpression experiments where the authors overexpressed these microRNAs in $Ink4c^{-/-}$;$Ptch1^{+/-}$ mice and found that overexpression induced loss of the wild-type $Ptch1$ allele. Subsequent work from the same group recently demonstrated that inhibiting expression of the miR-17–92 cluster in mouse models of medulloblastoma reduced tumor growth in the flank and intracranially [100]. What was especially interesting about this work is that the group used 8-mer seed-targeting locked nucleic acid (LNA)-modified anti-miR oligonucleotides to target the microRNAs miR-17–92 and miR-106b–25. These tiny LNAs were effective at increasing survival of animals bearing medulloblastoma, and thus should be considered as possible therapies in humans.

Additional therapies will likely include small molecule regulators of microRNA expression. For instance, Takwi et al. [101] identified miR-33b as a microRNA whose expression is low in medulloblastoma and is regulated by lovastatin [101]. This study described a small molecule screen with US FDA-approved compounds to search for regulators of miR-33b expression. MiR-33b was shown to negatively regulate C-MYC levels and thus mechanisms to upregulate miR-33b would lower C-MYC in Group 3 medulloblastoma (Figure 14.6). Lovastatin-mediated upregulation of miR-33b reduced

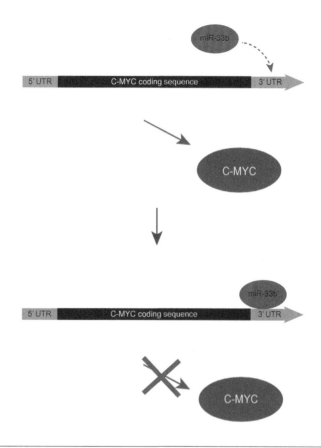

FIGURE 14.6

miR-33b regulates C-MYC expression. miR-33b directly binds to the 3′untranslated region (3′UTR) of C-MYC mRNA and represses its expression, leading to downregulation of C-MYC transactivation targets.

C-MYC levels and tumor growth *in vivo*. Further, lovastatin doses that were effective *in vivo* for reducing tumors were well tolerated in patients normally treated with lovastatin for cardiovascular disease. Finally, these exciting studies point to the possibility of repurposing FDA-approved compounds for medulloblastoma treatment.

10 DRUG DISCOVERY CONSIDERATIONS IN MEDULLOBLASTOMA

Several ongoing and completed clinical trials have targeted medulloblastoma (Tables 14.1 and 14.2). Many of these trials involve the DNA alkylating agent temozolomide (TMZ), which is approved for use in GBM. One important feature of TMZ is that it is brain penetrant, which is essential for a brain cancer compound. Further, combination therapies of brain penetrant small molecules such as the HDAC inhibitor vorinostat are also being tried with TMZ in medulloblastoma patients. However, many potentially

Table 14.1 Active, Recruiting, or Not Recruiting Yet Clinical Trials in Medulloblastoma

ClinicalTrials. Gov Identifier	Phase	Status	Drugs	Other Procedures
NCT02040376	3	Recruiting	Drug: Metformin Drug: Placebo	
NCT00336024	3	Active	Drug: Etoposide Drug: Cyclophosphamide Drug: Cisplatin Biological: Filgrastim Drug: Carboplatin Drug: Thiotepa Drug: Methotrexate Drug: Leucovorin calcium Drug: Vincristine sulfate	Procedure: Autologous hematopoietic stem cell transplantation Other: Laboratory biomarker analysis Procedure: Quality-of-life assessment
NCT01351870	3	Active		Radiation: Standard fractionation regimen Radiation: Hyperfractionated radiotherapy
NCT00085202	3	Active	Biological: Filgrastim Drug: Cisplatin Drug: Cyclophosphamide Drug: Vincristine	Procedure: Autologous hematopoietic stem cell transplantation Radiation: Radiation therapy
NCT02066220	3	Not recruiting yet	Drug: Reduced-intensity maintenance chemotherapy Drug: Maintenance chemotherapy	Radiation: Radiotherapy without carboplatin Radiation: Radiotherapy with carboplatin
NCT00085735	3	Recruiting	Drug: Cisplatin Drug: Cyclophosphamide Drug: Lomustine Drug: Vincristine sulfate	Radiation: Radiation therapy
NCT00716976	3	Active	Drug: Sodium thiosulfate	Procedure: Examination
NCT00749723	3	Active	Drug: Carboplatin Drug: Etoposide Drug: Temozolomide Drug: Thiotepa, carboplatin, etoposide Drug: Temozolomide, thiotepa Drug: Trofosfamide/etoposide	Procedure: Autologous stem cell transplantation
NCT01987596	3	Not recruiting yet	Drug: Etoposide Drug: Ifosfamide Drug: Carboplatin Drug: Topotecan hydrochloride Drug: Vincristine sulfate Drug: Cyclophosphamide Drug: Cisplatin Biological: Filgrastim	Other: Laboratory biomarker analysis
NCT01346267	3	Recruiting		Procedure: Real acupressure band Procedure: Placebo acupressure band

Table 14.2 Completed or Terminated Clinical Trials in Medulloblastoma

ClinicalTrials. Gov Identifier	Phase	Status	Drugs	Other Procedures
NCT00002875	3	Completed	Biological: Filgrastim Drug: Cisplatin Drug: Cyclophosphamide Drug: Lomustine Drug: Mesna Drug: Vincristine sulfate	Radiation: Low-LET electron therapy Radiation: Low-LET photon therapy
NCT00006461	3	Completed	Drug: Cisplatin Drug: Cyclophosphamide Drug: Vincristine sulfate Drug: Etoposide	Procedure: Therapeutic conventional surgery Radiation: Three-dimensional conformal radiation therapy
NCT01132547	3	Terminated	Drug: Cyproheptadine hydrochloride Other: Placebo	
NCT00538850	3	Completed	Drug: Fentanyl sublingual spray Drug: Placebo	
NCT00369785	3	Completed	Drug: Donepezil hydrochloride Other: Placebo	

effective therapeutics are not brain penetrant and thus efforts are underway to identify mechanisms to deliver these small molecules and microRNAs into the brain. One means of achieving this is through exosomes, which are small vesicles that are extruded from cell membranes. Studies in GBM have concentrated on marrow stromal cell (MSC) exosomes as a delivery vehicle for antitumor miRNAs. These exosomes reduce tumor cell growth *in vitro* and *in vivo* [102–104]. Therefore, delivering exosomes could be an interesting approach for reducing tumor recurrence [105–107]. Ideally, exosome therapy can be combined with small molecule, immunological, or viral approaches for reducing the incidence of GBM or possibly medulloblastoma tumor recurrence.

11 CONCLUSIONS

Epigenetic pathways have recently emerged as central to medulloblastoma cell proliferation and growth. These pathways contain epigenetic enzymes and microRNAs, which work in concert to induce cell cycle transitions during normal development. Mutation, deletion, or amplification of genomic regions coding for these microRNAs or enzymes induces changes associated with tumor growth. Thus, efforts to target both microRNAs and epigenetic enzymes in medulloblastoma may be therapeutically attractive since this would likely reduce the possibility of recurrence associated with many cancers. Similarly, strategies that modulate epigenetic enzymes, kinases, and ubiquitin ligases simultaneously may be equally therapeutically viable. Current and future studies should thus take a systems approach to identifying how various pathways intersect with epigenetic enzymes in order to identify effective combination therapies.

ACKNOWLEDGMENTS

We thank all members of the Center for Therapeutic Innovation, the Miami Project to Cure Paralysis for discussions and helpful suggestions. This work was supported by R21 NS056991 to Nagi Ayad, 1R01NS067289 to Nagi Ayad.

REFERENCES

[1] Hatten ME, Roussel MF. Development and cancer of the cerebellum. Trends Neurosci 2011;34:134–42.

[2] Roussel MF, Hatten ME. Cerebellum development and medulloblastoma. Curr Top Dev Biol 2011;94:235–82.

[3] Clifford SC, Lusher ME, Lindsey JC, Langdon JA, Gilbertson RJ, Straughton D, et al. Wnt/Wingless pathway activation and chromosome 6 loss characterize a distinct molecular sub-group of medulloblastomas associated with a favorable prognosis. Cell Cycle 2006;5:2666–70.

[3a] Tenzen T, Allen BL, Cole F, Kang JS, Krauss RS, McMAhon AP. The cell surface membrane proteins Cdo and Boc are components and targets of the Hedgehog signaling pathway and feedback network in mice. Dev Cell 2006;10:647–56.

[3b] Yao S, Lum L, Beachy P. The ihog cell-surface proteins bind Hedgehog and mediate pathway activation. Cell 2006;125:343–57.

[3c] Varjosalo M, Taipale J. Hedgehog: functions and mechanisms. Genes Dev 2008;22:254–72.

[4] Kool M, Korshunov A, Remke M, Jones DT, Schlanstein M, Northcott PA, et al. Molecular subgroups of medulloblastoma: an international meta-analysis of transcriptome, genetic aberrations, and clinical data of WNT, SHH, Group 3, and Group 4 medulloblastomas. Acta Neuropathol 2012;123:473–84.

[5] Northcott PA, Korshunov A, Witt H, Hielscher T, Eberhart CG, Mack S, et al. Medulloblastoma comprises four distinct molecular variants. J Clin Oncol 2011;29:1408–14.

[6] Taylor MD, Northcott PA, Korshunov A, Remke M, Cho YJ, Clifford SC, et al. Molecular subgroups of medulloblastoma: the current consensus. Acta Neuropathol 2012;123:465–72.

[7] Yang ZJ, Ellis T, Markant SL, Read TA, Kessler JD, Bourboulas M, et al. Medulloblastoma can be initiated by deletion of Patched in lineage-restricted progenitors or stem cells. Cancer Cell 2008;14:135–45.

[8] Schuller U, Heine VM, Mao J, Kho AT, Dillon AK, Han YG, et al. Acquisition of granule neuron precursor identity is a critical determinant of progenitor cell competence to form Shh-induced medulloblastoma. Cancer Cell 2008;14:123–34.

[9] Grammel D, Warmuth-Metz M, von Bueren AO, Kool M, Pietsch T, Kretzschmar HA, et al. Sonic hedgehog-associated medulloblastoma arising from the cochlear nuclei of the brainstem. Acta Neuropathol 2012;123:601–14.

[10] Pei Y, Moore CE, Wang J, Tewari AK, Eroshkin A, Cho YJ, et al. An animal model of MYC-driven medulloblastoma. Cancer Cell 2012;21:155–67.

[11] Gibson P, Tong Y, Robinson G, Thompson MC, Currle DS, Eden C, et al. Subtypes of medulloblastoma have distinct developmental origins. Nature 2010;468:1095–9.

[12] Jones DT, Jager N, Kool M, Zichner T, Hutter B, Sultan M, et al. Dissecting the genomic complexity underlying medulloblastoma. Nature 2012;488:100–5.

[13] Canettieri G, Di Marcotullio L, Greco A, Coni S, Antonucci L, Infante P, et al. Histone deacetylase and Cullin3-REN(KCTD11) ubiquitin ligase interplay regulates Hedgehog signalling through Gli acetylation. Nat Cell Biol 2010;12:132–42.

[14] De Smaele E, Di Marcotullio L, Moretti M, Pelloni M, Occhione MA, Infante P, et al. Identification and characterization of KCASH2 and KCASH3, 2 novel Cullin3 adaptors suppressing histone deacetylase and Hedgehog activity in medulloblastoma. Neoplasia 2011;13:374–85.

[15] Ecke I, Petry F, Rosenberger A, Tauber S, Monkemeyer S, Hess I, et al. Antitumor effects of a combined 5-aza-2'deoxycytidine and valproic acid treatment on rhabdomyosarcoma and medulloblastoma in Ptch mutant mice. Cancer Res 2009;69:887–95.

[16] Furchert SE, Lanvers-Kaminsky C, Juurgens H, Jung M, Loidl A, Fruhwald MC. Inhibitors of histone deacetylases as potential therapeutic tools for high-risk embryonal tumors of the nervous system of childhood. Int J Cancer 2007;120:1787–94.

[17] Lee SJ, Krauthauser C, Maduskuie V, Fawcett PT, Olson JM, Rajasekaran SA. Curcumin-induced HDAC inhibition and attenuation of medulloblastoma growth *in vitro* and *in vivo*. BMC Cancer 2011;11:144.

[18] Milde T, Oehme I, Korshunov A, Kopp-Schneider A, Remke M, Northcott P, et al. HDAC5 and HDAC9 in medulloblastoma: novel markers for risk stratification and role in tumor cell growth. Clin Cancer Res 2010;16:3240–52.

[19] Sredni ST, Halpern AL, Hamm CA, Bonaldo Mde F, Tomita T. Histone deacetylases expression in atypical teratoid rhabdoid tumors. Childs Nerv Syst 2013;29:5–9.

[20] Srivastava VK, Nalbantoglu J. The cellular and developmental biology of medulloblastoma: current perspectives on experimental therapeutics. Cancer Biol Ther 2010;9:843–52.

[21] Taylor P, Fangusaro J, Rajaram V, Goldman S, Helenowski IB, MacDonald T, et al. REST is a novel prognostic factor and therapeutic target for medulloblastoma. Mol Cancer Ther 2012;11:1713–23.

[22] Broide RS, Redwine JM, Aftahi N, Young W, Bloom FE, Winrow CJ. Distribution of histone deacetylases 1–11 in the rat brain. J Mol Neurosci 2007;31:47–58.

[23] Zhang CL, McKinsey TA, Chang S, Antos CL, Hill JA, Olson EN. Class II histone deacetylases act as signal-responsive repressors of cardiac hypertrophy. Cell 2002;110:479–88.

[24] Chang S, McKinsey TA, Zhang CL, Richardson JA, Hill JA, Olson EN. Histone deacetylases 5 and 9 govern responsiveness of the heart to a subset of stress signals and play redundant roles in heart development. Mol Cell Biol 2004;24:8467–76.

[25] McKinsey TA, Zhang CL, Lu J, Olson EN. Signal-dependent nuclear export of a histone deacetylase regulates muscle differentiation. Nature 2000;408:106–11.

[26] Roy S, Shor AC, Bagui TK, Seto E, Pledger WJ. Histone deacetylase 5 represses the transcription of cyclin D3. J Cell Biochem 2008;104:2143–54.

[27] Weichert W, Roske A, Gekeler V, Beckers T, Ebert MP, Pross M, et al. Association of patterns of class I histone deacetylase expression with patient prognosis in gastric cancer: a retrospective analysis. Lancet Oncol 2008;9:139–48.

[28] Liu WD, Wang HW, Muguira M, Breslin MB, Lan MS. INSM1 functions as a transcriptional repressor of the neuroD/beta2 gene through the recruitment of cyclin D1 and histone deacetylases. Biochem J 2006;397:169–77.

[29] Ma JX, Li H, Chen XM, Yang XH, Wang Q, Wu ML, et al. Expression patterns and potential roles of SIRT1 in human medulloblastoma cells *in vivo* and *in vitro*. Neuropathology 2013;33:7–16.

[30] Sheikh BN. Crafting the brain—role of histone acetyltransferases in neural development and disease. Cell Tissue Res 2014;356:553–73.

[31] Pfister S, Rea S, Taipale M, Mendrzyk F, Straub B, Ittrich C, et al. The histone acetyltransferase hMOF is frequently downregulated in primary breast carcinoma and medulloblastoma and constitutes a biomarker for clinical outcome in medulloblastoma. Int J Cancer 2008;122:1207–13.

[32] Malatesta M, Steinhauer C, Mohammad F, Pandey DP, Squatrito M, Helin K. Histone acetyltransferase PCAF is required for Hedgehog-Gli-dependent transcription and cancer cell proliferation. Cancer Res 2013;73:6323–33.

[33] Coni S, Antonucci L, D'Amico D, Di Magno L, Infante P, De Smaele E, et al. Gli2 acetylation at lysine 757 regulates hedgehog-dependent transcriptional output by preventing its promoter occupancy. PLoS One 2013;8:e65718.

[34] Kim HJ, Bae SC. Histone deacetylase inhibitors: molecular mechanisms of action and clinical trials as anticancer drugs. Am J Transl Res 2011;3:166–79.

[35] Li XN, Shu Q, Su JM, Perlaky L, Blaney SM, Lau CC. Valproic acid induces growth arrest, apoptosis, and senescence in medulloblastomas by increasing histone hyperacetylation and regulating expression of p21Cip1, CDK4, and CMYC. Mol Cancer Ther 2005;4:1912–22.

[36] Nor C, de Farias CB, Abujamra AL, Schwartsmann G, Brunetto AL, Roesler R. The histone deacetylase inhibitor sodium butyrate in combination with brain-derived neurotrophic factor reduces the viability of DAOY human medulloblastoma cells. Childs Nerv Syst 2011;27:897–901.

[37] Valdora F, Banelli B, Stigliani S, Pfister SM, Moretti S, Kool M, et al. Epigenetic silencing of DKK3 in Medulloblastoma. Int J Mol Sci 2013;14:7492–505.

[38] Vibhakar R, Foltz G, Yoon JG, Field L, Lee H, Ryu GY, et al. Dickkopf-1 is an epigenetically silenced candidate tumor suppressor gene in medulloblastoma. Neurooncology 2007;9:135–44.

[39] Spiller SE, Ditzler SH, Pullar BJ, Olson JM. Response of preclinical medulloblastoma models to combination therapy with 13-cis retinoic acid and suberoylanilide hydroxamic acid (SAHA). J Neurooncol 2008;87:133–41.

[40] Wilhelm SM, Adnane L, Newell P, Villanueva A, Llovet JM, Lynch M. Preclinical overview of sorafenib, a multikinase inhibitor that targets both Raf and VEGF and PDGF receptor tyrosine kinase signaling. Mol Cancer Ther 2008;7:3129–40.

[41] Tang Y, Yacoub A, Hamed HA, Poklepovic A, Tye G, Grant S, et al. Sorafenib and HDAC inhibitors synergize to kill CNS tumor cells. Cancer Biol Ther 2012;13:567–74.

[42] Alimova I, Birks DK, Harris PS, Knipstein JA, Venkataraman S, Marquez VE, et al. Inhibition of EZH2 suppresses self-renewal and induces radiation sensitivity in atypical rhabdoid teratoid tumor cells. Neuro Oncol 2013;15:149–60.

[43] Alimova I, Venkataraman S, Harris P, Marquez VE, Northcott PA, Dubuc A, et al. Targeting the enhancer of zeste homologue 2 in medulloblastoma. Int J Cancer 2012;131:1800–9.

[44] Dubuc AM, Remke M, Korshunov A, Northcott PA, Zhan SH, Mendez-Lago M, et al. Aberrant patterns of H3K4 and H3K27 histone lysine methylation occur across subgroups in medulloblastoma. Acta Neuropathol 2013;125:373–84.

[45] Boulay G, Rosnoblet C, Guerardel C, Angrand PO, Leprince D. Functional characterization of human Polycomb-like 3 isoforms identifies them as components of distinct EZH2 protein complexes. Biochem J 2011;434:333–42.

[46] Smits M, van Rijn S, Hulleman E, Biesmans D, van Vuurden DG, Kool M, et al. EZH2-regulated DAB2IP is a medulloblastoma tumor suppressor and a positive marker for survival. Clin Cancer Res 2012;18:4048–58.

[47] Bunt J, Hasselt NA, Zwijnenburg DA, Koster J, Versteeg R, Kool M. OTX2 sustains a bivalent-like state of OTX2-bound promoters in medulloblastoma by maintaining their H3K27me3 levels. Acta Neuropathol 2013;125:385–94.

[48] Kim W, Bird GH, Neff T, Guo G, Kerenyi MA, Walensky LD, et al. Targeted disruption of the EZH2–EED complex inhibits EZH2-dependent cancer. Nat Chem Biol 2013;9:643–50.

[49] Knutson SK, Wigle TJ, Warholic NM, Sneeringer CJ, Allain CJ, Klaus CR, et al. A selective inhibitor of EZH2 blocks H3K27 methylation and kills mutant lymphoma cells. Nat Chem Biol 2012;8:890–6.

[50] Konze KD, Ma A, Li F, Barsyte-Lovejoy D, Parton T, Macnevin CJ, et al. An orally bioavailable chemical probe of the lysine methyltransferases EZH2 and EZH1. ACS Chem Biol 2013;8:1324–34.

[51] Lee JM, Lee JS, Kim H, Kim K, Park H, Kim JY, et al. EZH2 generates a methyl degron that is recognized by the DCAF1/DDB1/CUL4 E3 ubiquitin ligase complex. Mol Cell 2012;48:572–86.

[51a] Pugh TJ, Weeraratne SD, Archer TC, Pomeranz Krummels DA, Auclair D, Bochicchio J, et al. Medulloblastoma exome sequencing uncovers subtype-specific somatic mutations. Nature 2012;488:106–10.

[51b] Robinson G, Parker M, Kranenburg TA, Lu C, Chen X, Ding L, et al. Novel mutations target distinct subgroups of medulloblastoma. Nature 2012;488:43–8.

[52] Guo C, Chen LH, Huang Y, Chang CC, Wang P, Pirozzi CJ, et al. KMT2D maintains neoplastic cell proliferation and global histone H3 lysine 4 monomethylation. Oncotarget 2013;4:2144–53.

[53] Jones PA, Baylin SB. The epigenomics of cancer. Cell 2007;128:683–92.

[54] Esteller M. Epigenetics in cancer. N Engl J Med 2008;358:1148–59.

[55] Diede SJ, Guenthoer J, Geng LN, Mahoney SE, Marotta M, Olson JM, et al. DNA methylation of developmental genes in pediatric medulloblastomas identified by denaturation analysis of methylation differences. Proc Natl Acad Sci USA 2010;107:234–9.

[56] Harada K, Toyooka S, Maitra A, Maruyama R, Toyooka KO, Timmons CF, et al. Aberrant promoter methylation and silencing of the RASSF1A gene in pediatric tumors and cell lines. Oncogene 2002;21:4345–9.

[57] Lusher ME, Lindsey JC, Latif F, Pearson AD, Ellison DW, Clifford SC. Biallelic epigenetic inactivation of the RASSF1A tumor suppressor gene in medulloblastoma development. Cancer Res 2002;62:5906–11.

[58] Donninger H, Vos MD, Clark GJ. The RASSF1A tumor suppressor. J Cell Sci 2007;120:3163–72.

[59] Henssen A, Thor T, Odersky A, Heukamp L, El-Hindy N, Beckers A, et al. BET bromodomain protein inhibition is a therapeutic option for medulloblastoma. Oncotarget 2013;4:2080–95.

[60] Bandopadhayay P, Bergthold G, Nguyen B, Schubert S, Gholamin S, Tang Y, et al. BET bromodomain inhibition of MYC-amplified medulloblastoma. Clin Cancer Res 2014;20:912–25.

[61] Belkina AC, Denis GV. BET domain co-regulators in obesity, inflammation and cancer. Nat Rev Cancer 2012;12:465–77.

[62] Cheng Z, Gong Y, Ma Y, Lu K, Lu X, Pierce LA, et al. Inhibition of BET bromodomain targets genetically diverse glioblastoma. Clin Cancer Res 2013;19:1748–59.

[63] Pastori C, Daniel M, Penas C, Volmar CH, Johnstone AL, Brothers SP, et al. BET bromodomain proteins are required for glioblastoma cell proliferation. Epigenetics 2014;9:611–20.

[64] Ciceri P, Muller S, O'Mahony A, Fedorov O, Filippakopoulos P, Hunt JP, et al. Dual kinase-bromodomain inhibitors for rationally designed polypharmacology. Nat Chem Biol 2014;10:305–12.

[65] Dittmann A, Werner T, Chung CW, Savitski MM, Falth Savitski M, Grandi P, et al. The commonly used PI3-kinase probe LY294002 is an inhibitor of BET bromodomains. ACS Chem Biol 2014;9:495–502.

[66] Ember SW, Zhu JY, Olesen SH, Martin MP, Becker A, Berndt N, et al. Acetyl-lysine binding site of bromodomain-containing protein 4 (BRD4) interacts with diverse kinase inhibitors. ACS Chem Biol 2014;9:1160–71.

[67] Hambardzumyan D, Becher OJ, Rosenblum MK, Pandolfi PP, Manova-Todorova K, Holland EC. PI3K pathway regulates survival of cancer stem cells residing in the perivascular niche following radiation in medulloblastoma *in vivo*. Genes Dev 2008;22:436–48.

[68] Markant SL, Esparza LA, Sun J, Barton KL, McCoig LM, Grant GA, et al. Targeting sonic hedgehog-associated medulloblastoma through inhibition of Aurora and Polo-like kinases. Cancer Res 2013;73:6310–22.

[69] Mohan AL, Friedman MD, Ormond DR, Tobias M, Murali R, Jhanwar-Uniyal M. PI3K/mTOR signaling pathways in medulloblastoma. Anticancer Res 2012;32:3141–6.

[70] Harris PS, Venkataraman S, Alimova I, Birks DK, Balakrishnan I, Cristiano B, et al. Integrated genomic analysis identifies the mitotic checkpoint kinase WEE1 as a novel therapeutic target in medulloblastoma. Mol Cancer 2014;13:72.

[71] Mahajan K, Fang B, Koomen JM, Mahajan NP. H2B Tyr37 phosphorylation suppresses expression of replication-dependent core histone genes. Nat Struct Mol Biol 2012;19:930–7.

[72] Mahajan K, Mahajan NP. WEE1 tyrosine kinase, a novel epigenetic modifier. Trends Genet 2013;29:394–402.

[73] Parker LL, Atherton-Fessler S, Lee MS, Ogg S, Falk JL, Swenson KI, et al. Cyclin promotes the tyrosine phosphorylation of p34cdc2 in a wee1+ dependent manner. EMBO J 1991;10:1255–63.

[74] Parker LL, Atherton-Fessler S, Piwnica-Worms H. p107wee1 is a dual-specificity kinase that phosphorylates p34cdc2 on tyrosine 15. Proc Natl Acad Sci USA 1992;89:2917–21.

[75] Rowley R, Hudson J, Young PG. The wee1 protein kinase is required for radiation-induced mitotic delay. Nature 1992;356:353–5.

[76] Ayad NG, Rankin S, Murakami M, Jebanathirajah J, Gygi S, Kirschner MW. Tome-1, a trigger of mitotic entry, is degraded during G1 via the APC. Cell 2003;113:101–13.

[77] Madoux F, Mishra J, Mercer BA, Ayad N, Roush W, Hodder P, et al.. Small molecule inhibitors of wee1 degradation and mitotic entry. Probe Reports from the NIH Molecular Libraries Program (Bethesda, MD); 2010.

[78] Penas C, Ramachandran V, Simanski S, Lee C, Madoux F, Rahaim RJ, et al. Casein kinase 1delta dependent wee1 degradation. J Biol Chem 2014;289:18893–903.

[79] Simanski S, Madoux F, Rahaim RJ, Chase P, Schurer S, Cameron M, et al.. Identification of small molecule inhibitors of wee1 degradation and mitotic entry. In Probe Reports from the NIH Molecular Libraries Program (Bethesda, MD); 2010.

[80] Watanabe N, Arai H, Iwasaki J, Shiina M, Ogata K, Hunter T, et al. Cyclin-dependent kinase (CDK) phosphorylation destabilizes somatic Wee1 via multiple pathways. Proc Natl Acad Sci USA 2005;102:11663–8.

[81] Watanabe N, Arai H, Nishihara Y, Taniguchi M, Watanabe N, Hunter T, et al. M-phase kinases induce phospho-dependent ubiquitination of somatic Wee1 by SCFbeta-TrCP. Proc Natl Acad Sci USA 2004;101:4419–24.

[82] Wei Y, Chen YH, Li LY, Lang J, Yeh SP, Shi B, et al. CDK1-dependent phosphorylation of EZH2 suppresses methylation of H3K27 and promotes osteogenic differentiation of human mesenchymal stem cells. Nat Cell Biol 2011;13:87–94.

[83] Wu SC, Zhang Y. Cyclin-dependent kinase 1 (CDK1)-mediated phosphorylation of enhancer of zeste 2 (Ezh2) regulates its stability. J Biol Chem 2011;286:28511–19.

[84] Zeng X, Chen S, Huang H. Phosphorylation of EZH2 by CDK1 and CDK2: a possible regulatory mechanism of transmission of the H3K27me3 epigenetic mark through cell divisions. Cell Cycle 2011;10:579–83.

[85] Chen S, Bohrer LR, Rai AN, Pan Y, Gan L, Zhou X, et al. Cyclin-dependent kinases regulate epigenetic gene silencing through phosphorylation of EZH2. Nat Cell Biol 2010;12:1108–14.

[86] Kaneko S, Li G, Son J, Xu CF, Margueron R, Neubert TA, et al. Phosphorylation of the PRC2 component Ezh2 is cell cycle-regulated and up-regulates its binding to ncRNA. Genes Dev 2010;24:2615–20.

[87] Sharif J, Endoh M, Koseki H. Epigenetic memory meets G2/M: to remember or to forget? Dev Cell 2011;20:5–6.

[88] Canettieri G, Di Marcotullio L, Greco A, Coni S, Antonucci L, Infante P, et al. Histone deacetylase and Cullin3-REN(KCTD11) ubiquitin ligase interplay regulates Hedgehog signalling through Gli acetylation. Nat Cell Biol 2010;12:132–42.

[89] Pierson J, Hostager B, Fan R, Vibhakar R. Regulation of cyclin dependent kinase 6 by microRNA 124 in medulloblastoma. J Neurooncol 2008;90:1–7.

[90] Mendrzyk F, Radlwimmer B, Joos S, Kokocinski F, Benner A, Stange DE, et al. Genomic and protein expression profiling identifies CDK6 as novel independent prognostic marker in medulloblastoma. J Clin Oncol 2005;23:8853–62.

[91] Silber J, Hashizume R, Felix T, Hariono S, Yu M, Berger MS, et al. Expression of miR-124 inhibits growth of medulloblastoma cells. Neuro Oncol 2013;15:83–90.

[92] Wu J, Qian J, Li C, Kwok L, Cheng F, Liu P, et al. miR-129 regulates cell proliferation by downregulating Cdk6 expression. Cell Cycle 2010;9:1809–18.

[93] Li Y, Guessous F, Zhang Y, Dipierro C, Kefas B, Johnson E, et al. MicroRNA-34a inhibits glioblastoma growth by targeting multiple oncogenes. Cancer Res 2009;69:7569–76.

[94] Venkataraman S, Birks DK, Balakrishnan I, Alimova I, Harris PS, Patel PR, et al. MicroRNA 218 acts as a tumor suppressor by targeting multiple cancer phenotype-associated genes in medulloblastoma. J Biol Chem 2013;288:1918–28.

[95] Lv SQ, Kim YH, Giulio F, Shalaby T, Nobusawa S, Yang H, et al. Genetic alterations in microRNAs in medulloblastomas. Brain Pathol 2012;22:230–9.

[96] Ferretti E, De Smaele E, Miele E, Laneve P, Po A, Pelloni M, et al. Concerted microRNA control of Hedgehog signalling in cerebellar neuronal progenitor and tumour cells. EMBO J 2008;27:2616–27.

[97] Friggi-Grelin F, Lavenant-Staccini L, Therond P. Control of antagonistic components of the hedgehog signaling pathway by microRNAs in Drosophila. Genetics 2008;179:429–39.

[98] Luo X, Liu J, Cheng SY. The role of microRNAs during the genesis of medulloblastomas induced by the hedgehog pathway. J Biomed Res 2011;25:42–8.

[99] Uziel T, Karginov FV, Xie S, Parker JS, Wang YD, Gajjar A, et al. The miR-17–92 cluster collaborates with the Sonic Hedgehog pathway in medulloblastoma. Proc Natl Acad Sci USA 2009;106:2812–17.

[100] Murphy BL, Obad S, Bihannic L, Ayrault O, Zindy F, Kauppinen S, et al. Silencing of the miR-17–92 cluster family inhibits medulloblastoma progression. Cancer Res 2013;73:7068–78.

[101] Takwi AA, Li Y, Becker Buscaglia LE, Zhang J, Choudhury S, Park AK, et al. A statin-regulated microRNA represses human c-Myc expression and function. EMBO Mol Med 2012;4:896–909.

[102] Katakowski M, Buller B, Zheng X, Lu Y, Rogers T, Osobamiro O, et al. Exosomes from marrow stromal cells expressing miR-146b inhibit glioma growth. Cancer Lett 2013;335:201–4.

[103] Katakowski M, Zheng X, Jiang F, Rogers T, Szalad A, Chopp M. MiR-146b-5p suppresses EGFR expression and reduces *in vitro* migration and invasion of glioma. Cancer Invest 2010;28:1024–30.

[104] Munoz JL, Bliss SA, Greco SJ, Ramkissoon SH, Ligon KL, Rameshwar P. Delivery of functional anti-miR-9 by mesenchymal stem cell-derived exosomes to glioblastoma multiforme cells conferred chemosensitivity. Mol Ther Nucleic Acids 2013;2:e126.

[105] Chen Y, Liu L. Modern methods for delivery of drugs across the blood-brain barrier. Adv Drug Deliv Rev 2012;64:640–65.

[106] Alvarez-Erviti L, Seow Y, Yin H, Betts C, Lakhal S, Wood MJ. Delivery of siRNA to the mouse brain by systemic injection of targeted exosomes. Nat Biotechnol 2011;29:341–5.

[107] Grapp M, Wrede A, Schweizer M, Huwel S, Galla HJ, Snaidero N, et al. Choroid plexus transcytosis and exosome shuttling deliver folate into brain parenchyma. Nat Commun 2013;4:2123.

CLINICAL SIGNIFICANCE OF EPIGENETIC ALTERATIONS IN GLIOBLASTOMA

15

Fumiharu Ohka[1,2], Atsushi Natsume[2], and Yutaka Kondo[1]

[1]*Department of Epigenomics, Nagoya City University Graduate School of Medical Sciences, Mizuho-cho, Mizuho-ku, Nagoya, Japan,* [2]*Department of Neurosurgery, Nagoya University Graduate School of Medicine, Nagoya, Japan*

CHAPTER OUTLINE

1 INTRODUCTION

Glioblastoma multiforme (GBM) is one of the most malignant brain tumors [1]. Combined therapy, such as tumor removal surgery, subsequent chemotherapy plus radiation therapy, is a global standard for GBM treatment. However, in spite of these aggressive treatments, median overall survival of GBM patients is about 15 months [2,3]. One of characteristic features, which link to the dismal prognosis of GBM, is that tumor cells tend to infiltrate into the surrounding normal brain tissues extensively. This feature sometimes disables to distinguish tumor from normal brain tissue during surgical treatment: thus intraoperative damages of surrounding normal brain result in permanent dysfunction for patients. Therefore, total tumor removal is theoretically difficult in most GBM cases. Additionally, in some cases, tumors arise from the brain stem where surgical approach is not applicable.

Although pediatric GBMs are less frequent than adult GBMs, GBM is the most common in pediatric malignant brain tumors. In pediatric GBMs, many tumors arise from midline region in brain, including thalamus, brain stem. Therefore, surgery is sometimes not attempted due to this location, resulting in dismal prognosis of this disease [4–6]. Therefore, in order to develop a novel effective therapy and

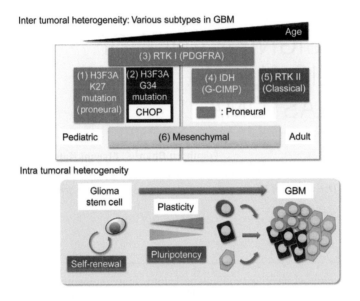

FIGURE 15.1

Intertumoral and intratumoral heterogeneity in GBM. *Intertumoral heterogeneity*: GBM can be divided into six subtypes, (1) K27, (2) G34, (3) RTK I, (4) IDH, (5) RTK II, and (6) mesenchymal, based on their transcriptional and DNA methylation status. Two of these clusters are closely associated with K27 and G34 mutations of *H3F3A*. Interestingly, GBM with G34 mutations shows widespread hypomethylation across the whole genome, defined as CpG hypomethylator phenotype (CHOP), while GBM with *IDH1* mutations shows hypermethylator phenotype (G-CIMP). The other cluster is enriched with *PDGFRA* amplification, termed as receptor tyrosine kinase 1 (RTK I). *Intratumoral heterogeneity*: GBM consists of various cell types (intratumoral heterogeneity). Such multiple distinct subpopulations of cancer cells within tumors may derive from a limited source of cancer cells (GSC) that have plasticity and respond to signals they receive from their microenvironment.

produce better patient outcomes, a more thorough understanding of the molecular pathways of pediatric GBM is required.

Recently, The Cancer Genome Atlas (TCGA) groups identified aberrant accumulation of DNA methylation in a subset of GBM, defined as glioma-CpG island methylator phenotype (G-CIMP) [7]. Additionally, TCGA have analyzed comprehensive genetic and epigenetic alterations and identified a large number of inactivating mutations in genes that control the epigenome. One study reported that 46% of cases in 291 GBM had at least one somatic mutation in genes associated with chromatin modification [8]. These indicate that interplay between genetic alterations and subsequent accumulation of epigenetic alterations may play pivotal roles in GBM formation (Figure 15.1) [9]. Thus, elucidating the mechanisms how epigenetic alterations are induced and contribute to tumor formation may be a clue for potential therapeutic development. This chapter outlines recent advances in genetic and epigenetic research with respect to gliomas and discusses the clinical implications of the development of novel therapies for this devastating disease.

2 INTERTUMORAL AND INTRATUMORAL HETEROGENEITY OF GBM

Although temozolomide (TMZ), alkylating agent, is widely used for the treatment of GBM, the effect of this drug is limited and tumor progression still occurs in >40% of patients [2]. This difficulty in treatment may be due to the existence of inter- and intratumoral heterogeneity in most of GBMs. As the term glioblastoma "multiforme" suggests, GBM consists of various cells. For example, a subset of cells shows glial fibrillary acidic protein (GFAP) and platelet-derived growth factor receptor alpha (PDGFRA), which are the characteristic markers of differentiated glial cells. By contrasts, another set of cells show high expression of Nestin or Musashi-1, which is generally expressed in neural stem cell population. After GBM patient undergoes treatment, although a large number of tumor cells are eradicated, residual treatment-resistant cells regenerate tumors. Thus, "intratumoral" heterogeneity offers a possible reason as to why certain treatment strategies fail in GBM patients.

In order to elucidate how GBM establishes intratumoral heterogeneity, glioma stem cells (GSCs) theory is proposed [10–12]. GSC displays stem cell properties which are self-renewal capacity and pluripotency (Figure 15.1). GSC can differentiate into different cell lineages, such as glial cell type and neuronal cell type, in response to signals from their microenvironment. In addition to conversion from GSCs into non-GSCs, the reverse process is now also being considered [13]. Studies indicated that this plasticity of GSC is regulated by epigenetic mechanisms and may associate with establishment of intratumoral heterogeneity [13–15]. Epigenetic regulation of GSCs is further discussed later.

3 ABERRANT DNA METHYLATION IN GBM

Epigenetic alterations have emerged as common hallmarks of various cancers, including GBM [7–9,16,17]. Recent genome-wide DNA methylation analyses reveal that cancer cells exhibit a distinct DNA methylation profile compared with normal cells. Cancer cells exhibit global DNA hypomethylation and hypermethylation of several specific gene promoters. Several cancers show a characteristic subtype, in which a large number of CpG islands in specific genes are methylated, termed as CpG island methylator phenotype (CIMP) group. In colon cancer, CIMP tumors exhibit characteristic features, frequent *BRAF* and *KRAS* mutations and low incidence of *TP53* mutations [18,19]. These indicate that interplay between specific genetic alterations and aberrant DNA methylation may affect clinical features.

It is demonstrated that GBM can be divided into several subtypes, including glioma-CIMP (G-CIMP) subtype and CpG hypomethylator phenotype (CHOP) subtype based on their DNA methylation status. These subtypes exhibit characteristic clinical features. G-CIMP subtype mostly consists of young adult patients, while CHOP consists of pediatric patients. G-CIMP subtype is associated with isocitrate dehydrogenase 1 (*IDH1*) mutation and CHOP subtype is associated with *H3F3A* mutation [9]. These indicate that interplay between genetic alterations and aberrant DNA methylation may contribute to GBM formation.

Recent studies explained the mechanistic link between genetic alterations and epigenetic alterations in tumor cells. IDH1 is a metabolic enzyme in tricarboxylic acid (TCA) cycle. *IDH1* mutation, which is a recurrent point mutation, is frequently seen in hematological malignancy and glioma [20,21]. IDH1 catalyzes conversion of isocitrate to alpha-ketoglutarate (α-KG), while mutated IDH1 produces 2-hydroxyglutarate (2-HG) from isocitrate instead of α-KG [22,23]. Aberrantly accumulated 2-HG

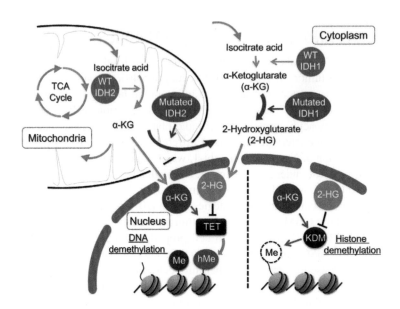

FIGURE 15.2

IDH mutation contributes to epigenetic alterations. Mutated IDH1 and IDH2 gain the ability to produce the metabolite, 2-HG, which inhibits α-KG-dependent dioxygenases, including histone demethylases and the TET protein family. Therefore, mutation of IDH1 is the mechanistic cause of G-CIMP through inhibition of the TET-mediated production of 5-hmC, which is a primary mode of DNA demethylation.

inhibits the function of α-KG-dependent enzymes, such as Ten-eleven translocation (TET) or lysine (K) specific histone demethylase (KDM) [24–26]. TET catalyzes the conversion of 5-methylcytosine (5-mC) to 5-hydroxymethylcytosine (5-hmC) in α-KG dependent manner. The conversion of 5-mC to 5-hmC has been proposed as an initial step of DNA demethylation. However, accumulation of 2-HG inhibits the function of TET, results in accumulation of DNA methylation (Figure 15.2). Taken together, *IDH1* mutation may contribute to formation of G-CIMP subtype in GBM. Indeed, a study demonstrated that overexpression of mutated *IDH1* in human astrocytes induced accumulation of aberrant DNA methylation [27]. In addition to DNA methylation, accumulation of 2-HG is also considered to be associated with aberrant histone modifications through inhibition of KDMs. KDM4C, an α-KG dependent enzyme, demethylates histone H3K9 and K27 methylation. Inhibition of KDM4C results in aberrantly accumulated methylation of histone tail (i.e., increase of H3K9me3, H3K27me3) in *IDH1*-mutated GBM [25].

4 ABERRANT HISTONE MODIFICATIONS AND CHROMATIN REMODELING IN GBM

It has been known that histone modification regulates gene transcription since 1960s. To date, more than 16 modifications have been identified and their regulation are associated with various biological

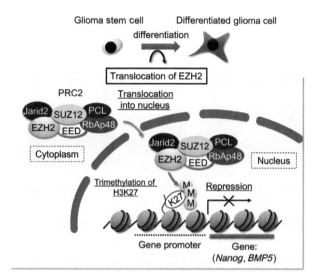

FIGURE 15.3

Translocation of EZH2 into nucleus during GSC differentiation. When GSC differentiates into non-GSC, EZH2 translocates into nucleus and methylates H3K27 in specific loci (e.g., *Nanog, BMP5*) as an early step of differentiation. This reveals that PRC2-mediated epigenetic mechanism plays an important role in GSC differentiation through aberrant H3K27me3 modifications of specific genes.

processes. Polycomb group proteins (PcG) are known for silencing *HOX* genes, which are responsible for assigning specific regional identities on body parts during development via regulation of histone modifications [28,29]. In GBM, Enhancer of Zeste Homolog 2 (EZH2) is frequently overexpressed. EZH2, a catalytic subunit of polycomb repressive complex 2 (PRC2), has a histone methyltransferase activity (Figure 15.3). EZH2 generally acts as a gene repressor through H3K27me3. B lymphoma Mo-MLV insertion region 1 homolog (BMI1), a component of another PcG complex (PRC1), is frequently overexpressed in GBM [30].

Recently, studies showed that recurrent mutation of *H3F3A*, which encodes histone H3.3, is frequently found in pediatric GBM [31,32]. *H3F3A* mutation exhibits two hotspots, which affect two critical amino acids of histone tail, K27 and Glycine 34 (G34). Intriguingly, genome-wide transcriptome analysis revealed that GBM with mutations of K27 and G34 exhibits distinct pattern of gene expression and DNA methylation [9]. G34 mutations are closely associated with CHOP subtypes. Methionine 27 (M27), which is encoded by K27 mutation, aberrantly binds to PRC2 and impairs the catalytic function of EZH2, results in reduction of H3K27me3 level (Figure 15.4). Since *H3F3A* mutation is mutually exclusive with *IDH1* mutation, altered histone modifications by *H3F3A* mutations may drive tumor formation in distinct epigenetic pathway than G-CIMP.

Recently, somatic mutation of *ATRX* (α-thalassemia/mental retardation syndrome X-linked) is identified in 6% of adult GBMs and in 14–29% of pediatric GBMs [17]. Mutation of *ATRX* is associated with an X-linked mental retardation (XLMR) syndrome and most often accompanied by ATRX syndrome. ATRX protein is an important regulator of chromatin structure through its ATRX-DNMT3-DNMT3L

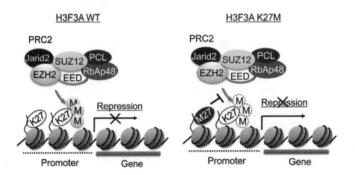

FIGURE 15.4

K27M mutation of H3F3A inhibits PRC2 function. K27M mutation of *H3F3A* is heterozygous mutation and impaired its function. H3M27 aberrantly binds to PRC2 and interferes enzymatic activity of EZH2. K27M tumors reveal that global downregulation of H3K27me3 level and aberrant gene expression profiling.

(ADD) domain and chromatin remodeling domain. In ATRX syndrome, mutations of *ATRX* genes are within these two domains and affect chromatin structure, histone modifications and DNA methylation [33,34]. Intriguingly, mutations of *ATRX* genes also impair H3.3 loading at telomeres and disrupt their heterochromatic state, affecting alternative lengthening of telomeres (ALT) [31]. ALT is defined as telomere length maintenance that is not dependent on telomerase activity. Activation of ALT may also contribute to GBM formation in ATRX-mutated cells.

Together, various epigenetic alterations, which are induced by genetic alterations in epigenetic modifiers, appear to contribute to GBM formation.

5 CONTRIBUTION OF GSC TO GBM FORMATION AND PROGRESSION

As aforementioned above, epigenetic alterations regulate stem cell properties of GSC, which is closely associated with intratumoral heterogeneity. GSCs are considered able to aberrantly differentiate into diverse cell types in response to oncogenic cues [35,36]. Whereas GSCs can differentiate into non-GSCs, the reverse process is now also being considered regulated by epigenetic alterations. Recently, it is reported that epigenetic regulation by EZH2 is a key mediator of tumor cell plasticity, which is required for the adaptation of glioblastoma cells to their microenvironment. At an early step of GSC differentiation, EZH2 translocates from cytoplasm into nucleus and changes distribution of H3K27me3 (Figure 15.3) [13]. In addition, inhibition of EZH2 results in dysregulation of differentiation as well as dedifferentiation and impairs tumor cell invasion. Thus, bidirectional conversion of GSC in response to the extrinsic cues is regulated by plastic EZH2-mediated epigenetic mechanism, which is important for the tumor expansion and may contribute to establishment of tissue heterogeneity.

Tumor cell plasticity also evokes another plausible therapeutic problem. Aberrant self-renewal may be a mechanism that mediates tumorigenicity/recurrence by replenishing the cancer stem cell pool; therefore, developing cancer stem cell-specific therapies should be required. However, interconversion between GSCs and non-GSCs exists, which may raise the possibility that therapeutic elimination of GSCs is followed by their regeneration from residual non-GSCs, allowing tumor regrowth and clinical relapse. As aforementioned, intratumoral heterogeneity is linked to therapeutic resistance, tumor recurrence, and metastatic progression. Targeting this plasticity may be a novel strategy for cancer treatment that attacks molecular mechanisms underlying the tumor heterogeneity, in which tumor cells exist in multiple states of differentiation that show distinct tumor-seeding properties.

6 MECHANISM OF ESTABLISHMENT OF H3K27ME3 LANDSCAPE DURING GSC DIFFERENTIATION

Knowledge about how the epigenome is altered specifically at certain loci and how this affects the phenotypes of GBM is still limited. Following the discovery of GSCs, it is important to elucidate the epigenetic mechanisms by which environmental cues control the differentiation of GSCs into the diverse array of cell types that form GBMs.

Recent genome-wide studies have shown that the large number of noncoding RNAs (ncRNAs) that are transcribed from the human genome include a group termed long ncRNAs (lncRNAs). LncRNA is generally greater than 200 nucleotides and up to 100 kb in length. LncRNA binds to protein including chromatin modifying proteins, such as methylases, demethylases, and deacetylases, and plays important roles in formation of protein complexes, serving as scaffold [37]. A subset of lncRNA is reported to bind to PcG and regulates specific gene transcription via recruiting these proteins to target loci [38]. HOTAIR binds to PRC2 and recruits PRC2 to Homeobox D (*HOXD*) genes cluster, resulting in repression of *HOXD* [39]. Additionally, HOTAIR also binds to LSD1, a demethyltransferase of H3K4, and inhibits the methylation of H3K4 [40]. Thus, lncRNA serves as scaffold through which several proteins bind to each other and contributes to form repressive chromatin structure efficiently at certain specific loci. Similarly, other lncRNAs are identified and are thought to play various roles [41,42]. However, the scaffold function has been identified in only a few lncRNAs. Further investigations are required to clarify the functional roles of lncRNAs in order to elucidate the gene regulatory mechanisms that are important in gliomagenesis. If specific lncRNAs are identified, which play important roles in GBM formation, those may be a potent target for novel treatment strategy.

7 NOVEL TREATMENT STRATEGY FOR GBM

Recently, many selective inhibitors targeting aberrant molecular pathway are used for clinical trial in GBM, such as targeting receptor tyrosine kinase (RTK) signaling (erlotinib, gefitinib, cetuximab, imatinib) and targeting angiogenesis (cediranib) (Table 15.1) [43–47]. However, these drugs did not provide a significant prolonged survival. Bevacizumab is a recombinant humanized monoclonal antibody that binds to human vascular endothelial growth factor (VEGF). Two randomized trials for newly diagnosed GBM patients have been tested, however, combination of Bevacizumab plus radiotherapy-TMZ did not improve overall survival (Table 15.1) [48,49].

Accumulating studies showed the importance of various epigenetic alterations during GBM formation. Until now, a few epigenetic drugs have been successfully used for cancer treatment. 5-aza-2'-deoxycytidine, a DNA demethylating agent, was the first epigenetic drug approved for the treatment of myelodysplastic syndromes (MDS). Its survival benefit over best conventional care was confirmed

Table 15.1 List of Several Clinical Trials, Evaluating the Efficiency of Novel Drugs

Treatment	Target	Drug	Trial Identifier[a]	Trial Name	Phase	Reference
RTK inhibitor	EGFR	Erlotinib	NCT00187486	Phase II study of tarceva plus temodar during and following radiation therapy in patients with newly diagnosed glioblastoma multiforme and gliosarcoma	II	[43]
		Gefitinib	NCT00025675	ZD1839 for treatment of recurrent or progressive malignant astrocytoma or glioblastoma and recurrent or progressive meningioma: A Phase II study with a phase i component for patients receiving EIAEDs	II	[44]
		Cetuximab	Not available in ClinicalTrials.gov		II	[45]
		Imatinib	NCT00039364	Open-label Phase II study on STI571 (Glivec) administered as a daily oral treatment in gliomas	II	[46]
	VEGF	Cediranib	NCT00777153	A Phase III, randomized, parallel group, multicenter study in recurrent glioblastoma patients to compare the efficacy of cediranib (RECENTIN™, AZD2171) monotherapy and the combination of cediranib with lomustine to the efficacy of lomustine alone	III	[47]
Monoclonal antibody	VEGF-A	Bevacizumab	NCT00943826	A study of Avastin® (Bevacizumab) in combination with TMZ and radiotherapy in patients with newly diagnosed glioblastoma	III	[48]
			NCT00884741	Phase III, double-blind, placebo-controlled trial of conventional concurrent chemoradiation and adjuvant TMZ plus bevacizumab vs. conventional concurrent chemoradiation and adjuvant TMZ in patients with newly diagnosed glioblastoma	III	[49]
Inhibitor of epigenetic modifiers	HDAC	Vorinostat	NCT00238303	Phase II trial of suberoylanilide hydroxamic acid (SAHA) in patients with recurrent glioblastoma	II	[55]

RTK, Receptor tyrosine kinase; *EGFR*, epidermal growth factor receptor; *VEGF*, vascular endothelial growth factor; *HDAC*, histone deacetylase.
[a]*Trial Identifier shows ClinicalTrials.gov number.*

[50]. Inhibitors of EZH2 have been identified independently from multiple groups [51–53]. Among them, Phase 1/2 study using E7438 for diffuse large B-cell lymphoma patients has been started in 2013. Regarding epigenetic treatments for GBM, a recent study showed that AGI-5198, an inhibitor of R132H-mutant IDH1 activity, results in decrease of 2-HG level and growth inhibition of glioma cells in both *in vitro* and *in vivo* [54]. This indicates that targeting epigenetic alterations might be an attractive option for a subset of GBM. Several clinical trials targeting epigenetic modifiers are also ongoing in GBM. Histone deacetylase (HDAC) inhibitors, such as valproic acid (VPA) or vorinostat, are investigated in Phase 1 or 2 studies (Table 15.1) [55,56].

A rationale for epigenetic treatment for cancers is based on pharmacologic targeting of various core transcriptional programs that sustains cancer cell identity. This also indicates that targeting aberrant epigenetic modifier may be effective for multiple processes in comparison with selective inhibitor (e.g., RTK inhibitor) of aberrant single signaling pathway. Since subset of GBMs displayed aberrant DNA hypermethylation and dysregulation of EZH2, which appear to play an important role in tumor formation, progression and establishment of tissue heterogeneity, targeting such aberrant epigenome might be a promising and effective therapeutic strategy for GBM treatment. Notably, K27 heterozygous mutation of *H3F3A* gene substitutes the K27 with a methionine (K27M). M27 aberrantly binds to PRC2 and impairs the catalytic function of EZH2, resulting in reduction of H3K27me3 level (Figure 15.4) [57]. These suggest that in pediatric GBM with K27M mutation of *H3F3A*, targeting EZH2 is not appropriate. In order to develop effective epigenetic therapeutic strategy for GBM, further investigations are required to clarify the interplay between genetic and epigenetic alterations during gliomagenesis in each GBM subtype. Based on the knowledge of genetic and epigenetic profiling of each GBM, targeting specific epigenetic alterations may be effective strategy for GBM treatment.

ACKNOWLEDGMENTS

This work was supported by the P-Direct from the Ministry of Education, Culture, Sports, Science and Technology of Japan and Grant-in-Aid for Scientific Research from the Japan Society for the Promotion of Science (25290048).

REFERENCES

[1] Louis DN, Oghaki H, Wiestler OD, Cavenee WK. WHO classification of tumors of the central nervous system. Lyon: IARC; 2007.

[2] Stupp R, Mason WP, van den Bent MJ, Weller M, Fisher B, Taphoorn MJ, et al. Radiotherapy plus concomitant and adjuvant temozolomide for glioblastoma. N Engl J Med 2005;352(10):987–96.

[3] Stupp R, Hegi ME, Mason WP, van den Bent MJ, Taphoorn MJ, Janzer RC, et al. Effects of radiotherapy with concomitant and adjuvant temozolomide versus radiotherapy alone on survival in glioblastoma in a randomised phase III study: 5-year analysis of the EORTC-NCIC trial. Lancet Oncol 2009;10(5):459–66.

[4] Broniscer A, Gajjar A. Supratentorial high-grade astrocytoma and diffuse brainstem glioma: two challenges for the pediatric oncologist. Oncologist 2004;9(2):197–206.

[5] Hargrave D, Bartels U, Bouffet E. Diffuse brainstem glioma in children: critical review of clinical trials. Lancet Oncol 2006;7(3):241–8.

[6] Jansen MH, van Vuurden DG, Vandertop WP, Kaspers GJ. Diffuse intrinsic pontine gliomas: a systematic update on clinical trials and biology. Cancer Treat Rev 2012;38(1):27–35.

[7] Noushmehr H, Weisenberger DJ, Diefes K, Phillips HS, Pujara K, Berman BP, et al. Identification of a CpG island methylator phenotype that defines a distinct subgroup of glioma. Cancer Cell 2010;17(5): 510–22.

[8] Brennan CW, Verhaak RG, McKenna A, Campos B, Noushmehr H, Salama SR, et al. The somatic genomic landscape of glioblastoma. Cell 2013;155(2):462–77.

[9] Sturm D, Witt H, Hovestadt V, Khuong-Quang DA, Jones DT, Konermann C, et al. Hotspot mutations in H3F3A and IDH1 define distinct epigenetic and biological subgroups of glioblastoma. Cancer Cell 2012;22(4):425–37.

[10] Gupta PB, Chaffer CL, Weinberg RA. Cancer stem cells: mirage or reality? Nat Med 2009;15(9):1010–12.

[11] Polyak K, Weinberg RA. Transitions between epithelial and mesenchymal states: acquisition of malignant and stem cell traits. Nat Rev Cancer 2009;9(4):265–73.

[12] Gupta PB, Fillmore CM, Jiang G, Shapira SD, Tao K, Kuperwasser C, et al. Stochastic state transitions give rise to phenotypic equilibrium in populations of cancer cells. Cell 2011;146(4):633–44.

[13] Natsume A, Ito M, Katsushima K, Ohka F, Hatanaka A, Shinjo K, et al. Chromatin regulator PRC2 is a key regulator of epigenetic plasticity in glioblastoma. Cancer Res 2013;73(14):4559–70.

[14] Katsushima K, Shinjo K, Natsume A, Ohka F, Fujii M, Osada H, et al. Contribution of microRNA-1275 to Claudin11 protein suppression via a polycomb-mediated silencing mechanism in human glioma stem-like cells. J Biol Chem 2012;287(33):27396–406.

[15] Roesch A, Fukunaga-Kalabis M, Schmidt EC, Zabierowski SE, Brafford PA, Vultur A, et al. A temporarily distinct subpopulation of slow-cycling melanoma cells is required for continuous tumor growth. Cell 2010;141(4):583–94.

[16] Jones PA, Baylin SB. The epigenomics of cancer. Cell 2007;128(4):683–92.

[17] Sturm D, Bender S, Jones DT, Lichter P, Grill J, Becher O, et al. Paediatric and adult glioblastoma: multiform (epi)genomic culprits emerge. Nat Rev Cancer 2014;14(2):92–107.

[18] Toyota M, Ahuja N, Ohe-Toyota M, Herman JG, Baylin SB, Issa JP. CpG island methylator phenotype in colorectal cancer. Proc Natl Acad Sci USA 1999;96(15):8681–6.

[19] Issa JP. CpG island methylator phenotype in cancer. Nat Rev Cancer 2004;4(12):988–93.

[20] Parsons DW, Jones S, Zhang X, Lin JC, Leary RJ, Angenendt P, et al. An integrated genomic analysis of human glioblastoma multiforme. Science 2008;321(5897):1807–12.

[21] Wen PY, Kesari S. Malignant gliomas in adults. N Engl J Med 2008;359(5):492–507.

[22] Dang L, White DW, Gross S, Bennett BD, Bittinger MA, Driggers EM, et al. Cancer-associated IDH1 mutations produce 2-hydroxyglutarate. Nature 2009;462(7274):739–44.

[23] Ward PS, Patel J, Wise DR, Abdel-Wahab O, Bennett BD, Coller HA, et al. The common feature of leukemia-associated IDH1 and IDH2 mutations is a neomorphic enzyme activity converting alpha-ketoglutarate to 2-hydroxyglutarate. Cancer Cell 2010;17(3):225–34.

[24] Figueroa ME, Abdel-Wahab O, Lu C, Ward PS, Patel J, Shih A, et al. Leukemic IDH1 and IDH2 mutations result in a hypermethylation phenotype, disrupt TET2 function, and impair hematopoietic differentiation. Cancer Cell 2010;18(6):553–67.

[25] Lu C, Ward PS, Kapoor GS, Rohle D, Turcan S, Abdel-Wahab O, et al. IDH mutation impairs histone demethylation and results in a block to cell differentiation. Nature 2012;483(7390):474–8.

[26] Xu W, Yang H, Liu Y, Yang Y, Wang P, Kim SH, et al. Oncometabolite 2-hydroxyglutarate is a competitive inhibitor of alpha-ketoglutarate-dependent dioxygenases. Cancer Cell 2011;19(1):17–30.

[27] Turcan S, Rohle D, Goenka A, Walsh LA, Fang F, Yilmaz E, et al. IDH1 mutation is sufficient to establish the glioma hypermethylator phenotype. Nature 2012;483(7390):479–83.

[28] Boyer LA, Plath K, Zeitlinger J, Brambrink T, Medeiros LA, Lee TI, et al. Polycomb complexes repress developmental regulators in murine embryonic stem cells. Nature 2006;441(7091):349–53.

[29] Lee TI, Jenner RG, Boyer LA, Guenther MG, Levine SS, Kumar RM, et al. Control of developmental regulators by Polycomb in human embryonic stem cells. Cell 2006;125(2):301–13.

[30] Hayry V, Tanner M, Blom T, Tynninen O, Roselli A, Ollikainen M, et al. Copy number alterations of the polycomb gene BMI1 in gliomas. Acta Neuropathol 2008;116(1):97–102.

[31] Schwartzentruber J, Korshunov A, Liu XY, Jones DT, Pfaff E, Jacob K, et al. Driver mutations in histone H3.3 and chromatin remodelling genes in paediatric glioblastoma. Nature 2012;482(7384):226–31.

[32] Wu G, Broniscer A, McEachron TA, Lu C, Paugh BS, Becksfort J, et al. Somatic histone H3 alterations in pediatric diffuse intrinsic pontine gliomas and non-brainstem glioblastomas. Nat Genet 2012;44(3):251–3.

[33] Higgs DR, Garrick D, Anguita E, De Gobbi M, Hughes J, Muers M, et al. Understanding alpha-globin gene regulation: aiming to improve the management of thalassemia. Ann N Y Acad Sci 2005;1054:92–102.

[34] Iwase S, Xiang B, Ghosh S, Ren T, Lewis PW, Cochrane JC, et al. ATRX ADD domain links an atypical histone methylation recognition mechanism to human mental-retardation syndrome. Nat Struct Mol Biol 2011;18(7):769–76.

[35] Chen J, Li Y, Yu TS, McKay RM, Burns DK, Kernie SG, et al. A restricted cell population propagates glioblastoma growth after chemotherapy. Nature 2012;488(7412):522–6.

[36] Cheng L, Huang Z, Zhou W, Wu Q, Donnola S, Liu JK, et al. Glioblastoma stem cells generate vascular pericytes to support vessel function and tumor growth. Cell 2013;153(1):139–52.

[37] Gupta RA, Shah N, Wang KC, Kim J, Horlings HM, Wong DJ, et al. Long non-coding RNA HOTAIR reprograms chromatin state to promote cancer metastasis. Nature 2010;464(7291):1071–6.

[38] Katsushima K, Kondo Y. Non-coding RNAs as epigenetic regulator of glioma stem-like cell differentiation. Front Genet 2014;5:14.

[39] Rinn JL, Kertesz M, Wang JK, Squazzo SL, Xu X, Brugmann SA, et al. Functional demarcation of active and silent chromatin domains in human HOX loci by noncoding RNAs. Cell 2007;129(7):1311–23.

[40] Tsai MC, Manor O, Wan Y, Mosammaparast N, Wang JK, Lan F, et al. Long noncoding RNA as modular scaffold of histone modification complexes. Science 2010;329(5992):689–93.

[41] Khalil AM, Guttman M, Huarte M, Garber M, Raj A, Rivea Morales D, et al. Many human large intergenic noncoding RNAs associate with chromatin-modifying complexes and affect gene expression. Proc Natl Acad Sci USA 2009;106(28):11667–72.

[42] Kaneko S, Li G, Son J, Xu CF, Margueron R, Neubert TA, et al. Phosphorylation of the PRC2 component Ezh2 is cell cycle-regulated and up-regulates its binding to ncRNA. Genes Dev 2010;24(23):2615–20.

[43] Prados MD, Chang SM, Butowski N, DeBoer R, Parvataneni R, Carliner H, et al. Phase II study of erlotinib plus temozolomide during and after radiation therapy in patients with newly diagnosed glioblastoma multiforme or gliosarcoma. J Clin Oncol 2009;27(4):579–84.

[44] Lassman AB, Rossi MR, Raizer JJ, Abrey LE, Lieberman FS, Grefe CN, et al. Molecular study of malignant gliomas treated with epidermal growth factor receptor inhibitors: tissue analysis from North American Brain Tumor Consortium Trials 01-03 and 00-01. Clin Cancer Res 2005;11(21):7841–50.

[45] Neyns B, Sadones J, Joosens E, Bouttens F, Verbeke L, Baurain JF, et al. Stratified phase II trial of cetuximab in patients with recurrent high-grade glioma. Ann Oncol 2009;20(9):1596–603.

[46] Raymond E, Brandes AA, Dittrich C, Fumoleau P, Coudert B, Clement PM, et al. Phase II study of imatinib in patients with recurrent gliomas of various histologies: a European Organisation for Research and Treatment of Cancer Brain Tumor Group Study. J Clin Oncol 2008;26(28):4659–65.

[47] Batchelor TT, Mulholland P, Neyns B, Nabors LB, Campone M, Wick A, et al. Phase III randomized trial comparing the efficacy of cediranib as monotherapy, and in combination with lomustine, versus lomustine alone in patients with recurrent glioblastoma. J Clin Oncol 2013;31(26):3212–18.

[48] Chinot OL, Wick W, Mason W, Henriksson R, Saran F, Nishikawa R, et al. Bevacizumab plus radiotherapy-temozolomide for newly diagnosed glioblastoma. N Engl J Med 2014;370(8):709–22.

[49] Gilbert MR, Dignam JJ, Armstrong TS, Wefel JS, Blumenthal DT, Vogelbaum MA, et al. A randomized trial of bevacizumab for newly diagnosed glioblastoma. N Engl J Med 2014;370(8):699–708.

[50] Kantarjian H, Issa JP, Rosenfeld CS, Bennett JM, Albitar M, DiPersio J, et al. Decitabine improves patient outcomes in myelodysplastic syndromes: results of a phase III randomized study. Cancer 2006;106(8):1794–803.

[51] McCabe MT, Ott HM, Ganji G, Korenchuk S, Thompson C, Van Aller GS, et al. EZH2 inhibition as a therapeutic strategy for lymphoma with EZH2-activating mutations. Nature 2012;492(7427):108–12.

[52] Qi W, Chan H, Teng L, Li L, Chuai S, Zhang R, et al. Selective inhibition of Ezh2 by a small molecule inhibitor blocks tumor cells proliferation. Proc Natl Acad Sci USA 2012;109(52):21360–5.

[53] Knutson SK, Wigle TJ, Warholic NM, Sneeringer CJ, Allain CJ, Klaus CR, et al. A selective inhibitor of EZH2 blocks H3K27 methylation and kills mutant lymphoma cells. Nat Chem Biol 2012;8(11):890–6.

[54] Rohle D, Popovici-Muller J, Palaskas N, Turcan S, Grommes C, Campos C, et al. An inhibitor of mutant IDH1 delays growth and promotes differentiation of glioma cells. Science 2013;340(6132):626–30.

[55] Galanis E, Jaeckle KA, Maurer MJ, Reid JM, Ames MM, Hardwick JS, et al. Phase II trial of vorinostat in recurrent glioblastoma multiforme: a North Central Cancer Treatment Group Study. J Clin Oncol 2009;27(12):2052–8.

[56] Friday BB, Anderson SK, Buckner J, Yu C, Giannini C, Geoffroy F, et al. Phase II trial of vorinostat in combination with bortezomib in recurrent glioblastoma: a North Central Cancer Treatment Group Study. Neuro Oncol 2012;14(2):215–21.

[57] Bender S, Tang Y, Lindroth AM, Hovestadt V, Jones DT, Kool M, et al. Reduced H3K27me3 and DNA hypomethylation are major drivers of gene expression in K27M mutant pediatric high-grade gliomas. Cancer Cell 2013;24(5):660–72.

ESOPHAGEAL CANCER

16

Zhe Jin[1,2,3], Xiaojing Zhang[1,3], and Xinmin Fan[1]

[1]*Department of Pathology, The Shenzhen University School of Medicine, Shenzhen, Guangdong, People's Republic of China,* [2]*Shenzhen Key Laboratory of Micromolecule Innovatal Drugs, The Shenzhen University School of Medicine, Shenzhen, Guangdong, People's Republic of China,* [3]*Shenzhen Key Laboratory of Translational Medicine of Tumor, The Shenzhen University School of Medicine, Shenzhen, Guangdong, People's Republic of China*

CHAPTER OUTLINE

1 INTRODUCTION

Esophageal cancer ranks as the eighth most frequent and sixth most fatal cancer type worldwide, with an estimated 480,000 new cases diagnosed and 400,000 deaths globally in 2008 [1]. This malignancy exists in two major histologic subtypes: esophageal adenocarcinoma (EAC), which is more prevalent in Western countries, with a rapidly increasing incidence; and esophageal squamous cell carcinoma (ESCC), which is more frequent in developing countries, especially in Asia [2]. Although significant advances have been made in treating esophageal cancers, these aggressive malignancies usually present as locally advanced disease with a very poor prognosis (approximately 14% 5-year survival) [1].

EAC and ESCC possess distinct pathological characteristics: the striking variety of geographic distribution of these two subtypes possibly reflects the differences in exposure to specific environmental factors. Risk factors for EAC include smoking, overweight and obesity, male sex, Caucasian race, and chronic gastroesophageal reflux disease (GERD), which is thought to trigger Barrett's esophagus (BE). Risk factors for ESCC are thought to include tobacco smoking, alcohol consumption, poor nutritional status, and low intake of fruits and vegetables. BE is the premalignant condition for EAC where the normal squamous epithelium is replaced by a columnar-lined epithelium in the distal portion of the esophagus [3]. This condition is most frequently associated with long-standing GERD. BE can proceed through low-grade dysplasia (LBE), high-grade dysplasia (HBE), intramucosal carcinoma, and eventually becomes invasive carcinoma at an overall rate of progression to EAC of 0.33% per person per year [3]. ESCC appears to develop via a hyperplasia–dysplasia (low, intermediate, and high-grade)–carcinoma *in situ*-invasive carcinoma sequence [2]. The currently accepted marker for esophageal cancer risk is histologic dysplasia, with high-grade dysplasia (HGD) being considered more accurate than low-grade dysplasia (LGD). As a result, there has been interest in screening to detect asymptomatic precursor lesions namely BE and squamous cell dysplasia. BE is usually found in middle-aged adults, and EAC risk in BE is increased 30- to 125-fold relative to the general population [4]. Endoscopic surveillance in BE patients is recommended at intervals of 2–3 years [5]. Esophageal cancer detected by endoscopic surveillance at an early stage prior to lymph node spread and when the disease is confined to the mucosa or submucosa had better prognoses (approximately 80% 5-year survival) [6,7]. However, endoscopic surveillance suffers from high cost, inconvenience, patient anxiety, low yield, and procedure-related risks. In addition, the current marker of EAC risk in BE, dysplasia, is plagued by high interobserver variability and limited predictive accuracy [8]. Because neoplastic progression is infrequent (approximately 1/200 patient-years) in BE, the merits and appropriate interval for endoscopic surveillance in BE have led to frequent debate [5]. This process would benefit greatly from effective biomarkers to stratify patients according to their level of neoplastic progression risk.

Of note, intensive molecular biological studies have revealed the molecular mechanisms involved in the development and progression of esophageal cancer. Progression of esophageal cancer is a multistep process that begins with the accumulation of genetic and epigenetic alterations, and leads to the inactivation of tumor suppressor genes (TSGs) and the activation of oncogenes. To date, DNA hypermethylation and miRNA deregulation are the best-documented epigenetic alterations in human esophageal cancer. DNA methylation and miRNA deregulation are recognized as common molecular alterations and contribute to the neoplastic development and progression by different mechanisms in human esophageal cancer. Furthermore, hypermethylation of TSG promoter regions and deregulation of miRNA occur not only in advanced cancer but also in premalignant lesions, and are related to prognosis and response to chemoradiotherapy in esophageal cancer. This chapter is mainly focused on these epigenetic changes in esophageal cancer and the value of early detection, risk stratification, prognosis, predicting response to treatment and potential targeting therapy, with special attention to DNA methylation and miRNA deregulation.

2 DNA METHYLATION

Esophageal cancer cells exhibit two types of alterations of DNA methylation: global DNA hypomethylation and promoter CpG island hypermethylation of TSGs. During the past decade, the list of TSGs inactivated by promoter hypermethylation has grown considerably to include essentially most of the pathways vital for tumorigenesis such as cell cycle regulation, apoptosis, signaling transduction, adhesion, motility, tumor cell invasion and metastasis, angiogenesis, and immune recognition. In contrast, few studies regarding global hypomethylation have been reported in esophageal cancer. Numerous reports have described DNA methylation in EAC and ESCC, the representative genes are summarized in Table 16.1 and discussed later. The wide range of distribution in the frequencies of TSG hypermethylation shown in the table might be due to differences in the patient cohorts (e.g., sample size), the methods used to assess the TSG hypermethylation, or from differential contamination by nonneoplastic cells. In this section, we will mainly discuss the promoter CpG island hypermethylation of TSG in esophageal cancer.

2.1 HYPERMETHYLATED GENES IN BE AND EAC

To date, the hypermethylation of TSGs is well documented in EAC and its premalignant condition BE. Among these studies, hypermethylation of the promoter CpG island of *CDKN2A/p16* (*cyclin-dependent kinase inhibitor2A*), *APC* (*adenomatous polyposis coli*), and *MGMT* (*O-methylguanine-DNA methyltransferase*), which induces transcriptional silencing, has been investigated in both BE and EAC by a number of research groups. Hypermethylation of these TSGs usually occurs in a large contiguous field, suggesting either a concerted methylation change associated with metaplasia or a clonal expansion of hypermethylated cells [16,34].

The tumor suppressor *CDKN2A/p16* encodes a cell cycle regulatory protein CDK1 that inhibits cyclin-dependant kinases 4 and 6, blocks the phosphorylation of pRb protein, and subsequently inhibits cell cycle progression from G1 to S phase. Inactivation of *CDKN2A/P16* gene is one of the most common genetic abnormalities in human neoplasia, including EAC. Several studies have provided evidence that loss of heterozygosity (LOH) of 9p21 (which contains the *p16* locus) combined with mutation or hypermethylation of *CDKN2A/p16* leads to *CDKN2A/p16* inactivation in some individuals with EAC or BE with dysplasia [35–37]. *CDKN2A/p16* is generally unmethylated in normal esophagus (NE) and hypermethylated in BE varied from 3% to 77%, in LBE from 20% to 56%, in HBE from 60% to 75%, and in EAC from 39% to 85% [9].

The tumor suppressor *APC* encodes a protein that acts as an antagonist of the Wnt signaling pathway. It is also involved in other processes including cell migration and adhesion, transcriptional activation, and apoptosis. Defects in this gene cause familial adenomatous polyposis, an autosomal dominant premalignant disease that usually progresses to malignancy. Similar to *CDKN2A/p16*, *APC* is generally unmethylated in NE and hypermethylated in BE varied from 40% to 85%, in 83% LBE, in 66% HBE, and in EAC from 42% to 92% [9].

The tumor suppressor *MGMT* encodes a DNA repair enzyme that removes methyl or alkyl groups at position O6 of guanine and prevents a G:A mutation or a DNA strand break. Unlike *p16* and *APC*, *MGMT* is hypermethylated in NE ranged from 17% to 55%, BE from 25% to 88.9%, dysplastic BE (DBE, including LBE and HBE) from 71.4% to 100%, and EAC from 23% to 78.9% [10–14].

Table 16.1 Representative Hypermethylated Genes in NE, Precursor, EAC and ESCC

Gene	Function	NE	Premalignant Lesion	Frequency of Hypermethylation (%)		References
				EAC	ESCC	
CDKN2A/p16	Cell cycle regulation	1.3%	BE (3–77%); LBE (20–56%); HBE (60–75%); LGD (31%); IGD (42%); HGD (33%)	39–85%	40–62%	[9] (review), [10]
APC	Wnt signaling	6.3%	BE (40–85%); LBE (83%); HBE (66%); LGD (3%); IGD (0%); HGD (0%)	42–92%	50%	[9] (review), [10]
MGMT	DNA repair	17–55%	BE (25–88.9%); DBE (71.4–100%); LGD (23%); IGD (17%); HGD (11%)	23–78.9%	33–39%	[10–15]
E-cadherin	Cell adhesion	3–19.4%	BE (8%); LBE (0%); HBE (0%); LGD (10%); IGD (17%); HGD (33%)	0–83.9%	34%	[10,11,16–18]
DAPK	Apoptosis	0–20%	BE (50%); DBE (53%); LGD (28%); IGD (25%); HGD (11%)	19–60%	26%	[9] (review)
TAC1	Neurotransmitter	7.5%	BE (63.3%); LBE (63.2%); HBE (52.4%)	61.2%	50%	[19]
NELL1	Cell growth and differentiation	0%	BE (46.7%); LBE (42.1%); HBE (61.9%)	47.8%	11.5%	[20]
AKAP12	Signal transduction	0%	BE (48.3%); LBE (52.6%); HBE (52.4%)	52.2%	7.7%	[21]
SST	Tumor suppressor	9%	BE (70%); LBE (63.2%); HBE (71.4%)	71.6%	53.8%	[22]
CDH13	Tumor suppressor	0%	BE (70%); LBE (78.9%); HBE (76.2%)	76.1%	19.2%	[23]
MAL	Tumor suppressor	0%	BE (53.3%); LBE (63.2%); HBE (57.1%)	40.3%	7.7%	[24]
ENG	Angiogenesis	11.9%	BE (13.3%); DBE (25%)	26.9%	46.2%	[25]
Reprimo	Cell cycle regulation	0%	BE (36%); DBE (63.6)	62.6%		[26]
TIMP3	MMP inhibitor	0–19.3%	BE (54.1–88%); DBE (78%)	19.5–86.3%		[11,12,15, 18,27]
hMLH1	Mismatch repair	6%	LGD (8%); IGD (17%); HGD (33%)		23–62%	[28,29]
RRAB	Nuclear receptor	0–39%	Hyperplasia (48%); LGD (13%); IGD (33%); HGD (44%)		36–70%	[25,28,30]
FHIT	Signal transduction	0%	DSE (78%)		14–85%	[10,31,32]
HIN1	Tumor suppressor	0%	LGD (31%); IGD (33%); HGD (44%)		50%	[9] (review)
RASSF1A	Signal transduction	4%			51%	[33]

NE, Normal esophagus; BE, Barrett's esophagus; LBE, BE with low-grade dysplasia; IGD, BE with intermediate-grade dysplasia; HBE, BE with high-grade dysplasia; DBE, dysplastic BE; LGD, squamous epithelium with low-grade dysplasia; IGD, squamous epithelium with intermediate-grade dysplasia; HGD, squamous epithelium with high-grade dysplasia; DSE, dysplastic squamous epithelium; EAC, esophageal adenocarcinoma; ESCC, esophageal squamous cell carcinoma.

E-cadherin (*CDH1*) encodes a transmembrane glycoprotein that mediates calcium-dependent intercellular adhesion that is essential for normal tissue homeostasis. Loss of E-cadherin occurs in a variety of epithelial tumors and is correlated with invasion and metastasis. *E-cadherin* is hypermethylated in NE rated from 3% to 19.4%, in 8% BE, and in EAC from 0% to 83.9% [10,15–17].

TIMP3 (*Tissue inhibitor of metalloproteinase 3*) encodes a multifunctional secreted protein with tumor suppressor properties including the inhibition of matrix metalloproteinases, reduction of metastasis, induction of apoptosis, and inhibition of tumor growth. *TIMP3* is hypermethylated in NE from 0% to 19.3%, in BE from 54.1% to 88%, in 78% DBE, and in EAC from 19.5% to 86.3% [10,11,14,17,18].

2.2 HYPERMETHYLATED GENES IN ESCC

Although hypermethylation of TSG in ESCC is not as well characterized as in EAC, especially for the precursor lesions (i.e., hyperplasia and dysplasia), accumulating evidences have been shown that hypermethylation of TSG is a critical step in carcinogenesis of ESCC. In two recent reviews, as many as 75 hypermethylated genes with tumor-suppressive function are listed, though most of these genes are investigated only by one research group [9,27]. As in EAC, ESCC is characterized by frequent methylation and a lower mutation rate of *CDKN2A/p16*, *APC*, and *E-cadherin* compared with frequent allelic alteration at these loci. *CDKN2A/p16*, *APC*, and *E-cadherin* are hypermethylated in LGD (31%, 3%, and 10%, respectively), intermediate-grade dysplasia (IGD; 42%, 0%, and 17%, respectively), HGD (33%, 0%, and 33%, respectively), and ESCC (40–62%, 50%, and 34%, respectively) [9], suggesting that hypermethylation of *CDKN2A/p16* and *E-cadherin* is an early event, and hypermethylation of *APC*, however, occurs late in ESCC carcinogenesis.

MGMT promoter methylation has been reported in about 30% of human cancers and is the key mechanism responsible for *MGMT* gene silencing [38]. *MGMT* hypermethylation is increased along with ESCC tumorigenesis, and the frequency is reported to be in 23% LGD, 17% IGD, 11% HGD, and 33–39% ESCC and is associated with a reduction in MGMT protein levels [9].

The tumor suppressor *FHIT* (*fragile histidine triad*), which plays an essential role in chromosomal abnormality and DNA damage, is frequently hypermethylated in ESCC but rarely in EAC. Hypermethylation of *FHIT* is reported in 0% NE, 78% dysplasia, and 14–85% ESCC, and is responsible for reduction of its protein [9,31,39].

The tumor suppressor *RARB* (*retinoic acid receptor β*) modulates the activity of retinoids in their regulation of cellular proliferation, differentiation, and apoptosis. *RARB* has been implicated as a tumor suppressor and has been shown to induce growth arrest and apoptosis in cancer cells. *RARB* hypermethylation is detected in NE from 0% to 39%, 48% hyperplasia, 13% LGD, 33% IGD, 44% HGD, and EAC from 36% to 70%, and is related with ESCC grade, suggesting that *RARB* may play an important role in ESCC development [25,30,32].

The *RASSF1A* (*RAS association domain family 1A*) protein inhibits the anaphase-promoting complex and prevents cyclin A and cyclin B degradation until the spindle checkpoint becomes fully operational. Hypermethylation of *RASSF1A* is found in 4% NE and 51% ESCC, and is correlated with reduction of *RASSF1A* mRNA and advanced tumor stage [28].

The *hMLH1* (*mismatch repair gene mutL homolog 1*) plays an important role in the DNA mismatch repair system, and hypermethylation of this gene is the main cause of *hMLH1* silencing, leading to frequent microsatellite instability (MSI) in a subset of human cancers. *hMLH1* hypermethylation has been reported in 6% NE, 8% LGD, 17% IGD, 33% HGD, and ESCC from 23% to 62%, and is associated with MSI in ESCC patients, indicating that *hMLH1* plays a critical role in ESCC [25,29]. *hMLH1* hypermethylation is uncommon and is consistent with the lower MSI prevalence in EAC.

2.3 TSG HYPERMETHYLATION AS BIOMARKER IN ESOPHAGEAL CANCER

As mentioned before, esophageal cancer is often not diagnosed until a late stage when it is incurable and the prognosis of affected patients is unsatisfactory. Furthermore, endoscopy-based screening is insufficient because of poor reliability of the histological grading of precursor lesion and poor specificity to predict the progression to esophageal cancer. Biomarkers are needed to screen multiple stages in the clinical pathway of BE and squamous dysplasia patients; from early diagnosis to risk stratification to prognosis, and therapeutic strategies to prevent and treat esophageal cancer. Although the clear epigenetic events, mainly in the form of TSG hypermethylation, are still being sought after, several potential hypermethylated gene biomarkers have been described in both EAC and ESCC, as well as in the precursor lesion. A subset of these hypermethylated TSGs is predicted to play an important role in the pathogenesis of esophageal cancer. Furthermore, some of these hypermethylated genes might be useful biomarkers as they appear to precede and thus predict the progression of BE to EAC or squamous dysplasia to ESCC.

2.3.1 TSG hypermethylation as biomarker in EAC

In the majority of EAC patients, TSG hypermethylation is acquired very early during EAC carcinogenesis, hence these hypermethylated genes could be used as early diagnostic biomarkers. As summarized in Table 16.1, the frequencies of *CDKN2A/p16*, *APC*, *Reprimo*, *E-cadherin*, and *TIMP3* hypermethylation were relatively low in NE but increased at the very early preneoplastic stage of BE, while being maintained in DBE and EAC. In a series of reports, Jin et al. have systematically investigated hypermethylation frequency and the normalized methylation value (NMV) of 7 genes (*NELL1*, *TAC1*, *AKAP12*, *SST*, *CDH13*, *MAL* and *ENG*) in approximately 260 esophageal tissue specimens of differing histologic types and grades, as well as in esophageal carcinoma cell lines using real-time quantitative methylation-specific PCR [19–24,33]. In all of these studies, both hypermethylation frequency and NMV of these genes are significantly higher in EAC as well as in BE and DBE as compared to NE. Meanwhile, incremental increases in the frequency of *TAC1*, *AKAP12*, *SST*, *CDH13*, and *ENG* hypermethylation are observed during EAC progression (i.e., metaplasia–dysplasia–carcinoma sequence). Treatment esophageal cancer cell lines with the demethylating agent 5-aza-2'-deoxycytidine revealed that reversal of methylation and restoration of mRNA expression, a negative correlation between NMV and mRNA expression was also observed for *AKAP12*, *SST*, *MAL*, and *NELL1* genes in EAC specimens with differing methylation status (i.e., unmethylated vs. hypermethylated). These results are consistent with the interpretation that DNA promoter hypermethylation silences these genes.

Ultimately, the biomarkers will be more useful if they can be detected in the remote specimens from the site of the tumor. Whereas endoscopic biopsies are utilized primarily as the source of tissue for biomarker studies in EAC patients, there are increasing attempts to identify informative biomarkers in alternative biological samples, such as serum or plasma, urine, and stool. The advantages of testing biomarkers utilizing a less invasive method are low cost, ease of performance, and greater acceptability. This would be particularly beneficial for the cancer screening in the primary care. Tumor DNA has been reported in the circulation of cancer patients from a number of primary tumors, including esophageal cancer. Kawakami et al. detected methylated *APC* DNA in the plasma of 25% of EAC patients, which was statistically associated with reduced patient survival [40]. In a study by Jin et al., *TAC1* methylation has been evaluated in 126 plasma samples from 35 control subjects, 10 additional patients with BE, 20 with DBE, and 61 with EAC, and both mean NMV and frequency of *TAC1* hypermethylation in plasma were significantly higher in EAC patients than in control subjects [20]. Recently, Hoffmann et al. reported that 36 of 59 patients (61.0%) with esophageal cancer had detectable levels of methylated *DAPK* or *APC* promoter DNA in plasma and preoperative detection was significantly associated with an unfavorable prognosis [41]. The combination of both markers significantly increased sensitivity and specificity for discriminating short (<2.5 years) and long survivors. Postoperative *APC* detection was significantly different if residual tumor was apparent. These studies indicate that TSG hypermethylation occurs early during EAC development and is detectable in plasma or serum, suggesting that hypermethylation of these genes could be useful as early diagnostic biomarkers, though large-scale validation studies are still needed.

As mentioned before, only a low percentage of BE patients develop EAC, and the main challenge is to identify the subset of BE patients with the greatest propensity to develop EAC and then target them for more intensive surveillance. The surveillance programs would benefit greatly from effective biomarkers to stratify patients according to individual neoplastic progression risk. Thus, the search for biomarkers for risk stratification of EAC in those with BE is currently an area of active research. A number of predictive markers have been reported in the literature, but have frequently been assessed in studies of limited power, or have lacked sufficient sensitivity or specificity when assessed in wider population-based studies. Hence, biomarker panels may be a better solution. Hypermethylation APC, TIMP3, and TERT genes were compared in 12 progressors (i.e., individuals who progressed to HBE/EAC) to those in 16 nonprogressors (i.e., individuals who did not progress to HBE/EAC), and an increased frequency of methylation of these three genes was found in the progressors versus the nonprogressors (*APC*: 100% vs. 36%; *TIMP3*: 91% vs. 23%; *TERT*: 92% vs. 17%, respectively) [18]. Hypermethylation of *APC* and *CDKN2A/p16* is also associated with an increased risk of BE ***progression to*** HBE or EAC [34]. A retrospective study demonstrated that hypermethylation of *CDKN2A/p16* (OR 1.74), *RUNX3* (OR 1.80), and *HPP1* (OR 1.77) in patients with non-DBE or LBE is independently associated with an increased risk of malignant transformation to HBE and EAC. Age, BE segment length, and hypermethylation of other genes (*TIMP3*, *APC*, or *CRBP1*) were not found to be independent risk factors [11]. In a follow-up study performed by the same research group, a three-tiered risk stratification model has been developed using three hypermethylated markers (*CDKN2A/p16*, *RUNX3*, and *HPP1*) combined with three clinical parameters (gender, BE segment length, and pathologic assessment) in order to generate receiver–operator characteristic curves stratifying BE patients into high, intermediate, and low risks for progression to HBE or EAC. This model might have impact upon the accuracy and efficiency of BE surveillance but has not been adopted into routine clinical practice to date [42]. In addition, hypermethylation of *SST*, *TAC1*, *CDH13*, *AKAP12*, and *NELL1* showed a strong relationship ***with*** BE segment length, a known clinical risk factor for neoplastic progression [19–22,33], indicating that hypermethylation of these genes may constitute both moleculars correlate of BE segment length and potential biomarkers for the prediction of BE progression. Furthermore, it has been suggested that hypermethylation of *SST*, *TAC1*, and *CDH13* may be useful as screening marker since both hypermethylation frequency and NMV of these genes were higher in BE with than without accompanying EAC (83.3% vs. 61.1% and 0.2763 vs. 0.2431; 75% vs. 55.6% and 0.2313 vs. 0.2145; 87.5% vs. 58.3% and 0.3871 vs. 0.2623, respectively) [20–22]. In light of these findings, Jin et al. performed a retrospective, multicenter, double-blinded validation study of eight methylation biomarkers (i.e., *p16*, *RUNX3*, *HPP1*, *NELL1*, *TAC1*, *SST*, *AKAP12*, and *CDH13*) for their accuracy in predicting neoplastic progression in BE in 145 nonprogressors and 50 progressors [43]. This study showed that the sensitivities of progression prediction approached 50% with specificity at 0.9, and the 8-marker panel was more objective, quantifiable, and possessed higher predictive sensitivity and specificity than clinical features, including age. This model is expected to reduce endoscopic procedures performed in BE surveillance while simultaneously increasing detection *rates* at earlier stages. Future studies should explore additional potentially predictive methylation targets, along with alternative means of assessing methylation biomarkers (such as immunohistochemical staining for reduced biomarker expression). Thus, these findings suggest that a methylation biomarker panel offers promise as a very useful tool clinically in the risk stratification of BE patients.

Prognostic biomarkers are useful to determine which patients can benefit from aggressive therapies, as measurement of survival is usually the clinical endpoint. Recent studies suggested that DNA hypermethylation may serve as a prognostic factor for EAC patient survival. Brocks et al. showed that patients with EAC whose tumors had more than 50% of a seven gene panel (*APC*, *E-cadherin*, *MGMT*, *ER*, *p16*, *DAPK*, *TIMP3*) methylated had significantly poorer survival and earlier tumor recurrence [17]. Hypermethylation of *NELL1* was significantly associated with shortened survival in stages I–II EAC patients [33]. As discussed before, methylated DNAs of *APC* and *DAPK* genes in the plasma are also linked with reduced survival for EAC patients [40,41].

Since most patients with esophageal cancer have a poor clinical outcome with surgical treatment alone, neoadjuvant chemoradiotherapy is frequently applied to improve survival of patients with locally advanced disease. Identification of biomarkers which predicts poor response to the therapy would be highly valuable to clinicians planning personalized treatment to maximize benefit while limiting the toxicity associated with these therapies. The tumor suppressor *Reprimo* encodes a cytoplasmic protein that is involved in regulation of p53-mediated cell cycle arrest at the G_2 phase. *Reprimo* hypermethylation is found to 0% in NE, 36% in BE, 63.6% in DBE, and 62.6% in EAC [44]. A nine gene methylation panel (*Reprimo*, *p57*, *p73*, *CDKN2A/p16*, *RUNX3*, *CHFR*, *MGMT*, *TIMP3*, and *HPP1*) showed a significantly higher number of methylated genes in chemoradiation nonresponders than in responders (2.4 vs. 1.4 genes per patient), and the frequency (64% vs. 15%) and level (0.313 vs. 0.078) of *Reprimo* methylation were also significantly higher in nonresponders than in responders [26].

2.3.2 TSG hypermethylation as biomarker in ESCC

As mentioned above, numerous numbers of genes have been shown to be hypermethylated in ESCC. These methylated genes have the potential to be used as molecular biomarkers for early detection, patient prognosis, and responsiveness to chemotherapy treatment. Here, we will introduce some of the well-characterized epigenetic biomarkers of ESCC.

Given the high mortality, early diagnosis is important for successful treatment of ESCC patients. TSG hypermethylation of *CDKN2A/p16*, *HIN1*, *hMLH1*, *E-cadherin*, *DAPK*, *FHIT*, *RARB*, and *MGMT* is detected in precancerous basal cell hyperplasia or dysplastic lesions, indicating their early diagnostic values in ESCC [9]. Adams et al. reported that a panel of four methylated genes (*AHRR*, *p16*, *MT1G*, and *CLDN3*) was used to successfully screen esophageal balloon cytology specimens with much better specificity and sensitivity for early detection of ESCC as compared to a single gene methylation [45]. Just as in EAC, hypermethylated genes have also been detected in the plasma or serum of patients with ESCC. The study done by Hibi et al. showed that 23% of ESCC patients who had methylated *CDKN2A/p16* in ESCC tissues also had this same methylation change detected in DNA isolated from their serum [46].

Furthermore, a panel of five methylated genes (*RARB*, *DAPK*, *CDH11*, *p16*, and *RASSF1A*) had a diagnostic sensitivity of 82% and specificity of 100% for ESCC in serum DNA of ESCC patients [47]. These data suggest that detection of DNA methylation in the plasma or serum could work as a tumor marker, and a panel of methylated genes would be more efficient for early detection of ESCC.

The TSG hypermethylation has also been evaluated as prognostic biomarkers following a curative resection of ESCC. Hypermethylation of *APC*, *TAC1*, *FHIT*, and *UCHL1* has been associated with reduced survival in ESCC patients after esophagectomy [20,48–50]. *MSH2* promoter hypermethylation in circulating tumor DNA has been suggested as a valuable predictor of disease-free survival after esophagectomy for ESCC patients [51]. Hypermethylation of *FHIT*, *CDH1*, and *ITGA4* has been linked to ESCC recurrence [49,52]. Liu et al. evaluated the methylation status of three Wnt antagonists (*SFRP1*, *DKK3*, and *RUNX3*) in circulating tumor DNA of ESCC patients and showed that patients with hypermethylation of 2 out of 3 genes had an elevated risk of recurrence with an odds ratio of 15.69 compared to those with no methylated genes [53]. ESCC patients with the hypermethylation of *UCHL1* and *CDH1* had an increased incidence of lymph nodes metastasis [50,54]. Recently, two CpG loci were identified as promising methylation markers to predict lymph node metastasis of ESCC by genome-wide methylation analysis [55]. Ling et al. investigated methylation status of nine cell cycle-associated genes and showed that 54% of the ESCC patients showed a CpG island methylator phenotype (CIMP) which was defined when more than 5 out of 9 genes were methylated in each case but that CIMP was not detected in any NE. Significant differences between CIMP status and metastasis and poor prognosis were found in this study [56].

3 DYSREGULATION OF miRNA

Although the functions of miRNAs still remain largely unknown, accumulating evidences suggest that miRNAs are deregulated in a wide variety of human cancers, including esophageal cancer. miRNAs can act as either tumor suppressor (ts-miR) by inhibiting the expression of oncogenes or tumor promoters (onco-miR) by suppressing the expression of target TSGs. In esophageal cancer, miRNA deregulation plays a significant role in the molecular oncogenic pathway, cancer prognosis, and patients' responsiveness to neoadjuvant and adjuvant therapies. Therefore, miRNAs have gathered a special attention as a new diagnostic and therapeutic tool for human esophageal cancer. Deregulated miRNAs, their functions and targets validated by luciferase reporter assay in EAC and ESCC, are summarized in Table 16.2. In addition, numerous studies have found deregulated miRNAs in esophageal cancer, but their targets have not been identified, in-depth studies of these miRNAs will be important.

3.1 miRNAs PROFILING AND THEIR TARGET GENES IN ESOPHAGEAL CANCER

3.1.1 onco-miR

A large number of onco-miRs have been identified and target important TSGs in esophageal cancer. The miR-25–93-106b polycistron is progressively upregulated during NE-BE-EAC progression and is associated with genomic amplification of the *MCM7* locus at chromosome 7q22.1. miRs-93 and -106b appear to contribute to EAC carcinogenesis by targeting *p21* and impacting cell cycle progression, meanwhile miR-25 targets *Bim* and reduces apoptosis [57]. miR-223 is also significantly upregulated during the BE–dysplasia–EAC sequence. miR-223 promotes migration and invasion, and modulates

Table 16.2 Deregulated miRNAs and their Targets Validated by Luciferase Reporter Assay in EAC and/or ESCC

miRNA	Histology	Deregulation	Target Genes	Functions	References
onco-miR					
106b	EAC	Up	p21	Promote proliferation	[57]
196a	EAC	Up	ANXA1	Promote proliferation and inhibit apoptosis	[58] (review)
196a	EAC	Up	KRT5, SPRR2C, S100A9	Promote BE to EAC	[58] (review)
25	EAC	Up	Bim	Promote proliferation	[57]
223	EAC	Up	PARP1	Promote migration and invasion	[59]
93	EAC	Up	p21	Promote proliferation	[57]
10b	ESCC	Up	KLF4	Promote migration and invasion	[58] (review)
19a	ESCC	Up	TNF-α	Promote proliferation	[58] (review)
141	ESCC	up	YAP1	Resistance to cisplatin-induced apoptosis	[58] (review)
21	ESCC	Up	PDCD4, FASL, TIMP3, RECK	Promote proliferation and invasion; inhibit apoptosis	[58] (review), [60]
25	ESCC	Up	E-cadherin, DSC2	Promote migration and invasion	[61] (review), [62]
200c	ESCC	Up	PPP2R1B	Induce chemoresistance of cisplatin	[58] (review)
373	ESCC	Up	LAST2	Promote proliferation	[58] (review)
577	ESCC	Up	TSGA10	Promote proliferation	[63]
92a	ESCC	Up	E-cadherin	Promote migration and invasion	[58] (review)
ts-miR					
26a	EAC	Down	Rb1	Regulation of anoikis-resistance	[64]
100	ESCC	Down	MTOR	Inhibit proliferation and promote apoptosis	[65]
101	ESCC	Down	EZH2	Inhibit migration and invasion	[66]
133a/b	ESCC	Down	FSCN1	Inhibit proliferation and invasion	[58] (review)

133a	ESCC	Down	CD47, FSCN1, MMP14	Inhibit proliferation and invasion	[58] (review), [67]
138	ESCC	Down	FLOT1, FLOT2, CAV1	Inhibit NF-κB activation	[68]
143	ESCC	Down	FSCN1	Inhibit proliferation	[69]
145	ESCC	Down	FSCN1	Inhibit proliferation and invasion	[58] (review), [69]
150	ESCC	Down	ZEB1	Inhibit proliferation and EMT	[70]
195	ESCC	Down	CDC42	Inhibit proliferation and invasion	[71]
27a	ESCC	Down	KRAS	Inhibit proliferation and invasion	[72]
29c	ESCC	Down	Cyclin E	Cell cycle arrest	[58] (review)
200b	ESCC	Down	CDK2, PAF, Kindlin-2	Inhibit migration and invasion	[73]
203	ESCC	Down	ΔNp63, LASP1	Inhibit proliferation, migration, and invasion	[58] (review), [74]
210	ESCC	Down	FGFRL1	Inhibit proliferation	[58] (review)
214	ESCC	Down	EZH2	Inhibit migration and invasion	[66]
223	ESCC	Down	ARTN	Inhibit migration and invasion	[58] (review)
375	ESCC	Down	PDK1, IGF1R	Inhibit proliferation and metastasis; promote apoptosis	[61] (review)
429	ESCC	Down	BCL-2, SP-1	Inhibit invasion and promote apoptosis	[75]
518b	ESCC	Down	Rap1b	Inhibit proliferation and invasion; promote apoptosis	[76]
593	ESCC	Down	PLK1	Inhibit proliferation	[58] (review)
655	ESCC	Down	PTTG1	Inhibit migration and invasion	[77]
98	ESCC	Down	EZH2	Inhibit migration and invasion	[66]
99a	ESCC	Down	MTOR	Inhibit proliferation and promote apoptosis	[65]

sensitivity to chemotherapy by targeting PARP1 (*poly ADP-ribose polymerase 1*) in EAC [59]. Upregulation of miR-196a has been confirmed in a progressive manner from BE to DBE and finally to EAC by two independent reports; miR-196a targets *ANXA1* (*annexin A1*), *KRT5* (*keratin 5*), *SPRR2C* (*small proline-rich protein 2C*), and *S100A9* (*S100 calcium-binding protein A9*), thereby exerting anti-poptotic effects and promoting BE to EAC [58].

miR-21 functions as an important onco-miR because it is upregulated in many types of cancers, including esophageal cancer. miR-21 promotes progression of ESCC via targeting multiple TSGs: *PDCD4* (*programmed cell death 4*), *FASL* (*Fas ligand*), *TIMP3* (*tissue inhibitor of metalloproteinase 3*), and *RECK* (*reversion-inducing-cysteine-rich protein with kazal motifs*) [58,60]. miR-21 should be explored as a therapeutic target for preclinical or clinical trials for subset of patients with high expression levels of this miRNA. Both miR-25 and miR-92a are upregulated and target *E-cadherin* whose down-regulation of this gene can lead to epithelial–mesenchymal transition (EMT) and metastasis in ESCC [61]. It has also been reported that miR-25-mediated downregulation of *DSC2* (*desmocollin-2*) promotes ESCC cell aggressiveness through redistributing adherens junctions and activating beta-catenin signaling [62]. Other onco-miRs in ESCC include: miR-10b, which targets *KLF4* [58,61]; miR-19a, which targets *TNF-α* [58]; miR-373, which targets *LAST2* [58]; and miR-577, which targets *TSGA10* [63].

3.1.2 ts-miR

Similarly, many ts-miRs and their targets have been identified in esophageal cancer. Downregulation of miR-26a is related to tumorigenesis and metastasis of EAC. miR-26a directly participates in the regulation of cell cycle and anoikis of OE33 EAC cells by targeting *Rb1* (*retinoblastoma 1*) [64]. miR-375 is consistently downregulated either in tumor tissues or plasma from ESCC patients and functions as a ts-miR. Hypermethylation of miR-375 promoter is the reported mechanism of its downregulation. miR-375 targets *IGFR1* (*insulin-like growth factor 1 receptor*) and *PDK1* (*3'-phosphoinositide-dependent protein kinase 1*), which are frequently overexpressed in many malignancies and have a crucial role in promoting cell proliferation, survival, and metastasis [58,61]. Downregulation of miR-133a has been reported in ESCC by multiple studies. Suzuki et al. reported that miR-133a expression is significantly lower in ESCC compared to adjacent noncancerous tissues and targets *CD47*; miR-133a significantly inhibits tumorigenesis and growth *in vivo*; overexpression of *CD47* is associated with lymph node metastasis and is an independent prognostic factor in ESCC patients [58,61]. Another study suggests that miR-133a in concert with miR-145 and miR-133b target actin-binding oncogenic protein, *FSCN1* (*Fascin homolog 1*), and thereby inhibit cell proliferation and invasion in ESCC [58,61]. Recently, it has been reported that miR-133a inhibits cell proliferation and invasion in ESCC by targeting both *FSCN1* and *MMP14* (matrix metalloprotease 14) [67]. It has been demonstrated that both miR-143 and miR-145 are downregulated and coregulates *FSCN1* in ESCC [69]. miR-203 inhibits cell proliferation, migration, and invasion in ESCC by targeting *ΔNp63* and *LASP1* (*LIM and SH3 protein 1*) [58,74]. Both miR-99a and miR-100 are downregulated and inhibit cell proliferation by inducing apoptosis via their cotarget *MTOR* (*mechanistic target of rapamycin*) in the ESCC [65]. More ts-miRNAs in ESCC are shown in Table 16.2 [66,68,70–73,75–77].

3.2 miRNA AS BIOMARKER IN ESOPHAGEAL CANCER

Dysregulation of miRNAs is observed during EAC progression (i.e., metaplasia–dysplasia–carcinoma sequence) in several studies. In a study of 22 NE, 24 BE, and 22 EAC tissues, Kan et al. found that 3 (miRs-25, -93 and -106b) and 4 miRNAs (miRs-100, -125b, -205 and -31) were upregulated and downregulated in BE and EAC as compared to NE, respectively [57]. Yang et al. reported that nine (miRs-126, -143, -145, -181a, -181b, -199a, -28, and -30a-5p) and seven miRNAs (miRs-149, -203, -210, -27b, -513, -617, and -99a) were sequentially upregulated and downregulated from NE to HBE to EAC, respectively [78]. Fassan et al. reported that miR-215 and miR-192 were significantly overexpressed, whereas miR-205, miR-203, and let-7c were significantly underexpressed during BE progression [78]. Leidner et al. identified that miR-31 and miR-31* were downregulated in HBE and EAC only, suggesting an association with transition from LBE to HBE; miR-375 was downregulated in EAC but not in LBE or HBE. They proposed miR-31 and -375 as novel candidate miRNAs specifically associated with early- and late-stage malignant progression in EAC carcinogenesis, respectively [78]. Recently, Wu et al. reported that 10 miRNAs (miR-21, -25, -223, -205, -203, let-7c, -133a, -301b, -618, and -23b) showed progressively altered expression from BE to EAC [79]. In a long-term follow-up study, Revilla-Nuin et al. found that miR-192, -194, -196a, and -196b showed a significantly higher expression in BE samples from the progressors as compared with the nonprogressors [80]. These results suggest that similar to DNA methylation, miRNAs are also promising in categorizing BE-associated lesions (including dysplasia) into subgroups, stratifying the risk of susceptibility to developing EAC. Though further work is required in order to utilize miRNAs for risk stratification in BE patients, it is clear that currently available studies have found significant differences in expression, and the use of miRNAs for this purpose shows promise.

Same as DNA methylation, miRNAs expression can be exploited in serum, plasma, urine, saliva, and other bodily fluids, and this property would be beneficial in clinical practice as they can be obtained noninvasively. In ESCC, the expression levels of miRs-21, -1322 are higher in serum than in the healthy controls, so these miRNAs could serve as biomarkers [61]. In addition, miR-21 levels are reduced postoperatively and postchemotherapy treatment, and this reduction is more significant in responders compared to nonresponders [78]. Komatsu et al. found that the plasma level of miR-21 tended to be higher in ESCC patients, while that of miR-375 was significantly lower and the miR-21/miR-375 ratio was significantly higher in ESCC patients than in controls [58,61].

In addition to early detection, miRNA deregulation may also be an important marker in determining prognosis, as well as predicting response to treatment. Many studies found the relationship between miR-21 upregulation and poor survival, thereby miR-21 seems to be a reliable poor prognostic marker with clinical potential in both EAC and ESCC [61,78]. Downregulation of miR-375 was associated with worse prognosis in both EAC and ESCC [78]. miRNAs overexpression (miR-103, -107, -145, -200c, -18a, -17, -19a, and -92a) and underexpression (miR-145 and -150) are correlated with poor survival in ESCC patients [58,61,70,81]. High levels of miRNAs (miR-18a, -31, -223, -1246) and low levels of miR-375 in serum from ESCC patients portend poor prognosis [58,82–84]. Overexpression of miR-92a, -99b, and -199a in EAC patients are poor prognosticators [61]. In addition, miR-16-2, -30e, and -200a expression are associated with shorter overall and disease-free survival in EAC patients [58].

Chemotherapy remains an important treatment strategy for patients with esophageal cancer. Chemotherapy could downstage tumors ahead of surgery, inhibit tumor recurrence, and kill metastatic tumor cells. Adjuvant and neoadjuvant chemotherapy have been applied clinically in patients with esophageal cancer; however, a great number of studies have indicated chemotherapy is unable to improve overall survival of patients, and multiple drug resistance (MDR) of cancer cells is a major clinical obstacle to curative effect and leads to poor prognosis of the patient [85]. Emerging evidence have suggested that dysregulation of miRNAs regulates MDR-related genes expression and plays key roles in MDR of patients with esophageal cancer. miR-200c is upregulated and induces chemoresistance of cisplatin in ESCC by targeting *PPP2R1B* (*protein phosphatase 2, regulatory subunit A, beta*), a subunit of phosphatase 2A, and reduction of PPP2R1B causes a decrease in phospho-AKT expression, which is involved in cell survival [58,78]. miR-141 confers cisplatin resistance in ESCC cells by targeting *YAP1* (*Yes-associated protein 1*), which is known to have a crucial role in apoptosis induced by DNA-damaging agents [61,85]. Though the exact mechanisms involved are unclear, miR-148a upregulation increases sensitivity to cisplatin and 5-fluorouracil in both ESCC and EAC cells, including cell lines that had originally been resistant to treatment [58,61,78,85]. miR-296 and miR-27a are capable of increasing sensitivity to both P-glycoprotein-related and P-glycoprotein-nonrelated drugs, in turn promoting adriamycin-induced apoptosis in esophageal cancer cells [58]. Therefore, these miRNAs might play important roles in the pathogenesis of esophageal cancer and might be considered as markers for response to treatment and potential targets for intervention of the drug resistance. Further studies on involvement of miRNA in MDR are required to discover effective strategies to attenuate or reverse MDR of esophageal cancer.

4 OPPORTUNITIES TO EPIGENETICALLY TARGET ESOPHAGEAL CANCER

Epigenetic reagents intended to reactivate epigenetically silenced TSGs are currently under preclinical and clinical investigations in the treatment of various human cancers. Nucleoside analogues 5-azacytidine (azacytidine) or its deoxy derivative, 5-aza-2'-deoxycytidine (decitabine), can effectively reverse silencing of TSGs via blocking the activity of DNA methyltransferase (DNMT) in tumor cells. These drugs have been approved by the US Food and Drug Administration (FDA) and are clinically successful for treating myelodysplastic syndrome, a preleukemia disease [86]. However, these drugs block DNA methylation at the enzymatic level and lack specificity for target genes, possibly resulting in global DNA hypomethylation. This may lead to the activation of oncogenes and/or increased genomic instability, causing the opposite effects such as acceleration of proliferation. Thereby, development of target-specific DNMT inhibitors is required [86].

To date, although there have been no published clinical trials specifically evaluating the therapeutic efficacy of these drugs in patient with esophageal cancer, combining these drugs with conventional chemotherapeutic drugs should be a promising prospect for treatment of esophageal cancer patients in the future. Clinical trials with epigenetic drugs in patients with BE or esophageal cancer from the clinicaltrials.gov website are summarized in Table 16.3.

Onco-miRs have the potential to serve as molecular therapeutic targets, since their inhibition should result in increased levels of TSG proteins. Conversely, reintroduction of ts-miRs into tumor cells will result in the downregulation of target oncogenes, causing tumor suppression. Targeting miRNAs has a unique advantage that one miRNA targets multiple genes simultaneously, powerfully inactivating entire pathways at once. Intensive efforts have been made to develop miRNA-based therapeutics in preclinical models of breast, pancreatic, and prostate cancers to modify oncogene or tumor suppressor functions [68]. The published literature dealing with miRNAs as therapeutic targets in esophageal cancer does not abound and relies mostly on results obtained from cell lines. To date, miRNA-based therapeutics have not yet chosen for clinical trials in patients with esophageal cancer, but this is a promising area of research.

Table 16.3 Clinical Trials with Epigenetic Drugs in Patients with Barrett Esophagus or Esophageal Cancer

Drug	Conditions	Status	NCT number	Phases
Decitabine	Esophageal lung cancer, malignant mesothelioma	Completed	NCT00019825	Phase 1
Decitabine	Small cell carcinoma, mesothelioma, nonsmall-cell lung carcinoma, advanced esophageal cancer	Completed	NCT00037817	Phase 1
Azacitidine	Esophageal cancer, malignant neoplasm of cardioesophageal junction of stomach	Unknown	NCT01386346	Phase 1
Entinostat	Multiple cancers including esophageal cancer	Completed	NCT00020579	Phase 1
Vorinostat	Multiple cancers including recurrent esophageal cancer	Recruiting	NCT01249443	Phase 1
Vorinostat	Esophageal cancer, gastric cancer, liver cancer		NCT00537121	Phase 1
Defined green tea catechin extract	Barrett esophagus	Completed	NCT00233935	Phase 1

5 OTHER EPIGENETIC EVENTS IN ESOPHAGEAL CANCER

5.1 HISTONE ACETYLASES/DEACETYLASES AND HISTONE METHYLTRANSFERASES/DEMETHYLASES

Alterations to histone-modifying proteins such as histone deacetylases (HDACs) and histone demethyltransferases (HDMTs) could lead to mutations in oncogenes, TSGs, or DNA repair genes which would result in genomic instability, oncogenic transformation, and development of cancer. Relatively limited information is available concerning the direct clinical implications of aberrant expression of histone acetyltransferases (HATs) and HDACs, as well as histone methyltransferases (HMTs) and HDMTs in esophageal cancer. Toh et al. demonstrated that histone H4 was significantly hyperacetylated in the early stage of cancer, and thereafter changed into a hypoacetylated state according to the degree of cancer progression in ESCC. The cases in which HDAC1 was less expressed in esophageal carcinoma cells than in the normal mucosa were significantly increased as the carcinoma invaded into the deeper layers of the esophageal wall. Furthermore, both the hyperacetylation of histone H4 and the high expression of HDAC1 were shown to topologically colocalize in the same tumor. These results suggested that a dynamic equilibrium between the HATs and HDACs activities is disrupted in ESCC, thus implying that a certain interaction may exist between the hyperacetylation of histone H4 and the HDAC1 expression [86]. In addition, the acetylation levels of histone H4 were inversely correlated to the depth of cancer invasion and pathological stage, and the ESCC patients with higher level of histone H4 acetylation had a better prognosis [86]. Similarly, Tzao et al. reported that the aberrant expression of acetylated histone H4 correlated positively with tumor stage, nodal involvement, and distant metastasis [86]. Langer et al. showed that approximately 50% of EACs had no detectable or very low HDAC1 expression, whereas 30% of EACs had no detectable HDAC2 expression [87]. High HDAC2 expression correlated with poor differentiation and lymph node metastases, and no correlation was observed between HDAC1 or HDAC2 expression and response to therapy or survival. HADC2 was overexpressed in ESCC, and HDAC inhibitor trichostatin A suppressed metastasis of ESCC cells through HADC2 reduced MMP-2/9 [88].

Recent studies have also revealed that aberrant patterns of HMTs and HDMTs are implicated in the pathogenesis of esophageal cancer. Knockdown of plant homeodomain finger protein 8 (PHF8), a histone lysine demethylases, inhibited cell proliferation, migration and invasion, and promoted apoptosis in ESCC cells *in vitro* [89]. Furthermore, downregulation of PHF8 attenuated the tumorigenicity of ESCC cells *in vivo*. Hudlebush et al. reported multiple myeloma SET (MMSET) histone lysine methyltransferase, which mediates di- and trimethylation of H3K36, was overexpressed in 38.9% ESCCs and 56.4% EACs [87]. The lysine-specific demethylase 1 (LSD1), a member of the histone demethylases, was elevated in cancerous tissue and correlated with lymph node metastasis and poorer overall survival in patients with ESCC [90]. Knockdown of LSD1 using lentivirus delivery of LSD1-specific shRNA abrogated the migration and invasion of ESCC cells *in vitro*. The Jumonji AT-rich interactive domain 1B (JARID1B) and PHD finger protein 2 (PHF2), members of the histone demethylases, were overexpressed on tissue microarrays of ESCC samples in 120 cases using immunohistochemical staining [91]. Sun et al. detected expression of histone demethylase GASC1 (gene amplified in squamous cell carcinoma 1) on tissue microarrays of ESCC samples in 185 cases using immunohistochemical staining [92]. The nuclear expression of GASC1 was significantly associated with lymph node metastasis and tumor-node metastasis stages. Chen et al. found that both H3 hypoacetylation and H3K27 hypermethylation were correlated with the severity and histological differentiation of the tumor, and H3K4 hypermethylation also correlated with tumor differentiation in ESCC [87]. Study done by Tzao et al. showed a significant positive relationship between tumor differentiation and acetylated histone 3 lysine 18, demethylated histone 4 arginine 3, and trimethylated histone 3 lysine 27 (H3K27triMe). The expression of H3K27triMe significantly correlated with lymph node metastasis and stage, and was one of the independent prognostic markers in ESCC patients [86]. ESCC patients whose tumors expressed high global levels of H3K18Ac and H4R3Me2 had significantly reduced recurrence free survivals relative to comparably staged patients with tumors exhibiting low global levels of these histone marks [86].

5.2 LONG NONCODING RNA

Long noncoding RNAs (lncRNAs) have been recently recognized as a major class of regulators in mammalian systems and play critical roles in the development and progression of cancer. Accumulating evidence has revealed the participation of lncRNAs in tumorigenesis of esophageal cancer. LncRNA *91H* is downregulated and contributes to the occurrence and progression of ESCC by inhibiting IGF2 expression [93]. LncRNA *HNF1A-AS1* is upregulated and promotes cell proliferation and migration in EAC [94]. LncRNA *PlncRNA-1* is upregulated and correlates with advanced clinical stage and lymph node metastasis in ESCC [95]. Knockdown of lncRNA *PlncRNA-1* reduces cell proliferation and increased the apoptosis *in vitro*. Li et al. identified a three-lncRNA signature (including the lncRNAs ENST00000435885.1, XLOC_013014, and ENST00000547963.1) which classified the ESCC patients into two groups with significantly different overall survival (median survival 19.2 months vs. >60 months, $P<0.0001$) [96]. The signature was applied to the test group (median survival 21.5 months vs. > 60 months, $P = 0.0030$) and independent cohort (median survival 25.8 months vs. >48 months, $P = 0.0187$) and showed similar prognostic values in both. Recent studies have also emerged linking lncRNA alterations with methylation changes in esophageal cancer. LncRNA *AFAP1-AS1* is extremely hypomethylated and overexpressed in BE and EAC [97]. Its silencing by small interfering

RNA inhibits proliferation and colony-forming ability, induces apoptosis, and reduces EAC cell migration and invasion without altering the expression of its protein-coding counterpart, AFAP1. LncRNA *POU3F3* is upregulated and contributes to the development of ESCC by interacting with EZH2 to promote methylation of *POU3F3* [98]. LncRNA *HOTAIR* is upregulated and correlates with cancer metastasis, elevated TNM stage classification and lowered overall survival rates in ESCC [99]. Furthermore, knockdown of *HOTAIR* reduces cell invasiveness and migration while increasing the response of cells to apoptosis *in vitro*. Mechanistically, lncRNA *HOTAIR* directly decreases WIF-1 expression by promoting its histone H3K27 methylation in the promoter region and then activates the Wnt/β-catenin signaling pathway in ESCC cells [100]. These data indicates that lncRNAs play an important role in tumorigenesis of esophageal cancer and represent a novel clinically relevant event to identify potential prognostic biomarkers and therapeutic targets for patients with esophageal cancer.

REFERENCES

[1] Jemal A, Bray F, Center MM, Ferlay J, Ward E, Forman D. Global cancer statistics. CA Cancer J Clin 2011;61:69–90.

[2] Zhang XM, Guo MZ. The value of epigenetic markers in esophageal cancer. Front Med China 2010;4:378–84.

[3] Desai TK, Krishnan K, Samala N, Singh J, Cluley J, Perla S, et al. The incidence of oesophageal adenocarcinoma in non-dysplastic Barrett's oesophagus: a meta-analysis. Gut 2012;61:970–6.

[4] Hameeteman W, Tytgat GN, Houthoff HJ, van den Tweel JG. Barrett's esophagus: development of dysplasia and adenocarcinoma. Gastroenterology 1989;96:1249–56.

[5] Wang KK, Sampliner RE. Updated guidelines 2008 for the diagnosis, surveillance and therapy of Barrett's esophagus. Am J Gastroenterol 2008;103:788–97.

[6] Pennathur A, Farkas A, Krasinskas AM, Ferson PF, Gooding WE, Gibson MK, et al. Esophagectomy for T1 esophageal cancer: outcomes in 100 patients and implications for endoscopic therapy. Ann Thorac Surg 2009;87:1048–54. Discussion 1054–1045.

[7] Barbour AP, Jones M, Brown I, Gotley DC, Martin I, Thomas J, et al. Risk stratification for early esophageal adenocarcinoma: analysis of lymphatic spread and prognostic factors. Ann Surg Oncol 2010;17:2494–502.

[8] Alikhan M, Rex D, Khan A, Rahmani E, Cummings O, Ulbright TM. Variable pathologic interpretation of columnar lined esophagus by general pathologists in community practice. Gastrointest Endosc 1999;50:23–6.

[9] Kaz AM, Grady WM. Epigenetic biomarkers in esophageal cancer. Cancer Lett 2014;342:193–9.

[10] Eads CA, Lord RV, Wickramasinghe K, Long TI, Kurumboor SK, Bernstein L, et al. Epigenetic patterns in the progression of esophageal adenocarcinoma. Cancer Res 2001;61:3410–8.

[11] Schulmann K, Sterian A, Berki A, Yin J, Sato F, Xu Y, et al. Inactivation of p16, RUNX3, and HPP1 occurs early in Barrett's-associated neoplastic progression and predicts progression risk. Oncogene 2005;24:4138–48.

[12] Baumann S, Keller G, Puhringer F, Napieralski R, Feith M, Langer R, et al. The prognostic impact of O6-methylguanine-DNA methyltransferase (MGMT) promotor hypermethylation in esophageal adenocarcinoma. Int J Cancer 2006;119:264–8.

[13] Kuester D, El-Rifai W, Peng D, Ruemmele P, Kroeckel I, Peters B, et al. Silencing of MGMT expression by promoter hypermethylation in the metaplasia-dysplasia-carcinoma sequence of Barrett's esophagus. Cancer Lett 2009;275:117–26.

[14] Smith E, De Young NJ, Pavey SJ, Hayward NK, Nancarrow DJ, Whiteman DC, et al. Similarity of aberrant DNA methylation in Barrett's esophagus and esophageal adenocarcinoma. Mol Cancer 2008;7:75.

[15] Corn PG, Heath EI, Heitmiller R, Fogt F, Forastiere AA, Herman JG, et al. Frequent hypermethylation of the 5' CpG island of E-cadherin in esophageal adenocarcinoma. Clin Cancer Res 2001;7:2765–9.

[16] Eads CA, Lord RV, Kurumboor SK, Wickramasinghe K, Skinner ML, Long TI, et al. Fields of aberrant CpG island hypermethylation in Barrett's esophagus and associated adenocarcinoma. Cancer Res 2000;60:5021–6.

[17] Brock MV, Gou M, Akiyama Y, Muller A, Wu TT, Montgomery E, et al. Prognostic importance of promoter hypermethylation of multiple genes in esophageal adenocarcinoma. Clin Cancer Res 2003;9:2912–9.

[18] Clement G, Braunschweig R, Pasquier N, Bosman FT, Benhattar J. Methylation of APC, TIMP3, and TERT: a new predictive marker to distinguish Barrett's oesophagus patients at risk for malignant transformation. J Pathol 2006;208:100–7.

[19] Jin Z, Hamilton JP, Yang J, Mori Y, Olaru A, Sato F, et al. Hypermethylation of the AKAP12 promoter is a biomarker of Barrett's-associated esophageal neoplastic progression. Cancer Epidemiol Biomarkers Prev 2008;17:111–7.

[20] Jin Z, Olaru A, Yang J, Sato F, Cheng Y, Kan T, et al. Hypermethylation of tachykinin-1 is a potential biomarker in human esophageal cancer. Clin Cancer Res 2007;13:6293–300.

[21] Jin Z, Mori Y, Hamilton JP, Olaru A, Sato F, Yang J, et al. Hypermethylation of the somatostatin promoter is a common, early event in human esophageal carcinogenesis. Cancer 2008;112:43–9.

[22] Jin Z, Cheng Y, Olaru A, Kan T, Yang J, Paun B, et al. Promoter hypermethylation of CDH13 is a common, early event in human esophageal adenocarcinogenesis and correlates with clinical risk factors. Int J Cancer 2008;123:2331–6.

[23] Jin Z, Wang L, Zhang Y, Cheng Y, Gao Y, Feng X, et al. MAL hypermethylation is a tissue-specific event that correlates with MAL mRNA expression in esophageal carcinoma. Sci Rep 2013;3:2838.

[24] Jin Z, Zhao Z, Cheng Y, Dong M, Zhang X, Wang L, et al. Endoglin promoter hypermethylation identifies a field defect in human primary esophageal cancer. Cancer 2013;119:3604–9.

[25] Guo M, Ren J, House MG, Qi Y, Brock MV, Herman JG. Accumulation of promoter methylation suggests epigenetic progression in squamous cell carcinoma of the esophagus. Clin Cancer Res 2006;12:4515–22.

[26] Hamilton JP, Sato F, Greenwald BD, Suntharalingam M, Krasna MJ, Edelman MJ, et al. Promoter methylation and response to chemotherapy and radiation in esophageal cancer. Clin Gastroenterol Hepatol 2006;4:701–8.

[27] Li JS, Ying JM, Wang XW, Wang ZH, Tao Q, Li LL. Promoter methylation of tumor suppressor genes in esophageal squamous cell carcinoma. Chin J Cancer 2013;32:3–11.

[28] Kuroki T, Trapasso F, Yendamuri S, Matsuyama A, Alder H, Mori M, et al. Promoter hypermethylation of RASSF1A in esophageal squamous cell carcinoma. Clin Cancer Res 2003;9:1441–5.

[29] Tzao C, Hsu HS, Sun GH, Lai HL, Wang YC, Tung HJ, et al. Promoter methylation of the hMLH1 gene and protein expression of human mutL homolog 1 and human mutS homolog 2 in resected esophageal squamous cell carcinoma. J Thorac Cardiovasc Surg 2005;130:1371.

[30] Wang Y, Fang MZ, Liao J, Yang GY, Nie Y, Song Y, et al. Hypermethylation-associated inactivation of retinoic acid receptor beta in human esophageal squamous cell carcinoma. Clin Cancer Res 2003;9:5257–63.

[31] Baba Y, Watanabe M, Baba H. A review of the alterations in DNA methylation in esophageal squamous cell carcinoma. Surg Today 2013;43:1355–64.

[32] Kuroki T, Trapasso F, Yendamuri S, Matsuyama A, Alder H, Mori M, et al. Allele loss and promoter hypermethylation of VHL, RAR-beta, RASSF1A, and FHIT tumor suppressor genes on chromosome 3p in esophageal squamous cell carcinoma. Cancer Res 2003;63:3724–8.

[33] Jin Z, Mori Y, Yang J, Sato F, Ito T, Cheng Y, et al. Hypermethylation of the nel-like 1 gene is a common and early event and is associated with poor prognosis in early-stage esophageal adenocarcinoma. Oncogene 2007;26:6332–40.

[34] Wang JS, Guo M, Montgomery EA, Thompson RE, Cosby H, Hicks L, et al. DNA promoter hypermethylation of p16 and APC predicts neoplastic progression in Barrett's esophagus. Am J Gastroenterol 2009;104:2153–60.

[35] Wong DJ, Barrett MT, Stoger R, Emond MJ, Reid BJ. p16INK4a promoter is hypermethylated at a high frequency in esophageal adenocarcinomas. Cancer Res 1997;57:2619–22.

[36] Klump B, Hsieh CJ, Holzmann K, Gregor M, Porschen R. Hypermethylation of the CDKN2/p16 promoter during neoplastic progression in Barrett's esophagus. Gastroenterology 1998;115:1381–6.

[37] Bian YS, Osterheld MC, Fontolliet C, Bosman FT, Benhattar J. p16 inactivation by methylation of the CDKN2A promoter occurs early during neoplastic progression in Barrett's esophagus. Gastroenterology 2002;122:1113–21.

[38] Liu L, Gerson SL. Targeted modulation of MGMT: clinical implications. Clin Cancer Res 2006;12:328–31.

[39] Tanaka H, Shimada Y, Harada H, Shinoda M, Hatooka S, Imamura M, et al. Methylation of the 5' CpG island of the FHIT gene is closely associated with transcriptional inactivation in esophageal squamous cell carcinomas. Cancer Res 1998;58:3429–34.

[40] Kawakami K, Brabender J, Lord RV, Groshen S, Greenwald BD, Krasna MJ, et al. Hypermethylated APC DNA in plasma and prognosis of patients with esophageal adenocarcinoma. J Natl Cancer Inst 2000;92:1805–11.

[41] Hoffmann AC, Vallbohmer D, Prenzel K, Metzger R, Heitmann M, Neiss S, et al. Methylated DAPK and APC promoter DNA detection in peripheral blood is significantly associated with apparent residual tumor and outcome. J Cancer Res Clin Oncol 2009;135:1231–7.

[42] Sato F, Jin Z, Schulmann K, Wang J, Greenwald BD, Ito T, et al. Three-tiered risk stratification model to predict progression in Barrett's esophagus using epigenetic and clinical features. PLoS One 2008;3:e1890.

[43] Jin Z, Cheng Y, Gu W, Zheng Y, Sato F, Mori Y, et al. A multicenter, double-blinded validation study of methylation biomarkers for progression prediction in Barrett's esophagus. Cancer Res 2009;69:4112–5.

[44] Hamilton JP, Sato F, Jin Z, Greenwald BD, Ito T, Mori Y, et al. Reprimo methylation is a potential biomarker of Barrett's-associated esophageal neoplastic progression. Clin Cancer Res 2006;12:6637–42.

[45] Adams L, Roth MJ, Abnet CC, Dawsey SP, Qiao YL, Wang GQ, et al. Promoter methylation in cytology specimens as an early detection marker for esophageal squamous dysplasia and early esophageal squamous cell carcinoma. Cancer Prev Res (Phila) 2008;1:357–61.

[46] Hibi K, Taguchi M, Nakayama H, Takase T, Kasai Y, Ito K, et al. Molecular detection of p16 promoter methylation in the serum of patients with esophageal squamous cell carcinoma. Clin Cancer Res 2001;7:3135–8.

[47] Li B, Wang B, Niu LJ, Jiang L, Qiu CC. Hypermethylation of multiple tumor-related genes associated with DNMT3b up-regulation served as a biomarker for early diagnosis of esophageal squamous cell carcinoma. Epigenetics 2011;6:307–16.

[48] Zare M, Jazii FR, Alivand MR, Nasseri NK, Malekzadeh R, Yazdanbod M. Qualitative analysis of adenomatous polyposis coli promoter: hypermethylation, engagement and effects on survival of patients with esophageal cancer in a high risk region of the world, a potential molecular marker. BMC Cancer 2009;9:24.

[49] Lee EJ, Lee BB, Kim JW, Shim YM, Hoseok I, Han J, et al. Aberrant methylation of Fragile Histidine Triad gene is associated with poor prognosis in early stage esophageal squamous cell carcinoma. Eur J Cancer 2006;42:972–80.

[50] Mandelker DL, Yamashita K, Tokumaru Y, Mimori K, Howard DL, Tanaka Y, et al. PGP9.5 promoter methylation is an independent prognostic factor for esophageal squamous cell carcinoma. Cancer Res 2005;65:4963–8.

[51] Ling ZQ, Zhao Q, Zhou SL, Mao WM. MSH2 promoter hypermethylation in circulating tumor DNA is a valuable predictor of disease-free survival for patients with esophageal squamous cell carcinoma. Eur J Surg Oncol 2012;38:326–32.

[52] Lee EJ, Lee BB, Han J, Cho EY, Shim YM, Park J, et al. CpG island hypermethylation of E-cadherin (CDH1) and integrin alpha4 is associated with recurrence of early stage esophageal squamous cell carcinoma. Int J Cancer 2008;123:2073–9.

[53] Liu JB, Qiang FL, Dong J, Cai J, Zhou SH, Shi MX, et al. Plasma DNA methylation of Wnt antagonists predicts recurrence of esophageal squamous cell carcinoma. World J Gastroenterol 2011;17:4917–21.

[54] Maesawa C, Tamura G, Nishizuka S, Ogasawara S, Ishida K, Terashima M, et al. Inactivation of the CDKN2 gene by homozygous deletion and *de novo* methylation is associated with advanced stage esophageal squamous cell carcinoma. Cancer Res 1996;56:3875–8.

[55] Gyobu K, Yamashita S, Matsuda Y, Igaki H, Niwa T, Oka D, et al. Identification and validation of DNA methylation markers to predict lymph node metastasis of esophageal squamous cell carcinomas. Ann Surg Oncol 2011;18:1185–94.

[56] Ling Y, Huang G, Fan L, Wei L, Zhu J, Liu Y, et al. CpG island methylator phenotype of cell-cycle regulators associated with TNM stage and poor prognosis in patients with oesophageal squamous cell carcinoma. J Clin Pathol 2011;64:246–51.

[57] Kan T, Sato F, Ito T, Matsumura N, David S, Cheng Y, et al. The miR-106b-25 polycistron, activated by genomic amplification, functions as an oncogene by suppressing p21 and Bim. Gastroenterology 2009;136:1689–700.

[58] Li SQ, Chen FJ, Cao XF. Distinctive microRNAs in esophageal tumor: early diagnosis, prognosis judgment, and tumor treatment. Dis Esophagus 2013;26:288–98.

[59] Streppel MM, Pai S, Campbell NR, Hu C, Yabuuchi S, Canto MI, et al. MicroRNA 223 is upregulated in the multistep progression of Barrett's esophagus and modulates sensitivity to chemotherapy by targeting PARP1. Clin Cancer Res 2013;19:4067–78.

[60] Wang N, Zhang CQ, He JH, Duan XF, Wang YY, Ji X, et al. MiR-21 down-regulation suppresses cell growth, invasion and induces cell apoptosis by targeting FASL, TIMP3, and RECK genes in esophageal carcinoma. Dig Dis Sci 2013;58:1863–70.

[61] Song S, Ajani JA. The role of microRNAs in cancers of the upper gastrointestinal tract. Nat Rev Gastroenterol Hepatol 2013;10:109–18.

[62] Fang WK, Liao LD, Li LY, Xie YM, Xu XE, Zhao WJ, et al. Down-regulated desmocollin-2 promotes cell aggressiveness through redistributing adherens junctions and activating beta-catenin signalling in oesophageal squamous cell carcinoma. J Pathol 2013;231:257–70.

[63] Yuan X, He J, Sun F, Gu J. Effects and interactions of MiR-577 and TSGA10 in regulating esophageal squamous cell carcinoma. Int J Clin Exp Pathol 2013;6:2651–67.

[64] Zhang YF, Zhang AR, Zhang BC, Rao ZG, Gao JF, Lv MH, et al. MiR-26a regulates cell cycle and anoikis of human esophageal adenocarcinoma cells through Rb1-E2F1 signaling pathway. Mol Biol Rep 2013;40:1711–20.

[65] Sun J, Chen Z, Tan X, Zhou F, Tan F, Gao Y, et al. MicroRNA-99a/100 promotes apoptosis by targeting mTOR in human esophageal squamous cell carcinoma. Med Oncol 2013;30:411.

[66] Huang SD, Yuan Y, Zhuang CW, Li BL, Gong DJ, Wang SG, et al. MicroRNA-98 and microRNA-214 post-transcriptionally regulate enhancer of zeste homolog 2 and inhibit migration and invasion in human esophageal squamous cell carcinoma. Mol Cancer 2012;11:51.

[67] Akanuma N, Hoshino I, Akutsu Y, Murakami K, Isozaki Y, Maruyama T, et al. MicroRNA-133a regulates the mRNAs of two invadopodia-related proteins, FSCN1 and MMP14, in esophageal cancer. Br J Cancer 2014;110:189–98.

[68] Gong H, Song L, Lin C, Liu A, Lin X, Wu J, et al. Downregulation of miR-138 sustains NF-kappaB activation and promotes lipid raft formation in esophageal squamous cell carcinoma. Clin Cancer Res 2013;19:1083–93.

[69] Liu R, Liao J, Yang M, Sheng J, Yang H, Wang Y, et al. The cluster of miR-143 and miR-145 affects the risk for esophageal squamous cell carcinoma through co-regulating fascin homolog 1. PLoS One 2012;7:e33987.

[70] Yokobori T, Suzuki S, Tanaka N, Inose T, Sohda M, Sano A, et al. MiR-150 is associated with poor prognosis in esophageal squamous cell carcinoma via targeting the EMT inducer ZEB1. Cancer Sci 2013;104:48–54.

[71] Fu MG, Li S, Yu TT, Qian LJ, Cao RS, Zhu H, et al. Differential expression of miR-195 in esophageal squamous cell carcinoma and miR-195 expression inhibits tumor cell proliferation and invasion by targeting of Cdc42. FEBS Lett 2013;587:3471–9.

[72] Zhu L, Wang Z, Fan Q, Wang R, Sun Y. microRNA-27a functions as a tumor suppressor in esophageal squamous cell carcinoma by targeting KRAS. Oncol Rep 2014;31:280–6.

[73] Zhang HF, Zhang K, Liao LD, Li LY, Du ZP, Wu BL, et al. miR-200b suppresses invasiveness and modulates the cytoskeletal and adhesive machinery in esophageal squamous cell carcinoma cells via targeting Kindlin-2. Carcinogenesis 2014.

[74] Takeshita N, Mori M, Kano M, Hoshino I, Akutsu Y, Hanari N, et al. miR-203 inhibits the migration and invasion of esophageal squamous cell carcinoma by regulating LASP1. Int J Oncol 2012;41:1653–61.

[75] Wang Y, Li M, Zang W, Ma Y, Wang N, Li P, et al. MiR-429 up-regulation induces apoptosis and suppresses invasion by targeting Bcl-2 and SP-1 in esophageal carcinoma. Cell Oncol (Dordr) 2013;36:385–94.

[76] Zhang M, Zhou S, Zhang L, Zhang J, Cai H, Zhu J, et al. miR-518b is down-regulated, and involved in cell proliferation and invasion by targeting Rap1b in esophageal squamous cell carcinoma. FEBS Lett 2012;586:3508–21.

[77] Wang Y, Zang W, Du Y, Ma Y, Li M, Li P, et al. Mir-655 up-regulation suppresses cell invasion by targeting pituitary tumor-transforming gene-1 in esophageal squamous cell carcinoma. J Transl Med 2013;11:301.

[78] Sakai NS, Samia-Aly E, Barbera M, Fitzgerald RC. A review of the current understanding and clinical utility of miRNAs in esophageal cancer. Semin Cancer Biol 2013;23:512–21.

[79] Wu X, Ajani JA, Gu J, Chang DW, Tan W, Hildebrandt MA, et al. MicroRNA expression signatures during malignant progression from Barrett's esophagus to esophageal adenocarcinoma. Cancer Prev Res (Phila) 2013;6:196–205.

[80] Revilla-Nuin B, Parrilla P, Lozano JJ, de Haro LF, Ortiz A, Martinez C, et al. Predictive value of microRNAs in the progression of barrett esophagus to adenocarcinoma in a long-term follow-up study. Ann Surg 2013;257:886–93.

[81] Xu XL, Jiang YH, Feng JG, Su D, Chen PC, Mao WM. MicroRNA-17, microRNA-18a, and microRNA-19a are prognostic indicators in esophageal squamous cell carcinoma. Ann Thorac Surg 2014.

[82] Wu C, Li M, Hu C, Duan H. Clinical significance of serum miR-223, miR-25 and miR-375 in patients with esophageal squamous cell carcinoma. Mol Biol Rep 2014;41:1257–66.

[83] Takeshita N, Hoshino I, Mori M, Akutsu Y, Hanari N, Yoneyama Y, et al. Serum microRNA expression profile: miR-1246 as a novel diagnostic and prognostic biomarker for oesophageal squamous cell carcinoma. Br J Cancer 2013;108:644–52.

[84] Hirajima S, Komatsu S, Ichikawa D, Takeshita H, Konishi H, Shiozaki A, et al. Clinical impact of circulating miR-18a in plasma of patients with oesophageal squamous cell carcinoma. Br J Cancer 2013;108:1822–9.

[85] Hong L, Han Y, Lu Q, Zhang H, Zhao Q, Wu K, et al. Drug resistance-related microRNAs in esophageal cancer. Expert Opin Biol Ther 2012;12:1487–94.

[86] Toh Y, Egashira A, Yamamoto M. Epigenetic alterations and their clinical implications in esophageal squamous cell carcinoma. Gen Thorac Cardiovasc Surg 2013;61:262–9.

[87] Schrump DS. Targeting epigenetic mediators of gene expression in thoracic malignancies. Biochim Biophys Acta 2012;1819:836–45.

[88] Wang F, Qi Y, Li X, He W, Fan QX, Zong H. HDAC inhibitor trichostatin A suppresses esophageal squamous cell carcinoma metastasis through HADC2 reduced MMP-2/9. Clin Invest Med 2013;36:E87–94.

[89] Sun X, Qiu JJ, Zhu S, Cao B, Sun L, Li S, et al. Oncogenic features of PHF8 histone demethylase in esophageal squamous cell carcinoma. PLoS One 2013;8:e77353.

[90] Yu Y, Wang B, Zhang K, Lei Z, Guo Y, Xiao H, et al. High expression of lysine-specific demethylase 1 correlates with poor prognosis of patients with esophageal squamous cell carcinoma. Biochem Biophys Res Commun 2013;437:192–8.

[91] Sun LL, Sun XX, Xu XE, Zhu MX, Wu ZY, Shen JH, et al. Overexpression of Jumonji AT-rich interactive domain 1B and PHD finger protein 2 is involved in the progression of esophageal squamous cell carcinoma. Acta Histochem 2013;115:56–62.

[92] Sun LL, Holowatyj A, Xu XE, Wu JY, Wu ZY, Shen JH, et al. Histone demethylase GASC1, a potential prognostic and predictive marker in esophageal squamous cell carcinoma. Am J Cancer Res 2013;3:509–17.

[93] Gao T, He B, Pan Y, Xu Y, Li R, Deng Q, et al. Long non-coding RNA 91H contributes to the occurrence and progression of esophageal squamous cell carcinoma by inhibiting IGF2 expression. Mol Carcinog 2015;54:359–67.

[94] Yang X, Song JH, Cheng Y, Wu W, Bhagat T, Yu Y, et al. Long non-coding RNA HNF1A-AS1 regulates proliferation and migration in oesophageal adenocarcinoma cells. Gut 2014;63:881–90.

[95] Wang CM, Wu QQ, Li SQ, Chen FJ, Tuo L, Xie HW, et al. Upregulation of the long non-coding RNA PlncRNA-1 promotes esophageal squamous carcinoma cell proliferation and correlates with advanced clinical stage. Dig Dis Sci 2014;59:591–7.

[96] Li J, Chen Z, Tian L, Zhou C, He MY, Gao Y, et al. LncRNA profile study reveals a three-lncRNA signature associated with the survival of patients with oesophageal squamous cell carcinoma. Gut 2014;63:1700–10.

[97] Wu W, Bhagat TD, Yang X, Song JH, Cheng Y, Agarwal R, et al. Hypomethylation of noncoding DNA regions and overexpression of the long noncoding RNA, AFAP1-AS1, in Barrett's esophagus and esophageal adenocarcinoma. Gastroenterology 2013;144:956–66. e954.

[98] Li W, Zheng J, Deng J, You Y, Wu H, Li N, et al. Increased levels of the long intergenic non-protein coding RNA POU3F3 promote DNA methylation in esophageal squamous cell carcinoma cells. Gastroenterology 2014;146:1714–26.

[99] Chen FJ, Sun M, Li SQ, Wu QQ, Ji L, Liu ZL, et al. Upregulation of the long non-coding RNA HOTAIR promotes esophageal squamous cell carcinoma metastasis and poor prognosis. Mol Carcinog 2013;52:908–15.

[100] Ge XS, Ma HJ, Zheng XH, Ruan HL, Liao XY, Xue WQ, et al. HOTAIR, a prognostic factor in esophageal squamous cell carcinoma, inhibits WIF-1 expression and activates Wnt pathway. Cancer Sci 2013;104:1675–82.

NASOPHARYNGEAL CANCER

17

Li-Xia Peng and Chao-Nan Qian

State Key Laboratory of Oncology in South China, Sun Yat-sen University Cancer Center,
Guangzhou, Guangdong, China

CHAPTER OUTLINE

1 INTRODUCTION

Nasopharyngeal carcinoma (NPC) is a common malignancy in southern China and Southeast Asia [1–4] and resulted in 65,000 deaths in 2010 globally, representing an increase from 45,000 in 1990 [5]. NPC has a unique geographic and ethnic distribution as there is a very low incidence in Caucasians from North America and other Western countries (under 1 per 100,000 persons per year) in contrast to a high incidence in southern China, Southeast Asia, and other Asian countries. In Sihui city in southern China, the NPC incidence is 27.2 for men and 11.3 for women per 100,000 persons per year. In fact, more than 80% of NPCs have been reported in China, Southeast Asia, and other Asian countries [6,7]. Southern Chinese immigrants also have a higher risk of NPC compared to the local Western population. Men are two- to threefold more frequently affected than women, independent of race or ethnicity [8,9]. Therapy for NPC includes radiotherapy, chemoradiotherapy, and salvage surgery [10,11]. Radiotherapy is the main treatment of choice for this cancer because of the high radiosensitivity of NPC cells.

The initiation of NPC is believed to be a process of multiple-step carcinogenesis as there are strong associations with genetic, environmental, and viral factors [12,13]. The linkage analysis conducted at the Sun Yat-sen University Cancer Center using highly polymorphic microsatellite markers identified two susceptibility loci on chromosomes 4p15.1-q12 and 3p21, but not on the MHC region, from the

S.G. Gray (Ed): Epigenetic Cancer Therapy. DOI: http://dx.doi.org/10.1016/B978-0-12-800206-3.00017-3

Guangdong and Hunan provinces in China, respectively [14]. Environmental factors, such as salted fish, are related to an increased risk of NPC in a southern Chinese population [8]. The association of NPC with other non-dietary factors, such as cigarette smoking, is still under investigation. Unique features of NPC, including its strong association with Epstein–Barr virus (EBV) and higher EBV antibody titers, especially the IgA class, are observed in most NPC patients compared to normal controls [15,16].

The accumulation of genetic and epigenetic alterations may play an important role in the initiation and progression of NPC. Epigenetics is defined as inheritable changes in gene expression without DNA sequence changes. The known epigenetic proteins include DNA methyltransferase (DNMT), histone deacetylase (HDAC), histone acetylase (HAT), methyl-CpG-binding protein (MBP), and histone methyltransferase (HMT) [17–19]. In this chapter, we mainly focus on epigenetic alterations in NPC, including DNA methylation, histone modification, miRNA, lncRNA, and the potential utility of epigenetic targeting therapies in the treatment of NPC.

2 DNA HYPERMETHYLATION IN NPC

DNA methylation is a normal process used to maintain normal patterns of gene expression in mammalian cells [20]. The process involves the regulation of gene imprinting, X-chromosome inactivation, and other biological activities [17,21]. DNMT1 predominantly methylates hemimethylated CpG dinucleotides in the mammalian genome [22,23] and maintains methylation status, whereas DNMT3a and DNMT3b are *de novo* methyltransferases that mainly establish methylation patterns [24]. DNMT3L is required for the methylation of imprinted genes in germ cells and interacts with DNMT3a and DNMT3b in *de novo* activity. The function of DNMT2 remains unclear [25]. MBPs may block transcription factors from binding to DNA by recruiting chromatin remodeling corepressor complexes [26,27]. HATs, HADCs, and HMTs are responsible for histone modification and chromatin remodeling [28,29].

DNMT1 is elevated in many types of tumors, but it is unknown whether DNMTs are relevant to NPC [30–32]. Latent membrane protein 1(LMP1) is an EBV-encoded oncoprotein that activates DNMT1 by c-Jun NH(2)-terminal kinase signaling, resulting in hypermethylation and silencing of E-cadherin and enhancing cell migration [33].

Aberrant DNA methylation patterns are found in many types of tumors [20,34–37]. As shown in Table 17.1, a series of tumor suppressor genes (TSGs) silenced by promoter methylation have been reported in NPC, indicating that the aberrant methylation pattern is a key factor in NPC carcinogenesis. Many of these genes are involved in cell cycle regulation (p14, p15, p16,14-3-3-δ, and Fez family zinc finger 2 (FEZF2)), apoptosis (CASP8, death-associated protein kinase (DAPK), GSTP1, tissue factor pathway inhibitor-2 (TFPI-2), ubiquitin carboxy-terminal hydrolase L1 (UCHL1), and WW domain-containing oxidoreductase (WWOX)), signaling pathways (WNT inhibitory factor-1 (WIF-1), retinoic acid receptor-β2 (RARβ2), Ras association domain family 1 isoform A (RASSF1A), high in normal-1 (HIN-1), and DAPK), DNA repair (glutathione S-transferase P (GSTP1), O6-methyguanine-DNA-methyltransferase (MGMT), and mutL homolog 1 (MLH1)), angiogenesis (thrombospondin 1 (THBS1)), cell adhesion, and migration and invasion (E-cadherin, deleted in liver cancer-1 (DLC-1), H-cadherin, matrix metalloprotease 19 (MMP19), popioid binding protein/cell adhesion molecule-like gene (OPCML), and tumor suppressor in lung cancer 1 (TSLC1)) [38]. These changes may cause chromosomal instability and mutations in signaling pathways and may ultimately contribute to NPC carcinogenesis.

Table 17.1 Summary of Major Tumor Suppressor Genes Methylated in NPC

Gene Name	Chromosome Location	Methylation Percentage in NPC Tumors (%)	Methylation Percentage in NPC cell Lines (%)	Major Function	References
BLU	3p21.3	66	–	Unknown	[39,40]
BRD7	16q12	100	71.4	Transcription regulation, cell cycle regulation	[41]
CASP8	2q33-34	7	–	Apoptosis	[42]
CDH4	20q13.33	94.3	–	Inhibiting cell proliferation, migration	[43]
CHFR	12q24.33	61.1	–	Mitotic checkpoint control	[44]
COX2	1q25.2-3	20	25	Inhibition of apoptosis	[45]
CRBPs (I + IV)	3q23	93.9	100	Transcription factors	[46]
CYB5R2	11p15.4	84	–	Unknown	[47]
DAB2	5p13.1	65.2	–	Growth inhibition	[48]
DAPK	9q21.33	76.1	–	Apoptosis	[49–51]
DLC-1	8P22.3	79	50	Signal transduction, cell adhesion, invasion	[52,53]
DLEC1	3p21-22	86.3	–	Unknown	[40,54]
E-cadherin	16q22.1	47	–	Cell–cell adhesion	[40]
EDRNB	19q22	84	–	G-protein coupled receptor	[55]
FEZF2	3p14.2	75.5	–	Cell cycle arrest, apoptosis, migration and stemness	[56]
GADD45G	9q22	16	73	Growth arrest, DNA repair	[57]
GNAT1	3p21.3	80	–	Tumor suppressor	[58]
GSTP1	11q13	-	–	Apoptosis, metabolism, energy pathways	[51]
H-cadherin	16q24.2-3	89.7	20	Cell adhesion, proliferation, metastasis	[59]
HIN-1	5q35-qter	77	100	Cell communication, Akt signaling pathway	[60]
hMLH1	3p21.3	21	–	DNA repair, cell cycle arrest	[61]
HOXA2	7p15.2	-	–	Transcription regulator	[62]
Kpm	13q12.11	36.7	78	Cell cycle arrest	[63]
LARS2	3p21.3	64	–	Tumor suppressor	[64]
LTF	3p21.3	63.6	100	Cell cycle control, inhibition of cell growth	[65,66]

(Continued)

Table 17.1 (Continued)

Gene Name	Chromosome Location	Methylation Percentage in NPC Tumors (%)	Methylation Percentage in NPC cell Lines (%)	Major Function	References
MIPOL1	14q13.3	–	–	Cell cycle regulation	[67]
MGMT	10q26	20	–	DNA repair, cell cycle	[42,51]
MMP19	12q14	–	–	Matrix metalloproteinase, anti-angiogenesis	[68]
NOR1	1p34.3	61.9	–	Unknown	[69]
OPCML	11q25	98	75	Cell adhesion	[70]
P14	9p21	20	–	ARF, stabilize p53	[51,71]
P15/ CDKN2B	9p21	50	–	Kinase inhibitor, cell cycle regulation	[42,72,73]
p16/ CDKN2A	9p21	22	–	Cell cycle arrest	[40,51,74]
P73	1p36.3	20	–	Cell cycle, apoptosis, transcription factor	[42]
PCDH10	4q28.3	82	92	Cell growth, migration, invasion, apoptosis	[75]
PCDH8	13q14.3	85.3	100	Tumor suppressor	[76]
PRDM2	1p36.21	60	–	Cell cycle arrest	[77]
PTEN	10q23.3	82.2	–	Cell cycle arrest, apoptosis, migration	[78]
PTPRG	3p14-21	41	86	Cell cycle regulation	[79]
RARRES1	3q25.32	90.7	20	Unknown	[80]
RARβ2	3p21	78.7	–	Tumor-suppressive activity	[49,51]
RASAL1	12q23-24	53	100	Ras GTPase–activating protein	[81]
RASSF1A	3p21.3	66.7	75	Cell cycle control, apoptosis, DNA repair	[49,51,82,83]
RASSF2A	20p13	50.9	80	Growth inhibition	[84]
RRAD	16q22.1	74.3	–	Unknown	[85]
Sky	9q22	–	–	Growth and metastasis inhibition	[86]
SOX11	2p25.2	–	–	Inhibiting cell growth and invasion	[87]
TFPI-2	7q22	88.6	67	Apoptosis, metastasis	[88]
THBS1	15q15	50	–	Cell adhesion, cell signaling	[42]
THY1	11q22-23	65	50	Proliferation, differentiation. apoptosis	[89]

(Continued)

Table 17.1 (Continued)

Gene Name	Chromosome Location	Methylation Percentage in NPC Tumors (%)	Methylation Percentage in NPC cell Lines (%)	Major Function	References
TSLC1	11q22-23	68	–	Cell–cell adhesion, apoptosis	[82]
UCHL1	4p14	–	100	Apoptosis, activates p53 pathway	[90]
WIF-1	12q14.3	89.7	–	Wnt signaling inhibitor	[91]
WWOX	16q23.1-23.2	87	–	Apoptosis	[92]
14-3-3-δ	2p25.1	84	–	Cell cycle arrest, metastasis	[93]

Note:– information not available.

3 HISTONE MODIFICATIONS IN NPC

Histone modifications and related proteins include histone acetylation (HATs)/deacetylation (HDACs), histone acetylation readers (BRDs), histone lysine methylation (KMTs)/demethylation (helminth defense molecules (HDMs)), histone methylation readers, and histone phosphorylation (kinases) [94]. Histone modifications have a major influence not just on transcription, but on all DNA-templated processes [95]. We have learned that coexisting histone modifications may be activators, but some may be repressors. These modification patterns change dynamically and have been termed "histone crosstalk." These patterns are widespread and have great biological significance [96]. Bromodomain-containing protein 7 (BRD7), a bromodomain gene, was the first bromodomain gene described in NPC. The overexpression of BRD7 inhibits NPC cell growth and the cell cycle by transcriptional regulation, which is achieved by BRD7 binding to lysine-acetylated peptides derived from histone H3 with K9 or K14 acetylated and from histone H4 with K8, K12, or K16 acetylated. Chromatin remodeling, not chromatin modification, is the major mechanism of BRD7-mediated gene transcription [97,98]. P300 is one of the histone acetyltransferases (HATs). Upregulated expression of p300 has been detected in NPC tissues compared to nasopharyngeal mucosal tissues. This protein might be important in conferring a more aggressive behavior and may be an independent molecular marker for a shortened survival in NPC patients [99]. Enhancer of zeste homolog 2 (EZH2) is the catalytic component of the PRC2 complex responsible for the methylation of H3K2. Excessive EZH2 expression by the inhibition of GSK-3β activity enhances NPC tumor invasion [100]. EZH2 may cause the reduction of p16INK4a expression by hypermethylating its promoter, which contributes to NPC tumorigenesis [101]. EZH2 supports NPC cell aggressiveness by forming a corepressor complex with HDAC1/HDAC2 and snail to inhibit E-cadherin [102]. Recently, researchers at the Sun Yat-Sen University Cancer Center discovered an epigenetic mechanism in the differentiation of NPC. The mechanism reveals that EZH2 directs IKKα transcriptional repression via H3K27 histone methylation on the IKKα promoter, while reduced IKKα expression is responsible for the undifferentiated phenotype of NPC [103]. Recently, it has been recognized that some kinases may also have nuclear functions, including the phosphorylation of histones [104–106].

One such enzyme is Janus kinase 2 (JAK2), which specifically phosphorylates H3Y41, disrupts the binding of the chromatin repressor HP1a, and activates hematopoietic oncogenes, such as Lmo2, in hematological malignancies [106]. However, JAK2 has not been studied in NPC. Metastasis-associated gene 1 (MTA1) is part of the chromatin-remodeling family. Aberrant expression of the MTA1 gene may be involved in the cell proliferation, tumor invasion, and metastasis of NPC. MTA1 overexpression correlates significantly with poor survival in NPC [107–111].

4 MicroRNA ALTERATIONS IN NPC

MicroRNAs (miRNAs) are endogenous; they are small noncoding RNA transcripts that are single-stranded and approximately 18–22 nucleotides long. They bind to the 3' UTR of target mRNA, resulting in the inhibition of target gene translation or degradation. The expression levels and patterns of miRNAs in NPC are significantly different from normal cells and tissues. MiRNAs play an important role in carcinogenesis and progression, including in NPC [112].

As shown in Table 17.2, miRNAs may act as oncomiRs or suppressors. They can affect NPC cell proliferation, apoptosis, and the cell cycle and regulate migration, invasion, and NPC metastasis. EBV mRNA BamHI A rightward transcripts (BARTs) regulate several key proteins that are responsible for facilitating latency maintenance, immune suppression, and tumor promotion [161]. Four miRNAs (miR-22, miR-572, miR-638, and miR-1234) in NPC serum are altered when comparing shorter and longer survival in patients; thus, these miRNAs may serve as potential biomarkers [162]. MiRNAs, such as miR-26a, miR-101, and miR-98, may interact with histone methylation proteins. These miR-NAs have been validated as regulators of EZH2 expression, which is a factor that may affect NPC aggressiveness [163].

5 LONG NONCODING RNAs IN NPC

Long noncoding RNAs (lncRNAs) are poorly conserved in different species. The mechanism of action in transcriptional regulation is varied. lncRNAs appear to have a key function at chromatin and may act as scaffolds or molecular chaperones, and the function may be subverted in NPC. In primary NPC, high levels of lnc-C22orf32-1 and lnc-AL355149.1-1 are significantly associated with male patients, and increased expression of lnc-C22orf32-1 and lnc-ZNF674-1 is associated with advanced tumors. Lnc-BCL2L11-3 is significantly increased in recurrent NPC tissues, while a significant reduction of lnc-AL355149.1-1 and lnc-ZNF674-1 has been observed. The association of lncRNA with gender, tumor size, and recurrent NPC implies that lncRNA may play a role in the pathogenesis of primary NPC and the mechanism of recurrence [164].

6 CLINICAL TRIALS

Several epigenetics-regulating inhibitors are under investigation alone or in combination with other anticancer drugs in different phases of clinical trials. Table 17.3 lists the registered trials in this field. More efforts are ongoing to develop more specific targeted agents for altered cancer epigenetics.

Table 17.2 Summary of miRNAs in NPC Cells or Tissues

Name	Role in NPC	Reference(s)
let-7	Suppresses cell proliferation	[113,114]
let-7a	Inhibits proliferation and induces apoptosis	[115]
let-7c	Tumor-suppressive factor	[116]
miR-1	Induces apoptosis	[117]
miR-10b	Promotes metastasis	[118]
miR-125a-5p	A regulator and predictor of gefitinib's effect on NPC	[119]
miR-138	Suppresses NPC growth and tumorigenesis	[120]
miR-141	Upregulated, cell cycle, apoptosis, cell growth, migration and invasion	[121]
miR-144	Promotes cell proliferation, migration and invasion	[122]
miR-146a	Overexpression in NPC specimens	[123]
miR-149	Promotes EMT and invasion	[124]
miR-155	Upregulated, promotes cell proliferation, migration and invasion	[125,126]
miR-184	Suppresses cell proliferation and survival	[127]
miR-18a	Promotes malignant progression	[128]
miR-18b	Widely overexpressed	[129]
miR-200a	Inhibits cell growth, migration and invasion	[130,131]
miR-200b	Suppresses cell growth, migration and invasion	[132]
miR-203	Downregulated in NPC	[133]
miR-204	Inhibits NPC invasion and metastasis	[134]
miR-205	Determines radioresistance	[135]
miR-214	Promotes tumorigenesis by targeting lactotransferrin	[136]
miR-216b	Suppresses tumor growth and invasion	[137]
miR-26a	Inhibits cell growth, invasion, and metastasis	[138,139]
miR-29c	Suppresses invasion and metastasis, enhances the sensitivities to cisplatin-based chemotherapy and radiotherapy	[140,141]
miR-324-3p	Regulates radioresistance	[142]
miR-34b/c	May singly or collaboratively contribute to the risk of NPC	[143]
miR-378	Functions as an oncomiR	[144]
miR-421	Induces cell proliferation and apoptosis resistance	[145]
miR-483-5p	High expression levels	[146]
miR-548q	High expression levels	[146]
miR-663	Promotes proliferation and tumorigenesis	[147]
miR-9	Tumor suppressor, shows decreased trend during progression	[148–150]
miR-93	Promotes uncontrolled cell growth, invasion, metastasis, and EMT	[151]
miR-BART1	Has been found to be highly expressed in NPC	[152]
miR-BART17	Was significantly more abundant in plasma samples from NPC patients	[153]
miR-BART22.	Facilitates NPC carcinogenesis by evading the host immune response	[154]
miR-BART3	Promotes cellular growth and transformation	[155]
miR-BART7	A potential biomarker for undifferentiated NPC	[156,157]
miR-BART9	Promotes tumor metastasis by targeting E-cadherin	[158,159]
miR-BHRF1-1	Potentiates viral lytic replication in NPC	[160]

Table 17.3 The Epigenetics-Regulating Inhibitors under Clinical Trials Targeting NPC Alone or in Combination with Other Anticancer Drugs

Status	Phase	Drug	Study
Unknown	Phase I	LBH589 and RAD001	Safety and Tolerability Study of RAD001 and LBH589 in All Solid Tumors with Enrichment for EBV Driven Tumors
Completed	Phase 1	Entinostat	MS-275 in Treating Patients with Advanced Solid Tumors or Lymphoma
Active, not recruiting	Phase 2	Vorinostat capecitabine	Capecitabine and Vorinostat in Treating Patients with Recurrent and/or Metastatic Head and Neck Cancer
Active, not recruiting	Phase 1	Azacitidine vorinostat	Vorinostat and Azacitidine in Treating Patients with Locally Recurrent or Metastatic Nasopharyngeal Cancer or Nasal Natural Killer T-Cell Lymphoma
Recruiting	Phase 1	Vorinostat, carboplatin, paclitaxel	Vorinostat in Combination with Paclitaxel and Carboplatin in Treating Patients with Metastatic or Recurrent Solid Tumors and HIV Infection
Completed	Phase 1	Vorinostat	Suberoylanilide Hydroxamic Acid in Treating Patients with Advanced Cancer

7 FUTURE PERSPECTIVES

Aberrant methylation of tumor suppressor genes is an important event in the formation and progression of cancer. Tumor-specific aberrant DNA methylations have been identified in different solid tumor types. Many studies have reported that genes silenced by DNA methylation may be established at the early stages of carcinogenesis. The methylation of specific genes or groups of genes is associated with chemotherapy response and survival. Therefore, an aberrant DNA methylation pattern may be used for cancer diagnostics in three ways: (1) as a marker for detecting cancer cells or cancer DNA; (2) as a marker for predicting prognosis; (3) as a biomarker for the assessment of the therapeutic response, which would be the most useful because gene methylation detection is easier than gene mutation detection, and gene methylation can be detected in noninvasive bodily fluids, such as blood, urine, saliva, and stool [165]. Recently, it was reported that NPC methylation analysis by a single PCR reaction with a small amount of tumor DNA derived from bodily fluid is similar to that of the primary tumor [166]. The detection of tumor suppressor gene methylation, such as for retinoblastoma protein-interacting zinc finger gene 1 (RIZ1), in bodily fluid samples of NPC patients indicates that further evaluation of the clinical application of screening for NPC is worthwhile [72,77,167]. Methylated genes are predicted to be a new generation of cancer biomarkers; however, all of these findings require validation in prospective clinical studies prior to clinical application.

Epigenetic change is heritable, but it can be changed. During the last few decades, a number of drugs targeting DNA methylation have been improved to increase efficacy and decrease toxicity. Demethylation agents and HDAC inhibitors are two major types of epigenetic agents. 5-Azacytidine, the most investigated demethylation drug to date, is currently recommended as the first-line treatment of high-risk myelodysplastic syndromes (MDS). However, global demethylation treatment using 5-azacytidine has also been challenged, as this treatment might also activate the expression of other genes responsible for cancer metastasis. Several classes of HDAC inhibitors have been isolated and are currently undergoing evaluation as potential therapeutic modalities in the treatment of NPC using cell lines [168]. Therefore, new methods for gene-specific epigenetic modification are being developed and tested in solid tumors [169].

REFERENCES

[1] Cao SM, Simons MJ, Qian CN. The prevalence and prevention of nasopharyngeal carcinoma in China. Chin J Cancer 2011;30(2):114–19.

[2] Adham M, Kurniawan AN, Muhtadi AI, Roezin A, Hermani B, Gondhowiardjo S, et al. Nasopharyngeal carcinoma in Indonesia: epidemiology, incidence, signs, and symptoms at presentation. Chin J Cancer 2012;31(4):185–96.

[3] Kataki AC, Simons MJ, Das AK, Sharma K, Mehra NK. Nasopharyngeal carcinoma in the Northeastern states of India. Chin J Cancer 2011;30(2):106–13.

[4] Sarmiento MP, Mejia MB. Preliminary assessment of nasopharyngeal carcinoma incidence in the Philippines: a second look at published data from four centers. Chin J Cancer 2014;33(3):159–64.

[5] Lozano R, Naghavi M, Foreman K, Lim S, Shibuya K, Aboyans V, et al. Global and regional mortality from 235 causes of death for 20 age groups in 1990 and 2010: a systematic analysis for the Global Burden of Disease Study 2010. Lancet 2012;380(9859):2095–128.

[6] Parkin DM, Bray F, Ferlay J, Pisani P. Estimating the world cancer burden: Globocan 2000. Int J Cancer 2001;94(2):153–6.

[7] Stiller CA. International variations in the incidence of childhood carcinomas. Cancer Epidemiol Biomarkers Prev 1994;3(4):305–10.

[8] Yu MC, Yuan JM. Epidemiology of nasopharyngeal carcinoma. Semin Cancer Biol 2002;12(6):421–9.

[9] Lu X, Wang FL, Guo X, Wang L, Zhang HB, Xia WX, et al. Favorable prognosis of female patients with nasopharyngeal carcinoma. Chin J Cancer 2013;32(5):283–8.

[10] Brennan B. Nasopharyngeal carcinoma. Orphanet J Rare Dis 2006;1:23.

[11] Chen MY, Jiang R, Guo L, Zou X, Liu Q, Sun R, et al. Locoregional radiotherapy in patients with distant metastases of nasopharyngeal carcinoma at diagnosis. Chin J Cancer 2013;32(11):604–13.

[12] Gu AD, Zeng MS, Qian CN. The criteria to confirm the role of epstein-barr virus in nasopharyngeal carcinoma initiation. Int J Mol Sci 2012;13(10):13737–47.

[13] Wee JT, Ha TC, Loong SL, Qian CN. Is nasopharyngeal cancer really a Cantonese cancer? Chin J Cancer 2010;29(5):517–26.

[14] Feng BJ, Huang W, Shugart YY, Lee MK, Zhang F, Xia JC, et al. Genome-wide scan for familial nasopharyngeal carcinoma reveals evidence of linkage to chromosome 4. Nat Genet 2002;31(4):395–9.

[15] Raab-Traub N, Flynn K. The structure of the termini of the Epstein-Barr virus as a marker of clonal cellular proliferation. Cell 1986;47(6):883–9.

[16] Raab-Traub N. Epstein-Barr virus in the pathogenesis of NPC. Semin Cancer Biol 2002;12(6):431–41.

[17] Csankovszki G, Nagy A, Jaenisch R. Synergism of Xist RNA, DNA methylation, and histone hypoacetylation in maintaining X chromosome inactivation. J Cell Biol 2001;153(4):773–84.

[18] Copeland RA, Olhava EJ, Scott MP. Targeting epigenetic enzymes for drug discovery. Curr Opin Chem Biol 2010;14(4):505–10.

[19] Marks P, Rifkind RA, Richon VM, Breslow R, Miller T, Kelly WK. Histone deacetylases and cancer: causes and therapies. Nat Rev Cancer 2001;1(3):194–202.

[20] Li JS, Ying JM, Wang XW, Wang ZH, Tao Q, Li LL. Promoter methylation of tumor suppressor genes in esophageal squamous cell carcinoma. Chin J Cancer 2013;32(1):3–11.

[21] Jones PA, Takai D. The role of DNA methylation in mammalian epigenetics. Science 2001;293(5532):1068–70.

[22] Jair KW, Bachman KE, Suzuki H, Ting AH, Rhee I, Yen RW, et al. De novo CpG island methylation in human cancer cells. Cancer Res 2006;66(2):682–92.

[23] Ting AH, Jair KW, Schuebel KE, Baylin SB. Differential requirement for DNA methyltransferase 1 in maintaining human cancer cell gene promoter hypermethylation. Cancer Res 2006;66(2):729–35.

[24] Hsieh CL. In vivo activity of murine de novo methyltransferases, Dnmt3a and Dnmt3b. Mol Cell Biol 1999;19(12):8211–18.

[25] Chedin F, Lieber MR, Hsieh CL. The DNA methyltransferase-like protein DNMT3L stimulates de novo methylation by Dnmt3a. Proc Natl Acad Sci USA 2002;99(26):16916–21.

[26] Meehan RR, Lewis JD, Bird AP. Characterization of MeCP2, a vertebrate DNA binding protein with affinity for methylated DNA. Nucleic Acids Res 1992;20(19):5085–92.

[27] Free A, Wakefield RI, Smith BO, Dryden DT, Barlow PN, Bird AP. DNA recognition by the methyl-CpG binding domain of MeCP2. J Biol Chem 2001;276(5):3353–60.

[28] Hassig CA, Schreiber SL. Nuclear histone acetylases and deacetylases and transcriptional regulation: HATs off to HDACs. Curr Opin Chem Biol 1997;1(3):300–8.

[29] Rice JC, Allis CD. Histone methylation versus histone acetylation: new insights into epigenetic regulation. Curr Opin Cell Biol 2001;13(3):263–73.

[30] Lee PJ, Washer LL, Law DJ, Boland CR, Horon IL, Feinberg AP. Limited up-regulation of DNA methyltransferase in human colon cancer reflecting increased cell proliferation. Proc Natl Acad Sci USA 1996;93(19):10366–70.

[31] Robertson KD, Uzvolgyi E, Liang G, Talmadge C, Sumegi J, Gonzales FA, et al. The human DNA methyltransferases (DNMTs) 1, 3a and 3b: coordinate mRNA expression in normal tissues and overexpression in tumors. Nucleic Acids Res 1999;27(11):2291–8.

[32] Kautiainen TL, Jones PA. DNA methyltransferase levels in tumorigenic and nontumorigenic cells in culture. J Biol Chem 1986;261(4):1594–8.

[33] Tsai CL, Li HP, Lu YJ, Hsueh C, Liang Y, Chen CL, et al. Activation of DNA methyltransferase 1 by EBV LMP1 Involves c-Jun NH(2)-terminal kinase signaling. Cancer Res 2006;66(24):11668–76.

[34] Xiang TX, Yuan Y, Li LL, Wang ZH, Dan LY, Chen Y, et al. Aberrant promoter CpG methylation and its translational applications in breast cancer. Chin J Cancer 2013;32(1):12–20.

[35] Jia Y, Guo M. Epigenetic changes in colorectal cancer. Chin J Cancer 2013;32(1):21–30.

[36] Hu XT, He C. Recent progress in the study of methylated tumor suppressor genes in gastric cancer. Chin J Cancer 2013;32(1):31–41.

[37] Yang HJ. Aberrant DNA methylation in cervical carcinogenesis. Chin J Cancer 2013;32(1):42–8.

[38] Li LL, Shu XS, Wang ZH, Cao Y, Tao Q. Epigenetic disruption of cell signaling in nasopharyngeal carcinoma. Chin J Cancer 2011;30(4):231–9.

[39] Qiu GH, Tan LK, Loh KS, Lim CY, Srivastava G, Tsai ST, et al. The candidate tumor suppressor gene BLU, located at the commonly deleted region 3p21.3, is an E2F-regulated, stress-responsive gene and inactivated by both epigenetic and genetic mechanisms in nasopharyngeal carcinoma. Oncogene 2004;23(27):4793–806.

[40] Ayadi W, Karray-Hakim H, Khabir A, Feki L, Charfi S, Boudawara T, et al. Aberrant methylation of p16, DLEC1, BLU and E-cadherin gene promoters in nasopharyngeal carcinoma biopsies from Tunisian patients. Anticancer Res 2008;28(4b):2161–7.

[41] Liu H, Zhang L, Niu Z, Zhou M, Peng C, Li X, et al. Promoter methylation inhibits BRD7 expression in human nasopharyngeal carcinoma cells. BMC Cancer 2008;8:253.

[42] Wong TS, Tang KC, Kwong DL, Sham JS, Wei WI, Kwong YL, et al. Differential gene methylation in undifferentiated nasopharyngeal carcinoma. Int J Oncol 2003;22(4):869–74.

[43] Du C, Huang T, Sun D, Mo Y, Feng H, Zhou X, et al. CDH4 as a novel putative tumor suppressor gene epigenetically silenced by promoter hypermethylation in nasopharyngeal carcinoma. Cancer Lett 2011;309(1):54–61.

[44] Cheung HW, Ching YP, Nicholls JM, Ling MT, Wong YC, Hui N, et al. Epigenetic inactivation of CHFR in nasopharyngeal carcinoma through promoter methylation. Mol Carcinog 2005;43(4):237–45.

[45] Soo R, Putti T, Tao Q, Goh BC, Lee KH, Kwok-Seng L, et al. Overexpression of cyclooxygenase-2 in nasopharyngeal carcinoma and association with epidermal growth factor receptor expression. Arch Otolaryngol Head Neck Surg 2005;131(2):147–52.

[46] Kwong J, Lo KW, Chow LS, To KF, Choy KW, Chan FL, et al. Epigenetic silencing of cellular retinolbinding proteins in nasopharyngeal carcinoma. Neoplasia 2005;7(1):67–74.

[47] Xiao X, Zhao W, Tian F, Zhou X, Zhang J, Huang T, et al. Cytochrome b5 reductase 2 is a novel candidate tumor suppressor gene frequently inactivated by promoter hypermethylation in human nasopharyngeal carcinoma. Tumour Biol 2014;35(4):3755–63.

[48] Tong JH, Ng DC, Chau SL, So KK, Leung PP, Lee TL, et al. Putative tumour-suppressor gene DAB2 is frequently down regulated by promoter hypermethylation in nasopharyngeal carcinoma. BMC Cancer 2010;10:253.

[49] Fendri A, Masmoudi A, Khabir A, Sellami-Boudawara T, Daoud J, Frikha M, et al. Inactivation of RASSF1A, RARbeta2 and DAP-kinase by promoter methylation correlates with lymph node metastasis in nasopharyngeal carcinoma. Cancer Biol Ther 2009;8(5):444–51.

[50] Kong WJ, Zhang S, Guo CK, Wang YJ, Chen X, Zhang SL, et al. Effect of methylation-associated silencing of the death-associated protein kinase gene on nasopharyngeal carcinoma. Anticancer Drugs 2006;17(3):251–9.

[51] Kwong J, Lo KW, To KF, Teo PM, Johnson PJ, Huang DP. Promoter hypermethylation of multiple genes in nasopharyngeal carcinoma. Clin Cancer Res 2002;8(1):131–7.

[52] Peng D, Ren CP, Yi HM, Zhou L, Yang XY, Li H, et al. Genetic and epigenetic alterations of DLC-1, a candidate tumor suppressor gene, in nasopharyngeal carcinoma. Acta Biochim Biophys Sin (Shanghai) 2006;38(5):349–55.

[53] Seng TJ, Low JS, Li H, Cui Y, Goh HK, Wong ML, et al. The major 8p22 tumor suppressor DLC1 is frequently silenced by methylation in both endemic and sporadic nasopharyngeal, esophageal, and cervical carcinomas, and inhibits tumor cell colony formation. Oncogene 2007;26(6):934–44.

[54] Kwong J, Chow LS, Wong AY, Hung WK, Chung GT, To KF, et al. Epigenetic inactivation of the deleted in lung and esophageal cancer 1 gene in nasopharyngeal carcinoma. Genes Chromosomes Cancer 2007;46(2):171–80.

[55] Zhou L, Feng X, Shan W, Zhou W, Liu W, Wang L, et al. Epigenetic and genetic alterations of the EDNRB gene in nasopharyngeal carcinoma. Oncology 2007;72(5–6):357–63.

[56] Shu XS, Li L, Ji M, Cheng Y, Ying J, Fan Y, et al. FEZF2, a novel 3p14 tumor suppressor gene, represses oncogene EZH2 and MDM2 expression and is frequently methylated in nasopharyngeal carcinoma. Carcinogenesis 2013;34(9):1984–93.

[57] Ying J, Srivastava G, Hsieh WS, Gao Z, Murray P, Liao SK, et al. The stress-responsive gene GADD45G is a functional tumor suppressor, with its response to environmental stresses frequently disrupted epigenetically in multiple tumors. Clin Cancer Res 2005;11(18):6442–9.

[58] Yi HM, Ren CP, Peng D, Zhou L, Li H, Yao KT. [Expression, loss of heterozygosity, and methylation of GNAT1 gene in nasopharyngeal carcinoma]. Ai Zheng 2007;26(1):9–14.

[59] Sun D, Zhang Z, Van do N, Huang G, Ernberg I, Hu L. Aberrant methylation of CDH13 gene in nasopharyngeal carcinoma could serve as a potential diagnostic biomarker. Oral Oncol 2007;43(1):82–7.

[60] Wong TS, Kwong DL, Sham JS, Tsao SW, Wei WI, Kwong YL, et al. Promoter hypermethylation of high-innormal 1 gene in primary nasopharyngeal carcinoma. Clin Cancer Res 2003;9(8):3042–6.

[61] Tan SH, Ida H, Goh BC, Hsieh W, Loh M, Ito Y. Analyses of promoter hypermethylation for RUNX3 and other tumor suppressor genes in nasopharyngeal carcinoma. Anticancer Res 2006;26(6b):4287–92.

[62] Li HP, Peng CC, Chung IC, Huang MY, Huang ST, Chen CC, et al. Aberrantly hypermethylated Homeobox A2 derepresses metalloproteinase-9 through TBP and promotes invasion in Nasopharyngeal carcinoma. Oncotarget 2013;4(11):2154–65.

[63] Zhang Y, Hu CF, Chen J, Yan LX, Zeng YX, Shao JY. LATS2 is de-methylated and overexpressed in nasopharyngeal carcinoma and predicts poor prognosis. BMC Cancer 2010;10:538.

[64] Zhou W, Feng X, Li H, Wang L, Zhu B, Liu W, et al. Inactivation of LARS2, located at the commonly deleted region 3p21.3, by both epigenetic and genetic mechanisms in nasopharyngeal carcinoma. Acta Biochim Biophys Sin (Shanghai) 2009;41(1):54–62.

[65] Zhang H, Feng X, Liu W, Jiang X, Shan W, Huang C, et al. Underlying mechanisms for LTF inactivation and its functional analysis in nasopharyngeal carcinoma cell lines. J Cell Biochem 2011;112(7):1832–43.

[66] Yi HM, Li H, Peng D, Zhang HJ, Wang L, Zhao M, et al. Genetic and epigenetic alterations of LTF at 3p21.3 in nasopharyngeal carcinoma. Oncol Res 2006;16(6):261–72.

[67] Cheung AK, Lung HL, Ko JM, Cheng Y, Stanbridge EJ, Zabarovsky ER, et al. Chromosome 14 transfer and functional studies identify a candidate tumor suppressor gene, mirror image polydactyly 1, in nasopharyngeal carcinoma. Proc Natl Acad Sci USA 2009;106(34):14478–83.

[68] Chan KC, Ko JM, Lung HL, Sedlacek R, Zhang ZF, Luo DZ, et al. Catalytic activity of matrix metalloproteinase-19 is essential for tumor suppressor and anti-angiogenic activities in nasopharyngeal carcinoma. Int J Cancer 2011;129(8):1826–37.

[69] Li W, Li X, Wang W, Li X, Tan Y, Yi M, et al. NOR1 is an HSF1- and NRF1-regulated putative tumor suppressor inactivated by promoter hypermethylation in nasopharyngeal carcinoma. Carcinogenesis 2011;32(9):1305–14.

[70] Cui Y, Ying Y, van Hasselt A, Ng KM, Yu J, Zhang Q, et al. OPCML is a broad tumor suppressor for multiple carcinomas and lymphomas with frequently epigenetic inactivation. PLoS One 2008;3(8):e2990.

[71] Lo KW, Huang DP. Genetic and epigenetic changes in nasopharyngeal carcinoma. Semin Cancer Biol 2002;12(6):451–62.

[72] Chang HW, Chan A, Kwong DL, Wei WI, Sham JS, Yuen AP. Evaluation of hypermethylated tumor suppressor genes as tumor markers in mouth and throat rinsing fluid, nasopharyngeal swab and peripheral blood of nasopharygeal carcinoma patient. Int J Cancer 2003;105(6):851–5.

[73] Wong TS, Kwong DL, Sham JS, Wei WI, Kwong YL, Yuen AP. Quantitative plasma hypermethylated DNA markers of undifferentiated nasopharyngeal carcinoma. Clin Cancer Res 2004;10(7):2401–6.

[74] Lo KW, Cheung ST, Leung SF, van Hasselt A, Tsang YS, Mak KF, et al. Hypermethylation of the p16 gene in nasopharyngeal carcinoma. Cancer Res 1996;56(12):2721–5.

[75] Ying J, Li H, Seng TJ, Langford C, Srivastava G, Tsao SW, et al. Functional epigenetics identifies a protocadherin PCDH10 as a candidate tumor suppressor for nasopharyngeal, esophageal and multiple other carcinomas with frequent methylation. Oncogene 2006;25(7):1070–80.

[76] He D, Zeng Q, Ren G, Xiang T, Qian Y, Hu Q, et al. Protocadherin8 is a functional tumor suppressor frequently inactivated by promoter methylation in nasopharyngeal carcinoma. Eur J Cancer Prev 2012;21(6):569–75.

[77] Chang HW, Chan A, Kwong DL, Wei WI, Sham JS, Yuen AP. Detection of hypermethylated RIZ1 gene in primary tumor, mouth, and throat rinsing fluid, nasopharyngeal swab, and peripheral blood of nasopharyngeal carcinoma patient. Clin Cancer Res 2003;9(3):1033–8.

[78] Li J, Gong P, Lyu X, Yao K, Li X, Peng H. Aberrant CpG island methylation of PTEN is an early event in nasopharyngeal carcinoma and a potential diagnostic biomarker. Oncol Rep 2014;31(5):2206–12.

[79] Cheung AK, Lung HL, Hung SC, Law EW, Cheng Y, Yau WL, et al. Functional analysis of a cell cycle-associated, tumor-suppressive gene, protein tyrosine phosphatase receptor type G, in nasopharyngeal carcinoma. Cancer Res 2008;68(19):8137–45.

[80] Kwong J, Lo KW, Chow LS, Chan FL, To KF, Huang DP. Silencing of the retinoid response gene TIG1 by promoter hypermethylation in nasopharyngeal carcinoma. Int J Cancer 2005;113(3):386–92.

[81] Jin H, Wang X, Ying J, Wong AH, Cui Y, Srivastava G, et al. Epigenetic silencing of a Ca(2+)-regulated Ras GTPase-activating protein RASAL defines a new mechanism of Ras activation in human cancers. Proc Natl Acad Sci USA 2007;104(30):12353–8.

[82] Zhou L, Jiang W, Ren C, Yin Z, Feng X, Liu W, et al. Frequent hypermethylation of RASSF1A and TSLC1, and high viral load of Epstein-Barr Virus DNA in nasopharyngeal carcinoma and matched tumor-adjacent tissues. Neoplasia 2005;7(9):809–15.

[83] Lo KW, Kwong J, Hui AB, Chan SY, To KF, Chan AS, et al. High frequency of promoter hypermethylation of RASSF1A in nasopharyngeal carcinoma. Cancer Res 2001;61(10):3877–81.

[84] Zhang Z, Sun D, Van do N, Tang A, Hu L, Huang G. Inactivation of RASSF2A by promoter methylation correlates with lymph node metastasis in nasopharyngeal carcinoma. Int J Cancer 2007;120(1):32–8.

[85] Mo Y, Midorikawa K, Zhang Z, Zhou X, Ma N, Huang G, et al. Promoter hypermethylation of Ras-related GTPase gene RRAD inactivates a tumor suppressor function in nasopharyngeal carcinoma. Cancer Lett 2012;323(2):147–54.

[86] Yan C, Liu C, Jin Q, Li Z, Tao B, Cai Z. The promoter methylation of the Syk gene in nasopharyngeal carcinoma cell lines. Oncol Lett 2012;4(3):505–8.

[87] Zhang S, Li S, Gao JL. Promoter methylation status of the tumor suppressor gene SOX11 is associated with cell growth and invasion in nasopharyngeal carcinoma. Cancer Cell Int 2013;13(1):109.

[88] Wang S, Xiao X, Zhou X, Huang T, Du C, Yu N, et al. TFPI-2 is a putative tumor suppressor gene frequently inactivated by promoter hypermethylation in nasopharyngeal carcinoma. BMC Cancer 2010;10:617.

[89] Lung HL, Bangarusamy DK, Xie D, Cheung AK, Cheng Y, Kumaran MK, et al. THY1 is a candidate tumour suppressor gene with decreased expression in metastatic nasopharyngeal carcinoma. Oncogene 2005;24(43):6525–32.

[90] Li L, Tao Q, Jin H, van Hasselt A, Poon FF, Wang X, et al. The tumor suppressor UCHL1 forms a complex with p53/MDM2/ARF to promote p53 signaling and is frequently silenced in nasopharyngeal carcinoma. Clin Cancer Res 2010;16(11):2949–58.

[91] Fendri A, Khabir A, Hadri-Guiga B, Sellami-Boudawara T, Daoud J, Frikha M, et al. Epigenetic alteration of the Wnt inhibitory factor-1 promoter is common and occurs in advanced stage of Tunisian nasopharyngeal carcinoma. Cancer Invest 2010;28(9):896–903.

[92] Yang Z, Lan H, Chen X, Li P, Li S, Mo W, et al. Molecular alterations of the WWOX gene in nasopharyngeal carcinoma. Neoplasma 2014;61(2):170–6.

[93] Yi B, Tan SX, Tang CE, Huang WG, Cheng AL, Li C, et al. Inactivation of 14-3-3 sigma by promoter methylation correlates with metastasis in nasopharyngeal carcinoma. J Cell Biochem 2009;106(5):858–66.

[94] Dawson MA, Kouzarides T. Cancer epigenetics: from mechanism to therapy. Cell 2012;150(1):12–27.

[95] Kouzarides T. Chromatin modifications and their function. Cell 2007;128(4):693–705.

[96] Lee JS, Smith E, Shilatifard A. The language of histone crosstalk. Cell 2010;142(5):682–5.

[97] Peng C, Zhou J, Liu HY, Zhou M, Wang LL, Zhang QH, et al. The transcriptional regulation role of BRD7 by binding to acetylated histone through bromodomain. J Cell Biochem 2006;97(4):882–92.

[98] Sun H, Liu J, Zhang J, Shen W, Huang H, Xu C, et al. Solution structure of BRD7 bromodomain and its interaction with acetylated peptides from histone H3 and H4. Biochem Biophys Res Commun 2007;358(2):435–41.

[99] Liao ZW, Zhou TC, Tan XJ, Song XL, Liu Y, Shi XY, et al. High expression of p300 is linked to aggressive features and poor prognosis of nasopharyngeal carcinoma. J Transl Med 2012;10:110.

[100] Ma R, Wei Y, Huang X, Fu R, Luo X, Zhu X, et al. Inhibition of GSK 3beta activity is associated with excessive EZH2 expression and enhanced tumour invasion in nasopharyngeal carcinoma. PLoS One 2013;8(7):e68614.

[101] Zhong J, Min L, Huang H, Li L, Li D, Li J, et al. EZH2 regulates the expression of p16 in the nasopharyngeal cancer cells. Technol Cancer Res Treat 2013;12(3):269–74.

[102] Tong ZT, Cai MY, Wang XG, Kong LL, Mai SJ, Liu YH, et al. EZH2 supports nasopharyngeal carcinoma cell aggressiveness by forming a co-repressor complex with HDAC1/HDAC2 and Snail to inhibit E-cadherin. Oncogene 2012;31(5):583–94.

[103] Yan M, Zhang Y, He B, Xiang J, Wang ZF, Zheng FM, et al. IKKalpha restoration via EZH2 suppression induces nasopharyngeal carcinoma differentiation. Nat Commun 2014;5:3661.

[104] Baek SH. When signaling kinases meet histones and histone modifiers in the nucleus. Mol Cell 2011;42(3):274–84.

[105] Bungard D, Fuerth BJ, Zeng PY, Faubert B, Maas NL, Viollet B, et al. Signaling kinase AMPK activates stress-promoted transcription via histone H2B phosphorylation. Science 2010;329(5996):1201–5.

[106] Dawson MA, Bannister AJ, Gottgens B, Foster SD, Bartke T, Green AR, et al. JAK2 phosphorylates histone H3Y41 and excludes HP1alpha from chromatin. Nature 2009;461(7265):819–22.

[107] Deng Y, Zhou D, Zeng L. [Expression and significance of MTA1 and RECK gene in nasopharyngeal carcinoma]. Lin Chung Er Bi Yan Hou Tou Jing Wai Ke Za Zhi 2011;25(12):534–8.

[108] Song Q, Li Y, Zheng X, Fang Y, Chao Y, Yao K, et al. MTA1 contributes to actin cytoskeleton reorganization and metastasis of nasopharyngeal carcinoma by modulating Rho GTPases and Hedgehog signaling. Int J Biochem Cell Biol 2013;45(7):1439–46.

[109] Song Q, Zhang H, Wang M, Song W, Ying M, Fang Y, et al. MTA1 promotes nasopharyngeal carcinoma growth in vitro and in vivo. J Exp Clin Cancer Res 2013;32(1):54.

[110] Deng YF, Zhou DN, Ye CS, Zeng L, Yin P. Aberrant expression levels of MTA1 and RECK in nasopharyngeal carcinoma: association with metastasis, recurrence, and prognosis. Ann Otol Rhinol Laryngol 2012;121(7):457–65.

[111] Li WF, Liu N, Cui RX, He QM, Chen M, Jiang N, et al. Nuclear overexpression of metastasis-associated protein 1 correlates significantly with poor survival in nasopharyngeal carcinoma. J Transl Med 2012;10:78.

[112] Lo AK, Dawson CW, Jin DY, Lo KW. The pathological roles of BART miRNAs in nasopharyngeal carcinoma. J Pathol 2012;227(4):392–403.

[113] Wong TS, Man OY, Tsang CM, Tsao SW, Tsang RK, Chan JY, et al. MicroRNA let-7 suppresses nasopharyngeal carcinoma cells proliferation through downregulating c-Myc expression. J Cancer Res Clin Oncol 2011;137(3):415–22.

[114] Pan XM, Jia J, Guo XM, Li ZH, Zhang Z, Qin HJ, et al. Lack of association between let-7 binding site polymorphism rs712 and risk of nasopharyngeal carcinoma. Fam Cancer 2014;13(1):93–7.

[115] Cai K, Wan Y, Sun G, Shi L, Bao X, Wang Z. Let-7a inhibits proliferation and induces apoptosis by targeting EZH2 in nasopharyngeal carcinoma cells. Oncol Rep 2012;28(6):2101–6.

[116] Liu Z, Long X, Chao C, Yan C, Wu Q, Hua S, et al. Knocking down CDK4 mediates the elevation of let-7c suppressing cell growth in nasopharyngeal carcinoma. BMC Cancer 2014;14(1):274.

[117] Wu CD, Kuo YS, Wu HC, Lin CT. MicroRNA-1 induces apoptosis by targeting prothymosin alpha in nasopharyngeal carcinoma cells. J Biomed Sci 2011;18:80.

[118] Li G, Wu Z, Peng Y, Liu X, Lu J, Wang L, et al. MicroRNA-10b induced by Epstein-Barr virus-encoded latent membrane protein-1 promotes the metastasis of human nasopharyngeal carcinoma cells. Cancer Lett 2010;299(1):29–36.

[119] Liu Y, Li Z, Wu L, Wang Z, Wang X, Yu Y, et al. MiRNA-125a-5p: a regulator and predictor of gefitinib's effect on nasopharyngeal carcinoma. Cancer Cell Int 2014;14(1):24.

[120] Liu X, Lv XB, Wang XP, Sang Y, Xu S, Hu K, et al. MiR-138 suppressed nasopharyngeal carcinoma growth and tumorigenesis by targeting the CCND1 oncogene. Cell Cycle 2012;11(13):2495–506.

[121] Zhang L, Deng T, Li X, Liu H, Zhou H, Ma J, et al. microRNA-141 is involved in a nasopharyngeal carcinoma-related genes network. Carcinogenesis 2010;31(4):559–66.

[122] Zhang LY, Ho-Fun Lee V, Wong AM, Kwong DL, Zhu YH, Dong SS, et al. MicroRNA-144 promotes cell proliferation, migration and invasion in nasopharyngeal carcinoma through repression of PTEN. Carcinogenesis 2013;34(2):454–63.

[123] Zhao Y, Chen X, Jing M, Du H, Zeng Y. Expression of miRNA-146a in nasopharyngeal carcinoma is upregulated by Epstein-Barr virus latent membrane protein 1. Oncol Rep 2012;28(4):1237–42.

[124] Luo Z, Zhang L, Li Z, Jiang C, Dai Y, Liu X, et al. miR-149 promotes epithelial-mesenchymal transition and invasion in nasopharyngeal carcinoma cells. Zhong Nan Da Xue Xue Bao Yi Xue Ban 2011;36(7):604–9.

[125] Zhu X, Wang Y, Sun Y, Zheng J, Zhu D. MiR-155 up-regulation by LMP1 DNA contributes to increased nasopharyngeal carcinoma cell proliferation and migration. Eur Arch Otorhinolaryngol 2014;271(7):1939–45.

[126] Du ZM, Hu LF, Wang HY, Yan LX, Zeng YX, Shao JY, et al. Upregulation of MiR-155 in nasopharyngeal carcinoma is partly driven by LMP1 and LMP2A and downregulates a negative prognostic marker JMJD1A. PLoS One 2011;6(4):e19137.

[127] Zhen Y, Liu Z, Yang H, Yu X, Wu Q, Hua S, et al. Tumor suppressor PDCD4 modulates miR-184-mediated direct suppression of C-MYC and BCL2 blocking cell growth and survival in nasopharyngeal carcinoma. Cell Death Dis 2013;4:e872.

[128] Luo Z, Dai Y, Zhang L, Jiang C, Li Z, Yang J, et al. miR-18a promotes malignant progression by impairing microRNA biogenesis in nasopharyngeal carcinoma. Carcinogenesis 2013;34(2):415–25.

[129] Yu X, Zhen Y, Yang H, Wang H, Zhou Y, Wang E, et al. Loss of connective tissue growth factor as an unfavorable prognosis factor activates miR-18b by PI3K/AKT/C-Jun and C-Myc and promotes cell growth in nasopharyngeal carcinoma. Cell Death Dis 2013;4:e634.

[130] Xia H, Ng SS, Jiang S, Cheung WK, Sze J, Bian XW, et al. miR-200a-mediated downregulation of ZEB2 and CTNNB1 differentially inhibits nasopharyngeal carcinoma cell growth, migration and invasion. Biochem Biophys Res Commun 2010;391(1):535–41.

[131] Xia H, Cheung WK, Sze J, Lu G, Jiang S, Yao H, et al. miR-200a regulates epithelial-mesenchymal to stem-like transition via ZEB2 and beta-catenin signaling. J Biol Chem 2010;285(47):36995–7004.

[132] Yang X, Ni W, Lei K. miR-200b suppresses cell growth, migration and invasion by targeting Notch1 in nasopharyngeal carcinoma. Cell Physiol Biochem 2013;32(5):1288–98.

[133] Yu H, Lu J, Zuo L, Yan Q, Yu Z, Li X, et al. Epstein-Barr virus downregulates microRNA 203 through the oncoprotein latent membrane protein 1: a contribution to increased tumor incidence in epithelial cells. J Virol 2012;86(6):3088–99.

[134] Ma L, Deng X, Wu M, Zhang G, Huang J. Down-regulation of miRNA-204 by LMP-1 enhances CDC42 activity and facilitates invasion of EBV-associated nasopharyngeal carcinoma cells. FEBS Lett 2014;588(9):1562–70.

[135] Qu C, Liang Z, Huang J, Zhao R, Su C, Wang S, et al. MiR-205 determines the radioresistance of human nasopharyngeal carcinoma by directly targeting PTEN. Cell Cycle 2012;11(4):785–96.

[136] Deng M, Ye Q, Qin Z, Zheng Y, He W, Tang H, et al. miR-214 promotes tumorigenesis by targeting lactotransferrin in nasopharyngeal carcinoma. Tumour Biol 2013;34(3):1793–800.

[137] Deng M, Tang H, Zhou Y, Zhou M, Xiong W, Zheng Y, et al. miR-216b suppresses tumor growth and invasion by targeting KRAS in nasopharyngeal carcinoma. J Cell Sci 2011;124(Pt 17):2997–3005.

[138] Lu J, He ML, Wang L, Chen Y, Liu X, Dong Q, et al. MiR-26a inhibits cell growth and tumorigenesis of nasopharyngeal carcinoma through repression of EZH2. Cancer Res 2011;71(1):225–33.

[139] Yu L, Lu J, Zhang B, Liu X, Wang L, Li SY, et al. miR-26a inhibits invasion and metastasis of nasopharyngeal cancer by targeting EZH2. Oncol Lett 2013;5(4):1223–8.

[140] Liu N, Tang LL, Sun Y, Cui RX, Wang HY, Huang BJ, et al. MiR-29c suppresses invasion and metastasis by targeting TIAM1 in nasopharyngeal carcinoma. Cancer Lett 2013;329(2):181–8.

[141] Zhang JX, Qian D, Wang FW, Liao DZ, Wei JH, Tong ZT, et al. MicroRNA-29c enhances the sensitivities of human nasopharyngeal carcinoma to cisplatin-based chemotherapy and radiotherapy. Cancer Lett 2013;329(1):91–8.

[142] Li G, Liu Y, Su Z, Ren S, Zhu G, Tian Y, et al. MicroRNA-324-3p regulates nasopharyngeal carcinoma radioresistance by directly targeting WNT2B. Eur J Cancer 2013;49(11):2596–607.

[143] Li L, Wu J, Sima X, Bai P, Deng W, Deng X, et al. Interactions of miR-34b/c and TP-53 polymorphisms on the risk of nasopharyngeal carcinoma. Tumour Biol 2013;34(3):1919–23.

[144] Yu BL, Peng XH, Zhao FP, Liu X, Lu J, Wang L, et al. MicroRNA-378 functions as an onco-miR in nasopharyngeal carcinoma by repressing TOB2 expression. Int J Oncol 2014;44(4):1215–22.

[145] Chen L, Tang Y, Wang J, Yan Z, Xu R. miR-421 induces cell proliferation and apoptosis resistance in human nasopharyngeal carcinoma via downregulation of FOXO4. Biochem Biophys Res Commun 2013;435(4):745–50.

[146] Zheng XH, Cui C, Ruan HL, Xue WQ, Zhang SD, Hu YZ, et al. Plasma microRNA profiles of nasopharyngeal carcinoma patients reveal miR-548q and miR-483-5p as potential biomarkers. Chin J Cancer 2014;33(7):330–8.

[147] Yi C, Wang Q, Wang L, Huang Y, Li L, Liu L, et al. MiR-663, a microRNA targeting p21(WAF1/CIP1), promotes the proliferation and tumorigenesis of nasopharyngeal carcinoma. Oncogene 2012;31(41):4421–33.

[148] Lu J, Luo H, Liu X, Peng Y, Zhang B, Wang L, et al. miR-9 targets CXCR4 and functions as a potential tumor suppressor in nasopharyngeal carcinoma. Carcinogenesis 2014;35(3):554–63.

[149] Gao F, Zhao ZL, Zhao WT, Fan QR, Wang SC, Li J, et al. miR-9 modulates the expression of interferon-regulated genes and MHC class I molecules in human nasopharyngeal carcinoma cells. Biochem Biophys Res Commun 2013;431(3):610–16.

[150] Lu J, Xu X, Liu X, Peng Y, Zhang B, Wang L, et al. Predictive value of miR-9 as a potential biomarker for nasopharyngeal carcinoma metastasis. Br J Cancer 2014;110(2):392–8.

[151] Lyu X, Fang W, Cai L, Zheng H, Ye Y, Zhang L, et al. TGFbetaR2 is a major target of miR-93 in nasopharyngeal carcinoma aggressiveness. Mol Cancer 2014;13:51.

[152] Ye Y, Zhou Y, Zhang L, Chen Y, Lyu X, Cai L, et al. EBV-miR-BART1 is involved in regulating metabolism-associated genes in nasopharyngeal carcinoma. Biochem Biophys Res Commun 2013;436(1):19–24.

[153] Gourzones C, Ferrand FR, Amiel C, Verillaud B, Barat A, Guerin M, et al. Consistent high concentration of the viral microRNA BART17 in plasma samples from nasopharyngeal carcinoma patients--evidence of non-exosomal transport. Virol J 2013;10:119.

[154] Lung RW, Tong JH, Sung YM, Leung PS, Ng DC, Chau SL, et al. Modulation of LMP2A expression by a newly identified Epstein-Barr virus-encoded microRNA miR-BART22. Neoplasia 2009;11(11):1174–84.

[155] Lei T, Yuen KS, Xu R, Tsao SW, Chen H, Li M, et al. Targeting of DICE1 tumor suppressor by Epstein-Barr virus-encoded miR-BART3* microRNA in nasopharyngeal carcinoma. Int J Cancer 2013;133(1):79–87.

[156] Chan JY, Gao W, Ho WK, Wei WI, Wong TS. Overexpression of Epstein-Barr virus-encoded microRNA-BART7 in undifferentiated nasopharyngeal carcinoma. Anticancer Res 2012;32(8):3201–10.

[157] Chang PL, Chang YS, Chen JH, Chen SJ, Chen HC. Analysis of BART7 microRNA from Epstein-Barr virus-infected nasopharyngeal carcinoma cells by capillary electrophoresis. Anal Chem 2008;80(22):8554–60.

[158] Hsu CY, Yi YH, Chang KP, Chang YS, Chen SJ, Chen HC. The Epstein-Barr virus-encoded microRNA MiR-BART9 promotes tumor metastasis by targeting E-cadherin in nasopharyngeal carcinoma. PLoS Pathog 2014;10(2):e1003974.

[159] Ramakrishnan R, Donahue H, Garcia D, Tan J, Shimizu N, Rice AP, et al. Epstein-Barr virus BART9 miRNA modulates LMP1 levels and affects growth rate of nasal NK T cell lymphomas. PLoS One 2011;6(11):e27271.

[160] Li Z, Chen X, Li L, Liu S, Yang L, Ma X, et al. EBV encoded miR-BHRF1-1 potentiates viral lytic replication by downregulating host p53 in nasopharyngeal carcinoma. Int J Biochem Cell Biol 2012;44(2):275–9.

[161] Zhu JY, Pfuhl T, Motsch N, Barth S, Nicholls J, Grasser F, et al. Identification of novel Epstein-Barr virus microRNA genes from nasopharyngeal carcinomas. J Virol 2009;83(7):3333–41.

[162] Liu N, Cui RX, Sun Y, Guo R, Mao YP, Tang LL, et al. A four-miRNA signature identified from genome-wide serum miRNA profiling predicts survival in patients with nasopharyngeal carcinoma. Int J Cancer 2014;134(6):1359–68.

[163] Alajez NM, Shi W, Hui AB, Bruce J, Lenarduzzi M, Ito E, et al. Enhancer of Zeste homolog 2 (EZH2) is overexpressed in recurrent nasopharyngeal carcinoma and is regulated by miR-26a, miR-101, and miR-98. Cell Death Dis 2010;1:e85.

[164] Gao W, Chan JY, Wong TS. Differential expression of long noncoding RNA in primary and recurrent nasopharyngeal carcinoma. Biomed Res Int 2014;2014:404567.

[165] Imperiale TF, Ransohoff DF, Itzkowitz SH, Levin TR, Lavin P, Lidgard GP, et al. Multitarget stool DNA testing for colorectal-cancer screening. N Engl J Med 2014;370(14):1287–97.

[166] Zhang Z, Sun D, Hutajulu SH, Nawaz I, Nguyen Van D, Huang G, et al. Development of a non-invasive method, multiplex methylation specific PCR (MMSP), for early diagnosis of nasopharyngeal carcinoma. PLoS One 2012;7(11):e45908.

[167] Johnson PJ, Lo YM. Plasma nucleic acids in the diagnosis and management of malignant disease. Clin Chem 2002;48(8):1186–93.

[168] Gray SG, Qian CN, Furge K, Guo X, Teh BT. Microarray profiling of the effects of histone deacetylase inhibitors on gene expression in cancer cell lines. Int J Oncol 2004;24(4):773–95.

[169] Yang X, Lay F, Han H, Jones PA. Targeting DNA methylation for epigenetic therapy. Trends Pharmacol Sci 2010;31(11):536–46.

TARGETING
ABERRANT
EPIGENETICS

NUTRITIONAL EPIGENETIC REGULATORS IN THE FIELD OF CANCER: NEW AVENUES FOR CHEMOPREVENTIVE APPROACHES

18

Michael Schnekenburger[1] and Marc Diederich[2]

[1]*Laboratoire de Biologie Moléculaire et Cellulaire du Cancer, Hôpital Kirchberg, Luxembourg, Luxembourg,*
[2]*College of Pharmacy, Seoul National University, Gwanak-gu, Seoul, Korea*

CHAPTER OUTLINE

S.G. Gray (Ed): Epigenetic Cancer Therapy. DOI: http://dx.doi.org/10.1016/B978-0-12-800206-3.00018-5

1 INTRODUCTION

Cancer remains a major burden of disease worldwide and the number of cancer-related cell death is still rising. However, cancer incidence and mortality frequencies largely fluctuate worldwide with highest rates identified in westernized countries and lower ones in Asian and African countries [1]. Remarkably, data from migration population studies indicate that ethnic groups present increased cancer incidence rate when they are migrating from a historically low- to high-risk countries, to ultimately reach either comparable or higher rates to the living country [2]. Although incidence and mortality rates for most cancers are declining in most developed countries, they are significantly rising in less developed and developing areas. These observations are related to the adoption of unhealthy, westernized lifestyles such as smoking, physical inactivity, and consumption of calorie-dense and highly processed food lacking fruit and vegetable uptake [2,3]. Furthermore, prospective epidemiological studies and randomized prevention trials clearly pinpoint that besides genetic disparities, environmental exposure and lifestyle factors play a major role in the ethology of many cancer subtypes including the most widespread ones in the developed world, which are lung, colorectal cancers, and prostate or breast cancer for men and women, respectively (Figure 18.1). Altogether, these observations suggest that many cancer cases are not a fatality of the modern society but are preventable with a healthy diet, the avoidance of sedentary lifestyle and obesity, and a controlled exposure to toxic compounds (e.g., smoking), radiations (e.g., sun exposure), and infections [2,4,5].

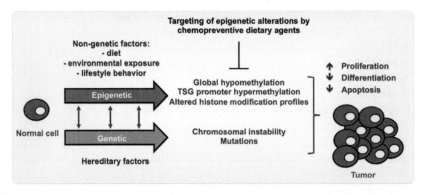

FIGURE 18.1

Influence of environment factors on epigenetic mechanisms in the context of tumorigenesis. Tumorigenesis is a multistep process associated with both genetic and epigenetic alterations. Epigenetic mechanisms, such as DNA methylation and histone modifications, are strongly impacted by environmental factors. Conversely, they represent promising targets for cancer prevention by dietary agents.

It is becoming clear that one major way by which environmental exposure and food active compounds act on our health is by targeting epigenetic mechanisms (Figure 18.1). Epigenetic events correspond to functionally relevant genomic changes without changes in DNA sequence, and include DNA methylation, histone modifications, and noncoding RNA functions. Altogether, these mechanisms regulate gene expression and therefore are essential for cell homeostasis and functions [6–9]. Epigenetic aberrations governing tumor suppressor gene (TSG) inactivation, oncogene activation, and chromosomal instability play a fundamental role in all critical pathways and steps of carcinogenesis including tumor initiation, and some events occur usually before neoplastic transformation. Interestingly, unlike genetic changes, epigenetic aberrations are potentially reversible. This fascinating feature of the epigenetic regulatory system has provided an attractive opportunity to the research community to develop new anticancer strategies using molecules able to target the epigenetic machinery, called epigenetic drugs, to remodel and potentially restore to a "normal" state epigenetic landscapes and the expression of epigenetically silenced genes found in tumor cells [6,8–10].

Furthermore, a growing body of evidence demonstrates that various dietary compounds including micro- and macronutrients such as methyl donors (folate, choline, and various vitamins) and phytochemicals such as thiosulfonates, polyphenols, glucosinolates, or terpenoids can interfere or influence epigenetic mechanisms and therefore affect the expression of genes involved in tumor development. The ability of such dietary components to modulate the epigenetic machinery is discussed below.

2 OVERVIEW OF THE CANCER EPIGENOME

Chromatin structure and activity are highly dependent on chromatin compaction and accessibility, which largely depend on two main epigenetic mechanisms including DNA methylation and covalent histone modifications. A third epigenetic layer represented by noncoding RNAs further controls gene expression. All mechanisms were described to be altered in cancer leading to aberrant gene expression profiles, associated with carcinogenesis [6–9].

2.1 HISTONE MODIFICATIONS

The N-terminal tails of core histones are subjected to various covalent modifications from which the most studied events are acetylation of lysine residues and methylation of arginine and lysine residues. Multiple and distinct combinations of histone modifications, called "histone code," are involved in the regulation of chromatin packaging and gene expression. These modifications are catalyzed by several specific families of enzymes categorized as "writers" and "erasers" families: acetyl groups are transferred by histone acetyltransferases (HATs) and removed by histone deacetylases (HDACs), whereas methyl groups are added by histone methyltransferases (HMTs) and removed by histone demethylases (HDMs) (Figure 18.2). More recently, additional histone modifiers were discovered but the significance of these modifications remains unclear even though links to cancer could be established. As an example, histone citrullination by the noncoding amino acid citrullin is conferred by peptidyl arginine deiminase (PADI)4. Such modifications were recently shown to displace histone H1 from chromatin leading to global chromatin decondensation [11]. To what extend such modifications can be conferred by dietary agents remains to be elucidated.

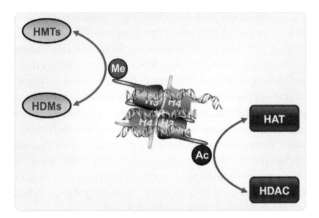

FIGURE 18.2

Schematic representation of posttranslational modifications of histones. Various families of histone modifying enzymes are involved in the regulation of histone acetylation: histone deacetylases (HDACs), histone acetyltransferases (HATs), and methylation: histone demethylases (HDMs) and histone methyltransferases (HTMs).

HDAC isoenzymes are divided into four classes. Class I comprises HDAC1, 2, 3, and 8, which are located mainly in the nucleus. Class II is subdivided into two subclasses: subclass IIa contains HDAC4, 5, 7, and 9 and subclass IIb comprises HDAC6 and 10. Class IV consists of only HDAC11, which is localized in the nucleus and has a catalytic domain in the N-terminal region. Class III HDAC, also called sirtuins (SIRTs), is composed of seven members SIRT1 to 7. SIRT1, 6, and 7 are predominantly found in the nucleus, SIRT2 is mainly in the cytoplasm, and SIRT3, 4 and, 5 are localized in the mitochondria. Remarkably, while non-sirtuin HDACs use zinc as a cofactor, SIRTs depend on the cofactor nicotinamide adenine dinucleotide (NAD^+) [12,13].

HATs are classified in five families from which three have been extensively studied: the p300/cAMP response element-binding protein (CREB)-binding protein (CBP) family, the general control nonderepressible 5 (Gcn5)–related N-acetyltransferase (GNAT) family including Gcn5 and p300/CBP-associated factor (PCAF) and the monocytic leukemia zinc finger protein (MOZ), Ybf2/Sas3, Sas2, TAT-interacting protein 60 (Tip60) (MYST) family. All HAT members use acetyl-coenzyme A (acetyl-CoA) as a cofactor of acetylation reactions [12,13].

HDMs are distinguished by their catalytic activity mechanisms that require various cofactors to catalyze protein demethylation. The members of the Jumonji C (JmjC) domain class of HDMs are α-ketoglutarate (α-KG)-, and iron (Fe^{2+})-dependent dioxygenases [14], whereas the members of the lysine-specific demethylase (LSD1) class are flavin adenine dinucleotide (FAD)-dependent amine oxidase HDMs [15].

The HMT family comprises many proteins that catalyze the transfer of one, two, or three methyl groups to lysine and arginine residues of histone proteins classified in three families: Su(var)3,9, Enhancer of zest, Trithorax (SET) domain lysine methyltransferases, non-SET domain lysine methyltransferases, and arginine methyltransferases; also known as protein arginine methyltransferases

(PRMTs) or protein lysine methyltransferases (PKMTs). *S*-adenosyl methionine (SAM) serves as a cofactor and methyl donor group for all three families [15].

Noteworthy, besides targeting histones, all these epigenetic effectors target and regulate functions of numerous nonhistone proteins including enzymes, transcription factors, structural proteins, and many more. Therefore, these epigenetic enzymes regulate many cellular processes. Accordingly, it is easy to conceive that the slightest changes in the expression and/or activity profiles of such epigenetic enzymes are provoking profound changes in epigenetic modifications and gene expression profiles which are frequently found associated with carcinogenesis. Furthermore, disruption of histone modification profiles is implicated in genomic instability, impaired DNA repair, and altered cell cycle checkpoints, causing tumor initiation and progression [12,13,15].

2.2 DNA METHYLATION

DNA methylation corresponds to the catalytic addition of a methyl group from the cofactor SAM on cytosine residues within CpG dinucleotides to generate 5-methylcytosine (5mC). The reaction is catalyzed by the DNA methyltransferase (DNMT) family composed of DNMT1, DNMT3A, and DNMT3B. DNA methylation is considered as a relatively stable epigenetic mark; however, under certain physiological or pathological circumstances such as during development and tumorigenesis, respectively, significant loss of DNA methylation is detected. DNA demethylation is the consequence of either an absence of maintenance of methylation profiles during replication (i.e., passive demethylation) or by active DNA demethylation mechanisms through multistep reactions involving oxidative and deamination processes followed by DNA repair pathways (Figure 18.3). The oxidative steps are catalyzed by ten-eleven translocation (TET) proteins and deamination is controlled by activation-induced deaminase (AID)/apolipoprotein B mRNA editing enzyme, catalytic polypeptide-like (APOBEC)-family of cytidine deaminases. Notably, TET1, 2, and 3 are α-KG- and Fe^{2+}-dependent enzymes capable of converting 5mC to 5-hydroxymethylcytosine (5hmC), and then to 5-formylcytosine (5fC) and 5-carboxylcytosine (5caC) by successive oxidative steps [16].

The DNA methylome of cancer cells is characterized by a global hypomethylation, which is associated with genomic instability, chromosomal aberrations, and oncogenic activation. Nevertheless, this cancer hallmark occurs concomitantly to site-specific DNA hypermethylation of CpG-rich promoters region (i.e., CpG islands) within TSGs that are usually unmethylated in healthy individuals leading to their transcriptional repression.

2.3 EPIGENETIC READERS

Besides the writers and erasers of the epigenetic landscape, a third category of proteins called epigenetic or chromatin "readers" is implicated in translating the epigenetic code of chromatin into functional gene regulation. These proteins can be subdivided into two subgroups: readers of the different state of methylation and oxidation of cytosine (5mc, 5hmC, 5fC, 5caC; Figure 18.4) such as methyl-binding domain (MBD) proteins [17,18] and readers of histone marks. Among this latter group, readers of methyl lysine are constituted of proteins with specific domains: alpha thalassemia/mental retardation syndrome X-linked (ATRX)-DNMT3-DNMT3L (ADD), ankyrin, bromo-adjacent homology (BAH), chromo-barrel, chromo-domain, double chromodomain (DCD), malignant brain tumor (MBT), plant homeodomain (PHD), PWWP (Pro-Trp-Trp-Pro), tandem Tudor domain (TTD), Tudor, WD40, and zinc finger CW (zf-CW); readers of

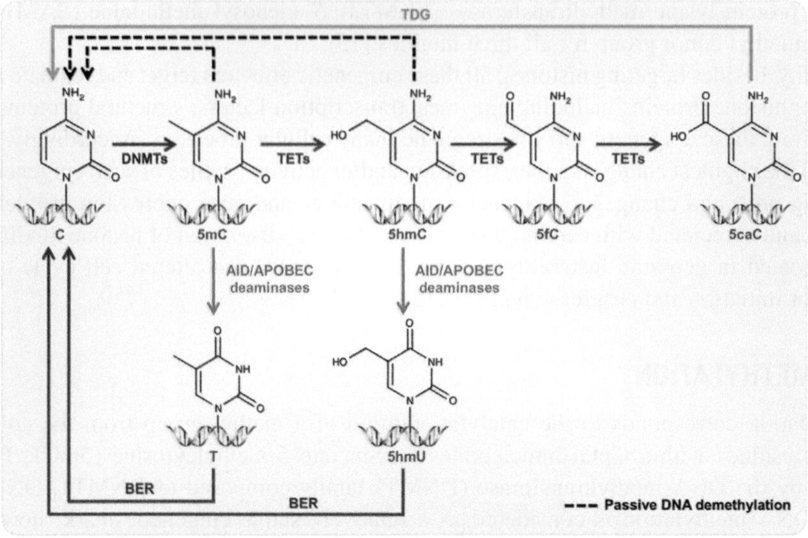

FIGURE 18.3

DNA methylation and demethylation pathways. DNA methylation occurs on cytosines (C) within CpG dinucleotides and is catalyzed by DNA methyltransferases (DNMTs) to generate 5-methylcytosine (5mC). Three pathways have been implicated in active DNA demethylation: (i) 5mCs are successively hydroxylated by ten-eleven translocation (TET) enzymes to be converted into 5-hydroxymethylcytosine (5hmC), 5-formylcytosine (5fC), and 5-carboxylcytosine (5caC); (ii) 5mC and 5hmC are deaminated by the activation-induced deaminase (AID)/apolipoprotein B mRNA editing enzyme, catalytic polypeptide-like (APOBEC)-family members to form 5-methyluracil (5mU), and 5-hydroxymethyluracil (5hmU), respectively; and (iii) 5mU/5hmU and 5caC can be excised by base-excision repair (BER) and by thymine DNA glycosylase (TDG) pathways, respectively. Besides active pathways, DNA demethylation is achieved in a passive manner in the absence of maintenance of 5mC and 5hmC during replication.

methylarginine with ADD•, Tudor•, and WD40 domains; and reader of acetyllysine with bromodomain, double bromodomain (DBD), double PHD finger (DPF), and double pleckstrin homology (PH) motifs [19].

Remarkably, aberrant activity due to mutations or altered expression of a number of these epigenetic effectors with reader domains is recurrently observed in cancer [20].

2.4 MicroRNAs

MicroRNAs (miRNAs) are a family of single-stranded 19- to 24-nucleotide non-protein-coding RNAs that posttranscriptionally regulate mRNA functions. MiRNAs target DNA regions located essentially in 5′ and 3′ untranslated regions (UTRs) of target genes degrading mRNA or blocking translation. In the human genome, this highly evolutionarily conserved RNA family represents approximately 800 members targeting probably more than half of all coding genes and largely contributes to the complexity of gene expression regulation (see Ref. [7] for review).

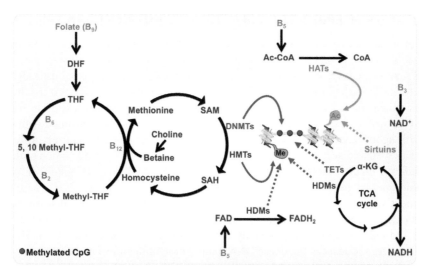

FIGURE 18.4

The activity of epigenetic enzymes dependent on various cofactors produce by cell metabolism. α-ketogutarate (α-KG), acetyl-coenzyme A (Ac-CoA), dihydrofolate (DHF), DNA methyltransferases (DNMTs), flavin adenine dinucleotide (FAD), histone acetyltransferases (HATs), histone deacetylases (HDACs), histone demethylases (HDMs), histone methyltransferases (HTMs), nicotinamide adenine dinucleotide (NAD), S-adenosyl homocysteine (SAH), S-adenosyl methionine (SAM), tricarboxylic acid (TCA), ten-eleven translocation (TET), tetrahydrofolate (THF).

Accumulating evidence suggests that the deregulation of miRNA expression signatures is associated with tumorigenesis. These altered miRNA expression profiles may result from alterations of other epigenetic marks targeting miRNA genes or mutations affecting proteins involved in miRNA biogenesis and processing [6,7].

3 DIETARY FACTORS AND THEIR INFLUENCE ON EPIGENETICS: AVENUE FOR DIETARY INTERVENTION TO PREVENT CANCER

As exposed above, it is now clearly established that epigenetic alterations are associated with the ethology of cancer due to unbalanced activities and overexpression of epigenetic writers, erasers, or readers [6,8,9,21]. Accordingly, the development of agents named epigenetic drugs has provided the ability to fight against cancer by reversing these alterations [9,12,13]. Besides representing a therapeutic potential, epigenetic alterations gained much attention in recent years as a promising molecular target for cancer prevention. Considering that epigenetic mechanisms are greatly influenced by environment and especially diet, we discuss below various classes of compounds found in our diet including nutrients such as methyl donors (folate, choline, and various vitamins) and phytochemicals and how they can influence cancer development or have cancer prevention potential.

3.1 MODULATION OF EPIGENETIC COFACTORS BY NUTRIENTS AND METABOLISM

Epigenetic enzymes require cofactors that are generated from dietary vitamins and minerals through various metabolic pathways including fatty acid oxidation, glycolysis, one-carbon metabolism/methionine pathways, and oxidative phosphorylation via the tricarboxylic (TCA) cycle, also called citric acid cycle or Krebs cycle (Table 18.1, Figure 18.4). Thus, availability of nutrients and cell metabolism preconditions epigenetic activities and therefore the regulation of gene expression to finally control phenotypes. One striking example of such regulatory mechanism can be illustrated with experiments conducted with agouti mice, in which genetically identical twins can look entirely different in both color and size as well as disease susceptibility based on the level of DNA methylation of their genome, which depends on dietary supplementation of their mothers with methyl donors (folate, choline, betaine, methionine, vitamin B_{12}, and methionine) and zinc during gestation [22]. These animal data are in agreement with those observed in human studies suggesting that adult disease risk is associated with adverse environmental conditions early in development and persistent epigenetic changes. For instance, individuals who were prenatally exposed to famine during the Dutch Hunger Winter (1944-45) presented, six decades later, less DNA methylation in the imprinted insulin-like growth factor 2 (IGF2) gene compared with unexposed individuals [23].

3.1.1 S-adenosyl methionine

SAM is the methylation donor for methylation reactions catalyzed either by DNMTs or HMTs generating S-adenosyl homocysteine (SAH), which is subsequently hydrolyzed into adenosine and homocysteine. SAH in turn acts as an inhibitor of methylation reactions. Therefore, the SAM:SAH ratio is critical in regulating the activity of essential cellular methylating enzymes. Accordingly, several studies reported that dietary methyl deficiency leading to SAM depletion is associated with both global and gene-specific DNA hypomethylation leading to activation of oncogenes concomitantly to DNA hypermethylation–mediated TSG silencing and subsequently tumorigenesis in rodent models [24,25]. In the

Table 18.1 Metabolic Cofactors and Epigenetic Reactions

Cofactor	Enzyme	Mechanism	Metabolic Process
α-KG	TETs	Actives DNA demethylation by oxidation	TCA cycle, glutamine catabolism
Acetyl-CoA	HATs	Histone acetylation	β oxidation, TCA cycle, lipid biosynthesis
FAD	HDMs (Jumonji family)	Histone demethylation	TCA cycle, glutamine catabolism
	HDMs (LSD1 family)	Histone demethylation	TCA cycle, glutamine catabolism
NAD+	Sirtuins	Histone deacetylation	TCA cycle, redox
SAM	DNMTs	DNA methylation	One-carbon metabolism and methionine pathway
	HMTs	Histone methylation	One-carbon metabolism and methionine pathway

α-KG, α-ketoglutarate; acetyl-CoA, acetyl coenzyme A; DNMTs, DNA methyltransferases; FAD, flavin adenine dinucleotide; HATs, histone acetyltransferases; HDAC, histone deacetylases; HDMs, histone demethylases; HTMs, histone methyltransferases; LSD1, lysine-specific demethylase; NAD, nicotinamide adenine dinucleotide; SAM, S-adenosyl methionine; TCA, tricarboxylic; TET, ten-eleven translocation.

same line, it has been shown that abundance of SAM upregulates tissue inhibitor of metalloproteinase-2 in colorectal cancer [26] and influences histone methylation, especially trimethylation of histone H3 on lysine 4 (H3K4me3), a mark of active and poised promoters [27].

The availability of several dietary factors involved in one-carbon metabolism and methionine pathways is regulating the SAM:SAH ratio. The most important is folate and its partners: vitamin B_6, and B_{12}. Alternatively, when folate is not available in sufficient amounts to assure adequate methylation, choline, betaine, and methionine are critical to contribute to maintenance of adequate SAM levels to allow methylation (Figure 18.4).

3.1.1.1 Folate

Folate or vitamin B_9 is not synthetized by human; therefore it has to be supplied through the diet. Folate has many bodily functions and plays an important role in one-carbon metabolism and synthesis of SAM. Foods with the highest levels include spinaches, asparagus, broccolis, offal, and liver. In addition, folate is found in a wide variety of dietary sources including dark green leafy vegetables, fruits, various nuts, beans, and peas but also poultry and meats.

Several epidemiological studies indicate that folate deficiency may represent a risk factor for several cancer subtypes [28,29]. Notably, large alcohol consumption is associated with folate deficiency and increased cancer risk [30]. Accordingly, folate deficiency or restriction has been reported to induce global loss of DNA methylation [31,32] and genomic instability [33]. In contrast, SAM repletion is increasing folate levels and leads to increased global DNA methylation [32]. Similarly, *in vitro* SAM supplementation in a glioblastoma cell line or in patients suffering of colorectal cancer was correlated with increased DNA methylation levels [34,35]. These evidences suggest that folate intake and global DNA methylation are positively correlated. Nevertheless, additional nutri-epigenomics studies have shown that excessive supplementation levels may also be associated with increased cancer risk and promote cancer progression in humans. Furthermore, high folate intake was significantly associated with a lower level of global methylation of leukocyte DNA in a group of healthy Japanese females [36]. Finally, additional studies show no effect of dietary folate on the risk of breast and colorectal cancer [37,38]. All together, these findings suggest that the link between folate status, DNA methylation, and human cancer risk is highly complex and displays a certain extend of fluctuability, which may depend on many parameters including the tissue considered, timing, and intensity of depletion/supplementation, polymorphism in genes implicated in SAM metabolism [39].

There is also evidence that folate intake can influence miRNA expression. Rodent fed with a folate- and methionine-deficient diet develop liver tumors after 54 weeks, a downregulation of liver-specific miR-122 is observed, while replenishment of folate starting at 36 weeks restores expression of miR-122 and prevents tumorigenesis. In cultured cells, folate deficiency leads to a global increase in miRNA expression profile, which is reversible by culturing cells in regular medium. Among miRNAs overexpressed in folate-deficient conditions, miR-222 is also upregulated in human peripheral blood cells of individuals with low folate intake [30].

These observations suggest that more investigations are required to better understand the role of folate levels and dietary supplementation for beneficial effects and potential therapeutic interventions.

3.1.1.2 Choline and betaine

Choline is found mainly in eggs, meat, and fish products; important amounts are found in many plant foods such as collard greens, Brussels sprouts, broccoli, Swiss chard, cauliflower, spinaches, and asparagus. Betaine is mainly found in seafood and certain plants (spinaches, beets, and cereals).

Choline is a major source of methyl groups as it is metabolized by oxidation to form betaine (trimethylglycine), which then contributes to methionine synthesis through the transfer of a methyl group to homocysteine (Figure 18.4). Therefore, both molecules are dietary factors that are important to maintain SAM:SAH ratio by participating in SAM synthesis pathway [28,40]. Conversely, several studies investigated the impact of choline and/or betaine levels on DNA methylation and cancer susceptibility. A population-based study revealed high intakes of choline and betaine to reduce breast cancer mortality [41]. Furthermore, animals injected with betaine present a dose-dependent increase in SAM levels in erythrocytes, whereas a low choline diet results in decreased folate levels [40]. In rodents, choline deficiency results in global hypomethylation of hepatic DNA and aberrant DNA hypermethylation at targeted TSG promoters such as glutathione S-transferase pi (GSTP1) gene involved in carcinogen detoxification: this prototypical hypermethylated biomarker gene becomes progressively methylated in prostate cancer [42,43] leading to gene silencing [40]. Remarkably, choline deficiency triggers spontaneous tumorigenesis and sensitizes rodents to carcinogens. Conversely, dietary supplementation with 0.8% choline completely prevents development of cancer in experimental models [40]. Besides affecting DNA methylation, choline deficiency also changes expression and activity of the HMT G9a associated to global and gene-specific changes in histone methylation levels [40]. Although the impact of such changes has not been thoroughly investigated, it is legitimate to assume it may also contribute to carcinogenesis depending on the duration and intensity of such deficiency and therefore may also represent an alternative to modulate epigenetic regulatory mechanisms as a preventive approach.

3.1.1.3 Methionine

Methionine is one of the nine essential amino acids and the only sulfurcontaining one that serves as a precursor for all other sulfur-containing amino acids and their derivatives. Methionine is continuously regenerated from homocysteine via one-carbon metabolism (Figure 18.4). Nevertheless, dietary methionine being converted into SAM through an ATP-driven reaction has potential effects on methylation reactions. Methionine is found in high amounts in eggs, chicken breast, some fishes (e.g., salmon, halibut, and mahi-mahi or dolphinfish), dairy products, and many vegetables including spinaches, zucchini, mushrooms, and asparagus.

Although methionine is a key methyl donor for methylation reactions, there is only limited evidence about its direct effect on epigenetic-dependent regulatory mechanisms. In rats receiving a methionine-supplemented diet, there was no evidence for any change in SAM:SAH ration nor exacerbated p53 gene methylation [44]. Conversely, high methionine intake cannot be associated with esophageal and gastric cancer [29]. Nonetheless, a study suggests that dietary methionine supplementation in rodents significantly increases methylation levels and prevents gamma-radiation-induced-global DNA hypomethylation [45]. In humans, methionine was even associated with decreased risks of proximal colon cancer in men and rectal cancer as well as ovarian cancer risk in women [38,46]. A methionine-deficient diet modulates the growth of gastric tumor cells and *in vitro* deficiency of methionine increases apoptosis rate and decreases cellular adhesion and migration associated with decreased E-cadherin promoter methylation and increased gene expression, *in vitro* and *in vivo* [47]. Further investigations deciphering the role of methionine uptake on cancer risk are required for potential dietary interventions.

3.1.1.4 Vitamins B$_2$, B$_6$, and B$_{12}$

Additional micronutrients such as vitamin B$_2$ (or riboflavin), B$_6$, and B$_{12}$ may influence both DNA methylation and related carcinogenic processes due to their role as catalytic cofactors of enzymes involved in the conversion of folate via the one-carbon metabolism pathway (Figure 18.4). Vitamin B_2 is a precursor for FAD, which is a cofactor of methylene tetrahydrofolatereductase, the enzyme responsible in the reduction of 5,10-methylene tetrahydrofolate (THF) to 5-methyl THF. Pyridoxine or vitamin B_6 is the cofactor of the enzyme serine hydroxymethyltransferase, which converts THF to 5,10-methylene THF. Cobalamin or vitamin B_{12} is the coenzyme of methionine synthase, the enzyme converting homocysteine to methionine [28].

The main sources of vitamin B_2 are dairy products, bananas, green beans, and asparagus. Vitamin B_6 is found in high amounts in many vegetables including lentils, soybeans, spinach, potatoes, green peas, yams, broccoli and asparagus, various nuts and seeds, and various meats and fishes. Vitamin B_{12} is found in seafood (fish, crustacean, and shellfish), meat and dairy products. Thus, depending on our diet, these vitamins potentially affect the efficiency of the one-carbon metabolism pathway.

Several studies initially failed to report an association between vitamin B_6 intake and ovarian cancer risk [48–50]. However, other studies reported an inverse association between dietary vitamin B_6 and ovarian cancer risk and esophageal adenocarcinoma [46,51]. In contrast, vitamin B_6 and vitamin B_{12} intake was associated with increased risk of sporadic colorectal tumors in women [38]. Conversely, intake of vitamin B_{12} is correlated to esophageal adenocarcinoma [51]. Xiao and collaborators reported that the intake of vitamins B_6 and B_{12} does not appear associated with the development of esophageal and gastric cancer [29]. Vitamin B_2 appears to be associated with decreased proximal colon cancer risk in women [38].

Despite the role of these vitamins in methylation reactions, there are only few studies investigating the relationship between these micronutrients and DNA methylation. For instance, it has been shown in rodents that a diet lacking vitamin B_{12} in presence of folate-enhanced placental global hypomethylation compared to a diet exclusively supplemented with folate [52]. Interestingly, in woman at parturition time, a study has shown that IGF2 methylation at the P3 promoter was matching in maternal and cord blood and was correlated to vitamin B_{12} levels in maternal blood serum. These findings suggest that environment controls methylation patterns in maternal blood, and then the maternal patterns influence DNA methylation and vitamin B_{12} levels in cord blood [53]. In another study, it has been reported that in former smokers an increase in vitamin B_{12} levels in serum was associated with a decrease in methylation in the promoter of the TSG Ras association (RalGDS/AF-6) domain family member 1A (RASSF1A), found frequently methylated in several cancer subtypes [54].

The relationship between vitamin uptake and cancer risk is rather complex and difficult to apprehend. This complexity probably relies on multiple parameters, which are not fully evaluated and understood yet. Considering that probably the most difficult part remains to control and determine the effect of only one parameter.

3.1.2 NAD$^+$

In cells, NAD is present under two forms: NAD$^+$ and NADH being the oxidized and reduced forms of NAD, respectively. When ATP is abundantly available, NAD$^+$ biosynthesis mostly consists of the formation of the pyridine mononucleotide, either nicotinamide mononucleotide (NMN) or nicotinic acid mononucleotide (NAMN) from exogenous precursors nicotinamide (NAM) and nicotinic acid (NA), respectively, both present in our diet as vitamin B_3. Vitamin B_3 or niacin, is an essential vitamin found essentially in various fishes and meats, peanuts, mushrooms, and avocado. These pathways, also known as the salvage or Preiss–Handler pathway, are important for NAD$^+$ homeostasis. Otherwise, NAD is synthesized *de novo* from tryptophan or regenerated by recycling degraded NAD products such as nicotinamide [55,56].

NAD+ is a metabolic cofactor for numerous enzymatic reactions. NAD^+ was initially identified for its involvement in redox reactions, transferring electrons from one reaction to another. Later, NAD^+ was identified to also act as cofactor of class III SIRT HDACs. Since, NAD^+:NADH redox ratio is a rate-limiting metabolite for protein deacetylation mediated by SIRT HDAC enzymatic activities, SIRTs represent metabolic sensors translating NAD^+ levels into a transcriptional signal for bioenergetic homeostasis by deacetylating transcription factors such as members of the O subgroup of forkhead box proteins (FOX) and histones at specific gene loci. Furthermore, SIRT1 plays a key role in regulating mitochondrial biogenesis and stress response through deacetylation of peroxisome proliferator–activated receptor-gamma coactivator (PGC)1α and a series of key transcription factors including forkhead box O (FOXO)1, p53, hypoxia inducible factor (HIF)1α and nuclear factor-kappa B (NF-κB) [56]. Interestingly, SIRTs mediate tumor suppression by antagonizing HIF-1α effects. Indeed, SIRT1 and SIRT6 inhibit the nuclear translocation and transcriptional activity of HIF-1α, respectively. In addition, SIRT3 suppresses mitochondrial reactive oxygen species production leading to HIF-1α destabilization. Finally, several SIRTs directly regulate the catalytic activity of key metabolic enzymes such as cytosolic and mitochondrial acetyl-CoA synthetases [57].

Conversely, enhancing SIRT1 activity has many beneficial effects and is now recognized as a potential preventive strategy against cancer development and for healthy ageing. Increased SIRT1 activity can be achieved through different types of interventions including: (i) caloric restriction, (ii) consumption of certain phytochemicals like resveratrol (see Section 3.3.3.7), (iii) enhancement of NAD^+ biosynthesis using supplementation with NAD^+ precursors (NA, NAM, NMN, and NAMN), and (iv) inhibition of other NAD^+-consuming reactions, such as the one catalyzed by poly(ADP-ribose) polymerase [55–57].

3.1.3 Acetyl-CoA

Acetyl-CoA is a thioester between the acyl group carrier, acetic acid and a thiol, coenzyme A. Acetyl-CoA, as a carrier of acyl groups, is an essential cofactor in the posttranslational acetylation reactions of histone and nonhistone proteins catalyzed by HATs.

De novo synthesis of CoA is a well-conserved enzymatic pathway, in which the first and rate-limiting step corresponds to phosphorylation of vitamin B_5 (or pantothenic acid). Vitamin B_5 is found in high amounts in mushrooms, dairy products, oily fish, avocado, and various meats. Accordingly, levels of vitamin B_5 are associated with the subsequent metabolite CoA levels, which affect the status of protein acetylation [58].

Acetyl-CoA is generated either by oxidative decarboxylation of pyruvate from glycolysis, which occurs in mitochondrial matrix, by oxidation of long-chain fatty acids, or by oxidative degradation of certain amino acids. Acetyl-CoA then enters in the TCA cycle where it is oxidized for energy production. In the cytosol, the initial step of *de novo* lipid biogenesis consists in conversion of citrate to acetyl-CoA and oxaloacetate by the enzyme ATP-citrate lyase using the energy of ATP hydrolysis [59]. Cytosolic/nuclear acetyl-CoA is also produced by two acetyl-CoA synthetase enzymes that condense acetate and thiol. Furthermore, downregulation of enzymes required for the synthesis of acetyl-CoA from acetate or citrate reduces acetylation of specific protein and histone substrates [58,59]. Accordingly, serum starvation or glucose deprivation induces a marked decrease in global histone acetylation levels [59].

Thus, dietary intake and energy balance through modulation of acetyl-CoA-driven HAT activities influence many cellular processes by regulating gene expression and increasing acetylation of a myriad of nonhistone proteins such as metabolic enzymes, transcription factors, or cell cycle regulators [12,13].

3.1.4 FAD

FAD is a redox cofactor of several important reactions in metabolism. This cofactor exists in two different redox states, with FAD and $FADH_2$ being the oxidized and reduced forms, respectively. FAD is formed of a riboflavin moiety (vitamin B_2), coupled to a phosphate group of an ADP molecule. The reaction starts by the conversion of riboflavin into flavin mononucleotide catalyzed by riboflavin kinase. Flavin mononucleotide is subsequently transformed into FAD by addition of an AMP moiety from ATP catalyzed by FAD-synthase [15]. Therefore, FAD availability is tightly depending on vitamin B_2 and energy metabolism (see Section 3.1.1.4).

Although the activity of certain HDMs depends on FAD, whose availability is driven by the intracellular energy content, these effectors of histone (or protein) demethylation appear to be good candidates to reprogram gene expression and therefore to be modulated in preventive approaches against cancer development.

3.1.5 α-ketogutarate

α-ketogutarate (α-KG) (2-oxoglutarate) plays an important role in oxidative metabolism as a key intermediate of the aerobic TCA cycle: α-KG derives from isocitrate and is concerted into succinyl CoA. Remarkably, conversion of α-KG by α-KG dehydrogenase into succinyl-CoA requires FAD and NAD, therefore the level of these two other epigenetic cofactors potentially regulate the pool of α-KG. This reaction is a critical and highly regulated step of the TCA cycle mainly controlled by feedback regulation. Notably, this conversion is also inhibited by high intracellular ATP levels [15]. Furthermore, anaplerotic reactions allow replenishment of TCA cycle intermediates by synthesizing α-KG through the action of glutamate dehydrogenase from glutamine, which is an abundant amino acid with important functions in cellular energetics. α-KG pool is also restored from isocitrate by isocitrate dehydrogenase (IDH).

In cancer cells, the pool of α-KG is elevated by the increased utilization of glutamine as an energy source. Moreover, mutations of *IDH1* and *IDH2* were shown in secondary glioblastoma and in acute myeloid leukemia (AML). These mutations are associated with an accumulation of metabolites that competitively inhibit α-KG such as 2-hydroxyglutarate, fumarate, and succinate; the latter two being TCA cycle metabolites [6,15,60]. In addition, IDH1/2 mutations confer a gain-of-function, allowing enzymes to process α-KG to the oncometabolite 2-hydroxyglutarate (2-HG), which inhibits TET2 activity and ultimately induces alterations of the methylation phenotype thus inhibiting hematopoietic differentiation [60]. Additional oncometabolites remain to be discovered, one can hypothesize that yet other epigenetic events can be regulated by such compounds generated by neomorphic enzymes.

α-KG is available as a dietary supplement; however, little is known about the safety and biological consequences of long-term use of such dietary supplements. Nevertheless, It has been reported that α-KG displays antitumor effects by inhibiting angiogenesis [61] and may have chemopreventive properties due to its ability to positively modulate transaminase activities and oxidant–antioxidant imbalance during N-nitrosodiethylamine-induced hepatocarcinogenesis [62].

α-KG is a cofactor of TET proteins and JmjC HDMs. Modulation of its cellar levels by metabolic cofactors, whose availability depends on intracellular energy levels, could become a good target for the reprogramming of gene expression by modulating demethylation processes in preventive interventions.

3.2 OTHER DIET-DERIVED NUTRIENTS AFFECTING EPIGENETIC MECHANISMS

3.2.1 Selenium

Selenium is an essential mineral found in Brazil nuts, seafood, meat, and fish products. Based on both epidemiological and preclinical studies, selenium has been reported to possess anticancer and chemopreventive activities. Increasing evidence suggests that it may have a propensity to interfere with various epigenetic processes leading to reexpression of silenced TSG in cancer cells [63]. Notably, the biological activity of selenium is tightly dependent on its chemical form (i.e., inorganic or organoselenium compounds). First, it has been shown that DNA from rodent fed with a selenium-rich diet was hypomethylated, whereas low-selenium diet increases DNA methylation of the von Hippel–Lindau TSG [64]. These findings were linked to the potential of selenium to inhibit DNMT1 activity and protein expression in colon cells [65]. Furthermore, selenium modulates SAM:SAH ratio and therefore DNA methylation by its action on 1-carbon metabolism [66]. On the other hand, the organoselenium dietary metabolites α-keto-γ-methylselenobutyrate and methylselenopyruvate trigger apoptosis in cancer cells by acting as HDAC inhibitors [67]. The use of such findings for dietary intervention requires deeper and careful investigations as they may depend on multiple parameters as observed for instance for folate.

3.2.2 Zinc

Zinc is an essential mineral found in large amount in oysters, meat, spinach, cashew nuts, chocolate, beans, and mushrooms.

Zinc plays a key role in cellular functions as it is a key constituent of many proteins or serves as cofactor of a myriad of mammalian enzymes. Conversely, zinc deficiency may lead to various cell dysfunctions and several studies suggest that it impairs DNA damage response resulting in DNA damage and loss of DNA integrity. Furthermore, dietary deficiencies in zinc intake have been associated with increased cancer risk [68]. Remarkably, zinc deficiency can cause a methyl deficiency similar to the one observed with other methyl donors, resulting in abnormal gene expression and developmental defects. These effects are most likely explained by the fact that zinc is the cofactor of several enzymes involved in the methionine pathway. Accordingly, zinc deficiency decreased SAM turnover and diminished both DNA and histone methylation in rats [69]. Based on these findings, it has been suggested that zinc possesses chemopreventive properties [68].

3.2.3 Butyrate

Butyrate is a major short-chain fatty acid produced during gut flora-mediated fermentation of dietary fibers. Legumes (beans, peas, and soybeans), fruits, nuts, cereals, and whole grains are good sources of dietary fibers. Butyrate is also found in butter and cheese.

Animal and epidemiological studies suggest that high-fiber diet may lower colorectal cancer risk. The beneficial effects of butyrate appear to be mediated by its HDAC inhibitory activity, which promotes histone acetylation, leading to expression of genes involved in cellular differentiation and apoptosis in several cancer models [70,71]. In this context, it has been proposed that butyrate can act in primary and secondary chemoprevention by its pleiotropic potential [71]. Therefore, butyrate is evaluated for the prevention of inflammation-mediated ulcerative colitis and colorectal cancer by lifelong continuous exposure to dietary fiber. Besides its HDAC inhibitory activity, butyrate was also reported to trigger DNA demethylation by a mechanism most likely dependent on changes in chromatin structure that could also contribute to chemopreventive properties of butyrate [72].

Surprisingly, there are only a very limited number of reports about the effect of butyrate on miRNA. Hu and collaborators have shown that the overexpression of the cyclin-dependent kinase inhibitor (CDKN)1A gene, coding for p21, a negative regulator of cell cycle induced by most HDAC inhibitors, is not only dependent on its HDAC inhibitory potential to increase the transcription via promoter acetylation, but also depends on the inhibition of several members of the miR-106b family, which control p21 translation [73]. These data suggest that the preventive role of butyrate may rely on the modulation of several epigenetic mechanisms.

3.2.4 Ursodeoxycholic acid

Ursodeoxycholic acid is a tertiary bile acid formed in the liver by epimerization of the lithocholic acid, which is one of the secondary bile acids formed in the colon by the action of intestinal bacteria on primary bile acids. Primary bile acids are derivatives of dietary lipid cholesterol synthetized in the liver and secreted in response to dietary fats [74].

Ursodeoxycholic acid has shown chemopreventive potential in preclinical and animal models of colon cancer by counteracting the tumor-promoting effects of either chemicals or secondary bile acids such as deoxycholic acid [74]. Akare and collaborators are the only one who suggested that the chemopreventive properties of ursodeoxycholic acid could be mediated by the induction of cell differentiation and senescence in colon carcinoma cells [75]. Differentiation was revealed by the upregulation of cytokeratin 8, 18, and 19 and E-cadherin, cytokeratin remodeling, and accumulation of lipid droplets. Senescence induction was accompanied by a decrease of telomerase activity. Authors also demonstrated that these cellular outcomes were accompanied by HDAC6 overexpression and subsequent reduced global histone acetylation. The regulation of HDAC6 by ursodeoxycholic acid might play a key role in its preventive activity as HDAC6 overexpression is sufficient to induce senescence in colon cancer cells [75].

3.3 PHYTOCHEMICALS TARGETING EPIGENETIC ENZYMES

Targeting epigenetic alterations associated with tumorigenesis is recognized as a promising anticancer strategy [6,9]. In this context, modulation of epigenetic processes such as DNA methylation and histone modifications by dietary phytochemicals represent a relevant approach for the modulation of pathways controlling, among others, proliferation, differentiation, and cell death that are altered during tumorigenesis: altogether these compounds are excellent candidates that mediate epigenetic cancer prevention [8,30,76,77]. Nowadays, a long list of these dietary molecules including glucosinates and thiosulfonates is being studied. An overview of phytochemicals from food interfering with DNA methylation and histone modifications is provided in Table 18.2.

In this context, we discuss the role of the most thoroughly investigated dietary compounds from fruits, vegetables, and herbs as nutraceuticals that have potential chemotherapeutic and chemopreventive properties (Figure 18.5).

3.3.1 Thiosulfonates

Thiosulfonates are allylsulfide compounds including allicin, diallyldisulfide, diallyltrisulfide, S-allylmercaptocysteine, S-allylcysteine [78]. This class of phytochemical organosulfurs is found essentially in garlic and other vegetables from *Allium* species. Epidemiological evidence indicates an inverse correlation between the consumption of *Allium*-derived compounds and the risk of stomach

Table 18.2 Food Compounds Reported to Act as Epigenetic Modulators

Compound	Class	Food Source	Potential Epigenetic Target
6-methoxy-2E,9E-humuladien-8-one	Sesquiterpene	Ginger	Histone modifications
Allicin	Thiosulfonate	*Allium* family	Histone modifications
Allylisothiocyanate	Glucosinate	Wasabi, mustard, radish	Histone modifications, miRNA
Anacardic acid	Polyphenol	Cashew nuts	Histone modifications
Apigenin	Polyphenol	Parsley, celery	DNA methylation, histone modifications
Baicalein	Polyphenol	*Oroxylumindicum*	DNA methylation
Betanin	Betalain	Beetroot red	DNA methylation
Biochanin A	Polyphenol	Soy	DNA methylation
Caffeic acid	Polyphenol	Coffee	DNA methylation
Catechin/epicatechin	Polyphenol	Green tea	DNA methylation
Chlorogenic acid	Polyphenol	Coffea	DNA methylation
Clusianone	Polyisoprenylatedbenzophenone	*Garcinia sp.*	Histone modifications
Curcumin	Polyphenol	Turmeric	DNA methylation, histone modifications, miRNA
Cyanidin	Anthocyanidin	Pigment found in many red fruits, vegetables, berries	DNA methylation
Daidzein	Polyphenol	Soy	DNA methylation, miRNA
Delphinidin	Polyphenol	Cranberries, concord grapes, pomegranates	Histone modifications
Diallyl disulfide	Thiosulfonate	*Allium* family	Histone modifications, miRNA
Diallyltrisulfide	Thiosulfonate	*Allium* family	Histone modifications, miRNA
EGCG	Polyphenol	Green tea	DNA methylation, histone modifications, miRNA
Ellagic acid	Polyphenol	Found in numerous fruits and vegetables	
Emodin	Polyphenol	Rhubarb	Histone modifications, miRNA
Epicatechingallate	Polyphenol	Green tea	DNA methylation
Epigallocatechin	Polyphenol	Green tea	DNA methylation
Fisetin	Polyphenol	Strawberries	DNA methylation, histone modifications, miRNA
Flavone	Polyphenol	Mandarin	Histone modifications
Galangin	Polyphenol	*Alpiniaofficinarum, Alpiniagalanga*	DNA methylation
Gallic acid	Polyphenol	Mango, blackberry	Histone modifications
Garcinol	Polyisoprenylatedbenzophenone	Kokum	Histone modifications, miRNA
Genistein	Polyphenol	Soy	DNA methylation, histone modifications, miRNA

(Continued)

Table 18.2 (Continued)

Compound	Class	Food Source	Potential Epigenetic Target
Guttiferone A	Polyisoprenylatedbenzophenone	*Garcinia sp.*	Histone modifications
Guttiferone E	Polyisoprenylatedbenzophenone	*Garcinia sp.*	Histone modifications
Hesperetin	Polyphenol	Citrus	DNA methylation
Indole-3-carbinol	Glucosinolate	Cruciferous vegetables	Histone modifications, miRNA
Isoliquiritigenin	Polyphenol	Liquorice	Histone modifications, miRNA
Kaempferol	Polyphenol	Apples, nuts, onions	Histone modifications
Luteolin	Polyphenol	Celery, parsley	DNA methylation
Lycopene	Terpenoid	Tomato	DNA methylation
Mahanine	Alkaloid	Curry tree	DNA methylation
MCP30	Protein	Better melon	Histone modifications
Myricetin	Polyphenol	Walnuts and various berries, fruits,vegetables	DNA methylation
Naringenin	Polyphenol	Citrus	DNA methylation
Phenethyl isothiocyanate	Glucosinate	Cruciferous vegetables	DNA methylation, histone modifications, miRNA
Phloretin (dihydrochalcone)	Polyphenol	Apples	DNA methylation
Piceatannol	Polyphenol	Grape	DNA methylation, histone modifications, miRNA
Protocatechuic acid	Polyphenol	Olives	DNA methylation
Quercetin	Polyphenol	Apple, tea, onion, nuts, berries	DNA methylation, histone modifications, miRNA
Resveratrol	Polyphenol	Grapes	DNA methylation, histone modifications, miRNA
Rosmarinic acid	Polyphenol	Rosemary	DNA methylation
S-allyl mercaptocysteine	Thiosulfonate	*Allium* family	Histone modifications
S-allylcysteine	Thiosulfonate	*Allium* family	Histone modifications
Sinapic acid	Polyphenol	Rapeseed	DNA methylation, histone modifications
Sulforaphane	Glucosinate	Cruciferous vegetables	DNA methylation, histone modifications, miRNA
Theaflavin 3,3′-digallate	Polyphenol	Black tea	DNA methylation
Theophylline	Terpenoid	Cocoa beans, tea	Histone modifications
Thymoquinone	Terpenoid	*Nigella sativa*	Histone modifications
Ursolic acid	Terpenoid	Basil	Histone modifications, miRNA

Compounds undergoing clinical trials (or in recruiting phase) in the field of cancer are marked in bold (http://www.clinicaltrial.gov/). EGCG, (-)-epigallocatechin-3-gallate; miRNA, microRNA.

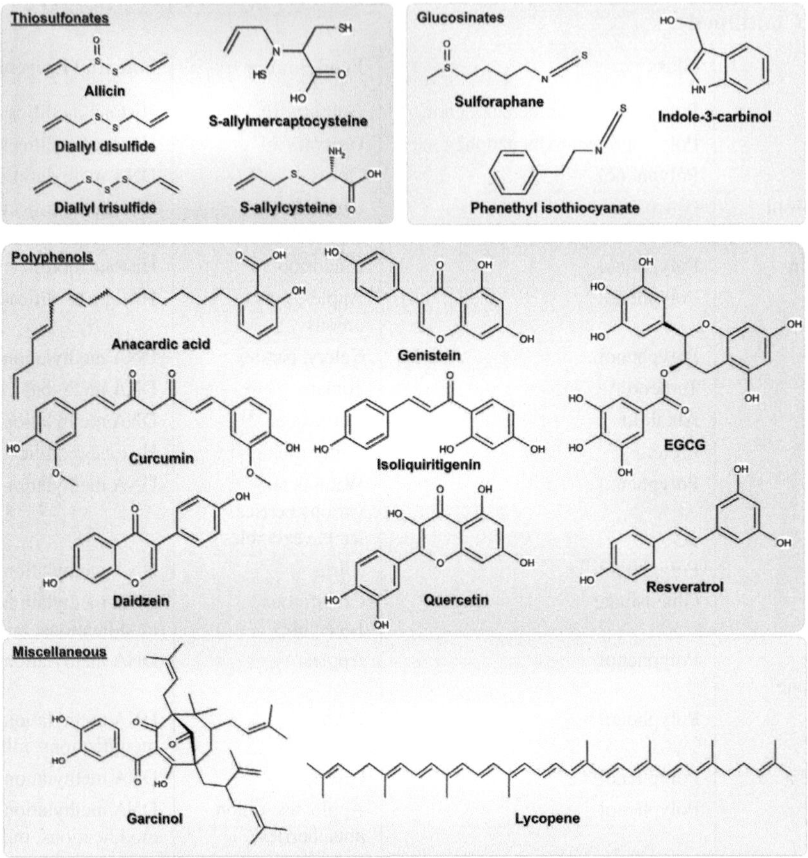

FIGURE 18.5

Chemical structures of cancer chemopreventive agents targeting epigenetic mechanisms. EGCG = (-)-epigallocatechin-3-gallate.

and colon cancers [79]. These compounds display a common feature: they induce histone acetylation as a consequence of the inhibition of HDAC activities both in cancer cells and in animal models [30,80]. Nonetheless, several reports have demonstrated that the inhibitory potential of this family of compounds is in fact mediated by the metabolite derivative allyl mercaptan. Accordingly, allyl mercaptan is among all allyl derivatives and precursors, the most potent HDAC inhibitor in *in vitro* cell–free HDAC activity assays, and also the most effective to induce histone acetylation in various models. Furthermore, it has been demonstrated in rodents that diallyl disulfide is metabolized to allyl mercaptan within 30 min [80]. As most HDAC inhibitors, allyl mercaptan increases global histone H3 and H4 acetylation and recruits positive histone marks to the CDKN1A (p21) promoter, leading to increased expression levels. Consequently, allyl mercaptan induces a G_1 cell cycle arrest [80]. Conversely, it is now well accepted that metabolic conversion of allyl organosulfurs into the HDAC inhibitor allyl mercaptan is in part responsible for the chemopreventive properties of allyl vegetables [30,80], among other properties that those compound families display [81,82].

A few reports investigated the effect of allyl compounds on miRNA regulatory networks. Diallyl disulfide suppresses proliferation and induces apoptosis in human gastric cancer through Wnt-1 signaling pathway by upregulation of miR-22 and -200b [83]. Diallyltrisulfide induces tumor suppressor miRNAs such as miR-34a, -143, -145, and -200b/c that are typically lost in osteosarcoma leading to an inactivation of Notch-1 signaling associated with an inhibition of proliferation, invasion, and angiogenesis in osteosarcoma cells [84]. These data provide additional evidence that modulation of miRNAs by allyl compounds may be implicated in their preventive properties against cancer.

3.3.2 Glucosinates

3.3.2.1 Isothiocyanates

Sulforaphane and phenethyl isothiocyanate are isothiocyanates found in large amounts in cruciferous vegetables. Accumulating data suggests that such vegetables possess interesting *in vitro* and *in vivo* cancer-preventive potential [85]. Furthermore, a growing body of evidence is connecting their chemopreventive properties against cancer with epigenetic mechanisms [80]. In particular, both compounds are reported to display potent HDAC inhibitory activity increasing total histone acetylation as well as gene-specific histone acetylation at certain promoters such as the p21 promoter in various cancer cell lines, which coincident with cell growth inhibition and tumor-suppressive effects [80,86]. Remarkably, 3 hours after an intake of a portion of 68 g of broccoli sprouts, a transient increase of global histone acetylation is detectable in peripheral blood mononuclear cells from human subjects [80]. *In silico* and *in vitro* data reveal that the HDAC inhibitory activity of sulforaphane is mediated by its metabolite sulforaphane–cysteine. More recently, sulforaphane was identified as a DNA demethylating agent due to its proficiency to downregulate the expression of DNMT1 and DNMT3B leading subsequently to the induction of cyclin D2 gene promoter demethylation and expression in cancer cells [87].

There are a few studies investigating the role of sulforaphane on miRNA expression levels. For instance, sulforaphane is implicated in the modulation of the expression of miR-23b, -27b, and miR-155 [88]. Interestingly, in another study, authors demonstrated that sulforaphane upregulates miR-140, which is a negative regulator of cancer stem cell formation in basal-like early stage breast cancer. These results highlight its potential preventive properties for breast cancer [89].

Similarly, in prostate cancer cells, phenethyl isothiocyanate was reported to restore expression of silenced GSTP1 by a mechanism involving promoter demethylation and increased histone acetylation. These effects are associated with increased expression of the CDKNs p21 and p27, which are negative cell cycle regulators [30]. Furthermore, phenethyl isothiocyanate prevents alterations of miRNA expression profiles associated with cigarette smoke exposure in rats [7].

3.3.2.2 Indole-3-carbinol

Indole-3-carbinol is found in cruciferous vegetables, a family of vegetables that displays valuable cancer-preventive properties [85]. Conversely, indole-3-carbinol is a well-recognized chemopreventive agent with many biological activities including inhibition of inflammation and angiogenesis, decreased proliferation, and promotion of tumor cell death [90]. In the acidic environment of the stomach, indole-3-carbinol is rapidly converted by extensive and rapid self-condensation into the digestive product 3,3′-diindolylmethane (formed from the dimerization of two molecules of indole-3-carbinol). Accumulating data suggests that the preventive effects of indole-3-carbinol might be due, at least partially, to the inhibitory effect of 3,3′-diindolylmethane on HDAC activity by inducing proteasome-mediated downregulation of class I HDAC isoenzymes (HDAC1, HDAC2, HDAC3, and HDAC8) both in cultured cells and in tumor-xenografted animals. Decreased HDAC expression was accompanied by increased expression of the pro-apoptotic Bcl-2 (B-cell lymphoma 2)–associated X (Bax) protein and the CDKNs p21 and p27 leading to cell cycle arrest and increased rate of apoptosis [91].

Besides affecting HDAC expression, indole-3-carbinol treatments prevent alterations of miRNA signatures observed in rodent models following cigarette smoke exposure or in chemically induced tumorigenesis [7]. Interestingly, indole-3-carbinol sensitizes gemcitabine-resistant pancreatic cancer cells by downregulating miR-21, a key regulator of cell proliferation and apoptosis in various cancer models [92].

3.3.3 Polyphenols

3.3.3.1 Anacardic acid

Anacardic acid has been initially isolated from cashew nuts. Anacardic acid was the first natural product inhibitor of HAT activities reported: in *in vitro* cell-free assays it inhibits the activity of p300, PCAF, and Tip60 HATs in a low micromolar range [93,94]. Tips60 activity plays a key role in DNA damage repair pathway by activating ataxia telangiectasia mutated (ATM) and DNA-dependent protein kinase catalytic subunit (DNA-PKc), two kinases implicated in DNA repair pathway, which are recruited and activated by DNA double-strand breaks. Conversely, anacardic acid–treated cells are sensitized to ionizing radiation [94]. Anacardic acid also inhibits both inducible and constitutive NF-κB activation, which is dependent upon p300 HAT activity by impairing activation of the inhibitor of kappa B alpha kinase (IKK) associated with an inhibition of the acetylation and nuclear translocation of NF-κB p65 subunit, which in turn suppresses NF-κB-dependent gene expression. In this context, anacardic acid presents anti-inflammatory and anti-invasive properties by suppressing tumor necrosis factor (TNF)-α-induced overexpression of anti-apoptotic proteins (e.g., Bcl-2, Bcl-xl, and survivin) and proteins involved in invasion and angiogenesis (e.g., intercellular adhesion molecule (ICAM)1 and matrix metalloproteinase (MMP)9), respectively. Finally, it has been reported that anacardic acid prevents UV-induced tumorigenesis [95]. Collectively, these data support a potential preventive role of anacardic acid against cancer development.

3.3.3.2 Chalcones

Chalcones are found in many fruits (citrus, apple, tomato, etc), vegetables (shallots, bean sprouts, potatoes, etc.), and also some edible plants (licorice) from our daily diet. This group of natural occurring compounds possesses a broad spectrum of biological activities consistent with anticarcinogenic activity including antioxidant, anti-inflammatory, and promotion of cell cycle arrest and apoptosis [96]. Although the effects of chalcones related to epigenetic mechanisms have not been thoroughly investigated, some members of this family such as isoliquiritigenin were reported to inhibit HDAC activity and to modulate miRNA signatures in lymphoblastoid cell lines which may account for their chemopreventive activities [97,98]. Similarly, the synthetic chalcone derivative 2′,4′,6′-tris(methoxymethoxy) was shown to induce histone hyperacetylation by inhibiting HDAC activities [99]. In agreement with this inhibitory activity, many chalcones are reported to induce the expression of p21, which is consistently induced by HDAC inhibitors [13]. Furthermore, the dihydrochalcone phloretin, found in apples and in apple-derived products, has also been reported to inhibit DNMT activity [100]. Based on their chemopreventive properties, chalcones probably deserve a better investigation of their potential effect on the epigenetic machinery.

3.3.3.3 Curcumin

Curcumin (diferuloylmethane) is a flavonoid from the rhizome of *Curcuma longa* with well-recognized anticancer and chemopreventive properties [101]. Indeed, this spice possesses antioxidant, anti-inflammatory, anti-angiogenic, anti-proliferative, and pro-apoptotic activities against several cancers [102]. Many evidences suggest that these anticancer properties might result from epigenetic changes triggered by curcumin and leading to a modulation of the expression of genes controlling these pathways (reviewed in [103]). Curcumin was reported to decrease both global DNA methylation and local promoter methylation. Furthermore, depending on models, curcumin decreases HAT activities and histone acetylation or reduces the expression of several HDAC isoenzymes accompanied by increased histone acetylation [103]. Finally, additional evidence suggests that curcumin may also interfere with histone methylation patterns as it decreases the expression of the HMT EZH2 and the level of the repressive histone mark H3K27me3 [104].

Curcumin is one of the most thoroughly studied phytochemicals related to its effect on miRNAs. An initial study in pancreatic cancer cells demonstrated that curcumin downregulates 18 miRNAs including miR-199a that leads to a decreased expression of its downstream target MET proto-oncogenes and downregulates 11 miRNA including miR-22 known to target the expression of estrogen receptor (ER)1 and specificity protein (Sp)1 transcription factor. Later, several studies highlighted a continuously growing list of miRNAs that are differentially regulated following curcumin treatment. Among upregulated miRNAs, miR-15a and -16 are associated with decreased expression of anti-apoptotic *Bcl-2* gene and downregulation of Wilm's tumor 1 (*WT1*) gene. Examples of downregulated miRs are represented by the oncomirsmiR- 21, -136, -186*, and -200 (see for review Ref. [103,105]). Remarkably, other reports highlighted a link between the potential DNA demethylating properties of curcumin to specific miRNA expression profiles. For instance, curcumin was reported to induce the expression of the tumor suppressor miR-203 in bladder cancer through miR-203 promoter demethylation [103]. Taking together these data suggest that the anticancer activity of curcumin has also been linked to changes in the expression profile of many miRNAs involved in the regulation of proliferation, apoptosis, metastasis, and chemoresistance.

3.3.3.4 EGCG

(-)-Epigallocatechin-3-gallate (EGCG), a major polyphenol of green tea, has been extensively reported for its *in vitro* and *in vivo* anticancer activities. Several studies have shown that consumption of green tea could decrease cancer risk and that EGCG prevents carcinogenesis in various models of induced tumors [106]. However, other studies failed to support these findings [107]. Several reports have highlighted the effects of EGCG on the epigenetic machinery that might account for its anticancer activities. EGCG was reported to inhibit DNMT1 and therefore acts as a DNA demethylating agent. Treatments with EGCG lead to the demethylation and gene reactivation of various TSG [108,109]. For instance, it was established that GSTP1 gene was demethylated and reactivated following exposure to green tea polyphenols in prostate cancer cells [110]. Nevertheless, several reports also show lack of DNMT inhibition by EGCG in selected cellular models [111].

Besides targeting DNA methylation, EGCG impacts histone modifications. Indeed, EGCG inhibits HAT activity leading to a decrease of histone and nonhistone protein acetylation levels in epidermal tumor models [112]. Furthermore, EGCG was reported to decrease the expression of the HMT EZH2 accompanied by reduced H3K27me3 levels [113].

Interestingly, EGCG also affects miRNA expression, which adds an additional layer of chemopreventive mechanisms related to epigenetics. For instance, a downregulation of oncogenic miRNAs miR-21, -92, -93, -106b, and -125b [7,105] and an upregulation of tumor suppressor miRNAs miR-7-1, -16, -20a, -34a, -99a, -210, -221, -330, and let-7 family [7,105] was reported. Analyzing the pathways affected by these miRNA, their expression profiles after EGCG exposure, at least at high concentrations, correspond to a reduction of proliferation and increase of apoptosis rate in cancer cells. Remarkably, it has been shown recently that EGCG binds directly to specific miRNAs (miR-33a and -122) leading to a downregulation of their expression [114]. Finally, An and collaborators identified miRNA expression signatures associated with EGCG-mediated UVB protection in human dermal fibroblasts. Functional analysis revealed that these miRNAs control expression of genes involved in transcription regulation and inhibition of apoptosis [115].

Therefore, health-promoting and cancer-preventing effects of EGCG remain to be unraveled and may be dependent on the modulation of several epigenetic mechanisms; nevertheless, the beneficial effect of green tea consumption may also rely on other catechins present in this common beverage or their combinatory action as they were reported to have weak DNMT inhibitory activity [30].

3.3.3.5 Genistein and daidzein

Genistein and daidzein, two phytochemicals abundantly present in soybeans, are modulators of oxidative stress, angiogenesis, cell growth, and apoptosis [116]. Epidemiological studies indicate an inverse correlation between soy product–rich diet and cancer risk in Asian populations [116]. Furthermore, the preventive role of genistein and soy isoflavones has been demonstrated in various animal cancer models. In line with *in vitro* data, genistein and daidzein are undergoing clinical trials for chemotherapeutic and chemopreventive purposes in several cancer subtypes [30]. Interestingly, both compounds were shown to decrease DNMT1 activity leading to DNA demethylation of TSGs and gene reactivation in cancer cells. Furthermore, genistein induces histone acetylation by inhibiting HDACs and activating HATs [30,117–120]. Over the past 2 years, genistein has been one of the most widely studied phytochemicals regarding its impact on miRNA signatures in multiple cancer-cell models. For instance, genistein decreases the expression of miR-23b-3p, -27a, -151, -221, -222, -223, and 1260b [7,120–123] and upregulates miR-34a, 574-3p, and 1296 [7,124] in cancer cells leading to diminished cell proliferation, migration, invasion, and migration activities. Genistein was also reported to downregulate the oncogenic long noncoding RNA homeobox antisense intergenic RNA (HOTAIR) [125]. Remarkably, Rabiau and collaborators have observed a similar effect on miRNA expression profiles in three prostate cancer cell lines after a treatment with genistein, daidzein, or the DNA demethylating agent 5-azacytidine [126]. These data suggest that these natural soy products act similarly to therapeutically used demethylating agents.

3.3.3.6 Quercetin

Quercetin is an abundant plant flavonoid with strong antioxidant, anti-inflammatory, and anti-proliferative activities. Quercetin displays a broad spectrum of anticancer activities and accumulating evidence also suggests cancer-preventive properties [127]. Interestingly, quercetin was reported to enhance the activity of the NAD^+-dependent HDAC SIRT1 in yeast and to decrease DNMT activity leading to p16 promoter demethylation associated with gene reactivation [128,129]. Nevertheless, whether these epigenetic targets are involved in the cancer-preventive properties of quercetin remains to be investigated. Several recent studies suggest that chemopreventive properties of quercetin could be associated with its capacity to regulate expression of miRNAs involved in inflammatory response. Quercetin downregulates the pro-inflammatory miR-155 and the oncomir miR-21 and upregulates miR-125b, a negative regulator of inflammation [130–132].

3.3.3.7 Resveratrol

Resveratrol is a stilbene found in various fruits and berries [133,134] and mainly in grape and grape-derived products (e.g., red wine). This polyphenol is recognized for its properties to modulate various tumorigenesis pathways and thereby possesses anticancer properties mediated by antioxidant and anti-inflammatory activities, inhibition of cell growth, induction of anti-angiogenic response, and increased apoptosis rate [135]. Furthermore, a growing body of evidence demonstrates that resveratrol displays pleiotropic health promoting effects, and therefore prevents the development of various pathologies such as cancer [136]. Molecular mechanisms underpinning chemopreventive effects of resveratrol were recently unraveled. Among these findings, the potential of resveratrol to increase lifespan in yeast and *Caenorhabditis elegans* by inducing SIRT1 activity is one remarkable feature of resveratrol; however, later this observation was reported to be an artifact due to *in vitro* fluorimetric-based HDAC activity assays [137]. Nonetheless, nowadays several reports confirmed that resveratrol acts as an SIRT inhibitor and mimics caloric restriction. Furthermore, resveratrol poorly protects SIRT1-null mice bearing the adenomatous polyposis coli (multiple intestinal neoplasia) (Apc(min)) mutation of tumorigenesis [138]. Collectively, these findings suggest that the preventive properties of resveratrol against cancer are SIRT-dependent [139]. Recently, resveratrol has also been reported to induce weak DNMT inhibitory activity by reducing DNMT1 expression leading to DNA demethylation of retinoic acid receptor (RAR)β2 and phosphatase and tensin homolog (PTEN) genes [25,100,140]. Moreover, resveratrol was reported to also decrease DNMT3B expression and to demethylate RASSF1A in women with increased breast cancer risk and in tumors from rodents. Interestingly, resveratrol increases DNMT3B expression in normal tissue. These changes in DNMT3B expression in rodents were inversely correlated to the expression of miR-129, -204, and -489 in normal and tumor tissues [141]. Additional reports demonstrated that resveratrol modulates the expression of some miRNAs that could account for its chemopreventive properties as this compound decreases the expression of various oncomirs including miR-17, -21, -25, and -92a-2, known to be upregulated in colon cancer. Resveratrol also induces the expression of the tumor suppressor miR-663 [7,30]. Remarkably, it has been shown recently that resveratrol binds directly to miR-33a and -122 leading to their upregulation [114].

3.3.4 Other phytochemicals

3.3.4.1 Garcinol

Garcinol is a polyisoprenylated benzophenone isolated from tree *Garcinia indica*. Several studies reported inhibitory activity of garcinol against PCAF and p300 HAT activities associated with significant histone H3 and H4 deacetylating activities [142,143]. Accordingly, garcinol modulates multiple pro-inflammatory signaling cascades including constitutively activated signal transducers and activators of transcription (STAT)3 and NF-κB pathways associated with reduced expression of several anti-apoptotic proteins, including survivin and Bcl-2, leading to growth suppression and cell death [30,143]. Furthermore, garcinol prevents chemically induced tumorigenesis in various animal models [30]. Recently, it has been reported that garcinol induces a specific miRNA signature associated with the sensitization of pancreatic cancer cells to gemcitabine treatment [144]. Altogether these findings suggest that garcinol may represent an interesting lead compound for further cancer prevention strategies.

3.3.4.2 Lycopene

Lycopene is a carotenoid terpenoid mainly found in tomatoes [145]. Epidemiological data suggests that a regular intake of tomatoes or tomato-derived products may be associated with reduced prostate cancer risk [30]. Lycopene modulates expression of numerous genes relevant to cell cycle control, DNA repair, and cell death. As a result, this compound inhibits cell proliferation and induces apoptosis in cell culture and was found to reduce tumor growth in animal models of breast, lung, and prostate cancer [30]. Interestingly, lycopene reactivated GSTP1 gene expression through reduced promoter methylation in MDA-MB-468 breast cancer cells [118]. Furthermore, lycopene induces DNA demethylation in RARβ2 and high in normal (HIN)1 genes in noncancerous human breast epithelial MCF-10A cells. However, lycopene does not induce any DNA demethylation in RARβ2 promoter in MDA-MB-468 or MCF-7 breast cancer cells. Although the exact mechanism underlying lycopene-induced DNA demethylation has not been further investigated, these data suggest that lycopene could potentially modulate tumorigenic processes via the modulation of epigenetic mechanisms.

4 CONCLUSIONS AND CRITICAL CONSIDERATIONS

The use of plants for medicinal purposes is a traditional practice since decades. Over the past years, there is a growing interest for preventive approaches in addition to curative approaches in many diseases such as cancer. With the identification of epigenetic mechanisms as an interface between our genome and the environment, including our diet and our genome, preventive methods even gain further attention. Accordingly, many epidemiological and experimental data are supporting beneficial effect of a healthy diet with at least seven fruits and vegetables rich in phytochemicals a day.

Here we reviewed the current knowledge on dietary compounds including nutrients and various phytochemical classes able to interfere or influence epigenetic mechanisms including DNA methylation or histone modifications (essentially histone acetylation) and therefore affect the expression of genes involved in tumor development, which accordingly could be of main importance for the preventive properties of such compounds. Although the list of dietary phytochemicals and nutrients impacting the regulation of epigenetic mechanisms is continuously growing (Table 18.2), for most of them, it is currently unclear whether the observed inhibitory potential has relevance for cancer-preventive activities. Thus, there are still a number of questions that need to be addressed to fully apprehend the contribution of epigenetic mechanisms in their preventive potential against different cancer subtypes in order to improve and optimize the rational use of potential dietary interventions.

First, even now, many food compounds are reported to be able to modulate epigenetic targets only on the basis of cell-free *in vitro* data. Therefore, there is a need to further investigate these compounds to get a better understanding of the molecular mechanisms underlying the chemopreventive effect of phytochemicals and nutrients from our diet.

Next, a major body of evidence was obtained in *in cellulo* systems with sometimes concentrations of compounds that are far beyond physiologically achievable concentrations, at least from a preventive perspective, and that fall most likely in the range of inducing more directly cell death mechanisms; in other words, how relevant are the observations related to epigenetic mechanisms based on a balanced and healthy daily diet? Hence, there is still a lack of *in vivo* evidence for the majority of these dietary compounds and even less data in human. Accordingly, there is a need for collecting more preclinical and clinical data on the epigenetic changes induced by many of the dietary phytochemicals. Currently, several phytochemicals are undergoing clinical trials (Table 18.2); however, in most cases, these are not epigenetically driven clinical trials. Indeed, most of these compounds display preventive effects that might as well be independent of any epigenetic regulation of gene expression.

Furthermore, we have to be really cautious about the interpretation of data obtained in animals, as many of these epigenetic modulations and signatures are species-specific and tightly dependent on metabolism; they are therefore not necessarily transposable to human.

Another point to consider is to clarify the effective and safety doses and concentrations of bioactive food components as well as the time of exposure relevant for cancer-prevention concomitant to the analysis of tissue distribution and bioavailability of these dietary compounds to reach the targeted tissues and to observe epigenetic changes in order to gain maximum beneficial effects, while keeping in mind that the protective effect is unlikely to rely on a single dietary agent. Accordingly, the identification of the relevant metabolites responsible of the beneficial effect is needed for many compounds.

An additional critical point consists in the observation that epigenetic modifications are tissue- specific and are involved in cell differentiation. Furthermore, epigenetic alterations are considered to be cancer- specific; therefore, active dietary agents may trigger different epigenetic changes in different tissues and even in different cell types of the same tissue. These considerations may explain some of the discrepancies observed between different studies.

Most data about the ability of chemopreventive phytochemicals to modulate the epigenome landscape were obtained by focusing on DNA methylation and histone modifications, essentially acetylation. However, to fully apprehend the chemopreventive potential of dietary compounds through the regulation of epigenetic machinery, we need to take in consideration all layers of epigenetic gene regulation including also the effect of phytochemicals on miRNA and epigenetic readers. Up to now the number of studies investigating the effect of dietary compounds on miRNA remains rather limited. Nevertheless, with the development of technologies allowing high throughput analyses of DNA methylation associated with histone modification profiles and miRNA transcriptomic tools at a genome-wide scale, coupled to an increased power of analysis by bioinformatics, there is no doubt that in the future more integrative data will be obtained, which would help in the design and application of further chemopreventive strategies. Furthermore, targeting epigenetic readers, which allow the epigenetic code readout into biological functions, could represent an attractive alternative of intervention for preventive purposes. In this context, modulators of specific epigenetic protein–protein interactions would probably provide new mechanistic insights into chromatin regulation and unravel new therapeutic opportunities. To the best of our knowledge, this aspect was so far not much investigated and there is no molecule from food reported to act in such a manner.

Besides their role on DNA methylation, only a few studies have been investigating the effect of dietary nutrients affecting cofactors of epigenetic reactions. Nevertheless, considering the impact of methyl donor deficiency to alter DNA methylation, we can hypothesize that it may also alter, for instance, histone methylation patterns as observed during tumorigenesis. A better comprehension of the effect of dietary nutrients on histone modifications and eventually miRNA regulation would be of importance to determine the feasibility of dietary interventions.

The epidemiological data regarding the effects of nutrients derived from studies in which they were evaluated based on a diet composed of several nutrients that may affect, for instance, DNA methylation in a similar way. Furthermore, the levels of intake of the different nutrients are evaluated by scores reflecting intake through diet and/or plasma measurements, which may not be accurate and altered by several other parameters or such other lifestyle factors or genetic variations in metabolic enzymes. Therefore, it is sometimes difficult to raise conclusions about the effect of individual nutrients. Furthermore, some studies attempt to investigate the role of some nutrients by dietary supplementation or depletion; however, looking at individual nutrients may be too simplistic. Nevertheless, improved characterization of micronutrients found in plant- and fruit-rich diet is of absolute necessity by bearing in mind that some other food co-contaminants reported to interfere with epigenetic mechanisms might counteract the beneficial effects of an otherwise healthy diet.

Finally, it seems also of importance to consider long-term and eventually trans-generational effects of sustained preventive/dietary interventions. In this regard, for an implementation of preventive interventions by epigenetic modulators as a personalized medical approach in the future either for primary prevention or eventually for secondary and tertiary prevention, there is an urgent need to identify and develop robust biomarkers to predict the need of such interventions, to monitor their efficiency and potentially measure their adverse health effects.

ACKNOWLEDGMENTS

MS is supported by a "Waxweiler grant for cancer prevention research" from the Action Lions "Vaincre le Cancer." This work was supported by the "Recherche Cancer et Sang" foundation, the "Recherches Scientifiques Luxembourg" association, the "EenHäerz fir kriibskrank Kanner" association, the Action LIONS "Vaincre le Cancer" association, and Télévie Luxembourg. MD is supported by the National Research Foundation of Korea (NRF) grant for the Global Core Research Center (GCRC) funded by the Korean government, Ministry of Science, ICT & Future Planning (MSIP) (No.2011-0030001).

REFERENCES

[1] Jemal A, Bray F, Center MM, Ferlay J, Ward E, Forman D. Global cancer statistics. CA Cancer J Clin 2011;61(2):69–90.
[2] Anand P, Kunnumakkara AB, Sundaram C, Harikumar KB, Tharakan ST, Lai OS, et al. Cancer is a preventable disease that requires major lifestyle changes. Pharm Res 2008;25(9):2097–116.
[3] Jemal A, Center MM, DeSantis C, Ward EM. Global patterns of cancer incidence and mortality rates and trends. Cancer Epidemiol Biomarkers Prev 2010;19(8):1893–907.
[4] Thun MJ, DeLancey JO, Center MM, Jemal A, Ward EM. The global burden of cancer: priorities for prevention. Carcinogenesis 2010;31(1):100–10.
[5] Cogliano VJ, Baan R, Straif K, Grosse Y, Lauby-Secretan B, El Ghissassi F, et al. Preventable exposures associated with human cancers. J Natl Cancer Inst 2011;103(24):1827–39.
[6] Florean C, Schnekenburger M, Grandjenette C, Dicato M, Diederich M. Epigenomics of leukemia: from mechanisms to therapeutic applications. Epigenomics 2011;3(5):581–609.
[7] Karius T, Schnekenburger M, Dicato M, Diederich M. MicroRNAs in cancer management and their modulation by dietary agents. Biochem Pharmacol 2012;83(12):1591–601.
[8] Schnekenburger M, Diederich M. Epigenetics offer new horizons for colorectal cancer prevention. Curr Colorectal Cancer Rep 2012;8(1):66–81.
[9] Seidel C, Florean C, Schnekenburger M, Dicato M, Diederich M. Chromatin-modifying agents in anti-cancer therapy. Biochimie 2012;94(11):2264–79.

[10] Schnekenburger M, Grandjenette C, Ghelfi J, Karius T, Foliguet B, Dicato M, et al. Sustained exposure to the DNA demethylating agent, 2'-deoxy-5-azacytidine, leads to apoptotic cell death in chronic myeloid leukemia by promoting differentiation, senescence, and autophagy. Biochem Pharmacol 2011;81(3):364–78.

[11] Christophorou MA, Castelo-Branco G, Halley-Stott RP, Oliveira CS, Loos R, Radzisheuskaya A, et al. Citrullination regulates pluripotency and histone H1 binding to chromatin. Nature 2014;507(7490): 104–8.

[12] Folmer F, Orlikova B, Schnekenburger M, Dicato M, Diederich M. Naturally occurring regulators of histone acetylation/deacetylation. Curr Nutr Food Sci 2010;6:78–99.

[13] Seidel C, Schnekenburger M, Dicato M, Diederich M. Histone deacetylase modulators provided by Mother Nature. Genes Nutr 2012;7(3):357–67.

[14] Tsukada Y, Fang J, Erdjument-Bromage H, Warren ME, Borchers CH, Tempst P, et al. Histone demethylation by a family of JmjC domain-containing proteins. Nature 2006;439(7078):811–16.

[15] Teperino R, Schoonjans K, Auwerx J. Histone methyl transferases and demethylases; can they link metabolism and transcription? Cell Metab 2010;12(4):321–7.

[16] Wu YC, Ling ZQ. The role of TET family proteins and 5-hydroxymethylcytosine in human tumors. Histol Histopathol 2014;29(8):991–7.

[17] Spruijt CG, Gnerlich F, Smits AH, Pfaffeneder T, Jansen PW, Bauer C, et al. Dynamic readers for 5-(hydroxy) methylcytosine and its oxidized derivatives. Cell 2013;152(5):1146–59.

[18] Iurlaro M, Ficz G, Oxley D, Raiber EA, Bachman M, Booth MJ, et al. A screen for hydroxymethylcytosine and formylcytosine binding proteins suggests functions in transcription and chromatin regulation. Genome Biol 2013;14(10):R119.

[19] Musselman CA, Lalonde ME, Cote J, Kutateladze TG. Perceiving the epigenetic landscape through histone readers. Nat Struct Mol Biol 2012;19(12):1218–27.

[20] Dawson MA, Kouzarides T, Huntly BJ. Targeting epigenetic readers in cancer. N Engl J Med 2012;367(7):647–57.

[21] Rodriguez-Paredes M, Esteller M. Cancer epigenetics reaches mainstream oncology. Nat Med 2011;17(3):330–9.

[22] Waterland RA, Jirtle RL. Transposable elements: targets for early nutritional effects on epigenetic gene regulation. Mol Cell Biol 2003;23(15):5293–300.

[23] Heijmans BT, Tobi EW, Stein AD, Putter H, Blauw GJ, Susser ES, et al. Persistent epigenetic differences associated with prenatal exposure to famine in humans. Proc Natl Acad Sci USA 2008;105(44):17046–9.

[24] James SJ, Pogribny IP, Pogribna M, Miller BJ, Jernigan S, Melnyk S. Mechanisms of DNA damage, DNA hypomethylation, and tumor progression in the folate/methyl-deficient rat model of hepatocarcinogenesis. J Nutr 2003;133(11 Suppl. 1):3740S–3747SS.

[25] Stefanska B, Karlic H, Varga F, Fabianowska-Majewska K, Haslberger A. Epigenetic mechanisms in anti-cancer actions of bioactive food components--the implications in cancer prevention. Br J Pharmacol 2012;167(2):279–97.

[26] Hussain Z, Khan MI, Shahid M, Almajhdi FN. S-adenosylmethionine, a methyl donor, up regulates tissue inhibitor of metalloproteinase-2 in colorectal cancer. Genet Mol Res 2013;12(2):1106–18.

[27] Shyh-Chang N, Locasale JW, Lyssiotis CA, Zheng Y, Teo RY, Ratanasirintrawoot S, et al. Influence of threonine metabolism on S-adenosylmethionine and histone methylation. Science 2013;339(6116):222–6.

[28] Anderson OS, Sant KE, Dolinoy DC. Nutrition and epigenetics: an interplay of dietary methyl donors, one-carbon metabolism and DNA methylation. J Nutr Biochem 2012;23(8):853–9.

[29] Xiao Q, Freedman ND, Ren J, Hollenbeck AR, Abnet CC, Park Y. Intakes of folate, methionine, vitamin B6, and vitamin B12 with risk of esophageal and gastric cancer in a large cohort study. Br J Cancer 2014;110(5):1328–33.

[30] Huang J, Plass C, Gerhauser C. Cancer chemoprevention by targeting the epigenome. Curr Drug Targets 2011;12(13):1925–56.

[31] Rampersaud GC, Kauwell GP, Hutson AD, Cerda JJ, Bailey LB. Genomic DNA methylation decreases in response to moderate folate depletion in elderly women. Am J Clin Nutr 2000;72(4):998–1003.

[32] Shelnutt KP, Kauwell GP, Gregory III JF, Maneval DR, Quinlivan EP, Theriaque DW, et al. Methylenetetrahydrofolate reductase 677C–>T polymorphism affects DNA methylation in response to controlled folate intake in young women. J Nutr Biochem 2004;15(9):554–60.

[33] Bistulfi G, Vandette E, Matsui S, Smiraglia DJ. Mild folate deficiency induces genetic and epigenetic instability and phenotype changes in prostate cancer cells. BMC Biol 2010;8:6.

[34] Pufulete M, Al-Ghnaniem R, Khushal A, Appleby P, Harris N, Gout S, et al. Effect of folic acid supplementation on genomic DNA methylation in patients with colorectal adenoma. Gut 2005;54(5):648–53.

[35] Hervouet E, Debien E, Campion L, Charbord J, Menanteau J, Vallette FM, et al. Folate supplementation limits the aggressiveness of glioma via the remethylation of DNA repeats element and genes governing apoptosis and proliferation. Clin Cancer Res 2009;15(10):3519–29.

[36] Ono H, Iwasaki M, Kuchiba A, Kasuga Y, Yokoyama S, Onuma H, et al. Association of dietary and genetic factors related to one-carbon metabolism with global methylation level of leukocyte DNA. Cancer Sci 2012;103(12):2159–64.

[37] Liu M, Cui LH, Ma AG, Li N, Piao JM. Lack of effects of dietary folate intake on risk of breast cancer: an updated meta-analysis of prospective studies. Asian Pac J Cancer Prev 2014;15(5):2323–8.

[38] de Vogel S, Dindore V, van Engeland M, Goldbohm RA, van den Brandt PA, Weijenberg MP. Dietary folate, methionine, riboflavin, and vitamin B-6 and risk of sporadic colorectal cancer. J Nutr 2008;138(12):2372–8.

[39] Ong TP, Moreno FS, Ross SA. Targeting the epigenome with bioactive food components for cancer prevention. J Nutrigenet Nutrigenomics 2011;4(5):275–92.

[40] Zeisel SH. Dietary choline deficiency causes DNA strand breaks and alters epigenetic marks on DNA and histones. Mutat Res 2012;733(1–2):34–8.

[41] Xu X, Gammon MD, Zeisel SH, Bradshaw PT, Wetmur JG, Teitelbaum SL, et al. High intakes of choline and betaine reduce breast cancer mortality in a population-based study. FASEB J 2009;23(11):4022–8.

[42] Duvoix A, Schmitz M, Schnekenburger M, Dicato M, Morceau F, Galteau MM, et al. Transcriptional regulation of glutathione S-transferase P1-1 in human leukemia. Biofactors 2003;17(1–4):131–8.

[43] Karius T, Schnekenburger M, Ghelfi J, Walter J, Dicato M, Diederich M. Reversible epigenetic fingerprint-mediated glutathione-S-transferase P1 gene silencing in human leukemia cell lines. Biochem Pharmacol 2011;81(11):1329–42.

[44] Amaral CL, Bueno Rde B, Burim RV, Queiroz RH, Bianchi Mde L, Antunes LM. The effects of dietary supplementation of methionine on genomic stability and p53 gene promoter methylation in rats. Mutat Res 2011;722(1):78–83.

[45] Batra V, Verma P. Dietary L-methionine supplementation mitigates gamma-radiation induced global DNA hypomethylation: enhanced metabolic flux towards S-adenosyl-L-methionine (SAM) biosynthesis increases genomic methylation potential. Food Chem Toxicol 2014;7.

[46] Harris HR, Cramer DW, Vitonis AF, DePari M, Terry KL. Folate, vitamin B(6), vitamin B(12), methionine and alcohol intake in relation to ovarian cancer risk. Int J Cancer 2012;131(4):E518–29.

[47] Graziosi L, Mencarelli A, Renga B, D'Amore C, Bruno A, Santorelli C, et al. Epigenetic modulation by methionine deficiency attenuates the potential for gastric cancer cell dissemination. J Gastrointest Surg 2013;17(1):39–49.

[48] Bidoli E, La Vecchia C, Talamini R, Negri E, Parpinel M, Conti E, et al. Micronutrients and ovarian cancer: a case-control study in Italy. Ann Oncol 2001;12(11):1589–93.

[49] Tworoger SS, Hecht JL, Giovannucci E, Hankinson SE. Intake of folate and related nutrients in relation to risk of epithelial ovarian cancer. Am J Epidemiol 2006;163(12):1101–11.

[50] Webb PM, Ibiebele TI, Hughes MC, Beesley J, van der Pols JC, Chen X, et al. Folate and related micronutrients, folate-metabolising genes and risk of ovarian cancer. Eur J Clin Nutr 2011;65(10):1133–40.

[51] Sharp L, Carsin AE, Cantwell MM, Anderson LA, Murray LJ, Group FS. Intakes of dietary folate and other B vitamins are associated with risks of esophageal adenocarcinoma, Barrett's esophagus, and reflux esophagitis. J Nutr 2013;143(12):1966–73.

[52] Kulkarni A, Dangat K, Kale A, Sable P, Chavan-Gautam P, Joshi S. Effects of altered maternal folic acid, vitamin B12 and docosahexaenoic acid on placental global DNA methylation patterns in Wistar rats. PLoS One 2011;6(3):e17706.

[53] Ba Y, Yu H, Liu F, Geng X, Zhu C, Zhu Q, et al. Relationship of folate, vitamin B12 and methylation of insulin-like growth factor-II in maternal and cord blood. Eur J Clin Nutr 2011;65(4):480–5.

[54] Vineis P, Chuang SC, Vaissiere T, Cuenin C, Ricceri F, Genair EC, et al. DNA methylation changes associated with cancer risk factors and blood levels of vitamin metabolites in a prospective study. Epigenetics 2011;6(2):195–201.

[55] Chiarugi A, Dolle C, Felici R, Ziegler M. The NAD metabolome--a key determinant of cancer cell biology. Nat Rev Cancer 2012;12(11):741–52.

[56] Mouchiroud L, Houtkooper RH, Auwerx J. NAD(+) metabolism: a therapeutic target for age-related metabolic disease. Crit Rev Biochem Mol Biol 2013;48(4):397–408.

[57] Morris BJ. Seven sirtuins for seven deadly diseases of aging. Free Radic Biol Med 2013;56:133–71.

[58] Siudeja K, Srinivasan B, Xu L, Rana A, de Jong J, Nollen EA, et al. Impaired Coenzyme A metabolism affects histone and tubulin acetylation in Drosophila and human cell models of pantothenate kinase associated neurodegeneration. EMBO Mol Med 2011;3(12):755–66.

[59] Donohoe DR, Bultman SJ. Metaboloepigenetics: interrelationships between energy metabolism and epigenetic control of gene expression. J Cell Physiol 2012;227(9):3169–77.

[60] Madzo J, Vasanthakumar A, Godley LA. Perturbations of 5-hydroxymethylcytosine patterning in hematologic malignancies. Semin Hematol 2013;50(1):61–9.

[61] Matsumoto K, Obara N, Ema M, Horie M, Naka A, Takahashi S, et al. Antitumor effects of 2-oxoglutarate through inhibition of angiogenesis in a murine tumor model. Cancer Sci 2009;100(9):1639–47.

[62] Dakshayani KB, Subramanian P, Manivasagam T, Mohamed Essa M. Metabolic normalization of alpha-ketoglutarate against N-nitrosodiethylamine-induced hepatocarcinogenesis in rats. Fundam Clin Pharmacol 2006;20(5):477–80.

[63] Xiang N, Zhao R, Song G, Zhong W. Selenite reactivates silenced genes by modifying DNA methylation and histones in prostate cancer cells. Carcinogenesis 2008;29(11):2175–81.

[64] Uthus E, Begaye A, Ross S, Zeng H. The von Hippel-Lindau (VHL) tumor-suppressor gene is down-regulated by selenium deficiency in Caco-2 cells and rat colon mucosa. Biol Trace Elem Res 2011;142(2):223–31.

[65] Davis CD, Uthus EO, Finley JW. Dietary selenium and arsenic affect DNA methylation in vitro in Caco-2 cells and in vivo in rat liver and colon. J Nutr 2000;130(12):2903–9.

[66] Uthus EO, Ross SA, Davis CD. Differential effects of dietary selenium (se) and folate on methyl metabolism in liver and colon of rats. Biol Trace Elem Res 2006;109(3):201–14.

[67] Pinto JT, Lee JI, Sinha R, MacEwan ME, Cooper AJ. Chemopreventive mechanisms of alpha-keto acid metabolites of naturally occurring organoselenium compounds. Amino Acids 2011;41(1):29–41.

[68] Dhawan DK, Chadha VD. Zinc: a promising agent in dietary chemoprevention of cancer. Indian J Med Res 2010;132:676–82.

[69] Wallwork JC, Duerre JA. Effect of zinc deficiency on methionine metabolism, methylation reactions and protein synthesis in isolated perfused rat liver. J Nutr 1985;115(2):252–62.

[70] Schnekenburger M, Morceau F, Henry E, Blasius R, Dicato M, Trentesaux C, et al. Transcriptional and post-transcriptional regulation of glutathione S-transferase P1 expression during butyric acid-induced differentiation of K562 cells. Leuk Res 2006;30(5):561–8.

[71] Scharlau D, Borowicki A, Habermann N, Hofmann T, Klenow S, Miene C, et al. Mechanisms of primary cancer prevention by butyrate and other products formed during gut flora-mediated fermentation of dietary fibre. Mutat Res 2009;682(1):39–53.

[72] Spurling CC, Suhl JA, Boucher N, Nelson CE, Rosenberg DW, Giardina C. The short chain fatty acid butyrate induces promoter demethylation and reactivation of RARbeta2 in colon cancer cells. Nutr Cancer 2008;60(5):692–702.

[73] Hu S, Dong TS, Dalal SR, Wu F, Bissonnette M, Kwon JH, et al. The microbe-derived short chain fatty acid butyrate targets miRNA-dependent p21 gene expression in human colon cancer. PLoS One 2011;6(1):e16221.

[74] Solimando R, Bazzoli F, Ricciardiello L. Chemoprevention of colorectal cancer: a role for ursodeoxycholic acid, folate and hormone replacement treatment? Best Pract Res Clin Gastroenterol 2011;25(4–5):555–68.

[75] Akare S, Jean-Louis S, Chen W, Wood DJ, Powell AA, Martinez JD. Ursodeoxycholic acid modulates histone acetylation and induces differentiation and senescence. Int J Cancer 2006;119(12):2958–69.

[76] Vanden Berghe W. Epigenetic impact of dietary polyphenols in cancer chemoprevention: lifelong remodeling of our epigenomes. Pharmacol Res 2012;65(6):565–76.

[77] Schnekenburger M, Dicato M, Diederich M. Plant-derived epigenetic modulators for cancer treatment and prevention. Biotechnol Adv 2014;31

[78] Cerella C., Kelkel M., Viry E., Dicato M., Jacob C., Diederich M. Naturally occurring organic sulfur compounds: an example of a multitasking class of phytochemicals in anti-cancer research. In: Rasooli PI, editor. Phytochemicals - Bioactivities and Impact on Health 2011.

[79] Powolny AA, Singh SV. Multitargeted prevention and therapy of cancer by diallyl trisulfide and related Allium vegetable-derived organosulfur compounds. Cancer Lett 2008;269(2):305–14.

[80] Nian H, Delage B, Ho E, Dashwood RH. Modulation of histone deacetylase activity by dietary isothiocyanates and allyl sulfides: studies with sulforaphane and garlic organosulfur compounds. Environ Mol Mutagen 2009;50(3):213–21.

[81] Cerella C, Scherer C, Cristofanon S, Henry E, Anwar A, Busch C, et al. Cell cycle arrest in early mitosis and induction of caspase-dependent apoptosis in U937 cells by diallyltetrasulfide (Al2S4). Apoptosis 2009;14(5):641–54.

[82] Kelkel M, Cerella C, Mack F, Schneider T, Jacob C, Schumacher M, et al. ROS-independent JNK activation and multisite phosphorylation of Bcl-2 link diallyl tetrasulfide-induced mitotic arrest to apoptosis. Carcinogenesis 2012;33(11):2162–71.

[83] Tang H, Kong Y, Guo J, Tang Y, Xie X, Yang L, et al. Diallyl disulfide suppresses proliferation and induces apoptosis in human gastric cancer through Wnt-1 signaling pathway by up-regulation of miR-200b and miR-22. Cancer Lett 2013;340(1):72–81.

[84] Li Y, Zhang J, Zhang L, Si M, Yin H, Li J. Diallyl trisulfide inhibits proliferation, invasion and angiogenesis of osteosarcoma cells by switching on suppressor microRNAs and inactivating of Notch-1 signaling. Carcinogenesis 2013;34(7):1601–10.

[85] Verkerk R, Schreiner M, Krumbein A, Ciska E, Holst B, Rowland I, et al. Glucosinolates in Brassica vegetables: the influence of the food supply chain on intake, bioavailability and human health. Mol Nutr Food Res 2009;53(Suppl. 2):S219.

[86] Myzak MC, Karplus PA, Chung FL, Dashwood RH. A novel mechanism of chemoprotection by sulforaphane: inhibition of histone deacetylase. Cancer Res 2004;64(16):5767–74.

[87] Hsu A, Wong CP, Yu Z, Williams DE, Dashwood RH, Ho E. Promoter de-methylation of cyclin D2 by sulforaphane in prostate cancer cells. Clin Epigenet 2011;3:3.

[88] Slaby O, Sachlova M, Brezkova V, Hezova R, Kovarikova A, Bischofova S, et al. Identification of microRNAs regulated by isothiocyanates and association of polymorphisms inside their target sites with risk of sporadic colorectal cancer. Nutr Cancer 2013;65(2):247–54.

[89] Li Q, Yao Y, Eades G, Liu Z, Zhang Y, Zhou Q. Downregulation of miR-140 promotes cancer stem cell formation in basal-like early stage breast cancer. Oncogene 2013 Jun 10. PubMed PMID:23752191. Pubmed Central PMCID: 3883868.

[90] Acharya A, Das I, Singh S, Saha T. Chemopreventive properties of indole-3-carbinol, diindolylmethane and other constituents of cardamom against carcinogenesis. Recent Pat Food Nutr Agric 2010;2(2):166–77.

[91] Li Y, Li X, Guo B. Chemopreventive agent 3,3'-diindolylmethane selectively induces proteasomal degradation of class I histone deacetylases. Cancer Res 2010;70(2):646–54.

[92] Paik WH, Kim HR, Park JK, Song BJ, Lee SH, Hwang JH. Chemosensitivity induced by down-regulation of microRNA-21 in gemcitabine-resistant pancreatic cancer cells by indole-3-carbinol. Anticancer Res 2013;33(4):1473–81.

[93] Balasubramanyam K, Swaminathan V, Ranganathan A, Kundu TK. Small molecule modulators of histone acetyltransferase p300. J Biol Chem 2003;278(21):19134–40.

[94] Sun Y, Jiang X, Chen S, Price BD. Inhibition of histone acetyltransferase activity by anacardic acid sensitizes tumor cells to ionizing radiation. FEBS Lett 2006;580(18):4353–6.

[95] Kim MK, Shin JM, Eun HC, Chung JH. The role of p300 histone acetyltransferase in UV-induced histone modifications and MMP-1 gene transcription. PLoS One 2009;4(3):e4864.

[96] Orlikova B, Tasdemir D, Golais F, Dicato M, Diederich M. Dietary chalcones with chemopreventive and chemotherapeutic potential. Genes Nutr 2011;6(2):125–47.

[97] Orlikova B, Schnekenburger M, Zloh M, Golais F, Diederich M, Tasdemir D. Natural chalcones as dual inhibitors of HDACs and NF-kappaB. Oncol Rep 2012;28(3):797–805.

[98] Lee JE, Hong EJ, Nam HY, Hwang M, Kim JH, Han BG, et al. Molecular signatures in response to Isoliquiritigenin in lymphoblastoid cell lines. Biochem Biophys Res Commun 2012;427(2):392–7.

[99] Lee SH, Zhao YZ, Park EJ, Che XH, Seo GS, Sohn DH. 2',4',6'-Tris(methoxymethoxy) chalcone induces apoptosis by enhancing Fas-ligand in activated hepatic stellate cells. Eur J Pharmacol 2011;658(1):9–15.

[100] Paluszczak J, Krajka-Kuzniak V, Baer-Dubowska W. The effect of dietary polyphenols on the epigenetic regulation of gene expression in MCF7 breast cancer cells. Toxicol Lett 2010;192(2):119–25.

[101] Teiten MH, Eifes S, Dicato M, Diederich M. Curcumin-the paradigm of a multi-target natural compound with applications in cancer prevention and treatment. Toxins 2010;2(1):128–62.

[102] Duvoix A, Blasius R, Delhalle S, Schnekenburger M, Morceau F, Henry E, et al. Chemopreventive and therapeutic effects of curcumin. Cancer Lett 2005;223(2):181–90.

[103] Teiten MH, Dicato M, Diederich M. Curcumin as a regulator of epigenetic events. Mol Nutr Food Res 2013;57(9):1619–29.

[104] Hua WF, Fu YS, Liao YJ, Xia WJ, Chen YC, Zeng YX, et al. Curcumin induces down-regulation of EZH2 expression through the MAPK pathway in MDA-MB-435 human breast cancer cells. Eur J Pharmacol 2010;637(1–3):16–21.

[105] Milenkovic D, Jude B, Morand C. miRNA as molecular target of polyphenols underlying their biological effects. Free Radic Biol Med 2013;64:40–51.

[106] Shirakami Y, Shimizu M, Moriwaki H. Cancer chemoprevention with green tea catechins: from bench to bed. Curr Drug Targets 2012;13(14):1842–57.

[107] Khan N, Mukhtar H. Cancer and metastasis: prevention and treatment by green tea. Cancer Metastasis Rev 2010;29(3):435–45.

[108] Fang MZ, Wang Y, Ai N, Hou Z, Sun Y, Lu H, et al. Tea polyphenol (-)-epigallocatechin-3-gallate inhibits DNA methyltransferase and reactivates methylation-silenced genes in cancer cell lines. Cancer Res 2003;63(22):7563–70.

[109] Lee WJ, Shim JY, Zhu BT. Mechanisms for the inhibition of DNA methyltransferases by tea catechins and bioflavonoids. Mol Pharmacol 2005;68(4):1018–30.

[110] Pandey M, Shukla S, Gupta S. Promoter demethylation and chromatin remodeling by green tea polyphenols leads to re-expression of GSTP1 in human prostate cancer cells. Int J Cancer 2010;126(11):2520–33.

[111] Chuang JC, Yoo CB, Kwan JM, Li TW, Liang G, Yang AS, et al. Comparison of biological effects of non-nucleoside DNA methylation inhibitors versus 5-aza-2'-deoxycytidine. Mol Cancer Ther 2005;4(10):1515–20.

[112] Choi KC, Jung MG, Lee YH, Yoon JC, Kwon SH, Kang HB, et al. Epigallocatechin-3-gallate, a histone acetyltransferase inhibitor, inhibits EBV-induced B lymphocyte transformation via suppression of RelA acetylation. Cancer Res 2009;69(2):583–92.

[113] Balasubramanian S, Adhikary G, Eckert RL. The Bmi-1 polycomb protein antagonizes the (-)-epigallocatechin-3-gallate-dependent suppression of skin cancer cell survival. Carcinogenesis 2010;31(3):496–503.

[114] Baselga-Escudero L, Blade C, Ribas-Latre A, Casanova E, Suarez M, Torres JL, et al. Resveratrol and EGCG bind directly and distinctively to miR-33a and miR-122 and modulate divergently their levels in hepatic cells. Nucleic Acids Res 2014;42(2):882–92.

[115] An IS, An S, Park S, Lee SN, Bae S. Involvement of microRNAs in epigallocatechin gallate-mediated UVB protection in human dermal fibroblasts. Oncol Rep 2013;29(1):253–9.

[116] Rietjens IM, Sotoca AM, Vervoort J, Louisse J. Mechanisms underlying the dualistic mode of action of major soy isoflavones in relation to cell proliferation and cancer risks. Mol Nutr Food Res 2013;57(1):100–13.

[117] Fang MZ, Chen D, Sun Y, Jin Z, Christman JK, Yang CS. Reversal of hypermethylation and reactivation of p16INK4a, RARbeta, and MGMT genes by genistein and other isoflavones from soy. Clin Cancer Res 2005;11(19 Pt 1):7033–41.

[118] King-Batoon A, Leszczynska JM, Klein CB. Modulation of gene methylation by genistein or lycopene in breast cancer cells. Environ Mol Mutagen 2008;49(1):36–45.

[119] Kikuno N, Shiina H, Urakami S, Kawamoto K, Hirata H, Tanaka Y, et al. Genistein mediated histone acetylation and demethylation activates tumor suppressor genes in prostate cancer cells. Int J Cancer 2008;123(3):552–60.

[120] Hirata H, Hinoda Y, Shahryari V, Deng G, Tanaka Y, Tabatabai ZL, et al. Genistein downregulates oncomiR-1260b and upregulates sFRP1 and Smad4 via demethylation and histone modification in prostate cancer cells. Br J Cancer 2014;110(6):1645–54.

[121] Ma J, Cheng L, Liu H, Zhang J, Shi Y, Zeng F, et al. Genistein down-regulates miR-223 expression in pancreatic cancer cells. Curr Drug Targets 2013;14(10):1150–6.

[122] Zaman MS, Thamminana S, Shahryari V, Chiyomaru T, Deng G, Saini S, et al. Inhibition of PTEN gene expression by oncogenic miR-23b-3p in renal cancer. PLoS One 2012;7(11):e50203.

[123] Chiyomaru T, Yamamura S, Zaman MS, Majid S, Deng G, Shahryari V, et al. Genistein suppresses prostate cancer growth through inhibition of oncogenic microRNA-151. PLoS One 2012;7(8):e43812.

[124] Chiyomaru T, Yamamura S, Fukuhara S, Hidaka H, Majid S, Saini S, et al. Genistein up-regulates tumor suppressor microRNA-574-3p in prostate cancer. PLoS One 2013;8(3):e58929.

[125] Chiyomaru T, Yamamura S, Fukuhara S, Yoshino H, Kinoshita T, Majid S, et al. Genistein inhibits prostate cancer cell growth by targeting miR-34a and oncogenic HOTAIR. PLoS One 2013;8(8):e70372.

[126] Rabiau N, Trraf HK, Adjakly M, Bosviel R, Guy L, Fontana L, et al. miRNAs differentially expressed in prostate cancer cell lines after soy treatment. In Vivo 2011;25(6):917–21.

[127] Murakami A, Ashida H, Terao J. Multitargeted cancer prevention by quercetin. Cancer Lett 2008;269(2):315–25.

[128] Howitz KT, Bitterman KJ, Cohen HY, Lamming DW, Lavu S, Wood JG, et al. Small molecule activators of sirtuins extend Saccharomyces cerevisiae lifespan. Nature 2003;425(6954):191–6.

[129] Tan S, Wang C, Lu C, Zhao B, Cui Y, Shi X, et al. Quercetin is able to demethylate the p16INK4a gene promoter. Chemotherapy 2009;55(1):6–10.

[130] Sheth S, Jajoo S, Kaur T, Mukherjea D, Sheehan K, Rybak LP, et al. Resveratrol reduces prostate cancer growth and metastasis by inhibiting the Akt/MicroRNA-21 pathway. PLoS One 2012;7(12):e51655.

[131] Boesch-Saadatmandi C, Wagner AE, Wolffram S, Rimbach G. Effect of quercetin on inflammatory gene expression in mice liver in vivo—role of redox factor 1, miRNA-122 and miRNA-125b. Pharmacol Res 2012;65(5):523–30.

[132] Boesch-Saadatmandi C, Loboda A, Wagner AE, Stachurska A, Jozkowicz A, Dulak J, et al. Effect of quercetin and its metabolites isorhamnetin and quercetin-3-glucuronide on inflammatory gene expression: role of miR-155. J Nutr Biochem 2011;22(3):293–9.

[133] Kelkel M, Jacob C, Dicato M, Diederich M. Potential of the dietary antioxidants resveratrol and curcumin in prevention and treatment of hematologic malignancies. Molecules 2010;15(10):7035–74.

[134] Folmer F, Basavaraju U, Jaspars M, Hold G, El-Omar E, Dicato M, et al. Anticancer effects of bioactive berry compounds. Phytochem Rev 2014;13(1):295–322.

[135] Aluyen JK, Ton QN, Tran T, Yang AE, Gottlieb HB, Bellanger RA. Resveratrol: potential as anticancer agent. J Diet Suppl 2012;9(1):45–56.

[136] Scott E, Steward WP, Gescher AJ, Brown K. Resveratrol in human cancer chemoprevention--choosing the 'right' dose. Mol Nutr Food Res 2012;56(1):7–13.

[137] Pacholec M, Bleasdale JE, Chrunyk B, Cunningham D, Flynn D, Garofalo RS, et al. SRT1720, SRT2183, SRT1460, and resveratrol are not direct activators of SIRT1. J Biol Chem 2010;285(11):8340–51.

[138] Boily G, He XH, Pearce B, Jardine K, McBurney MW. SirT1-null mice develop tumors at normal rates but are poorly protected by resveratrol. Oncogene 2009;28(32):2882–93.

[139] Farghali H, Kutinova Canova N, Lekic N. Resveratrol and related compounds as antioxidants with an allosteric mechanism of action in epigenetic drug targets. Physiol Res 2013;62(1):1–13.

[140] Stefanska B, Rudnicka K, Bednarek A, Fabianowska-Majewska K. Hypomethylation and induction of retinoic acid receptor beta 2 by concurrent action of adenosine analogues and natural compounds in breast cancer cells. Eur J Pharmacol 2010;638(1–3):47–53.

[141] Qin W, Zhang K, Clarke K, Weiland T, Sauter ER. Methylation and miRNA effects of resveratrol on mammary tumors vs. normal tissue. Nutr Cancer 2014;66(2):270–7.

[142] Balasubramanyam K, Altaf M, Varier RA, Swaminathan V, Ravindran A, Sadhale PP, et al. Polyisoprenylated benzophenone, garcinol, a natural histone acetyltransferase inhibitor, represses chromatin transcription and alters global gene expression. J Biol Chem 2004;279(32):33716–26.

[143] Li F, Shanmugam MK, Chen L, Chatterjee S, Basha J, Kumar AP, et al. Garcinol, a polyisoprenylated benzophenone modulates multiple proinflammatory signaling cascades leading to the suppression of growth and survival of head and neck carcinoma. Cancer Prev Res (Phila) 2013;6(8):843–54.

[144] Parasramka MA, Ali S, Banerjee S, Deryavoush T, Sarkar FH, Gupta S. Garcinol sensitizes human pancreatic adenocarcinoma cells to gemcitabine in association with microRNA signatures. Mol Nutr Food Res 2013;57(2):235–48.

[145] Kelkel M, Schumacher M, Dicato M, Diederich M. Antioxidant and anti-proliferative properties of lycopene. Free Radic Res 2011;45(8):925–40.

EMERGING EPIGENETIC THERAPIES—LYSINE METHYLTRANSFERASE/PRC COMPLEX INHIBITORS

19

Vineet Pande[1], David J. Pocalyko[2], and Dashyant Dhanak[2]

[1]*Janssen Research & Development, Beerse, Belgium,* [2]*Janssen Research & Development, Spring House, PA, USA*

CHAPTER OUTLINE

1 INTRODUCTION

Two classes of protein complexes—trithorax and polycomb group (PcG) proteins, are responsible for deposition of histone marks correlating with gene activation or repression, respectively [1]. The latter is associated with trimethylation of lysine 27 residue on histone 3 (H3K27). In mammals, there are two polycomb complexes—polycomb repressive complex (PRC)1 and PRC2. Enhancer of zeste homolog 2 (EZH2) is a histone lysine methyltransferase that catalyzes the transfer of a methyl group from the cofactor *S*-adenosyl methionine (SAM) to H3K27 and functions as part of the PRC2 complex [2]. PRC2 is further thought to facilitate the recruitment of PRC1 to methylated histones to repress the target gene expression [1]. In addition to EZH2, the PRC2 complex contains other PcG proteins including SUZ12 and embryonic ectoderm development protein (EED). Together these three proteins constitute the core holoenzyme to which accessory proteins may complex to regulate PRC2 recruitment and activity [3]. The methylation of H3K27 is catalyzed by the suppressor of variegation 3–9 (Suv (3–9)), enhancer of zeste (E(z)) and trithorax (SET) domain of EZH2 and in cell-free systems minimally requires the presence of both SUZ12 and EED [4]. PRC2 catalyzes a three-step sequential methylation (Figure 19.1) of H3K27 resulting in the formation of mono-, di-, and trimethylated product.

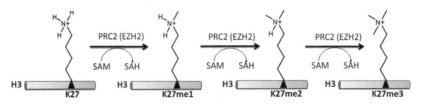

FIGURE 19.1

Molecular details of sequential methylation by PRC2 of lysine 27 side chain in a putative histone 3 tail.

2 PRC2 STRUCTURE–FUNCTION AND DRUGABILITY

The PRC2 core complex (excluding other cofactor proteins) found in humans is conserved across all multicellular organisms and is composed of four proteins (summing up to about 230 kDa in molecular weight): EED (different isoforms), EZH1 or EZH2, Suz12, and RbAp46 or RbAp48 [3]. Both EZH2 and its close homolog, EZH1, have catalytic methyltransferase activity within the C-terminal SET domain. Generally, SET domains are approximately 130 amino acids long and possess catalytic activity toward the ε-amino group of lysine residues and are able to mono-, di-, or trimethylate lysine residues by using SAM as a cofactor [5]. PRC2 complexes containing EZH1 typically have lower enzymatic activity [6] than those containing EZH2 and the activity of EZH1/2 depends on interaction with both Suz12 and the WD40 domain in EED. The WD40 beta propeller domain of EED interacts, in turn with H3K27me3 repressive marks and promotes allosteric activation of the methyltransferase activity of PRC2 [7,8]. Mutations in EED preventing the recognition of trimethyl-lysine marks abolish the activation of PRC2 *in vitro* and reduce global methylation in Drosophila [8]. RbAp48 also contains a WD40 propeller required for interaction with both Suz12 and the first 10 residues of unmodified histone 3 [9,10]. It is known that additional cofactors stabilize this core PRC2 complex in a functional sense. One such important cofactor is AEBP2 and Ciferri et al. [11] shed light on the structure of the PRC2 complex by using electron microscopy to produce a low-resolution 3D image of the human PRC2 complex bound to AEBP2. This study showed that the cofactor stabilizes the architecture of the PRC2 complex by binding it to a central hinge point. In addition, this study proposed a model for the binding of PRC2–AEBP2 complex to a di-nucleosome, spatially explaining the collaborative molecular mechanisms of the different components of the complex (e.g., EZH2 and EED) described above.

The catalytic activity of PRC2 is of particular interest from a drug-discovery perspective. The EZH2–SET domain recruits SAM as a cofactor to transfer methyl groups to histone substrates. Computational studies on the physicochemical features (e.g., hydrophobicity, buriedness, similarity to well-established drugable pockets) of several SAM binding sites [12] predict this site is generally chemically tractable for binding by small molecules and indeed as described below, a number of drug-like small molecules targeting the SAM binding pocket of EZH2 have been discovered. Antonysamy et al. [13] reported the first crystal structure (Figure 19.2) at a resolution of 2Å of the SET domain of EZH2, albeit in an inactive, *apo* form. The structure reveals that in its uncomplexed form, the C-terminus folds back into the active site occluding substrate access. This structure suggests a conformational change in the SET domain is required upon complex formation for the binding of both cofactor and histone substrate. The SET domain of EZH2 has moderate sequence homology (40–55% amino acid sequence similarity upon

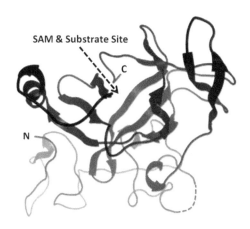

SAM & Substrate Site

C

N

FIGURE 19.2

Crystal structure of the *apo* form of human EZH2 SET domain (Protein Data Bank accession ID: 4MI5). N- and C-terminal ends are indicated. The C-terminal domain occludes the putative SAM and substrate binding site.

pairwise alignment) with other SET domain containing proteins such as GLP/EHMT1, mixed lineage lymphoma (MLL), EZH1, SETD1A, and SETD1B, and the availability of *holo* structures of MLL [14] and GLP/EHMT1 [15] has proven to be useful in comparative structural modeling of the EZH2 active site for *in silico* drug design [16,17].

3 ROLE OF PRC2 IN CANCER AND EZH2 AS A DRUG TARGET

PRC2 mediates transcriptional repression at least in part by H3K27 methylation although the exact mechanism has not been fully elucidated [6,18]. Studies in Drosophila [19] indicate PRC2 is recruited to target genes and catalyzes the methylation of H3K27. PRC1 is subsequently recruited and catalyzes the ubiquitination of H2AK119 leading to a more condensed chromatin configuration. This model would predict that PRC2 and PRC1 have largely overlapping target genes, which has been confirmed in human embryonic fibroblasts [20], although there is evidence for opposing roles in hematopoietic stem cells [21,22]. In addition, PRC2 cooperates with other epigenetic silencing enzymes. For example, EZH2 can bind the DNA methyltransferases DNMT1, DMNT3A, and DMNT3B [23] as well as recruit several HDACs suggesting additional mechanisms for transcriptional repression [24].

EZH2 is overexpressed and associated with poor clinical outcome in metastatic prostate [25] and breast cancer [26]. Overexpression has also been described in other cancers including lung [27], gastric [28], renal [29], bladder [30], and hepatocellular [31]. In addition, inactivating mutations in ubiquitously transcribed tetratricopeptide repeat, X chromosome (UTX), the corresponding H3K27 demethylase, have been correlated to increased levels of H3K27me3 and poor outcomes in metastatic prostate cancer [25]. More recently, various heterozygous mutations of tyrosine 641 (Y641) in the C-terminal SET domain have been reported in 7% of follicular lymphoma and 22% of diffuse large B-cell lymphoma (DLBCL) leading to increased H3K27me3 [32–34]. A variety of Y641 mutations in these lymphomas (e.g., Y641F, Y641N, Y641S, Y641H, and Y641C) [35] result in a gain of function in which the mutant enzymes have enhanced catalytic activity toward aH3K27me2 substrate and a relatively limited ability to convert H3K27me to H3K27me2. This is in contrast to the wild-type enzyme which has highest catalytic activity for the first methylation reaction [35,36]. These findings suggest that deregulation of EZH2 and hypermethylation of H3K27 result in the silencing of key genes involved in tumor growth and survival such as the p130-dependent pathway [37] and retinoblastoma (RB)-dependent pathways [37,38]. Furthermore, the above observations provide a strong rationale for targeting EZH2 therapeutically in a number of these malignancies.

Importantly, mutations in EZH2 have also been described in myelodysplasia–myeloproliferation neoplasms, myelofibrosis, and various subtypes of myelodysplastic syndromes [39,40]. In contrast to the gain of function mutations found in lymphoma, the mutations in myeloid neoplasms are consistent with a loss of function of the enzyme. The fact that EZH2 activating and inactivating mutations are both associated with tumor growth and maintenance highlights the complex role of PcG proteins and the exquisite context dependency of the enzyme and its product(s).

4 DISCOVERY OF INHIBITORS OF EZH2/PRC2

The mounting evidence of H3K27 hypermethylation and its role in multiple cancers have prompted research groups to develop potent and selective small molecule inhibitors of EZH2. Among the first compounds reported to effect EZH2 activity was 3-deazaneplanocin A (DZNep) (1), an S-adenosyl homocysteine hydrolase (SAH) inhibitor, which indirectly inhibits EZH2 by interfering with SAM and SAH metabolisms [41]. Inhibition by DZNep induces depletion of the PRC2 proteins and the associated H3K27 methylation although the effect is not specific to EZH2 catalytic inhibition and is rather an effect of reduced cellular PRC2 levels. DZNep has been reported to induce apoptosis in breast and colon cancer cells [41]; however, due to the mechanism of DZNep inhibition, attribution of the apoptotic effect in these cells specifically to EZH2 inhibition is not possible.

Recently several selective, SAM-competitive inhibitors of EZH2 have been reported following high-throughput screening (HTS) of small molecule libraries accompanied by various orthogonal assays and subsequent medicinal chemistry optimization. One of the most potent of these is GSK126 (2) which inhibits both wild-type and mutant EZH2 with similar potencies (K_i^{app}=0.5–3 nM) [42]. GSK126 is highly selective for EZH2, showing > 1000-fold selectivity over 20 other SET- and non-SET-domain-containing protein methyltransferases (PMTs) and > 150-fold selectivity over EZH1 (K_i^{app}=89 nM) despite the high sequence similarity of their SET domains (96% identical). GSK126 induced loss of H3K27me3 in both EZH2 mutant and wild-type DLBCL cell lines and inhibited proliferation in EZH2 mutant DLBCL cell lines. Robust transcriptional activation was observed in mutant sensitive cell lines while conversely, minimal transcriptional activation was noted in wild-type cell lines where proliferation was unaffected by EZH2 inhibition. GSK126 demonstrated complete inhibition of tumor growth in KARPAS-422 and Pfeiffer (a lymphoma cell line containing A677G mutant EZH2) xenograft models at 50 mg/kg administered intraperitoneally and induced tumor regression at higher doses. The compound was well tolerated at doses and schedules tested and improved survival of mice carrying KARPAS-422 xenografts. Further studies have shown that EZH2/PRC2 is activated by H3K27me3 binding [43] and interestingly, GSK126 has a greater affinity for this activated form and a significantly longer residence time relative to the inactivated EZH2/PRC2 [43]. These data suggest that activation of PRC2 by H3 peptide may allow the enzyme to adopt a distinct conformation possessing greater affinity for the inhibitor (Figure 19.3).

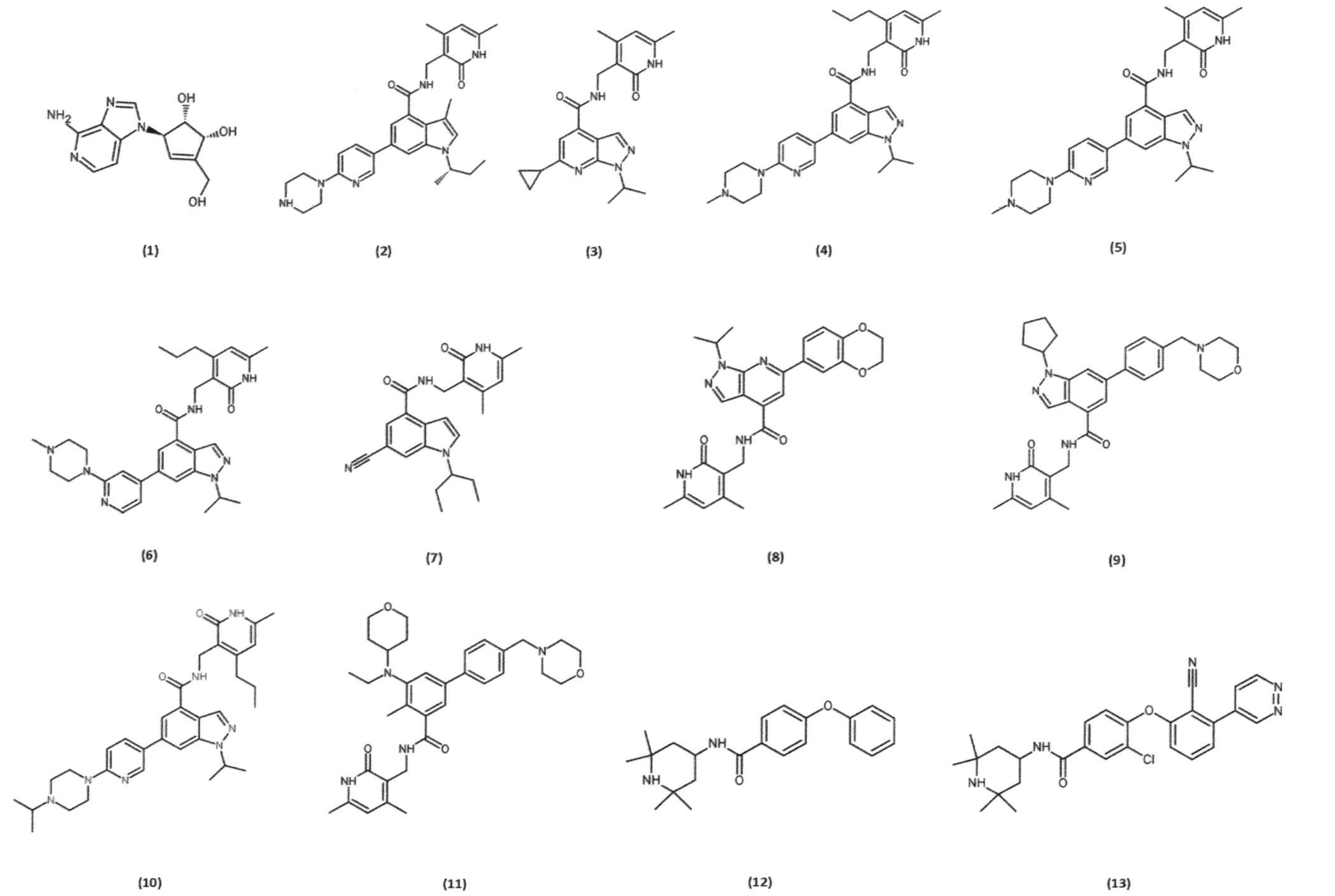

FIGURE 19.3

Chemical structures of EZH2 inhibitors.

From a similar screening and medicinal chemistry effort, Verma et al. [16] reported a structure activity relationship (SAR) study around the screening hit (**3**) that showed a modest K_i of 149 ± 28 nM against EZH2. The cyclopropyl group at the six-position when replaced by a piperazinylpyridine moiety provided a significant boost in potency as exemplified by (**4**), (**5**), and (**6**). A combination of replacing the central pyrazolopyridine scaffold with an indazole and substituting at the pyridine C4 with an *n*-propyl group resulted in compounds with Ki values of 0.6, 7.9, and 1.2 nM, respectively against EZH2. All these inhibitors are reportedly around 100-fold selective over several other PMTs.

The SAM-competitive inhibitor, EI1 (**7**) was identified following screening and medicinal chemistry optimization. The compound has similar potencies for the wild-type ($IC_{50} = 15 \pm 2$ nM) and the Y641F mutant EZH2 ($IC_{50} = 13 \pm 3$ nM) [44]. EI1 showed > 10,000-fold selectivity over ten other PMTs tested and ~90-fold selectivity over EZH1. EI1 dramatically inhibited H3K27me3 and H3K27me2 levels in DLBCL cell lines containing wild-type and EZH2 mutations. Proliferation of the EZH2 mutant DLBCL cells was strongly affected by EI1 inhibition while in contrast growth of the EZH2 wild-type DLBCL cells was only weakly inhibited. Inhibition led to transcriptional activation in EZH2 mutant KARPAS-422 cells, downregulation of cyclin A and B1, reactivation of PRC2 target gene expression, and corresponding G1 arrest [44].

At Epizyme, an HTS campaign followed by hit triaging led to the discovery of compound (**8**) with an IC_{50} of 620nM for wild type PRC2 complex. This compound needed further optimization in terms of physicochemical properties and increasing the size of the lipophilic group emanating from the indazole1-position led to the discovery of EPZ005687 (**9**). This analogue displayed SAM competitive inhibition of PRC2 enzymatic activity with a K_i of 24 ± 7 nM [45]. It showed > 500-fold selectivity over related PMTs and ~50-fold selectivity over EZH1. The affinity of EPZ005687 was similar for wild-type and Y641-containing mutant EZH2 but had a greater affinity for the A677G mutant (5.4-fold). As with the other EZH2 inhibitors above, EPZ005687 inhibited H3K27me3 in both wild-type and mutant EZH2 lymphoma cell lines but only showed inhibition of proliferation in mutant EZH2-containing cells. Pfeiffer cell line was particularly sensitive to inhibition by EPZ005687. Compound treatment led to a G1 accumulation in EZH2 Y641 mutant lymphoma cell lines and a negative enrichment for cell cycle genes from the Kyoto Encyclopedia of Genes and Genomes (KEGG) gene set such as CDK2, CCNA2, CCNB1, and E2F2. [45]

While GSK126, EI1, and EPZ005687 are all potent, selective inhibitors of EZH2, the compounds lacked suitable pharmacokinetic properties for oral administration. UNC1999 (**10**), derived from EPZ005687 using molecular modeling, is the first orally bioavailable inhibitor of EZH2 to be reported [17]. The compound, however, showed only 10-fold selectivity over EZH1 and is thus considered a dual EZH2/1 inhibitor. Similarly, the compound displayed less than five-fold selectivity for wild-type EZH2 over mutant enzymes (Y641N, Y641F). However, UNC1999 was > 10,000-fold more selective for EZH2 over 15 other DNA, lysine, and arginine methyltransferases tested. The compound reduced H3K27me3 in cell-based assays and selectively inhibited the growth of DLBCL cell harboring the Y641N mutation. Oral bioavailability was demonstrated in mice but the effect on tumors in xenograft models was not examined.

EPZ-6438/E7438 (**11**) is an improved version of EPZ005687 and shares similar *in vitro* properties (mechanism of action, cellular activity, and selectivity) but has better oral bioavailability and pharmacokinetic properties [46]. Inhibition of EZH2 by EPZ-6438 in SMARCB1-deleted malignant rhabdoid tumor (MRT) cell lines led to strong anti-proliferative effects with IC_{50} values in the nanomolar range

Table 19.1 Clinical Studies Evaluating EZH2 Inhibitors

Compound	Company	Clinical Trial Identifier	Clinical Phase	Description
EPZ-6438/E7438	Epizyme	NCT01897571	Phase I/II	Single agent in subjects with advanced solid tumors or with B-cell lymphomas
GSK2816126	GlaxoSmithKline	NCT02082977	Phase I	Single agent in subjects with relapsed/ refractory diffuse large B-cell and transformed follicular lymphoma

[47] demonstrating for the first time activity in a non-hematological malignancy. Oral, twice-daily dosing of 250 or 500 mg/kg EPZ-6438 led to near complete regression in MRT xenografts with no tumor regrowth observed 30 days after inhibitor dosing was stopped. Inhibition of EZH2 in these tumors led to derepression of neuronal differentiation genes, cell cycle regulatory, and tumor repressors while reducing the expression of GLI1, PTCH1, MYC, and EZH2. The combination of these effects is likely the cause of the pronounced and durable antitumor effects induced by EPZ-6438 in these tumors. The antitumor activity of EPZ-6438 was also recently reported in EZH2 mutant non-Hodgkin lymphoma (NHL) [46]. Oral dosing of EPZ-6348 in mutant NHL xenografts led to a range of antitumor activity depending on the cell line. The most robust response was again observed in Pfeiffer xenografts. Treatment with 114 mg/kg of EPZ-6438 once daily led to complete tumor regression with no tumor regrowth observed 36 days after cessation of dosing. EPZ-6438 is now in Phase I clinical studies in cancer patients with lymphomas and solid tumors with high levels of H3K27me3 (NCT01897571, Table 19.1).

Researchers at Constellation Pharmaceuticals reported SAM-competitive EZH2 inhibitors having distinct tetramethyl piperidinyl benzamide scaffolds structurally unrelated to the pyridone amide–motif-containing compounds described above. Following an HTS campaign, compound (**12**) was identified as a SAM competitive hit with moderate affinity of 51 μM [48]. Following a comprehensive medicinal chemistry program [49] compound (**13**) was obtained as a potent SAM competitive inhibitor. This compound, was approximately six fold more potent for wild-type EZH2 ($IC_{50} = 32$ nM) than for the Y641N mutant EZH2 ($IC_{50} = 197$ nM) or EZH1 ($IC_{50} = 213$ nM) and showed good selectivity against the small panel of PMTs tested. The compound was active in cellular assays and reduced global H3K27me3 and H3K27me2 levels in HT and SUDHL6 cells. As with the other EZH2 inhibitors above, inhibition of proliferation was observed upon compound treatment in DLBCL cell lines harboring EZH2 mutations while wild-type DLBCL cell lines remained insensitive.

In a conceptually different approach to inhibiting the PRC2 complex, Kim et al. [50] reported development of cell-penetrant, hydrocarbon stapled peptides based on an alpha-helix portion of EZH2 interacting with EED as a part of the PRC2 complex. These stapled peptides (SAH-EZH2a & SAH EZH2b- Figure 19.4) selectively inhibited H3K27 trimethylation (H3K27me3) by dose dependently disrupting the EZH2:EED protein–protein interaction (EED-binding affinity around 300 nM for various lengths and modifications) and reducing EZH2 protein levels. PRC2-dependent MLL-AF9 cells underwent growth arrest and monocyte-macrophage differentiation upon treatment with SAH-EZH2, consistent with changes observed in expression of lineage-specific marker genes, regulated by PRC2.

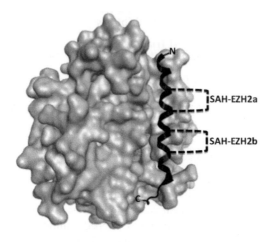

FIGURE 19.4

Crystal structure (Protein Data Bank accession ID: 2QXV) of EZH2 peptide (alpha helical representation) and EED (protein surface representation) used to design hydrocarbon stapled (shown as dotted lines) peptide inhibitors of various lengths. SAH–EZH2a peptides are chemically stapled at residues 47 and 51 and SAH–EZH2a at residues 54 and 58 to stabilize the helical peptide. The full length helix in the complex is 30 residues long and the various stapled peptides span lengths between 24 and 28 residues.

5 CONCLUSIONS

The misregulation of H3K27me3 levels, through EZH2 overexpression or somatic alteration, silences target genes important in tumor development. EZH2 as the catalytic piece of the multi-protein PRC2 complex is an emerging drug target in the post-HDAC inhibitor and hypomethylating agents' era. At a molecular level, the SET domain containing the methyltransferase catalytic activity of EZH2 has been the target of interest for a number of drug-discovery programs. Screening campaigns followed by extensive medicinal chemistry based optimization has led to the discovery and development of drug-like small molecules selectively targeting the catalytic activity of EZH2 in context of the PRC2 complex. Two of these inhibitors have entered clinical trials in oncology (Table 19.1). In addition to efficacy in the clinic, the safety of these novel agents in humans remains to be seen. It is anticipated that the demonstrated role of EZH2 in multiple cancer types will spur future clinical trials in non-hematological malignancies such as prostate and breast cancers. As an emerging target EZH2 is also being investigated in other settings such as viral latency, where Friedman et al. [51] have shown HIV-1 proviral reactivation is enhanced with shRNA knockdown of EZH2. Despite much recent progress, a detailed understanding of the function and extensive biology of PRC2 function continues to evolve and deepen. In addition to offering potential novel targeted therapeutics in cancer, small molecule inhibitors of EZH2/PRC2 complex provide much needed chemical tools to help decipher the complex epigenetic regulation of cellular transcription and function. Such agents will continue to be invaluable in the discovery and development of novel therapeutics in a variety of indications.

REFERENCES

[1] Cao R, Wang LJ, Wang HB, et al. Role of histone H3 lysine 27 methylation in polycomb-group silencing. Science 2002;298(5595):1039–43.

[2] Morey L, Helin K. Polycomb group protein-mediated repression of transcription. Trends Biochem Sci 2010;35(6):323–32.

[3] Margueron R, Reinberg D. The Polycomb complex PRC2 and its mark in life. Nature 2011;469(7330):343–9.

[4] Simon JA, Lange CA. Roles of the EZH2 histone methyltransferase in cancer epigenetics. Mutat Res 2008;647(1–2):21–9.

[5] Herz HM, Garruss A, Shilatifard A. SET for life biochemical activities and biological functions of SET domain-containing proteins. Trends Biochem Sci 2013;38(12):621–39.

[6] Margueron R, Li GH, Sarma K, et al. Ezh1 and Ezh2 maintain repressive chromatin through different mechanisms. Mol Cell 2008;32(4):503–18.

[7] Xu C, Bian CB, Yang W, et al. Binding of different histone marks differentially regulates the activity and specificity of polycomb repressive complex 2 (PRC2). Proc Natl Acad Sci USA 2010;107(45):19266–71.

[8] Margueron R, Justin N, Ohno K, et al. Role of the polycomb protein EED in the propagation of repressive histone marks. Nature 2009;461(7265) 762-U11.

[9] Ketel CS, Andersen EF, Vargas ML, Su J, Strome S, Simon JA. Subunit contributions to histone methyltransferase, activities of fly and worm Polycomb group complexes. Mol Cell Biol 2005;25(16):6857–68.

[10] Yamamoto K, Sonoda M, Inokuchi J, Shirasawa S, Sasazuki T. Polycomb group suppressor of zeste 12 links heterochromatin protein 1 alpha and enhancer of zeste 2. J Biol Chem 2004;279(1):401–6.

[11] Ciferri C, Lander GC, Maiolica A, Herzog F, Aebersold R, Nogales E. Molecular architecture of human polycomb repressive complex 2. Elife 2012:1.

[12] Campagna-Slater V, Mok MW, Nguyen KT, Feher M, Najmanovich R, Schapira M. Structural chemistry of the histone methyltransferases cofactor binding site. J Chem Inf Model 2011;51(3):612–23.

[13] Antonysamy S, Condon B, Druzina Z, et al. Structural context of disease-associated mutations and putative mechanism of autoinhibition revealed by X-Ray crystallographic analysis of the EZH2-SET domain. PLoS One 2013;8(12).

[14] Southall SM, Wong PS, Odho Z, Roe SM, Wilson JR. Structural basis for the requirement of additional factors for MLL1 SET domain activity and recognition of epigenetic marks. Mol Cell 2009;33(2):181–91.

[15] Wu H, Min JR, Lunin VV, et al. Structural biology of human H3K9 methyltransferases. PLoS One 2010;5(1).

[16] Verma SK, Tian XR, LaFrance LV, et al. Identification of potent, selective, cell-active inhibitors of the histone lysine methyltransferase EZH2. Acs Med Chem Lett 2012;3(12):1091–6.

[17] Konze KD, Ma A, Li F, et al. An orally bioavailable chemical probe of the Lysine Methyltransferases EZH2 and EZH1. ACS Chem Biol 2013;8(6):1324–34.

[18] Hansen KH, Bracken AP, Pasini D, et al. A model for transmission of the H3K27me3 epigenetic mark. Nat Cell Biol 2008;10(11):1291–300.

[19] Wang L, Brown JL, Cao R, Zhang Y, Kassis JA, Jones RS. Hierarchical recruitment of polycomb group silencing complexes. Mol Cell 2004;14(5):637–46.

[20] Bracken AP, Dietrich N, Pasini D, Hansen KH, Helin K. Genome-wide mapping of polycomb target genes unravels their roles in cell fate transitions. Genes Dev 2006;20(9):1123–36.

[21] Majewski IJ, Ritchie ME, Phipson B, et al. Opposing roles of polycomb repressive complexes in hematopoietic stem and progenitor cells. Blood 2010;116(5):731–9.

[22] Lessard J, Schumacher A, Thorsteinsdottir U, van Lohuizen M, Magnuson T, Sauvageau G. Functional antagonism of the Polycomb-Group genes eed and Bmi1 in hemopoietic cell proliferation. Genes Dev 1999;13(20):2691–703.

[23] Vire E, Brenner C, Deplus R, et al. The polycomb group protein EZH2 directly controls DNA methylation. Nature 2006;439(7078):871–4.

[24] van der Vlag J, Otte AP. Transcriptional repression mediated by the human polycomb-group protein EED involves histone deacetylation. Nat Genet 1999;23(4):474–8.

[25] Varambally S, Dhanasekaran SM, Zhou M, et al. The polycomb group protein EZH2 is involved in progression of prostate cancer. Nature 2002;419(6907):624–9.

[26] Kleer CG, Cao Q, Varambally S, et al. EZH2 is a marker of aggressive breast cancer and promotes neoplastic transformation of breast epithelial cells. Proc Natl Acad Sci USA 2003;100(20):11606–11.

[27] Watanabe H, Soejima K, Yasuda H, et al. Deregulation of histone lysine methyltransferases contributes to oncogenic transformation of human bronchoepithelial cells. Cancer Cell Int 2008;8:15.

[28] Matsukawa Y, Semba S, Kato H, Ito A, Yanagihara K, Yokozaki H. Expression of the enhancer of zeste homolog 2 is correlated with poor prognosis in human gastric cancer. Cancer Sci 2006;97(6):484–91.

[29] Wagener N, Macher-Goeppinger S, Pritsch M, et al. Enhancer of zeste homolog 2 (EZH2) expression is an independent prognostic factor in renal cell carcinoma. BMC Cancer 2010;10:524.

[30] Arisan S, Buyuktuncer ED, Palavan-Unsal N, Caskurlu T, Cakir OO, Ergenekon E. Increased expression of EZH2, a polycomb group protein, in bladder carcinoma. Urol Int 2005;75(3):252–7.

[31] Sudo T, Utsunomiya T, Mimori K, et al. Clinicopathological significance of EZH2 mRNA expression in patients with hepatocellular carcinoma. Br J Cancer 2005;92(9):1754–8.

[32] Morin RD, Johnson NA, Severson TM, et al. Somatic mutations altering EZH2 (Tyr641) in follicular and diffuse large B-cell lymphomas of germinal-center origin. Nat Genet 2010;42(2):181–5.

[33] Pasqualucci L, Trifonov V, Fabbri G, et al. Analysis of the coding genome of diffuse large B-cell lymphoma. Nat Genet 2011;43(9):830–7.

[34] Ryan RJ, Nitta M, Borger D, et al. EZH2 codon 641 mutations are common in BCL2-rearranged germinal center B cell lymphomas. PLoS One 2011;6(12):e28585.

[35] Sneeringer CJ, Scott MP, Kuntz KW, et al. Coordinated activities of wild-type plus mutant EZH2 drive tumor-associated hypertrimethylation of lysine 27 on histone H3 (H3K27) in human B-cell lymphomas. Proc Natl Acad Sci USA 2010;107(49):20980–5.

[36] Yap DB, Chu J, Berg T, et al. Somatic mutations at EZH2 Y641 act dominantly through a mechanism of selectively altered PRC2 catalytic activity, to increase H3K27 trimethylation. Blood 2011;117(8):2451–9.

[37] Bohrer LR, Chen S, Hallstrom TC, Huang H. Androgens suppress EZH2 expression via retinoblastoma (RB) and p130-dependent pathways: a potential mechanism of androgen-refractory progression of prostate cancer. Endocrinology 2010;151(11):5136–45.

[38] Bracken AP, Pasini D, Capra M, Prosperini E, Colli E, Helin K. EZH2 is downstream of the pRB-E2F pathway, essential for proliferation and amplified in cancer. EMBO J 2003;22(20):5323–35.

[39] Ernst T, Chase AJ, Score J, et al. Inactivating mutations of the histone methyltransferase gene EZH2 in myeloid disorders. Nat Genet 2010;42(8):722–6.

[40] Nikoloski G, Langemeijer SM, Kuiper RP, et al. Somatic mutations of the histone methyltransferase gene EZH2 in myelodysplastic syndromes. Nat Genet 2010;42(8):665–7.

[41] Tan J, Yang X, Zhuang L, et al. Pharmacologic disruption of Polycomb-repressive complex 2-mediated gene repression selectively induces apoptosis in cancer cells. Genes Dev 2007;21(9):1050–63.

[42] McCabe MT, Ott HM, Ganji G, et al. EZH2 inhibition as a therapeutic strategy for lymphoma with EZH2-activating mutations. Nature 2012;492(7427):108–12.

[43] Van Aller GS, Pappalardi MB, Ott HM, et al. Long residence time inhibition of EZH2 in activated polycomb repressive complex 2. ACS Chem Biol 2013.

[44] Qi W, Chan H, Teng L, et al. Selective inhibition of Ezh2 by a small molecule inhibitor blocks tumor cells proliferation. Proc Natl Acad Sci USA 2012;109(52):21360–5.

[45] Knutson SK, Wigle TJ, Warholic NM, et al. A selective inhibitor of EZH2 blocks H3K27 methylation and kills mutant lymphoma cells. Nat Chem Biol 2012;8(11):890–6.

[46] Knutson SK, Kawano S, Minoshima Y, et al. Selective inhibition of EZH2 by EPZ-6438 leads to potent anti-tumor activity in EZH2 mutant non-hodgkin Lymphoma. Mol Cancer Ther 2014.

[47] Knutson SK, Warholic NM, Wigle TJ, et al. Durable tumor regression in genetically altered malignant rhab-doid tumors by inhibition of methyltransferase EZH2. Proc Natl Acad Sci USA 2013;110(19):7922–7.

[48] Garapaty-Rao S, Nasveschuk C, Gagnon A, et al. Identification of EZH2 and EZH1 small molecule inhibitors with selective impact on diffuse large B cell lymphoma cell growth. Chem Biol 2013;20(11):1329–39.

[49] Nasveschuk CG, Gagnon A, Garapaty-Rao S, et al. Discovery and optimization of tetramethylpiperidinyl benzamidesas inhibitors of EZH2. ACS Med Chem Lett 2014;5:378–83.

[50] Kim W, Bird GH, Neff T, et al. Targeted disruption of the EZH2-EED complex inhibits EZH2-dependent cancer. Nat Chem Biol 2013;9(10):643–50.

[51] Friedman J, Cho WK, Chu CK, et al. Epigenetic silencing of HIV-1 by the Histone H3 Lysine 27 Methyltransferase Enhancer of Zeste 2. J Virol 2011;85(17):9078–89.

INHIBITORS OF JUMONJI C-DOMAIN HISTONE DEMETHYLASES

20

Peter Staller

EpiTherapeutics ApS, Copenhagen, Denmark

CHAPTER OUTLINE

1 INTRODUCTION: HISTONE METHYLATION—NORMAL AND PATHOLOGICAL FUNCTIONS

Genomic DNA in eukaryotic cells is compacted into chromatin, whose basic building blocks are the nucleosomes. In these, the DNA is wrapped around an octamer of the core histones H2A, H2B, H3, and H4. The linker histone H1 and its isoforms sit at the base of the nucleosome near the DNA entry and exit, bind to the linker region of the DNA, and serve to compact the chromatin. Covalent modifications of histones and the chromosomal DNA form the basis of an epigenetic code that plays a critical role in controlling DNA replication, gene transcription, and DNA repair. These patterns of posttranslational chromatin modifications are set up by a group of enzymes known as "writer" proteins, removed by another group ("eraser" proteins) and "read" by a third group of effectors that contain protein-binding modules to recognize specific modifications. By influencing both global and local gene expression, epigenetic mechanisms influence the establishment and the maintenance of cellular identity. On the flipside, the system of epigenetic writers, readers, and erasers can contribute to diseases such as cancer, chronic inflammation, and neurological disorders when its components are deregulated by, e.g., gene mutations, deletions, amplifications, or changes in protein expression levels.

Among a plethora of histone modifications, methylation of lysine and arginine residues plays a key role in the regulation of transcription and genomic stability. These modifications are catalyzed by several protein families, which are often referred to as histone methyltransferases. The enzymes use

S-adenosyl methionine (SAM) as a methyl donor. The Su(var)3-9, Enhancer-of-zeste and Trithorax (SET) domain family and the DOT1L protein both methylate lysines, whereas a second group of protein arginine methyltransferases exclusively targets arginine residues at different substrate proteins. Specific lysines in histone H3 and H4 can be mono-, di-, or trimethylated (me1, me2, or me3). Arginine residues can be monomethylated (me1), symmetrically dimethylated (me2s), or asymmetrically dimethylated (me2a). In contrast to acetylation, lysine methylation does not change the net charge of the modified residue and hence does not directly affect the interaction of histones with DNA. However, different families of reader proteins recognize the methylated amino acids. These readers often serve as subunits of larger protein complexes that contain enzymatic activities, which modify the chromatin and thereby alter the local structure and affect gene transcription. Mapping of the distribution of methylated lysines on genome-wide scale indicates that modified histones occupy distinct regions of the genome. Actively transcribed areas are enriched for H3K36me3 (trimethylated lysine 36 on histone H3) and H3K79me2/3 (the di- and trimethylated state of lysine 79 on histone H3). In contrast, H3K9me2/3, H3K27me3, and H4K20me3 are primarily found in transcriptionally silenced parts. A third class of chromatin marks, such as H3K4me3, is located in both silenced and actively transcribed areas. The importance of methylation of H3K4, H3K9, H3K27, and H4K20 is underscored by the fact that they are required for normal embryonic development in mice. Please refer to several recent reviews that cover the important roles of the different histone methylation marks in stem cells, cellular differentiation, and organismal development [1–3]. In addition, work performed in model organisms such as *Caenorhabditis elegans* and *Drosophila melanogaster* indicates that lysine methylation also contributes to the regulation of organismal lifespan and aging-related processes (reviewed in Ref. [4]).

Many histone methyltransferases are involved in the etiology of neurological disorders and cancer. Inactivating or activating mutations or changes in protein expression levels cause alterations of histone methylation patterns and a subsequent deregulation of gene transcription. Examples include the overexpression of the H3K27 methyltransferase EZH2, which is the catalytic subunit of the polycomb repressive complex 2 (PRC2), in a range of human cancers [5,6]. Interestingly, somatic mutations of EZH2 that increase H3K27 trimethylation activity were found in a subset of B-cell lymphomas, and the recent development of specific EZH2 inhibitors offers therapeutics potential in this context [7–11]. In contrary, in pediatric glioblastoma recurrent mutations target lysine 27 of histone H3.3 (H3.3K27M), inhibit EZH2 methylation activity, and lead to a global loss of H3K27me3 [12–14]. Although H3.3 mutations remain to be fully characterized with respect to a driver function in glioblastoma, their occurrence indicates that first of all normal development requires tightly controlled H3K27me3 levels and secondly that a given methyltransferase can act both as an oncogene and a tumor suppressor in a context-specific manner. In addition to EZH2, this has recently been demonstrated for the PRC2 complex component SUZ12, which is required for EZH2 enzymatic activity but also acts as a tumor suppressor in tumors of the nervous system and in melanomas [15].

Further examples of the importance of balanced histone methylations comprise the H3K36 histone methyltransferase NSD2/MMSET. NSD2 is overexpressed in a subset of multiple myelomas due to a recurrent chromosomal translocation and triggers increased H3K36me2 levels and the activation of oncogenes [16]. The H3K79 (unmethylated lysine 79 on histone H3) methyltransferase DOT1L is itself not changed in cancers. However, in acute myeloid leukemia (AML) cells, a range of MLL fusion proteins that originate from chromosomal translocations involving the Mixed Lineage Leukemia gene (*MLL*) increase the recruitment of DOT1L to target genes and thereby promote a local gain of H3K79 methylation. This mechanism is required for the induction of critical target genes of MLL fusion proteins and the survival of leukemic cells in experimental systems [17–19]. Consequently, a small molecule inhibitor of DOT1L enzymatic activity is currently used in a clinical Phase 1 trial for the treatment of MLL-rearranged leukemias.

2 THE FIRST HISTONE DEMETHYLASES: LSD1 AND LSD2

In 2004, the discovery that the lysine-specific demethylase 1 (LSD1, also known as KDM1A and AOF2) demethylates H3K4me1 (monomethylated lysine 4 on histone H3) and H3K4me2 (dimethylated lysine 4 on histone H3) changed the prevailing view that histone methylation was a stable mark that could only be removed by histone exchange and DNA replication [20]. LSD1 and the closely related LSD2 contain an amine oxidase-like (AOL) domain that is also present in several metabolic enzymes, such as the closest relatives, the monoamine oxidases MAO-A and MAO-B. The catalytic mechanism of the demethylation is dependent on both flavine adenine dinucleotide (FAD) as a cofactor and a lone electron pair on the lysine ε-nitrogen atom (Figure 20.1A). Hence, the LSD enzymes cannot demethylate trimethylated lysines. Moreover, partner proteins influence the catalytic activity of LSD1. This was demonstrated by a switch to H3K9me1 and H3K9me2 specificity induced by complex formation with the androgen receptor [21]. In addition, nonhistone protein substrates for LSD1 have been reported, such as the transcription factors p53 and E2F1 as well as the DNA methyltransferase DNMT1 [22–24]. LSD2 (also known as KDM1B and AOF1) targets H3K4me2 and H3K4me1 [25,26] and LSD2 catalytic activity is stimulated by binding to its cofactor GLYR1 through stabilization of the interaction with the histone substrate [27].

Deletion of LSD1 is embryonic lethal in mice around day E5.5 and induces a progressive loss of DNA methylation by destabilizing DNMT1 [23]. As part of different activating and repressive chromatin-modifying complexes, LSD1 is a central regulator of chromatin (recently reviewed in [28]) with crucial functions in both stem cell maintenance and cellular differentiation programs such as hematopoiesis [29]. LSD2 is required for maternal imprinting in oocytes, and embryos derived from LSD2-deficient female mice die before mid-gestation [30]. In the context of human disease, LSD1 is overexpressed in several malignancies, among them prostate cancer, neuroblastoma, estrogen-receptor negative breast cancer, lung adenocarcinoma, bladder cancer, and colorectal carcinoma [31–35]. Notably, two recent studies demonstrated that the inhibition of LSD1 enzymatic activity in AML cells abrogated oncogenic potential and triggered cellular differentiation [36,37]. These findings highlighted the potential of a targeted anti-LSD1 therapy and encouraged the development of specific inhibitors. The nonselective and irreversible monoamine oxidase inhibitor tranylcypromine (TCP, **1**, Figure 20.1B) that has been used for the treatment of depression inhibits LSD1 with a half-maximal inhibitory concentration (IC_{50}) of below $2\,\mu M$ and induces H3K4 methylation in cultured cells and *in vivo* [38]. TCP has proven to be valuable tool in functional studies, however, more specific inhibitors that do not inactivate the monoamine oxidases MAO-A and MAO-B are desired for targeted therapy. To achieve this, derivatives of TCP have been tested and they indeed show enhanced potency and target selectivity for LSD1 [39,40]. A patent application filed by Oryzon Genomics, a Spanish biopharmaceutical company, describes many TCP derivatives with substitutions on both the phenyl ring and the nitrogen with greatly improved selectivity and potency for LSD1 [41]. Some of these derivatives show brain penetrance and an antileukemic effect in animal models (an example structure **2** is depicted in Figure 20.1B) [36,42]. In 2013, Oryzon Genomics received orphan designation from the European Medicines Agency (EMA) for their TCP-based LSD1 inhibitor ORY-1001 for the treatment of AML. Subsequently, clinical studies have

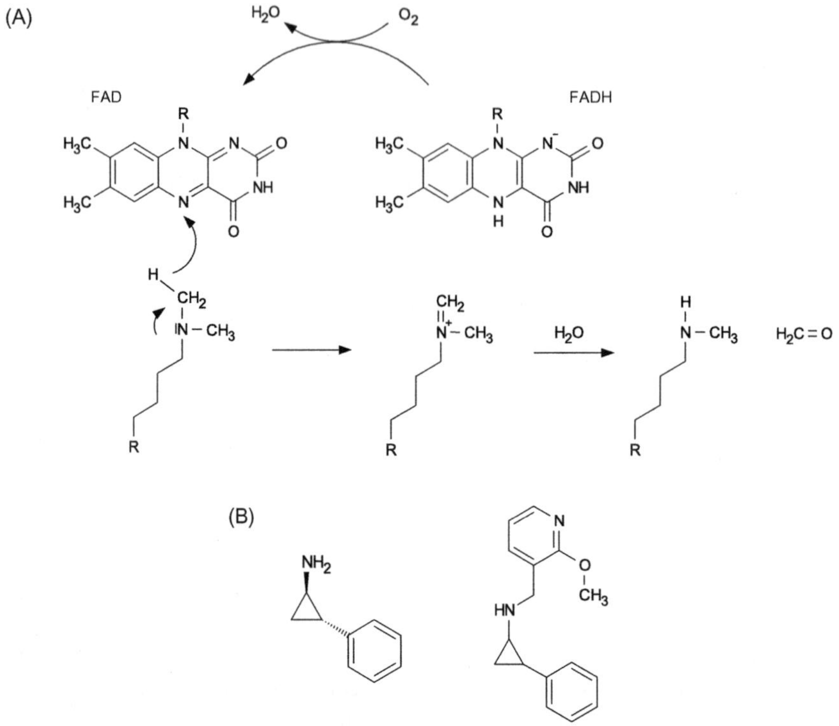

FIGURE 20.1

Lysine demethylation by LSD enzymes. (A) Lysine demethylation by LSD1 and LSD2 with flavin adenine dinucleotide (FAD) acting as cofactor is likely to occur through a hydride transfer mechanism. (B) The monoamine oxidase inhibitor tranylcypromine **1** was the first reported inhibitor of LSD1, which led to the development of LSD1-selective analogues such as **2**.

been started. In addition, the irreversible LSD1 inhibitor GSK2879552 developed by GlaxoSmithKline has entered clinical trials in the United States for the treatment of AML and refractory small cell lung cancer, respectively. The chemical structures of GSK2879552 and ORY-1001 have not been disclosed.

3 THE JUMONJI C-DOMAIN HISTONE DEMETHYLASES

In 2006, several publications described the histone lysine demethylases activity of jumonji C (JMJC)-domain proteins [43–47]. The catalytic JMJC domain contains a double-stranded β-helix (DSBH) fold, a structural motif that is shared with a large family of 2-oxoglutarate-dependent oxygenases [48]. Three conserved residues in the DSBH fold (consensus sequence HXXD/E…H), termed the facial triad, coordinate a catalytic Fe(II). The second cofactor is 2-oxoglutarate, which is held in place by two

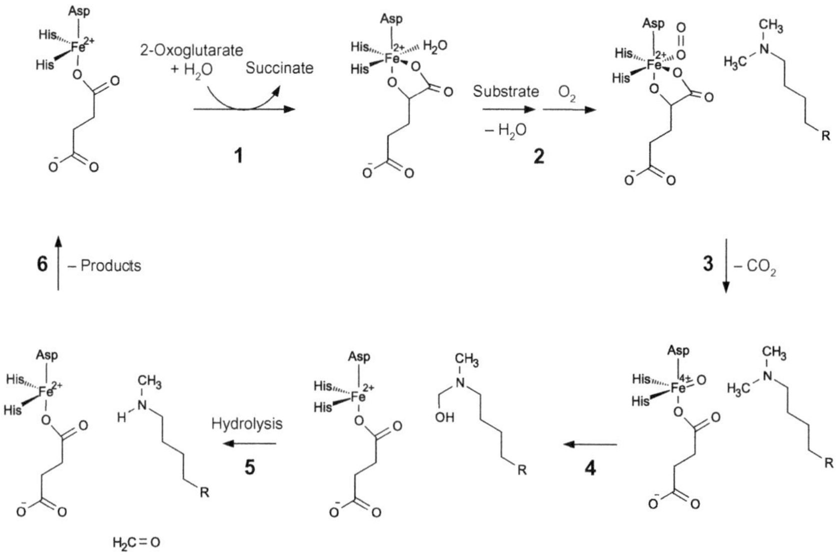

FIGURE 20.2

Lysine demethylation by the JMJC domain-containing enzymes. Catalytic mechanism of the 2-oxoglutarate-dependent hydroxylases. Step 1: 2-Oxoglutarate binds to the Fe^{2+}-bound enzyme. Step 2: Substrate binding destabilizes a Fe^{2+}-coordinated water molecule and allows binding of dioxygen. Step 3: Dioxygen reacts with 2-oxoglutarate to produce succinate, carbon dioxide, and an oxoferryl species ($Fe^{4+}=O$). Step 4: The oxoferryl species hydroxylates the methyl-lysine. Step 5: The lysyl hemiaminal is hydrolyzed to generate demethylated lysine and formaldehyde. Step 6: The demethylated lysine is released.

residues that are also conserved [49,50]. During the enzymatic reaction, Fe(II) and 2-oxoglutarate react with dioxygen to form an oxoferryl (Fe(IV)=O) intermediate that is highly reactive and hydroxylates the ζ-methyl group of the methylated lysine substrate. The resulting lysyl hemiaminal spontaneously decomposes and releases the methyl group from nitrogen as formaldehyde (Figure 20.2). Due to this enzymatic mechanism, JMJC proteins can target all three lysine methylation states (mono-, di-, and trimethylated lysines).

The human genome encodes 32 JMJC proteins, and so far, histone demethylase activity has been documented for 18 members. Figure 20.3 shows a phylogenetic tree of the JMJC proteins. In general, binding of the substrate to these enzymes is mediated by a broad hydrophobic cleft that leads to a polar binding pocket for the methylated nitrogen of the target lysine residue [51–53]. Lack of conservation of the three Fe(II)-coordinating residues in some JMJC family members indicates the absence of enzymatic activity, e.g. in the case of JARID2 (jumonji/ARID domain-containing protein2) [50]. The characterization of the JMJC proteins is a field of active research and conflicting results on the enzymatic activities of, e.g., PHF2 (PHD finger protein 2), KDM8, and JMJD6 (jumonji domain-containing protein 6) have been obtained [54–60].

The JMJC demethylases can be grouped into seven subfamilies based on the sequence of the catalytic domains. The members of a family have largely overlapping substrate specificities, which are

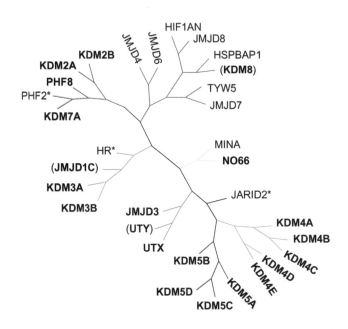

FIGURE 20.3

Phylogenetic tree of JMJC containing human proteins in the SMART database. Distinct phylogenetic branches are highlighted with colors. Proteins for which histone lysine demethylase activity has been demonstrated are indicated in bold. Proteins for which a controversial enzymatic activity has been reported are in brackets. Proteins with a noncanonical amino acid at a position involved in cofactor binding are marked with an asterisk and may be catalytically inactive. *Abbreviations*: HIF1AN, hypoxia-inducible factor 1α inhibitor; HR, hairless homolog; HSPBAP1, HSPB1 (27 kDa heat shock protein)-associated protein 1; MINA, MYC-induced nuclear antigen; NO66, nucleolar protein 66; PHF2, PHD finger protein 2; PHF8, PHD finger protein 8; TYW5, tRNA wybutosine-synthesizing protein 5; UTX, ubiquitously transcribed TPR protein on the X chromosome; UTY, ubiquitously transcribed TPR protein on the Y chromosome.

determined by amino acid variations in the catalytic center or in nearby domains [61]. Many JMJC proteins contain chromatin-binding motifs such as plant homeobox domain (PHD) fingers or Tudor domains. Interestingly, the binding of methylated histone residues proximal to the targeted lysine by these recognition units can modulate both the demethylase activity and specificity, as it was demonstrated for PHF8 (PHD finger protein 8) and KDM7A [62].

4 ROLE OF JMJC HISTONE DEMETHYLASES IN HUMAN CANCER

The current knowledge about the biological functions of the JMJC proteins is still limited. For an overview, please see a collection of excellent recent reviews [63–65]. Mouse models with constitutive and conditional deletions of several JMJC family genes have been generated. The resulting phenotypes underscore the importance of the JMJC enzymes for not only stem cell maintenance but also early organismal

Table 20.1 JMJC Histone Demethylases and Associated Knockout Phenotypes

Histone demethylase	Other names	Histone substrates	Phenotype
KDM2A	FBXL11, JHDM1A	H3K36me1, H3K36me2	–
KDM2B	FBXL10, JHDM1B	H3K36me, H3K36me2 (H3K4me3)	Partial peri- or postnatal lethality, neural tube closure defects, exencephaly, and reduced sperm count [66]
KDM3A	JMJD1A, JHDM2A, TSGA	H3K9me1, H3K9me2	Male infertility and adult-onset obesity [67–70]
KDM3B	JMJD1B	H3K9me1, H3K9me2	–
KDM4A	JMJD2A, JHDM3A	H3K9me2, H3K9me3 H3K36me2, H3K36me3 H1.4K26me2, H1.4K26me3	Altered response to cardiac stress [71]
KDM4B	JMJD2B, JHDM3B	H3K9me2, H3K9me3 H3K36me2, H3K36me3 H1.4K26me2, H1.4K26me3	No gross abnormality and delayed mammary gland development when targeted in breast epithelium [72]
KDM4C	JMJD2C, JHDM3C, GASC1	H3K9me2, H3K9me3 H3K36me2, H3K36me3 H1.4K26me2, H1.4K26me3	No gross abnormality [73]
KDM4D	JMJD2D, JHDM3D	H3K9me3	No obvious phenotype [74]
KDM4E	KDM4DL, JMJD2E	H3K9me3	–
KDM5A	KDM5A, JARID1A, RBP2	H3K4me2, H3K4me3	Aberrant behavior when held by the tail and hematological abnormalities [75]
KDM5B	JARID1B, PLU1	H3K4me2, H3K4me3	Embryonic lethality before E7.5 [76] Neonatal lethality due to respiratory failure [77]
KDM5C	JARID1C, SMCX	H3K4me2, H3K4me3	Neurulation and cardiac looping defects [78]
KDM5D	JARID1D, SMCY	H3K4me2, H3K4me3	–
JMJD3	KDM6B	H3K27me2, H3K27me3	Perinatal lethality, premature lung development and respiratory failure [79]
UTX	KDM6A	H3K27me2, H3K27me3	Neural tube defects at E9.5 [78]. Female embryonic lethality at E10.5, partial male embryonic lethality, defects in cardiac development, increased tumor formation [80]
UTY	KDM6C	(H3K27me2, H3K27me3)	Males viable, no phenotype. UTY/UTX negative males phenocopy UTX negative females [81]
JMJD1C	–	(H3K9me1, H3K9me2)	–
JMJD4	–		–
KDM8	JMJD5	(H3K36me2)	Embryonic lethality between E6.5 and E11.5 [82]
JMJD6	–	(H3R2) (H4R3)	–

(Continued)

Table 20.1 (Continued)

Histone demethylase	Other names	Histone substrates	Phenotype
JMJD7	–	–	–
JMJD8	–	–	–
KDM7A	KIAA1718, JHDM1D	H3K9me1, H3K9me2 H3K27me1, H3K27me2	–
PHF2	JHDM1E	(H3K9me2)	Partial neonatal death, growth retardation, less adipose tissue, and reduced adipocyte numbers [83]
PHF8	JHDM1F	H3K9me1, H3K9me2 (H4K20me1)	–
JARID2	–	–	Embryonic lethality between EE11.5 and E15.5 [84]
HIF1AN	FIH1	–	Viable, elevated metabolic rate, increased insulin sensitivity, and decreased adiposity [85]
HSPBAP1	PASS1	–	–
HR	–	(H3K9me1, H3K9me2)	Hair loss, skin defects [86]
NO66	C14orf169	H3K4me2, H3K4me3	–
MINA	–	–	–
TYW5	C2orf60	–	–

Histone substrates in brackets are controversial and have not been confirmed by laboratories.

development and tissue specification. Table 20.1 provides an overview of phenotypes observed in knockout studies. However, compensatory effects by other subfamily members are likely to occur in some cases and may explain weak or absent phenotypes. For other demethylases, investigators obtained divergent knockout phenotypes, which are believed to result from the use of different mouse strains. In addition, these studies do not address the relevance of the enzymatic activity of JMJC demethylases versus their scaffolding and recruitment functions in larger chromatin-associated complexes. For example, the H3K36 demethylase FBXL10 (F-box and leucine-rich repeat protein 10)/KDM2B recruits the PRC1 complex to CpG islands in the promoters of lineage-specific genes. Thereby it mediates monoubiquitylation of lysine 119 of histone H2A, which is required for stable transcriptional repression [87–90]. The demethylase activity of FBXL10 is not required for this function. This example illustrates the importance of "rescue experiments" in the context of knockdown or knockout approaches. Reconstitution with catalytically inactive mutants at comparable expression levels to the endogenous proteins is necessary to gain insight into the relevance of enzymatic activity. Hence, small molecule inhibitors that specifically target single JMJC demethylases or at least subfamilies would serve as valuable research tools to complement these studies.

Tumor suppressor genes are frequently target of recurrent inactivating mutations, copy number loss or epigenetic silencing in cancer cells, whereas oncogenes can be activated by mutations or gene rearrangements as well as by amplification/polysomy or overexpression. Recurrent mutations are well documented for the gene encoding ubiquitously transcribed TPR protein on the X chromosome (*UTX/KDM6A*). UTX demethylase activity targets H3K27me2 and H3K27me3 methylations. Consequently, the inactivation of

UTX disturbs the regulation of these repressive histone marks, which are generated by the methyltransferase EZH2. Inactivating somatic mutations of *UTX* occur in multiple myeloma, esophageal cancers, renal cell carcinomas, breast cancer, colorectal cancer, prostate cancer, glioblastomas, and medulloblastomas [91–95]. Loss of UTX leads to a gain of H3K27me3 as it is also observed upon enhanced activity of EZH2 in a variety of human tumors such as prostate cancers and B-cell lymphomas [5,6,96,97]. Germline mutations of *UTX* are associated with Kabuki syndrome, a rare congenital abnormality syndrome that confers an increased risk of cancer development (reviewed in [98]). In addition to *UTX*, inactivating mutations in the gene encoding the H3K4me2 and H3K4me3 demethylase JARID1C/KDM5C were identified in a large-scale sequencing approach of renal cell carcinoma tumors [91]. KDM5C plays a complex role in gene regulation. A recent paper describes that at gene enhancer elements KDM5C helps to maintain H3K4 monomethylation and thereby stimulates enhancer activity and transcription. However, at transcriptional start sites, KDM5C restricts the transcriptional output through removal of the (in this context) activating H3K4me3 mark [99]. An important role of *KDM5C* in kidney cancer is supported by the finding that it is a target gene of the hypoxia-inducible factor (HIF)/von Hippel Lindau tumor suppressor protein (pVHL) pathway. A study using RNA interference against KDM5C in *VHL*-deficient cells showed accelerated tumor growth in xenograft studies, suggesting that the inactivation of both *VHL* and *KDM5C* promotes carcinogenesis in this setting [100]. As *UTX*, *KDM5C* is encoded on the X chromosome, and germline mutations of *KDM5C* are causing X-linked mental retardation in males [101–103]. An essential function of KDM5C for neuronal survival and dendritic outgrowth is supported by several studies that report increased expression levels of *KDM5C* transcripts in normal fetal brain tissue [101] and neuronal cell death and reduced dendrites upon knockdown [104].

There is no strong evidence for the existence of recurrent activating mutations or rearrangements of genes encoding JMJC demethylases in cancer so far. However, gene amplifications and overexpression phenotypes have been reported for several subfamilies. A prominent example are the KDM4/JMJD2 enzymes that remove in context-dependent way the repressive histone marks H3K9me2 and H3K9me3 (trimethylated lysine 9 on histone H3) as well as H3K36me2 and H3K36me3. Mouse studies have provided genetic evidence for a link between the loss of H3K9me3 and the development of cancer. Double knockout mice for the histone methyltransferases Suv39h1 and Suv39h2 that catalyze H3K9 trimethylation display genomic instability and an increased risk of late onset B-cell lymphomas [105]. In a mouse model for T-cell lymphoma, the ablation of Suv39h1 and concomitant lower levels of H3K9me3 accelerated carcinogenesis [106]. Given these results, a current hypothesis proposes that a pathologically increased activity of the KDM4 demethylases may disturb the balance of H3K9me3 with similar consequences. Along these lines the locus encoding KDM4C/JMJD2C was identified as a gene amplified in esophageal squamous cell carcinoma (GASC1) [107], and high expression levels of KDM4C/JMJD2C/GASC1 correlate with poor patient prognosis [108]. Amplifications of KDM4C have also been reported for breast cancer and medulloblastoma [109,110]. Moreover, functional studies have shown that KDM4C is required for proliferation in cancer cell lines derived from squamous cell carcinoma, prostate carcinoma, breast cancer, and diffuse large B-cell lymphoma among others [43,110–112]. The closely related enzyme JMJD2A/KDM4A is overexpressed in gastric cancer [113]. Expression levels of KDM4A correlate with tumor stage and nodal status and predict poor survival. The authors propose that KDM4A negatively regulates apoptosis in gastric cancer cells. Overexpression of *KDM4A* was also found in ovarian cancer [114], breast cancer [115], and squamous cell carcinoma in which levels were associated with lymph node metastasis [116]. Interestingly, elevated expression of KDM4A leads to a rereplication phenotype with a gain of specific chromosomal regions in ovarian cancer cells [114]. However, this might be a tissue-specific effect and it should be mentioned that *KDM4A* was also reported to be downregulated in bladder cancer [35]. The third KDM4 subgroup member *KDM4B/JMJD2B* was found to be amplified in malignant peripheral nerve sheath tumors [117], and increased expression levels were detected in breast cancer [118] and colorectal cancer [119]. *JMJD2B* is regulated by hypoxia, and knockdown approaches have shown that depletion of the protein reduces the growth rates of multiple cancer cell lines [72,120] or in rare occasions can trigger apoptosis [121]. Based on these findings, the inhibition of the enzymatic activity of the KDM4 subfamily could have therapeutic potential in human cancer treatment.

The KDM5/JARID family targets H3K4me2 and H3K4me3 for demethylation. Genome-wide studies on the distribution of H3K4me3 and H3K4me2 indicate that these marks peak at promoters and transcriptional start sites at the 5'-end of genes. In addition, H3K4me3 overlaps with H3K9 and H3K14 acetylation and its abundance correlates with transcriptional output [122–124]. Given the correlation of H3K4me3 with transcriptional activity, removal of this modification by histone demethylases is believed to trigger transcriptional repression, and there is growing evidence that the catalytic activities of the KDM5 family members are involved [99,125,126].

KDM5A/JARID1A was initially identified as a binding partner of the tumor suppressor retinoblastoma protein (pRB) [127]. The demethylase is involved in the control of cell proliferation and differentiation exerted by the pRB pathway as demonstrated both in human tumor cells and in mouse embryo fibroblasts (MEFs). Tumor suppression by pRB may be partially mediated through its ability to sequester KDM5A, thereby directly inhibiting an oncogenic function [128]. Accordingly, the catalytic activity of Kdm5a suppresses hallmarks of senescence in MEFs and is required for their proliferation *in vitro* [129]. A recent mouse model supports a role of *Kdm5a* in cancer: the ablation of *Kdm5a* in either $Rb^{+/-}$ mice or in mice with a targeted deletion of the tumor suppressor *Menin1* delayed tumor development and decreased tumor burden [129]. Interestingly, tumors that eventually arose in the $Rb^{+/-}$ background had elevated levels of *Kdm5b* mRNA, indicating a compensatory function of the closely related protein during tumor onset or progression. In human cancer, several tumor types show elevated expression levels of KDM5A. Liang et al. described an upregulation of KDM5A in hepatocellular carcinoma (HCC) [130]. In HCC cell lines, KDM5A suppresses the expression of the cyclin-dependent kinase inhibitors p21Cip1 and p27Kip1, and siRNA knockdown of KDM5A induces markers of cellular senescence. A similar mechanism was proposed for gastric cancer [131], in which the majority of investigated tumor samples displayed an upregulation of KDM5A. An increased expression of KDM5A was also found in human lung cancer biopsies [132], and a role for KDM5A in epithelial–mesenchymal transition in nonsmall cell lung cancer cells has recently been described [133]. For human breast cancer, contradictory results were reported. Whereas a meta-analysis of available expression data yielded a positive correlation of *KDM5A* mRNA levels with absence of recurrence after mastectomy and a positive response to docetaxel [134], a second study described a correlation between high levels of *KDM5A* mRNA and breast tumor metastasis [135]. Moreover, increased resistance to EGFR inhibitors has been documented for breast cancer cell lines with *KDM5A* gene amplifications and increased protein expression [136]. Similarly, the development of *in vitro* tolerance to EGFR inhibitors and cisplatin in a lung cancer cell line required *KDM5A* expression and was accompanied by an increase of KDM5A protein and a global reduction of cellular H3K4me3 levels [137].

KDM5B was initially identified as a gene upregulated by the tyrosine kinase HER2 in breast cancer cells [138]. A subsequent study verified an increased expression of *KDM5B* in human primary breast cancer samples compared to normal tissue and in human breast cancer-derived cell lines [139]. A recent study identified *KDM5B* as an oncogene in luminal breast cancer and an association with worse patient prognosis [140]. RNA interference targeting KDM5B in breast cancer cells induces tumor suppressor genes, reduces cell proliferation, and inhibits anchorage independent growth in soft agar [126]. In *in vivo* models, the depletion of KDM5B strongly reduces tumor growth as demonstrated for a syngeneic mouse mammary tumor model [126] and in xenografts using human breast cancer cells [76].

KDM5B is also upregulated in bladder cancer and lung cancer [141,142] and it is required for proliferation and survival of cancer cell lines of this origin. Similar results were obtained in colorectal tumors and derived cell lines thereof [143]. Moreover, KDM5B expression increases during the progression of prostate cancer and promotes gene activation by the androgen receptor via an unknown mechanism [144]. Roesch et al. suggest that KDM5B regulates stem cell-like properties in melanoma cells [145]. Treatment of cultured melanoma cells with cytotoxic agents or with the BRAF inhibitor vemurafenib leads to an enrichment of a drug-resistant subpopulation characterized by high expression of KDM5B [146] and elevated levels of KDM5B were also found in patients' melanomas that relapsed under vemurafenib treatment. In conclusion, there is emerging evidence for elevated expression levels of KDM5A and KDM5B and important phenotypic consequences in various types of human cancer.

The KDM2 family comprises only two members, KDM2A/FBXL11 and KDM2B/FBXL10, that target H3K36me2 and H3K36me1. H3K36me2 is present in actively transcribed chromatin regions, especially in gene bodies, and numerous studies support a function of H3K36 methylation in gene activation and control of transcription initiation. Whereas a connection between KDM2A and human cancer is only beginning to emerge, such as the fact that it is overexpressed in some lung cancers [147], there are some interesting findings on KDM2B/FBXL10. The catalytic activity of KDM2B is required to inhibit premature senescence in MEFs [148,149] and ectopic expression of KDM2B immortalizes these cells. *KDM2B* expression is elevated in human leukemias and pancreatic ductal adenocarcinomas [150,151] and it is essential for cell proliferation. In triple negative breast cancer, Kottakis et al. found elevated KDM2B levels and a correlation with relapse [152]. In addition, KDM2B enzymatic activity is required in a mouse model of AML driven by the oncogenes HoxA9/Meis1 [150], and human AML-derived cells depend on KDM2B expression for proliferation, as demonstrated by shRNA experiments [153]. These findings indicate that besides DOT1L (see above), KDM2B might be the second epigenetic enzyme to play an important role in AML cells. However, KDM2B also exerts functions that are independent of its histone demethylase activity. It also recruits Polycomb group proteins to a subset of their target genes [87–90]. Thereby KDM2B both maintains embryonic stem cells and ensures proper expression of homeotic genes and normal vertebral development [90]. To what extent a deregulation of Polycomb targets by overexpression of KDM2B contributes to human cancer is currently investigated [151,152], and a specific inhibitor targeting the demethylase activity of KDM2 would surely help to address this question, and might bear great therapeutic potential.

In addition to KDM4, KDM5, and KDM2 family members, further JMJC demethylases are overexpressed in cancer. The Myc target gene MINA/Mina53 is induced in several human malignancies, such as colon cancer [154], esophageal cancer (here Mina expression levels correlated to shorter survival [155]), renal cell carcinoma [156], breast carcinoma [157], and lung cancer [158,159]. Although knockdown of MINA causes an antiproliferative effect in several of these studies, no enzymatic activity has been detected so far, which complicates a targeted approach. A similar scenario exists for JMJD1C. This gene was identified as essential in a shRNA screening approach in a mouse model for human AML driven by the MLL-AF9 fusion gene, and as required in human AML-derived cell lines to suppress apoptosis [160]. An initial publication describes H3K9me1 and H3K9me2 demethylase activity of JMJD1C [161], but these results await confirmation. However, JMJD1C might target nonhistone substrates instead [162]. Please see Table 20.2 for an overview of additional JMJC demethylases with a link to cancer.

Table 20.2 JMJC Histone Demethylases in Human Cancer

Histone demethylase	Association with cancer
KDM2A	Overexpressed in lung cancer [147,163]
KDM2B	Overexpressed in leukemias [150], bladder carcinoma [164], pancreatic cancer [151], and breast cancer [152]
KDM3A	Overexpressed in malignant colorectal cancer [165], metastasized prostate carcinoma [166], renal cell carcinoma [167], and hepatocellular carcinoma [168]
KDM4A	Overexpressed in breast cancer [115], squamous cell carcinoma [116], ovarian cancer [114], and gastric cancer [113] Silenced or downregulated in bladder cancer [35]
KDM4B	Amplified in malignant peripheral nerve sheath tumor [117] Overexpressed breast [118] and colorectal cancer [119]
KDM4C	Amplified in esophageal cancer [107,108], breast cancer [110], and medulloblastoma [109] Translocated in lymphoma [169]
KDM5A	Silenced or deleted in melanoma [170] Translocated in acute leukemia [171,172] Overexpressed in hepatocellular carcinoma [130], gastric cancer [131,173], and lung cancer [132,133]
KDM5B	Amplified and/or overexpressed in breast cancer [138,139,174] Overexpressed in bladder cancer [141,142], lung cancer [141], colorectal cancer [143], prostate cancer [144], and malignant melanoma [146]
KDM5C	Mutated (inactivated) in renal cell carcinoma [91]
KDM5D	Deleted in prostate cancer [175]
JMJD3	Overexpressed in various cancers including lung and liver carcinomas and several hematological malignancies [176–178] and myelodysplastic syndrome [179] Overexpressed in primary Hodgkin's lymphoma [180]
UTX	Mutated in multiple tumor types including multiple myeloma, esophageal squamous cell carcinoma [95], renal clear cell carcinoma [91,95], transitional cell carcinoma [181], and chronic myelomonocytic leukemia [182] Overexpressed in breast cancer [115]
JMJD1C	Target of MLL–AF4 in acute leukemia [160] Mutated in intracranial germ cell tumors [183]
JMJD5	Upregulated in breast cancer [184] Downregulated in lung cancer [185]
PHF2	Mutated, silenced, or downregulated in breast carcinoma [186] as well as head and neck squamous cell carcinoma [187] and colon cancer [188] Overexpressed in esophageal cancer [189]
PHF8	Overexpressed in prostate [190] and lung cancer [191] Involved in retinoic acid response in leukemia [192]
NO66	Overexpressed in nonsmall-cell lung cancer [159]
MINA	Overexpressed in colon cancer [154], esophageal cancer [155], renal cell carcinoma [156], breast carcinoma [157], and lung cancer [159]

5 SMALL COMPOUND INHIBITORS OF JMJC HISTONE DEMETHYLASES

As illustrated by the examples given above, JMJC histone demethylases are considered interesting putative drug targets. This is based on many findings that document increased expression levels in cancer and encouraging results in functional assays, which rely primarily on suppression or ablation of protein expression. The relevance of the enzymatic activity remains to be investigated in many cases, but the well-defined active side cavity of the demethylases provides as structural feature to support the development of high affinity and selective small compound inhibitors. However, research in this direction is in early stages, and of the JMJC demethylase inhibitors reported so far, only very few have sufficient potency and selectivity to serve as leads for drug development. In addition, it is currently unclear if overexpression of the demethylases in cancer is a sufficient criterion to identify patients that could benefit from treatment with an inhibitor. We are currently lacking reliable predictive biomarkers for patient stratification. These could be gene expression signatures in cancer cells or specific rearrangements or mutations in other cancer-relevant genes that would induce a dependency on histone demethylase activity. Examples for the latter are emerging for histone methyltransferases, such as the already mentioned MLL rearrangement that render leukemic cells dependent on DOT1L activity [18,19,193]. Another example are mutations of *SMARCB1* (which encodes a subunit of a histone-remodeling complex) that predict sensitivity to inhibition of EZH2 in malignant rhabdoid tumors [194].

Historically, the first small molecules that were found to inhibit JMJC enzymes belong to a class of general inhibitors of 2-oxoglutarate oxygenases. These compounds were used to target the HIF prolyl-hydroxylases (PHD or EGLN enzymes, which regulate the stability of the transcription factor HIF by posttranslational prolyl-hydroxylation [195]) but also inhibit the JMJC enzymes. Compounds such as *N*-oxalylglycine (NOG, **3**, Figure 20.4A), one representative member of this class, compete with the 2-oxoglutarate (2OG) cofactor and coordinate to the active site ferrous iron. This holds true for intermediates of the tricarboxylic acid (TCA) cycle such as succinate and fumarate (**4** and **5**, Figure 20.4A), as well as 2-hydroxyglutarate (2HG, **6**, Figure 20.4A). Accumulation of 2HG can result from point mutations of isocitrate dehydrogenase (IDH) type 1 and type 2 enzymes, which are found in several human tumors. These mutations create neomorphic enzymatic activity allowing the conversion of 2OG to 2HG, which competitively inhibits the activity of multiple αKG-dependent dioxygenases, among them the JMJC enzymes, though rather weakly with IC_{50} values in the micromolar to millimolar range [196,198]. In fact, most of the reported JMJC inhibitors to date are metal chelators and 2OG competitors.

An increasing amount of structural information on the 2OG dioxygenases has become available. Among others, the crystal structures of the catalytic domains of the disease-associated demethylases KDM4A, KDM4B, KDM4D, UTX, KDM2A as well as PHD2 and PHF8 have been solved [51,53,54,61,199–203]. These structures indicate both a substantial general conservation in the Fe(II) and 2OG binding sites, but also subfamily-specific differences that can be exploited for the development of specific inhibitors. The variations in the substrate-binding elements between the different JMJC enzyme subfamilies offer additional advantages. However, our knowledge on conformational changes during catalysis is limited, and recent studies with inhibitors indicate compound-induced movements of the metal within the active site in the range of more than one angstrom [204]. Many of the reported crystal structures actually use the above-mentioned NOG as a stable 2OG analogue. They revealed that the active site metal ion is bound by NOG via its C-2 oxo and C-1 carboxylate groups, whereas its glycine carboxylate occupies the same position as the C-5 carboxylate of 2OG. Several derivatives of NOG have been designed, some of them with substituents that mimic the interaction

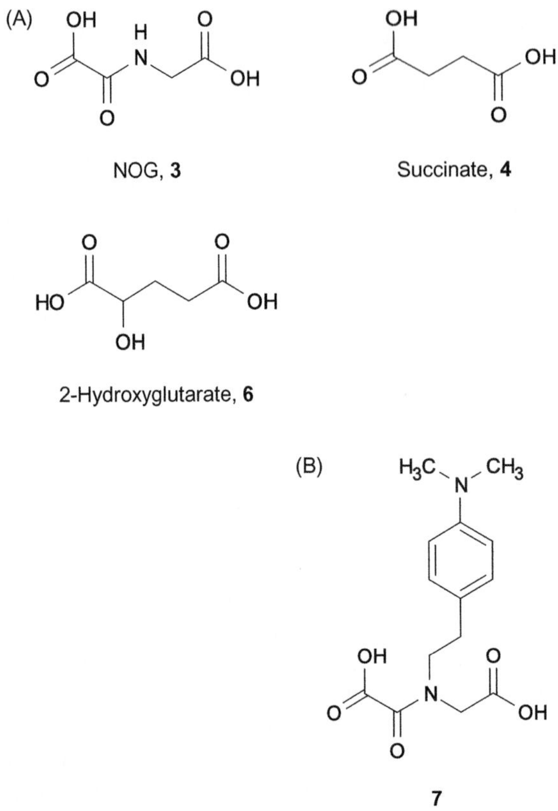

NOG, 3

Succinate, 4

Fumarate, 5

2-Hydroxyglutarate, 6

7

FIGURE 20.4

Representative 2-oxoglutarate mimetics and TCA cycle intermediates that inhibit JMJC demethylases. (A) Structure of the 2-oxoglutarate mimetic N-oxalylglycine (NOG, 3), a close isostere of 2OG that competes for binding to the catalytic center. Most crystal structures of JMJC proteins have been obtained with bound NOG. The TCA cycle intermediates succinate 4 and fumarate 5 are weak inhibitors of JMJC proteins. Another weak inhibitor is 2-hydroxyglutarate (2HG, 6), which is produced by mutant IDH enzymes in gliomas or myeloid leukemias. Elevated levels of TCA intermediates or 2HG may change histone methylation levels in cancer cells [196,197]. (B) One representative compound 7 of a class of NOG derivatives substituted with an alkyl-linked dimethylaniline group. These compounds occupy the 2OG and peptide substrate binding sites of JMJC enzymes.

of the methylated peptide with the enzyme (a representative compound (7) is shown in Figure 20.4B) [205]. These compounds inhibit KDM4 enzymes in enzymatic assays with a potency in the millimolar concentration range.

The observation that the histone deacetylase inhibitor suberanilohydroxamic acid (SAHA, **8**, marketed as Vorinostat, Figure 20.5) exerted an inhibitory effect on JMJC demethylases gave rise to a series of hydroxamic acids and resulted in the discovery of Methylstat (**9**, Figure 20.5). Methylstat inhibits

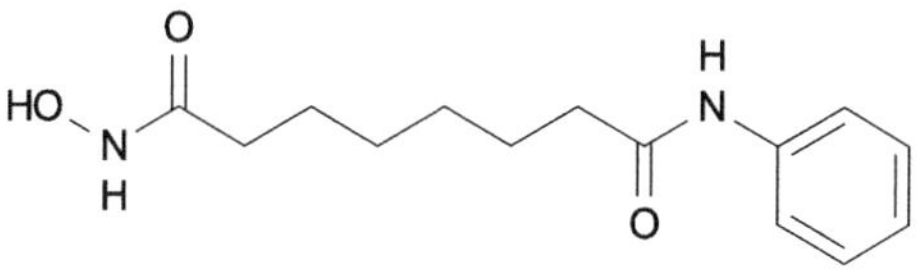

SAHA, **8**

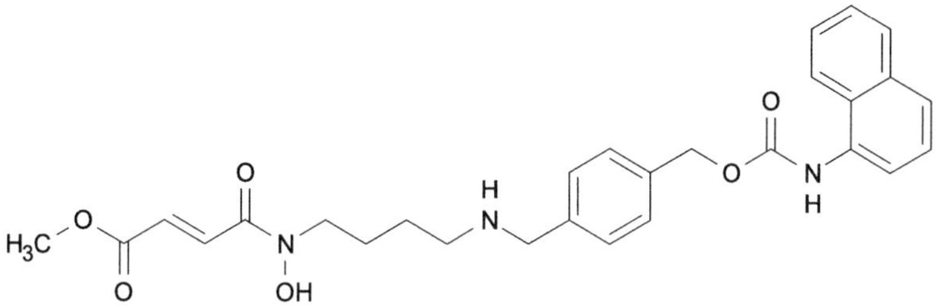

Methylstat, **9**

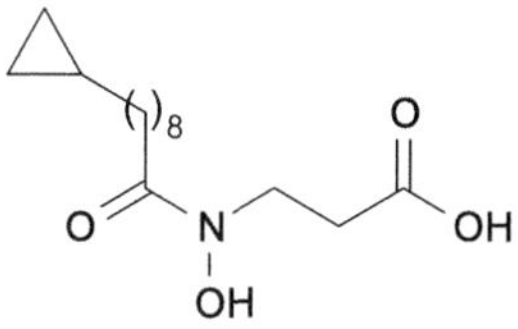

10

FIGURE 20.5

Structures of representative hydroxamic acid-derived inhibitors. Hydroxamic acids are metal chelators and can inhibit JMJC demethylase with potencies in the micromolar range. The observation that the clinically used histone deacetylase (HDAC) inhibitor SAHA **8** can inhibit KDM4E in enzymatic assays gave rise to the development of several derivatives such as Methylstat **9**. The cyclopropane ring of compound **10** confers selectivity for the KDM2/7 enzymes over the KDM4 subfamily.

KDM4C in the micromolar range and has cellular activity in methyl ester form [206]. A subsequent report by Suzuki et al. describes a hydroxamate with specificity for the KDM2 and PHF subfamilies (**10**, Figure 20.5) that shows a micromolar potency on PHF8 enzymatic activity and inhibits the proliferation of cancer cell lines at a concentration of 40 μM [207].

A high-throughput screening approach using a biochemical assay identified 8-hydroxyquinolines as pan-specific low micromolar JMJC demethylase inhibitors [208]. These molecules act as iron chelators via their quinoline nitrogen and phenyl oxygen atoms and the most potent compounds such as 5-carboxy-8-hydroxyquinoline (IOX1, **12**, Figure 20.6A) display half-maximal inhibitory concentrations (IC$_{50}$) of 200 nM in enzymatic assays. An 8-hydroxyquinoline derivative (**13**, Figure 20.6A) can block reactivation of herpes simplex virus via inhibition of KDM4 in a mouse model [209]. Another

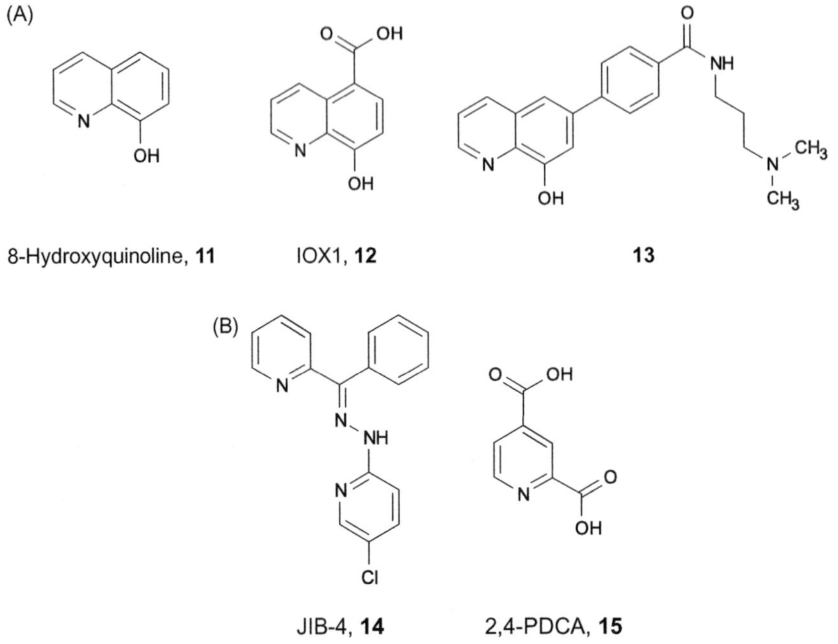

(A)

8-Hydroxyquinoline, **11**   IOX1, **12**   **13**

(B)

JIB-4, **14**   2,4-PDCA, **15**

FIGURE 20.6

8-Hydroxyquinoline and pyridine derivatives. (A) 8-Hydroxyquinoline **11** and derivatives chelate iron via their pyridyl nitrogen and phenolic oxygen atoms. IOX1 **12** was identified in a high-throughput screen against recombinant KDM4E is a broad spectrum 2OG oxygenase inhibitor. Substitutions such as in compound **13** have improved selectivity for KDMs. (B) 2,4-PDCA **15** is another metal chelating unselective 2OG oxygenase inhibitor. The carboxyl groups are not required for inhibition of JMJC proteins by pyridines, and the pyridylhydrazone derivative JIB-4 **14** was identified in a high-throughput cellular screen. JIB-4 is active against the KDM4, KDM5, and KDM6/UTX families of JMJC demethylases.

pan-specific JMJC inhibitor was identified in a cell-based screening exercise that employed a stably integrated reporter gene. This reporter was silenced in cells and its reactivation was used as readout for active compounds [210]. The molecule, termed JIB-4 (**14**, Figure 20.6C), belongs to the class of pyridine hydrazones and inhibits KDM4, KDM6, and KDM5 enzymes at submicromolar to micromolar concentrations. It is cell permeable, diminished tumor growth, and prolonged survival in xenograft cancer models. Although chelation of iron and competition with 2OG have been demonstrated for other pyridine derivatives such as pyridine-2,4-dicarboxylic acid (2,4-PDCA, **15**, Figure 20.6D, a broad spectrum 2OG oxygenase inhibitor), the exact mode of action of JIB-4 has not been resolved and might involve iron binding and a mixed competition with the histone peptide and 2OG [210].

A variety of other compounds have been reported to inhibit the JMJC demethylases, among those natural products such as catechols and flavonoids, the agrochemical daminozide [211], and 2-4(4-methylphenyl)-1,2-benzisothiazol-3(2H)-one (PBIT, **16**, Figure 20.7A) as a micromolar inhibitor of KDM5B that suppresses KDM5B-mediated demethylation of H3K4me3 and breast cancer cell proliferation [212]. However, the fact that the PBIT analogue ebselen (**17**, Figure 20.7A), a zinc rejecting

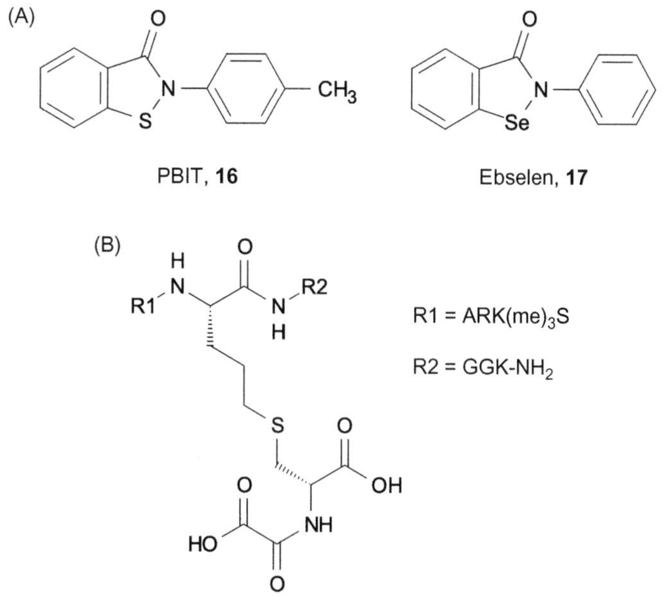

(A)

PBIT, **16**   Ebselen, **17**

(B)

R1 = ARK(me)$_3$S

R2 = GGK-NH$_2$

18

FIGURE 20.7

Compounds with unknown mechanism of action and peptide-derived inhibitors. (A) PBIT **16** and ebselen **17** inhibit both the KDM4 and KDM5 demethylase subfamilies. The mechanism of action of these compounds is unknown, however, it is possible that the ejection of a structural zinc ion underlies their activity. (B) An example of a histone-substrate competitive inhibitor that uses both a histone H3 fragment and an iron-chelating group (*N*-oxalyl-D-cysteine). Compound **18** displays a more than 300-fold selectivity for KDM4 over KDM2, KDM3, and KDM6/UTX family members.

compound [213], also inhibits KDM5B indicates the need for additional research addressing the mechanism of inhibition. As most compounds show only limited specificity between subfamilies of demethylases, some research groups have started a peptidomimetic approach for inhibitor design. Lohse and co-workers obtained proof of principle with a compound with fourfold selectivity for KDM4C over KDM4A, albeit with low potency in enzymatic assays [214]. A similar approach by Woon et al. who used a histone peptide (H3(7–14)K9me3) coupled with *N*-oxalyl-D-cysteine resulted in a KDM4A-specific inhibitor (IC$_{50}$ of 270 nM) with only little inhibition of other JMJC demethylases (**18**, Figure 20.7B) [215]. A combination of structural information on both the substrate and cofactor binding by the JMJC demethylases might provide an interesting future avenue for inhibitor design. Please refer to recent excellent reviews for a broader overview on the different inhibitors that have been reported [216,217].

Given the general problems to combine high selectivity and potency with cell permeability, the compounds described so far are unlikely to be useful as leads for pharmaceutical development. In contrast, a highly encouraging lead compound is GSK-J1 and its cell-active ethyl ester prodrug GSK-J4, an inhibitor of the KDM6/JMJD3 subfamily that demethylates the H3K27me3 mark (**19** and **20**,

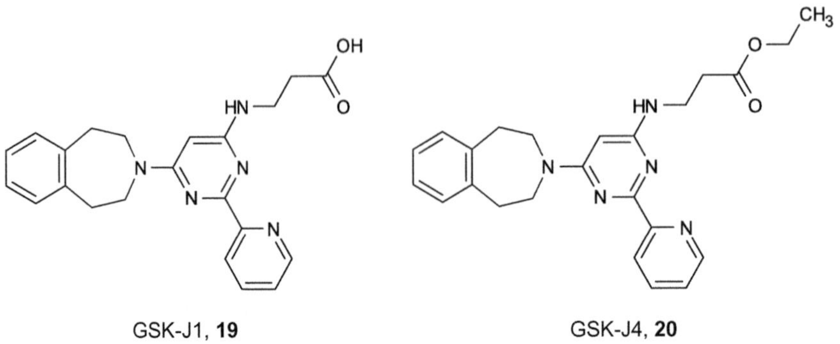

FIGURE 20.8

Structures of the reported KDM6 inhibitor GSK-J1 **19** and its ethyl ester prodrug GSK-J4 **20** which is active in cells. GSK-J1 is an iron chelator and was shown to induce metal movement within the active site of KDM6B. It inhibits KDM6B with an IC_{50} of 60 nM and shows up to 10-fold selectivity over KDM5B and above 1000-fold selectivity over KDM2, KDM3, and KDM4 demethylases.

Figure 20.8) [204]. GSK-J1 is an active site binder, which chelates Fe(II) and competes with 2OG and very potently inhibits human KDM6 demethylases with an IC_{50} of about 60 nM. As mentioned above, structural analysis indicates that GSK-J1 induces a movement in the active site metal and that one of the interactions with a residue of the facial triade is made through a water molecule. This flexibility is an encouraging finding since it may allow for a wider set of inhibitor scaffolds. GSK-J1 is more than 100-fold selective over the KDM4 subfamily, and thermal shift analysis of further JMJC demethylases with GSK-J1 showed only little binding compared to JMJD3. However, it has recently been reported that GSK-J1 also inhibits KDM5B and KDM5C, although it is 5–10 times less potent on KDM5 enzymes compared to KDM6 [218]. The prodrug GSK-J4 reduced lipopolysaccharide-induced production of proinflammatory cytokines by human macrophages at concentrations below 10 nM and confirmed results obtained with RNAi-mediated knockdown of KDM6A and KDM6B [204], providing evidence that the catalytic activity of the KDM6 demethylases is crucial for their role in inflammation.

The biotech company EpiTherapeutics has developed a series of highly potent and selective KDM5 inhibitors. The most potent compounds show subnanomolar potencies in enzymatic assays and are at least 20-fold selective for KDM5B over the KDM4 demethylases, more than 100-fold selective over KDM2, and more than 1000-fold selective over KDM1, KDM3, KDM6, and KDM7 subfamily members. In cells, these compounds inhibit H3K4me3 and H3K4me2 demethylation at single digit nanomolar concentrations and show antiproliferative potency at concentrations below 1 μM in different cancer cell lines. Moreover, they inhibit the onset of drug tolerance to kinase inhibitors in several cellular models, such as nonsmall cell lung cancer cell lines exposed to EGFR kinase and MEK inhibitors. Similar synergistic effects were found in B-Raf mutated melanoma cell lines treated with B-Raf inhibitors. Although the underlying mechanisms are incompletely understood, a causal role of KDM5A in the onset of drug tolerance has recently been demonstrated [137]. The compounds are well tolerated in animals, and in several mouse xenograft models of multiple myeloma, a dose-dependent reduction of tumor growth was demonstrated (P. Staller, unpublished results).

6 CONCLUSION

Extensive research efforts have demonstrated that histone demethylases can be targeted by drug-like small molecules to inhibit their enzymatic activity. These inhibitors will enable us to achieve a better understanding of the biological functions of the demethylases and their roles in human disease. In addition and most importantly, they may offer a therapeutic perspective. In an ideal scenario, targeting of epigenetic factors (e.g., histone methyltransferases and demethylases) that influence cancer cell identity would be complementary to more common therapeutic approaches such as the induction of apoptosis and the inhibition of cell proliferation.

Our currently limited knowledge about the cellular functions of the histone demethylases remains a major challenge in their validation as targets for cancer therapy. Studies of knockout mice have revealed complex phenotypes that are difficult to interpret. The enzymes are involved in multiple biological processes and functional compensation by closely related family members seems to occur. These results raise the question as to whether pan-family versus specific inhibitors might be advantageous for therapy, perhaps on the expense of an increased risk for side effects. In addition, our understanding of the noncatalytic domains of the demethylases is limited. These domains are involved in substrate recognition and specificity, target recruitment and interaction with other epigenetic factors. Many JMJC proteins contain chromatin or DNA recognition modules such as chromodomains, Tudor domains, PHD fingers, CXXC motifs as well as protein–protein interaction domains. These might serve as targets for inhibitors in addition to the catalytic domains in the future. A first step in this direction is the discovery of small molecule inhibitors that target the interaction between H3K4me3 and the C-terminal PHD finger of KDM5A by Wagner et al. [219]. In rare cases in AML, the C-terminal PHD finger of KDM5A is fused to the nucleoporin NUP98 due to a chromosomal translocation event and recruits the oncogenic NUP98/KDM5A fusion protein to chromatin. NUP98/KDM5A prevents the differentiation-associated removal of H3K4me3 at loci encoding lineage-specific transcription factors and maintains their inappropriate activity, which ultimately blocks cellular differentiation [171]. In addition to inhibitors that target chromatin binding, other concepts such as agents that promote the activity of demethylases with tumor suppressor function or that target allosteric interactions may be conceivable in future, albeit challenging.

The complex biology of histone demethylation was also highlighted in recent study that used an affinity-tag purification/mass spectrometry approach to establish a protein–protein interaction map of chromatin-associated factors including 25 JMJC proteins, KDM1A and 41 lysine methyltransferases. This approach confirmed existing data on KDM2B, whose major role in development is believed to be the recruitment of polycomb complexes to CpG island-rich promoters, a function that does not require KDM2B demethylase activity. However, whether polycomb recruitment or the enzymatic activity of KDM2B is more relevant in the context of cancer cell immortalization and proliferation remains to be established. The authors reported evidence for 36 different complexes, among them at least 7 potentially novel ones containing demethylases [220]. Several components with different enzymatic activities are present in these complexes, and the interdependencies of the individual activities are incompletely understood and might be regulated by development stage and tissue identity. As an important consequence of this functional diversity, no consistent gene expression signatures or biomarkers for patient stratification have been published for the histone demethylases so far. Although gene amplifications or overexpression in tumor cells may help to identify patients who could benefit from treatment with a demethylase inhibitor, it is currently unclear if these parameters are sufficient for stratification. Therefore, predictive markers for indications and biomarkers for response are needed. An important step to achieve this knowledge will be the use of specific inhibitors in large cell line panels with subsequent transcriptional profiling. Association studies of response with the mutational status of tumor suppressors and oncogenes as well as larger chromosomal aberrations in cell lines may be used to define predictive markers, and patient-derived xenograft models offer the possibility to test responder hypotheses in primary tumor material.

A discussed above, the JMJC family comprises both putative oncogenes and tumor suppressors. Therefore, subclass-specific inhibitors are required to minimize risks and to create a strong safety profile for these compounds. This should also enable the use of demethylase inhibitors in combination therapy. Targeted cancer therapy directed against only one pathway typically does not yield sustained tumor eradication, but after an initial strong effect, drug-resistant clones often give rise to remission of disease. The use of multiple drugs to target different deregulated signaling pathways in the tumor can trigger synergistic effects and a sustained antitumor response. Some potential combination strategies are emerging for the histone demethylases. Rui et al. demonstrated that the amplification of the chromosome band 9p24 in a subset of B-cell lymphomas leads to oncogenic cooperation between KDM4C and the Janus tyrosine kinase 2 (JAK2), both of which are encoded within the same amplicon. Inhibition of JAK2 combined with a knockdown of KDM4C led to a cooperative killing of the treated lymphoma cells, pointing to an interesting therapeutic option once KDM4-specific inhibitors will be available [112]. In addition, the finding that the KDM5A demethylases can associate with SIN3B histone deacetylase and NURD complexes [221,222] indicates the possibility of a synergistic effect of a combination of histone deacetylase and demethylase inhibitors.

Initial results obtained in tissue culture experiments indicate an involvement of the JMJC demethylases in the development of drug tolerance, a cellular phenotype that is characterized by a dramatically reduced responsiveness to various anticancer agents, increased viability, activation of growth factor signaling pathways, and an altered chromatin state. The H3K4 demethylases KDM5A is upregulated in lung cancer cells exposed to EGFR inhibitors or cisplatin, whereas global H3K4me3 levels are decreased, and shRNA against *KDM5A* inhibits the emergence of drug tolerant cells [137]. This phenotype is transiently acquired and reversible, and thereby resembles the transient nature of *KDM5B* upregulation in a subpopulation of melanoma cells that were required for tumor maintenance in serial transplantation experiments performed by Roesch et al. [145]. Of note, a combination of *KDM5B* knockdown and the proteasome inhibitor bortezomib or the B-Raf inhibitor vemurafenib enhanced the antitumor effect of the targeted drugs in animal experiments [146]. Currently, these findings are restricted to model systems and there are no published data available to what extent the histone demethylases contribute to acquired drug tolerance in human cancers. However, many clinical examples document that patients respond to re-introduction of the same therapy (drug rechallenge) after a drug holiday following disease relapse or progression during treatment (recently reviewed in Ref. [223]). Currently, the basis of this effect is not understood, and it remains to be demonstrated that KDM5 enzymes play a role in drug sensitivity in human disease. However, it is tempting to speculate that small molecule inhibitors of KDM5 may be beneficial in delaying the onset of drug tolerance or in supporting the resensitization to therapy. Obviously, this will be limited to cases where resistance is not based on genomic alterations such as gene amplifications or mutations as, e.g., the recurrent T790M mutation in EGFR, which renders the receptor insensitive to several kinase inhibitors used in the clinic.

The question as to whether demethylase inhibitors may raise specific safety concerns awaits further studies. The ongoing clinical Phase 1 studies with the LSD1 inhibitors are of great interest. In general, the functions of KDMs in stem cells as well as in cellular differentiation indicate that the hematopoietic system and organs with a high rate of homeostatic self-renewal such as the intestine might be affected, but in this respect, epigenetic drugs do not differ from already existing treatment for cancer. In addition, it has been discussed if compounds that influence histone marks might elicit effects on the germline that could be transmitted to the next generation and even beyond ("transgenerational epigenetic inheritance," a concept that is the matter of a controversial scientific debate [224,225]). Of course, any *in utero* exposure of a developing embryo should be avoided; however, the risk of induced changes in germ cells in younger patients might exist, even if this is shared with a large fraction of clinical cancer drugs and if the naturally occurring extensive reprogramming of the embryonic epigenome might help to "reset" drug-induced changes.

In conclusion, small molecule inhibitors have been generated for LSD1, the KDM5/JARID family and KDM6/UTX so far and these have shown target engagement in *in vivo* studies. The fact that there are ongoing clinical studies with the LSD1 inhibitors in leukemia and lung cancer patients encourages the search for specific inhibitors with drug-like properties for the remaining histone demethylases and offers hope that they bear a great potential in the continuous fight against cancer.

REFERENCES

[1] Greer EL, Shi Y. Histone methylation: a dynamic mark in health, disease and inheritance. Nat Rev Genet 2012;13(5):343–57.

[2] Herz HM, Garruss A, Shilatifard A. SET for life: biochemical activities and biological functions of SET domain-containing proteins. Trends Biochem Sci 2013;38(12):621–39.

[3] Thiagarajan RD, Morey R, Laurent LC. The epigenome in pluripotency and differentiation. Epigenomics 2014;6(1):121–37.

[4] McCauley BS, Dang W. Histone methylation and aging: lessons learned from model systems. Biochim Biophys Acta 2014;1839(12):1454–62.

[5] Bracken AP, Pasini D, Capra M, Prosperini E, Colli E, Helin K. EZH2 is downstream of the pRB-E2F pathway, essential for proliferation and amplified in cancer. Embo J 2003;22(20):5323–35.

[6] Varambally S, Dhanasekaran SM, Zhou M, Barrette TR, Kumar-Sinha C, Sanda MG, et al. The polycomb group protein EZH2 is involved in progression of prostate cancer. Nature 2002;419(6907):624–9.

[7] Sneeringer CJ, Scott MP, Kuntz KW, Knutson SK, Pollock RM, Richon VM, et al. Coordinated activities of wild-type plus mutant EZH2 drive tumor-associated hypertrimethylation of lysine 27 on histone H3 (H3K27) in human B-cell lymphomas. Proc Natl Acad Sci USA 2010;107(49):20980–5.

[8] McCabe MT, Ott HM, Ganji G, Korenchuk S, Thompson C, Van Aller GS, et al. EZH2 inhibition as a therapeutic strategy for lymphoma with EZH2-activating mutations. Nature 2012;492(7427):108–12.

[9] McCabe MT, Graves AP, Ganji G, Diaz E, Halsey WS, Jiang Y, et al. Mutation of A677 in histone methyltransferase EZH2 in human B-cell lymphoma promotes hypertrimethylation of histone H3 on lysine 27 (H3K27). Proc Natl Acad Sci USA 2012;109(8):2989–94.

[10] Knutson SK, Wigle TJ, Warholic NM, Sneeringer CJ, Allain CJ, Klaus CR, et al. A selective inhibitor of EZH2 blocks H3K27 methylation and kills mutant lymphoma cells. Nat Chem Biol 2012;8(11):890–6.

[11] Beguelin W, Popovic R, Teater M, Jiang Y, Bunting KL, Rosen M, et al. EZH2 is required for germinal center formation and somatic EZH2 mutations promote lymphoid transformation. Cancer Cell 2013;23(5):677–92.

[12] Schwartzentruber J, Korshunov A, Liu XY, Jones DT, Pfaff E, Jacob K, et al. Driver mutations in histone H3.3 and chromatin remodelling genes in paediatric glioblastoma. Nature 2012;482(7384):226–31.

[13] Wu G, Broniscer A, McEachron TA, Lu C, Paugh BS, Becksfort J, et al. Somatic histone H3 alterations in pediatric diffuse intrinsic pontine gliomas and non-brainstem glioblastomas. Nat Genet 2012;44(3):251–3.

[14] Lewis PW, Muller MM, Koletsky MS, Cordero F, Lin S, Banaszynski LA, et al. Inhibition of PRC2 activity by a gain-of-function H3 mutation found in pediatric glioblastoma. Science 2013;340(6134):857–61.

[15] De Raedt T, Beert E, Pasmant E, Luscan A, Brems H, Ortonne N, et al. PRC2 loss amplifies ras-driven transcription and confers sensitivity to BRD4-based therapies. Nature 2014.

[16] Kuo AJ, Cheung P, Chen K, Zee BM, Kioi M, Lauring J, et al. NSD2 links dimethylation of histone H3 at lysine 36 to oncogenic programming. Mol Cell 2011;44(4):609–20.

[17] Bernt KM, Zhu N, Sinha AU, Vempati S, Faber J, Krivtsov AV, et al. MLL-rearranged leukemia is dependent on aberrant H3K79 methylation by DOT1L. Cancer Cell 2011;20(1):66–78.

[18] Daigle SR, Olhava EJ, Therkelsen CA, Majer CR, Sneeringer CJ, Song J, et al. Selective killing of mixed lineage leukemia cells by a potent small-molecule DOT1L inhibitor. Cancer Cell 2011;20(1):53–65.

[19] Daigle SR, Olhava EJ, Therkelsen CA, Basavapathruni A, Jin L, Boriack-Sjodin PA, et al. Potent inhibition of DOT1L as treatment of MLL-fusion leukemia. Blood 2013;122(6):1017–25.

[20] Shi Y, Lan F, Matson C, Mulligan P, Whetstine JR, Cole PA, et al. Histone demethylation mediated by the nuclear amine oxidase homolog LSD1. Cell 2004;119(7):941–53.

[21] Metzger E, Wissmann M, Yin N, Muller JM, Schneider R, Peters AH, et al. LSD1 demethylates repressive histone marks to promote androgen-receptor-dependent transcription. Nature 2005;437(7057):436–9.

[22] Huang J, Sengupta R, Espejo AB, Lee MG, Dorsey JA, Richter M, et al. p53 is regulated by the lysine demethylase LSD1. Nature 2007;449(7158):105–8.

[23] Wang J, Hevi S, Kurash JK, Lei H, Gay F, Bajko J, et al. The lysine demethylase LSD1 (KDM1) is required for maintenance of global DNA methylation. Nat Genet 2009;41(1):125–9.

[24] Kontaki H, Talianidis I. Lysine methylation regulates E2F1-induced cell death. Mol Cell 2010;39(1):152–60.

[25] Karytinos A, Forneris F, Profumo A, Ciossani G, Battaglioli E, Binda C, et al. A novel mammalian flavin-dependent histone demethylase. J Biol Chem 2009;284(26):17775–82.

[26] Fang R, Barbera AJ, Xu Y, Rutenberg M, Leonor T, Bi Q, et al. Human LSD2/KDM1b/AOF1 regulates gene transcription by modulating intragenic H3K4me2 methylation. Mol Cell 2010;39(2):222–33.

[27] Fang R, Chen F, Dong Z, Hu D, Barbera AJ, Clark EA, et al. LSD2/KDM1B and its cofactor NPAC/GLYR1 endow a structural and molecular model for regulation of H3K4 demethylation. Mol Cell 2013;49(3):558–70.

[28] Rudolph T, Beuch S, Reuter G. Lysine-specific histone demethylase LSD1 and the dynamic control of chromatin. Biol Chem 2013;394(8):1019–28.

[29] Kerenyi MA, Shao Z, Hsu YJ, Guo G, Luc S, O'Brien K, et al. Histone demethylase Lsd1 represses hematopoietic stem and progenitor cell signatures during blood cell maturation. Elife 2013;2:e00633.

[30] Ciccone DN, Su H, Hevi S, Gay F, Lei H, Bajko J, et al. KDM1B is a histone H3K4 demethylase required to establish maternal genomic imprints. Nature 2009;461(7262):415–18.

[31] Kahl P, Gullotti L, Heukamp LC, Wolf S, Friedrichs N, Vorreuther R, et al. Androgen receptor coactivators lysine-specific histone demethylase 1 and four and a half LIM domain protein 2 predict risk of prostate cancer recurrence. Cancer Res 2006;66(23):11341–7.

[32] Schulte JH, Lim S, Schramm A, Friedrichs N, Koster J, Versteeg R, et al. Lysine-specific demethylase 1 is strongly expressed in poorly differentiated neuroblastoma: implications for therapy. Cancer Res 2009;69(5):2065–71.

[33] Lim S, Janzer A, Becker A, Zimmer A, Schule R, Buettner R, et al. Lysine-specific demethylase 1 (LSD1) is highly expressed in ER-negative breast cancers and a biomarker predicting aggressive biology. Carcinogenesis 2010;31(3):512–20.

[34] Hayami S, Kelly JD, Cho HS, Yoshimatsu M, Unoki M, Tsunoda T, et al. Overexpression of LSD1 contributes to human carcinogenesis through chromatin regulation in various cancers. Int J Cancer 2011;128(3):574–86.

[35] Kauffman EC, Robinson BD, Downes MJ, Powell LG, Lee MM, Scherr DS, et al. Role of androgen receptor and associated lysine-demethylase coregulators, LSD1 and JMJD2A, in localized and advanced human bladder cancer. Mol Carcinog 2011;50(12):931–44.

[36] Harris WJ, Huang X, Lynch JT, Spencer GJ, Hitchin JR, Li Y, et al. The histone demethylase KDM1A sustains the oncogenic potential of MLL-AF9 leukemia stem cells. Cancer Cell 2012;21(4):473–87.

[37] Schenk T, Chen WC, Gollner S, Howell L, Jin L, Hebestreit K, et al. Inhibition of the LSD1 (KDM1A) demethylase reactivates the all-trans-retinoic acid differentiation pathway in acute myeloid leukemia. Nat Med 2012;18(4):605–11.

[38] Lee MG, Wynder C, Schmidt DM, McCafferty DG, Shiekhattar R. Histone H3 lysine 4 demethylation is a target of nonselective antidepressive medications. Chem Biol 2006;13(6):563–7.

[39] Ueda R, Suzuki T, Mino K, Tsumoto H, Nakagawa H, Hasegawa M, et al. Identification of cell-active lysine specific demethylase 1-selective inhibitors. J Am Chem Soc 2009;131(48):17536–7.

[40] Mimasu S, Umezawa N, Sato S, Higuchi T, Umehara T, Yokoyama S. Structurally designed trans-2-phenylcyclopropylamine derivatives potently inhibit histone demethylase LSD1/KDM1. Biochemistry 2010;49(30):6494–503.

[41] Oryzon Genomics SA Phenylcyclopropamine derivatives and their medical use. WO2010084160A1. 2010.

[42] Neelamegam R, Ricq EL, Malvaez M, Patnaik D, Norton S, Carlin SM, et al. Brain-penetrant LSD1 inhibitors can block memory consolidation. ACS Chem Neurosci 2012;3(2):120–8.

[43] Cloos PA, Christensen J, Agger K, Maiolica A, Rappsilber J, Antal T, et al. The putative oncogene GASC1 demethylates tri- and dimethylated lysine 9 on histone H3. Nature 2006;442(7100):307–11.

[44] Fodor BD, Kubicek S, Yonezawa M, O'Sullivan RJ, Sengupta R, Perez-Burgos L, et al. Jmjd2b antagonizes H3K9 trimethylation at pericentric heterochromatin in mammalian cells. Genes Dev 2006;20(12):1557–62.

[45] Klose RJ, Yamane K, Bae Y, Zhang D, Erdjument-Bromage H, Tempst P, et al. The transcriptional repressor JHDM3A demethylates trimethyl histone H3 lysine 9 and lysine 36. Nature 2006;442(7100):312–16.

[46] Tsukada Y, Fang J, Erdjument-Bromage H, Warren ME, Borchers CH, Tempst P, et al. Histone demethylation by a family of JmjC domain-containing proteins. Nature 2006;439(7078):811–16.

[47] Whetstine JR, Nottke A, Lan F, Huarte M, Smolikov S, Chen Z, et al. Reversal of histone lysine trimethylation by the JMJD2 family of histone demethylases. Cell 2006;125(3):467–81.

[48] McDonough MA, Loenarz C, Chowdhury R, Clifton IJ, Schofield CJ. Structural studies on human 2-oxoglutarate dependent oxygenases. Curr Opin Struct Biol 2010;20(6):659–72.

[49] Hegg EL, Que Jr. L. The 2-His-1-carboxylate facial triad—an emerging structural motif in mononuclear non-heme iron(II) enzymes. Eur J Biochem 1997;250(3):625–9.

[50] Klose RJ, Kallin EM, Zhang Y. JmjC-domain-containing proteins and histone demethylation. Nat Rev Genet 2006;7(9):715–27.

[51] Ng SS, Kavanagh KL, McDonough MA, Butler D, Pilka ES, Lienard BM, et al. Crystal structures of histone demethylase JMJD2A reveal basis for substrate specificity. Nature 2007;448(7149):87–91.

[52] Chen Z, Zang J, Kappler J, Hong X, Crawford F, Wang Q, et al. Structural basis of the recognition of a methylated histone tail by JMJD2A. Proc Natl Acad Sci USA 2007;104(26):10818–23.

[53] Couture JF, Collazo E, Ortiz-Tello PA, Brunzelle JS, Trievel RC. Specificity and mechanism of JMJD2A, a trimethyllysine-specific histone demethylase. Nat Struct Mol Biol 2007;14(8):689–95.

[54] Horton JR, Upadhyay AK, Hashimoto H, Zhang X, Cheng X. Structural basis for human PHF2 Jumonji domain interaction with metal ions. J Mol Biol 2011;406(1):1–8.

[55] Stender JD, Pascual G, Liu W, Kaikkonen MU, Do K, Spann NJ, et al. Control of proinflammatory gene programs by regulated trimethylation and demethylation of histone H4K20. Mol Cell 2012;48(1):28–38.

[56] Wen H, Li J, Song T, Lu M, Kan PY, Lee MG, et al. Recognition of histone H3K4 trimethylation by the plant homeodomain of PHF2 modulates histone demethylation. J Biol Chem 2010;285(13):9322–6.

[57] Baba A, Ohtake F, Okuno Y, Yokota K, Okada M, Imai Y, et al. PKA-dependent regulation of the histone lysine demethylase complex PHF2-ARID5B. Nat Cell Biol 2011;13(6):668–75.

[58] Del Rizzo PA, Krishnan S, Trievel RC. Crystal structure and functional analysis of JMJD5 indicate an alternate specificity and function. Mol Cell Biol 2012;32(19):4044–52.

[59] Webby CJ, Wolf A, Gromak N, Dreger M, Kramer H, Kessler B, et al. Jmjd6 catalyses lysyl-hydroxylation of U2AF65, a protein associated with RNA splicing. Science 2009;325(5936):90–3.

[60] Chang B, Chen Y, Zhao Y, Bruick RK. JMJD6 is a histone arginine demethylase. Science 2007;318(5849):444–7.

[61] Krishnan S, Trievel RC. Structural and functional analysis of JMJD2D reveals molecular basis for site-specific demethylation among JMJD2 demethylases. Structure 2013;21(1):98–108.

[62] Horton JR, Upadhyay AK, Qi HH, Zhang X, Shi Y, Cheng X. Enzymatic and structural insights for substrate specificity of a family of jumonji histone lysine demethylases. Nat Struct Mol Biol 2010;17(1):38–43.

[63] Kooistra SM, Helin K. Molecular mechanisms and potential functions of histone demethylases. Nat Rev Mol Cell Biol 2012;13(5):297–311.

[64] Mosammaparast N, Shi Y. Reversal of histone methylation: biochemical and molecular mechanisms of histone demethylases. Annu Rev Biochem 2010;79:155–79.

[65] Pedersen MT, Helin K. Histone demethylases in development and disease. Trends Cell Biol 2010;20(11):662–71.

[66] Fukuda T, Tokunaga A, Sakamoto R, Yoshida N. Fbxl10/Kdm2b deficiency accelerates neural progenitor cell death and leads to exencephaly. Mol Cell Neurosci 2011;46(3):614–24.

[67] Inagaki T, Tachibana M, Magoori K, Kudo H, Tanaka T, Okamura M, et al. Obesity and metabolic syndrome in histone demethylase JHDM2a-deficient mice. Genes Cells 2009;14(8):991–1001.

[68] Liu Z, Zhou S, Liao L, Chen X, Meistrich M, Xu J. Jmjd1a demethylase-regulated histone modification is essential for cAMP-response element modulator-regulated gene expression and spermatogenesis. J Biol Chem 2010;285(4):2758–70.

[69] Okada Y, Scott G, Ray MK, Mishina Y, Zhang Y. Histone demethylase JHDM2A is critical for Tnp1 and Prm1 transcription and spermatogenesis. Nature 2007;450(7166):119–23.

[70] Tateishi K, Okada Y, Kallin EM, Zhang Y. Role of Jhdm2a in regulating metabolic gene expression and obesity resistance. Nature 2009;458(7239):757–61.

[71] Zhang QJ, Chen HZ, Wang L, Liu DP, Hill JA, Liu ZP. The histone trimethyllysine demethylase JMJD2A promotes cardiac hypertrophy in response to hypertrophic stimuli in mice. J Clin Invest 2011;121(6):2447–56.

[72] Kawazu M, Saso K, Tong KI, McQuire T, Goto K, Son DO, et al. Histone demethylase JMJD2B functions as a co-factor of estrogen receptor in breast cancer proliferation and mammary gland development. PLoS One 2011;6(3):e17830.

[73] Pedersen MT, Agger K, Laugesen A, Johansen JV, Cloos PA, Christensen J, et al. The demethylase JMJD2C localizes to H3K4me3-positive transcription start sites and is dispensable for embryonic development. Mol Cell Biol 2014;34(6):1031–45.

[74] Iwamori N, Zhao M, Meistrich ML, Matzuk MM. The testis-enriched histone demethylase, KDM4D, regulates methylation of histone H3 lysine 9 during spermatogenesis in the mouse but is dispensable for fertility. Biol Reprod 2011;84(6):1225–34.

[75] Klose RJ, Yan Q, Tothova Z, Yamane K, Erdjument-Bromage H, Tempst P, et al. The retinoblastoma binding protein RBP2 is an H3K4 demethylase. Cell 2007;128(5):889–900.

[76] Catchpole S, Spencer-Dene B, Hall D, Santangelo S, Rosewell I, Guenatri M, et al. PLU-1/JARID1B/KDM5B is required for embryonic survival and contributes to cell proliferation in the mammary gland and in ER+ breast cancer cells. Int J Oncol 2011;38(5):1267–77.

[77] Albert M, Schmitz SU, Kooistra SM, Malatesta M, Morales Torres C, Rekling JC, et al. The histone demethylase Jarid1b ensures faithful mouse development by protecting developmental genes from aberrant H3K4me3. PLoS Genet 2013;9(4):e1003461.

[78] Cox BJ, Vollmer M, Tamplin O, Lu M, Biechele S, Gertsenstein M, et al. Phenotypic annotation of the mouse X chromosome. Genome Res 2010;20(8):1154–64.

[79] Satoh T, Takeuchi O, Vandenbon A, Yasuda K, Tanaka Y, Kumagai Y, et al. The Jmjd3-Irf4 axis regulates M2 macrophage polarization and host responses against helminth infection. Nat Immunol 2010;11(10):936–44.

[80] Lee S, Lee JW, Lee SK. UTX, a histone H3-lysine 27 demethylase, acts as a critical switch to activate the cardiac developmental program. Dev Cell 2012;22(1):25–37.

[81] Shpargel KB, Sengoku T, Yokoyama S, Magnuson T. UTX and UTY demonstrate histone demethylase-independent function in mouse embryonic development. PLoS Genet 2012;8(9):e1002964.

[82] Ishimura A, Minehata K, Terashima M, Kondoh G, Hara T, Suzuki T. Jmjd5, an H3K36me2 histone demethylase, modulates embryonic cell proliferation through the regulation of Cdkn1a expression. Development 2012;139(4):749–59.

[83] Okuno Y, Ohtake F, Igarashi K, Kanno J, Matsumoto T, Takada I, et al. Epigenetic regulation of adipogenesis by PHF2 histone demethylase. Diabetes 2013;62(5):1426–34.

[84] Takeuchi T, Kojima M, Nakajima K, Kondo S. Jumonji gene is essential for the neurulation and cardiac development of mouse embryos with a C3H/He background. Mech Dev 1999;86(1–2):29–38.

[85] Zhang N, Fu Z, Linke S, Chicher J, Gorman JJ, Visk D, et al. The asparaginyl hydroxylase factor inhibiting HIF-1alpha is an essential regulator of metabolism. Cell Metab 2010;11(5):364–78.

[86] Zarach JM, Beaudoin III GM, Coulombe PA, Thompson CC. The co-repressor hairless has a role in epithelial cell differentiation in the skin. Development 2004;131(17):4189–200.

[87] Farcas AM, Blackledge NP, Sudbery I, Long HK, McGouran JF, Rose NR, et al. KDM2B links the Polycomb Repressive Complex 1 (PRC1) to recognition of CpG islands. Elife 2012;1:e00205.

[88] He J, Shen L, Wan M, Taranova O, Wu H, Zhang Y. Kdm2b maintains murine embryonic stem cell status by recruiting PRC1 complex to CpG islands of developmental genes. Nat Cell Biol 2013;15(4):373–84.

[89] Wu X, Johansen JV, Helin K. Fbxl10/Kdm2b recruits polycomb repressive complex 1 to CpG islands and regulates H2A ubiquitylation. Mol Cell 2013;49(6):1134–46.

[90] Blackledge NP, Farcas AM, Kondo T, King HW, McGouran JF, Hanssen LL, et al. Variant PRC1 complex-dependent H2A ubiquitylation drives PRC2 recruitment and polycomb domain formation. Cell 2014;157(6):1445–59.

[91] Dalgliesh GL, Furge K, Greenman C, Chen L, Bignell G, Butler A, et al. Systematic sequencing of renal carcinoma reveals inactivation of histone modifying genes. Nature 2010;463(7279):360–3.

[92] Grasso CS, Wu YM, Robinson DR, Cao X, Dhanasekaran SM, Khan AP, et al. The mutational landscape of lethal castration-resistant prostate cancer. Nature 2012;487(7406):239–43.

[93] Jones DT, Jager N, Kool M, Zichner T, Hutter B, Sultan M, et al. Dissecting the genomic complexity underlying medulloblastoma. Nature 2012;488(7409):100–5.

[94] Robinson G, Parker M, Kranenburg TA, Lu C, Chen X, Ding L, et al. Novel mutations target distinct subgroups of medulloblastoma. Nature 2012;488(7409):43–8.

[95] van Haaften G, Dalgliesh GL, Davies H, Chen L, Bignell G, Greenman C, et al. Somatic mutations of the histone H3K27 demethylase gene UTX in human cancer. Nat Genet 2009;41(5):521–3.

[96] Morin RD, Johnson NA, Severson TM, Mungall AJ, An J, Goya R, et al. Somatic mutations altering EZH2 (Tyr641) in follicular and diffuse large B-cell lymphomas of germinal-center origin. Nat Genet 2010;42(2):181–5.

[97] Yap DB, Chu J, Berg T, Schapira M, Cheng SW, Moradian A, et al. Somatic mutations at EZH2 Y641 act dominantly through a mechanism of selectively altered PRC2 catalytic activity, to increase H3K27 trimethylation. Blood 2010;117(8):2451–9.

[98] Van der Meulen J, Speleman F, Van Vlierberghe P. The H3K27me3 demethylase UTX in normal development and disease. Epigenetics 2014;9(5):658–68.

[99] Outchkourov NS, Muino JM, Kaufmann K, van Ijcken WF, Groot Koerkamp MJ, van Leenen D, et al. Balancing of histone H3K4 methylation states by the Kdm5c/SMCX histone demethylase modulates promoter and enhancer function. Cell Rep 2013;3(4):1071–9.

[100] Niu X, Zhang T, Liao L, Zhou L, Lindner DJ, Zhou M, et al. The von Hippel–Lindau tumor suppressor protein regulates gene expression and tumor growth through histone demethylase JARID1C. Oncogene 2012;31(6):776–86.

[101] Jensen LR, Amende M, Gurok U, Moser B, Gimmel V, Tzschach A, et al. Mutations in the JARID1C gene, which is involved in transcriptional regulation and chromatin remodeling, cause X-linked mental retardation. Am J Hum Genet 2005;76(2):227–36.

[102] Santos C, Rodriguez-Revenga L, Madrigal I, Badenas C, Pineda M, Mila M. A novel mutation in JARID1C gene associated with mental retardation. Eur J Hum Genet 2006;14(5):583–6.

[103] Tzschach A, Lenzner S, Moser B, Reinhardt R, Chelly J, Fryns JP, et al. Novel JARID1C/SMCX mutations in patients with X-linked mental retardation. Hum Mutat 2006;27(4):389.

[104] Iwase S, Lan F, Bayliss P, de la Torre-Ubieta L, Huarte M, Qi HH, et al. The X-linked mental retardation gene SMCX/JARID1C defines a family of histone H3 lysine 4 demethylases. Cell 2007;128(6):1077–88.

[105] Peters AH, O'Carroll D, Scherthan H, Mechtler K, Sauer S, Schofer C, et al. Loss of the Suv39h histone methyltransferases impairs mammalian heterochromatin and genome stability. Cell 2001;107(3):323–37.

[106] Braig M, Lee S, Loddenkemper C, Rudolph C, Peters AH, Schlegelberger B, et al. Oncogene-induced senescence as an initial barrier in lymphoma development. Nature 2005;436(7051):660–5.

[107] Yang ZQ, Imoto I, Fukuda Y, Pimkhaokham A, Shimada Y, Imamura M, et al. Identification of a novel gene, GASC1, within an amplicon at 9p23-24 frequently detected in esophageal cancer cell lines. Cancer Res 2000;60(17):4735–9.

[108] Sun LL, Holowatyj A, Xu XE, Wu JY, Wu ZY, Shen JH, et al. Histone demethylase GASC1, a potential prognostic and predictive marker in esophageal squamous cell carcinoma. Am J Cancer Res 2013;3(5):509–17.

[109] Ehrbrecht A, Muller U, Wolter M, Hoischen A, Koch A, Radlwimmer B, et al. Comprehensive genomic analysis of desmoplastic medulloblastomas: identification of novel amplified genes and separate evaluation of the different histological components. J Pathol 2006;208(4):554–63.

[110] Liu G, Bollig-Fischer A, Kreike B, van de Vijver MJ, Abrams J, Ethier SP, et al. Genomic amplification and oncogenic properties of the GASC1 histone demethylase gene in breast cancer. Oncogene 2009;28(50):4491–500.

[111] Wissmann M, Yin N, Muller JM, Greschik H, Fodor BD, Jenuwein T, et al. Cooperative demethylation by JMJD2C and LSD1 promotes androgen receptor-dependent gene expression. Nat Cell Biol 2007;9(3):347–53.

[112] Rui L, Emre NC, Kruhlak MJ, Chung HJ, Steidl C, Slack G, et al. Cooperative epigenetic modulation by cancer amplicon genes. Cancer Cell 2010;18(6):590–605.

[113] Hu CE, Liu YC, Zhang HD, Huang GJ. JMJD2A predicts prognosis and regulates cell growth in human gastric cancer. Biochem Biophys Res Commun 2014;449(1):1–7.

[114] Black JC, Manning AL, Van Rechem C, Kim J, Ladd B, Cho J, et al. KDM4A lysine demethylase induces site-specific copy gain and rereplication of regions amplified in tumors. Cell 2013;154(3):541–55.

[115] Patani N, Jiang WG, Newbold RF, Mokbel K. Histone-modifier gene expression profiles are associated with pathological and clinical outcomes in human breast cancer. Anticancer Res 2011;31(12):4115–25.

[116] Ding X, Pan H, Li J, Zhong Q, Chen X, Dry SM, et al. Epigenetic activation of AP1 promotes squamous cell carcinoma metastasis. Sci Signal 2013;6(273) ra28 1–13, S0–5.

[117] Pryor JG, Brown-Kipphut BA, Iqbal A, Scott GA. Microarray comparative genomic hybridization detection of copy number changes in desmoplastic melanoma and malignant peripheral nerve sheath tumor. Am J Dermatopathol 2011;33(8):780–5.

[118] Yang J, Jubb AM, Pike L, Buffa FM, Turley H, Baban D, et al. The histone demethylase JMJD2B is regulated by estrogen receptor alpha and hypoxia, and is a key mediator of estrogen induced growth. Cancer Res 2010;70(16):6456–66.

[119] Fu L, Chen L, Yang J, Ye T, Chen Y, Fang J. HIF-1alpha-induced histone demethylase JMJD2B contributes to the malignant phenotype of colorectal cancer cells via an epigenetic mechanism. Carcinogenesis 2012;33(9):1664–73.

[120] Shi L, Sun L, Li Q, Liang J, Yu W, Yi X, et al. Histone demethylase JMJD2B coordinates H3K4/H3K9 methylation and promotes hormonally responsive breast carcinogenesis. Proc Natl Acad Sci USA 2011;108(18):7541–6.

[121] Sun BB, Fu LN, Wang YQ, Gao QY, Xu J, Cao ZJ, et al. Silencing of JMJD2B induces cell apoptosis via mitochondria-mediated and death receptor-mediated pathway activation in colorectal cancer. J Dig Dis 2014.

[122] Barski A, Cuddapah S, Cui K, Roh TY, Schones DE, Wang Z, et al. High-resolution profiling of histone methylations in the human genome. Cell 2007;129(4):823–37.

[123] Bernstein BE, Kamal M, Lindblad-Toh K, Bekiranov S, Bailey DK, Huebert DJ, et al. Genomic maps and comparative analysis of histone modifications in human and mouse. Cell 2005;120(2):169–81.

[124] Heintzman ND, Stuart RK, Hon G, Fu Y, Ching CW, Hawkins RD, et al. Distinct and predictive chromatin signatures of transcriptional promoters and enhancers in the human genome. Nat Genet 2007;39(3):311–18.

[125] Tahiliani M, Mei P, Fang R, Leonor T, Rutenberg M, Shimizu F, et al. The histone H3K4 demethylase SMCX links REST target genes to X-linked mental retardation. Nature 2007;447(7144):601–5.

[126] Yamane K, Tateishi K, Klose RJ, Fang J, Fabrizio LA, Erdjument-Bromage H, et al. PLU-1 is an H3K4 demethylase involved in transcriptional repression and breast cancer cell proliferation. Mol Cell 2007;25(6):801–12.

[127] Fattaey AR, Helin K, Dembski MS, Dyson N, Harlow E, Vuocolo GA, et al. Characterization of the retinoblastoma binding proteins RBP1 and RBP2. Oncogene 1993;8(11):3149–56.

[128] Benevolenskaya EV, Murray HL, Branton P, Young RA, Kaelin Jr. WG. Binding of pRB to the PHD protein RBP2 promotes cellular differentiation. Mol Cell 2005;18(6):623–35.

[129] Lin W, Cao J, Liu J, Beshiri ML, Fujiwara Y, Francis J, et al. Loss of the retinoblastoma binding protein 2 (RBP2) histone demethylase suppresses tumorigenesis in mice lacking Rb1 or Men1. Proc Natl Acad Sci USA 2011;108(33):13379–86.

[130] Liang X, Zeng J, Wang L, Fang M, Wang Q, Zhao M, et al. Histone demethylase retinoblastoma binding protein 2 is overexpressed in hepatocellular carcinoma and negatively regulated by hsa-miR-212. PLoS One 2013;8(7):e69784.

[131] Jiping Z, Ming F, Lixiang W, Xiuming L, Yuqun S, Han Y, et al. MicroRNA-212 inhibits proliferation of gastric cancer by directly repressing retinoblastoma binding protein 2. J Cell Biochem 2013;114(12):2666–72.

[132] Teng YC, Lee CF, Li YS, Chen YR, Hsiao PW, Chan MY, et al. Histone demethylase RBP2 promotes lung tumorigenesis and cancer metastasis. Cancer Res 2013;73(15):4711–21.

[133] Wang S, Wang Y, Wu H, Hu L. RBP2 Induces Epithelial–Mesenchymal transition in non-small cell lung Cancer. PLoS One 2013;8(12):e84735.

[134] Paolicchi E, Crea F, Farrar WL, Green JE, Danesi R. Histone lysine demethylases in breast cancer. Crit Rev Oncol Hematol 2013;86(2):97–103.

[135] Cao J, Liu Z, Cheung WK, Zhao M, Chen SY, Chan SW, et al. Histone demethylase RBP2 is critical for breast cancer progression and metastasis. Cell Rep 2014;6(5):868–77.

[136] Hou J, Wu J, Dombkowski A, Zhang K, Holowatyj A, Boerner JL, et al. Genomic amplification and a role in drug-resistance for the KDM5A histone demethylase in breast cancer. Am J Transl Res 2012;4(3):247–56.

[137] Sharma SV, Lee DY, Li B, Quinlan MP, Takahashi F, Maheswaran S, et al. A chromatin-mediated reversible drug-tolerant state in cancer cell subpopulations. Cell 2010;141(1):69–80.

[138] Lu PJ, Sundquist K, Baeckstrom D, Poulsom R, Hanby A, Meier-Ewert S, et al. A novel gene (PLU-1) containing highly conserved putative DNA/chromatin binding motifs is specifically up-regulated in breast cancer. J Biol Chem 1999;274(22):15633–45.

[139] Barrett A, Madsen B, Copier J, Lu PJ, Cooper L, Scibetta AG, et al. PLU-1 nuclear protein, which is upregulated in breast cancer, shows restricted expression in normal human adult tissues: a new cancer/testis antigen? Int J Cancer 2002;101(6):581–8.

[140] Yamamoto S, Wu Z, Russnes HG, Takagi S, Peluffo G, Vaske C, et al. JARID1B is a luminal lineage-driving oncogene in breast cancer. Cancer Cell 2014;25(6):762–77.

[141] Hayami S, Yoshimatsu M, Veerakumarasivam A, Unoki M, Iwai Y, Tsunoda T, et al. Overexpression of the JmjC histone demethylase KDM5B in human carcinogenesis: involvement in the proliferation of cancer cells through the E2F/RB pathway. Mol Cancer 2010;9:59.

[142] Li X, Su Y, Pan J, Zhou Z, Song B, Xiong E, et al. Connexin 26 is down-regulated by KDM5B in the progression of Bladder Cancer. Int J Mol Sci 2013;14(4):7866–79.

[143] Ohta K, Haraguchi N, Kano Y, Kagawa Y, Konno M, Nishikawa S, et al. Depletion of JARID1B induces cellular senescence in human colorectal cancer. Int J Oncol 2013;42(4):1212–18.

[144] Xiang Y, Zhu Z, Han G, Ye X, Xu B, Peng Z, et al. JARID1B is a histone H3 lysine 4 demethylase up-regulated in prostate cancer. Proc Natl Acad Sci USA 2007;104(49):19226–31.

[145] Roesch A, Fukunaga-Kalabis M, Schmidt EC, Zabierowski SE, Brafford PA, Vultur A, et al. A temporarily distinct subpopulation of slow-cycling melanoma cells is required for continuous tumor growth. Cell 2010;141(4):583–94.

[146] Roesch A, Vultur A, Bogeski I, Wang H, Zimmermann KM, Speicher D, et al. Overcoming intrinsic multidrug resistance in melanoma by blocking the mitochondrial respiratory chain of slow-cycling JARID1B(high) cells. Cancer Cell 2013;23(6):811–25.

[147] Wagner KW, Alam H, Dhar SS, Giri U, Li N, Wei Y, et al. KDM2A promotes lung tumorigenesis by epigenetically enhancing ERK1/2 signaling. J Clin Invest 2013;123(12):5231–46.

[148] He J, Kallin EM, Tsukada Y, Zhang Y. The H3K36 demethylase Jhdm1b/Kdm2b regulates cell proliferation and senescence through p15(Ink4b). Nat Struct Mol Biol 2008;15(11):1169–75.

[149] Pfau R, Tzatsos A, Kampranis SC, Serebrennikova OB, Bear SE, Tsichlis PN. Members of a family of JmjC domain-containing oncoproteins immortalize embryonic fibroblasts via a JmjC domain-dependent process. Proc Natl Acad Sci USA 2008.

[150] He J, Nguyen AT, Zhang Y. KDM2b/JHDM1b, an H3K36me2-specific demethylase, is required for initiation and maintenance of acute myeloid leukemia. Blood 2011;117(14):3869–80.

[151] Tzatsos A, Paskaleva P, Ferrari F, Deshpande V, Stoykova S, Contino G, et al. KDM2B promotes pancreatic cancer via Polycomb-dependent and -independent transcriptional programs. J Clin Invest 2013;123(2):727–39.

[152] Kottakis F, Foltopoulou P, Sanidas I, Keller P, Wronski A, Dake BT, et al. NDY1/KDM2B functions as a master regulator of polycomb complexes and controls self-renewal of breast cancer stem cells. Cancer Res 2014;74(14):3935–46.

[153] Nakamura S, Tan L, Nagata Y, Takemura T, Asahina A, Yokota D, et al. JmjC-domain containing histone demethylase 1B-mediated p15(Ink4b) suppression promotes the proliferation of leukemic progenitor cells through modulation of cell cycle progression in acute myeloid leukemia. Mol Carcinog 2013;52(1):57–69.

[154] Teye K, Tsuneoka M, Arima N, Koda Y, Nakamura Y, Ueta Y, et al. Increased expression of a Myc target gene Mina53 in human colon cancer. Am J Pathol 2004;164(1):205–16.

[155] Tsuneoka M, Fujita H, Arima N, Teye K, Okamura T, Inutsuka H, et al. Mina53 as a potential prognostic factor for esophageal squamous cell carcinoma. Clin Cancer Res 2004;10(21):7347–56.

[156] Ishizaki H, Yano H, Tsuneoka M, Ogasawara S, Akiba J, Nishida N, et al. Overexpression of the myc target gene Mina53 in advanced renal cell carcinoma. Pathol Int 2007;57(10):672–80.

[157] Thakur C, Lu Y, Sun J, Yu M, Chen B, Chen F. Increased expression of mdig predicts poorer survival of the breast cancer patients. Gene 2014;535(2):218–24.

[158] Komiya K, Sueoka-Aragane N, Sato A, Hisatomi T, Sakuragi T, Mitsuoka M, et al. Mina53, a novel c-Myc target gene, is frequently expressed in lung cancers and exerts oncogenic property in NIH/3T3 cells. J Cancer Res Clin Oncol 2010;136(3):465–73.

[159] Suzuki C, Takahashi K, Hayama S, Ishikawa N, Kato T, Ito T, et al. Identification of Myc-associated protein with JmjC domain as a novel therapeutic target oncogene for lung cancer. Mol Cancer Ther 2007;6(2):542–51.

[160] Sroczynska P, Cruickshank VA, Bukowski JP, Miyagi S, Bagger FO, Walfridsson J, et al. shRNA screening identifies JMJD1C as being required for leukemia maintenance. Blood 2014;123(12):1870–82.

[161] Kim SM, Kim JY, Choe NW, Cho IH, Kim JR, Kim DW, et al. Regulation of mouse steroidogenesis by WHISTLE and JMJD1C through histone methylation balance. Nucleic Acids Res 2010;38(19):6389–403.

[162] Watanabe S, Watanabe K, Akimov V, Bartkova J, Blagoev B, Lukas J, et al. JMJD1C demethylates MDC1 to regulate the RNF8 and BRCA1-mediated chromatin response to DNA breaks. Nat Struct Mol Biol 2013;20(12):1425–33.

[163] Dhar SS, Alam H, Li N, Wagner KW, Chung J, Ahn YW, et al. Transcriptional repression of histone deacetylase 3 by the histone demethylase KDM2A is coupled to tumorigenicity of lung cancer cells. J Biol Chem 2014;289(11):7483–96.

[164] Kottakis F, Polytarchou C, Foltopoulou P, Sanidas I, Kampranis SC, Tsichlis PN. FGF-2 regulates cell proliferation, migration, and angiogenesis through an NDY1/KDM2B-miR-101-EZH2 pathway. Mol Cell 2011;43(2):285–98.

[165] Uemura M, Yamamoto H, Takemasa I, Mimori K, Hemmi H, Mizushima T, et al. Jumonji domain containing 1A is a novel prognostic marker for colorectal cancer: *in vivo* identification from hypoxic tumor cells. Clin Cancer Res 2010;16(18):4636–46.

[166] Qi J, Nakayama K, Cardiff RD, Borowsky AD, Kaul K, Williams R, et al. Siah2-dependent concerted activity of HIF and FoxA2 regulates formation of neuroendocrine phenotype and neuroendocrine prostate tumors. Cancer Cell 2010;18(1):23–38.

[167] Guo X, Shi M, Sun L, Wang Y, Gui Y, Cai Z, et al. The expression of histone demethylase JMJD1A in renal cell carcinoma. Neoplasma 2011;58(2):153–7.

[168] Yamada D, Kobayashi S, Yamamoto H, Tomimaru Y, Noda T, Uemura M, et al. Role of the hypoxia-related gene, JMJD1A, in hepatocellular carcinoma: clinical impact on recurrence after hepatic resection. Ann Surg Oncol 2011;19(Suppl. 3):S355–64.

[169] Vinatzer U, Gollinger M, Mullauer L, Raderer M, Chott A, Streubel B. Mucosa-associated lymphoid tissue lymphoma: novel translocations including rearrangements of ODZ2, JMJD2C, and CNN3. Clin Cancer Res 2008;14(20):6426–31.

[170] Vogt T, Kroiss M, McClelland M, Gruss C, Becker B, Bosserhoff AK, et al. Deficiency of a novel retinoblastoma binding protein 2-homolog is a consistent feature of sporadic human melanoma skin cancer. Lab Invest 1999;79(12):1615–27.

[171] Wang GG, Song J, Wang Z, Dormann HL, Casadio F, Li H, et al. Haematopoietic malignancies caused by dysregulation of a chromatin-binding PHD finger. Nature 2009;459(7248):847–51.

[172] van Zutven LJ, Onen E, Velthuizen SC, van Drunen E, von Bergh AR, van den Heuvel-Eibrink MM, et al. Identification of NUP98 abnormalities in acute leukemia: JARID1A (12p13) as a new partner gene. Genes Chromosomes Cancer 2006;45(5):437–46.

[173] Zeng J, Ge Z, Wang L, Li Q, Wang N, Bjorkholm M, et al. The histone demethylase RBP2 Is overexpressed in gastric cancer and its inhibition triggers senescence of cancer cells. Gastroenterology 2010;138(3):981–92.

[174] Network TCGA Comprehensive molecular portraits of human breast tumours. Nature 2012;490(7418):61–70.

[175] Perinchery G, Sasaki M, Angan A, Kumar V, Carroll P, Dahiya R. Deletion of Y-chromosome specific genes in human prostate cancer. J Urol 2000;163(4):1339–42.

[176] Agger K, Cloos PA, Rudkjaer L, Williams K, Andersen G, Christensen J, et al. The H3K27me3 demethylase JMJD3 contributes to the activation of the INK4A-ARF locus in response to oncogene- and stress-induced senescence. Genes Dev 2009;23(10):1171–6.

[177] Barradas M, Anderton E, Acosta JC, Li S, Banito A, Rodriguez-Niedenfuhr M, et al. Histone demethylase JMJD3 contributes to epigenetic control of INK4a/ARF by oncogenic RAS. Genes Dev 2009;23(10):1177–82.

[178] Shen Y, Guo X, Wang Y, Qiu W, Chang Y, Zhang A, et al. Expression and significance of histone H3K27 demethylases in renal cell carcinoma. BMC Cancer 2012;12:470.

[179] Wei Y, Chen R, Dimicoli S, Bueso-Ramos C, Neuberg D, Pierce S, et al. Global H3K4me3 genome mapping reveals alterations of innate immunity signaling and overexpression of JMJD3 in human myelodysplastic syndrome CD34[+] cells. Leukemia 2013;27(11):2177–86.

[180] Anderton JA, Bose S, Vockerodt M, Vrzalikova K, Wei W, Kuo M, et al. The H3K27me3 demethylase, KDM6B, is induced by Epstein–Barr virus and over-expressed in Hodgkin's Lymphoma. Oncogene 2011;30(17):2037–43.

[181] Gui Y, Guo G, Huang Y, Hu X, Tang A, Gao S, et al. Frequent mutations of chromatin remodeling genes in transitional cell carcinoma of the bladder. Nat Genet 2011;43(9):875–8.

[182] Jankowska AM, Makishima H, Tiu RV, Szpurka H, Huang Y, Traina F, et al. Mutational spectrum analysis of chronic myelomonocytic leukemia includes genes associated with epigenetic regulation: UTX, EZH2, and DNMT3A. Blood 2011;118(14):3932–41.

[183] Wang L, Yamaguchi S, Burstein MD, Terashima K, Chang K, Ng HK, et al. Novel somatic and germline mutations in intracranial germ cell tumours. Nature 2014;511(7508):241–5.

[184] Hsia DA, Tepper CG, Pochampalli MR, Hsia EY, Izumiya C, Huerta SB, et al. KDM8, a H3K36me2 histone demethylase that acts in the cyclin A1 coding region to regulate cancer cell proliferation. Proc Natl Acad Sci USA 2010;107(21):9671–6.

[185] Wang Z, Wang C, Huang X, Shen Y, Shen J, Ying K. Differential proteome profiling of pleural effusions from lung cancer and benign inflammatory disease patients. Biochim Biophys Acta 2012;1824(4):692–700.

[186] Sinha S, Singh RK, Alam N, Roy A, Roychoudhury S, Panda CK. Alterations in candidate genes PHF2, FANCC, PTCH1 and XPA at chromosomal 9q22.3 region: pathological significance in early- and late-onset breast carcinoma. Mol Cancer 2008;7:84.

[187] Ghosh A, Ghosh S, Maiti GP, Mukherjee S, Mukherjee N, Chakraborty J, et al. Association of FANCC and PTCH1 with the development of early dysplastic lesions of the head and neck. Ann Surg Oncol 2011;19(Suppl. 3):S528–38.

[188] Lee KH, Park JW, Sung HS, Choi YJ, Kim WH, Lee HS, et al. PHF2 histone demethylase acts as a tumor suppressor in association with p53 in cancer. Oncogene 2014.

[189] Sun LL, Sun XX, Xu XE, Zhu MX, Wu ZY, Shen JH, et al. Overexpression of Jumonji AT-rich interactive domain 1B and PHD finger protein 2 is involved in the progression of esophageal squamous cell carcinoma. Acta Histochem 2013;115(1):56–62.

[190] Bjorkman M, Ostling P, Harma V, Virtanen J, Mpindi JP, Rantala J, et al. Systematic knockdown of epigenetic enzymes identifies a novel histone demethylase PHF8 overexpressed in prostate cancer with an impact on cell proliferation, migration and invasion. Oncogene 2012;31(29):3444–56.

[191] Shen Y, Pan X, Zhao H. The histone demethylase PHF8 is an oncogenic protein in human non-small cell lung cancer. Biochem Biophys Res Commun 2014.

[192] Arteaga MF, Mikesch JH, Qiu J, Christensen J, Helin K, Kogan SC, et al. The histone demethylase PHF8 governs retinoic acid response in acute promyelocytic leukemia. Cancer Cell 2013;23(3):376–89.

[193] Bernt KM, Armstrong SA. A role for DOT1L in MLL-rearranged leukemias. Epigenomics 2011;3(6):667–70.

[194] Knutson SK, Warholic NM, Wigle TJ, Klaus CR, Allain CJ, Raimondi A, et al. Durable tumor regression in genetically altered malignant rhabdoid tumors by inhibition of methyltransferase EZH2. Proc Natl Acad Sci USA 2013;110(19):7922–7.

[195] Schofield CJ, Ratcliffe PJ. Oxygen sensing by HIF hydroxylases. Nat Rev Mol Cell Biol 2004;5(5):343–54.

[196] Lu C, Ward PS, Kapoor GS, Rohle D, Turcan S, Abdel-Wahab O, et al. IDH mutation impairs histone demethylation and results in a block to cell differentiation. Nature 2012;483(7390):474–8.

[197] Turcan S, Rohle D, Goenka A, Walsh LA, Fang F, Yilmaz E, et al. IDH1 mutation is sufficient to establish the glioma hypermethylator phenotype. Nature 2012;483(7390):479–83.

[198] Chowdhury R, Yeoh KK, Tian YM, Hillringhaus L, Bagg EA, Rose NR, et al. The oncometabolite 2-hydroxyglutarate inhibits histone lysine demethylases. EMBO Rep 2011;12(5):463–9.

[199] Cheng Z, Cheung P, Kuo AJ, Yukl ET, Wilmot CM, Gozani O, et al. A molecular threading mechanism underlies Jumonji lysine demethylase KDM2A regulation of methylated H3K36. Genes Dev 2014;28(16):1758–71.

[200] Chu CH, Wang LY, Hsu KC, Chen CC, Cheng HH, Wang SM, et al. KDM4B as a target for prostate cancer: structural analysis and selective inhibition by a novel inhibitor. J Med Chem 2014;57(14):5975–85.

[201] Sengoku T, Yokoyama S. Structural basis for histone H3 Lys 27 demethylation by UTX/KDM6A. Genes Dev 2011;25(21):2266–77.

[202] Yu L, Wang Y, Huang S, Wang J, Deng Z, Zhang Q, et al. Structural insights into a novel histone demethylase PHF8. Cell Res 2010;20(2):166–73.

[203] Yue WW, Hozjan V, Ge W, Loenarz C, Cooper CD, Schofield CJ, et al. Crystal structure of the PHF8 Jumonji domain, an Nepsilon-methyl lysine demethylase. FEBS Lett 2010;584(4):825–30.

[204] Kruidenier L, Chung CW, Cheng Z, Liddle J, Che K, Joberty G, et al. A selective jumonji H3K27 demethylase inhibitor modulates the proinflammatory macrophage response. Nature 2012;488(7411):404–8.

[205] Hamada S, Kim TD, Suzuki T, Itoh Y, Tsumoto H, Nakagawa H, et al. Synthesis and activity of *N*-oxalylglycine and its derivatives as Jumonji C-domain-containing histone lysine demethylase inhibitors. Bioorg Med Chem Lett 2009;19(10):2852–5.

[206] Luo X, Liu Y, Kubicek S, Myllyharju J, Tumber A, Ng S, et al. A selective inhibitor and probe of the cellular functions of Jumonji C domain-containing histone demethylases. J Am Chem Soc 2011;133(24):9451–6.

[207] Suzuki T, Ozasa H, Itoh Y, Zhan P, Sawada H, Mino K, et al. Identification of the KDM2/7 histone lysine demethylase subfamily inhibitor and its antiproliferative activity. J Med Chem 2013;56(18):7222–31.

[208] King ON, Li XS, Sakurai M, Kawamura A, Rose NR, Ng SS, et al. Quantitative high-throughput screening identifies 8-hydroxyquinolines as cell-active histone demethylase inhibitors. PLoS One 2010;5(11):e15535.

[209] Liang Y, Vogel JL, Arbuckle JH, Rai G, Jadhav A, Simeonov A, et al. Targeting the JMJD2 histone demethylases to epigenetically control herpesvirus infection and reactivation from latency. Sci Transl Med 2013;5(167) 167ra5.

[210] Wang L, Chang J, Varghese D, Dellinger M, Kumar S, Best AM, et al. A small molecule modulates Jumonji histone demethylase activity and selectively inhibits cancer growth. Nat Commun 2013;4:2035.

[211] Rose NR, Woon EC, Tumber A, Walport LJ, Chowdhury R, Li XS, et al. Plant growth regulator daminozide is a selective inhibitor of human KDM2/7 histone demethylases. J Med Chem 2012;55(14):6639–43.

[212] Sayegh J, Cao J, Zou MR, Morales A, Blair LP, Norcia M, et al. Identification of small molecule inhibitors of Jumonji AT-rich interactive domain 1B (JARID1B) histone demethylase by a sensitive high throughput screen. J Biol Chem 2013;288(13):9408–17.

[213] Sekirnik R, Rose NR, Thalhammer A, Seden PT, Mecinovic J, Schofield CJ. Inhibition of the histone lysine demethylase JMJD2A by ejection of structural Zn(II). Chem Commun (Camb) 2009;14(42):6376–8.

[214] Lohse B, Nielsen AL, Kristensen JB, Helgstrand C, Cloos PA, Olsen L, et al. Targeting histone lysine demethylases by truncating the histone 3 tail to obtain selective substrate-based inhibitors. Angew Chem Int Ed Engl 2011;50(39):9100–3.

[215] Woon EC, Tumber A, Kawamura A, Hillringhaus L, Ge W, Rose NR, et al. Linking of 2-oxoglutarate and substrate binding sites enables potent and highly selective inhibition of JmjC histone demethylases. Angew Chem Int Ed Engl 2012;51(7):1631–4.

[216] Thinnes CC, England KS, Kawamura A, Chowdhury R, Schofield CJ, Hopkinson RJ. Targeting histone lysine demethylases-Progress, challenges, and the future. Biochim Biophys Acta 2014.

[217] Wang Z, Patel DJ. Small molecule epigenetic inhibitors targeted to histone lysine methyltransferases and demethylases. Q Rev Biophys 2013;46(4):349–73.

[218] Heinemann B, Nielsen JM, Hudlebusch HR, Lees MJ, Larsen DV, Boesen T, et al. Inhibition of demethylases by GSK-J1/J4. Nature 2014;514(7520):E1–E2.

[219] Wagner EK, Nath N, Flemming R, Feltenberger JB, Denu JM. Identification and characterization of small molecule inhibitors of a plant homeodomain finger. Biochemistry 2012;51(41):8293–306.

[220] Marcon E, Ni Z, Pu S, Turinsky AL, Trimble SS, Olsen JB, et al. Human-chromatin-related protein interactions identify a demethylase complex required for chromosome segregation. Cell Rep 2014;8(1):297–310.

[221] Hayakawa T, Ohtani Y, Hayakawa N, Shinmyozu K, Saito M, Ishikawa F, et al. RBP2 is an MRG15 complex component and down-regulates intragenic histone H3 lysine 4 methylation. Genes Cells 2007;12(6):811–26.

[222] Nishibuchi G, Shibata Y, Hayakawa T, Hayakawa N, Ohtani Y, Shinmyozu K, et al. Physical and functional interactions between the histone H3K4 demethylase KDM5A and the nucleosome remodeling and deacetylase (NuRD) complex. J Biol Chem 2014.

[223] Kuczynski EA, Sargent DJ, Grothey A, Kerbel RS. Drug rechallenge and treatment beyond progression—implications for drug resistance. Nat Rev Clin Oncol 2013;10(10):571–87.

[224] Heard E, Martienssen RA. Transgenerational epigenetic inheritance: myths and mechanisms. Cell 2014;157(1):95–109.

[225] Lim JP, Brunet A. Bridging the transgenerational gap with epigenetic memory. Trends Genet 2013;29(3):176–86.

EMERGING EPIGENETIC THERAPIES: LYSINE ACETYLTRANSFERASE INHIBITORS

21

Stephanie Kaypee[1], Somnath Mandal[1,2], Snehajyoti Chatterjee[1,3], and Tapas K. Kundu[1]

[1]*Transcription and Disease Laboratory, Molecular Biology and Genetics Unit, Jawaharlal Nehru Centre for Advanced Scientific Research, Bangalore, Karnataka, India,* [2]*Department of Biochemistry, Faculty of Agriculture, Uttar Banga Krishi Viswavidyalaya, Pundibari, Cooch-Behar, West Bengal, India,* [3]*Laboratoire de Neurosciences Cognitives et Adaptatives, Université de Strasbourg-CNRS, GDR CNRS, Strasbourg, France*

CHAPTER OUTLINE

1 INTRODUCTION

The wrapping of DNA around the histone octamer (two H2A-H2B dimers and a H3-H4 tetramer) forms the nucleosome which is the basic unit of chromatin. The nucleosomal DNA is further compacted in the presence of H1 linker histone to form chromatosomes [1,2]. The extent of compaction of chromatin is orchestrated by the interplay between the histone post translational modifications (PTM) and chromatin-associated proteins such as modifiers and remodelers.

Histones undergo a vast array of PTMs which are mainly clustered on their N-terminal unstructured tails. Histone PTMs direct the recruitment of these chromatin-associated factors and form a regulatory platform on which various nuclear events such as transcription, replication and DNA damage response are executed. Combinatorial effect of these modifications lead to specific epigenetic readouts interpreted in the form of chromatin compaction, differential gene expression patterns, replication regulation, and DNA damage foci recognition. These modification moieties are "read" by proteins possessing specialized domains which can bind and in turn influence chromatin dynamicity [3]. This combination of histone modifications, their crosstalk and its functional interpretation establishes the "language" of histone crosstalk [4].

Since the discovery of histone acetylation, a predominant PTM present on chromatin, lysine N-ε-acetylation has been implicated in various nuclear functions mainly pertaining to transcription activation. Acetylation of the positively charged lysine (K) residue presumably neutralizes the charge which leads to the alteration of the histone octamer and DNA interaction [5]. The presence of histone acetylation dictates the degree of compaction of chromatin, which directly translates into the level of gene expression [6–9]. Apart from altering the degree of DNA compaction, acetylation serves another important purpose; the acetylation on histone tails provides a platform for the recruitment of various chromatin modifiers and transcription factors [10,11]. These proteins are recruited to the chromatin via their bromodomain that specifically recognizes acetylated lysine residues. These PTM "readers" help in the integration of various nuclear signals and thereby allow extensive crosstalk between histone modifications.

Over the past decade, lysine acetylation has gained importance as an indispensible non-histone posttranslational modification. The presence of histone acetylation was reported as early as 1964. In this pioneering study, it was first shown that acetylated histones had a reduced inhibition toward messenger RNA (mRNA) synthesis compared to non-acetylated histones *in vitro* [8]. In 1979, the first non-histone protein reported to be acetylated was high mobility group proteins, but the functional role of acetylation was appreciated only after the tumor suppressor Tp53 was reported to be acetylated on its C-terminal domain 20 years later [12,13]. Acetylation of p53 has been shown to be indispensible for its function as a tumor suppressor during DNA damage response. CBP/p300 and p300/CBP associated factor (PCAF) mediated acetylation of p53 enhances its sequence specific DNA binding and increases protein stability by interfering with MDM2 binding, while acetylation of p53 by Tip60 is important for p53 mediated apoptosis [13–18]. Since the discovery of p53 as an acetylation substrate, a surge of studies began to identify other such proteins undergoing this particular PTM.

In the first study toward identifying the global acetylome, Kim et al. demonstrated for the first time that proteins involved in diverse cellular functions like transcription, chromatin remodeling, RNA splicing and translation, chaperones, cytoskeletal proteins were substrates for lysine acetylation. Their group enriched 388 acetylated peptides on 195 proteins, present in different cellular compartments using antibody against the acetylated lysine modification. Interestingly, their study identified 133 mitochondrial proteins and abundant metabolic enzymes in the acetylome [19]. This study revealed that the function of lysine acetylation is not limited to the nucleus.

Adopting a similar approach, Choudhary et al. advanced the study using the stable-isotope labeling by amino acid in cell culture method to identify a staggering 3500 acetylation sites on over 1700 proteins. Again the distribution of these acetylated proteins provided confounding evidence that lysine acetylation plays a key regulatory role in diverse cellular pathways. The astounding numbers of acetylation sites are comparable to the phosphoproteome. The interesting observations made by this study demonstrated that lysine acetylation is well conserved through evolution and there is a predominant presence of lysine acetylation on structured regions of proteins compared to unstructured regions. It is apparent that lysine acetylation plays an important role in protein structure and protein-protein interaction. Components of numerous macromolecular protein complexes involved in functions ranging from chromatin remodeling to DNA damage repair and replication have been found to be acetylated [20].

The "writers" of lysine acetylation are lysine acetyltransferases (KATs) and the "erasers" are Lysine Deacetylases (KDACs). Chemically, the transfer of an acetyl functional group from Acetyl-Coenzyme A (acetyl-CoA) onto the ε-N on the lysine residue is catalyzed by KATs and the reverse reaction is catalyzed by KDACs. The first KAT was discovered as p55 or HAT A from the macronuclei in *Tetrahymena*, which was identified as a close homolog of the yeast transcription adapter protein Gcn5p a year later [21,22]. The first KAT to be cloned and bacterially expressed was the yeast KAT, HAT1 (or KAT1) [23]. Subsequently, several proteins were found to possess intrinsic acetyltransferase activity, and they are now broadly classified into Type A and Type B based on their subcellular localization. Type-B KATs are predominantly located in the cytoplasm and catalyze the acetylation on H4K5 and -K12 of nascent histones before they are transported to the nucleus and assembled into the chromatin. Hat1 is the catalytic subunit of the Type-B KAT complex [24]. Earlier studies in yeast have shown that the Hat1 complex is involved in nuclear functions such as telomeric silencing, mismatch repair [25]. Another cytoplasmic KAT, HAT4, has be shown to acetylate free H4 including its globular lysine sites, H4K79ac and H4K91ac [26]. Type-B HATs share homology with general control non-derepressible 5 (Gcn5) and belong to the GNAT superfamily [26,27]. In contrast to the Hat1 complex which cannot acetylate histones assembled into chromatin, the Type-A KATs are nuclear acetyltransferases which are capable of acetylating chromatinized histones [28]. Nuclear KATs belong to five major families based on their structural homology, namely, Gcn5-related N-acetyltransferase (GNAT) family consisting of Gcn5/KAT2A), PCAF /KAT2B) and related proteins, the p300/CBP family consisting of p300 (E1A-associated 300 kDa protein/KAT3B) and CBP (CREB binding protein/KAT3A) and the MYST family where MYST is an acronym of the four founding members MOZ (monocytic leukemia zinc finger protein/KAT6A), Ybf2 (KAT6), Sas2 (something about silencing 2/ KAT8), and TIP60 (TAT interacting 60 kDa protein/ KAT5). General transcription factors related KATs such as TAF1 (KAT4), TFIIIC subunits 90 (KAT12) and 110, and TAFII250, and nuclear receptors such as SRC1 (KAT13A), ACTR (KAT13B) possess acetyltransferase activity [29,30].

The ubiquitous and dynamic nature of lysine acetylation demands for a tight regulation of acetylation levels for the maintenance of cellular homeostasis. Thus, the deregulation of the KATs or KDACs can alter the fine balance between acetylation and deacetylation resulting in oncogenesis, metabolic and neurodegenerative disorders [31–33].

In this chapter, we will discuss in details the roles played by different KATs in the manifestation of cancer. We will also shed light on the advancement in the field of epigenetic therapeutics with respect to the development of natural or synthetic KAT inhibitors which are emerging as useful modalities to combat several diseases.

2 ABERRANT HISTONE ACETYLATION PATTERNS IN DISEASES

Alteration in global histone "language" is often associated with disease conditions which are mainly the result of deregulation of chromatin modifying enzymes and associated chromatin remodelers. Hence, certain chromatin modifications have been observed as potential biomarker of diseases like cancer. Altered histone acetylation patterns are often predictive of the progression of the disease. Global loss of histone H4 lysine 16 acetylation (H4K16ac) with a concomitant decrease in histone H4 lysine 20 trimethylation (H4K20me^3) levels have been considered a hallmark of cancer [34]. In prostate cancer, global hypoacetylation, particularly in H3K9ac, H3K18ac, and H3K16ac levels are indicative of poor prognosis and recurrence [35]. Increase in the level of H3K56ac, a globular histone mark associated with DNA damage, has been observed in multiple cancers and is elevated in undifferentiated cells [36]. Study from our group has shown a global histone hyperacetylation which strongly correlated with oral cancer manifestation [37]. In nonsmall cell lung carcinoma (NSCLC), the loss of H2AK5ac correlates with poor survival, while loss of H3K9ac is associated with better survival of patients [38]. It has also been reported that loss of acetylation on H3K9 and H3K18 is linked to better prognostic outcome in glioma, whereas loss of H3K18ac is indicative of better survival in esophageal carcinoma [39–41]. Distinct histone acetylation patterns have been observed in inflammatory disorders as well; the acetylation on H4K8 and -K12 was associated with interleukin-1β-induced gene expression [42]. In polyglutamine (poly(Q)) diseases, such as Huntington disease, CBP is sequestered by unfolded poly(Q) protein aggregates, resulting in decreased acetyltransferase activity which culminates in global reduction in histone acetylation levels [43,44]. Hence, it is evident that the state of the epigenetic landscape plays a pivotal role in the progression and manifestation of cancer and other diseases (Figure 21.1), thereby justifying the need for therapeutic intervention which may help reverse disease phenotypes.

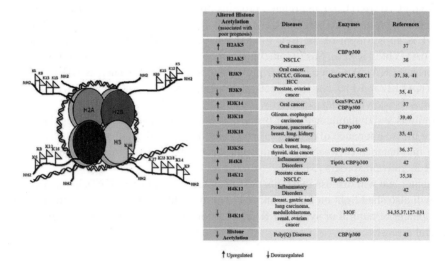

Altered Histone Acetylation (associated with poor prognosis)	Diseases	Enzymes	References
↑ H2AK5	Oral cancer	CBP/p300	37
↓ H2AK5	NSCLC		38
↑ H3K9	Oral cancer, NSCLC, Glioma, HCC	GcnS/PCAF, SRC1	37, 38, 41
↓ H3K9	Prostate, ovarian cancer		35, 41
↑ H3K14	Oral cancer	GcnS/PCAF, CBP/p300	37
↑ H3K18	Glioma, esophageal carcinoma	CBP/p300	39,40
↓ H3K18	Prostate, pancreatic, breast, lung, kidney cancer		35, 41
↑ H3K56	Oral, breast, lung, thyroid, skin cancer	CBP/p300, Gcn5	36, 37
↑ H4K8	Inflammatory Disorders	Tip60, CBP/p300	42
↓ H4K12	Prostate cancer, NSCLC	Tip60, CBP/p300	35,38
↑ H4K12	Inflammatory Disorders		42
↓ H4K16	Breast, gastric and lung carcinoma, medulloblastoma, renal, ovarian cancer	MOF	34,35,37,127-131
↓ Histone Acetylation	Poly(Q) Diseases	CBP/p300	43

↑ Upregulated   ↓ Downregulated

FIGURE 21.1

Histone acetylation marks and their link to diseases. The nucleosome cartoon depicts the positions of histone acetylation marks on the N-terminal tails of the four core histones: H2A, H2B, H3, and H4. The table alongside summarizes the deregulation of histone acetylation in different diseases.

3 ROLE OF KATs IN CANCER

There are numerous evidences that implicate the role of KATs or their acetylated substrates in tumorigenesis. Functional alteration of KATs in malignancies is a well documented phenomenon. Mutations in KATs, such as point mutations, gene/gene segment amplification or deletions, translocations have been observed in many cancers. Alteration at the expression levels of KATs have also been linked to the progression of cancers. This perturbation in KAT functions has a rather broad consequence, manifested through the altered functions of their substrates, as an example it is known that aberrant histone acetylation patterns have a widespread effect on the expression profile of genes in different pathological conditions.

3.1 p300/CBP FAMILY

CBP and its paralog p300 are known transcriptional coactivators possessing intrinsic acetyltransferase activity [45,46]. These large proteins facilitate the transcription of active genes by acting as adapter proteins bridging DNA binding transcription factors to the basal transcription machinery and by loosening chromatin through histone tail acetylation [47]. CBP and p300 share an overall 63% amino acid sequence identity and ~90% sequence homology at the histone acetyltransferase domain leading to a degree of target redundancy [48]. CBP/p300 bind to a myriad of transcription factors through its four transactivation domains namely, TAZ1, KIX, TAZ2, and IBiD. Moreover, p300/CBP can function both as tumor suppressors and oncogenes, depending on the cellular context. This dual nature, in part, is due to the vast interactome of these proteins. CBP/p300 interacts with tumor suppressors like p53, E2F, SMAD, FOXO-1, -3a and -4, BRCA1, RUNX while they also act as coactivators for oncogenes such as myb, myc, c-Fos, adenoviral E1A, and HPV E6 protein [49]. By virtue of its interaction, CBP/p300 is involved in an array of cellular functions such as DNA repair, differentiation, cell cycle, and apoptosis. Heterozygous germline mutations in the CBP (rarely p300) gene results in a genetic disorder known as the Rubinstein-Tyabi syndrome. This disorder is marked by craniofacial abnormalities, mental retardation, and a predisposition toward tumorigenesis [50–52]. CBP/p300 is an important component of hematopoiesis, hence deregulation of these enzymes have been frequently linked to hematological malignancies [53]. Mice heterozygous for CBP (CBP$^{+/-}$) have a higher incidence of hematological cancers, moreover chimera mice generated by injecting p300$^{-/-}$ CBP$^{-/-}$ cells develop tumors originating from these null cells [54,55]. Several inactivating somatic mutations associated with loss of wild type allele, that is, loss of heterozygosity have been reported in solid cancers such as colorectal, breast, ovarian, hepatocellular, and oral carcinomas [56]. These observations indicate the role of p300/CBP as a putative tumor suppressor.

On the flip-side, CBP/p300 has been reported to be involved in malignant transformations. Frequent chromosomal translocations disrupt the CBP and p300 genes and form gain-of-function (GOF) fusion proteins which contribute to tumorigenesis in leukemia and lymphoma. In acute myeloid leukemia (AML), fusions between MOZ and CBP (t(8,16)) or p300 (t(8, 22)), and mixed-lineage leukemia (MLL) and CBP (t(11, 16)) or p300 (t(11,22)) through translocations, result in chimeric proteins possessing in-frame bromo and HAT domains of CBP/p300 [57–59]. These fusion proteins are constitutively active resulting in chromatin alterations and a higher overall gene expression level which is a consequence of aberrant histone acetylation. The oncogenicity of CBP and p300 may also be conferred by virtue of their interaction with oncoproteins such as c-myb, c-myc, c-jun, c-fos, which are associated with cell proliferation and tumorigenesis [60–63]. Acute myeloid leukemia1-eleven twenty one oncoproteins induce myeloid tumorigenesis in an myb-p300-dependent manner and the interaction between myb and p300 is important for the cellular transformation [64]. Likewise, it has been observed that targeting p300/CBP or disrupting the interaction between p300/CBP and their oncogenic partners such as HIF1-α, β-catenin, NFκB, and AR can inhibit cell proliferation and tumor progression [65–70]. CBP/p300 is important for proper cell cycle progression, as their inhibition by microinjected antibodies result in G1/S arrest and embryonic fibroblast null for CBP/p300 have proliferation defects and have a higher rate of attaining senescence [71]. Hence, it is not surprising that overexpression of p300 has been reported in several cancers including prostate cancer, breast cancer, nonsmall cell lung carcinoma, and hepatocellular carcinoma (HCC) [72–77]. Cancer cell proliferation in AR positive prostate cancer and doxorubicin-resistance in bladder cancer has been linked to p300 protein levels. The higher expression levels of p300 are also indicative of poor prognosis in these cancers [78,79].

Recent studies have shown the importance of PTMs in the regulation of CBP/p300 activity. p300 phosphorylation by cyclin-dependent kinase 1 (CDK1) and ERK1/2 on Ser1038 and Ser2039 induces the degradation of p300 which in turn enhanced the progression in lung cancer [80]. CBP/p300 is capable of enhancing its own activity by autoacetylation on an unstructured lysine-rich loop present at its active site [81]. Work from our group has shown the role of histone chaperone, nucleophosmin (NPM1), in conjunction with GAPDH, in the induction of p300 autoacetylation and the positive correlation between hyperactivated p300 and oral cancer manifestation [37]. Therefore, it is evident that the proper functioning of CBP/p300 is required for normal cellular functioning and deviation from it may lead to adverse pathological conditions including cancer (Figure 21.2).

3.2 GNAT FAMILY

Gcn5 and PCAF are homologous transcription co-activators present in metazoans as part of large multisubunit complexes like the human SPT3-TAF9-GCN5 acetyltransferase (STAGA), TATA binding protein (TBP)–free–TAF complex (TFTC), and PCAF complexes [82–85]. These complexes have a role in UV-damage DNA repair, the acetyltransferase catalytic subunit acetylates the histones around the damaged DNA and facilitates the loosening of chromatin and recruitment of nucleotide excision repair machinery [86]. Since Gcn5 and PCAF containing complexes are involved in maintenance of the integrity of DNA, it is likely that these lysine acetyltransferases have a role in tumorigenesis, a condition where the integrity of the genome is perturbed. PCAF is a co-activator of p53-target genes and moreover it regulates p53 activity by acetylating K320, which promotes p53-dependent p21 transcription and cell cycle arrest during DNA damage. PCAF also possesses ubiquitin E3 ligase activity and it can negatively regulate p53 levels after DNA damage, by promoting its degradation in conjunction with the oncoprotein HDM2 [87]. PCAF has been observed to be downregulated in HCC and HCC cell lines while overexpression of PCAF in these cell lines were shown to induce apoptosis through inactivation of AKT pathway and an increase in the levels of H4 acetylation [88]. Studies have shown that PCAF- and Gcn5-containing complexes associate with the oncoprotein, Myc, and this association promotes the recruitment of Gcn5/PCAF to chromatin at the Myc-target promoters [89]. Moreover, truncated Myc protein that loses the ability to interact with the STAGA complex has been shown to possess reduced transactivation ability and cannot transform cells. These results suggest the indispensible role of Gcn5 and PCAF in Myc-mediated oncogenic transformations [90].

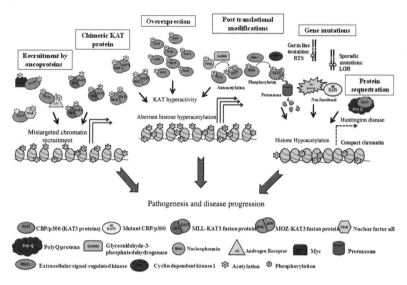

FIGURE 21.2

Deregulation of CBP/p300 in diseases. Three scenarios of CBP/p300 (KAT3 proteins) deregulation have been illustrated: first, the targeted chromatin recruitment of CBP/p300 by oncoproteins and fusion proteins; second, the aberrant histone acetylation by hyperactive CBP/p300 due to in-frame MOZ-CBP/p300 chimeric protein, p300/CBP overexpression and enhancement of autoacetylation by NPM1and GAPDH; third, loss of KAT activity by degradation in the presence of CDK1 or ERK1/2 mediated phosphorylation, gene mutations in the case of Rubinstein-Tyabi Syndrome (RTS) or Loss of Heterozygosity (LOH) in different malignancies and the sequestration of CBP by poly(Q) proteins in Huntington Disease.

PCAF is also a key positive cofactor in the Hedgehog-Gli pathway and is essential for the proliferation of medulloblastoma and glioblastoma cells [91]. PCAF endows drug resistance in a Twist1-dependent or E2F-dependent manner and by upregulating DNA damage genes [92,93]. It has also been reported that PCAF-mediated H3K9 acetylation is elevated manifold at MDR1 promoters conferring multidrug resistance in breast cancer, while knockdown of PCAF/Gcn5 effectively sensitizes these cells to therapeutics [94].

3.3 MYST FAMILY

MYST family KATs are evolutionarily conserved acetyltransferases. In humans, MYST family comprises of Tip60, HAT bound to ORC1 (HBO1), MOZ, MOZ-related factor (MORF), and males absent on first (MOF). These KATs share a conserved MYST domain consisting of an acetyl-CoA-binding domain and a Zinc finger domain [95]. These KATs are members of multi-protein complexes involved in crucial cellular functions such as gene expression, DNA damage response and replication. Dysregulation of these enzymes have been linked to a number of pathologies including carcinogenesis [96].

3.3.1 Tip60

HIV-1 Tat interacting protein, 60kDa (Tip60) is intimately linked to DNA repair pathway [97]. It is responsible for the acetylation and activation of ATM kinase involved in the double-strand break (DBS) repair pathway [98]. Activated ATM kinase then rapidly phosphorylates downstream targets such as p53 and the histone variant H2AX (γH2AX) around the DBS demarcating the damage foci for the recruitment of damage repair proteins [99–101]. Tip60 acetylates p53 on Lys120 and activates p53 in concert with ATM [16,17]. Depending on the persistence and severity of the damage the cell either undergoes cell cycle arrest or apoptosis. In colon and lung carcinoma, there is a significant downregulation in the expression levels of Tip60, which could possibly lead to hypoacetylation and reduced activation of p53 and subsequently p53 downstream gene expression [102]. Tip60 is a transcription co-activator and can be linked to cancer through its interacting partners. Studies have shown that Tip60 can enhance Myc-dependent transformation and can contribute to cancer progression [103]. Tip60 has also been linked to hormone-independent prostate cancer progression, where in advanced forms of the cancer there is a direct correlation between nuclear Tip60 accumulation and ligand-independent AR activation in prostate cancer progression [104].

3.3.2 HBO1

HBO1/KAT7 has a close association with replication. It was first identified as a lysine acetyltransferase associated with ORC1 protein, the largest subunit of the origin recognition complex [105]. HBO1 aids in the pre-replication complex assembly and replication initiation [106–108]. HBO1 exists in two distinct complexes; ING5-associated HBO1 which co-purifies with MCM proteins and may be involved in replication and ING4-associated HBO1 [107]. HBO1 is linked to cancer by its association with tumor suppressors ING4/5 and JADE 1/2/3. ING 4/5 specifically recognizes H3K4 trimethylation through its PHD domain and targets HBO1 to its target genes. The interaction of ING4/5 and H3K4 trimethylation augments HBO1 activity which is required for ING-specific tumor suppressor activity such as apoptosis to genotoxic stress and inhibition of anchorage-independent growth of tumor cells [109]. Alternate splicing of JADE 1/2/3 regulates this association of HBO1 with ING 4/5 and in turn modulates the binding of the complex to chromatin [110]. Tumor suppressor p53 directly binds to HBO1 during cellular stress and negatively regulates its activity to stall replication [111]. There have also been reports indicating the role of HBO1 in tumor progression and drug resistance. Phosphorylation of HBO1 by Polo-like kinase 1 transcriptionally elevates c-Fos expression which consequently results in the higher expression of MDR1 thereby conferring resistance to drugs such as the deoxycytidine-analogue, gemcitabine [112]. Cyclin E/cdk2 phosphorylates HBO1 at Tyr88, and this modified form of HBO1 is important for promoting the enrichment of breast cancer stem cells [113]. HBO1 has also been found to be a magnitude more abundant in cancer cell lines such as Saos-2 and MCF7 as well as primary tumors of the breast, stomach, testis, ovary, esophagus, and bladder in comparison to normal tissue [114].

3.3.3 MOZ and MORF

MOZ/KAT6A and MORF/KAT6B) are highly homologous transcription co-activators belonging to the MYST family [115,116]. MOZ was first identified as a chimeric protein fused with CBP by chromosomal translocation t (8;16) (p11;p13) in AML, which is characterized by poor response to chemotherapeutics [117]. Fusions between MOZ and CBP homolog p300 t (8;22) (p11;q13) is also observed in AML [118]. Similarly, MORF-CBP fusions t (10;16) (q22;p13) mirror the MOZ-CBP in terms of

GOF, aberrant histone acetylation and gene expression [119,120]. Transcription intermediary factor 2 (TIF2), a member of the p160 co-activator family of proteins, is yet another MOZ fusion partner in AML [121,122]. The MOZ-TIF2 fusion protein contains an intact CBP/p300 interacting domain (CID) which associates with CBP/p300 forming a hyperactive chimera which results in deregulated acetylation and depletion of CBP/p300 from PML bodies preventing p53 transactivation. Moreover, the CID domain of MOZ-TIF2 fusion protein is essential for transformation and leukemogenesis suggesting the contribution of CBP/p300 in the GOF of MOZ-TIF2. MOZ-TIF2 can also confer self-renewable properties to hematopoietic progenitor cells [123].

3.3.4 MOF

MOF/KAT8) is the human homolog of the drosophila catalytic subunit of the Male specific lethal (dMSL) complex involved in dosage compensation [124]. hMOF associates with two complexes hMSL and non-specific lethal (hNSL) which have distinct substrate specificities. hMSL is responsible for H4K16ac, a key acetylation mark associated with higher chromatin structures, while hNSL has a broader specificity for K5, K8, and K16 on histone H4 [125,126]. Hence, hMOF is responsible for the majority of H4K16 acetylation in human cells [125]. Loss of H4K16ac is considered a hallmark of cancer and is associated with chromosomal aberration, cell cycle abnormalities, and genomic instability [34]. Downregulation of hMOF is a common feature of many cancers including colorectal cancer, gastric cancer, renal cell carcinoma, ovarian cancer, primary breast carcinoma, and medulloblastoma [127–130]. Conversely, hMOF has been reported to be overexpressed in NSCLC and mediates drug-resistance in an Nrf2-dependent manner. The overexpression of hMOF is associated with large tumor size, metastasis, and poor prognosis of the disease [131].

4 LYSINE ACETYLTRANSFERASES AS POTENTIAL THERAPEUTIC TARGETS

KATs (EC No. 2.3.1.48) were the first family of histone modification enzymes characterized at the molecular, biochemical and structural levels. Since, the identification and characterization of the *Tetrahymena* Gcn5 KAT in 1996 [22], a large number of KATs have now been identified and characterized in metazoans. The last decade and half has witnessed tremendous progress in the field of small molecule inhibition of KATs. Based on their synthesis and source they can be divided into four broad categories, namely, bisubstrate inhibitors, natural products identified as KAT inhibitors, synthetic derivatives, and analogues of natural products, and synthetic small molecules (Table 21.1).

4.1 BISUBSTRATE INHIBITORS

This class of inhibitors is the pioneer in the KAT inhibitor field, since the discovery of a structural mimic of acetyl-CoA, covalently linked to peptide substrate with various chain lengths. The first reported KAT inhibitor, Lysyl CoA showed high selectivity toward p300. By the same rational-design approach H3CoA20 was synthesized which exhibited specificity toward PCAF [132]. In spite of having distinct advantages like specificity and potency (works in the nanomolar range), these molecules suffer from a number of limitations like poor cell permeability, metabolic instability and poor pharmacokinetic properties. To improve these shortcomings, a "tat" peptide was conjugated along with the molecule [133] which improves the cell permeability. This compound has been used extensively as biological probe

Table 21.1 Lysine Acetyltransferase Inhibitors

Name	Source	Class of compound	Class of KAT inhibition	IC$_{50}$	Nature of inhibition
Natural					
Anacardic acid	Cashew nut Shell Liquid (CNSL)	6-pentadecylsalicylic acid	p300, PCAF, Tip60	8.5 μM for p300, 5 μm for PCAF and 64 ± 15 μM Tip60	Non-competitive for Ac-CoA (p300 and PCAF), Competitive for Ac-CoA (Tip60)
Garcinol	Garcinia indica (kokum)	Polyisoprenylated benzophone(PBD)	p300, PCAF	5 μM for p300 and 7 μm for PCAF	Competitive for histone and uncompetitive for Ac-CoA
Curcumin	Curcuma longa (turmeric)	Polyphenols (diferuloyl methane)	p300	25 μm for p300	Mixed inhibition
Plumbagin	Plumbago rosea (chitraka)	1,4 napthoxy hydroquinone	p300	20–25 μm for p300	Noncompetitive
Epigallocatechin-3-gallate (EGCG)	Green tea	Polyphenol	p300, PCAF, Tip60	30 μm for p300, 60 μm for PCAF and 70 μm for Tip60.	Uncompetitive
Sanguinarine	Sanguinaria Canadensis and Argemone Mexicana	Quaternary benzo-phenanthridine alkaloids (QBA)	p300, PCAF	10 μm	Not determined
Procyanidin B3	Grape seeds	Flavonoids	Predominantly p300.	Around 25 μm	Uncompetitive
Delphinidin	Punica granatum	Polyphenol	P300 and CBP	30 μm	Not determined
Gallic acid	Rosa rugosa	(3,4,5 trihydroxy benzoic acid) polyphenol	P300 and CBP	20 μm	Uncompetitive
Embellin	Embelia ripes	Hydroxybenzoquinone	PCAF	7.2 μm	Non-competitive for H3 and competitive for acetyl CoA
Semi-Synthetic					
LTK-13, LTK-14, LTK-19	Isogarcinol as lead compound	Polyisoprenylated benzophone (PBD)	p300	5–7 μm	Non-competitive
CTK7A	Curcumin as lead molecule	Hydrazinocucurmin	p300 and PCAF	25 μm	Mixed inhibition for both the substrates.
PTK1	Methyl derivative of Plumbagin as lead molecule	1,4 naphthoquinone analogue	p300	20–25 μm	Noncompetitive for both acetyl CoA and core histones

Compound	Lead/Type	Structure	Target	K_i/IC$_{50}$	Mechanism
Compound 6d	Anacardic acid as lead structure	(Z)-2-Hydroxy-6-(4-(pentyloxy)styryl) benzoic acid	PCAF	662 ± 64 μM	Not determined
MG149	Anacardic acid as lead molecule	6-alkylsalicylate analogue	Tip60	74 ± 20 μM	Competitive
4-cyano-3-trifluoro methylphenylbenza mides	2,6-dihydroxybenzoic acid	Benzamides	p300	50 μM	Not determined
Cyclohexanone derivatives	Curcumin as lead structures	2,6-Bis(3-bromo-4-hydroxybenzylidene) cyclohexanone	p300	5 μM	Not determined
Synthetic					
Lysyl CoA	Bi-substrate analogue	Peptide	p300	50 nM	Competitive
H3-CoA	Bi-substrate analogue	Peptide	PCAF	300 nM	Competitive
Spd CoA	Bi-substrate analogue	Peptide	p300 and PCAF	20 nM for p300 and 20 μM for PCAF	Competitive
γ-butyrolactone	Synthetic	α-methylene-γ-butyrolactones	GCN5	100 μM	Nonirreversible
CCT000791 and CCT000792	Synthetic	isothiazolones	p300 and PCAF	7.3 and 15 μM for PCAF and 15 μM for p300	Irreversible covalent binding via thiol interactions
C646	Synthetic	Pyrazolone	p300	400 nM	Competitive
L002	Synthetic		p300	1.98 μM	Not determined
NU9056	Synthetic	isothiazolone	Tip60	2 μM	Not determined
TH1834	Synthetic	pentamidine	Tip60	Not determined	Not determined

to elucidate the role of p300/CBP in physiological processes [134]. In the similar line of research an arginine-rich peptide sequence attached to lysyl CoA, where the sulfur atom is linked to spermidine had been designed. The resultant molecule (Spd-CoA) proved to be a p300/CBP acetyltransferase inhibitor, which enhances the sensitivity to cisplatin, fluracil, UV-C indication, impeded acetylation-dependent DNA repair [135]. Due to its polyamine nature its uptake was markedly facilitated in cell via polyamine transporter. Similar potent bisubstrate inhibitors with peptide nature were developed against Tip60 and Esa1 (H4K16CoA) [136].

4.2 KAT INHIBITORS: NATURAL PRODUCTS

Natural small molecules, traditionally known for their multiple medicinal properties emerged as an invaluable resource for development of lysine acetyltransferase enzyme inhibitors. They provide tremendous molecular diversity and perfect template for further derivatization.

Work from our group reported the first natural KAT inhibitor, anacardic acid which was isolated from cashew nut shell liquid. This salicylic acid derivative (nonadecyl salicylic acid) resulted to be a potent inhibitor of p300 and PCAF lysine acetyltransferases and has an IC50 value in the micromolar range [137]. AA blocks the activation of the p65 subunit of NFκB. Human cancer cells treated with AA exhibits sensitivity to ionizing radiation, and imparts anti-inflammatory anti-obese and anti-parasitic activity [138].

The polyphenolic compound isolated from *Curcuma longa* rhizome, is a p300-specific KAT inhibitor with no inhibitory effect on PCAF [139]. This molecule shows tremendous possibility for lead molecule, as it posses broad spectrum cancer chemo-preventive activity in preclinical animal models. Curcumin exhibited strong anticancer activity by inhibiting lung cancer, prostate cancer, breast cancer and colorectal cancer [140]. The anticancer potential of curcumin results from its ability to modulate transcription factors like NFκB, AP-1, and EGR-1 which leads to the perturbation of downstream signaling pathways [141]. In addition to this, curcumin is also a potent antioxidant and anti-inflammation agent. In recent studies, curcumin has been found to be influencing adult hippocampal neurogenesis and neural plasticity indicating its potential application in neurological disorders like Alzheimer's disease [142]. Furthermore, the ability of curcumin to inhibit viral Tat protein acetylation by p300 presents itself as a promising drug for the treatment of AIDS [143]. Pharmacologically, curcumin has been found to be nontoxic and safe.

Garcinol, a naturally occurring polyisoprenylated benzophenone isolated from *Garcinia indica* or kokum fruit rind (indigenous to southern parts of India) acts as a potent inhibitor of both PCAF and p300 acetyltransferase activity [144]. This molecule shows KAT inhibitory activity at sub micromolar level, and is a potent inducer of apoptosis in HeLa cells. Garcinol also exhibits strong antioxidative and free radical scavenging activity, antiangiogenic (by downregulating VEGF, MMP-9, PGE-2), antiproliferative (downregulating cyclinD1), anti-inflammatory activity (by downregulating COX, NOS, NFκB expression), activation of apoptotic pathways (activation of caspase-3). Garcinol treatment alters expression of chromatin modifying enzymes in breast cancer cell line (MCF7) resulting in severe growth arrest and highlights its potential for cancer chemotherapeutic agent [145]. It has also been tested in several cancer cell types *in vitro* and *in vivo* [146]. Constitutive activation of pro-inflammatory transcription factors like STAT3 and NFκB is associated with tumorigenesis and metastasis. In a recent study, Sethi et al. demonstrated the antiproliferative and pro-apoptotic effects of garcinol on HCC cells by the inhibition of the transcription factor STAT3 dimerization and sequence-specific

DNA binding [147]. The same group also reported the antitumor effect of garcinol in head and neck squamous cell carcinoma (HNSCC cells) by demonstrating its ability to suppress multiple pro-inflammatory signaling cascades [148].

In search of new scaffold for KAT inhibitors, Ravindra et al. in 2009 isolated and identified a natural hydroxynaphthoquinone, plumbagin from the roots of *Plumbago rosea* (an Indian ayurvedic medicinal herb known as chitraka) as yet another naturally occurring p300 specific KAT inhibitor [149]. The remarkable part of study was its structure-activity analysis, where for the first time a chemical entity (a single hydroxyl group functional group) was shown to be responsible for inhibition of p300 acetyltransferase activity. The anti-cancerous property of plumbagin has been attributed to its effect on multiple signaling pathways and its ability to undergo redox cycling properties generating reactive oxygen species (ROS) [150].

Sanguinarine is a DNA intercalator alkaloid obtained from the root of *Sanguinaria canadensis* and *Argemone mexicana* is known to possess anti-microbial, anti-oxidant, anti-inflammatory, and pro-apoptotic properties. It targets a variety of cellular components including NFκB, blocks cell cycle, and induces apoptosis in multiple cancerous cell lines [151]. Sanguinarine is a potent inhibitor of p300 and PCAF *in vitro* and *in vivo* [152]. Sanguinarine treatment potently inhibits histone acetylation in cell lines and mice liver with minimal toxicity. Apart from KAT inhibition, sanguinarine can also potently inhibit G9a and CARM1 mediated methylation, conclusively exhibiting a global chromatin modulation property.

Epigallocatechin-3-gallate (EGCG), a major polyphenol found in green tea also adds to the list of growing naturally occurring KAT inhibitors [153]. It exhibited anti-NFκB activity in multiple human malignant cell types. EGCG is a pan-acetyltransferase inhibitor with micromolar specificity toward p300, PCAF, and Tip60 but it does not inhibit KDACs or methyltransferases. EGCG abrogates p300-induced p65 acetylation, enhanced cytosolic IκBα expression levels and subsequently suppressed TNF-α-induced NFκB activation and downstream gene expression.

ProcyanidinB3, a flavonoid richly found in grape seed extract was found to inhibit lysine acetyl-transferase p300 activity. This molecule is a potent anti-cancerous molecule which inhibits p300 mediated acetylation of androgen receptor [154].

Another flavonoid delphinidin isolated from *Punica granatum*, identified as novel p300 KAT inhibitor, does not affect other epigenetic enzymes like KDACs or DNMTs. This molecule suppresses inflammatory signaling via prevention of NFκB acetylation in fibroblast-like synoviocyte MH7A cells [155].

Rosa rugosa Thunb. (Rosaceae) has been traditionally used for treatments of chronic inflammatory disorders, diabetes, pain, and anticancer in Korea. Gallic acid, a polyphenol isolated from *Rosa rugosa* is a potent KAT inhibitor and possessed specificity for majority of KAT enzymes. Gallic acid induced hypoacetylation of p65, by directly inhibiting the activity of KAT enzymes resulting into suppression of NFκB transactivation. Additionally, it interferes with lipopolysaccharide and cytokine-induced inflammatory responses *in vivo*. In short, this study showed selective modulation of NFκB acetylation by KAT inhibition could be a potential mechanism for a new class of anti-inflammatory and chemotherapeutic drugs [156].

Embelin (hydroxybenzoquinone), a cell permeable small molecule isolated from *Embelia ripes* berries is traditionally known for its various biological activities. The pro-apoptotic, anti-tumorogenic, and anti-cancer activities of embelin have been attributed to its inhibitory effect on X-linked Inhibitor of Apoptosis Protein (XIAP). The anti-inflammatory and pro-apoptotic activities of embelin are also mediated through the inhibition of TLR2-mediated activation of NFκB. Embelin treatment leads to the downregulation of genes involved in cell proliferation (COX-2, cyclin-D1, c-Myc) and tumor metastasis (VEGF, ICAM-1, MMP-9) as well as XIAP [157,158]. This molecule has also been associated with antiviral, antibacterial, and antihelmintic activities. In a recent study, our group has reported that embelin specifically inhibits H3K9ac in mice and inhibits recombinant PCAF-mediated acetylation with near complete specificity *in vitro*. Furthermore, the use of embelin made it possible to identify the gene networks that are regulated by PCAF during muscle differentiation, predominantly via PCAF-mediated MyoD acetylation [159].

4.3 KAT INHIBITORS: SYNTHETIC DERIVATIVES AND ANALOGS OF NATURAL PRODUCTS

Although there have been tremendous progress in identifying natural small molecule inhibitors of KATs from the diverse "natural chemical library" present, these molecules suffer from limitations like lack of specificity, pleiotropic effects, toxicity inside mammalian cells, limited bioavailability due to poor cell permeability, promiscuous substrates, limited potency, and poor pharmacokinetic properties. Nevertheless, these molecules provide an excellent parent scaffold on which further "tweaking" is possible. Attempts have been made to further derivatize these molecules to improve the general druggability of these molecules. In the process, these resultant molecules also served as biological probes to delineate the cellular physiology and pathophysiology.

The approach of studying the structure-function relationship of molecules like in the case of garcinol, a nonspecific and toxic KAT inhibitor, lead to the derivatization and synthesis of isogarcinol (IG) which is a product of internal cyclization of garcinol. Controlled modification and monosubstitution at "14" position yields 14-isopropoxy IG (LTK-13) and 14-methoxy IG (LTK-14). Di-substitution of Isogarcinol resulted in 13, 14 disulfoxy IG (LTK-19). The mono substituted IG derivatives LTK-13, -14, and di-substituted LTK-19 specifically inhibit the p300 KAT activity but not the PCAF KAT activity. Selective nature of one such derivative LTK-14 has been used in understanding the regulation of gene expression by p300 KAT. The remarkable property of LTK-14 is that it retains the anti-HIV properties while it is nontoxic to the host T cells [160].

Cellular toxicity is a limitation in case of another natural anticancer p300 inhibitor plumbagin. By exploiting the chemical structure and functional properties of this 1,4-naphthoquinone class of molecules, recently a nontoxic yet equally potent p300 KAT inhibitor, PTK1 was reported. This molecule retains all the other properties of plumbagin but they generate negligible ROS via redox cycling [161].

Semi-synthetic derivatives of AA have been identified with KAT inhibitory properties. *In vitro* KAT inhibition has been shown to correlate with their anti-proliferative effects in multiple cancer cell lines including breast cancer, cervical carcinoma, T-cell lymphoma and prostate adenocarcinoma. By exploring the PCAF enzyme crystal structure conjugated with Coenzyme A, Ghizzoni et al. provided a binding mechanism mode for AA [162]. 4-Cyano-3-trifluoromethylphenylbenzamides have been reported as p300 inhibitors *in vitro* with activity comparable to that of AA [163].

Inhibition of Tip60-dependent activation of ATM by AA sensitized cancer cells to ionization therapy, hence warranting the need for specific Tip60 inhibitors [164]. Ghizzoni et al. derivatized a highly specific Tip60 inhibitor, MG 149, which can effectively inhibit histone acetylation at micromolar concentrations [165].

Curcumin-derived cinnamoyl analogs have been reported to inhibit p300 acetyltransferase activity *in vitro*. The bromo-substituted 2,6-arylidene cyclohexanone was proven to be the most potent compound which led to a decrease in histone H3 acetylation in mammalian cells [166]. Notably, these bromo-substituted phenolic compounds also inhibits methyltransferases (PRMT1, CARM1, and SET7) and sirtuins, and hence are referred to as epigenetic multiple ligands (epi-MLs) [167]. Recently, water-soluble KAT inhibitor CTK7A (hydrazinocurcumin) was synthesized using curcumin as a parent molecule shows p300 and PCAF inhibitory activities. Treatment of CTK7a showed substantial reduction of xenografted oral tumor growth in mice [37]. These results, therefore, not only establish an epigenetic target for oral cancer, but also implicate a KAT inhibitor as a potential therapeutic molecule.

4.4 KAT INHIBITORS: SYNTHETIC SMALL MOLECULES

The first report of development of a synthetic small molecule inhibitor for GNAT family KAT (Gcn5) came from Gianni's group. Based on published enzyme-kinetic mechanisms, they explored the structure-activity relationship of a biologically abundant natural scaffold named γ-butyrolactone. Among all the synthesized candidate derivatives, only α-methylene- γ-butyrolactone MB-3 was found to specifically inhibit Gcn5 [168].

High-throughput screening (HTS) of a large chemical compound library led to the discovery of isothiazolones as potent p300 and PCAF inhibitors. Moreover, these analogs have exhibited growth inhibition phenotype in human colon carcinoma cells in a time and concentration-dependent manner. Treatment with isothiazolones caused a global reduction in the acetylation levels of histones H3-H4 and α-tubulin have been observed. The proposed mechanism of inhibition implies a plausible reaction with cysteine thiol groups with cleavage of the sulfur-nitrogen bond and concomitant disulfide formation [169]. In another similar kind of study a series of N-aliphatic substituted isothiazolones have been investigated to explore their structure–activity relationships. N-aromatic and N-benzylic substituted isothiazolones have also been shown to inhibit p300 and PCAF enzymatic activities. Based on the generic structure of isothiazolones, a virtual screening of the National Cancer Institute Library was successfully applied in order to identify related KAT inhibitors [170].

Recently, in search of a potent specific p300 inhibitor, Philip Cole's group screened commercially available small-molecule library based on the *in silico* docking studies directed toward the substrate binding pocket of the p300 KAT domain crystal structure. Using this approach, the molecule C646 was identified, which is a novel competitive, highly specific p300 inhibitor with a Ki of 400 nM and the only synthetic KAT inhibitor known to work in a nanomolar range. Structure-activity investigations demonstrated the importance of carboxylic and the nitro group for inhibition of p300 due to their involvement in hydrogen bonds within the enzyme active site. The conjugated enone seems to be crucial for inhibitory activity as well. Importantly, C646 suppressed histone H3 and H4 acetylation in mouse fibroblast cell line and inhibited melanoma and lung cancer cell growth *in vitro* [171]. HTS of small molecule libraries have also led to identification of another potent p300 acetyltransferase inhibitor called L002 which is a new chemical scaffold. This molecule could effectively suppress the growth of MDA-MB-231, a triple negative breast cancer cell line, and reduce histone acetylation in xenografted mice *in vivo* [172].

Recently significant progress has been made in small molecule inhibitor development for Tip60. Extensive virtual screening based on Esa1 crystal structure has lead to the identification of a few promising candidates [173]. Similar HTS screening led to identification of isothiazole compound (NU9056) exhibiting promising effects against prostate cancer [174]. Structure-based rational drug designing using computational tools have identified TH1834, a Tip60-specific inhibitor, as candidate therapeutics against breast cancer [175].

5 CONCLUSION AND PERSPECTIVE

In the cell, a proper stoichiometry is always maintained between lysine acetylation and deacetylation, and perturbation of this stoichiometry is hallmark of congenital diseases like cancer. The importance of lysine acetylation in pathobiology and disease progression is well perceived among the scientific communities. Remarkable advancement in the inhibitor development targeting KDACs was also achieved (two drugs were approved by FDA) but the development of therapeutics, targeting certain lysine acetyltransferases was neglected until the last decade. The stupendous efforts of the scientists have led to tremendous enhancement of knowledge regarding lysine acetyltransferase inhibitors and their molecular targets within the cell. Several bisubstrate, natural, semi-synthetic analogs, and synthetic compounds have been identified as KAT inhibitors. But unfortunately till now none of them have entered clinical trials. Nonspecificity or pleiotropic nature and the lack of potency are the two major bottlenecks in the development of these compounds as potential drug candidates. Regardless of these setbacks, the requisite for KAT inhibitors as potential epigenetic therapeutics is now emerging. The regulatory role of KATs in cellular pathways, the ubiquitous presence of KATs in diversified multicellular organisms, and the important role KATs play in the progression of diseases have led to the need for the development of nontoxic, potent, and specific KAT inhibitors. The knowledge and understanding of evolution of this enzyme [176] can further lead to species specific inhibitors without off-target inhibition.

REFERENCES

[1] Luger K, Mader AW, Richmond RK, Sargent DF, Richmond TJ. Crystal structure of the nucleosome core particle at 2.8 A resolution. Nature 1997;389(6648):251–60.

[2] Bednar J, Horowitz RA, Grigoryev SA, Carruthers LM, Hansen JC, Koster AJ, et al. Nucleosomes, linker DNA, and linker histone form a unique structural motif that directs the higher-order folding and compaction of chromatin. Proc Natl Acad Sci USA 1998;95(24):14173–8.

[3] Seet BT, Dikic I, Zhou MM, Pawson T. Reading protein modifications with interaction domains. Nat Rev Mol Cell Biol 2006;7(7):473–83.

[4] Lee JS, Smith E, Shilatifard A. The language of histone crosstalk. Cell 2010;142(5):682–5.

[5] Garcia-Ramirez M, Rocchini C, Ausio J. Modulation of chromatin folding by histone acetylation. J Biol Chem 1995;270(30):17923–8.

[6] Kouzarides T. Chromatin modifications and their function. Cell 2007;128(4):693–705.

[7] Grunstein M. Histone acetylation in chromatin structure and transcription. Nature 1997;389(6649):349–52.

[8] Allfrey VG, Faulkner R, Mirsky AE. Acetylation and methylation of histones and their possible role in the regulation of RNA synthesis. Proc Natl Acad Sci USA 1964;51:786–94.

[9] Hebbes TR, Thorne AW, Crane-Robinson C. A direct link between core histone acetylation and transcriptionally active chromatin. EMBO J 1988;7(5):1395–402.

[10] Struhl K. Histone acetylation and transcriptional regulatory mechanisms. Genes Dev 1998;12(5):599–606.

[11] Jenuwein T, Allis CD. Translating the histone code. Science (New York, NY) 2001;293(5532):1074–80.

[12] Sterner R, Vidali G, Allfrey VG. Studies of acetylation and deacetylation in high mobility group proteins. Identification of the sites of acetylation in HMG-1. J Biol Chem 1979;254(22):11577–83.

[13] Gu W, Roeder RG. Activation of p53 sequence-specific DNA binding by acetylation of the p53 C-terminal domain. Cell 1997;90(4):595–606.

[14] Sakaguchi K, Herrera JE, Saito S, Miki T, Bustin M, Vassilev A, et al. DNA damage activates p53 through a phosphorylation-acetylation cascade. Genes Dev 1998;12(18):2831–41.

[15] Liu L, Scolnick DM, Trievel RC, Zhang HB, Marmorstein R, Halazonetis TD, et al. p53 sites acetylated in vitro by PCAF and p300 are acetylated in vivo in response to DNA damage. Mol Cell Biol 1999;19(2):1202–9.

[16] Tang Y, Luo J, Zhang W, Gu W. Tip60-dependent acetylation of p53 modulates the decision between cell-cycle arrest and apoptosis. Mol Cell 2006;24(6):827–39.

[17] Sykes SM, Mellert HS, Holbert MA, Li K, Marmorstein R, Lane WS, et al. Acetylation of the p53 DNA-binding domain regulates apoptosis induction. Mol Cell 2006;24(6):841–51.

[18] Tang Y, Zhao W, Chen Y, Zhao Y, Gu W. Acetylation is indispensable for p53 activation. Cell 2008;133(4):612–26.

[19] Kim SC, Sprung R, Chen Y, Xu Y, Ball H, Pei J, et al. Substrate and functional diversity of lysine acetylation revealed by a proteomics survey. Mol Cell 2006;23(4):607–18.

[20] Choudhary C, Kumar C, Gnad F, Nielsen ML, Rehman M, Walther TC, et al. Lysine acetylation targets protein complexes and co-regulates major cellular functions. Science (New York, NY) 2009;325(5942): 834–40.

[21] Brownell JE, Allis CD. An activity gel assay detects a single, catalytically active histone acetyltransferase subunit in Tetrahymena macronuclei. Proc Natl Acad Sci USA 1995;92(14):6364–8.

[22] Brownell JE, Zhou J, Ranalli T, Kobayashi R, Edmondson DG, Roth SY, et al. Tetrahymena histone acetyltransferase A: a homolog to yeast Gcn5p linking histone acetylation to gene activation. Cell 1996;84(6):843–51.

[23] Kleff S, Andrulis ED, Anderson CW, Sternglanz R. Identification of a gene encoding a yeast histone H4 acetyltransferase. J Biol Chem 1995;270(42):24674–7.

[24] Parthun MR, Widom J, Gottschling DE. The major cytoplasmic histone acetyltransferase in yeast: links to chromatin replication and histone metabolism. Cell 1996;87(1):85–94.

[25] Kelly TJ, Qin S, Gottschling DE, Parthun MR. Type B histone acetyltransferase Hat1p participates in telomeric silencing. Mol Cell Biol 2000;20(19):7051–8.

[26] Yang X, Yu W, Shi L, Sun L, Liang J, Yi X, et al. HAT4, a Golgi apparatus-anchored B-type histone acetyltransferase, acetylates free histone H4 and facilitates chromatin assembly. Mol Cell 2011;44(1):39–50.

[27] Dutnall RN, Tafrov ST, Sternglanz R, Ramakrishnan V. Structure of the histone acetyltransferase Hat1: a paradigm for the GCN5-related N-acetyltransferase superfamily. Cell 1998;94(4):427–38.

[28] Brownell JE, Allis CD. Special HATs for special occasions: linking histone acetylation to chromatin assembly and gene activation. Curr Opin Genet Dev 1996;6(2):176–84.

[29] Roth SY, Denu JM, Allis CD. Histone acetyltransferases. Annu Rev Biochem 2001;70:81–120.

[30] Allis CD, Berger SL, Cote J, Dent S, Jenuwien T, Kouzarides T, et al. New nomenclature for chromatin-modifying enzymes. Cell 2007;131(4):633–6.

[31] Di Cerbo V, Schneider R. Cancers with wrong HATs: the impact of acetylation. Brief Funct Genomic 2013;12(3):231–43.

[32] Arif M, Senapati P, Shandilya J, Kundu TK. Protein lysine acetylation in cellular function and its role in cancer manifestation. Biochim Biophys Acta 2010;1799(10–12):702–16.

[33] Schneider A, Chatterjee S, Bousiges O, Selvi BR, Swaminathan A, Cassel R, et al. Acetyltransferases (HATs) as targets for neurological therapeutics. Neurotherapeutics 2013;10(4):568–88.

[34] Fraga MF, Ballestar E, Villar-Garea A, Boix-Chornet M, Espada J, Schotta G, et al. Loss of acetylation at Lys16 and trimethylation at Lys20 of histone H4 is a common hallmark of human cancer. Nat Genet 2005;37(4):391–400.

[35] Seligson DB, Horvath S, Shi T, Yu H, Tze S, Grunstein M, et al. Global histone modification patterns predict risk of prostate cancer recurrence. Nature 2005;435(7046):1262–6.

[36] Das C, Lucia MS, Hansen KC, Tyler JK. CBP/p300-mediated acetylation of histone H3 on lysine 56. Nature 2009;459(7243):113–17.

[37] Arif M, Vedamurthy BM, Choudhari R, Ostwal YB, Mantelingu K, Kodaganur GS, et al. Nitric oxide-mediated histone hyperacetylation in oral cancer: target for a water-soluble HAT inhibitor, CTK7A. Chem Biol 2010;17(8):903–13.

[38] Barlesi F, Giaccone G, Gallegos-Ruiz MI, Loundou A, Span SW, Lefesvre P, et al. Global histone modifications predict prognosis of resected non small-cell lung cancer. J Clin Oncol 2007;25(28):4358–64.

[39] Liu BL, Cheng JX, Zhang X, Wang R, Zhang W, Lin H, et al. Global histone modification patterns as prognostic markers to classify glioma patients. Cancer Epidemiol Biomarkers Prev 2010;19(11):2888–96.

[40] Tzao C, Tung HJ, Jin JS, Sun GH, Hsu HS, Chen BH, et al. Prognostic significance of global histone modifications in resected squamous cell carcinoma of the esophagus. Mod Pathol 2009;22(2):252–60.

[41] Fullgrabe J, Kavanagh E, Joseph B. Histone onco-modifications. Oncogene 2011;30(31):3391–403.

[42] Ito K, Barnes PJ, Adcock IM. Glucocorticoid receptor recruitment of histone deacetylase 2 inhibits interleukin-1beta-induced histone H4 acetylation on lysines 8 and 12. Mol Cell Biol 2000;20(18):6891–903.

[43] Yeh HH, Young D, Gelovani JG, Robinson A, Davidson Y, Herholz K, et al. Histone deacetylase class II and acetylated core histone immunohistochemistry in human brains with Huntington's disease. Brain Res 2013;1504:16–24.

[44] McCampbell A, Taylor JP, Taye AA, Robitschek J, Li M, Walcott J, et al. CREB-binding protein sequestration by expanded polyglutamine. Hum Mol Genet 2000;9(14):2197–202.

[45] Ogryzko VV, Schiltz RL, Russanova V, Howard BH, Nakatani Y. The transcriptional coactivators p300 and CBP are histone acetyltransferases. Cell 1996;87(5):953–9.

[46] Bannister AJ, Kouzarides T. The CBP co-activator is a histone acetyltransferase. Nature 1996;384(6610):641–3.

[47] Kundu TK, Palhan VB, Wang Z, An W, Cole PA, Roeder RG. Activator-dependent transcription from chromatin in vitro involving targeted histone acetylation by p300. Mol Cell 2000;6(3):551–61.

[48] Liu X, Wang L, Zhao K, Thompson PR, Hwang Y, Marmorstein R, et al. The structural basis of protein acetylation by the p300/CBP transcriptional coactivator. Nature 2008;451(7180):846–50.

[49] Goodman RH, Smolik S. CBP/p300 in cell growth, transformation, and development. Genes Dev 2000;14(13):1553–77.

[50] Rubinstein JH, Taybi H. Broad thumbs and toes and facial abnormalities. A possible mental retardation syndrome. Am J Dis Child 1963(105):588–608.

[51] Miller RW, Rubinstein JH. Tumors in Rubinstein-Taybi syndrome. Am J Med Genet 1995;56(1):112–15.

[52] Blough RI, Petrij F, Dauwerse JG, Milatovich-Cherry A, Weiss L, Saal HM, et al. Variation in microdeletions of the cyclic AMP-responsive element-binding protein gene at chromosome band 16p13.3 in the Rubinstein-Taybi syndrome. Am J Med Genet 2000;90(1):29–34.

[53] Oike Y, Takakura N, Hata A, Kaname T, Akizuki M, Yamaguchi Y, et al. Mice homozygous for a truncated form of CREB-binding protein exhibit defects in hematopoiesis and vasculo-angiogenesis. Blood 1999;93(9):2771–9.

[54] Kung AL, Rebel VI, Bronson RT, Ch'ng LE, Sieff CA, Livingston DM, et al. Gene dose-dependent control of hematopoiesis and hematologic tumor suppression by CBP. Genes Dev 2000;14(3):272–7.

[55] Rebel VI, Kung AL, Tanner EA, Yang H, Bronson RT, Livingston DM. Distinct roles for CREB-binding protein and p300 in hematopoietic stem cell self-renewal. Proc Natl Acad Sci USA 2002;99(23):14789–94.

[56] Iyer NG, Ozdag H, Caldas C. p300/CBP and cancer. Oncogene 2004;23(24):4225–31.

[57] Sobulo OM, Borrow J, Tomek R, Reshmi S, Harden A, Schlegelberger B, et al. MLL is fused to CBP, a histone acetyltransferase, in therapy-related acute myeloid leukemia with a t(11;16)(q23;p13.3). Proc Natl Acad Sci USA 1997;94(16):8732–7.

[58] Ida K, Kitabayashi I, Taki T, Taniwaki M, Noro K, Yamamoto M, et al. Adenoviral E1A-associated protein p300 is involved in acute myeloid leukemia with t(11;22)(q23;q13). Blood 1997;90(12):4699–704.

[59] Taki T, Sako M, Tsuchida M, Hayashi Y. The t(11;16)(q23;p13) translocation in myelodysplastic syndrome fuses the MLL gene to the CBP gene. Blood 1997;89(11):3945–50.

[60] Dai P, Akimaru H, Tanaka Y, Hou DX, Yasukawa T, Kanei-Ishii C, et al. CBP as a transcriptional coactivator of c-Myb. Genes Dev 1996;10(5):528–40.

[61] Vervoorts J, Luscher-Firzlaff JM, Rottmann S, Lilischkis R, Walsemann G, Dohmann K, et al. Stimulation of c-MYC transcriptional activity and acetylation by recruitment of the cofactor CBP. EMBO Rep 2003;4(5):484–90.

[62] Bannister AJ, Oehler T, Wilhelm D, Angel P, Kouzarides T. Stimulation of c-Jun activity by CBP: c-Jun residues Ser63/73 are required for CBP induced stimulation in vivo and CBP binding in vitro. Oncogene 1995;11(12):2509–14.

[63] Bannister AJ, Kouzarides T. CBP-induced stimulation of c-Fos activity is abrogated by E1A. EMBO J 1995;14(19):4758–62.

[64] Pattabiraman DR, McGirr C, Shakhbazov K, Barbier V, Krishnan K, Mukhopadhyay P, et al. Interaction of c-Myb with p300 is required for the induction of acute myeloid leukemia (AML) by human AML oncogenes. Blood 2014;123(17):2682–90.

[65] Wu D, Zhang R, Zhao R, Chen G, Cai Y, Jin J. A novel function of novobiocin: disrupting the interaction of HIF1alpha and p300/CBP through direct binding to the HIF1alpha C-terminal activation domain. PLoS One 2013;8(5):e62014.

[66] Burslem GM, Kyle HF, Breeze AL, Edwards TA, Nelson A, Warriner SL, et al. Small-molecule proteomimetic inhibitors of the HIF-1alpha-p300 protein-protein interaction. Chembiochem 2014;15(8):1083–7.

[67] Reece KM, Richardson ED, Cook KM, Campbell TJ, Pisle ST, Holly AJ, et al. Epidithiodiketopiperazines (ETPs) exhibit in vitro antiangiogenic and in vivo antitumor activity by disrupting the HIF-1alpha/p300 complex in a preclinical model of prostate cancer. Mol Cancer 2014;13(1):91.

[68] Zhou B, Liu Y, Kahn M, Ann DK, Han A, Wang H, et al. Interactions between beta-catenin and transforming growth factor-beta signaling pathways mediate epithelial-mesenchymal transition and are dependent on the transcriptional co-activator cAMP-response element-binding protein (CREB)-binding protein (CBP). J Biol Chem 2012;287(10):7026–38.

[69] Yoshida T, Hashimura M, Mastumoto T, Tazo Y, Inoue H, Kuwata T, et al. Transcriptional upregulation of HIF-1α by NF-κB/p65 and its associations with β-catenin/p300 complexes in endometrial carcinoma cells. Lab Invest 2013;93(11):1184–93.

[70] Santer FR, Hoschele PP, Oh SJ, Erb HH, Bouchal J, Cavarretta IT, et al. Inhibition of the acetyltransferases p300 and CBP reveals a targetable function for p300 in the survival and invasion pathways of prostate cancer cell lines. Mol Cancer Ther 2011;10(9):1644–55.

[71] Ait-Si-Ali S, Polesskaya A, Filleur S, Ferreira R, Duquet A, Robin P, et al. CBP/p300 histone acetyltransferase activity is important for the G1/S transition. Oncogene 2000;19(20):2430–7.

[72] Heemers HV, Debes JD, Tindall DJ. The role of the transcriptional coactivator p300 in prostate cancer progression. Adv Exp Med Biol 2008;617:535–40.

[73] Debes JD, Sebo TJ, Lohse CM, Murphy LM, Haugen DA, Tindall DJ. p300 in prostate cancer proliferation and progression. Cancer Res 2003;63(22):7638–40.

[74] Xiao XS, Cai MY, Chen JW, Guan XY, Kung HF, Zeng YX, et al. High expression of p300 in human breast cancer correlates with tumor recurrence and predicts adverse prognosis. Chin J Cancer Res 2011;23(3):201–7.

[75] Hou X, Li Y, Luo RZ, Fu JH, He JH, Zhang LJ, et al. High expression of the transcriptional co-activator p300 predicts poor survival in resectable non-small cell lung cancers. Eur J Surg Oncol 2012;38(6):523–30.

[76] Yokomizo C, Yamaguchi K, Itoh Y, Nishimura T, Umemura A, Minami M, et al. High expression of p300 in HCC predicts shortened overall survival in association with enhanced epithelial mesenchymal transition of HCC cells. Cancer Lett 2011;310(2):140–7.

[77] Ishihama K, Yamakawa M, Semba S, Takeda H, Kawata S, Kimura S, et al. Expression of HDAC1 and CBP/p300 in human colorectal carcinomas. J Clin Pathol 2007;60(11):1205–10.

[78] Ianculescu I, Wu DY, Siegmund KD, Stallcup MR. Selective roles for cAMP response element-binding protein binding protein and p300 protein as coregulators for androgen-regulated gene expression in advanced prostate cancer cells. J Biol Chem 2012;287(6):4000–13.

[79] Takeuchi A, Shiota M, Tatsugami K, Yokomizo A, Tanaka S, Kuroiwa K, et al. p300 mediates cellular resistance to doxorubicin in bladder cancer. Mol Med Rep 2012;5(1):173–6.

[80] Wang SA, Hung CY, Chuang JY, Chang WC, Hsu TI, Hung JJ. Phosphorylation of p300 increases its protein degradation to enhance the lung cancer progression. Biochim Biophys Acta 2014;1843(6):1135–49.

[81] Thompson PR, Wang D, Wang L, Fulco M, Pediconi N, Zhang D, et al. Regulation of the p300 HAT domain via a novel activation loop. Nat Struct Mol Biol 2004;11(4):308–15.

[82] Martinez E, Kundu TK, Fu J, Roeder RG. A human SPT3-TAFII31-GCN5-L acetylase complex distinct from transcription factor IID. J Biol Chem 1998;273(37):23781–5.

[83] Wieczorek E, Brand M, Jacq X, Tora L. Function of TAF(II)-containing complex without TBP in transcription by RNA polymerase II. Nature 1998;393(6681):187–91.

[84] Brand M, Yamamoto K, Staub A, Tora L. Identification of TATA-binding protein-free TAFII-containing complex subunits suggests a role in nucleosome acetylation and signal transduction. J Biol Chem 1999;274(26):18285–9.

[85] Ogryzko VV, Kotani T, Zhang X, Schiltz RL, Howard T, Yang XJ, et al. Histone-like TAFs within the PCAF histone acetylase complex. Cell 1998;94(1):35–44.

[86] Ura K, Araki M, Saeki H, Masutani C, Ito T, Iwai S, et al. ATP-dependent chromatin remodeling facilitates nucleotide excision repair of UV-induced DNA lesions in synthetic dinucleosomes. EMBO J 2001;20(8):2004–14.

[87] Linares LK, Kiernan R, Triboulet R, Chable-Bessia C, Latreille D, Cuvier O, et al. Intrinsic ubiquitination activity of PCAF controls the stability of the oncoprotein Hdm2. Nat Cell Biol 2007;9(3):331–8.

[88] Zheng X, Gai X, Ding F, Lu Z, Tu K, Yao Y, et al. Histone acetyltransferase PCAF up-regulated cell apoptosis in hepatocellular carcinoma via acetylating histone H4 and inactivating AKT signaling. Mol Cancer 2013;12(1):96.

[89] Frank SR, Schroeder M, Fernandez P, Taubert S, Amati B. Binding of c-Myc to chromatin mediates mitogen-induced acetylation of histone H4 and gene activation. Genes Dev 2001;15(16):2069–82.

[90] Spotts GD, Patel SV, Xiao Q, Hann SR. Identification of downstream-initiated c-Myc proteins which are dominant-negative inhibitors of transactivation by full-length c-Myc proteins. Mol Cell Biol 1997;17(3):1459–68.

[91] Malatesta M, Steinhauer C, Mohammad F, Pandey DP, Squatrito M, Helin K. Histone acetyltransferase PCAF is required for Hedgehog-Gli-dependent transcription and cancer cell proliferation. Cancer Res 2013;73(20):6323–33.

[92] Hirano G, Izumi H, Kidani A, Yasuniwa Y, Han B, Kusaba H, et al. Enhanced expression of PCAF endows apoptosis resistance in cisplatin-resistant cells. Mol Cancer Res 2010;8(6):864–72.

[93] Shiota M, Yokomizo A, Tada Y, Uchiumi T, Inokuchi J, Tatsugami K, et al. P300/CBP-associated factor regulates Y-box binding protein-1 expression and promotes cancer cell growth, cancer invasion and drug resistance. Cancer Sci 2010;101(8):1797–806.

[94] Toth M, Boros IM, Balint E. Elevated level of lysine 9-acetylated histone H3 at the MDR1 promoter in multidrug-resistant cells. Cancer Sci 2012;103(4):659–69.

[95] Utley RT, Cote J. The MYST family of histone acetyltransferases. Curr Top Microbiol Immunol 2003;274:203–36.

[96] Yang XJ. The diverse superfamily of lysine acetyltransferases and their roles in leukemia and other diseases. Nucleic Acids Res 2004;32(3):959–76.

[97] Ikura T, Ogryzko VV, Grigoriev M, Groisman R, Wang J, Horikoshi M, et al. Involvement of the TIP60 histone acetylase complex in DNA repair and apoptosis. Cell 2000;102(4):463–73.

[98] Sun Y, Jiang X, Chen S, Fernandes N, Price BD. A role for the Tip60 histone acetyltransferase in the acetylation and activation of ATM. Proc Natl Acad Sci USA 2005;102(37):13182–7.

[99] Shroff R, Arbel-Eden A, Pilch D, Ira G, Bonner WM, Petrini JH, et al. Distribution and dynamics of chromatin modification induced by a defined DNA double-strand break. Curr Biol 2004;14(19):1703–11.

[100] Rogakou EP, Boon C, Redon C, Bonner WM. Megabase chromatin domains involved in DNA double-strand breaks in vivo. J Cell Biol 1999;146(5):905–16.

[101] Bewersdorf J, Bennett BT, Knight KL. H2AX chromatin structures and their response to DNA damage revealed by 4Pi microscopy. Proc Natl Acad Sci USA 2006;103(48):18137–42.

[102] Lleonart ME, Vidal F, Gallardo D, Diaz-Fuertes M, Rojo F, Cuatrecasas M, et al. New p53 related genes in human tumors: significant downregulation in colon and lung carcinomas. Oncol Rep 2006;16(3):603–8.

[103] Awasthi S, Sharma A, Wong K, Zhang J, Matlock EF, Rogers L, et al. A human T-cell lymphotropic virus type 1 enhancer of Myc transforming potential stabilizes Myc-TIP60 transcriptional interactions. Mol Cell Biol 2005;25(14):6178–98.

[104] Halkidou K, Gnanapragasam VJ, Mehta PB, Logan IR, Brady ME, Cook S, et al. Expression of Tip60, an androgen receptor coactivator, and its role in prostate cancer development. Oncogene 2003;22(16):2466–77.

[105] Iizuka M, Stillman B. Histone acetyltransferase HBO1 interacts with the ORC1 subunit of the human initiator protein. J Biol Chem 1999;274(33):23027–34.

[106] Aggarwal BD, Calvi BR. Chromatin regulates origin activity in Drosophila follicle cells. Nature 2004;430(6997):372–6.

[107] Doyon Y, Cayrou C, Ullah M, Landry AJ, Cote V, Selleck W, et al. ING tumor suppressor proteins are critical regulators of chromatin acetylation required for genome expression and perpetuation. Mol Cell 2006;21(1):51–64.

[108] Iizuka M, Matsui T, Takisawa H, Smith MM. Regulation of replication licensing by acetyltransferase Hbo1. Mol Cell Biol 2006;26(3):1098–108.

[109] Hung T, Binda O, Champagne KS, Kuo AJ, Johnson K, Chang HY, et al. ING4 mediates crosstalk between histone H3 K4 trimethylation and H3 acetylation to attenuate cellular transformation. Mol Cell 2009;33(2):248–56.

[110] Saksouk N, Avvakumov N, Champagne KS, Hung T, Doyon Y, Cayrou C, et al. HBO1 HAT complexes target chromatin throughout gene coding regions via multiple PHD finger interactions with histone H3 tail. Mol Cell 2009;33(2):257–65.

[111] Iizuka M, Sarmento OF, Sekiya T, Scrable H, Allis CD, Smith MM. Hbo1 Links p53-dependent stress signaling to DNA replication licensing. Mol Cell Biol 2008;28(1):140–53.

[112] Song B, Liu XS, Rice SJ, Kuang S, Elzey BD, Konieczny SF, et al. Plk1 phosphorylation of orc2 and hbo1 contributes to gemcitabine resistance in pancreatic cancer. Mol Cancer Ther 2013;12(1):58–68.

[113] Duong MT, Akli S, Macalou S, Biernacka A, Debeb BG, Yi M, et al. Hbo1 is a cyclin E/CDK2 substrate that enriches breast cancer stem-like cells. Cancer Res 2013;73(17):5556–68.

[114] Iizuka M, Takahashi Y, Mizzen CA, Cook RG, Fujita M, Allis CD, et al. Histone acetyltransferase Hbo1: catalytic activity, cellular abundance, and links to primary cancers. Gene 2009;436(1–2):108–14.

[115] Champagne N, Pelletier N, Yang XJ. The monocytic leukemia zinc finger protein MOZ is a histone acetyltransferase. Oncogene 2001;20(3):404–9.

[116] Champagne N, Bertos NR, Pelletier N, Wang AH, Vezmar M, Yang Y, et al. Identification of a human histone acetyltransferase related to monocytic leukemia zinc finger protein. J Biol Chem 1999;274(40):28528–36.

[117] Borrow J, Stanton Jr. VP, Andresen JM, Becher R, Behm FG, Chaganti RS, et al. The translocation t(8;16) (p11;p13) of acute myeloid leukaemia fuses a putative acetyltransferase to the CREB-binding protein. Nat Genet 1996;14(1):33–41.

[118] Kitabayashi I, Aikawa Y, Yokoyama A, Hosoda F, Nagai M, Kakazu N, et al. Fusion of MOZ and p300 histone acetyltransferases in acute monocytic leukemia with a t(8;22)(p11;q13) chromosome translocation. Leukemia 2001;15(1):89–94.

[119] Panagopoulos I, Fioretos T, Isaksson M, Samuelsson U, Billstrom R, Strombeck B, et al. Fusion of the MORF and CBP genes in acute myeloid leukemia with the t(10;16)(q22;p13). Hum Mol Genet 2001;10(4):395–404.

[120] Kojima K, Kaneda K, Yoshida C, Dansako H, Fujii N, Yano T, et al. A novel fusion variant of the MORF and CBP genes detected in therapy-related myelodysplastic syndrome with t(10;16)(q22;p13). Br J Haematol 2003;120(2):271–3.

[121] Carapeti M, Aguiar RC, Goldman JM, Cross NC. A novel fusion between MOZ and the nuclear receptor coactivator TIF2 in acute myeloid leukemia. Blood 1998;91(9):3127–33.

[122] Leo C, Chen JD. The SRC family of nuclear receptor coactivators. Gene 2000;245(1):1–11.

[123] Deguchi K, Ayton PM, Carapeti M, Kutok JL, Snyder CS, Williams IR, et al. MOZ-TIF2-induced acute myeloid leukemia requires the MOZ nucleosome binding motif and TIF2-mediated recruitment of CBP. Cancer Cell 2003;3(3):259–71.

[124] Hilfiker A, Hilfiker-Kleiner D, Pannuti A, Lucchesi JC. mof, a putative acetyl transferase gene related to the Tip60 and MOZ human genes and to the SAS genes of yeast, is required for dosage compensation in Drosophila. EMBO J 1997;16(8):2054–60.

[125] Smith ER, Cayrou C, Huang R, Lane WS, Cote J, Lucchesi JC. A human protein complex homologous to the Drosophila MSL complex is responsible for the majority of histone H4 acetylation at lysine 16. Mol Cell Biol 2005;25(21):9175–88.

[126] Cai Y, Jin J, Swanson SK, Cole MD, Choi SH, Florens L, et al. Subunit composition and substrate specificity of a MOF-containing histone acetyltransferase distinct from the male-specific lethal (MSL) complex. J Biol Chem 2010;285(7):4268–72.

[127] Cao L, Zhu L, Yang J, Su J, Ni J, Du Y, et al. Correlation of low expression of hMOF with clinicopathological features of colorectal carcinoma, gastric cancer and renal cell carcinoma. Int J Oncol 2014;44(4):1207–14.

[128] Wang Y, Zhang R, Wu D, Lu Z, Sun W, Cai Y, et al. Epigenetic change in kidney tumor: downregulation of histone acetyltransferase MYST1 in human renal cell carcinoma. J Exp Clin Cancer Res 2013;32:8.

[129] Liu N, Zhang R, Zhao X, Su J, Bian X, Ni J, et al. A potential diagnostic marker for ovarian cancer: involvement of the histone acetyltransferase, human males absent on the first. Oncol Lett 2013;6(2):393–400.

[130] Pfister S, Rea S, Taipale M, Mendrzyk F, Straub B, Ittrich C, et al. The histone acetyltransferase hMOF is frequently downregulated in primary breast carcinoma and medulloblastoma and constitutes a biomarker for clinical outcome in medulloblastoma. Int J Cancer 2008;122(6):1207–13.

[131] Chen Z, Ye X, Tang N, Shen S, Li Z, Niu X, et al. The histone acetylranseferase hMOF acetylates Nrf2 and regulates anti-drug responses in human non-small cell lung cancer. Br J Pharmacol 2014;171(13):3196–211.

[132] Lau OD, Kundu TK, Soccio RE, Ait-Si-Ali S, Khalil EM, Vassilev A, et al. HATs off: selective synthetic inhibitors of the histone acetyltransferases p300 and PCAF. Mol Cell 2000;5(3):589–95.

[133] Zheng Y, Balasubramanyam K, Cebrat M, Buck D, Guidez F, Zelent A, et al. Synthesis and evaluation of a potent and selective cell-permeable p300 histone acetyltransferase inhibitor. J Am Chem Soc 2005;127(49):17182–3.

[134] Cole PA. Chemical probes for histone-modifying enzymes. Nat Chem Biol 2008;4(10):590–7.

[135] Bandyopadhyay K, Baneres JL, Martin A, Blonski C, Parello J, Gjerset RA. Spermidinyl-CoA-based HAT inhibitors block DNA repair and provide cancer-specific chemo-and radiosensitization. Cell Cycle (Georgetown, Tex) 2009;8(17):2779–88.

[136] Wu J, Xie N, Wu Z, Zhang Y, Zheng YG. Bisubstrate inhibitors of the MYST HATs Esa1 and Tip60. Bioorg Med Chem 2009;17(3):1381–6.

[137] Balasubramanyam K, Swaminathan V, Ranganathan A, Kundu TK. Small molecule modulators of histone acetyltransferase p300. J Biol Chem 2003;278(21):19134–40.

[138] Hemshekhar M, Sebastin Santhosh M, Kemparaju K, Girish KS. Emerging roles of anacardic acid and its derivatives: a pharmacological overview. Basic Clin Pharmacol Toxicol 2011.

[139] Balasubramanyam K, Varier RA, Altaf M, Swaminathan V, Siddappa NB, Ranga U, et al. Curcumin, a novel p300/CREB-binding protein-specific inhibitor of acetyltransferase, represses the acetylation of histone/nonhistone proteins and histone acetyltransferase-dependent chromatin transcription. J Biol Chem 2004;279(49):51163–71.

[140] Palve YP, Nayak PL. Curcumin: a wonder anticancer drug. Int J Biomed Pharm Sci 2012;3(2):60–9.

[141] Shishodia S, Singh T, Chaturvedi MM. Modulation of transcription factors by curcumin. Adv Exp Med Biol 2007;595:127–48.

[142] Tiwari SK, Agarwal S, Seth B, Yadav A, Nair S, Bhatnagar P, et al. Curcumin-loaded nanoparticles potently induce adult neurogenesis and reverse cognitive deficits in Alzheimer's disease model via canonical Wnt/beta-catenin pathway. ACS Nano 2014;8(1):76–103.

[143] Barthelemy S, Vergnes L, Moynier M, Guyot D, Labidalle S, Bahraoui E. Curcumin and curcumin derivatives inhibit Tat-mediated transactivation of type 1 human immunodeficiency virus long terminal repeat. Res Virol 1998;149(1):43–52.

[144] Balasubramanyam K, Altaf M, Varier RA, Swaminathan V, Ravindran A, Sadhale PP, et al. Polyisoprenylated benzophenone, garcinol, a natural histone acetyltransferase inhibitor, represses chromatin transcription and alters global gene expression. J Biol Chem 2004;279(32):33716–26.

[145] Collins HM, Abdelghany MK, Messmer M, Yue B, Deeves SE, Kindle KB, et al. Differential effects of garcinol and curcumin on histone and p53 modifications in tumour cells. BMC Cancer 2013;13:37.

[146] Saadat N, Gupta SV. Potential role of garcinol as an anticancer agent. J Oncol 2012;2012:647206.

[147] Sethi G, Chatterjee S, Rajendran P, Li F, Shanmugam MK, Wong KF, et al. Inhibition of STAT3 dimerization and acetylation by garcinol suppresses the growth of human hepatocellular carcinoma in vitro and in vivo. Mol Cancer 2014;13:66.

[148] Li F, Shanmugam MK, Chen L, Chatterjee S, Basha J, Kumar AP, et al. Garcinol, a polyisoprenylated benzophenone modulates multiple proinflammatory signaling cascades leading to the suppression of growth and survival of head and neck carcinoma. Cancer Prev Res (Phila Pa) 2013;6(8):843–54.

[149] Ravindra KC, Selvi BR, Arif M, Reddy BA, Thanuja GR, Agrawal S, et al. Inhibition of lysine acetyltransferase KAT3B/p300 activity by a naturally occurring hydroxynaphthoquinone, plumbagin. J Biol Chem 2009;284(36):24453–64.

[150] Padhye S, Dandawate P, Yusufi M, Ahmad A, Sarkar FH. Perspectives on medicinal properties of plumbagin and its analogs. Med Res Rev 2012;32(6):1131–58.

[151] Slaninová I, Pěnčíková K, Urbanová J, Slanina J, Táborská E. Antitumor activities of sanguinarine and related alkaloids. Phytochemistry Rev 2013;13(1):51–68.

[152] Selvi BR, Pradhan SK, Shandilya J, Das C, Sailaja BS, Shankar GN, et al. Sanguinarine interacts with chromatin, modulates epigenetic modifications, and transcription in the context of chromatin. Chem Biol 2009;16(2):203–16.

[153] Choi KC, Jung MG, Lee YH, Yoon JC, Kwon SH, Kang HB, et al. Epigallocatechin-3-gallate, a histone acetyltransferase inhibitor, inhibits EBV-induced B lymphocyte transformation via suppression of RelA acetylation. Cancer Res 2009;69(2):583–92.

[154] Choi KC, Park S, Lim BJ, Seong AR, Lee YH, Shiota M, et al. Procyanidin B3, an inhibitor of histone acetyltransferase, enhances the action of antagonist for prostate cancer cells via inhibition of p300-dependent acetylation of androgen receptor. Biochem J 2011;433(1):235–44.

[155] Seong AR, Yoo JY, Choi K, Lee MH, Lee YH, Lee J, et al. Delphinidin, a specific inhibitor of histone acetyltransferase, suppresses inflammatory signaling via prevention of NF-kappaB acetylation in fibroblast-like synoviocyte MH7A cells. Biochem Biophys Res Commun 2011;410(3):581–6.

[156] Choi KC, Lee YH, Jung MG, Kwon SH, Kim MJ, Jun WJ, et al. Gallic acid suppresses lipopolysaccharide-induced nuclear factor-kappaB signaling by preventing RelA acetylation in A549 lung cancer cells. Mol Cancer Res 2009;7(12):2011–21.

[157] Huang M, Tang SN, Upadhyay G, Marsh JL, Jackman CP, Shankar S, et al. Embelin suppresses growth of human pancreatic cancer xenografts, and pancreatic cancer cells isolated from KrasG12D mice by inhibiting Akt and Sonic hedgehog pathways. PLoS One 2014;9(4):e92161.

[158] Poojari R. Embelin-a drug of antiquity: shifting the paradigm towards modern medicine. Expert Opin Investig Drugs 2014;23(3):427–44.

[159] Modak R, Basha J, Bharathy N, Maity K, Mizar P, Bhat AV, et al. Probing p300/CBP associated factor (PCAF)-dependent pathways with a small molecule inhibitor. ACS Chem Biol 2013;8(6):1311–23.

[160] Mantelingu K, Reddy BA, Swaminathan V, Kishore AH, Siddappa NB, Kumar GV, et al. Specific inhibition of p300-HAT alters global gene expression and represses HIV replication. Chem Biol 2007;14(6):645–57.

[161] Vasudevarao MD, Mizar P, Kumari S, Mandal S, Siddhanta S, Swamy MM, et al. Naphthoquinone-mediated inhibition of lysine acetyltransferase KAT3B/p300, basis for non-toxic inhibitor synthesis. J Biol Chem 2014;289(11):7702–17.

[162] Ghizzoni M, Boltjes A, Graaf C, Haisma HJ, Dekker FJ. Improved inhibition of the histone acetyltransferase PCAF by an anacardic acid derivative. Bioorg Med Chem 2010;18(16):5826–34.

[163] Souto JA, Conte M, Alvarez R, Nebbioso A, Carafa V, Altucci L, et al. Synthesis of benzamides related to anacardic acid and their histone acetyltransferase (HAT) inhibitory activities. ChemMedChem 2008;3(9):1435–42.

[164] Sun Y, Jiang X, Chen S, Price BD. Inhibition of histone acetyltransferase activity by anacardic acid sensitizes tumor cells to ionizing radiation. FEBS Lett 2006;580(18):4353–6.

[165] Ghizzoni M, Wu J, Gao T, Haisma HJ, Dekker FJ, George Zheng Y. 6-alkylsalicylates are selective Tip60 inhibitors and target the acetyl-CoA binding site. Eur J Med Chem 2012;47(1):337–44.

[166] Costi R, Di Santo R, Artico M, Miele G, Valentini P, Novellino E, et al. Cinnamoyl compounds as simple molecules that inhibit p300 histone acetyltransferase. J Med Chem 2007;50(8):1973–7.

[167] Mai A, Cheng D, Bedford MT, Valente S, Nebbioso A, Perrone A, et al. Epigenetic multiple ligands: mixed histone/protein methyltransferase, acetyltransferase, and class III deacetylase (sirtuin) inhibitors. J Med Chem 2008;51(7):2279–90.

[168] Biel M, Kretsovali A, Karatzali E, Papamatheakis J, Giannis A. Design, synthesis, and biological evaluation of a small-molecule inhibitor of the histone acetyltransferase Gcn5. Angew Chem Int Ed Engl 2004;43(30):3974–6.

[169] Stimson L, Rowlands MG, Newbatt YM, Smith NF, Raynaud FI, Rogers P, et al. Isothiazolones as inhibitors of PCAF and p300 histone acetyltransferase activity. Mol Cancer Ther 2005;4(10):1521–32.

[170] Ghizzoni M, Haisma HJ, Dekker FJ. Reactivity of isothiazolones and isothiazolone-1-oxides in the inhibition of the PCAF histone acetyltransferase. Eur J Med Chem 2009;44(12):4855–61.

[171] Bowers EM, Yan G, Mukherjee C, Orry A, Wang L, Holbert MA, et al. Virtual ligand screening of the p300/CBP histone acetyltransferase: identification of a selective small molecule inhibitor. Chem Biol 2010;17(5):471–82.

[172] Yang H, Pinello CE, Luo J, Li D, Wang Y, Zhao LY, et al. Small-molecule inhibitors of acetyltransferase p300 identified by high-throughput screening are potent anticancer agents. Mol Cancer Ther 2013;12(5):610–20.

[173] Wu J, Wang J, Li M, Yang Y, Wang B, Zheng YG. Small molecule inhibitors of histone acetyltransferase Tip60. Bioorg Chem 2011;39(1):53–8.

[174] Coffey K, Blackburn TJ, Cook S, Golding BT, Griffin RJ, Hardcastle IR, et al. Characterisation of a Tip60 specific inhibitor, NU9056, in prostate cancer. PLoS One 2012;7(10):e45539.

[175] Gao C, Bourke E, Scobie M, Famme MA, Koolmeister T, Helleday T, et al. Rational design and validation of a Tip60 histone acetyltransferase inhibitor. Sci Rep 2014;4:5372.

[176] Wang L, Tang Y, Cole PA, Marmorstein R. Structure and chemistry of the p300/CBP and Rtt109 histone acetyltransferases: implications for histone acetyltransferase evolution and function. Curr Opin Struct Biol 2008;18(6):741–7.

EMERGING EPIGENETIC THERAPIES—BROMODOMAIN LIGANDS

22

David S. Hewings, Timothy P.C. Rooney, and Stuart J. Conway

Department of Chemistry, Chemistry Research Laboratory, University of Oxford, Oxford, UK

CHAPTER OUTLINE

S.G. Gray (Ed): Epigenetic Cancer Therapy. DOI: http://dx.doi.org/10.1016/B978-0-12-800206-3.00022-7

1 INTRODUCTION

Lysine acetylation has been recognized as a protein posttranslational modification (PTM) since the 1960s, but it is only in the last two decades that the molecular mechanisms by which it exerts is effects have been revealed. Allfrey, who first identified acetyl lysine (KAc) as a PTM of histones [1,2], suggested a connection between KAc and transcription, proposing that "acetylation, by neutralizing the positive charges associated with the ε-amino groups of lysine residues, would be expected to influence the interactions between histones and DNA... and this may offer a molecular basis for the pronounced changes in histone acetylation and RNA synthesis during the course of gene activation in many cell types" [2]. The identification, in the 1990s, of histone acetyltransferases (HATs) and histone deacetylases (HDACs) with homology to known yeast transcriptional coactivators [3–5] appeared to support Allfrey's early hypothesis by providing a functional link between lysine acetylation and gene expression.

Changes in electrostatic interactions between histones and DNA are not the only mechanism by which lysine acetylation can influence gene expression. Significantly, a specialized KAc recognition domain, the bromodomain, is present in many transcriptional coactivators, acting as a key mediator between acetylated histones and components of the transcriptional machinery. Recent years have seen a rapid advance in our understanding of these protein domains, which recognize KAc in a sequence-selective manner on both histone and nonhistone proteins. Notably, it has emerged that bromodomain-containing proteins (BCPs) are involved in regulating important genes involved in cell proliferation and inflammation, leading to the identification of BCPs as therapeutic targets for diverse cancer types. The study of BCPs in disease states has been greatly assisted by the development of small molecule inhibitors of the bromodomain–KAc interaction, notably for bromodomains of the bromodomain and extraterminal domain (BET) family. This chapter will discuss the advances in the development of BET bromodomain ligands and their application in biological systems related to cancer. In addition, progress in the identification of non-BET bromodomain ligands will be described. Detailed discussions of structure–activity relationships are beyond the scope of this chapter, and the reader is directed to several recent reviews that deal with this topic thoroughly [6–10].

1.1 BROMODOMAINS—A KAc RECOGNITION DOMAIN

The bromodomain was first identified by Tamkun et al. in 1992 as a structural motif present in the *Drosophila* protein brahma and was named after this protein by analogy with the chromodomain [11,12]. The structure of the cyclic adenosine monophosphate (cAMP) response element-binding protein (CREB) binding protein (CREBBP) bromodomain is shown in Figure 22.1 (PDB ID: 3P1C) [13]. As with all human bromodomains, it exists as part of a much larger, multidomain protein (Figure 22.1A). Figure 22.1B illustrates the characteristic four-helix bundle structure. Two loop regions are also present (shown in green, Figure 22.1B): the short BC loop and the longer ZA loop. These two loops are diverse in primary sequence and allow bromodomains to discriminate between KAc in different sequence contexts [13]. KAc binds in a well-defined pocket at the end of the four-helix bundle. The

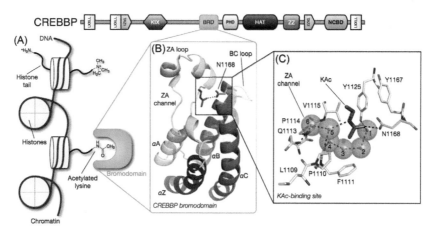

FIGURE 22.1

Bromodomains bind to acetylated lysine residues. (A) Histone tails are subjected to multiple PTMs, including lysine acetylation. Lysine acetylation state is "read" by bromodomains, protein modules that invariably exist as part of a more complex protein architecture. The example shown is cAMP response element-binding protein (CREB) binding protein (CREBBP or CBP). (B) X-ray crystal structure of the CREBBP bromodomain in complex with KAc (carbon = purple, PDB code 3P1C). The bromodomain fold comprises four helices αZ (blue), αA (yellow), αB (orange), and αC (red) and two loop regions known as the ZA loop and the BC loop (light green). (C) The KAc residue binds in a well-defined pocket and, in CREBBP, forms interactions with N1168 and a second interaction with Y1125, via structured water molecule 1 (red sphere). The pocket is substantially hydrophobic, but four structured water molecules form its base (waters 1 – 4). Many bromodomains have been crystallized with one or two additional water molecules bound in the ZA channel. In the CREBBP above, water molecules 5 and 6 are observed in the ZA channel. Water 5 binds to the NH of the acetylated lysine residue, P1110, and water 6. Water 6 binds to Q1113 and water 5 [13]. *Adapted with permission from Ref. [6]. Copyright 2012 American Chemical Society.*

pocket is largely hydrophobic in nature, but is lined by a network of water molecules that are observed in most bromodomain structures determined to date. A direct hydrogen bond between the carbonyl oxygen and an asparagine residue at the C-terminal end of helix B (N1168 in CREBBP) anchors KAc in the binding site. This residue is present in 48 out the 61 human bromodomains identified by Filippakopoulos et al. in their study of bromodomain structure. In 12 bromodomains, it is replaced by other hydrogen bond donors such as threonine or tyrosine, however, the KAc-binding ability is demonstrably maintained in some cases [13]. One isolated example (MLL, mixed lineage leukemia) exists of an aspartate at this position suggesting that the domain might not bind KAc. A second hydrogen bond is formed between the carbonyl group and an even more highly conserved tyrosine residue in the ZA loop (58/61 human bromodomains; Y1125 in CREBBP), via a structured water molecule (Water 1, Figure 22.1C). The KAc N–H interacts with the protein backbone via another structured water molecule (Water 5, Figure 22.1C), which is located further away from the base of the pocket. A second binding mode is observed in some bromodomains, including the BET family, and allows the simultaneous recognition of two KAc residues on one peptide by a single bromodomain.

1.2 CLASSIFICATION OF HUMAN BROMODOMAINS

The human genome contains 61 distinct bromodomains that are found in 46 different proteins [13,14]. Most proteins contain only one bromodomain, though 10 contain 2, and 1 (polybromo-1 or PB1) contains 5. BCPs are primarily transcriptional coregulators or histone-modifying enzymes and include HATs, ATP-dependent chromatin remodeling complexes, helicases, methyltransferases, transcriptional coregulators, and nuclear scaffolding proteins [15]. The bromodomains have been classified into eight families on the basis of three-dimensional structure-based alignment (Figure 22.2) [13]. The precise biological roles of many of the proteins are poorly understood, and the role of the bromodomain in protein function frequently less so.

1.3 FUNCTIONS OF BET BROMODOMAINS

Arguably the best characterized family of BCPs is the BET proteins bromodomain comprising 2, 3, 4, and testis-specific (BRD2, BRD3, BRD4 and BRDT; Family II, Figure 22.2). Each protein contains two N-terminal bromodomains and a long C-terminal region containing the extraterminal (ET) protein–protein interaction domain. The first bromodomains are more closely related to each other than to the second bromodomain in the same protein (Figure 22.2). BET proteins associate with hyperacetylated histone tails, particularly histone H4, and possibly to a lesser extent H3 [13,16–20]. The N-terminal bromodomains can recognize two adjacent acetylation marks simultaneously on the same histone tail, one binding as a "canonical" KAc, the other interacting primarily through hydrophobic contacts [13,20]. The interactions of the N-terminal bromodomains with tetra-acetylated histone H4 (H4K5AcK8AcK12AcK16Ac) are some of the strongest bromodomain–peptide interactions reported, with dissociation constants in the low micromolar range [13]. Interactions between acetylated histone peptides and the second bromodomain are weaker.

BRD2 was the first BET protein to be studied in human cells. Originally named RING3, it was identified as a nuclear kinase [21,22], with a role in cell cycle control through transactivation of E2F transcription factor responsive genes [23]. Only later was it shown to bind to acetylated chromatin [16,24], where it interacts with transcription factors (including E2F) [25], HATs [26], chromatin remodeling proteins [25], and essential components of the transcriptional machinery such as the Mediator complex [24]. Furthermore, BRD2 allows RNA polymerase II (Pol II) to transcribe through acetylated histones which otherwise hinder its passage. BRD2 may also play a role in hematopoiesis: forced overexpression of BRD2 in transgenic mice leads to B-cell lymphomas and leukemias, in part through upregulation of cyclin A [27,28].

BRD4 is also a transcriptional coactivator, recruiting general transcription factors [29], notably the positive transcription elongation factor (P-TEFb), to acetylated chromatin [30–33]. BRD4 binds to the regulatory subunit of P-TEFb, cyclin T, via an alpha-helical C-terminal region referred to as the P-TEFb interaction domain (PID) [34]. This promotes the phosphorylation of the C-terminal domain (CTD) of the Pol II subunit retinol-binding protein 1 (RBP1) on serine 2 by the catalytic P-TEFb subunit cyclin-dependent kinase 9 (CDK9), releasing the polymerase from promoter-proximal pausing and stimulating transcriptional elongation [30,31]. The BRD4-dependent switch from initiation to elongation plays a particularly important role in the induction of primary response genes (i.e., genes that do not require *de novo* protein synthesis for their expression) following stimulation [35,36]. Intriguingly, both the C-terminal PID and the second bromodomain appear to bind to P-TEFb, the latter by interacting with KAc on cyclin T [34,37]. Given the weaker affinity of the second bromodomain for histones,

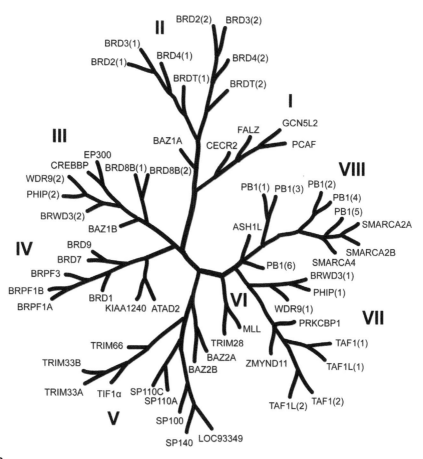

FIGURE 22.2

Phylogenetic tree of the human bromodomain family constructed on a structure-based alignment [13]. The eight bromodomain families are highlighted with Roman numerals. *Reprinted with permission from Ref. [6]. Copyright 2012 American Chemical Society.*

it is possible that the first bromodomain is primarily responsible for interactions with histones, while the second interacts significantly with nonhistone substrates. BRD4 may also activate Pol II-dependent transcription independently of P-TEFb. This may occur through the recruitment of other effectors such as the histone methyltransferase NSD3 (nuclear SET domain-containing protein 3) or the histone dem-ethylase JMJD6 (JmjC domain-containing protein 6), via the ET domain of BRD4 [38]. Additionally, the bromodomain-containing region of BRD4 has been reported to possess intrinsic kinase activity and is able to phosphorylate serine 2 of the Pol II CTD, though the relative importance of direct versus P-TEFb-mediated phosphorylation of Pol II by BRD4 *in vivo* is unclear [39].

As suggested above, the bromodomains of BRD4 might form functional associations with acety-lated lysines on nonhistone proteins. In addition to the aforementioned interaction with cyclin T, the bromodomains interact with acetylated RELA (transcription factor p65 (alternative name: NF-κB p65 subunit)), a component of the NF-κB transcription factor that controls the expression of many proinflammatory genes. Acetylation of RELA recruits BRD4 to the promoters of NF-κB-dependent genes where it acts as a coactivator by recruiting P-TEFb and activating Pol II [40].

Unlike most chromatin-associated proteins, BRD4 remains bound to chromatin during mitosis [41,42], and is therefore proposed to act as a "mitotic bookmark" for actively transcribed genes [43,44]. After the near-shutdown of transcription during mitosis, transcription must be activated on entry into G_1 in order to maintain cell lineage and continued cell cycle progression [45,46]. BRD4 appears to play a key role in accelerating transcription initiation after mitosis at the genes where it has remained bound by recruiting P-TEFb and promoting chromatin decompaction after mitosis [47]. BRD2 also binds to mitotic chromosomes, but the functional significance of the interaction remains unclear, as BRD2 lacks the PID necessary for P-TEFb recruitment [48].

BRD3 is much less studied than BRD2 or BRD4. It binds to acetylated nucleosomes and might play a similar role to BRD2 in coupling histone acetylation to transcription [18]. It also binds to erythroid transcription factor (GATA1) in an acetylation-dependent manner, which appears to be critical for erythroid maturation [49,50].

While BRD2, BRD3, and BRD4 are ubiquitously expressed, BRDT expression is limited to the testes [51]. It is essential in multiple stages of spermatogenesis [52–54], and consequently BRDT inhibition has been proposed for male contraception [55].

1.4 THE ROLES OF BET BCPS IN CANCER

BRD2 and more especially BRD4 have been implicated in various disease states, including cancer and viral infection. In a rare but aggressive form of squamous cell carcinoma known as NUT (nuclear protein in testis) midline carcinoma, *BRD4*, or more occasionally *BRD3*, are fused in frame to the *NUT* gene of unknown function [56]. The resultant fusion is under the control of the *BRD4* promoter and encodes the two bromodomains and ET domain of BRD4, and almost the entire coding region of *NUT*, which includes a binding domain for the HATs CREBBP and p300 [57]. As a result, NUT, which is normally expressed only in testes, becomes expressed in cells harboring the rearrangement. Reynoird et al. have proposed an intriguing mechanism for the oncogenic effect of this translocation: the NUT region of the fusion recruits CREBBP/p300 to acetylated chromatin via the CREBBP/p300 TAZ2 (transcription adaptor putative zinc finger 2) domain, bringing about further acetylation, binding of BRD4-NUT and recruitment of CREBBP/p300. The resultant sequestration of CREBBP/p300 on chromatin prevents it from activating p53, reducing the activity of this critical tumor suppressor [57].

BET proteins, particularly BRD4, play a critical role in many hematological and solid tumors, by acting as coactivators for the expression of proproliferative genes. Notably, BRD4 is required for the expression of *MYC*, an oncogenic driver in many cancers [58,59]. This observation has attracted considerable attention, as therapeutic inhibition of BRD4–histone interactions represent a novel strategy to target *MYC*-dependent cancers, which have otherwise proved difficult to drug.

Many of the roles of BET proteins in disease have only become apparent following the development of chemical probes to inhibit the bromodomain–KAc interaction, described later. These discoveries have opened up new experimental approaches to the treatment of various diseases, testifying to the power of chemical probe development.

2 THE DEVELOPMENT OF BET BROMODOMAIN LIGANDS

2.1 DIAZEPINE-BASED BET BROMODOMAIN LIGANDS

2.1.1 I-BET762 (1)

The first potent and selective bromodomain ligands were methyltriazolodiazepines (**1–2**, Figure 22.3), published simultaneously in late 2010 by Filippakopoulos et al. [60] and Nicodeme et al. [61]. The studies from GSK [61,62] that eventually led to the discovery of I-BET762 (**1**, Figure 22.3) were based on a phenotypic screening approach to discover small molecule upregulators of apolipoprotein A1 (ApoA1). It was not initially known that these compounds functioned by binding to the BET bromodomains. ApoA1 upregulation is involved in protection from atherosclerosis progression and with anti-inflammatory effects; however, when these studies commenced, no method existed to achieve this upregulation in a therapeutic manner. A luciferase reporter-based screen was used to identify small molecule enhancers of ApoA1 expression. Subsequent medicinal chemistry efforts led to the development of I-BET762, with an EC_{50} value of 700 nM in the reporter gene assay. The benzodiazepine core was essential for activity, as was the aryl group extending from the 6-position [63]. It was noted that the C(4) stereochemistry of the molecule had a large influence on its potency, with only the (*S*)-enantiomer showing activity. A chemoproteomic approach was employed to identify the target in which an affinity matrix of the active compound was generated and exposed to a lysate of HepG2 cells. This technique identified the bromodomains of BRD2, BRD3, and BRD4, members of the BET family, as potential cellular targets of I-BET762; and the proposed interaction was confirmed by *in vitro* ITC (isothermal titration calorimetry) and SPR (surface plasmon resonance) experiments. I-BET762 bound to the tandem bromodomains of BRD2–4 [K_D(BRD4) = 55 nM by ITC] and displaced a tetra-acetylated histone H4 peptide [IC_{50}(BRD4) = 36 nM by FRET], while showing no affinity for bromodomains outside the BET family [61,62].

2.1.2 Mitsubishi compounds and (+)-JQ1 (2)

A second triazolodiazepine-based BET ligand, (+)-JQ1 (**2**, Figure 22.3), was independently developed by a collaboration between scientists at the Structural Genomics Consortium (SGC), Harvard Medical School and Dana-Farber Cancer Institute [60] based on compounds disclosed in a patent from Mitsubishi Tanabe Pharma Corporation [64]. A previous patent from this company indicates that related thienotriazolodiazepine compounds were identified from phenotypic studies investigating potential therapies for transplant rejection or autoimmune diseases [65]. The authors were encouraged by the synthetic availability of the lead compounds and the fact that they each contain a "privileged" structural core found in drug molecules such as alprazolam and triazolam. Using docking studies based on the X-ray crystal structure of *apo*-BRD4(1), **2** was designed as a putative ligand. Like I-BET762, only a single enantiomer was active, providing a useful negative control for cellular studies. (+)-JQ1 bound to BRD4(1) with a K_D value of 49 nM (ITC) and showed excellent selectivity for the BET family of bromodomains. (+)-JQ1 displaced a tetra-acetylated histone H4 peptide (H4$_{1-21}$K5AcK8AcK12AcK16Ac) from BRD4(1) with an IC_{50} of 77 nM, as determined by a bead-based amplified luminescent proximity homogeneous assay (AlphaScreen®) [66]. Furthermore, (+)-JQ1 inhibited the interaction of BRD4 with nuclear chromatin in human cells, as demonstrated by fluorescence recovery after photobleaching (FRAP) experiments, which measure the mobility of green fluorescent protein (GFP)-labeled BRD4 [60].

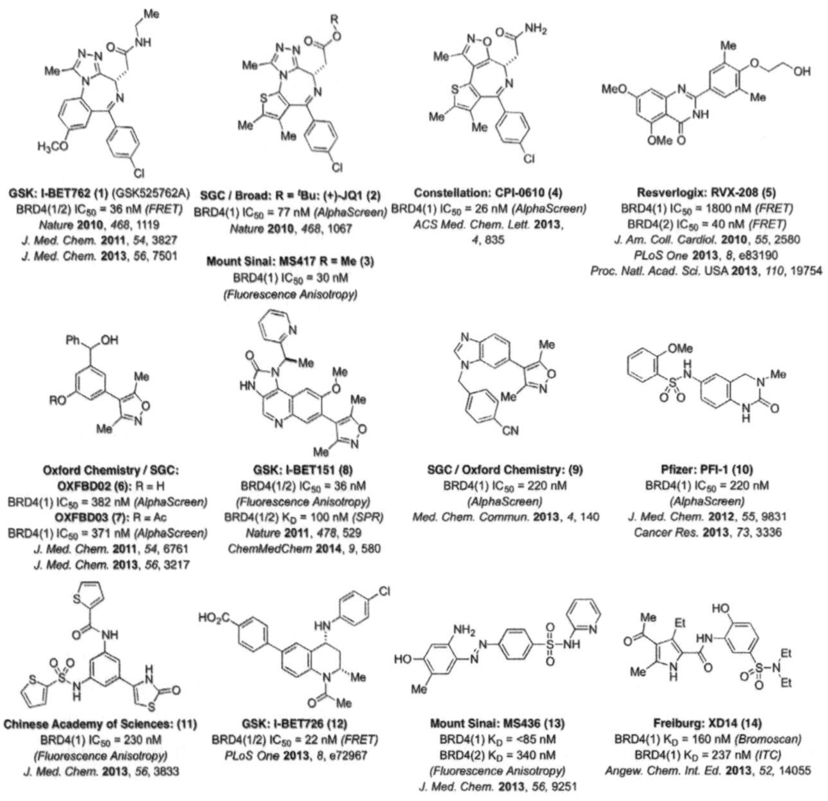

FIGURE 22.3

BET bromodomain inhibitors. The structures of key BET bromodomain inhibitors and their IC_{50} or K_D values with the method used to obtain these values.

MS417 (**3**, Figure 22.3) is a structurally related BET bromodomain ligand that was developed by the laboratory of Ming-Ming Zhou. This compound contains a methyl ester in place of the *tert*-butyl ester of (+)-JQ1 and binds to BRD4(1) and BRD4(2) with similar affinities to (+)-JQ1 [67].

2.2 3,5-DIMETHYLISOXAZOLE-BASED BROMODOMAIN LIGANDS

The 3,5-dimethylisoxazole motif has been independently discovered by a number of groups and has emerged as an excellent KAc mimic that can form the basis of ligands for the BET [68–74] and CREBBP [68,75,76] bromodomains.

Development of an AlphaScreen-based peptide displacement assay by Philpott et al. identified NMP (**15**, Figure 22.4A) and the dihydroquinazolinone (**16**, DHQ) core as promising KAc mimics. The 3,5-dimethylisoxazole-containing DHQ derivative (**18**) displayed an unexpectedly low IC_{50} value of ~7 μM against BRD4(1) in the initial assay, leading to the identification of the 3,5-dimethylisoxazole-based compound **19** (Figure 22.4B) as a BET bromodomain ligand [64,68].

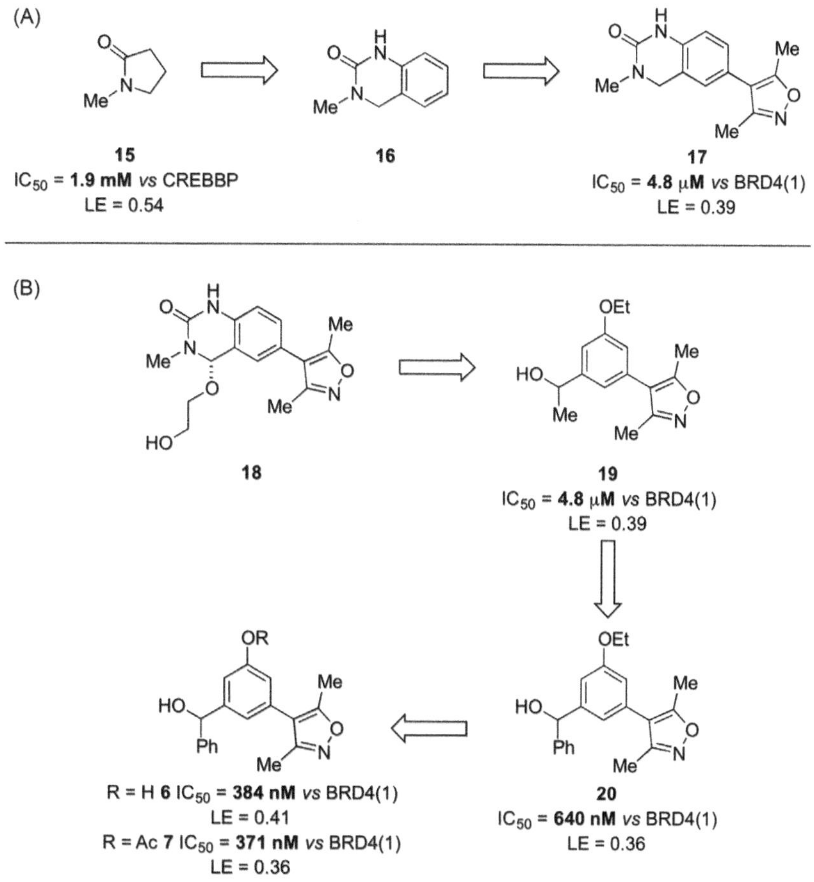

FIGURE 22.4

(A) The fragment-based identification of the DHQ (**16**) and 3,5-dimethylisoxazole KAc mimics. (B) Optimization of the lead compound **18** to give the BET bromodomain ligands **6** and **7** reported by Hewings et al. [68,73]. IC_{50} values were determined using an AlphaScreen assay. Ligand efficiency (LE) was calculated using the equation $(pIC_{50} \times 1.4\,kcal/mol)/heavy\,atom\,count$ [77].

Optimization of the methyl group on compound **19** to a phenyl ring improved the interaction with the WPF shelf region of BRD4(1), giving **20**. Removing the ethyl group to give the phenol (**6**) allowed hydrogen bonding with one of the ZA channel water molecules and provided a BET bromodomain ligand with an IC_{50} value of 384 nM (average value of the (*R*)- and (*S*)-enantiomers). Addition of an acetate group on the phenol (**7**) gave a similar IC_{50} value of 371 nM for BRD4(1) and moderately improved the selectivity over CREBBP [73]. Compounds **6** and **7** were evaluated for their cytotoxic effects on MV4;11 cells and shown to have IC_{50} values of 794 nM and 616 nM, respectively.

The 3,5-dimethylisoxazole moiety was independently identified by Bamborough et al. [70], ultimately leading to the development of a range of potent and selective BET bromodomain ligands. The

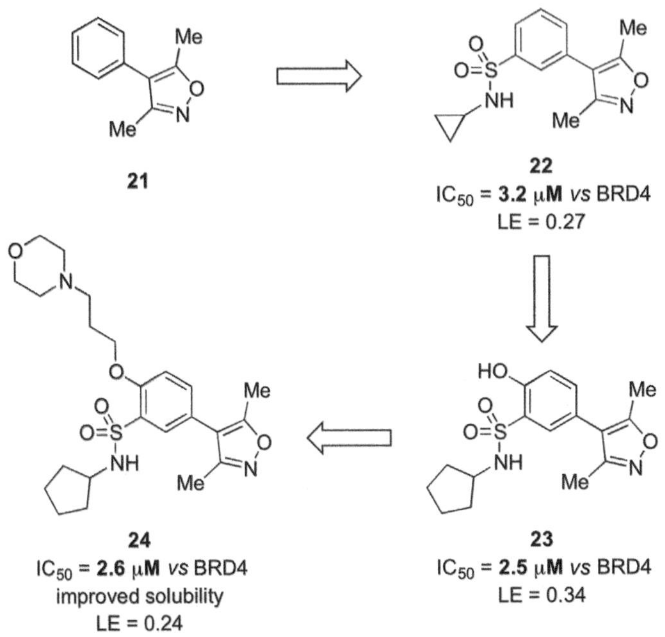

FIGURE 22.5

The optimization of the lead fragment **21** to give the BET bromodomain ligand **24** reported by Bamborough et al. [70]. IC$_{50}$ values were determined using a time-resolved fluorescence resonance energy transfer (TR-FRET) assay.

3,5-dimethyl-4-phenylisoxazole fragment (**21**) showed 32% inhibition of BRD3 and 26% inhibition of BRD4 at 200 μM. A comparison of the X-ray crystal structures of the fragment **21** (Figure 22.5) and I-BET762 led to the development of a 3D-pharmacophore model for BET bromodomain binding.

A search of commercially available compounds using the pharmacophore gave a number of hits based on the 3,5-dimethylisoxazole with a sulfonamide substituent on the phenyl ring *meta* to the 3,5-dimethylisoxazole [70]. Both the pharmacophore and docking models indicated that the sulfonamide was effective at directing the attached lipophilic substituent into the WPF shelf region of the bromodomain. X-ray crystal structures of these ligands bound to BRD2(1) confirmed that the sulfonamide and its substituent were oriented as expected. Initial hits possessed a cyclopropyl ring binding to the WPF shelf (**22**) but this was optimized to a cyclopentyl ring (**23**). A substituent *para* to the 3,5-dimethylisoxazole was found to be beneficial, with methyl, methoxy, and hydroxyl all tolerated. A limitation of the sulfonamide series was their low solubility. This problem was addressed by the incorporation of a morpholine solubilizing group attached to an oxygen atom *para* to the 3,5-dimethylisoxazole giving compound **24**. This compound has low μM IC$_{50}$ values for the BET bromodomains (determined using a TR-FRET assay). DSF (differential scanning fluorimetry) analysis indicated reasonable selectivity for the BET bromodomains over other BCPs evaluated. Compound **24** showed an IC$_{50}$ value of 3.0 μM in a cellular assay based on inhibiting release of interleukin 6 (IL-6), a cytokine shown to be regulated by BET ligands [61].

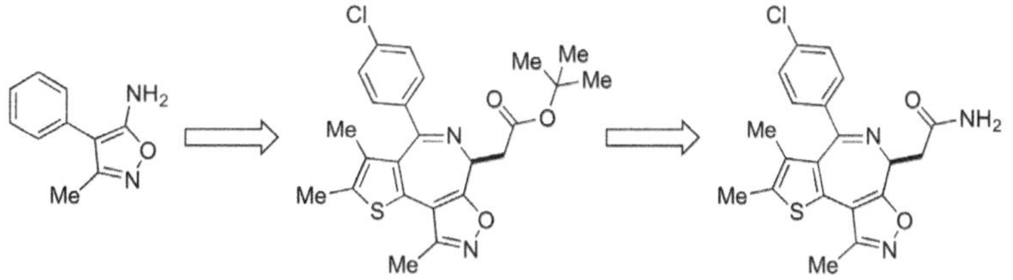

25
IC_{50} = **33 μM** *vs* BRD4(1)
LE = 0.48

26
IC_{50} = **290 nM** *vs* BRD4(1)
LE = 0.30

4
IC_{50} = **26 nM** *vs* BRD4(1)
LE = 0.39

FIGURE 22.6

The optimization of the aminoisoxazole (**25**) fragment to give the isoxazole azepine BET bromodomain ligand **4**, reported by Gehling et al. [74]. IC_{50} values were determined using an AlphaScreen assay.

I-BET151 (**8**, Figure 22.3) is another 3,5-dimethylisoxazole-based BET bromodomain ligand, which was discovered by phenotypic screening, in a manner similar to I-BET762 [69]. Subsequent to the identification of the 3,5-dimethylisoxazole KAc mimic being essential for activity, the initial hit compound was optimized to display favorable pharmacokinetic and metabolism properties [71].

Researchers at Constellation Pharmaceuticals have reported an isoxazole-based BET bromodomain ligand that was developed using hits from a fragment screen [74]. This screen identified several compounds with potencies in the micromolar range. Analysis of cocrystals of these compounds with BRD4(1) led to the selection of **26** (Figure 22.6), as the lead to be progressed, as it mimicked the key interactions formed between BRD4(1) and KAc. Based on the structure of (+)-JQ1 [60], compound **26** was synthesized, in which the triazole is replaced with the isoxazole ring. This compound (**26**) possessed a higher IC_{50} value for BRD4(1) than (+)-JQ1 when evaluated in the same assay. Changing the ^{t}Bu ester to a carboxamide (**4**) resulted in an IC_{50} value of 26 nM, which was similar to the values displayed by (+)-JQ1 and I-BET151 in the same assay. A subsequent analysis of the SAR around the aryl ring that occupies the WPF shelf did not improve the compound potency, indicating that the original 4-chlorophenyl group was optimal. An X-ray crystal structure of **4** bound to BRD4(1) confirmed that the compound bound as expected. The isoxazole resides in the KAc-binding pocket and acts as a KAc mimic and the 4-chlorophenyl group binds to the WPF shelf. The carboxamide NH_2 group forms a water-mediated hydrogen bond with the carbonyl oxygen atom of N140. Compound **4** was shown to be a potent BET bromodomain probe in cellular assays and pharmacokinetic profiling in rat and dog demonstrated suitable characteristics for further *in vivo* experiments.

2.3 RVX-208 (5)

RXV-208 (**5**) is currently undergoing clinical development as a treatment for atherosclerosis [78,79]. As with I-BET762 and I-BET151, it was originally identified as an upregulator of ApoA1, and two recent studies have shown that it binds to BET bromodomains [80,81]. In contrast to previously described BET inhibitors, RVX-208 (**5**) shows some selectivity for the second over the first bromodomain of the BETs. For example, ITC experiments carried out by Picaud et al. gave a K_D of 140 nM for BRD4(2), but 1.1 μM for BRD4(1) [80].

2.4 PFI-1 (10)

A collaboration between the SGC and Pfizer led to the development of BET bromodomain inhibitors based around the DHQ-based 3-methyl-3,4-dihydroquinazolin-2(1*H*)-one scaffold [82,83]. The carbonyl group of the cyclic urea mimics that of KAc, while the *N*-methyl group binds in the same position as the methyl group of KAc. Introducing sulfonamide-linked aromatic substituents at the 6-position led to a substantial increase in potency, consistent with a crystal structure of the *ortho*-methoxyphenyl-substituted compound, PFI-1 (**10**), in which the pendant aromatic group was directed into the WPF shelf. PFI-1 (**10**) showed an IC_{50} value of 220 nM against BRD4(1) in the AlphaScreen® peptide displacement assay, and an EC_{50} value of 1.89 μM for the inhibition of IL-6 production in LPS-stimulated PBMCs [82]. Additionally the compound displays antileukemic activity, downregulating *MYC* expression and inducing differentiation [83]. However, pharmacokinetic properties are suboptimal with the low oral bioavailability in rats being attributed to poor solubility in the gut [82].

2.5 THIAZOL-2-ONE-BASED BROMODOMAIN LIGANDS

A publication from Zhao et al. at the State Key Laboratory of Drug Research in the Chinese Academy of Sciences describes a fragment-based approach to identifying new BET bromodomain inhibitor chemotypes [84]. Docking was used to select 41 putative BRD4-binding fragments, which were progressed to crystallographic studies. Nine fragments were identified, of which four were disclosed in the 2013 paper. Structure-based optimization of a 4-phenyl-thiazol-2-one hit identified several compounds with submicromolar IC_{50} values, including **11**, which had an IC_{50} value of 230 nM in a fluorescence anisotropy (FA) assay for the displacement of fluorescein-tagged (+)-JQ1 from BRD4(1). No data were presented on the selectivity of such compounds over other bromodomains. Activity in cellular assays, including growth inhibition and *MYC* mRNA downregulation in HT-29 colon cancer cells, correlated poorly with inhibitory activity in the FA assay, possibly as a result of limited membrane permeability. The most potent of the compounds in the cellular assay was only moderately active (GI_{50} of 37 μM in HT-29 proliferation assay, *c.f.* 2.3 μM for (+)-JQ1) [84].

2.6 DIAZOBENZENE-BASED BROMODOMAIN LIGANDS

Ming-Ming Zhou and coworkers identified a diazobenzene compound **43**, referred to as ischemin, as a ligand for the CREBBP bromodomain (Figure 22.9) [85], and have subsequently used a similar scaffold to develop potent BET bromodomain inhibitors [86]. MS436 (**13**) inhibited the binding of fluorescein-tagged MS417 (**3**) to BRD4(1) in a fluorescence anisotropy assay with a K_i value of <85 nM (limited by the affinity of the assay probe). The compound also showed activity against the bromodomains of CREBBP, PCAF (p300-CREBBP associated factor), BRD7, BPTF (bromodomain PHD finger transcription factor), BAZ2B, and SMARCA4, with K_i values between 2 and 8 μM in FA assays. The compound also showed activity in several cellular assays, including suppressing the release of proinflammatory cytokine IL-6 after LPS stimulation, with an IC_{50} value of 4.9 μM in an enzyme-linked immunosorbent assay (ELISA). Data published in a patent from the same laboratory revealed that the diazo linker can be replaced with an alkene without loss of affinity [87].

2.7 4-ACYLPYRROLES—XD14 (14)

Lucas et al. used virtual screening followed by X-ray crystallography and ITC measurements to identify new inhibitors of BRD4(1) [88]. By far the most active hit was **14**, referred to as XD14, which had a K_D of 237 nM against BRD4(1) by ITC and a K_D value of 160 nM in the BROMOscan ligand displacement assay. It also interacted significantly with the CREBBP bromodomain (K_D value of 1.6 μM in BROMOscan assay), but showed good selectivity over other bromodomains from outside the BET family. The compound was crystallized with BRD4(1), and the X-ray structure revealed that the ketone carbonyl group mimics the carbonyl of KAc in its hydrogen bonding interactions, while a methyl group at C(5) occupies a similar position as the terminal methyl group of KAc. The amide carbonyl interacts with a water molecule in the ZA channel, as with I-BET151 and compound **6**, and the pyrrole N–H forms a hydrogen bond to the carbonyl group of P82. The compound does not occupy the WPF shelf, but the phenyl ring forms an edge-face π-stacking interaction with W81 [89]. Interestingly, an analogue with a methyl group in place of the substituted phenyl ring bound to BRD4(1) with an almost identical pose but significantly reduced affinity (16 μM by ITC), suggesting that this interaction contributes significantly to the binding of XD14.

2.8 I-BET726 (GSK1324726A, 12)

The 1-acetyl tetrahydroquinoline KAc mimic, originally disclosed in a GSK patent filed in 2010 [90], was subsequently published by Wyce et al. [91]. Compound **12**, referred to as I-BET726, binds to the tandem bromodomains of BRD4 with a K_D of 4 nM by ITC (IC$_{50}$ value of 22 nM in a TR-FRET assay), making it the most potent BET ligand in the published literature to date. It shows excellent selectivity against the closely related CREBBP bromodomain (K_D value of 6.3 μM by ITC) and no appreciable activity against a panel of other bromodomains. The compound has been extensively profiled in cells and shows potent antiproliferative activity in many, but not all, solid tumor cell lines tested, with IC$_{50}$ values in a 6-day growth inhibition assay of less than 50 nM in some neuroblastoma cell lines.

2.9 *IN SILICO* APPROACHES TO THE DISCOVERY OF BET BROMODOMAIN-BINDING FRAGMENTS

The interest in developing bromodomain ligands has prompted a number of computational approaches to investigating bromodomain structure and ligand design. This work is briefly summarized here, but comprehensive description of this computational research is beyond the scope of this chapter, however, it has been discussed in more detail in a recent review [94]. The laboratory of Caflisch has been particularly active in this area. This group has applied computational techniques to study the influence of bromodomain flexibility on histone recognition [95], the mechanism and kinetics of KAc binding to bromodomains [96], and an analysis of whether the structured water molecules in the KAc-binding pockets of CREBBP and BAZ2B can be displaced [97]. Vidler et al. took a computational approach to determining the predicted druggability of all human bromodomains and hence provided useful information for future ligand development programs [98].

Of more relevance here is the use of computational techniques to identify bromodomain ligands or bromodomain-binding fragments. Vidler et al. took a virtual screening approach to identify predicted

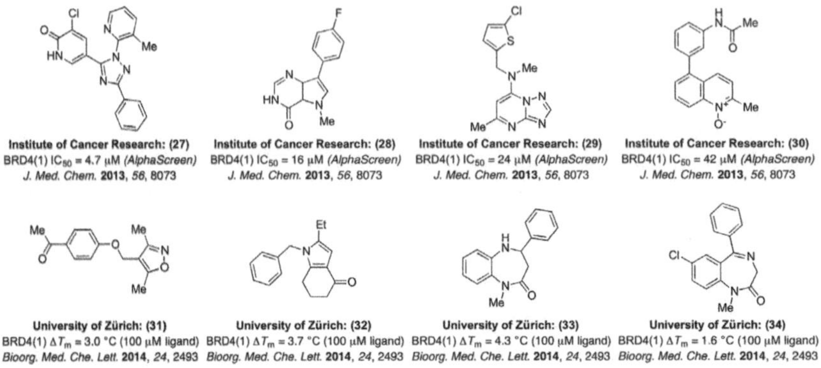

Institute of Cancer Research: (27)
BRD4(1) IC$_{50}$ = 4.7 µM (AlphaScreen)
J. Med. Chem. **2013**, *56*, 8073

Institute of Cancer Research: (28)
BRD4(1) IC$_{50}$ = 16 µM (AlphaScreen)
J. Med. Chem. **2013**, *56*, 8073

Institute of Cancer Research: (29)
BRD4(1) IC$_{50}$ = 24 µM (AlphaScreen)
J. Med. Chem. **2013**, *56*, 8073

Institute of Cancer Research: (30)
BRD4(1) IC$_{50}$ = 42 µM (AlphaScreen)
J. Med. Chem. **2013**, *56*, 8073

University of Zürich: (31)
BRD4(1) ΔT_m = 3.0 °C (100 µM ligand)
Bioorg. Med. Che. Lett. **2014**, *24*, 2493

University of Zürich: (32)
BRD4(1) ΔT_m = 3.7 °C (100 µM ligand)
Bioorg. Med. Che. Lett. **2014**, *24*, 2493

University of Zürich: (33)
BRD4(1) ΔT_m = 4.3 °C (100 µM ligand)
Bioorg. Med. Che. Lett. **2014**, *24*, 2493

University of Zürich: (34)
BRD4(1) ΔT_m = 1.6 °C (100 µM ligand)
Bioorg. Med. Che. Lett. **2014**, *24*, 2493

FIGURE 22.7

BET bromodomain-binding fragments. The structures of BET bromodomain-binding fragments reported by Vidler (**27–30**) [92] and Zhao (**31–34**) [93].

BRD4(1)-binding fragments. In this work, they tested 143 compounds that were predicted to bind to the KAc-binding pocket of BRD4(1). In an AlphaScreen analysis of these compounds six novel hit compounds were identified, four of which (**27–30**) were based on unprecedented KAc-mimicking groups (Figure 22.7). These four compounds have RMMs (relative molecular mass) of between 245 and 363; they showed low micromolar IC$_{50}$ values against BRD4(1) and possess ligand efficiencies of 0.27–0.31. Zhao et al. based their *in silico* screening on the ZINC all-now library. Nine million compounds were filtered to 238,408 fragments based on parameters including RMM, number of rings, and number of rotatable bonds. Following docking studies and molecular dynamics (MD) simulations 24 compounds were evaluated biologically to identify 4 compounds (**31–34**) that were observed to bind BRD4(1) using DSF (Figure 22.7) [93]. Muvva et al. used an *in silico* approach to screening compounds from the NCI Diversity, Drug Bank, and Toslab databases for predicted BRD4(1) affinity [99]. Although a number of predicted ligands were identified, no experimental evidence for BRD4(1) binding was presented.

2.10 DUAL KINASE-BROMODOMAIN INHIBITORS

It has recently been shown that a number of kinase inhibitors are also able to bind to the KAc-binding pocket of some bromodomains. The first such report was from Martin et al. who showed that Dinaciclib (**35**, Figure 22.8), a CDK inhibitor currently in Phase III trials for chronic lymphocytic leukemia, was observed to bind to BRDT(1), crystallographically [100]. Analysis in the BROMOscan assay reveals only weak binding, with a K_D value of 37 µM.

The phosphoinositide 3-kinase (PI3K) inhibitor LY294002 (**36**), widely used as a chemical probe, also interacts with BET bromodomains. Dissociation constants were estimated by titrating in LY284002 to compete bromodomains off an immobilized LY294002 derivative, giving an apparent K_D value of 2 µM for BRD4. LY294002 also showed cellular effects consistent with BET bromodomain inhibition such as suppression of cytokine release after LPS stimulation [101]. Neither study investigated whether these compounds could inhibit the reported kinase activity of BRD4. Remarkably, BRD4 had been identified as a cellular target of LY294002 in 2007 by a similar chemoproteomic approach using

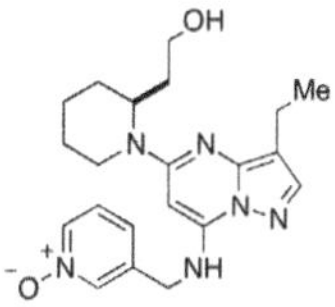

Dinaciclib (35)
BRDT K_D = 37 μM *(BROMOscan)*
ACS Chem. Biol. **2013**, *8*, 2360

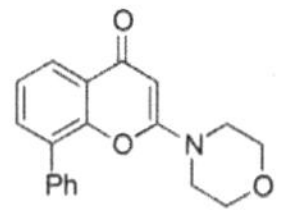

LY294002 (36)
BRD4 app K_D = 2 μM
(chemoproteomic MS/MS)
ACS Chem. Biol. **2014**, *9*, 495

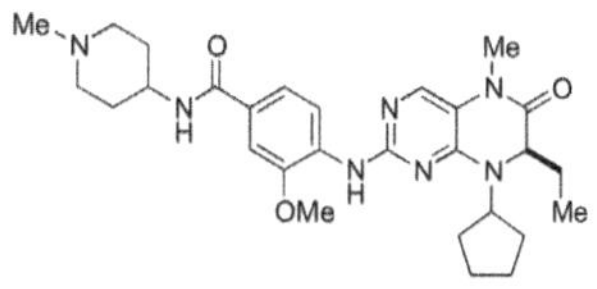

BI2536 (37)
BRD4(1) K_D = 37 nM *(ITC)*
BRD4(1) IC_{50} = 25 nM *(AlphaScreen)*
Nature Chem. Biol. **2014**, *10*, 305
ACS Chem. Biol. **2014**, *9*, 1160

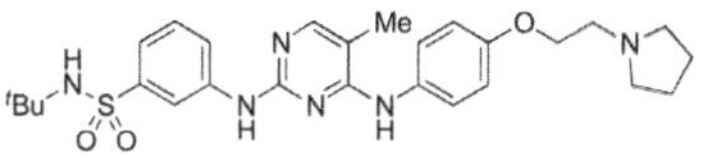

TG101348 (38)
BRD4(1) K_D = 164 nM *(ITC)*
Nature Chem. Biol. **2014**, *10*, 305

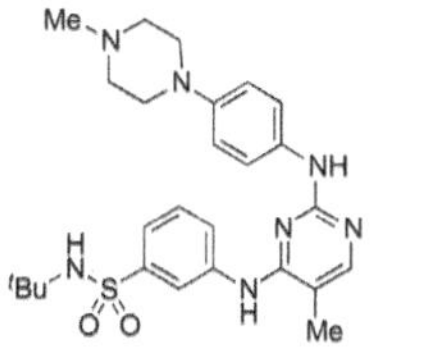

TG101209 (39)
BRD4(1) K_D = 123 nM *(ITC)*
Nature Chem. Biol. **2014**, *10*, 305

FIGURE 22.8

Dual kinase-bromodomain inhibitors. The structures of reported dual kinase-bromodomain inhibitors (**35–39**) and their IC_{50} or K_D values with the method used to obtain these values.

immobilized inhibitor, predating all reported BET inhibitors by several years [102]. Furthermore, a study in 2005 showed that LY294002 inhibited NF-κB-dependent transcription and nitric oxide production while another PI3K inhibitor, wortmannin, did not [103]. These effects now appear likely to have resulted from BRD4 inhibition [40,104].

To identify kinase inhibitors that displayed more potent BET bromodomain affinity, with the aim of developing single agent dual kinase-bromodomain inhibitors for cancer therapy, Ciceri et al. screened 628 kinase inhibitors against BRD4(1), using an AlphaScreen assay [105]. The compound collection contained more than 200 clinical kinase inhibitors. A number of these compounds were found to be potent BRD4(1) ligands, with the PLK1 inhibitor BI2536 (**37**) having K_D value of 37 nM for BRD4(1) and the JAK (Janus kinase) inhibitors TG101209 (**39**) and TG-101348 (**38**) having K_D values of 123 and 164 nM for BRD4(1), respectively. BI-2536 and TG-101348 were shown to displace BRD4 from chromatin in a FRAP assay in U2OS cells. It is suggested that the BET bromodomains, and not the JAK kinases, are the key targets for these compounds, at least in certain cell lines. Based on the analysis in this paper, the authors propose a method for the rational design of dual kinase-bromodomain inhibitors that possess complementary pharmacology for use in cancer therapy. In related work, Ember et al. identified 14 kinase inhibitors as BRD4(1) ligands that occupy the KAc-binding site [106]. This study also identified BI2536 and TG101209 as among the most potent BRD4(1) ligands.

3 BET BROMODOMAIN LIGANDS IN CLINICAL TRIALS

An important endorsement of the utility of the BET bromodomain ligands, described above, is their application in exploring the roles of these proteins *in vivo*. A remarkable number of studies employing

BET inhibitors have been published since 2010, and we do not attempt to review them all. Instead, we will cover some of the early studies that used these compounds and attempt to give an indication of the therapeutic potential of BET inhibition.

3.1 BET BROMODOMAIN LIGANDS AND CANCER

Much attention has focused on the use of BET inhibitors in hematological malignancies. In one of the first studies, (+)-JQ1 was used to corroborate the findings of an RNAi screen that identified BRD4 as a therapeutic target for acute myeloid leukemia (AML) [58]. In addition to reproducing the antiproliferative and differentiation-inducing effects of BRD4 short hairpin RNA (shRNA) in AML cells, (+)-JQ1 showed single agent antileukemic effects in an AML mouse model. Intriguingly, the effect of (+)-JQ1 on leukemia cells was due, at least in part, to downregulation of the *MYC* oncogene, an attractive but challenging therapeutic target. A reduction in *MYC* expression was also shown to be partly responsible for the activity of (+)-JQ1 in multiple myeloma (MM) [107]. Downregulation of *MYC* and its target genes was accompanied by reduced proliferation and cell cycle arrest in cell lines and primary samples. Similar findings have been reported by Mertz et al. who independently demonstrated the MYC-dependent antiproliferative effects of (+)-JQ1 in leukemia and MM cell lines and showed activity in xenograft models of AML and Burkitt's lymphoma [59]. Encouragingly, (+)-JQ1 showed selectivity for malignant cells in all studies; however, the reason underpinning the observed selectivity for transformed cells remains to be fully elucidated.

I-BET151 showed efficacy against leukemia cell lines and xenograft leukemia models containing MLL gene fusions [69]. The authors suggest that BET BCPs recruit members of the superelongation complex (SEC) or polymerase-associated factor complex (PAFc). SEC and PAFc are regulators of transcriptional elongation, and since components of these complexes form fusions with MLL proteins in leukemias, they are crucial to the transforming effects of the translocations. Known MLL target genes, including *BCL2*, *CDK6*, and *MYC*, were downregulated by I-BET151 treatment, consistent with the observed phenotype of the compound. These findings strengthen the association between the inhibition of BET bromodomains and *MYC* downregulation, suggesting that the inhibition of BET bromodomains is a promising therapeutic strategy for treating MLL-fusion leukemias [58,69].

Due to the potential for a MYC-targeted agent to act as a cancer therapy [107], there has been considerable interest in establishing the effects of a *MYC*-downregulating compound, such as a BET bromodomain inhibitor, in other tumor types [108]. Indeed, the antiproliferative effects of BET bromodomain inhibitors in diverse leukemias [109–113], lymphomas [114–117], and myelomas [118] are at least in part dependent on MYC. Three BET inhibitors (I-BET762 from GSK, OTX015, a triazolodiazepine licensed by OncoEthix from Mitsubishi Tanabe, and CPI-0610, a compound from Constellation Pharmaceuticals) are currently in early stage clinical trials for hematological malignancies (*vide infra*).

Recent work has shown that a range of solid tumors is also sensitive to BET inhibition. In many cases, the mode of action appears to be MYC dependent. Cancers in which *MYC* or *MYCN* are frequently amplified or overexpressed are often particularly sensitive. These include neuroblastomas [91,119], medulloblastomas [120,121], lung adenocarcinomas [122], and prostate cancer [123,124]. However, in some cancers, the effects of BET inhibition are largely *MYC* independent [125–129]. It will be interesting to see whether *MYC* status is a useful indicator of a patient's responsiveness to BET inhibitors.

In the aforementioned malignancies, BET proteins are required to express key genes involved in cancer growth and progression, but are not themselves oncogenic drivers. There is little correlation between BRD4 expression and sensitivity to BET inhibitors, and in fact high BRD4 levels are associated with reduced invasiveness and improved survival in breast cancers [130]. NUT midline carcinoma (NMC) is unusual in that BET bromodomains are fundamental to the etiology of the disease. Consequently, BET inhibitors are being investigated as potential therapies. Filippakopoulos et al. observed that (+)-JQ1 led to differentiation, growth arrest, and apoptosis in patient-derived NMC cell lines, phenocopying the effect of BRD4-NUT siRNA (Small interfering RNA) knock down. (+)-JQ1 treatment also led to the downregulation of BRD4-dependent genes. In patient-derived NMC mouse xenografts, (+)-JQ1 led to tumor shrinkage and increased overall survival rate [60]. Two clinical trials are now underway to investigate BET inhibitors in NMC and other solid tumors: GSK are studying I-BET726, while Tensha therapeutics (a spin-off from the Bradner laboratory at the Dana Faber Cancer Centre) is using an undisclosed compound referred to as TEN-010.

3.2 BET BROMODOMAIN LIGANDS AND INFLAMMATION

Inhibitors of BET bromodomains have shown potential as anti-inflammatory agents. Phenotypic screening identified lead compounds as upregulators of ApoA1, resulting in I-BET762 and I-BET151 [61–63,71,72]. ApoA1 is the major component of high-density lipoprotein and is associated with anti-inflammatory effects. Nicodeme et al. investigated the effects of I-BET762 on the inflammatory response in LPS-stimulated macrophages [61]. I-BET762 downregulated the expression of LPS-inducible cytokines in stimulated macrophages but had little effect on unstimulated cells. The susceptibility of a gene to BET inhibition depended on its pattern of chromatin modifications, suggesting that the anti-inflammatory effects of I-BET762 arose through inhibiting BET–chromatin interactions. Anti-inflammatory activity was also demonstrated *in vivo* with I-BET762 improving survival and downregulating proinflammatory genes in mouse models of endotoxic shock and bacterial sepsis. I-BET151 also showed activity as both a prophylactic and a therapeutic agent in the same mouse model of sepsis [72].

BET inhibitors may exert a second, independent anti-inflammatory effect by inhibiting the BRD4–NF-κB interaction [40]. Several BET inhibitors suppress the expression of NF-κB target genes following stimulation [63,67,86,104]. Treatment with BET inhibitors has therefore been suggested to reduce inflammatory kidney damage in HIV-associated kidney disease [67], and as a treatment for cancers in which constitutive NF-κB activity drives proliferation [104].

3.3 BET BROMODOMAIN LIGANDS IN CLINICAL TRIALS

It is a testament to the rapid progress in this field that there are currently five BET bromodomain inhibitors in clinical trials (Table 22.1). Following promising results in animal models two Phase I clinical trials have been initiated focusing on NMC, one using I-BET762 (ClinicalTrials.gov identifier: NCT01587703) and the other using the triazolothienodiazepine TEN-010 (ClinicalTrials.gov identifier: NCT01987362). The structurally related compound OTX015 is the subject of a clinical trial focusing on acute leukemia and other hematological malignancies (ClinicalTrials.gov identifier: NCT01713582).

A Phase I study using CPI-0610, developed by Constellation Pharmaceuticals, has also been initiated for the treatment of patients with progressive lymphoma (ClinicalTrials.gov identifier: NCT01949883).

Table 22.1 Bromodomain Ligands in Clinical Trials

Compound	Company	Phase	Indications	ClinicalTrials.gov identifier (status)
RVX-208	Resverlogix	II	Atherosclerosis	NCT01058018 (completed)
		II	Coronary artery disease	NCT01067820 (completed)
		II	Dyslipidemia	NCT01423188 (completed)
			Type II diabetes	NCT01728467 (completed)
I-BET762 (GSK525762)	GSK	I/II	NMC	NCT01587703 (recruiting)
		I/II	Relapsed refractory hematological malignancies	NCT01943851 (recruiting)
OTX015	OncoEthix	I	Acute leukemia and other hematological malignancies	NCT01713582 (recruiting)
CPI-0610	Constellation Pharmaceuticals	I	Progressive lymphoma	NCT01949883 (recruiting)
		I	Multiple myeloma	NCT02157636 (recruiting)
		I	Acute leukemia, myelodysplastic syndrome, or myelodysplastic/ myeloproliferative neoplasms	NCT02158858 (recruiting)
TEN-010	Tensha Therapeutics	I	NMC	NCT01987362 (recruiting)

More recently, two new Phase 1 clinical trials of CPI-0610 commenced. The first evaluates CPI-0610 in patients with previously treated MM (ClinicalTrials.gov identifier: NCT02157636), while the second trial focuses on patients with acute leukemia, myelodysplastic syndrome, or myelodysplastic/myeloproliferative neoplasms (ClinicalTrials.gov identifier: NCT02158858).

Clinical trials on RVX-208 were initiated based on its ability to modulate ApoA1 levels before it was shown to be a BET bromodomain ligand. The first indications investigated the effect of the compound on atherosclerosis (ClinicalTrials.gov identifier: NCT01058018) and coronary artery disease (ClinicalTrials.gov identifier: NCT01067820). This trial failed to meet its primary end point, but it did show that RVX-208 caused a modest regression of coronary atherosclerosis and that a synergistic effect was observed when given in combination with Rosuvastatin. It is not entirely clear that the effects of BET bromodomain ligands modulate ApoA1 levels by direct interaction with BET bromodomains, and further studies are needed in this area [8]. Trials focusing on the effect of RVX-208 on dyslipidemia (ClinicalTrials.gov identifier: NCT01423188) and type II diabetes (ClinicalTrials.gov identifier: NCT01728467) have also been completed. Interestingly, analysis of the pooled data collected in these trials revealed a statistically significant reduction of major adverse cardiovascular events (MACE) in patients with diabetes mellitus. Analysis of the pooled patient data showed again that RVX-208 treatment led to a statistically significant relative risk reduction in MACE of >65% compared to placebo. This finding is important as the majority of diabetes mellitus patients die from cardiovascular diseases. Diabetes mellitus patients given RVX-208 tended to have lower blood glucose compared to those patients given a placebo. In diabetes mellitus patients who had low levels of high-density lipoprotein, the blood glucose was significantly lower following treatment with RVX-208 compared to placebo. It is interesting to note that the time required for RVX-208 to reduce blood glucose was at least 12 weeks following initiation of treatment.

4 THE DEVELOPMENT OF CREBBP BROMODOMAIN LIGANDS

The BCP CREBBP has emerged as another important target for small molecule bromodomain inhibitors. CREBBP was first discovered as a nuclear protein that binds to phosphorylated CREB and was shown to regulate protein kinase A activity as a component of a multiprotein complex [131]. Its paralogue, p300 (or EP300), was identified as a binding partner of the adenovirus early region 1A protein (E1A) [132]. Both CREBBP and p300 were later revealed to possess HAT catalytic activity [133,134]. The two proteins are known to have at least 400 interacting protein partners making them key nodes in the known mammalian protein–protein interactome [135]. One such interacting partner is p53, and the specific recognition of KAc382 of p53 by the CREBBP/p300 bromodomain is required for acetylation-dependent coactivator recruitment after DNA damage.

4.1 N-ACETYLATED FRAGMENTS

The first attempts to identify for small molecule ligands for the CREBBP bromodomain were reported by the Zhou group in 2006 [136]. Having recently solved the nuclear magnetic resonance (NMR) structure of CREBBP bound to p53 [137], a structure-based screening program was initiated. A focused library of 200 compounds containing an anticipated acetyl lysine mimic was screened by NMR spectroscopy, looking for perturbations in the 2D ^{1}H-^{15}N HSQC (heteronuclear single quantum correlation) spectra. From this screen, 14 bromodomain-binding compounds were identified, of which 2, **40** (MS2126) and **41** (MS7972) (Figure 22.9), were shown to block the interaction between p53 and CREBBP in a concentration-dependent manner. In addition, in cells treated with doxorubicin to stimulate DNA damage, both **40** and **41** were shown to suppress the levels of p53 and the expression of p21 (a target of p53). Further characterization using a fluorescence assay determined **41** to bind with a K_D of approximately 20 μM. An NMR structure of **41** bound to CREBBP was solved showing that while the compound partly resided in the KAc-binding site, no specific interactions were formed with the conserved asparagine.

4.2 AZOBENZENE-BASED CREBBP BROMODOMAIN LIGANDS

In a subsequent NMR screen, on a larger set of compounds, the same group identified a number of azobenzene-based ligands of the CREBBP bromodomain [85]. A number of analogues based on this framework were synthesized and screened in a cell-based assay measuring p53-dependent expression of p21 in response to doxorubicin. The optimized compound **43** (Figure 22.9) had an IC$_{50}$ value of 5 μM in the cellular assay and a K_D value of 19 μM in a fluorescence assay counterscreen, and was up to fivefold selective over the PCAF, BRD4(1), and BAZ2B bromodomains. Treatment of cardiomyocytes with **43** prevented p53-induced apoptosis upon doxorubicin treatment (a model for ischemic stress), giving **43** the name ischemin.

4.3 A CREBBP BROMODOMAIN-BINDING CYCLIC PEPTIDE

Aided by structural information, the Zhou group developed a cyclic peptide inhibitor of CREBBP, based on the p53 sequence [138]. MD simulations were employed to identify which residues in p53 could be converted to cysteine and chemically linked, while still maintaining the β-turn adopted by the bound

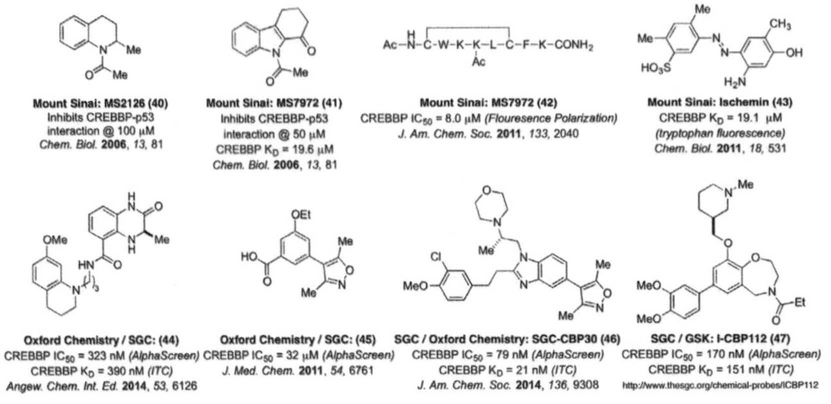

FIGURE 22.9

CREBBP bromodomain-binding fragment and ligands. The structures of CREBBP bromodomain-binding fragments and ligands reported and their IC$_{50}$ or K_D values with the method used to obtain these values.

peptide. Optimization of the amino acid sequence and linker strategy resulted in the cyclic peptide **42** (Figure 22.9) that bound CREBBP with 8 μM affinity, compared to 190 μM for the native p53 peptide. Treatment of colorectal carcinoma and embryonic kidney cells with **42** showed a reduction in p53-induced p21 expression, suggesting the importance of CREBBP bromodomain binding for p53 activity.

4.4 DIHYDROQUINOXALINONE-BASED CREBBP BROMODOMAIN LIGANDS

The first CREBBP bromodomain inhibitors with submicromolar affinity were developed by Rooney et al. [139]. The DHQ fragment was selected at the KAc mimic on which the inhibitors would be based. However, as work in the same group had shown that the DHQ moiety is susceptible to oxidation [68], modification of the KAc group to the more stable benzoxazinone was explored. The compounds in this series only showed micromolar affinity for the CREBBP bromodomain and hence the KAc mimic was altered to the dihydroquinoxalinone motif. The dihydroquinoxalinone-based compounds were ligands for the CREBBP bromodomain, displaying submicromolar affinity in an AlphaScreen assay. X-ray crystallography revealed that the tetrahydroquinoline group (**44**, Figure 22.9) formed a cation–π interaction with R1173. Computational studies indicated that this interaction can form spontaneously in water and contributes toward the binding affinity of the ligand for the bromodomain. The highest affinity compound (**44**) displays modest selectivity for the CREBBP bromodomain over BRD4(1) but showed good activity in a FRAP assay in U2OS cells.

4.5 3,5-DIMETHYLISOXAZOLE-BASED CREBBP BROMODOMAIN LIGANDS

In the course of early work on the development of 3,5-dimethylisoxazole-based BET bromodomain inhibitors, it was observed that a number of these compounds also showed some affinity for the CREBBP bromodomain [68]. The most potent CREBBP analogue, **45** (Figure 22.9), had an IC$_{50}$ value of 32 μM and similar potencies for BET bromodomains. A crystal structure of **45** bound to CREBBP confirmed that the 3,5-dimethylisoxazole moiety acted as the KAc mimic in the same manner as in the BRD4(1) bromodomain. A carboxylate substitution directed toward the BC loop, which might contribute to the affinity of **45** for CREBBP, potentially formed a long range interaction with R1173. A similar interaction between a sulfonate and R1173 was observed by Borah et al. [85] giving an early indication that this residue might be important for CREBBP bromodomain affinity.

Benzimidazole substituted 3,5-dimethylisoxazole-based compounds have also been identified as promiscuous BET and CREBBP bromodomain inhibitors [75]. An extensive optimization of this fragment has recently resulted in the development of SGC-CBP30 (**46**), which has a K_D value of 21 nM for CREBBP and 40-fold selectivity over BRD4(1) [76]. Throughout the development of **46**, selectivity over BRD4(1) remained an issue and it was only through numerous substitutions at the N(1) and C(2) positions in the final compound that resulted in significant selectivity for the CREBBP bromodomain over the BET BCPs. A morpholine group at N(1) was deemed optimum, while a stereogenic methyl group in the linker was integral to the eventual selectivity. Aryl substituents at C(2) were screened and electron rich analogues found to be favored. A cocrystal structure of the final compound **46** with CREBBP revealed that the dimethylisoxazole group bound, as expected, in the KAc pocket. The morpholine group was directed into the ZA channel, while the aryl substituent formed a cation-π interaction with R1173, consistent with the findings of Rooney et al. Using an FRAP assay, **46** was shown to disrupt the binding of the CREBBP bromodomain with chromatin in a cellular setting. In a luciferase-based reporter assay, **46** was shown to inhibit the CREBBP coactivation of p53 target genes. This compound is likely to be a valuable probe in determining the role of the bromodomain in the complex functions of CREBBP and EP300, although it remains to be seen whether 40-fold selectivity over the BET bromodomains is sufficient to suppress the strong BET phenotype in a cellular setting.

Recently, the SGC and GSK have developed a benzoxazepine-based CREBBP inhibitor, I-CBP112 (**47**), which binds to the bromodomains of CREBBP and EP300 with K_D values of 151 and 625 nM, respectively. Selectivity over other bromodomains appears satisfactory and the compound was active in an FRAP assay in U2OS cells, demonstrating an on target effect in a cellular setting. Although the design of **47** is yet to be published, a structure of **47** bound to the CREBBP bromodomain is available in the PDB that highlights a number of interesting features. The benzoxazepine core binds across the entrance to the KAc pocket, while the acyl group attached to N(4) acts as the acetyl lysine mimic forming the typical direct hydrogen bond to N1168 and water-mediated hydrogen bond to Y1125. The aryl group at C(7) forms a cation-π interaction to R1173 and the two methoxy substituents suggest that this interaction was optimized during the design process. A unique feature of **47** is the presence of a propionate rather than acetate group in the KAc mimic. The longer ethyl group extended toward, but did not displace, the conserved water molecules at the base of the KAc pocket.

5 THE DEVELOPMENT OF LIGANDS FOR OTHER BROMODOMAINS

The success in developing BET and CREBBP bromodomain ligands and their utility in elucidating the complex biology of the BCPs has inspired the search for small molecules that bind to other, phylogenetically diverse, bromodomains. The SGC has recently reported the [1,2,4]triazolo[4,3-*a*] phthalazines, which bind to multiple bromodomain classes including the BET bromodomains. For example, **51** (Figure 22.10) displaced a tetra-acetylated H4 peptide from BRD4(1) with an IC$_{50}$ value of 160 nM (AlphaScreen) and also showed submicromolar IC$_{50}$ values against the BRD9 and CREBBP

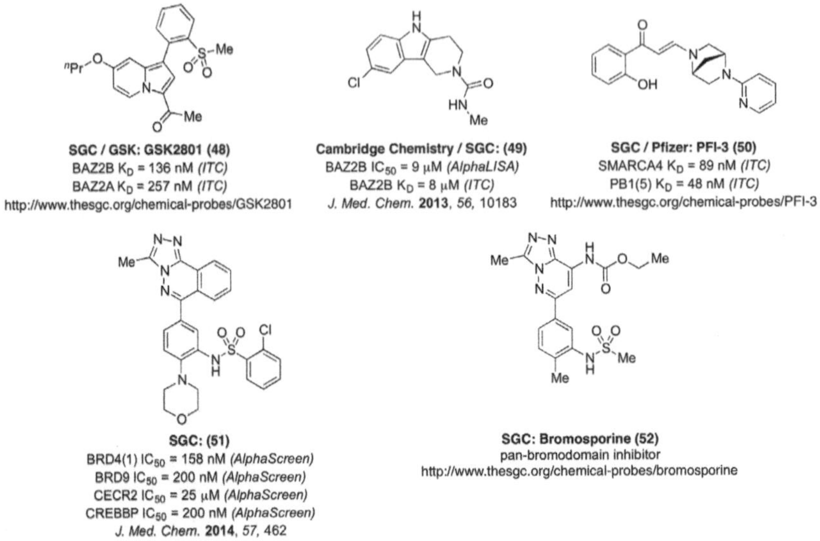

FIGURE 22.10

Ligands and fragments for other non-BET bromodomains. The structures of bromodomain-binding ligands and fragments reported their IC_{50} or K_D values with the method used to obtain these values.

bromodomains [140]. An X-ray crystal structure of **51** with BRD4(1) showed the methyltriazole acts as a KAc mimic. The related fused bicyclic triazole, bromosporine (**52**), is a broad-spectrum bromodomain inhibitor, which shows affinity for a large number of bromodomains and is named with reference to the nonselective kinase inhibitor, staurosporine [141]. This compound was developed as a tool for probing bromodomains for which no other ligands are known.

The bromodomain adjacent to zinc finger domain (BAZ) family of BCPs contains four members: BAZ1A, BAZ1B, BAZ2A, and BAZ2B. All four proteins possess a carboxy-terminal bromodomain next to a PHD finger located toward the amino terminus. While BAZ1A and BAZ1B are involved in chromatin remodeling, the function of BAZ2B is less clear [142]. BAZ2A is constitutively expressed in a range of tissues protein plays a central role in the nucleolar remodeling complex (NoRC) [143]. In order to further understand the function of the BAZ2A and BAZ2B proteins, three groups of researchers have developed ligands for the BAZ2A/B bromodomains. Ferguson et al. identified a number of BAZ2B bromodomain-binding fragments, the highest affinity of which is shown in Figure 22.10 (**49**) [144]. X-ray crystal structures of four fragments bound to the BAZ2B bromodomain were obtained, providing insight into the development of higher affinity ligands for this bromodomain. GSK and the SGC have developed a high-affinity probe for the BAZ2A/B bromodomains, GSK2801 (**48**, Figure 22.10), which is reported on the SGC website (http://www.thesgc.org/chemical-probes/GSK2801). This compound has K_D values of 136 and 257 nM (ITC) for the BAZ2B and BAZ2A bromodomains, respectively. The SGC website also reports another BAZ2 bromodomain probe that was developed in collaboration with the Institute of Cancer Research. This compound, known as BAZ2-ICR, binds to BAZ2A with a K_D value of 109 nM (ITC) and to BAZ2B with a K_D value of 170 nM (ITC).

SWI/SNF-related, matrix-associated, actin-dependent regulator of chromatin, subfamily A, member 4 (SMARCA4) is a member of the switch/sucrose non-fermentable (SWI/SNF) family of proteins. Loss of SMARCA4 function and components of SWI/SNF have been linked to cancer development suggesting a tumor suppressor function for SMARCA4. This protein also interacts with the breast cancer gene breast cancer 1, early onset (BRCA1), and regulates the expression of the tumorigenic protein CD44. Pfizer and the SGC have collaborated to develop a ligand (PFI-3, **50**, Figure 22.10) for the SMARCA4 bromodomain, with a K_D value of 89 nM (ITC). Compound **50** also binds to the polybromodomain-containing protein 1 (PB1) with a K_D value of 48 nM (ITC).

6 CONCLUSION

There has been rapid and significant progress on the development of BET bromodomain inhibitors. The readily available structural information on the binding of these ligands to members of the BET bromodomain family has been key in accelerating the development of new ligands. There are now a significant number of potent BET ligands and these are derived from a range of structurally distinct chemotypes. The availability of structurally diverse ligands give confidence that the phenotypes that these compounds evoke are likely to be predominantly as a result of BET bromodomain inhibition, rather than off-target effects. Despite the intense focus on the development of BET bromodomain ligands, only modest progress has been made on developing ligands that show selectivity for the individual BCPs within the family, or for the first bromodomains over the second bromodomains. Whether these developments are required for the development of clinically useful compounds will be revealed by the clinical trials that are being conducted on five BET bromodomain ligands currently.

The success in developing BET bromodomain ligands as chemical probes, and the success of these compounds in underpinning biological studies on these targets, has promoted the development of small molecule ligands for non-BET bromodomains. This endeavor is progressing rapidly, with potent and selective ligands for a number of bromodomains being reported recently.

The work described here clearly demonstrates that bromodomains play fundamental roles in a number of cancer types. The development of drug-like small molecule ligands for these protein modules has given rise to the prospect that bromodomains, particularly those in the BET family, are new therapeutic targets for treating certain forms of cancer. Ongoing clinical trials will determine whether this promise translates into clinically useful drugs.

REFERENCES

[1] Allfrey V, Faulkner R, Mirsky A. Acetylation and methylation of histones and their possible role in regulation of RNA synthesis. Proc Natl Acad Sci USA 1964;51:786–94.

[2] Gershey EL, Vidali G, Allfrey VG. Chemical studies of histone acetylation. The occurrence of epsilon-*N*-acetyllysine in the F2a1 histone. J Biol Chem 1968;243:5018–22.

[3] Kleff S, Andrulis ED, Anderson CW, Sternglanz R. Identification of a gene encoding a yeast histone H4 acetyltransferase. J Biol Chem 1995;270:24674–7.

[4] Brownell JE, Zhou J, Ranalli T, Kobayashi R, Edmondson DG, Roth SY, et al. Tetrahymena histone acetyltransferase a: a homolog to yeast Gcn5p linking histone acetylation to gene activation. Cell 1996;84:843–51.

[5] Taunton J, Hassig CA, Schreiber SL. A mammalian histone deacetylase related to the yeast transcriptional regulator Rpd3p. Science 1996;272:408–11.

[6] Hewings DS, Rooney TPC, Jennings LE, Hay DA, Schofield CJ, Brennan PE, et al. Progress in the development and application of small molecule inhibitors of bromodomain-acetyl-lysine interactions. J Med Chem 2012;55:9393–413.

[7] Jennings LE, Measures AR, Wilson BG, Conway SJ. Phenotypic screening and fragment-based approaches to the discovery of small-molecule bromodomain ligands. Future Med Chem 2014;6:179–204.

[8] Filippakopoulos P, Knapp S. Targeting bromodomains: epigenetic readers of lysine acetylation. Nat Rev Drug Disc 2014;13:337–56.

[9] Gallenkamp D, Gelato KA, Haendler B, Weinmann H. Bromodomains and their pharmacological inhibitors. ChemMedChem 2014;9:438–64.

[10] Garnier J-M, Sharp PP, Burns CJ. BET bromodomain inhibitors: a patent review. Expert Opin Ther Pat 2014;24:185–99.

[11] Haynes SR, Dollard C, Winston F, Beck S, Trowsdale J, Dawid IB. The bromodomain: a conserved sequence found in human, drosophila and yeast proteins. Nucleic Acids Res 1992;20:2603.

[12] Tamkun JW, Deuring R, Scott MP, Kissinger M, Pattatucci AM, Kaufman TC, et al. Brahma: a regulator of drosophila homeotic genes structurally related to the yeast transcriptional activator SNF2/SWI2. Cell 1992;68:561–72.

[13] Filippakopoulos P, Picaud S, Mangos M, Keates T, Lambert J-P, Barsyte-Lovejoy D, et al. Histone recognition and large-scale structural analysis of the human bromodomain family. Cell 2012;149:214–31.

[14] Arrowsmith CH, Bountra C, Fish PV, Lee K, Schapira M. Epigenetic protein families: a new frontier for drug discovery. Nat Rev Drug Disc 2012;11:384–400.

[15] Filippakopoulos P, Knapp S. The bromodomain interaction module. FEBS Lett 2012;586:2692–704.

[16] Kanno T, Kanno Y, Siegel RM, Jang MK, Lenardo MJ, Ozato K. Selective recognition of acetylated histones by bromodomain proteins visualized in living cells. Mol Cell 2004;13:33–43.

[17] Huang H, Zhang J, Shen W, Wang X, Wu J, Wu J, et al. Solution structure of the second bromodomain of Brd2 and its specific interaction with acetylated histone tails. BMC Struct Biol 2007;7:57.

[18] LeRoy G, Rickards B, Flint SJ. The double bromodomain proteins Brd2 and Brd3 couple histone acetylation to transcription. Mol Cell 2008;30:51–60.

[19] Liu Y, Wang X, Zhang J, Huang H, Ding B, Wu J, et al. Structural basis and binding properties of the second bromodomain of Brd4 with acetylated histone tails. Biochemistry 2008;47:6403–17.

[20] Morinière J, Rousseaux S, Steuerwald U, Soler-López M, Curtet S, Vitte A-L, et al. Cooperative binding of two acetylation marks on a histone tail by a single bromodomain. Nature 2009;461:664–8.

[21] Rachie NA, Seger R, Valentine MA, Ostrowski J, Bomsztyk K. Identification of an inducible 85-Kda nuclear-protein kinase. J Biol Chem 1993;268:22143–9.

[22] Denis GV, Green MR. A novel, mitogen-activated nuclear kinase is related to a drosophila developmental regulator. Genes Dev 1996;10:261–71.

[23] Denis GV, Vaziri C, Guo N, Faller DV. RING3 kinase transactivates promoters of cell cycle regulatory genes through E2F. Cell Growth Differ 2000;11:417–24.

[24] Crowley TE, Kaine EM, Yoshida M, Nandi A, Wolgemuth DJ. Reproductive cycle regulation of nuclear import, euchromatic localization, and association with components of Pol II mediator of a mammalian double-bromodomain protein. Mol Endocrinol 2002;16:1727–37.

[25] Denis GV, McComb ME, Faller DV, Sinha A, Romesser PB, Costello CE. Identification of transcription complexes that contain the double bromodomain protein Brd2 and chromatin remodeling machines. J Proteome Res 2006;5:502–11.

[26] Sinha A, Faller DV, Denis GV. Bromodomain analysis of Brd2-dependent transcriptional activation of cyclin A. Biochem J 2005;387:257–69.

[27] Greenwald RJ, Tumang JR, Sinha A, Currier N, Cardiff RD, Rothstein TL, et al. E Mu-BRD2 transgenic mice develop B-cell lymphoma and leukemia. Blood 2004;103:1475–84.

[28] Belkina AC, Blanton WP, Nikolajczyk BS, Denis GV. The double bromodomain protein Brd2 promotes B cell expansion and mitogenesis. J Leukoc Biol 2014;95:451–60.

[29] Wu S-Y, Chiang C-M. The double bromodomain-containing chromatin adaptor Brd4 and transcriptional regulation. J Biol Chem 2007;282:13141–5.

[30] Jang MK, Mochizuki K, Zhou MS, Jeong HS, Brady JN, Ozato K. The bromodomain protein Brd4 is a positive regulatory component of P-TEFb and stimulates RNA polymerase II-dependent transcription. Mol Cell 2005;19:523–34.

[31] Yang ZY, Yik J, Chen RC, He NH, Jang MK, Ozato K, et al. Recruitment of P-TEFb for stimulation of transcriptional elongation by the bromodomain protein Brd4. Mol Cell 2005;19:535–45.

[32] Zippo A, Serafini R, Rocchigiani M, Pennacchini S, Krepelova A, Oliviero S. Histone crosstalk between H3S10ph and H4K16ac generates a histone code that mediates transcription elongation. Cell 2009;138:1122–36.

[33] Zhang W, Prakash C, Sum C, Gong Y, Li Y, Kwok JJT, et al. Bromodomain-containing protein 4 (BRD4) regulates RNA polymerase II serine 2 phosphorylation in human CD4+ T Cells. J Biol Chem 2012;287:43137–55.

[34] Bisgrove DA, Mahmoudi T, Henklein P, Verdin E. Conserved P-TEFb-interacting domain of BRD4 inhibits HIV transcription. Proc Natl Acad Sci USA 2007;104:13690–5.

[35] Hargreaves DC, Horng T, Medzhitov R. Control of inducible gene expression by signal-dependent transcriptional elongation. Cell 2009;138:129–45.

[36] Patel MC, Debrosse M, Matthew S, Dey A, Huynh W, Sarai N, et al. BRD4 coordinates recruitment of pause release factor P-TEFb and the pausing complex NELF/DSIF to regulate transcription elongation of interferon-stimulated genes. Mol Cell Biol 2013;33:2497–507.

[37] Schroeder S, Cho S, Zeng L, Zhang Q, Kaehlcke K, Mak L, et al. Two-pronged binding with bromodomain-containing protein 4 liberates positive transcription elongation factor B from inactive ribonucleoprotein complexes. J Biol Chem 2012;287:1090–9.

[38] Rahman S, Sowa ME, Ottinger M, Smith JA, Shi Y, Harper JW, et al. The Brd4 extraterminal domain confers transcription activation independent of pTEFb by recruiting multiple proteins, including NSD3. Mol Cell Biol 2011;31:2641–52.

[39] Devaiah BN, Lewis BA, Cherman N, Hewitt MC, Albrecht BK, Robey PG, et al. BRD4 is an atypical kinase that phosphorylates serine2 of the RNA polymerase II carboxy-terminal domain. Proc Natl Acad Sci USA 2012;109:6927–32.

[40] Huang B, Yang X-D, Zhou M-M, Ozato K, Chen L-F. Brd4 coactivates transcriptional activation of NFf-Kappa B via specific binding to acetylated RelA. Mol Cell Biol 2009;29:1375–87.

[41] Dey A, Ellenberg J, Farina A, Coleman AE, Maruyama T, Sciortino S, et al. A bromodomain protein, MCAP, associates with mitotic chromosomes and affects G(2)-to-M transition. Mol Cell Biol 2000;20:6537–49.

[42] Toyama R, Rebbert ML, Dey A, Ozato K, Dawid IB. Brd4 associates with mitotic chromosomes throughout early zebrafish embryogenesis. Dev Dyn 2008;237:1636–44.

[43] Dey A, Chitsaz F, Abbasi A, Misteli T, Ozato K. The double bromodomain protein Brd4 binds to acetylated chromatin during interphase and mitosis. Proc Natl Acad Sci USA 2003;100:8758–63.

[44] Dey A, Nishiyama A, Karpova T, McNally J, Ozato K. Brd4 marks select genes on mitotic chromatin and directs postmitotic transcription. Mol Biol Cell 2009;20:4899–909.

[45] Yang Z, He N, Zhou Q. Brd4 recruits P-TEFb to chromosomes at late mitosis to promote G1 gene expression and cell cycle progression. Mol Cell Biol 2008;28:967–76.

[46] Devaiah BN, Singer DS. Two faces of Brd4: mitotic bookmark and transcriptional lynchpin. Transcription 2013;4:13–17.

[47] Zhao R, Nakamura T, Fu Y, Lazar Z, Spector DL. Gene bookmarking accelerates the kinetics of post-mitotic transcriptional re-activation. Nat Cell Biol 2011;13:1295–304.

[48] Belkina AC, Denis GV. BET domain co-regulators in obesity, inflammation and cancer. Nat Rev Cancer 2012;12:465–77.

[49] Lamonica J, Deng W, Kadauke S, Campbell A, Gamsjaeger R, Wang H, et al. Bromodomain protein Brd3 associates with acetylated GATA1 to promote its chromatin occupancy at erythroid target genes. Proc Natl Acad Sci USA 2011;108:E159–68.

[50] Gamsjaeger R, Webb SR, Lamonica JM, Billin A, Blobel GA, Mackay JP. Structural basis and specificity of acetylated transcription factor GATA1 recognition by BET family bromodomain protein Brd3. Mol Cell Biol 2011;31:2632–40.

[51] Jones MH, Numata M, Shimane M. Identification and characterization of BRDT: a testis-specific gene related to the bromodomain genes RING3 and drosophila Fsh. Genomics 1997;45:529–34.

[52] Shang E, Nickerson HD, Wen D, Wang X, Wolgemuth DJ. The first bromodomain of Brdt, a testis-specific member of the BET sub-family of double-bromodomain-containing proteins, is essential for male germ cell differentiation. Development 2007;134:3507–15.

[53] Berkovits BD, Wang L, Guarnieri P, Wolgemuth DJ. The testis-specific double bromodomain-containing protein BRDT forms a complex with multiple spliceosome components and is required for mRNA splicing and 3′-UTR truncation in round spermatids. Nucleic Acids Res 2012;40:7162–75.

[54] Aston KI, Krausz C, Laface I, Ruiz-Castane E, Carrell DT. Evaluation of 172 candidate polymorphisms for association with oligozoospermia or azoospermia in a large cohort of men of European descent. Hum Reprod 2010;25:1383–97.

[55] Matzuk MM, McKeown MR, Filippakopoulos P, Li Q, Ma L, Agno JE, et al. Small-molecule inhibition of BRDT for male contraception. Cell 2012;150:673–84.

[56] French CA. Pathogenesis of NUT midline carcinoma. Annu Rev Pathol 2012;7:247–65.

[57] Reynoird N, Schwartz BE, Delvecchio M, Sadoul K, Meyers D, Mukherjee C, et al. Oncogenesis by sequestration of CBP/P300 in transcriptionally inactive hyperacetylated chromatin domains. EMBO J 2010;29:2943–52.

[58] Zuber J, Shi J, Wang E, Rappaport AR, Herrmann H, Sison EA, et al. RNAi screen identifies Brd4 as a therapeutic target in acute myeloid leukaemia. Nature 2011;478:524–8.

[59] Mertz JA, Conery AR, Bryant BM, Sandy P, Balasubramanian S, Mele DA, et al. Targeting MYC dependence in cancer by inhibiting BET bromodomains. Proc Natl Acad Sci USA 2011;108:16669–74.

[60] Filippakopoulos P, Qi J, Picaud S, Shen Y, Smith WB, Fedorov O, et al. Selective inhibition of BET bromodomains. Nature 2010;468:1067–73.

[61] Nicodeme E, Jeffrey KL, Schaefer U, Beinke S, Dewell S, Chung C-W, et al. Suppression of inflammation by a synthetic histone mimic. Nature 2010;468:1119–23.

[62] Chung C-W, Coste H, White JH, Mirguet O, Wilde J, Gosmini RL, et al. Discovery and characterization of small molecule inhibitors of the BET family bromodomains. J Med Chem 2011;54:3827–38.

[63] Mirguet O, Gosmini R, Toum J, Clément CA, Barnathan M, Brusq J-M, et al. Discovery of epigenetic regulator I-BET762: lead optimization to afford a clinical candidate inhibitor of the BET bromodomains. J Med Chem 2013;56:7501–15.

[64] Miyoshi S, Ooike S, Iwata K, Hikwa H, Sugahara K, Hikwawa H, et al. Antitumor agent. WO2009084693; 2009.

[65] Adachi K, Hikwawa H, Hamada M, Endoh J, Ishibuchi S, Fujie N, et al. Thienotriazolodiazepine compound and a medicinal use thereof. WO2006129623; 2006.

[66] Philpott M, Yang J, Tumber T, Fedorov O, Uttarkar S, Filippakopoulos P, et al. Bromodomain-peptide displacement assays for interactome mapping and inhibitor discovery. Mol BioSyst 2011;7:2899–908.

[67] Zhang G, Liu R, Zhong Y, Plotnikov AN, Zhang W, Zeng L, et al. Down-regulation of NF-κB transcriptional activity in HIV-associated kidney disease by BRD4 inhibition. J Biol Chem 2012;287:28840–51.

[68] Hewings DS, Wang M, Philpott M, Fedorov O, Uttarkar S, Filippakopoulos P, et al. 3,5-Dimethylisoxazoles act as acetyl-lysine-mimetic bromodomain ligands. J Med Chem 2011;54:6761–70.

[69] Dawson MA, Prinjha RK, Dittmann A, Giotopoulos G, Bantscheff M, Chan W-I, et al. Inhibition of BET recruitment to chromatin as an effective treatment for MLL-fusion leukaemia. Nature 2011;478:529–33.

[70] Bamborough P, Diallo H, Goodacre JD, Gordon L, Lewis A, Seal JT, et al. Fragment-based discovery of bromodomain inhibitors Part 2: optimization of phenylisoxazole sulfonamides. J Med Chem 2012;55:587–96.

[71] Mirguet O, Lamotte Y, Donche F, Toum J, Gellibert F, Bouillot A, et al. From ApoA1 upregulation to BET family bromodomain inhibition: discovery of I-BET151. Bioorg Med Chem Lett 2012;22:2963–7.

[72] Seal J, Lamotte Y, Donche F, Bouillot A, Mirguet O, Gellibert F, et al. Identification of a novel series of BET family bromodomain inhibitors: binding mode and profile of I-BET151 (GSK1210151A). Bioorg Med Chem Lett 2012;22:2968–72.

[73] Hewings DS, Fedorov O, Filippakopoulos P, Martin S, Picaud S, Tumber A, et al. Optimization of 3,5-dimethylisoxazole derivatives as potent bromodomain ligands. J Med Chem 2013;56:3217–27.

[74] Gehling VS, Hewitt MC, Vaswani RG, Leblanc Y, Côté A, Nasveschuk CG, et al. Discovery, design, and optimization of isoxazole azepine BET inhibitors. ACS Med Chem Lett 2013;4:835–40.

[75] Hay D, Fedorov O, Filippakopoulos P, Martin S, Philpott M, Picaud S, et al. The design and synthesis of 5 and 6-isoxazolylbenzimidazoles as selective inhibitors of the BET bromodomains. Med Chem Commun 2013;4:140–4.

[76] Hay DA, Fedorov O, Martin S, Singleton DC, Tallant C, Wells C, et al. Discovery and optimization of small-molecule ligands for the CBP/P300 bromodomains. J Am Chem Soc 2014;136:9308–19.

[77] Hopkins AL, Groom CR, Alex A. Ligand efficiency: a useful metric for lead selection. Drug Discov Today 2004;9:430–1.

[78] Bailey D, Jahagirdar R, Gordon A, Hafiane A, Campbell S, Chatur S, et al. RVX-208: a small molecule that increases apolipoprotein a-I and high-density lipoprotein cholesterol *in vitro* and *in vivo*. J Am Coll Cardiol 2010;55:2580–9.

[79] Nicholls SJ, Gordon A, Johansson J, Wolski K, Ballantyne CM, Kastelein JJP, et al. Efficacy and safety of a novel oral inducer of apolipoprotein a-I synthesis in statin-treated patients with stable coronary artery disease a randomized controlled trial. J Am Coll Cardiol 2011;57:1111–19.

[80] Picaud S, Wells C, Felletar I, Brotherton D, Martin S, Savitsky P, et al. RVX-208, an Inhibitor of BET transcriptional regulators with selectivity for the second bromodomain. Proc Natl Acad Sci USA 2013;110:19754–9.

[81] McLure KG, Gesner EM, Tsujikawa L, Kharenko OA, Attwell S, Campeau E, et al. RVX-208, an inducer of ApoA-I in humans, is a BET bromodomain antagonist. PLoS One 2013;8:e83190.

[82] Fish PV, Filippakopoulos P, Bish G, Brennan PE, Bunnage ME, Cook AS, et al. Identification of a chemical probe for bromo and extra C-terminal bromodomain inhibition through optimization of a fragment-derived hit. J Med Chem 2012;55:9831–7.

[83] Picaud S, Da Costa D, Thanasopoulou A, Filippakopoulos P, Fish PV, Philpott M, et al. PFI-1, a highly selective protein interaction inhibitor, targeting BET bromodomains. Cancer Res 2013;73:3336–46.

[84] Zhao L, Cao D, Chen T, Wang Y, Miao Z, Xu Y, et al. Fragment-based drug discovery of 2-thiazolidinones as inhibitors of the histone reader BRD4 bromodomain. J Med Chem 2013;56:3833–51.

[85] Borah JC, Mujtaba S, Karakikes I, Zeng L, Muller M, Patel J, et al. A small molecule binding to the coactivator CREB-binding protein blocks apoptosis in cardiomyocytes. Chem Biol 2011;18:531–41.

[86] Zhang G, Plotnikov AN, Rusinova E, Shen T, Morohashi K, Joshua J, et al. Structure-guided design of potent diazobenzene inhibitors for the BET bromodomains. J Med Chem 2013;56:9251–64.

[87] Zhou MM, Ohlmeyer M, Mujtaba S, Plotnikov A, Kastrinsky D, Zhang, G. Inhibitors of bromodomains as modulators of gene expression. WO2012116170; 2012.

[88] Lucas X, Wohlwend D, Hügle M, Schmidtkunz K, Gerhardt S, Schüle R, et al. 4-Acyl pyrroles: mimicking acetylated lysines in histone code reading. Angew Chem Int Ed 2013;52:14055–9.

[89] Bissantz C, Kuhn B, Stahl M. A medicinal chemist's guide to molecular interactions. J Med Chem 2010;53:5061–84.

[90] Tetrahydroquinolines derivatives as bromodomain inhibitors; May 12, 2011.

[91] Wyce A, Ganji G, Smitheman KN, Chung C-W, Korenchuk S, Bai Y, et al. BET inhibition silences expression of MYCN and BCL2 and induces cytotoxicity in neuroblastoma tumor models. PLoS One 2013;8:e72967.

[92] Vidler LR, Filippakopoulos P, Fedorov O, Picaud S, Martin S, Tomsett M, et al. Discovery of novel small-molecule inhibitors of BRD4 using structure-based virtual screening. J Med Chem 2013;56:8073–88.

[93] Zhao H, Gartenmann L, Dong J, Spiliotopoulos D, Caflisch A. Discovery of BRD4 bromodomain inhibitors by fragment-based high-throughput docking. Bioorg Med Chem Lett 2014;24:2493–6.

[94] Brand M, Measures AM, Wilson BG, Cortopassi WA, Alexander R, Höss M, et al. Small molecule inhibitors of bromodomain-acetyl-lysine interactions. ACS Chem Biol 2015;10:22–39.

[95] Steiner S, Magno A, Huang D, Caflisch A. Does bromodomain flexibility influence histone recognition? FEBS Lett 2013;587:2158–63.

[96] Magno A, Steiner S, Caflisch A. Mechanism and kinetics of acetyl-lysine binding to bromodomains. J Chem Theory Comput 2013;9:4225–32.

[97] Huang D, Rossini E, Steiner S, Caflisch A. Structured water molecules in the binding site of bromodomains can be displaced by cosolvent. ChemMedChem 2014;9:573–9.

[98] Vidler LR, Brown N, Knapp S, Hoelder S. Druggability analysis and structural classification of bromodomain acetyl-lysine binding sites. J Med Chem 2012;55:7346–59.

[99] Muvva C, Singam ERA, Raman SS, Subramanian V. Structure-based virtual screening of novel, high-affinity BRD4 inhibitors. Mol BioSyst 2014.

[100] Martin MP, Olesen SH, Georg GI, Schönbrunn E. Cyclin-dependent kinase inhibitor dinaciclib interacts with the acetyl-lysine recognition site of bromodomains. ACS Chem Biol 2013;8:2360–5.

[101] Dittmann A, Werner T, Chung C-W, Savitski MM, Fälth Savitski M, Grandi P, et al. The commonly used PI3-kinase probe LY294002 is an inhibitor of BET bromodomains. ACS Chem Biol 2014;9:495–502.

[102] Gharbi SI, Zvelebil MJ, Shuttleworth SJ, Hancox T, Saghir N, Timms JF, et al. Exploring the specificity of the PI3K family inhibitor LY294002. Biochem J 2007;404:15–21.

[103] Kim YH, Choi K-H, Park J-W, Kwon TK. LY294002 inhibits LPS-induced NO production through a inhibition of NF-kappaB activation: independent mechanism of phosphatidylinositol 3-kinase. Immunol Lett 2005;99:45–50.

[104] Zou Z, Huang B, Wu X, Zhang H, Qi J, Bradner J, et al. Brd4 maintains constitutively active NF-κB in cancer cells by binding to acetylated RelA. Oncogene 2014;33:2395–404.

[105] Ciceri P, Muller S, O'Mahony A, Fedorov O, Filippakopoulos P, Hunt JP, et al. Dual kinase-bromodomain inhibitors for rationally designed polypharmacology. Nat Chem Biol 2014;10:305–12.

[106] Ember SWJ, Zhu J-Y, Olesen SH, Martin MP, Becker A, Berndt N, et al. Acetyl-lysine binding site of bromodomain-containing protein 4 (BRD4) interacts with diverse kinase inhibitors. ACS Chem Biol 2014;9:1160–71.

[107] Delmore JE, Issa GC, Lemieux ME, Rahl PB, Shi J, Jacobs HM, et al. BET bromodomain inhibition as a therapeutic strategy to target C-Myc. Cell 2011;146:904–17.

[108] Watson JD. Curing "Incurable" cancer. Cancer Discov 2011;1:477–80.

[109] Da Costa D, Agathanggelou A, Perry T, Weston V, Petermann E, Zlatanou A, et al. BET inhibition as a single or combined therapeutic approach in primary paediatric B-precursor acute lymphoblastic leukaemia. Blood Cancer J 2013;3:e126.

[110] Cheng Z, Gong Y, Ma Y, Lu K, Lu X, Pierce LA, et al. Inhibition of BET bromodomain targets genetically diverse glioblastoma. Clin Cancer Res 2013;19:1748–59.

[111] Ott CJ, Kopp N, Bird L, Paranal RM, Qi J, Bowman T, et al. BET bromodomain inhibition targets both C-Myc and IL7R in high-risk acute lymphoblastic leukemia. Blood 2012;120:2843–52.

[112] King B, Trimarchi T, Reavie L, Xu L, Mullenders J, Ntziachristos P, et al. The ubiquitin ligase FBXW7 modulates leukemia-initiating cell activity by regulating MYC stability. Cell 2013;153:1552–66.

[113] Roderick JE, Tesell J, Shultz LD, Brehm MA, Greiner DL, Harris MH, et al. C-Myc inhibition prevents leukemia initiation in mice and impairs the growth of relapsed and induction failure pediatric T-ALL cells. Blood 2014;123:1040–50.

[114] Chapuy B, McKeown MR, Lin CY, Monti S, Roemer MGM, Qi J, et al. Discovery and characterization of super-enhancer-associated dependencies in diffuse large B cell lymphoma. Cancer Cell 2013;24:777–90.

[115] Zhao X, Lwin T, Zhang X, Huang A, Wang J, Marquez VE, et al. Disruption of the MYC-miRNA-EZH2 loop to suppress aggressive B-cell lymphoma survival and clonogenicity. Leukemia 2013;27:2341–50.

[116] Tolani B, Gopalakrishnan R, Punj V, Matta H, Chaudhary PM. Targeting Myc in KSHV-associated primary effusion lymphoma with BET bromodomain inhibitors. Oncogene 2014;33:2928–37.

[117] Emadali A, Rousseaux S, Bruder-Costa J, Rome C, Duley S, Hamaidia S, et al. Identification of a novel BET bromodomain inhibitor-sensitive cancers, gene regulatory circuit that controls rituximab response and tumour growth in aggressive lymphoid cancers. EMBO Mol Med 2013;5:1180–95.

[118] Chaidos A, Caputo V, Gouvedenou K, Liu B, Marigo L, Chaudhry MS, et al. Potent antimyeloma activity of the novel bromodomain inhibitors I-BET151 and I-BET762. Blood 2014;123:697–705.

[119] Puissant A, Frumm SM, Alexe G, Bassil CF, Qi J, Chantery YH, et al. Targeting MYCN in neuroblastoma by BET bromodomain inhibition. Cancer Discov 2013;3:308–23.

[120] Henssen A, Thor T, Odersky A, Heukamp L, El-Hindy N, Beckers A, et al. BET bromodomain protein inhibition is a therapeutic option for medulloblastoma. Oncotarget 2013;4:2080–95.

[121] Bandopadhayay P, Bergthold G, Nguyen B, Schubert S, Gholamin S, Tang Y, et al. BET bromodomain inhibition of MYC-amplified medulloblastoma. Clin Cancer Res 2014;20:912–25.

[122] Shimamura T, Chen Z, Soucheray M, Carretero J, Kikuchi E, Tchaicha JH, et al. Efficacy of BET bromodomain inhibition in Kras-mutant non-small cell lung cancer. Clin Cancer Res 2013;19:6183–92.

[123] Gao L, Schwartzman J, Gibbs A, Lisac R, Kleinschmidt R, Wilmot B, et al. Androgen receptor promotes ligand-independent prostate cancer progression through C-Myc upregulation. PLoS One 2013;8.

[124] Wyce A, Degenhardt Y, Bai Y, Le B, Korenchuk S, Crouthamel M-C, et al. Inhibition of BET bromodomain proteins as a therapeutic approach in prostate cancer. Oncotarget 2013;4:2419–29.

[125] Stewart HJS, Horne GA, Bastow S, Chevassut TJT. BRD4 associates with P53 in DNMT3A-mutated leukemia cells and is implicated in apoptosis by the bromodomain inhibitor JQ1. Cancer Med 2013;2:826–35.

[126] Dawson MA, Gudgin EJ, Horton SJ, Giotopoulos G, Meduri E, Robson S, et al. Recurrent mutations, including NPM1c, activate a BRD4-dependent core transcriptional program in acute myeloid leukemia. Leukemia 2014;28:311–20.

[127] Patel AJ, Liao C-P, Chen Z, Liu C, Wang Y, Le LQ. BET bromodomain inhibition triggers apoptosis of NF1-associated malignant peripheral nerve sheath tumors through Bim induction. Cell Rep 2014;6:81–92.

[128] Segura MF, Fontanals-Cirera B, Gaziel-Sovran A, Guijarro MV, Hanniford D, Zhang G, et al. BRD4 sustains melanoma proliferation and represents a new target for epigenetic therapy. Cancer Res 2013;73:6264–76.

[129] Lockwood WW, Zejnullahu K, Bradner JE, Varmus H. Sensitivity of human lung adenocarcinoma cell lines to targeted inhibition of BET epigenetic signaling proteins. Proc Natl Acad Sci USA 2012;109:19408–13.

[130] Crawford NPS, Alsarraj J, Lukes L, Walker RC, Officewala JS, Yang HH, et al. Bromodomain 4 activation predicts breast cancer survival. Proc Natl Acad Sci USA 2008;105:6380–5.

[131] Chrivia J, Kwok R, Lamb N, Hagiwara M, Montminy M, Goodman R. Phosphorylated CREB binds specifically to the nuclear-protein CBP. Nature 1993;365:855–9.

[132] Wang L, Tang Y, Cole PA, Marmorstein R. Structure and chemistry of the P300/CBP and Rtt109 histone acetyltransferases: implications for histone acetyltransferase evolution and function. Curr Opin Struct Biol 2008;18:741–7.

[133] Bannister AJ, Kouzarides T. The CBP co-activator is a histone acetyltransferase. Nature 1996;384:641–3.

[134] Ogryzko VV, Schiltz RL, Russanova V, Howard BH, Nakatani Y. The transcriptional coactivators P300 and CBP are histone acetyltransferases. Cell 1996;87:953–9.

[135] Bedford DC, Kasper LH, Fukuyama T, Brindle PK. Target gene context influences the transcriptional requirement for the KAT3 family of CBP and P300 histone acetyltransferases. Epigenetics 2010;5:9–15.

[136] Sachchidanand Resnick-Silverman L, Yan S, Mutjaba S, Liu Zeng L, et al. Target structure-based discovery of small molecules that block human P53 and CREB binding protein association. Chem Biol 2006;13:81–90.

[137] Mujtaba S, He Y, Zeng L, Yan S, Plotnikova O, Sachchidanand Structural mechanism of the bromodomain of the coactivator CBP in P53 transcriptional activation. Mol Cell 2004;13:251–63.

[138] Gerona-Navarro G, Yoel-Rodríguez Mujtaba S, Frasca A, Patel J, Zeng L, et al. Rational design of cyclic peptide modulators of the transcriptional coactivator CBP: a new class of P53 inhibitors. J Am Chem Soc 2011;133:2040–3.

[139] Rooney TPC, Filippakopoulos P, Fedorov O, Picaud S, Cortopassi WA, Hay DA, et al. A series of potent CREBBP bromodomain ligands reveals an induced-fit pocket stabilized by a cation-π interaction. Angew Chem Int Ed 2014;53:6126–30.

[140] Fedorov O, Lingard H, Wells C, Monteiro OP, Picaud S, Keates T, et al. [1,2,4]Triazolo[4,3-α]Phthalazines: inhibitors of diverse bromodomains. J Med Chem 2014;57:462–76.

[141] Ruegg UT, Burgess GM. Staurosporine, K-252 and Ucn-01 - Potent but nonspecific inhibitors of protein-kinases. Trends Pharmacol Sci 1989;10:218–20.

[142] Jones MH, Hamana N, Nezu JI, Shimane M. A novel family of bromodomain genes. Genomics 2000;63:40–5.

[143] Guetg C, Lienemann P, Sirri V, Grummt I, Hernandez-Verdun D, Hottiger MO, et al. The NoRC complex mediates the heterochromatin formation and stability of silent rRNA genes and centromeric repeats. EMBO J 2010;29:2135–46.

[144] Ferguson FM, Fedorov O, Chaikuad A, Philpott M, Muniz JRC, Felletar I, et al. Targeting low-druggability bromodomains: fragment based screening and inhibitor design against the BAZ2B bromodomain. J Med Chem 2013;56:10183–7.

CLINICAL TRIALS

23

Wei Zhu and Jiaqi Qian

Department of Oncology, First Affiliated Hospital of Nanjing Medical University, Nanjing, People's Republic of China

CHAPTER OUTLINE

1 INTRODUCTION

"Epigenetics" is a term coined by Conrad Waddington to describe heritable changes in a cellular phenotype that were independent of alterations in the DNA sequence. Although the consensus definition of epigenetics remains contentious, it is well known that epigenetic modification includes DNA methylation, histone modifications, and noncoding RNAs regulation. These mechanisms are interrelated and

S.G. Gray (Ed): Epigenetic Cancer Therapy. DOI: http://dx.doi.org/10.1016/B978-0-12-800206-3.00023-9

Table 23.1 Drugs in Trials Mentioned in This Chapter

Inhibitors	Drug Classification	Drug name	Targets
DNMTis	Nucleoside analogues	5-Azacytidine	DNMTs
		5-Aza-2'-deoxycytidine (Decitabine)	DNMTs
	Nonnucleosides	EGCG	DNMTs
		Hydralazine	DNMT1
		Procainamide	DNMT1
		Procaine	DNMTs
		Curcumin	DNMTs
HDACis	Short-chain fatty acids	Valproic acid	HDAC I, II a
		Sodium butyrate	HDAC I, II a
	Hydroxamates	Suberoylanilide hydroxamic acid (SAHA, Vorinostat)	HDAC 1,2,3,4,6,7,9
		Trichostatin A (TSA)	HDAC 1,2,3,4,6,7,9
		Belinostat (PXD-101)	HDAC 1,2,3,6
		Panobinostat (LBH589)	HDAC 1,2,3,6
	Benzamides	MS-275 (Entinostat)	HDAC 1,2
		Mocetinostat (MGCD0103)	HDAC 1,2
	Cyclic peptides	Romidepsin (FR901228/FK228, Istodax)	HDAC 1,2,3,8

need to be stably maintained during cell divisions to conserve cellular identity but also react to cell intrinsic signals during development or to external factors to adopt to altered environmental cues [1]. The epigenetic modifications are involved in various diseases including cancer, diabetes, autoimmune diseases, Parkinson's and Alzheimer's diseases, etc. Taking cancer as an example, cumulative epigenetic changes is observed during the process of cancer development, such as initial, progress, metastasis, and multi-drug resistance in both hematological malignancies and solid tumors. This shows epigenetic drugs may have potential efficacy in cancer treatment. In the past three decades, various epigenome-targeted drugs have been developed (Table 23.1). In this chapter, clinical trials found on the ClinicalTrials .gov website of epigenetic drugs in various human cancers are summarized (Table 23.2). Meanwhile the efficacy of the treatment through targeting aberrant epigenetics is presented in Table 23.3.

2 SINGLE AGENT THERAPY

2.1 DNA METHYLTRANSFERASE INHIBITORS

DNA is methylated by addition of a methyl group to the $5'$ position of cytosine residues in the cytosine-phosphoguanine (CpG) dinucleotide. This process is common throughout the genome. In cancer cells, where DNA methylation was the first epigenetic alteration to be observed, abnormal methylation patterns like regional promoter hypermethylation which led to the silencing of cancer suppressor genes are familiar phenomenons. Unlike mutations or deletions, these silenced genes typically have intact DNA sequences, which cause chemical reversal of gene silencing has become an exciting approach for

Table 23.2 Trials in the ClinicalTrials.gov Website with Epigenetic Drugs in Cancer Patients

Initiated Time	Drug	Cancer	Status	Single or in Combination	Phases	NCT Number
1999	Azacitidine	Head and neck cancer	Completed	in combination	Phase I	NCT00004062
2006	Azacitidine	Prostate cancer	Completed	single	Phase II	NCT00384839
2007	Azacitidine	Advanced cancers	Completed	in combination	Phase I	NCT00496444
2010	Azacitidine	Colorectal cancer	Completed	in combination	Phase II	NCT01105377
2012	Azacitidine	MDS, AML	Completed	in combination	Phase I	NCT01575691
2000	Azacitidine	Leukemia, MDSs	Completed	in combination	Phase I	NCT00004871
2000	Azacitidine	Lymphoma, small intestine cancer	Completed	in combination	Phase I	NCT00005639
2000	Azacitidine	Hematologic and solid tumors	Completed	in combination	Phase II	NCT00006019
2005	Azacitidine	Leukemia, MDSs	Completed	in combination	Phase I	NCT00101179
2005	Azacitidine	Leukemia, MDSs	Completed	in combination	Phase II	NCT00118196
2006	Azacitidine	MDS, AML	Completed	in combination	Phase II	NCT00326170
2006	Azacitidine	AML, MDS	Completed	in combination	Phase II	NCT00339196
2006	Azacitidine	MDS, leukemia	Completed	single	Phase I	NCT00350818
2006	Azacitidine	AML, MDS, leukemia	Completed	in combination	Phase II	NCT00382590
2007	Azacitidine	AML, MDS	Completed	in combination	Phase III	NCT00422890
2007	Azacitidine	AML, MDS, leukemia	Completed	in combination	Phase IIII	NCT00569010
2008	Azacitidine	Leukemia	Completed	single	Phase II	NCT00739388
2008	Azacitidine	MDS, AML	Completed	single	Phase II	NCT00795548
2009	Azacitidine	Advanced solid tumor malignancies	Completed	in combination	Phase I	NCT00996515
2006	Azacitidine	Lung cancer	Recruiting	in combination	Phase IIII	NCT00387465
2010	Azacitidine	Hematologic cancers	Suspended	in combination	Phase IIII	NCT01093573
2005	Azacitidine	Leukemia, MDSs	Terminated	in combination	Phase I	NCT00234000
2006	Azacitidine	Leukemia, MDSs	Terminated	in combination	N/A	NCT00398047
2007	Azacitidine	SCLC	Terminated	in combination	Phase I	NCT00443261
2007	Azacitidine	AML, MDS	Terminated	single	Phase II	NCT00446303
2007	Azacitidine	HL and NHL	Terminated	in combination	Phase II	NCT00543582
2008	Azacitidine	AML, MDS	Terminated	in combination	Phase II	NCT00666497

(*Continued*)

Table 23.2 (Continued)

Initiated Time	Drug	Cancer	Status	Single or in Combination	Phases	NCT Number
2009	Azacitidine	Ovarian cancer	Terminated	single	Phase I	NCT00842582
2004	Azacitidine	CML	Active, not recruiting	in combination	Phase I	NCT00101179
2006	Azacitidine	AML	Active, not recruiting	in combination	Phase I	NCT00275080
2006	Azacitidine	CML	Active, not recruiting	in combination	Phase II	NCT00313586
2006	Azacitidine	Adult nasal type extranodal NK/T-cell lymphoma	Active, not recruiting	in combination	Phase I	NCT00336063
2006	Azacitidine	ALL	Active, not recruiting	single	Phase I	NCT00349596
2006	Azacitidine	CML	Active, not recruiting	in combination	Phase I	NCT00351975
2006	Azacitidine	Leukemia	Active, not recruiting	single	Phase II	NCT00387647
2006	Azacitidine	Adult acute basophilic leukemia	Active, not recruiting	in combination	Phase III	NCT00392353
2006	Azacitidine	CLL	Active, not recruiting	single	Phase II	NCT00413478
2006	Azacitidine	AML	Active, not recruiting	in combination	Phase II	NCT00414310
2007	Azacitidine	Ovarian cancer	Active, not recruiting	single	Phase IIII	NCT00477386
2007	Azacitidine	AML	Active, not recruiting	single	Phase II	NCT00492401
2007	Azacitidine	Prostate cancer	Active, not recruiting	in combination	Phase IIII	NCT00503984
2007	Azacitidine	CMML	Active, not recruiting	in combination	Phase I	NCT00528983
2007	Azacitidine	Mesothelioma	Active, not recruiting	in combination	Phase I	NCT00629343
2008	Azacitidine	Leukemia	Active, not recruiting	single	NA	NCT00660400
2008	Azacitidine	Leukemia	Active, not recruiting	in combination	Phase IIII	NCT00691938
2008	Azacitidine	AML	Active, not recruiting	in combination	Phase I	NCT00703300
2008	Azacitidine	MDS	Active, not recruiting	single	Phase II	NCT00721214
2008	Azacitidine	AML	Active, not recruiting	single	Phase I	NCT00761722
2008	Azacitidine	Melanoma	Active, not recruiting	in combination	Phase IIII	NCT00791271
2008	Azacitidine	Leukemia	Active, not recruiting	in combination	Phase II	NCT00813124
2008	Azacitidine	MDS	Active, not recruiting	single	Phase II	NCT00897130
2009	Azacitidine	Lymphoma	Active, not recruiting	in combination	Phase I	NCT00901069
2009	Azacitidine	RCC	Active, not recruiting	in combination	Phase IIII	NCT00934440
2009	Azacitidine	Leukemia	Active, not recruiting	in combination	Phase II	NCT00948064

Year	Drug	Condition	Status	Combination	Phase	NCT number
2009	Azacitidine	Leukemia	Active, not recruiting	in combination	Phase III	NCT01038635
2009	Azacitidine	Hematopoietic/lymphoid cancer	Active, not recruiting	in combination	Phase I	NCT01039155
2010	Azacitidine	AML	Active, not recruiting	in combination	Phase III	NCT01074047
2010	Azacitidine	Follicular lymphoma	Active, not recruiting	in combination	Phase II	NCT01121757
2010	Azacitidine	Adult acute myeloid leukemia with Del(5q)	Active, not recruiting	in combination	Phase I	NCT01130506
2010	Azacitidine	AML	Active, not recruiting	in combination	Phase II	NCT01168219
2010	Azacitidine	Acute myeloid leukemia (AML)	Active, not recruiting	in combination	Phase II	NCT01180322
2010	Azacitidine	MDS	Active, not recruiting	single	NA	NCT01192945
2010	Azacitidine	Colorectal cancer	Active, not recruiting	in combination	Phase III	NCT01193517
2010	Azacitidine	MDS	Active, not recruiting	single	Phase VI	NCT01201811
2011	Azacitidine	Leukemia	Active, not recruiting	in combination	Phase III	NCT01202877
2009	Azacitidine	Lung cancer	Active, not recruiting	in combination	NA	NCT01209520
2010	Azacitidine	CML	Active, not recruiting	single	Phase II	NCT01241500
2011	Azacitidine	AML	Active, not recruiting	in combination	Phase I	NCT01249430
2011	Azacitidine	Leukemia	Active, not recruiting	in combination	Phase III	NCT01254890
2010	Azacitidine	Adult acute basophilic leukemia	Active, not recruiting	in combination	Phase I	NCT01260714
2010	Azacitidine	MDS	Active, not recruiting	in combination	Phase III	NCT0130513
2011	Azacitidine	MDS	Active, not recruiting	single	Phase III	NCT01305460
2011	Azacitidine	Estrogen receptor-negative breast cancer	Active, not recruiting	in combination	Phase II	NCT01349959
2011	Azacitidine	CML	Active, not recruiting	single	Phase II	NCT01350947
2011	Azacitidine	AML	Active, not recruiting	in combination	Phase I	NCT01390311
2011	Azacitidine	AML	Active, not recruiting	in combination	Phase II	NCT01420926
2012	Azacitidine	Leukemia	Active, not recruiting	in combination	Phase III	NCT01460498
2010	Azacitidine	AML	Active, not recruiting	in combination	Phase II	NCT01488565
2012	Azacitidine	Leukemia	Active, not recruiting	single	Phase I	NCT01519011
2012	Azacitidine	CMML	Active, not recruiting	in combination	Phase I	NCT01613976
2012	Azacitidine	Leukemia	Active, not recruiting	in combination	Phase III	NCT01636609
2013	Azacitidine	AML	Active, not recruiting	single	Phase IIIII	NCT01809392

(Continued)

Table 23.2 (Continued)

Initiated Time	Drug	Cancer	Status	Single or in Combination	Phases	NCT Number
2006	Azacitidine	Adult acute basophilic leukemia	Active, not recruiting	in combination	Phase II	NCT00416598
2011	Azacitidine	NSCLC	Active, not recruiting	single	Phase II	NCT01281124
2008	Azacitidine	AML	Active, not recruiting has results	in combination	Phase II	NCT00658814
2011	Azacitidine	AML	Not yet recruiting	in combination	Phase III	NCT01369368
2013	Azacitidine	AML	Not yet recruiting	in combination	Phase II	NCT01794169
2013	Azacitidine	NSCLC	Not yet recruiting	single	Phase I	NCT02009436
2014	Azacitidine	Adult acute monoblastic leukemia (M5a)	Not yet recruiting	in combination	Phase II	NCT02029417
2014	Azacitidine	MDS	Not yet recruiting	in combination	Phase II	NCT02038816
2014	Azacitidine	AML	Not yet recruiting	in combination	Phase II	NCT02073838
2014	Azacitidine	AML	Not yet recruiting	in combination	NA	NCT02121418
2006	Azacitidine	Recurrent non–small cell lung cancer	Recruiting	in combination	Phase III	NCT00387465
2008	Azacitidine	Advanced or metastatic solid tumors	Recruiting	in combination	Phase III	NCT00748553
2015	Azacitidine	AML	Recruiting	in combination	Phase III	NCT00766116
2009	Azacitidine	Leukemia	Recruiting	single	Phase III	NCT00887068
2009	Azacitidine	CML	Recruiting	in combination	Phase II	NCT00946647
2010	Azacitidine	Leukemia	Recruiting	in combination	Phase III	NCT01016600
2010	Azacitidine	MM	Recruiting	in combination	NA	NCT01050790
2011	Azacitidine	MDS	Recruiting	in combination	Phase II	NCT01053806
2010	Azacitidine	MDS	Recruiting	in combination	Phase I	NCT01065129
2010	Azacitidine	MDS	Recruiting	in combination	Phase II	NCT01088373
2009	Azacitidine	Hematopoietic/ lymphoid cancer	Recruiting	in combination	Phase III	NCT01093573
2011	Azacitidine	MDS	Recruiting	in combination	Phase I	NCT01152346
2010	Azacitidine	MM	Recruiting	in combination	Phase III	NCT01155583
2010	Azacitidine	Pancreatic cancer	Recruiting	single	Phase I	NCT01167816
2011	Azacitidine	Relapsed or refractory solid tumors	Recruiting	in combination	Phase I	NCT01241500

Year	Drug	Condition	Status	Type	Phase	NCT Number
2011	Azacitidine	Thrombocytopenia	Recruiting	single	Phase III	NCT01286038
2011	Azacitidine	AML	Recruiting	in combination	Phase II	NCT01305499
2011	Azacitidine	Acute myeloid leukemia	Recruiting	in combination	Phase II	NCT01358734
2011	Azacitidine	CML	Recruiting	in combination	Phase II	NCT01404741
2011	Azacitidine	AML	Recruiting	single	Phase II	NCT01462578
2012	Azacitidine	Leukemia	Recruiting	in combination	Phase III	NCT01498445
2012	Azacitidine	CML	Recruiting	in combination	Phase II	NCT01522976
2012	Azacitidine	NSCLC	Recruiting	in combination	Phase I	NCT01537744
2011	Azacitidine	AML	Recruiting	in combination	Phase II	NCT01541280
2012	Azacitidine	NSCLC	Recruiting	in combination	Phase I	NCT01545947
2012	Azacitidine	AML	Recruiting	in combination	Phase II	NCT01556477
2012	Azacitidine	MDS	Recruiting	in combination	Phase III	NCT01566695
2012	Azacitidine	CML	Recruiting	single	NA	NCT01595295
2012	Azacitidine	MDS	Recruiting	single	Phase II	NCT01599325
2012	Azacitidine	Leukemia	Recruiting	in combination	Phase II	NCT01617226
2011	Azacitidine	Adult acute basophilic leukemia	Recruiting	in combination	Phase II	NCT01627041
2012	Azacitidine	MDS	Recruiting	single	Phase II	NCT01652781
2012	Azacitidine	Leukemia	Recruiting	single	Phase II	NCT01687400
2012	Azacitidine	AML	Recruiting	in combination	Phase II	NCT01700673
2012	Azacitidine	Leukemia	Recruiting	in combination	Phase II	NCT01720225
2012	Azacitidine	Adult acute basophilic leukemia	Recruiting	in combination	Phase III	NCT01729845
2012	Azacitidine	AML	Recruiting	in combination	Phase II	NCT01743859
2012	Azacitidine	AML	Recruiting	single	Phase III	NCT01747499
2012	Azacitidine	MDS	Recruiting	in combination	Phase II	NCT01748240
2012	Azacitidine	AML	Recruiting	in combination	Phase III	NCT01757535
2012	Azacitidine	AML	Recruiting	single	Phase III	NCT01758367
2013	Azacitidine	Leukemia	Recruiting	in combination	Phase III	NCT01787487
2013	Azacitidine	AML	Recruiting	in combination	Phase I	NCT01798901
2013	Azacitidine	CML	Recruiting	in combination	NA	NCT01812252
2013	Azacitidine	AML	Recruiting	in combination	Phase I	NCT01814826
2013	Azacitidine	MDS	Recruiting	in combination	Phase III	NCT01828346

(Continued)

Table 23.2 (Continued)

Initiated Time	Drug	Cancer	Status	Single or in Combination	Phases	NCT Number
2013	Azacitidine	CML	Recruiting	in combination	Phase I	NCT01834248
2012	Azacitidine	AML	Recruiting	single	Phase III	NCT01835587
2012	Azacitidine	Adult AML	Recruiting	in combination	Phase I	NCT01839240
2013	Azacitidine	Resected pancreatic adenocarcinoma	Recruiting	in combination	Phase II	NCT01845805
2013	Azacitidine	AML	Recruiting	in combination	Phase II	NCT01846624
2013	Azacitidine	Lymphoblastic leukemia	Recruiting	in combination	Phase I	NCT01861002
2013	Azacitidine	AML	Recruiting	in combination	Phase II	NCT01869114
2013	Azacitidine	MDS	Recruiting	in combination	Phase II	NCT01873703
2013	Azacitidine	NSCLC	Recruiting	in combination	NA	NCT01886573
2013	Azacitidine	Leukemia	Recruiting	in combination	Phase II	NCT01892371
2013	Azacitidine	Leukemia	Recruiting	in combination	Phase II	NCT01893372
2013	Azacitidine	AML	Recruiting	single	Phase II	NCT01912274
2013	Azacitidine	MDS	Recruiting	in combination	Phase I	NCT01913951
2013	Azacitidine	AML	Recruiting	in combination	Phase III	NCT01926587
2013	Azacitidine	CML	Recruiting	single	Phase III	NCT01928537
2013	Azacitidine	NSCLC	Recruiting	in combination	Phase II	NCT01928576
2013	Azacitidine	NSCLC	Recruiting	in combination	Phase II	NCT01935947
2013	Azacitidine	CML	Recruiting	in combination	Phase I	NCT01957644
2013	Azacitidine	Advanced cancers	Recruiting	in combination	Phase III	NCT01983969
2013	Azacitidine	MDS	Recruiting	in combination	Phase II	NCT01993641
2013	Azacitidine	AML	Recruiting	single	Phase II	NCT0199557
2013	Azacitidine	HL	Recruiting	in combination	Phase III	NCT01998035
2013	Azacitidine	AML	Recruiting	in combination	Phase II	NCT02017457
2013	Azacitidine	MDS	Recruiting	in combination	Phase I	NCT02018926
2010	Azacitidine	Adult acute minimally differentiated myeloid leukemia (M0)	Recruiting	in combination	Phase III	NCT02085408
2014	Azacitidine	AML	Recruiting	in combination	Phase II	NCT02088541
2014	Azacitidine	AML	Recruiting	in combination	Phase I	NCT02093403

Year	Drug	Cancer	Status	Type	Phase	NCT number
2014	Azacitidine	Leukemia	Recruiting	in combination	Phase III	NCT02096042
2013	Azacitidine	AML	Recruiting	single	Phase III	NCT02109744
2001	Curcumin	Colorectal cancer	Completed	single	Phase I	NCT00027495
2005	Curcumin	MM	Completed	in combination	N/A	NCT00113841
2005	Curcumin	Pancreatic cancer	Completed	in combination	Phase II	NCT00192842
2010	Curcumin	Breast cancer	Completed	single	Phase II	NCT01042938
1999	Curcumin	Colorectal cancer	Terminated	in combination	N/A	NCT00003365
2005	Curcumin	FAP	Terminated	single	Phase II	NCT00248053
2004	Decitabine	Recurrent thyroid cancer	Active, not recruiting	in combination	Phase II	NCT00085293
2006	Decitabine	MDS	Active, not recruiting	in combination	Phase III	NCT00382200
2010	Decitabine	AML	Active, not recruiting	single	Phase II	NCT01149408
2008	Decitabine	AML	Active, not recruiting	in combination	Phase II	NCT00778375
2010	Decitabine	AML	Active, not recruiting	in combination	Phase III	NCT01211457
2012	Decitabine	AML	Active, not recruiting	single	Phase III	NCT01633099
2010	Decitabine	CML	Active, not recruiting	single	Phase II	NCT01251627
2009	Decitabine	Leukemia	Active, not recruiting	in combination	Phase I	NCT01001143
2009	Decitabine	AML	Active, not recruiting	single	Phase VI	NCT01806116
2014	Decitabine	AML	Active, not recruiting	in combination	Phase III	NCT02059720
2010	Decitabine	MDS	Active, not recruiting	single	Phase VI	NCT01400633
1999	Decitabine	Leukemia	Completed	in combination	Phase III	NCT00002831
1999	Decitabine	MDS, leukemia, lymphoma, MM	Completed	single	Phase I	NCT00002980
1999	Decitabine	Leukemia, MDS	Completed	single	Phase II	NCT00003361
2001	Decitabine	Esophageal, lung cancer, mesothelioma	Completed	single	Phase I	NCT00019825
2002	Decitabine	SCC, mesothelioma, NSCLC	Completed	in combination	Phase I	NCT00037817
2002	Decitabine	Bladder cancer, breast cancer, melanoma	Completed	single	Phase I	NCT00030615
2002	Decitabine	CML	Completed	single	Phase II	NCT00041990
2002	Decitabine	CML	Completed	single	Phase II	NCT00042003
2002	Decitabine	CML	Completed	single	Phase II	NCT00042016
2003	Decitabine	Leukemia, MDS	Completed	in combination	Phase III	NCT00075010
2003	Decitabine	Leukemia	Completed	in combination	Phase II	NCT00054431

(Continued)

Table 23.2 (Continued)

Initiated Time	Drug	Cancer	Status	Single or in Combination	Phases	NCT Number
2003	Decitabine	MDS, CML	Completed	single	Phase II	NCT00067808
2004	Decitabine	Neuroblastoma	Completed	in combination	Phase I	NCT00075634
2004	Decitabine	Lymphoma, intestinal neoplasms	Completed	single	Phase I	NCT00089089
2005	Decitabine	Lymphoma	Completed	in combination	Phase I	NCT00109824
2005	Decitabine	Leukemia, MDS	Completed	in combination	Phase I	NCT00114257
2005	Decitabine	AML	Completed	in combination	Phase III	NCT00260832
2006	Decitabine	Leukemia	Completed	in combination	Phase I	NCT00357708
2006	Decitabine	AML	Completed	single	Phase II	NCT00358644
2006	Decitabine	AML	Completed	single	Phase IIIII	NCT00398983
2007	Decitabine	Leukemia, MDS	Completed	in combination	Phase I	NCT00479232
2008	Decitabine	Leukemia	Completed	single	Phase I	NCT01378416
2008	Decitabine	AML, MDS	Completed	single	Phase II	NCT00760084
2009	Decitabine	Leukemia, MDS	Completed	in combination	Phase IIII	NCT00002832
2009	Decitabine	AML	Completed	single	Phase II	NCT00866073
2010	Decitabine	CML	Completed	single	Phase II	NCT01098084
2011	Decitabine	AML	Not yet recruiting	in combination	Phase I	NCT01483274
2010	Decitabine	Breast cancer	Recruiting	in combination	Phase III	NCT01194908
2012	Decitabine	NSCLC	Recruiting	in combination	Phase III	NCT01628471
2013	Decitabine	Colon cancer	Recruiting	single	NA	NCT01882660
2012	Decitabine	Solid tumors	Recruiting	in combination	Phase III	NCT01799083
2012	Decitabine	AML	Recruiting	in combination	Phase II	NCT02084563
2010	Decitabine	Ewings sarcoma	Recruiting	in combination	Phase I	NCT01241162
2012	Decitabine	Leukemia	Recruiting	in combination	Phase II	NCT01515527
2013	Decitabine	Leukemia	Recruiting	in combination	Phase III	NCT01794702
2013	Decitabine	Leukemia	Recruiting	single	Phase II	NCT01786343
2013	Decitabine	AML	Recruiting	in combination	Phase II	NCT01829503
2013	Decitabine	Leukemia	Recruiting	in combination	Phase III	NCT01893320
2013	Decitabine	Leukemia	Recruiting	in combination	Phase II	NCT02010645
2009	Decitabine	MDS	Recruiting	in combination	Phase II	NCT00903760

Year	Drug	Cancer	Status		Phase	NCT number
2012	Decitabine	MDS	Recruiting	in combination	Phase II	NCT01593670
2011	Decitabine	AML	Recruiting	in combination	Phase III	NCT01303796
2012	Decitabine	MDS; AML	Recruiting	in combination	Phase IIII	NCT01690507
2012	Decitabine	AML	Recruiting	in combination	Phase IIII	NCT01546038
2011	Decitabine	ALL	Recruiting	in combination	Phase IIII	NCT01483690
2011	Decitabine	AML	Recruiting	single	Phase I	NCT01277484
2009	Decitabine	AML	Recruiting	in combination	Phase II	NCT00867672
2009	Decitabine	Leukemia	Recruiting	single	Phase I	NCT00986804
2013	Decitabine	AML	Recruiting	in combination	Phase II	NCT01853228
2013	Decitabine	Leukemia	Recruiting	in combination	Phase I	NCT02003573
2011	Decitabine	AML	Recruiting	in combination	Phase I	NCT01352650
2012	Decitabine	AML	Recruiting	in combination	Phase II	NCT01707004
2013	Decitabine	MDS	Recruiting	single	Phase VI	NCT02013102
2014	Decitabine	MDS	Recruiting	single	NA	NCT02045654
2002	Decitabine	Leukemia	terminated	single	Phase I	NCT00042796
2002	Decitabine	Leukemia, MDS	terminated	single	Phase I	NCT00049582
2005	Decitabine	MDS, CML	Terminated	single	Phase II	NCT00113321
2007	Decitabine	RCC	Terminated	in combination	Phase II	NCT00561912
2008	Decitabine	Unspecified adult solid tumor	terminated	in combination	Phase I	NCT00701298
2008	Decitabine	Fallopian tube, ovarian, peritoneal cavity cancer	Terminated	in combination	Phase II	NCT00748527
2009	Decitabine	Cancer	Terminated	in combination	Phase I	NCT00886457
2009	Decitabine	AML	Terminated	in combination	Phase II	NCT00943553
2004	Entinostat	CML	Active, not recruiting	in combination	Phase I	NCT00101179
2006	Entinostat	CML	Active, not recruiting	in combination	Phase II	NCT00313586
2009	Entinostat	HL	Active, not recruiting	single	Phase II	NCT00866333
2010	Entinostat	Recurrent adult ALL	Active, not recruiting	in combination	Phase I	NCT01132573
2011	Entinostat	ERBC	Active, not recruiting	in combination	Phase II	NCT01349959
2012	Entinostat	Lung Cancer	Active, not recruiting	in combination	Phase I	NCT01594398
2001	Entinostat	Leukemia, MM and plasma cell neoplasm, MDS	Completed	single	Phase I	NCT00015925
2001	Entinostat	Cancer	Completed	single	Phase I	NCT00020579

(*Continued*)

Table 23.2 (Continued)

Initiated Time	Drug	Cancer	Status	Single or in Combination	Phases	NCT Number
2004	Entinostat	Lymphoma, small intestine cancer	Completed	in combination	Phase I	NCT00098891
2005	Entinostat	Leukemia, MDS	Completed	in combination	Phase I	NCT00101179
2005	Entinostat	Melanoma	Completed	single	Phase II	NCT00185302
2010	Entinostat	Colorectal cancer	Completed	in combination	Phase II	NCT01105377
2014	Entinostat	Breast cancer	not yet recruiting	in combination	Phase II	NCT02115594
2009	Entinostat	Clear cell RCC	Recruiting	in combination	Phase III	NCT01038778
2011	Entinostat	HER2-positive breast cancer	Recruiting	in combination	Phase I	NCT01434303
2011	Entinostat	AML	Recruiting	in combination	Phase II	NCT01305499
2013	Entinostat	NSCLC	Recruiting	in combination	NA	NCT01886573
2013	Entinostat	NSCLC	Recruiting	in combination	Phase II	NCT01928576
2013	Entinostat	NSCLC	Recruiting	in combination	Phase II	NCT01935947
2006	Entinostat	NSCLC	Recruiting	in combination	Phase III	NCT00387465
2014	Entinostat	Estrogen receptor-positive breast cancer	Recruiting	in combination	Phase III	NCT02115282
2007	Epigallocatechin gallate	Prostate cancer	Completed	single	Phase I	NCT00459407
2011	Epigallocatechin gallate	Prostate cancer	Terminated	single	Phase II	NCT01340599
2006	Hydralazine	Refractory solid tumors	Completed	in combination	Phase II	NCT00404508
2006	Hydralazine	Cervical cancer	Completed	in combination	Phase III	NCT00404326
2007	Hydralazine	Rectal cancer	Withdrawn	single	Phase III	NCT00575640
2007	Hydralazine	Breast cancer	Withdrawn	single	Phase III	NCT00575978
2006	MGCD0103	Tumors	Completed	in combination	Phase III	NCT00372437
2006	MGCD0103	Tumors, NHL	Completed	single	Phase I	NCT00323934
2006	MGCD0103	Leukemia, MDS	Completed	single	Phase I	NCT00324129
2006	MGCD0103	Leukemia, MDS	Completed	single	Phase I	NCT00324194
2006	MGCD0103	AML, MDS	Completed	single	Phase IIII	NCT00324220
2006	MGCD0103	Lymphoma	Completed	single	Phase II	NCT00359086
2007	MGCD0103	CLL	Completed	single	Phase II	NCT00431873
2013	MGCD0103	MDS	Recruiting	in combination	Phase I	NCT02018926
2006	MGCD0103	HL	Terminated	single	Phase II	NCT00358982

Year	Drug	Indication	Status	Type	Phase	NCT
2006	MGCD0103	AML, MDS	Terminated	single	Phase II	NCT00374296
2007	MGCD0103	Solid cancer	Terminated	in combination	Phase I	NCT00511576
2007	MGCD0103	HL	Terminated	in combination	Phase II	NCT00543582
2008	MGCD0103	AML, MDS	Terminated	in combination	Phase II	NCT00666497
2009	Panobinostat (LBH589)	Thyroid carcinoma	Active, not recruiting	single	Phase II	NCT01013597
2006	Panobinostat (LBH589)	Tumors, CTCL	Completed	single	Phase I	NCT00412997
2007	Panobinostat (LBH589)	Breast cancer	Completed	in combination	Phase I	NCT00567879
2007	Panobinostat (LBH589)	NHL, neoplasms	Completed	single	Phase I	NCT00503451
2007	Panobinostat (LBH589)	Cancer	Completed	single	Phase I	NCT00570284
2007	Panobinostat (LBH589)	CTCL	Completed	single	Phase IIIII	NCT00490776
2008	Panobinostat (LBH589)	HER-2 positive breast cancer, metastatic breast cancer	Completed	in combination	Phase I	NCT00788931
2008	Panobinostat (LBH589)	Cancer, advanced solid tumor	Completed	single	Phase I	NCT00739414
2008	Panobinostat (LBH589)	Lymphoma, leukemia, MM	Completed	single	Phase III	NCT00621244
2008	Panobinostat (LBH589)	Leukemia	Completed	single	Phase II	NCT00723203
2010	Panobinostat (LBH589)	SCLC	Completed	single	Phase I	NCT01222936
2007	Panobinostat (LBH589)	CTCL	Terminated	single	Phase II	NCT00699296
2009	Panobinostat (LBH589)	Recurrent malignant gliomas	Terminated	single	Phase II	NCT00848523
2010	Panobinostat (LBH589)	NHL	Terminated	single	Phase II	NCT01090973
2009	Panobinostat (LBH589)	NDCLC	Recruiting	in combination	Phase III	NCT00907179
2009	Panobinostat (LBH589)	Renal cancer	Recruiting	in combination	Phase I	NCT01005797
2009	Panobinostat (LBH589)	Neuroendocrine tumors	Active, not recruiting	single	Phase II	NCT00985946
2009	Panobinostat (LBH589)	Advanced solid tumors	Active, not recruiting	single	Phase I	NCT00997399
2011	Panobinostat (LBH589)	Solid tumors	Recruiting	in combination	Phase I	NCT01336842
2010	Panobinostat (LBH589)	Clear cell RCC	Recruiting	in combination	Phase III	NCT01582009
2009	Panobinostat (LBH589)	Prostate cancer	Recruiting	in combination	Phase III	NCT00878436
2008	Panobinostat (LBH589)	Lung cancer	Active, not recruiting	in combination	Phase I	NCT00738751
2010	Panobinostat (LBH589)	Breast cancer	Recruiting	in combination	Phase III	NCT01194908
2009	Panobinostat (LBH589)	Unspecified adult solid tumor	Active, not recruiting	in combination	Phase I	NCT00877904

(Continued)

Table 23.2 (Continued)

Initiated Time	Drug	Cancer	Status	Single or in Combination	Phases	NCT Number
2012	Panobinostat (LBH589)	Nodal lymphoma	Recruiting	single	Phase II	NCT01658241
2011	Panobinostat (LBH589)	Lymphoblastic leukemia	Recruiting	in combination	Phase I	NCT01321346
2014	Panobinostat (LBH589)	Skin cancer	Recruiting	in combination	Phase I	NCT02032810
2008	Panobinostat (LBH589)	Breast cancer	Active, not recruiting	single	Phase II	NCT00777049
2009	Panobinostat (LBH589)	Hematopoietic/ lymphoid cancer	Recruiting	in combination	Phase III	NCT00918333
2009	Panobinostat (LBH589)	Malignant glioma	Active, not recruiting	in combination	Phase III	NCT00859222
2011	Panobinostat (LBH589)	Myeloma	Recruiting	in combination	Phase III	NCT01440582
2010	Panobinostat (LBH589)	Melanoma	Active, not recruiting	single	Phase I	NCT01065467
2011	Panobinostat (LBH589)	Recurrent glioma	Recruiting	in combination	Phase I	NCT01324635
2013	Panobinostat (LBH589)	Hematologic neoplasms	Recruiting	single	Phase II	NCT01802879
2011	Panobinostat (LBH589)	Diffuse LBCL	Recruiting	in combination	Phase II	NCT01282476
2009	Panobinostat (LBH589)	HL	Recruiting	single	Phase I	NCT01032148
2012	Panobinostat (LBH589)	MM	Recruiting	in combination	Phase I	NCT01549431
2014	Panobinostat (LBH589)	MM	not yet recruiting	in combination	Phase I	NCT02057640
2011	Panobinostat (LBH589)	Myeloma	Active, not recruiting	in combination	Phase I	NCT01301807
2013	Panobinostat (LBH589)	MM	Recruiting	in combination	Phase I	NCT01965353
2011	Panobinostat (LBH589)	HL	Recruiting	in combination	Phase II	NCT01460940
2010	Panobinostat (LBH589)	HL	Recruiting	in combination	Phase III	NCT01169636
2010	Panobinostat (LBH589)	Anaplastic large cell lymphoma	Recruiting	in combination	Phase II	NCT01261247
2011	Panobinostat (LBH589)	MM	Recruiting	in combination	Phase III	NCT01496118
2009	Panobinostat (LBH589)	MM	Active, not recruiting	in combination	Phase III	NCT01023308
2009	Panobinostat (LBH589)	Metastatic melanoma	Recruiting	in combination	Phase III	NCT00925132
2010	Panobinostat (LBH589)	Relapsed and bortezomib refractory MM	Active, not recruiting	in combination	Phase IV	NCT01083602
2007	Panobinostat (LBH589)	MM	Active, not recruiting	single	Phase I	NCT00532675
2010	Panobinostat (LBH589)	Chordoma	Active, not recruiting	in combination	Phase I	NCT01175109
2010	Panobinostat (LBH589)	AML	Active, not recruiting	single	Phase I	NCT01242774
2009	Panobinostat (LBH589)	CML	Recruiting	in combination	Phase II	NCT00946647
2011	Panobinostat (LBH589)	AML	Recruiting	in combination	Phase I	NCT01463046

Year	Drug	Indication	Status	Single/Combination	Phase	NCT Number
2010	Panobinostat (LBH589)	Diffuse LBCL	Recruiting	in combination	Phase II	NCT01238692
2012	Panobinostat (LBH589)	MM	Recruiting	in combination	Phase II	NCT01651039
2011	Panobinostat (LBH589)	AML	Recruiting	single	Phase III	NCT01451268
2008	Panobinostat (LBH589)	Leukemia	Active, not recruiting	in combination	Phase III	NCT00691938
2012	Panobinostat (LBH589)	CMML	Active, not recruiting	single	Phase I	NCT01613976
2009	Panobinostat (LBH589)	AML	Active, not recruiting	single	Phase III	NCT00840346
2011	Panobinostat (LBH589)	HL	Recruiting	in combination	Phase I	NCT01184428
2009	Panobinostat (LBH589)	Diffuse LBCL	Recruiting	in combination	Phase II	NCT00978432
2004	Romidepsin	Adult alveolar soft-part sarcoma	Active, not recruiting	single	Phase II	NCT00112463
2000	Romidepsin	CTCL	Active, not recruiting	single	Phase II	NCT00007345
2007	Romidepsin	Peripheral T-cell lymphoma	Active, not recruiting has results	single	Phase II	NCT00426764
2001	Romidepsin	Lymphoma	Completed	single	Phase I	NCT00019318
2001	Romidepsin	Leukemia, lymphoma	Completed	single	Phase I	NCT00024180
2002	Romidepsin	SCLC	Completed	in combination	Phase I	NCT00037817
2002	Romidepsin	Neoplasms	Completed	in combination	Phase I	NCT00048334
2002	Romidepsin	Leukemia, lymphoma, MDS syndrome	Completed	single	Phase II	NCT00042822
2003	Romidepsin	Nervous system tumors, leukemia	Completed	single	Phase I	NCT00053963
2004	Romidepsin	Lymphoma	Completed	in combination	Phase II	NCT00079443
2004	Romidepsin	Lymphoma	Completed	single	Phase II	NCT00077194
2004	Romidepsin	Colorectal cancer	Completed	single	Phase II	NCT00077337
2004	Romidepsin	Brain and nervous system tumors	Completed	single	Phase II	NCT00085540
2004	Romidepsin	Lung cancer	Completed	single	Phase II	NCT00086827
2004	Romidepsin	Ovarian, primary peritoneal cavity cancer	Completed	single	Phase II	NCT00091195
2004	Romidepsin	Breast cancer	Completed	single	Phase II	NCT00098397
2004	Romidepsin	Head and neck cancer	Completed	single	Phase II	NCT00098813
2004	Romidepsin	Peritoneal, epithelial, ovarian cancer	Completed	single	Phase II	NCT01645670
2005	Romidepsin	Leukemia, lymphoma, MDS syndrome	Completed	in combination	Phase I	NCT00114257

(Continued)

Table 23.2 (Continued)

Initiated Time	Drug	Cancer	Status	Single or in Combination	Phases	NCT Number
2005	Romidepsin	Prostate cancer, metastases	Completed	single	Phase II	NCT00106418
2005	Romidepsin	CTCL	Completed	single	Phase II	NCT00106431
2005	Romidepsin	RCC, metastases	Completed	single	Phase II	NCT00106613
2006	Romidepsin	Pancreatic cancer	Completed	in combination	Phase III	NCT00379639
2006	Romidepsin	Lymphoma	Completed	single	Phase II	NCT00383565
2007	Romidepsin	SCLC	Completed	single	Phase II	NCT00020202
2011	Romidepsin	Hematologic cancer	Completed	in combination	Phase I	NCT01324310
2011	Romidepsin	Hematologic cancer	Completed	in combination	Phase I	NCT01324323
2013	Romidepsin	Lymphoma	not yet recruiting	in combination	Phase I	NCT01902225
2009	Romidepsin	Leukemia	Recruiting	in combination	Phase I	NCT00963274
2011	Romidepsin	Peripheral T cell lymphoma	Recruiting	in combination	Phase III	NCT01280526
2011	Romidepsin	Relapsed peripheral T-cell lymphoma	Recruiting	in combination	Phase III	NCT01482962
2011	Romidepsin	Lymphoma	Recruiting	single	Phase III	NCT01456039
2012	Romidepsin	Solid tumors	Recruiting	in combination	Phase I	NCT01537744
2012	Romidepsin	Lymphoma	Recruiting	in combination	Phase I	NCT01590732
2012	Romidepsin	Lymphoma	Recruiting	in combination	Phase I	NCT01638533
2012	Romidepsin	Recurrent cutaneous T-cell NHL	Recruiting	in combination	Phase I	NCT01738594
2012	Romidepsin	HL	Recruiting	in combination	Phase III	NCT01742793
2012	Romidepsin	MM	Recruiting	in combination	Phase III	NCT01755975
2012	Romidepsin	Peripheral T-cell lymphoma	Recruiting	in combination	Phase II	NCT01822886
2013	Romidepsin	Peripheral T-cell lymphoma	Recruiting	in combination	Phase I	NCT01846390
2013	Romidepsin	B-cell adult ALL	Recruiting	in combination	Phase I	NCT01897012
2013	Romidepsin	HER2-negative breast cancer	Recruiting	in combination	Phase III	NCT01938833
2013	Romidepsin	MM	Recruiting	in combination	Phase III	NCT01947140

Year	Drug	Indication	Status	Type	Phase	NCT Number
2013	Romidepsin	MM	Recruiting	in combination	Phase III	NCT01979276
2013	Romidepsin	Lymphoma	Recruiting	in combination	Phase III	NCT01998035
2013	Romidepsin	T Cell NHL	Recruiting	in combination	Phase II	NCT01908777
2013	Romidepsin	Peripheral T-cell lymphoma	Recruiting	in combination	Phase III	NCT01796002
2014	Romidepsin	Cutaneous T-cell lymphoma	Recruiting	in combination	Phase I	NCT02061449
2004	Romidepsin	Esophageal cancer, gastric cancer	Suspended	single	Phase II	NCT00098527
2004	Romidepsin	Bladder, renal pelvis and urethral cancer	Terminated	single	Phase II	NCT00087295
2008	Romidepsin	MM	Terminated	in combination	Phase II	NCT00765102
2011	Romidepsin	Mycosis fungoides, CTCL, neoplasms	Terminated	single	Phase I	NCT01445340
2006	Valproic acid	Brain tumors	Active, not recruiting	in combination	Phase II	NCT00302159
2008	Valproic acid	Lung cancer	Active, not recruiting	in combination	Phase I	NCT00996060
2009	Valproic acid	MDS	Active, not recruiting	in combination	Phase I	NCT00903422
2006	Valproic acid	AML	Active, not recruiting	in combination	Phase II	NCT00414310
2003	Valproic acid	Carcinoma, NSCLC	Completed	single	Phase II	NCT00073385
2006	Valproic acid	Cervical cancer	Completed	in combination	Phase II	NCT00404326
2006	Valproic acid	Refractory solid tumors	Completed	in combination	Phase II	NCT00404508
2007	Valproic acid	Advanced cancers	Completed	in combination	Phase I	NCT00496444
2011	Valproic acid	AML	not yet recruiting	in combination	Phase IIII	NCT01369368
2011	Valproic acid	MDS	not yet recruiting	in combination	Phase II	NCT01356875
2011	Valproic acid	Pancreatic cancer	not yet recruiting	single	Phase II	NCT01333631
2013	Valproic acid	Breast cancer	not yet recruiting	in combination	Phase II	NCT01900730
2014	Valproic acid	Ciliary body and choroid melanoma	not yet recruiting	in combination	Phase II	NCT02068586
2008	Valproic acid	SCLC	Recruiting	in combination	Phase II	NCT00759824
2009	Valproic acid	AML	Recruiting	in combination	Phase II	NCT00867672
2009	Valproic acid	Glial cell tumors	Recruiting	in combination	Phase II	NCT00879437
2009	Valproic acid	Cancer	Recruiting	single	Phase I	NCT01007695
2009	Valproic acid	Sarcoma	Recruiting	single	Phase I	NCT01010958
2010	Valproic acid	Sarcoma	Recruiting	in combination	Phase IIII	NCT01106872

(Continued)

Table 23.2 (Continued)

Initiated Time	Drug	Cancer	Status	Single or in Combination	Phases	NCT Number
2010	Valproic acid	CLL	Recruiting	in combination	Phase III	NCT01295593
2010	Valproic acid	Locally advanced inoperable NSCLC	Recruiting	single	Phase III	NCT01203735
2010	Valproic acid	Thyroid neoplasm	Recruiting	single	Phase II	NCT01182285
2012	Valproic acid	Diffuse LBCL	Recruiting	in combination	Phase III	NCT01622439
2012	Valproic acid	Colorectal cancer	Recruiting	in combination	Phase III	NCT01898104
2012	Valproic acid	Head and neck cancer	Recruiting	single	Phase II	NCT01695122
2008	Valproic acid	Breast cancer	Recruiting	in combination	Phase II	NCT01010854
2009	Valproic acid	AML	Recruiting	in combination	Phase III	NCT00995332
2012	Valproic acid	Advanced cancers	Recruiting	in combination	Phase I	NCT01552434
2004	Valproic acid	Malignant melanoma	Terminated	single	Phase III	NCT00087477
2006	Valproic acid	Malignant melanoma	Terminated	in combination	Phase III	NCT00358319
2006	Valproic acid	Locally advanced breast cancer	Terminated	in combination	Phase II	NCT00395655
2007	Valproic acid	Brain metastases	Terminated	in combination	Phase I	NCT00437957
2007	Valproic acid	CLL	Terminated	in combination	Phase II	NCT00524667
2013	Valproic acid	Pediatric brain tumor; glioma; anaplastic astrocytoma; Medulloblastoma; glioblastoma	Withdrawn	single	Phase I	NCT01861990
2005	Vorinostat	Recurrent adult brain tumor	Active, not recruiting	in combination	Phase I	NCT00268385
2005	Vorinostat	Recurrent marginal zone lymphoma	Active, not recruiting	in combination	Phase II	NCT00253630
2005	Vorinostat	Ciliary body and choroid melanoma	Active, not recruiting	single	Phase II	NCT00121225
2006	Vorinostat	Recurrent adult Burkitt lymphoma	Active, not recruiting	in combination	Phase I	NCT00275080
2006	Vorinostat	Adult nasal type extranodal NK/T-cell lymphoma	Active, not recruiting	in combination	Phase I	NCT00336063

Year	Drug	Condition	Status		Phase	NCT Number
2006	Vorinostat	Recurrent RCC	Active, not recruiting	in combination	Phase III	NCT00324740
2006	Vorinostat	Clear cell RCC	Active, not recruiting	in combination	Phase III	NCT00324870
2006	Vorinostat	Male breast cancer	Active, not recruiting	in combination	Phase III	NCT00368875
2006	Vorinostat	Adult acute basophilic leukemia	Active, not recruiting	in combination	Phase III	NCT00392353
2006	Vorinostat	Breast cancer	Active, not recruiting	in combination	Phase II	NCT00262834
2006	Vorinostat	Breast cancer	Active, not recruiting	single	Phase III	NCT00416130
2007	Vorinostat	Stage I prostate cancer	Active, not recruiting	in combination	Phase II	NCT00589472
2008	Vorinostat	NSCLC	Active, not recruiting	in combination	Phase I	NCT00702572
2008	Vorinostat	Breast cancer	Active, not recruiting	in combination	Phase I	NCT00788112
2008	Vorinostat	Lymphoma	Active, not recruiting	in combination	Phase III	NCT00667615
2008	Vorinostat	SCLC	Active, not recruiting	in combination	Phase III	NCT00702962
2008	Vorinostat	Adult giant cell glioblastoma	Active, not recruiting	in combination	Phase III	NCT00731731
2008	Vorinostat	Leukemia	Active, not recruiting	in combination	Phase III	NCT00764517
2008	Vorinostat	Breast cancer	Active, not recruiting	in combination	Phase II	NCT00616967
2008	Vorinostat	Lymphoma	Active, not recruiting	in combination	Phase II	NCT00720876
2008	Vorinostat	MM and plasma cell neoplasm	Active, not recruiting	in combination	Phase II	NCT00744354
2009	Vorinostat	Leukemia	Active, not recruiting	in combination	Phase I	NCT00875745
2009	Vorinostat	Neuroblastoma	Active, not recruiting	in combination	Phase I	NCT01019850
2009	Vorinostat	MM	Active, not recruiting	in combination	Phase I	NCT01038388
2009	Vorinostat	B-cell CLL	Active, not recruiting	in combination	Phase III	NCT00918723
2009	Vorinostat	Contiguous stage II adult diffuse large cell lymphoma	Active, not recruiting	in combination	Phase III	NCT00972478
2009	Vorinostat	MM	Active, not recruiting	in combination	Phase II	NCT00839956
2009	Vorinostat	Leukemia	Active, not recruiting	in combination	Phase II	NCT00948064
2009	Vorinostat	Lymphoma	Active, not recruiting	single	Phase II	NCT00875056
2010	Vorinostat	Male breast cancer	Active, not recruiting	in combination	Phase I	NCT01084057
2010	Vorinostat	Lymphoma	Active, not recruiting	in combination	Phase I	NCT01169532
2010	Vorinostat	Disseminated neuroblastoma	Active, not recruiting	in combination	Phase I	NCT01208454
2010	Vorinostat	Lymphoma	Active, not recruiting	in combination	Phase I	NCT01276717

(Continued)

Table 23.2 (Continued)

Initiated Time	Drug	Cancer	Status	Single or in Combination	Phases	NCT Number
2010	Vorinostat	Recurrent salivary gland cancer	Active, not recruiting	in combination	Phase II	NCT01175980
2010	Vorinostat	Tongue cancer	Active, not recruiting	in combination	Phase II	NCT01267240
2010	Vorinostat	NSCLC	Active, not recruiting	in combination	Phase II	NCT01413750
2011	Vorinostat	Adult anaplastic astrocytoma	Active, not recruiting	in combination	Phase I	NCT01378481
2011	Vorinostat	Lymphoma	Active, not recruiting	in combination	Phase I	NCT01421173
2007	Vorinostat	Anaplastic glioma	Active, not recruiting	in combination	Phase IIII	NCT00555399
2007	Vorinostat	Adult nasal type extranodal NK/T-cell lymphoma	Active, not recruiting	in combination	Phase IIII	NCT00601718
2008	Vorinostat	MM	Active, not recruiting	in combination	Phase I	NCT00729118
2008	Vorinostat	NSCLC	Active, not recruiting	in combination	Phase II	NCT00798720
2008	Vorinostat	MM	Active, not recruiting	in combination	Phase III	NCT00773747
2009	Vorinostat	Pancreatic cancer	Active, not recruiting	in combination	Phase I	NCT00983268
2010	Vorinostat	AML	Active, not recruiting	in combination	Phase I	NCT01130506
2010	Vorinostat	NSCLC	Active, not recruiting	in combination	Phase IIII	NCT01027676
2009	Vorinostat	Brain metastases	Active, not recruiting has results	in combination	Phase I	NCT00838929
2005	Vorinostat	Recurrent RCC	Active, not recruiting has results	single	Phase II	NCT00278395
2000	Vorinostat	Hematologia and solid cancer	Completed	single	Phase I	NCT00005634
2002	Vorinostat	Cancer	Completed	single	Phase I	NCT00045006
2004	Vorinostat	CTCL, mycosis fungoides	Completed	in combination	Phase II	NCT00091559
2004	Vorinostat	B-cell lymphoma	Completed	in combination	Phase II	NCT00097929
2005	Vorinostat	Advanced cancer	Completed	in combination	Phase I	NCT00106626
2005	Vorinostat	Multiple myeloma	Completed	in combination	Phase I	NCT00111813
2005	Vorinostat	Colorectal cancer	Completed	in combination	Phase I	NCT00138177
2005	Vorinostat	Brain, neuroblastoma, CNS, hematologic tumors	Completed	in combination	Phase I	NCT00217412

Year	Drug	Indication	Status	Type	Phase	NCT
2005	Vorinostat	Unspecified adult solid tumor, protocol specific	Completed	in combination	Phase I	NCT00227513
2005	Vorinostat	Unspecified adult solid tumor, protocol specific	Completed	in combination	Phase I	NCT00243100
2005	Vorinostat	Tumors	Completed	single	Phase I	NCT00127127
2005	Vorinostat	Lymphoma	Completed	single	Phase I	NCT00127140
2005	Vorinostat	Breast cancer	Completed	single	Phase II	NCT00132002
2005	Vorinostat	Ovarian cancer, primary peritoneal cavity cancer	Completed	single	Phase II	NCT00132067
2005	Vorinostat	Head and neck cancer	Completed	single	Phase II	NCT00134043
2005	Vorinostat	Lung cancer	Completed	single	Phase II	NCT00138203
2005	Vorinostat	Brain and central nervous system tumors	Completed	single	Phase II	NCT00238303
2005	Vorinostat	Glioblastoma, gliosarcoma, brain tumor	Completed	single	Phase II	NCT01647100
2005	Vorinostat	Mesothelioma lung cancer	Completed	single	Phase III	NCT00128102
2006	Vorinostat	Leukemia, MDS	Completed	in combination	Phase I	NCT00278330
2006	Vorinostat	Unspecified adult solid tumor, protocol specific	Completed	in combination	Phase I	NCT00287937
2006	Vorinostat	MM and plasma cell neoplasm	Completed	in combination	Phase I	NCT00310024
2006	Vorinostat	Unspecified adult solid tumor, protocol specific	Completed	in combination	Phase I	NCT00324480
2006	Vorinostat	Leukemia, MDS	Completed	in combination	Phase I	NCT00331513
2006	Vorinostat	Unspecified adult solid tumor, protocol specific	Completed	in combination	Phase I	NCT01645514
2006	Vorinostat	Colorectal cancer	Completed	single	Phase I	NCT00336141
2006	Vorinostat	Leukemia, MDS	Completed	single	Phase II	NCT00305773
2006	Vorinostat	Kidney cancer	Completed	single	Phase II	NCT00354250
2007	Vorinostat	NSCLC	Completed	in combination	Phase I	NCT00423449
2007	Vorinostat	Lung cancer	Completed	in combination	Phase II	NCT00481078
2007	Vorinostat	Pelvic cancer radiotherapy	Completed	single	Phase I	NCT00455351
2007	Vorinostat	Brain, CNS, intestine tumors, lymphoma	Completed	single	Phase I	NCT00499811

(Continued)

Table 23.2 (Continued)

Initiated Time	Drug	Cancer	Status	Single or in Combination	Phases	NCT Number
2008	Vorinostat	Cancer,	Completed	in combination	Phase I	NCT00750178
2008	Vorinostat	Leukemia	Completed	in combination	Phase I	NCT00816283
2008	Vorinostat	Relapsed or refractory MM	Completed	in combination	Phase II	NCT00773838
2008	Vorinostat	Advanced cancer relapsed and refractory	Completed	single	Phase I	NCT00632931
2008	Vorinostat	Lymphoma	Completed	single	Phase I	NCT00771472
2008	Vorinostat	Glioblastoma, gliosarcoma, brain tumor	Completed	single	Phase II	NCT00641706
2009	Vorinostat	MM	Completed	in combination	Phase I	NCT00858234
2009	Vorinostat	Hematologic and solid cancer	Completed	in combination	Phase II	NCT00942266
2009	Vorinostat	Advanced cancer	Completed	single	Phase II	NCT00907738
2010	Vorinostat	Lymphoma, sarcoma, Wilms tumor, neuroblastoma	Completed	in combination	Phase I	NCT01132911
2006	Vorinostat	Breast cancer	Completed with result	in combination	Phase II	NCT00365599
2012	Vorinostat	MM	not yet recruiting	in combination	Phase II	NCT01720875
2014	Vorinostat	Advanced cancers	not yet recruiting	in combination	Phase I	NCT02042989
2014	Vorinostat	Neuroblastoma	not yet recruiting	in combination	Phase II	NCT02035137
2012	Vorinostat	Mantle cell lymphoma	not yet recruiting	in combination	Phase II	NCT01578343
2014	Vorinostat	Myeloma	not yet recruiting	in combination	Phase II	NCT02114502
2008	Vorinostat	NHL	Recruiting	in combination	Phase I	NCT00691210
2008	Vorinostat	Recurrent adult diffuse large cell lymphoma	Recruiting	in combination	Phase II	NCT00703664
2009	Vorinostat	Untreated childhood supratentorial primitive neuroectodermal tumor	Recruiting	in combination	NA	NCT00867178
2009	Vorinostat	Malignant solid tumor	Recruiting	in combination	Phase I	NCT01023737
2009	Vorinostat	Stage I cutaneous T-cell non-Hodgkin lymphoma	Recruiting	in combination	Phase II	NCT00958074

Year	Drug	Condition	Status	Combination	Phase	NCT Number
2009	Vorinostat	Anaplastic large cell lymphoma	Recruiting	in combination	Phase II	NCT00992446
2009	Vorinostat	Locally advanced non–small cell lung cancer	Recruiting	single	Phase I	NCT01059552
2010	Vorinostat	Stage IV squamous cell carcinoma of the oropharynx	Recruiting	in combination	Phase I	NCT00946673
2010	Vorinostat	Advanced cancer	Recruiting	in combination	Phase I	NCT01087554
2010	Vorinostat	Prostate cancer	Recruiting	in combination	Phase I	NCT01174199
2010	Vorinostat	Advanced cancers	Recruiting	in combination	Phase I	NCT01266057
2010	Vorinostat	Brain cancer	Recruiting	in combination	Phase I/II	NCT01110876
2010	Vorinostat	Noncontiguous stage II mantle cell lymphoma	Recruiting	in combination	Phase I/II	NCT01193842
2010	Vorinostat	Brain cancer	Recruiting	in combination	Phase I/II	NCT01266031
2010	Vorinostat	MM	Recruiting	in combination	Phase I/II	NCT01554852
2011	Vorinostat	Unspecified adult solid tumor	Recruiting	in combination	NA	NCT01281176
2011	Vorinostat	Solid tumors	Recruiting	in combination	Phase I/II	NCT01294670
2011	Vorinostat	MM	Recruiting	in combination	Phase I/II	NCT01297764
2011	Vorinostat	MM	Recruiting	in combination	Phase I/II	NCT01394354
2011	Vorinostat	Children with relapsed solid tumor	Recruiting	in combination	Phase I/II	NCT01422499
2011	Vorinostat	ALL	Recruiting	in combination	Phase I/II	NCT01483690
2011	Vorinostat	MM	Recruiting	in combination	Phase I/II	NCT01502085
2012	Vorinostat	Peripheral T-cell lymphoma	Recruiting	in combination	Phase I	NCT01567709
2012	Vorinostat	AML	Recruiting	in combination	Phase I/II	NCT01534260
2012	Vorinostat	AML	Recruiting	in combination	Phase II	NCT01550224
2012	Vorinostat	MDS	Recruiting	in combination	Phase II	NCT01593670
2012	Vorinostat	Adult brain tumor	Recruiting	in combination	Phase II	NCT01738646
2012	Vorinostat	MDS	Recruiting	in combination	Phase II	NCT01748240
2013	Vorinostat	Advanced cancers	Recruiting	in combination	Phase I/II	NCT01983969
2013	Vorinostat	AML	Recruiting	in combination	Phase I/II	NCT01802333
2014	Vorinostat	Leukemia	Recruiting	in combination	Phase I	NCT02083250

(Continued)

Table 23.2 (Continued)

Initiated Time	Drug	Cancer	Status	Single or in Combination	Phases	NCT Number
2010	Vorinostat	Liver cancer	Recruiting	in combination	Phase I	NCT01075113
2010	Vorinostat	Recurrent anal cancer	Recruiting	in combination	Phase I	NCT01249443
2011	Vorinostat	Advanced cancer	Recruiting	in combination	Phase I	NCT01339871
2012	Vorinostat	Breast carcinoma	Recruiting	in combination	NA	NCT01655004
2012	Vorinostat	Male breast cancer	Recruiting	in combination	NA	NCT01720602
2012	Vorinostat	CML	Recruiting	in combination	Phase II	NCT01522976
2012	Vorinostat	Metastatic intraocular melanoma	Recruiting	in combination	Phase II	NCT01587352
2012	Vorinostat	Leukemia	Recruiting	in combination	Phase II	NCT01617226
2012	Vorinostat	CTCL	Recruiting	in combination	Phase III	NCT01728805
2013	Vorinostat	Sarcoma	Recruiting	in combination	Phase III	NCT01879085
2013	Vorinostat	Accelerated phase CML	Recruiting	in combination	Phase II	NCT01789255
2006	Vorinostat	Leukemia, MDS	Suspended	in combination	Phase III	NCT00392353
2005	Vorinostat	Breast cancer, colorectal cancer, NSCL	Terminated	in combination	Phase II	NCT00126451
2007	Vorinostat	Neoplasms	Terminated	in combination	Phase I	NCT00424775
2007	Vorinostat	NSCL, prostate, bladder, urothelial cancers	Terminated	in combination	Phase I	NCT00565227
2007	Vorinostat	Stage IIIB or IV non–small cell lung cancer	Terminated	in combination	Phase IIIII	NCT00473889
2008	Vorinostat	Malignant solid tumor	Terminated	in combination	Phase I	NCT00801151

Year	Drug	Condition	Status		Phase	NCT number
2008	Vorinostat	Malignant solid tumor	Terminated	in combination	Phase I	NCT00801151
2008	Vorinostat	SCLC	Terminated	in combination	Phase III	NCT00697476
2008	Vorinostat	Hematologic cancer	Terminated	in combination	Phase II	NCT00673153
2008	Vorinostat	Lymphoma	Terminated	in combination	Phase II	NCT00810576
2009	Vorinostat	Pancreatic cancer	Terminated	in combination	Phase III	NCT00831493
2009	Vorinostat	Leukemia, myelodysplastic syndromes	Terminated	in combination	Phase II	NCT00818649
2010	Vorinostat	Breast cancer	Terminated	in combination	Phase II	NCT01194427
2010	Vorinostat	Hematologic and solid cancer	Terminated	single	Phase I	NCT01116154
2005	Vorinostat	NSCLC	Terminated with result	single	Phase III	NCT00251589
2013	GSK2879552	SCLC	Recruiting	single	Phase I	NCT02034123
2013	E7438	Diffuse LBCL	Recruiting	single	Phase III	NCT01897571
2012	OTX015	Acute Leukemia; other HM	Recruiting	single	Phase I	NCT01713582
2013	CPI-0610	Lymphoma	Recruiting	single	Phase I	NCT01949883
2005	Vorinostat	Lymphoma	Terminated with with result	in combination	Phase I	NCT00127101

Table 23.3 Related Phase II and Phase III Trials with Results Reported in PubMed

Study (year)	Country	Phase	Total Number	Cancer Types	Prior treatment	Treatment Regimen	CR	PR	SD	PD
Lübbert et al. (2011)	Germany	Phase II	227	AML	No	decitabine	30	29	57	19
Nand et al. (2013)	America	Phase II	142	AML	No	azacitidine+ gemtuzumab ozogamicin	35/19	0/1	41/NA	NA
Kantarjian et al. (2012)	France	Phase III	485	AML	No	decitabine / cytarabine	38/17	6/8	67/52	50/69
Schroeder et al. (2013)	Germany	Phase II	30	AML or MDS	Yes	azacitidine + donor lymphocyte infusions	7	2	5	NA
Sonpavde et al. (2011)	America	Phase II	36	Prostate	Yes	azacitidine + combined androgen blockade	0	0	5	7
Platzbecker et al. (2012)	Germany	Phase II	20	AML or MDS	Yes	azacitidine	0	0	NA	NA
Malik et al. (2013)	America	Phase II	9	CLL	Yes	azacitidine	0	0	0	NA
Al-Ali et al. (2012)	Germany	Phase I/II	40	AML	Yes	azacitidine	2	3	15	2
Fenaux et al. (2010)	France	Phase III	113	AML	Yes	azacitidine/ conventional care regimens	10/9	0/0	NA	NA
Raffoux et al. (2010)	France	Phase II	65	AML or MDS	Yes	azacitidine + valproic acid + all-*trans* retinoic acid	14	3	NA	NA
Juergens et al. (2011)	America	Phase II	42	NSCL	Yes	azacitidine + Entinostat	1	1	10	21
Matei et al. (2012)	America	Phase II	17	Ovarian	Yes	decitabine + cytarabine	1	5	6	5
Braun et al. (2014)	France	Phase II	39	CML	Yes	decitabine	4	8	3	NA
Cashen et al. (2009)	America	Phase II	55	AML	No	decitabine	13	0	NA	21

Author (year)	Country	Phase	N	Disease		Drug				
Ravandi et al. (2013)	America	Phase II	43	AML	Yes	azacitidine + sorafenib	16	1	NA	NA
Fu et al. (2013)	America	Phase I/II	29	Ovarian	Yes	azacitidine + carboplatin	1	3	10	NA
Pohlmann et al. (2002)	America	Phase II	24	SCC	Yes	decitabine + Cisplatin	0	8	5	8
Scott et al. (2010)	America	Phase II	22	MDS or CMML	NA	azacitidine + etanercept	9	2	7	2
Borthakur et al. (2010)	America	Phase I/II	34	AML	Yes	azacitidine + cytarabine	2	0	NA	NA
Tawbi et al. (2013)	America	Phase I/II	35	metastatic melanoma	Yes	decitabine + temozolomide	2	4	14	13
Pollyea et al. (2012)	America	Phase I/II	42	AML	No	azacitidine + lenalidomide	7	4	NA	NA
Garcia-Manero et al. (2006)	America	Phase I/II	54	AML or MDS	No	decitabine + valproic acid	10	0	NA	NA
Jean-Pierre et al. (2005)	America	Phase II	35	CML	Yes	decitabine	12	7	NA	NA
Coronel et al. (2010)	America	Phase III	36	cervical	Yes	hydralazine + valproate /placebo	0/0	4/1	5/6	8/12
Candelaria et al. (2007)	Mexico	Phase II	17	solid tumors	Yes	hydralazine+ magnesium valproate	0	4	8	NA
Trudel et al. (2013)	Canada	Phase II	16	ovarian	Yes	EGCG-enriched green tea drink	0	0	0	NA
Shanafelt et al. (2013)	America	Phase II	42	CLL	Yes	Polyphenon E	0	0	NA	NA
Modesitt et al. (2008)	America	Phase II	27	ovarian / peritonea	No	vorinostat	0	1	9	14
Vansteenkiste et al. (2008)	Belgium	Phase II	16	breast/ colorectal/ lung	Yes	vorinostat	0	0	8	2
Woyach et al. (2009)	America	Phase II	19	thyroid/ DTC /MTC	Yes	vorinostat	0	0	9	10
Blumenschein et al. (2008)	America	Phase II	12	head and neck	Yes	vorinostat	0	1	3	8

(Continued)

Table 23.3 (Continued)

Study (year)	Country	Phase	Total Number	Cancer Types	Prior treatment	Treatment Regimen	CR	PR	SD	PD
Galanis et al. (2009)	America	Phase II	66	glioblastoma multiforme	Yes	vorinostat	0	0	10	54
Traynor et al. (2009)	America	Phase II	16	NSCLC	Yes	vorinostat	0	0	8	6
Bradley et al. (2009)	America	Phase II	27	prostate	Yes	vorinostat	0	0	2	25
Munster et al. (2011)	America	Phase II	43	breast	Yes	vorinostat + tamoxifen	0	8	9	26
Ramalingam et al. (2010)	America	Phase II	62	NSCLC	Yes	vorinostat + carboplatin + paclitaxel	1	20	NA	NA
Kirschbaum et al. (2012)	America	Phase II	25	HL	Yes	vorinostat	0	1	12	NA
Schaefer et al. (2009)	America	Phase II	37	AML	No	vorinostat	1	0	NA	NA
Duvic et al. (2007)	America	Phase II	33	CTCL	Yes	vorinostat	0	8	NA	NA
Kirschbaum et al. (2010)	America	Phase II	35	NHL / MCL	Yes	vorinostat	5	5	1	NA
Crump et al. (2008)	Canada	Phase II	18	DLBCL	Yes	vorinostat	1	0	1	16
Friday et al. (2012)	America	Phase II	37	GBM	Yes	vorinostat + bortezomib	0	1	0	34
Olsen et al. (2007)	America	Phase II	74	CTCL	Yes	vorinostat	1	29	10	NA
Dana et al. (2013)	Germany	Phase II	35	prostate	Yes	panobinostat	0	0	4	NA
Cassier et al. (2013)	France	Phase II	47	STS	Yes	panobinostat	0	0	17	NA
Younes et al. (2011)	America	Phase II	129	HL	Yes	panobinostat	5	30	14	71
Wolff et al. (2012)	America	Phase II	38	MM	Yes	panobinostat	0	1	9	NA

Reference	Country	Phase	N	Cancer		Treatment				
Daud et al. (2009)	America	Phase I/II	39	melanoma	Yes	valproic acid + Karenitecin	0	0	7	NA
Wolff et al. (2008)	America	Phase II	44	melanoma	Yes	valproic acid radiation + Cisplatin + Etoposide + Vincristine + Ifosfamide	0	1	26	9
Rocca et al. (2009)	Italy	Phase I/II	29	melanoma	Yes	vorinostat + chemoimmunotherapy	1	2	3	NA
Scherpereel et al. (2011)	France	Phase II	45	mesothelioma	Yes	valproic acid + doxorubicin	0[a]/0[b]	6[a]/7[b]	10[a]/4[b]	20[a]/25[b]
Hauschild et al. (2008)	Germany	Phase II	28	melanoma	Yes	entinostat	0	0	7	21
Denise et al. (2013)	America	Phase II	130	breast	Yes	entinostat/ exemestane	0	0	NA	NA
Stadler et al. (2006)	America	Phase II	29	renal	Yes	romidepsin	1	1	2	22
Schrump et al. (2008)	America	Phase II	18	NSCLC/ SCLC	Yes	romidepsin	0	0	9	9
Whitehead et al. (2009)	America	Phase II	25	colorectal	Yes	romidepsin	0	0	4	20
Molife et al. (2010)	America	Phase II	35	prostate	Yes	romidepsin	0	2	11	22
Iwamoto et al. (2011)	America	Phase I/II	40	Anaplastic glioma/ Glioblastoma	Yes	romidepsin	0	0	10	25
Otterson et al. (2010)	America	Phase II	16	SCLC	Yes	romidepsin	0	0	3	12
Shah et al. (2006)	America	Phase II	15	neuroendo- crine	Yes	romidepsin	NA	NA	NA	3
Harrison et al. (2014)	Australia	Phase II	25	MM	Yes	romidepsin + bortezomib + dexamethasone	2	13	2	1
Whittaker et al. (2010)	United Kingdom	Phase II	96	CTCL	Yes	romidepsin	6	NA	NA	NA

(Continued)

Table 23.3 (Continued)

Study (year)	Country	Phase	Total Number	Cancer Types	Prior treatment	Treatment Regimen	CR	PR	SD	PD
Niesvizky et al. (2011)	America	Phase II	13	MM	Yes	romidepsin	0	0	0	13
Piekarz et al. (2008)	America	Phase II	71	CTCL	Yes	romidepsin	4	20	26	15
Kanai et al. (2010)	Germany	Phase II	21	pancreatic	Yes	curcumin + gemcitabine-based chemotherapy	0	0	5	NA
Epelbaum et al. (2010)	America	Phase II	16	pancreatic	Yes	Curcumin + Gemcitabine	0	1	4	6
Dhillon et al. (2014)	America	Phase II	25	pancreatic	Yes	curcumin	0	0	1	NA

[a]Response at three cycles.
[b]Response at six cycles.

cancer treatment [2]. The transference of methyl group from donor S-adenosyl methionine (SAM) to the $5'$ position of cytosine is catalyzed by DNA methyltransferases (DNMTs). In mammals, three DNMTs have so far been identified, including two "de novo" methyltransferases (DNMT3A and DNMT3B) and "maintenance" methyltransferase (DNMT1), generally the most abundant active of the three [3].

Due to the pivotal role of DNMTs, intense interest has focused on developing drugs able to interfere with aberrant DNMT activities. Various compounds have been identified with inhibitory effects on DNA methylation and it is possible to reactivate a hypermethylated tumor suppressor gene by the use of DNA methyltransferase inhibitors (DNMTi). Some of the inhibitors has been studied in Phase II and/or Phase III trials. These tested drugs could be classified into three categories (Table 23.1). Most are nucleoside analogues, such as azacitidine (5-azacitidine, 5-Aza-CR) and decitabine (5-aza-2'-deoxycitidine, 5-Aza-CdR). Once incorporated into DNA, they form a covalent bond with the DNMTs trapping the enzyme and making it unavailable for further methylation, thus resulting in demethylation of replicating nascent DNA [4]. Another category is small molecular compounds including hydralazine and procainamide, which have been approved for the treatment of hypertension and cardiac arrhythmia [5]. Hydralazine was tested as a DNMT inhibitor due to its capability to induce (as a side effect) a lupus-like syndrome known to be related to disorders associated with DNA methylation [6]. In addition, some natural molecules such as Epigallocatechin-3-gallate (EGCG), the major polyphenol from green tea, exhibited a very strong DNMT inhibitory action in preclinical studies, although its mechanism was unclear [7,8].

2.1.1 Azacitidine

Azacitidine, a pyrimidine nucleoside analogue of cytidine, which can be incorporated into the DNA and interfering with its metabolism, primarily affecting cells in the DNA synthesis phase (S-phase), was initially synthesized as a cytotoxic agent in 1970s [9]. Several years later, however, it was discovered to have DNA demethylation activity at low doses [10]. This improved understanding of azacitidine influenced the following studies. From 1993 to 2002, the Cancer and Leukemia Group B (CALGB) conducted three sequential trials (CALGB Protocols 8421, 8921, and 9221), which involved 291 evaluated patients affected by myelodysplasia syndrome (MDS) or acute myelocytic leukemia (AML) (according to the new WHO classification) receiving low dose azacitidine as a continuous intravenous infusion. A significantly prolonged survival (19.3 months) was observed, suggesting that azacitidine might affect the natural course of MDS [11]. These positive results finally resulted in the approval of azacitidine by the FDA for the treatment of MDS in 2004.

In 2012, a multicenter Phase I/II study conducted by Al-Ali et al. further validated the efficiencies of azacitidine in AML. In this study, 36 patients with AML medically unfit for or resistant to conventional chemotherapy were treated with azacitidine 75 mg/m²/day subcutaneously for 5 consecutive days every 4 weeks. 67.5% patients were benefited by azacitidine on different degree including complete response (CR), partial response (PR), stable disease (SD), and hematologic improvement (HI) [12]. Meanwhile, Fenaux et al. also reported that azacitidine increased the overall survival (OS) of elder AML patients from 16.0 to 24.5 months while reducing side effects [13]. Despite the considerable effect in MDS and AML, newly released results of a national Phase II Study in Fludarabine-Refractory Chronic Lymphocytic Leukemia (CLL) patients showed limitations of azacitidine. Nine patients with recurrent fludarabine-refractory CLL were enrolled. Azacitidine (75 mg/m²) was administered by subcutaneous injection daily for 7 consecutive days every 3–8 weeks, no partial or complete responses occurred in these patients [14]. The only attempt of azacitidine for solid tumors in 1977 met with the same failure.

One hundred seventy-seven patients were evaluated after therapy with 5-azacytidine using a dose of 1.6 mg/kg/day for 10 days followed by a maintenance regimen. Antitumor effect was seen in 17% of the evaluable patients with carcinoma of the breast and 21% of the patients with malignant lymphomas. It is believed that the drug is only of minimal value as a single agent in solid tumors [15].

In summary, azacitidine monotherapy is effective and safe in the treatment of MDS and AML. However, there is insufficient evidence to support azacitidine in solid tumors or other hematological malignancies like CLL at the present time. Twenty-six trials of single azacitidine trials of cancer are ongoing, including 10 AML, 4 CML, 2 nonsmall cell lung cancer (NSCLC), and 3 other solid tumors (Table 23.2).

2.1.2 Decitabine

Decitabine, as deoxy analogue of azacitidine, was also approved by FDA for treatment of MDS and chronic myelomonocytic leukemia (CMML) in 2006. Early studies of decitabine, however, demonstrated minimal-to-moderate activity, with substantial toxicity as the development and improvement of this drug was achieved by markedly reducing the doses to overcome limiting toxicities, which also likely enhanced their targeted effects on DNA methyltransferases [16]. In 2000, however, a pivotal Phase III study was initiated to compare low-dose decitabine with supportive care in patients with MDS, after which the optimizing dosing schedules of decitabine (low dose, high dose intensity, and multiple cycles) improved results further and further, suggesting that decitabine is an active therapy that alters the natural course of MDS [17].

In recent decades, clinicians have been seeking more effective and safe drugs for elderly AML, as older patients have difficulty tolerating chemotherapy due to comorbidities, concomitant end-organ dysfunction, and poor performance status. Treatment candidates included cytarabine, azacitidine, and decitabine. The recent study indicated that decitabine has certain advantages. Following two studies suggested that decitabine was well tolerated, with encouraging responses noted in elderly patients. First, Cashen et al. treated 55 newly diagnosed older AML patients with intermediate or poor-risk cytogenetics with decitabine at $20 \, \text{mg/m}^2$/day for 5 days, repeated on a 4-week cycle. After a median of three cycles of decitabine, the overall response rate was 25% and an additional 29% had stable disease. The drug was fairly well tolerated, with myelosuppression, febrile neutropenia, and fatigue as the major toxicities [18]. Another Phase II study in Europe treated 227 elderly AML patients with decitabine at $15 \, \text{mg/m}^2$ every 8 hours for 3 days as part of a 6-week cycle. The CR rate in this study was 13%, a little lower than that in the study mentioned above. This minor difference might be reflective of the study design, which called for application of four courses of therapy and then optional maintenance therapy without systematic recording of improvement in response rate while on maintenance.

These promising response rate and toxicity profile set the stage for the use of this decitabine dose in the following Phase III study. In this open-label Phase III trial, 485 patients in 15 countries were randomized to decitabine or their treatment choice (TC) of either low-dose cytarabine (LDAC) or supportive care. The decitabine dose was the same as that used in the Phase II study by Cashen et al., at $20 \, \text{mg/m}^2$/day intravenous drip (IV) for 5 days as a 4-week cycle. Regarding the secondary end points, patients on decitabine had significantly improved rates of CR and CR rate without platelet recovery (CRp) (17.8% vs. 7.8% in the TC group) at the prespecified cutoff. The safety profile of decitabine was similar to LDAC, with the majority of patients in both arms experiencing some grade 3 or 4 toxicity [19]. Finally, decitabine administered at a dose of $20 \, \text{mg/m}^2$ by a 1-hour IV for 5 consecutive days of a 4-week cycle was approved by the European Medicines Agency (EMA) for treatment in adult patients aged ≥ 65 years with de novo or secondary AML who are not candidates for standard induction therapy in 2012. Ongoing trials are focused on further defining subgroups of elderly AML patients who may derive more benefit from decitabine and combining it with other agents for AML.

On the contrary, the therapeutic value of single agent decitabine in solid tumors has not been seen by far. Only toxicity test and safety evaluation were carried out in three Phase I trials. However single drug treatment of decitabine was applied to the treatment for MDS and AML. Clinical research of both drugs for elderly AML is underway, which has become a research hotspot in the world now.

2.1.3 EGCG in clinical trials

Epigallocatechin gallate (EGCG)-enriched tea drink, the double-brewed green tea (DBGT), is the most focused natural DNMT inhibitor. A Phase II study was conducted to assess its effectiveness and safety as a maintenance treatment in women with advanced stage serous or endometrioid ovarian cancer. This trial was terminated due to only 5 of the 16 women remained free of recurrence 18 months after CR [20]. Although failed in solid tumor therapy, Tait D et al. reported that Polyphenon E (epigallocatechin-3-gallate-rich) seemed to be able to delay disease progression in CLL patients at a dose of 1000 mg orally twice per day for the first 7 days of cycle 1 and then increased the dose to 2000 mg orally twice per day [21]. More research is needed to evaluate the anticancer effect of EGCG.

2.2 HISTONE DEACETYLASE INHIBITORS

Eukaryotic DNA is packed in a high-level structure called chromatin, resulting from the assembly of an elementary unit, the nucleosome, an octameric structure obtained from eight proteins called histones. The interactions between the DNA and the histones terminal tails control the activation or repression of gene transcription and several chemical modifications can change the status of histones with impact on gene transcription. In particular the acetylation of lysine residues found in histones is equilibrated by two enzymes: the histone acetyl transferases (HAT) and the histone deacetylases (HDACs) [22]. Acetylation of histones in chromatin is one mechanism involved in the epigenetic alterations of gene expression and is tightly controlled by the balance of these two enzymes. HDAC inhibitor (HDACi) activities resulted in an increase of the acetylated level of histones, promoting in turn the re-expression of silenced regulatory genes. HDACis also have immunomodulatory activity that may contribute to mediating their anticancer effects. Furthermore, in contrast to most cancer therapy agents, HDACis can induce the death of transformed cells in both proliferative and non-proliferative phases of cell cycle [23]. The mechanisms of action of HDACi are complex and not completely clear. The term "HDAC inhibitors" is commonly used for compounds that target the classical class I (HDAC1, 2, 3, and 8), II (HDAC4, 5, 6, 7, 8, 9, and 10), and IV (HDAC11). A number of structurally diverse HDACis have been identified, many of which are or derived from natural products. HDACi includes a variety of compounds belonging to several structural classes: hydroxamic acids (trichostatin A, panobinostat, or vorinostat), short-chain fatty acids (butyric acid and valproic acid), benzamides, and cyclic peptides (trapoxin and depsipeptide) [24] (Table 23.1). Some of them are currently being evaluated in clinical trials. In this part we will present and preliminarily discuss the results obtained from clinical trials that used HDACis as single agent therapy in solid cancers and hematological malignancies.

2.2.1 Hydroxamic acids

Hydroxamic acids are the most widely explored class of HDACis. Essential characteristics of hydroxamic acid-based molecules are the polar hydroximic group, a four–six carbon hydrophobic methylene spacer (CU, polar connection unit), a second polar site, and a terminal hydrophobic group. Present in most HDACis, the CU can interact with amino acids in the tunnel, and a four- or six-carbon unit hydrophobic spacer (linker), allowing the following zinc-binding group to reach and complex with the zinc ion inhibiting the enzyme [25]. Suberoylanilide hydroxamic acid (SAHA, Vorinostat), the prototype of this kind of HDACi, efficiently inhibits HDAC1, 2, 3, 4, 6, 7, and 9, and displays lower potency against HDAC8 [26]. At the molecular level the antiproliferative effects of SAHA involve the accumulation of acetylated. histones, resulting in transcriptional activation of p21, c/ebpα, rarα, and e-cadherin, and decrease of cyclin B1, c-myc, and cyclin D1 levels (independent of the active β-catenin pathway), leading to induction of apoptosis, G2/M cell cycle arrest, and cell differentiation [27–30]. Trichostatin A (TSA) was the first natural product discovered to display HDACi activity [31,32]. TSA inhibits HDAC1, 2, and 3, and HDAC4, 6, 7, and 9 HDACs at single-digit nanomolar level, being less efficient against HDAC8 [33,34]. Data from a wide spectrum of studies suggest that TSA causes apoptosis in tumor cells, whereas in normal cells it predominantly arrests cell cycle progression. Besides SAHA and TSA, additional hydroxamate-based compounds have been designed, such as Belinostat (PXD-101), Panobinostat (LBH589), and Dacinostat (LAQ824).

As the first approved drugs of HDACis, vorinostat has been studied in a variety of clinical trials. The most powerful effect was obtained in refractory cutaneous T-cell lymphoma (CTCL). In a Phase IIb multicenter trial of vorinostat in patients with persistent, progressive, or treatment refractory CTCL, the objective response rate was 29.7% overall, and 29.5% in patients with stage IIB or higher disease [35]. Vorinostat also showed efficiency in other hematological malignancies. In a Phase II study reported in 2011, 35 patients with relapsed/refractory follicular lymphoma (FL), marginal zone lymphoma (MZL), or mantle cell lymphoma (MCL) were treated with oral vorinostat administered at a dose of 200 mg twice daily. ORR was 29% (five complete responses and five partial responses) [36]. On the contrary, there remained a lack of activity of vorinostat for some solid tumors. Totally 95 patients were treated in trials for prostate cancer, head and neck cancer, ovarian cancer, and NSCLC. Only two partial and no complete responses occurred in these heavily pretreated patients [37–40]. However, in a Phase II study of vorinostat monotherapy in patients with GBM, 66 patients who had received ≤ 1 prior chemotherapy regimen for progressive/recurrent glioblastoma multiforme (GBM), and who were not undergoing surgery, were treated with 200 mg vorinostat bid on Days 1–14 every 3 weeks [41]. The primary efficacy endpoint was met; 9 of the first 52 patients were progression-free at 6 months. This unexpected result suggested that the dose and schedule of vorinostat employed in this Phase II trial had a biologic effect on glioblastoma tumors, may concern the ability of vorinostat to cross the blood–brain barrier, affecting target pathways in GBM. The authors of this study concluded that vorinostat has single-agent activity in GBM and is well tolerated.

Panobinostat (LBH589) is a structurally novel cinnamic hydroxamic acid analogue, and both intravenous and oral formulations are being investigated. Based on the hypothesis that leukemic cells might require an extended dosing period for disease control, a two-arm, dose-escalation Phase IA/II study in patients with advanced hematologic malignancies was initiated with a 7-consecutive-day dose schedule of intravenous panobinostat. Fifteen patients were treated but asymptomatic grade 3 QT interval corrected with Fridericia's formula (QTCF) prolongation was reported in four patients, resulting in premature discontinuation of the study, and all subsequent studies have utilized an intermittent dosing schedule with minimal cardiac effects observed [42].

2.2.2 Short-chain fatty acids

The mechanism of action of short-chain fatty acids consists in a zinc-binding function of the carboxylic group or in a competition of the same group with the acetate released in the deacetylation reaction, with the carboxylic group occupying the acetate escaping tunnel [43]. The only drug in clinical trials of this group is valproic acid (VPA); VPA inhibits class I/IIa HDACs in the low millimolar/submillolar range, being weakly active against class IIb HDACs [44]. Unlike other short-chain fatty acids, VPA seems to act more efficiently inducing differentiation in carcinoma cells, transformed hematopoietic progenitor cells, and leukemic blasts from acute myeloid leukemia (AML) patients [45]. Due to the weak inhibitory of HDAC, VPA is currently used mostly in combination with other drugs. No Phase II monotherapy trial of VPA has been reported.

2.2.3 Benzamides

Belinostat (PXD101) is a hydroxamic acid derivative, which has been administered as an infusion on days 1 to 5 of a 21-day cycle in a Phase I study in patients with advanced B-cell malignancies refractory to standard therapy [46]. A benzamide derivative, Ms-275 (entinostat), inhibits HDACs by binding the zinc-activated catalytic site. Entinostat has been investigated in patients with advanced refractory acute leukemias, mainly AML [47]. Fatigue and gastrointestinal symptoms were reported: however, no CR or PR was seen, despite 12 patients having a transient reduction in peripheral blood blasts.

2.2.4 Cyclic peptides

Cyclic peptides are a group of HDACis composed of cyclic tetrapeptide containing a 2-amino-8-oxo-9, 10-epoxy-decanoyl (AOE) moiety and cyclic peptides without AOE moiety. Romidepsin (FR901228/FK228, Istodax) is a cyclic peptide without the AOE moiety, isolated from *Chromobacterium viola-ceum*. Romidepsin predominantly inhibits HDAC1 and 2 and weakly HDAC4. Romidepsin induces differentiation, growth arrest, and apoptosis as well as inhibiting metastasis and angiogenesis [48–51]. A dose-escalation study in CLL and AML patients was done by Byrd and colleagues, with the aim of achieving an in vivo dose that increased acetylation of histone proteins H3 and H4 by 100% in vitro [52]. Although no formal CR or PR was seen in 10 CLL patients, antitumor activity was noted. Of 10 patients with AML none achieved CR or PR, although one patient experienced tumor lysis syndrome. Another study with 11 AML/ MDS patients had one CR, with SD in six patients [53]. In 2001, responses in four patients with T-cell lymphoma were reported in a Phase I trial conducted at the National Cancer Institute [54]. Bates et al. reported the final results of 71 patients with CTCL treated on the multicenter NCI study of romidepsin administered as a 4-hour infusion on days 1, 8, and 15 of a 28-day cycle with a starting dose of $14\,mg/m^2$ [55]. The ORR was 34%, with a CR observed in four patients, a PR in 20, and SD in 26. The duration of responses improved with increased depth of response and the median time to progression for patients with a major response (CR or PR) was 15.1 months. CR was achieved even in patients with Sézary syndrome. Favorable responses in CTCL also have been confirmed in a European-US study of 96 patients, with remarkably similar outcomes to that of Bates and colleagues [56]. The ORR was 32%, as measured by SWAT, with CR observed in six patients. One recent Phase II trial of the HDACI romidepsin in androgen-independent prostate cancer, although well tolerated, likewise showed minimal antineoplastic activity [57].

2.3 HISTONE ACETYLATION TRANSFERASE (HAT) INHIBITORS

The only HAT inhibitor in clinical trials is curcumin, the major curcuminoid of turmeric (*Curcuma longa*). The first indication of curcumin's anticancer activities in human participants was reported in 1987. In a clinical trial involving 62 patients with external cancerous lesions, topical curcumin was found to produce remarkable symptomatic relief as evidenced by reductions in smell, itching, lesion size, and pain. Although the effect continued for several months in many patients, only one patient had an adverse reaction [58]. Since then, curcumin, either alone or in combination with other agents, has demonstrated potential against colorectal cancer (CRC), pancreatic cancer, breast cancer, prostate cancer, multiple myeloma, lung cancer, oral cancer, and head and neck squamous cell carcinoma (HNSCC).

3 COMBINATION THERAPY

3.1 IN COMBINATION WITH OTHER ONCOLOGY DRUGS

Four epigenetic derepressive agents are now FDA approved for two hematologic malignancies, MDS treatment with DNMTIs azacitidine and decitabine, and cutaneous T-cell lymphoma therapy using the HDACis vorinostat and romidepsin; moreover, other malignancies such as peripheral T-cell lymphoma and Hodgkin's disease will likely gain approval for monotherapy DNMTis and HDACi. However, single-agent clinical studies of various solid tumors have proved fairly disappointing. For the latter, epigenetic drugs will likely prove more beneficial when combined with long-established approaches such as conventional cytotoxic chemotherapies, endocrine therapies, molecular targeted therapy, and radiotherapy.

3.1.1 Combination of DNMTis with other oncology drugs

While single-agent azacitidine or decitabine demonstrated significant efficacy for AML and MDS, solid tumor studies have been fairly disappointing, causing studies of DNMTis in combination with other conventional agents.

The combination of azacitidine and donor lymphocyte infusions (DLI) was recommended as first salvage therapy for relapse after allogeneic transplantation in a Phase II study of 28 AML patients. Overall response rate was 30%, including seven complete remissions (CRs, 23%) and two partial remissions (7%). Five patients remain in CR for a median of 777 days (range 461–888). Patients with MDS or AML with myelodysplasia-related changes were more likely to respond ($P=0.011$), and a lower blast count ($P=0.039$) as well as high-risk cytogenetics ($P=0.035$) correlated with the likelihood to achieve CR [59]. Forty-three AML patients treated with azacitidine and sorafenib (small molecule tyrosine kinase inhibitor) also responded in a encouraging rate (46%), including 10 (27%) complete response with incomplete count recovery (CRi), 6 (16%) complete responses (CR), and 1 (3%) partial response [60].

The same as single agent therapy, early combination trials of decitabine also turned out to be a failure, because of its exorbitant dosages. In a Phase II study of NSCLC, a maximum tolerated decitabine dose of $67\,mg/m^2$, given concurrently with $33\,mg/m^2$ cisplatin over a 2-h period for 3 consecutive days of a 21-day cycle, resulted in no objective responses and significant hematologic toxicity [61]. Similarly, a Phase II trial of squamous cell cervical cancer, administering decitabine at $50\,mg/\,m^2/$ day for 3 days, concurrent with $30\,mg/m^2$ cisplatin, resulted in one patient death [62]. Luckily, based on the better understanding of its hypomethylating effects, and optimization of dose schedules, newer trials have examined lower doses of decitabine in various combined regimens. First, there seems to be no improvement. In a Phase I/II ovarian cancer study combinating decitabine with carboplatin, no significant improvement over carboplatin alone was obtained [63], while a separate Phase IIa clinical trial of azacitidine and carboplatin resulted in one complete, three partial, and 10 stable disease responses.

[64]. However, the combination of decitabine and temozolomide, which is widely used for chemotherapy of metastatic melanoma, was proved safe and led to 18% ORR and 61% clinical benefit rate (CR + PR + SD) in 35 metastatic melanoma patients [65]. In addition, a 13-patient AML Phase I study, decitabine combined with arsenic trioxide and/or ascorbic acid resulted in one complete remission and five patients with stable disease [66].

3.1.2 Combination of HDACis with other oncology drugs

Like DNMTIs, despite successful studies of hematologic malignances, solid tumor clinical trials of monotherapy HDACis suggest similarly limited clinical activity. It is now widely believed that these agents will be most effective in combination with conventional chemotherapies. A Phase II study of the HDACI vorinostat combined with the antiestrogen tamoxifen, in hormone-refractory breast cancer patients, yielded a clinical benefit rate (response or stable disease for over 24 weeks) of 40%, although toxicity necessitated dose adjustment in several patients [67]. Similarly, a 12-Patient Phase I trial combining the HDACI panobinostat with the angiogenesis inhibitor bevacizumab resulted in three partial responses and seven cases of stable disease [68]. Daud et al. also conducted a Phase I/II trial of valproic acid and topoisomerase I inhibitor Karenitecin in patients with stage IV melanoma. The results showed that the addition of HDACi appeared to be a promising modality to increase sensitivity to topoisomerase I inhibition in preclinical models in melanoma [69]. Based on the fact that most patients were receiving their second, third, or fourth line of chemotherapy and progressing after chemotherapy, the antitumor effects of HDACi combined with other drugs were probably achieved by overcoming resistance to chemotherapy. Compared with single-arm trials, evidence obtained from properly randomized controlled trials are of a higher quality. In a 2011 published double-blind, placebo-controlled, randomized Phase III trial, Coronel et al. compared hydralazine valproate versus placebo when added to the cisplatin topotecan regimen in cervical cancer [70]. The results demonstrated a significant advantage in progression-free survival for epigenetic therapy over that of the current standard combination chemotherapy. These results were supported by another randomized, double-blinded, placebo-controlled Phase II trial [71], which suggested that vorinostat enhances the efficacy of carboplatin and paclitaxel in patients with advanced NSCLC. Results of a randomized, Phase II, double-blind study of exemestane, with or without entinostat, in postmenopausal women with locally recurrent or metastatic estrogen receptor-positive breast cancer progressing on a nonsteroidal aromatase inhibitor was also a large positive trial with 130 patients enrolled. It confirmed positive results of entinostat in combination with antihormone therapy in breast cancers [72]. However, Richards et al. revealed that gemcitabine plus oral HDACi CI-994 offered no advantage over gemcitabine alone in the treatment of patients with advanced pancreatic cancer in a Phase II randomized, double-blind, placebo-controlled, multicenter study [73]. These conflicting results call for further high-quality trials.

3.2 IN COMBINATION WITH OTHER EPIGENETIC AGENTS

3.2.1 In hematologic malignancies

DNA methylation and modification of histone tails both contribute to the regulation of gene expression. Preclinical studies support the point that pharmacologic targeting of both DNMT and HDAC may cause interrelated anticancer activity. Recently, a series of Phase I–II studies have reported encouraging results. Garcia-Manero et al. treated 54 patients (AML or high-risk MDS) with a modified dose of decitabine (at 15 mg/m², IV for 10 days) administered concomitantly with increasing doses of VPA (orally,

10 days). Dosage changed to 50 mg/kg/day for Phase II part of the study. The overall response rate (CR plus PR) was 22%, including a 19% CR rate. Survival was 15.3 months in responders [74]. Parallel results were observed in a study on azacitidine in combination with VPA and All Trans Retinoic Acid (ATRA) which had already been the first-line therapy for acute promyelocytic leukemia. [75]. This study had a Phase I/II design as well with azacitidine administered at a fixed dose of 75 mg/m^2/day for 7 days, and VPA was again concomitantly dose escalated (orally, 7 days). ATRA was given at a dose of 45 mg/m^2/day (days 3–7). On the 7-day schedule as well, a dose of 50 mg/kg was the maximum tolerated dose for VPA. Fifty-three patients with AML or high-risk MDS were included. Overall response rate was 42%, including 22% CRs, 5% CRps, and 13% bone marrow responses. Dose-limiting toxicity was neurotoxicity in both trials. Evidence for a beneficial effect of VPA in this combination can be derived from a higher response rate in patients with higher VPA levels in the second study. In the first study, a trend for a higher response rate was observed in patients with higher VPA levels in the subgroup of untreated patients. There also was an association of responses with higher VPA doses, and responses were attained already after a median of only one cycle, which is earlier than expected with decitabine monotherapy. A smaller Phase I study on decitabine plus VPA in 25 AML patients could not verify this beneficial effect of VPA, although responses appeared to occur earlier with the combination treatment versus single agent decitabine as well. In this trial, 14 patients received decitabine alone to determine the optimal biologic dose, which was 20 mg/m^2/day (days 1–10). Only 11 patients received the combination with dose-escalating VPA (days 5–21). Dose-limiting encephalopathy already occurred in two patients at 25 mg/kg/day. Responses included two CRs, two CRis, and two PRs (ORR 54%), but the authors conclude that VPA might be associated with too much toxicity in this elderly patient population [76]. Two further studies were conducted with panobinostat plus azacitidine. In a pilot trial, 10 patients (eight AML, two MDS) were treated with azacitidine (75 mg/m^2, 7 days) followed by 5 days of sodium panobinostat (200 mg/kg) [77]. Five patients achieved a clinical response (PR or SD) and one patient received subsequent allogeneic stem cell transplantation. Another study by Gore et al. investigated the optimal dosing schedule for azacitidine in this combination [78]. The overall response rate was 38% (11/29), but it was 56% (5/9) and 50% (3/6) in the dose cohorts receiving prolonged azacitidine schedules (50 mg/m^2/day for 10 days and 25 mg/m^2 for 14 days, respectively). Major toxicity for panobinostat was neurotoxicity in both studies.

3.2.2 In solid tumors

Unlike the striking activity of combination therapy in some hematologic malignancies, the outcomes of multi-agents trials in solid tumors are conflicted. Researchers added inhibitors of DNA methylation hydralazine plus the HDACi magnesium valproate to chemotherapy in three studies [79–81]. The response and disease stabilization rates observed suggest that hydralazine and valproate overcome epigenetic changes, increasing the efficacy of antitumor therapy.

4 CONCLUSION AND FUTURE PERSPECTIVES

The recent acknowledgment of cancer as an epigenetic disease that aberrant epigenetic changes lead to inactivation of pivotal genes involved in correct cell growth, contributing to cancer development and progression, has resulted in several initiatives implementing epigenetics in cancer treatment. The reversibility of these processes has led to the development of so-called epigenetic drugs. In that scope, both demethylating and histone acetylating drugs have been recognized as potential therapeutics. Several reports have reported on the in vitro and in vivo anticancer potential of these epigenetic drugs. These positive results have been extended to cancer clinical trials and made a series of achievements. Four epigenetic drugs have been FDA approved for treatment of hematopoietic malignancies by now.

Although epigenetic drugs in monotherapy treatment of solid tumors showed limited efficacy, combinational epigenetic therapies for the treatment of cancers after chemotherapy provided clinical benefit, suggesting that chemotherapy might cause epigenetic alterations that are modulated by epigenetic drugs. Nevertheless, this modulation is weak and subtle, which might be the reason why antitumor efficacy only existed in combination therapy. Ongoing Phase II trials are launched for expanding applications of traditional epigenetic drugs like using low dose azacitidine and/or decitabine in the therapy of elderly AML, large B-cell lymphoma and CML, while newly Phase I trials busy evaluating safety and efficacy of administration schedule for newly epigenetic agents such as lysine demethylases inhibitors, lysine methyltransferase inhibitors, and bromodomain containing proteins [82–85]. With the development of new techniques and the knowledge of epigenetic mechanisms, better application of these drugs is expectable. Braun et al. used gene detection technology in their trials of decitabine in chronic myelomonocytic leukemia patients and successfully predicted the improved overall survival [86]; we have reasons to believe that examination of related gene expression might be helpful for better analysis of experimental results and will exert more important role in further epigenetic clinical trials. There might also be applications of epigenetic drugs as radiosensitizers. In preclinical studies, HDACis act as radiation protectants of normal tissue. Demethylating drugs also hold potential as radiosensitizers. Clinical investigations on demethylating agents in combination with irradiation are lacking but seem promising. One step further may be the combination of demethylating agents plus HDACis and irradiation.

REFERENCES

[1] Jones PA, Baylin SB. The epigenomics of cancer. Cell 2007;128(4):683–92.
[2] Bhalla KN. Epigenetic and chromatin modifiers as targeted therapy of hematologic malignancies. J Clin Oncol 2005;23(17):3971–93.
[3] Goll MG, Bestor TH. Eukaryotic cytosine methyltransferases. Annu Rev Biochem 2005;74:481–514.
[4] Epigenetics SM. DNA methylation, and chromatin modifying drugs. Annu Rev Pharmacol Toxicol 2009;49:243–63.
[5] Amatori S, Bagaloni I, Donati B, Fanelli M. DNA Demethylating Antineoplastic Strategies A Comparative Point of View. Genes Cancer 2010;1(3):197–209.
[6] Lu Q, Wu A, Richardson BC. Demethylation of the same promoter sequence increases CD70 expression in lupus T cells and T cells treated with lupus-inducing drugs. J Immunol 2005;174(10):6212–9.
[7] Nandakumar V, Vaid M, Katiyar SK. (−)-Epigallocatechin-3-gallate reactivates silenced tumor suppressor genes, Cip1/p21 and p16INK4a, by reducing DNA methylation and increasing histones acetylation in human skin cancer cells. Carcinogenesis 2011;32(4):537–44.
[8] Gu B, Ding Q, Xia G, Fang Z. EGCG inhibits growth and induces apoptosis in renal cell carcinoma through TFPI-2 overexpression. Oncol Rep 2009;21(3):635.
[9] Notari RE, Deyoung JL. Kinetics and mechanisms of degradation of the antileukemic agent 5-azacytidine in aqueous solutions. J Pharm Sci 1975;64(7):1148–57.
[10] Jones PA, Taylor SM. Cellular differentiation, cytidine analogs and DNA methylation[J]. Cell 1980;20(1):85–93.

[11] Cheson BD, Bennett JM, Kantarjian H, Pinto A, Schiffer CA, Nimer SD, et al. Report of an international working group to standardize response criteria for myelodysplastic syndromes. Blood 2000;96(12):3671–4.

[12] Al-Ali HK, Jaekel N, Junghanss C, Maschmeyer G, Krahl R, Cross M, et al. Azacitidine in patients with acute myeloid leukemia medically unfit for or resistant to chemotherapy: a multicenter phase I/II study. Leuk Lymphoma 2012;53(1):110–7.

[13] Fenaux P, Mufti GJ, Hellström-Lindberg E, Santini V, Gattermann N, Germing U, et al. Azacitidine prolongs overall survival compared with conventional care regimens in elderly patients with low bone marrow blast count acute myeloid leukemia. J Clin Oncol 2010;28(4):562–9.

[14] Malik A, Shoukier M, Garcia-Manero G, Wierda W, Cortes J, Bickel S, et al. Azacitidine in fludarabine-refractory chronic lymphocytic leukemia: a phase II study. Clin. Lymphoma Myeloma Leuk 2013;13(3):292–5.

[15] Weiss AJ, Metter GE, Nealon TF, Keanan JP, Ramirez G, Swaiminathan A, et al. Phase II study of 5-azacytidine in solid tumors[J]. Cancer Treat Rep 1976;61(1):55–8.

[16] Yoo CB, Jones PA. Epigenetic therapy of cancer: past, present and future. Nat Rev Drug Discov 2006;5(1):37–50.

[17] Jabbour E, Issa JP, Garcia-Manero G, Kantarjian H. Evolution of decitabine development. Cancer 2008:2341–51.

[18] Cashen AF, Schiller GJ, O'Donnell MR, DiPersio JF. Multicenter, phase II study of decitabine for the first-line treatment of older patients with acute myeloid leukemia. J Clin Oncol 2010;28(4):556–61.

[19] Lübbert M, Rüter BH, Claus R, Schmoor C, Schmid M, Germing U, et al. A multicenter phase II trial of decitabine as first-line treatment for older patients with acute myeloid leukemia judged unfit for induction chemotherapy. Haematologica 2012;97(3):393–401.

[20] Trudel D, Labbé DP, Araya-Farias M, Doyen A, Bazinet L, Duchesne T, et al. A two-stage, single-arm, phase II study of EGCG-enriched green tea drink as a maintenance therapy in women with advanced stage ovarian cancer. Gynecol Oncol 2013;131(2):357–61.

[21] Shanafelt TD, Call TG, Zent CS, LaPlant B, Bowen DA, Roos M, et al. Phase I trial of daily oral Polyphenon E in patients with asymptomatic Rai stage 0 to II chronic lymphocytic leukemia. J Clin Oncol 2009;27(23):3808–14.

[22] Bertrand P. Inside HDAC with HDAC inhibitors. Eur J Med Chem 2010;45(6):2095–116.

[23] Burgess A, Ruefli A, Beamish H, Warrener R, Saunders N, Johnstone R, et al. Histone deacetylase inhibitors specifically kill nonproliferating tumour cells. Oncogene 2004;23(40):6693–701.

[24] Carey N, La Thangue NB. Histone deacetylase inhibitors: gathering pace. Curr Opin Pharmacol 2006;6(4):369–75.

[25] Mai A, Massa S, Pezzi R, Valente S, Loidl P, Brosch G. Synthesis and Biological Evaluation of 2-, 3-, and 4-Acylaminocinnamyl-Nhydroxyamides as Novel Synthetic HDAC Inhibitors. Med Chem 2005;1(3):245–54.

[26] Khan O, La Thangue NB. Drug Insight: histone deacetylase inhibitor-based therapies for cutaneous T-cell lymphomas. Nat Clin Pract Oncol 2008;5(12):714–26.

[27] Marks PA. Discovery and development of SAHA as an anticancer agent. Oncogene 2007;26(9):1351–6.

[28] Zhang C, Richon V, Ni X, Talpur R, Duvic M. Selective induction of apoptosis by histone deacetylase inhibitor SAHA in cutaneous T-cell lymphoma cells: relevance to mechanism of therapeutic action. J Invest Dermatol 2005;125(5):1045–52.

[29] Huang L, Sowa Y, Sakai T, Ke CC, Chen FD, Wang HE, et al. Activation of the p21 WAF1/CIP1 promoter independent of p53 by the histone deacetylase inhibitor suberoylanilide hydroxamic acid (SAHA) through the Sp1 sites. Oncogene 2000;19(50).

[30] Li D, Marchenko ND, Moll UM. SAHA shows preferential cytotoxicity in mutant p53 cancer cells by destabilizing mutant p53 through inhibition of the HDAC6-Hsp90 chaperone axis. Cell Death Differ. 2011;18(12):1904–13.

[31] Tsuji N, Kobayashi M, Nagashima K. A new antifungal antibiotic, trichostatin. J Antibiot (Tokyo) 1976;29(1):1–6.

[32] Tsuji N, Kobayashi M, Trichostatin C. A glucopyranosyl hydroxamate. J Antibiot (Tokyo) 1978;31(10):939–44.

[33] Vanhaecke T, Papeleu P, Elaut G, Rogiers V. Trichostatin A-like hydroxamate histone deacetylase inhibitors as therapeutic agents: toxicological point of view. Curr Med Chem 2004;11(12):1629–43.

[34] Khan ANH, Gregorie CJ, Tomasi TB. Histone deacetylase inhibitors induce TAP, LMP, Tapasin genes and MHC class I antigen presentation by melanoma cells. Cancer Immunol Immunother 2008;57(5):647–54.

[35] Olsen EA, Kim YH, Kuzel TM, Pacheco TR, Foss FM, Parker S, et al. Phase IIb multicenter trial of vorinostat in patients with persistent, progressive, or treatment refractory cutaneous T-cell lymphoma. J Clin Oncol 2007;25(21):3109–15.

[36] Kirschbaum M, Frankel P, Popplewell L, Zain J, Delioukina M, Pullarkat V, et al. Phase II study of vorinostat for treatment of relapsed or refractory indolent non-Hodgkin's lymphoma and mantle cell lymphoma. J Clin Oncol 2011;29(9):1198–203.

[37] Modesitt SC, Sill M, Hoffman JS, Bender DP, Gynecologic Oncology Group A phase II study of vorinostat in the treatment of persistent or recurrent epithelial ovarian or primary peritoneal carcinoma: a Gynecologic Oncology Group study. Gynecol Oncol 2008;109(2):182–6.

[38] Vansteenkiste J, Van Cutsem E, Dumez H, Chen C, Ricker JL, Randolph SS, et al. Early phase II trial of oral vorinostat in relapsed or refractory breast, colorectal, or non-small cell lung cancer. Invest New Drugs 2008;26(5):483–8.

[39] Woyach JA, Kloos RT, Ringel MD, Arbogast D, Collamore M, Zwiebel JA, et al. Lack of therapeutic effect of the histone deacetylase inhibitor vorinostat in patients with metastatic radioiodine-refractory thyroid carcinoma. J Clin Endocrinol Metab 2009;94(1):164–70.

[40] Blumenschein Jr GR, Kies MS, Papadimitrakopoulou VA, Lu C, Kumar AJ, Ricker JL, et al. Phase II trial of the histone deacetylase inhibitor vorinostat (Zolinza™, suberoylanilide hydroxamic acid, SAHA) in patients with recurrent and/or metastatic head and neck cancer. Invest New Drugs 2008;26(1):81–7.

[41] Kantarjian HM, Thomas XG, Dmoszynska A, Wierzbowska A, Mazur G, Mayer J, et al. Multicenter, randomized, open-label, phase III trial of decitabine versus patient choice, with physician advice, of either supportive care or low-dose cytarabine for the treatment of older patients with newly diagnosed acute myeloid leukemia. J Clin Oncol 2012;30(21):2670–7.

[42] Giles F, Fischer T, Cortes J, Garcia-Manero G, Beck J, Ravandi F, et al. A phase I study of intravenous LBH589, a novel cinnamic hydroxamic acid analogue histone deacetylase inhibitor, in patients with refractory hematologic malignancies. Clin Cancer Res 2006;12(15):4628–35.

[43] Mai A, Altucci L. Epi-drugs to fight cancer: From chemistry to cancer treatment, the road ahead. Int J Biochem Cell Biol 2009;41(1):199–213.

[44] De Felice L, Tatarelli C, Mascolo MG, Gregorj C, Agostini F, Fiorini R, et al. Histone deacetylase inhibitor valproic acid enhances the cytokine-induced expansion of human hematopoietic stem cells. Cancer Res 2005;65(4):1505–13.

[45] Raffoux E, Cras A, Recher C, Boëlle PY, de Labarthe A, Turlure P, et al. Phase 2 clinical trial of 5-azacitidine, valproic acid, and all-trans retinoic acid in patients with high-risk acute myeloid leukemia or myelodysplastic syndrome. Oncotarget 2010;1(1):34–42.

[46] Gimsing P, Hansen M, Knudsen LM, Knoblauch P, Christensen IJ, Ooi CE, et al. A phase I clinical trial of the histone deacetylase inhibitor belinostat in patients with advanced hematological neoplasia. Eur J Haematol 2008;81(3):170–6.

[47] Gojo I, Jiemjit A, Trepel JB, Sparreboom A, Figg WD, Rollins S, et al. Phase 1 and pharmacologic study of Ms-275, a histone deacetylase inhibitor, in adults with refractory and relapsed acute leukemias. Blood 2007;109:2781–90.

[48] Furumai R, Matsuyama A, Kobashi N, Lee KH, Nishiyama M, Nakajima H, et al. FK228 (depsipeptide) as a natural prodrug that inhibits class I histone deacetylases. Cancer Res 2002;62(17):4916–21.

[49] Bertino EM, Otterson GA. Romidepsin: a novel histone deacetylase inhibitor for cancer. Expert Opin Investig Drugs 2011;20(8):1151–8.

[50] Sato N, Ohta T, Kitagawa H, Kayahara M, Ninomiya I, Fushida S, et al. FR901228, a novel histone deacety-lase inhibitor, induces cell cycle arrest and subsequent apoptosis in refractory human pancreatic cancer cells. Int J Oncol 2004;24(3):679–85.

[51] VanOosten RL, Moore JM, Ludwig AT, Griffith TS. Depsipeptide (FR901228) enhances the cytotoxic activity of TRAIL by redistributing TRAIL receptor to membrane lipid rafts. Mol Ther 2005;11(4): 542–52.

[52] Byrd JC, Marcucci G, Parthun MR, Xiao JJ, Klisovic RB, Moran M, et al. A phase 1 and pharmacody-namic study of depsipeptide (FK228) in chronic lymphocytic leukemia and acute myeloid leukemia. Blood 2005;105(3):959–67.

[53] Klimek VM, Fircanis S, Maslak P, Guernah I, Baum M, Wu N, et al. Tolerability, pharmacodynamics, and pharmacokinetics studies of depsipeptide (romidepsin) in patients with acute myelogenous leukemia or advanced myelodysplastic syndromes. Clin Cancer Res 2008;14(3):826–32.

[54] Piekarz RL, Robey R, Sandor V, Bakke S, Wilson WH, Dahmoush L, et al. Inhibitor of histone deacetylation, depsipeptide (FR901228), in the treatment of peripheral and cutaneous T-cell lymphoma: a case report. Blood 2001;98(9):2865–8.

[55] Bates S., Piekarz R., Wright J. Final clinical results of a phase 2 NCI multicenter study of romidepsin in recurrent cutaneous T-cell lymphoma (molecular analyses included). ASH Annual Meeting Abstracts 2008, 112: 1568.

[56] Kim Y., Whittaker S., Demierre M.F., Lerner A., Duvic M., Reddy S., et al. Clinically significant responses achieved with romidepsin in treatment-refractory cutaneous T-cell lymphoma: final results from a phase 2B, international, multicenter, registration study. ASH Annual Meeting Abstracts 2008;112(11):263.

[57] Molife LR, Attard G, Fong PC, Karavasilis V, Reid AH, Patterson S, et al. Phase II, two-stage, single-arm trial of the histone deacetylase inhibitor (HDACi) romidepsin in metastatic castration-resistant prostate cancer (CRPC). Ann Oncol 2010;21(1):109–13.

[58] Kuttan R, Sudheeran PC, Josph CD. Turmeric and curcumin as topical agents in cancer therapy. Tumori 1987;73(1):29–31.

[59] Schroeder T, Czibere A, Platzbecker U, Bug G, Uharek L, Luft T, et al. Azacitidine and donor lymphocyte infusions as first salvage therapy for relapse of AML or MDS after allogeneic stem cell transplantation. Leukemia 2013.

[60] Ravandi F, Alattar ML, Grunwald MR, Rudek MA, Rajkhowa T, Richie MA, et al. Phase 2 study of azacyti-dine plus sorafenib in patients with acute myeloid leukemia and FLT-3 internal tandem duplication mutation. Blood 2013;121(23):4655–62.

[61] Pohlmann P1, DiLeone LP, Cancella AI, Caldas AP, Dal Lago L, Campos Jr O, et al. A phase I trial of cis-platin plus decitabine, a new DNA-hypomethylating agent, in patients with advanced solid tumors and a follow-up early phase II evaluation in patients with inoperable non-small cell lung cancer. Invest New Drugs 2000;18(1):83–91.

[62] Pohlmann P1, DiLeone LP, Cancella AI, Caldas AP, Dal Lago L, Campos Jr O, et al. Phase II trial of cisplatin plus decitabine, a new DNA hypomethylating agent, in patients with advanced squamous cell carcinoma of the cervix. Am J Clin Oncol 2002;25(5):496–501.

[63] Issa JPJ, Gharibyan V, Cortes J, Jelinek J, Morris G, Verstovsek S, et al. Phase II study of low-dose decitabine in patients with chronic myelogenous leukemia resistant to imatinib mesylate. J Clin Oncol 2005;23(17):3948–56.

[64] Board EABEA. Randomized phase II study of decitabine in combination with carboplatin compared with carboplatin alone in patients with recurrent advanced ovarian cancer.. J Clin Oncol (Meeting Abstracts) 2009;27(15S):5562.

[65] Tawbi HA, Beumer JH, Tarhini AA, Moschos S, Buch SC, Egorin MJ, et al. Safety and efficacy of decitabine in combination with temozolomide in metastatic melanoma: a phase I/II study and pharmacokinetic analysis. Ann Oncol 2013;24(4):1112–9.

[66] Welch JS, Klco JM, Gao F, Procknow E, Uy GL, Stockerl-Goldstein KE, et al. Combination decitabine, arsenic trioxide, and ascorbic acid for the treatment of myelodysplastic syndrome and acute myeloid leukemia: a phase I study. Am J Hematol 2011;86(9):796–800.

[67] Munster PN, Lacevic M, Thomas S, Irwin D, Paroly W, Natale R, et al. Phase II trial of the histone deacetylase inhibitor, vorinostat, to restore hormone sensitivity to the antiestrogen tamoxifen in patients with advanced breast cancer who progressed on prior hormone therapy. J Clin Oncol 2009;27(15S May 20 Supplement):1075.

[68] Drappatz J, Lee EQ, Hammond S, Grimm SA, Norden AD, Beroukhim R, et al. Phase I study of panobinostat in combination with bevacizumab for recurrent high-grade glioma. J Neurooncol 2012;107(1):133–8.

[69] Daud AI, Dawson J, DeConti RC, Bicaku E, Marchion D, Bastien S, et al. Potentiation of a topoisomerase I inhibitor, karenitecin, by the histone deacetylase inhibitor valproic acid in melanoma: translational and phase I/II clinical trial. Clin Cancer Res 2009;15(7):2479–87.

[70] Coronel J, Cetina L, Pacheco I, et al. A double-blind, placebo-controlled, randomized phase III trial of chemotherapy plus epigenetic therapy with hydralazine valproate for advanced cervical cancer. Preliminary results. Med Oncol 2011;28(1):540–6.

[71] Ramalingam SS, Maitland ML, Frankel P, Trejo-Becerril C, González-Fierro A, de la Cruz-Hernandez E, et al. Carboplatin and Paclitaxel in combination with either vorinostat or placebo for first-line therapy of advanced non–small-cell lung cancer. J Clin Oncol 2010;28(1):56–62.

[72] Yardley DA1, Ismail-Khan RR, Melichar B, Lichinitser M, Munster PN, Klein PM, et al. Randomized phase II, double-blind, placebo-controlled study of exemestane with or without entinostat in postmenopausal women with locally recurrent or metastatic estrogen receptor-positive breast cancer progressing on treatment with a nonsteroidal aromatase inhibitor. J Clin Oncol 2013;31(17):2128–35.

[73] Pauer LR, Cunningham C, Williams A, Grove W, Kraker A, et al. Phase I study of oral CI-994 in combination with carboplatin and paclitaxel in the treatment of patients with advanced solid tumors. Cancer Invest 2004:886–96. 22.6.

[74] Garcia-Manero G, Kantarjian HM, Sanchez-Gonzalez B, Yang H, Rosner G, Verstovsek S, et al. Phase 1/2 study of the combination of 5-aza-2'-deoxycytidine with valproic acid in patients with leukemia. Blood 2006;108(10):3271–9.

[75] Soriano AO1, Yang H, Faderl S, Estrov Z, Giles F, Ravandi F, et al. Safety and clinical activity of the combination of 5-azacytidine, valproic acid, and all-trans retinoic acid in acute myeloid leukemia and myelodysplastic syndrome. Blood 2007;110(7):2302–8.

[76] Blum W, Klisovic RB, Hackanson B, Liu Z, Liu S, Devine H, et al. Phase I study of decitabine alone or in combination with valproic acid in acute myeloid leukemia. J Clin Oncol 2007;25(25):3884–91.

[77] Ungewickell A, Medeiros BC. Novel agents in acute myeloid leukemia. Int J Hematol 2012;96(2):178–85.

[78] Kuendgen A, Lübbert M. Current status of epigenetic treatment in myelodysplastic syndromes. Ann Hematol 2008;87(8):601–11.

[79] Candelaria M, Gallardo-Rincón D, Arce C, et al. A phase II study of epigenetic therapy with hydralazine and magnesium valproate to overcome chemotherapy resistance in refractory solid tumors. Ann Oncol 2007;18(9):1529–38.

[80] Coronel J, Cetina L, Pacheco I, Trejo-Becerril C, González-Fierro A, de la Cruz-Hernandez E, et al. A double-blind, placebo-controlled, randomized phase III trial of chemotherapy plus epigenetic therapy with hydralazine valproate for advanced cervical cancer. Preliminary results. Med Oncol 2011;28(1):540–6.

[81] Arce C, Pérez-Plasencia C, González-Fierro A, de la Cruz-Hernández E, Revilla-Vázquez A, Chávez-Blanco A, et al. A proof-of-principle study of epigenetic therapy added to neoadjuvant doxorubicin cyclophosphamide for locally advanced breast cancer. PLoS One 2006;1(1):e98.

[82] Garcia-Manero G, Jabbour E, Borthakur G, Faderl S, Estrov Z, Yang H, et al. Randomized open-label phase II study of decitabine in patients with low-or intermediate-risk myelodysplastic syndromes. J Clin Oncol 2013;31(20):2548–53.

[83] Clozel T, Yang SN, Elstrom RL, Tam W, Martin P, Kormaksson M, et al. Mechanism-based epigenetic chemosensitization therapy of diffuse large B-cell lymphoma. Cancer Discov 2013;3(9):1002–19.

[84] Phillips CL, Davies SM, McMasters R, Absalon M, O'Brien M, Mo J, et al. Low dose decitabine in very high risk relapsed or refractory acute myeloid leukaemia in children and young adults. Br J Haematol 2013;161(3):406–10.

[85] Krug U, Koschmieder A, Schwammbach D, Gerss J, Tidow N, Steffen B, et al. Feasibility of azacitidine added to standard chemotherapy in older patients with acute myeloid leukemia—a randomised SAL pilot study. PloS one 2012;7(12):e52695.

[86] Braun T, Itzykson R, Renneville A, de Renzis B, Dreyfus F, Laribi K, et al. Molecular predictors of response to decitabine in advanced chronic myelomonocytic leukemia: a phase 2 trial. Blood 2011;118(14):3824–31.

ISSUES TO OVERCOME/AREAS OF CONCERN

PART

4

ISSUES TO OVERCOME AREAS OF CONCERN

GENETIC INTRATUMOR HETEROGENEITY

24

Donat Alpar, Louise J. Barber, and Marco Gerlinger

Centre for Evolution and Cancer, The Institute of Cancer Research, London, UK

CHAPTER OUTLINE

1 INTRODUCTION: INTER- AND INTRATUMOR HETEROGENEITY

Disease outcomes can vary significantly even between patients with cancers of the same type, histological characteristics, and stage. Recent large-scale sequencing approaches of various cancer types identified a large catalogue of genetic cancer driver aberrations, including somatic copy number alterations (SCNAs), rearrangements, and point mutations. However, most individual driver genes were only altered in a small percentage of samples from the same histological cancer type and driver mutation profiles showed high inter-individual variability [1]. This genetic intertumor heterogeneity is thought to be a major reason for different outcomes of patients with otherwise similar cancers. The identification of substantial intertumor heterogeneity for most cancer types further suggests that drugs need to be targeted to the specific genetic alterations present in an individual tumor in order to maximize efficacy. These concepts are the basis of personalized or precision cancer medicine, an approach which has led to major breakthroughs in oncology. For example, patients with nonsmall cell lung cancers (NSCLCs) only benefit from epidermal growth factor receptor (EGFR) tyrosine kinase inhibitor (TKI) therapy if their tumors harbor an activating mutation in *EGFR* [2]. In colorectal cancer, anti-EGFR antibody therapy is only active in patients with *RAS* wild-type tumors whereas *RAS* mutant tumors show primary resistance [3]. Treatment stratification approaches based on these data have increased treatment

S.G. Gray (Ed): Epigenetic Cancer Therapy. DOI: http://dx.doi.org/10.1016/B978-0-12-800206-3.00024-0

efficacy and avoid unnecessary side effects from the administration of inactive therapies. However, many other attempts to increase therapeutic efficacy by stratifying therapy to the driver aberrations identified in individual cancers have failed [4] and prognostic and predictive genetic biomarkers have not been identified for the majority of tumor types.

Genetic intratumor heterogeneity (ITH), defined as the variability of genetic aberrations between cancer cells within an individual patient, has also been discovered in many cancer types. Single biopsies and analysis methods that miss potential subclonal structures in human cancers may fail to elucidate the most aggressive and drug resistant subclones in heterogeneous tumors. Treatment stratification based on single biopsies may target genetic alterations that are only present in minor subclones and this is likely to be clinically futile [5]. Thus, ITH is a potential explanation for the overall moderate success of personalized cancer medicine approaches. Considering the potential impact of ITH from the outset is critical to improve the success rate of biomarker and drug target discovery research. This review provides an overview of ITH in human cancers and identifies strategies to optimize biomarker and drug target discovery and the development of effective therapies in heterogeneous cancers.

2 GENETIC ITH AS A RESULT OF POPULATION EXPANSION

Most cancers are of monoclonal origin [6], arising from a single cell that has acquired the genetic driver aberrations for cancerous transformation. However, monoclonality is unlikely to be maintained in expanding cancer cell populations as further genetic alterations can be acquired by individual cancer cells. Genomic instability, which increases the rate at which mutations occur in the genome, may promote ITH even in small tumors whose cells have only undergone a small number of cell doublings. However, clinically detectable tumors usually contain billions of cancer cells and the background mutation rate, estimated for non-cancerous human tissues to be in the order of magnitude of 10^{-8} to 10^{-9} per cell division [7], may suffice to generate significant ITH even in newly diagnosed malignancies.

Conceptually, each somatic genetic alteration is the result of a specific cell extrinsic or intrinsic mutational process [8]. Tobacco smoke and ultraviolet radiation are examples of extrinsic influences and intrinsic mutational processes include the spontaneous deamination of methylated cytosines which contributes to the background mutation rate in mammalian genomes [9]. Genomic instability mechanisms occur frequently in cancers and enhance the generation of genomic alterations. An example of this is chromosomal instability (CIN) which is characterized by an increased rate of gain or loss of whole chromosomes or large portions of chromosomes [10]. Genomic instability can also result from smaller scale changes such as base substitutions and small nucleotide insertions or deletions. One example of this is microsatellite instability (MIN), which defines a DNA mismatch repair deficiency commonly caused by the hypermethylation-mediated inactivation of *MLH1* or germline mutations in *MLH1*, *MSH2*, *MSH6*, or *PMS2* [11]. Microsatellite instability results in a hypermutator phenotype, leading to some of the highest mutational loads observed in human cancers [8]. A recent analysis of the catalogue of somatic mutations in more than 7000 tumors across 30 different cancer types identified as many as 21 distinct mutational signatures. Some signatures were identified across a variety of cancer types whereas others were confined to a small number or a single histological cancer type. Each of these is thought to reflect the fingerprint left in the genome by a specific DNA damage or repair process operative in cancer cells and their precursors [8]. Less than half of these signatures could be attributed to well-characterized mutational processes such as MIN but the remaining mechanisms still need to be investigated.

Somatic genetic aberrations can be accumulated gradually over time or suddenly, in single catastrophic events. For example, chromothripsis describes complex structural rearrangements resulting from multiple simultaneous chromosome breaks followed by the re-ligation of fragments in the wrong order or orientation [12]. Chromoplexy also leads to complex structural aberrations through the erroneous repair of fractures occurring simultaneously in multiple chromosomes. This results in chain-like translocations which can alter several cancer genes in a single catastrophic event, representative of punctuated rather than gradual and stepwise cancer evolution [13]. These mechanisms can potentially generate multiple driver aberrations simultaneously, permitting cells to acquire novel and complex phenotypes that may not be achievable in a stepwise fashion. Kataegis describes a further process generating multiple localized single nucleotide substitutions within a short time period. Kataegis was originally found in breast cancer and is driven by the AID/APOBEC proteins [14,15].

Genomic instability can clearly contribute to cancer development [16] but the contribution to cancer progression and therapeutic resistance development is less well defined. Associations of genomic instability with poor clinical outcome in various cancer types including breast, ovarian, colon, endometrial, and lung cancers, and some types of hematological malignancies [17] suggest an ongoing influence in established cancers. In the most extreme scenario, genomic instability may lead to cancers in which no two tumor cells are exactly alike [18]. Interestingly, breast, ovarian, gastric, and NSCLC with very low or very high CIN had a better prognosis than those with intermediate instability levels [19]. These results are in line with the concept of an optimal level of genomic instability for tumor development and progression [20]. Beyond a certain level, instability may provide no further growth advantage and may even be detrimental for cancer cells. This may provide therapeutic opportunities: for example, through drugs that cause DNA damage in cancer cells, which are unable to tolerate any further increase in DNA error rates [21,22].

3 CANCER FROM AN EVOLUTIONARY PERSPECTIVE

Since initially proposed [23], substantial evidence shows that neoplasms can adapt to changing environments through the Darwinian evolution of subclones harboring advantageous mutations [24]. The generation of a genetically heterogeneous cancer cell population through new genetic aberrations is a basic requirement for evolutionary adaptation. Beneficial mutations can then expand through Darwinian selection. These two processes, mutation and selection, are commonly thought of as the key drivers of evolutionary adaptation. However, genetic drift is a third force influencing cancer evolution [25]. Drift defines changes in the frequency of an allele in a population resulting from random processes. For example, a small clone can be eliminated by untoward chance events even if it is the fittest clone within a tumor.

Although the mutation rate can vary across the cancer genome [26], mutations are essentially generated randomly. The majority of randomly generated new mutations are likely to be neutral or even deleterious and only a small proportion of mutations will establish advantageous phenotypes which can be positively selected. Importantly, the fitness effect of a genetic aberration depends on the cellular context. Interactions among genetic variants, generally referred to as epistasis, and cell-extrinsic features such as the cellular microenvironment or drug therapy determine whether a new genetic alteration provides a growth advantage or not and variation over time can promote or diminish fitness effects of existing mutations [27,28].

Based on their fitness effects and contribution to oncogenesis, somatic aberrations can be categorized as "drivers" or "passengers." Driver aberrations are positively selected in a particular environmental context and have key roles in initiating and promoting oncogenesis. In contrast, passenger mutations are not under positive selection. Passenger mutations can increase in abundance if they occur in clones which are expanding due to advantageous mutations. This so-called hitchhiking leads to large numbers of clonally expanded passenger mutations in cancer genomes. The analysis of large numbers of predominantly passenger aberrations permitted insights into the ancestral relationships of cancer subclone evolution during progression and therapy resistance [29–34], and into the temporal changes in mutation spectra during cancer progression [14,30]. Although passenger mutations support the analysis of cancer evolution, the identification of driver mutations is critical for drug target discovery and to define the relevance of ITH for patient outcomes. For example, ITH may be irrelevant for clinical management if all heterogeneous mutations are passenger mutations but personalized cancer medicine feasibility may be compromised if many of the driver aberrations within an individual cancer are heterogeneous.

Separating driver from passenger aberrations has proved challenging [1,35]. The lack of functional models to test whether an individual aberration is a driver or a passenger and the high mutational burdens of some cancer genomes, leading to excessive passenger mutation loads, are among the reasons for slow progress. Most studies apply statistical approaches to identify genes which are more frequently mutated than expected by chance, characteristic of a driver gene under positive selection [36]. Estimates suggest that thousands of samples need to be sequenced per tumor type in order to identify all driver genes which are mutated in at least 2% of samples with these statistical approaches [37], yet only a few hundred cases have been sequenced at best per tumor type to date. Furthermore, current large-scale sequencing approaches focus mainly on treatment-naïve primary tumors but therapy and metastases to different organs may lead to the selection of additional driver aberrations which are likely to be absent in untreated primary tumors. Thus, many further driver genes may remain to be discovered through the expansion of sequencing projects and inclusion of clinically more relevant specimens. Twenty to fifty percent of silent mutations in oncogenes have also been suggested to have driver roles through as yet unknown mechanisms [38], challenging the assumption that most silent mutations are passengers. Furthermore, the majority of cancer genome studies have focused on only approximately 1.5% of the genome which encodes proteins, but additional driver aberrations may affect genetic features located within the 80% of the genome which is expected to have functional roles in human cells [39]. The recent discovery of mutations in the promoter of human telomerase reverse transcriptase provides an example of driver mutations in intergenic DNA regions [40]. The tremendous progress in cancer genome analysis which started through the next-generation sequencing (NGS) revolution will most likely reveal further insights into cancer driver mutation landscapes.

Similar to asexually reproducing organisms such as microorganisms, cancer cells are unable to remove deleterious mutations from the genome through recombination. This irreversible accumulation of deleterious mutations, called Muller's ratchet [41], may gradually impair the fitness of cancer cell populations and may even lead to their extinction. Although these considerations highlight the potential negative effects of high mutation rates, it is conceivable that genomic instability could be particularly advantageous for the acquisition of phenotypes which require the alteration of multiple genes within a cell. Such cooperating mutations may be highly unlikely to occur in cancer cells through the background mutation rate alone (Figure 24.1A,B). The acquisition of mutations through single catastrophic events may be relevant if mutational alteration of single genes contributing to such polygenic phenotypes has deleterious effects and if only the combination is advantageous (Figure 24.1C). Finally, high

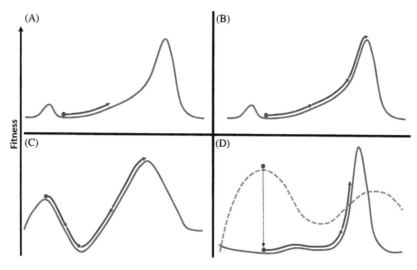

FIGURE 24.1

Evolutionary adaptation on the fitness landscape. Red arrows represent the fitness effect of individual genetic aberrations. **A**. The gain of an advantageous aberration confers a movement toward a fitness maximum. Single aberrations may fail to reach the fitness maximum if this requires multiple cooperating aberrations. **B**. Genomic instability increases the probability for multiple cooperating aberrations to co-occur within a single cell and allow cancer cells to reach these fitness peaks. **C**. Fitness maxima requiring multiple driver aberrations may be inaccessible if single aberrations occurring in isolation are detrimental and only the combination is advantageous, unless multiple drivers are acquired simultaneously in a single catastrophic event (e.g., chromothripsis or chromoplexy). **D**. Drug treatment can re-shape the fitness landscape. Cancer cell evolution toward new fitness peaks (e.g., drug-resistant phenotypes) on the altered fitness landscape could be accelerated by genomic instability. Orange broken line: fitness landscape before drug therapy; blue line: fitness landscape after drug therapy.

mutation rates may also be critical for rapid adaptation when fitness landscapes change dramatically through environmental changes (Figure 24.1D).

Subclones, which have undergone significant clonal expansion can be detected through standard NGS but the unselected population heterogeneity, including single cells harboring newly acquired genetic aberrations or small clones which have only expanded for a few generations, are usually below the detection limit. Recently developed single-cell sequencing technologies allow interrogation of population heterogeneity down to the most fundamental level of the single cell [42–44]. Whether cancer evolution is predictable based on the increasingly detailed assessments of cancer subclonal landscapes is of major interest for personalized cancer medicine endeavors. As discussed, cancer evolution is governed by three different forces: mutation generation, drift, and clonal selection. Mutation and drift are stochastic processes whereas clonal selection is deterministic in nature. The influence of stochastic elements is likely to limit the long-term predictability of evolutionary trajectories of individual cancers. However, the interrogation of clones which have already expanded to a level where deterministic selective forces dominate may allow short and medium term predictability. Assessing other fundamental population genetic parameters such as mutation rates and cancer population sizes may permit proba-bilistic approaches to estimate future events such as the acquisition of drug resistance or of cancer recurrence. Ongoing evolutionary adaptation of cancers is likely to require regular re-assessments of cancer genomic landscapes in order to maintain an up-to-date record of the subclonal cancer composi-tion. Emerging liquid biopsy techniques based on circulating tumor cells [45] and cell-free tumor DNA might facilitate this without the need for re-biopsies [46,47].

4 PATTERNS OF ITH

The evolutionary nature of cancer is likely to be a key driver of cancer progression and therapy failure. In contrast to the evolution of species in the wild, cancer evolution occurs over a relatively short time period and can be observed repeatedly in thousands of patients in order to understand the rules govern-ing cancer evolution. Studying the patterns of heterogeneity and the biology of the identified subclones will be critically important for the development of rational strategies to treat heterogeneous cancers. Solid tumors and leukemias will be considered separately as spatial heterogeneity is likely to add an additional layer of complexity to the study of the former.

4.1 ITH IN SOLID TUMORS

Multi-region exome sequencing of 10 stage II–IV clear cell renal cell carcinomas (ccRCC) found that two-thirds of somatic mutations and SCNAs were heterogeneous on average across each tumor [30,31]. Reconstruction of the ancestral relationships of the detected subclones through phylogenetic analysis revealed branched rather than linear evolutionary patterns for all 10 tumors (Figure 24.2). Mutations or hypermethylation of one copy of the *von Hippel Lindau* (*VHL*) tumor suppressor gene and loss of chro-mosome 3p, which inactivates the second copy of *VHL*, were the only driver aberrations which were present in every analyzed region of all 10 tumors. Mutations in the SWI/SNF-chromatin remodeling complex member *PBRM1* and gain of chromosome 5q and losses of 4q, 8p, and 14q were the only other driver aberrations located on the trunk of the phylogenetic tree in some cases [30] (Figure 24.2). Thus, these were the likely founder mutations of these tumors. In contrast, the vast majority of driver aberra-tions were heterogeneous and therefore located on the branches of the phylogenetic trees (Figure 24.2). These included mutations in all other common ccRCC driver genes such as *BAP1*, *SETD2*, *KDM5C*, *TP53*, *PIK3CA*, and *MTOR*, as well as driver SCNAs. Due to the spatial separation of heterogeneous subclones, these complex subclonal architectures would have been missed by single biopsies. Many individual biopsies lacked any obvious subclonal structures, generating an illusion of monoclonality despite ITH in every tumor when assessed by multi-region analysis. Importantly, the number of het-erogeneous mutations detected in most of these tumors increased with each additional biopsy that was taken [30]. Thus, an optimal number of biopsies to reliably detect the majority of somatic mutations within a ccRCC tumor could not be defined. This suggests that ccRCC subclonal architectures may be even more complex than estimated, based on the limited number of biopsies used. Assessment of signaling pathway activity and of gene expression profiles further demonstrated that driver mutation heterogeneity conferred phenotypic heterogeneity. In one tumor, regions harboring an activating muta-tion in the kinase domain of the mammalian target of rapamycin (mTOR) showed phosphorylation of the downstream mTOR pathway targets S6-kinase and 4EBP, which have been suggested as biomarkers

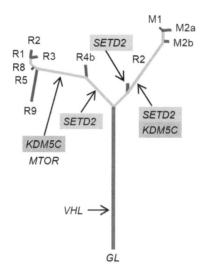

FIGURE 24.2

Phylogenetic tree reconstructed from multi-region exome sequencing data from a clear cell renal cell carcinoma. The tree shows branched evolution with multiple distinct regionally separated subclones which are evolving simultaneously. Branch lengths are proportional to the number of non-synonymous mutations separating the branching points. Driver mutations were acquired by the indicated genes in the branches the arrows point to. Driver mutations defining parallel evolution events are highlighted by colors. Tree is rooted at the "germ line" (GL) DNA sequence, determined by exome sequencing of DNA from peripheral blood. *Adapted from [30].*

for mTOR-inhibition response in ccRCC [31]. In contrast, phosphorylation was absent in cancer cells with wild-type mTOR within the same tumor. Thus, depending on the region analyzed, biomarkers would have predicted a high or a low probability of responding to an mTOR inhibitor [31]. Overall, actionable driver mutations in the PI3K-mTOR pathway, such as in *PTEN*, *PIK3CA*, or *TSC2*, were subclonal by multi-region sequencing in all six tumors where these were detected [30]. An *ARID1A* mutation, which might be targetable through synthetic lethal interactions with *ARID1B* [48], was also heterogeneous [30]. Spatial heterogeneity might also hamper the accurate identification of prognostic signatures, as demonstrated by the detection of both good and poor prognosis gene expression signatures in different regions of the primary tumor and metastatic sites from one patient [31]. Some tumors harbored multiple distinct mutations in a specific driver gene, including *BAP1*, *KDM5C*, *PIK3CA*, *PBRM1*, *PTEN*, or *SETD2* [30]. This parallel evolution of independent but functionally convergent genetic alterations within tumors suggests that the evolutionary trajectories of individual cancers may differ. What determines these trajectories remains unknown but epistatic interactions between genetic or epigenetic alterations, or selection pressure from the tumor microenvironment are possible explanations. Taken together, the frequent occurrence of branched evolution with spatially separated actionable driver mutations questions the benefit of using single biopsies for treatment stratification approaches in this tumor type. These results also raise important questions about the impact of subclonal compositions on treatment outcomes and the design of the next generation of biomarker detection approaches to reliably assess subclonality of genetic aberrations.

Genetic ITH characterized by spatially separated subclones was also identified by multi-region exome sequencing of six high-grade serous ovarian cancers. Alterations in cancer driver genes such as *PIK3CA*, *CTNNB1*, and *NF1* were heterogeneous in some patients [49] whereas five out of six cases harboring mutations in *TP53* showed identical mutations across all tumor regions from that individual. Thus, *TP53* mutations were early truncal drivers in the majority of these tumors, whereas several actionable driver mutations had been acquired as subclonal aberrations during cancer progression.

Multi-region copy number profiling of glioblastomas, the most lethal type of adult brain tumor, identified spatially heterogeneous driver SCNAs in eight of nine patients [50]. *CDKN2A/B* losses and *EGFR* amplifications were detected in the majority of these cases and were always homogenous, suggesting that they are often truncal alterations. In contrast, *CDK6*, *MET*, and *PDGFRA* amplifications and *PTEN* and *TP53* deletions were heterogeneous in some cases and truncal in others [50]. *RB1*, *AKT3*, and *MDM4* SCNAs were heterogeneous in all cases where they were identified, suggesting they usually occurred late during cancer evolution [50]. Gene expression signatures, which had been described to define glioblastoma subtypes with different survival and treatment responses, were also heterogeneous within individual tumors from 6 out of 10 patients [50]. Fluorescence in situ hybridization (FISH) analysis demonstrated heterogeneous *EGFR*, *MET*, and *PDGFRA* amplifications in distinct and intermixed cancer cell subclones within the same glioblastoma [51]. Cooperation of *EGFR* mutated and *EGFR* wild-type intratumoral glioblastoma subclones has been demonstrated, suggesting that co-dependencies may drive this pattern of intermingling subclones [52]. Such cooperating ecosystems may be vulnerable to the eradication of only one subclone, which may then impair the fitness of a dependent clone.

Neuroblastomas are aggressive pediatric cancers. Tumors with *MYCN* gene amplifications define a high-risk group with poor outcomes despite stratification to intensive therapy regimens [53,54]. FISH studies found heterogeneity of *MYCN* amplification status in 7–8% of *MYCN*-amplified neuroblastomas [55]. Although *MYCN* heterogeneity is infrequent, this may cause difficulties in therapy stratification in this group of patients. DNA content and microsatellite analyses aiming to scrutinize tumor heterogeneity in spatially or temporally separated tumors from the same neuroblastoma patient revealed genetic ITH between multiple specimens in 66% of cases, where samples were obtained without intervening chemotherapy [56]. Comparison of DNA content in paired pre- and post-chemotherapy samples detected a difference in 60% of cases, with pre-chemotherapy samples displaying co-existence of distinct near-triploid and diploid clones whereas all post-chemotherapy samples solely contained diploid clone(s). This clonal genetic evolution from triploidy toward diploidy was found to be associated with disease progression in patients with locoregional neuroblastoma [57]. Combined analysis of multiple loci, including known driver aberrations such as *MYCN* and 17q gains and 1p and 11q losses, revealed ITH in 8 out of 18 neuroblastoma patients. These findings suggest that neuroblastoma exhibits clonal heterogeneity, and clonal evolution occurs during the course of treatment and clinical progression.

Examination of SCNAs in medulloblastomas, the most common type of malignant pediatric brain tumor, identified genetic heterogeneity between each of seven matched primary and metastatic samples [58]. This involved driver aberrations: for example, *MYCN* amplification was only found in the primary tumors but not in the metastases in three cases. Thus, anti-MYCN therapeutics may lack efficacy in the metastatic compartments. Analysis of promoter hypermethylation further revealed heterogeneous methylation patterns between primary tumors and metastatic sites and *MLH1* sequencing identified heterogeneous somatic mutations in three cases.

Deep sequencing of single biopsies from 65 triple negative breast cancers found that mutations in driver genes such as *TP53*, *PIK3CA*, and *PTEN* were clonally dominant in some tumors, but they were also detected as subclonal mutations in other cases [59]. These findings are highly relevant for treatment personalization efforts: for example, the targeting of actionable *PIK3CA* mutations may be clinically futile in cases where this mutation is subclonal. Whole-genome sequencing of biopsies from 21 breast cancers of various subtypes and computational reconstruction of their evolutionary histories detected subclonal structures in all tumors [34]. Intratumoral subclones showed distinct mutation, copy number, and translocation profiles. Several aberrations including *TP53* and *PIK3CA* mutations and *ERBB2*, *MYC*, and *CCND1* amplifications appeared to be early drivers, located on the trunk of the phylogenetic trees. SCNAs continued to evolve over the lifetime of the tumor until the biopsy was taken and whole genome doubling events, which have been associated with poor prognosis in other cancer types [60,61], were late events during breast cancer evolution [34]. Changes in the mutational spectra suggested that different mutational processes were operative at different time points during breast cancer evolution. Genome-wide copy number profiling of single cancer cells from two high-grade triple-negative breast cancers identified a small number of predominant subclones in each tumor which differed significantly in their genomic profiles [42]. This has been interpreted as tumor evolution through phases of punctuated clonal expansions as opposed to gradual evolution which would be expected to leave many more intermediate subclones detectable. ITH may vary between different breast cancer subtypes as shown by the comparably low discordance rates of less than 5% for mutations detected by whole-genome sequencing in two separate biopsies taken from each of five estrogen receptor (ER)-positive primary breast cancers [62]. Further studies are necessary to define subtype-specific subclonal structures and to investigate the possibility of spatial heterogeneity in breast cancers. The association of *HER2* amplification heterogeneity between primary breast tumors and their metastases with poor prognosis demonstrates the clinical relevance of ITH [63].

Spatially separated subclones with heterogeneous ploidy profiles and chromosomal aberrations have been identified in primary gastric cancers [64]. *ERBB2* amplification heterogeneity was detected by FISH in 54% of gastric cancers, which had been classified as *ERBB2*-amplified tumors [65]. This is clinically relevant, as tumors with *ERBB2* amplifications are routinely targeted through the addition of trastuzumab to chemotherapy. To date, the clinical benefits of trastuzumab therapy in cases where *ERBB2* amplification is subclonal are unknown.

Multi-region analysis of key driver mutations suggested limited ITH in some cancer types. For example, *RAS/RAF*, *PIK3CA*, and *TP53* driver mutation status was concordant between primary colorectal cancers and metastatic sites in more than 93% of patients [66]. The concordance rate of *RAS/RAF* mutations was even higher, reaching nearly 98%. Thus, *RAS/RAF* mutations can occur as early truncal alterations but they may not confer a selective advantage during tumor progression in initially *RAS/RAF* wild-type tumors, explaining the infrequent detection of subclonal mutations. Similarly high *KRAS* mutation concordance rates of up to 96.4% were identified in other studies [67–70]. The homogeneity of *RAS/RAF* mutation status most likely contributes to the success of these markers as negative predictors of benefit from EGFR-targeted therapies in colorectal cancer patients. Activating *EGFR* mutations detected to date in nonsmall cell lung cancer (NSCLC) have high concordance between primary tumors and matched metastases [4,71–74]. Thus, the detection of an activating *EGFR* mutation in a single biopsy is likely to be predictive of the presence of this mutation in all NSCLC cells within an individual patient.

Epigenetic deregulation ranging from changes in DNA methylation to histone modifications are increasingly recognized across different cancer types and may offer new therapeutic opportunities.

However, epigenetic alterations can be subject to ITH, and therapeutic approaches exploiting these alterations may be constrained by heterogeneity. For example, driver mutations in the chromatin remodeling gene *PBRM1* were truncal and hence homogenous in a subgroup of ccRCC cases, and inactivation of the tumor suppressor gene *VHL* through DNA methylation was truncal in one case [30]. Targeting the epigenetically mediated VHL inactivation or potential therapeutic vulnerabilities arising from inactivation of the SWI/SNF complex member PBRM1 may be an excellent strategy in these cases. In contrast, driver mutations in other histone-modifying genes including *SETD2*, *KDM5C*, *ARID1A*, *SMARCA4*, and *BAP1* were heterogeneous in all tumors in which they occurred [30]. Even if these tumor suppressor genes were targetable through synthetic lethal approaches, as recently identified for *ARID1A* loss which confers ARID1B dependency [48], targeting of these drivers may be clinically futile due to their heterogeneous nature. Biomarker approaches which can reliably identify cancers with truncal alterations may define patients suitable for targeted therapeutic intervention.

Follicular lymphoma (FL) also has significant potential for epigenetic targeting. Driver mutations in several epigenetic regulator genes including *CREBBP*, *EZH2*, and *KMT2D* have been identified as clonally dominant mutations in most FL cases [75]. However, FL relapses after effective therapy often lacked the original driver mutations in *CREBBP*, *EZH2*, or *KMT2D* but had acquired novel mutations in the same genes. The repeated evolution of driver mutations in identical genes shows that relapses from an immature precursor clone may be highly dependent on the acquisition of specific genetic alterations. Identifying the specific epigenetic drivers that are necessary for progression in individual FLs may allow effective treatment personalization despite the potential presence of immature precursor clones lacking the target (Table 24.1).

As metastases are the ultimate cause of death and the main target of systemic therapy in most cancer patients, the genetic, epigenetic, and functional differences between cancer subclones in the primary

Table 24.1 Examples of Genes Involved in Epigenetic Regulation That Have Been Found to Exhibit Intratumor and Clonal Heterogeneity

Gene	Function	Observed ITH	Reference
PBRM1	Chromatin remodeling	ccRCC (truncal)	[30]
		Metastatic castration-resistant prostate cancer	[77]
ARID1A	Chromatin remodeling	ccRCC	[30]
SMARCA4	Chromatin remodeling	ccRCC	[30]
		Glioma	[33]
SETD2	Histone methyltransferase	ccRCC	[30]
BAP1	Histone deubiquitinase	ccRCC	[30]
KDM5C	Histone demethylase	ccRCC	[30]
CREBBP	Histone acetyltransferase	FL	[75]
EZH2	Histone methyltransferase	FL	[75]
		Myelodysplastic syndrome	[87]
KMT2D	Histone methyltransferase	FL	[75]
TET2	DNA demethylase	Myelodysplastic syndrome	[87]
		Metastatic castration-resistant prostate cancer	[77]
ASXL1	Chromatin remodeling	Myelodysplastic syndrome	[87]

tumor and metastases are critically important for the personalization of cancer medicine. Single case studies in prostate cancers and ccRCC suggested that all metastatic sites studied in each patient derived from a single subclone from an otherwise heterogeneous primary tumor [31,76,77]. Unsupervised hierarchical clustering of primary medulloblastomas and multiple metastatic sites based on promoter CpG hypermethylation, SCNA profiles, and point mutation data demonstrated higher epigenetic and genetic similarity between the different metastatic sites than between the metastatic and the primary tumors in two patients analyzed [58]. Although larger numbers of cases need to be studied, this may suggest that metastatic seeding could be a rate-limiting step in these cancer types that can only be accomplished by a selected sub-population of tumor cells. Such evolutionary bottlenecks may promote the acquisition of genetic alterations that are shared by all metastases, providing an opportunity for therapeutic targeting. Importantly, the subclones seeding the metastases in both cases were small minority clones within the primary tumors and they would most likely have been missed by single biopsies. Multi-region assessment of primary tumors could increase the ability to detect such clones. However, obtaining biopsies directly from metastatic sites may be a more robust and clinically feasible approach to define the driver landscape of metastatic disease for therapeutic targeting. Autopsy studies of 14 prostate cancer patients with multiple metastases revealed that liver metastases were more closely related to each other than to metastases in other organ sites [78]. The identification of such organ-specific SCNA signatures suggests that metastases to diverse sites can be indicative of ITH and organ site-specific genotypes of cancer metastases could perhaps be exploitable for therapeutic benefit. In pancreatic cancer, patterns of chromosomal rearrangements and point mutations suggest that metastases are seeded from multiple different primary clones [79,80]. Although it remains unknown, this may be the result of the high metastatic potential of pancreatic cancers, reflected in the tendency toward early incidence of metastasis when the primary tumor is still small. Further, a subset of unique metastatic variants in pancreatic and prostate cancer is suggestive of continued evolution in individual metastases after the initiation of metastatic expansion [77,79,80].

4.2 ITH IN LEUKEMIAS

ITH has been observed in leukemias of lymphoid cell origin. Single cell (FISH) analysis of *ETV6/RUNX1*-positive childhood B-cell acute lymphoblastic leukemia (ALL) samples identified subclones which were heterogeneous for several driver SCNAs in most cases [81]. Phylogenetic reconstruction revealed a nonlinear, branched clonal architecture with subclones evolving in parallel. The clonal composition of ALL populations changed after xenografting into immunodeficient mice, suggesting fitness differences of the identified subclones. Xenografting of *BCR-ABL* fusion gene-positive B-cell ALLs provided further evidence of multiple genetically and functionally heterogeneous subclones [82]. Whole-exome sequencing of chronic lymphocytic leukemia (CLL) samples demonstrated subclonal driver mutations in 46% of 149 cases. Driver aberrations in genes which are relatively specific for CLL or B cell malignancies (e.g., del(13q), *MYD88* mutations, trisomy 12) were mostly clonally dominant [83] whereas cancer genes which are mutated in a wide variety of neoplasms such as *ATM*, *TP53*, and *RAS* frequently harbored subclonal mutations. Thus, aberrations specifically affecting B cells may drive cancer initiation whereas mutations in more generic cancer driver genes may foster ongoing evolutionary adaptation and disease progression [83]. The detection of subclonal driver mutations in CLL was an independent poor prognostic marker, suggesting a link between the ability to evolve and cancer outcome. CLL is a rare example of a cancer in which the interplay between genetic and DNA methylation heterogeneity has been assessed [84]. The detection of genetic CLL driver aberration ITH was positively correlated with DNA methylation heterogeneity. Increased heterogeneity was associated with shorter time to treatment and these samples were less likely to harbor good prognosis genetic markers, such as *IGHV* mutations. Longitudinal analysis also showed that the evolution of novel genetically distinct subclones is associated with a change in DNA methylation patterns. Although the study did not investigate whether the evolving DNA methylation aberrations had driver potential, it is possible that genetic and epigenetic aberrations contribute to cancer evolution and ITH.

NGS of paired bone marrow aspirates and peripheral blood samples from acute myeloid leukemia (AML) patients identified subclonal structures and ITH in most of 19 cases analyzed [85]. Importantly, the subclonal composition of bone marrow aspirates and peripheral blood were highly similar in 68% of these patients and all subclones detected in the bone marrow were represented in the blood in all cases. Thus, AML subclones readily circulate from the bone marrow into the peripheral blood and simple blood samples may allow a representative assessment of the heterogeneous AML genomic landscapes. These genetically distinct AML subclones differed in morphology and immunophenotype and in their ability to engraft in different mouse strains, suggesting functional differences. AML subclonal heterogeneity revealed by conventional cytogenetics correlates with a poor patient prognosis and has been proposed as an additional prognostic marker particularly in the cytogenetically adverse-risk subgroup [86]. ITH was also identified in myelodysplastic syndrome [87]. In contrast to AML and CLL where higher ITH was associated with poor outcomes, the clonal dominance or subclonality of driver mutations with known survival effects (*TET2*, *ASXL1*, *SRSF2*, *EZH2*, *CBL*, and *RUNX1*) was not associated with leukemia-free survival in these malignancies, suggesting cancer type specific differences [87].

5 ITH AND THE EVOLUTION OF DRUG RESISTANCE

With multiple subclones existing in many solid and hematological malignancies, actionable therapeutic targets may not be present homogeneously throughout the cancer cell population in an individual patient. Subclones lacking the targeted driver mutation may rapidly progress and lead to primary resistance or early therapy failure. Deep sequencing of cancers with intermixing subclones such as leukemias or multi-region profiling of solid tumors may detect such therapeutically important subclonal compositions. Combination therapies which are simultaneously targeting the individual drivers of multiple subclones could be employed to address such heterogeneous cancers, but high toxicities of targeted therapy combinations have hampered this approach to date. Alternatively, patients could be prioritized for a specific drug therapy if their tumor harbors the targeted genetic aberrations on the trunk of the cancer's phylogenetic tree, present in every cancer cell within the tumor [5,88]. Sophisticated detection methods would of course be obsolete for targetable driver aberrations which are usually homogeneous and truncal whenever they are present in a specific tumor type, such as activating *EGFR* mutations in NSCLC [4,71–74,89].

Parallel evolution may foster functional homogeneity of specific biological processes in individual cancers. For example, the *SETD2* histone methyltransferase was inactivated through three distinct somatic mutations in a single ccRCC, with one mutation present in every single region analyzed [31]. Such functional convergence of genetically heterogeneous subclones could facilitate effective therapeutic targeting, as supported by an analysis of ccRCCs which had responded exceptionally well to mTOR inhibitors. One of these cases harbored a truncal mutation in *TSC1*, activating the PI3-kinase-mTOR pathway, and multiple distinct mTOR pathway-activating mutations were identified in two further tumors, affecting every single tumor region analyzed [90].

Biopsies of multiple metastatic sites are unlikely to be clinically feasible for the identification of targetable drivers activated through distinct mutations in every single subclone of a cancer. The identification of the mechanisms driving such remarkable parallel evolution in individual cancers may provide surrogate markers to predict which tumors show this behavior. Alternatively, "liquid biopsies" based on the genetic analysis of circulating tumor cells or of circulating free DNA (cfDNA) from blood samples [91] may be applicable for the minimally invasive identification of dominant and subclonal targetable aberrations. Such techniques would also permit regular re-assessments of dynamic changes and could foster adaptive therapy approaches.

Dynamic evolutionary adaptation to targeted therapy has been demonstrated for metastatic colorectal cancers through sensitive methods which detect tumor-derived mutations in cfDNA isolated from blood samples. Activating *RAS* and *BRAF* mutations had evolved after the failure of EGFR therapy in 23 out of 24 patients with initially *RAS* and *BRAF* wild-type tumors [92]. Notably, multiple independent drug resistant clones evolved in parallel in 67% of these patients, with an average of 2.9 distinct mutations and a maximum of 12 per patient. Mathematical modeling based on the kinetics of *KRAS* mutations in cfDNA suggested that *KRAS* mutant subclones had most likely been present before therapy initiation in most patients [47]. Population sizes of the resistant subclones were estimated to have been in the range of 2000–3000 cells before therapy initiation. Biopsies would most likely have missed these small clones due to sampling biases and even if they had been present in a biopsy, they were probably far below the detection limit of standard NGS techniques. These modeling results suggests that apparent *RAS/BRAF* mutation homogeneity between primary tumors and metastases may be an illusion resulting from technical limitations in the detection of multiple small *RAS/BRAF* mutant subclones before therapy-driven clonal selection [66,68–70,93].

Even the targeting of truncal mutations does not always guarantee high response rates and acquired resistance eventually occurs in the majority of cancers. In NSCLC harboring targetable activating EGFR mutations, resistance to TKIs frequently arises through the second site *EGFR* mutation, T790M [94,95]. Several lines of evidence suggest that acquired resistance often results from rare drug resistant cancer cell subclones that were already present prior to treatment initiation. Subclonal *EGFR* T790M mutations were detectable in over 25% of treatment-naïve NSCLCs [96–99]. Patients with detectable T790M mutations had poorer progression-free survival [98,99] and lower response rates to *EGFR* TKIs, as well as shorter overall survival compared to those in which T790M mutations could not be identified [100]. Based on these data, it is conceivable that the pre-treatment quantity of cancer cell subclones harboring the T790M mutation influences the magnitude and duration of targeted drug responses. Multiple sequential re-biopsies of 17 NSCLCs found that five tumors in which T790M mutations were detected post-TKI failure appeared negative for the T790M mutation when re-tested at a later time point [101]. These highly dynamic changes are evidence of ongoing cancer evolutionary adaptation resulting in temporal heterogeneity of driver mutation landscapes. Furthermore, combinations of different EGFR-TKI resistance mechanisms including *EGFR* T790M mutations, *MET* amplification, *HER2* amplification, and small cell histologic transformation were identified in 4% of biopsies taken from a cohort of 155 *EGFR*-mutant lung cancers after the failure of EGFR TKI therapy [102]. Not all samples were analyzed for aberrations other than T790M, and multiple-region samples were not available from metastases. Hence, it is possible that heterogeneity of resistance mechanisms may be even more prevalent which could significantly complicate the ability to rationally identify second-line targeted therapy.

Ultra-deep sequencing of sequential CML and Philadelphia-chromosome positive AML samples collected before BCR-ABL inhibitor therapy and at relapse showed similar subclonal complexities. The *ABL1* kinase domain revealed multiple distinct resistance mutations and up to 13 different subclonal populations in the majority of patients [103]. Many of these would have remained undetected by conventional Sanger sequencing due to the low sensitivity of this technology. A longitudinal study evaluating one high-risk multiple myeloma patient during four sequential lines of therapy revealed the existence of two major subclones which repeatedly switched dominance during the course of therapy [104]. Although the clonal composition changed with each line of therapy, even sensitive clones were never completely eradicated. Whole-genome sequencing of eight paired AML samples collected before cytarabine and anthracyclin treatment and at relapse [105] revealed that the dominant clone at relapse had evolved from the dominant clone before therapy in three patients, and from a minority clone present before therapy in five patients. More than one minority subclone survived chemotherapy in two of these cases leading to polyclonal relapses. All relapse clones harbored mutations that had been undetectable before therapy but specific drivers of drug resistance could not be identified, most likely due to small numbers of patients. Many of the mutations that were only detectable at relapse displayed a signature of chemotherapy-induced mutagenesis [105], suggesting that chemotherapy exposure directly contributed to the generation of resistance drivers. This has also been shown for glioblastomas, in which mutational inactivation of the mismatch repair gene *MSH6* drives acquired resistance to the alkylating agent temozolomide [106–108]. The patterns of *MSH6* inactivating mutations were highly similar to those resulting from exposure to alkylating agents in experimental systems suggesting that chemotherapy facilitated the acquisition of drug resistance [106–108]. Similarly, a recent study of glioma recurrences after treatment with temozolomide detected driver mutations in the *RB* gene and Akt-mTOR pathway that displayed the characteristic mutagenic signature of temozolomide exposure [33]. In addition, the relapse clones had branched off early from a common ancestor clone and displayed distinct driver mutation profiles from the primary tumors that had evolved independently from the same common ancestor clone [33]. Consequentially, the genetic landscapes of the pre-treatment clones provided limited insight into the genetic landscape and putative targets within the relapse clones.

The changes in driver mutation profiles, the selection of genomic instability mechanisms, and the alteration of subclonal composition through drug therapy demonstrate the profound impact effective therapy can have on cancer genomic landscapes. Such changes in cancer genotypes and phenotypes after the acquisition of drug resistance suggest that re-biopsies may be necessary to effectively personalize subsequent therapy. Drug treatment has also been suggested to alter ITH through the pruning of drug-sensitive tumor subclones [88,105] (Figure 24.3). Such evolutionary bottlenecks theoretically reduce ITH temporarily before expansion of the resistant clones re-establishes ITH. This may provide a therapeutic opportunity for efficient targeting of genetic alterations driving the resistant phenotype or hitchhiking with these drivers of resistance in a relatively small and homogenous cancer cell population. However, polyclonal resistance such as that observed through cfDNA analysis in colorectal cancer patients after EGFR-targeted therapy failure [92] is testament to the extensive heterogeneity and evolvability of advanced cancers. The parallel evolution of several resistance clones may prohibit effective pruning of ITH (Figure 24.3) and impair strategies aiming to target residual cancer populations after bottleneck events. Yet, this data shows dramatic functional convergence of multiple resistance clones toward mutations activating the RAS/RAF pathway downstream of EGFR [92]. This potentially indicates that the number of readily accessible resistance pathways to a specific therapy and in a specific

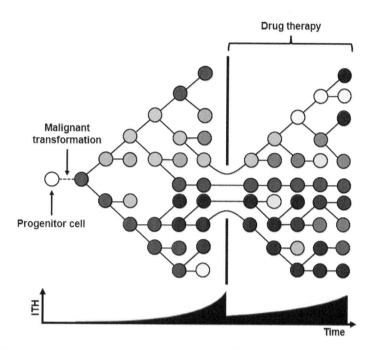

FIGURE 24.3

Branched cancer evolution and intratumor heterogeneity (ITH) during cancer development and drug therapy. Most cancers arise from a single progenitor cell that undergoes malignant transformation through the acquisition of genetic driver aberrations. Further driver aberrations can increase the growth advantage of cancer cells and accelerate clonal expansion while fitness-reducing alterations lead to negative selection. The ongoing acquisition of aberrations increases the degree of ITH which provides substrates for clonal selection. Drug therapy confers selective survival of resistant subclones thus generating an evolutionary bottleneck which temporarily reduces genetic ITH. Recent data suggests that multiple drug-resistant subclones are likely to occur in advanced cancers and pass through the bottleneck, reducing the ability of effective therapy to homogenize the cancer temporarily. The subsequent expansion of drug-resistant subclones allows the acquisition of new aberrations and re-establishment of ITH. Circles with different colors represent cancer subclones with a unique genomic landscape.

cancer type may be limited. Thus, the mechanisms of resistance to any one targeted therapy may be highly predictable.

Rational approaches to prevent or delay the evolution of therapy resistance based on the anticipation and targeting of the most prevalent resistance mechanisms have indeed been successful. For example, second-generation BCR-ABL kinase inhibitors, designed to retain activity against some of the mutant *BCR-ABL* variants evolving during and conferring resistance to standard first-line imatinib therapy, have improved outcomes of CML patients when used as first-line therapy [109]. Importantly, third-generation BCR-ABL inhibitors which are even effective against *BCR-ABL* with the T315I resistance mutation can achieve very high response rates after the failure of these second-generation inhibitors [110]. Similarly, the ALK inhibitor ceritinib was designed to be effective even in the presence of resistance mutations in the ALK tyrosine kinase domain, achieving response rates of 56% in patients with advanced NSCLC that progressed during first-generation ALK-inhibitor crizotinib therapy [111,112].

DNA methylation is a heritable epigenetic mechanism which can contribute to the evolution of drug resistance through clonal selection. In estrogen receptor-positive breast cancer, silencing of *HOXC10* through promoter hypermethylation leads to aromatase inhibitor resistance [113]. Further, silencing of genes which are part of the Akt/mTOR-, p53-, redox-, and homologous recombination repair pathways through aberrant DNA methylation can lead to platinum resistance in ovarian cancer [114]. It remains unknown whether several of these mechanisms can evolve simultaneously in individual patients, conferring epigenetic heterogeneity and further subclonal complexities. The reversibility of DNA methylation aberrations, for example through DNA methyltransferase inhibitors, may be a decisive advantage for the targeting of these drug resistance drivers. However, DNA methylation contributes to the suppression of endogenous transposable elements throughout the genome [115]. An increase in somatic insertion frequency has been observed in hypomethylated colorectal cancer genomes [116], suggesting that drugs resulting in global hypomethylation may also reactivate endogenous retroviruses which might lead to detrimental genomic instability [117]. Drugs targeting specific DNA methylation aberrations could be necessary to prevent such untoward effects.

Histone modifications can also contribute to cancer drug resistance. Gefitinib resistance was found in a small subpopulation of NSCLC cell lines harboring activating *EGFR* mutations. Resistance was transiently maintained and driven by chromatin alterations, modulated by the histone demethylase KDM5A [118]. This study suggested that a reversible drug resistant state could provide a mechanism that allows a small cancer cell subpopulation to survive drug therapy until permanent resistance mechanisms, for example genetic resistance, have been established. Histone deacetylase inhibitor treatment reversed this resistant chromatin state, suggesting that epigenetic targeting can reduce such non-genetic heterogeneity and maintain NSCLC dependency on activating EGFR mutations [118]. Activating *NOTCH1* mutations in T-cell ALL can be effectively targeted with γ-secretase inhibitors, though the response is transient [119]. γ-Secretase inhibitor-tolerant "persister" cells were found to have an altered chromatin state resulting in sensitivity to inhibitors of the chromatin regulatory factor BRD4 [120]. Combination treatment with γ-secretase and BRD4 inhibitors were effective in patient-derived xenograft models, and may prove to be effective in the clinic [120].

Epigenetic mechanisms are important in controlling tissue type specific cell differentiation and may be implicated in non-genetic resistant states manifesting in morphological trans-differentiation. In prostate cancer, castration resistance can occur through trans-differentiation into a neuroendocrine/small-cell phenotype which confers loss of tissue type-specific androgen receptor signaling dependency [121]. Similar grand-differentiation of NSCLCs to a small-cell lung cancer phenotype has also been observed as a mechanism of EGFR-inhibitor resistance through the loss of addiction to activating *EGFR* mutations [72]. It may be possible to exploit epigenetic targeting to maintain cancer cells in the differentiated or functional state that optimizes the efficacy of available drug therapies and increases therapeutic success.

An increasing number of studies into the genetics of acquired drug resistance suggest that the clonal evolution of genetically resistant cancer cell subclones is a common mechanism of resistance development to a wide range of cancer therapies [88]. These observations raise important questions about the molecular and evolutionary mechanisms which allow cancers to readily evolve, about the changes in cancer genomic landscapes resulting from the outgrowth of drug resistant clones and whether evolution can be prevented.

6 CLINICAL IMPLICATIONS OF ITH

ITH hinders the accurate genetic profiling of many cancers and impairs the appropriate selection of targeted therapies and the identification of prognostic and predictive biomarkers. The identification of ITH across many cancer types requires novel approaches to detect ITH and to incorporate ITH into biomarker analysis approaches. A combination of multi-region analysis and deep sequencing of single biopsies and circulating tumor DNA "liquid biopsies" [91] are likely to be necessary in order to delineate subclonal complexities in individual patients and across different tumor types. The development of biomarkers based on ITH is a further important task. The clinical potential of genetic heterogeneity metrics have been provided in premalignant Barrett's esophagus lesions where greater heterogeneity was an independent predictor of progression to adenocarcinoma [122], as well as in AML and CLL where subclonal heterogeneity confers poor prognosis [83,86]. Standardized biopsy approaches will be necessary to robustly assess ITH for biomarker development purposes and to compare heterogeneity patterns across different solid tumors. Identification of targetable truncal driver mutations, including truncal DNA methylation driver events, should help to improve treatment efficiency and increase the therapeutic armamentarium. The assessment of subclonal mutation compositions may help to determine the time to drug resistance evolution as a function of the pre-existing drug-resistant subclonal population. Ultimately, this could lead to pre-emptive targeting of the most likely resistance mechanisms to evolve in the recurrence tumor in order to prolong time to treatment failure. Development of drugs targeting recurrent genetic and epigenetic drivers of drug resistance is a further attractive strategy to prevent or reverse drug resistance evolution. Also, as genomic instability mechanisms foster the generation of population heterogeneity and potentially the evolvability of cancers during challenging environmental conditions such as drug therapy, re-stabilization approaches may improve therapeutic outcomes. For example, it may be possible to reverse MIN driven by hypermethylation of *MLH1* by epigenetic targeting approaches [123]. In addition, assessing ITH in individual cancers could inform rational combination therapy strategies to address the multiple dominant harboring distinct driver mutations. Further consideration should also be given to non-heritable heterogeneity, such as that resulting from the interaction between tumor cells and their microenvironment, as this may also influence therapeutic efficacy.

ACKNOWLEDGMENTS

MG was funded by grants from Cancer Research UK, Prostate Cancer UK, the Prostate Cancer Foundation, the Royal Marsden NIHR Biomedical Research Centre for Cancer and the Wellcome Trust (Grant number: 105104/Z/14/Z). DA and The Institute of Cancer Research acknowledge the support of the European Commission under the Marie Curie Intra-European Fellowship Programme (No. 299946).

REFERENCES

[1] Vogelstein B, Papadopoulos N, Velculescu VE, Zhou S, Diaz Jr. LA, Kinzler KW. Cancer genome landscapes. Science 2013;339(6127):1546–58.

[2] Mok TS, Wu YL, Thongprasert S, Yang CH, Chu DT, Saijo N, et al. Gefitinib or carboplatin-paclitaxel in pulmonary adenocarcinoma. N Engl J Med 2009;361(10):947–57.

[3] Douillard J-Y, Oliner KS, Siena S, Tabernero J, Burkes R, Barugel M, et al. Panitumumab–FOLFOX4 treatment and RAS Mutations in colorectal cancer. N Engl J Med 2013;369(11):1023–34.

[4] Dienstmann R, Rodon J, Barretina J, Tabernero J. Genomic medicine frontier in human solid tumors: prospects and challenges. J Clin Oncol 2013;31(15):1874–84.

[5] Yap TA, Gerlinger M, Futreal PA, Pusztai L, Swanton C. Intratumor heterogeneity: seeing the wood for the trees. Sci Transl Med 2012;4(127) 127ps10.

[6] Fialkow PJ. Clonal origin of human tumors. Annu Rev Med 1979;30:135–43.

[7] Tomlinson IP, Novelli MR, Bodmer WF. The mutation rate and cancer. Proc Natl Acad Sci USA 1996;93(25):14800–3.

[8] Alexandrov LB, Nik-Zainal S, Wedge DC, Aparicio SAJR, Behjati S, Biankin AV, et al. Signatures of mutational processes in human cancer. Nature 2013;500(7463):415–21.

[9] Hodgkinson A, Eyre-Walker A. Variation in the mutation rate across mammalian genomes. Nat Rev Genet 2011;12(11):756–66.

[10] Burrell RA, McGranahan N, Bartek J, Swanton C. The causes and consequences of genetic heterogeneity in cancer evolution. Nature 2013;501(7467):338–45.

[11] Vilar E, Gruber SB. Microsatellite instability in colorectal cancer-the stable evidence. Nat Rev Clin Oncol 2010;7(3):153–62.

[12] Stephens PJ, Greenman CD, Fu B, Yang F, Bignell GR, Mudie LJ, et al. Massive genomic rearrangement acquired in a single catastrophic event during cancer development. Cell 2011;144(1):27–40.

[13] Baca SC, Prandi D, Lawrence MS, Mosquera JM, Romanel A, Drier Y, et al. Punctuated evolution of prostate cancer genomes. Cell 2013;153(3):666–77.

[14] Nik-Zainal S, Alexandrov LB, Wedge DC, Van Loo P, Greenman CD, Raine K, et al. Mutational processes molding the genomes of 21 breast cancers. Cell 2012;149(5):979–93.

[15] Taylor BJ, Nik-Zainal S, Wu YL, Stebbings LA, Raine K, Campbell PJ, et al. DNA deaminases induce break-associated mutation showers with implication of APOBEC3B and 3A in breast cancer kataegis. Elife 2013;2:e00534.

[16] Negrini S, Gorgoulis VG, Halazonetis TD. Genomic instability--an evolving hallmark of cancer. Nat Rev Mol Cell Biol 2010;11(3):220–8.

[17] McGranahan N, Burrell RA, Endesfelder D, Novelli MR, Swanton C. Cancer chromosomal instability: therapeutic and diagnostic challenges. EMBO Rep 2012;13(6):528–38.

[18] Lengauer C, Kinzler KW, Vogelstein B. Genetic instabilities in human cancers. Nature 1998;396(6712):643–9.

[19] Birkbak NJ, Eklund AC, Li Q, McClelland SE, Endesfelder D, Tan P, et al. Paradoxical relationship between chromosomal instability and survival outcome in cancer. Cancer Res 2011;71(10):3447–52.

[20] Cahill DP, Kinzler KW, Vogelstein B, Lengauer C. Genetic instability and darwinian selection in tumours. Trends Cell Biol 1999;9(12):M57–60.

[21] Janssen A, Kops GJ, Medema RH. Elevating the frequency of chromosome mis-segregation as a strategy to kill tumor cells. Proc Natl Acad Sci USA 2009;106(45):19108–13.

[22] Bryant HE, Schultz N, Thomas HD, Parker KM, Flower D, Lopez E, et al. Specific killing of BRCA2-deficient tumours with inhibitors of poly(ADP-ribose) polymerase. Nature 2005;434(7035):913–17.

[23] Nowell PC. The clonal evolution of tumor cell populations. Science 1976;194(4260):23–8.

[24] Greaves M, Maley CC. Clonal evolution in cancer. Nature 2012;481(7381):306–13.

[25] Merlo LMF, Pepper JW, Reid BJ, Maley CC. Cancer as an evolutionary and ecological process. Nat Rev Cancer 2006;6(12):924–35.

[26] Lawrence MS, Stojanov P, Polak P, Kryukov GV, Cibulskis K, Sivachenko A, et al. Mutational heterogeneity in cancer and the search for new cancer-associated genes. Nature 2013;499(7457):214–18.

[27] Hashimoto K, Rogozin IB, Panchenko AR. Oncogenic potential is related to activating effect of cancer single and double somatic mutations in receptor tyrosine kinases. Hum Mutat 2012;33(11):1566–75.

[28] Junttila MR, de Sauvage FJ. Influence of tumour micro-environment heterogeneity on therapeutic response. Nature 2013;501(7467):346–54.

[29] Bolli N, Avet-Loiseau H, Wedge DC, Van Loo P, Alexandrov LB, Martincorena I, et al. Heterogeneity of genomic evolution and mutational profiles in multiple myeloma. Nat Commun 2014;5:2997.

[30] Gerlinger M, Horswell S, Larkin J, Rowan AJ, Salm MP, Varela I, et al. Genomic architecture and evolution of clear cell renal cell carcinomas defined by multiregion sequencing. Nat Genet 2014;46(3):225–33.

[31] Gerlinger M, Rowan AJ, Horswell S, Larkin J, Endesfelder D, Gronroos E, et al. Intratumor heterogeneity and branched evolution revealed by multiregion sequencing. N Engl J Med 2012;366(10):883–92.

[32] Welch JS, Ley TJ, Link DC, Miller CA, Larson DE, Koboldt DC, et al. The origin and evolution of mutations in acute myeloid leukemia. Cell 2012;150(2):264–78.

[33] Johnson BE, Mazor T, Hong C, Barnes M, Aihara K, McLean CY, et al. Mutational analysis reveals the origin and therapy-driven evolution of recurrent glioma. Science 2014;343(6167):189–93.

[34] Nik-Zainal S, Van Loo P, Wedge DC, Alexandrov LB, Greenman CD, Lau KW, et al. The life history of 21 breast cancers. Cell 2012;149(5):994–1007.

[35] Parmigiani G, Boca S, Lin J, Kinzler KW, Velculescu V, Vogelstein B. Design and analysis issues in genome-wide somatic mutation studies of cancer. Genomics 2009;93(1):17–21.

[36] McDonald JH, Kreitman M. Adaptive protein evolution at the Adh locus in Drosophila. Nature 1991;351(6328):652–4.

[37] Lawrence MS, Stojanov P, Mermel CH, Robinson JT, Garraway LA, Golub TR, et al. Discovery and saturation analysis of cancer genes across 21 tumour types. Nature 2014;505(7484):495–501.

[38] Supek F, Minana B, Valcarcel J, Gabaldon T, Lehner B. Synonymous mutations frequently Act as driver mutations in human cancers. Cell 2014;156(6):1324–35.

[39] Barretina J, Caponigro G, Stransky N, Venkatesan K, Margolin AA, Kim S, et al. The cancer cell line encyclopedia enables predictive modelling of anticancer drug sensitivity. Nature 2012;483(7391):603–7.

[40] Huang FW, Hodis E, Xu MJ, Kryukov GV, Chin L, Garraway LA. Highly recurrent TERT promoter mutations in human melanoma. Science 2013;339(6122):957–9.

[41] Muller HJ. The relation of recombination to mutational advance. Mutat Res 1964;106:2–9.

[42] Navin N, Kendall J, Troge J, Andrews P, Rodgers L, McIndoo J, et al. Tumour evolution inferred by single-cell sequencing. Nature 2011;472(7341):90–4.

[43] Ke R, Mignardi M, Pacureanu A, Svedlund J, Botling J, Wahlby C, et al. In situ sequencing for RNA analysis in preserved tissue and cells. Nat Methods 2013;10(9):857–60.

[44] Voet T, Kumar P, Van Loo P, Cooke SL, Marshall J, Lin ML, et al. Single-cell paired-end genome sequencing reveals structural variation per cell cycle. Nucleic Acids Res 2013;41(12):6119–38.

[45] Ni X, Zhuo M, Su Z, Duan J, Gao Y, Wang Z, et al. Reproducible copy number variation patterns among single circulating tumor cells of lung cancer patients. Proc Natl Acad Sci USA 2013;110(52):21083–8.

[46] Dawson SJ, Tsui DW, Murtaza M, Biggs H, Rueda OM, Chin SF, et al. Analysis of circulating tumor DNA to monitor metastatic breast cancer. N Engl J Med 2013;368(13):1199–209.

[47] Diaz Jr. LA, Williams RT, Wu J, Kinde I, Hecht JR, Berlin J, et al. The molecular evolution of acquired resistance to targeted EGFR blockade in colorectal cancers. Nature 2012;486(7404):537–40.

[48] Helming KC, Wang X, Wilson BG, Vazquez F, Haswell JR, Manchester HE, et al. ARID1B is a specific vulnerability in ARID1A-mutant cancers. Nat Med 2014;20(3):251–4.

[49] Bashashati A, Ha G, Tone A, Ding J, Prentice LM, Roth A, et al. Distinct evolutionary trajectories of primary high-grade serous ovarian cancers revealed through spatial mutational profiling. J Pathol 2013;231(1):21–34.

[50] Sottoriva A, Spiteri I, Piccirillo SG, Touloumis A, Collins VP, Marioni JC, et al. Intratumor heterogeneity in human glioblastoma reflects cancer evolutionary dynamics. Proc Natl Acad Sci USA 2013;110(10):4009–14.

[51] Snuderl M, Fazlollahi L, Le LP, Nitta M, Zhelyazkova BH, Davidson CJ, et al. Mosaic amplification of multiple receptor tyrosine kinase genes in glioblastoma. Cancer Cell 2011;20(6):810–17.

[52] Inda MM, Bonavia R, Mukasa A, Narita Y, Sah DW, Vandenberg S, et al. Tumor heterogeneity is an active process maintained by a mutant EGFR-induced cytokine circuit in glioblastoma. Genes Dev 2010; 24(16):1731–45.

[53] Brodeur GM, Seeger RC, Schwab M, Varmus HE, Bishop JM. Amplification of N-myc in untreated human neuroblastomas correlates with advanced disease stage. Science 1984;224(4653):1121–4.

[54] Seeger RC, Brodeur GM, Sather H, Dalton A, Siegel SE, Wong KY, et al. Association of multiple copies of the N-myc oncogene with rapid progression of neuroblastomas. N Engl J Med 1985;313(18):1111–16.

[55] Theissen J, Boensch M, Spitz R, Betts D, Stegmaier S, Christiansen H, et al. Heterogeneity of the MYCN oncogene in neuroblastoma. Clin Cancer Res 2009;15(6):2085–90.

[56] Mora J, Cheung NK, Gerald WL. Genetic heterogeneity and clonal evolution in neuroblastoma. Br J Cancer 2001;85(2):182–9.

[57] Mora J, Lavarino C, Alaminos M, Cheung NK, Rios J, de Torres C, et al. Comprehensive analysis of tumoral DNA content reveals clonal ploidy heterogeneity as a marker with prognostic significance in locoregional neuroblastoma. Genes Chromosomes Cancer 2007;46(4):385–96.

[58] Wu X, Northcott PA, Dubuc A, Dupuy AJ, Shih DJ, Witt H, et al. Clonal selection drives genetic divergence of metastatic medulloblastoma. Nature 2012;482(7386):529–33.

[59] Shah SP, Roth A, Goya R, Oloumi A, Ha G, Zhao Y, et al. The clonal and mutational evolution spectrum of primary triple-negative breast cancers. Nature 2012;486(7403):395–9.

[60] Dewhurst SM, McGranahan N, Burrell RA, Rowan AJ, Gronroos E, Endesfelder D, et al. Tolerance of whole-genome doubling propagates chromosomal instability and accelerates cancer genome evolution. Cancer Discov 2014;4(2):175–85.

[61] Carter SL, Cibulskis K, Helman E, McKenna A, Shen H, Zack T, et al. Absolute quantification of somatic DNA alterations in human cancer. Nat Biotechnol 2012;30(5):413–21.

[62] Ellis MJ, Ding L, Shen D, Luo J, Suman VJ, Wallis JW, et al. Whole-genome analysis informs breast cancer response to aromatase inhibition. Nature 2012;486(7403):353–60.

[63] Niikura N, Liu J, Hayashi N, Mittendorf EA, Gong Y, Palla SL, et al. Loss of human epidermal growth factor receptor 2 (HER2) expression in metastatic sites of HER2-overexpressing primary breast tumors. J Clin Oncol 2012;30(6):593–9.

[64] Furuya T, Uchiyama T, Murakami T, Adachi A, Kawauchi S, Oga A, et al. Relationship between chromosomal instability and intratumoral regional DNA ploidy heterogeneity in primary gastric cancers. Clin Cancer Res 2000;6(7):2815–20.

[65] Lee HE, Park KU, Yoo SB, Nam SK, Park do J, Kim HH, et al. Clinical significance of intratumoral HER2 heterogeneity in gastric cancer. Eur J Cancer 2013;49(6):1448–57.

[66] Vakiani E, Janakiraman M, Shen R, Sinha R, Zeng Z, Shia J, et al. Comparative genomic analysis of primary versus metastatic colorectal carcinomas. J Clin Oncol 2012;30(24):2956–62.

[67] Baldus SE, Schaefer KL, Engers R, Hartleb D, Stoecklein NH, Gabbert HE. Prevalence and heterogeneity of KRAS, BRAF, and PIK3CA mutations in primary colorectal adenocarcinomas and their corresponding metastases. Clin Cancer Res 2010;16(3):790–9.

[68] Knijn N, Mekenkamp LJ, Klomp M, Vink-Borger ME, Tol J, Teerenstra S, et al. KRAS mutation analysis: a comparison between primary tumours and matched liver metastases in 305 colorectal cancer patients. Br J Cancer 2011;104(6):1020–6.

[69] Oltedal S, Aasprong OG, Moller JH, Korner H, Gilje B, Tjensvoll K, et al. Heterogeneous distribution of K-ras mutations in primary colon carcinomas: implications for EGFR-directed therapy. Int J Colorectal Dis 2011;26(10):1271–7.

[70] Kim MJ, Lee HS, Kim JH, Kim YJ, Kwon JH, Lee JO, et al. Different metastatic pattern according to the KRAS mutational status and site-specific discordance of KRAS status in patients with colorectal cancer. BMC Cancer 2012;12:347.

[71] Vignot S, Frampton GM, Soria JC, Yelensky R, Commo F, Brambilla C, et al. Next-generation sequencing reveals high concordance of recurrent somatic alterations between primary tumor and metastases from patients with non-small-cell lung cancer. J Clin Oncol 2013;31(17):2167–72.

[72] Sequist LV, Waltman BA, Dias-Santagata D, Digumarthy S, Turke AB, Fidias P, et al. Genotypic and histological evolution of lung cancers acquiring resistance to EGFR inhibitors. Sci Transl Med 2011;3(75):75ra26.

[73] Takahashi K, Kohno T, Matsumoto S, Nakanishi Y, Arai Y, Yamamoto S, et al. Clonal and parallel evolution of primary lung cancers and their metastases revealed by molecular dissection of cancer cells. Clin Cancer Res 2007;13(1):111–20.

[74] Jakobsen JN, Sorensen JB. Intratumor heterogeneity and chemotherapy-induced changes in EGFR status in non-small cell lung cancer. Cancer Chemother Pharmacol 2012;69(2):289–99.

[75] Okosun J, Bodor C, Wang J, Araf S, Yang CY, Pan C, et al. Integrated genomic analysis identifies recurrent mutations and evolution patterns driving the initiation and progression of follicular lymphoma. Nat Genet 2014;46(2):176–81.

[76] Haffner MC, Mosbruger T, Esopi DM, Fedor H, Heaphy CM, Walker DA, et al. Tracking the clonal origin of lethal prostate cancer. J Clin Invest 2013;123(11):4918–22.

[77] Nickerson ML, Im KM, Misner KJ, Tan W, Lou H, Gold B, et al. Somatic alterations contributing to metastasis of a castration-resistant prostate cancer. Hum Mutat 2013;34(9):1231–41.

[78] Letouze E, Allory Y, Bollet MA, Radvanyi F, Guyon F. Analysis of the copy number profiles of several tumor samples from the same patient reveals the successive steps in tumorigenesis. Genome Biol 2010;11(7):R76.

[79] Campbell PJ, Yachida S, Mudie LJ, Stephens PJ, Pleasance ED, Stebbings LA, et al. The patterns and dynamics of genomic instability in metastatic pancreatic cancer. Nature 2010;467(7319):1109–13.

[80] Yachida S, Jones S, Bozic I, Antal T, Leary R, Fu B, et al. Distant metastasis occurs late during the genetic evolution of pancreatic cancer. Nature 2010;467(7319):1114–17.

[81] Anderson K, Lutz C, van Delft FW, Bateman CM, Guo Y, Colman SM, et al. Genetic variegation of clonal architecture and propagating cells in leukaemia. Nature 2011;469(7330):356–61.

[82] Notta F, Mullighan CG, Wang JC, Poeppl A, Doulatov S, Phillips LA, et al. Evolution of human BCR-ABL1 lymphoblastic leukaemia-initiating cells. Nature 2011;469(7330):362–7.

[83] Landau DA, Carter SL, Stojanov P, McKenna A, Stevenson K, Lawrence MS, et al. Evolution and impact of subclonal mutations in chronic lymphocytic leukemia. Cell 2013;152(4):714–26.

[84] Oakes CC, Claus R, Gu L, Assenov Y, Hullein J, Zucknick M, et al. Evolution of DNA methylation is linked to genetic aberrations in chronic lymphocytic leukemia. Cancer Discov 2014;4(3):348–61.

[85] Klco JM, Spencer DH, Miller CA, Griffith M, Lamprecht TL, O'Laughlin M, et al. Functional heterogeneity of genetically defined subclones in acute myeloid leukemia. Cancer Cell 2014;25(3):379–92.

[86] Bochtler T, Stolzel F, Heilig CE, Kunz C, Mohr B, Jauch A, et al. Clonal heterogeneity as detected by metaphase karyotyping is an indicator of poor prognosis in acute myeloid leukemia. J Clin Oncol 2013;31(31):3898–905.

[87] Papaemmanuil E, Gerstung M, Malcovati L, Tauro S, Gundem G, Van Loo P, et al. Clinical and biological implications of driver mutations in myelodysplastic syndromes. Blood 2013;122(22):3616–27.

[88] Gerlinger M, Swanton C. How darwinian models inform therapeutic failure initiated by clonal heterogeneity in cancer medicine. Br J Cancer 2010;103(8):1139–43.

[89] Govindan R, Ding L, Griffith M, Subramanian J, Dees ND, Kanchi KL, et al. Genomic landscape of non-small cell lung cancer in smokers and never-smokers. Cell 2012;150(6):1121–34.

[90] Voss MH, Hakimi AA, Pham CG, Brannon AR, Chen YB, Cunha LF, et al. Tumor genetic analyses of patients with metastatic renal cell carcinoma and extended benefit from mTOR inhibitor therapy. Clin Cancer Res 2014;20(7):1955–64.

[91] Crowley E, Di Nicolantonio F, Loupakis F, Bardelli A. Liquid biopsy: monitoring cancer-genetics in the blood. Nat Rev Clin Oncol 2013;10(8):472–84.

[92] Bettegowda C, Sausen M, Leary RJ, Kinde I, Wang Y, Agrawal N, et al. Detection of circulating tumor DNA in early- and late-stage human malignancies. Sci Transl Med 2014;6(224):224ra24.

[93] Thierry AR, Mouliere F, El Messaoudi S, Mollevi C, Lopez-Crapez E, Rolet F, et al. Clinical validation of the detection of KRAS and BRAF mutations from circulating tumor DNA. Nat Med 2014;20(4):430–5.

[94] Engelman JA, Zejnullahu K, Mitsudomi T, Song Y, Hyland C, Park JO, et al. MET amplification leads to gefitinib resistance in lung cancer by activating ERBB3 signaling. Science 2007;316(5827):1039–43.

[95] Kobayashi S, Boggon TJ, Dayaram T, Janne PA, Kocher O, Meyerson M, et al. EGFR mutation and resistance of non-small-cell lung cancer to gefitinib. N Engl J Med 2005;352(8):786–92.

[96] Bell DW, Gore I, Okimoto RA, Godin-Heymann N, Sordella R, Mulloy R, et al. Inherited susceptibility to lung cancer may be associated with the T790M drug resistance mutation in EGFR. Nat Genet 2005;37(12):1315–16.

[97] Inukai M, Toyooka S, Ito S, Asano H, Ichihara S, Soh J, et al. Presence of epidermal growth factor receptor gene T790M mutation as a minor clone in non-small cell lung cancer. Cancer Res 2006;66(16):7854–8.

[98] Maheswaran S, Sequist LV, Nagrath S, Ulkus L, Brannigan B, Collura CV, et al. Detection of mutations in EGFR in circulating lung-cancer cells. N Engl J Med 2008;359(4):366–77.

[99] Su KY, Chen HY, Li KC, Kuo ML, Yang JC, Chan WK, et al. Pretreatment epidermal growth factor receptor (EGFR) T790M mutation predicts shorter EGFR tyrosine kinase inhibitor response duration in patients with non-small-cell lung cancer. J Clin Oncol 2012;30(4):433–40.

[100] Lee Y, Lee GK, Lee YS, Zhang W, Hwang JA, Nam BH, et al. Clinical outcome according to the level of preexisting epidermal growth factor receptor T790M mutation in patients with lung cancer harboring sensitive epidermal growth factor receptor mutations. Cancer 2014.

[101] Kuiper JL, Heideman DA, Thunnissen E, Paul MA, van Wijk AW, Postmus PE, et al. Incidence of T790M mutation in (sequential) rebiopsies in EGFR-mutated NSCLC-patients. Lung Cancer 2014;85(1):19–24.

[102] Yu HA, Arcila ME, Rekhtman N, Sima CS, Zakowski MF, Pao W, et al. Analysis of tumor specimens at the time of acquired resistance to EGFR-TKI therapy in 155 patients with EGFR-mutant lung cancers. Clin Cancer Res 2013;19(8):2240–7.

[103] Soverini S, De Benedittis C, Machova Polakova K, Brouckova A, Horner D, Iacono M, et al. Unraveling the complexity of tyrosine kinase inhibitor-resistant populations by ultra-deep sequencing of the BCR-ABL kinase domain. Blood 2013;122(9):1634–48.

[104] Keats JJ, Chesi M, Egan JB, Garbitt VM, Palmer SE, Braggio E, et al. Clonal competition with alternating dominance in multiple myeloma. Blood 2012;120(5):1067–76.

[105] Ding L, Ley TJ, Larson DE, Miller CA, Koboldt DC, Welch JS, et al. Clonal evolution in relapsed acute myeloid leukaemia revealed by whole-genome sequencing. Nature 2012;481(7382):506–10.

[106] Cahill DP, Levine KK, Betensky RA, Codd PJ, Romany CA, Reavie LB, et al. Loss of the mismatch repair protein MSH6 in human glioblastomas is associated with tumor progression during temozolomide treatment. Clin Cancer Res 2007;13(7):2038–45.

[107] Hunter C, Smith R, Cahill DP, Stephens P, Stevens C, Teague J, et al. A hypermutation phenotype and somatic MSH6 mutations in recurrent human malignant gliomas after alkylator chemotherapy. Cancer Res 2006;66(8):3987–91.

[108] Cancer Genome Atlas Research N Comprehensive genomic characterization defines human glioblastoma genes and core pathways. Nature 2008;455(7216):1061–8.

[109] Kantarjian H, Shah NP, Hochhaus A, Cortes J, Shah S, Ayala M, et al. Dasatinib versus imatinib in newly diagnosed chronic-phase chronic myeloid leukemia. N Engl J Med 2010;362(24):2260–70.

[110] Cortes JE, Kim DW, Pinilla-Ibarz J, le Coutre P, Paquette R, Chuah C, et al. A phase 2 trial of ponatinib in Philadelphia chromosome-positive leukemias. N Engl J Med 2013;369(19):1783–96.

[111] Shaw AT, Kim DW, Mehra R, Tan DS, Felip E, Chow LQ, et al. Ceritinib in ALK-rearranged non-small-cell lung cancer. N Engl J Med 2014;370(13):1189–97.

[112] Friboulet L, Li N, Katayama R, Lee CC, Gainor JF, Crystal AS, et al. The ALK inhibitor ceritinib overcomes crizotinib resistance in non-small cell lung cancer. Cancer Discov 2014;4(6):662–73.

[113] Pathiraja TN, Nayak SR, Xi Y, Jiang S, Garee JP, Edwards DP, et al. Epigenetic reprogramming of HOXC10 in endocrine-resistant breast cancer. Sci Transl Med 2014;6(229):229ra41.

[114] Dai W, Zeller C, Masrour N, Siddiqui N, Paul J, Brown R. Promoter CpG island methylation of genes in key cancer pathways associates with clinical outcome in high-grade serous ovarian cancer. Clin Cancer Res 2013;19(20):5788–97.

[115] Howard G, Eiges R, Gaudet F, Jaenisch R, Eden A. Activation and transposition of endogenous retroviral elements in hypomethylation induced tumors in mice. Oncogene 2008;27(3):404–8.

[116] Lee E, Iskow R, Yang L, Gokcumen O, Haseley P, Luquette III LJ, et al. Landscape of somatic retrotransposition in human cancers. Science 2012;337(6097):967–71.

[117] Ehrlich M. DNA methylation in cancer: too much, but also too little. Oncogene 2002;21(35):5400–13.

[118] Sharma SV, Lee DY, Li B, Quinlan MP, Takahashi F, Maheswaran S, et al. A chromatin-mediated reversible drug-tolerant state in cancer cell subpopulations. Cell 2010;141(1):69–80.

[119] Palomero T, Ferrando A. Therapeutic targeting of NOTCH1 signaling in T-cell acute lymphoblastic leukemia. Clin Lymphoma Myeloma 2009;9(Suppl. 3):S205–10.

[120] Knoechel B, Roderick JE, Williamson KE, Zhu J, Lohr JG, Cotton MJ, et al. An epigenetic mechanism of resistance to targeted therapy in T cell acute lymphoblastic leukemia. Nat Genet 2014;46(4):364–70.

[121] Aparicio AM, Harzstark AL, Corn PG, Wen S, Araujo JC, Tu SM, et al. Platinum-based chemotherapy for variant castrate-resistant prostate cancer. Clin Cancer Res 2013;19(13):3621–30.

[122] Maley CC, Galipeau PC, Finley JC, Wongsurawat VJ, Li X, Sanchez CA, et al. Genetic clonal diversity predicts progression to esophageal adenocarcinoma. Nat Genet 2006;38(4):468–73.

[123] Fox EJ, Prindle MJ, Loeb LA. Do mutator mutations fuel tumorigenesis? Cancer Metastasis Rev 2013;32(3–4):353–61.

EPIGENETICS UNDERPINNING DNA DAMAGE REPAIR

25

Derek J. Richard[1], Emma Bolderson[1], Anne-Marie Baird[1,2], and Kenneth J. O'Byrne[1,2]

[1]*Genome Stability Laboratory, Cancer and Ageing Research Program, Institute of Health and Biomedical Innovation, Queensland University of Technology, Woolloongabba, Queensland, Australia,* [2]*Thoracic Oncology Research Group, Institute of Molecular Medicine, St. James's Hospital, Dublin, Ireland*

CHAPTER OUTLINE

1 INTRODUCTION

Epigenetics is the study of heritable changes in gene expression that are not the result of genetic alterations. These changes include DNA methylation, histone modifications, or indeed microRNA (miRNA) expression [1,2]. Chromatin is a tightly compacted DNA–protein complex that allows approximately two meters of DNA to be packaged inside a cell, only a few micrometers across. Although the resulting DNA structure is very stable, it is not very amiable to DNA-dependent processes, so mechanisms have to exist to allow processes such as transcription, replication, and DNA repair to occur.

Two diverse groups of enzymes are responsible for changes in chromatin structure, these include histone-modifying enzymes that can add a variety of chemical residues to covalently alter chromatin. The second group of enzymes is the ATP-dependent chromatin remodelers, these enzymes use energy generated from ATP hydrolysis to displace or move histones. The new nomenclature for these enzymes will be used in this chapter [3].

This chapter will look at how a cell responds to and deals with genomic instability at the epigenetic level and highlight how critical chromatin remodeling is for correct DNA repair and cell survival following DNA damage. This chapter will initially look at the DNA repair pathways that function in human cells and then at how the repair of DNA damage is controlled by epigenetics.

2 EPIGENETIC CHANGES IN CHROMATIN

2.1 HISTONE MODIFICATIONS

The average human cell contains approximately two meters of DNA. This DNA must be packed into a cell that is only a few micrometers across, however, cells have evolved an elegant mechanism to overcome this problem. Chromatin represents a cellular structure composed of DNA that has been highly condensed through interactions with specific condensation proteins. The basic unit of chromatin is the nucleosome, which is formed by an octamer of histones, containing two copies of each H2A, H2B, H3, and H4. From the histones extend N-terminal tails that are highly regulated by posttranslational modifications. Each nucleosome is wrapped with 146 bp of DNA and form linear 10 nm "beads on a string" structures that further pack together in 30 nm arrays [4]. These nucleosomes interact to form higher order structures making the chromatin. Chromatin itself can either be classified as euchromatin, which represents more open areas of chromatin, often the center of transcriptional activity, or it can be highly compacted into heterochromatin, regions that remain relatively dormant in that particular cell type. The cell has the ability to regulate how compact its DNA is by chemically modifying the core histones around which the DNA is wrapped. It is important that the cell can regulate the compaction of DNA so that it may carry out cellular transactions such as transcription, DNA replication, DNA repair, and cell division. The primary mechanism by which chromatin structure is regulated, is through posttranslational modification of the core histones, including phosphorylation, ubquitination, acetylation, and methylation. These modified histones can then be remodeled by chromatin remodeling enzymes that are large multisubunit complexes that effectively function around an ATP-driven motor. There are four families of these remodeling complexes in eukaryotes, namely SWI/SNF, KAT5 (Tip60)/p400 and the NuA4 complex, INO80, and ISWI [5]. Chromatin remodeling complexes are driven by ATP hydrolysis to move along DNA. As they progress they can remove nucleosomes thereby opening up stretches of DNA, they can slide nucleosomes along the DNA and can assist in the exchange of histones. Histone modifications often work with these remodeling complexes allowing the recruitment of repair proteins [4,6]. Chromatin can be seen as a very dynamic state within the cells, one that can be changed with a multitude of regulatory modalities in order that cellular transactions may occur without hindrance.

The critical nature of genome organization is highlighted by human genome sequencing projects, particularly in cancers. These demonstrate that mutations are not evenly spread through the genome, but are clustered to regions of heterochromatin (as marked by H3K9me3) [7]. This is consistent with the fact that DNA repair rates in heterochromatin are much lower than for euchromatin, even when negative selection of mutations in coding regions is taken into account [8,9]. However, other errors such as gene amplification and deletion are more common in open DNA regions associated with lower nucleosome compaction [10–13]. Epigenetic markers present on the histones also influence these mutation rates [7,13]. Taken together the literature strongly suggests that compaction of DNA, as influenced by epigenetic factors, plays an important role in maintaining genomic stability.

2.2 DNA METHYLATION

DNA methylation occurs when a methyl group is added to the 5'-position of cytosine (CpG). CpG dinucleotides are common in both intergenic and repetitive regions of the genome with methylation affecting approximately 60–90% of these sites. Approximately 60% of gene promoters contain CpG islands and methylation of these promoters is less frequent and results in transcriptional silencing, a feature common in many cancers [14]. Although little evidence exists to suggest DNA methylation regulates DNA repair directly, DNA methylation of DNA repair gene promoters is common in cancer [15,16]. There is also growing evidence to suggest that DNA repair-mediated proofreading occurs, where repair systems check DNA methylation patterns and that this process is as important as the DNA repair itself [17].

2.3 MICRORNAS

The third form of epigenetic regulation is via miRNAs. miRNAs function to bind the 3' untranslated region of a target mRNA and negatively regulate the mRNA function. miRNAs are small molecules of 20–24 nucleotides that bind in a sequence-dependent manner to their target sequence. There is growing evidence that epigenetic mechanisms regulate not only gene promoter activity but also the expression of miRNAs such as let-7a, miR-9, miR-34a, miR-124, miR-137, miR-148, and miR-203. However, studies have also shown that conversely miRNAs function to control important epigenetic regulators, these include histone deacetylases (HDACs), DNA methyltransferases, and polycomb group genes. This regulation and counterregulation form a feedback loop and when regulation of this loop is disrupted then so is the epigenetic state of the cell. Disruption of this loop is clearly linked with disease [18].

3 DNA DAMAGE

It is intrinsically important that cells are able to detect and repair genomic instability, even in areas of heterochromatin, in order that gene coding potential is maintained. DNA itself is extremely fragile and highly reactive and is subject to spontaneous damage through the intracellular generation of reactive oxygen species, the spontaneous hydrolysis of nucleotides, through the introduction of mistakes during S phase of the cell cycle and the generation of cytotoxic double-strand DNA breaks (DSBs) during DNA replication. External environmental factors such as UV light, exposure to many chemical carcinogens and ionizing radiation (IR) all impact on genomic integrity. These factors combined result in the introduction of tens of thousands of DNA lesions in any 1 day [19]. If left unrepaired, these lesions can impair transcriptional machinery, stop cell cycle progression, induce cellular senescence, cause cell death/apoptosis, or loss of cellular programming and the development of disease.

4 DNA REPAIR PATHWAYS

To cope with the vast array of DNA damage types and volume of damage, cells have dedicated a large armament of proteins to detect, signal, and repair DNA damage. There are four distinct DNA repair mechanisms that exist to repair the different types of lesions generated. These are the DNA mismatch repair (MMR), base excision repair (BER), nucleotide excision repair (NER), and double-strand break repair pathways. Each one of these pathways can themselves be subdivided further into subpathways that deal with a particular type of lesion.

4.1 MISMATCH REPAIR

DNA MMR is a highly conserved pathway in eukaryotes and is designed to find and replace mismatched nucleotides within the genome. This pathway detects and repairs erroneous insertions, deletions, and base substitutions that have not been detected by the proofreading function of DNA polymerase during DNA replication [20]. The process of MMR has been relatively well described and the fundamental pathway is highly conserved [21,22]. Like other repair pathways, MMR is a multistep process involving recognition of the error, excision of the incorrect base, filling the gap with the correct base and ligation of the DNA backbone. The repair process is initially catalyzed by the heterodimeric complexes such as MSH2–MSH6, MSH2–MSH3, MLH1–PMS2, each one specific for a particular type of damage [23,24]. PMS2 then functions to introduce a nick into the DNA, with this being stimulated by ATP, PCNA (proliferating cell nuclear antigen), and RFC (recombinant factor C). The nuclease, EXO1 (exonuclease 1), then functions to remove the base and the exposed single-stranded DNA that can then be coated with RPA (replication protein A). The last stages of repair involve resynthesis by replicative polδ and ligation of the phosphate backbone by DNA ligase [25].

4.2 BASE EXCISION REPAIR

BER is a relatively simple pathway designed to remove chemically modified bases that do not generate significant distortion to the DNA structure. The primary lesions removed by the BER pathway are oxidized DNA bases. These are also the most common form of DNA damage a cell encounters and there are over 100 different types of oxidized base modifications that have been identified in human cells. By far the most common of these is the 7,8-dihydro-8-oxo-guanine (8-oxo-G) [26]. Failure to correctly repair these oxidized lesions can result in G to T substitutions resulting in G:C to A:T transversions, a feature common in many cancers [27]. There are more than 20 proteins involved in the BER pathways and repair complexes vary depending on the type of oxidized lesion present within the genome [28]. BER consists of five distinct steps: lesion recognition, strand scission, gap processing, DNA synthesis, and ligation [28]. The initial step of the BER process is catalyzed by bifunctional DNA glycosylases such as OGG1 (8-oxoguanine glycosylase a), NTHL1 (endonuclease III-like protein 1), and NEIL3 (Nei endonuclease VIII-like 3). These DNA glycosylases function to recognize and excise the damaged base, then to hydrolyze the DNA backbone [28,29]. The DNA glycosylase generates a gap that is then further processed by the 3′ phosphodiesterase enzyme APE1 (AP endonuclease), this then allows DNA synthesis to occur mediated by Polβ. Finally the phosphate backbone is ligated by the XRCC1/LigIII heterodimer [29].

4.3 NUCLEOTIDE EXCISION REPAIR

NER is primarily involved in the repair of bulky DNA lesions or adducts. For example cyclobutane pyrimidine dimers (CPDs) and pyrimidine-(6,4)-pyrimidone products (6-4PP) distort the DNA helix [30]. The NER pathway consists of over 30 different proteins and consists of 2 distinct subpathways. Transcription-coupled NER (TC-NER) is highly active and functions in regions of transcribed DNA, while global genome repair (GG-NER) is slower and functions by continuously scanning and repairing DNA damage throughout the genome [31]. These distinct pathways are, however, composed of similar processing steps. Initially, the DNA damage lesion is recognized, following this the preincision protein complex is recruited and the DNA is unwound, nucleases then function to excise the damaged DNA fragment and the repair is then completed with the recruitment of a DNA polymerase and finally the phosphate backbone is ligated by a DNA ligase. However, the enzymes involved in these processing steps differ between TC-NER and GC-NER [32]. In TC-NER, the stalling of RNA polymerase II at the lesion site results in the recruitment of the Cockayne syndrome complementation groups A and B, initiating the repair process [30]. In GG-NER, XPC/hHR23B complex are initially recruited to the lesion along with DDB1 and DDB2/XPE complex [30]. However, following this initial DNA damage recognition stage both pathways then repair their damaged DNA in the same way. The TFIIH complex (part of RNA polymerase II) functions to unwind the DNA, the damage is then bound by the preincision complex which consists of XPD, XPB, and XPA, prior to being cleaved by the proteases XPG and XPF/ERCC1 [30]. The cleavage event generates an approximately 24–32 bp nucleotide fragment and this gap is then filled by the DNA polymerases Pol δ, Pol ε, and Pol κ. Finally the phosphate backbone is ligated [30].

4.4 DOUBLE-STRAND BREAK REPAIR

The most cytotoxic of all forms of DNA damage is the DSB. If not repaired correctly these lesions can result in severe chromosomal aberrations and failure to repair even a single DSB may result in cell death. Two pathways exist that function to repair these DSBs; namely the nonhomologous end joining (NHEJ) pathway and the homologous recombination pathway (HR). NHEJ occurs in all phases of the cell cycle and is a relatively simple process that ligates the two broken ends together. Because the DNA ends are not always compatible for ligation, a degree of processing of the ends may occur, with the resultant loss of genetic code [33]. Homologous recombination can only occur in the S and G_2 phases of the cell cycle, where a sister chromatid is available to be copied. Since genetic information is copied from the undamaged sister chromatid, HR has a high fidelity [34]. Following induction of a DSB the Ataxia Telangiectasia Mutated kinase (ATM) autoactivates, a process marked with a Serine 1981 phosphorylation. ATM then phosphorylates H2AX at Serine 139, which is subsequently referred to as γH2AX [35]. In NHEJ, γH2AX promotes the recruitment of one Ku heterodimer (Ku70/Ku80) at each DNA double-strand break termini. Once bound, Ku, which is a ring-shaped complex, acts as a docking site for DNA protein kinase (DNA-PKc) [33,36]. Once bound, DNA-PKc forms a bridge between the

DNA ends and this triggers its autophosphorylation on Threonine 2609 [37], catalyzing the connection of both DNA termini protecting the DNA from aberrant processing [38]. Following this the Artemis nuclease is recruited to process the DNA ends (if required). Artemis activity requires its phosphorylation by DNA-PKc (or ATM) [39,40]. To further process the ends, DNA polymerase μ and/or λ can be recruited. To complete the repair process, XRCC4 is recruited and this then recruits DNA ligase IV [38].

As mentioned previously, HR only occurs during the S and G_2 phases of the cell cycle when a sister chromatid is available to use as a template [34,41]. HR is initiated by the recruitment of hSSB1 to the DSB [42]. hSSB1 then functions to recruit the MRN (Mre11/Rad50/Nbs1) complex [43,44]. The MRN complex then functions to further activate ATM kinase and to initiate the 5'–3' resection at the DSB end, generating single-stranded DNA [45]. This single-stranded DNA is then bound by the RAD51 recombinase, which generates a nucleoprotein filament. This Rad51 nucleoprotein filament facilitates homology searching and the subsequent strand invasion into the sister chromatin [46,47].

5 EPIGENETIC MODIFICATIONS OF DOUBLE-STRAND BREAK REPAIR

5.1 DNA REPAIR-INDUCED HISTONE MODIFICATIONS

Changes in DNA structure are required following DNA damage, these changes are required to allow repair of the DNA damage to occur and to allow initiation of the cellular checkpoints. Changes to chromatin structure also influence gene expression and in particular alter the expression of DNA repair proteins. Remodeling of the chromatin surrounding areas containing damaged DNA is required to allow access to DNA repair proteins. It is generally believed that the more "open" the chromatin is, the easier DNA damage is to repair. Certainly depletion of the linker histone, H1, which is responsible for the high compaction of chromatin through the linking of neighboring histones, leads to cellular hyperresistance to DNA damage. Treatment of cells with HDAC inhibitors, which lead to global chromatin relaxation throughout the genome, also results in a hyperresistant phenotype to DNA damaging agents [48].

5.2 PHOSPHORYLATION

There are a number of endogenous and exogenous factors that will generate a DSB within the genome. It is vital that this DSB is marked quickly and repair initiated. At the same time, the cell must fire checkpoints so that the cell cycle is halted to prevent catastrophic events. As mentioned previously, it is thought that hSSB1 is the first protein factor to arrive at the DSB site and indeed cells lacking hSSB1 can neither repair DNA breaks nor signal their presence [42–44,49,50]. The first epigenetic signal is initiated by the ATM kinase and this involves the phosphorylation of the histone H2AX on Serine 139 (SQ motif) with this variant of H2AX called γ-H2AX [51,52]. Although the ATM kinase is the principal kinase targeting H2AX, two other phosphatidylinositol 3-kinase-related kinases (PIKKs) can also phosphorylate H2AX on Serine 139, these are the ATM- and Rad3-related (ATR), and the DNA-dependent protein kinase (DNA-PK). Whereas ATM and DNA-PK function primarily following IR, ATR mainly responds to replication stress and ultraviolet (UV) irradiation [53–55]. Once initiated the γ-H2AX mark extends from the break site in a MDC1-dependent manner [56], spreading over more than 1 Mb on each side of the break [52,57–60]. MDC1 itself binds to γ-H2AX and recruits activated ATM, allowing the extension and amplification of the signal [56]. Consistent with the requirement for γ-H2AX, cells from mice defective for γ-H2AX show chromosomal aberrations and are radiosensitive [61].

After the phosphorylation of H2AX, a number of repair proteins are recruited to the DSB site, these include the MRN complex, which is composed of three subunits, Mre11, NBS1, and Rad50. NBS1 itself binds to γ-H2AX directly through the FHA/BRCT domain [62,63]. As mentioned previously, MDC1 is then recruited to the break site and amplifies the ATM signal. MDC1 binds to both γ-H2AX and to NBS1 via its N-terminal forkhead-associated domain in a phosphorylation-dependent manner [64]. Other repair proteins that localize to the break site include BRCA1, 53BP1, UBC13/RNF8, RNF168, and chromatin-remodeling complexes INO80, SWR1, KAT5-p400 [61,65–67]. H2AX itself does not appear to be required for chromatin organization around the break site, nor is it required for the initial recruitment of DNA repair and signaling factors, it is however required for their retention and accumulation at the break site [61,65,66,68,69]. γ-H2AX is also thought to function with cohesins around the sites of DSBs, maintaining chromosome stability by holding DNA ends in close proximity, allowing repair to occur [60,70–73].

In response to DNA DSBs, in addition to γ-H2AX, histone H4 is also phosphorylated by Casein Kinase II (CKII) [74,75]. This histone modification is enhanced around DNA DSBs and has been implicated in restoration of chromatin structure following DNA repair.

Another histone phosphorylation event implicated in DNA repair is the phosphorylation of H3 on threonine 11 by the DNA repair and checkpoint kinase Chk1. This chromatin modification is implicated in transcriptional transactivation of the Cyclin B and Cdk1 genes. Following DNA damage the Chk1 kinase dissociates from chromatin, leading to a rapid reduction in H3 threonine 11 phosphorylation. This reduction in H3 threonine 11 phosphorylation leads to reduced binding of the acetyltransferase KAT2A (GCN5), and a subsequent reduction in Cyclin B and Cdk1 expression [76].

5.3 UBIQUITINATION

As well as being phosphorylated, γ-H2AX is also ubiquitinated in a DNA damage responsive manner. Indeed ubiquitination has a major regulatory impact on the DNA damage response proteins, both functioning to control the stability of these repair proteins and to modulate their cellular functions.

γ-H2AX itself functions with MDC1 to recruit the UBC13/RNF8 ubiquitin ligase complex [77–80]. Once recruited, RNF8 then polyubiquitinates both γ-H2AX and H2A at the DSB and functions with HERC2 to coordinate the ubiquitination of other targets including 53BP1, RAP80, and BRCA1 [81]. The polyubiquitinated H2A is required for the recruitment of 53BP1 and the BRCA1-Abraxas-RAP80 complex, while monoubiquitinated H2A is associated with more localized chromatin remodeling [82–86]. In addition to RNF8, RNF168 was also identified as having ubiquitin ligase activity against H2A and H2AX [87]. Here, RNF168 recruitment is dependent on prior RNF8-mediated substrate ubiquitination. The E2 ubiquitin-conjugating enzyme, UBC13, functions with RNF8 to promote the initial formation of K63-linked ubiquitin chains on histones. Whereas RNF168 is required to amplify the ubiquitin signal, enabling the recruitment of 53BP1 and BRCA1 [87,88]. H2AX is also constitutively phosphorylated on Y142 by the WSTF tyrosine kinase, in a process required for the DNA damage response [89]. The regulation of H2AX Y142 phosphorylation appears to be a balance between phosphorylation by WSTF and dephosphorylation by EYA1 (eyes absent homolog 1). The dephosphorylation of H2AX Y142 potentially enhances MDC1 and ATM recruitment to the damage site, promoting the phosphorylation of H2AX to γ-H2AX [89,90]. H2B has also been shown to be phosphorylated rapidly on Serine 14 following induction of DSBs and H2B phosphorylated on Ser14 is present in DSB foci at late stages of repair. The phosphorylation of H2B on Serine 14 is dependent on the presence of γ-H2AX [91].

5.4 METHYLATION

As discussed above, histone function is altered by both phosphorylation and ubiquitination, however, histones are also methylated to regulate their function. It is generally accepted that methylation of histones promotes a condensed chromatin structure. It is well understood that methylation and demethylation of histones function like an "off" and "on" switch for transcription, however, methylation of histones H3K79 and H4K20 is also important for DSB repair [92,93]. H4K20me and H3K79me are not induced by DNA damage, but these residues can be exposed due to distortion of the chromatin structure induced by DNA damage. H4K20me and H3K79me modifications are required for the recruitment of DSB repair factors and have been suggested to act as a potential mechanism for DNA repair pathway activation. For instance, H4K20me is required for the recruitment of 53BP1 homologue in fission yeast. 53BP1 interacts with H4K20me through its Tudor domain while also binding γ-H2AX through its BRCT domain [94,95]. Like in yeast, human 53BP1 can also bind directly to H4K20me2 and γ-H2AX [96].

Another protein, BBAP, functions to monoubiquitinate histone H4 on lysine 91. This monoubiquitination is required for the retention of the chromatin-associated H4K20 methylase and the mono- and dimethylation of H4K20. Cells depleted of BBAP are sensitive to DNA damaging agents and have impaired 53BP1 recruitment. The recruitment of BBAP and its partner BAL1 to the DNA damage site is PARP1 dependent [97–99]. Interestingly, lysine 91 of H4 is also acetylated and this is important for chromatin assembly following the repair of the damage [100].

5.5 ACETYLATION AND DEACETYLATION

In order for the DNA break site to be repaired, the cell must relax chromatin around the site. This process is associated with an increase in the acetylation of histones H2A and H4 around the site of the damage. Like γ-H2AX, this acetylation mark extends for hundreds of kilobases from either side of the break site [57,101–104]. Acetylation of histones is performed by histone acetyltransferases (HATs). HATs function to transfer an acetyl group onto the histone, causing relaxation of the chromatin. HDACs remove the acetyl groups, restoring order to the chromatin. Dynamic alterations in chromatin structure have been correlated with the process of homologous recombination and in terms of cellular viability it is essential that acetylation of histones is tightly controlled during this process [105]. Sensitivity to DNA damaging agents was observed in yeast with K5, K8, K12, and K16 mutated lysine residues in Histone H4, implicating acetylation in DNA repair [106]. The activity of HATs and HDACs is highly regulated to control the turnover of histone acetylation. Following DSB induction, HATs are recruited to the break site and this correlates with the acetylation of K9, K14, K18, K23, and K27 on Histone H3 and K5, K8, K12, and K16 on Histone H4 [105].

One of the best characterized HATs in DNA repair is KAT5, part of the NuA4 chromatin remodeling complex (reviewed in [107]). This HAT acetylates both H2A and H4 and is recruited to sites of DSBs. KAT5 is required for recruitment of repair proteins, including the MRN complex and subsequent DNA repair via homologous recombination [108,109].

A recent study has implicated histone acetyltransferase 1 in DNA repair. Histone acetyltransferase 1 functions to acetylate the newly synthesized N-terminal tails of H4, facilitating histone transfer, with this allowing homologous recombination repair to occur [110]. The KAT3A (CBP) and KAT3B (p300) HATs have been shown to transcriptionally regulate the DNA repair proteins BRCA1 and Rad51. Depletion of KAT3A or KAT3B led to decreased H3 and H4 acetylation, and this led to a reduction in

E2F1 binding to the Rad51 and BRCA1 promoters and resulted in the downregulation of transcript and protein [111].

Following the repair of the DNA break, the cell must compact/restore the chromatin. This is catalyzed by the removal of the acetyl groups from the histones by HDACs. There are several HDACs implicated in the repair of DSBs [112,113]. Much of the work around genome stability and HDACs has been done in yeast where Sin3/Rpd3, Sir2, and Hst1 have all been implicated in DSB repair [112,114,115]. In budding yeast, the HDACs, Rpd3 and Hda1 have been shown to target the nucleases Sae2 and Exo1, instead of nucleosomes, implicating HDACs in homologous recombination repair. Deacetylation of Sae1 and Exo1 leads to protein stabilization via protection from degradation through the autophagy pathway [116]. CtIP, the mammalian homologue of Sae2, is also deacetylated by a HDAC, SIRT6 and this event also promotes DNA end resection [113].

In mammalian cells, HDAC3 has been linked to DNA damage repair, but its exact role is still unknown [117]. HDAC4 has also been implicated in double-strand break repair and colocalizes with 53BP1 at sites of DSBs [118]. Yeast [119] and human [120] cells exhibit radiosensitivity when HDACs are inhibited, suggesting that HDACs may be involved in DSB repair or transcriptional regulation of DNA repair genes, but the precise mechanism again remains unknown. In addition, HDAC1 and 2 have been implicated in DSB repair via NHEJ and are rapidly recruited to sites of DSBs [121]. HDAC1 and HDAC2 are required for global deacetylation of histones at H3K56 and H4K16. It is suggested that HDAC1 and HDAC 2 may regulate the accessibility and activity of NHEJ factors at sites of DSBs or prevent transcription from interfering with DNA repair [121].

Chromatin and nucleosome assembly following DNA damage is clearly an important factor in the DNA damage repair process. Critical chromatin remodeling factors CAF1 (Chromatin Assembly Factor 1), Asf1, or FACT are recruited to the sites of DNA damage where they have been shown to mediate nucleosome disassembly and reassembly [122–125]. CAF1 is recruited to both UV-induced DNA damage and to sites of DSBs. CAF1 participates with Asf1 in the loading of H3 and H4 onto the DNA and is also required for chromatin restoration following UV DNA damage, through the hRing1b-dependant monoubiquitination of H2AK119 [122–124,126,127]. In addition, H3K56 acetylation is dependent on Asf1 with KAT3A/KAT3B or the KAT2A HAT, and in yeast this acetylation is required for the DNA damage response [125,128–132].

6 CHROMATIN REMODELING FACTORS RECRUITED TO SITES OF DNA DAMAGE

In addition to direct modification of DNA histones, there is also a separate group of enzymes known as the ATP-dependent chromatin modelers. These enzymes use energy from ATP hydrolysis to disrupt histone–DNA interactions leading to changes in nucleosome structure.

Within seconds of DSB induction, chromatin close to the break was found to undergo energy-dependent expansion suggesting the involvement of ATP-dependent chromatin remodelers [133]. These data led to the formation of a model whereby following DSB induction, ATP-dependent chromatin remodeling enzymes rapidly decondense chromatin, allowing recruitment of DNA repair proteins and subsequent DNA repair.

Several ATP-dependent chromatin remodeling enzymes have been shown to be required for DNA repair. For example in yeast, SWR1 is required for recruitment of Ku proteins in the early stages of

NHEJ [134]. In addition, in mammalian cells, the BRG1 component of the chromatin remodeling SWI/SNF complex is required for recruitment of Ku70/80 and NHEJ proficiency [135]. Another study also showed that depletion of BRG1 reduced H2AX phosphorylation and led to cellular sensitivity to IR [136]. In yeast, mutations that disrupt the catalytic activity of the SWI/SNF complex have been shown to inhibit homologous recombination repair [137].

The chromatin remodeling INO80 complex is also required for DNA repair and was found to be recruited to chromatin as far as 10 kb from a DSB, which correlates with nucleosome loss which also extends a few kilobases from a DSB [138,139]. Cells depleted of the INO80 protein were found to be deficient in resection of DNA double-strand breaks and subsequently homologous recombination [140]. In support of this, in yeast the INO80 complex has been shown to be required for the recruitment of Rad51 and Rad52, which are required for strand exchange during the process of homologous recombination [139]. INO80 deficient cells have been shown to have defective nucleosome removal surrounding a double-strand break in both yeast [139] and mammalian [133] cells. Yeast mutants in INO80 are also unable to recruit Mre11, the major nuclease involved in homologous recombination [134]. Thus, in light of the above it has been suggested that INO80 is responsible for eliminating nucleosomes from around a double-strand break, promoting resection and homologous recombination [134]. INO80 also facilitates the recruitment of NER factors to the sites of UV-induced DNA damage [141].

The NuA4 complex comprises two types of chromatin remodelers, as discussed earlier: an enzyme with HAT activity (KAT5) and an ATP-dependent chromatin remodeler (p400) in higher eukaryotes. Mammalian KAT5 has both ATP-dependent and HAT activities. Irradiated cells expressing HAT-defective KAT5 have been shown to accumulate DSBs, implicating the HAT activity of KAT5 in DNA repair [142]. KAT5 has been suggested to be required for the removal of γH2AX and in support of this, depletion of KAT5 from human cells led to accumulation of γH2AX on chromatin [143]. Thus KAT5 is postulated to facilitate the removal and dephosphorylation of H2AX, attenuating the DNA damage signal and the restructuring and restoration of the chromatin. KAT5 is also implicated in the sensing and signaling of DNA damage and supporting this, has been shown to be required for activation of the ATM kinase through acetylation [144].

7 DNA REPAIR IN HETEROCHROMATIN

The highly compact structure of heterochromatin contains repressive histone modifications such as trimethylation of histone H3 at lysine 9 (H3K9me3). In addition to histone modifications, heterochromatin also contains H3K9me3-associated histone interacting proteins such as Heterochromatin Protein 1 (HP1). Repression of repeat sequences in DNA is proposed to protect it from aberrant homologous recombination, therefore when DNA repair in heterochromatin is required, restructuring of chromatin is necessary. Consistent with this, repair of IR-induced DSBs is slower in heterochromatin [145,146]. Supporting the idea that heterochromatin impedes DNA repair, depletion of two of the major constituents of heterochromatin, HP1 and its partner protein KAP-1 enhances the rate of DNA repair [146,147]. Following DNA damage, HP1β is phosphorylated by Casein kinase II and rapidly removed from chromatin surrounding DSBs and then can be detected to reaccumulate on chromatin [147–149]. Phosphorylation of KAP-1 by the ATM kinase leads to its removal from chromatin promoting global chromatin relaxation [150]. The removal of HP1β from heterochromatin is dependent upon KAP-1 phosphorylation by Chk2 [151]. Loss of HP1 from the heterochromatin allows the chromodomain of the acetyltransferase KAT5 to bind to H3K9me3, promoting recruitment of other DNA repair factors [152].

8 CONCLUSION

It is clear that DNA damage induces significant chromatin restructuring around the damage site, in order for its efficient repair to occur. These epigenetic changes appear to occur very early in the repair process and even proceed resection of DNA ends. The changes to the chromatin structure are primarily based on the organization of histones and the subsequent posttranslational modification of these histones to allow recruitment of repair proteins and to relax the DNA structures. Once repair has occurred in the chromatin structure, it must be reorganized and DNA methylation reintroduced. Indeed there is growing evidence to suggest the DNA repair pathways also function to insure epigenetic changes to the DNA are maintained.

One major challenge for investigators studying the "histone code" in the future is to determine whether specific histone modifications correlate with distinct types of DNA damage.

REFERENCES

[1]   Baylin SB, Jones PA. A decade of exploring the cancer epigenome-biological and translational implications. Nat Rev Cancer 2011;11:726–34.

[2]   Klaunig JE, Wang Z, Pu X, Zhou S. Oxidative stress and oxidative damage in chemical carcinogenesis. Toxicol Appl Pharmacol 2011;254:86–99.

[3]   Allis CD, Berger SL, Cote J, Dent S, Jenuwien T, Kouzarides T, et al. New nomenclature for chromatin-modifying enzymes. Cell 2007;131:633–6.

[4]   Campos EI, Reinberg D. Histones: annotating chromatin. Annu Rev Genet 2009;43:559–99.

[5]   Clapier CR, Cairns BR. The biology of chromatin remodeling complexes. Annu Rev Biochem 2009;78:273–304.

[6]   Cairns BR. Chromatin remodeling complexes: strength in diversity, precision through specialization. Curr Opin Genet Dev 2005;15:185–90.

[7]   Schuster-Bockler B, Lehner B. Chromatin organization is a major influence on regional mutation rates in human cancer cells. Nature 2012;488:504.

[8]   Goodarzi AA, Noon AT, Deckbar D, Ziv Y, Shiloh Y, Löbrich M, et al. ATM signaling facilitates repair of DNA double-strand breaks associated with heterochromatin. Mol Cell 2008;31:167–77.

[9]   Noon AT, Shibata A, Rief N, Löbrich M, Stewart GS, Jeggo PA, et al. 53BP1-dependent robust localized KAP-1 phosphorylation is essential for heterochromatic DNA double-strand break repair. Nat Cell Biol 2010;12:177–91.

[10] Chen X, Chen Z, Chen H, Su Z, Yang J, Lin F, et al. Nucleosomes suppress spontaneous mutations base-specifically in eukaryotes. Science 2012;335:1235–8.

[11] Sasaki S, Mello CC, Shimada A, Nakatani Y, Hashimoto S, Ogawa M, et al. Chromatin-associated periodicity in genetic variation downstream of transcription start sites. Medaka: A Model for Organogenesis, Human Disease, and Evolution 2011:39–47.

[12] Sasaki S, Mello CC, Shimada A, Nakatani Y, Hashimoto S, Ogawa M, et al. Chromatin-associated periodicity in genetic variation downstream of transcription start sites. Science 2009;323:401–4.

[13] Tolstorukov MY, Volfovsky N, Stephens RM, Park PJ. Impact of chromatin structure on sequence variability in the human genome. Nat Struct Mol Biol 2011;18:510–15.

[14] Deaton AM, Bird A. CpG islands and the regulation of transcription. Genes Dev 2011;25:1010–22.

[15] DiNardo DN, Butcher DT, Robinson DP, Archer TK, Rodenhiser DI. Functional analysis of CpG methylation in the BRCA1 promoter region. Oncogene 2001;20:5331–40.

[16] Moelans CB, Verschuur-Maes AH, van Diest PJ. Frequent promoter hypermethylation of BRCA2, CDH13, MSH6, PAX5, PAX6 and WT1 in ductal carcinoma *in situ* and invasive breast cancer. J Pathol 2011;225:222–31.

[17] Schar P, Fritsch O. DNA repair and the control of DNA methylation. Prog Drug Res 2011;67:51–68.

[18] Sato F, Tsuchiya S, Meltzer SJ, Shimizu K. MicroRNAs and epigenetics. FEBS J 2011;278:1598–609.

[19] Lindahl T. Instability and decay of the primary structure of DNA. Nature 1993;362:709–15.

[20] Hsieh P, Yamane K. DNA mismatch repair: molecular mechanism, cancer, and ageing. Mech Ageing Dev 2008;129:391–407.

[21] Fukui K. DNA mismatch repair in eukaryotes and bacteria. J Nucleic Acids 2010;2010.

[22] Morita R, Nakane S, Shimada A, Inoue M, Iino H, Wakamatsu T, et al. Molecular mechanisms of the whole DNA repair system: a comparison of bacterial and eukaryotic systems. J Nucleic Acids 2010;2010:179594.

[23] Iyer RR, Pluciennik A, Burdett V, Modrich PL. DNA mismatch repair: functions and mechanisms. Chem Rev 2006;106:302–23.

[24] Li GM. Mechanisms and functions of DNA mismatch repair. Cell Res 2008;18:85–98.

[25] Jiricny J. Mediating mismatch repair. Nat Genet 2000;24:6–8.

[26] Nyaga SG, Jaruga P, Lohani A, Dizdaroglu M, Evans MK. Accumulation of oxidatively induced DNA damage in human breast cancer cell lines following treatment with hydrogen peroxide. Cell Cycle 2007;6:1472–8.

[27] Greenman C, Stephens P, Smith R, Dalgliesh GL, Hunter C, Bignell G, et al. Patterns of somatic mutation in human cancer genomes. Nature 2007;446:153–8.

[28] Svilar D, Goellner EM, Almeida KH, Sobol RW. Base excision repair and lesion-dependent subpathways for repair of oxidative DNA damage. Antioxid Redox Signal 2011;14:2491–507.

[29] David SS, O'Shea VL, Kundu S. Base-excision repair of oxidative DNA damage. Nature 2007;447:941–50.

[30] Melis JP, van Steeg H, Luijten M. Oxidative DNA damage and nucleotide excision repair. Antioxid Redox Signal 2013;18:2409–19.

[31] Kamileri I, Karakasilioti I, Garinis GA. Nucleotide excision repair: new tricks with old bricks. Trends Genet 2012;28:566–73.

[32] Vermeulen W. Dynamics of mammalian NER proteins. DNA Repair (Amst) 2011;10:760–71.

[33] Lieber MR. The mechanism of human nonhomologous DNA end joining. J Biol Chem 2008;283:1–5.

[34] Khanna KK, Jackson SP. DNA double-strand breaks: signaling, repair and the cancer connection. Nat Genet 2001;27:247–54.

[35] Burma S, Chen BP, Murphy M, Kurimasa A, Chen DJ. ATM phosphorylates histone H2AX in response to DNA double-strand breaks. J Biol Chem 2001;276:42462–7.

[36] Lieber MR, Yu K, Raghavan SC. Roles of nonhomologous DNA end joining, V(D)J recombination, and class switch recombination in chromosomal translocations. DNA Repair (Amst) 2006;5:1234–45.

[37] Chan DW, Chen BP, Prithivirajsingh S, Kurimasa A, Story MD, Qin J, et al. Autophosphorylation of the DNA-dependent protein kinase catalytic subunit is required for rejoining of DNA double-strand breaks. Genes Dev 2002;16:2333–8.

[38] Weterings E, Chen DJ. The endless tale of non-homologous end-joining. Cell Res 2008;18:114–24.

[39] Dahm K. Functions and regulation of human artemis in double strand break repair. J Cell Biochem 2007;100:1346–51.

[40] Ma Y, Pannicke U, Schwarz K, Lieber MR. Hairpin opening and overhang processing by an Artemis/DNA-dependent protein kinase complex in nonhomologous end joining and V(D)J recombination. Cell 2002;108:781–94.

[41] Shrivastav M, De Haro LP, Nickoloff JA. Regulation of DNA double-strand break repair pathway choice. Cell Res 2008;18:134–47.

[42] Richard DJ, Bolderson E, Cubeddu L, Wadsworth RI, Savage K, Sharma GG, et al. Single-stranded DNA-binding protein hSSB1 is critical for genomic stability. Nature 2008;453:677–81.

[43] Richard DJ, Cubeddu L, Urquhart AJ, Bain A, Bolderson E, Menon D, et al. hSSB1 interacts directly with the MRN complex stimulating its recruitment to DNA double-strand breaks and its endonuclease activity. Nucleic Acids Res 2011;39:3643–51.

[44] Richard DJ, Savage K, Bolderson E, Cubeddu L, So S, Ghita M, et al. hSSB1 rapidly binds at the sites of DNA double-strand breaks and is required for the efficient recruitment of the MRN complex. Nucleic Acids Res 2011;39:1692–702.

[45] Lamarche BJ, Orazio NI, Weitzman MD. The MRN complex in double-strand break repair and telomere maintenance. FEBS Lett 2010;584:3682–95.

[46] Ashton NW, Bolderson E, Cubeddu L, O'Byrne KJ, Richard DJ. Human single-stranded DNA binding proteins are essential for maintaining genomic stability. BMC Mol Biol 2013;14:9.

[47] Zhang J, Ma Z, Treszezamsky A, Powell SN. MDC1 interacts with Rad51 and facilitates homologous recombination. Nat Struct Mol Biol 2005;12:902–9.

[48] Murga M, Jaco I, Fan Y, Soria R, Martinez-Pastor B, Cuadrado M, et al. Global chromatin compaction limits the strength of the DNA damage response. J Cell Biol 2007;178:1101–8.

[49] Li Y, Bolderson E, Kumar R, Muniandy PA, Xue Y, Richard DJ, et al. HSSB1 and hSSB2 form similar multiprotein complexes that participate in DNA damage response. J Biol Chem 2009;284:23525–31.

[50] Skaar JR, Richard DJ, Saraf A, Toschi A, Bolderson E, Florens L, et al. INTS3 controls the hSSB1-mediated DNA damage response. J Cell Biol 2009;187:25–32.

[51] Downs JA, Lowndes NF, Jackson SP. A role for *Saccharomyces cerevisiae* histone H2A in DNA repair. Nature 2000;408:1001–4.

[52] Rogakou EP, Pilch DR, Orr AH, Ivanova VS, Bonner WM. DNA double-stranded breaks induce histone H2AX phosphorylation on serine 139. J Biol Chem 1998;273:5858–68.

[53] An J, Huang YC, Xu QZ, Zhou LJ, Shang ZF, Huang B, et al. DNA-PKcs plays a dominant role in the regulation of H2AX phosphorylation in response to DNA damage and cell cycle progression. BMC Mol Biol 2010;11:18.

[54] Stiff T, O'Driscoll M, Rief N, Iwabuchi K, Löbrich M, Jeggo PA. ATM and DNA-PK function redundantly to phosphorylate H2AX after exposure to ionizing radiation. Cancer Res 2004;64:2390–6.

[55] Ward IM, Chen J. Histone H2AX is phosphorylated in an ATR-dependent manner in response to replicational stress. J Biol Chem 2001;276:47759–62.

[56] Lou Z, Minter-Dykhouse K, Franco S, Gostissa M, Rivera MA, Celeste A, et al. MDC1 maintains genomic stability by participating in the amplification of ATM-dependent DNA damage signals. Mol Cell 2006;21:187–200.

[57] Downs JA, Allardn S, Jobin-Robitaille O, Javaheri A, Auger A, Bouchard N, et al. Binding of chromatin-modifying activities to phosphorylated histone H2A at DNA damage sites. Mol Cell 2004;16:979–90.

[58] Rogakou EP, Boon C, Redon C, Bonner WM. Megabase chromatin domains involved in DNA double-strand breaks *in vivo*. J Cell Biol 1999;146:905–16.

[59] Shroff R, Arbel-Eden A, Pilch D, Ira G, Bonner WM, Petrini JH, et al. Distribution and dynamics of chromatin modification induced by a defined DNA double-strand break. Curr Biol 2004;14:1703–11.

[60] Unal E, Arbel-Eden A, Sattler U, Shroff R, Lichten M, Haber JE, et al. DNA damage response pathway uses histone modification to assemble a double-strand break-specific cohesin domain. Mol Cell 2004;16:991–1002.

[61] Celeste A, Petersen S, Romanienko PJ, Fernandez-Capetillo O, Chen HT, Sedelnikova OA, et al. Genomic instability in mice lacking histone H2AX. Science 2002;296:922–7.

[62] Kobayashi J. Molecular mechanism of the recruitment of NBS1/hMRE11/hRAD50 complex to DNA double-strand breaks: NBS1 binds to gamma-H2AX through FHA/BRCT domain. J Radiat Res 2004;45:473–8.

[63] Kobayashi J, Tauchi H, Sakamoto S, Nakamura A, Morishima K, Matsuura S, et al. NBS1 localizes to gamma-H2AX foci through interaction with the FHA/BRCT domain. Curr Biol 2002;12:1846–51.

[64] Spycher C, Miller ES, Townsend K, Pavic L, Morrice NA, Janscak P, et al. Constitutive phosphorylation of MDC1 physically links the MRE11–RAD50–NBS1 complex to damaged chromatin. J Cell Biol 2008;181:227–40.

[65] Celeste A, Fernandez-Capetillo O, Kruhlak MJ, Pilch DR, Staudt DW, Lee A, et al. Histone H2AX phosphorylation is dispensable for the initial recognition of DNA breaks. Nat Cell Biol 2003;5:675–9.

[66] Paull TT, Rogakou EP, Yamazaki V, Kirchgessner CU, Gellert M, Bonner WM, et al. A critical role for histone H2AX in recruitment of repair factors to nuclear foci after DNA damage. Curr Biol 2000;10:886–95.

[67] van Attikum H, Gasser SM. Crosstalk between histone modifications during the DNA damage response. Trends Cell Biol 2009;19:207–17.

[68] Bassing CH, Chua KF, Sekiguchi J, Suh H, Whitlow SR, Fleming JC, et al. Increased ionizing radiation sensitivity and genomic instability in the absence of histone H2AX. Proc Natl Acad Sci USA 2002;99:8173–8.

[69] Fink M, Imholz D, Thoma F. Contribution of the serine 129 of histone H2A to chromatin structure. Mol Cell Biol 2007;27:3589–600.

[70] Bauerschmidt C, Arrichiello C, Burdak-Rothkamm S, Woodcock M, Hill MA, Stevens DL, et al. Cohesin promotes the repair of ionizing radiation-induced DNA double-strand breaks in replicated chromatin. Nucleic Acids Res 2010;38:477–87.

[71] Bekker-Jensen S, Lukas C, Kitagawa R, Melander F, Kastan MB, Bartek J, et al. Spatial organization of the mammalian genome surveillance machinery in response to DNA strand breaks. J Cell Biol 2006;173:195–206.

[72] Kim BJ, Li Y, Zhang J, Xi Y, Li Y, Yang T, et al. Genome-wide reinforcement of cohesin binding at pre-existing cohesin sites in response to ionizing radiation in human cells. J Biol Chem 2010;285:22782–90.

[73] Kim JS, Krasieva TB, LaMorte V, Taylor AMR, Yokomori K. Specific recruitment of human cohesin to laser-induced DNA damage. J Biol Chem 2002;277:45149–53.

[74] Cheung WL, Turner FB, Krishnamoorthy T, Wolner B, Ahn SH, Foley M, et al. Phosphorylation of histone H4 serine 1 during DNA damage requires casein kinase II in *S. cerevisiae*. Curr Biol 2005;15:656–60.

[75] Utley RT, Lacoste N, Jobin-Robitaille O, Allard S, Cote J. Regulation of NuA4 histone acetyltransferase activity in transcription and DNA repair by phosphorylation of histone H4. Mol Cell Biol 2005;25:8179–90.

[76] Shimada M, Niida H, Zineldeen DH, Tagami H, Tanaka M, Saito H, et al. Chk1 is a histone H3 threonine 11 kinase that regulates DNA damage-induced transcriptional repression. Cell 2008;132:221–32.

[77] Huen MS, Grant R, Manke I, Minn K, Yu X, Yaffe MB, et al. RNF8 transduces the DNA-damage signal via histone ubiquitylation and checkpoint protein assembly. Cell 2007;131:901–14.

[78] Jackson SP, Durocher D. Regulation of DNA damage responses by ubiquitin and SUMO. Mol Cell 2013;49:795–807.

[79] Kolas NK, Chapman JR, Nakada S, Ylanko J, Chahwan R, Sweeney FD, et al. Orchestration of the DNA-damage response by the RNF8 ubiquitin ligase. Science 2007;318:1637–40.

[80] Mailand N, Bekker-Jensen S, Faustrup H, Melander F, Bartek J, Lukas C, et al. RNF8 ubiquitylates histones at DNA double-strand breaks and promotes assembly of repair proteins. Cell 2007;131:887–900.

[81] Bekker-Jensen S, Rendtlew Danielsen J, Fugger K, Gromova I, Nerstedt A, Lukas C, et al. HERC2 coordinates ubiquitin-dependent assembly of DNA repair factors on damaged chromosomes. Nat Cell Biol 2010;12:80–6.

[82] Marteijn JA, Bekker-Jensen S, Mailand N, Lans H, Schwertman P, Gourdin AM, et al. Nucleotide excision repair-induced H2A ubiquitination is dependent on MDC1 and RNF8 and reveals a universal DNA damage response. J Cell Biol 2009;186:835–47.

[83] Sobhian B, Shao G, Lilli DR, Culhane AC, Moreau LA, Xia B, et al. RAP80 targets BRCA1 to specific ubiquitin structures at DNA damage sites. Science 2007;316:1198–202.

[84] Wang B, Elledge SJ. Ubc13/Rnf8 ubiquitin ligases control foci formation of the Rap80/Abraxas/Brca1/Brcc36 complex in response to DNA damage. Proc Natl Acad Sci USA 2007;104:20759–63.

[85] Wu J, Liu C, Chen J, Yu X. RAP80 protein is important for genomic stability and is required for stabilizing BRCA1-A complex at DNA damage sites *in vivo*. J Biol Chem 2012;287:22919–26.

[86] Bergink S, Salomons FA, Hoogstraten D, Groothuis TA, de Waard H, Wu J, et al. DNA damage triggers nucleotide excision repair-dependent monoubiquitylation of histone H2A. Genes Dev 2006;20:1343–52.

[87] Doil C, Mailand N, Bekker-Jensen S, Menard P, Larsen DH, Pepperkok R, et al. RNF168 binds and amplifies ubiquitin conjugates on damaged chromosomes to allow accumulation of repair proteins. Cell 2009;136:435–46.

[88] Stewart GS, Panier S, Townsend K, Al-Hakim AK, Kolas NK, Miller ES, et al. The RIDDLE syndrome protein mediates a ubiquitin-dependent signaling cascade at sites of DNA damage. Cell 2009;136:420–34.

[89] Cook PJ, Ju BG, Telese F, Wang X, Glass CK, Rosenfeld MG. Tyrosine dephosphorylation of H2AX modulates apoptosis and survival decisions. Nature 2009;458:591–6.

[90] Xiao A, Li H, Shechter D, Ahn SH, Fabrizio LA, Erdjument-Bromage H, et al. WSTF regulates the H2A.X DNA damage response via a novel tyrosine kinase activity. Nature 2009;457:57–62.

[91] Fernandez-Capetillo O, Allis CD, Nussenzweig A. Phosphorylation of histone H2B at DNA double-strand breaks. J Exp Med 2004;199:1671–7.

[92] Huyen Y, Zgheib O, Ditullio Jr RA, Gorgoulis VG, Zacharatos P, Petty TJ, et al. Methylated lysine 79 of histone H3 targets 53BP1 to DNA double-strand breaks. Nature 2004;432:406–11.

[93] Schotta G, Sengupta R, Kubicek S, Malin S, Kauer M, Callén E, et al. A chromatin-wide transition to H4K20 monomethylation impairs genome integrity and programmed DNA rearrangements in the mouse. Genes Dev 2008;22:2048–61.

[94] Nakamura TM, Moser BA, Du LL, Russell P. Cooperative control of Crb2 by ATM family and Cdc2 kinases is essential for the DNA damage checkpoint in fission yeast. Mol Cell Biol 2005;25:10721–30.

[95] Sanders SL, Portoso M, Mata J, Bähler J, Allshire RC, Kouzarides T, et al. Methylation of histone H4 lysine 20 controls recruitment of Crb2 to sites of DNA damage. Cell 2004;119:603–14.

[96] Botuyan MV, Lee J, Ward IM, Kim JE, Thompson JR, Chen J, et al. Structural basis for the methylation state-specific recognition of histone H4-K20 by 53BP1 and Crb2 in DNA repair. Cell 2006;127:1361–73.

[97] Yan Q, Dutt S, Xu R, Graves K, Juszczynski P, Manis JP, et al. The BBAP E3 ligase monoubiquitylates histone H4 at lysine 91 and selectively modulates the DNA damage response in chemotherapy-resistant lymphomas. Blood 2009;114:1522–1522.

[98] Yan Q, Dutt S, Xu R, Graves K, Juszczynski P, Manis JP, et al. BBAP monoubiquitylates histone H4 at lysine 91 and selectively modulates the DNA damage response. Mol Cell 2009;36:110–20.

[99] Yan Q, Xu R, Zhu L, Cheng X, Wang Z, Manis J, et al. BAL1 and its partner E3 ligase, BBAP, link Poly(ADP-ribose) activation, ubiquitylation, and double-strand DNA repair independent of ATM, MDC1, and RNF8. Mol Cell Biol 2013;33:845–57.

[100] Ye J, Ai X, Eugeni EE, Zhang L, Carpenter LR, Jelinek MA, et al. Histone H4 lysine 91 acetylation: a core domain modification associated with chromatin assembly. Mol Cell 2005;18:123–30.

[101] Jha S, Shibata E, Dutta A. Human Rvb1/Tip49 is required for the histone acetyltransferase activity of Tip60/NuA4 and for the downregulation of phosphorylation on H2AX after DNA damage. Mol Cell Biol 2008;28:2690–700.

[102] Kusch T, Florens L, Macdonald WH, Swanson SK, Glaser RL, Yates III JR. Acetylation by Tip60 is required for selective histone variant exchange at DNA lesions. Science 2004;306:2084–7.

[103] Murr R, Loizou JI, Yang YG, Cuenin C, Li H, Wang ZQ, et al. Histone acetylation by Trrap-Tip60 modulates loading of repair proteins and repair of DNA double-strand breaks. Nat Cell Biol 2006;8:91–9.

[104] Xu Y, Sun Y, Jiang X, Ayrapetov MK, Moskwa P, Yang S, et al. The p400 ATPase regulates nucleosome stability and chromatin ubiquitination during DNA repair. J Cell Biol 2010;191:31–43.

[105] Tamburini BA, Tyler JK. Localized histone acetylation and deacetylation triggered by the homologous recombination pathway of double-strand DNA repair. Mol Cell Biol 2005;25:4903–13.

[106] Bird AW, Yu DY, Pray-Grant MG, Qiu Q, Harmon KE, Megee PC, et al. Acetylation of histone H4 by Esa1 is required for DNA double-strand break repair. Nature 2002;419:411–15.

[107] Sun Y, Jiang X, Price BD. Tip60: connecting chromatin to DNA damage signaling. Cell Cycle 2010;9:930–6.

[108] Robert F, Hardy S, Nagy Z, Baldeyron C, Murr R, Déry U, et al. The transcriptional histone acetyltransferase cofactor TRRAP associates with the MRN repair complex and plays a role in DNA double-strand break repair. Mol Cell Biol 2006;26:402–12.

[109] Murr R, Loizou JI, Yang YG, Cuenin C, Li H, Wang ZQ, et al. Histone acetylation by Trrap-Tip60 modulates loading of repair proteins and repair of DNA double-strand breaks. Nat Cell Biol 2006;8:91–9.

[110] Yang X, Li L, Liang J, Shi L, Yang J, Yi X, et al. Histone acetyltransferase 1 promotes homologous recombination in DNA repair by facilitating histone turnover. J Biol Chem 2013;288:18271–82.

[111] Ogiwara H, Kohno T. CBP and p300 histone acetyltransferases contribute to homologous recombination by transcriptionally activating the BRCA1 and RAD51 genes. PLoS One 2012;7:e52810.

[112] Tamburini BA, Tyler JK. Localized histone acetylation and deacetylation triggered by the homologous recombination pathway of double-strand DNA repair. Mol Cell Biol 2005;25:4903–13.

[113] Kaidi A, Weinert BT, Choudhary C, Jackson SP. Human SIRT6 promotes DNA end resection through CtIP deacetylation. Science 2010;329:1348–53.

[114] Jazayeri A, McAinsh AD, Jackson SP. *Saccharomyces cerevisiae* Sin3p facilitates DNA double-strand break repair. Proc Natl Acad Sci USA 2004;101:1644–9.

[115] Utley RT, Lacoste N, Jobin-Robitaille O, Allard S, Cote J. Regulation of NuA4 histone acetyltransferase activity in transcription and DNA repair by phosphorylation of histone H4. Mol Cell Biol 2005;25:8179–90.

[116] Robert T, Vanoli F, Chiolo I, Shubassi G, Bernstein KA, Rothstein R, et al. HDACs link the DNA damage response, processing of double-strand breaks and autophagy. Nature 2011;471:74–9.

[117] Bhaskara S, Chyla BJ, Amann JM, Knutson SK, Cortez D, Sun ZW, et al. Deletion of histone deacetylase 3 reveals critical roles in S phase progression and DNA damage control. Mol Cell 2008;30:61–72.

[118] Kao GD, McKenna WG, Guenther MG, Muschel RJ, Lazar MA, Yen TJ. Histone deacetylase 4 interacts with 53BP1 to mediate the DNA damage response. J Cell Biol 2003;160:1017–27.

[119] Nicolas E, Yamada T, Cam HP, Fitzgerald PC, Kobayashi R, Grewal SI. Distinct roles of HDAC complexes in promoter silencing, antisense suppression and DNA damage protection. Nat Struct Mol Biol 2007;14:372–80.

[120] Cerna D, Camphausen K, Tofilon PJ. Histone deacetylation as a target for radiosensitization. Curr Top Dev Biol 2006;73:173–204.

[121] Miller KM, Tjeertes JV, Coates J, Legube G, Polo SE, Britton S, et al. Human HDAC1 and HDAC2 function in the DNA-damage response to promote DNA nonhomologous end-joining. Nat Struct Mol Biol 2010;17:1144–51.

[122] Green CM, Almouzni G. Local action of the chromatin assembly factor CAF-1 at sites of nucleotide excision repair *in vivo*. EMBO J 2003;22:5163–74.

[123] Moggs JG, Grandi P, Quivy JP, Jónsson ZO, Hübscher U, Becker PB, et al. A CAF-1-PCNA-mediated chromatin assembly pathway triggered by sensing DNA damage. Mol Cell Biol 2000;20:1206–18.

[124] Nabatiyan A, Szuts D, Krude T. Induction of CAF-1 expression in response to DNA strand breaks in quiescent human cells. Mol Cell Biol 2006;26:1839–49.

[125] Ransom M, Dennehey BK, Tyler JK. Chaperoning histones during DNA replication and repair. Cell 2010;140:183–95.

[126] Polo SE, Roche D, Almouzni G. New histone incorporation marks sites of UV repair in human cells. Cell 2006;127:481–93.

[127] Zhu Q, Wani G, Arab HH, El-Mahdy MA, Ray A, Wani AA. Chromatin restoration following nucleotide excision repair involves the incorporation of ubiquitinated H2A at damaged genomic sites. DNA Repair (Amst) 2009;8:262–73.

[128] Collins SR, Miller KM, Maas NL, Roguev A, Fillingham J, Chu CS, et al. Functional dissection of protein complexes involved in yeast chromosome biology using a genetic interaction map. Nature 2007;446:806–10.

[129] Das C, Lucia MS, Hansen KC, Tyler JK. CBP/p300-mediated acetylation of histone H3 on lysine 56 (vol 459, pg 113, 2009). Nature 2009;460.

[130] Driscoll R, Hudson A, Jackson SP. Yeast Rtt109 promotes genome stability by acetylating histone H3 on lysine 56. Science 2007;315:649–52.

[131] Han J, Zhou H, Horazdovsky B, Zhang K, Xu RM, Zhang Z. Rtt109 acetylates histone H3 lysine 56 and functions in DNA replication. Science 2007;315:653–5.

[132] Tsubota T, Berndsen CE, Erkmann JA, Smith CL, Yang L, Freitas MA, et al. Histone H3-K56 acetylation is catalyzed by histone chaperone-dependent complexes. Mol Cell 2007;25:703–12.

[133] Kruhlak MJ, Celeste A, Dellaire G, Fernandez-Capetillo O, Müller WG, McNally JG, et al. Changes in chromatin structure and mobility in living cells at sites of DNA double-strand breaks. J Cell Biol 2006;172:823–34.

[134] van Attikum H, Fritsch O, Gasser SM. Distinct roles for SWR1 and INO80 chromatin remodeling complexes at chromosomal double-strand breaks. Embo J 2007;26:4113–25.

[135] Ogiwara H, Ui A, Otsuka A, Satoh H, Yokomi I, Nakajima S, et al. Histone acetylation by CBP and p300 at double-strand break sites facilitates SWI/SNF chromatin remodeling and the recruitment of non-homologous end joining factors. Oncogene 2011;30:2135–46.

[136] Lee HS, Park JH, Kim SJ, Kwon SJ, Kwon J. A cooperative activation loop among SWI/SNF, gamma-H2AX and H3 acetylation for DNA double-strand break repair. EMBO J 2010;29:1434–45.

[137] Chai B, Huang J, Cairns BR, Laurent BC. Distinct roles for the RSC and Swi/Snf ATP-dependent chromatin remodelers in DNA double-strand break repair. Genes Dev 2005;19:1656–61.

[138] Chen CC, Carson JJ, Feser J, Tamburini B, Zabaronick S, Linger J, et al. Acetylated lysine 56 on histone H3 drives chromatin assembly after repair and signals for the completion of repair. Cell 2008;134:231–43.

[139] Tsukuda T, Fleming AB, Nickoloff JA, Osley MA. Chromatin remodelling at a DNA double-strand break site in *Saccharomyces cerevisiae*. Nature 2005;438:379–83.

[140] Gospodinov A, Vaissiere T, Krastev DB, Legube G, Anachkova B, Herceg Z. Mammalian Ino80 mediates double-strand break repair through its role in DNA end strand resection. Mol Cell Biol 2011;31:4735–45.

[141] Jiang Y, Wang X, Bao S, Guo R, Johnson DG, Shen X, et al. INO80 chromatin remodeling complex promotes the removal of UV lesions by the nucleotide excision repair pathway. Proc Natl Acad Sci USA 2010;107:17274–9.

[142] Ikura T, Ogryzko VV, Grigoriev M, Groisman R, Wang J, Horikoshi M, et al. Involvement of the TIP60 histone acetylase complex in DNA repair and apoptosis. Cell 2000;102:463–73.

[143] Jha S, Shibata E, Dutta A. Human Rvb1/Tip49 is required for the histone acetyltransferase activity of Tip60/NuA4 and for the downregulation of phosphorylation on H2AX after DNA damage. Mol Cell Biol 2008;28:2690–700.

[144] Sun Y, Jiang X, Chen S, Fernandes N, Price BD. A role for the Tip60 histone acetyltransferase in the acetylation and activation of ATM. Proc Natl Acad Sci USA 2005;102:13182–7.

[145] Cowell IG, Sunter NJ, Singh PB, Austin CA, Durkacz BW, Tilby MJ. gammaH2AX foci form preferentially in euchromatin after ionising-radiation. PLoS One 2007;2:e1057.

[146] Goodarzi AA, Noon AT, Deckbar D, Ziv Y, Shiloh Y, Löbrich M, et al. ATM signaling facilitates repair of DNA double-strand breaks associated with heterochromatin. Mol Cell 2008;31:167–77.

[147] Baldeyron C, Soria G, Roche D, Cook AJ, Almouzni G. HP1alpha recruitment to DNA damage by p150CAF-1 promotes homologous recombination repair. J Cell Biol 2011;193:81–95.

[148] Ayoub N, Jeyasekharan AD, Bernal JA, Venkitaraman AR. HP1-beta mobilization promotes chromatin changes that initiate the DNA damage response. Nature 2008;453:682–6.

[149] Ayoub N, Jeyasekharan AD, Venkitaraman AR. Mobilization and recruitment of HP1: a bimodal response to DNA breakage. Cell Cycle 2009;8:2945–50.

[150] Ziv Y, Bielopolski D, Galanty Y, Lukas C, Taya Y, Schultz DC, et al. Chromatin relaxation in response to DNA double-strand breaks is modulated by a novel ATM- and KAP-1 dependent pathway. Nat Cell Biol 2006;8:870–6.

[151] Bolderson E, Savage KI, Mahen R, Pisupati V, Graham ME, Richard DJ, et al. Kruppel-associated Box (KRAB)-associated co-repressor (KAP-1) Ser-473 phosphorylation regulates heterochromatin protein 1beta (HP1-beta) mobilization and DNA repair in heterochromatin. J Biol Chem 2012;287:28122–31.

[152] Sun Y, Jiang X, Xu Y, Ayrapetov MK, Moreau LA, Whetstine JR, et al. Histone H3 methylation links DNA damage detection to activation of the tumour suppressor Tip60. Nat Cell Biol 2009;11:1376–82.

EPIGENETICS OF CISPLATIN RESISTANCE

26

Steven G. Gray[1,2]

[1]Hope Directorate, St James' Hospital, Dublin, Ireland, [2]Thoracic Oncology Research Group, Institute of Molecular Medicine, St. James's Hospital, Dublin, Ireland

CHAPTER OUTLINE

S.G. Gray (Ed): Epigenetic Cancer Therapy. DOI: http://dx.doi.org/10.1016/B978-0-12-800206-3.00026-4

1 INTRODUCTION

Currently, the standard of care for many cancer types involves treatment regimens which contain platinum. Many patients, however, gain no benefit from this treatment, due to an innate intrinsic resistance to this agent. For example, the first line treatment for patients with malignant pleural mesothelioma (MPM) receives either cisplatin/pemetrexed (or cisplatin/raltitrexed). All patients presenting with MPM will receive this treatment but only 41% of patients will demonstrate an initial response to this therapy. This means that a large proportion of the patient's tumors' had an intrinsic resistance to the platinum-based therapy.

Another common issue that emerges when patients are treated with cisplatin is that while they may initially respond to treatment, over time this response fails. In other words, their tumor gains an acquired resistance to cisplatin.

Epigenetics is currently defined as a specialized form of gene regulation whereby stable and heritable changes are effected on gene expression which are not due to changes in the primary DNA sequence. The current established epigenetic mechanisms identified to date involve the following: DNA CpG methylation, histone posttranslational modifications (PTMs), histone variants, and noncoding RNA (ncRNA).

In this chapter using non-small cell lung cancer (NSCLC) and ovarian cancer as examples, I shall discuss how epigenetics and the cellular machinery involved with this regulation may be of critical importance in the development of both innate and acquired resistance to cisplatin. An overview is provided in Figure 26.1.

2 DNA METHYLATION

In a previous chapter, we have discussed the general issue of DNA CpG methylation and hydroxymethylation in cancer (see Chapter 3). It is now well established that aberrant DNA CpG methylation is a common element in the pathogenesis of cancer including both ovarian [1] and NSCLC [2].

2.1 DNA METHYLATION CHANGES ASSOCIATED WITH CISPLATIN RESISTANCE

Global Methyl-Capture sequencing (MethylCap-seq) combines the use of recombinant methyl-CpG-binding domain of methyl-CpG-binding domain protein 2, MBD2 protein to immunoprecipitate methylated DNA which is then sequenced using next-generation sequencing (NGS) to provide a global and unbiased analysis of DNA methylation patterns. Using this technique, Yu et al. conducted a genome-wide DNA methylation profile of both the cisplatin-sensitive ovarian cancer cell line A2780 and its isogenic derivative resistant line A2780CP. From a comparison of the related profiles, 224 hypermethylated and 1216 hypomethylated DMRs (differentially methylated regions) were observed between the resistant cell line compared to its sensitive parent [3]. A similar type of study again using the same cell lines used genome-wide methylation profiling across 27,578 CpG sites, resulting in the identification of loci at 4092 genes which had become hypermethylated in the chemoresistant A2780/cp70 cell line compared with the parental-sensitive A2780 cell line [4].

In a similar vein, Zhang et al. used a combined integrated DNA methylation and gene profiling analysis of the NSCLC lung adenocarcinoma cisplatin-sensitive parent A549 versus its isogenic

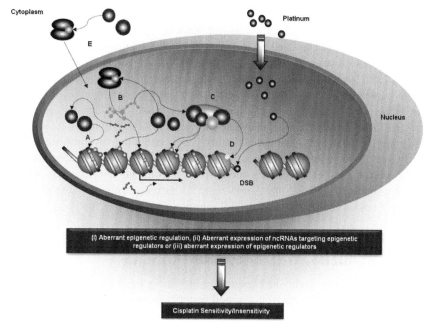

FIGURE 26.1

Epigenetics underpinning cisplatin resistance. Diagram summarizing the available evidence linking the various forms of epigenetic regulation to how cancer cells respond to cisplatin-based therapies. (A) DNA methylation; (B) ncRNA (including miRNAs and lncRNAs); (C) epigenetic readers, writers, and erasers (including multisubunit complexes); (D) histone variants; and (E) nonhistone PTMs.

resistant clone A549/DDP, and identified 372 genes that showed both DNA CpG hypermethylation and reduced expression in cisplatin-resistant A549/DDP cells. From this they validated 10 candidate genes. Treatment of the A549/DDP cell line with a low dose combination regimen of 5-aza-2'-deoxycytidine (5-Aza-dC) and trichostatin A (TSA) was found to reverse drug resistance of A549/DDP cells both *in vitro* and *in vivo*, along with demethylation and restoration of expression of candidate genes (GAS1, TIMP4, ICAM1, and WISP2). Validation of these candidate genes was demonstrated when GAS1 was overexpressed in the A549/DDP cells resulting in increased sensitivity to DDP, inhibition of cellular proliferation, with associated cell cycle arrest, enhanced apoptosis, and growth retardation of xenografts *in vivo* [5].

2.2 METHYLATION IN CANCER CELL LINES MAY NOT TRULY REFLECT THE METHYLATION FROM THE PRIMARY TUMOR

It is now an established fact that *de novo* DNA methylation can occur during the cell-culturing process. In an elegant study using experimental and bioinformatic approaches, Jones and coworkers unveiled genomic regions that require DNA methylation for survival of cancer cells. Initially, the authors compared the DNA methylation profiles of HCT-116 cells to those of subclones of HCT-116 cells which had been engineered to have deficient levels of DNA methyltransferases *DNMT3B* and *DNMT1*. The resulting profiles were then clustered according to their DNA methylation status in primary normal and tumor tissues. Finally, using gene expression meta-analysis the authors identified genomic regions that are dependent on DNA methylation-mediated gene silencing. Critically, the authors found a subset of a group of genes whose silencing by DNA methylation was required for cells to survive in culture. This group of genes was found to be highly methylated in colon, bladder, and breast cell lines and unmethylated in primary-matched tissue analyzed, independent of the tumorigenic state. Intriguingly, this group encompasses many nucleosome assembly genes, including several histone variants about which currently very little is known. These results therefore suggest that during the cell-culturing process, extensive changes in expression of nucleosome constituents are necessary for cell survival. Furthermore it also highlights the fact that data derived from experiments investigating epigenetic phenomena using cell lines should invoke very careful interpretation [6].

3 EPIGENETIC READERS, WRITERS, AND ERASERS

The PTM of histones is a well-established epigenetic regulatory mechanism. The addition of modifications such as acetylation, methylation, in defined patterns represents a "code" used by the cellular machinery to regulate transcription. To utilize this code, various families of proteins have emerged with different functions based on their ability to either "read," "write," or "erase" these modifications. Evidence is now emerging that many of these proteins may also play roles in cisplatin resistance.

3.1 LYSINE ACETYLTRANSFERASES

Of the many epigenetic writers, lysine acetyltransferases are emerging as major elements in responses to cisplatin and the development of resistance to this agent.

For instance in lung cancer, these include the lysine acetyltransferases KAT13D (Clock) [7], KAT5 (Tip60) [8], KAT2B (PCAF) [9], and SRC-3 [10].

Expression of CBP/p300-Interacting Transactivator with GLU/ASP-Rich C-Terminal Domain 2 (CITED2) has also been shown to be involved with cisplatin resistance in cancer cell lines by a process dependent upon p53. Chao and coworkers demonstrated that knockdown of CITED2 sensitized cells in p53 positive cells, whereas H1299 cells which are p53 defective had negligible responses to cisplatin. Knockdown of CITED2 induced KAT3A-mediated p53 acetylation (Lys373) preventing ubiquitination and turnover of p53. This resulted in increased levels of the p53 target Bax and was further increased following cisplatin treatment [11].

It is well established that the lysine acetyltransferase KAT3B/p300 is involved with the stabilization of E2F1 in response to DNA damage [12] and its acetyltransferase activity has been linked to nucleotide excision repair (NER) of DNA damage [13,14]. More recently, KAT3B has been shown to be recruited to DNA damage sites located within heterochromatin allowing damage recognition factors access to damaged DNA [15]. KAT3B has also been associated with regulating base excision repair (BER) through acetylation of DNA polymerase beta [16], and the NER factor xeroderma pigmentosum complementation group G (XPG) [17].

3.2 TIP60/KAT5—A MASTER REGULATOR OF CISPLATIN RESISTANCE

Methylation at Histone H3 links DNA damage detection at DNA double-strand breaks (DSBs) to activation of the KAT5/Tip60 [18].

A recent study has linked both KAT5 (Tip60) and HDAC6 as important regulators of lung cancer cell responses to cisplatin. The acetyltransferase Tip60 acetylates an important splicing factor SRSF2 on its lysine 52 residue promoting its proteasomal degradation, while HDAC6 abrogates this. In response to cisplatin, an acetylation/phosphorylation signaling network regulates both the accumulation of SRSF2 and splicing of caspase-8 pre-mRNA and determines whether cells undergo apoptosis or G(2)/M cell cycle arrest [19]. KAT5 (Tip60) has also been shown to regulate expression of excision repair cross-complementation group 1 (ERCC1) protein expression [20], an enzyme involved in the repair of cisplatin-induced DNA lesions further underscoring the importance of this lysine acetyltransferase in resistance mechanisms to cisplatin.

3.3 EZH2/PRC2 COMPLEXES AND CISPLATIN RESISTANCE

One of the first indicators that the lysine methyltransferase EZH2 (enhancer of zeste homolog 2) could be involved with cisplatin resistance came from a study in ovarian cancer which found that EZH2 was overexpressed in cisplatin-resistant ovarian cancer cells compared with cisplatin-sensitive cells [21]. Targeting EZH2 resulted in enhanced sensitivity to platinum-based therapy in a cisplatin-resistant cell line both *in vitro* and *in vivo* [21,22]. This observation was also seen in NSCLC lung adenocarcinoma, with overexpression noted in primary tissues and cisplatin-resistant cell lines. Targeting EZH2 in both cisplatin-resistant and cisplatin-sensitive cell lines resulted in significantly inhibited cell line proliferation and migration, leading to G(2)/M cell cycle arrest and apoptosis. Moreover, knockdown of EZH2 was found to enhance cisplatin sensitivity of cisplatin-resistant cells and was associated with reduced expression of multidrug resistance-related protein 1 (MDR1) levels [23].

In lung cancer, the VEGF signaling pathway has been linked to regulation of EZH2. In a cell line-based approach, VEGF signaling through VEGFR2 was found to upregulate EZH2 [24], and when EZH2 was therapeutically targeted using 3-deazaneplanocin A, an inhibitor of *S*-adenosyl methionine-dependent methyltransferases that targets the degradation of EZH2, cells were found to have increased sensitivity to treatment with cisplatin or carboplatin [24].

3.4 EPIGENETIC ERASERS ASSOCIATED WITH CISPLATIN RESISTANCE

Evidence linking histone deacetylases to cisplatin resistance is emerging. For instance, SIRT1 expression has been linked to cisplatin resistance in epidermoid and hepatoma cells [25]. In ovarian cancer cell lines, Kim and coworkers found that the expression level of acetylated histone 3 protein in SKOV3 cells increased after exposure to cisplatin, which was associated with decreased expression of HDAC1 and HDAC4. Forced overexpression of HDACs 1–4 in these cells was associated with an increased cell viability in the presence of cisplatin [26].

SIRT6 has recently been shown to recruit the chromatin remodeler SNF2H to DSBs and focally deacetylates histone H3K56 [27]. The link between epigenetic erasers and chromatin remodeling complexes in DNA repair and potentially mediating cisplatin resistance is becoming stronger as loss of SWI/SNF chromatin remodeling factor subunits has been shown to modulate cisplatin cytotoxicity [28].

3.5 EPIGENETIC READERS ASSOCIATED WITH CISPLATIN RESISTANCE

Epigenetic readers can be thought of as effector proteins that contain domains which recognize and are recruited to specific marks on histones or nucleotides. It must be noted that proteins which "write" or "erase" epigenetic marks may also contain such reader domains, thus leading coordinated "reading/ writing" and "reading/erasing" modifications for the regulation or setting of epigenetic events. Proteins that contain these reader domains can loosely be classified into four groups: chromatin architectural proteins, chromatin remodeling enzymes, chromatin modifiers, and adaptor proteins that function to recruit/assemble the protein complexes involved in gene expression. Some of the best recognized reader domains are the chromo domains, WD repeat domains, Tudor domains, plant homeodomain (PHD) domains, and bromodomains [29], and evidence is emerging to link several of these domain containing proteins with cisplatin resistance.

For instance the WD repeat protein WDR82, which recognizes histone H2B ubiquitination-dependent manner [30] and is a regulatory component of the Setd1a and Setd1b histone H3K4 methyltransferase complexes [31], has been found to be associated in a Tox4 protein complex which recognizes DNA adducts generated by cisplatin [32]. Likewise another WD repeat containing protein F-box/WD repeat-containing protein 7 (FBW7) has been shown to be involved with sensitivity to cisplatin [33].

3.6 BRCA1 COMPLEXES CONTAINING EPIGENETIC READERS/WRITERS AND ERASERS AS A CRITICAL ELEMENT IN CISPLATIN RESISTANCE

Epigenetic readers, writers, and erasers are often found in large multiprotein complexes. Complexes involving BRCA1 and the epigenetic regulatory machinery are now emerging as playing important roles in cisplatin sensitivity and resistance.

3.6.1 BRCA1 complexes, the epigenetic machinery, and DNA damage responses

The Breast Cancer 1 gene (BRCA1) has two important functions namely (i) the regulation of gene transcription and (ii) the regulation of the cellular response to DNA damage (DNA Repair) [34]. In this regard, BRCA1 acts mainly as a tumor suppressor through transcriptionally regulating genes involved with DNA repair [35].

Loss of BRCA1 expression is a frequent event in NSCLC [36,37], and loss of expression due to epigenetic inactivation via DNA CpG methylation is a factor in 18–30% of tumors [36,38]. Likewise BRCA1 is also commonly lost in ovarian cancer, and methylation at the BRCA1 promoter is a causative event in a proportion of patients [39]. Following the functional isolation of the first HDACs in 1996, it was soon shown that BRCA1 was able to associate with HDACs, indicating that chromatin remodeling may be an integral element in BRCA1 regulation of genome integrity [40]. In response to DNA damage, the BRCA1 protein forms several complexes, containing epigenetic readers, writers, and erasers to both execute and coordinate various aspects of the DNA damage response [41]. In this regard, it has recently been shown that the lysine demethylase KDM5B is required for efficient DSB repair and for the recruitment of Ku70 and BRCA1 [42]. Another critical component of DSB is the mediator of DNA damage checkpoint 1 (MDC1) protein, and its function is regulated by posttranslational phosphorylation, ubiquitylation, and sumoylation [43], and it has recently been shown that the lysine demethylase JMJD1C plays a crucial role in this process, and in response to DSBs JMJD1C demethylates MDC1 to regulate BRCA1-mediated chromatin responses to DNA breaks [44].

With respect to cisplatin, in response to DNA damage BRCA1 forms a heterodimer with the protein BRCA1-associated RING domain-1 (BARD1) generating a complex with a ubiquitin E3 ligase activity [45] that ubiquitylates both H2A and H2B histones [46] and effecting chromatin remodeling [47]. Cisplatin has been shown to directly bind to BRCA1 and its transcriptional transactivation activity is dramatically diminished in the presence of multiple cisplatin-damaged DNA sites [48]. Furthermore, when complexed with BARD1, cisplatin treatment results in a significantly reduced E3 ligase activity [49].

BRCA1 is also associated with the Mi-2/NuRD (nucleosome remodeling and deacetylase) complex. This complex has been shown to assemble on DSBs in breast cancer cells treated with cisplatin. Critically, phosphorylation of HDAC2 would appear to be essential for the formation of the NuRD complex at the DSB [50]. Other histone PTMs would appear to play essential roles in the repair of DSBs by NuRD. Upon induction of DSBs, the catalytic subunit of NuRD, CHD4 stimulates the accrual of RFN168 and RFN8, two proteins which together mediate ubiquitylation of histone H2A, resulting in a chromatin environment at the DSB that is permissive for the assembly of checkpoint and repair proteins including BRCA1 [51–53].

The effects of BRCA1 mutation in human and mouse cells are mitigated by concomitant deletion of the protein p53BP1. This protein binds histone H4 dimethylated at position lysine 20 (H4K20me2) to promote nonhomologous end joining. This therefore suggests that a balance between BRCA1 and 53BP1 is responsible for regulating DNA DSB repair mechanism choice. Greenberg and coworkers have shown that the activities of KATs and HDACs are essential to this balance. In particular, hypoacetylation due to loss of KAT5 (TIP60) acetyltransferase deficiency mimicked BRCA1 mutation, resulting in reduced BRCA1 and Rad51 DSB localization, and enhanced radial chromosome formation associated with increases in 53BP1, whereas HDAC inhibition yielded the opposite effect [54]. The data suggests that acetylation of H4K16 plays a central role in determining the balance of BRCA1 and 53BP1 at DSB chromatin by reducing, albeit not eliminating 53BP1 Tudor domain's binding affinity for H4K20me2 when present on the same H4 tail [54]. Critically, loss or impairment of KAT5 (TIP60) activity reduced homologous recombination and conferred sensitivity to PARP inhibition, suggesting that inhibitors directed at KAT5 may be useful in this context.

3.6.2 BRCA1 is linked with sensitivity to cisplatin

The critical role that BRCA1 plays in determining sensitivity to cisplatin originally came from murine studies of breast cancer. In these studies cells that were deficient for BRCA1 were found to be cisplatin sensitive, whereas restoring BRCA1 in these cells caused increased resistance. Xenograft studies confirmed these initial observations [55–57].

Such results have been confirmed in ovarian cancer. In ovarian cancer, of 115 primary sporadic ovarian carcinomas, 39 (34%) had low BRCA1 protein and 49 (42%) had low BRCA2 expression. Restoration of BRCA1 and BRCA2 mediated resistance to platinum chemotherapy in recurrent BRCA1 and BRCA2 mutated hereditary ovarian carcinomas [58]. Inactivation of BRCA1 by DNA CpG methylation predicts enhanced sensitivity to platinum-derived drugs to the same extent as BRCA1 mutations in breast and ovarian cancer. Most importantly, BRCA1 hypermethylation proved to be a predictor of longer time to relapse and improved overall survival in ovarian cancer patients undergoing chemotherapy with cisplatin [59].

Perhaps the first indicator within the clinical setting that levels of BRCA1 could predict response to cisplatin in NSCLC came from a study of patients treated with gemcitabine/cisplatin in the neoadjuvant setting. This study found that patients whose tumors had low levels of BRCA1 mRNA had a better outcome than those whose tumors high levels of BRCA1 mRNA [37,60]. Further evidence for this came from a study within the second-line setting, where low BRCA1 expression levels were significantly correlated with higher response rates, longer progression free survival, and median overall survival [61]. Finally in more advanced metastatic disease, Baorui Liu and coworkers confirmed that BRCA1 mRNA expression levels measured in pleural effusions from patients with metastatic NSCLC were inversely correlated with sensitivity to cisplatin [62].

To test the possibility that BRCA1 levels could be used to customize therapy of NSCLC patients, Rosell and coworkers conducted a Phase II clinical trial where patients were segregated and treated based on epidermal growth factor receptor (EGFR) mutation status and BRCA1 expression. EGFR-mutated patients received erlotinib, whereas nonmutated patients received chemotherapy with or without cisplatin based on their BRCA1 mRNA expression as follows: low, cisplatin plus gemcitabine; intermediate, cisplatin plus docetaxel; high, docetaxel alone. To determine whether any BRCA1 interacting partners could provide additional prognostic value, levels of interacting partner proteins (RAP80 and Abraxas) were also examined. It was found that patients with both low BRCA1 and RAP80 levels had a median survival exceeding 26 months compared to 11 months for patients with low BRCA1 alone. RAP80 was a significant factor for survival in patients treated according to BRCA1 levels (hazard ratio, 1.3 [95% CI, 1–1.7]; P=0.05) [63]. Supporting evidence for the role of BRCA1 expression in predicting sensitivity to cisplatin in NSCLC has come from additional studies [64–68]. In addition, BRCA1 expression also has predictive value for patients with metastatic NSCLC undergoing second-line cisplatin-based chemotherapy [61]. For those patients who have high expression of BRCA1, anti-tubulin-containing regimens are emerging as a potential therapeutic intervention [35], NSCLC patients with high BRCA1 mRNA expression were found to benefit more from this type of treatment (8.7 vs. 13.0 months) [69].

3.6.3 The link between BRCA1, K-methyltransferases, and acquired cisplatin resistance

KDMs are also linked to BRCA1 and response to cisplatin. The first indicators for this came from studies in murine BRCA1-deficient mammary tumor cells, which were found to be selectively sensitive to EZH2 (KMT6) inhibitors [70]. This KMT-added methyl groups to lysine 27 of histone H3 and loss of trimethyl lysine 27 (H3K27me3) is prognostic for poor outcome in ovarian cancer [71]. In a similar manner, high levels of H3K27me3 were found to have better prognosis in NSCLC [72]. Indeed across all pathological stages of NSCLC, patients with high levels of EZH2 have significantly poorer prognosis than patients with low EZH2 [73].

Ovarian or NSCLC patients with either high BRCA1 or EZH2 levels might potentially benefit from treatments with EZH2 inhibitors such as DZNep (3-Deazaneplanocin A). Indeed as mentioned earlier targeting EZH2 in NSCLC cell lines with DZNep was found to sensitize lung cancer cells to cisplatin [24]. As mentioned in a previous section, loss or downregulation of BRCA1 is common in ovarian cancer, and indeed it has now been shown that overexpression of EZH2 is linked to reduced expression of BRCA1 in manner similar to that seen in lung cancer [22].

4 NONCODING RNAs

ncRNAs have been explored in detail in earlier chapters of this volume. Two types of ncRNAs micro-RNAs (miRNAs) and long noncoding RNAs (lncRNAs) play important roles in epigenetic regulation. Furthermore, there is increasing evidence that both of these forms of ncRNA also play functional roles in cisplatin sensitivity/resistance.

4.1 miRNAs ASSOCIATED WITH CISPLATIN RESISTANCE/SENSITIVITY

Intriguingly, various miRNAs associated with resistance to cisplatin have been found to be downregulated by DNA CpG methylation [74]. For instance, in ovarian cancer the let-7e miRNA was found to be significantly reduced in a cisplatin-resistant human epithelial ovarian cancer (EOC) cell line A2780/CP compared with the parental A2780 cell [75]. Likewise, levels of this miRNA have been shown to be reduced in NSCLC and associated with poorer overall survival [76].

When A2780, SKOV3, and ES2 ovarian cancer cells were treated with cisplatin the levels of this miRNA decreased in a concentration-dependent manner. Overexpression of let-7e by transfection of agomir could resensitize A2780/CP and reduce the expression of cisplatin-resistant-related proteins EZH2 and cyclin D1 (CCND1), whereas let-7e inhibitors increased resistance to cisplatin in parental A2780 cells [75]. Quantitative methylation-specific PCR analysis showed hypermethylation of the CpG island adjacent to let-7e in A2780/CP cells, and demethylation treatment with 5-aza-CdR or transfection of pYr-let-7e-shRNA plasmid containing unmethylated let-7e DNA sequence could restore let-7e expression and partly reduce the chemoresistance. In addition, cisplatin combined with let-7e agomirs inhibited the growth of A2780/CP xenograft more effectively than cisplatin alone [75].

Overexpression of EZH2 has been shown to contribute to the development of acquired cisplatin resistance in ovarian cancer cells *in vitro* and *in vivo* [21]. The miRNA miR-101 is a critical regulator of EZH2 and loss or depletion of this miRNA results in elevated levels of EZH2 [77]. In both NSCLC and ovarian cancer, levels of miR-101 are downregulated [78,79], indicating that this miRNA may be an effective target for potentially sensitizing/resensitizing NSCLC and ovarian cancer to platinum-based therapy.

Using an integrative approach in lymphoblastoid cell lines (LCLs), Huang and coworkers exposed cells to the IC_{50} for cisplatin treatment (48-h treatment) and identified genes and miRNAs whose expression changed. They then integrated transcriptional gene expression and miRNA expression with their known single nucleotide polymorphism (SNP) genotypes to identify robust SNP associations with the IC_{50}. These were then mapped to genomic loci as quantitative trait loci (QTLs), and the associations between these QTLs and platinum sensitivity were subsequently evaluated. From the analysis for cisplatin, one 1 SNP associated with the expression of 1 miRNA and 1 gene was found (rs11138019 genotype, miR-30d expression, and ABCD2 expression), and functional characterization of this association was confirmed in the SKOV3 ovarian cancer cell line [80]. Intriguingly, miR-31 is associated with cisplatin resistance in NSCLC through regulating ABCD2 [81].

Other examples of miRNAs which are associated with cisplatin resistance in NSCLC and ovarian cancer through their activities on particular genes include miR-17/TGFβR2 [82]; miR-489/Akt3 [83]; let-7c/SMC1A, mir-133a/SUMO-1 [84]; miR-519d/XIAP-1 [85]; and miR-224-5p/PRKCD [86].

4.2 EPI-miRNAs AND CISPLATIN SENSITIVITY

A subset of miRNAs has been shown to regulate the expression of epigenetic effector proteins, which has led to them being described as epi-miRNAs. Several of these epi-miRNAs have been shown to be aberrantly expressed in NSCLC [87]. The importance of an epi-miRNA network in cisplatin resistance has been demonstrated by Edward Ratvitski and coworkers who have identified a network of miRNAs associated with the Tumor Protein p63 as a major element in the development of chemoresistance to cisplatin in laryngeal squamous cell carcinoma. Many of these include miRNAs that directly target the epigenetic machinery including miR-297 (DNMT3A), miR-92b-3p (HDAC9), and miR-485-5p (KDM4C) [88], while in a response to cisplatin a network of miRNA promoters was found to be regulated by p63 in complexes with other transcriptional and chromatin-associated factors including HDACs, KATs, KMTs, and KDMs [89].

Clearly a role between miRNAs and the epigenetic machinery exists to influence cellular responses to cisplatin and bigger networks will emerge in the future.

4.3 lncRNAs ASSOCIATED WITH RESISTANCE/SENSITIVITY

The role of lncRNAs in epigenetic regulation has been discussed in depth in an earlier chapter, and the reader is also directed to general overviews in the following reviews [90,91]. lncRNAs are also emerging as epigenetic elements involved with resistance to cisplatin. Initially, Yang et al. showed that a well-established lncRNA called HOTAIR could be involved with mediating sensitivity/resistance to cisplatin in hepatocellular carcinoma (HCC) cells. In HCC cell lines, siRNA suppression of HOTAIR reduced cell viability and cell invasion, sensitized cells to TNF-α induced apoptosis, and increased the chemotherapeutic sensitivity of the cancer cells to cisplatin and doxorubicin [92]. Indeed in NSCLC cell lines, expression of HOTAIR was found to be significantly upregulated in cisplatin-resistant A549/DDP cells compared with parental A549 cells and HOTAIR was observed to be significantly downregulated in adenocarcinoma samples from patients who were responsive to cisplatin therapy [93].

HOTAIR is a lncRNA that is known to associate with the PRC2 complex to regulate transcription by altered histone H3 lysine 27 (H3K27) methylation [94]. One of the pathways that HOTAIR regulates via H3K27me3 is the wingless-type MMTV integration site family (Wnt) canonical pathway. Increased H3K27me3 levels at the WIF-1 promoter result in decreased WIF-1 expression and concomitant activation of the Wnt signaling pathway [95], suggesting that HOTAIR when bound by PRC2 (and the associated KMT, EZH2) inhibits the expression of WIF-1. Critically, additional links between other lncRNAs and the Wnt signaling and cisplatin resistance are emerging.

Using microarray expression profiling of mRNAs, lncRNA and miRNA in A549 cells and cisplatin-resistant A549/CDDP cells, Yang et al. [96] identified and validated 8 mRNAs, 8 lncRNAs, and 5 miRNAs that were differentially expressed between the cells. By using gene coexpression network analysis, a link between one of these lncRNAs (AK126698) and the Wnt pathway was identified. Ablation of AK126698 not only greatly decreased expression of NKD2, a negative regulator of the Wnt/β-catenin signaling, but it also increased the accumulation and nuclear translocation of β-catenin and significantly decreased the rate of apoptosis induced by cisplatin in A549 cells [96]. Another lncRNA UCA1 has also been observed to be associated with resistance to cisplatin in human bladder cancer, and this also is linked to the Wnt signaling pathway [97].

Taken together, the data emerging suggests that lncRNAs may be important elements modulating the epigenetic landscape in the cellular response to cisplatin, and therefore playing central roles in the development of resistance to this chemotherapy.

5 HISTONE VARIANTS

Originally discovered in 1998, one of the commonest responses to DNA damage is the marking of the damaged DNA by the histone H2 variant gamma-H2AX [98]. Despite being a marker for DSBs, murine models with either absence or mutation of H2Ax were found to have no influence on sensitivity to cisplatin [99]. Nevertheless, Khabele and coworkers used gamma-H2AX phosphorylation (pH2AX) as a mark to identify novel cytotoxic histone deacetylase inhibitors (HDACi) for ovarian cancer. A panel of HDACi was tested in combination with cisplatin to determine if the HDACi enhanced the effects of cisplatin in ovarian cancer cells. The authors found that high levels of pH2AX in HDACi-treated ovarian cancer cells were tightly associated with decreased cell viability and increased apoptosis. From their analysis, a ketone-based HDACi was subsequently chosen for further development and found to enhance the effects of cisplatin, even in ovarian cancer cells resistant to this agent [100].

6 CANCER STEM CELLS AND CISPLATIN RESISTANCE

The cancer stem cell (CSC) hypothesis suggests that there is a small subset of cancer cells that are responsible for tumor initiation and growth, possessing properties such as indefinite self-renewal, slow replication, intrinsic resistance to chemotherapy and radiotherapy, and an ability to give rise to differentiated progeny [101]. Critically, CSCs or side populations may be an emerging central element in cisplatin resistance. For instance, we have shown that in NSCLC cell lines during the development of resistance a subpopulation of NSCLC cells emerge with a putative stem-like signature including significantly elevated Nanog, Oct-4, and SOX-2, along with the epithelial–mesenchymal transition (EMT) markers, c-Met, and beta-catenin [102].

In this regard, it is well established that in the regulation embryonic stem cells (ESCs) identity, KAT5 (Tip60) plays critical role by an associated Tip60-p400 complex [103]. However, it has now been shown that Hdac6 regulates Tip60-p400 function in stem cells [104]. In contrast to differentiated cells, where Hdac6 is mainly cytoplasmic, HDAC6 is largely nuclear in ESCs, neural stem cells (NSCs), and some cancer cell lines, and interacts with Tip60-p400 in each. Given that in a previous section, we have discussed the evidence linking resistance to cisplatin was found to involve both KAT5 and HDAC6 [19], it is tempting to speculate that this may link the resistance mechanism with the subpopulation of CSCs present.

Members of the miRNA processing machinery may also be linked to CSC development and the tumoral response to cisplatin. The PIWI/Argonaut family plays key roles in germ cell and stem cell self-renewal [105]. One of these genes the PIWI-like 2 (Piwil2) gene has been shown to be involved with chromatin remodeling in response to cisplatin. This remodeling involves histone H3 acetylation by p300 (KAT3B) [106], and additional studies determined that the Piwil2-mediated chromatin relaxation is essential for DNA damage repair, and may therefore act as a critical gatekeeper in DNA damage induced tumorigenesis [107]. Given the role of Piwil2 in stem cell self-renewal, and that fact that it is induced in response to DNA damage it has been suggested that this may contribute to drug resistance of tumors through enhanced activity in CSCs [107]

7 PROTEIN PTMs ASSOCIATED WITH DEVELOPMENT OF CISPLATIN RESISTANCE

Much of the current focus in the previous sections has discussed the role of the epigenetic machinery at the level of chromatin. However, it must be noted that many members of the cellular epigenetic machinery can also effect PTMs on nonhistone proteins which has also led to the suggestion that there may be protein code (in addition to the classical "Histone Code") [108]. There is increasing evidence that PTMs on nonhistone proteins play crucial roles in the regulation not only of their function and activity but also recent evidence indicates that protein PTMs may be involved with the development of resistance to cisplatin.

Foxo3a is a member of the Forkhead O transcription factors (FOXO) which are critical for the regulation of cell cycle arrest, cell death, and DNA damage repair [109]. In lung cancer patients a poor prognosis is associated with low FOXO3a expression [110], and loss or deletion of Foxo3a is a frequent event in lung cancer [111].

Acetylation of Foxo3a is essential to its basal activity through the activities of KATs and HDACs [112,113], and the acetylation of Foxo3a is emerging as a central element in cancer cell responses to various chemotherapies including paclitaxel, epirubicin [114], and imatinib [115]. Indeed acetylation of Foxo3a may be a central element in cisplatin resistance. In cells treated with cisplatin, levels of acetylated Foxo3a were reduced, which was also a feature in cisplatin-resistant cells [116].

The nuclear factor kappa B (NF-kB) pathway is well established as a major element in many cellular responses including DNA damage. Furthermore it is also well established that the regulation of NF-kB also involves PTMs particularly through lysine acetylation of the p65/RelA subunit through the activities of KATs/HDACs [117,118]. NF-kB can also be modulated through lysine methylation by the lysine methyltransferases Set9 (KMT7) [119] and NSD1 (KMT3B) [120], and by arginine methylation via protein arginine methyltransferase 5 (PRMT5) [121]. Interestingly, cytoplasmic expression of PRMT5 is correlated with a poor prognosis [122], while CRAM1 and PRMT1 are found to be significantly overexpressed in NSCLC, although neither correlate with survival [123].

A common problem for patients being treated with cisplatin is the development of cisplatin-induced nephrotoxicity. Cisplatin has been shown to reduce the levels of SIRT1 [124], and it was shown that the cisplatin-mediated toxicity involved acetylation of p65/RelA by this histone deacetylase SIRT1 [125]. while activators of SIRT1 or forced overexpression of this HDAC result in amelioration of cisplatin-induced renal apoptotic injury in cell line models and animal experiments [124,125].

Targeting NF-kB is therefore emerging as a promising potential therapeutic avenue for chemoresistance [126]. In a panel of isogenic parent/resistant NSCLC overexpression of Nuclear Factor Of Kappa Light Polypeptide Gene Enhancer In B-Cells Inhibitor, Alpha (NFKBI or IKBA) was observed in the cisplatin-resistant cells compared to the parent, and treating cells with a NF-kB-specific inhibitor (DHMEQ) led to significantly enhanced effects on viability and proliferation in cisplatin-resistant cells compared with parent cells, indicating that targeting the NF-kB pathway may be an important potential therapeutic avenue in treating cisplatin-resistant patients. In this regard, it was recently shown that the epigenetic reader Brd4 maintains the activated form of NF-kB (involving acetylated p65/RelA at position lysine 310) and therefore inhibitors of Brd4 may be able to sensitize cells to cisplatin therapy. Inhibitors of BRD4 are being actively pursued and several are currently in clinical trials.

8 TARGETING CISPLATIN RESISTANCE EPIGENETICALLY

In the previous sections, I have set out the data linking epigenetics as a key mechanism underpinning cisplatin resistance. It would seem that potentially targeting the enzymes involved could be used as a therapeutic approach to enhancing cisplatin activity and/or resensitizing resistant tumors' to cisplatin-based chemotherapy regimens. In the following sections, I shall describe some of the data showing that targeting the epigenetic machinery may have clinical potential.

For instance, expression of the tumor suppressor Regulator of G protein signaling 10 (RGS10) is suppressed in cell models of ovarian cancer chemoresistance. Hooks and coworkers demonstrated that this was due to both DNA methylation and chromatin compaction via HDACs [127]. Subsequent studies demonstrated that pharmacological inhibition of DNMT or HDAC enzymatic activity resulted in significantly increased RGS10 expression and this was associated with enhanced cisplatin-mediated cell death [128].

Another study by Wang and coworkers developed a novel HDAC inhibitor called YCW1. This compound was shown to exert cancer-specific cytotoxicity via mitochondria-mediated apoptosis in a model of lung cancer. Moreover, this antitumor effect was found to be synergistic when combined with cisplatin. In both orthotopic and subcutaneous implanted models of lung cancer, YCW1 significantly inhibited tumor growth and was found to prevent lung metastases via inhibition of focal adhesion complex [129].

A study examining the efficacy of a histone deacetylase inhibitor and a DNA demethylating agent in combination with low dose cisplatin in ovarian cancer cell lines and xenografts was found to potentiate the anticancer effects of cisplatin [130].

It may not be necessary to use inhibitors that target the epigenetic machinery directly. The family of lysine demethylases is known to be FE++ dependent. Therefore the potential exists to use of iron chelators in therapeutic regimens using cisplatin. Beland and coworkers examined the potential utility of the iron chelator desferrioxamine (DFO) to target cancer. In this paper, the authors showed that DFO caused significant changes in epigenetic alterations. Specifically, DFO treatment decreased the protein levels of the histone H3 lysine 9 demethylase, KDM4A (JMJD2A), and the histone H3 lysine 4 demethylase, KDM1A (LSD1), in breast cancer cell lines. DFO treatment activates apoptotic programs in cancer cells and enhances their sensitivity to cisplatin. In p53 wild-type cells this was found to be due to the downregulation of KDM4A whereas in p53 mutant cells, the activation of the p53-independent apoptotic program was driven predominantly by the epigenetic upregulation of p21 [131].

Given the potential significance of KAT5/Tip60 in cisplatin resistance, it will be interesting to see if agents that specifically target this protein have efficacy. Several inhibitors targeting KAT5/Tip60 have recently emerged including MG-149 [132], Nu9056 [133], and TH1834 [134]. I am unaware of any clinical trials currently running for these inhibitors, so it remains to be seen if they could have any efficacy in the clinical setting.

8.1 NATURAL BIOACTIVES

The natural bioactive polyphenol epigallocatechin gallate (EGCG) has been shown to enhance the efficacy of cisplatin by both NSCLC cell lines [135–137], while a combination of EGCG and sulforaphane, or EGCG and curcumin enhanced cisplatin mediated responses in ovarian cancer cell lines

[138,139]. EGCG is a compound which can inhibit DNA methyltransferases, sulforaphane inhibits HDACs, and curcumin can inhibit KATs. Sulforaphane itself has also been shown independently to enhance cisplatin sensitivity in ovarian and lung cancers [140,141]. In this regard, sulforaphane has also been shown to protect from cisplatin induced liver and kidney toxicity [142,143], suggesting that this compound could potentially have additional benefit to the patient. A more detailed description of natural epigenetic inhibitors is provided in Chapter 20.

9 CLINICAL TRIALS

Many clinical trials involving epigenetic therapies and cisplatin have been conducted. Initially, the results of such trials were discouraging. For example, a Phase I trial of cisplatin plus decitabine, in patients with advanced solid tumors and a follow-up early Phase II evaluation in patients with inoperable NSCLC, came to the conclusion that the cisplatin plus decitabine combination did not exhibit significant antitumor activity in patients with NSCLC at the dose and schedule applied to justify its further evaluation [144]. Nevertheless, trials are ongoing and a selection is shown in Table 26.1. Several are showing promise and are discussed in more detail.

One combination of epigenetic therapy inhibitors which is beginning to show some promise is hydralazine/valproate (valproic acid). Hydralazine acts as a DNMT inhibitor [145], while valproic acid functions to inhibit HDACs [146].

In a Phase III clinical trial of Cervical Cancer (Phase III) designed to test the superiority of hydralazine and valproic acid plus standard cisplatin topotecan against placebo plus cisplatin topotecan upon progression free survival. At a median follow-up time of 7 months (1–22), the median PFS (progression free survival) is 6 months for chemotherapy + placebo and 10 months for chemotherapy + hydralazine/valproic acid ($P=0.0384$, two-tailed) [147].

An earlier Phase II clinical trial using hydralazine/valproic acid was tested in combination with various chemotherapies in refractory solid tumors' to see if this combination could overcome chemotherapy resistance including cisplatin. The results of this small clinical trial found a clinical benefit in 12 (80%) patients: 4 PR and 8 SD. Particularly noteworthy from this study is that of seven ovarian cancer patients included in the trial all achieved either partial responses ($n=3$) or disease stabilization ($n=4$) [148].

A large number of studies have shown that valproate has efficacy in resensitizing or enhancing the effects of cisplatin in cell line models. These include studies in ovarian cancer [149], lung cancer (both NSCLC and SCLC) [150,151], and mesothelioma [152]. The standard first line therapy for mesothelioma is cisplatin/pemetrexed (or cisplatin/raltitrexed). Only 41% of patients have an objective response and the other 59% are refractory. In a Phase II study of valproate plus doxorubicin in patients ($n=45$) having failed cisplatin/pemetrexed, 16% had a partial response. As there is no recommended second-line therapy for mesothelioma this represents a potential new intervention in cisplatin refractory mesothelioma [153].

The data would suggest that histone deacetylase inhibitors might have a potentially strong role to play in targeting cisplatin refractory tumors'. However, the largest Phase III, randomized, double-blind, placebo-controlled clinical trial to date in mesothelioma, which tested the efficacy of the histone deacetylase inhibitor Vorinostat in Patients With Advanced MPM who had Failed Prior Pemetrexed and Either Cisplatin or Carboplatin Therapy, failed. Vorinostat did not significantly extend the overall

Table 26.1 Some of the Clinical Trials Involving Epigenetic Targeting Agents with Links to Cisplatin

ClinicalTrials.Gov Identifier	Drug/regimen	Target	Phase	Data
NCT00404326	Hydralazine; valproate; cisplatin	Cervical cancer	II	Completed
NCT00532818	Hydralazine; valproate; cisplatin; topotecan	Cervical cancer	III	[147]
NCT00404508	Hydralazine; valproate; chemotherapy	Solid tumors'	II	[148]
NCT00634205	Valproate; doxorubicin	Malignant mesothelioma	II	[153]
NCT01203735	Valproate; chemoradiotherapy	NSCLC	I/II	Recruiting
Vantage 014 NCT00128102 (?)	Vorinostat	Mesothelioma	III	[154]
NCT01336842	Panobinostat	Solid tumors' NSCLC	I	Recruiting
NCT00387465	Entinostat; azacitidine	NSCLC	I/II	[161]
NCT00106626	Vorinostat; pemetrexed; cisplatin	Solid tumors'	I	Completed
NCT00907738 (continuation of NCT00106626)	Vorinostat; pemetrexed; cisplatin	Solid tumors'	II	Completed
NCT01353482	Cisplatin; pemetrexed; vorinostat; placebo	Mesothelioma	I/II	Withdrawn
NCT01045538	Vorinostat; capecitabine; cisplatin	Gastric cancer	I/II	Active not recruiting
NCT01064921	Vorinostat; cisplatin; radiation	Oropharyngeal squamous cell carcinoma	I	Recruiting
NCT00662311	Vorinostat; paclitaxel; radiation therapy	NSCLC	I	Terminated
NCT01059552	Dose escalation of vorinostat, cisplatin, pemetrexed, and radiation	NSCLC	I	Recruiting
NCT00901537	Azacitidine; cisplatin	Lung cancer head and neck cancer	I	Terminated
NCT00443261	Azacitidine; cisplatin	Head and neck squamous cell carcinoma	I	Terminated
NCT01209520	Cisplatin; carboplatin; paclitaxel; vidaza	NSCLC	Pilot study	Active, not recruiting
NCT01846390	Gemcitabine; dexamethasone; cisplatin; romidepsin	Peripheral T-cell lymphoma diffuse large B-cell lymphoma	I	Recruiting
NCT00842582	Azacytidine; carboplatin; paclitaxel	Ovarian cancer	I	Terminated
NCT00529022	Azacitidine; valproic acid; carboplatin	Ovarian cancer	I	Completed [156]
NCT00748527	Carboplatin; decitabine	Ovarian cancer	II	Terminated
NCT00477386	Carboplatin; decitabine	Ovarian cancer	I/II	Active, not recruiting
NCT01478685	CC-486 (oral azacitidine) carboplatin; ABI-007	Relapsed or refractory solid tumor	I	Recruiting
NCT01928576	Nivolumab (anti-PD1); azacitidine; entinostat CC-486	Recurrent metastatic NSCLC which must have received at least one platinum-based chemotherapy	II	Recruiting

survival of patients with advanced MPM who had failed prior chemotherapy compared to placebo [154]. Nevertheless new HDAC inhibitors are being developed which have improved efficacy in combination with cisplatin in both NSCLC and MPM cell lines [155].

A recent Phase I clinical trial in ovarian cancer examined the effect of Azacitidine and Valproate in combination with carboplatin in patients with ovarian cancer. Minor responses or stable disease lasting ≥ 4 months were achieved by six patients (18.8%), including three with platinum-resistant or platinum refractory ovarian cancer, suggesting that regimens including epigenetic targeting agents may have a role to play in the management of platinum refractory ovarian cancer [156].

Returning to natural bioactives, the efficacy of EGCG and cisplatin has recently been examined in a Phase I study of concurrent chemotherapy and thoracic radiotherapy with oral EGCG protection in patients with locally advanced stage III NSCLC (NCT01481818). Whilst the purpose of this study was to examine the effect of EGCG on acute radiation-induced esophagitis (ARIE), the patients in the trial ($n=24$), overall response rate for the standard chemoradiation regimen was 66.7%, which was measured by a CT scan 6–8 weeks after completion of treatment. There were no complete responses, partial responses (PR) were observed in 16 patients, stable disease (SD) in 3 patients, and 5 patients had progressive disease (PD) in 5. Strikingly, EGCG showed dramatic regression of esophagitis to grade 0/1 was observed in 22 of 24 patients and has subsequently led to an extension of this trial into Phase II [157].

9.1 LOW DOSE THERAPIES AS "EPIGENETIC PRIMING" EVENTS

Data is emerging to show that low dose treatments with epigenetic targeting agents may prove to be the next breakthrough in the treatment of cancer. Using clinically relevant doses that do not cause any cytotoxicity, tumors may become "epigenetically primed" for standard chemotherapy or targeted therapy. This has implications for targeting patients whose tumors have become resistant to cisplatin.

For example, nontoxic dose of the histone deacetylase inhibitor chidamide has been shown to synergistically enhances platinum-induced DNA damage responses and apoptosis in NSCLC cells [158]. There are also indications that such a strategy may also be useful for treatment of ovarian cancer. Lung and coworkers demonstrated that the addition of a low dose of a histone deacetylase inhibitor (LBH589 or Panobinostat) resensitized a cisplatin refractory ovarian cancer cell line [159].

In probably the most important development to date in this regard, transient treatments with clinically relevant nanomolar doses of DNA-demethylating agents exert durable antitumor effects on hematological and epithelial tumor cells. Treatments generated an antitumor "memory" response, including inhibition of subpopulations of cancer stem-like cells, accompanied by sustained decreases in genome-wide promoter DNA methylation, gene reexpression, and antitumor changes in key cellular regulatory pathways [160]. Based on the data that emerged from the clinical trial involving low dose entinostat/azacitidine combination in NSCLC (NCT00387465) [161], and the initial results from the anti-PD1 therapies (which included patients with NSCLC-NCT00730639) [162], NCT01928576 is a Phase II clinical trial in NSCLC examining whether epigenetic priming with low dose azacitidine/entinostat prior to therapy with an anti-PD1 agent (Nivolumab) will improve objective response rates. In this study of patients with recurrent metastatic NSCLC, to be eligible, patients must have received at least one prior platinum-based chemotherapy, and not more than three for stage IIIB/IV disease, or alternatively if they have received adjuvant or neoadjuvant platinum-doublet chemotherapy (after surgery and/or radiation therapy) and developed recurrent or metastatic disease within 6 months of completing therapy. It will be very interesting to see if such a treatment regimen will enhance objective response and indicates a new therapeutic avenue for treating patients which have developed cisplatin refractory cancer.

While to my knowledge no studies involving epigenetic priming have yet completed for cisplatin, a recent Phase I/II trial (NCT01004991) using epigenetic priming in patients newly diagnosed with diffuse large B-cell lymphoma (DLBCL) demonstrated very strong responses. Eleven of 12 patients achieved a complete response and 10 remained in remission with a median follow-up of 13 months (range 5–28 months) at the time of publication [163]. Another Phase I/II clinical trial (NCT01799083) with refractory advanced solid tumors using just a low dose demethylating agent (Decitabine) coupled with chemotherapy and decitabine-based immunotherapy (Cytokine-Induced Killer Cells (CIK) isolated from patient blood and expanded) has reported some responses. These include partial response and stable disease in patients with malignant pleural disease, ovarian cancer, and lung adenocarcinoma [164].

Another factor indicating the potential usefulness of low dose epigenetic therapies concerns the treatment of elderly patients. One recent report has shown the successful treatment of elderly patients aged over 80 suffering from acute myelogenous leukemia with low dose decitabine [165].

10 CONCLUSION

It is clear from the previous sections that epigenetics plays important roles in both the cellular response to cisplatin, and strong evidence exists linking epigenetic mechanisms to both the development and maintenance of resistance to platinum-based chemotherapy regimens.

The emerging clinical data for combinatorial low dose epigenetic therapies is compelling and future work will determine if such therapies will revolutionize the way chemotherapy-resistant tumors are treated. One concern, however, is the issue of intratumoral heterogeneity, and if this may affect how epigenetic therapies (both standard and epigenetic priming) will affect the long-term outcome of patients treated to resensitize to cisplatin-based regimens.

REFERENCES

[1] Nguyen HT, Tian G, Murph MM. Molecular epigenetics in the management of ovarian cancer: are we investigating a rational clinical promise? Front Oncol 2014;4:71.

[2] Langevin SM, Kratzke RA, Kelsey KT. Epigenetics of lung cancer. Transl Res 2015;165(1):74–90.

[3] Yu W, Jin C, Lou X, Han X, Li L, He Y, et al. Global analysis of DNA methylation by methyl-capture sequencing reveals epigenetic control of cisplatin resistance in ovarian cancer cell. PLoS One 2011;6(12):e29450.

[4] Zeller C, Dai W, Steele NL, Siddiq A, Walley AJ, Wilhelm-Benartzi CS, et al. Candidate DNA methylation drivers of acquired cisplatin resistance in ovarian cancer identified by methylome and expression profiling. Oncogene 2012;31(42):4567–76.

[5] Zhang YW, Zheng Y, Wang JZ, Lu XX, Wang Z, Chen LB, et al. Integrated analysis of DNA methylation and mRNA expression profiling reveals candidate genes associated with cisplatin resistance in non-small cell lung cancer. Epigenetics 2014;9(6).

[6] De Carvalho DD, Sharma S, You JS, Su SF, Taberlay PC, Kelly TK, et al. DNA methylation screening identifies driver epigenetic events of cancer cell survival. Cancer Cell 2012;21(5):655–67.

[7] Igarashi T, Izumi H, Uchiumi T, Nishio K, Arao T, Tanabe M, et al. Clock and ATF4 transcription system regulates drug resistance in human cancer cell lines. Oncogene 2007;26(33):4749–60.

[8] Miyamoto N, Izumi H, Noguchi T, Nakajima Y, Ohmiya Y, Shiota M, et al. Tip60 is regulated by circadian transcription factor clock and is involved in cisplatin resistance. J Biol Chem 2008;283(26):18218–26.

[9] Hirano G, Izumi H, Kidani A, Yasuniwa Y, Han B, Kusaba H, et al. Enhanced expression of PCAF endows apoptosis resistance in cisplatin-resistant cells. Mol Cancer Res 2010;8(6):864–72.

[10] Cai D, Shames DS, Raso MG, Xie Y, Kim YH, Pollack JR, et al. Steroid receptor coactivator-3 expression in lung cancer and its role in the regulation of cancer cell survival and proliferation. Cancer Res 2010;70(16):6477–85.

[11] Wu ZZ, Sun NK, Chao CC. Knockdown of CITED2 using short-hairpin RNA sensitizes cancer cells to cisplatin through stabilization of p53 and enhancement of p53-dependent apoptosis. J Cell Physiol 2011;226(9):2415–28.

[12] Ianari A, Gallo R, Palma M, Alesse E, Gulino A. Specific role for p300/CREB-binding protein-associated factor activity in E2F1 stabilization in response to DNA damage. J Biol Chem 2004;279(29):30830–5.

[13] Ogiwara H, Ui A, Otsuka A, Satoh H, Yokomi I, Nakajima S, et al. Histone acetylation by CBP and p300 at double-strand break sites facilitates SWI/SNF chromatin remodeling and the recruitment of non-homologous end joining factors. Oncogene 2011;30(18):2135–46.

[14] Rubbi CP, Milner J. p53 is a chromatin accessibility factor for nucleotide excision repair of DNA damage. EMBO J 2003;22(4):975–86.

[15] Wang QE, Han C, Zhao R, Wani G, Zhu Q, Gong L, et al. p38 MAPK-and Akt-mediated p300 phosphorylation regulates its degradation to facilitate nucleotide excision repair. Nucleic Acids Res 2013;41(3):1722–33.

[16] Hasan S, El-Andaloussi N, Hardeland U, Hassa PO, Burki C, Imhof R, et al. Acetylation regulates the DNA end-trimming activity of DNA polymerase beta. Mol Cell 2002;10(5):1213–22.

[17] Tillhon M, Cazzalini O, Nardo T, Necchi D, Sommatis S, Stivala LA, et al. p300/CBP acetyl transferases interact with and acetylate the nucleotide excision repair factor XPG. DNA Repair (Amst) 2012;11(10):844–52.

[18] Sun Y, Jiang X, Xu Y, Ayrapetov MK, Moreau LA, Whetstine JR, et al. Histone H3 methylation links DNA damage detection to activation of the tumour suppressor Tip60. Nat Cell Biol 2009;11(11):1376–82.

[19] Edmond V, Moysan E, Khochbin S, Matthias P, Brambilla C, Brambilla E, et al. Acetylation and phosphorylation of SRSF2 control cell fate decision in response to cisplatin. Embo J 2011;30(3):510–23.

[20] Van Den Broeck A, Nissou D, Brambilla E, Eymin B, Gazzeri S. Activation of a Tip60/E2F1/ERCC1 network in human lung adenocarcinoma cells exposed to cisplatin. Carcinogenesis 2012;33(2):320–5.

[21] Hu S, Yu L, Li Z, Shen Y, Wang J, Cai J, et al. Overexpression of EZH2 contributes to acquired cisplatin resistance in ovarian cancer cells *in vitro* and *in vivo*. Cancer Biol Ther 2010;10(8):788–95.

[22] Li T, Cai J, Ding H, Xu L, Yang Q, Wang Z. EZH2 participates in malignant biological behavior of epithelial ovarian cancer through regulating the expression of BRCA1. Cancer Biol Ther 2014;15(3):271–8.

[23] Lv Y, Yuan C, Xiao X, Wang X, Ji X, Yu H, et al. The expression and significance of the enhancer of zeste homolog 2 in lung adenocarcinoma. Oncol Rep 2012;28(1):147–54.

[24] Riquelme E, Suraokar M, Behrens C, Lin HY, Girard L, Nilsson MB, et al. VEGF/VEGFR-2 upregulates EZH2 expression in lung adenocarcinoma cells and EZH2 depletion enhances the response to platinum-based and VEGFR-2-targeted therapy. Clin Cancer Res 2014;20(14):3849–61.

[25] Liang XJ, Finkel T, Shen DW, Yin JJ, Aszalos A, Gottesman MM. SIRT1 contributes in part to cisplatin resistance in cancer cells by altering mitochondrial metabolism. Mol Cancer Res 2008;6(9):1499–506.

[26] Kim MG, Pak JH, Choi WH, Park JY, Nam JH, Kim JH. The relationship between cisplatin resistance and histone deacetylase isoform overexpression in epithelial ovarian cancer cell lines. J Gynecol Oncol 2012;23(3):182–9.

[27] Toiber D, Erdel F, Bouazoune K, Silberman DM, Zhong L, Mulligan P, et al. SIRT6 recruits SNF2H to DNA break sites, preventing genomic instability through chromatin remodeling. Mol Cell 2013;51(4):454–68.

[28] Kothandapani A, Gopalakrishnan K, Kahali B, Reisman D, Patrick SM. Downregulation of SWI/SNF chromatin remodeling factor subunits modulates cisplatin cytotoxicity. Exp Cell Res 2012;318(16):1973–86.

[29] Dawson MA, Kouzarides T. Cancer epigenetics: from mechanism to therapy. Cell 2012;150(1):12–27.

[30] Wu M, Wang PF, Lee JS, Martin-Brown S, Florens L, Washburn M, et al. Molecular regulation of H3K4 trimethylation by Wdr82, a component of human Set1/COMPASS. Mol Cell Biol 2008;28(24):7337–44.

[31] Lee JH, You J, Dobrota E, Skalnik DG. Identification and characterization of a novel human PP1 phosphatase complex. J Biol Chem 2010;285(32):24466–76.

[32] Bounaix Morand du Puch C, Barbier E, Kraut A, Coute Y, Fuchs J, Buhot A, et al. TOX4 and its binding partners recognize DNA adducts generated by platinum anticancer drugs. Arch Biochem Biophys 2011;507(2):296–303.

[33] Yu HG, Wei W, Xia LH, Han WL, Zhao P, Wu SJ, et al. FBW7 upregulation enhances cisplatin cytotoxicity in non-small cell lung cancer cells. Asian Pac J Cancer Prev 2013;14(11):6321–6.

[34] Foulkes WD, Shuen AY. In brief: BRCA1 and BRCA2. J Pathol 2013;230(4):347–9.

[35] Price M, Monteiro AN. Fine tuning chemotherapy to match BRCA1 status. Biochem Pharmacol 2010; 80(5):647–53.

[36] Lee MN, Tseng RC, Hsu HS, Chen JY, Tzao C, Ho WL, et al. Epigenetic inactivation of the chromosomal stability control genes BRCA1, BRCA2, and XRCC5 in non-small cell lung cancer. Clin Cancer Res 2007;13(3):832–8.

[37] Rosell R, Skrzypski M, Jassem E, Taron M, Bartolucci R, Sanchez JJ, et al. BRCA1: a novel prognostic factor in resected non-small-cell lung cancer. PLoS One 2007;2(11):e1129.

[38] Wang Y, Zhang D, Zheng W, Luo J, Bai Y, Lu Z. Multiple gene methylation of nonsmall cell lung cancers evaluated with 3-dimensional microarray. Cancer 2008;112(6):1325–36.

[39] Catteau A, Harris WH, Xu CF, Solomon E. Methylation of the BRCA1 promoter region in sporadic breast and ovarian cancer: correlation with disease characteristics. Oncogene 1999;18(11):1957–65.

[40] Yarden RI, Brody LC. BRCA1 interacts with components of the histone deacetylase complex. Proc Natl Acad Sci USA 1999;96(9):4983–8.

[41] Roy R, Chun J, Powell SN. BRCA1 and BRCA2: different roles in a common pathway of genome protection. Nat Rev Cancer 2012;12(1):68–78.

[42] Li X, Liu L, Yang S, Song N, Zhou X, Gao J, et al. Histone demethylase KDM5B is a key regulator of genome stability. Proc Natl Acad Sci USA 2014;111(19):7096–101.

[43] Lu J, Matunis MJ. A mediator methylation mystery: JMJD1C demethylates MDC1 to regulate DNA repair. Nat Struct Mol Biol 2013;20(12):1346–8.

[44] Watanabe S, Watanabe K, Akimov V, Bartkova J, Blagoev B, Lukas J, et al. JMJD1C demethylates MDC1 to regulate the RNF8 and BRCA1-mediated chromatin response to DNA breaks. Nat Struct Mol Biol 2013;20(12):1425–33.

[45] Baer R, Ludwig T. The BRCA1/BARD1 heterodimer, a tumor suppressor complex with ubiquitin E3 ligase activity. Curr Opin Genet Dev 2002;12(1):86–91.

[46] Thakar A, Parvin J, Zlatanova J. BRCA1/BARD1 E3 ubiquitin ligase can modify histones H2A and H2B in the nucleosome particle. J Biomol Struct Dyn 2010;27(4):399–406.

[47] Zhao Y, Brickner JR, Majid MC, Mosammaparast N. Crosstalk between ubiquitin and other post-translational modifications on chromatin during double-strand break repair. Trends Cell Biol 2014;24(7):426–34.

[48] Ratanaphan A, Wasiksiri S, Canyuk B, Prasertsan P. Cisplatin-damaged BRCA1 exhibits altered thermostability and transcriptional transactivation. Cancer Biol Ther 2009;8(10):890–8.

[49] Atipairin A, Canyuk B, Ratanaphan A. The RING heterodimer BRCA1–BARD1 is a ubiquitin ligase inactivated by the platinum-based anticancer drugs. Breast Cancer Res Treat 2011;126(1):203–9.

[50] Sun JM, Chen HY, Davie JR. Differential distribution of unmodified and phosphorylated histone deacetylase 2 in chromatin. J Biol Chem 2007;282(45):33227–36.

[51] Larsen DH, Poinsignon C, Gudjonsson T, Dinant C, Payne MR, Hari FJ, et al. The chromatin-remodeling factor CHD4 coordinates signaling and repair after DNA damage. J Cell Biol 2010;190(5):731–40.

[52] Luijsterburg MS, Acs K, Ackermann L, Wiegant WW, Bekker-Jensen S, Larsen DH, et al. A new non-catalytic role for ubiquitin ligase RNF8 in unfolding higher-order chromatin structure. EMBO J 2012;31(11):2511–27.

[53] Smeenk G, Wiegant WW, Vrolijk H, Solari AP, Pastink A, van Attikum H. The NuRD chromatin-remodeling complex regulates signaling and repair of DNA damage. J Cell Biol 2010;190(5):741–9.

[54] Tang J, Cho NW, Cui G, Manion EM, Shanbhag NM, Botuyan MV, et al. Acetylation limits 53BP1 association with damaged chromatin to promote homologous recombination. Nat Struct Mol Biol 2013;20(3):317–25.

[55] Bhattacharyya A, Ear US, Koller BH, Weichselbaum RR, Bishop DK. The breast cancer susceptibility gene BRCA1 is required for subnuclear assembly of Rad51 and survival following treatment with the DNA cross-linking agent cisplatin. J Biol Chem 2000;275(31):23899–903.

[56] Tassone P, Tagliaferri P, Perricelli A, Blotta S, Quaresima B, Martelli ML, et al. BRCA1 expression modulates chemosensitivity of BRCA1-defective HCC1937 human breast cancer cells. Br J Cancer 2003;88(8):1285–91.

[57] Tassone P, Di Martino MT, Ventura M, Pietragalla A, Cucinotto I, Calimeri T, et al. Loss of BRCA1 function increases the antitumor activity of cisplatin against human breast cancer xenografts in vivo. Cancer Biol Ther 2009;8(7):648–53.

[58] Swisher EM, Gonzalez RM, Taniguchi T, Garcia RL, Walsh T, Goff BA, et al. Methylation and protein expression of DNA repair genes: association with chemotherapy exposure and survival in sporadic ovarian and peritoneal carcinomas. Mol Cancer 2009;8:48.

[59] Stefansson OA, Villanueva A, Vidal A, Marti L, Esteller M. BRCA1 epigenetic inactivation predicts sensitivity to platinum-based chemotherapy in breast and ovarian cancer. Epigenetics 2012;7(11):1225–9.

[60] Taron M, Rosell R, Felip E, Mendez P, Souglakos J, Ronco MS, et al. BRCA1 mRNA expression levels as an indicator of chemoresistance in lung cancer. Hum Mol Genet 2004;13(20):2443–9.

[61] Papadaki C, Sfakianaki M, Ioannidis G, Lagoudaki E, Trypaki M, Tryfonidis K, et al. ERCC1 and BRAC1 mRNA expression levels in the primary tumor could predict the effectiveness of the second-line cisplatin-based chemotherapy in pretreated patients with metastatic non-small cell lung cancer. J Thorac Oncol 2012;7(4):663–71.

[62] Wang L, Wei J, Qian X, Yin H, Zhao Y, Yu L, et al. ERCC1 and BRCA1 mRNA expression levels in metastatic malignant effusions is associated with chemosensitivity to cisplatin and/or docetaxel. BMC Cancer 2008;8:97.

[63] Rosell R, Perez-Roca L, Sanchez JJ, Cobo M, Moran T, Chaib I, et al. Customized treatment in non-small-cell lung cancer based on EGFR mutations and BRCA1 mRNA expression. PLoS One 2009;4(5):e5133.

[64] Bonanno L, Costa C, Majem M, Sanchez JJ, Gimenez-Capitan A, Rodriguez I, et al. The predictive value of 53BP1 and BRCA1 mRNA expression in advanced non-small-cell lung cancer patients treated with first-line platinum-based chemotherapy. Oncotarget 2013;4(10):1572–81.

[65] Li Z, Qing Y, Guan W, Li M, Peng Y, Zhang S, et al. Predictive value of APE1, BRCA1, ERCC1 and TUBB3 expression in patients with advanced non-small cell lung cancer (NSCLC) receiving first-line platinum-paclitaxel chemotherapy. Cancer Chemother Pharmacol 2014;74(4):777–86.

[66] Qin X, Yao W, Li W, Feng X, Huo X, Yang S, et al. ERCC1 and BRCA1 mRNA expressions are associated with clinical outcome of non-small cell lung cancer treated with platinum-based chemotherapy. Tumour Biol 2014;35(5):4697–704.

[67] Wang TB, Zhang NL, Wang SH, Li HY, Chen SW, Zheng YG. Expression of ERCC1 and BRCA1 predict the clinical outcome of non-small cell lung cancer in patients receiving platinum-based chemotherapy. Genet Mol Res 2014;13(2):3704–10.

[68] Xian-Jun F, Xiu-Guang Q, Li Z, Hui F, Wan-Ling W, Dong L, et al. ERCC1 and BRCA1 mRNA expression predicts the clinical outcome of non-small cell lung cancer receiving platinum-based chemotherapy. Pak J Med Sci 2014;30(3):488–92.

[69] Su C, Zhou S, Zhang L, Ren S, Xu J, Zhang J, et al. ERCC1, RRM1 and BRCA1 mRNA expression levels and clinical outcome of advanced non-small cell lung cancer. Med Oncol 2011;28(4):1411–7.

[70] Puppe J, Drost R, Liu X, Joosse SA, Evers B, Cornelissen-Steijger P, et al. BRCA1-deficient mammary tumor cells are dependent on EZH2 expression and sensitive to polycomb repressive complex 2-inhibitor 3-deazaneplanocin A. Breast Cancer Res 2009;11(4):R63.

[71] Wei Y, Xia W, Zhang Z, Liu J, Wang H, Adsay NV, et al. Loss of trimethylation at lysine 27 of histone H3 is a predictor of poor outcome in breast, ovarian, and pancreatic cancers. Mol Carcinog 2008;47(9):701–6.

[72] Chen X, Song N, Matsumoto K, Nanashima A, Nagayasu T, Hayashi T, et al. High expression of trimethylated histone H3 at lysine 27 predicts better prognosis in non-small cell lung cancer. Int J Oncol 2013;43(5):1467–80.

[73] Kikuchi J, Kinoshita I, Shimizu Y, Kikuchi E, Konishi J, Oizumi S, et al. Distinctive expression of the polycomb group proteins Bmi1 polycomb ring finger oncogene and enhancer of zeste homolog 2 in nonsmall cell lung cancers and their clinical and clinicopathologic significance. Cancer 2010;116(12):3015–24.

[74] Drayton RM. The role of microRNA in the response to cisplatin treatment. Biochem Soc Trans 2012;40(4):821–5.

[75] Cai J, Yang C, Yang Q, Ding H, Jia J, Guo J, et al. Deregulation of let-7e in epithelial ovarian cancer promotes the development of resistance to cisplatin. Oncogenesis 2013;2:e75.

[76] Zhu WY, Luo B, An JY, He JY, Chen DD, Xu LY, et al. Differential expression of miR-125a-5p and let-7e predicts the progression and prognosis of non-small cell lung cancer. Cancer Invest 2014;32(8):394–401.

[77] Varambally S, Cao Q, Mani RS, Shankar S, Wang X, Ateeq B, et al. Genomic loss of microRNA-101 leads to overexpression of histone methyltransferase EZH2 in cancer. Science 2008;322(5908):1695–9.

[78] Semaan A, Qazi AM, Seward S, Chamala S, Bryant CS, Kumar S, et al. MicroRNA-101 inhibits growth of epithelial ovarian cancer by relieving chromatin-mediated transcriptional repression of p21(waf(1)/cip(1)). Pharm Res 2011;28(12):3079–90.

[79] Yang Y, Li X, Yang Q, Wang X, Zhou Y, Jiang T, et al. The role of microRNA in human lung squamous cell carcinoma. Cancer Genet Cytogenet 2010;200(2):127–33.

[80] Lacroix B, Gamazon ER, Lenkala D, Im HK, Geeleher P, Ziliak D, et al. Integrative analyses of genetic variation, epigenetic regulation, and the transcriptome to elucidate the biology of platinum sensitivity. BMC Genomics 2014;15(1):292.

[81] Dong Z, Zhong Z, Yang L, Wang S, Gong Z. MicroRNA-31 inhibits cisplatin-induced apoptosis in non-small cell lung cancer cells by regulating the drug transporter ABCB9. Cancer Lett 2014;343(2):249–57.

[82] Jiang Z, Yin J, Fu W, Mo Y, Pan Y, Dai L, et al. miRNA 17 family regulates cisplatin-resistant and metastasis by targeting TGFbetaR2 in NSCLC. PLoS One 2014;9(4):e94639.

[83] Wu H, Xiao Z, Zhang H, Wang K, Liu W, Hao Q. MiR-489 modulates cisplatin resistance in human ovarian cancer cells by targeting Akt3. Anticancer Drugs 2014;25(7):799–809.

[84] Liu M, Zhang X, Hu CF, Xu Q, Zhu HX, Xu NZ. MicroRNA-mRNA functional pairs for cisplatin resistance in ovarian cancer cells. Chin J Cancer 2014;33(6):285–94.

[85] Pang Y, Mao H, Shen L, Zhao Z, Liu R, Liu P. MiR-519d represses ovarian cancer cell proliferation and enhances cisplatin-mediated cytotoxicity *in vitro* by targeting XIAP. Onco Targets Ther 2014;7:587–97.

[86] Zhao H, Bi T, Qu Z, Jiang J, Cui S, Wang Y. Expression of miR-224-5p is associated with the original cisplatin resistance of ovarian papillary serous carcinoma. Oncol Rep 2014;32(3):1003–12.

[87] O'Byrne KJ, Barr MP, Gray SG. The role of epigenetics in resistance to Cisplatin chemotherapy in lung cancer. Cancers (Basel) 2011;3(1):1426–53.

[88] Ratovitski EA. Phospho-DeltaNp63alpha/microRNA network modulates epigenetic regulatory enzymes in squamous cell carcinomas. Cell Cycle 2014;13(5):749–61.

[89] Huang Y, Kesselman D, Kizub D, Guerrero-Preston R, Ratovitski EA. Phospho-DeltaNp63alpha/microRNA feedback regulation in squamous carcinoma cells upon cisplatin exposure. Cell Cycle 2013;12(4):684–97.

[90] Morlando M, Ballarino M, Fatica A, Bozzoni I. The role of long noncoding RNAs in the epigenetic control of gene expression. ChemMedChem 2014;9(3):505–10.

[91] Orom UA, Shiekhattar R. Long noncoding RNAs usher in a new era in the biology of enhancers. Cell 2013;154(6):1190–3.

[92] Yang Z, Zhou L, Wu LM, Lai MC, Xie HY, Zhang F, et al. Overexpression of long non-coding RNA HOTAIR predicts tumor recurrence in hepatocellular carcinoma patients following liver transplantation. Ann Surg Oncol 2011;18(5):1243–50.

[93] Liu Z, Sun M, Lu K, Liu J, Zhang M, Wu W, et al. The long noncoding RNA HOTAIR contributes to cisplatin resistance of human lung adenocarcinoma cells via downregualtion of p21(WAF1/CIP1) expression. PLoS One 2013;8(10):e77293.

[94] Gupta RA, Shah N, Wang KC, Kim J, Horlings HM, Wong DJ, et al. Long non-coding RNA HOTAIR reprograms chromatin state to promote cancer metastasis. Nature 2010;464(7291):1071–6.

[95] Ge XS, Ma HJ, Zheng XH, Ruan HL, Liao XY, Xue WQ, et al. HOTAIR, a prognostic factor in esophageal squamous cell carcinoma, inhibits WIF-1 expression and activates Wnt pathway. Cancer Sci 2013; 104(12):1675–82.

[96] Yang Y, Li H, Hou S, Hu B, Liu J, Wang J. The noncoding RNA expression profile and the effect of lncRNA AK126698 on cisplatin resistance in non-small-cell lung cancer cell. PLoS One 2013;8(5):e65309.

[97] Fan Y, Shen B, Tan M, Mu X, Qin Y, Zhang F, et al. Long non-coding RNA UCA1 increases chemoresistance of bladder cancer cells by regulating Wnt signaling. FEBS J 2014;281(7):1750–8.

[98] Lukas J, Lukas C, Bartek J. More than just a focus: the chromatin response to DNA damage and its role in genome integrity maintenance. Nat Cell Biol 2011;13(10):1161–9.

[99] Revet I, Feeney L, Bruguera S, Wilson W, Dong TK, Oh DH, et al. Functional relevance of the histone gammaH2Ax in the response to DNA damaging agents. Proc Natl Acad Sci USA 2011;108(21):8663–7.

[100] Wilson AJ, Holson E, Wagner F, Zhang YL, Fass DM, Haggarty SJ, et al. The DNA damage mark pH2AX differentiates the cytotoxic effects of small molecule HDAC inhibitors in ovarian cancer cells. Cancer Biol Ther 2011;12(6):484–93.

[101] O'Flaherty JD, Barr M, Fennell D, Richard D, Reynolds J, O'Leary J, et al. The cancer stem-cell hypothesis: its emerging role in lung cancer biology and its relevance for future therapy. J Thorac Oncol 2012;7(12):1880–90.

[102] Barr MP, Gray SG, Hoffmann AC, Hilger RA, Thomale J, O'Flaherty JD, et al. Generation and characterisation of cisplatin-resistant non-small cell lung cancer cell lines displaying a stem-like signature. PLoS One 2013;8(1):e54193.

[103] Fazzio TG, Huff JT, Panning B. An RNAi screen of chromatin proteins identifies Tip60-p400 as a regulator of embryonic stem cell identity. Cell 2008;134(1):162–74.

[104] Chen PB, Hung JH, Hickman TL, Coles AH, Carey JF, Weng Z, et al. Hdac6 regulates Tip60-p400 function in stem cells. Elife 2013;2:e01557.

[105] Juliano C, Wang J, Lin H. Uniting germline and stem cells: the function of Piwi proteins and the piRNA pathway in diverse organisms. Annu Rev Genet 2011;45:447–69.

[106] Wang QE, Han C, Milum K, Wani AA. Stem cell protein Piwil2 modulates chromatin modifications upon cisplatin treatment. Mutat Res 2011;708(1–2):59–68.

[107] Yin DT, Wang Q, Chen L, Liu MY, Han C, Yan Q, et al. Germline stem cell gene PIWIL2 mediates DNA repair through relaxation of chromatin. PLoS One 2011;6(11):e27154.

[108] Sims III RJ, Reinberg D. Is there a code embedded in proteins that is based on post-translational modifications? Nat Rev Mol Cell Biol 2008;9(10):815–20.

[109] Yang JY, Hung MC. Deciphering the role of forkhead transcription factors in cancer therapy. Curr Drug Targets 2011;12(9):1284–90.

[110] Yang YC, Tang YA, Shieh JM, Lin RK, Hsu HS, Wang YC. DNMT3B overexpression by deregulation of FOXO3a-mediated transcription repression and MDM2 overexpression in lung cancer. J Thorac Oncol 2014;9(9):1305–15.

[111] Mikse OR, Blake Jr. DC, Jones NR, Sun YW, Amin S, Gallagher CJ, et al. FOXO3 encodes a carcinogen-activated transcription factor frequently deleted in early-stage lung adenocarcinoma. Cancer Res 2010;70(15):6205–15.

[112] Beharry AW, Sandesara PB, Roberts BM, Ferreira LF, Senf SM, Judge AR. HDAC1 activates FoXO and is both sufficient and required for skeletal muscle atrophy. J Cell Sci 2014;127(Pt 7):1441–53.

[113] Senf SM, Sandesara PB, Reed SA, Judge AR. p300 Acetyltransferase activity differentially regulates the localization and activity of the FOXO homologues in skeletal muscle. Am J Physiol Cell Physiol 2011;300(6):C1490–501.

[114] Khongkow M, Olmos Y, Gong C, Gomes AR, Monteiro LJ, Yague E, et al. SIRT6 modulates paclitaxel and epirubicin resistance and survival in breast cancer. Carcinogenesis 2013;34(7):1476–86.

[115] Corrado P, Mancini M, Brusa G, Petta S, Martinelli G, Barbieri E, et al. Acetylation of FOXO3a transcription factor in response to imatinib of chronic myeloid leukemia. Leukemia 2009;23(2):405–6.

[116] Shiota M, Yokomizo A, Kashiwagi E, Tada Y, Inokuchi J, Tatsugami K, et al. Foxo3a expression and acetylation regulate cancer cell growth and sensitivity to cisplatin. Cancer Sci 2010;101(5):1177–85.

[117] Lawless MW, O'Byrne KJ, Gray SG. Oxidative stress induced lung cancer and COPD: opportunities for epigenetic therapy. J Cell Mol Med 2009;13(9A):2800–21.

[118] Mankan AK, Lawless MW, Gray SG, Kelleher D, McManus R. NF-kappaB regulation: the nuclear response. J Cell Mol Med 2009;13(4):631–43.

[119] Ea CK, Baltimore D. Regulation of NF-kappaB activity through lysine monomethylation of p65. Proc Natl Acad Sci USA 2009;106(45):18972–7.

[120] Lu T, Jackson MW, Wang B, Yang M, Chance MR, Miyagi M, et al. Regulation of NF-kappaB by NSD1/FBXL11-dependent reversible lysine methylation of p65. Proc Natl Acad Sci USA 2010;107(1):46–51.

[121] Wei H, Wang B, Miyagi M, She Y, Gopalan B, Huang DB, et al. PRMT5 dimethylates R30 of the p65 subunit to activate NF-kappaB. Proc Natl Acad Sci USA 2013;110(33):13516–21.

[122] Ibrahim R, Matsubara D, Osman W, Morikawa T, Goto A, Morita S, et al. Expression of PRMT5 in lung adenocarcinoma and its significance in epithelial–mesenchymal transition. Hum Pathol 2014;45(7):1397–405.

[123] Elakoum R, Gauchotte G, Oussalah A, Wissler MP, Clement-Duchene C, Vignaud JM, et al. CARM1 and PRMT1 are dysregulated in lung cancer without hierarchical features. Biochimie 2014;97:210–8.

[124] Kim DH, Jung YJ, Lee JE, Lee AS, Kang KP, Lee S, et al. SIRT1 activation by resveratrol ameliorates cisplatin-induced renal injury through deacetylation of p53. Am J Physiol Renal Physiol 2011;301(2):F427–35.

[125] Jung YJ, Lee JE, Lee AS, Kang KP, Lee S, Park SK, et al. SIRT1 overexpression decreases cisplatin-induced acetylation of NF-kappaB p65 subunit and cytotoxicity in renal proximal tubule cells. Biochem Biophys Res Commun 2012;419(2):206–10.

[126] Godwin P, Baird AM, Heavey S, Barr MP, O'Byrne KJ, Gately K. Targeting nuclear factor-kappa B to overcome resistance to chemotherapy. Front Oncol 2013;3:120.

[127] Ali MW, Cacan E, Liu Y, Pierce JY, Creasman WT, Murph MM, et al. Transcriptional suppression, DNA methylation, and histone deacetylation of the regulator of G-protein signaling 10 (RGS10) gene in ovarian cancer cells. PLoS One 2013;8(3):e60185.

[128] Cacan E, Ali MW, Boyd NH, Hooks SB, Greer SF. Inhibition of HDAC1 and DNMT1 modulate RGS10 expression and decrease ovarian cancer chemoresistance. PLoS One 2014;9(1):e87455.

[129] Huang WJ, Tang YA, Chen MY, Wang YJ, Hu FH, Wang TW, et al. A histone deacetylase inhibitor YCW1 with antitumor and antimetastasis properties enhances cisplatin activity against non-small cell lung cancer in preclinical studies. Cancer Lett 2014;346(1):84–93.

[130] Meng F, Sun G, Zhong M, Yu Y, Brewer MA. Anticancer efficacy of cisplatin and trichostatin A or 5-aza-2'-deoxycytidine on ovarian cancer. Br J Cancer 2013;108(3):579–86.

[131] Pogribny IP, Tryndyak VP, Pogribna M, Shpyleva S, Surratt G, Gamboa da Costa G, et al. Modulation of intracellular iron metabolism by iron chelation affects chromatin remodeling proteins and corresponding epigenetic modifications in breast cancer cells and increases their sensitivity to chemotherapeutic agents. Int J Oncol 2013;42(5):1822–32.

[132] Ghizzoni M, Wu J, Gao T, Haisma HJ, Dekker FJ, George Zheng Y. 6-alkylsalicylates are selective Tip60 inhibitors and target the acetyl-CoA binding site. Eur J Med Chem 2012;47(1):337–44.

[133] Coffey K, Blackburn TJ, Cook S, Golding BT, Griffin RJ, Hardcastle IR, et al. Characterisation of a Tip60 specific inhibitor, NU9056, in prostate cancer. PLoS One 2012;7(10):e45539.

[134] Gao C, Bourke E, Scobie M, Famme MA, Koolmeister T, Helleday T, et al. Rational design and validation of a Tip60 histone acetyltransferase inhibitor. Sci Rep 2014;4:5372.

[135] Kim KC, Lee C. Reversal of cisplatin resistance by epigallocatechin gallate is mediated by downregulation of axl and tyro 3 expression in human lung cancer cells. Korean J Physiol Pharmacol 2014;18(1):61–6.

[136] Zhou DH, Wang X, Feng Q. EGCG enhances the efficacy of cisplatin by downregulating hsa-miR-98-5p in NSCLC A549 cells. Nutr Cancer 2014;66(4):636–44.

[137] Deng PB, Hu CP, Xiong Z, Yang HP, Li YY. Treatment with EGCG in NSCLC leads to decreasing interstitial fluid pressure and hypoxia to improve chemotherapy efficacy through rebalance of Ang-1 and Ang-2. Chin J Nat Med 2013;11(3):245–53.

[138] Chen H, Landen CN, Li Y, Alvarez RD, Tollefsbol TO. Enhancement of cisplatin-mediated apoptosis in ovarian cancer cells through potentiating G2/M arrest and p21 upregulation by combinatorial epigallocatechin gallate and sulforaphane. J Oncol 2013;2013:872957.

[139] Yunos NM, Beale P, Yu JQ, Huq F. Synergism from sequenced combinations of curcumin and epigallocatechin-3-gallatewithcisplatininthekillingofhumanovariancancercells.AnticancerRes2011;31(4): 1131–40.

[140] Di Pasqua AJ, Hong C, Wu MY, McCracken E, Wang X, Mi L, et al. Sensitization of non-small cell lung cancer cells to cisplatin by naturally occurring isothiocyanates. Chem Res Toxicol 2010;23(8):1307–9.

[141] Hunakova L, Gronesova P, Horvathova E, Chalupa I, Cholujova D, Duraj J, et al. Modulation of cisplatin sensitivity in human ovarian carcinoma A2780 and SKOV3 cell lines by sulforaphane. Toxicol Lett 2014;230(3):479–86.

[142] Guerrero-Beltran CE, Calderon-Oliver M, Tapia E, Medina-Campos ON, Sanchez-Gonzalez DJ, Martinez-Martinez CM, et al. Sulforaphane protects against cisplatin-induced nephrotoxicity. Toxicol Lett 2010;192(3):278–85.

[143] Guerrero-Beltran CE, Mukhopadhyay P, Horvath B, Rajesh M, Tapia E, Garcia-Torres I, et al. Sulforaphane, a natural constituent of broccoli, prevents cell death and inflammation in nephropathy. J Nutr Biochem 2012;23(5):494–500.

[144] Schwartsmann G, Schunemann H, Gorini CN, Filho AF, Garbino C, Sabini G, et al. A phase I trial of cisplatin plus decitabine, a new DNA-hypomethylating agent, in patients with advanced solid tumors and a follow-up early phase II evaluation in patients with inoperable non-small cell lung cancer. Invest New Drugs 2000;18(1):83–91.

[145] Cornacchia E, Golbus J, Maybaum J, Strahler J, Hanash S, Richardson B. Hydralazine and procainamide inhibit T cell DNA methylation and induce autoreactivity. J Immunol 1988;140(7):2197–200.

[146] Phiel CJ, Zhang F, Huang EY, Guenther MG, Lazar MA, Klein PS. Histone deacetylase is a direct target of valproic acid, a potent anticonvulsant, mood stabilizer, and teratogen. J Biol Chem 2001;276(39):36734–41.

[147] Coronel J, Cetina L, Pacheco I, Trejo-Becerril C, Gonzalez-Fierro A, de la Cruz-Hernandez E, et al. A double-blind, placebo-controlled, randomized phase III trial of chemotherapy plus epigenetic therapy with hydralazine valproate for advanced cervical cancer. Preliminary results. Med Oncol 2011;28(Suppl. 1): S540–6.

[148] Candelaria M, Gallardo-Rincon D, Arce C, Cetina L, Aguilar-Ponce JL, Arrieta O, et al. A phase II study of epigenetic therapy with hydralazine and magnesium valproate to overcome chemotherapy resistance in refractory solid tumors. Ann Oncol 2007;18(9):1529–38.

[149] Lin CT, Lai HC, Lee HY, Lin WH, Chang CC, Chu TY, et al. Valproic acid resensitizes cisplatin-resistant ovarian cancer cells. Cancer Sci 2008;99(6):1218–26.

[150] Tesei A, Brigliadori G, Carloni S, Fabbri F, Ulivi P, Arienti C, et al. Organosulfur derivatives of the HDAC inhibitor valproic acid sensitize human lung cancer cell lines to apoptosis and to cisplatin cytotoxicity. J Cell Physiol 2012;227(10):3389–96.

[151] Hubaux R, Vandermeers F, Crisanti MC, Kapoor V, Burny A, Mascaux C, et al. Preclinical evidence for a beneficial impact of valproate on the response of small cell lung cancer to first-line chemotherapy. Eur J Cancer 2010;46(9):1724–34.

[152] Vandermeers F, Hubert P, Delvenne P, Mascaux C, Grigoriu B, Burny A, et al. Valproate, in combination with pemetrexed and cisplatin, provides additional efficacy to the treatment of malignant mesothelioma. Clin Cancer Res 2009;15(8):2818–28.

[153] Scherpereel A, Berghmans T, Lafitte JJ, Colinet B, Richez M, Bonduelle Y, et al. Valproate-doxorubicin: promising therapy for progressing mesothelioma. A phase II study. Eur Respir J 2011;37(1):129–35.

[154] Krug LM, Kindler HL, Calvert H, Manegold C, Tsao AS, Fennell D, et al. Vorinostat in patients with advanced malignant pleural mesothelioma who have progressed on previous chemotherapy (VANTAGE-014): a phase 3, double-blind, randomised, placebo-controlled trial. Lancet Oncol 2015;16(4):447–56.

[155] Gueugnon F, Cartron PF, Charrier C, Bertrand P, Fonteneau JF, Gregoire M, et al. New histone deacetylase inhibitors improve cisplatin antitumor properties against thoracic cancer cells. Oncotarget 2014;5(12):4504–15.

[156] Falchook GS, Fu S, Naing A, Hong DS, Hu W, Moulder S, et al. Methylation and histone deacetylase inhibition in combination with platinum treatment in patients with advanced malignancies. Invest New Drugs 2013;31(5):1192–200.

[157] Zhao H, Zhu W, Xie P, Li H, Zhang X, Sun X, et al. A phase I study of concurrent chemotherapy and thoracic radiotherapy with oral epigallocatechin-3-gallate protection in patients with locally advanced stage III non-small-cell lung cancer. Radiother Oncol 2014;110(1):132–6.

[158] Zhou Y, Pan DS, Shan S, Zhu JZ, Zhang K, Yue XP, et al. Non-toxic dose chidamide synergistically enhances platinum-induced DNA damage responses and apoptosis in Non-Small-Cell lung cancer cells. Biomed Pharmacother 2014;68(4):483–91.

[159] Ma YY, Lin H, Moh JS, Chen KD, Wang IW, Ou YC, et al. Low-dose LBH589 increases the sensitivity of cisplatin to cisplatin-resistant ovarian cancer cells. Taiwan J Obstet Gynecol 2011;50(2):165–71.

[160] Tsai HC, Li H, Van Neste L, Cai Y, Robert C, Rassool FV, et al. Transient low doses of DNA-demethylating agents exert durable antitumor effects on hematological and epithelial tumor cells. Cancer Cell 2012;21(3):430–46.

[161] Juergens RA, Wrangle J, Vendetti FP, Murphy SC, Zhao M, Coleman B, et al. Combination epigenetic therapy has efficacy in patients with refractory advanced non-small cell lung cancer. Cancer Discov 2011;1(7):598–607.

[162] Topalian SL, Hodi FS, Brahmer JR, Gettinger SN, Smith DC, McDermott DF, et al. Safety, activity, and immune correlates of anti-PD-1 antibody in cancer. N Engl J Med 2012;366(26):2443–54.

[163] Clozel T, Yang S, Elstrom RL, Tam W, Martin P, Kormaksson M, et al. Mechanism-based epigenetic chemosensitization therapy of diffuse large B-cell lymphoma. Cancer Discov 2013;3(9):1002–19.

[164] Fan H, Lu X, Wang X, Liu Y, Guo B, Zhang Y, et al. Low-dose decitabine-based chemoimmunotherapy for patients with refractory advanced solid tumors: a phase I/II report. J Immunol Res 2014;2014:371087.

[165] Lin J, Zhu H, Li S, Fan H, Lu X, Chang C, et al. Successful treatment with low-dose decitabine in acute myelogenous leukemia in elderly patients over 80 years old: five case reports. Oncol Lett 2013;5(4):1321–4.

THERAPEUTICALLY TARGETING EPIGENETIC REGULATION OF CANCER STEM CELLS

27

Brendan Ffrench[1], John J. O'Leary[1], and Michael F. Gallagher[2]

[1]Department of Pathology, Coombe Women's and Infant's University Hospital, Dublin, Ireland, [2]Department of Histopathology, University of Dublin, Trinity College, Trinity Centre, St James Hospital, Dublin, Ireland

CHAPTER OUTLINE

S.G. Gray (Ed): Epigenetic Cancer Therapy. DOI: http://dx.doi.org/10.1016/B978-0-12-800206-3.00027-6

1 PRINCIPLES OF STEM CELL BIOLOGY

1.1 INTRODUCTION

To understand the nature of stem cells (SCs) in malignancy it is important to first understand the biology of the nonmalignant stem cell. The term stem cell refers to cells that can (i) replenish themselves in a process termed self-renewal (SR), (ii) produce different types of more mature cells in a process termed "differentiation," and (iii) regenerate their tissue(s) of origin. Together, these properties facilitate the principle function of SCs, growth and repair of the various tissues in the body. In this section, each of these properties will be discussed before describing the role of SCs in malignancy in Section 2.

1.2 POTENCY

The term potency refers to the number of cell types (lineages) a SC can produce by differentiation and is the principle factor by which SCs are classified. "Totipotent" SCs can produce all the cells that make up an individual, including the trophoblast and placenta in mammals. "Pluripotent" SCs can differentiate to produce cells representative of all three germ layers (endoderm, mesoderm, and ectoderm). The most studied pluripotent SCs are embryonic stem cells (ESCs), which are derived from the inner cell mass of the developing embryo. "Multipotent" SCs can produce several types of related cells by differentiation. Most SCs from adult tissues (non-ESCs) are multipotent and are collectively referred to as "Adult SCs." The most studied multipotent SCs are the hematopoietic SCs (HSCs) and mesenchymal SCs (MSCs) that are responsible for the production of all blood and structural (adipocytes, osteoblasts, and chondrocytes) cells in the bone marrow.

1.3 MAINTENANCE OF THE SR STATE

SCs undergo cell division by mitosis similarly to other somatic cells. However, when discussing SC biology, the term SR is used to describe how SC mitosis involves additional mechanisms that prevent differentiation. SR confers SCs with an unlimited proliferative potential and allows for the amplification of the SC pool, providing sufficient cells for growth and development. For example, the upscaling of pluripotent SCs in the blastocyst prior to germ layer formation is achieved via SR [1]. Additionally, SR maintains lifelong pools of somatic SCs in each tissue for growth and repair [2]. As such, the failure of SC SR would result in the depletion of the SC pool, after which no further growth and repair of the tissue would be possible.

SR is not a default state of SCs. In fact, SR is actively maintained by complex mechanisms. The main function of this maintenance is to ensure that the SC pool is not depleted in an environment where SCs constantly encounter differentiation and other stimuli. SR must be maintained when SCs are in an active or inactive state and during differentiation, as described in subsections later. Maintenance of SR involves two mechanisms: (i) promotion of epigenetic regulators that maintain the SR state and (ii) inhibition of epigenetic regulators that facilitate SC differentiation. *In vivo*, maintenance of SR is achieved through four broad mechanisms. First SCs physically interact with surrounding cells, which regulate SR/differentiation. Second and thirdly, SCs are regulated by ligands secreted into their niche by both themselves and surrounding cells. Finally, SC SR is maintained by a network of molecular mechanisms that are self-expressed within the SC in response to the other three factors. These molecular mechanisms are described in detail in Section 3.

This four-part mechanism is well illustrated by the HSC–MSC system in the bone marrow, in which it has been very well characterized [3,4] (Figure 27.1). HSCs differentiate to produce the different populations of blood cell while MSCs produce other components such as bone, cartilage, and adipose tissue. HSCs are maintained in an inactive state in the MSC-rich osteoblastic (bone side) niche by physical interaction with chemokines secreted by MSCs. When required, HSCs are activated and released from the osteoblastic niche by MSCs and migrate to the perivascular (blood side) niche, where they produce blood cells by differentiation as required. The different factors present in each niche are primarily determined by the presence/absence of MSCs and result in a change in gene expression profile (described in Section 3) that facilitates maintenance of SR and differentiation.

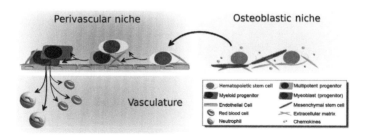

FIGURE 27.1

The bone marrow niche regulates HSC activity. Two types of SC reside in the bone marrow. MSCs are responsible for the production of adipocytes, osteoblasts, and chondrocytes while HSCs produce blood cells. HSCs primarily reside in an inactive state in the osteoblastic (bone-producing) niche, where HSC activity is principally regulated by physical interaction with MSCs. When required, chemokines and proteases mobilize HSCs to the perivascular (near-vascular) niche, where they self-renew and differentiate to produce blood cells via committed progenitors. In both locations, HSC activity is determined by physical interaction with the surrounding environment, chemokines secreted into the niche, and self-expressed internal molecular mechanisms. As such multiple factors interact in the stem niche to maintain SR and ensure appropriate differentiation and function.

1.4 ASYMMETRIC DIVISION AND DIFFERENTIATION

Differentiation refers to the process by which SCs produce different types of mature cells to facilitate embryonic development and growth and repair of tissues in the adult. Differentiation properties are determined by SC potency. For example, pluripotent SCs can produce cells representative of all three germ layers while multipotent SCs can produce a limited number of related cell types. While the main role of SR is to maintain the SC population of a tissue, differentiation produces the cells that maintain a tissue-specific function.

Uniquely, SCs can divide to produce (i) two identical daughter SCs by mitosis (symmetrical SR: Figure 27.2A) (ii) one SC and one more mature "differentiated" cell. More rarely, SCs will produce two differentiated daughter cells in a process known as symmetrical differentiation (Figure 27.2B). The mechanism by which SCs produce one SC and one differentiated cell simultaneously is termed "Asymmetric Division" (AD) and is a defining property of SCs (Figure 27.2C). AD allows the production of differentiated cells without depleting the SC pool. The precise mechanisms of AD are poorly understood but appear to involve two broad, nonmutually exclusive mechanisms. First, "Intrinsic AD" involves segregation of the modulators of differentiation, where SR mechanisms are passed on to the SR daughter cell and differentiation mechanisms activated in the differentiated daughter cell (Figure 27.3A). Second, "Niche-Dependent AD" involves orientation of the SC prior to cell division so that, post-cell division, the SC daughter cell is maintained in an SR niche while the differentiated daughter cell is deposited in a niche that promotes differentiation (Figure 27.3B). In both cases, the establishment of an axis of polarity during cell division determines the fate of each daughter cell after AD.

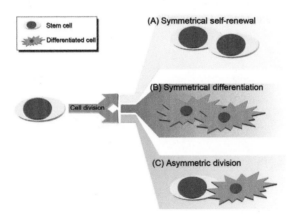

FIGURE 27.2

SCs self-renew and differentiate via AD. SCs are principally defined by their ability to self-renew and produce mature cells through differentiation. (A) SCs can divide by mitosis to produce two identical daughter SCs. This "symmetric SR" enhances the SC pool. (B) SCs can also differentiate to produce two mature cells in a process termed "symmetric differentiation." Differentiated cells, which must be replenished over time, are required for tissue function. (C) Significantly, SCs can combine these two mechanisms to "asymmetrically divide," producing one identical daughter SC and one differentiated cell simultaneously. AD allows SCs to produce mature cells when required while ensuring the SC pool is not exhausted.

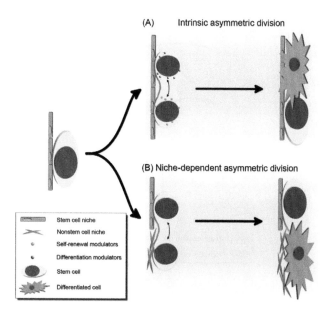

FIGURE 27.3

Mechanisms of SCAD. Two AD mechanisms have been described. Intrinsic AD involves the segregation of SR and differentiation modulators into the self-renewing and differentiating daughter cells, respectively. Niche-dependent differentiation involves the depositing of daughter cells into SR or differentiation promoting niches as appropriate. These are nonmutually exclusive and both share the theme of establishing an axis of polarity during cell division.

1.5 ESTABLISHMENT, GROWTH, AND REPAIR OF TISSUES

The third defining property of SCs is the ability to establish tissues *in vivo* via extensive SR and differentiation. In the developing embryo, ESCs can produce cells representative of all three germ layers, which are organized to produce an embryo. In the laboratory, newly isolated SCs can be shown to be pluripotent by demonstration that they can produce or contribute to all parts of the embryo via animal models. In another experimental approach, SCs can be xenografted into immune-compromised mice, which permit the establishment of teratomas. The tissues present in the teratomas produced by these SCs act as an indicator of potency. This latter approach was employed to demonstrate that human ESCs (hESCs) had been first isolated by Thompson et al. [5]. More recently, xenograft models were used to show that the overexpression of four transcription factors (Oct3/4, Sox2, Klf4, and c-Myc) was sufficient to epigenetically alter a somatic adult cell into a pluripotent cell [6]. The teratomas produced by these "induced pluripotent stem" (iPS) cells contained tissues from all three layers. This breakthrough demonstrated the key role of epigenetic regulation in SCs, which was acknowledged by the awarding of the Noble Prize to Yamanaka Shinya in 2012.

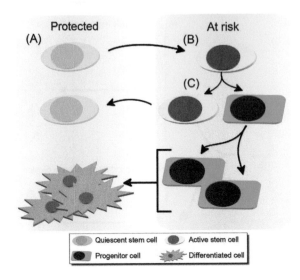

FIGURE 27.4

Stem, progenitor, and differentiated cells are arranged as hierarchies. SC populations are arranged as stem–progenitor–differentiated cell hierarchies. Both SCs and progenitors possess SC properties, the term "progenitor" reflecting a decreased potency. (A) Mechanistically, SCs reside in a quiescent state for long periods of time. (B) When required, SCs are activated and produce progenitor cells. (C) These progenitor cells undergo exhaustive SR/differentiation cell divisions, while SCs return to quiescence. As cell division increases exposure to acquired mutations, SCs (and differentiated cells) are largely protected in G_0, outside the cell cycle. In contrast, progenitor cells are largely exposed (at risk) to acquired mutations. Thus, SC hierarchies protect the longest living cells from genetic mutations through extensive periods of quiescence.

1.6 SC POPULATIONS ARE ORGANIZED AS STEM–PROGENITOR CELL HIERARCHIES

SCs tend not to produce differentiated cells directly. Rather, they produce a set of intermediate cells (progenitor cells) which in turn produce differentiated cells (Figure 27.4). Progenitor cells possess SR, AD, and tissue genesis properties but have a lower potency than their parent SC. In tissues, SCs are long lived, while progenitors are short lived cells with high levels of turnover in the tissue. As such, progenitor cells are the true functional power houses of SC hierarchies. SCs and progenitors work together to drive the growth, development, and repair of tissues. This hierarchical arrangement preserves the long-lived SC from excessive cell divisions and thus acquired mutations (Figure 27.4). Progenitors are transient, shorter lived members of the hierarchy, proliferating and differentiating until all are eventually "terminally differentiated," at which point they carry out a specialized, tissue-specific function. In another layer of complexity, many tissues appear to contain several parallel stem–progenitor cell hierarchies. From a research point of view, such hierarchical arrangements complicate the isolation and study of pure SC populations from progenitors and differentiated cells.

1.7 SC PROPERTIES AND REGULATION OF CELL CYCLE EXIT

Regulation of the cell cycle is a key mechanism for SC activity, SR, and differentiation. In Section 2, we will learn that cancer stem cells (CSCs) can only contribute to tumorigenesis in the SR state. As such, forcing CSCs out of their cell cycle is at the center of many CSC-targeting strategies. Before this can be discussed in later sections, it is important that SC cell cycle exit mechanisms are understood.

SC SR is characterized by a standard G_1-S and G_2-M phase cell cycle under epigenetic regulation by p53-p21 signaling, Retinoblastoma (Rb) activation, cyclin-dependent kinases (CDKs), and CDK inhibitors (CKIs). Maintenance of the SC SR state involves several layers of epigenetic regulation aimed at keeping SCs within the cell cycle. Additionally, SCs can transit between active and inactive (quiescent) states, which require programmed epigenetic regulation of cell cycle exit and reentry. Finally, older SCs are removed from the SC pool to prevent the passing on of potential DNA damage in a specialized cell cycle exit mechanism termed "cellular senescence." Cell cycle exit for all three processes involves transition from G_1 to G_0 via p53-p21 signaling, which is only reversible in quiescence. This postmitotic state is tightly regulated by epigenetic mechanisms (Section 3).

The hierarchical SC model indicates that SCs are longer lived than progenitors. Normal growth and repair functions are facilitated by progenitors while SCs enter long periods of dormancy termed "quiescence." When new cells are required in response to injury or stress, SCs are activated to SR and differentiate as required and subsequently return to dormancy. Thus SCs require coordinated cell cycle exit/reentry mechanisms to carry out their SR and AD roles during tissue homeostasis. While quiescence was originally believed to be a cell state absent of cell cycle regulation, it is now understood that epigenetic regulation (e.g., p53-p21 monitoring of DNA damage) is active during and required for quiescence [7].

Senescence is the process by which aging cells are removed from the cell cycle to prevent passing on of DNA damage. Aging is detected as a shortening of the telomeres that cap chromosomes, which eventually compromise cellular function. Mechanistically, telomere shortening is detected by the ATM/ATR protein, which recruits checkpoint "Chk" proteins to activate p53-mediated cell cycle exit to senescence, a process that involves several DNA methyltransferases [8–10]. Additionally, precise levels of p16 appear to define entry to and maintenance of senescence via the formation of senescence-associated heterochromatin foci (SHAF) [10]. Senescence is considered to be an energy-efficient alternative to apoptosis, where an old cell can perform a metabolic function while DNA damage is removed from the mitotic pool.

2 CANCER AND CANCER STEM CELLS

2.1 INTRODUCTION

For many years, it has been clear that only a subpopulation of cells within a tumor are capable of initiating and driving the growth of the tumor or "tumorigenesis." It is now understood that these tumor-initiating cells share some of the properties of SCs. Collectively, tumor-initiating cells have become known as CSCs. In this section, we will describe the history of CSC research toward a contemporary definition of CSCs and CSC Theory.

2.2 DEVELOPMENT OF CSC THEORY

CSC Theory originated with a rare gonadal malignancy termed an embryonal carcinoma (EC). As far back as the late 1800s, pathologists had described ECs as a disorganized caricature of the tissues (endoderm, mesoderm, and ectoderm) of an early stage embryo. This led to the postulation that ECs may develop from a pluripotent cell in a parallel fashion to the embryo. By the 1960s, single EC cells (ECCs) had been isolated and shown to be capable of reproducing their original malignancy in *xenograft* models [11]. By the 1980s, pluripotent ECCs had been developed into what was in effect the first CSC cell culture model [12–16], although the "CSC" term was not used at the time. Cells from a well-differentiated EC termed a "teratocarcinoma" could be maintained in the SR state in culture and readily differentiated by the addition of all-trans retinoic acid (ATRA) [16]. In parallel, bone marrow research had attracted substantial interest following the devastating effects of the 1945 nuclear attacks in Japan. Substantial progress was made in understanding bone marrow stem cells (HSCs and MSCs), which ultimately led to the identification of Leukemia SCs (LSCs) in 1997, when the term "CSC" was coined [17,18]. At this point, however, it was not clear whether CSCs were involved in other malignancies.

The "Coming of Age" for CSC Theory arrived in 2001 with the publication of the definitive Reya et al. review article [19]. For the first time, CSC Theory was combined into a single concept after which CSC Theory became widely respected if not universally accepted. CSC Theory is based upon the observation that some, if not all, malignancies are initiated and driven by a subpopulation of cells that behave akin to SCs [19,20]. The principle implication from CSC Theory was that targeting malignancy *en masse* must be complimented with specific targeting of CSC populations [19]. For example, where a single CSC remained postintervention, CSC Theory holds that this is sufficient to explain recurrence. By 2001, it was clear that CSC targeting had to be developed to complement existing cancer targeting strategies. However, before CSCs from specific malignancies could be understood, they first had to be identified, isolated, and studied.

2.3 DEFINING PRINCIPLES OF CSC THEORY

Our understanding of CSCs has developed rapidly in the years following the Reya et al. article. Recently, lineage tracing studies have demonstrated that CSCs are the initial targets of malignant transformation in at least some cancers [21]. Today, CSCs are defined by the SC processes of SR, differentiation, and tumorigenesis. CSCs have been successfully identified and isolated for analysis from several malignancies including breast [22], brain [23], and prostate [24] (reviewed in [25]). To date, a malignancy devoid of CSCs has not been described. It is now well established that low numbers of CSC are sufficient to reestablish the original malignancy from which they were derived. Although experimentally challenging, a single CSC has in fact been shown to be sufficient for tumorigenesis [11,22]. It is now clear that many, if not all, tumors are generated by one or more CSC stem–progenitor cell hierarchies [26,27]. Stem and progenitor cells are capable of reconstituting the malignancy in animal models, a property that is not shared by differentiated cells. This clearly identifies cancer stem and progenitor cells as targets for anticancer therapeutics.

Alarmingly, CSCs from several malignancies have been shown to be highly resistant to conventional therapies. For example, CSCs are known to tolerate standard chemotherapies and radiation therapies [25]. The precise nature of these resistance mechanisms is as yet poorly understood. CSCs may possess inherent resistance or an adaptive plasticity that allows reconstitution of the tumor postintervention.

It appears likely that CSC resilience is related to their hierarchical organization. CSCs clearly enter periods of quiescence, which are resistant to conventional chemotherapy targeting of rapidly dividing cells. Additionally, CSCs are highly tolerant of the low oxygen (hypoxic) conditions found in tumors, which are highly toxic to many other cell types [28]. Our understanding of these mechanisms is hindered by challenges associated with establishing CSC models, as discussed in Section 4. However, an improved understanding of CSC resistance to current cancer therapies is vital if CSC targeting is to be coupled to existing clinical approaches.

2.4 CSCs AND TUMOR DEVELOPMENT

A single CSC is sufficient to generate a malignancy through extensive AD [11,22] (Figure 27.5). Hierarchically, CSCs are initially active and produce subpopulations of cancer progenitor cells that employ AD mechanisms to produce the bulk of the tumor. Once sufficient progenitors have been produced, CSCs appear to enter a state of quiescence [7]. The role of differentiated cells in the tumor is easily underappreciated but these cells carry out many of the functions required for tumorigenesis.

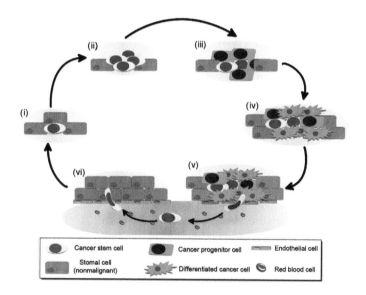

FIGURE 27.5

Single CSCs are sufficient to generate tumors. (i) Single CSCs are sufficient to generate a tumor *in vivo* through extensive SR and differentiation. (ii) It appears that tumor expansion is initially suppressed to permit establishment of the malignancy from the CSC. (iii) The first step in tumor expansion is the production of progenitor cells. (iv) While progenitor cells continue tumorigenesis through extensive SR and differentiation, CSCs have the opportunity to return to a quiescent state, which appears to be resistant to chemotherapeutics and radiation. (v) Differentiated tumor cells form leaky vasculature, which facilitates intravasation of cancer cells to begin metastasis. Escaping cancer cells undergo EMT to enter the vasculature. (vi) Upon reaching the secondary site, these "CTCs" undergo MET, exiting the vasculature to establish a metastasis and perpetuate the cycle. Each of these processes has been linked to SC properties.

For example, the vasculature produced through differentiation produces a leaky network of blood vessels that provides oxygen and nutrients to some parts of the tumor and promotes hypoxia-driven tumorigenesis in other parts [29]. Once the primary tumor is established, this leaky vasculature facilitates intravasation of cancer (stem) cells into the circulation. These "circulating tumor cells" (CTCs) ultimately extravasate at a secondary site where metastatic disease is established [30]. The epigenetically regulated SC-like differentiation mechanism "epithelial to mesenchymal transition" (EMT) is required for intravasation while the reverse dedifferentiation process "MET (mesenchymal to epithelial transition)" is required for extravasation during metastasis. As such, CSC SR, AD, and tumorigenic properties drive the progress of malignant disease [31,32] (Figure 27.5). The importance of epigenetic regulation in tumorigenesis from CSCs is clear when it is considered that self-renewing (tumorigenic) CSCs and their differentiated (nontumorigenic) progeny differ only in terms of epigenetic regulation and not DNA compliment. Each of the CSC processes described here requires regulation by epigenetic mechanisms, as will be described in Section 3.

3 EPIGENETIC REGULATION OF CSC SR AND DIFFERENTIATION

3.1 INTRODUCTION

In their SR and differentiated states, SCs and CSCs possess an identical DNA complement. As such, SC state is determined by several layers of epigenetic regulation, which balance promotion and inhibition of SR and differentiation mechanisms. These epigenetic regulation mechanisms culminate in oversight of a number of stemness signal transduction pathways. These signaling pathways are the primary determinant of SC regulation and are themselves regulated by other epigenetic mechanisms such as microRNAs (miRNAs) and DNA remodeling, as well as self-regulating and regulating one another. Many of these epigenetic mechanisms were first described in nonmalignant SCs, where they were shown to be necessary for maintenance of the SR state and/or differentiation. Unsurprisingly, many of these mechanisms were shown to behave aberrantly when investigated in CSCs. So crucial is their role that these mechanisms have become cornerstones of CSC-targeting strategies. These mechanisms are described in detail later and their exploitation in CSC targeting in Section 4. These pathways all target key transcription factors, which are the main determinants of SC state. The best studied of these, Oct4–Sox2–Nanog regulation of pluripotency, will be used as an example to illustrate specific mechanisms.

3.2 REGULATION OF PLURIPOTENCY BY OCT4, SOX2, AND NANOG

The pluripotent SC and CSC state is regulated by a trinity of transcription factors known as Oct4, Sox2, and Nanog, each of which is necessary for maintenance of the pluripotent state [33–35]. Collectively, Oct4–Sox2–Nanog regulation is the best studied example of epigenetic regulation of SCs. Additionally, the regulation of nonpluripotent processes by one or two of these proteins is widely described. All three are highly expressed in the pluripotent SR state and downregulated during early differentiation. Nanog is the most important of the three, which is reflected in it being named after the "Tir na n'Og" (Land of Eternal Youth) Irish legend [36,37]. While all three determine pluripotency in hESCs and hECCs, their regulation is different [38,39]. For example, hECCs can differentiate without loss of Sox2 expression. As the best studied example of regulation of SC function, regulation of Oct4, Sox2, and Nanog will be used to illustrate the mechanisms of other epigenetic regulators of SC function.

There are two main models proposed as mechanisms through which Oct4, Sox2, and Nanog expression facilitates the balance between pluripotency and differentiation. In mouse ESCs (mESCs), Austin Smith and coworkers have proposed a two-step "Ground State" model, where Fgf (fibroblast growth factor) and Mapk (mitogen-activated protein kinase) signaling control transition of mESCs between a "naïve" Ground State and a more mature state, which is primed for differentiation [34]. In the second "Competition Model," Bing Lim and coworkers have proposed that in hESCs, Oct4, Sox2, Nanog and other transcription factors each inhibit and promote certain differentiation lineages [35]. Thus, hESCs are proposed to maintain the pluripotent state via a perfectly balanced network of promotion and inhibition of lineage differentiation [35]. Differences between the models are debated and most likely relate to mESCs and hESCs representing slightly earlier and later stages of pluripotency respectively.

3.3 EPIGENETIC REGULATION OF SCs AND CSCs BY STEMNESS SIGNALING PATHWAYS

The basic mechanism of a signal transduction pathway is the detection of a ligand by a membrane-bound receptor, which transfers a signal through a number of intermediate proteins toward the alteration of a key pathway modulator, which enters the nucleus to influence the expression of target genes. Several signal transduction pathways are so closely associated with SCs as to be commonly referred to as "stemness signaling pathways." Four of the principle stemness signaling pathways will be discussed in detail in this section, namely "Wnt," "Hedgehog," "TGF-β," and "Notch" signaling. In both their inactive and active states, these pathways regulate maintenance of SR or lineage differentiation.

Stemness signaling pathways are characterized by families of ligands, receptors, coreceptors, and modulators, which allows for a large matrix of possible interactions within each pathway. This complexity permits each pathway to regulate several different processes and provides redundancy to ensure functionality is maintained. Alarmingly, this complexity is often hijacked by CSCs to drive tumorigenesis. In this regard, tumorigenesis is a CSC AD process, which is highly regulated toward promoting tumor growth.

3.4 STEMNESS SIGNALING PATHWAYS AND EPIGENETIC REGULATION BY miRNAs AND DNA REMODELING

Stemness signaling pathways are regulated by other epigenetic mechanisms such as miRNAs and DNA remodeling. Similarly, these pathways employ miRNAs and DNA remodeling mechanisms to achieve their roles. A comprehensive discussion of these roles is beyond the scope of this chapter but has been well reviewed in previous chapters. Here, the overall SC roles of miRNAs and DNA remodeling mechanisms will be discussed and specific examples detailed in the subsections later. As we learned in previous chapters, miRNAs posttranscriptionally regulate protein production. It is well established that different populations of miRNA are expressed in the SR and differentiated states of all types of SC/CSC studied [40,41]. Many miRNAs have been shown to regulate and be regulated by key SC/CSC genes such as Oct4, Sox2, and Nanog [40,41]. The precise mechanisms through which miRNAs target mRNAs are complex and only partially understood. For example, a single miRNA may target one mRNA at one time and a different mRNA at another. This complicates miRNA analysis and, in particular, their exploitation as cancer therapy targets, as will be discussed in Section 4.

DNA remodeling mechanisms have been described in detail in earlier chapters. As one of the most prominent epigenetic regulation mechanisms it is unsurprising that SC/CSC SR–differentiation mechanisms are often regulated by DNA remodeling [42–44]. Specific examples of DNA remodeling are discussed for each of the stemness signaling pathways below. In overview, ESCs possess a dynamic chromatin structure, which is characterized by global DNA hypomethylation, where chromatin is relatively condensed. ESCs appear to be protected, to some extent, from DNA methylation, as specific histone modifications appear to be vital for ESC SR [43,45]. During differentiation, chromatin remodeling of differentiation genes by Polycomb Group (PcG) proteins is relaxed, facilitating lineage differentiation. From early pluripotency studies, a regulatory role for DNA methylation in the SR differentiation switch was apparent. For example, Oct4 and Nanog promoters are known to be hypomethylated in ESCs and hypermethylated during differentiation [43,46]. There is growing certainty that change in promoter methylation of specific genes, rather than global changes, is a key component of lineage differentiation. Epigenetic mechanisms operate synergistically with stemness signaling pathways as is described below.

Polycomb repressive complexes (PRCs) are important epigenetic repressors that were described in detail in earlier chapters. The catalytic component of PRC2, "Enhancer of Zeste Homologue 2" (EZH2), has emerged as a potential anticancer target due to its key role as an epigenetic regulator in SCs, CSCs, and cancer. Mechanistically, EZH2 is a histone methyltransferases that targets genes via H3K27me (histone 3, lysine 27 methylation) marks, which facilitates recruitment of PRC1 and DNA methyltransferases to ultimately epigenetically silence gene expression [47,48]. Although EZH2 is best described for its gene silencing role, gene activation and bivalent roles have also been described. 3000 bivalent domains, where a gene promoter carries repressive (H3K27me) and activator (H3K4me, histone 3, lysine 4 methylation) marks simultaneously, have been described in ESCs, a number that is reduced to 1000 during early differentiation [49]. This complex mechanism has been described in ESCs, NSCs, HSCs, and hair follicle SCs (HFSCs) and is thought to facilitate the rapid alterations in gene expression required during the SR differentiation switch [48]. Broadly, PRCs are essential for SC and CSC proliferation and the switch from SR to differentiation [50,51]. EZH2 has a prominent role in embryonic development and is known to directly regulate Nanog and Sox2 [52,53] as well as several of the signaling pathways discussed later. EZH2 is highly expressed in many cancers and has emerged as a potential diagnostic, prognostic, and therapeutic target (see Section 4).

3.5 WNT/β-CATENIN SIGNALING

The term "Wnt" is an amalgamation reflecting how Wnt signaling was originally identified in two parallel systems, where the same protein was termed "Integration1" (Int1) in breast cancer research and "Wingless" (Wg) in developmental biology [54,55]. Decades of research have identified a family if 19 Wnts as secreted glycoproteins that activate three parallel pathways, only one of which, "Canonical" Wnt (Wnt/β-catenin) signaling is discussed here. Mechanistically, Wnt ligands are received by a family of 7 "Frizzled" (Fzd) receptors, which cooperate with a number of coreceptors (Lrp5/6, R-Spondins, etc.) to transduce signaling through an array of protein intermediaries. The key step in Wnt/β-catenin signaling is the stabilization of "β-catenin." In the absence of Wnt ligand, β-catenin is marked for degradation by the "β-catenin Destruction Complex," which is composed of the "Axin," "Apc," and "Gsk3" proteins. Upon Wnt ligand binding, the Destruction Complex is inhibited, which permits stabilization and accumulation of cytoplasmic "β-catenin," which enters the nucleus to affect the expression of "Tcf/Lef" transcription factors, which directly regulated target genes such as Oct4–Sox2–Nanog [54,55].

Various differentiation processes are facilitated by differing levels of Wnt signaling and different combinations of Wnts, Fzds, etc. This complexity is well illustrated by Wnt regulation of ESCs. Wnts are required to maintain mESC SR, which likely relates to direct binding between "Tcf3" and Oct4, Sox2 and Nanog. In hESCs, Wnts appear to have no role in SR but Wnt activation leads to mesodermal differentiation. ESCs and the feeder layer cells on which they are grown in culture are known to secrete different populations of Wnts and the inclusion of Wnts enhances ESC isolation efficiency. Additionally, Wnt signaling is known to regulate the SR state in HSCs, HFSCs, and mammary SCs (reviewed in [54,55]). For example, β-catenin knockout mice have been shown to produce HSCs with increased proliferation rates and lower levels of apoptosis and differentiation. Wnt has been prominently studied in intestine development, where gradients of Wnts facilitate proliferation and differentiation of intestinal SCs [56].

Anomalies, such as constitutive Wnt signaling, are unsurprisingly associated with several cancers [54]. Malignant Wnt signaling has been most studied in colorectal cancer, where almost all patients possess a "β-catenin on" phenotype due to mutations in APC (loss of function) or β-catenin (gain of function). In a parallel fashion to intestinal SCs, high/low Wnt niches in the colon appear to promote CSC and differentiated cell proliferation, respectively. Aberrant Wnt signaling is described in neural and breast CSCs [54]. In contrast to the Wnt role in HSCs, Wnt loss does not affect LSCs from chronic myeloid leukemia. However, β-catenin is active in LSCs from mixed lineage leukemia, where β-catenin inhibition reverts LSCs to a pre-LSC state. As such, the role of Wnts in leukemia is complex and disease specific.

Wnt signaling is regulated by and employs DNA modification and miRNA mechanisms. β-catenin is known to recruit histone methyltransferases, including EZH2, and acetylases to alter target gene expression [57]. EZH2 has been shown to directly repress several Wnt genes [58]. In ovarian follicle SCs, Wnt expression is restricted by PcG proteins. In breast cancer, several Wnts display methylation, particularly EHZ2-driven H3K27me3 [59]. Additionally, Wnt signaling is known to be controlled by and alter the expression of a population of miRNAs in mechanisms that are only partially described. For example, several miRNAs directly target β-catenin and APC, which results in enhanced growth of many cancers [54]. As such, Wnt signaling is a complex cascading pathway, which is regulated by other epigenetic mechanisms as well as activating other epigenetic mechanisms to facilitate its regulatory role. This complexity is the principle challenge for the development of Wnt-targeting anticancer therapies, as will be discussed in Section 4.

3.6 HEDGEHOG SIGNALING

The term "Hedgehog" (Hh) reflects a role in developmental biology, where Hh-mutated *Drosophila* larvae appeared hedgehog-like due to abnormal appendages. Subsequently, organization in the developing embryo was shown to be determined by Hh signaling gradients and duration [60,61]. Three Hh pathways have been described, the most prominent of which, "Sonic Hedgehog" (Shh), will be described here. Mechanistically, Shh ligands bind to and inhibit the cell surface receptor "Patched" (Ptch), which permits accumulation of "Smoothened" (Smo). Accumulated Smo prevents destruction of "Gli" proteins, which translocate to the nucleus of the cell to influence expression of target genes. As Gli targets include Glis and other Hh promoters, Hh signaling is an example of a self-promoting cascade. While the role of Gli1 is largely unknown, Gli2 and Gli3 are responsible for activity and repression, respectively, of targets including Nanog [60,62]. Hh signaling regulates SC homeostasis in many adult tissues but has been principally associated with embryonic central nervous system (CNS) development. In the CNS, Hh signaling maintains neural SCs (NSCs) and progenitors. Mechanistically, Hh signaling results in Gli2 being recruited to Nanog promoters in both NSCs and medullablastoma (MB) CSCs [62].

Persistent Hh signaling is associated with many cancers, particularly in the brain (MB) and skin (basal cell carcinoma, BCC) [60,61]. Hh plays a key role in SR/differentiation of MB and glioma CSCs. Hh signaling in tumors is due to Hh being secreted by tumor cells containing mutations in pathway components, overreaction to Hh secretion by the surrounding tissues or both. Hh signals are known to promote tumor growth, the tumor microenvironment or both. Shh is known to cooperate with and/or regulate Wnt and TGF-β signaling, which may reflect a role in early fate decisions. As the main modulators of Hh signaling, it is unsurprising that Gli genes are reported to display modifications such as acetylation in MB. As an adaptation, MB expresses high levels of Hh, which promotes increased HDAC expression to promote tumorigenesis. Glis are also common miRNA targets in cancer, although these roles are poorly characterized. Hh signaling has attracted much attention as a potential anticancer target, particularly in CNS malignancies, which will be discussed in Section 5.

3.7 NOTCH SIGNALING

Notch signaling also acquired its name through *Drosophila* studies, where mutations in this developmental regulator resulted in flies with wings displaying a notch pattern. Subsequently, Notch signaling has been observed in several types of SC and CSC and is a key area of CSC-targeting research [63]. Notch signaling is a simple pathway containing no secondary messengers or cascade. Notch signaling facilitates cell to cell communication, where "Jagged" and "Delta" receptors on one cell interact with Notch transmembrane receptors on an adjacent cell. This interaction results in cleavage of the "Notch intercellular domain" (NICD), which translocates to the nucleus of the cell to regulate target genes [64]. In the nucleus, NICD binding alters "RBP-Jκ" from a gene expression repressor to activator role. The principle Notch target genes are the "Hes" and "Hey" gene families, which control SC SR and cell fate decisions during differentiation in many adult tissues, often via regulation of Nanog [65]. The key molecule in Notch signaling is the enzyme "γ-secretase," which facilitates cleavage of NICD. As well as cell-to-cell communication, recent studies suggest a parallel self-signaling role via secreted Jagged. Notch signaling is primarily known for maintaining SC pools in adult tissues and determining cell fate decisions in the developing embryo [65].

Notch signaling is generally overexpressed in cancers (leukemia, breast, lung, glioma, head and neck, MB, colorectal, pancreatic, melanoma), although some exceptions with downregulated Notch levels have been described [63,66]. High Notch expression has been correlated with strong survival of malignancies and poor patient outcomes in various malignancies. Specifically, Notch signaling has been shown to maintain the CSC pool and to be one of the main regulators of EMT and chemoresistance [66]. Notch signaling is so vital to EMT, and thus to metastasis, that Notch targeting is a key area of antimetastatic research (Section 4). Notch inhibition reduces CSC function in leukemia, breast (via direct Notch targeting of Nanog), glioma, colon, ovary, lung, and liver cancers. In different malignancies, overexpression of Notch signaling often occurs via the same pathway component but through various different mechanisms. For example, Notch1 overexpression appears to act via Myc in leukemia, breast and lung cancers while in glioma, aberrant expression of Notch1 is controlled by miR-525-5p regulation of Jagged [63]. Various Notch pathway genes are epigenetically regulated in cancers. For example, aberrant acetylation of Jagged in multiple myeloma results in amplified Notch signaling. In pancreatic CSCs, miRs-34 and -181c control the Notch-mediated SR/differentiation switch, while miR-146a aberrantly regulates "Numb" in many breast carcinomas [63]. Recently, EZH2 has been shown to regulate Notch1 in breast CSCs [67]. Although EZH2-Notch mechanisms are as yet poorly described, it is tempting to speculate that Notch signaling is associated with EZH2 silencing of anti-metastatic genes as dysregulation of both mechanisms generates similar prometastatic phenotypes [57]. This is discussed in detail in Section 4.

3.8 TGF-β SIGNALING

Transforming growth factor (TGF)-β signaling is one of the most studied pathways in molecular biology and has several growth and development roles in nonmalignant and malignant cells and SCs [68]. In overview, precise levels of TGF-β signaling vary during different processes within the same tissue to regulate the precise downstream mechanisms expressed. TGF-β signaling involves three parallel pathways (bone morphogenic protein (BMP), TGF-β, and activin pathways), all of which are key SC regulators. TGF-β signaling is transduced within the cell by a number of "Smad" protein modulators, which ultimately enter the nucleus of the cell to influence the expression of gene targets including Nanog. As all three pathways are composed of families of ligands and receptors, the different signaling combinations permit regulation of a large number of growth and development processes in a highly specific and redundant manner.

From the earliest establishment of ESCs in culture, TGF-β signaling components were identified as being required for maintenance of the SR state. TGF-β signaling components such as BMP (mESCs) and activin (hESCs) are essential for SR through direct Smad–Nanog interaction [34,35]. TGF-β signaling has been cited as controlling "most, if not all, cell differentiation events" [68]. Different TGF-β signaling events control initial ESC differentiation decisions while others control later lineage differentiation mechanisms. TGF-β signaling has also been described as a key regulator of adult SCs (mammary, intestinal, skin, MSCs, HSCs, and NSCs). For example, TGF-β signaling is the principle regulatory factor determining the interaction between HSCs and MSCs in the bone marrow, where low and high levels stimulate or inhibit proliferation, respectively [69]. TGF-β signaling also controls the quiescence activation switch in HSCs [70].

During early tumorigenesis, low levels of TGF-β signaling inhibit tumor expansion, which appears to facilitate establishment of the initial tumor by CSCs [71]. Subsequently, tumorigenesis is accelerated by medium levels of TGF-β signaling and tumor maturation achieved by higher levels of TGF-β signaling. As such, most advanced tumors express high levels of TGF-β signaling while TGF-β pathway genes are mutation hotspots in many malignancies. TGF-β signaling plays a key role in genome stability, regulates cell cycle progression at G_1, and is sufficient for EMT during metastasis [71]. TGF-β signaling plays a key role in EMT in classical breast CD44[high]/CD24[Low] CSCs and CD133[+] colon CSCs, while activated Nodal and Activin are required for pancreatic CSC SR. LSCs secrete TGF-β to promote self-survival and proliferation. TGF-β signaling facilitates EMT so readily that TGF-β-driven EMT models are commonly employed. A link between TGF-β signaling and CSCs was clearly established when it was shown that TGF-β-driven EMT produces cells displaying CSC markers and properties [72].

Epigenetic regulation of TGF-β signaling has been shown to enhance tumorigenesis in several malignancies (breast, glioblastoma). The mir-17-92 family is a complex upstream and downstream regulator of Smad activity. TGF-β signaling is an obvious anticancer target due to its prominent role in malignancy and CSCs specifically. One potential target is EZH2, which has been shown to be an essential upstream regulator of TGF-β signaling during EMT-driven migration and invasion [73,74]. However, as we will learn in Section 4, targeting TGF-β signaling in a clinical setting is particularly challenging.

4 THERAPEUTIC TARGETING OF CSCs

4.1 INTRODUCTION

In this section, we will discuss how CSCs have proved much more difficult to clinically target than originally hoped. In spite of this, CSC targeting remains an area of great hope in cancer therapeutics. Broadly, targeting strategies have focused on forcing CSCs into a differentiated nontumorigenic state. Additionally, there is increasing evidence indicating that CSC targeting greatly enhances responses to chemotherapy and radiotherapy, particularly in therapy refractory disease. As we will describe, CSC targeting shows promise but further studies are required before CSCs can be efficiently targeted in the clinic.

4.2 CHALLENGES FOR CSC THERAPEUTICS

CSCs present specific challenges for development of targeted therapies, which are related to their inherent CSC biology [75]. The first of these relates to our limited understanding of CSC biology. Without a precise characterization of the epigenetic regulation of CSC SR and differentiation, it is impossible to target CSCs without unwanted side effects. For example, we have previously discussed how SCs and CSCs from the same tissue employ very similar SR/differentiation mechanisms. As such it is difficult to target CSCs without diminishing the SC pool, which would severely compromise the growth and repair of the tissue. The main reason for our limited understanding of CSC biology is the difficulty in establishing cell culture models. While this may appear to be easily achievable, SCs will in fact generally spontaneously differentiate when removed from their *in vivo* niche. Establishing cell culture models, therefore, requires considerable study to identify the specific types and levels of growth factors required to prevent SCs from differentiating. While CSCs appear to be more resistant in cell culture and thus models more easily established, without a comparable SC cell culture model it is difficult to assess the effect of potential therapies on nonmalignant SCs in surrounding tissues. The complexity is compounded by SC hierarchies, the implications of which are that multiple malignant and nonmalignant stem and progenitor cells may be found together in the same tissue.

The description of SC hierarchies suggested that progenitor cells are attractive targets, which appear to be the most active members of the hierarchy. However, we have previously discussed how elimination of progenitor cells will likely result in an awakening of dormant CSCs, which are ideally primed to drive tumorigenesis. As such, standard cancer interventions, by removing the active CSC pool, may awaken the dormant CSC pool to produce a new population of cells that are now resistant to therapies. As such, strategies that target CSCs and progenitor cells in parallel may ultimately prove successful. As the different cells in the CSC hierarchy appear to differ only in terms of epigenetic regulation of gene expression, much attention has focused on targeting CSCs epigenetically. As we will now learn, such strategies are often targeted toward stemness signaling pathways.

4.3 TARGETING CSCs VIA FORCED DIFFERENTIATION

CSC tumorigenic properties are linked to the SR state and are lost upon differentiation. As such, forcing differentiation upon CSCs has attracted considerable attention as a potential anticancer therapy. The best studied differentiation morphogen is ATRA. In culture, ATRA readily forces hESCs and hEC cells into a neural differentiation lineage, which is characterized by loss of stemness properties including tumorigenicity. The addition of ATRA has been shown to have promising, if limited, pre-clinical and clinical effects against leukemia [76,77]. Subsequently, ATRA was studied extensively as an anticancer therapy [78]. Results indicate that ATRA treatment is only successful in some cancers. For example, lung cancer has been shown to be resistant to ATRA treatment, which results in cancer cell promotion [79]. Several other approaches toward forced differentiation of CSCs target the stemness signaling mechanisms that are crucial to SR maintenance, as described in the following subsections.

4.4 TARGETING CSCs VIA WNT SIGNALING

Wnt signaling has been shown to be targetable in many experimental studies but has not progressed to strong clinical data to date (reviewed in [56]). This is primarily due to the complicated role that different levels and combinations of Wnts play in different malignancies and their surrounding tissues. As such, Wnt-targeting strategies must address the precise level of Wnt signaling within the malignancy. Additionally, the effectiveness of Wnt inhibitors has been shown to be proportional to the precise mutations found in APC and β-catenin [56]. We have previously described how Wnt signaling levels and mutations are prominently associated with colorectal cancer (Section 3.4). However, Wnt signaling gradients also control SC proliferation and differentiation in intestinal tissues. Thus it is unsurprising that Wnt-inhibition studies have observed extensive negative effects on the intestine in animal models. Wnt-inhibition strategies include targeting "Porcupine" and "Axin" stability/activity [80,81]. Porcupine is an acetyltransferase that is essential for processing Wnts for secretion and Axin is a scaffold protein that is important for the stability of the β-catenin destruction complex [81]. These studies are currently showing promising results in early clinical trials. However, further investigation is required before Wnt inhibition can be safely and effectively achieved in the clinic.

4.5 TARGETING CSCs VIA HEDGEHOG SIGNALING

Targeting Hh signaling has been primarily investigated in cancers of the skin, CNS and pancreas, where Hh expression is well characterized [61]. Hh is targeted through Smo inhibitors and one such drug, vismodegib has received FDA approval for treatment of locally advanced and metastatic BCC [61]. Smo inhibition was unsuccessful in colorectal and ovarian cancer studies [82,83] but Hh inhibition via another drug, sulforaphane, has been shown to specifically target pancreatic CSCs [84]. Additional lessons have been learned from Hh targeting studies. Hh targeting is almost exclusively successful in tumors with active Hh signaling and unsuccessful in Hh-inactive tumors [61]. Demonstration that essentially all positive BCC responders to Hh targeting were positive for expression of a 5-gene Hh signaling panel strongly supports patient prescreening. Epigenetic targeting of CSCs, it seems, is about individualization: delivering the correct treatment to the appropriate patient.

4.6 TARGETING CSCs VIA NOTCH SIGNALING

Notch signaling has attracted interest as a potential anticancer target due to its seemingly universal role in EMT during metastasis and in specific cancers where it is highly expressed such as pancreatic cancer [63]. The principle strategy used is inhibition of γ-secretase, which activates NICD, via γ-secretase inhibitors (GSIs). While GSIs do not appear to be effective in isolation, there has been very promising data for the use of GSIs in combination with standard chemotherapies. In pancreatic cancer, the combination of a GSI with chemotherapy drug gemcitabine was found to dramatically reduce metastasis, tumor progression, CSCs and tumor angiogenesis while reducing induction of apoptosis, when compared to GSIs or gemcitabine alone [85]. In ovarian cancer, GSIs were found to deplete CSCs and reduce resistance to platinum-based chemotherapy [86]. Notably, GSIs and platinum treatment targeted both the CSCs and tumor bulk in combination. Despite these successes, GSIs have caused substantial side effects in animal models, such as extensive remodeling of intestinal structures [87]. As such, Notch-specific antibodies are being assessed currently, an approach that is proving challenging due to the similarities between Notch proteins [63]. Finally, Notch signaling is a very promising biomarker, particularly for metastasis.

4.7 TARGETING CSCs VIA TGF-β SIGNALING

Targeting TGF-β signaling in tumor cells has attracted much attention due to the high levels of TGF-β found in many advanced tumors [68]. However, the level of TGF-β signaling in the tumor niche is developmentally regulated, which complicates targeting strategies [71]. During early tumorigenesis, low levels of TGF-β signaling facilitate a somewhat suppressive state, which appears to facilitate the initial establishment of the tumor before TGF-β-driven tumor expansion. Later, tumorigenesis is driven by very high levels of TGF-β signaling. Additionally, the effect on different downstream molecular pathways regulated by TGF-β signaling during each stage of tumorigenesis must be considered. As such, stage-specific TGF-β signaling targeting strategies are being developed. Broadly, targeting strategies aim to inhibit oncogenic TGF-β signaling while restoring autonomous TGF-β signaling [71]. This presents clinical challenges such as the specific dose required to alter TGF-β signaling levels precisely and the length of time TGF-β signaling targeting protocols should be followed. Incorrect doses or time frames will not simply fail to diminish the malignancy but may in fact drive enhanced tumorigenesis. As such, stage-specific combinational approaches are seen as the most likely successful future approach.

TGF-β signaling has been targeted via ligand traps, antisense oligonucleotides (ASOs), and small molecule receptor kinase inhibitors (RKIs). Ligand traps are antiligand antibodies that act as sinks for excess TGF-β and have been successful in breast cancer metastasis preclinical studies [71,88]. ASOs act to block TGF-β synthesis by degrading TGF-β mRNAs. ASO targeting is challenging to administer and is complicated by off-target effects but offers the benefit of large-scale blockade of the target protein [71]. ASOs for TGF-β have had limited results in mouse models for breast metastasis but showed significant anticancer effects in a mouse model of pancreatic metastasis [89,90]. RKIs are small molecule drugs that offer competitive inhibition of the catalytic activity of TGF-β receptors. RKIs for TGF-β receptors showed inhibition of motility and invasion in breast metastasis models [91]. However, other studies have shown strong short-term but weak long-term effects of TGF-β RKIs [92], reflecting the complex expression pattern of TGF-β signaling during tumorigenesis. Additionally, in murine breast cancer models, metastasis is alleviated by BMP treatment, which specifically decreases CSC via induced differentiation

[93]. In pancreatic cancer patients, TGF-β inhibition reduces CSC tumorigenicity, while increasing chemosensitivity [94]. As these strategies are quite new, limited clinical data has been produced for TGF-β targeting to date. These few studies indicate that TGF-β targeting in the clinic must address the complex stage- and patient-specific nature of TGF-β signaling during tumorigenesis to prove beneficial.

4.8 TARGETING CSCs VIA PRC2 COMPONENT EZH2

EZH2 inhibition has received much attention due to its high expression in many cancers, association with tumorigenic and invasive properties of CSCs, and the fact that EZH2 can be blocked by several inhibitor drugs. In malignancy, EZH2 is principally known to silence specific tumor suppressor genes and, to a lesser extent, promote expression of selected oncogenes [47,95,96]. In line with this, the anticancer strategy aims to alleviate EZH2-driven gene silencing to restore tumor suppressor gene expression. However, PRCs display bivalent properties of promoting and silencing target gene expression. As such, EZH2 inhibition is only suitable for specific types of malignancy. Additionally, the nature of the EZH2 defect, such as the precise mutation involved, may be key to generating efficient targeting strategies. Mechanistically, EZH2 appears to repress the expression of antimetastatic genes in many cancers and CSCs [47,48]. As such, EZH2 targeting may represent a strong opportunity to target metastasis, which is responsible for most cancer deaths and is untreatable in most cases. High EZH2 expression is an indicator of malignancy, poor prognosis, and aggressive (potentially metastatic) disease in several cancers including breast, prostate, ovary, liver, brain, and head and neck [47,48]. EZH2 is known to regulate proliferation and the SR–differentiation switch in many types of CSCs [47]. EZH2 components do not appear to be essential for stand-out CSC properties such as resistance to therapies. Instead, EZH2 acts to repress the expression of key antimetastatic genes in CSCs [47]. Where EZH2 contributes to key CSC properties it tends to be an indirect mechanism. For example, EZH2 mediates chemo- and radiotherapy resistance through regulation of DNA damage mechanisms [47].

To date, the most successful and widely studied "EZH2 inhibitor" is DNZeP (3-Deazaneplanocin), which is a global methylation inhibitor that has been shown to rescue the expression of many EZH2 targets. EZH2-dependent methylation requires S-adenosyl-L-homocysteine (SAH) as a cofactor: DNZeP is a SAH-hydrolase inhibitor [47,95–97]. DNZeP has been shown to confer antitumor activity in breast, lung, prostate, and liver malignancies and leukemia. DNZeP has not entered clinical trials but has been shown to have preclinical anticancer effects in several cancers [47,95–97]. DNZeP is 20 times better at targeting BRCA$^-$ than BRCA$^+$ cells, which indicates potential utility as a new treatment for poorly differentiated, more stem-like breast cancers [98]. DNZeP can specifically target colon CSCs, resulting in an increased apoptotic rate that does not affect nonmalignant SCs [97]. As such, DNZeP has attracted much excitement and the outcome of presumed clinical trials is highly anticipated. There are two other EZH2 inhibitors under clinical trial, namely, "GSK2816126" (NCT02082977) and "E7438" (NCT01897571), but these trials are at too early a stage for conclusions on their efficacy to be made.

4.9 TARGETING CSCs VIA miRNA MECHANISMS

Targeting a single miRNA has been shown to affect multiple downstream pathways, which is a major attraction. However, targeting single miRNAs can, therefore, have multiple unwanted downstream effects. Thus, miRNA targeting has great potential but is highly complex. Broadly, therapeutic strategies aim to inhibit oncogenic miRNAs and/or restore expression of tumor suppressor miRNAs [99–101].

Prolonged inhibition and restoration of specific miRNAs has been achieved in animal studies, which opens up the possibility of miRNA targeting in the clinic [102,103: reviewed in Ref. 100]. In parallel, research has focused upon identification of cancer stage-specific miRNAs. For example, some success has been observed in pancreatic cancer, where aberration of miRs-21 and -221 and restoration of let-7 sensitizes cells to chemotherapy drug gemcitabine [reviewed in Refs. 101,104,105]. Additionally, miR-27a and miR-200 targeting specifically diminished CSC properties in separate pancreatic cancer studies [106]. While miRNA targeting has strong potential as a future therapy, the strategy is immediately hindered by technical difficulties associated with identification of the direct targets regulated by specific miRNAs. As target identification strategies improve, the likelihood of clinically relevant miRNA targeting of CSCs grows stronger. In the interim, miRNA profiles have important prognostic and diagnostic uses, which are being actively studied.

5 PERSPECTIVE: CSCs AND THE FUTURE OF CANCER THERAPEUTICS

To successfully treat cancer, one must be able to detect the presence of a malignancy and remove its tumorigenic potential (via CSC targeting), while preserving the integrity of the surrounding healthy tissue. In this section, the future of CSC targeting will be discussed with respect to these three principles.

5.1 EPIGENETIC CSC SIGNATURES AS BIOMARKERS

Although it is often overlooked, CSC-linked epigenetic mechanisms have been successfully exploited as diagnostic, prognostic, and/or treatment-management biomarkers. At one level, the expression of specific components of stemness signaling pathways is already being used to inform clinicians as to the precise nature of the disease so that the most appropriate treatment can be given to each patient. While many cancers are now treated with high levels of efficiency, such screening strategies will be key components of efforts to tackle cancers that are highly resistant to current treatment regimes. At another level, it is likely that CTCs express specific CSC-like epigenetic signatures (e.g., stemness signaling pathways, DNA methylation, miRNAs) can be exploited to screen patients for early detection [107–111]. Such an approach is particularly applicable to the "silent killer" malignancies, which can develop and even metastasize before the primary malignancy has been detected. Additionally, CTC epigenetic signatures may hold the key to pinpointing the location of otherwise undetected primary tumors. Looking forward, it may soon be possible to routinely screen for CTCs, using specific epigenetic signatures to identify the primary tumor's origin long before clinical symptoms are apparent. With modern technology, such analysis can be carried out at clinically relevant speeds and costs. Such epigenetically enabled early detection would have a great impact on the survival of patients who would otherwise present with incurable late stage disease.

5.2 FIGHTING FIRE WITH FIRE: AWAKENING CSCs FOR ENHANCED CHEMORESPONSE

The hierarchical CSC model indicates that CSCs spend long periods of time in a quiescent state, where they are protected from cancer therapeutics designed to target rapidly growing cells. Addressing this, a pretreatment that stimulates CSCs to leave the quiescent state, reenter the cell cycle, and hyperproliferate may enhance CSC sensitivity to current treatments. This was long-ago demonstrated in principle. A single dose of chemotherapy drug 5-fluorouracil (5-FU) ablated the active HSC system, awakening quiescent HSCs. However, a second 5-FU dose targeted these newly awakened HSCs, removing their regenerative potential [112]. While the concept of CSC stimulation as an anticancer measure seems counterintuitive, this strategy could expose a large population of hidden CSCs to current cancer therapeutics, with obvious benefits for patients. We have described how targeting stemness signaling pathways have been successful in enhancing response to chemotherapy drugs. It is tempting to speculate that stemness signaling pathway targeting awakens CSCs from quiescence, which would expose CSCs to the conventional chemotherapeutic. Regardless of the mechanism involved or the approach chosen, stimulation followed by ablation strategies may represent the future of CSC targeting.

5.3 ZERO-COLLATERAL CSC TARGETING

While still in its infancy, CSC targeting clearly has immense potential in the future of cancer therapeutics. It is widely acknowledged that eliminating CSCs could dramatically enhance first line therapies for currently refractory disease. CSC targeting will likely form part of a combined strategy that synergistically enhances, rather than replacing, current therapies. The scientific and clinical challenge is development of CSC-targeting strategies with few negative side effects. For example, CSC-directed therapies can have deleterious effects on healthy adult SCs, due to similarities between CSCs and SCs. Targeting of LSCs by rapamycin, an inhibitor of PI3-kinase signaling, has been achieved in a manner that did not negatively affect the HSC pool, demonstrating that CSC-specific strategies can be achieved [113]. In this regard, it is important that a comprehensive understanding of the multiple SC and CSC hierarchies involved in specific tissues and malignancies is generated. The more information that is known about these mechanisms, the more likely it is that a CSC-targeting strategy, which does not adversely affect the SC pool, will be developed. In this way, CSC targeting can be realized in a fashion that does not adversely affect the patient and is specifically tailored to the precise biology of their malignancy.

REFERENCES

[1] Ito K, Suda T. Metabolic requirements for the maintenance of self-renewing stem cells. Nat Rev Mol Cell Biol 2014;15:243–56.

[2] Simons BD, Clevers H. Strategies for homeostatic stem cell self-renewal in adult tissues. Cell 2011;145(6):851–62.

[3] Copley MR, Beer PA, Eaves CJ. Hematopoietic stem cell heterogeneity takes centre stage. Cell Stem Cell 2012;10:690–7.

[4] Noll JE, Williams SA, Purton LE, Zannettino AC. Tug of war in the haematopoietic stem cell niche: do myeloma plasma cells compete for the HSE niche? Blood Cancer J 2012;2:e91.

[5] Thompson JA, Itskovitz-Eldor J, Sharipo SS, Waknitz MA, Swiergiel JJ, Marshall VS, et al. Embryonic stem cell lines derived from human blastocysts. Science 1998;282:1145–8.

[6] Takahashi K, Tanabe K, Ohnuki M, Narita M, Ichisaka T, Tomada K, et al. Induction of pluripotent stem cells from adult human fibroblasts by defined factors. Cell 2007;131:1–12.

[7] Cheung TH, Rando TA. Molecular regulation of stem cell quiescence. Nat Rev Mol Cell Biol 2013;14:329–40.

[8] Beltrami AP, Cesselli D, Beltrami CA. At the stem of youth and health. Pharmacol Ther 2011;129:3–20.

[9] Beltrami AP, Cesselli D, Beltrami CA. Stem cell senescence and regenerative paradigms. Clin Pharmac Ther 2012;91(1):21–9.

[10] Fiorentino FP, Symonds CE, Macaluso M, Giordano A. Senescence and p130/Rbl2: a new beginning to the end. Cell Res 2009;19:1044–51.

[11] Kleinsmith L, Pierce GB. Multipotency of single embryonal carcinoma cells. Cancer Res 1964;24:1544–51.

[12] Andrews PW, Bronson DL, Benham F, Strickland S, Knowles BB. A comparative study of eight cell lines derived from human testicular teratocarcinoma. Intl J Cancer 1980;26:269–80.

[13] Andrews PW, Goodfellow PN, Shevinsky LH, Bronson DL, Knowles BB. Cell-surface antigens of a clonal human embryonal carcinoma cell line: morphological and antigenic differentiation in culture. Intl J Cancer 1982;29:523–31.

[14] Andrews PW, Damjanov I, Simon D, Banting GS, Carlin C, Dracopoli C, et al. Pluripotent embryonal carcinoma clones derived from the human teratocarcinoma cell line Tera-2. Labor Invest 1984;50(2):147–62.

[15] Andrews PW, Banting G, Damjanov I, Arnaud D, Avner P. Three monoclonal antibodies defining distinct differentiation antigens associated with different high molecular weight polypeptides on the surface of human embryonal carcinoma cells. Hybridoma 1984;3(4):347–61.

[16] Andrews PW. Retinoic acid induces neuronal differentiation of a cloned human embryonal carcinoma cell line *in vitro*. Dev Biol 1984;103:285–93.

[17] Bonnet D, Dick JE. Human acute myeloid leukemia is organised as a hierarchy that originates from a primitive hematopoietic cell. Nat Med 1997;3(7):730–7.

[18] Dick JE. Acute myeloid leukaemia stem cells. Ann NY Acad Sci 2005;1044:1–5.

[19] Reya T, Morrison SJ, Clarke MF, Weissman IL. Stem cells, cancer and cancer stem cells. Nature 2001;414:105–11.

[20] Hanahan D, Weinberg RA. Hallmarks of cancer: the next generation. Cell 2011;144:646–74.

[21] Driessiens G, Beck B, Caauwe A, Simons BD, Blanpain C. Defining the mode of tumor growth by clonal analysis. Nature 2012;488(7412):527–30.

[22] Al-Hajj M, Wicha MS, Benito-Hernandez A, Morrison SJ, Clarke MF. Prospective identification of tumorigenic breast cancer cells. PNAS 2003;100(7):3983–8.

[23] Hemmati HD, Nakano I, Lazareff Masterman-Smith M, Greschwind DH, Bonner-Fraser M, et al. Cancerous stem cells can arise from paediatric brain tumors. PNAS 2001;100:15178–83.

[24] Richardson GD, Robson CN, Lang SH, Neal DE, Maitland NJ, Colline AT. CD133, a novel marker of human prostatic epithelial stem cells. J Cell Sci 2004;117(16):3539–45.

[25] Chen K, Huang Y, Chen J. Understanding and targeting cancer stem cells: therapeutic implications and challenges. Acta Pharma Sin 2013;34:732–40.

[26] Huntley BJ, Gilliland Leukaemia stem cells and the evolution of cancer stem cell research. Nat Rev Cancer 2005;5:311–21.

[27] Oishi N, Wang XW. Novel therapeutic strategies for targeting liver cancer stem cells. Intl J Biol Sci 2011;7(5):517–35.

[28] Keith B, Simon MC. Hypoxia-inducible factors, stem cells and cancer. Cell 2007;129(3):465–72.

[29] Hillen F, Griffioen AW. Tumor vascularisation: sprouting angiogenesis and beyond. Cancer Mets Rev 2007;26:489–502.

[30] Valastayan S, Weinburg RA. Tumor metastasis: molecular insights and evolving paradigms. Cell 2011;147:275–92.

[31] Brabletz T. EMT and MET in metastasis: where are the cancer stem cells? Cancer Cell 2012;22:699–700.

[32] Baccelli I, Trumpp A. The evolving concept of cancer and metastasis stem cells. J Cell Biol 2012;198(3):281–93.

[33] Chambers I, Smith A. Self-renewal of teratocarcinoma and embryonic stem cells. Oncogene 2004;23:7150–60.

[34] Silva J, Smith A. Capturing pluripotency. Cell 2008;132:532–8.

[35] Loh KM, Lim B. A precarious balance: pluripotency factors as lineage specifiers. Cell Stem Cell 2011;8:363–9.

[36] Mitsui K, Tokuzawa Y, Itoh H, Segawa K, Murakami M, Takahashi K, et al. The homeoprotein Nanog is required for maintenance of pluripotency in mouse epiblast and ES cells. Cell 2003;113:631–42.

[37] Chambers I, Colby D, Robertson M, Nichols J, Lee S, Tweedie A, et al. Functional expression cloning of Nanog, a pluripotency sustaining factor in embryonic stem cells. Cell 2003;113:643–55.

[38] Andrews PW, Matin MM, Bahrami AR, Damjanov I, Gokhale P, Draper JS. Embryonic stem (ES) and embryonal carcinoma (EC) cells: opposite sides of the same coin. Biochem Soc Trans 2005;33:1526–30.

[39] Andrews PW. From teratocarcinomas to embryonic stem cells. Phil Trans R Soc Lon B 2002;357:405–17.

[40] Takahashi R, Miyazaki H, Ochiya T. The role of microRNAs in the regulation of cancer stem cells. Front Genet 2014;4(295):1–11.

[41] Greve TS, Judson RL, Blelloch R. microRNA control of mouse and human pluripotent stem cell behaviour. Annu Rev Cell Dev Biol 2013;29:213–39.

[42] Munoz P, Iliou MS, Esteller M. Epigenetic alterations involved in cancer stem cell reprogramming. Mol Oncol 2012;6:620–36.

[43] Berdasco M, Esteller M. DNA methylation in stem cell renewal and multipotency. J Stem Cell Res Ther 2011;2:42–51.

[44] Baylin SB. Stem cells, cancer and epigenetics. StemBook, ed., 2009 October.

[45] Sørensen AL, Jacobsen BM, Reiner AH, Andersen IS, Collas P. Promoter DNA methylation patterns of differentiated cells are largely programmed at the progenitor stage. Mol Cell Biol 2010;21:2066–77.

[46] Fouse SD, Shen Y, Pellegrini M, Cole S, Meissner A, Van Neste L, et al. Promoter CpG methylation contributes to ES cell gene regulation in parallel with Oct4/Nanog, PcG complex and histone H3 K4/27. Cell Stem Cell 2008;2:160–9.

[47] Crea F, Paolicchi E, Marquez VE, Danesi R. Polycomb genes and cancer: time for clinical application? Crit Rev Oncol Hematol 2012;83:184–93.

[48] Aloia L, Di Stefano B, Di Croce L. Polycomb complexes in stem cells and embryonic development. Development 2013;140:2525–34.

[49] Marks H, Kalkan T, Menafra R, Denissov S, Jones K, Hofemeister H, et al. The transcriptional and epigenetic foundations of ground state pluripotency. Cell 2012;149:590–604.

[50] Sauvageau M, Sauvageau G. Polycomb group proteins: multi-faceted regulators of somatic stem cells and cancer. Cell Stem Cell 2010;7:299–313.

[51] Richly H, Aloia L, Di Croce L. Roles of polycomb group proteins in stem cells and cancer. Cell Death Dis 2011;2:e204.

[52] Bardot ES, Valdes VJ, Zhang J, Perdigoto CN, Nicolis S, Silva JM, et al. Polycomb subunits Ezh1 and Ezh2 regulate the Merkel cell differentiation program in skin stem cells. EMBO J 2013;32(14):1990–2000.

[53] Villasante A, Piazzolla D, Li H, Gomez-Lopez G, Djabail M, Serrano M. Epigenetic regulation of Nanog expression by Ezh2 in pluripotent stem cells. Cell Cycle 2011;10(9):1488–9.

[54] Holland JD, Klaus A, Garratt AN, Birchmeier W. Wnt signalling in stem and cancer stem cells. Curr Opin Cell Biol 2013;25:254–64.

[55] Kuhl JS, Kuhl M. On the role of Wnt/β-catenin signalling in stem cells. Biochim Biophys Acta 2013;1830:2297–306.

[56] De Sousa EM, Veumeulen L, Richel D, Medema JP. Targeting Wnt signalling in colon cancer stem cells. Clin Cancer Res 2010;17:647–53.

[57] Tsang DPF, Cheng ASL. Epigenetic regulation of signalling pathways in cancer: role of the histone methyltransferases EZH2. J Gastroenterol Hepatol 2011;26:19–27.

[58] Wang L, Jin Q, Lee J-E, Su I-H, Ge K. Histone H3K27 methyltransferase Ezh2 represses Wnt genes to facilitate adipogenesis. PNAS 2010;107(16):7317–22.

[59] Shi B, Liang J, Yang X, Wang Y, Zhao Y, Wu H, et al. Integration of estrogen and wnt signalling circuits by the polycomb group protein EZH2 in breast cancer cells. Mol Cell Biol 2007;27(14):5105–19.

[60] Briscoe J, Therond PP. The mechanisms of Hedgehog signalling and its roles in development and disease. Nat Rev Mol Cell Biol 2013;14:416–29.

[61] Amakye D, Jagani Z, Dorsch M. Unravelling the therapeutic potential of the hedgehog pathway in cancer. Nat Med 2013;19(11):1410–22.

[62] Po A, Ferretti E, Miele E, De Smaele E, Paganelli A, Canettieri G, et al. Hedgehog controls neural stem cells through p53-independent regulation of Nanog. EMBO J 2010;29:2646–58.

[63] Capaccione KM, Pine SR. The Notch signalling pathway as a mediator of tumor survival. Carcinogen 2013;34(7):1420–30.

[64] Fiuza UM, Arias AM. Cell and molecular biology of Notch. J Endocrin 2007;194:459–74.

[65] Noisa P, Lund C, Kanduri K, Lund R, Lahdesmaki H, Lahesmaa R, et al. Notch signaling regulates neural crest differentiation from human pluripotent stem cells. J Cell Sci 2014 25 ePub.

[66] Espinoza E, Pochampally R, Xing F, Watabe K, Miele L. Notch signaling: targeting cancer stem cells and epithelial-to-mesenchymal transition. OncoTargets and Ther 2013;6:1249–59.

[67] Gonzalez ME, Moore HM, Li X, Toy KA, Huang W, Sabel MS, et al. EZH2 expands breast stem cells through activation of notch1 signaling. PNAS 2014;111(8):3098–103.

[68] Sakaki-Yumoto M, Katsuno Y, Derynck R. TGF-β family signalling in stem cells. Biochim Biophys Acta 2013;1830:2280–96.

[69] Kale VP, Vaidya AA. Molecular mechanisms behind the dose-dependent differential activation of MAPK pathways induced by transforming growth factos-beta1 in hematopoietic cells. Stem Cells Dev 2004:536–47.

[70] Yamazaki S, Iwama A, Takayanagi S, Eto K, Ema H, Nakauchi H. TGF-beta as a candidate bone marrow niche signal to induce hematopoietic stem cell hibernation. Blood 2009;113:1250–6.

[71] Connolly EC, Freimuth J, Nkhurst RJ. Complexities of TGF-β targeted cancer therapy. Intl J Biol Sci 2012;8:964–78.

[72] Mani SA, Guo W, Liao MJ, Eaton EN, Ayyanan A, Zhou AY, et al. The epithelial-mesenchymal transiation generates cells with properties of stem cells. Cell 2008;133:704–15.

[73] Rao Z-Y, Cai M-Y, Yang G-F, He LR, Mai SJ, Hua WF, et al. EZH2 supports ovarian carcinoma cell invasion and/or metastasis via regulation of TGF-β1 and is a predictor of outcome in ovarian carcinoma patients. Carcinogen 2010;31(9):1576–83.

[74] Tiwari N, Tiwari VK, Waldmeier L, Balwierz PJ, Arnold P, Meyer-Schaller N, et al. Sox4 is a master regulator of epithelial–mesenchymal transition by controlling Ezh2 expression and epigenetic reprogramming. Cancer Cell 2013;23:768–83.

[75] Han L, Shi S, Gong T, Zhang Z, Sun X. Cancer stem cells: therapeutic implications and perspective in cancer therapy. Acta Pharma Sinica B 2013;3(2):65–75.

[76] Castaigne S, Chomienne C, Daniel MT, Ballerini P, Berger R, Fenauc P, et al. All-trans retinoic acid as a differentiation therapy for acute promyelocytic leukaemia. I. Clinical results. Blood 1990;76:1704–9.

[77] Chomienne C, Ballerune P, Bailtrand N, Daniel MT, Fenaux P, Castaigne S, et al. All-trans retinoic acid in acute prolyelocytic leukaemias. II. *In vitro* studies: structure–function relationship. Blood 1990;76:1710–17.

[78] Connolly R, Nguyen NK, Sukumar S. Molecular pathways: current role and future directions of the retinoic acid pathway in cancer prevention and treatment. Clin Cancer Res 2013;19(7):1651–9.

[79] Garcia-Regalado A, Vargas M, Garcia-Carranca A, Arechaga-Ocampo E, Gonzalez-De la Rosa CH. Activation of Akt pathway by transcription-independent mechanisms of retinoic acid promotes survival and invasion in lung cancer cells. Mol Cancer 2013;12:44–54.

[80] Lepourcelet M, Chen YN, France DS, Wang H, Crews P, Petersen F, et al. Small-molecular antagonists of the oncogenic Tcf/beta-catenin protein complex. Cancer Cell 2004;5:91–102.

[81] Chen B, Dodge ME, Tang W, Lu J, Ma Z, Fan CW, et al. Small molecule-mediated disruption of Wnt-dependent signalling in tissue regeneration and cancer. Nat Chem Viol 2009;5:100–7.

[82] Berlin J, Bendell JC, Firdaus I, Gore I, Hermann RC, Mulcahy MF, et al. A randomized phase II trial of vismodegib versus placebo with FOLFOX or FOLFIRI and bevacizumab in patients with previously untreated colorectal cancer. Clin Cancer Res 2013;19:258–67.

[83] Kaye SB, Fehrenbacher L, Holloway R, Amit A, Karlan B, Slomovitz B, et al. A phase II, randomized, placebo-controlled study of vismodegib as maintenance therapy in patients with ovarian cancer in the second or third complete remission. Clin Cancer Res 2012;18:6509–18.

[84] Rodova M, Fu J, Watkins DN, Srivastava RK, Shankar S. Sonic hedgehog signalling inhibition provides opportunities for targeted therapy by sulforaphane in regulating pancreatic cancer stem cell self-renewal. PLoS One 2012;7(9):e46083.

[85] Yabuuchi S, Pai SG, Campbell NR, de Wilde RF, De Oliveira E, Korangath P, et al. Notch signal pathway targeted therapy suppresses tumor progression and metastatic spread in pancreatic cancer. Cancer Lett 2013;335:41–51.

[86] McAuliffe SM, Morgan SL, Wyant GA, Tran LT, Muto KW, Chen YS, et al. Targeting Notch, a key pathway for ovarian cancer stem cells, sensitizes tumors to platinum therapy. PNAS 2012;109(43):E2939–48.

[87] Wong GT, Manfra D, Poulet FM, Zhang Q, Josien H, Bara T, et al. Chronic treatment with the gamma-secretase inhibitor LY-411,575 inhibits beta-amyloid peptide production and alters lymphopoiesis and intestinal cell differentiation. J Cell Biol 2004;279:12876–82.

[88] Nam JS, Terabe M, Mamura M, Kang MJ, Chae H, Stuelten C, et al. An anti-transforming growth factor beta antibody suppresses metastasis via cooperative effects on multiple cell compartments. Cancer Res 2008;68:3835–43.

[89] Muraoka-Cook RS, Kurokawa H, Koh Y, Forbes JT, Roebuck LR, Barcellos-Hoff MH, et al. Conditional overexpression of active transforming growth factor beta1 *in vivo* accelerates metastases of transgenic mammary tumours. Cancer Res 2004;64:9002–11.

[90] Schlingensiepen KH, Jaschinski F, Lang SA, Moser C, Geissler EK, Schlitt HJ, et al. Transforming growth factor-beta2 gene silencing with trabedersen (AP 12009) in pancreatic cancer. Cancer Sci 2011;102:1193–200.

[91] Ehata S, Hanyu A, Fujime M, Katsuno Y, Fukunaga E, Goto K, et al. Ki26894, a novel transforming growth factor-beta type 1 receptor kinase inhibitor, inhibits *in vitro* invasion and *in vivo* metastasis of a human breast cancer cell line. Cancer Sci 2007;98:127–33.

[92] Yang L, Huang J, Ren X, Gorska AE, Chytil A, Carbone DP, et al. Abrogation of TGF beta signalling in mammary carcinomas recruits Gr-1 + CD11b+ myeloid cells that promote metastasis. Cancer Cell 2008;13:23–35.

[93] Scheel C, Eaton EN, Li SH, Chaffer CL, Reinhardt F, Kah KJ, et al. Paracrine and autocrine signals induce and maintain mesenchymal and stem cell states in the breast. Cell 2011;145:926–40.

[94] Lonardo E, Herman PC, Mueller MT, Huber S, Balic A, Miranda-Lorenzo I, et al. Nodal/Activin signalling drives self-renewal and tumorigenicity of pancreatic cancer stem cells and provides a target for combined drug therapy. Cell Stem Cell 2011;9:433–46.

[95] Crea F, Fornaro L, Bocci G, Sun L, Farrar WL, Falcone A, et al. EZH2 inhibition: targeting the crossroad of tumor invasion and angiogenesis. Cancer Mets Rev 2012;31(3–4):753–61.

[96] Deb G, Thanur VS, Gupta S. Multifaceted role of EZH2 in breast and prostate tumorigenesis. Epigenetics 2013;8(5):464–76.

[97] Benoit YD, Witherspoon MS, Laursen KB, Guezquez A, Beausejour M, Beaulieu JF, et al. Pharmacological inhibition of Polycomb repressive complex-2 activity induces apoptosis in human colon cancer stem cells. Expt Cell Res 2013;319:1463–70.

[98] Puppe J, Drost R, Liu X, Joosse SA, Evers B, Cornelissen-Steijger P, et al. BRCA1-deficient mammary tumor cells are dependent on EZH2 expression and sensitive to Polycomb repressive complex 2-inhibitor 3-deazaneplanocin A. Breast Cancer Res 2009;11(4):1–12.

[99] Garzan R, Marcucci G, Croce CM. Targeting microRNAs in cancer: rationale, strategies and challenges. Nat Rev Drug Disc 2010;9:275–89.

[100] Khan S, Anasrullah Kumar D, Jaqqi M, Chauhan SC. Targeting microRNAs in pancreatic cancer: micro-players in the big game. Cancer Res 2013;73(22):6541–7.

[101] Liu C, Tang DG. MicroRNA regulation of cancer stem cells. Cancer Res 2011;71(18):5950–4.

[102] Landford RE, Hildebrandt-Eriksen ES, Petri A, Persson R, Lindow M, Munk ME, et al. Therapeutic silencing of microRNA-122 in primates with chronic hepatitis C virus infection. Science 2010;327:198–201.

[103] Jansson MD, Lund AH. MicroRNA an cancer. Mol Oncol 2012;6:590–610.

[104] Moriyama T, Ohuchida K, Mizumoto K, Yu J, Sato N, Nabae T, et al. MicroRNA-21 modulates biological functions of pancreatic cancer cells including their proliferation, invasion and chemoresistance. Mol Cancer Ther 2009;8(5):1067–74.

[105] Oh JS, Kim JJ, Byun JY, et al. Lin-28/Let-7 modulates radiosensity of human cancer cells with activation of K-Ras. Int J Radiat Oncol Biol Phys 2010;76(1):5–8.

[106] Cheng HE, Shi SI, Cai X, et al. MicroRNA signature for human pancreatic cancer invasion and metastasis (review). Exp Ther Med 2012;4:181–7.

[107] Raimondi C, Gradilone A, Gazzaniga P. Circulating tumour cells in early bladder cancer: insight into micrometastatic disease. Expert Rev Mol Diagn 2014 epub.

[108] Wendel M, Barzhenova L, Boshuizen R, et al. Fluid biopsy for circulating tumor cell identification in patients with early-and late-stage non-small cell lung cancer: a glimpse into lung cancer biology. Phys Biol 2012;9(1) epub.

[109] Wulfing P, Borchard J, Buerger H, et al. HER2-positive circulating tumor cells indicate poor clinical outcome in stage I to III breast cancer patients. Clin Cancer Res 2006;12(6):1715–20.

[110] Geiman TM, Muegge K. DNA methylation in early development. Mol Reprod Dev 2010;77(2):105–13.

[111] Sheaffer KL, Kim R, Aoki R, et al. DNA methylation is required for the control of stem cell differentiation in the small intestine. Genes Dev 2014;28(6):652–64.

[112] Harrison DE, Lerner CP. Most primitive hematopoietic stem cells are stimulated to cycle rapidly after treatment with 5-fluorouracil. Blood 1991;78(5):1237–40.

[113] Yilmaz OH, Vladez R, Theisen Pten dependence distinguishes haematopoietic stem cells from leukaemia-initiating cells. Nature 2006;441(7092):475–82.

FUTURE DIRECTIONS: TRANSLATION TO THE CLINIC

PART

5

FUTURE
DIRECTIONS:
TRANSLATION
TO THE CLINIC

PERSONALIZED EPIGENETIC THERAPY— CHEMOSENSITIVITY TESTING

28

Benet Pera and Leandro Cerchietti

*Hematology and Oncology Division, Medicine Department, Weill Cornell
Medical College of Cornell University, New York, NY, USA*

CHAPTER OUTLINE

1 INTRODUCTION

Diffuse large B-cell lymphomas (DLBCLs) are common non-Hodgkin lymphoid (NHL) malignancies that for most patients follow an aggressive clinical course. Contrasting with most other "solid" tumors, DLBCLs are potentially chemotherapy curable diseases. Chemotherapy is, therefore, the cornerstone of DLBCL treatment. It has been known for more than 40 years that some patients with NHL can be cured using chemotherapy in combination [1]. Since then, a great deal of effort has gone to identify the best combinatorial treatment for DLBCL patients. The best cost-effective regimen, now part of the standard in most countries, resulted to be CHOP, a combination of four chemotherapy agents (cyclophosphamide, doxorubicin, vincristine, and prednisolone) [2,3].

The majority of these drugs target the DNA and/or mitosis, common strategies in the 1960s and 1970s to kill highly proliferating tumor cells. Cyclophosphamide, a nitrogen mustard, mainly acts as a DNA-cross-linker attaching into the nitrogen 7 of the guanine base and interfering with DNA replication by forming intrastrand and interstrand DNA cross-links [4]. Vincristine is a semisynthetic natural product with a well-characterized action mechanism that involves the disruption of the microtubules and the subsequent cell death by apoptosis [5]. The anthracycline doxorubicin intercalates into DNA and stabilizes the DNA topoisomerase II (Topo II) complex resulting in replication stopping [6]. For three decades, there were no further substantial advances on the treatment of DLBCL until rituximab (an anti-CD20 monoclonal antibody) was adopted as part of the standard treatment (R-CHOP) [7]. As result, the overall response rate for patients with advanced disease increased to 65–70% from about 50–55% with CHOP. Currently, there is no superior regimen for the treatment of DLBLC, however, approximately 25% of the patients will succumb to the disease either due to primary refractory DLBCL or because of a relapsing lymphoma.

2 CHEMORESISTANCE IN LYMPHOMAS

The most consisting finding from a series of randomized controlled clinical trials for NHL patients is the superiority of an anthracycline-containing regimen over regimens that do not contain an anthracycline. Therefore, anthracycline resistance, in particular to doxorubicin, remains the primary obstacle for improving the curability rate of DLBCL with chemotherapy. Chemoresistance may be intrinsic or acquired (a.k.a. adaptive) and may occur through multiple mechanisms. These mechanisms may be classified into two major groups. The first group is decreased drug accumulation and/or increased drug inactivation, e.g., via upregulation of the MDR/TAP (multidrug resistance/antigen peptide transporter) genes. Although this mechanism has been well characterized in cell and animal models [8], its relevance to clinical resistance remains unclear [9]. A second mechanism involves biological mechanisms that prevent cells from dying after chemotherapy-induced damage. The capacity of most cancers to resist the cytotoxic effects of chemotherapy is more closely connected to biological abnormalities that affect critical cell cycle checkpoints and cell death pathways than to specific mechanisms of resistance unique to each drug. One approach to improve treatments in DLBCL would be the addition of adjuvant drugs to "chemosensitize" the tumor cells by interfering with aberrant biological pathways that make cells to resist dead from DNA damage. Thus, and in the light of many decades of successful curative chemotherapy in DLBCL, chemosensitization emerges as an attractive strategy to defeat relapsed and refractory disease, and epigenetic therapy promises to be a valid approach to achieve it.

3 EPIGENETICALLY ENCODED CHEMORESISTANCE

Gene expression is coordinated by transcription factors (that can activate or repress genes) and by the regulatory state of the chromatin, both at the DNA and histone components. This regulatory state is orchestrated by chromatin modifications such as DNA cytosine methylation and histone acetylation. Several enzymes, brought to regulatory transcriptional complexes by transcription factors and cofactors, impose these chemical modifications in the DNA and histones. Mutations in these enzymes or alteration in substrates and cofactors may induce aberrant DNA methylation patterns and histone modifications, a common feature in cancer [10–16]. This aberrant patterning is used by malignant cells to alter the expression of specific genes that will allow them to sustain the oncogenic state and survive extrinsic changes such as those imposed by chemotherapy agents.

Decreased expression of tumor suppressor genes and checkpoints is one of the most common mechanisms of lymphomagenesis. This can be achieved, among others mechanisms, by genetic lesions (e.g., mutations in TP53) as well as by epigenetic repression of genes (gene silencing) [17–19]. From the therapeutic point of view, epigenetic tumor suppressor silencing offers the opportunity of reversibility. Thus, pharmacological target of enzymes imposing gene silencing allows the restoring of nonfunctional pathways in tumor cells that may increase their susceptibility to cancer chemotherapy.

4 IMPLEMENTING EPIGENETIC THERAPY TO CHEMOSENSITIZE LYMPHOMA

Although new strategies are being tested, in the current clinical setting epigenetic therapy is currently represented by approaches that tiling the balance in favor of histone acetylation or induce DNA hypomethylation (Figure 28.1). Histone acetylation status is part of the so-called histone code [20] and is regulated through a group of covalent histone-modifying enzymes grouped into histone acetyltransferases (HATs) that increase histone acetylation and histone deacetylases (HDACs) that decrease

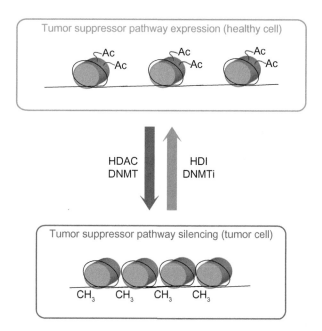

FIGURE 28.1

In cancer cells, the epigenetic balance is tilted to the silencing of tumor suppressor pathways by DNA hypermethylation of genes in the pathway through the action of the DNA methyltransferase and histone deacetylases. Histones deacetylase inhibitors and/or DNMT inhibitors can reverse the balance to reexpression of tumor suppressor pathways.

histone acetylation. Available drugs include HDAC inhibitors (HDI) that upon administration will tilt the balance to HATs, with a net increase in histone acetylation. Histone acetylation induces a relaxation in the chromatin making it more accessible to transcription activation and reexpression of genes.

A similar transcriptional effect to HDI is obtained upon DNA demethylation. The enzymes that methylate DNA are grouped into DNA methyltransferases (DNMTs) that introduce a methyl group into cytosines in CpG dinucleotides. Inhibition of DNMT can reverse gene silencing, restoring the expression of pathways ultimately affecting DLBCL survival [21]. Currently, there are two DNMT inhibitors (DNMTis) under clinical investigation in DLBCL; 5-aza-2′-deoxycitidine (decitabine) and azacitidine. Both molecules are cytidine nucleoside analogues that incorporate into DNA inactivating irreversibly DNMTs by forming a covalent bond between their 5-azacytosine ring and the DNMT enzyme. Thus, DNMTs are not able to incorporate methyl groups in newly synthesized DNA, resulting in the reduction of DNA methylation.

5 WHAT ANTITUMORAL EFFECT TO EXPECT FROM EPIGENETIC DRUGS?

Although histone acetylation and DNA demethylation can result in tumor suppressor reactivation with consequent triggering of cell death in cell cultures [22], this is not a typical effect in the clinical setting. Several clinical trials employing HDI in DLBCL, such as vorinostat and panobinostat, as well as demethylating agents have shown modest antilymphoma effect as single agents in relapsed DLBCLs [22]. Majority of these trials focused in tumor burden as primary goal to test the antilymphoma effect. However, recent animal studies and clinical trials suggest that tumor shrinkage is a rare event, while phenotypic and molecular changes, such as senescence or differentiation, are more common. Therefore, prospective trials should focus in delivering an amount of drug necessary to maximally achieve a biological effect rather than in tumor burden decrease. A recent report on panobinostat in relapsed DLBCL showed a modest tumor response measured with the standard criteria for tumor response, however, up to 50% of the patients in the trial exhibited a change in their tumor phenotype. Interestingly, cell lines mimicking this effect *in vivo* have shown a dramatic increase in their response to proteasome inhibitors, prompted researchers to pursue this idea in a new clinical trial combining an HDI with a proteasome inhibitor.

Since DNMTIs are incorporated into DNA, they may induce dose-dependent DNA damage. When this DNA damage exceeds a critical threshold it may cause apoptosis or any other form of cell death, but typically does not result in DNA hypomethylation, an effect requiring surviving cells. Compared to other cytosine analogues, such as cytarabine, DNMTis are poor inducers of DNA damage and may create adaptive resistance to other DNA damaging agents. Therefore, DNMTis must be used at doses that induce DNA hypomethylation while minimizing the DNA damaging effect when are aimed to restore lymphoma chemosensitivity to chemotherapeutic drugs. Recent data suggest that doses require to induce pathway reactivation and phenotypic changes are significantly lower than maximum tolerated doses for DNMTis [23].

6 SELECTING THE RIGHT DRUG FOR THE RIGHT PATIENT AND VICE VERSA

Although the clinical prognostic factors can stratify populations of patients with DLBCL into broad groups with respect to disease progression, it is clear that deeper understanding of the molecular disease heterogeneity, both between patients and within lymphomas, is required to a personalized therapeutic approach. In this regard, molecular profiling of lymphomas can serve to select candidates for epigenetic chemosensitizing therapies. Patient selection and mechanism of epigenetic sensitization are sides of the same coin. There are at least two complimentary approaches toward tumor-selective personalized epigenetic chemosensitization: cellular reprogramming and synthetic lethality.

6.1 CELLULAR REPROGRAMMING

During carcinogenesis certain apoptotic and differentiation pathways are epigenetically inactivated in order to assure tumor cell survival, being many of these pathways also required for the antineoplastic effect of most chemotherapeutic agents. By reversing epigenetic changes, epigenetic therapy can result in cell reprogramming that, in turn, may trigger cell differentiation, cell death, and/or senescence among other possible phenotypes. Cellular reprogramming may also restate pathways that lead to increased chemotherapy sensitivity.

For example, it has been reported that sequential treatment of drug-resistant breast cancer cells first with decitabine followed by doxorubicin reverses this resistance in more than 80% of treated cells [24]. Decitabine pretreatment induces a depletion in the DNMT1 protein levels that relieves transcriptional repression of *p21* (which is responsible for cell cycle regulation), thereby increasing its expression [24,25]. This DNMT1 reduction plays also an indirect role in *p21* induction via reexpression of methylation inactivated transcription factors such as *EGR1*, *SMAD3*, and *HES6*, which interact with *p21* promoter and increase its expression in tumor cells [25–27]. Induction of *p21* causes cells to undergo G2/M arrest that might result in accumulation of Topoisomerase II. Since stabilization of Topoisomerase II is also an effect of doxorubicin, this pathway reactivation results in increasing doxorubicin tumor cytotoxic effects [28,29].

It is worth nothing that reactivation of tumor cell-specific pathways may be more therapeutically crucial than reaching global epigenetic changes (i.e., increase in histone acetylation or DNA hypomethylation). In DLBCL, this is the case for the transforming growth factor β (TGFβ) pathway, specifically for its intracellular transducer SMAD1. TGFβ belongs to the superfamily of growth factors, involves in the control of cell growth, proliferation, differentiation, apoptosis, and homeostasis of normal B cells. These functions are mediated through proteins from the SMAD family that transduce the extracellular signals from the TGFβ ligands to the nucleus where they activate target genes [30]. We recently found that SMAD1 is epigenetically silenced by DNA hypermethylation in malignant B cells, allowing these cells to sustain the oncogenic phenotype. Moreover, we also demonstrated that DNA methylation-mediated silencing of SMAD1 results in chemoresistance in DLBCL. Treatment of DLBCL patients with the DNMTi azacitdine hypomethylates and reactivates SMAD1 expression. This makes the TGFβ pathway to become functional again, similarly to what is seen in B cells. Lymphoma cells exhibiting this gained functionality are now responsive to growth inhibitory signals through the TGFβ pathway, and more importantly, to chemotherapeutic drugs [23] (Figure 28.2). Therefore, cellular epigenetic reprogramming restores a "normal" tumor suppressor pathway in lymphoma cells that causes higher vulnerability to chemotherapy agents. Although speculative, patients with genetic lesions affecting TGFβ and/or SMAD1 could less likely benefit from this approach and will not be selected for such therapy.

6.2 SYNTHETIC LETHALITY

Originally described in the fruit fly [31] and later introduced to cancer biology [32], the concept of synthetic lethality represents a promising approach to be considered in the field of epigenetic therapy.

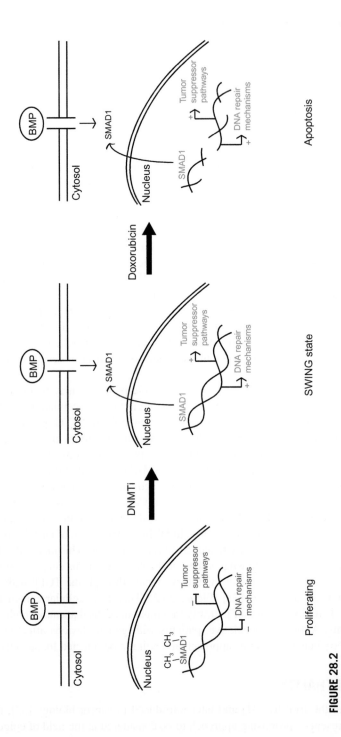

FIGURE 28.2

A depiction of the model proposed for the mechanism-based epigenetic chemosensitization of aggressive lymphomas upon low-dose DNMT inhibitor treatment. SWING: Senescence with incomplete growth arrest.

Synthetic lethality originally refers to the relationship between two genes where loss or inhibition of either gene is compatible with cell viability, but simultaneous loss or inhibition of both genes ends in cell death. This concept expands to synthetic lethal relationships among pathways that ultimately regulate essential survival functions.

In the context of epigenetics, mutations generated in the cancer cell can result in *de novo* dependences on the activity of certain gene products that in normal cells are not essential. Then, activation or inactivation of these new essential genes by epigenetic drugs could lead to cell death on cancer cells, leaving normal cells largely unaffected [33]. In this line, we can speculate the following scenario leading to synthetic lethality: genes *A* and *B* form a synthetic lethal pair in normal cells. In turn, gene *B* expression is negatively regulated by a third gene, gene *C*, normally hypermethylated. A DNMTi will hypomethylate and reactivate C that in turn will inhibit B. This situation will be lethal only for cells lacking the synthetic lethal pair A, which may occur exclusively in cancer cells by mutations or any other mechanism (Figure 28.3). Patients bearing inactivating mutations in gene *A* will be eligible for a therapy with DNMTis. Thus, epigenetic therapy personalization in lymphoma must address the identification of the mutations present in the patient that confer synthetic lethality. Currently, the only clinical application of synthetic lethality is the use of poly (ADP-ribose) polymerase (PARP) inhibitors in patients bearing tumors with *BRCA1* or *BRCA2* mutations [34]. PARP detects single strand DNA breaks and orchestrates its reparation [35]. When PARP activity is inhibited, an alternative form of DNA repair, homologous recombination (HR), buffers the PARP lack of effect [36]. HR is reliant on functional *BRCA1* and *BRCA2* genes, and when these genes are mutated this form of DNA repair cannot occur and cells die when exposed to PARP inhibitors.

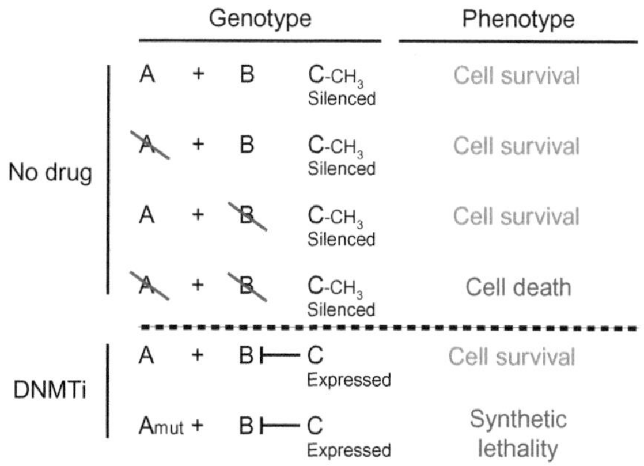

FIGURE 28.3

Genes A and B form a synthetic lethal pair. Loss or inhibition of either gene is compatible with cell viability, but loss or inhibition of both genes results in cell death. Gene C inhibits gene B. Gene C is normally hypermethylated and not expressed. Treatment with a DNA methyltransferase inhibitor restores gene C expression that will inhibit gene B. This approach will be lethal only to cell lacking gene A, for example by a mutation (Amut).

7 RESISTANCE TO EPIGENETIC-ACTING DRUGS AND CHEMOSENSITIVITY TESTING

Mutations in chromatin regulators are not rare events in lymphoma and other tumors. Gain-of-function mutations may lead to highly cancer-specific inhibitors such as in cases with EZH2 and IDH1/2 mutations present in several malignancies. On the other hand, mutations affecting HATs could decrease the effect of HDI as reported in preclinical models. Pharmacological cross-resistance among chemical-related compounds such as nucleotides analogues (e.g., Ara-C, azacitidine, and 5′deoxy-azacitidine) started to emerge as a clinical problem [37]. Other resistance mechanisms to epigenetic drugs are likely to arise, since perturbation of epigenetic pattering are frequent events in B-cell lymphomas [38].

In this context, it is relevant to highlight the utility of chemosensitivity testing to identify effective personalized epigenetic treatments. At least two complementary strategies can be adopted; one involving patient-derived material, and the other employing the Cancer Cell Line Encyclopedia (CCLE). The first approach involves the establishment of long-term cultures of patient tumor cells in which drug activity can be assayed. Although more challenging for hematological malignances and low-proliferating tumors, researchers used conditional reprogramming [39] to culture normal and cancer cells from a patient with recurrent respiratory papillomatosis, and identified vorinostat as an effective agent in killing the patient-derived tumor cells *in vitro*. On the basis of this data, the patient was subsequently treated with vorinostat resulting in tumor size shrinkage and stabilization [40]. On the other hand, the CCLE compiles large-scale genomic data sets of human cancer cell lines that include gene expression, chromosomal copy number, and massively parallel sequencing data, together with pharmacological profiling of drugs across most of these cell lines [41]. This tool represents a promising example for chemosensitivity identification (including to epigenetic drugs), as well as for identification of biomarkers associated with drug activity (or lack of). For example, investigators integrated the genetic profiles deposited in CCLE with sensitivity data of a novel fibroblast growth factor receptor (FGFR) inhibitor, and they identified that FGFR genetic alterations are the most significant predictors of sensitivity [42]. This data was later incorporated in the clinical trials for this FGFR inhibitor as a biomarker for patient selection.

Looking forward, exposing patient-derived chemotherapy-resistant cells to epigenetic agents and analyzing the reprogramming effect (e.g., by gene expression profiling), could help to determine whether a newly acquired chemotherapy sensitive state is reached, helping clinicians to select most likely effective drugs to use in combinatorial treatment with the epigenetic agent.

8 CONCLUSION

Pharmacologic regulation of epigenetic pathways represents a promising therapeutic strategy in DLBCL. In light of the preclinical and early clinical data, several epigenetic agents such as DNMTis and HDI have demonstrated encouraging pharmacodynamic effects. Importantly, epigenetic chemosensitization is emerging as a key mechanism of action underlying the effect of these drugs. This could probably be the most likely clinical scenario for translation of pleiotropic epigenetic agents such as DNMTis and HDIs.

The exploitation of cancer-specific vulnerabilities through the identification of synthetic lethal pairs will remain a key issue for the development of personalized epigenetic therapies. It will be crucial to identify and characterize epigenetic lesions present in each patient to select the best strategy to activate or inactivate the lethal counterpart. More importantly, the concept of biologically driven treatments should be included from the early beginning when designing clinical trials of epigenetic drugs. Research aimed to characterize tumor genetic and epigenetic features, as well as to identify biomarkers that account for the molecular effect of these drugs will be necessary to successfully translate epigenetic therapies to patients.

REFERENCES

[1] Levitt M, Marsh JC, DeConti RC, Mitchell MS, Skeel RT, Farber LR, et al. Combination sequential chemotherapy in advanced reticulum cell sarcoma. Cancer 1972;29:630–6.

[2] Gottlieb JA, Gutterman JU, McCredie KB, Rodriguez V, Frei III E. Chemotherapy of malignant lymphoma with adriamycin. Cancer Res 1973;33:3024–8.

[3] McKelvey EM, Gottlieb JA, Wilson HE, Haut A, Talley RW, Stephens R, et al. Hydroxyldaunomycin (Adriamycin) combination chemotherapy in malignant lymphoma. Cancer 1976;38:1484–93.

[4] Dong Q, Barsky D, Colvin ME, Melius CF, Ludeman SM, Moravek JF, et al. A structural basis for a phosphoramide mustard-induced DNA interstrand cross-link at 5′-d(GAC). Proc Natl Acad Sci USA 1995;92:12170–4.

[5] Gidding CE, Kellie SJ, Kamps WA, de Graaf SS. Vincristine revisited. Crit Rev Oncol Hematol 1999;29:267–87.

[6] Gewirtz DA. A critical evaluation of the mechanisms of action proposed for the antitumor effects of the anthracycline antibiotics adriamycin and daunorubicin. Biochem Pharmacol 1999;57:727–41.

[7] Coiffier B, Haioun C, Ketterer N, Engert A, Tilly H, Ma D, et al. Rituximab (anti-CD20 monoclonal antibody) for the treatment of patients with relapsing or refractory aggressive lymphoma: a multicenter phase II study. Blood 1998;92:1927–32.

[8] Binkhathlan Z, Lavasanifar A. P-glycoprotein inhibition as a therapeutic approach for overcoming multidrug resistance in cancer: current status and future perspectives. Curr Cancer Drug Targets 2013;13:326–46.

[9] Libby E, Hromas R. Dismounting the MDR horse. Blood 2010;116:4037–8.

[10] Cerchietti LC, Hatzi K, Caldas-Lopes E, Yang SN, Figueroa ME, Morin RD, et al. BCL6 repression of EP300 in human diffuse large B cell lymphoma cells provides a basis for rational combinatorial therapy. J Clin Invest 2010;120(12):4569–82.

[11] Palomero T, Couronne L, Khiabanian H, Kim MY, Ambesi-Impiombato A, Perez-Garcia A, et al. Recurrent mutations in epigenetic regulators, RHOA and FYN kinase in peripheral T cell lymphomas. Nat Genet 2014;46:166–70.

[12] Abdel-Wahab O, Levine RL. Mutations in epigenetic modifiers in the pathogenesis and therapy of acute myeloid leukemia. Blood 2013;121:3563–72.

[13] Heyn H, Esteller M. EZH2: an epigenetic gatekeeper promoting lymphomagenesis. Cancer Cell 2013;23:563–5.

[14] Pasqualucci L, Dominguez-Sola D, Chiarenza A, Fabbri G, Grunn A, Trifonov V, et al. Inactivating mutations of acetyltransferase genes in B-cell lymphoma. Nature 2011;471:189–95.

[15] Morin RD, Mendez-Lago M, Mungall AJ, Goya R, Mungall KL, Corbett RD, et al. Frequent mutation of histone-modifying genes in non-Hodgkin lymphoma. Nature 2011;476:298–303.

[16] Aumann S, Abdel-Wahab O. Somatic alterations and dysregulation of epigenetic modifiers in cancers. Biochem Biophys Res Commun 2014.

[17] Taylor KH, Briley A, Wang Z, Cheng J, Shi H, Caldwell CW. Aberrant epigenetic gene regulation in lymphoid malignancies. Semin Hematol 2013;50:38–47.

[18] Jiang Y, Hatzi K, Shaknovich R. Mechanisms of epigenetic deregulation in lymphoid neoplasms. Blood 2013;121:4271–9.

[19] Mullighan CG. Genome sequencing of lymphoid malignancies. Blood 2013;122:3899–907.

[20] Strahl BD, Allis CD. The language of covalent histone modifications. Nature 2000;403:41–5.

[21] Li Y, Nagai H, Ohno T, Yuge M, Hatano S, Ito E, et al. Aberrant DNA methylation of p57(KIP2) gene in the promoter region in lymphoid malignancies of B-cell phenotype. Blood 2002;100:2572–7.

[22] Cerchietti L, Leonard JP. Targeting the epigenome and other new strategies in diffuse large B-cell lymphoma: beyond R-CHOP. Hematology Am Soc Hematol Educ Program 2013;2013:591–5.

[23] Clozel T, Yang S, Elstrom RL, Tam W, Martin P, Kormaksson M, et al. Mechanism-based epigenetic chemo-sensitization therapy of diffuse large B-cell lymphoma. Cancer Discov 2013;3:1002–19.

[24] Vijayaraghavalu S, Dermawan JK, Cheriyath V, Labhasetwar V. Highly synergistic effect of sequential treatment with epigenetic and anticancer drugs to overcome drug resistance in breast cancer cells is mediated via activation of p21 gene expression leading to G2/M cycle arrest. Mol Pharm 2013;10:337–52.

[25] Young JI, Sedivy JM, Smith JR. Telomerase expression in normal human fibroblasts stabilizes DNA 5-meth-ylcytosine transferase I. J Biol Chem 2003;278:19904–8.

[26] Drenzek JG, Seiler NL, Jaskula-Sztul R, Rausch MM, Rose SL. Xanthohumol decreases Notch1 expression and cell growth by cell cycle arrest and induction of apoptosis in epithelial ovarian cancer cell lines. Gynecol Oncol 2011;122:396–401.

[27] Ijichi H, Otsuka M, Tateishi K, Ikenoue T, Kawakami T, Kanai F, et al. Smad4-independent regulation of p21/WAF1 by transforming growth factor-beta. Oncogene 2004;23:1043–51.

[28] Walker JV, Nitiss JL. DNA topoisomerase II as a target for cancer chemotherapy. Cancer Invest 2002;20:570–89.

[29] Quan ZW, Yue JN, Li JY, Qin YY, Guo RS, Li SG. Somatostatin elevates topoisomerase II alpha and enhances the cytotoxic effect of doxorubicin on gall bladder cancer cells. Chemotherapy 2008;54:431–7.

[30] Massague J. TGFbeta in. Cancer Cell 2008;134:215–30.

[31] Lucchesi JC. Synthetic lethality and semi-lethality among functionally related mutants of Drosophila melan-fgaster. Genetics 1968;59:37–44.

[32] Hartwell LH, Szankasi P, Roberts CJ, Murray AW, Friend SH. Integrating genetic approaches into the discovery of anticancer drugs. Science 1997;278:1064–8.

[33] Mair B, Kubicek S, Nijman SM. Exploiting epigenetic vulnerabilities for cancer therapeutics. Trends Pharmacol Sci 2014;35:136–45.

[34] Fong PC, Boss DS, Yap TA, Tutt A, Wu P, Mergui-Roelvink M, et al. Inhibition of poly(ADP-ribose) polymerase in tumors from BRCA mutation carriers. N Engl J Med 2009;361:123–34.

[35] Ame JC, Spenlehauer C, de Murcia G. The PARP superfamily. BioEssays 2004;26:882–93.

[36] Farmer H, McCabe N, Lord CJ, Tutt AN, Johnson DA, Richardson TB, et al. Targeting the DNA repair defect in BRCA mutant cells as a therapeutic strategy. Nature 2005;434:917–21.

[37] Zahreddine HA, Culjkovic-Kraljacic B, Assouline S, Gendron P, Romeo AA, Morris SJ, et al. The sonic hedgehog factor GLI1 imparts drug resistance through inducible glucuronidation. Nature 2014;511:90–3.

[38] Shaknovich R, Melnick A. Epigenetics and B-cell lymphoma. Curr Opin Hematol 2011;18:293–9.

[39] Liu X, Ory V, Chapman S, Yuan H, Albanese C, Kallakury B, et al. ROCK inhibitor and feeder cells induce the conditional reprogramming of epithelial cells. Am J Pathol 2012;180:599–607.

[40] Yuan H, Myers S, Wang J, Zhou D, Woo JA, Kallakury B, et al. Use of reprogrammed cells to identify therapy for respiratory papillomatosis. N Engl J Med 2012;367:1220–7.

[41] Barretina J, Caponigro G, Stransky N, Venkatesan K, Margolin AA, Kim S, et al. The cancer cell line encyclopedia enables predictive modelling of anticancer drug sensitivity. Nature 2012;483:603–7.

[42] Guagnano V, Kauffmann A, Wohrle S, Stamm C, Ito M, Barys L, et al. FGFR genetic alterations predict for sensitivity to NVP-BGJ398, a selective pan-FGFR inhibitor. Cancer Discov 2012;2:1118–33.

PERSONALIZED THERAPY— EPIGENETIC PROFILING AS PREDICTORS OF PROGNOSIS AND RESPONSE

29

Holger Heyn

Cancer Epigenetics and Biology Program, Bellvitge Biomedical Research Institute, Barcelona, Catalonia, Spain

CHAPTER OUTLINE

1 EPIGENETIC BIOMARKERS FOR PRECISION MEDICINE

Personalized medicine is defined as an emerging practice of medicine that uses an individual's genetic profile to guide decisions made in regard to the prevention, diagnosis, and treatment of disease. Knowledge of a patient's genetic profile can help doctors select the proper medication or therapy and administer it using the proper dose or regimen (Genetics Home Reference, NIH). Nevertheless, patients being diagnosed with cancer are currently treated with less well-defined therapeutic strategies, based on cancer type or tumor histology. In rare cases, molecular or genetic biomarkers, such as cell surface receptors and mutation status, are assessed to realize specifically tailored therapies. In this respect, an improved stratification of current patient classification and particularly novel targeted therapeutics could enable an advanced personalized patient therapy, improving treatment efficiencies and reducing side effects. Therefore, the objectives of current research in personalized cancer medicine center on the identification of novel biomarkers to allow individually designed and efficient therapy decisions, applicable for daily clinical use.

Epigenetics emerged into the research field by presenting a stable and easy-to-access modification of the genetic code, with implication in cancerogenesis and well-suited properties as disease biomarker [1]. Consequently, epigenetic biomarkers for diagnosis, prognosis, and therapeutic intervention are currently enlarging the arsenal of genetic and molecular markers, with high potential for rapid translation into clinical use. Herein, screening efforts mainly focus on the identification of DNA methylation alterations, as the covalent modification of the DNA sequence has a direct regulatory impact on transcriptional activities and, more importantly, it is highly variant in virtually all cancer types [2]. Furthermore, DNA methylation represents a stable modification, which can be assessed using standardized methodologies, supporting its value as clinically applied biomarker. In contrary, the diagnostic potential of other epigenetic modification, such as histone marks, faces more technical challenges, as they are less stable and have been shown to be more dynamic [3,4]. Additionally, the assessment of histone modifications is more difficult to apply and standardize, as it relies on the use of antibodies, which vary in performance.

Most strikingly, in addition to being a potent biomarker with potential clinical value, alterations in DNA methylation actively participate in disease formation, wherein the silencing of tumor suppressor genes displays the best-studied biological implication. Promoter hypermethylation and subsequent transcriptional silencing of genes implemented in processes described as *Hallmarks of Cancer* underscores the functional importance of the epigenetic aberration and its value as biomarker or therapeutic target (Figure 29.1) [5,6]. Moreover, the gain of DNA methylation at respective promoters presents a frequent and cancer-specific event, properties required for clinically suitable biomarkers. Consequently, a number of epigenetic biomarker have proven their translational value, with some of these being currently evaluated in ongoing clinical trials [7].

From a diagnostic point of view, early tumor detection is one of the major factors for a successful cancer treatment. In this respect, the analysis of aberrant DNA methylation signatures can be performed using biopsies or noninvasive methods. Herein, the use of biological fluids, such as blood, stool, saliva, or urine samples, has clear advantages over invasive techniques in the diagnosis at early stage cancers or even preoplastic lesions. Due to its stable character, DNA methylation profiles can be accessed from various biological fluids. Moreover, besides being a valuable biomarker for cancer diagnostics, alterations of DNA methylation were shown to be informative for patient prognosis or to guide therapy decision [8].

From a therapeutic perspective, DNA methylation depicts a directly targetable modification, with two epigenetic drugs that modify the methyltransferase activity of DNA methylating enzymes (DNMTs) being approved by the Food and Drug Administration (FDA) for the treatment of hematopoietic neoplasms. In addition, epigenetic drugs targeting histone modifiers or chromatin remodelers are approved or under evaluation for leukemia and lymphoma treatment. Considering the tightly regulated network of epigenetic marks, wherein DNA methylation is closely associated to specific histone

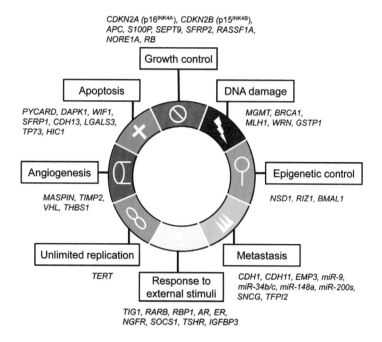

FIGURE 29.1

The implication of aberrant DNA methylation in the hallmarks of cancer.

modifications and nucleosome positioning, a perturbation of this system is likely to affect cell homeostasis and to favor cellular transformation. Accordingly, a growing number of proposed driver mutations in epigenetic modifiers are observed in different cancer types, underlining the importance of gene regulation in cancerogenesis [9,10]. That this is also of therapeutic interest is reflected by the number of epigenetic modification suitable as therapeutic target [11] and the number of epigenetic drugs that are currently evaluated in ongoing clinical trials [12]. In addition to variants of the FDA-approved DNMT and HDAC inhibitors, these involve novel candidates for epigenetic-based targeted therapies such as BED inhibitors [13].

In the following sections, single aspects of epigenetic biomarker discovery and clinical application are highlighted, with emphasis on ongoing research and future perspectives in the field. Considering the relatively recent advance of epigenetic biomarker into translational valuable tools and particularly the latest boost in detection technologies (Box 29.1), epigenetic biomarkers might still be in their infancy, however, they present an enormous potential to contribute to the realization of the longstanding promise of a *Personalized Cancer Medicine*.

In addition to sequencing-based strategies, high density array-based techniques are capable to unravel DNA methylation alterations with high accuracy [19] and provide a comprehensive and cost-effective tool for the epigenetic screening of high sample numbers. Specifically, the Infinium HM450 DNA methylation BeadChip (Illumina) analyzes DNA methylation levels of more than 450,000 CpG sites (dinucleotides that are the main target for methylation) and was previously established as reliable

---

BOX 29.1 FROM SINGLE LOCI TO GENOME-WIDE PROFILING

Owing to a boost in detection technologies, numerous genetic and epigenetic variations have been related to phenotypic variability, including cancer [8,14]. Their interpretation gave an informative insight into aberrant biological processes and determined key factors that represent driver events directly associated to phenotype formation. DNA methylation screening technology recently advances from single loci-based approaches to a genome-wide characterization at high resolution. In particular, the immense decrease of sequencing costs enabled comprehensive analysis of epigenetic variation in natural and pathological contexts [15–18]. Comprehensive screening of the DNA methylation landscape using whole-genome bisulfite shotgun sequencing (WGBS) did not only reveal dramatic changes in the DNA methylome of diseased cells overall but also enabled a paradigm shift from hypothesis-driven to data-driven screening approaches to identify aberrantly methylated loci.

---

technology to detect epigenetic alterations [16,20–23]. Considering gene promoters to be the main region of transcriptional regulation, the platform covers CpG sites around the transcription start site of virtually all protein-coding genes (99%), but also of noncoding RNAs such as microRNAs. In addition, regions related to enhancer activity and DNaseI-sensitive sites, with probable regulative potential, were considered in the array design. Furthermore, the platform gives a comprehensive insight in DNA methylation levels of CpG islands (96%), and also flanking regions (CpG shores) and CpG poorer loci.

Together, sequencing and array-based technologies enable an unbiased view of epigenetic variation, eventually resulting in the comprehensive assessment of epigenetic alterations in cancer. Importantly, these techniques will provide a wealth of potential biomarkers for translational efforts [1].

2 SYSTEMS EPIGENOMICS FOR BIOMARKER DISCOVERY

Screening the human cancer DNA methylation landscape genome wide [18] or at genome scale [19] revealed a comparable magnitude of epigenetic variation to alterations observed at genetic levels with thousands of nucleotides being affected [24,25]. DNA methylation profiles of specific tumors even out-ranged this number, leading to the definition of an epigenetic tumor subtype: the CpG island methylator phenotype (CIMP) [26,27]. Interestingly, excessive hypermethylation events could be mechanistically traced back to genetic alteration with mutations in, for example, *IDH* or *TET* family members being responsible for the increased DNA methylation levels [28].

In general, aberrations of the DNA methylation landscape in cancers are related to large-scale blocks loosing and focal loci gaining DNA methylation [29–31]. While the functional implication of the former event remains largely elusive (although an increased transcriptional variability has been suggested), the regional gain, particularly in promoter regions, was associated to gene silencing. Herein, the repression of tumor suppressor genes has been suggested as an epigenetic driver event that contributes to cancer onset and progression [32]. However, as the majority of alterations are located outside the promoter context, with a direct regulatory effect challenging to identify, determining the impact of epigenetic aberrations on cancer is far more complex and novel strategies are required to fully elucidate their involvement in cancerogenesis. Accordingly, similar to their genetic counterparts, epigenetic studies need to distinguish driver from passenger events and herein cause from consequences of the disease [33]. Unraveling the complexity of epigenetic regulation [34,35] could present a key toward the meaningful interpretation of aberration observed in cancer [36,37] and the determination of novel epigenetic biomarkers.

2.1 FUNCTIONAL EPIGENETIC ALTERATIONS

Systematic integration of multidimensional cellular information and molecular traits is a powerful strategy to identify functional or clinically relevant changes in the DNA methylation landscape. Assuming that functional relevant alterations, both genetic and epigenetic, are affecting downstream mechanisms and resulting in measurable changes, functional events are thought to leave molecular footprints or quantitative traits [38]. These traits not only include directly involved mechanisms, such as transcriptional or gene splicing activities, but can also represent clinical features, including patient survival and treatment response. However, to pinpoint the functional impact and to assign driver characteristics, the underlying chain of events, associating epigenetic alterations to their respective traits, have to be entirely elucidated. Herein, a current focus lies on the association between DNA methylation and gene expression changes as directly related factors; however, the spectrum of quantitative traits can be enlarged using additional regulatory mediator traits, such as histone modifications, nucleosome positioning or noncoding RNAs, and a diverse spectrum of clinical traits, including metabolites.

In this regard, the epigenetic silencing of the glutathione *S*-transferase *GSTP1* presents a well-established examples, with consequences of the alteration affecting downstream mechanisms, contributing to cancerogenesis and being clinical relevant. The glutathione *S*-transferase pi 1 (*GSTP1*), frequently hypermethylated in prostate cancer, is involved in the cellular detoxification of xenobiotics and carcinogens [39,40]. From a systems epigenetics perspective, the gain of promoter methylation of *GSTP1* is associated to transcriptional silencing and the accumulation of potentially harmful molecules within affected cells. Due to the impaired detoxification, *GSTP1* hypermethylation favors carcinogenesis, being particularly frequent in prostate tissue, even at early lesions [41]. Using loci-specific methods, the aberration is an established epigenetic biomarker for prostate cancer diagnosis, detectable with high specificity and sensitivity from biopsies and biological fluids, such as serum [42–44] or urine [45,46]. *GSTP1* hypermethylation perfectly illustrates how causalities can be identified through the systematic integration of several layers of information (cellular and clinical), eventually resulting in a reliable biomarker and a powerful clinical tool.

Novel analysis strategies, particularly tailored for genome-wide epigenomic analysis, will enable to determine functional impacts of regulatory alterations and to define causal relationships between cancer gene deregulation and pathologic phenotypes. Being a widely hypothesis-free strategy, deregulation of established cancer genes and novel candidates will fall within the scope of the approach. Similar to genetic mutations, epigenetic deregulation is likely to improve our mechanistic understanding of cancerogenesis and could be implemented in precision oncology as epigenetic markers or therapeutic targets. Hence, functional alterations in the DNA methylation landscape represent a basic principle in tumorigenesis, however, with high translational potential.

2.2 EPIGENETIC SURROGATE BIOMARKER

Moving from hypothesis-driven to data-driven identification strategies, a wealth of novel markers with potential clinical value rapidly enlarged the arsenal of biomarker for diagnostic or prognostic use. However, taking into account that most of the novel markers represent alterations of single CpG sites or a combination at multiple loci (epigenetic signatures), their functional implication in cancer formation is challenging to assess in a comprehensive manner. Although, their value as clinical cancer biomarker is evident, they might rather represent a surrogate mark of the phenotype than biological functional alterations.

In this regard, genome-wide association studies (GWAS) represent a prototype for such clinical translational variability, as despite the majority of genetic risk variants are located in an noncoding context (and thus not directly affecting a gene product or protein function), the observed associations are significant and guide current research [47]. GWAS traits include natural phenotypes, but also cancer risk, suggesting the genetic polymorphism to be an interesting candidate for future risk assessment and disease management strategies. In order to complement genetic information in risk assessment, epigenetic variance represents a particular interesting mark, as its stable, however, reversible character reflects external and internal influences that might contribute to disease susceptibility. In this respect, DNA methylation not only represents a potent surrogate mark for aberrant mechanism occurring at the time of sampling but also reflects earlier events within the process of cancerogenesis. In particular, these screenshots of tumorigenesis represent a valuable source of information for epigenome-wide association studies and the identification of epigenetic biomarkers.

Furthermore, due to its plasticity, epigenetics reflects variation of multiple sources. As some of these are challenging to be assessed directly, associated surrogate marker represents valuable factors for clinical transfer. Exemplary for this scenario, the frequent hypermethylation of regions within the homeobox gene family, enriched in genes associated to developmental processes, is a potential epigenetic marker and surrogate mark for the differentiation state of the cancer cell [48].

3 EPIGENETIC BIOMARKERS FOR DIAGNOSIS, PROGNOSIS, AND DRUG RESPONSE

3.1 DIAGNOSTIC EPIGENETIC BIOMARKERS

DNA methylation represents a stable and accessible epigenetic mark, which specific alterations are a reliable indicator of disease, with best-studied examples described for cancer diagnostics. Particularly, due to its localized occurrence and high frequencies, it presents certain advantages compared to genetic markers, which often occur scattered over multiple loci and mostly display rather low recurrence rates. Examples of epigenetic biomarkers for cancer diagnostics include a broad range of frequent and likely functional alterations, often detectable in tumor biopsies and biological fluids, such as serum or urine and are outlined in the following sections.

One of the first and probably the most advanced DNA methylation marker for cancer diagnostics is the hypermethylation event at the promoter of *GSTP1*, mainly observed in prostate cancers [39,49–52], however, also detectable at lower frequencies in other cancer types [49]. Following its initial description, multiple studies and large-scale meta-analysis confirmed the hypermethylation event as sensitive and specific epigenetic biomarker for prostate cancer diagnosis [40], even challenging established molecular markers such as PSA diagnostics. Although being considerable specific as a single biomarker, *GSTP1*-based diagnostics benefited from joint analysis using additional frequent hypermethylation events. In particular, *APC, RASSF1A, PTGS2, RARβ, TIG1*, and *MDR1* were confirmed to gain promoter methylation to similar extents [50,52–54], accompanied by a number of genes involved in DNA repair (*MGMT*), cell adhesion (*CD44, EDNRB, ECADHERIN, LGALS3*), invasion and metastasis (*TIMP2, TIMP3, LGALS3*), apoptosis (*DAPK1*), cell cycle control (*CDKN2A, CDKN1A*), and hormonal responses (*AR, ER*) that completed the list of potential biomarker candidates [55]. Either alone or in combination these markers were able to reliably distinguish between primary prostate cancer and benign tissue [50,52,54].

3.1.1 Biomarkers in biological fluids

Early cancer detection presents a critical factor associated to patient survival. Once the primary tumor is visible by routine cancer screenings or symptomatic, the prognosis for the cancer patient is often fatal. In order to improve the overall patient survival, a critical step presents the tumor detection at early stages or even in preneoplastic states. Herein, biological fluids, such as blood, urine, saliva, or stool, are easy accessible sample sources that often contain the required information in early phases of cancer development. Specifically, circulating tumor cells or cell-free tumor DNA enters biological fluids in different ways: tumor cells can be directly released from the tissue of origin; necrotic tumor cells enter macrophages and subsequently the blood or urethra by phagocytosis or free tumor DNA directly reaches the fluids by cell lysis.

Epigenetic biomarkers from liquid biopsies could be established for tissue types being in direct contact to the biological fluid, such as prostate (urine), head and neck (saliva) or colon (stool), but particularly blood (serum) serves as universal source for epigenetic biomarker, including those for lung or brain tumors. Moreover, due to its cancer and tissue specificity, epigenetic alterations are often capable to point to the origin of the primary tumor. The following examples provide an overview of the value of epigenetic biomarkers from biological fluids and specifically highlight the sensitivity and specificity that can be achieved using such approaches.

Hypermethylation of *GSTP1* in prostate cancer was detectable in urine, ejaculates as well as serum and presents the best-studied epigenetic biomarker in fluids. Initial studies confirmed the presence of *GSTP1* hypermethylation in urine and ejaculates from prostate cancer samples [56]. Later, *GSTP1* hypermethylation was detected in serum with sensitivities ranging from 21% to 42% [42–44]. *GSTP1* hypermethylation in urine was further confirmed [45,46] and technically improved by prostatic massage [57] and the combination of different markers [58,59]. Although, the use of *GSTP1* hypermethylation as clinical biomarker has yet to be approved, there are diagnostic assays already commercially available.

Colorectal cancer (CRC) reveals frequent hypermethylation of *APC*, *MGMT*, *RASSF2A*, and *WIF1*. Combing all markers enabled the detection of CRC in patient plasma samples with a sensitivity of 87% and a specificity of 92% [60]. Further, a restriction enzyme-based testing 56 candidates identified *TMEFF2*, *NGFR*, and *SEPT9* as potential blood-based biomarker for CRC [61]. Stool-based studies for CRC revealed hypermethylation of *TFPI2* as potential epigenetic biomarker [62]. Here, *TFPI2* hypermethylation detected CRC with a sensitivity of 76–89%. Further, epigenetics biomarkers (*MGMT*, *CDKN2A*, and *DAPK1*) in saliva from patients suffering from head and neck cancers were shown to detect cancer incidence with 65% sensitivity [63]. Subsequent studies confirmed this association and suggested additional biomarkers to further improve sensitivity, when tested in combination [64]. The value of DNA hypermethylation as biomarker in biological fluids was also established for glioblastoma patients [65]. Here the use of patient serum was highly correlating with tissue methylation level and even predicting progression-free and overall survival. Especially for tumor types for which primary material is unavailable or difficult to obtain, such as brain, the use of biological fluids presents a sensitive alternative.

Taking into account the increasing number of potential biomarkers, allowing combinatory tests, a further improvement of sensitivities and specificities is likely. Particularly the combinatory application detecting mutations, circulating microRNAs, and epigenetic alteration could be applicable as routinely applied screening method of biological fluids for the early detection of tumor incidence, even in premalignant stages [66].

3.1.2 Cancer of unknown primary

Following the detection of metastatic diseases, it is crucial for the success of the treatment, and hence the prognosis, to identify the origin of the primary tumor. Failing to do so leads to a dramatic decrease in median survival for patients with cancers of unknown primary (CUP), due to the lack of specific treatments [67]. Conversely, the identification of the primary tumor, followed by tumor type-specific treatment, could greatly improve patients' prognosis [68]. In a comprehensive study, analyzing DNA methylation of 1505 CpG sites in 1628 human samples, Fernandez et al. determined tissue- and cancer-specific fingerprints for a variety of human cancers. Subsequently, they were able to assign a tissue of origin for 69% of the CUPs analyzed. Blinded-type pathological analyses confirmed the sensitivity in most cases. Considering the rapidly increasing assay resolution (from 1505 to more than 450,000 CpG sites, currently), suggests that there will be almost complete efficiency in the detection of tumor origin in the near future. As a considerable amount of patient and normal sample data will be publically available soon (reference set), the costs per CUP sample are minor in relation to the potential gain in patients' quality of life and survival.

3.2 PROGNOSTIC EPIGENETIC BIOMARKERS

The fact that altered DNA methylation correlates with the transcriptional silencing of tumor suppressor genes, supports the implication of epigenetic events in tumor progression and hence patient prognosis. Consistently, studies analyzing single loci reported the association between hypermethylation events of *NSD1*, *DAPK*, *EMP3*, and *CDKN2A* and patient survival for neuroblastoma, lung, brain, and colorectal cancers, respectively [2]. With an advance in screening technologies, allowing the analysis of multiple loci in single assays, several studies determined epigenetic signatures, significantly related to patient outcome in breast [69] and colorectal [26] cancers, leukemia [70], and gliomas [28]. In particular, nons-mall cell lung cancer (NSCLC) provides an illustrative example of the value of epigenetic biomarker in cancer patient prognosis, and also of the evolution from single loci approaches to genome-scale analysis methods and is therefore highlighted in the following section.

Using a single loci approach and 167 patient samples, four genes were able to predict tumor progression in NSCLC independently of other clinical features such as stage, age, and smoking history. In a multivariate model, promoter hypermethylation of *p16*, *CDH13*, *RASSF1A*, and *APC* was significantly associated with early tumor recurrence in stage I cancer patients [71]. The detection of novel biomarker in this specific tumor subtypes is of particular clinical interest, as no currently applied strategy is capable to indicate, which of these high-risk patients could benefit from adjuvant chemotherapy in additional to surgical intervention. A subsequent study, using genome-scale methods and profiling more than 450,000 CpG sites and 237 patient samples with stage I NSCLC, identified an epigenetic signature capable to determine patients with poor prognosis and proposed that hypermethylation of five genes was predictive for relapse-free survival in stage I NSCLC. Particularly, the gain of DNA methylation associated to HIST1H4F, PCDHGB6, NPBWR1, ALX1, and HOXA9 distinguished patients with high- and low-risk stage I NSCLC.

Importantly, both studies validated their findings in independent patient cohorts underlining the significance of their results. Conclusively, epigenetic biomarkers present a potent predictor of survival for patients with early stage NSCLC and could guide clinical decision to administer adjuvant chemotherapy. As prognostic biomarkers are not only highly informative for the patient and related disease management but also enable the stratification of patients to evaluate treatment efficacy in retroperspective studies, large-scale epigenomic profiling efforts will surely provide a valuable source of information driving future clinical guidelines.

3.3 BIOMARKER GUIDING THERAPEUTIC DECISION

Considering the fact that most clinically relevant epigenetic alterations present a loss of function event, mainly related to the transcriptional silencing of tumor suppressor genes, a direct therapeutic intervention is not applicable in the majority of cases. However, as tumor suppressor genes are incorporated into complex intracellular signaling cascades and multicomponent mechanisms, the epigenetic alteration can directly point to affected pathways that include targetable factors. In this regard, the use of epigenetic biomarker to predict treatment response can be interrogated from different viewpoints considering (i) a direct association between the alteration and the drug action or (ii) an effect based on mechanistic or functional interactions.

3.3.1 Direct implications of epigenetic silencing on drug response

Given a mechanistic relationship between drug action and gene deregulation, therapeutic efficiencies can be directly explained by the function of the protein and the effect of the drug treatment. Exemplary, epigenetic silencing of the DNA repair component *MGMT* leads to a reduced ability of tumor cells to remove alkyl groups introduced by carcinogens. Intriguingly, the epigenetic silencing of *MGMT* improves the efficacy of chemotherapeutic treatments using alkylating drugs such as carmustine and temozolomide [72,73]. Due to the impaired DNA repair ability, *MGMT* hypermethylated cells present an increased sensitivity not only to DNA modifying agents but also to cilengitide [74], cyclophosphamide [75], and radiation [76]. Importantly, epigenetic silencing is frequently observed in glioma patients (around 40% of cases [77]) for which temozolomide treatment displays the first-line therapeutic intervention. Hence, the mechanistic implication of *MGMT* in DNA repair directly points to a predictive potential for certain therapeutics and hence enables guidance for therapy decision. Accordingly, *MGMT* as epigenetic biomarker for the treatment of glioma patients is currently evaluated in clinical trials.

A similar direct epitype–drug association is represented by *BRCA1*, whose ability to repair DNA through homologous recombination is impaired in *BRCA1* hypermethylated cancer cells [78]. Consistently, epigenetic silencing of *BRCA1* results in hypersensitivity of cancer cells to DNA damaging agents as shown for cisplatin in an ovarian and breast cancer contexts [79]. Additional examples of direct associations between epigenetic alterations and drug sensitivity involve the hypermethylation of the *MLH1* promoter, which results in resistance of ovarian cancer cells to cisplatin treatment *in vitro* [80] and *in vivo* mouse models [81]. In addition to *BRCA1* and *MLH1*, further epigenetically deregulated genes in cancer were shown to alter drug treatment efficiencies and are summarized in Table 29.1.

3.3.2 Synthetic lethal interactions of epigenetic events

In addition to epigenetic alteration pointing directly to mechanisms that confer sensibility to specific drug treatments, indirect functional or mechanistic interactions present powerful associations to identify biomarker for drug treatment efficiencies or novel application windows for targeted drugs. Herein, a milestone in synthetic lethal interaction originated from the genetic alteration of *BRCA1* and *BRCA2*, whose impaired function resulted in sensitivity of cancer cell to the inhibition of the DNA

Table 29.1 Aberrant DNA Methylation Related to Drug Response		
Gene Name	**Therapeutic Agent**	**Effect**
APAF1	Adriamycin	Resistance
ER	Antiestrogens	Resistance
SULF2	Camptothecin	Sensitivity
PLK2	Carboplatin	Sensitivity
BRCA1	Cisplatin	Sensitivity
FANCF	Cisplatin	Sensitivity
IGFBP3	Cisplatin	Resistance
MLH1	Cisplatin	Sensitivity
MT1E	Cisplatin	Sensitivity
TGM2	Cisplatin	Resistance
TP73	Cisplatin	Sensitivity
BMP4	Cisplatin	Resistance
IGFBP3	Cisplatin	Resistance
FANCF	Cisplatin	Sensitivity
ERCC1	Cisplatin	Sensitivity
SFN	Cisplatin	Sensitivity
CHFR	Docetaxel	Sensitivity
ABCB1	Doxorubicin	Sensitivity
GSTP1	Doxorubicin	Sensitivity
TGM2	Doxorubicin	Resistance
LINE-1	Fluoropyrimidines	Resistance
TFAP2E	Fluorouracil	Sensitivity
SFN	Gemcitabine	Sensitivity
WRN	Irinotecan	Sensitivity
SLC19A1	Methotrexate	Resistance
ERCC5	Nemorubicin	Resistance
PLK2	Paclitaxel	Sensitivity
CHFR	Paclitaxel	Sensitivity
BRCA1	PARP inhibitors	Sensitivity
CDK10	Tamoxifen	Resistance
PITX2	Tamoxifen	Resistance
MGMT	Temozolomide	Sensitivity
PRKCDBP	TNFα	Resistance

repair enzyme PARP [82]. Functionally, the *BRCA* genes are involved in the homologous recombination-mediated DNA repair, whereas PARP is incorporated in the base excision repair system. Strikingly, inhibition of both systems is lethal to cancer cell, and thus *BRCA* mutant cells are primed to be specifically sensitive to *PARP* inhibitors. Hence, although the primary alteration is not directly targeted, its mechanistic implementation in multifactorial processes identified synthetic lethal interactions, resulting in the specific treatment of respective cancer cells [83]. Taking into account that germline mutation of *BRCA* genes is significantly associated to the incidence of inherited breast and ovarian cancers, provides a therapeutic window for personalized treatments of *BRCA* mutant patients with the PARP inhibitor, which is evaluated in ongoing clinical trials [84]. Despite the fact that *BRCA* genes are frequently mutated in inherited breast and ovarian cancers, alterations in sporadic cases are rare. However, epigenetic silencing of *BRCA1* presents a recurrent event in sporadic breast cancer cases, suggesting that hypermethylated tumor cells could be equally sensitive to PARP inhibition, as their genetically altered counterparts [85]. And indeed, breast cancer cells with epigenetically silenced *BRCA1* displayed increased sensitivity to PARP inhibitors [78]. Although, these results have to be validated in future studies, they suggest the value of PARP inhibitor treatment in *BRCA1* hypermethylated breast cancer patients. Certainly, the two independent repair mechanisms involving BRCA1 and PARP represent an illustrative example the synthetic lethal interactions specifically targetable by therapeutic intervention.

Likewise, a range of cellular processes, affected by frequent epigenetic perturbation, might be susceptive for synthetic lethal intervention. Herein, frequent hypermethylation events of negative regulators of cell proliferation could serve as potent surrogate marker for the use of therapeutics, specifically targeting molecules in downstream mitogenic cascades. Specifically, the epigenetic silencing of the *RAS* inhibitor *RASSF1*, one of the most frequent events observed in diverse cancer types, suggests the value of therapeutic intervention using inhibitors targeting RAF or MEK, signaling factors in the downstream cascade of RAS. Similarly, the hypermethylation of the cell cycle inhibitor *DAPK*, frequently observed in lung, head and neck, and bladder cancers, suggests a potential benefit of the application of CDK inhibitor, cell cycle components directly activated by the silencing of DAPK. Similarly, tumors with the highly recurrent hypermethylation of *CDKN2B* could benefit from the treatment with CDK4/6 inhibitors. In this regard, the epigenetic silencing of *CDKN2A*, a negative regulator of the TP53 inhibitor MDM2, could represent a valuable biomarker for the efficient treatment of TP53-repressed tumors; as MDM2 can be directly targeted using specific drugs.

Considering the implication of epigenetic regulation in crucial cellular processes and their perturbance through deregulation in cancer contexts, provides a wealth of possible therapeutic opportunities. However, the efficiency of such prediction has to be confirmed in biological assays. Herein, systematic screening efforts to determine such association need to be performed in a comprehensive manner to estimate the value of epigenetic biomarker in the prediction of drug response. Intriguingly, such efforts were recently undertaken for genetic studies, which could serve as prototype for future work in the epigenetic field [83,86].

3.3.3 Lessons learned from genetic biomarker screenings

Human cancer cell lines display a powerful tool to interrogate the efficiency of chemotherapeutic drugs for growth inhibition and tumor cell kill. Broad screening approaches of almost 500,000 different drugs, taking advantages of the well-characterized cancer cell lines of the NCI60 panel (representing 9 distinct tumor types), revealed valuable insights in anticancer drug action but also immunotherapy and viral and bacterial pathogenesis [1]. Recently, two independent studies published systematic drug screening approaches of 639 and 947 human cancer cell lines, respectively [2,3], resulting in profound knowledge of the efficiency of a variety of targeted agents and cytotoxic chemotherapeutics on cancer cell treatment. Strikingly, both studies not only validated known and clinically approved treatment strategies but also identified novel biomarkers suggested to be predictive for certain and before unrecognized drug applications. Combining an intense molecular characterization with drug sensitivity data and using computational modeling approaches, the authors describe gene mutations, copy number variations (CNVs), and gene expression to be predictive for drug resistance or sensitivity. Furthermore, they suggest novel and unexpected biomarkers with high translational potential to clinical use, such as the use of PARP inhibitors for Ewing's sarcoma cells harboring EWS-FLI1 translocations or AHR expression as predictor for the efficiency of MEK inhibitors in NRAS-mutant cells.

Using recent technology and characterizing a broad spectrum of alteration occurring in cancer, to date high-throughput drug screening studies focused on genetic aberration and gene expression changes but leaving out epigenetic alteration. Taking into account the value of epigenetic marks as biomarker for clinical application ranging from diagnostic to treatment decision, future model-based studies can benefit from including the comprehensive profiling of CpG methylation or other epigenetic marks. Combined the systematic screening of drug sensitivities and the subsequent integration in predictive computational model approaches will enable the identification of cancer cell-specific signatures and single biomarker genes that are capable to predict the sensitivity of cancer cells to certain drugs in the model system and are likely of translational value for patients currently suffering from the respective cancer types.

4 GENETIC ALTERATIONS IN EPIGENETIC MODIFIERS: POTENTIAL DRUG TARGETS

Tumors displaying global epigenomic alterations might also benefit from therapies that restore physiological patterns. A significant number of DNA and histone-modifying enzymes are mutated in disease (Table 29.2) and can be targeted by specific molecules, representing the first examples of personalized medicine therapies that combine genomic and epigenomic knowledge.

4.1 EPIGENETIC DRUGS IN CLINICAL USE AND TRIALS

To date, four epigenetic drugs have been approved by the FDA: two DNMT inhibitors and two HDAC inhibitors. 5-Azacytidine (5-Aza-CR) was approved in 2004 to specifically inhibit DNA methylation and 2 years later, so was its variant 5-Aza-2′-deoxycytidine (5-Aza-CdR). Both were approved for the treatment of higher risk myelodysplastic syndrome (MDS). 5-Aza-CR (Vidaza, Celgene) was confirmed to significantly increase the median overall survival (24.5 compared to 15 months) in a randomized study of 358 patients with higher risk MDS [87]. Most importantly, the therapeutic benefit was achieved at low drug dosage. Another MDS study of 309 patients reported complete remission in 10–17% of participants, while 23–36% of patients had hematological improvement [88]. Interestingly, 27 AML patients included in the study also showed improved median survival time (19.3 compared to 12.9 months). 5-Aza-CR is currently being tested in ongoing clinical trials for approval of AML and CML. In a study of 115 MDS patients, treated with the 5-Azacytidine variant 5-Aza-CdR (Dacogen, Eisai), 70% showed therapy benefits in terms of complete response (35%), partial response (2%), or hematologic improvement (33%) [89]. In addition, S110 [90], a dinucleotide containing 5-Aza-CdR (suggested to have enhanced stability and efficiency) complement the list of DNMT inhibitors that are in use or being tested in clinical trials. However, despite their defined molecular properties, the actual mechanism by which patients benefit from the treatment with epigenetic drugs is still uncertain [91].

Table 29.2 Disruption of Epigenetic Modifiers in Human Cancer

Gene family	Gene	Alteration	Cancer
DNMT	DNMT1	Overexpression	Solid tumors
	DNMT3A	Mutation, overexpression	AML, solid tumors
	DNMT3B	Overexpression, amplification	Solid tumors
DNDM	TET1	Translocation	ALL, AML
	TET2	Mutation	Leukemia
HAT	CREBBP	Translocation, mutation, deletion, E1A/SV40	Leukemia, solid cancers
	EP300	Translocation, mutation, deletion, E1A/SV40	Leukemia, solid cancers
	MYST3	Translocation	Leukemia
	MYST4	Translocation	Leukemia
	KAT5	Monoallelic loss	Lymphoma
	PCAF	Mutation	Epithelial cancer
	NCOA1	Translocation	Rhabdommyosarcoma
	NCOA3	Overexpression, amplification	Solid tumors
HDAC	HDAC1	Overexpression, repression, mutation	Solid tumors
	HDAC2	Overexpression, truncation	Solid tumors, CRC
	HDAC4	Overexpression, mutation	Solid tumors
	HDAC5	Repression	CRC
	HDAC6	Overexpression	Solid tumors
	HDAC7A	Overexpression	CRC
	SIRT1	Overexpression, repression, mutation	AML, solid tumors
	SIRT2	Repression	Solid tumors
	SIRT3	Overexpression	Solid tumors
	SIRT7	Overexpression	Solid tumors
HMT	MLL1	Translocation	Leukemia, lymphoma
	MLL3	Truncation, mutation, deletion	Leukemia, CRC
	MLL4	Amplification	Solid tumors
	EZH2	Overexpression, amplification, mutation	Solid tumors, lymphoma
	NSD1	Hypermethylation, mutation, LOH, translocation	Solid tumors
	NSD2	Translocation	Multiple myeloma
	NSD3	Translocation	AML, solid tumors
	PRDM1	Truncation	Lymphoma
	PRDM2	Mutation, repression	Solid tumors
	PRDM3	Translocation	Myeloid leukemia
	PRDM12	Deletion	CML
	PRDM16	Translocation	Leukemia
	SMYD3	Overexpression	Solid tumors
	SUV39H1	Mutation, overexpression	Solid tumors
	PRMT1	Repression	Breast cancer
	PRMT5	Overexpression	Gastric cancer

(Continued)

Table 29.2 (Continued)

Gene family	Gene	Alteration	Cancer
HDM	*KDM4C*	Amplification	Squamous cell carcinoma
	KDM5B	Overexpression	Breast cancer
	KDM1A	Overexpression	Solid tumors
	KDM2B	Overexpression	Leukemia
	KDM5A	Copy number variation, translocation	Glioblastoma, AML
	KDM5C	Mutation, overexpression	Solid tumors
	KDM6A	Mutation, overexpression, deletion	Solid tumors, AML
	KDM6B	Overexpression	Lymphoma

Variations in the efficiency of 5-Aza-CR and 5-Aza-CdR treatment were reported as a consequence of DNMT3B amplification in cancer cell line experiments [92]. Here, an elevated gene dosage of the methyltransferase led to a decreased sensitivity to demethylating agents. Vorinostat (Merck), an FDA-approved HDAC inhibitor for the treatment of cutaneous T-cell lymphoma (CTCL, 2006) [93,94], was also confirmed to induce complete response or hematological improvement in AML patients [95]. Besides Vorinostat, Romidepsin (Celgene), another HDAC inhibitor, revealed remarkable efficacies in the treatment of CTCL patients [96]. Two additional HDAC inhibitors—Panobinostat (Novartis) and CI-994 (Pfizer)—are currently being tested in clinical Phase III trials for the treatment of lymphomas and NSCLC, respectively.

The combination of Vorinostat with conventional drugs, particularly in solid tumors [97–99] is showing promise. Combination of DNMT inhibitors with conventional therapies is also showing promise [100,101], and so combinatorial therapies in general might lead to higher efficiencies for overcoming resistance and for administrating lower drug doses—reducing side effects and hence improve patients' quality of life. For example, the potential for the antitumor effects of low dose application of 5-Aza-CR and 5-Aza-CdR has been shown in hematological and epithelial tumor cell lines. However, so far these studies are based on model systems [102]. Furthermore, the combined low-dose therapy of the DNMT inhibitor 5-Aza-CR and the HDAC inhibitor Entinostat resulted in improved progression-free survival in an extensively pretreated patient with recurrent metastatic NSCLC, and so is likely to encourage further research into combinatorial usage [103]. Three additional patients experienced positive responses to anti-cancer treatment subsequent to the combinational epigenetic therapy, so this Phase I/II study suggests that administering epigenetic drugs in solid tumors could have clinical value [104]. Specifically, patients with hypermethylated biomarkers for NSCLC [105] detected in serum benefited from the treatment.

4.2 APPLICATION OF EPIGENETIC THERAPEUTICS IN A NONCANCER CONTEXT

Other diseases apart from cancer might benefit from epigenetic treatment. In particular, mutations in DNA methylation and histone-modifying enzymes have been reported for neurodevelopmental disorders and autoimmune diseases, and also aberrant gene promoter methylation is observed.

Triplet GAA/TCC expansion in the Frataxin (FXN) gene is associated with Friedrich's ataxia. Silencing of FXN by hypermethylation is accompanied by histone hypoacetylation, which was shown to be revertible by HDAC inhibitor treatment. Friedrich's ataxia presents the most advanced noncancer example for the use of epigenetic drugs, clinical trials are currently ongoing [106]. Patients with Rubinstein-Taybi syndrome are also likely to benefit from histone-related epigenetic treatment. In this syndrome, the histone acetyltransferases CREBBP and EP300 are mutated, contributing to an overall reduction of H2B acetylation in patients with this syndrome [107]. Therefore HDAC inhibitors could, at least in part, compensate for the loss of acetylation as their function in brain was shown to be CREBBP dependent [108].

Furthermore, age-related memory loss might be treated by drugs targeting histone-modifying enzymes [109]. The deregulation of H4K12 acetylation in the brains of old mice is suspected to cause memory-related gene expression alteration and HDAC inhibitor treatment in the mouse model has shown to reinstate the expression of learning-induced genes and the recovery of cognitive abilities. Furthermore, recent efforts in the exploration of novel therapeutic approaches to combat the neurodegeneration and cognitive impairment (e.g., Alzheimer's disease, AD) have increasingly revolved around epigenetic changes. Here, epigenetic studies assessed the consequences of increased histone acetylation, which causes chromatin remodeling and altered gene expression profiles, and showed that increased chromatin remodeling, as the result of treating mice with histone deacetylase inhibitors (HDAC inhibitors), restores the learning ability AD mouse models even after severe neuronal loss has already occurred [110]. This work hints at the possibility that cognitive abilities can be improved, even in patients with advanced stage of AD and age-related dementia. However, the genome-wide mode of action of HDAC inhibitors remains unknown and has to be systematically assessed in future studies.

It is of note that also cardiovascular diseases such as atherosclerosis might benefit from epigenetic drug treatments. Lipoproteins involved in disease onset were shown to trigger *de novo* DNA methylation as well as histone deacetylation, potential mechanisms that could be treated with already approved epigenetic drugs [111].

4.3 A NEW GENERATION OF TARGETED EPIGENETIC DRUGS

Due to an increased resolution of genomic and epigenetic profiling methods, targeted epigenetic treatments, such as the specific inhibition of enzymes by small molecules, are being developed, adding to the drug arsenal to improve drug performance further, while reducing toxicity to healthy tissue. Specific treatments targeting epigenetic modifiers are the subject of current investigation and the most recent examples are briefly considered in the following section.

Bromodomain and extra terminal domains (BET) are adapter molecules involved in chromatin-dependent signal transduction, resulting in transcriptional initiation and elongation. The bromodomain family member BRD4 has been shown to be involved in gene fusions in squamous carcinomas and leukemia. The small molecule JQ1 has been established as a potent inhibitor of BRD4 and JQ1 treatment showed strong antiproliferative effects in cell line and xenograft models harboring BRD4 fusion [112]. Chromosomal translocation involving *MLL* (mixed lineage leukemia) is an initiator of aggressive forms of leukemia with extremely poor prognosis. *MLL* fusion partners, especially members of the super elongation complex (SEC) such as AF4/6/9/10 and ENL, have been identified as interaction partners with BET proteins. Using the small molecule inhibitor of the BET family I-BET151 causes displacement of BRD3/4 and SEC from chromatin, followed by inhibition of coactivator function and subsequent repression of key oncogenes such as *MYC*, *BCL2*, and *CDK6* [113].

Fusion products of translocated *MLL* genes also initiate aberrant recruitment of DOT1L, a histone methyltransferase involved in H3K79 methylation, to MLL target promoters resulting in mislocalized enzymatic activity [114,115]. In patients harboring the translocated *MLL* genes, altered DOT1L activity functions as a downstream event, associated with the activation of leukemogenic genes such as *HOXA9* and *MEIS1*. Recently, Daigle and co-workers developed the small molecule, EPZ004777, which specifically inhibits the H3K79 methylation activity of DOT1L. EPZ004777 treatment results in selective killing of leukemia cells harboring the *MLL* translocation in mouse xenograft models [116].

5 FUTURE PERSPECTIVE OF BIOMARKER DISCOVERY AND APPLICATION

Taking into account the rather low resolution of epigenetic studies that were conducted to date suggests that we have only seen snapshots of the processes that actually happed during neoplastic transformation. However, the advent of affordable whole-genome profiles enables us to assess the complete picture of epigenetic variation occurring in cancer. Herein, future challenges will first and foremost be related to the separation of biological noise (passenger events) from functional alterations driving or accelerating the disease process.

Clinical valuable biomarkers, in particular those related to diagnosis and prognosis, are independent from this enigma. Being equally valuable when representing secondary events or surrogate marks, these biomarkers are not necessarily related to biological functional events. Accordingly, clinical biomarker can exploit the entire spectrum of variability observed in cancer sample cohorts. Henceforth, large-scale integrative studies will provide an estimate of the proportion of variable loci suitable as clinical biomarker and of the accuracy with which cancer events or patient outcome are predictable. Consequently, efforts in epigenome-wide association studies (EWAS) are likely to supplement genetically defined biomarkers and to open new avenues in *Personalized Medicine* [117,118]. However, particularly due to the plasticity and intraindividual variability of epigenetic modifications, EWAS have to cope with severe challenges, especially hidden confounding covariates such as cell composition heterogeneity [119].

Without any doubt, we currently see only the tip of the iceberg of epigenetic biomarkers valuable for clinical applications, such as diagnosis, prognosis, or therapy decision. Particularly, the use of comprehensive integrative analysis and multidimensional datasets will significantly increase the number, quality, and translational potential of future epigenetic biomarker for precision medicine.

REFERENCES

[1] Heyn H, Esteller M. DNA methylation profiling in the clinic: applications and challenges. Nat Rev Genet 2012;13(10):679–92.
[2] Esteller M. Epigenetics in cancer. N Engl J Med 2008;358:1148–59.
[3] Feinberg AP. Phenotypic plasticity and the epigenetics of human disease. Nature 2007;447(7143):433–40.
[4] Berdasco M, Esteller M. Aberrant epigenetic landscape in cancer: how cellular identity goes awry. Dev Cell 2010;19(5):698–711.
[5] Hanahan D, Weinberg RA. The hallmarks of cancer. Cell 2000;100(1):57–70.
[6] Hanahan D, Weinberg RA. Hallmarks of cancer: the next generation. Cell 2011;144(5):646–74.
[7] Heyn H, Méndez-González J, Esteller M. Epigenetic profiling joins personalized cancer medicine. Expert Rev Mol Diagn 2013;13(5):473–9.
[8] Baylin SB, Jones PA. A decade of exploring the cancer epigenome—biological and translational implications. Nat Rev Cancer 2011;11(10):726–34.

[9] Portela A, Esteller M. Epigenetic modifications and human disease. Nat Biotechnol 2010;28(10):1057–68.

[10] Gonzalez-Perez A, Jene-Sanz A, Lopez-Bigas N. The mutational landscape of chromatin regulatory factors across 4,623 tumor samples. Genome Biol 2013;14(9):r106.

[11] Kelly TK, De Carvalho DD, Jones PA. Epigenetic modifications as therapeutic targets. Nat Biotechnol 2010;28:1069–78.

[12] Rodríguez-Paredes M, Esteller M. Cancer epigenetics reaches mainstream oncology. Nat Med 2011; 17(3):330–9.

[13] Dawson MA, Kouzarides T. Cancer epigenetics: from mechanism to therapy. Cell 2012;150(1):12–27.

[14] Stratton MR. Exploring the genomes of cancer cells: progress and promise. Science 2011;331(6024):1553–8.

[15] Heyn H, Vidal E, Sayols S, Sanchez-Mut JV, Moran S, Medina I, et al. Whole-genome bisulfite DNA sequencing of a DNMT3B mutant patient. Epigenetics Off J DNA Methylation Soc [Internet]. 1 de junio de 2012 [citado 21 de mayo de 2012];7(6). Recuperado a partir de: http://www.ncbi.nlm.nih.gov/pubmed/22595875.

[16] Heyn H, Li N, Ferreira HJ, Moran S, Pisano DG, Gomez A, et al. Distinct DNA methylomes of newborns and centenarians. Proc Natl Acad Sci USA 2012;109(26):10522–7.

[17] Berman BP, Weisenberger DJ, Aman JF, Hinoue T, Ramjan Z, Liu Y, et al. Regions of focal DNA hypermethylation and long-range hypomethylation in colorectal cancer coincide with nuclear lamina-associated domains. Nat Genet 2011;44(1):40–6.

[18] Lister R, Pelizzola M, Dowen RH, Hawkins RD, Hon G, Tonti-Filippini J, et al. Human DNA methylomes at base resolution show widespread epigenomic differences. Nature 2009;462(7271):315–22.

[19] Bibikova M, Barnes B, Tsan C, Ho V, Klotzle B, Le JM, et al. High density DNA methylation array with single CpG site resolution. Genomics 2011;98(4):288–95.

[20] Heyn H, Carmona FJ, Gomez A, Ferreira HJ, Bell JT, Sayols S, et al. DNA methylation profiling in breast cancer discordant identical twins identifies DOK7 as novel epigenetic biomarker. Carcinogenesis 2013;34(1):102–8.

[21] Sandoval J, Heyn H, Moran S, Serra-Musach J, Pujana MA, Bibikova M, et al. Validation of a DNA methylation microarray for 450,000 CpG sites in the human genome. Epigenetics 2011;6(6):692–702.

[22] Sandoval J, Heyn H, Méndez-González J, Gomez A, Moran S, Baiget M, et al. Genome-wide DNA methylation profiling predicts relapse in childhood B-cell acute lymphoblastic leukaemia. Br J Haematol 2012.

[23] Sandoval J, Mendez-Gonzalez J, Nadal E, Chen G, Carmona FJ, Sayols S, et al. A prognostic DNA methylation signature for stage I non-small-cell lung cancer. J Clin Oncol 2013.

[24] Alexandrov LB, Nik-Zainal S, Wedge DC, Aparicio SAJR, Behjati S, Biankin AV, et al. Signatures of mutational processes in human cancer. Nature 2013;500(7463):415–21.

[25] Alexandrov LB, Nik-Zainal S, Wedge DC, Campbell PJ, Stratton MR. Deciphering signatures of mutational processes operative in human cancer. Cell Rep 2013;3(1):246–59.

[26] Hinoue T, Weisenberger DJ, Lange CPE, Shen H, Byun H-M, Van Den Berg D, et al. Genome-scale analysis of aberrant DNA methylation in colorectal cancer. Genome Res 2012;22(2):271–82..

[27] Noushmehr H, Weisenberger DJ, Diefes K, Phillips HS, Pujara K, Berman BP, et al. Identification of a CpG Island methylator phenotype that defines a distinct subgroup of glioma. Cancer Cell 2010;17(5):510–22.

[28] Turcan S, Rohle D, Goenka A, Walsh LA, Fang F, Yilmaz E, et al. IDH1 mutation is sufficient to establish the glioma hypermethylator phenotype. Nature 2012;483(7390):479–83.

[29] Berman BP. Regions of focal DNA hypermethylation and long-range 1 hypomethylation in colorectal cancer coincide with nuclear lamina-associated domains. Nat Genet 2011;44(1):40–6.

[30] Hon GC, Hawkins RD, Caballero OL, Lo C, Lister R, Pelizzola M, et al. Global DNA hypomethylation coupled to repressive chromatin domain formation and gene silencing in breast cancer. Genome Res 2012;22(2): 246–58.

[31] Hansen KD, Timp W, Bravo HC, Sabunciyan S, Langmead B, McDonald OG, et al. Increased methylation variation in epigenetic domains across cancer types. Nat Genet 2011;43(8):768–75.

[32] Jones PA, Baylin SB. The epigenomics of cancer. Cell 2007;128:683–92.

[33] Furney SJ, Gundem G, Lopez-Bigas N. Oncogenomics methods and resources. Cold Spring Harb Protoc 2012;2012(5).

[34] Jin F, Li Y, Dixon JR, Selvaraj S, Ye Z, Lee AY, et al. A high-resolution map of the three-dimensional chromatin interactome in human cells. Nature 2013;503(7475):290–4.

[35] Hon GC, Rajagopal N, Shen Y, McCleary DF, Yue F, Dang MD, et al. Epigenetic memory at embryonic enhancers identified in DNA methylation maps from adult mouse tissues. Nat Genet 2013;45(10):1198–206.

[36] Smith E, Shilatifard A. Enhancer biology and enhanceropathies. Nat Struct Mol Biol 2014;21(3):210–9.

[37] Herz H-M, Hu D, Shilatifard A. Enhancer malfunction in cancer. Mol Cell 2014;53(6):859–66.

[38] Civelek M, Lusis AJ. Systems genetics approaches to understand complex traits. Nat Rev Genet 2014; 15(1):34–48.

[39] Lee WH, Morton RA, Epstein JI, Brooks JD, Campbell PA, Bova GS, et al. Cytidine methylation of regulatory sequences near the pi-class glutathione *S*-transferase gene accompanies human prostatic carcinogenesis. Proc Natl Acad Sci USA 1994;91(24):11733–7.

[40] Van Neste L, Herman JG, Otto G, Bigley JW, Epstein JI, Van Criekinge W. The Epigenetic promise for prostate cancer diagnosis. The Prostate 2012;72(11):1248–61.

[41] Nakayama M, Bennett CJ, Hicks JL, Epstein JI, Platz EA, Nelson WG, et al. Hypermethylation of the human glutathione *S*-transferase-pi gene (GSTP1) CpG island is present in a subset of proliferative inflammatory atrophy lesions but not in normal or hyperplastic epithelium of the prostate: a detailed study using laser-capture microdissection. Am J Pathol 2003;163(3):923–33.

[42] Reibenwein J, Pils D, Horak P, Tomicek B, Goldner G, Worel N, et al. Promoter hypermethylation of GSTP1, AR, and 14-3-3σ in serum of prostate cancer patients and its clinical relevance. Prostate 2007;67(4): 427–32.

[43] Ellinger J, Haan K, Heukamp LC, Kahl P, Büttner R, Müller SC, et al. CpG Island hypermethylation in cell-free serum DNA identifies patients with localized prostate cancer. Prostate 2008;68(1):42–9.

[44] Sunami E, Shinozaki M, Higano CS, Wollman R, Dorff TB, Tucker SJ, et al. Multimarker circulating DNA assay for assessing blood of prostate cancer patients. Clin Chem 2009;55(3):559–67.

[45] Gonzalgo ML, Pavlovich CP, Lee SM, Nelson WG. Prostate cancer detection by GSTP1 methylation analysis of postbiopsy urine specimens. Clin Cancer Res 2003;9(7):2673–7.

[46] Cairns P, Esteller M, Herman JG, Schoenberg M, Jeronimo C, Sanchez-Cespedes M, et al. Molecular detection of prostate cancer in urine by GSTP1 hypermethylation. Clin Cancer Res 2001;7(9):2727–30.

[47] Hindorff LA, Gillanders EM, Manolio TA. Genetic architecture of cancer and other complex diseases: lessons learned and future directions. Carcinogenesis 2011;32(7):945–54.

[48] Heyn H, Esteller M. EZH2: an epigenetic gatekeeper promoting lymphomagenesis. Cancer Cell 2013;23(5): 563–5.

[49] Esteller M, Corn PG, Urena JM, Gabrielson E, Baylin SB, Herman JG. Inactivation of glutathione *S*-transferase P1 gene by promoter hypermethylation in human neoplasia. Cancer Res 1998;58(20):4515–8.

[50] Yegnasubramanian S, Kowalski J, Gonzalgo ML, Zahurak M, Piantadosi S, Walsh PC, et al. Hypermethylation of CpG islands in primary and metastatic human prostate cancer. Cancer Res 2004;64(6):1975–86.

[51] Jerónimo C, Usadel H, Henrique R, Oliveira J, Lopes C, Nelson WG, et al. Quantitation of GSTP1 methylation in non-neoplastic prostatic tissue and organ-confined prostate adenocarcinoma. J Natl Cancer Inst 2001;93(22):1747–52.

[52] Bastian PJ, Ellinger J, Wellmann A, Wernert N, Heukamp LC, Müller SC, et al. Diagnostic and prognostic information in prostate cancer with the help of a small set of hypermethylated gene loci. Clin Cancer Res 2005;11(11):4097–106.

[53] Padar A, Sathyanarayana UG, Suzuki M, Maruyama R, Hsieh J-T, Frenkel EP, et al. Inactivation of cyclin D2 gene in prostate cancers by aberrant promoter methylation. Clin Cancer Res 2003;9(13):4730–4.

[54] Ellinger J, Bastian PJ, Jurgan T, Biermann K, Kahl P, Heukamp LC, et al. CpG island hypermethylation at multiple gene sites in diagnosis and prognosis of prostate cancer. Urology 2008;71(1):161–7.

[55] Ahmed H. Promoter methylation in prostate cancer and its application for the early detection of prostate cancer using serum and urine samples. Biomark Cancer 2010;2010(2):17–33.

[56] Goessl C, Krause H, Müller M, Heicappell R, Schrader M, Sachsinger J, et al. Fluorescent methylation-specific polymerase chain reaction for DNA-based detection of prostate cancer in bodily fluids. Cancer Res 2000;60(21):5941–5.

[57] Goessl C, Müller M, Heicappell R, Krause H, Straub B, Schrader M, et al. DNA-based detection of prostate cancer in urine after prostatic massage. Urology 2001;58(3):335–8.

[58] Hoque MO, Topaloglu O, Begum S, Henrique R, Rosenbaum E, Van Criekinge W, et al. Quantitative methylation-specific polymerase chain reaction gene patterns in urine sediment distinguish prostate cancer patients from control subjects. J Clin Oncol 2005;23(27):6569–75.

[59] Rogers CG, Gonzalgo ML, Yan G, Bastian PJ, Chan DY, Nelson WG, et al. High concordance of gene methylation in post-digital rectal examination and post-biopsy urine samples for prostate cancer detection. J Urol 2006;176(5):2280–4.

[60] Lee BB, Lee EJ, Jung EH, Chun H-K, Chang DK, Song SY, et al. Aberrant methylation of APC, MGMT, RASSF2A, and Wif-1 genes in plasma as a biomarker for early detection of colorectal cancer. Clin Cancer Res 2009;15(19):6185–91.

[61] Lofton-Day C. DNA methylation biomarkers for blood-based colorectal cancer screening. Clin Chem 2008;54:414–23.

[62] Glockner SC. Methylation of TFPI2 in stool DNA: a potential novel biomarker for the detection of colorectal cancer. Cancer Res 2009;69:4691–9.

[63] Rosas SL, Koch W, da Costa Carvalho MG, Wu L, Califano J, Westra W, et al. Promoter hypermethylation patterns of p16, O6-methylguanine-DNA-methyltransferase, and death-associated protein kinase in tumors and saliva of head and neck cancer patients. Cancer Res 2001;61(3):939–42.

[64] Guerrero-Preston R, Soudry E, Acero J, Orera M, Moreno-López L, Macía-Colón G, et al. NID2 and HOXA9 promoter hypermethylation as biomarkers for prevention and early detection in oral cavity squamous cell carcinoma tissues and saliva. Cancer Prev Res (Phila) 2011;4(7):1061–72.

[65] Balaña C, Carrato C, Ramírez JL, Cardona AF, Berdiel M, Sánchez JJ, et al. Tumour and serum MGMT promoter methylation and protein expression in glioblastoma patients. Clin Transl Oncol 2011;13(9):677–85.

[66] Kahlert C, Melo SA, Protopopov A, Tang J, Seth S, Koch M, et al. Identification of double-stranded genomic DNA spanning all chromosomes with mutated KRAS and p53 DNA in the serum exosomes of patients with pancreatic cancer. J Biol Chem 2014;289(7):3869–75.

[67] Abbruzzese JL, Abbruzzese MC, Lenzi R, Hess KR, Raber MN. Analysis of a diagnostic strategy for patients with suspected tumors of unknown origin. J Clin Oncol 1995;13(8):2094–103.

[68] Greco FA, Pavlidis N. Treatment for patients with unknown primary carcinoma and unfavorable prognostic factors. Semin Oncol 2009;36(1):65–74.

[69] Van der Auwera I, Yu W, Suo L, Van Neste L, van Dam P, Van Marck EA, et al. Array-based DNA methylation profiling for breast cancer subtype discrimination. PLoS One 2010;5(9):e12616.

[70] Milani L, Lundmark A, Kiialainen A, Nordlund J, Flaegstad T, Forestier E, et al. DNA methylation for subtype classification and prediction of treatment outcome in patients with childhood acute lymphoblastic leukemia. Blood 2010;115(6):1214–25.

[71] Brock MV. DNA methylation markers and early recurrence in stage I lung cancer. N Engl J Med 2008;358:1118–28.

[72] Esteller M, Garcia-Foncillas J, Andion E, Goodman SN, Hidalgo OF, Vanaclocha V, et al. Inactivation of the DNA-repair gene MGMT and the clinical response of gliomas to alkylating agents. N Engl J Med 2000;343(19):1350–4.

[73] Hegi ME, Diserens A-C, Gorlia T, Hamou M-F, de Tribolet N, Weller M, et al. MGMT gene silencing and benefit from temozolomide in glioblastoma. N Engl J Med 2005;352(10):997–1003.

[74] Stupp R, Hegi ME, Neyns B, Goldbrunner R, Schlegel U, Clement PMJ, et al. Phase I/IIa study of cilengitide and temozolomide with concomitant radiotherapy followed by cilengitide and temozolomide maintenance therapy in patients with newly diagnosed glioblastoma. J Clin Oncol 2010;28(16):2712–18.

[75] Esteller M, Gaidano G, Goodman SN, Zagonel V, Capello D, Botto B, et al. Hypermethylation of the DNA repair gene O(6)-methylguanine DNA methyltransferase and survival of patients with diffuse large B-cell lymphoma. J Natl Cancer Inst 2002;94(1):26–32.

[76] Chakravarti A, Erkkinen MG, Nestler U, Stupp R, Mehta M, Aldape K, et al. Temozolomide-mediated radiation enhancement in glioblastoma: a report on underlying mechanisms. Clin Cancer Res 2006;12(15):4738–46.

[77] Esteller M, Hamilton SR, Burger PC, Baylin SB, Herman JG. Inactivation of the DNA repair gene O6-methylguanine-DNA methyltransferase by promoter hypermethylation is a common event in primary human neoplasia. Cancer Res 1999;59(4):793–7.

[78] Veeck J, Ropero S, Setien F, Gonzalez-Suarez E, Osorio A, Benitez J, et al. BRCA1 CpG island hyper-methylation predicts sensitivity to poly(adenosine diphosphate)-ribose polymerase inhibitors. J Clin Oncol 2010;28(29):e563–4. [author reply e565–e566].

[79] Stefansson OA, Villanueva A, Vidal A, Martí L, Esteller M. BRCA1 epigenetic inactivation predicts sensitivity to platinum-based chemotherapy in breast and ovarian cancer. Epigenetics 2012;7(11):1225–9.

[80] Strathdee G, MacKean MJ, Illand M, Brown R. A role for methylation of the hMLH1 promoter in loss of hMLH1 expression and drug resistance in ovarian cancer. Oncogene 1999;18(14):2335–41.

[81] Plumb JA, Strathdee G, Sludden J, Kaye SB, Brown R. Reversal of drug resistance in human tumor xenografts by 2′-deoxy-5-azacytidine-induced demethylation of the hMLH1 gene promoter. Cancer Res 2000;60(21):6039–44.

[82] Farmer H, McCabe N, Lord CJ, Tutt ANJ, Johnson DA, Richardson TB, et al. Targeting the DNA repair defect in BRCA mutant cells as a therapeutic strategy. Nature 2005;434(7035):917–21.

[83] Turner NC, Lord CJ, Iorns E, Brough R, Swift S, Elliott R, et al. A synthetic lethal siRNA screen identifying genes mediating sensitivity to a PARP inhibitor. EMBO J 2008;27(9):1368–77.

[84] Fong PC, Boss DS, Yap TA, Tutt A, Wu P, Mergui-Roelvink M, et al. Inhibition of poly(ADP-ribose) polymerase in tumors from BRCA mutation carriers. N Engl J Med 2009;361(2):123–34.

[85] Esteller M, Silva JM, Dominguez G, Bonilla F, Matias-Guiu X, Lerma E, et al. Promoter hypermethylation and BRCA1 inactivation in sporadic breast and ovarian tumors. J Natl Cancer Inst 2000;92(7):564–9.

[86] Bernards R. Finding effective cancer therapies through loss of function genetic screens. Curr Opin Genet Dev 2013;24C:23–9.

[87] Fenaux P. Efficacy of azacitidine compared with that of conventional care regimens in the treatment of higher-risk myelodysplastic syndromes: a randomised, open-label, phase III study. Lancet Oncol 2009;10:223–32.

[88] Silverman LR, McKenzie DR, Peterson BL, Holland JF, Backstrom JT, Beach CL, et al. Further analysis of trials with azacitidine in patients with myelodysplastic syndrome: studies 8421, 8921, and 9221 by the Cancer and Leukemia Group B. J Clin Oncol 2006;24(24):3895–903.

[89] Kantarjian HM, O'Brien S, Shan J, Aribi A, Garcia-Manero G, Jabbour E, et al. Update of the decitabine experience in higher risk myelodysplastic syndrome and analysis of prognostic factors associated with outcome. Cancer 2007;109(2):265–73.

[90] Yoo CB, Jeong S, Egger G, Liang G, Phiasivongsa P, Tang C, et al. Delivery of 5-aza-2′-deoxycytidine to cells using oligodeoxynucleotides. Cancer Res 2007;67(13):6400–8.

[91] Figueroa ME, Skrabanek L, Li Y, Jiemjit A, Fandy TE, Paietta E, et al. MDS and secondary AML display unique patterns and abundance of aberrant DNA methylation. Blood 2009;114(16):3448–58.

[92] Simó-Riudalbas L, Melo SA, Esteller M. DNMT3B gene amplification predicts resistance to DNA demethylating drugs. Genes Chromosomes Cancer 2011;50(7):527–34.

[93] Kelly WK, O'Connor OA, Krug LM, Chiao JH, Heaney M, Curley T, et al. Phase I study of an oral histone deacetylase inhibitor, suberoylanilide hydroxamic acid, in patients with advanced cancer. J Clin Oncol 2005;23(17):3923–31.

[94] Duvic M, Talpur R, Ni X, Zhang C, Hazarika P, Kelly C, et al. Phase 2 trial of oral vorinostat (suberoylanilide hydroxamic acid, SAHA) for refractory cutaneous T-cell lymphoma (CTCL). Blood 2007;109(1):31–9.

[95] Garcia-Manero G, Yang H, Bueso-Ramos C, Ferrajoli A, Cortes J, Wierda WG, et al. Phase 1 study of the histone deacetylase inhibitor vorinostat (suberoylanilide hydroxamic acid [SAHA]) in patients with advanced leukemias and myelodysplastic syndromes. Blood 2008;111(3):1060–6.

[96] Whittaker SJ, Demierre M-F, Kim EJ, Rook AH, Lerner A, Duvic M, et al. Final results from a multi-center, international, pivotal study of romidepsin in refractory cutaneous T-cell lymphoma. J Clin Oncol 2010;28(29):4485–91.

[97] Munster PN, Marchion D, Thomas S, Egorin M, Minton S, Springett G, et al. Phase I trial of vorinostat and doxorubicin in solid tumours: histone deacetylase 2 expression as a predictive marker. Br J Cancer 2009;101(7):1044–50.

[98] Kadia TM, Yang H, Ferrajoli A, Maddipotti S, Schroeder C, Madden TL, et al. A phase I study of vorinostat in combination with idarubicin in relapsed or refractory leukaemia. Br J Haematol 2010;150(1):72–82.

[99] Ramalingam SS, Parise RA, Ramanathan RK, Lagattuta TF, Musguire LA, Stoller RG, et al. Phase I and pharmacokinetic study of vorinostat, a histone deacetylase inhibitor, in combination with carboplatin and paclitaxel for advanced solid malignancies. Clin Cancer Res 2007;13(12):3605–10.

[100] Braiteh F, Soriano AO, Garcia-Manero G, Hong D, Johnson MM, De Padua Silva L, et al. Phase I study of epigenetic modulation with 5-azacytidine and valproic acid in patients with advanced cancers. Clin Cancer Res 2008;14(19):6296–301.

[101] Lin J, Gilbert J, Rudek MA, Zwiebel JA, Gore S, Jiemjit A, et al. A phase I dose-finding study of 5-azacytidine in combination with sodium phenylbutyrate in patients with refractory solid tumors. Clin Cancer Res 2009;15(19):6241–9.

[102] Tsai H-C, Li H, Van Neste L, Cai Y, Robert C, Rassool FV, et al. Transient low doses of DNA-demethylating agents exert durable antitumor effects on hematological and epithelial tumor cells. Cancer Cell 2012;21(3):430–46.

[103] Juergens RA, Wrangle J, Vendetti FP, Murphy SC, Zhao M, Coleman B, et al. Combination epigenetic therapy has efficacy in patients with refractory advanced non-small cell lung cancer. Cancer Discov 2011;1(7):598–607.

[104] Rodríguez-Paredes M, Esteller M. A combined epigenetic therapy equals the efficacy of conventional chemotherapy in refractory advanced non-small cell lung cancer. Cancer Discov 2011;1(7):557–9.

[105] Esteller M, Sanchez-Cespedes M, Rosell R, Sidransky D, Baylin SB, Herman JG. Detection of aberrant promoter hypermethylation of tumor suppressor genes in serum DNA from non-small cell lung cancer patients. Cancer Res 1999;59(1):67–70.

[106] Herman D, Jenssen K, Burnett R, Soragni E, Perlman SL, Gottesfeld JM. Histone deacetylase inhibitors reverse gene silencing in Friedreich's ataxia. Nat Chem Biol 2006;2(10):551–8.

[107] Alarcón JM, Malleret G, Touzani K, Vronskaya S, Ishii S, Kandel ER, et al. Chromatin acetylation, memory, and LTP are impaired in CBP +/− mice: a model for the cognitive deficit in Rubinstein–Taybi syndrome and its amelioration. Neuron 2004;42(6):947–59.

[108] Vecsey CG, Hawk JD, Lattal KM, Stein JM, Fabian SA, Attner MA, et al. Histone deacetylase inhibitors enhance memory and synaptic plasticity via CREB:CBP-dependent transcriptional activation. J Neurosci 2007;27(23):6128–40.

[109] Peleg S, Sananbenesi F, Zovoilis A, Burkhardt S, Bahari-Javan S, Agis-Balboa RC, et al. Altered histone acetylation is associated with age-dependent memory impairment in mice. Science 2010;328(5979):753–6.

[110] Gräff J, Rei D, Guan J-S, Wang W-Y, Seo J, Hennig KM, et al. An epigenetic blockade of cognitive functions in the neurodegenerating brain. Nature 2012;483(7388):222–6.

[111] Rangel-Salazar R, Wickström-Lindholm M, Aguilar-Salinas CA, Alvarado-Caudillo Y, Døssing KB, Esteller M, et al. Human native lipoprotein-induced *de novo* DNA methylation is associated with repression of inflammatory genes in THP-1 macrophages. BMC Genomics 2011;12:582.

[112] Filippakopoulos P, Qi J, Picaud S, Shen Y, Smith WB, Fedorov O, et al. Selective inhibition of BET bromodomains. Nature 2010;468(7327):1067–73.

[113] Dawson MA, Prinjha RK, Dittmann A, Giotopoulos G, Bantscheff M, Chan W-I, et al. Inhibition of BET recruitment to chromatin as an effective treatment for MLL-fusion leukaemia. Nature 2011;478(7370):529–33.

[114] Bernt KM, Zhu N, Sinha AU, Vempati S, Faber J, Krivtsov AV, et al. MLL-rearranged leukemia is dependent on aberrant H3K79 methylation by DOT1L. Cancer Cell 2011;20(1):66–78.

[115] Krivtsov AV, Feng Z, Lemieux ME, Faber J, Vempati S, Sinha AU, et al. H3K79 methylation profiles define murine and human MLL-AF4 leukemias. Cancer Cell 2008;14(5):355–68.

[116] Daigle SR. Selective killing of mixed lineage leukemia cells by a potent small-molecule DOT1L inhibitor. Cancer Cell 2011;20:53–65.

[117] Rakyan VK, Down TA, Balding DJ, Beck S. Epigenome-wide association studies for common human diseases. Nat Rev Genet 2011;12(8):529–41.

[118] Michels KB, Binder AM, Dedeurwaerder S, Epstein CB, Greally JM, Gut I, et al. Recommendations for the design and analysis of epigenome-wide association studies. Nat Methods 2013;10(10):949–55.

[119] Jaffe AE, Irizarry RA. Accounting for cellular heterogeneity is critical in epigenome-wide association studies. Genome Biol 2014;15(2):R31.

Index

Note: Page numbers followed by "*f*" and "*t*" refer to figures and tables, respectively.

Printed and bound by CPI Group (UK) Ltd, Croydon, CR0 4YY

13/05/2025

01870676-0001